Single Variable
CALCULUS
Early Transcendentals
Fourth Edition

THE JONES AND BARTLETT PUBLISHERS SERIES IN MATHEMATICS

Geometry

Geometry with an Introduction to Cosmic Topology
Hitchman (978-0-7637-5457-0) © 2009

Euclidean and Transformational Geometry: A Deductive Inquiry
Libeskind (978-0-7637-4366-6) © 2008

A Gateway to Modern Geometry: The Poincaré Half-Plane, Second Edition
Stahl (978-0-7637-5381-8) © 2008

Understanding Modern Mathematics
Stahl (978-0-7637-3401-5) © 2007

Lebesgue Integration on Euclidean Space, Revised Edition
Jones (978-0-7637-1708-7) © 2001

Precalculus

Precalculus: A Functional Approach to Graphing and Problem Solving, Sixth Edition
Smith (978-0-7637-5177-7) © 2012

Precalculus with Calculus Previews (Expanded Volume), Fourth Edition
Zill/Dewar (978-0-7637-6631-3) © 2010

Precalculus with Calculus Previews (Essentials Version), Fourth Edition
Zill/Dewar (978-0-7637-3779-5) © 2007

Calculus

Single Variable Calculus: Early Transcendentals, Fourth Edition
Zill/Wright (978-0-7637-4965-1) © 2011

Multivariable Calculus, Fourth Edition
Zill/Wright (978-0-7637-4966-8) © 2011

Calculus: Early Transcendentals, Fourth Edition
Zill/Wright (978-0-7637-5995-7) © 2011

Multivariable Calculus
Damiano/Freije (978-0-7637-8247-4) © 2011

Calculus: The Language of Change
Cohen/Henle (978-0-7637-2947-9) © 2005

Applied Calculus for Scientists and Engineers
Blume (978-0-7637-2877-9) © 2005

Calculus: Labs for Mathematica
O'Connor (978-0-7637-3425-1) © 2005

Calculus: Labs for MATLAB
O'Connor (978-0-7637-3426-8) © 2005

Linear Algebra

Linear Algebra with Applications, Seventh Edition
Williams (978-0-7637-8248-1) © 2011

Linear Algebra with Applications, Alternate Seventh Edition
Williams (978-0-7637-8249-8) © 2011

Linear Algebra: Theory and Applications
Cheney/Kincaid (978-0-7637-5020-6) © 2009

Advanced Engineering Mathematics

Advanced Engineering Mathematics, Fourth Edition
Zill/Wright (978-0-7637-7966-5) © 2011

An Elementary Course in Partial Differential Equations, Second Edition
Amaranath (978-0-7637-6244-5) © 2009

Complex Analysis

A First Course in Complex Analysis with Applications, Second Edition
Zill/Shanahan (978-0-7637-5772-4) © 2009

Complex Analysis for Mathematics and Engineering, Fifth Edition
Mathews/Howell (978-0-7637-3748-1) © 2006

Classical Complex Analysis
Hahn (978-0-8672-0494-0) © 1996

Real Analysis

Real Analysis
Denlinger (978-0-7637-7947-4) © 2011

An Introduction to Analysis, Second Edition
Bilodeau/Thie/Keough (978-0-7637-7492-9) © 2010

Basic Real Analysis
Howland (978-0-7637-7318-2) © 2010

Closer and Closer: Introducing Real Analysis
Schumacher (978-0-7637-3593-7) © 2008

The Way of Analysis, Revised Edition
Strichartz (978-0-7637-1497-0) © 2000

Topology

Foundations of Topology, Second Edition
Patty (978-0-7637-4234-8) © 2009

Discrete Math and Logic

Discrete Structures, Logic, and Computability, Third Edition
Hein (978-0-7637-7206-2) © 2010

Essentials of Discrete Mathematics
Hunter (978-0-7637-4892-0) © 2009

Logic, Sets, and Recursion, Second Edition
Causey (978-0-7637-3784-9) © 2006

Numerical Methods

Numerical Mathematics
Grasselli/Pelinovsky (978-0-7637-3767-2) © 2008

Exploring Numerical Methods: An Introduction to Scientific Computing Using MATLAB
Linz (978-0-7637-1499-4) © 2003

Advanced Mathematics

Mathematical Modeling with Excel®
Albright (978-0-7637-6566-8) © 2010

Clinical Statistics: Introducing Clinical Trials, Survival Analysis, and Longitudinal Data Analysis
Korosteleva (978-0-7637-5850-9) © 2009

Harmonic Analysis: A Gentle Introduction
DeVito (978-0-7637-3893-8) © 2007

Beginning Number Theory, Second Edition
Robbins (978-0-7637-3768-9) © 2006

A Gateway to Higher Mathematics
Goodfriend (978-0-7637-2733-8) © 2006

For more information on this series and its titles, please visit us online at http://www.jbpub.com/math. Qualified instructors, contact your Publisher's Representative at 1-800-832-0034 or info@jbpub.com to request review copies for course consideration.

THE JONES AND BARTLETT PUBLISHERS INTERNATIONAL SERIES IN MATHEMATICS

Mathematical Modeling with Excel®
Albright (978-0-7637-6566-8) © 2010

An Introduction to Analysis, Second Edition
Bilodeau/Thie/Keough (978-0-7637-7492-9) © 2010

Basic Real Analysis
Howland (978-0-7637-7318-2) © 2010

Advanced Engineering Mathematics, Fourth Edition, International Version
Zill/Wright (978-0-7637-7994-8) © 2011

Calculus: Early Transcendentals, Fourth Edition International Version
Zill/Wright (978-0-7637-8652-6) © 2011

Multivariable Calculus
Damiano/Freije (978-0-7637-8247-4) © 2011

Real Analysis
Denlinger (979-0-7637-7947-4) © 2011

Mathematical Modeling for the Scientific Method
Pravica (978-0-7637-7946-7) © 2011

A Journey into Partial Differential Equations
Bray (978-0-7637-7256-7) © 2011

Functions of Mathematics in Liberal Arts
Johnson (978-0-7637-8116-3) © 2012

For more information on this series and its titles, please visit us online at http://www.jbpub.com/math. Qualified instructors, contact your Publisher's Representative at 1-800-832-0034 or info@jbpub.com to request review copies for course consideration.

Formulas from Geometry

Area A, Circumference C, Volume V, Surface Area S

RECTANGLE

$A = lw, \ C = 2l + 2w$

PARALLELOGRAM

$A = bh$

TRAPEZOID

$A = \frac{1}{2}(a+b)h$

RIGHT TRIANGLE

Pythagorean Theorem:
$c^2 = a^2 + b^2$

TRIANGLE

$A = \frac{1}{2}bh, \ C = a + b + c$

EQUILATERAL TRIANGLE

$h = \frac{\sqrt{3}}{2}s, \ A = \frac{\sqrt{3}}{4}s^2$

CIRCLE

$A = \pi r^2, \ C = 2\pi r$

CIRCULAR RING

$A = \pi(R^2 - r^2)$

CIRCULAR SECTOR

$A = \frac{1}{2}r^2\theta, \ s = r\theta$

ELLIPSE
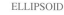
$A = \pi ab$

ELLIPSOID

$V = \frac{4}{3}\pi abc$

SPHERE

$V = \frac{4}{3}\pi r^3, \ S = 4\pi r^2$

RIGHT CYLINDER

$V = Bh, \ B$ area of base

RIGHT CIRCULAR CYLINDER

$V = \pi r^2 h, \ S = 2\pi rh$ (lateral side)

RECTANGULAR PARALLELEPIPED

$V = lwh, \ S = 2(hl + lw + hw)$

CONE

$V = \frac{1}{3}Bh, \ B$ area of base

RIGHT CIRCULAR CONE

$V = \frac{1}{3}\pi r^2 h, \ S = \pi r\sqrt{r^2 + h^2}$

FRUSTUM OF A CONE

$V = \frac{1}{3}\pi h(r_1^2 + r_1 r_2 + r_2^2)$

Single Variable
CALCULUS
Early Transcendentals
Fourth Edition

Dennis G. Zill
Loyola Marymount University

Warren S. Wright
Loyola Marymount University

JONES AND BARTLETT PUBLISHERS
Sudbury, Massachusetts
BOSTON   TORONTO   LONDON   SINGAPORE

World Headquarters
Jones and Bartlett Publishers
40 Tall Pine Drive
Sudbury, MA 01776
978-443-5000
info@jbpub.com
www.jbpub.com

Jones and Bartlett Publishers Canada
6339 Ormindale Way
Mississauga, Ontario L5V 1J2
Canada

Jones and Bartlett Publishers International
Barb House, Barb Mews
London W6 7PA
United Kingdom

Jones and Bartlett's books and products are available through most bookstores and online booksellers. To contact Jones and Bartlett Publishers directly, call 800-832-0034, fax 978-443-8000, or visit our website, www.jbpub.com.

Substantial discounts on bulk quantities of Jones and Bartlett's publications are available to corporations, professional associations, and other qualified organizations. For details and specific discount information, contact the special sales department at Jones and Bartlett via the above contact information or send an email to specialsales@jbpub.com.

Copyright © 2011 by Jones and Bartlett Publishers, LLC

All rights reserved. No part of the material protected by this copyright may be reproduced or utilized in any form, electronic or mechanical, including photocopying, recording, or by any information storage and retrieval system, without written permission from the copyright owner.

Production Credits
Chief Executive Officer: Clayton Jones
Chief Operating Officer: Don W. Jones, Jr.
President, Higher Education and Professional Publishing: Robert W. Holland, Jr.
V.P., Sales: William J. Kane
V.P., Design and Production: Anne Spencer
V.P., Manufacturing and Inventory Control: Therese Connell
Publisher: David Pallai
Acquisitions Editor: Timothy Anderson
Editorial Assistant: Melissa Potter
Production Director: Amy Rose
Production Manager: Jennifer Bagdigian
Production Assistant: Ashlee Hazeltine
Senior Marketing Manager: Andrea DeFronzo
Associate Marketing Manager: Lindsay Ruggiero
Cover and Title Page Design: Kristin E. Parker
Photo Research and Permissions Manager: Kimberly Potvin
Composition: Aptara®, Inc.
Front Cover Image: © Dennis Mammana, all rights reserved. Original image courtesy NASA/JPL-Caltech.
Back Cover Image: © Cenk E. Tezel & Tunç Tezel
Printing and Binding: Courier Kendallville
Cover Printing: Courier Kendallville

Library of Congress Cataloging-in-Publication Data
Zill, Dennis G., 1940-
 Single variable calculus: early transcendentals / Dennis Zill, Warren Wright.
  p. cm.
 Includes Index.
 ISBN-13: 978-0-7637-4965-1 (pbk.)
 ISBN-10: 0-7637-4965-6 (pbk.)
 1. Calculus. I. Wright, Warren S. II. Title.
 QA303.Z525 2009
 515—dc22
                    2008013286
6048
Printed in the United States of America
13 12 11 10 09 10 9 8 7 6 5 4 3 2 1

Preface

☰ To the Instructor

Philosophy

The **fourth edition** of *Single Variable Calculus: Early Transcendentals* represents a substantial revision of the last edition. Although there is much new in this edition, I have striven to keep intact my original goal of compiling a calculus text that is not just a collection of definitions and theorems, skills and formulas to be memorized, and problems to be solved, but rather a book that communicates with its primary audience, the students. It is my hope that these changes make the text more relevant and interesting for both student and instructor.

Features in This Revision

Precalculus Material The precalculus review material on the real number system, the Cartesian plane, lines, and circles that appeared in Appendix I in the last edition, has been moved to the *Student Resource Manual* (*SRM*).

Sections and Exercises Most of the material has been updated and, in some instances, reorganized. Many sections and exercise sets have been completely rewritten. Many new problems, especially applications, problems dealing with graphing calculators and computers, conceptual problems, and project problems have been added to the exercise sets. For the most part, the added applications are "real life" in that they have been thoroughly researched using original sources. Problems dealing with interpretation of graphs have also been added. Moreover, there is an increased emphasis on the trigonometric functions, both in the examples and in the exercise sets throughout the text. There are over 7300 problems in this Calculus Series.

To aid in assignment of problems, each exercise set is clearly partitioned into groups of problems using headings such as *Fundamentals*, *Applications*, *Mathematical Models*, *Projects*, *Calculator/CAS Problems*, and so on. I think most of the heads are self-explanatory, and so problems listed under the heading *Think About It* deal with conceptual aspects of the material covered in that section and are suitable either for assignment or for a classroom discussion. No answers are supplied for these problems in the text. Some problems are labeled *Mathematical Classics* and reflect the fact that they have been around for a long time, appear in almost every calculus text, or have an interesting twist, while other problems labeled *A Bit of History* naturally evince some historical content.

Chapter 1 is a review of functions and following the current prevailing fashion, functions are presented from the algebraic, graphical, numerical, or verbal points of view. In point of fact, the entire last section in Chapter 1 is entitled *From Words to Functions*. Because many students invariably encounter difficulties in solving related rate and applied optimization problems, I have included this section to give them a preview of how to set up, or construct, a function from a verbal description (with calculus content removed). Indeed, many of the problems in Section 1.7 reappear in a calculus context in Section 4.8.

Differential equations appear in Chapter 8. First-order equations are considered in Chapter 8 to benefit those students who encounter their applications in courses in physics and engineering. Proofs of some of the longer theorems are given in the appendix. Biographical sketches of some of the mathematicians who had a significant impact on the development of calculus appear at the end of appropriate sections under the rubric *Postscript—A Bit of History*.

Special Features Each chapter opens with its own table of contents and an introduction to the material covered in that chapter. The text ends with *Resource Pages*, which is a compact review of basic concepts from algebra, geometry, trigonometry, and calculus: the laws of exponents, factorization formulas, binomial expansions, Pascal's triangle, formulas from geometry, graphs and functions, trigonometric functions, exponential and logarithmic functions, and differentiation and integration formulas. Many of the topics summarized in the *Resource Pages* are discussed in greater depth in the *SRM*.

The popular feature introduced in the last edition, called *Test Yourself*, has been retained and expanded in this edition. The *Test Yourself* feature is a self-test consisting of 56 questions on four broad areas of precalculus mathematics. This test is intended to encourage students to review, perhaps on their own, some of the more essential prerequisite subjects, such as absolute values, the Cartesian plane, equations of lines, circles, and so on, that are used throughout the text. Answers to all questions in the test are given in the answer section.

Users of the previous three editions have been very receptive to the *Remarks* that often conclude a section. As a consequence their number has been increased and they have been renamed as *Notes from the Classroom*. These *Notes* are intended to be informal discussions that are aimed directly at the student. These discussions range from warnings about common algebraic, procedural, and notational errors; to misinterpretations of theorems; to advice; to questions asking the student to think about and possibly extend the ideas just presented.

Also, at the request of users, the number of marginal annotations and guidance annotations in the examples have been increased.

Student Resource Manual I believe that this manual can be of significant help to a student's success in a calculus course. Unlike the traditional student solutions manual, where a selected subset of the problems are worked out, the *SRM* is divided into four sections:

- ESSAYS • TOPICS IN PRECALCULUS • USE OF A CALCULATOR
- SELECTED SOLUTIONS

In the opening section of the *SRM*, Essays, there two interesting essays, one on the history of calculus and the other on its modern day applications:

The Story of Calculus, by Roger Cooke, University of Vermont, and
Calculus and Mathematical Modeling by Fred S. Roberts, Rutgers University.

(*The Story of Calculus* also appears on page xxiii of this book.) In Topics in Precalculus, selected topics from precalculus (such as sets, exponents, real numbers and inequalities, the Cartesian Plane, lines, matrices, and determinants) are reviewed because of their relevance to calculus. Because I feel that a mathematics textbook is not the appropriate place for a discussion of the use of technology, graphing calculator essentials are discussed in the section entitled Use of a Calculator. In Selected Solutions, a detailed solution of every odd problem in the exercise sets is given.

Figures, Definitions, Theorems A word about the numbering of the figures, definitions, and theorems is in order. Because of the great number of figures, definitions, and theorems in this text, I have switched to a double-decimal numeration system. For example, the interpretation of "Figure 1.2.3" is

Chapter Section of Chapter 1
↓ ↓
1.2.3 ← Third figure in Section 1.2

I feel that this type of numeration will make it easier to find, say, a theorem or figure when it is referred to in a later section or chapter. In addition, to better link a figure with the text, the *first* textual reference to each figure is done in the same font style and color as the figure number. For example, the first reference to the first figure in Section 7.5 is given as Figure 7.5.1 and all subsequent references are written in the traditional style of Figure 7.5.1. Also, in this revision each figure in the text has a brief explanatory caption.

Supplements

For Instructors

- An *Instructor's ToolKit* (*ITK*) offers instructors a computerized test bank that allows instructors to create customized tests and quizzes. The questions and answers are sorted by chapter and can be easily installed on a computer for accessibility. The *ITK* also includes *PowerPoint® Slides*, which feature all labeled figures as they appear in the text. This useful tool allows instructors to easily display and discuss figures and problems found within the text. The *ITK* is located on a secured website for qualified instructors only.

    http://www.jbpub.com/catalog/9780763496511/Instructor

- A *Complete Solutions Manual* (*CSM*) is also available on the website or on a CD. This manual contains detailed solutions of every problem in the text.
- *WebAssign* Developed by instructors for instructors, WebAssign is the premier independent online teaching and learning environment, guiding over 3 million students through their academic careers since 1997. With WebAssign, you can:
    - Create and distribute algorithmic assignments using questions specific to your textbook
    - Grade, record, and analyze student responses and performance instantly
    - Offer more practice exercises, quizzes, and homework
    - Upload resources to share and communicate with your students seamlessly

    For more detailed information and to sign up for free faculty access, please visit:

    www.webassign.net

    Please contact your Jones and Bartlett Publisher's Representative for information on how students can purchase access to WebAssign bundled with this textbook.
    Designated instructor's materials are for qualified instructors only. Jones and Bartlett reserves the right to evaluate all requests.

For Students

- A *Student Resource Manual* (*SRM*) prepared by Jeffrey M. Gervasi, EdD, of Porterville College. This printed manual can be ordered bundled with the text at substantial savings compared to buying the text and the *SRM* separately.
- WebAssign access code card

To the Student

You are enrolled in one of the most interesting courses in mathematics. Many years ago when I was a student in Calculus I, I was struck by the power and beauty of the material. It was unlike any mathematics that I had studied up to that point. It was fun, it was exciting, and it was a challenge. After teaching collegiate mathematics for many years, I have seen almost every type of student, from a budding genius who invented his own calculus, to students who struggled to master the most rudimentary mechanics of the subject. Over these years I have also witnessed a sad phenomenon: some students fail calculus, not because they find the subject matter impossibly difficult, but because they have weak algebra skills and an inadequate working knowledge of trigonometry. Calculus builds immediately on your prior knowledge and skills and there is much new ground to be covered. Consequently there is very little time to review basics in the formal classroom setting. So those of us who teach calculus must assume that you can factor, simplify and solve equations, solve inequalities, handle absolute values, use a calculator, apply the laws of exponents, find equations of lines, plot points, sketch basic graphs, and apply important logarithmic and trigonometric identities. The ability to do algebra and trigonometry, work with exponentials and logarithms, and sketch *by hand* basic graphs quickly and accurately are keys to success in a calculus course.

On page xix there is a list of 56 questions in a section called "Test Yourself." This "test" is an opportunity for you to check your knowledge on some of the topics that will be assumed in this text. Relax, take your time, read and work every question, and then compare your answers with those given on page ANS-1. Regardless of your "score" on this test, you are strongly encouraged to review precalculus material either given in the *Student Resource Manual* or from a precalculus text.

A word to those students who have taken calculus in high school: please do not assume that you can get by with minimal effort because you recognize some of the topics in differential and integral calculus. A sense of familiarity of the subject combined with an attitude of complacency is often the downfall of some students.

Learning mathematics is not like learning how to ride a bicycle, that once learned, the ability sticks for a lifetime. Mathematics is more like learning another language or learning to play a musical instrument; it requires time, effort, and a lot of practice to develop and maintain proficiency. Even experienced musicians still practice the fundamental scales. So ultimately, you the student can learn mathematics (that is, make it stick) only through the hard work of doing mathematics. Although I have tried to make *most* details in the solution of an example clear to the reader, inevitably you will have to fill in some missing steps. You cannot read a text such as this as you would a novel; you must work your way through the text with pencil and paper at the ready.

In conclusion, I wish you the best of luck in this course.

≡ Acknowledgments

Compiling a textbook of this complexity is a monumental task. Besides the authors, many people put much time and energy into this project. First and foremost, I would like to express my appreciation to the editorial, production, and marketing staffs at Jones and Bartlett Publishers and to the following reviewers of this and previous editions who contributed many suggestions, valid criticisms, and even an occasional word of support:

Scott Wilde, *Baylor University*
Salvatore Anastasio, *SUNY, New Paltz*
Thomas Bengston, *Penn State University, Delaware County*
Steven Blasberg, *West Valley College*
Robert Brooks, *University of Utah*
Dietrich Burbulla, *University of Toronto*
David Burton, *Chabot College*
Maurice Chabot, *University of Southern Maine*
H. Edward Donley, *Indiana University of Pennsylvania*
John W. Dulin, *GMI Engineering & Management Institute*
Arthur Dull, *Diablo Valley College*
Hugh Easler, *College of William and Mary*
Jane Edgar, *Brevard Community College*
Joseph Egar, *Cleveland State University*
Patrick J. Enright, *Arapahoe Community College*
Peter Frisk, *Rock Valley College*
Shirley Goldman, *University of California at Davis*
Joan Golliday, *Santa Fe Community College*
David Green, Jr., *GMI Engineering & Management Institute*
Harvey Greenwald, *California Polytechnic State University*
Walter Gruber, *Mercy College of Detroit*
Dave Hallenbeck, *University of Delaware*
Noel Harbetson, *California State University at Fresno*
Bernard Harvey, *California State University, Long Beach*
Christopher E. Hee, *Eastern Michigan University*
Jean Holton, *Tidewater Community College*
Rahim G. Karimpour, *Southern Illinois University*

Martin Kotler, *Pace University*
Carlon A. Krantz, *Kean College of New Jersey*
George Kung, *University of Wisconsin at Stevens Point*
John C. Lawlor, *University of Vermont*
Timothy Loughlin, *New York Institute of Technology*
Antonio Magliaro, *Southern Connecticut Slate University*
Walter Fred Martens, *University of Alabama at Birmingham*
William E. Mastrocola, *Colgate University*
Jill McKenney, *Lane Community College*
Edward T. Migliore, *Monterey Peninsula College*
Carolyn Narasimhan, *DePaul University*
Harold Olson, *Diablo Valley College*
Gene Ortner, *Michigan Technological University*
Aubrey Owen, *Community College of Denver*
Marvin C. Papenfuss, *Loras College*
Don Poulson, *Mesa Community College*
Susan Prazak, *College of Charleston*
James J. Reynolds, *Pennsylvania State University, Beaver Campus*
Susan Richman, *Penn State University, Harrisburg*
Rodd Ross, *University of Toronto*
Donald E. Rossi, *De Anza College*
Lillian Seese, *St. Louis Community College at Meramec*
Donald Sherbert, *University of Illinois*
Nedra Shunk, *Santa Clara University*
Phil R. Smith, *American River College*
Joseph Stemple, *CUNY Queens College*

Margaret Suchow, *Adirondack Community College*
John Suvak, *Memorial University of Newfoundland*
George Szoke, *University of Akron*
Hubert Walczak, *College of St. Thomas*
Richard Werner, *Santa Rosa Junior College*
Loyd V. Wilcox, *Golden West College*
Jack Wilson, *University of North Carolina, Asheville*

I would also like to extend an extra-special thank you to the following individuals:

- Jeff Dodd, Jacksonville State University, for the project in Problem 37 of Exercises 8.3,
- John David Dionisio, Loyola Marymount University, and Brian and Melanie Fulton, High Point University, for providing the solutions for the *Complete Solutions Manual* and the *Student Resource Manual*,
- Roger Cooke, University of Vermont, and Fred S. Roberts, Rutgers University, for graciously taking time out their busy schedules to write and contribute the excellent calculus essays,
- Carol Wright, for her help in the final stages of manuscript preparation for this and other texts,
- David Pallai, Publisher, and Tim Anderson, Editor, for putting up with all my verbal venting of frustrations,
- Jennifer Bagdigian, Production Manager, for smoothly coordinating the phases of production and for her patience in putting up with my never-ending changes, and
- Irving Drooyan and Charles Carico, for starting it all.

Even with all the help just listed, the accuracy of every letter, word, symbol, equation, and figure contained in this final product is the responsibility of the author. I would be very grateful to have any errors or "typos" called to my attention. Corrections can be sent to Tim Anderson at

tanderson@jbpub.com.

In conclusion, I welcome my long-time colleague Warren Scott Wright at Loyola Marymount University, and author of many of the supplements that accompany my texts, as a coauthor of this text.

Dennis G. Zill

Warren S. Wright

Contents

Preface   vii

About the Covers   xviii

Test Yourself   xix

Essay: The Story of Calculus   xxiii

1 Functions   1

1.1   Functions and Graphs   2
1.2   Combining Functions   10
1.3   Polynomial and Rational Functions   20
1.4   Transcendental Functions   30
1.5   Inverse Functions   37
1.6   Exponential and Logarithmic Functions   48
1.7   From Words to Functions   55
       Chapter 1 in Review   61

2 Limit of a Function   67

2.1   Limits—An Informal Approach   68
2.2   Limit Theorems   74
2.3   Continuity   81
2.4   Trigonometric Limits   88
2.5   Limits That Involve Infinity   94
2.6   Limits—A Formal Approach   103
2.7   The Tangent Line Problem   110
      Chapter 2 in Review   118

xiv Contents

3 The Derivative 121

3.1 The Derivative 122
3.2 Power and Sum Rules 130
3.3 Product and Quotient Rules 138
3.4 Trigonometric Functions 144
3.5 Chain Rule 149
3.6 Implicit Differentiation 156
3.7 Derivatives of Inverse Functions 162
3.8 Exponential Functions 167
3.9 Logarithmic Functions 172
3.10 Hyperbolic Functions 178
   Chapter 3 in Review 186

4 Applications of the Derivative 191

4.1 Rectilinear Motion 192
4.2 Related Rates 196
4.3 Extrema of Function 204
4.4 Mean Value Theorem 210
4.5 Limits Revisited—L'Hôpital's Rule 216
4.6 Graphing and the First Derivative 224
4.7 Graphing and the Second Derivative 230
4.8 Optimization 235
4.9 Linearization and Differentials 247
4.10 Newton's Method 254
   Chapter 4 in Review 260

5 Integrals 267

5.1 The Indefinite Integral 268
5.2 Integration by the u-Substitution 276
5.3 The Area Problem 286

5.4 The Definite Integral 295
5.5 Fundamental Theorem of Calculus 305
Chapter 5 in Review 316

6 Applications of the Integral 321

6.1 Rectilinear Motion Revisited 322
6.2 Area Revisited 325
6.3 Volumes of Solids: Slicing Method 333
6.4 Volumes of Solids: Shell Method 340
6.5 Length of a Graph 345
6.6 Area of a Surface of Revolution 348
6.7 Average Value of a Function 351
6.8 Work 355
6.9 Fluid Pressure and Force 362
6.10 Centers of Mass and Centroids 367
Chapter 6 in Review 373

7 Techniques of Integration 379

7.1 Integration—Three Resources 380
7.2 Integration by Substitution 382
7.3 Integration by Parts 386
7.4 Powers of Trigonometric Functions 393
7.5 Trigonometric Substitutions 399
7.6 Partial Fractions 406
7.7 Improper Integrals 415
7.8 Approximate Integration 423
Chapter 7 in Review 433

8 First-Order Differential Equations 439

8.1 Separable Equations 440
8.2 Linear Equations 445

8.3	Mathematical Models 450
8.4	Solution Curves without a Solution 459
8.5	Euler's Method 468
	Chapter 8 in Review 471

9 Sequences and Series   475

9.1	Sequences 476
9.2	Monotonic Sequences 485
9.3	Series 490
9.4	Integral Test 501
9.5	Comparison Tests 504
9.6	Ratio and Root Tests 509
9.7	Alternating Series 512
9.8	Power Series 519
9.9	Representing Functions by Power Series 523
9.10	Taylor Series 529
9.11	Binomial Series 540
	Chapter 9 in Review 544

10 Conics and Polar Coordinates   547

10.1	Conic Sections 548
10.2	Parametric Equations 560
10.3	Calculus and Parametric Equations 568
10.4	Polar Coordinate System 573
10.5	Graphs of Polar Equations 576
10.6	Calculus in Polar Coordinates 585
10.7	Conic Sections in Polar Coordinates 592
	Chapter 10 in Review 597

Appendix   APP-1

  Proofs of Selected Theorems   APP-1

Resource Pages   RP-1

  Review of Algebra   RP-2
  Formulas from Geometry   RP-2
  Graphs and Functions   RP-4
  Review of Trigonometry   RP-5
  Exponential and Logarithmic Functions   RP-7
  Differentiation   RP-8
  Integration Formulas   RP-9

Answers to Test Yourself   ANS-1

Answers to Selected Odd-Numbered Problems   ANS-2

Index   IND-1

Photo Credits   C-1

About the Covers

Front Cover: Martian Analemma

On Earth, the analemma is the figure-8 loop we get if we mark the Sun's position at the same time each day throughout the year. Its shape is determined by the tilt of our planet's axis and the variable speed at which Earth revolves around the Sun. If, however, we were to mark the Sun's position in the sky of Mars, we would discover a simpler, stretched, pear-shaped analemma. This digital illustration shows the late afternoon Sun that would have been seen from the Sagan Memorial Station once every 30 Martian days (sols) beginning on Sol 24 (July 29, 1997). Slightly less bright, the Martian Sun would appear only about one-third the size we see from Earth, while the Martian dust—responsible for Mars's pink sky—also scatters some blue light around the solar disk. This photo-illustration was created by first plotting the Martian analemma with Starry Night® Pro software (v6.2.3), and then using Adobe® Photoshop® CS2 (v9.0.2) to assemble it onto the correct celestial location on NASA's famous Presidential Panorama. © Dennis Mammana / dennismammana.com

Back Cover Inset: Total Solar Eclipse Analemma

This image is the world's first analemma (or *Tutulemma*, a term coined by the photographers based on the Turkish word for eclipse) photo that includes a total solar eclipse. This is a year-long image process, showing the Sun's motion in one frame. The photo was started in 2005 and ended in 2006 in Side, Turkey, about 500 km south of the photographer's home. Tunç Tezel, a leading amateur astronomer and night sky photographer, took the image with assistance from his brother, Cenk E. Tezel. Venus was also visible during the process, and can be seen toward the lower right portion of the photo. © Cenk E. Tezel & Tunç Tezel

Test Yourself

Answers to all questions are on page ANS-1.

In Preparation for Calculus

Basic Mathematics

1. (True/False) $\sqrt{a^2 + b^2} = a + b$. _____
2. (True/False) For $a > 0$, $(a^{4/3})^{3/4} = a$. _____
3. (True/False) For $x \neq 0$, $x^{-3/2} = \dfrac{1}{x^{2/3}}$. _____
4. (True/False) $\dfrac{2^n}{4^n} = \dfrac{1}{2^n}$. _____
5. (Fill in the blank) In the expansion of $(1 - 2x)^3$ the coefficient of x^2 is _____.
6. Without the aid of a calculator, evaluate $(-27)^{5/3}$.
7. Write as one expression without negative exponents:
$$x^2 \frac{1}{2}(x^2 + 4)^{-1/2} 2x + 2x\sqrt{x^2 + 4}.$$
8. Complete the square: $2x^2 + 6x + 5$.
9. Solve the equations:
   (a) $x^2 = 7x$ (b) $x^2 + 2x = 5$ (c) $\dfrac{1}{2x-1} - \dfrac{1}{x} = 0$ (d) $x + \sqrt{x-1} = 1$
10. Factor completely:
    (a) $10x^2 - 13x - 3$
    (b) $x^4 - 2x^3 - 15x^2$
    (c) $x^3 - 27$
    (d) $x^4 - 16$

Real Numbers

11. (True/False) If $a < b$, then $a^2 < b^2$. _____
12. (True/False) $\sqrt{(-9)^2} = -9$. _____
13. (True/False) If $a < 0$, then $\dfrac{-a}{a} < 0$. _____
14. (Fill in the blanks) If $|3x| = 18$, then $x = $ _____ or $x = $ _____.
15. (Fill in the blank) If $a - 5$ is a negative number, then $|a - 5| = $ _____.
16. Which of the following real numbers are rational numbers?
    (a) 0.25 (b) $8.131313\ldots$ (c) π
    (d) $\dfrac{22}{7}$ (e) $\sqrt{16}$ (f) $\sqrt{2}$
    (g) 0 (h) -9 (i) $1\dfrac{1}{2}$
    (j) $\dfrac{\sqrt{5}}{\sqrt{2}}$ (k) $\dfrac{\sqrt{3}}{2}$ (l) $\dfrac{-2}{11}$
17. Match the given interval with the appropriate inequality.
    (i) $(2, 4]$ (ii) $[2, 4)$ (iii) $(2, 4)$ (iv) $[2, 4]$
    (a) $|x - 3| < 1$ (b) $|x - 3| \leq 1$ (c) $0 \leq x - 2 < 2$ (d) $1 < x - 1 \leq 3$
18. Express the interval $(-2, 2)$ as
    (a) an inequality and   (b) an inequality involving absolute values.
19. Sketch the graph of $(-\infty, -1] \cup [3, \infty)$ on the number line.

20. Find all real numbers x that satisfy the inequality $|3x - 1| > 7$. Write your solution using interval notation.
21. Solve the inequality $x^2 \geq -2x + 15$ and write your solution using interval notation.
22. Solve the inequality $x \leq 3 - \dfrac{6}{x+2}$ and write your solution using interval notation.

≡ Cartesian Plane

23. (Fill in the blank) If (a, b) is a point in the third quadrant, then (a, b) is a point in the _____ quadrant.
24. (Fill in the blank) The midpoint of the line segment from $P_1(2, -5)$ to $P_2(8, -9)$ is _____.
25. (Fill in the blanks) If $(-2, 6)$ is the midpoint of the line segment from $P_1(x_1, 3)$ to $P_2(8, y_2)$, then $x_1 =$ _____ and $y_2 =$ _____.
26. (Fill in the blanks) The point $(1, 5)$ is on a graph. Give the coordinates of another point on the graph if the graph is:
    (a) symmetric with respect to the x-axis. _____
    (b) symmetric with respect to the y-axis. _____
    (c) symmetric with respect to the origin. _____
27. (Fill in the blanks) The x- and y-intercepts of the graph of $|y| = 2x + 4$ are, respectively, _____ and _____.
28. In which quadrants of the Cartesian plane is the quotient x/y negative?
29. The y-coordinate of a point is 2. Find the x-coordinate of the point if the distance from the point to $(1, 3)$ is $\sqrt{26}$.
30. Find an equation of the circle for which $(-3, -4)$ and $(3, 4)$ are endpoints of a diameter.
31. If the points P_1, P_2, and P_3 are collinear as shown in FIGURE TY.1, then find an equation relating the distances $d(P_1, P_2)$, $d(P_2, P_3)$, and $d(P_1, P_3)$.

FIGURE TY.1 Graph for Problem 31

32. Which of the following equations best describes the circle given in FIGURE TY.2? The symbols a, b, c, d, and e stand for nonzero constants.
    (a) $ax^2 + by^2 + cx + dy + e = 0$
    (b) $ax^2 + ay^2 + cx + dy + e = 0$
    (c) $ax^2 + ay^2 + cx + dy = 0$
    (d) $ax^2 + ay^2 + c = 0$
    (e) $ax^2 + ay^2 + cx + e = 0$

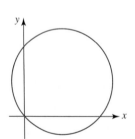

FIGURE TY.2 Graph for Problem 32

≡ Lines

33. (True/False) The lines $2x + 3y = 5$ and $-2x + 3y = 1$ are perpendicular. _____
34. (Fill in the blank) The lines $6x + 2y = 1$ and $kx - 9y = 5$ are parallel if $k =$ _____.
35. (Fill in the blank) A line with x-intercept $(-4, 0)$ and y-intercept $(0, 32)$ has slope _____.
36. (Fill in the blanks) The slope and the x- and y-intercepts of the line $2x - 3y + 18 = 0$ are, respectively, _____, _____, and _____.
37. (Fill in the blank) An equation of the line with slope -5 and y-intercept $(0, 3)$ is _____.
38. Find an equation of the line that passes through $(3, -8)$ and is parallel to the line $2x - y = -7$.

39. Find an equation of the line through the points $(-3, 4)$ and $(6, 1)$.
40. Find an equation of the line that passes through the origin and through the point of intersection of the graphs of $x + y = 1$ and $2x - y = 7$.
41. A tangent line to a circle at a point P on the circle is a line through P that is perpendicular to the line through P and the center of the circle. Find an equation of the tangent line L indicated in FIGURE TY.3.

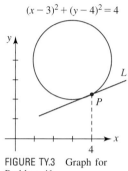

FIGURE TY.3 Graph for Problem 41

42. Match the given equation with the appropriate graph in FIGURE TY.4.
    (i) $x + y - 1 = 0$ (ii) $x + y = 0$ (iii) $x - 1 = 0$
    (iv) $y - 1 = 0$ (v) $10x + y - 10 = 0$ (vi) $-10x + y + 10 = 0$
    (vii) $x + 10y - 10 = 0$ (viii) $-x + 10y - 10 = 0$

(a)

(b)

(c)

(d)

(e)

(f)

(g)

(h)

FIGURE TY.4 Graphs for Problem 42

≡ Trigonometry

43. (True/False) $1 + \sec^2\theta = \tan^2\theta$. _____
44. (True/False) $\sin(2t) = 2\sin t$. _____
45. (Fill in the blank) The angle 240 degrees is equivalent to _____ radians.
46. (Fill in the blank) The angle $\pi/12$ radians is equivalent to _____ degrees.
47. (Fill in the blank) If $\tan t = 0.23$, $\tan(t + \pi) = $ _____.
48. Find $\cos t$ if $\sin t = \frac{1}{3}$ and the terminal side of the angle t lies in the second quadrant.
49. Find the values of the six trigonometric functions of the angle θ given in FIGURE TY.5.

FIGURE TY.5 Triangle for Problem 49

50. Express the lengths b and c in FIGURE TY.6 in terms of the angle θ.

FIGURE TY.6 Triangle for Problem 50

≡ Logarithms

51. Express the symbol k in the exponential statement $e^{(0.1)k} = 5$ as a logarithm.
52. Express the logarithmic statement $\log_{64} 4 = \frac{1}{3}$ as an equivalent exponential statement.
53. Express $\log_b 5 + 3\log_b 10 - \log_b 40$ as a single logarithm.
54. Use a calculator to evaluate $\dfrac{\log_{10} 13}{\log_{10} 3}$.
55. (Fill in the blank) $b^{3\log_b 10} = $ _____.
56. (True/False) $(\log_b x)(\log_b y) = \log_b(y^{\log_b x})$. _____

Essay

The Story of Calculus

by Roger Cooke
University of Vermont

Calculus is generally considered to be a creation of the seventeenth-century European mathematicians, with the main work having been done by Isaac Newton (1642–1727) and Gottfried Wilhelm Leibniz (1646–1711). This traditional view is correct in broad outline. Any large-scale theory, however, is a mosaic whose tiles were laid over a long period of time; and in any living theory new tiles are continually being laid. The strongest statement the historian dares to make is that a pattern became apparent at a certain time and place. Such is the case with calculus. We can say with some confidence that the main outlines of the subject appeared in the seventeenth century and that the pattern was made much clearer by the work of Newton and Leibniz. However, many of the essential principles of calculus were discovered as early as the time of Archimedes (287–211 BCE), and some of these same discoveries were made independently in China and Japan. Moreover if you dig deeply into the problems and methods of calculus, you will soon find yourself pursuing problems that lead into the modern areas of analytic function theory, differential geometry, and functions of a real variable. To change the metaphor from art to transportation, we can think of calculus as a large railroad station, where passengers arriving from many different places all come together for a brief time before setting out again for a variety of destinations. In the present essay we shall try to look in both directions from that station, to the sources, and to the destinations. Let us begin by describing the station itself.

Isaac Newton

Gottfried Leibniz

What Is Calculus? Calculus is traditionally divided into two parts, called *differential calculus* and *integral calculus*. Differential calculus investigates the properties of the comparative rates of change of variables that are linked by equations. For example, a fundamental result of differential calculus is that if $y = x^n$, then the rate of change of y with respect to x is nx^{n-1}. It turns out that when we think intuitively about certain phenomena—the motion of bodies, changes in temperature, growth of populations, and many others—we are led to postulate certain relations between these variables and their rates of change. These relations are written down in a form known as *differential equations*. Thus the primary purpose of studying differential calculus is to understand what rates of change are and how to write down differential equations. Integral calculus provides methods of recovering the original variables knowing their rates of change. The technique for doing so is called *integration*, and the primary purpose of studying integral calculus is to learn how to *solve* the differential equations that are provided by differential calculus.

These goals are often masked in calculus books, where differential calculus is used to find the maximum and minimum values of certain variables, and integral calculus is used to compute lengths, areas, and volumes. There are two reasons for emphasizing these applications in a textbook. First, the full use of calculus involving differential equations involves

xxiii

some rather elaborate theory that must be introduced gradually; meanwhile, the student must be shown *some* use for the techniques that are being put forth. Second, such problems were the source of the ideas that led to calculus; the uses we now make of the subject arose only after it was discovered.

In describing the problems that led to calculus and the problems that can be solved using calculus, we still have not pointed out the fundamental techniques that make calculus so much more powerful a tool of analysis than mere algebra and geometry. These techniques involve the use of what was once called *infinitesimal analysis.* The constructions and formulas of high school geometry and algebra all have a finite character. For example, to construct the tangent to a circle or to bisect an angle, you perform a finite number of operations with straightedge and compass. Although Euclid knew considerably more geometry than is found in modern high school courses, he, too, confined himself mostly to finite processes. Only in the limited context of the theory of proportion does he allow the infinite into his geometry, and even there it is surrounded by so much logical caution that the proofs involved are extraordinarily cumbersome and hard to read. This same situation occurs in algebra. In order to solve a polynomial equation, you perform a finite number of operations of addition, subtraction, multiplication, division, and root extraction. When the equation can be solved, the solution is expressed as a finite formula involving the coefficients.

These finite techniques, however, have a limited range of applicability. One cannot find the areas of most curved figures by a finite number of operations with straightedge and compass, nor can one solve most polynomial equations of degree five or higher using a finite number of algebraic operations. It was the desire to escape from the limitations of finite methods that led to the creation of calculus. We shall now look at some of the early attempts to develop techniques for handling the more difficult problems of geometry, after which we shall summarize the process by which calculus was worked out and finally exhibit some of the harvest it has provided.

The Geometric Sources of the Calculus One of the oldest mathematical problems is that of squaring the circle; that is, constructing a square equal in area to a given circle. It is now known that this problem cannot be solved by use of a finite number of applications of compass and straightedge. However, Archimedes discovered that if one could draw a spiral starting at the center of a circle that makes exactly one revolution before reaching the circle, then the tangent to that spiral at its point of intersection with the circle would form the hypotenuse of a right triangle with area exactly equal to the circle (see Figure 1). Thus if one could draw this spiral and its tangent, one could square the circle. Archimedes, however, was silent on the question of how one might draw this tangent.

We see here that one of the classical mathematical problems could be solved if only we could draw a certain curve and a tangent to it. This problem, and others like it, caused the purely mathematical problem of finding the tangent to a curve to become important. This

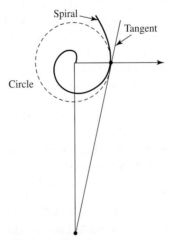

FIGURE 1 *The spiral of Archimedes.* The tangent at the end of the first turn and the two axes form a triangle with area equal to the circle about the origin through the point of tangency.

problem is the main source of differential calculus. The "infinitesimal" trick that allows the problem to be solved is to think of the tangent as the line determined by two points on the curve "infinitely close" together. Another way of saying the same thing is that an "infinitely short" piece of any curve is straight. The trouble is that it is hard to be precise about the meaning of the phrases "infinitely close" and "infinitely short."

Little progress was made on this problem until the invention of analytic geometry in the seventeenth century by Pierre de Fermat (1601–1665) and René Descartes (1596–1650). Once a curve could be represented by an equation, it became possible to say with more confidence what was meant by "infinitely close" points, at least for polynomial equations such as $y = x^2$. With algebraic symbolism to represent points on the curve, it was possible to consider two points on the curve with x-coordinates x_0 and x_1, so that $x_1 - x_0$ is the distance between the x-coordinates. When the equation of the curve was written at each of these points and one of the two equations subtracted from the other, one side of the resulting equation contained the factor $x_1 - x_0$, which could therefore be divided out. Thus if $y_0 = x_0^2$ and $y_1 = x_1^2$, then $y_1 - y_0 = x_1^2 - x_0^2 = (x_1 - x_0)(x_1 + x_0)$, and so $\frac{y_1 - y_0}{x_1 - x_0} = x_1 + x_0$. When $x_1 = x_0$, it follows that $y_1 = y_0$, and the expression $\frac{y_1 - y_0}{x_1 - x_0}$ has no meaning. However, the expression $x_1 + x_0$ has the perfectly definite value $2x_0$. Thus we can think of $2x_0$ as the ratio of the infinitely small difference in y, namely $y_1 - y_0$, to the infinitely small difference in x, namely $x_1 - x_0$, when the point (x_1, y_1) is infinitely close to the point (y_1, y_0) on the curve $y = x^2$. As you will learn in your study of calculus, this ratio gives enough information to draw the tangent line to the curve $y = x^2$.

The preceding argument is, except for small changes in notation, exactly the way Fermat found the tangent to a parabola. It was open to one logical objection, however: At one stage we divided both sides of an equation by $x_1 - x_0$, then at a later stage we decided that $x_1 - x_0 = 0$. Since division by zero is an illegal operation, we seem to be trying to eat our cake and have it, too. It took some time to find a convincing answer to this objection.

We have just seen that Archimedes was unable to solve the fundamental problem of differential calculus, drawing the tangent to a curve. Archimedes *was* able to solve some of the fundamental problems of integral calculus, however. In fact he found the volume of a sphere in an extremely ingenious way. He considered a cylinder containing a cone and a sphere and imagined this figure cut into infinitely thin slices. By looking at the areas of these sections of the cone, sphere, and cylinder, he was able to show how the cylinder would balance the cone and sphere if the figures were hung on opposite sides of a fulcrum. This balancing gave one relation among the three figures, and Archimedes already knew the volumes of the cone and cylinder; hence he was able to compute the volume of the sphere.

This argument illustrates the second infinitesimal technique that lies at the foundation of calculus: A volume can be regarded as a stack of plane figures, and an area can be regarded as a stack of line segments, in the sense that if every horizontal section of one region equals the same horizontal section of another region, then the two regions are equal. During the European Renaissance this principle came to be widely used under the name of the *method of indivisibles* for finding the areas and volumes of many figures. It is nowadays called *Cavalieri's Principle* after Bonaventura Cavalieri (1598–1647), who used it to prove many of the elementary formulas that now make up integral calculus. Cavalieri's Principle was also discovered in other lands where Euclid's work had never gone. The fifth-century Chinese mathematicians Zu Chongzhi and his son Zu Geng, for example, found the volume of a sphere using a technique very similar to Archimedes' method.

Thus we find mathematicians anticipating the integral calculus by using infinitesimal methods to find areas and volumes at a very early stage of geometry in both ancient Greece and China. Like the infinitesimal method of drawing tangents, however, this method of finding areas and volumes was open to objection. For example, the volume of each plane section of a figure is zero; how can a collection of zeros be put together to yield something that is not zero? Also, why doesn't the method work in one dimension? Consider the sections of a right triangle parallel to one of its legs. Each section intersects the hypotenuse and the other leg in congruent figures, namely one point each. Yet the hypotenuse and the other leg are not the same length. Objections like these were worrisome. The results obtained using these methods were so spectacular, however, that mathematicians preferred to take them on faith, continue to use them, and try to build the foundation under them later, just as a tree grows both roots and branches at the same time.

The Invention of the Calculus By the middle of the seventeenth century, many of the elementary techniques and facts of calculus were known, including methods for finding the tangents to simple curves and formulas for areas bounded by these curves. In other words, many of the formulas you will find in the early chapters of any calculus textbook were already known before Newton and Leibniz began their work. What was lacking until late in the seventeenth century, however, was the realization that these two kinds of problems are related to each other.

To see how the relation came to be discovered, we need to say a bit more about tangents. We mentioned above that to draw a tangent to a curve at a given point one needs to know how to find a second point on the line. In the early days of analytic geometry this second point was usually taken as the point at which the tangent intersects the x-axis. The projection onto the x-axis of the portion of the tangent between the point of tangency and the intersection with the axis was called the *subtangent*. In the study of tangents one very natural problem arose: *to reconstruct a curve, given the length of its subtangent at every point*. Through the study of this problem it came to be noticed that the ordinates of any curve are proportional to the area under a second curve whose ordinates are the lengths of the subtangents to the original curve. That result is the fundamental theorem of calculus. The honor of explicitly recognizing this relation goes to Isaac Barrow (1630–1677), who pointed it out in a book called *Lectiones Geometricae* in 1670. Barrow stated several theorems resembling the fundamental theorem of calculus. One of these is the following: *If a curve is drawn so that the ratio of its ordinate to its subtangent* [this ratio is precisely what is now called the derivative] *is proportional to the ordinate of a second curve, then the area under the second curve is proportional to the ordinate of the first.*

These relations provided a unifying principle for the large number of particular results on tangents and areas that had been obtained by the method of indivisibles in the early seventeenth century: To find the area under a curve, find a second curve for which the ratio of the ordinate to the subtangent equals the ordinate of the given curve. Then the ordinate of that second curve will give the area under the first curve.

At this point the calculus was ready to be born. It needed only someone to give systematic methods of computing tangents (actually subtangents) and inverting that process in order to find areas. This is the work actually performed by Newton and Leibniz. These two giants of mathematical creativity took quite different paths to their discoveries.

Newton's approach was algebraic and developed out of the problem of finding an efficient method of extracting the roots of a number. Although he had barely begun to study algebra in 1662, by 1665 Newton's reflections on the problem of extracting roots led him to the discovery of the infinite series now known as the binomial theorem; that is, the relation

$$(1 + x)^r = 1 + rx + \frac{r(r-1)}{2}x^2 + \frac{r(r-1)(r-2)}{1 \cdot 2 \cdot 3}x^3 + \cdots$$

By combining the binomial theorem with infinitesimal techniques, Newton was able to derive the basic formulas of differential and integral calculus. Central to Newton's approach was the use of infinite series to express the variables in question, and the fundamental problem that Newton did not solve was to establish that such series could be manipulated just like finite sums. Thus in a sense Newton drove the infinite away from one entrance to its burrow, only to find it showing its face in another.

Since he thought of variables as being physical quantities that change their value with time, Newton invented names for variables and their rates of change that reflected this intuition. According to Newton a *fluent* (x) is a moving or flowing quantity; its *fluxion* (x) is its rate of flow, what we now call its velocity or *derivative*. Newton expounded his results in 1671 in a treatise called *Fluxions* written in Latin; but this work was not published until an English version appeared in 1736. (The original Latin version was first published in 1742.)

Despite his notation and his arguments, which seem crude and inefficient today, the tremendous power of calculus shines through Newton's *Fluxions* in the solution of such difficult problems as finding the arc length of a curve. This "rectification" of a curve had been thought impossible, but Newton showed that one could find an infinite number of curves whose length could be expressed in finite terms.

Newton's approach to the calculus was algebraic, as we have just seen, and he inherited the fundamental theorem from Barrow. Leibniz, on the other hand, worked out the fundamental result on his own during the 1670s, and his approach was different from Newton's. Leibniz is considered the earliest pioneer of symbolic logic, and he had a much better appreciation than Newton of the importance of good symbolic notation. He invented the notation dx and dy that we still use today. For him dx was an abbreviation for "difference in x" and represented the difference between two infinitely close values of x. In other words, it expressed exactly what we had in mind above when we considered the infinitely small change $x_1 - x_0$. Leibniz thought of dx as an "infinitesimal" number, a number not zero, yet so small that no multiple of it could exceed any ordinary number. Not being zero, it could serve as the denominator in a fraction, and so dy/dx was the quotient of two infinitely small quantities. In this way he hoped to avoid the objections to the newly established method of finding tangents.

In the controversial technique of finding areas by adding up the sections, Leibniz also made a major contribution. Instead of thinking of an area [for example, the area under a curve $y = f(x)$] as a collection of line segments, he regarded it as the sum of the areas of "infinitely thin" rectangles of height $y = f(x)$ and infinitesimal base dx. Hence the difference between the area up to the point $x + dx$ and the area up to the point x was the infinitesimal difference in area $dA = f(x)\,dx$, and the total area was found by summing up these infinitesimal differences in area. Leibniz invented the elongated S (the integral sign $\int$) that is now universally used for expressing this summation process. Thus he would express the area under the curve $y = f(x)$ as $A = \int dA = \int f(x)\,dx$, and each part of this symbol expressed a simple and clear geometric idea.

With Leibniz's notation, Barrow's fundamental theorem of calculus merely says that the pair of equations

$$A = \int f(x)\,dx, \qquad dA = f(x)\,dx$$

are equivalent to each other. Because of what was just stated above, this equivalence is nearly obvious.

Both Newton and Leibniz had made huge advances in mathematics, and there was plenty of credit for both of them. It is unfortunate that the near coincidence of their work led to an acrimonious dispute over priority between their followers.

Some parts of the calculus, involving infinite series, had been invented in India in the fourteenth and fifteenth centuries. The late fifteenth-century Indian mathematician Jyesthadeva gave the series

$$\theta = r\left(\frac{\sin\theta}{\cos\theta} - \frac{\sin^3\theta}{3\cos^3\theta} + \frac{\sin^5\theta}{5\cos^5\theta} - \cdots\right)$$

for the length of an arc of a circle, proved this result, and explicitly stated that this series will converge only if θ is not larger than $45°$. If we write $\theta = \arctan x$, and use the fact that $\frac{\sin\theta}{\cos\theta} = \tan\theta = x$, this series becomes the standard series for $\arctan x$.

Likewise some infinite series were developed in Japan independently about the same time as in Europe. The Japanese mathematician Katahiro Takebe (1664–1739) found a series expansion equivalent to the series for the square of the arcsine function. He was considering the square of half the arc at height h in a circle of diameter d; this works out to be the function $f(h) = \left(\frac{d}{2}\arcsin\frac{h}{d}\right)^2$. Katahiro Takebe had no notation for the general term of a series, but he discovered patterns in the coefficients by computing the function geometrically at the particular value of $h = 0.000001$, $d = 10$ to a very large number of decimal places—more than fifty—and then using this extraordinary accuracy to refine the approximation by successively adding corrective terms. By proceeding in this way he was able to discern a pattern in the successive approximations, from which by extrapolation he was able to state the general term of the series:

$$f(h) = dh\left[1 + \sum_{n=1}^{\infty} \frac{2^{2n+1}(n!)^2}{(2n+2)!}\left(\frac{h}{d}\right)^n\right]$$

After Newton and Leibniz, there remained the problem of putting flesh on the skeleton these two geniuses had created. The majority of this work was completed by the Continental mathematicians, notably the circle around the Swiss mathematicians James and John Bernoulli ((1655–1705) and (1667–1748), respectively) and John Bernoulli's student the Marquis de l'Hôpital (1661–1704). These mathematicians and others worked out the familiar formulas for the derivatives and integrals of elementary functions that are found in textbooks today. The essential techniques of calculus were known by the early eighteenth century, and an eighteenth-century textbook such as Euler's *Introduction in analysin infinitorum* (1748), if translated into English, would look very much like a modern textbook.

The Legacy of the Calculus Having looked at the sources of calculus and the procedure by which it was constructed, let us now examine briefly the results it produced.

The calculus scored an amazing number of triumphs in its first two centuries. Dozens of previously obscure physical phenomena involving heat, fluid flow, celestial mechanics, elasticity, light, electricity, and magnetism turned out to have measurable properties whose relations could be described as differential equations. Physics was forever committed to speaking the language of calculus.

By no means were all of the mathematical problems arising from physics solved, however. For example, the area under a curve whose equation involved the square root of a cubic polynomial could not be found in terms of familiar elementary functions. Such integrals arose frequently in both geometry and physics, and came to be known as *elliptic integrals* because the problem of finding the length could be understood only when the real variable x is replaced by a complex variable $z = x + iy$. The reworking of the calculus in terms of complex variables led to many new and fascinating discoveries, which eventually came to be codified as a new branch of mathematics called analytic function theory.

The proper definition of integration remained a problem for some time. Integrals arose out of the use of infinitesimal processes to find areas and volumes. Should the integral be defined as a "sum of infinitesimal differences," or should it be defined as the reverse of differentiation? What functions can be integrated? Many definitions of integral were proposed in the nineteenth century, and the elaboration of these ideas has led to the subject now known as real analysis.

While the applications of calculus have moved on to more and more triumphs in an unending stream for the last three hundred years, its foundations lay in an unsatisfactory state for the first half of this period. The root of the difficulty was the meaning to be attached to Leibniz's dx. What was this quantity? How could it be neither positive nor zero? If zero, it could not be used as a denominator; if positive, then the equations in which it occurred were not truly equations. Leibniz believed that infinitesimals were real things, that areas and volumes could be synthesized by "adding up" their sections, as Zu Chongzhi, Archimedes, and others had done. Newton was less confident of the validity of infinitesimal methods and tried to justify his arguments in ways that would meet Euclidean standards of rigor. In his *Principia Mathematica* he wrote:

> These Lemmas are premised to avoid the tediousness of deducing involved demonstrations *ad absurdum,* according to the method of the ancient geometers. For demonstrations are shorter by the method of indivisibles; but because the hypothesis of indivisibles seems somewhat harsh, and therefore that method is reckoned less geometrical, I chose rather to reduce the demonstrations of the following Propositions to the first and last sums and ratios of evanescent quantities, that is, to the limits of those sums and rations... Therefore if hereafter I should happen to consider quantities as made up of particles, or should use little curved lines for right [straight] ones, I would not be understood to mean indivisibles, but evanescent divisible quantities...

... For those ultimate ratios with which quantities vanish are not truly the ratios of ultimate quantities, but limits towards which the ratios of quantities decreasing without limit do always converge; and to which they approach nearer than by any given difference, but never go beyond, nor in effect attain to, till the quantities are diminished *in infinitum*.

In this passage Newton was claiming that the lack of rigor involved in using infinitesimal arguments could be compensated for by the use of limits. His formulation of this concept in the passage just quoted is not so clear as one might wish, however. This lack of clarity led the philosopher Berkeley to refer contemptuously to fluxions as "ghosts of departed quantities." The advances achieved in physics using calculus, however, were so outstanding that for more than a century no one bothered to supply the extra rigor Newton alluded to (and physicists still don't bother with it!). A completely rigorous and systematic presentation of the calculus came only in the nineteenth century.

After the work of Augustin-Louis Cauchy (1789–1856) and Karl Weierstrass (1815–1896), the received view was that infinitesimals are merely heuristic in nature, and students were subjected to a rigorous "epsilon-delta" approach to limits. Somewhat surprisingly, however, in the twentieth century it was shown by Abraham Robinson (1918–1974) that a logically consistent model of the real numbers can be developed in which there are actual infinitesimals, just as Leibniz had believed. This new approach, called "nonstandard analysis," does not seem to be supplanting the now-traditional presentation of calculus, however.

Exercises

1. The kind of spiral considered by Archimedes in now named after him. An Archimedean spiral is the locus of a point moving at a constant speed along a ray rotating with constant angular speed about a fixed point. If the linear speed along the ray (the *radial* component of its velocity) is v, the point will be at a distance vt from the center of rotation (assuming that is where it starts) at time t. Suppose the angular speed of rotation of the ray is ω (radians per unit time). Given a circle of radius R, and a radial speed of u, what must ω be in order for the spiral to reach the circle at the end of its first turn? *Ans.* $\left(\frac{2\pi v}{R}\right)$

   The point will have a *circumferential* velocity $r\omega = vt\,\omega$. According to a principle enunciated in Aristotle's *Mechanics*, the actual velocity of the particle will be directed along the diagonal of a parallelogram (a rectangle in this case) having the two components as sides. Use this principle to show how to construct the tangent to the spiral (it will be the line containing the diagonal of this rectangle). Verify that sides of this rectangle are in the ratio $1 : 2\pi$. See Figure 1.

2. Figure 2 illustrates how Archimedes found the relation between the volumes of a sphere, cone, and cylinder. The diameter AB is doubled, making $BC = AB$. When the figure is revolved about this line, the circle generates a sphere, the triangle DBG generates a cone, and the rectangle $DEFG$ generates a cylinder. Prove the following facts.

   (a) If B is used as a fulcrum, the cylinder has the center K of the circle as its center of gravity, and therefore could all be concentrated there without changing the torque about B.

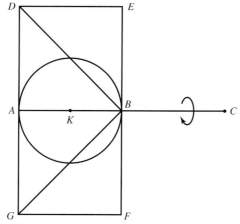

FIGURE 2 Section of Archimedes' sphere, cone, and cylinder

- **(b)** Each section of the cylinder perpendicular to the line *AB*, remaining in its present position, would exactly balance the same section of the cone plus the section of the sphere if both of the latter were moved to the point *C*.
- **(c)** Hence the cylinder concentrated at *K* would balance the cone and sphere concentrated at *C*.
- **(d)** Therefore the cylinder equals twice the sum of the cone and the sphere.
- **(e)** Since the cone is known to be one-third of the cylinder, it follows that the sphere must be one-sixth of it.
- **(f)** Since the volume of the cylinder is $8\pi r^3$.

3. The method by which Zu Chongzhi and Zu Geng found the volume of a sphere is as follows: Imagine the sphere as a ball tightly stuck inside the intersection of two cylinders at right angles to each other. The solid formed by the intersection of the two cylinders (called a *double umbrella* in Chinese) and containing the ball is then tightly fitted inside a cube whose edge equals the diameter of the sphere.

   From this description draw a section of the sphere within the double umbrella within the cube. Imagine this section being made parallel to the plane formed by the axes of the two cylinders and at a distance *h* below this plane. Verify the following facts.
   - **(a)** If the radius of the sphere is *r*, the circular section of it has diameter $2\sqrt{r^2 - h^2}$.
   - **(b)** Hence the square formed by the section of the double umbrella has area $4(r^2 - h^2)$, and so the area between the section of the cube and the section of the double umbrella is
     $$4r^2 - 4(r^2 - h^2) = 4h^2.$$
   - **(c)** The corresponding section of a pyramid whose base is the bottom of the cube and whose vertex is at the center of the sphere (or cube) would also have area $4h^2$. Hence the volume between the double umbrella and the cube is exactly the volume of such a pyramid plus its mirror image above the central plane. Conclude that the region between the double umbrella and the cube is one-third of the cube.
   - **(d)** Therefore the double umbrella occupies two-thirds of the volume of the cube; that is, its volume is $\frac{16}{3}r^3$.
   - **(e)** Each circular section of the sphere is inscribed in the corresponding square section of the double umbrella. Hence the circular section is $\frac{\pi}{4}$ of the section of the double umbrella.
   - **(f)** Therefore the volume of the sphere is $\frac{\pi}{4}$ of the volume of the double umbrella; that is, $\frac{4}{3}\pi r^3$.

4. Give an "infinitesimal" argument that the area of a sphere is three times its volume divided by its radius by imagining the sphere to be a collection of "infinitely thin" pyramids with vertices all stuck together at the origin. [*Hint*: Use the fact that the volume of a pyramid is one-third the area of its base times its altitude. Archimedes says that this is the reasoning that led him to discover the area of a sphere.]

Answers to the Essay's exercise questions can be found in the *Student Resource Manual* and the *Complete Solutions Manual*.

Single Variable
CALCULUS
Early Transcendentals
Fourth Edition

Chapter 1

Functions

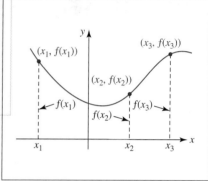

In This Chapter Have you ever heard remarks such as "Success is a function of hard work" and "Demand is a function of price"? The word *function* is often used to suggest a relationship or a dependence of one quantity on another. As you may already know, in mathematics the notion of a function has a similar but slightly more specialized interpretation.

Calculus is mostly about functions. Thus it is appropriate that we begin its study with a chapter devoted to a review of this important concept.

1.1 Functions and Graphs
1.2 Combining Functions
1.3 Polynomial and Rational Functions
1.4 Transcendental Functions
1.5 Inverse Functions
1.6 Exponential and Logarithmic Functions
1.7 From Words to Functions
    Chapter 1 in Review

1.1 Functions and Graphs

Introduction Using the objects and the persons around us, it is easy to make up a rule of correspondence that associates, or pairs, the members, or elements, of one set with the members of another set. For example, to each social security number there is a person, to each book there corresponds at least one author, to each state there is one governor, and so on. In mathematics we are interested in a special type of correspondence, a *single-valued correspondence*, called a **function**.

Definition 1.1.1 Function

A **function** from a set X to a set Y is a rule of correspondence that assigns to each element x in X exactly one element y in Y.

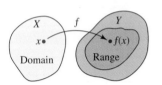

FIGURE 1.1.1 Domain and range of a function f

Terminology A function is usually denoted by a letter such as f, g, or h. We can then represent a function f from a set X to a set Y by the notation $f: X \to Y$. The set X is called the **domain** of f. The set of corresponding elements y in the set Y is called the **range** of the function. The unique element y in the range that corresponds to a selected element x in the domain X is called the **value** of the function at x, or the **image** of x, and is written $f(x)$. The latter symbol is read "f of x" or "f at x," and we write $y = f(x)$. It is also convenient at times to denote a function by $y = y(x)$. Notice in FIGURE 1.1.1 that the range of f need not be the entire set Y. Many instructors like to call an element x in the domain the *input* of the function and the corresponding element $f(x)$ in the range the *output* of the function. Since the value of y depends on the choice of x, y is called the **dependent variable**; x is called the **independent variable**. We will assume hereafter that the sets X and Y consist of real numbers; the function f is then called a **real-valued function of a real variable**.

Throughout the discussion and exercises of this text, functions are represented in several ways:

- *analytically*, that is, by a formula such as $f(x) = x^2$;
- *verbally*, that is, by a description in words;
- *numerically*, that is, by a table of numerical values; and
- *visually*, that is, by a graph.

EXAMPLE 1 Squaring Function

The rule for squaring a real number is given by the equation $f(x) = x^2$ or $y = x^2$. The values of f at $x = -5$ and $x = \sqrt{7}$ are obtained by replacing x, in turn, by the numbers -5 and $\sqrt{7}$:

$$f(-5) = (-5)^2 = 25 \quad \text{and} \quad f(\sqrt{7}) = (\sqrt{7})^2 = 7.$$ ■

EXAMPLE 2 Student and Desk Correspondence

A natural correspondence occurs between a set of 20 students and a set of, say, 25 desks in a classroom when each student selects and sits in a different desk. If the set of 20 students is the set X and the set of 25 desks is the set Y, then this correspondence is a function from the set X to the set Y provided no student sits in two desks at the same time. The set of 20 desks actually occupied by the students constitutes the range of the function. ■

Student/desk correspondence

Occasionally for emphasis we will write a function represented by a formula using parentheses in place of the symbol x. For example, we can write the squaring function $f(x) = x^2$ as

$$f(\ ) = (\ )^2. \tag{1}$$

Thus, if we wish to evaluate (1) at, say, $3 + h$, where h represents a real number, we put $3 + h$ into the parentheses and carry out the appropriate algebra:

See the *Resource Pages* for a review of binomial expansions. ▶

$$f(3 + h) = (3 + h)^2 = 9 + 6h + h^2.$$

If a function f is defined by means of a formula or an equation, then typically the domain of $y = f(x)$ is not expressly stated. We can usually deduce the domain of $y = f(x)$ either from the structure of the equation or from the context of the problem.

EXAMPLE 3 Domain and Range

In Example 1, since any real number x can be squared and the result x^2 is another real number, $f(x) = x^2$ is a function from R to R, that is, $f: R \to R$. In other words, the domain of f is the set R of real numbers. Using interval notation, we also write the domain as $(-\infty, \infty)$. Because $x^2 \geq 0$ for every real number x, it is easy to see that the range of f is the set of nonnegative real numbers or $[0, \infty)$. ∎

■ **Domain of a Function** As mentioned earlier, the domain of a function $y = f(x)$ that is defined by a formula is usually not specified. Unless stated or implied to the contrary, it is understood that

- *The **domain** of a function f is the largest subset of the set of real numbers for which $f(x)$ is a real number.*

This set is sometimes referred to as the **implicit domain** or **natural domain** of the function. For example, we cannot compute $f(0)$ for the **reciprocal function** $f(x) = 1/x$ since $1/0$ is not a real number. In this case we say that f is **undefined** at $x = 0$. Since every nonzero real number has a reciprocal, the domain of $f(x) = 1/x$ is the set of real numbers except 0. By the same reasoning, the function $g(x) = 1/(x^2 - 4)$ is not defined at either $x = -2$ or $x = 2$, and so its domain is the set of real numbers with -2 and 2 excluded. The **square root function** $h(x) = \sqrt{x}$ is not defined at $x = -1$ because $\sqrt{-1}$ is not a real number. In order for $h(x) = \sqrt{x}$ to be defined in the real number system we must require the **radicand**, in this case simply x, to be nonnegative. From the inequality $x \geq 0$ we see that the domain of the function h is the interval $[0, \infty)$. The domain of the **constant function** $f(x) = -1$ is the set of real numbers $(-\infty, \infty)$ and its range is the set consisting of the single number -1.

EXAMPLE 4 Domain and Range

Determine the domain and range of $f(x) = 4 + \sqrt{x - 3}$.

Solution The radicand $x - 3$ must be nonnegative. By solving the inequality $x - 3 \geq 0$ we get $x \geq 3$, and so the domain of f is $[3, \infty)$. Now, since the symbol $\sqrt{}$ denotes the nonnegative square root of a number, $\sqrt{x - 3} \geq 0$ for $x \geq 3$ and consequently $4 + \sqrt{x - 3} \geq 4$. The smallest value of $f(x)$ occurs at $x = 3$ and is $f(3) = 4 + \sqrt{0} = 4$. Moreover, because $x - 3$ and $\sqrt{x - 3}$ increase as x increases, we conclude that $y \geq 4$. Consequently the range of f is $[4, \infty)$. ∎

EXAMPLE 5 Domains of Two Functions

Determine the domain of

(a) $f(x) = \sqrt{x^2 + 2x - 15}$ (b) $g(x) = \dfrac{5x}{x^2 - 3x - 4}$.

Solution
(a) As in Example 4, the expression under the radical symbol—the radicand—must be nonnegative, that is, the domain of f is the set of real numbers x for which $x^2 + 2x - 15 \geq 0$ or $(x - 3)(x + 5) \geq 0$. The solution set of the inequality $(-\infty, -5] \cup [3, \infty)$ is also the domain of f.
(b) A function that is given by a fractional expression is not defined at the x-values for which its denominator is equal to 0. Since the denominator of $g(x)$ factors, $x^2 - 3x - 4 = (x + 1)(x - 4)$, we see that $(x + 1)(x - 4) = 0$ for $x = -1$ and $x = 4$. These are the *only* numbers for which g is not defined. Hence, the domain of the function g is the set of real numbers with $x = -1$ and $x = 4$ excluded. ∎

◂ In precalculus a quadratic inequality such as $(x - 3)(x + 5) \geq 0$ is often solved by means of a sign chart.

Using interval notation, the domain of g in part (b) of Example 5 can be written as $(-\infty, -1) \cup (-1, 4) \cup (4, \infty)$. As an alternative to this ungainly union of disjoint intervals, this domain can also be written using set-builder notation as $\{x \mid x \neq -1 \text{ and } x \neq 4\}$.

■ **Graphs** A function is often used to describe phenomena in fields such as science, engineering, and business. In order to interpret and utilize data, it is useful to display this data in the form of a graph. In the **rectangular** or **Cartesian coordinate system**, the graph of a function f is the graph of the set of ordered pairs $(x, f(x))$, where x is in the domain of f. In the xy-plane an ordered pair $(x, f(x))$ is a point, so that the graph of a function is a set of points. If a function is defined by an equation $y = f(x)$, then the graph of f is the graph of the equation. To obtain points on the graph of an equation $y = f(x)$, we judiciously choose numbers $x_1, x_2, x_3, \ldots$ in its domain, compute $f(x_1), f(x_2), f(x_3), \ldots$, plot the corresponding points $(x_1, f(x_1)), (x_2, f(x_2)), (x_3, f(x_3)), \ldots$, and then connect these points with a smooth curve (if possible). See FIGURE 1.1.2. Keep in mind that

- a value of x is a directed distance from the y-axis, and
- a function value $f(x)$ is a directed distance from the x-axis.

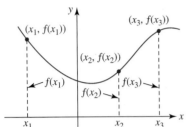

FIGURE 1.1.2 Points on the graph of an equation $y = f(x)$

A word about the figures in this text is in order. With a few exceptions, it is usually impossible to display the complete graph of a function, and so we often display only the more important features of the graph. In FIGURE 1.1.3(a), notice that the graph goes down on its left and right sides. Unless indicated to the contrary, we may assume that there are no major surprises beyond what we have shown and the graph simply continues in the manner indicated. The graph in Figure 1.1.3(a) indicates the so-called **end behavior** or **global behavior** of the function. If a graph terminates at either its right or left end, we will indicate this by a dot when clarity demands it. We will use a solid dot to represent the fact that the end point is included on the graph and an open dot to signify that the end point is not included on the graph.

■ **Vertical Line Test** From the definition of a function we know that for each x in the domain of f there corresponds only one value $f(x)$ in the range. This means a vertical line that intersects the graph of a function $y = f(x)$ (this is equivalent to choosing an x) can do so in at most one point. Conversely, if *every* vertical line that intersects a graph of an equation does so in at most one point, then the graph is the graph of a function. The last statement is called the **vertical line test** for a function. On the other hand, if *some* vertical line intersects a graph of an equation more than once, then the graph is not that of a function. See Figures 1.1.3(a)–(c). When a vertical line intersects a graph in several points, the same number x corresponds to different values of y in contradiction to the definition of a function.

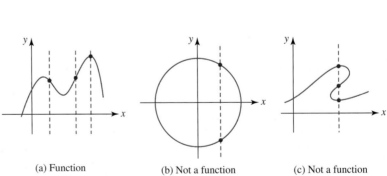

(a) Function (b) Not a function (c) Not a function

FIGURE 1.1.3 Vertical line test

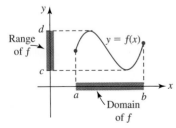

FIGURE 1.1.4 Domain and range interpreted graphically

If you have an accurate graph of a function $y = f(x)$, it is often possible to *see* the domain and range of f. In FIGURE 1.1.4 assume that the blue curve is the entire, or complete, graph of some function f. Then the domain of f is the interval $[a, b]$ on the x-axis and the range is the interval $[c, d]$ on the y-axis.

EXAMPLE 6 Example 4 Revisited

From the graph of $f(x) = 4 + \sqrt{x-3}$ given in FIGURE 1.1.5, we can see that the domain and range of f are, respectively, $[3, \infty)$ and $[4, \infty)$. This agrees with the results in Example 4. ∎

FIGURE 1.1.5 Graph of function f in Example 6

Intercepts To graph a function defined by an equation $y = f(x)$, it is usually a good idea to first determine whether the graph of f has any intercepts. Recall that all points on the y-axis are of the form $(0, y)$. Thus, if 0 is the domain of a function f, the **y-intercept** is the point on the y-axis whose y-coordinate is $f(0)$; in other words, $(0, f(0))$. See FIGURE 1.1.6(a). Similarly, all points on the x-axis have the form $(x, 0)$. This means that to find the **x-intercepts** of the graph of $y = f(x)$, we determine the values of x that make $y = 0$. That is, we must solve the equation $f(x) = 0$ for x. A number c for which $f(c) = 0$ is referred to as either a **zero** of the function f or a **root** (or **solution**) of the equation $f(x) = 0$. The *real* zeros of a function f are the x-coordinates of the x-intercepts of the graph of f. In Figure 1.1.6(b), we have illustrated a function that has three zeros x_1, x_2, and x_3 because $f(x_1) = 0$, $f(x_2) = 0$, and $f(x_3) = 0$. The corresponding three x-intercepts are the points $(x_1, 0)$, $(x_2, 0)$, and $(x_3, 0)$. Of course, the graph of the function may have no intercepts. This is illustrated in Figure 1.1.5.

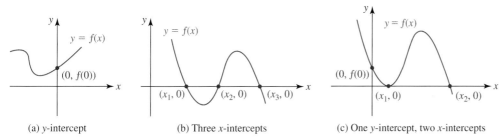

(a) y-intercept   (b) Three x-intercepts   (c) One y-intercept, two x-intercepts

FIGURE 1.1.6 Intercepts of the graph of a function f

A graph does not necessarily have to *cross* a coordinate axis at an intercept, a graph could simply touch, or be *tangent* to an axis. In Figure 1.1.6(c) the graph of $y = f(x)$ is tangent to the x-axis at $(x_1, 0)$.

EXAMPLE 7 Intercepts

Find, if possible, the x- and y-intercepts of the given function.

(a) $f(x) = x^2 + 2x - 2$ (b) $f(x) = \dfrac{x^2 - 2x - 3}{x}$

Solution

(a) Since 0 is in the domain of f, $f(0) = -2$ and so the y-intercept is the point $(0, -2)$. To obtain the x-intercepts we must determine whether f has any real zeros, that is, real solutions of the equation $f(x) = 0$. Since the left-hand side of the equation $x^2 + 2x - 2 = 0$ has no obvious factors, we use the quadratic formula to obtain $x = -1 \pm \sqrt{3}$. The x-intercepts are the points $(-1 - \sqrt{3}, 0)$ and $(-1 + \sqrt{3}, 0)$.

(b) Because 0 is not in the domain of f, the graph of f possesses no y-intercept. Now, since f is a fractional expression, the only way we can have $f(x) = 0$ is to have the numerator equal zero and the denominator not zero at the same number. Factoring the left-hand side of $x^2 - 2x - 3 = 0$ gives $(x + 1)(x - 3) = 0$. Therefore, the numbers -1 and 3 are the zeros of f. The x-intercepts are the points $(-1, 0)$ and $(3, 0)$. ∎

Piecewise-Defined Functions A function f may involve two or more expressions or formulas, with each formula defined on different parts of the domain of f. A function defined in this manner is called a **piecewise-defined function**. For example,

$$f(x) = \begin{cases} x^2, & x < 0 \\ x + 1, & x \geq 0 \end{cases}$$

is not two functions, but a single function in which the rule of correspondence is given in two pieces. In this case, one piece is used for the negative real numbers ($x < 0$) and the other part for the nonnegative numbers ($x \geq 0$); the domain of f is the union of the intervals $(-\infty, 0) \cup [0, \infty) = (-\infty, \infty)$. For example, since $-4 < 0$, the rule indicates that we square the number: $f(-4) = (-4)^2 = 16$; on the other hand, since $6 \geq 0$ we add 1 to the number: $f(6) = 6 + 1 = 7$.

EXAMPLE 8 Graph of a Piecewise-Defined Function

Consider the piecewise-defined function

$$f(x) = \begin{cases} -1, & x < 0 \\ 0, & x = 0 \\ x + 1, & x > 0. \end{cases} \qquad (2)$$

Although the domain of f consists of all real numbers $(-\infty, \infty)$, each piece of the function is defined on a different part of this domain. We draw

- the horizontal line $y = -1$ for $x < 0$,
- the point $(0, 0)$ for $x = 0$, and
- the line $y = x + 1$ for $x > 0$.

The graph is given in FIGURE 1.1.7.

FIGURE 1.1.7 Graph of piecewise-defined function in Example 8

Semicircles As shown in Figure 1.1.3(b), a circle is not the graph of a function. Actually, an equation such as $x^2 + y^2 = 9$ defines (at least) two functions of x. If we solve this equation for y in terms of x, we get $y = \pm\sqrt{9 - x^2}$. Because of the single-valued convention of the $\sqrt{}$ sign, both equations $y = \sqrt{9 - x^2}$ and $y = -\sqrt{9 - x^2}$ define functions. The first equation defines an **upper semicircle** and the second defines a **lower semicircle**. From the graphs shown in FIGURE 1.1.8, the domain of $y = \sqrt{9 - x^2}$ is $[-3, 3]$ and the range is $[0, 3]$; the domain and range of $y = -\sqrt{9 - x^2}$ are $[-3, 3]$ and $[-3, 0]$, respectively.

(a) Upper semicircle (b) Lower semicircle
FIGURE 1.1.8 These semicircles are graphs of functions

Absolute-Value Function The function $f(x) = |x|$, called the **absolute-value function**, appears frequently in the discussion of subsequent chapters. The domain of f is the set of all real numbers $(-\infty, \infty)$ and its range is $[0, \infty)$. In other words, for any real number x, the function values $f(x)$ are nonnegative. For example,

$$f(3) = |3| = 3, \quad f(0) = |0| = 0, \quad f\left(-\frac{1}{2}\right) = \left|-\frac{1}{2}\right| = -\left(-\frac{1}{2}\right) = \frac{1}{2}.$$

By the definition of the absolute value of x, we see that f is a piecewise-defined function consisting of two pieces

$$f(x) = |x| = \begin{cases} -x, & \text{if } x < 0 \\ x, & \text{if } x \geq 0. \end{cases} \qquad (3)$$

Its graph, shown in FIGURE 1.1.9(a), consists of two perpendicular half lines. Since $f(x) \geq 0$ for all x, another way of graphing (3) is simply to sketch the line $y = x$ and then reflect in the x-axis that portion of the line that is below the x-axis. See Figure 1.1.9(b).

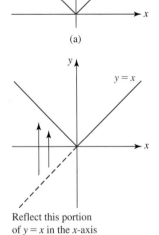

FIGURE 1.1.9 Absolute-value function (3)

■ **Greatest Integer Function** We consider next a piecewise-defined function f called the **greatest integer function**. This function, which has many notations, will be denoted here by $f(x) = \lfloor x \rfloor$ and is defined by the rule

$$\lfloor x \rfloor = n, \quad \text{where } n \text{ is an integer satisfying } n \leq x < n + 1. \tag{4}$$

◀ The greatest integer function is also written as $f(x) = [\![x]\!]$.

Translated into words, (4) means that

- *The function value $f(x)$ is the greatest integer n that is less than or equal to x.*

For example,

$$f(-1.5) = -2, \, f(0.4) = 0, \, f(\pi) = 3, \, f(5) = 5,$$

and so on. The domain of f is the set of real numbers and consists of the union of an infinite number of disjoint intervals; in other words, $f(x) = \lfloor x \rfloor$ is a piecewise-defined function given by

$$f(x) = \lfloor x \rfloor = \begin{cases} \vdots \\ -2, & -2 \leq x < -1 \\ -1, & -1 \leq x < 0 \\ 0, & 0 \leq x < 1 \\ 1, & 1 \leq x < 2 \\ 2, & 2 \leq x < 3 \\ \vdots \end{cases} \tag{5}$$

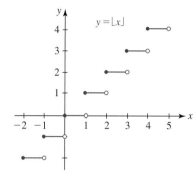

FIGURE 1.1.10 Greatest integer function

The range of f is the set of integers. The portion of the graph of f on the closed interval $[-2, 5]$ is given in FIGURE 1.1.10.

In computer science the greatest integer function is known as the **floor function**. A related function called the **ceiling function*** $g(x) = \lceil x \rceil$ is defined to be the least integer n that is greater than or equal to x. See Problems 57–59 in Exercises 1.1.

■ **A Mathematical Model** It is often desirable to describe the behavior of some real-life system or phenomenon, whether physical, sociological, or even economic, in mathematical terms. The mathematical description of a system or a phenomenon is called a **mathematical model** and can be as complicated as hundreds of simultaneous equations or as simple as a single function. We conclude this section with a real-world illustration of a piecewise-defined function called the *postage stamp function*. This function is similar to $f(x) = \lfloor x \rfloor$ in that both are examples of *step functions*; each function is constant on an interval and then jumps to another constant value on the next abutting interval.

As of this writing, the United States Postal Service (USPS) first-class mailing rates for a letter in a standard-size envelope depends on its weight in ounces:

$$\text{Postage} = \begin{cases} \$0.42, & 0 < \text{weight} \leq 1 \text{ ounce} \\ \$0.59, & 1 < \text{weight} \leq 2 \text{ ounces} \\ \$0.76, & 2 < \text{weight} \leq 3 \text{ ounces} \\ \vdots \\ \$2.87, & 12 < \text{weight} \leq 13 \text{ ounces.} \end{cases} \tag{6}$$

The rule in (6) is a function P consisting of 14 pieces (letters over 13 ounces are sent priority mail). A function value $P(w)$ is one of 14 constants; the constant changes depending on the weight w (in ounces) of the letter.† For example,

$$P(0.5) = \$0.42, \, P(1.7) = \$0.59, \, P(2.2) = \$0.76, \, P(2.9) = \$0.76, \text{ and } P(12.1) = \$2.87.$$

The domain of the function P is the union of the intervals:

$$(0, 1] \cup (1, 2] \cup (2, 3] \cup \cdots \cup (12, 13] = (0, 13].$$

---

*The floor and ceiling functions and their notation are due to the noted Canadian computer scientist Kenneth E. Iverson (1920–2004).

†Not shown in (6) is the fact that the postage of a letter whose weight falls in the interval (3, 4] is determined by whether its weight is in (3, 3.5] or in (3.5, 4]. This is the only interval subdivided in this manner.

$f(x)$ NOTES FROM THE CLASSROOM

When sketching the graph of a function, you should never resort to plotting a lot of points by hand. That is something a graphing calculator or a computer algebra system (CAS) does so well. On the other hand, you should not become dependent on a calculator to obtain a graph. Believe it or not, there are calculus instructors who do not allow the use of graphing calculators on quizzes or tests. Usually there is no objection to your using calculators or computers as an aid in checking homework problems, but in the classroom, instructors want to see the product of your own mind, namely, the ability to analyze. So you are strongly encouraged to develop your graphing skills to the point where you are able to quickly sketch by hand the graph of a function from a basic familiarity of types of functions and by plotting a minimum of well-chosen points.

Exercises 1.1   Answers to selected odd-numbered problems begin on page ANS-2.

≡ Fundamentals

In Problems 1–6, find the indicated function values.

1. If $f(x) = x^2 - 1$; $f(-5), f(-\sqrt{3}), f(3),$ and $f(6)$
2. If $f(x) = -2x^2 + x$; $f(-5), f(-\frac{1}{2}), f(2),$ and $f(7)$
3. If $f(x) = \sqrt{x+1}$; $f(-1), f(0), f(3),$ and $f(5)$
4. If $f(x) = \sqrt{2x+4}$; $f(-\frac{1}{2}), f(\frac{1}{2}), f(\frac{5}{2}),$ and $f(4)$
5. If $f(x) = \dfrac{3x}{x^2+1}$; $f(-1), f(0), f(1),$ and $f(\sqrt{2})$
6. If $f(x) = \dfrac{x^2}{x^3-2}$; $f(-\sqrt{2}), f(-1), f(0),$ and $f(\frac{1}{2})$

In Problems 7 and 8, find
$$f(x), f(2a), f(a^2), f(-5x), f(2a+1), f(x+h)$$
for the given function f and simplify as much as possible.

7. $f(\ ) = -2(\ )^2 + 3(\ )$
8. $f(\ ) = (\ )^3 - 2(\ )^2 + 20$
9. For what values of x is $f(x) = 6x^2 - 1$ equal to 23?
10. For what values of x is $f(x) = \sqrt{x-4}$ equal to 4?

In Problems 11–26, find the domain of the given function f.

11. $f(x) = \sqrt{4x-2}$
12. $f(x) = \sqrt{15-5x}$
13. $f(x) = \dfrac{10}{\sqrt{1-x}}$
14. $f(x) = \dfrac{2x}{\sqrt{3x-1}}$
15. $f(x) = \dfrac{2x-5}{x(x-3)}$
16. $f(x) = \dfrac{x}{x^2-1}$
17. $f(x) = \dfrac{1}{x^2-10x+25}$
18. $f(x) = \dfrac{x+1}{x^2-4x-12}$
19. $f(x) = \dfrac{x}{x^2-x+1}$
20. $f(x) = \dfrac{x^2-9}{x^2-2x-1}$
21. $f(x) = \sqrt{25-x^2}$
22. $f(x) = \sqrt{x(4-x)}$
23. $f(x) = \sqrt{x^2-5x}$
24. $f(x) = \sqrt{x^2-3x-10}$
25. $f(x) = \sqrt{\dfrac{3-x}{x+2}}$
26. $f(x) = \sqrt{\dfrac{5-x}{x}}$

In Problems 27–30, determine whether the graph in the figure is the graph of a function.

27.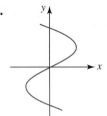

FIGURE 1.1.11   Graph for Problem 27

28.

FIGURE 1.1.12   Graph for Problem 28

29.

FIGURE 1.1.13   Graph for Problem 29

30.

FIGURE 1.1.14   Graph for Problem 30

In Problems 31–34, use the graph of the function f given in the figure to find its domain and range.

31.

FIGURE 1.1.15   Graph for Problem 31

32.

FIGURE 1.1.16   Graph for Problem 32

33.

FIGURE 1.1.17   Graph for Problem 33

34.

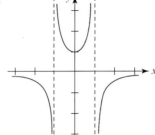

FIGURE 1.1.18 Graph for Problem 34

In Problems 35–44, find the *x*- and *y*-intercepts, if any, of the graph of the given function *f*. Do not graph.

35. $f(x) = \frac{1}{2}x - 4$

36. $f(x) = x^2 - 6x + 5$

37. $f(x) = 4(x - 2)^2 - 1$

38. $f(x) = (2x - 3)(x^2 + 8x + 16)$

39. $f(x) = x^3 - x^2 - 2x$

40. $f(x) = x^4 - 1$

41. $f(x) = \dfrac{x^2 + 4}{x^2 - 16}$

42. $f(x) = \dfrac{x(x + 1)(x - 6)}{x + 8}$

43. $f(x) = \dfrac{3}{2}\sqrt{4 - x^2}$

44. $f(x) = \dfrac{1}{2}\sqrt{x^2 - 2x - 3}$

In Problems 45 and 46, use the graph of the function f given in the figure to estimate the values of $f(-3), f(-2), f(-1), f(1)$, $f(2)$, and $f(3)$. Estimate the *y*-intercept.

45.

FIGURE 1.1.19 Graph for Problem 45

46.

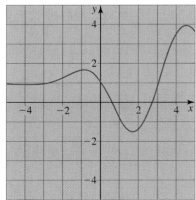

FIGURE 1.1.20 Graph for Problem 46

In Problems 47 and 48, use the graph of the function f given in the figure to estimate the values of $f(-2), f(-1.5), f(0.5)$, $f(1), f(2)$, and $f(3.2)$. Estimate the *x*-intercepts.

47.

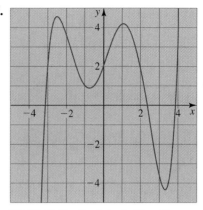

FIGURE 1.1.21 Graph for Problem 47

48.

FIGURE 1.1.22 Graph for Problem 48

In Problems 49 and 50, find two functions $y = f_1(x)$ and $y = f_2(x)$ defined by the given equation. Find the domain of the functions f_1 and f_2.

49. $x = y^2 - 5$

50. $x^2 - 4y^2 = 16$

51. Some of the functions that you will encounter later on in this text will have the set of positive integers n as their domain. The **factorial function** $f(n) = n!$ is defined as the product of the first n positive integers, that is,

$$f(n) = n! = 1 \cdot 2 \cdot 3 \cdots (n - 1) \cdot n.$$

(a) Evaluate $f(2), f(3), f(5)$, and $f(7)$.
(b) Show that $f(n + 1) = f(n) \cdot (n + 1)$.
(c) Simplify $f(5)/f(4)$ and $f(7)/f(5)$.
(d) Simplify $f(n + 3)/f(n)$.

52. Another function of a positive integer n gives the sum of the first n squared positive integers:

$$S(n) = \frac{1}{6}n(n + 1)(2n + 1).$$

(a) Find the value of the sum $1^2 + 2^2 + \cdots + 99^2 + 100^2$.
(b) Find n such that $300 < S(n) < 400$. [*Hint*: Use a calculator.]

Think About It

53. Determine an equation of a function $y = f(x)$ whose domain is
(a) $[3, \infty)$
(b) $(3, \infty)$.

54. Determine an equation of a function $y = f(x)$ whose range is
(a) $[3, \infty)$
(b) $(3, \infty)$.

55. From the graph of $f(x) = -x^2 + 2x + 3$ given in FIGURE 1.1.23 determine the range and domain of the function $g(x) = \sqrt{f(x)}$. Explain your reasoning in one or two sentences.

FIGURE 1.1.23 Graph for Problem 55

56. Let P denote any point $(x, f(x))$ on the graph of a function f. Suppose that the line segments PT and PS are perpendicular to the x- and y-axes, respectively. Let M_1, M_2, and M_3 be, in turn, the midpoints of PT, PS, and ST as shown in FIGURE 1.1.24. Find a function that describes the path of the points M_1. Repeat for the midpoints M_2 and M_3.

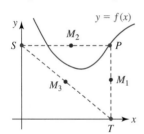

FIGURE 1.1.24 Graph for Problem 56

57. On page 7 we saw that the **ceiling function** $g(x) = \lceil x \rceil$ is defined to be the least integer n that is greater than or equal to x. Fill in the blanks.

$$g(x) = \lceil x \rceil = \begin{cases} \vdots & \\ \underline{\quad}, & -3 < x \leq -2 \\ \underline{\quad}, & -2 < x \leq -1 \\ \underline{\quad}, & -1 < x \leq 0 \\ \underline{\quad}, & 0 < x \leq 1 \\ \underline{\quad}, & 1 < x \leq 2 \\ \underline{\quad}, & 2 < x \leq 3 \\ \vdots & \end{cases}$$

58. Graph the ceiling function $g(x) = \lceil x \rceil$ defined in Problem 57.

59. The piecewise-defined function

$$\text{int}(x) = \begin{cases} \lfloor x \rfloor, & x \geq 0 \\ \lceil x \rceil, & x < 0 \end{cases}$$

is called the **integer function**. Graph $\text{int}(x)$.

60. Discuss how to graph the function $f(x) = |x| + |x - 3|$. Carry out your ideas.

In Problems 61 and 62, describe in words how the graphs of the given functions differ.

61. $f(x) = \dfrac{x^2 - 9}{x - 3}$,

$$g(x) = \begin{cases} \dfrac{x^2 - 9}{x - 3}, & x \neq 3 \\ 4, & x = 3 \end{cases}, \quad h(x) = \begin{cases} \dfrac{x^2 - 9}{x - 3}, & x \neq 3 \\ 6, & x = 3 \end{cases}$$

62. $f(x) = \dfrac{x^4 - 1}{x^2 - 1}$,

$$g(x) = \begin{cases} \dfrac{x^4 - 1}{x - 1}, & x \neq 1 \\ 0, & x = 1 \end{cases}, \quad h(x) = \begin{cases} \dfrac{x^4 - 1}{x^2 - 1}, & x \neq 1 \\ 2, & x = 1 \end{cases}$$

1.2 Combining Functions

Introduction Two functions f and g can be combined in several ways to create new functions. In this section we will examine two such ways in which functions can be combined: through arithmetic operations, and through the operation of function composition.

Power Functions A function of the form

$$f(x) = x^n \tag{1}$$

is called a **power function**. In this section we consider n to be a rational real number. The domain of a power function depends on the power n. For example, for $n = 2$, $n = \frac{1}{2}$, and $n = -1$, respectively,

- the domain of $f(x) = x^2$ is the set R of real numbers or $(-\infty, \infty)$,
- the domain of $f(x) = x^{1/2} = \sqrt{x}$ is $[0, \infty)$,
- the domain of $f(x) = x^{-1} = \dfrac{1}{x}$ is the set R of real numbers except $x = 0$.

Simple power functions, or modified versions of these functions, occur so often in problems in calculus that you do not want to spend valuable time plotting their graphs. We suggest that you know (memorize) the short catalogue of graphs of power functions given in FIGURE 1.2.1. You should recognize that the graph in part (a) of Figure 1.2.1 is a **line** and the graph in part (b) is a **parabola**.

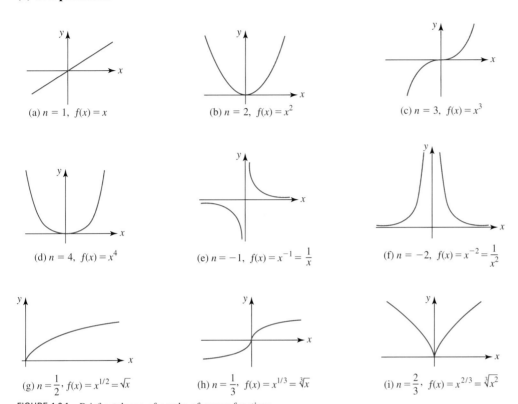

FIGURE 1.2.1 Brief catalogue of graphs of power functions

■ **Arithmetic Combinations** Two functions can be combined through the familiar four arithmetic operations of addition, subtraction, multiplication, and division.

Definition 1.2.1 Arithmetic Combinations

If f and g are two functions, then the **sum** $f + g$, the **difference** $f - g$, the **product** fg, and the **quotient** f/g are defined as follows:

$$(f + g)(x) = f(x) + g(x), \tag{2}$$

$$(f - g)(x) = f(x) - g(x), \tag{3}$$

$$(fg)(x) = f(x)g(x), \tag{4}$$

$$\left(\frac{f}{g}\right)(x) = \frac{f(x)}{g(x)}, \text{ provided } g(x) \neq 0. \tag{5}$$

■ **Domain of an Arithmetic Combination** When combining two functions arithmetically it is necessary that both f and g be defined at a same number x. Hence the **domain** of the functions $f + g$, $f - g$, and fg is the set of real numbers that are *common* to both domains, that is, the domain is the *intersection* of the domain of f with the domain of g. In the case of the quotient f/g, the domain is also the intersection of the two domains, *but* we must also exclude any values of x for which the denominator $g(x)$ is zero. In other words, if the domain of f is the set X_1 and the domain of g is the set X_2, then the domain of $f + g$, $f - g$, and fg is $X_1 \cap X_2$, and the domain of f/g is the set $\{x \mid x \in X_1 \cap X_2, g(x) \neq 0\}$.

EXAMPLE 1 Sum of Two Power Functions

We have already seen that the domain of $f(x) = x^2$ is the set R of real numbers or $(-\infty, \infty)$ and that the domain of $g(x) = \sqrt{x}$ is $[0, \infty)$. Therefore, the domain of the sum

$$f(x) + g(x) = x^2 + \sqrt{x}$$

is the intersection of the two domains: $(-\infty, \infty) \cap [0, \infty) = [0, \infty)$. ∎

Polynomial Functions Many of the functions that we work with in calculus are constructed by performing arithmetic operations on power functions. Of special interest are the power functions (1) where n is a nonnegative integer. For $n = 0, 1, 2, 3, \ldots$, the function $f(x) = x^n$ is called a **single-term polynomial function**. Using the arithmetic operations of addition, subtraction, and multiplication we can build polynomial functions with many terms. For example, if $f_1(x) = x^3$, $f_2(x) = x^2$, $f_3(x) = x$, and $f_4(x) = 1$, then

$$f_1(x) - f_2(x) + f_3(x) + f_4(x) = x^3 - x^2 + x + 1.$$

In general, a **polynomial function** $y = f(x)$ is a function of the form

$$f(x) = a_n x^n + a_{n-1} x^{n-1} + \cdots + a_2 x^2 + a_1 x + a_0, \tag{6}$$

where n is a nonnegative integer and the coefficients a_i, $i = 0, 1, \ldots, n$ are real numbers. The **domain** of any polynomial function f is the set of all real numbers $(-\infty, \infty)$. The following functions are *not* polynomials:

$$\underset{\underset{y = 5x^2 - 3x^{-1}}{\uparrow}}{\text{not a nonnegative integer}} \quad \text{and} \quad \underset{\underset{y = 2x^{1/2} - 4.}{\uparrow}}{\text{not a nonnegative integer}}$$

EXAMPLE 2 Sum, Difference, Product, and Quotient

Consider the polynomial functions $f(x) = x^2 + 4x$ and $g(x) = x^2 - 9$.

(a) From (2)–(4) of Definition 1.2.1 we can produce three new polynomial functions:

$$(f + g)(x) = f(x) + g(x) = (x^2 + 4x) + (x^2 - 9) = 2x^2 + 4x - 9,$$
$$(f - g)(x) = f(x) - g(x) = (x^2 + 4x) - (x^2 - 9) = 4x + 9,$$
$$(fg)(x) = f(x)g(x) = (x^2 + 4x)(x^2 - 9) = x^4 + 4x^3 - 9x^2 - 36x.$$

(b) Finally, from (5) of Definition 1.2.1,

$$\left(\frac{f}{g}\right)(x) = \frac{f(x)}{g(x)} = \frac{x^2 + 4x}{x^2 - 9}.$$
∎

Notice in Example 2, since $g(-3) = 0$ and $g(3) = 0$, the domain of the quotient $(f/g)(x)$ is $(-\infty, \infty)$ with $x = 3$ and $x = -3$ excluded, in other words, the domain of $(f/g)(x)$ is the union of three intervals: $(-\infty, -3) \cup (-3, 3) \cup (3, \infty)$.

Rational Functions The function in part (b) of Example 2 is an example of a rational functions. In general, a **rational function** $y = f(x)$ is a function of the form

▶ Polynomial and rational functions will be discussed in greater detail in Section 1.3.

$$f(x) = \frac{p(x)}{q(x)}, \tag{7}$$

where p and q are polynomial functions. For example, the functions

$$y = \frac{x}{x^2 + 5}, \quad y = \underset{\underset{\text{polynomial}}{\uparrow}}{\overset{\overset{\text{polynomial}}{\downarrow}}{\frac{x^3 - x + 7}{x + 3}}}, \quad y = \frac{1}{x},$$

are rational functions. The function

$$y = \frac{\sqrt{x}}{x^2 - 1} \quad \leftarrow \text{not a polynomial}$$

is not a rational function.

■ **Composition of Functions** Another method of combining functions f and g is called **function composition**. To illustrate the idea, let us suppose that for a given x in the domain of g the function value $g(x)$ is a number in the domain of the function f. This means we are able to evaluate f at $g(x)$, in other words, $f(g(x))$. For example, suppose $f(x) = x^2$ and $g(x) = x + 2$. Then for $x = 1$, $g(1) = 3$, and since 3 is the domain of f, we can write $f(g(1)) = f(3) = 3^2 = 9$. Indeed, for these two particular functions it turns out that we can evaluate f at any function value $g(x)$, that is,

$$f(g(x)) = f(x + 2) = (x + 2)^2.$$

The resulting function, called the **composition of f and g**, is defined next.

Definition 1.2.2 Function Composition

If f and g are two functions, then the **composition of f and g**, denoted by $f \circ g$, is the function defined by

$$(f \circ g)(x) = f(g(x)). \tag{8}$$

The **composition of g and f**, denoted by $g \circ f$, is the function defined by

$$(g \circ f)(x) = g(f(x)). \tag{9}$$

EXAMPLE 3 Two Compositions

If $f(x) = x^2 + 3x$ and $g(x) = 2x^2 + 1$ find

(a) $(f \circ g)(x)$ and   (b) $(g \circ f)(x)$.

Solution

(a) For emphasis we replace x by the set of parentheses ( ) and write f in the form $f(x) = (\ )^2 + 3(\ )$. Thus, to evaluate $(f \circ g)(x)$ we fill each set of parentheses with $g(x)$. We find

$$\begin{aligned}(f \circ g)(x) = f(g(x)) &= f(2x^2 + 1) \\ &= (2x^2 + 1)^2 + 3(2x^2 + 1) \\ &= 4x^4 + 4x^2 + 1 + 3 \cdot 2x^2 + 3 \cdot 1 \\ &= 4x^4 + 10x^2 + 4.\end{aligned}$$

(b) In this case write g in the form $g(x) = 2(\ )^2 + 1$. Then

$$\begin{aligned}(g \circ f)(x) = g(f(x)) &= g(x^2 + 3x) \\ &= 2(x^2 + 3x)^2 + 1 \\ &= 2(x^4 + 6x^3 + 9x^2) + 1 \\ &= 2x^4 + 12x^3 + 18x^2 + 1.\end{aligned}$$

Parts (a) and (b) of Example 3 illustrate that function composition is not commutative. That is, in general

$$f \circ g \neq g \circ f.$$

EXAMPLE 4 Writing a Function as a Composition

Express $F(x) = \sqrt{6x^3 + 8}$ as the composition of two functions f and g.

Solution If we define f and g as $f(x) = \sqrt{x}$ and $g(x) = 6x^3 + 8$, then

$$F(x) = (f \circ g)(x) = f(g(x)) = f(6x^3 + 8) = \sqrt{6x^3 + 8}.$$

There are other solutions to Example 4. For instance, if the functions f and g are defined by $f(x) = \sqrt{6x + 8}$ and $g(x) = x^3$, then observe $(f \circ g)(x) = f(x^3) = \sqrt{6x^3 + 8}$.

❚ **Domain of a Composition** To evaluate the composition $(f \circ g)(x) = f(g(x))$ the number $g(x)$ must be in the domain of f. For example, the domain of $f(x) = \sqrt{x}$ is $[0, \infty)$ and the domain of $g(x) = x - 2$ is the set of real numbers $(-\infty, \infty)$. Observe, we cannot evaluate $f(g(1))$ because $g(1) = -1$ and -1 is not in the domain of f. In order to substitute $g(x)$ into $f(x)$, $g(x)$ must satisfy the inequality that defines the domain of f, namely, $g(x) \geq 0$. This last inequality is the same as $x - 2 \geq 0$ or $x \geq 2$. The domain of the composition $f(g(x)) = \sqrt{g(x)} = \sqrt{x - 2}$ is $[2, \infty)$, which is only a portion of the original domain $(-\infty, \infty)$ of g. In general, the **domain of the composition** $f \circ g$ is the set of the numbers x in the domain of g such that $g(x)$ is in the domain of f.

For a constant $c > 0$, the functions defined by $y = f(x) + c$ and $y = f(x) - c$ are the *sum* and *difference* of the function $f(x)$ and the constant function $g(x) = c$. The function $y = cf(x)$ is the *product* of $f(x)$ and the constant function $g(x) = c$. The functions defined by $y = f(x + c)$, $y = f(x - c)$, and $y = f(cx)$ are *compositions* of $f(x)$ with the polynomial functions $g(x) = x + c$, $g(x) = x - c$, and $g(x) = cx$, respectively. As we see next, the graph of each of these is either a **rigid** or **nonrigid transformation** of the graph of $y = f(x)$.

❚ **Rigid Transformations** A **rigid transformation** of a graph is one that changes only the *position* of the graph in the xy-plane but not its shape. For the graph of a function $y = f(x)$ we examine four kinds of shifts or translations.

Translations

Suppose $y = f(x)$ is a function and c is a positive constant. Then the graph of

- $y = f(x) + c$ is the graph of f shifted vertically **up** c units,
- $y = f(x) - c$ is the graph of f shifted vertically **down** c units,
- $y = f(x + c)$ is the graph of f shifted horizontally to the **left** c units,
- $y = f(x - c)$ is the graph of f shifted horizontally to the **right** c units.

Consider the graph of a function $y = f(x)$ given in FIGURE 1.2.2. Vertical and horizontal shifts of this graph are the graphs in red in parts (a)–(d) of FIGURE 1.2.3. If (x, y) is a point on the graph of $y = f(x)$ and the graph of f is shifted, say, upward by $c > 0$ units, then $(x, y + c)$ is a point on the new graph. In general, the x-coordinates do not change as a result of a vertical shift. See Figures 1.2.3(a) and 1.2.3(b). Similarly, in a horizontal shift the y-coordinates of points on the shifted graph are the same as on the original graph. See Figures 1.2.3(c) and 1.2.3(d).

FIGURE 1.2.2 Graph of $y = f(x)$

(a) Vertical shift up

(b) Vertical shift down

(c) Horizontal shift left

(d) Horizontal shift right

FIGURE 1.2.3 Vertical and horizontal shifts of the graph of $y = f(x)$ by an amount $c > 0$

EXAMPLE 5 Shifted Graphs

The graphs of $y = x^2 + 1$, $y = x^2 - 1$, $y = (x + 1)^2$, and $y = (x - 1)^2$ are obtained from the graph of $f(x) = x^2$ in FIGURE 1.2.4(a) by shifting this graph, in turn, 1 unit up (Figure 1.2.4(b)), 1 unit down (Figure 1.2.4(c)), 1 unit to the left (Figure 1.2.4(d)), and 1 unit to the right (Figure 1.2.4(e)).

(a) Starting point   (b) Shift up   (c) Shift down   (d) Shift left   (e) Shift right

FIGURE 1.2.4 Shifted graphs in Example 5

■ **Combining Shifts** In general, the graph of a function

$$y = f(x \pm c_1) \pm c_2, \qquad (10)$$

◀ The order in which the shifts are done is irrelevant.

where c_1 and c_2 are positive constants, combines a horizontal shift (left or right) with a vertical shift (up or down). For example, the graph $y = (x + 1)^2 - 1$ is the graph of $f(x) = x^2$ shifted 1 unit to the left followed by a vertical shift 1 unit down. The graph is given in FIGURE 1.2.5.

Another way of rigidly transforming a graph of a function is by a **reflection** in a coordinate axis.

Reflections

Suppose $y = f(x)$ is a function. Then the graph of

- $y = -f(x)$ is the graph of f reflected in the **x-axis**,
- $y = f(-x)$ is the graph of f reflected in the **y-axis**.

FIGURE 1.2.5 Graph obtained by a horizontal and vertical shift

Reflection or mirror image

In FIGURE 1.2.6(a) we have reproduced the graph of a function $y = f(x)$ given in Figure 1.2.2. The reflections of this graph in the x- and y-axes are illustrated in Figures 1.2.6(b) and 1.2.6(c). Each of these reflections is a **mirror image** of the graph of $y = f(x)$ in the respective coordinate axis.

(a) Starting point   (b) Reflection in x-axis   (c) Reflection in y-axis

FIGURE 1.2.6 Reflections in the coordinate axes

EXAMPLE 6 Reflections

Graph

(a) $y = -\sqrt{x}$ (b) $y = \sqrt{-x}$.

Solution The starting point is the graph of $f(x) = \sqrt{x}$ given in FIGURE 1.2.7(a).
  (a) The graph of $y = -\sqrt{x}$ is the reflection of the graph of $f(x) = \sqrt{x}$ in the x-axis. Observe in Figure 1.2.7(b) that since $(1, 1)$ is on the graph of f, the point $(1, -1)$ is on the graph of $y = -\sqrt{x}$.
  (b) The graph of $y = \sqrt{-x}$ is the reflection of the graph of $f(x) = \sqrt{x}$ in the y-axis. Observe in Figure 1.2.7(c) that since $(1, 1)$ is on the graph of f the point $(-1, 1)$ is on the graph of $y = \sqrt{-x}$. The function $y = \sqrt{-x}$ looks a little strange, but bear in mind that its domain is determined by the requirement that $-x \geq 0$, or equivalently $x \leq 0$, and so the reflected graph is defined on the interval $(-\infty, 0]$.

(a) Starting point (b) Reflection in x-axis (c) Reflection in y-axis

FIGURE 1.2.7 Graphs in Example 6

■ **Nonrigid Transformations** If a function f is multiplied by a constant $c > 0$, the shape of the graph is changed but retains, *roughly*, its original shape. The graph of $y = cf(x)$ is the graph of $y = f(x)$ distorted vertically; the graph of f is either stretched (or elongated) vertically or is compressed (or flattened) vertically depending on the value of c. Put another way, a vertical stretch is a stretch of the graph of $y = f(x)$ away from the x-axis, whereas a vertical compression is a squeezing of the graph of $y = f(x)$ toward the x-axis. The graph of the function $y = f(cx)$ is distorted horizontally, either by a stretch of the graph of $y = f(x)$ away from the y-axis or a squeezing of the graph of $y = f(x)$ toward the y-axis. Stretching or compressing a graph are examples of **nonrigid transformations**.

Stretches and Compressions

Suppose $y = f(x)$ is a function and c is a positive constant. Then the graph of

- $y = cf(x)$ is the graph of f **vertically stretched** by a factor of c if $c > 1$,
- $y = cf(x)$ is the graph of f **vertically compressed** by a factor of $1/c$ if $0 < c < 1$,
- $y = f(cx)$ is the graph of f **horizontally stretched** by a factor of $1/c$ if $0 < c < 1$,
- $y = f(cx)$ is the graph of f **horizontally compressed** by a factor of c if $c > 1$.

EXAMPLE 7 Two Compressions

Given $f(x) = x^2 - x$. Compare the graphs of

(a) $y = \frac{1}{2}f(x)$ and   **(b)** $y = f(2x)$.

Solution The graph of the given polynomial function f is shown in **FIGURE 1.2.8**.
 (a) With the identification $c = \frac{1}{2}$, the graph of $y = \frac{1}{2}f(x)$ is the graph of f vertically compressed by a factor of 2. Of the three points shown on the graph of Figure 1.2.8(a), notice in Figure 1.2.8(b) the y-coordinates of the corresponding three points are one-half as large. The original graph is squeezed toward the x-axis.
 (b) With the identification $c = 2$, the graph of $y = f(2x)$ is the graph of f horizontally compressed by a factor of 2. Of the three points shown on the graph of Figure 1.2.8(a), in Figure 1.2.8(c) the x-coordinates of the corresponding three points are divided by 2. The original graph is squeezed toward the y-axis.

(a) $y = f(x)$ (b) $y = \frac{1}{2}f(x)$ (c) $y = f(2x)$

FIGURE 1.2.8 Graphs of functions in Example 7

The next example illustrates shifting, reflecting, and stretching of a graph.

EXAMPLE 8 Combining Transformations
Graph $y = 2 - 2\sqrt{x-3}$.

Solution You should recognize that the given function consists of four transformations of the basic function $f(x) = \sqrt{x}$:

$$y = \underset{\uparrow\ \uparrow}{\overset{\text{vertical shift up}\qquad\quad\text{horizontal shift to right}}{\underset{}{2 - 2\sqrt{x - 3}}}}.$$
$$\text{reflection in } x\text{-axis}\quad \text{vertical stretch}$$

We start with the graph of $f(x) = \sqrt{x}$ in FIGURE 1.2.9(a). The four transformations are illustrated in Figures 1.2.9(b)–(e).

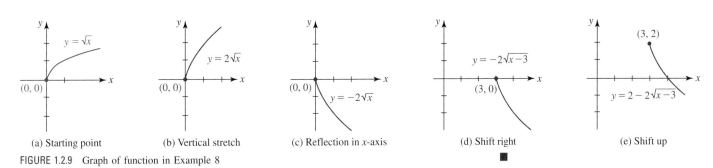

(a) Starting point (b) Vertical stretch (c) Reflection in x-axis (d) Shift right (e) Shift up

FIGURE 1.2.9 Graph of function in Example 8

■ **Symmetry** If the graph of a function is symmetric with respect to the y-axis, we say that f is an **even function**. A function whose graph is symmetric with respect to the origin is said to be an **odd function**. We have the following tests for symmetry.

Tests for Symmetry of the Graph of a Function

The graph of a function f with domain X is symmetric with respect to

- the **y-axis** if $f(-x) = f(x)$ for every x in X, or (11)
- the **origin** if $f(-x) = -f(x)$ for every x in X. (12)

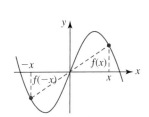

FIGURE 1.2.10 Even function; graph has y-axis symmetry

In FIGURE 1.2.10, observe that if f is an even function and

$$\underset{\downarrow}{f(x)} \qquad\qquad \underset{\downarrow}{f(-x)}$$
$$(x, y) \text{ is a point on its graph, then necessarily } (-x, y)$$

is also a point on its graph. Similarly, we see in FIGURE 1.2.11 that if f is an odd function and

$$\underset{\downarrow}{f(x)} \qquad\qquad \underset{\downarrow}{f(-x) = -f(x)}$$
$$(x, y) \text{ is a point on its graph, then necessarily } (-x, -y)$$

is a point on its graph.

FIGURE 1.2.11 Odd function; graph has origin symmetry

EXAMPLE 9 Odd and Even Functions

(a) $f(x) = x^3$ is an odd function since by (12),

$$f(-x) = (-x)^3 = (-1)^3 x^3 = -x^3 = -f(x).$$

Inspection of Figure 1.2.1(c) shows that the graph of f is symmetric with respect to the origin. For example, since $f(1) = 1$, $(1, 1)$ is a point on the graph of $y = x^3$. Because f is an odd function, $f(-1) = -f(1)$ implies $(-1, -1)$ is on the same graph.

(b) $f(x) = x^{2/3}$ is an even function since by (11) and the laws of exponents

$$f(-x) = (-x)^{2/3} = (-1)^{2/3}x^{2/3} = (\sqrt[3]{-1})^2 x^{2/3} = (-1)^2 x^{2/3} = x^{2/3} = f(x).$$
(cube root of -1 is -1)

In Figure 1.2.1(i), we see that the graph of f is symmetric with respect to the y-axis. For example, $(8, 4)$ and $(-8, 4)$ are points on the graph of $y = x^{2/3}$.

(c) $f(x) = x^3 + 1$ is neither even nor odd. From

$$f(-x) = (-x)^3 + 1 = -x^3 + 1$$

we see that $f(-x) \neq f(x)$, and $f(-x) \neq -f(x)$.

The graphs in Figure 1.2.1, with part (g) the only exception, possess either y-axis or origin symmetry. The functions in Figures 1.2.1(b), (d), (f), and (i) are even, whereas the functions in Figures 1.2.1(a), (c), (e), and (h) are odd.

Exercises 1.2
Answers to selected odd-numbered problems begin on page ANS-2.

≡ Fundamentals

In Problems 1–6, find $f + g, f - g, fg$, and f/g.

1. $f(x) = 2x + 5, g(x) = -4x + 8$

2. $f(x) = 5x^2, g(x) = 7x - 9$

3. $f(x) = \dfrac{x}{x+1}, g(x) = \dfrac{1}{x}$

4. $f(x) = \dfrac{2x-1}{x+3}, g(x) = \dfrac{x-3}{4x+2}$

5. $f(x) = x^2 + 2x - 3, g(x) = x^2 + 3x - 4$

6. $f(x) = x^2, g(x) = \sqrt{x}$

In Problems 7–10, let $f(x) = \sqrt{x-1}$ and $g(x) = \sqrt{2-x}$. Find the domain of the given function.

7. $f + g$ **8.** fg **9.** f/g **10.** g/f

In Problems 11–16, find $f \circ g$ and $g \circ f$.

11. $f(x) = 3x - 2, g(x) = x + 6$

12. $f(x) = 4x + 1, g(x) = x^2$

13. $f(x) = x^2, g(x) = x^3 + x^2$

14. $f(x) = 2x + 4, g(x) = \dfrac{1}{2x+4}$

15. $f(x) = \dfrac{3}{x}, g(x) = \dfrac{x}{x+1}$

16. $f(x) = x^2 + \sqrt{x}, g(x) = x^2$

In Problems 17 and 18, let $f(x) = \sqrt{x-3}$ and $g(x) = x^2 + 2$. Find the domain of the given function.

17. $f \circ g$ **18.** $g \circ f$

In Problems 19 and 20, let $f(x) = 5 - x^2$ and $g(x) = 2 - \sqrt{x}$. Find the domain of the given function.

19. $g \circ f$ **20.** $f \circ g$

In Problems 21 and 22, find $f \circ (2f)$ and $f \circ (1/f)$.

21. $f(x) = 2x^3$ **22.** $f(x) = \dfrac{1}{x-1}$

The composition of three functions f, g, and h is the function

$$(f \circ g \circ h)(x) = f(g(h(x))).$$

In Problems 23 and 24, find $f \circ g \circ h$.

23. $f(x) = x^2 + 6, g(x) = 2x + 1, h(x) = 3x - 2$

24. $f(x) = \sqrt{x-5}, g(x) = x^2 + 2, h(x) = \sqrt{2x+1}$

In Problems 25 and 26, find a function g.

25. $f(x) = 2x - 5, (f \circ g)(x) = -4x + 13$

26. $f(x) = \sqrt{2x+6}, (f \circ g)(x) = 4x^2$

In Problems 27 and 28, express the function F as a composition $f \circ g$ of two functions f and g.

27. $F(x) = 2x^4 - x^2$ **28.** $F(x) = \dfrac{1}{x^2+9}$

In Problems 29–36, the points $(-2, 1)$ and $(3, -4)$ are on the graph of the function $y = f(x)$. Find the corresponding points on the graph obtained by the given transformations.

29. the graph of f shifted up 2 units

30. the graph of f shifted down 5 units

31. the graph of f shifted to the left 6 units

32. the graph of f shifted to the right 1 unit

33. the graph of f shifted up 1 unit and to the left 4 units

34. the graph of f shifted down 3 units and to the right 5 units

35. the graph of f reflected in the y-axis

36. the graph of f reflected in the x-axis

In Problems 37–40, use the graph of the function $y = f(x)$ given in the figure to graph the following functions:

(a) $y = f(x) + 2$ (b) $y = f(x) - 2$
(c) $y = f(x + 2)$ (d) $y = f(x - 5)$
(e) $y = -f(x)$ (f) $y = f(-x)$

37.

FIGURE 1.2.12 Graph for Problem 37

38.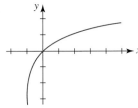

FIGURE 1.2.13 Graph for Problem 38

39.

FIGURE 1.2.14 Graph for Problem 39

40.

FIGURE 1.2.15 Graph for Problem 40

In Problems 41 and 42, use the graph of the function $y = f(x)$ given in the figure to graph the following functions:

(a) $y = f(x) + 1$ (b) $y = f(x) - 1$
(c) $y = f(x + \pi)$ (d) $y = f(x - \pi/2)$
(e) $y = -f(x)$ (f) $y = f(-x)$
(g) $y = 3f(x)$ (h) $y = -\frac{1}{2}f(x)$

41.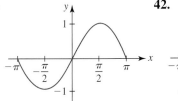

FIGURE 1.2.16 Graph for Problem 41

42.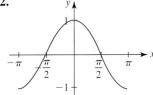

FIGURE 1.2.17 Graph for Problem 42

In Problems 43–46, find the equation of the final graph after the given transformations are applied to the graph of $y = f(x)$.

43. the graph of $f(x) = x^3$ shifted up 5 units and right 1 unit

44. the graph of $f(x) = x^{2/3}$ stretched vertically by a factor of 3 units, then shifted right 2 units

45. the graph of $f(x) = x^4$ reflected in the x-axis, then shifted left 7 units

46. the graph of $f(x) = \dfrac{1}{x}$ reflected in the y-axis, then shifted left 5 units and down 10 units

In Problems 47 and 48, complete the graph of the given function $y = f(x)$ if

(a) f is an even function   and   (b) f is an odd function.

47.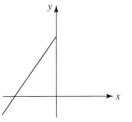

FIGURE 1.2.18 Graph for Problem 47

48.

FIGURE 1.2.19 Graph for Problem 48

49. Fill in the table where f is an even function.

x	0	1	2	3	4
$f(x)$	-1	2	10	8	0
$g(x)$	2	-3	0	1	-4
$(f \circ g)(x)$					

50. Fill in the table where g is an odd function.

x	0	1	2	3	4
$f(x)$	-2	-3	0	-1	-4
$g(x)$	9	7	-6	-5	13
$(g \circ f)(x)$					

A Mathematical Classic In the mathematical analysis of circuits or of signals it is convenient to define a special function that is 0 (off) up to a certain number and then the number 1 (on) after that. The **Heaviside function**

$$U(x - a) = \begin{cases} 0, & x < a \\ 1, & x \geq a, \end{cases}$$

is named after the brilliant and controversial English electrical engineer and mathematician **Oliver Heaviside** (1850–1925). The function U is also known as the **unit step function**.

In Problems 51 and 52, sketch the given function. The function in Problem 52 is sometimes called the **boxcar function**.

51. $y = 2U(x - 1) + U(x - 2)$

52. $y = U\left(x + \tfrac{1}{2}\right) - U\left(x - \tfrac{1}{2}\right)$

53. Find an equation for the function f illustrated in FIGURE 1.2.20 in terms of $U(x - a)$.

FIGURE 1.2.20 Graph for Problem 53

54. The Heaviside function $U(x - a)$ is frequently combined with other functions by addition and multiplication. Given that $f(x) = x^2$, compare the graphs of $y = f(x - 3)$ and $y = f(x - 3)U(x - 3)$.

In Problems 55 and 56, sketch the given function.

55. $y = (2x - 5)U(x - 1)$ **56.** $y = x - xU(x - 3)$

Think About It

57. Determine whether $f \circ (g + h) = f \circ g + f \circ h$ is true or false.

58. Suppose $[-1, 1]$ is the domain of $f(x) = x^2$. What is the domain of $y = f(x - 2)$?

59. Explain why the graph of a function cannot be symmetric with respect to the x-axis.

60. What points, if any, on the graph of $y = f(x)$ remain fixed, that is, the same on the resulting graph after a vertical stretch or compression? After a reflection in the x-axis? After a reflection in the y-axis?

61. Suppose the domain of f is $(-\infty, \infty)$. What is the relationship between the graphs of $y = f(x)$ and $y = f(|x|)$?

62. Review the graphs of $y = x$ and $y = 1/x$ in Figure 1.2.1. Then discuss how to obtain the graph of $y = 1/f(x)$ from the graph of $y = f(x)$. Sketch the graph of $y = 1/f(x)$ for the function f whose graph is given in Figure 1.2.15.

63. Suppose $f(x) = x$ and $g(x) = \lfloor x \rfloor$ is the greatest integer or floor function. The difference of f and g is the function $\text{frac}(x) = x - \lfloor x \rfloor$ called the **fractional part of x**. Explain the name and then graph $\text{frac}(x)$.

64. Using the notion of a reflection of a graph in an axis, express the ceiling function $g(x) = \lceil x \rceil$ in terms of the floor function $f(x) = \lfloor x \rfloor$ (see pages 7 and 15).

1.3 Polynomial and Rational Functions

Introduction In this section we continue our review of polynomial functions and rational functions. Functions such as $y = 2x - 1$, $y = 5x^2 - 2x + 4$, and $y = x^3$, in which the variable x is raised to a *nonnegative integer power*, are examples of polynomial functions. In the preceding section we saw that a general **polynomial function** $y = f(x)$ has the form

$$f(x) = a_n x^n + a_{n-1} x^{n-1} + \cdots + a_2 x^2 + a_1 x + a_0, \qquad (1)$$

where n is a nonnegative integer. A **rational function** is the quotient

$$f(x) = \frac{p(x)}{q(x)}, \qquad (2)$$

where p and q are polynomial functions.

Polynomial Functions The constants $a_n, a_{n-1}, \ldots, a_1, a_0$ in (1) are called **coefficients**; the number a_n is called the **leading coefficient** and a_0 is called the **constant term** of the polynomial. The highest power of x in a polynomial is said to be its **degree**. So if $a_n \neq 0$, then we say that $f(x)$ in (1) has **degree n**. For example,

$$f(x) = 3\overset{\text{degree 5}}{x^5} - 4x^3 - 3x + 8$$

(leading coefficient; constant term)

is a polynomial function of degree 5.

Polynomials of degrees $n = 0$, $n = 1$, $n = 2$, and $n = 3$ are respectively,

$f(x) = a$, constant function,
$f(x) = ax + b$, linear function,
$f(x) = ax^2 + bx + c$, quadratic function,
$f(x) = ax^3 + bx^2 + cx + d$, cubic function.

The constant function $f(x) = 0$ is called the **zero polynomial**.

Lines You are undoubtedly familiar with the fact that the graphs of a constant function and a linear function are **lines**. Since the notion of a line plays an important role in the study of differential calculus, it is appropriate that we review equations of lines. There are three types of lines in the xy-plane: horizontal lines, vertical lines, and slant or oblique lines.

Slope

We begin with the recollection from plane geometry that through any two distinct points (x_1, y_1) and (x_2, y_2) in the plane there passes only one line L. If $x_1 \neq x_2$, then the number

$$m = \frac{y_2 - y_1}{x_2 - x_1} \tag{3}$$

is called the **slope** of the line determined by these two points. It is customary to denote the **change in y** or the **rise** of the line by $\Delta y = y_2 - y_1$ and the **change in x** or the **run** of the line by $\Delta x = x_2 - x_1$, so that (3) is written $m = \Delta y / \Delta x$. See FIGURE 1.3.1. As indicated in FIGURE 1.3.2, any pair of distinct points on a line with slope, say, $(x_1, y_1), (x_2, y_2)$, and $(x_3, y_3), (x_4, y_4)$, will determine the same slope. In other words, the slope of a line is independent of the choice of points on the line.

In FIGURE 1.3.3 we compare the graphs of lines with positive, negative, zero, and undefined slopes. In Figure 1.3.3(a) we see, reading the graph left to right, that a line with positive slope ($m > 0$) rises as x increases. Figure 1.3.3(b) shows that a line with negative slope ($m < 0$) falls as x increases. If (x_1, y_1) and (x_2, y_2) are points on a horizontal line, then $y_1 = y_2$ and so its rise is $\Delta y = y_2 - y_1 = 0$. Hence from (3) the slope is zero ($m = 0$). See Figure 1.3.3(c). If (x_1, y_1) and (x_2, y_2) are points on a vertical line, then $x_1 = x_2$ and so its run is $\Delta x = x_2 - x_1 = 0$. In this case we say that the slope of the line is **undefined** or that the line has no slope. See Figure 1.3.3(d). Only lines with slope are the graphs of functions.

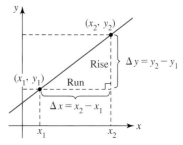

FIGURE 1.3.1 Slope of a line

FIGURE 1.3.2 Similar triangles

(a) $m > 0$

(b) $m < 0$

(c) $m = 0$

(d) m undefined

FIGURE 1.3.3 Lines with slope (a)–(c); line with no slope (d)

Equations of Lines

To find an equation of a line L with slope m let us suppose that (x_1, y_1) is on the line. If (x, y) represents any other point on L, then (3) gives

$$\frac{y - y_1}{x - x_1} = m.$$

Multiplying both sides of the last equality by $x - x_1$ gives an important equation. The **point–slope equation** of the line through (x_1, y_1) with slope m is

$$y - y_1 = m(x - x_1). \tag{4}$$

Any line that is not vertical must cross the y-axis. If the y-intercept is $(0, b)$, then with $x_1 = 0, y_1 = b$, (4) gives $y - b = m(x - 0)$. The last equation simplifies to the **slope-intercept equation** of the line

$$y = mx + b. \tag{5}$$

EXAMPLE 1 Point–Slope Equation

Find an equation of the line passing through the points $(4, 3)$ and $(-2, 5)$.

Solution First we compute the slope of the line through the points. From (3),

$$m = \frac{5 - 3}{-2 - 4} = \frac{2}{-6} = -\frac{1}{3}.$$

The point–slope equation (4) then gives $y - 3 = -\frac{1}{3}(x - 4)$ or $y = -\frac{1}{3}x + \frac{13}{3}$. ∎

An equation of *any* line in the plane is a special case of the general **linear equation**

$$Ax + By + C = 0, \qquad (6)$$

where A, B, and C are real constants. The characteristic that gives (6) its name *linear* is that the variables x and y appear only to the first power. The cases of special interest are

$$A = 0, B \neq 0, \text{ gives } y = -\frac{C}{B}, \qquad (7)$$

$$A \neq 0, B = 0, \text{ gives } x = -\frac{C}{A}, \qquad (8)$$

$$A \neq 0, B \neq 0, \text{ gives } y = -\frac{A}{B}x - \frac{C}{B}. \qquad (9)$$

The first and the third of these three equations define functions. By relabeling $-C/B$ in (7) as a we get a constant function $y = a$. By relabeling $-A/B$ and $-C/B$ in (9) as a and b, respectively, we get the form of a linear function $f(x) = ax + b$, which, except for symbols, is the same as (5). By relabeling $-C/A$ in (8) as a we get the equation of a vertical line $x = a$, which is not a function.

▌ **Increasing–Decreasing Functions** We have just seen in Figures 1.3.3(a) and 1.3.3(b) that if $a > 0$ (which, as we have just seen plays the part of m), the values of a linear function $f(x) = ax + b$ increase as x increases, whereas for $a < 0$, the values of $f(x)$ decrease as x increases. The notions of increasing and decreasing can be extended to *any* function. A function f is said to be

- **increasing** on an interval I if $f(x_1) < f(x_2)$, and (10)
- **decreasing** on an interval I if $f(x_1) > f(x_2)$. (11)

In FIGURE 1.3.4(a) the function f is increasing on the interval $[a, b]$, whereas f is decreasing on $[a, b]$ in Figure 1.3.4(b). A linear function $f(x) = ax + b$ increases on the interval $(-\infty, \infty)$ for $a > 0$ and decreases on the interval $(-\infty, \infty)$ for $a < 0$.

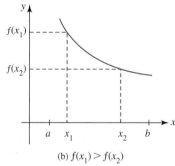

(a) $f(x_1) < f(x_2)$ (b) $f(x_1) > f(x_2)$

FIGURE 1.3.4 Increasing function in (a); decreasing function in (b)

This assumption means that L_1 and L_2 are nonvertical lines.

▌ **Parallel and Perpendicular Lines** If L_1 and L_2 are two distinct lines *with slope*, then necessarily L_1 and L_2 are either parallel or they intersect. If the lines intersect at a right angle, they are said to be perpendicular. We can determine whether two lines are parallel or are perpendicular by examining their slopes.

Parallel and Perpendicular Lines

Suppose L_1 and L_2 are lines with slopes m_1 and m_2, respectively. Then,

- L_1 is **parallel** to L_2 if and only if $m_1 = m_2$, and
- L_1 is **perpendicular** to L_2 if and only if $m_1 m_2 = -1$.

EXAMPLE 2 Parallel Lines

The linear equations $3x + y = 2$ and $6x + 2y = 15$ can be rewritten in the slope-intercept forms $y = -3x + 2$ and $y = -3x + \frac{15}{2}$, respectively. As noted in blue and red the slope of each line is -3. Therefore the lines are parallel. The graphs of these equations are shown in FIGURE 1.3.5. ∎

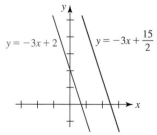

FIGURE 1.3.5 Parallel lines in Example 2

EXAMPLE 3 Perpendicular Lines

Find an equation of the line through $(0, -3)$ that is perpendicular to the graph of $4x - 3y + 6 = 0$.

Solution By solving for y, the given linear equation yields the slope-intercept form $y = \frac{4}{3}x + 2$. This line, whose graph is given in blue in FIGURE 1.3.6, has slope $\frac{4}{3}$. The slope of any line perpendicular to it is the negative reciprocal of $\frac{4}{3}$, namely, $-\frac{3}{4}$. Since $(0, -3)$ is the y-intercept of the required line, it follows from (5) that its equation is $y = -\frac{3}{4}x - 3$. The graph of the last equation is the red line in Figure 1.3.6. ∎

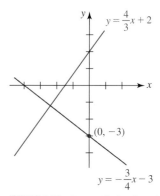

FIGURE 1.3.6 Perpendicular lines in Example 3

■ **Quadratic Functions** The squaring function $y = x^2$ that we have seen in Sections 1.1 and 1.2 is a member of a family of functions called **quadratic functions**, that is, polynomial functions of the form $f(x) = ax^2 + bx + c$, where $a \neq 0$, b, and c are constants. The graphs of quadratic functions, called **parabolas**, are simply rigid and nonrigid transformations of the graph of $y = x^2$ shown in FIGURE 1.3.7.

■ **Vertex and Axis** If the graph of a quadratic function opens upward $a > 0$ (or downward $a < 0$), the lowest (highest) point (h, k) on the parabola is called its **vertex**. All parabolas are symmetric with respect to a vertical line through the vertex (h, k). The line $x = h$ is called the **axis** of the parabola. See FIGURE 1.3.8.

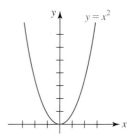

FIGURE 1.3.7 Graph of simplest parabola

■ **Standard Form** The vertex (h, k) of a parabola can be determined by recasting the equation $f(x) = ax^2 + bx + c$ into the **standard form**

$$f(x) = a(x - h)^2 + k. \qquad (12)$$

The form (12) is obtained from $f(x) = ax^2 + bx + c$ by completing the square in x. With the aid of differential calculus we will be able to find the vertex of a parabola without completing the square.

As the next example shows, a reasonable sketch of a parabola can be obtained by plotting the intercepts and the vertex. The form in (12) indicates that its graph is the graph of $y = ax^2$ shifted horizontally $|h|$ units and shifted vertically $|k|$ units.

EXAMPLE 4 Graph Using Intercepts and Vertex

Graph $f(x) = x^2 - 2x - 3$.

Solution Since $a = 1 > 0$ we know that the parabola will open upward. From $f(0) = -3$ we get the y-intercept $(0, -3)$. To see whether there are any x-intercepts we solve $x^2 - 2x - 3 = 0$ by factoring or by the quadratic formula. From $(x + 1)(x - 3) = 0$ we find the solutions $x = -1$ and $x = 3$. The x-intercepts are $(-1, 0)$ and $(3, 0)$. To locate the vertex we complete the square:

$$f(x) = (x^2 - 2x + 1) - 1 - 3 = (x^2 - 2x + 1) - 4.$$

Thus the standard form is $f(x) = (x - 1)^2 - 4$. By comparing the last equation with (12) we identify $h = 1$ and $k = -4$. We conclude that the vertex is $(1, -4)$. Using this information we draw a parabola through these four points as shown in FIGURE 1.3.9.

By finding the vertex of a parabola we automatically determine the range of the quadratic function. As Figure 1.3.9 clearly shows, the range of f is the interval $[-4, \infty)$ on the y-axis. Figure 1.3.9 also shows that f is decreasing on the interval $(-\infty, 1]$ but increasing on $[1, \infty)$. ∎

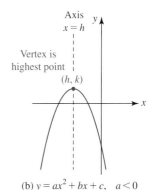

FIGURE 1.3.8 Vertex and axis of a parabola

■ **Higher-Degree Polynomial Functions** The graph of *every* linear function $f(x) = ax + b$ is a line and the graph of *every* quadratic function $f(x) = ax^2 + bx + c$ is a parabola. Such definitive descriptive statements cannot be made about the graph of a higher-degree polynomial function. What is the shape of the graph of a fifth-degree polynomial function? It turns out that the graph of a polynomial function of degree $n \geq 3$ can have several possible shapes. In general,

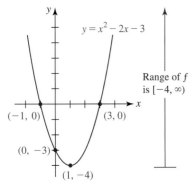

FIGURE 1.3.9 Parabola in Example 4

graphing a polynomial function f of degree $n \geq 3$ often demands the use of either calculus or a graphing utility. However, by keeping in mind shifting, end behavior, intercepts, and symmetry we can, in many instances, quickly sketch a reasonable graph of a higher-degree polynomial function while keeping point plotting to a minimum.

■ **End Behavior** In rough terms, the **end behavior** of *any* function f is simply how f behaves for very large values of $|x|$. In the case of a polynomial function f of degree n, its graph resembles the graph of $y = a_n x^n$ for large values of $|x|$. To see why the graph of a polynomial function such as $f(x) = -2x^3 + 4x^2 + 5$ resembles the graph of the single-term polynomial $y = -2x^3$ when $|x|$ is large, let us factor out the highest power of x, that is, x^3:

$$f(x) = x^3\left(-2 + \frac{4}{x} + \frac{5}{x^3}\right). \tag{13}$$

(both these terms become negligible when $|x|$ is large)

By letting $|x|$ increase without bound, both $4/x$ and $5/x^3$ can be made as close to 0 as we want. Thus, when $|x|$ is large, the values of the function f in (13) are closely approximated by the values of $y = -2x^3$. In general, there can be only four types of end behavior for polynomial functions. To interpret the arrows in FIGURE 1.3.10 let us examine the arrows in, say, Figure 1.3.10(c) where it is assumed that n is odd and $a_n > 0$. The position and direction of the left arrow (left arrow points down) indicates that as x becomes unbounded in the negative direction, the values $f(x)$ are decreasing. Stated another way, the graph is heading downward. Similarly, the position and direction of the right arrow (right arrow points up) indicates that as x becomes unbounded in the positive direction, the values $f(x)$ are increasing (the graph is heading upward). You can see the end behavior illustrated in Figures 1.3.10(a) and 1.3.10(c) in the graphs given in FIGURE 1.3.11 and FIGURE 1.3.12, respectively. The graphs of the functions $y = -x$, $y = -x^2$, $y = -x^3$, ..., $y = -x^8$ are the graphs in Figures 1.3.11 and 1.3.12 reflected in the x-axis, and so their end behavior is as shown in Figures 1.3.10(b) and 1.3.10(d).

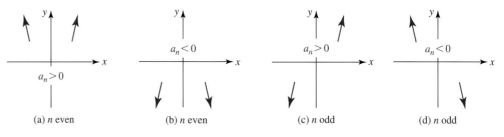

(a) n even (b) n even (c) n odd (d) n odd

FIGURE 1.3.10 End behavior of a polynomial function f depends on its degree n and on the sign of its leading coefficient

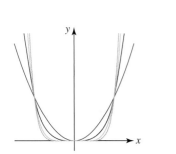

FIGURE 1.3.11 Graphs of $y = x^2$ (blue), $y = x^4$ (red), $y = x^6$ (green), $y = x^8$ (gold)

■ **Symmetry of Polynomial Functions** It is easy to tell by inspection those polynomial functions whose graphs possess **symmetry** with respect to either the y-axis or the origin. The words *even* and *odd* have special meaning for polynomial functions. The conditions $f(-x) = f(x)$ and $f(-x) = -f(x)$ hold for polynomial functions in which all the powers of x are even integers and odd integers, respectively. For example,

even powers: $f(x) = 5x^4 - 7x^2$ — even function
odd powers: $f(x) = 10x^5 + 7x^3 + 4x$ — odd function
mixed powers: $f(x) = -3x^7 + 2x^4 + x^3 + 2$ — neither even nor odd

A function such as $f(x) = 3x^6 - x^4 + 6$ is an even function because all powers are even integers; the constant term 6 is actually $6x^0$, and 0 is an even nonnegative integer.

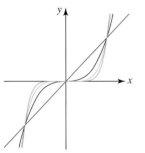

FIGURE 1.3.12 Graphs of $y = x$ (blue), $y = x^3$ (red), $y = x^5$ (green), $y = x^7$ (gold)

■ **Intercepts of Polynomial Functions** The graph of every polynomial function f passes through the y-axis since $x = 0$ is in the domain of the function. The y-intercept is the point

$(0, f(0))$. The real **zeros** of a polynomial function are the x-coordinates of the **x-intercepts** of its graph. A number c is a zero of a polynomial function f of degree n if and only if $x - c$ is a factor of f, that is, $f(x) = (x - c)q(x)$, where $q(x)$ is a polynomial of degree $n - 1$. If $(x - c)^m$ is a factor of f, where $m > 1$ is a positive integer, and $(x - c)^{m+1}$ is *not* a factor of f, then c is said to be a **repeated zero**, or a **zero of multiplicity m**. When $m = 1$, c is called a **simple zero**. For example, $-\frac{1}{3}$ and $\frac{1}{2}$ are simple zeros of $f(x) = 6x^2 - x - 1$ since f can be written as $f(x) = 6(x + \frac{1}{3})(x - \frac{1}{2})$, whereas 5 is a repeated zero or a zero of multiplicity 2 for $f(x) = x^2 - 10x + 25 = (x - 5)^2$. The behavior of the graph of f at an x-intercept $(c, 0)$ depends on whether c is a simple zero, or a zero of multiplicity $m > 1$, where m is either an even or an odd integer. See FIGURE 1.3.13.

(a) Simple zero

(b) Zero of odd multiplicity $m = 3, 5, \ldots$

(c) Zero of even multiplicity $m = 2, 4, \ldots$
FIGURE 1.3.13 x-intercepts of a polynomial function f

x-intercepts of Polynomials

- If c is a simple zero, then the graph of f passes directly through the x-axis at $(c, 0)$. See Figure 1.3.13(a).
- If c is a zero of odd multiplicity $m = 3, 5, \ldots$, then the graph of f passes through the x-axis but is flattened at $(c, 0)$. See Figure 1.3.13(b).
- If c is a zero of even multiplicity $m = 2, 4, \ldots$, then the graph of f does not pass through the x-axis but is tangent to, or touches, the x-axis at $(c, 0)$. See Figure 1.3.13(c).

In the case when c is either a simple zero or a zero of odd multiplicity, $f(x)$ changes sign at $(c, 0)$, whereas if c is a zero of even multiplicity, $f(x)$ does not change sign at $(c, 0)$. We note that depending on the sign of the leading coefficient of the polynomial, the graphs in Figure 1.3.13 could be reflected in the x-axis.

EXAMPLE 5 Graphs of Polynomial Functions

Graph

(a) $f(x) = x^3 - 9x$ (b) $g(x) = (1 - x)(x + 1)^2$ (c) $h(x) = -(x + 4)(x - 2)^3$.

Solution

(a) By ignoring all terms but the first, we see that the graph of f resembles the graph of $y = x^3$ for large $|x|$. This end behavior of f is shown in Figure 1.3.10(c). Since all the powers are odd integers, f is an odd function and its graph is symmetric with respect to the origin. Setting $f(x) = 0$ we see from

$$\underset{\underset{\downarrow}{\text{difference of two squares}}}{x(x^2 - 9) = 0} \quad \text{or} \quad x(x - 3)(x + 3) = 0$$

that the zeros of f are $x = 0$ and $x = \pm 3$. Since these numbers are simple zeros the graph passes directly through x-intercepts at $(0, 0)$, $(-3, 0)$, and $(3, 0)$ as shown in FIGURE 1.3.14.

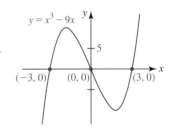
FIGURE 1.3.14 Graph of function in Example 5(a)

(b) Multiplying out, g is the same as $g(x) = -x^3 - x^2 + x + 1$ and so we see that the graph of g resembles the graph of $y = -x^3$ for large $|x|$, just the opposite of the end behavior of the function in part (a). Because there are both even and odd powers of x present, g is neither even nor odd; its graph possesses no y-axis or origin symmetry. Because -1 is a zero of multiplicity 2, the graph is tangent to the x-axis at $(-1, 0)$. Since 1 is a simple zero, the graph passes directly through the x-axis at $(1, 0)$. See FIGURE 1.3.15.

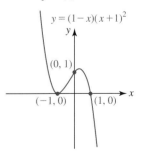
FIGURE 1.3.15 Graph of function in Example 5(b)

(c) Inspection of h shows that its graph resembles the graph of $y = -x^4$ for large $|x|$. This end behavior of h is shown in Figure 1.3.10(b). The function h is neither even nor odd. From the factored form of $h(x)$, we see that -4 is a simple zero and so the graph of h passes directly through the x-axis at $(-4, 0)$. Since 2 is a zero of multiplicity 3, its graph flattens as it passes through the x-intercept $(2, 0)$. See FIGURE 1.3.16.

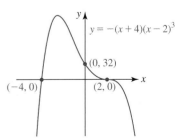
FIGURE 1.3.16 Graph of function in Example 5(c)

■ **Rational Functions** Graphing a rational function $f(x) = p(x)/q(x)$ is a little more complicated than graphing a polynomial function because in addition to paying attention to intercepts, symmetry, and shifting/reflecting/stretching of known graphs, you should also keep an eye on the domain of f and the degrees of $p(x)$ and $q(x)$. The last two items are important in determining whether a graph of a rational function possesses *asymptotes*.

■ **Intercepts of Rational Functions** The ***y*-intercept** of the graph of $f(x) = p(x)/q(x)$ is the point $(0, f(0))$, provided the number 0 is in the domain of f. For example, the graph of the rational function $f(x) = (1 - x)/x$ does not cross the *y*-axis since $f(0)$ is not defined. If the polynomials $p(x)$ and $q(x)$ have no common factors, then the ***x*-intercepts** of the graph of the rational function $f(x) = p(x)/q(x)$ are the points whose *x*-coordinates are the real zeros of the numerator $p(x)$. In other words, the only way we can have $f(x) = p(x)/q(x) = 0$ is to have $p(x) = 0$. Thus for $f(x) = (1 - x)/x$, $1 - x = 0$ gives $x = 1$ and so $(1, 0)$ is an *x*-intercept of the graph of f.

■ **Asymptotes** The graph of a rational function $f(x) = p(x)/q(x)$ can have asymptotes. For our purposes the asymptotes can be a horizontal line, a vertical line, or a slant line. On a practical level, vertical and horizontal asymptotes of the graph of a rational function f can be determined by inspection. So for the sake of discussion let us suppose that

$$f(x) = \frac{p(x)}{q(x)} = \frac{a_n x^n + a_{n-1} x^{n-1} + \cdots + a_1 x + a_0}{b_m x^m + b_{m-1} x^{m-1} + \cdots + b_1 x + b_0}, \quad a_n \neq 0, b_m \neq 0, \tag{14}$$

represents a general rational function. The degree of $p(x)$ is n and the degree of $q(x)$ is m.

Asymptotes of Graphs of Rational Functions

Suppose that the polynomial functions $p(x)$ and $q(x)$ in (14) *have no common factors*.

- If a is real zero of $q(x)$, then $x = a$ is a **vertical asymptote** for the graph of f.
- If $n = m$, then $y = a_n/b_m$ (the quotient of the leading coefficients) is a **horizontal asymptote** for the graph of f.
- If $n < m$, then $y = 0$ is a **horizontal asymptote** for the graph of f.
- If $n > m$, then the graph of f has *no* **horizontal asymptote**.
- If $n = m + 1$, then the quotient $y = mx + b$ of $p(x)$ and $q(x)$ is a **slant asymptote** for the graph of f.

We note from the foregoing bulleted list that horizontal and slant asymptotes are mutually exclusive. In other words, a graph of a rational function f cannot possess a slant asymptote *and* a horizontal asymptote.

EXAMPLE 6 Graphs of Rational Functions

Graph

(a) $f(x) = \dfrac{x}{1 - x^2}$ **(b)** $g(x) = \dfrac{x^2 - x - 6}{x - 5}$.

Solution
(a) We begin with the observation that the numerator $p(x) = x$ and denominator $q(x) = 1 - x^2$ of f have no common factors. Also, since $f(-x) = -f(x)$ the function f is odd. Therefore, its graph is symmetric with respect to the origin. Because $f(0) = 0$, the *y*-intercept is $(0, 0)$. Moreover, $p(x) = x = 0$ implies $x = 0$ and so the only intercept is $(0, 0)$. The zeros of the denominator $q(x) = 1 - x^2$ are ± 1. Therefore, the lines $x = -1$ and $x = 1$ are vertical asymptotes. Since the degree of the numerator x is 1 and the degree of the denominator $1 - x^2$ is 2 (and $1 < 2$) it follows that $y = 0$ is a horizontal asymptote for the graph of f. The graph consists of three distinct *branches*, one to the left of the line $x = -1$, one between the lines $x = -1$ and $x = 1$, and one to the right of the line $x = 1$. See FIGURE 1.3.17.

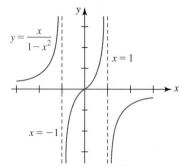

FIGURE 1.3.17 Graph of function in Example 6(a)

(b) Again, note that the numerator $p(x) = x^2 - x - 6$ and denominator $q(x) = x - 5$ of g have no common factors. Also, f is neither odd nor even. From $f(0) = \frac{6}{5}$ we get the y-intercept $\left(0, \frac{6}{5}\right)$. From $p(x) = x^2 - x - 6 = 0$ or $(x + 2)(x - 3) = 0$ we see that -2 and 3 are zeros of $p(x)$. The x-intercepts are $(-2, 0)$ and $(3, 0)$. The zero of $q(x) = x - 5$ is obviously 5 so that the line $x = 5$ is a vertical asymptote. Finally, from the fact that the degree of $p(x) = x^2 - x - 6$ (which is 2) *is exactly one greater* than the degree of $q(x) = x - 5$ (which is 1), the graph of $f(x)$ has a slant asymptote. To find it, we divide $p(x)$ by $q(x)$. By either long division or synthetic division, the result

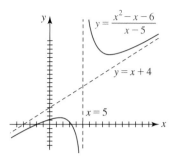

FIGURE 1.3.18 Graph of function in Example 6(b)

$$\frac{x^2 - x - 6}{x - 5} = x + 4 + \frac{14}{x - 5}$$

$y = x + 4$ is the slant asymptote

shows that the slant asymptote is $y = x + 4$. The graph consists of two branches, one to the left of the line $x = 5$ and one to the right of the line $x = 5$. See **FIGURE 1.3.18**. ∎

■ **Postscript—Graph with a Hole** We assumed throughout the discussion of asymptotes that the polynomial functions $p(x)$ and $q(x)$ in (14) had no common factors. We know that if $q(a) = 0$ and $p(x)$ and $q(x)$ have no common factors, then the line $x = a$ is necessarily a vertical asymptote for the graph of f. However, when $p(a) = 0$ *and* $q(a) = 0$, then $x = a$ *may not* be an asymptote; there may simply be a **hole** in the graph.

◀ If $p(a) = 0$ and $q(a) = 0$, then by the Factor Theorem of algebra $x - a$ is a factor of both p and q.

EXAMPLE 7 Graph with a Hole

Graph the function $f(x) = \dfrac{x^2 - 2x - 3}{x^2 - 1}$.

Solution Although the zeros of $x^2 - 1 = 0$ are ± 1, only $x = 1$ is a vertical asymptote. Note that the numerator $p(x)$ and denominator $q(x)$ have the common factor $x + 1$ which we cancel provided $x \neq -1$:

$$f(x) = \frac{(x + 1)(x - 3)}{(x + 1)(x - 1)} \overset{\text{equality is true for } x \neq -1}{=} \frac{x - 3}{x - 1}. \qquad (15)$$

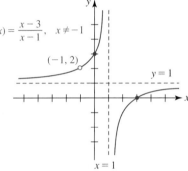

FIGURE 1.3.19 Graph of function in Example 7

We graph $y = \dfrac{x - 3}{x - 1}$, $x \neq -1$, by observing that the y-intercept is $(0, 3)$, an x-intercept is $(3, 0)$, a vertical asymptote is $x = 1$, and a horizontal asymptote is $y = 1$. Although $x = -1$ is not a vertical asymptote, we represent the fact that f is not defined at that number by drawing an open circle or hole in the graph at the point corresponding to $(-1, 2)$. See **FIGURE 1.3.19**. ∎

◀ The y-coordinate of the hole is the value of the reduced fraction (15) at $x = -1$.

$f(x)$ NOTES FROM THE CLASSROOM

In the last two sections we worked principally with polynomial functions. Polynomial functions are the fundamental building blocks of a class known as **algebraic functions**. In this section we saw that a rational function is the quotient of two polynomial functions. In general, an algebraic function f involves a finite number of additions, subtractions, multiplications, divisions, and roots of polynomial functions. Thus

$$y = 2x^2 - 5x, \quad y = \sqrt[3]{x^2}, \quad y = x^4 + \sqrt{x^2 + 5}, \quad \text{and} \quad y = \frac{\sqrt{x}}{x^3 - 2x^2 + 7}$$

are algebraic functions. Starting with the next section we consider functions that belong to a different class known as **transcendental functions**. A transcendental function f is defined to be one that is *not* algebraic. The six trigonometric functions and the exponential and logarithmic functions are examples of transcendental functions.

Exercises 1.3

≡ Fundamentals

In Problems 1–6, find an equation of the line through $(1, 2)$ with the indicated slope.

1. $\dfrac{2}{3}$
2. $\dfrac{1}{10}$
3. 0
4. -2
5. -1
6. undefined

In Problems 7–10, find the slope and the x- and y-intercepts of the given line. Graph the line.

7. $3x - 4y + 12 = 0$
8. $\dfrac{1}{2}x - 3y = 3$
9. $2x - 3y = 9$
10. $-4x - 2y + 6 = 0$

In Problems 11–16, find an equation of the line that satisfies the given conditions.

11. through $(2, 3)$ and $(6, -5)$
12. through $(5, -6)$ and $(4, 0)$
13. through $(-2, 4)$ parallel to $3x + y - 5 = 0$
14. through $(5, -7)$ parallel to the y-axis
15. through $(2, 3)$ perpendicular to $x - 4y + 1 = 0$
16. through $(-5, -4)$ perpendicular to the line through $(1, 1)$ and $(3, 11)$

In Problems 17 and 18, find a linear function $f(x) = ax + b$ that satisfies both of the given conditions.

17. $f(-1) = 5, f(1) = 6$
18. $f(-1) = 1 + f(2), f(3) = 4f(1)$

In Problems 19 and 20, find an equation of the red line L shown in the given figure.

19.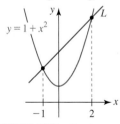

FIGURE 1.3.20 Graphs for Problem 19

20.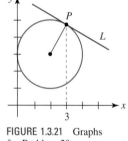

FIGURE 1.3.21 Graphs for Problem 20

In Problems 21–26, consider the quadratic function f.
(a) Find all intercepts of the graph of f.
(b) Express the function f in standard form.
(c) Find the vertex and axis of symmetry.
(d) Sketch the graph of f.
(e) What is the range of f?
(f) On what interval is f increasing? Decreasing?

21. $f(x) = x(x + 5)$
22. $f(x) = -x^2 + 4x$
23. $f(x) = (3 - x)(x + 1)$
24. $f(x) = (x - 2)(x - 6)$
25. $f(x) = x^2 - 3x + 2$
26. $f(x) = -x^2 + 6x - 5$

In Problems 27–32, describe in words how the graph of the given function can be obtained from the graph of $y = x^2$ by rigid or nonrigid transformations.

27. $f(x) = (x - 10)^2$
28. $f(x) = (x + 6)^2$
29. $f(x) = -\dfrac{1}{3}(x + 4)^2 + 9$
30. $f(x) = 10(x - 2)^2 - 1$
31. $f(x) = (-x - 6)^2 - 4$
32. $f(x) = -(1 - x)^2 + 1$

In Problems 33–42, proceed as in Example 5 and sketch the graph of the given polynomial function f.

33. $f(x) = x^3 - 4x$
34. $f(x) = 9x - x^3$
35. $f(x) = -x^3 + x^2 + 6x$
36. $f(x) = x^3 + 7x^2 + 12x$
37. $f(x) = (x + 1)(x - 2)(x - 4)$
38. $f(x) = (2 - x)(x + 2)(x + 1)$
39. $f(x) = x^4 - 4x^3 + 3x^2$
40. $f(x) = x^2(x - 2)^2$
41. $f(x) = -x^4 + 2x^2 - 1$
42. $f(x) = x^5 - 4x^3$

In Problems 43–48, match the given graph with one of the polynomial functions in (a)–(f).

(a) $f(x) = x^2(x - 1)^2$
(b) $f(x) = -x^3(x - 1)$
(c) $f(x) = x^3(x - 1)^3$
(d) $f(x) = -x(x - 1)^3$
(e) $f(x) = -x^2(x - 1)$
(f) $f(x) = x^3(x - 1)^2$

43.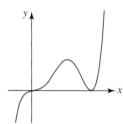

FIGURE 1.3.22 Graph for Problem 43

44.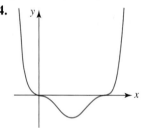

FIGURE 1.3.23 Graph for Problem 44

45.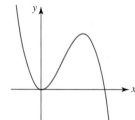

FIGURE 1.3.24 Graph for Problem 45

46.

FIGURE 1.3.25 Graph for Problem 46

47.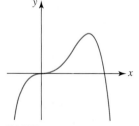

FIGURE 1.3.26 Graph for Problem 47

48.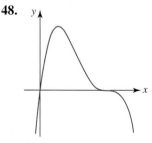

FIGURE 1.3.27 Graph for Problem 48

In Problems 49–62, find all asymptotes for the graph of the given rational function. Find x- and y-intercepts of the graph. Sketch the graph f.

49. $f(x) = \dfrac{4x - 9}{2x + 3}$

50. $f(x) = \dfrac{2x + 4}{x - 2}$

51. $f(x) = \dfrac{1}{(x - 1)^2}$

52. $f(x) = \dfrac{4}{(x + 2)^3}$

53. $f(x) = \dfrac{x}{x^2 - 1}$

54. $f(x) = \dfrac{x^2}{x^2 - 4}$

55. $f(x) = \dfrac{1 - x^2}{x^2}$

56. $f(x) = \dfrac{x(x - 5)}{x^2 - 9}$

57. $f(x) = \dfrac{x^2 - 9}{x}$

58. $f(x) = \dfrac{x^2 - 3x - 10}{x}$

59. $f(x) = \dfrac{x^2}{x + 2}$

60. $f(x) = \dfrac{x^2 - 2x}{x + 2}$

61. $f(x) = \dfrac{x^2 - 2x - 3}{x - 1}$

62. $f(x) = \dfrac{-(x - 1)^2}{x + 2}$

63. Determine whether the numbers -1 and 2 are in the range of the rational function $f(x) = \dfrac{2x - 1}{x + 4}$.

64. Determine the points where the graph of $f(x) = \dfrac{(x - 3)^2}{x^2 - 5x}$ crosses its horizontal asymptote.

≡ **Mathematical Models**

65. Related Temperatures The functional relationship between degrees Celsius T_C and degrees Fahrenheit T_F is linear. Express T_F as a function of T_C if $(0°C, 32°F)$ and $(60°C, 140°F)$ are on the graph of T_F. Show that $100°C$ is equivalent to the Fahrenheit boiling point $212°F$. See FIGURE 1.3.28.

66. Related Temperatures The functional relationship between degrees Celsius T_C and kelvin units T_K is linear. Express T_K as a function of T_C given that $(0°C, 273\,K)$ and $(27°C, 300\,K)$ are on the graph of T_K. Express the boiling point $100°C$ in kelvin units. Absolute zero is defined as $0\,K$. What is $0\,K$ in degrees Celsius? Express T_K as a linear function of T_F. What is $0\,K$ in degrees Fahrenheit? See Figure 1.3.28.

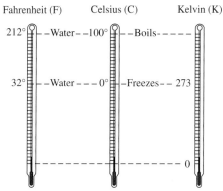

FIGURE 1.3.28 Thermometers in Problems 65 and 66

67. Simple Interest In simple interest, the amount A accrued over time is the linear function $A = P + Prt$, where P is the principal, t is measured in years, and r is the annual interest rate (expressed as a decimal). Compute A after 20 years if the principal is $P = 1000$, and the annual interest rate is 3.4%. At what time is $A = 2200$?

68. Linear Depreciation Straight line, or linear depreciation, consists of an item losing all its initial worth of A dollars over a period of n years by an amount A/n each year. If an item costing $20,000 when new is depreciated linearly over 25 years, determine a linear function giving its value V after x years, where $0 \le x \le 25$. What is the value of the item after 10 years?

69. A ball is thrown upward from ground level with an initial velocity of 96 ft/s. The height of the ball from the ground is given by the quadratic function $s(t) = -16t^2 + 96t$. At what times is the ball on the ground? Graph s over the time interval for which $s(t) \ge 0$.

70. In Problem 69, at what times is the ball 80 ft above the ground? How high does the ball go?

≡ **Think About It**

71. Consider the linear function $f(x) = \tfrac{5}{2}x - 4$. If x is changed by 1 unit, how many units will y change? If x is changed by 2 units? If x is changed by n (n a positive integer) units?

72. Consider the interval $[x_1, x_2]$ and the linear function $f(x) = ax + b$, $a \ne 0$. Show that

$$f\left(\frac{x_1 + x_2}{2}\right) = \frac{f(x_1) + f(x_2)}{2},$$

and interpret this result geometrically for $a > 0$.

73. How would you find an equation of the line that is the perpendicular bisector of the line segment through $\left(\tfrac{1}{2}, 10\right)$ and $\left(\tfrac{3}{2}, 4\right)$?

74. Using only the concepts of this section, how would you prove or disprove that the triangle with vertices $(2, 3)$, $(-1, -3)$, and $(4, 2)$ is a right triangle?

1.4 Transcendental Functions

Introduction In the first two sections of this chapter we examined various properties and graphs of **algebraic functions**. For the next three sections we examine **transcendental functions**. Basically, a transcendental function f is one that is not algebraic. A transcendental function could be as simple as the power function $y = x^\pi$, where the power is an irrational number, but the familiar transcendental functions from precalculus mathematics are the trigonometric functions, the inverse trigonometric functions, and the exponential and logarithmic functions. In this section we review the six trigonometric functions and their graphs. In Section 1.5 we consider the inverse trigonometric functions and in Section 1.6 we review exponential and logarithmic functions.

For a review of the basics of unit-circle and right-triangle trigonometry see the *Student Resource Manual*. Also see the *Resource Pages* at the end of the text.

Graphs of Sine and Cosine Recall from precalculus mathematics that the trigonometric sine and cosine functions have **period** 2π:

$$\sin(x + 2\pi) = \sin x \quad \text{and} \quad \cos(x + 2\pi) = \cos x. \tag{1}$$

The graph of *any* periodic function over an interval of length equal to its period is said to be one **cycle** of its graph. The graph of a periodic function is easily obtained by repeatedly drawing one cycle of its graph. FIGURE 1.4.1 shows one cycle of the graph of $f(x) = \sin x$ (in red); the graph of f on, say, the intervals $[-2\pi, 0]$ and $[2\pi, 4\pi]$ (in blue) is exactly the same as the graph on $[0, 2\pi]$. Because $f(-x) = \sin(-x) = -\sin x = -f(x)$ the sine function is an odd function and its graph is symmetric with respect to the origin.

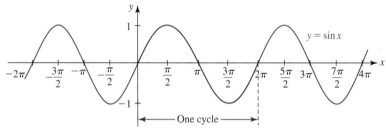

FIGURE 1.4.1 Graph of $y = \sin x$

FIGURE 1.4.2 shows one cycle (in red) of $g(x) = \cos x$ on $[0, 2\pi]$ along with the extension of that cycle (in blue) to the adjacent intervals $[-2\pi, 0]$ and $[2\pi, 4\pi]$. In contrast to the graph of $f(x) = \sin x$ where $f(0) = f(2\pi) = 0$, for the cosine function we have $g(0) = g(2\pi) = 1$. The cosine function is an even function: $g(-x) = \cos(-x) = \cos x = g(x)$, and so we see in Figure 1.4.2 its graph is symmetric with respect to the y-axis.

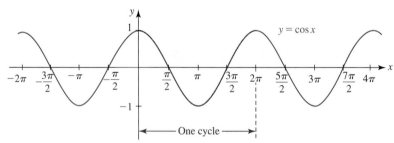

FIGURE 1.4.2 Graph of $y = \cos x$

The sine and cosine functions are defined for all real numbers x. Also, it is apparent in Figures 1.4.1 and 1.4.2 that

$$-1 \leq \sin x \leq 1 \quad \text{and} \quad -1 \leq \cos x \leq 1, \tag{2}$$

or equivalently, $|\sin x| \leq 1$ and $|\cos x| \leq 1$. In other words,

- the domain of $\sin x$ and $\cos x$ is $(-\infty, \infty)$, and the range of $\sin x$ and $\cos x$ is $[-1, 1]$.

■ **Intercepts** In this and subsequent courses in mathematics it is important that you know the x-coordinates of the x-intercepts of the sine and cosine graphs, in other words, the zeros of $f(x) = \sin x$ and $g(x) = \cos x$. From the sine graph in Figure 1.4.1 we see that the zeros of the sine function, or the numbers for which $\sin x = 0$, are $x = 0, \pm\pi, \pm 2\pi, \pm 3\pi, \ldots$. These numbers are integer multiples of π. From the cosine graph in Figure 1.4.2 we see that $\cos x = 0$ when $x = \pm\pi/2, \pm 3\pi/2, \pm 5\pi/2, \ldots$. These numbers are odd-integer multiples of $\pi/2$.

If n represents an integer, then $2n + 1$ is an odd integer. Therefore the **zeros** of $f(x) = \sin x$ and $g(x) = \cos x$ can be written in a compact form:

- $\sin x = 0$ for $x = n\pi$, n an integer, $\quad\quad\quad\quad\quad\quad\quad\quad\quad\quad\quad\quad\quad\quad\quad\quad$ (3)

- $\cos x = 0$ for $x = (2n + 1)\dfrac{\pi}{2}$, n an integer. $\quad\quad\quad\quad\quad\quad\quad\quad\quad\quad$ (4)

Additional important numerical values of the sine and cosine functions on the interval $[0, \pi]$ are given in the table that follows.

x	0	$\dfrac{\pi}{6}$	$\dfrac{\pi}{4}$	$\dfrac{\pi}{3}$	$\dfrac{\pi}{2}$	$\dfrac{2\pi}{3}$	$\dfrac{3\pi}{4}$	$\dfrac{5\pi}{6}$	π
$\sin x$	0	$\dfrac{1}{2}$	$\dfrac{\sqrt{2}}{2}$	$\dfrac{\sqrt{3}}{2}$	1	$\dfrac{\sqrt{3}}{2}$	$\dfrac{\sqrt{2}}{2}$	$\dfrac{1}{2}$	0
$\cos x$	1	$\dfrac{\sqrt{3}}{2}$	$\dfrac{\sqrt{2}}{2}$	$\dfrac{1}{2}$	0	$-\dfrac{1}{2}$	$-\dfrac{\sqrt{2}}{2}$	$-\dfrac{\sqrt{3}}{2}$	-1

(5)

You should be able to discern the values $\sin x$ and $\cos x$ on $[\pi, 2\pi]$ from this table using the concept of the unit circle and a reference angle. Of course, outside the interval $[0, 2\pi]$ we can determine corresponding function values using periodicity.

■ **Other Trigonometric Functions** Four additional trigonometric functions are defined in terms of quotients or reciprocals of the sine and cosine functions. The **tangent**, **cotangent**, **secant**, and **cosecant** functions are defined, respectively, as

$$\tan x = \frac{\sin x}{\cos x}, \quad\quad \cot x = \frac{\cos x}{\sin x}, \quad\quad\quad\quad\quad\quad (6)$$

$$\sec x = \frac{1}{\cos x}, \quad\quad \csc x = \frac{1}{\sin x}. \quad\quad\quad\quad\quad\quad (7)$$

The domain of each function in (6) and (7) is the set of real numbers except those numbers for which the denominator is zero. From (4) we see

- the domain of $\tan x$ and of $\sec x$ is $\{x \mid x \neq (2n + 1)\pi/2, n = 0, \pm 1, \pm 2, \ldots\}$.

Similarly, from (3) it follows that

- the domain of $\cot x$ and of $\csc x$ is $\{x \mid x \neq n\pi, n = 0, \pm 1, \pm 2, \ldots\}$.

Moreover, from (2)

$$|\sec x| = \left|\frac{1}{\cos x}\right| = \frac{1}{|\cos x|} \geq 1 \quad\quad\quad\quad\quad\quad (8)$$

and

$$|\csc x| = \left|\frac{1}{\sin x}\right| = \frac{1}{|\sin x|} \geq 1. \quad\quad\quad\quad\quad\quad (9)$$

Recall that an absolute-value inequality such as (8) means $\sec x \geq 1$ or $\sec x \leq -1$. Hence the range of the secant and the cosecant functions is $(-\infty, -1] \cup [1, \infty)$. The tangent and cotangent functions have the same range: $(-\infty, \infty)$. Using (5) we can determine some numerical values of $\tan x, \cot x, \sec x,$ and $\csc x$. For example,

$$\tan \frac{2\pi}{3} = \frac{\sin(2\pi/3)}{\cos(2\pi/3)} = \frac{\sqrt{3}/2}{-1/2} = -\sqrt{3}.$$

Graphs The numbers that make the denominators of $\tan x$, $\cot x$, $\sec x$, and $\csc x$ equal to zero correspond to vertical asymptotes of their graphs. In view of (4), the vertical asymptotes of the graphs of $y = \tan x$ and $y = \sec x$ are $x = \pm \pi/2, \pm 3\pi/2, \pm 5\pi/2, \ldots$. On the other hand, from (3), the vertical asymptotes for the graphs of $y = \cot x$ and $y = \csc x$ are $x = 0$, $\pm \pi, \pm 2\pi, \pm 3\pi, \ldots$. These asymptotes are the red dashed lines in FIGURES 1.4.3–1.4.6.

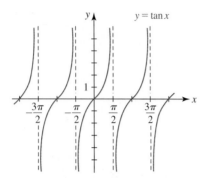

FIGURE 1.4.3 Graph of $y = \tan x$

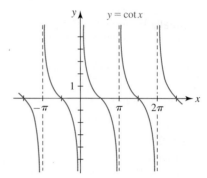

FIGURE 1.4.4 Graph of $y = \cot x$

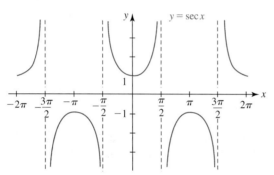

FIGURE 1.4.5 Graph of $y = \sec x$

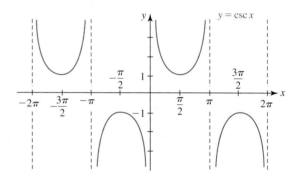

FIGURE 1.4.6 Graph of $y = \csc x$

Because the sine and cosine functions are 2π periodic, $\sec x$ and $\csc x$ are also 2π periodic. But it should be obvious from Figures 1.4.3 and 1.4.4 that tangent and cotangent are π periodic:

$$\tan(x + \pi) = \tan x \quad \text{and} \quad \cot(x + \pi) = \cot x. \tag{10}$$

Also, $\tan x$, $\cot x$, and $\csc x$ are odd functions; $\sec x$ is an even function.

Transformation and Graphs We can obtain variations of the graphs of the trigonometric functions through rigid and nonrigid transformations. Graphs of functions of the form

$$y = D + A\sin(Bx + C) \quad \text{or} \quad y = D + A\cos(Bx + C), \tag{11}$$

where $A, B > 0, C$, and D are real constants, represent shifts, compressions, and stretches of the basic sine and cosine graphs. For example,

$$y = \underset{\underset{\text{vertical shift}}{\uparrow}}{D} + \underset{\underset{\text{vertical stretch/compression/reflection}}{\uparrow}}{A} \sin(\underset{\underset{\substack{\text{horizontal stretch/compression} \\ \text{by changing period}}}{\uparrow}}{B}x + \underset{\underset{\text{horizontal shift}}{\uparrow}}{C}).$$

The number $|A|$ is called the **amplitude** of the functions or of their graphs. The amplitude of the basic functions $y = \sin x$ and $y = \cos x$ is $|A| = 1$. The **period** of each function in (11) is $2\pi/B$, $B > 0$, and the portion of the graph of each function in (11) over the interval $[0, 2\pi/B]$ is called one **cycle**.

EXAMPLE 1 Periods

(a) The period of $y = \sin 2x$ is $2\pi/2 = \pi$, and therefore one cycle of the graph is completed on the interval $[0, \pi]$.

(b) Before determining the period of $\sin(-\frac{1}{2}x)$ we must first rewrite the function as $\sin(-\frac{1}{2}x) = -\sin(\frac{1}{2}x)$ (the sine is an odd function). The period is now $2\pi/\frac{1}{2} = 4\pi$ and therefore one cycle of the graph is completed on the interval $[0, 4\pi]$. ∎

EXAMPLE 2 Vertically Transformed Graphs

Graph

(a) $y = -\dfrac{1}{2}\cos x$ (b) $y = 1 + 2\sin x$.

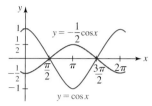

FIGURE 1.4.7 Graph of function in Example 2(a)

Solution

(a) The graph of $y = -\frac{1}{2}\cos x$ is the graph of $y = \cos x$ compressed vertically by a factor of 2 and the minus sign indicates that the graph is then reflected in the x-axis. With the identification $A = -\frac{1}{2}$ we see that the amplitude of the function is $|A| = |-\frac{1}{2}| = \frac{1}{2}$. The graph of $y = -\frac{1}{2}\cos x$ on the interval $[0, 2\pi]$ is shown in red in FIGURE 1.4.7.

(b) The graph of $y = 2\sin x$ is the graph of $y = \sin x$ stretched vertically by a factor of 2. The amplitude of the graph is $|A| = |2| = 2$. The graph of $y = 1 + 2\sin x$ is the graph of $y = 2\sin x$ shifted up 1 unit. See FIGURE 1.4.8. ∎

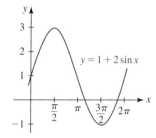

FIGURE 1.4.8 Graph of function in Example 2(b)

EXAMPLE 3 Horizontally Compressed Cosine Graph

Find the period of $y = \cos 4x$ and graph the function.

Solution With the identification that $B = 4$, we see that the period of $y = \cos 4x$ is $2\pi/4 = \pi/2$. We conclude that the graph of $y = \cos 4x$ is the graph of $y = \cos x$ compressed horizontally. To graph the function, we draw one cycle of the cosine graph with amplitude 1 on the interval $[0, \pi/2]$ and then use periodicity to extend the graph. FIGURE 1.4.9 shows four complete cycles of $y = \cos 4x$ (the basic cycle in red and the extended graph in blue) and one cycle of $y = \cos x$ (shown in green) on $[0, 2\pi]$. Notice that $y = \cos 4x$ attains its minimum at $x = \pi/4$ since $\cos 4(\pi/4) = \cos \pi = -1$ and its maximum at $x = \pi/2$ since $\cos 4(\pi/2) = \cos 2\pi = 1$. ∎

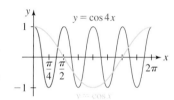

FIGURE 1.4.9 Graph of function in Example 3

From Section 1.2 we know that the graph of $y = \cos(x - \pi/2)$ is the basic cosine graph shifted to the right. In FIGURE 1.4.10 the graph of $y = \cos(x - \pi/2)$ (in red) on the interval $[0, 2\pi]$ is one cycle of $y = \cos x$ on the interval $[-\pi/2, 3\pi/2]$ (in blue) shifted horizontally $\pi/2$ units to the right. Similarly, the graphs of $y = \sin(x + \pi/2)$ and $y = \sin(x - \pi/2)$ are the basic sine graphs shifted $\pi/2$ units to the left and to the right, respectively. See FIGURE 1.4.11 and FIGURE 1.4.12.

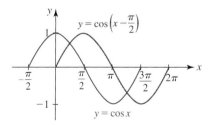

FIGURE 1.4.10 Horizontally shifted cosine graph

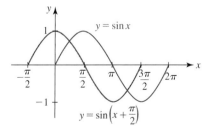

FIGURE 1.4.11 Horizontally shifted sine graph

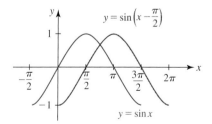

FIGURE 1.4.12 Horizontally shifted sine graph

By comparing the red graphs in Figures 1.4.10–1.4.12 with the graphs in Figures 1.4.1 and 1.4.2 we see that

- the cosine graph shifted $\pi/2$ units to the right is the sine graph,
- the sine graph shifted $\pi/2$ units to the left is the cosine graph, and
- the sine graph shifted $\pi/2$ units to the right is the cosine graph reflected in the x-axis.

In other words, we have graphically verified the identities

$$\cos\left(x - \frac{\pi}{2}\right) = \sin x, \quad \sin\left(x + \frac{\pi}{2}\right) = \cos x, \quad \text{and} \quad \sin\left(x - \frac{\pi}{2}\right) = -\cos x. \quad (12)$$

Suppose $f(x) = A \sin Bx$, then

$$f\left(x + \frac{C}{B}\right) = A \sin B\left(x + \frac{C}{B}\right) = A \sin(Bx + C). \quad (13)$$

The result in (13) shows that the graph of $y = A \sin(Bx + C)$ can be obtained by shifting the graph of $f(x) = A \sin Bx$ horizontally a distance $|C|/B$. If $C < 0$, the shift is to the right, whereas if $C > 0$, the shift is to the left. The number $|C|/B$ is called the **phase shift** of the graphs of the functions in (3).

EXAMPLE 4 Horizontally Shifted Cosine Graph

The graph of $y = 10 \cos 4x$ is shifted $\pi/12$ units to the right. Find its equation.

Solution By writing $f(x) = 10 \cos 4x$ and using (13), we find

$$f\left(x - \frac{\pi}{12}\right) = 10 \cos 4\left(x - \frac{\pi}{12}\right) \quad \text{or} \quad y = 10 \cos\left(4x - \frac{\pi}{3}\right).$$

In the last equation we would identify $C = -\pi/3$. The phase shift is $\pi/12$. ∎

Note: As a practical matter the phase shift for either $y = A \sin(Bx + C)$ or $y = A \cos(Bx + C)$ can be obtained by factoring the number B from $Bx + C$. For example,

$$y = A \sin(Bx + C) = A \sin B\left(x + \frac{C}{B}\right).$$

EXAMPLE 5 Horizontally Shifted Graphs

Graph

(a) $y = 3 \sin(2x - \pi/3)$ (b) $y = 2 \cos(\pi x + \pi)$.

Solution

(a) For purposes of comparison we will first graph $y = 3 \sin 2x$. The amplitude of $y = 3 \sin 2x$ is $|A| = 3$ and its period is $2\pi/2 = \pi$. Thus one cycle of $y = 3 \sin 2x$ is completed on the interval $[0, \pi]$. Then we extend this graph to the adjacent interval $[\pi, 2\pi]$ as shown in blue in FIGURE 1.4.13. Next, we rewrite $y = 3 \sin(2x - \pi/3)$ by factoring 2 from $2x - \pi/3$:

$$y = 3 \sin\left(2x - \frac{\pi}{3}\right) = 3 \sin 2\left(x - \frac{\pi}{6}\right).$$

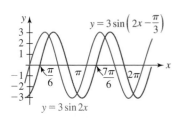

FIGURE 1.4.13 Graph of function in Example 5(a)

From the form of the last expression we see that the phase shift is $\pi/6$. The graph of the given function, shown in red in Figure 1.4.13, is obtained by shifting the graph of $y = 3 \sin 2x$ (in blue) to the right $\pi/6$ units.

(b) The amplitude of $y = 2 \cos \pi x$ is $|A| = 2$ and the period is $2\pi/\pi = 2$. Thus one cycle of $y = 2 \cos \pi x$ is completed on the interval $[0, 2]$. In FIGURE 1.4.14 two cycles of the graph of $y = 2 \cos \pi x$ (in blue) are shown. The x-intercepts of this graph correspond to the values of x for which $\cos \pi x = 0$. By (4), this implies $\pi x = (2n + 1)\pi/2$ or $x = (2n + 1)/2$, n an integer. In other words, for $n = 0, -1, 1, -2, 2, -3, \ldots$ we get $x = \pm\frac{1}{2}, \pm\frac{3}{2}, \pm\frac{5}{2}$, and so on. Now by rewriting the given function as

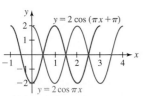

FIGURE 1.4.14 Graph of function in Example 5(b)

$$y = 2 \cos \pi(x + 1)$$

we see that the phase shift is 1. The graph of $y = 2 \cos(\pi x + \pi)$ shown in red in Figure 1.4.14, is obtained by shifting the graph of $y = 2 \cos \pi x$ (in blue) to the left 1 unit. This means that the x-intercepts are the same for both graphs. ∎

In applied mathematics, trigonometric functions serve as mathematical models for many periodic phenomena.

EXAMPLE 6 Alternating Current

A mathematical model for the current I (in amperes) in a wire of an alternating-current circuit is given by $I(t) = 30 \sin 120\pi t$, where t is time measured in seconds. Sketch one cycle of the graph. What is the maximum value of the current?

Solution The graph has amplitude 30 and period $2\pi/120\pi = \frac{1}{60}$. Therefore, we sketch one cycle of the basic sine curve on the interval $\left[0, \frac{1}{60}\right]$, as shown in FIGURE 1.4.15. From the figure it is evident that the maximum value of the current is $I = 30$ amperes and occurs in the interval $\left[0, \frac{1}{60}\right]$ at $t = \frac{1}{240}$ since

$$I\left(\frac{1}{240}\right) = 30 \sin\left(120\pi \cdot \frac{1}{240}\right) = 30 \sin \frac{\pi}{2} = 30.$$

FIGURE 1.4.15 Graph of current in Example 6 shows that there are 60 cycles in 1 second

■ **For Future Reference** Trigonometric identities are used throughout calculus, especially in the study of integral calculus. For convenience of reference we list next some identities that are of particular importance.

Pythagorean Identities

$$\sin^2 x + \cos^2 x = 1 \tag{14}$$

$$1 + \tan^2 x = \sec^2 x \tag{15}$$

$$1 + \cot^2 x = \csc^2 x \tag{16}$$

Sum and Difference Formulas

$$\sin(x_1 \pm x_2) = \sin x_1 \cos x_2 \pm \cos x_1 \sin x_2 \tag{17}$$

$$\cos(x_1 \pm x_2) = \cos x_1 \cos x_2 \mp \sin x_1 \sin x_2 \tag{18}$$

Double-Angle Formulas

$$\sin 2x = 2 \sin x \cos x \tag{19}$$

$$\cos 2x = \cos^2 x - \sin^2 x \tag{20}$$

Half-Angle Formulas

$$\sin^2 \frac{x}{2} = \frac{1}{2}(1 - \cos x) \tag{21}$$

$$\cos^2 \frac{x}{2} = \frac{1}{2}(1 + \cos x) \tag{22}$$

Additional identities can be found in the *Resource Pages* at the end of this text.

Exercises 1.4 Answers to selected odd-numbered problems begin on page ANS-5.

≡ Fundamentals

In Problems 1–6, use the techniques of shifting, stretching, compressing, and reflecting to sketch at least one cycle of the graph of the given function.

1. $y = \frac{1}{2} + \cos x$
2. $y = -1 + \cos x$
3. $y = 2 - \sin x$
4. $y = 3 + 3 \sin x$
5. $y = -2 + 4 \cos x$
6. $y = 1 - 2 \sin x$

In Problems 7–14, find the amplitude and period of the given function. Sketch at least one cycle of the graph.

7. $y = 4 \sin \pi x$
8. $y = -5 \sin \frac{x}{2}$
9. $y = -3 \cos 2\pi x$
10. $y = \frac{5}{2} \cos 4x$
11. $y = 2 - 4 \sin x$
12. $y = 2 - 2 \sin \pi x$
13. $y = 1 + \cos \frac{2x}{3}$
14. $y = -1 + \sin \frac{\pi x}{2}$

In Problems 15–18, the given figure shows one cycle of a sine or cosine graph. From the figure determine A and D and write an equation of the form $y = D + A\sin x$ or $y = D + A\cos x$ for the graph.

15.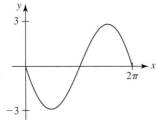

FIGURE 1.4.16   Graph for Problem 15

16.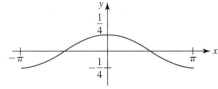

FIGURE 1.4.17   Graph for Problem 16

17.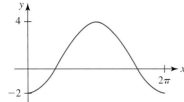

FIGURE 1.4.18   Graph for Problem 17

18.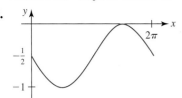

FIGURE 1.4.19   Graph for Problem 18

In Problems 19–24, the given figure shows one cycle of a sine or cosine graph. From the figure determine A and B and write an equation of the form $y = A\sin Bx$ or $y = A\cos Bx$ for the graph.

19.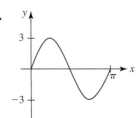

FIGURE 1.4.20   Graph for Problem 19

20.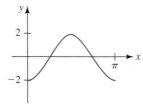

FIGURE 1.4.21   Graph for Problem 20

21.

FIGURE 1.4.22   Graph for Problem 21

22.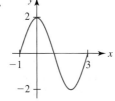

FIGURE 1.4.23   Graph for Problem 22

23.

FIGURE 1.4.24   Graph for Problem 23

24.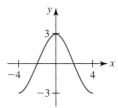

FIGURE 1.4.25   Graph for Problem 24

In Problems 25–34, find the amplitude, period, and phase shift of the given function. Sketch at least one cycle of the graph.

25. $y = \sin\left(x - \dfrac{\pi}{6}\right)$

26. $y = \sin\left(3x - \dfrac{\pi}{4}\right)$

27. $y = \cos\left(x + \dfrac{\pi}{4}\right)$

28. $y = -2\cos\left(2x - \dfrac{\pi}{6}\right)$

29. $y = 4\cos\left(2x - \dfrac{3\pi}{2}\right)$

30. $y = 3\sin\left(2x + \dfrac{\pi}{4}\right)$

31. $y = 3\sin\left(\dfrac{x}{2} - \dfrac{\pi}{3}\right)$

32. $y = -\cos\left(\dfrac{x}{2} - \pi\right)$

33. $y = -4\sin\left(\dfrac{\pi}{3}x - \dfrac{\pi}{3}\right)$

34. $y = 2\cos\left(-2\pi x - \dfrac{4\pi}{3}\right)$

In Problems 35 and 36, write an equation of the function whose graph is described in words.

35. The graph of $y = \sin \pi x$ is stretched vertically upward by a factor of 5 and is shifted $\tfrac{1}{2}$ unit to the right.

36. The graph of $y = 4\cos\dfrac{x}{2}$ is shifted downward 8 units and is shifted $2\pi/3$ units to the left.

In Problems 37 and 38, find the x-intercepts of the graph of the given function on the interval $[0, 2\pi]$. Then find all intercepts using periodicity.

37. $y = -1 + \sin x$

38. $y = 1 - 2\cos x$

In Problems 39–44, find the x-intercepts for the graph of the given function. Do not graph.

39. $y = \sin \pi x$

40. $y = -\cos 2x$

41. $y = 10\cos\dfrac{x}{2}$

42. $y = 3\sin(-5x)$

43. $y = \sin\left(x - \dfrac{\pi}{4}\right)$

44. $y = \cos(2x - \pi)$

In Problems 45–52, find the period, x-intercepts, and the vertical asymptotes of the given function. Sketch at least one cycle of the graph.

45. $y = \tan \pi x$

46. $y = \tan\dfrac{x}{2}$

47. $y = \cot 2x$

48. $y = -\cot\dfrac{\pi x}{3}$

49. $y = \tan\left(\dfrac{x}{2} - \dfrac{\pi}{4}\right)$

50. $y = \dfrac{1}{4}\cot\left(x - \dfrac{\pi}{2}\right)$

51. $y = -1 + \cot \pi x$

52. $y = \tan\left(x + \dfrac{5\pi}{6}\right)$

In Problems 53–56, find the period and the vertical asymptotes of the given function. Sketch at least one cycle of the graph.

53. $y = 3 \csc \pi x$
54. $y = -2 \csc \dfrac{x}{3}$
55. $y = \sec\left(3x - \dfrac{\pi}{2}\right)$
56. $y = \csc(4x + \pi)$

≡ **Mathematical Models**

57. **Depth of Water** The depth of water d at the entrance to a small harbor at time t is modeled by a function of the form

$$d(t) = D + A \sin B\left(t - \dfrac{\pi}{2}\right),$$

where A is one half the difference between the high- and low-tide depths, $2\pi/B$, $B > 0$, is the tidal period, and D is the average depth. Assume that the tidal period is 12 hours, the depth at high tide is 18 ft, and the depth at low tide is 6 ft. Sketch two cycles of the graph of d.

58. **Fahrenheit Temperature** Suppose that

$$T(t) = 50 + 10 \sin \dfrac{\pi}{12}(t - 8), \quad 0 \le t \le 24$$

is a mathematical model of the Fahrenheit temperature at t hours after midnight on a certain day of the week.
(a) What is the temperature at 8 A.M.?
(b) At what time(s) does $T(t) = 60$?
(c) Sketch the graph of T.
(d) Find the maximum and minimum temperatures and the times at which they occur.

≡ **Calculator/CAS Problems**

59. **Acceleration Due to Gravity** Because of Earth's rotation, its shape is not spherical but bulges at the equator and is flattened at the poles. As a result, the acceleration due to gravity is not a constant 980 cm/s², but varies with latitude θ. Satellite studies have suggested that the acceleration due to gravity g is approximated by the mathematical model

$$g = 978.0309 + 5.18552 \sin^2 \theta - 0.00570 \sin^2 2\theta.$$

Find g
(a) at the equator ($\theta = 0°$),
(b) at the north pole, and
(c) at 45° north latitude.

60. **Putting the Shot** The range of a shot put released from a height h above the ground with an initial velocity v_0 at an angle ϕ to the horizontal can be approximated by the mathematical model

$$R = \dfrac{v_0 \cos \phi}{g}\left[v_0 \sin \phi + \sqrt{v_0^2 \sin^2 \phi + 2gh}\right],$$

where g is the acceleration due to gravity. See FIGURE 1.4.26.
(a) If $v_0 = 13.7$ m/s, $\phi = 40°$, and $g = 9.81$ m/s², compare the ranges achieved for the release heights $h = 2.0$ m and $h = 2.4$ m.
(b) Explain why an increase in h yields an increase in the range R if the other parameters are held fixed.
(c) What does this imply about the advantage that height gives a shot-putter?

FIGURE 1.4.26 Projectile in Problem 60

≡ **Think About It**

61. The function $f(x) = \sin \tfrac{1}{2}x + \sin 2x$ is periodic. What is the period of f?

62. Discuss and then sketch the graphs of $y = |\sin x|$ and $y = |\cos x|$.

63. Discuss and then sketch the graphs of $y = |\sec x|$ and $y = |\csc x|$.

64. Can the given equation have any real-number solution x?
(a) $9 \csc x = 1$
(b) $7 + 10 \sec x = 0$
(c) $\sec x = -10.5$

In Problems 65 and 66, use the graphs of $y = \tan x$ and $y = \sec x$ to find numbers A and C for which the given equality is true.

65. $\cot x = A \tan(x + C)$
66. $\csc x = A \sec(x + C)$

1.5 Inverse Functions

▌**Introduction** In Section 1.1 we saw that a function f is a rule of correspondence that assigns to each value x in its domain X, a single or unique value y in its range. This rule does not preclude having the same number y associated with several *different* values of x. For example, for $f(x) = -x^2 + 2x + 4$, the value $y = 4$ in the range of f occurs at either $x = 0$ or $x = 2$ in the domain of f. On the other hand, for the function $f(x) = 2x + 3$, the value $y = 4$ occurs only at

$x = \frac{1}{2}$. Indeed, for every value y in the range of $f(x) = 2x + 3$, there corresponds only one value of x in the domain. Functions of this last kind are given the special name **one-to-one**.

Definition 1.5.1 One-to-One Function

A function f is said to be **one-to-one** if each number in the range of f is associated with exactly one number in its domain X.

■ **Horizontal Line Test** Interpreted geometrically, Definition 1.5.1 means that a horizontal line ($y = $ constant) can intersect the graph of a one-to-one function in at most one point. Furthermore, if *every* horizontal line that intersects the graph of a function does so in at most one point, then the function is necessarily one-to-one. A function is *not* one-to-one if *some* horizontal line intersects its graph more than once.

EXAMPLE 1 Horizontal Line Test

(a) The graph of the function $f(x) = x^2 + 1$ and a horizontal line $y = c$ intersecting the graph is shown in FIGURE 1.5.1(a). The figure clearly indicates that there are two numbers x_1 and x_2 in the domain of f for which $f(x_1) = f(x_2) = c$. Hence the function f is not one-to-one.
(b) Inspection of Figure 1.5.1(b) shows that for every horizontal line $y = c$ intersecting the graph of $f(x) = x^3$, there is only one number x_1 in the domain of f such that $f(x_1) = c$. The function f is one-to-one. ■

FIGURE 1.5.1 Two types of functions in Example 1

■ **Inverse of a One-to-One Function** Suppose f is a one-to-one function with domain X and range Y. Since every number y in Y corresponds to precisely one number x in X, the function f must actually determine a "reverse" function g whose domain is Y and range is X. As shown in FIGURE 1.5.2, f and g must satisfy

$$f(x) = y \quad \text{and} \quad g(y) = x. \tag{1}$$

The equations in (1) are actually the compositions of the functions f and g:

$$f(g(y)) = y \quad \text{and} \quad g(f(x)) = x. \tag{2}$$

The function g is called the **inverse** of f or the **inverse function** for f. Following convention that each domain element be denoted by the symbol x, the first equation in (2) is rewritten as $f(g(x)) = x$. We summarize the results given in (2).

FIGURE 1.5.2 A function f and its inverse function g

Definition 1.5.2 Inverse Function

Let f be a one-to-one function with domain X and range Y. The **inverse** of f is the function g with domain Y and range X for which

$$f(g(x)) = x \text{ for every } x \text{ in } Y, \tag{3}$$

and

$$g(f(x)) = x \text{ for every } x \text{ in } X. \tag{4}$$

Of course, if a function f is not one-to-one, then it has no inverse function.

■ **Notation** The inverse of a function f is usually written f^{-1} and is read "f inverse." This latter notation, although standard, is somewhat unfortunate. We hasten to point out that in the symbol $f^{-1}(x)$ the "-1" is *not* an exponent. In terms of the new notation, (3) and (4) become, respectively,

▶ In (3) and (4), the symbol g is playing the part of the symbol f^{-1}.

$$f(f^{-1}(x)) = x \quad \text{and} \quad f^{-1}(f(x)) = x. \tag{5}$$

■ **Properties** Before we examine a method for finding the inverse of a one-to-one function f, let us list some important properties about f and its inverse f^{-1}.

> **Theorem 1.5.1** Properties of Inverse Functions
>
> (i) The domain of f^{-1} = range of f.
> (ii) The range of f^{-1} = domain of f.
> (iii) An inverse function f^{-1} is one-to-one.
> (iv) The inverse of f^{-1} is f.
> (v) The inverse of f is unique.

■ **A Method for Finding f^{-1}** If f^{-1} is the inverse of a one-to-one function $y = f(x)$, then from (1), $x = f^{-1}(y)$. Thus we need only do the following two things to find f^{-1}.

Guidelines for Finding the Inverse Function

Suppose $y = f(x)$ is a one-to-one function. Then to find f^{-1}:

- Solve $y = f(x)$ for the symbol x in terms of y (if possible). This gives $x = f^{-1}(y)$.
- Relabel the variable x as y and the variable y as x. This gives $y = f^{-1}(x)$.

Note: It is sometimes convenient to interchange the steps in the foregoing guidelines:

- Relabel x and y in the equation $y = f(x)$, and solve (if possible) $x = f(y)$ for y. This gives $y = f^{-1}(x)$.

EXAMPLE 2 Inverse of a Function

Find the inverse of $f(x) = x^3$.

Solution In Example 1 we saw that this function was one-to-one. To begin, we rewrite the function as $y = x^3$. Solving for x then gives $x = y^{1/3}$. Next we relabel variables to obtain $y = x^{1/3}$. Thus $f^{-1}(x) = x^{1/3}$ or equivalently $f^{-1}(x) = \sqrt[3]{x}$. ■

Finding the inverse of a one-to-one function $y = f(x)$ is sometimes difficult and at times impossible. For example, FIGURE 1.5.3 suggests (and it can be shown) that the function $f(x) = x^3 + x + 3$ is one-to-one and so has an inverse f^{-1}. But solving the equation $y = x^3 + x + 3$ for x is difficult for everyone (including your instructor). Since f is a polynomial function its domain is $(-\infty, \infty)$ and, because its end behavior is that of $y = x^3$, the range of f is $(-\infty, \infty)$. Consequently the domain and range of f^{-1} are $(-\infty, \infty)$. Even though we do not know f^{-1} explicitly, it makes complete sense to talk about the values such as $f^{-1}(3)$ and $f^{-1}(5)$. In the case of $f^{-1}(3)$ note that $f(0) = 3$. This means that $f^{-1}(3) = 0$. Can you figure out the value of $f^{-1}(5)$?

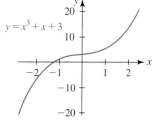

FIGURE 1.5.3 Graph suggests f is one-to-one

■ **Graphs of f and f^{-1}** Suppose that (a, b) represents any point on the graph of a one-to-one function f. Then $f(a) = b$ and

$$f^{-1}(b) = f^{-1}(f(a)) = a$$

implies that (b, a) is a point on the graph of f^{-1}. As shown in FIGURE 1.5.4(a), the points (a, b) and (b, a) are reflections of each other in the line $y = x$. This means that the line $y = x$ is the perpendicular bisector of the line segment from (a, b) to (b, a). Because each point on one graph is the reflection of a corresponding point on the other graph, we see in Figure 1.5.4(b) that the graphs of f^{-1} and f are **reflections** of each other in the line $y = x$. We also say that the graphs of f^{-1} and f are **symmetric** with respect to the line $y = x$.

FIGURE 1.5.4 Graphs of f and f^{-1} are reflections in the line $y = x$

EXAMPLE 3 Graphs of f and f^{-1}

In Example 2 we saw that the inverse of $y = x^3$ is $y = x^{1/3}$. In FIGURES 1.5.5(a) and 1.5.5(b) we show the graphs of these functions; in Figure 1.5.5(c) the graphs are superimposed on the same coordinate system to illustrate that the graphs are reflections of each other in the line $y = x$.

FIGURE 1.5.5 Graphs of f and f^{-1} in Example 3

Every linear function $f(x) = ax + b$, $a \neq 0$, is one-to-one.

EXAMPLE 4 Inverse of a Function

Find the inverse of the linear function $f(x) = 5x - 7$.

Solution Since the graph of $y = 5x - 7$ is a nonhorizontal line, it follows from the horizontal line test that f is a one-to-one function. To find f^{-1} solve $y = 5x - 7$ for x:

$$5x = y + 7 \quad \text{implies} \quad x = \frac{1}{5}y + \frac{7}{5}.$$

Relabeling variables in the last equation gives $y = \frac{1}{5}x + \frac{7}{5}$. Therefore $f^{-1}(x) = \frac{1}{5}x + \frac{7}{5}$. The graphs of f and f^{-1} are compared in FIGURE 1.5.6.

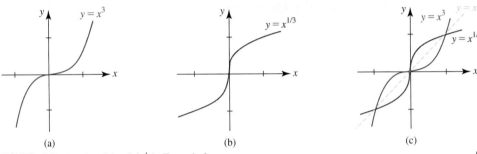

FIGURE 1.5.6 Graphs of f and f^{-1} in Example 4

Every quadratic function $f(x) = ax^2 + bx + c$, $a \neq 0$, is *not* one-to-one.

■ **Restricted Domains** For a function f that is not one-to-one, it may be possible to restrict its domain in such a manner so that the new function consisting of f defined on this restricted domain is one-to-one and so has an inverse. In most cases we want to restrict the domain so that the new function retains its original range. The next example illustrates this concept.

EXAMPLE 5 Restricted Domain

In Example 1 we showed graphically that the quadratic function $f(x) = x^2 + 1$ is not one-to-one. The domain of f is $(-\infty, \infty)$, and as seen in FIGURE 1.5.7(a), the range of f is $[1, \infty)$. Now by defining $f(x) = x^2 + 1$ only on the interval $[0, \infty)$, we see two things in Figure 1.5.7(b): the range of f is preserved and $f(x) = x^2 + 1$ confined to the domain $[0, \infty)$ passes the horizontal line test, in other words, is one-to-one. The inverse of this new one-to-one function is obtained in the usual manner. Solving $y = x^2 + 1$ for x and relabeling variables gives

$$x = \pm\sqrt{y - 1} \quad \text{and so} \quad y = \pm\sqrt{x - 1}.$$

The appropriate algebraic sign in the last equation is determined from the fact that the domain and range of f^{-1} are $[1, \infty)$ and $[0, \infty)$, respectively. This forces us to choose $f^{-1}(x) = \sqrt{x-1}$ as the inverse of f. See Figure 1.5.7(c).

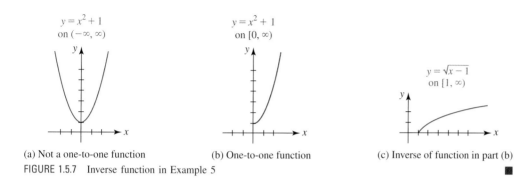

(a) Not a one-to-one function   (b) One-to-one function   (c) Inverse of function in part (b)

FIGURE 1.5.7 Inverse function in Example 5

■ **Inverse Trigonometric Functions** Although none of the trigonometric functions are one-to-one, by suitably restricting each of their domains we can define six inverse trigonometric functions.

■ **Inverse Sine Function** From FIGURE 1.5.8(a) we see that the function $y = \sin x$ on the closed interval $[-\pi/2, \pi/2]$ takes on all values in its range $[-1, 1]$. Notice that any horizontal line drawn to intersect the red portion of the graph can do so at most once. Thus the sine function on this restricted domain is one-to-one and has an inverse. There are two notations commonly used throughout mathematics to denote the inverse of the function shown in Figure 1.5.8(b):

$$\sin^{-1} x \quad \text{or} \quad \arcsin x,$$

and are read **inverse sine of x** and **arcsine of x**, respectively.

◀ The computer algebra system *Mathematica* uses the arcsine notation.

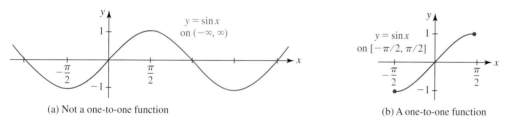

(a) Not a one-to-one function   (b) A one-to-one function

FIGURE 1.5.8 Restricting the domain of $y = \sin x$ to produce a one-to-one function

In FIGURE 1.5.9(a) we have reflected the portion of the graph of $y = \sin x$ on the interval $[-\pi/2, \pi/2]$ (the red graph in Figure 1.5.8(b)) in the line $y = x$ to obtain the graph of $y = \sin^{-1} x$ (in blue). For clarity, we have reproduced this blue graph in Figure 1.5.9(b). As this graph shows, the domain of the inverse sine function is $[-1, 1]$ and the range is $[-\pi/2, \pi/2]$.

(a)   (b)

FIGURE 1.5.9 Graph of the inverse sine function is the blue curve

> **Definition 1.5.3** Inverse Sine Function
>
> The **inverse sine function**, or **arcsine function**, is defined by
> $$y = \sin^{-1} x \quad \text{if and only if} \quad x = \sin y, \qquad (6)$$
> where $-1 \leq x \leq 1$ and $-\pi/2 \leq y \leq \pi/2$.

In words:

- *The inverse sine of the number x is that number y (or radian-measured angle) between $-\pi/2$ and $\pi/2$ whose sine is x.*

The symbols $y = \arcsin x$ and $y = \sin^{-1} x$ are used interchangeably throughout mathematics and its applications and so we will alternate their use so that you become comfortable with both notations.

EXAMPLE 6 Evaluating the Inverse Sine Function

Find

(a) $\arcsin \dfrac{1}{2}$ (b) $\sin^{-1}\left(-\dfrac{1}{2}\right)$ and   (c) $\sin^{-1}(-1)$.

Solution

(a) If we let $y = \arcsin \frac{1}{2}$, then by (6) we must find the number y (or radian-measured angle) that satisfies $\sin y = \frac{1}{2}$ and $-\pi/2 \leq y \leq \pi/2$. Since $\sin(\pi/6) = \frac{1}{2}$, and $\pi/6$ satisfies the inequality $-\pi/2 \leq y \leq \pi/2$ it follows that
$$y = \frac{\pi}{6}.$$

(b) If we let $y = \sin^{-1}(-\frac{1}{2})$, then $\sin y = -\frac{1}{2}$. Since we must choose y such that $-\pi/2 \leq y \leq \pi/2$, we find that $y = -\pi/6$.

(c) Letting $y = \sin^{-1}(-1)$, we have that $\sin y = -1$ and $-\pi/2 \leq y \leq \pi/2$. Hence $y = -\pi/2$. ∎

Read this paragraph several times. ▶   In parts (b) and (c) of Example 6 we were careful to choose y so that $-\pi/2 \leq y \leq \pi/2$. For example, it is a common error to think that because $\sin(3\pi/2) = -1$, then necessarily $\sin^{-1}(-1)$ can be taken to be $3\pi/2$. Remember: If $y = \sin^{-1} x$, then y is subject to the restriction $-\pi/2 \leq y \leq \pi/2$, and $3\pi/2$ does not satisfy this inequality.

EXAMPLE 7 Evaluating a Composition

Without using a calculator, find $\tan\left(\sin^{-1} \frac{1}{4}\right)$.

Solution We must find the tangent of the angle of t radians with sine equal to $\frac{1}{4}$, that is, $\tan t$ where $t = \sin^{-1}\frac{1}{4}$. The angle t is shown in FIGURE 1.5.10. Since
$$\tan t = \frac{\sin t}{\cos t} = \frac{1/4}{\cos t},$$
we want to determine the value of $\cos t$. From Figure 1.5.10 and the Pythagorean identity $\sin^2 t + \cos^2 t = 1$, we see that

FIGURE 1.5.10   The angle $t = \sin^{-1}\frac{1}{4}$ in Example 7

$$\left(\frac{1}{4}\right)^2 + \cos^2 t = 1 \quad \text{or} \quad \cos t = \frac{\sqrt{15}}{4}.$$

Hence
$$\tan t = \frac{1/4}{\sqrt{15}/4} = \frac{1}{\sqrt{15}} = \frac{\sqrt{15}}{15},$$

and so
$$\tan\left(\sin^{-1}\frac{1}{4}\right) = \tan t = \frac{\sqrt{15}}{15}.$$ ∎

The procedure that will be illustrated in Example 10 provides an alternative method for solving Example 7.

■ **Inverse Cosine Function** If we restrict the domain of the cosine function to the closed interval $[0, \pi]$, the resulting function is one-to-one and thus has an inverse. We denote this inverse by

$$\cos^{-1} x \quad \text{or} \quad \arccos x,$$

which gives us the following definition.

Definition 1.5.4 Inverse Cosine Function

The **inverse cosine function**, or **arccosine function**, is defined by

$$y = \cos^{-1} x \quad \text{if and only if} \quad x = \cos y, \tag{7}$$

where $-1 \leq x \leq 1$ and $0 \leq y \leq \pi$.

The graphs shown in FIGURE 1.5.11 illustrate how the function $y = \cos x$ restricted to the interval $[0, \pi]$ becomes a one-to-one function.

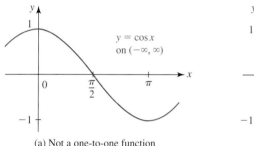
(a) Not a one-to-one function

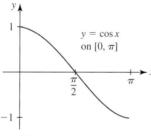
(b) A one-to-one function

FIGURE 1.5.11 Restricting the domain of $y = \cos x$ to produce a one-to-one function

FIGURE 1.5.12 Graph of the inverse cosine function

By reflecting the graph of the one-to-one function in Figure 1.5.11(b) in the line $y = x$ we obtain the graph of $y = \cos^{-1} x$ shown in FIGURE 1.5.12. The figure clearly shows that the domain and range of $y = \cos^{-1} x$ are $[-1, 1]$ and $[0, \pi]$, respectively.

EXAMPLE 8 Evaluating the Inverse Cosine Function

Evaluate $\arccos(-\sqrt{3}/2)$.

Solution If $y = \arccos(-\sqrt{3}/2)$, then $\cos y = -\sqrt{3}/2$. The only number in $[0, \pi]$ for which this is true is $y = 5\pi/6$. That is,

$$\arccos\left(-\frac{\sqrt{3}}{2}\right) = \frac{5\pi}{6}.$$

EXAMPLE 9 Evaluating the Compositions of Functions

Write $\sin(\cos^{-1} x)$ as an algebraic expression in x.

Solution In FIGURE 1.5.13 we have constructed an angle of t radians with cosine equal to x. Then $t = \cos^{-1} x$, or $x = \cos t$, where $0 \leq t \leq \pi$. Now to find $\sin(\cos^{-1} x) = \sin t$, we use the identity $\sin^2 t + \cos^2 t = 1$. Thus

$$\sin^2 t + x^2 = 1$$
$$\sin^2 t = 1 - x^2$$
$$\sin t = \sqrt{1 - x^2}$$
$$\sin(\cos^{-1} x) = \sqrt{1 - x^2}.$$

We use the positive square root of $1 - x^2$, since the range of $\cos^{-1} x$ is $[0, \pi]$, and the sine of an angle t in the first or second quadrant is positive.

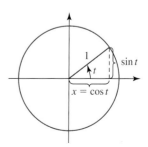
FIGURE 1.5.13 The angle $t = \cos^{-1} x$ in Example 9

■ **Inverse Tangent Function** If we restrict the domain of $\tan x$ to the open interval $(-\pi/2, \pi/2)$, then the resulting function is one-to-one and thus has an inverse. This inverse is denoted by

$$\tan^{-1} x \quad \text{or} \quad \arctan x.$$

Definition 1.5.5 Arctangent Function

The **inverse tangent function**, or **arctangent function**, is defined by

$$y = \tan^{-1} x \quad \text{if and only if} \quad x = \tan y, \tag{8}$$

where $-\infty < x < \infty$ and $-\pi/2 < y < \pi/2$.

The graphs shown in FIGURE 1.5.14 illustrate how the function $y = \tan x$ restricted to the open interval $(-\pi/2, \pi/2)$ becomes a one-to-one function. By reflecting the graph of the one-to-one function in Figure 1.5.14(b) in the line $y = x$ we obtain the graph of $y = \tan^{-1} x$ shown in FIGURE 1.5.15. We see in the figure that the domain and range of $y = \tan^{-1} x$ are, in turn, the intervals $(-\infty, \infty)$ and $(-\pi/2, \pi/2)$. For example, $y = \tan^{-1}(-1) = -\pi/4$ since $-\pi/4$ is the only number in the interval $(-\pi/2, \pi/2)$ for which $\tan(-\pi/4) = -1$.

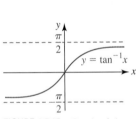

(a) Not a one-to-one function   (b) A one-to-one function

FIGURE 1.5.14 Restricting the domain of $y = \tan x$ to produce a one-to-one function.

FIGURE 1.5.15 Graph of the inverse tangent function

EXAMPLE 10 Evaluating Compositions of Functions

Without using a calculator, find $\cos(\arctan \frac{2}{3})$.

Solution If we let $y = \arctan \frac{2}{3}$, then $\tan y = \frac{2}{3}$. Using the right triangle in FIGURE 1.5.16 as an aid, we see that

$$\cos\left(\arctan \frac{2}{3}\right) = \cos y = \frac{3}{\sqrt{13}}. \quad \blacksquare$$

FIGURE 1.5.16 Triangle in Example 10

■ **Properties of the Inverses** Recall from (5) that $f^{-1}(f(x)) = x$ and $f(f^{-1}(x)) = x$ hold for any function f and its inverse under suitable restrictions on x. Thus for the inverse trigonometric functions, we have the following properties.

Theorem 1.5.2 Properties of Inverse Trig Functions

- (i) $\sin^{-1}(\sin x) = \arcsin(\sin x) = x$ if $-\pi/2 \leq x \leq \pi/2$
- (ii) $\sin(\sin^{-1} x) = \sin(\arcsin x) = x$ if $-1 \leq x \leq 1$
- (iii) $\cos^{-1}(\cos x) = \arccos(\cos x) = x$ if $0 \leq x \leq \pi$
- (iv) $\cos(\cos^{-1} x) = \cos(\arccos x) = x$ if $-1 \leq x \leq 1$
- (v) $\tan^{-1}(\tan x) = \arctan(\tan x) = x$ if $-\pi/2 < x < \pi/2$
- (vi) $\tan(\tan^{-1} x) = \tan(\arctan x) = x$ if $-\infty < x < \infty$

EXAMPLE 11 Applying the Inverse Properties

Without using a calculator, evaluate

(a) $\cos\left(\cos^{-1}\dfrac{1}{3}\right)$ (b) $\tan^{-1}\left(\tan\dfrac{3\pi}{4}\right)$.

Solution

(a) By Theorem 1.5.2(*iv*), $\cos\left(\cos^{-1}\dfrac{1}{3}\right) = \dfrac{1}{3}$.

(b) In this case we *cannot* apply property (*v*), since $3\pi/4$ is not in the interval $(-\pi/2, \pi/2)$. If we first evaluate $\tan(3\pi/4) = -1$, then we have

$$\tan^{-1}\left(\tan\dfrac{3\pi}{4}\right) = \tan^{-1}(-1) = -\dfrac{\pi}{4}.$$

■

Inverses of the Other Trigonometric Functions With the domains suitably restricted the remaining trigonometric functions $y = \cot x$, $y = \sec x$, and $y = \csc x$ also have inverses.

Definition 1.5.6 Other Inverse Trig Functions

(*i*) $y = \cot^{-1}x$ if and only if $x = \cot y$, $-\infty < x < \infty$ and $0 < y < \pi$
(*ii*) $y = \sec^{-1}x$ if and only if $x = \sec y$, $|x| \geq 1$ and $0 \leq y \leq \pi$, $y \neq \pi/2$
(*iii*) $y = \csc^{-1}x$ if and only if $x = \csc y$, $|x| \geq 1$ and $-\pi/2 \leq y \leq \pi/2$, $y \neq 0$

The graphs of $y = \cot^{-1}x$, $y = \sec^{-1}x$, and $y = \csc^{-1}x$ as well as their domains and ranges are summarized in FIGURE 1.5.17.

(a) $y = \cot^{-1}x$
domain: $(-\infty, \infty)$
range: $(0, \pi)$

(b) $y = \sec^{-1}x$
domain: $(-\infty, -1] \cup [1, \infty)$
range: $[0, \pi/2) \cup (\pi/2, \pi]$

(c) $y = \csc^{-1}x$
domain: $(-\infty, -1] \cup [1, \infty)$
range: $[-\pi/2, 0) \cup (0, \pi/2]$

FIGURE 1.5.17 Graphs of the inverse cotangent, inverse secant, and inverse cosecant functions

$f(x)$ NOTES FROM THE CLASSROOM

The ranges specified in Definitions 1.5.3, 1.5.4, 1.5.5, and 1.5.6(*i*) are universally agreed upon and grew out of the most logical and most convenient limitation of the original function. Thus, when we see $\arccos x$ or $\tan^{-1}x$ in any context, we know that $0 \leq \arccos x \leq \pi$ and $-\pi/2 < \tan^{-1}x < \pi/2$. These conventions are the same as those used in calculators when the $\boxed{\sin^{-1}}$, $\boxed{\cos^{-1}}$, and $\boxed{\tan^{-1}}$ keys are employed. However, there has been no universal agreement on the ranges of either $y = \sec^{-1}x$ or $y = \csc^{-1}x$. The ranges specified in (*ii*) and (*iii*) in Definition 1.5.6 are gaining in popularity because these are the ranges employed in computer algebra systems such as *Mathematica* and *Maple*. But you should be aware that there are popular calculus texts that define the domain and range of $y = \sec^{-1}x$ to be

domain: $(-\infty, -1] \cup [1, \infty)$,   range: $[0, \pi/2) \cup [\pi, 3\pi/2)$,

and the domain and range of $y = \csc^{-1}x$ to be

domain: $(-\infty, -1] \cup [1, \infty)$,   range: $(0, \pi/2] \cup (\pi, 3\pi/2]$.

Exercises 1.5   Answers to selected odd-numbered problems begin on page ANS-6.

≡ Fundamentals

In Problems 1 and 2, reread the introduction to this section. Then explain why the given function f is not one-to-one.

1. $f(x) = 1 + x(x - 5)$ **2.** $f(x) = x^4 + 2x^2$

In Problems 3–8, determine whether the given function is one-to-one by examining its graph.

3. $f(x) = 5$ **4.** $f(x) = 6x - 9$
5. $f(x) = \frac{1}{3}x + 3$ **6.** $f(x) = |x + 1|$
7. $f(x) = x^3 - 8$ **8.** $f(x) = x^3 - 3x$

In Problems 9–12, the given function f is one-to-one. Find f^{-1}.

9. $f(x) = 3x^3 + 7$
10. $f(x) = \sqrt[3]{2x - 4}$
11. $f(x) = \dfrac{2 - x}{1 - x}$
12. $f(x) = 5 - \dfrac{2}{x}$

In Problems 13 and 14, verify that $f(f^{-1}(x)) = x$ and $f^{-1}(f(x)) = x$.

13. $f(x) = 5x - 10,\ f^{-1}(x) = \dfrac{1}{5}x + 2$

14. $f(x) = \dfrac{1}{x + 1},\ f^{-1}(x) = \dfrac{1 - x}{x}$

In Problems 15–18, the given function f is one-to-one. Without finding the inverse, find the domain and range of f^{-1}.

15. $f(x) = \sqrt{x + 2}$
16. $f(x) = 3 + \sqrt{2x - 1}$
17. $f(x) = \dfrac{1}{x + 3}$
18. $f(x) = \dfrac{x - 1}{x - 4}$

In Problems 19 and 20, the given function f is one-to-one. Without finding the inverse, find the point on the graph of f^{-1} corresponding to the indicated value of x in the domain of f.

19. $f(x) = 2x^3 + 2x;\ x = 2$
20. $f(x) = 8x - 3;\ x = 5$

In Problems 21 and 22, the given function f is one-to-one. Without finding the inverse, find x in the domain of f^{-1} that satisfies the indicated equation.

21. $f(x) = x + \sqrt{x};\ f^{-1}(x) = 9$
22. $f(x) = \dfrac{4x}{x + 1};\ f^{-1}(x) = \dfrac{1}{2}$

In Problems 23 and 24, sketch the graph of f^{-1} from the graph of f.

23.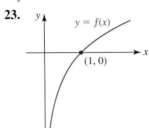

FIGURE 1.5.18   Graph for Problem 23

24.

FIGURE 1.5.19   Graph for Problem 24

In Problems 25 and 26, sketch the graph of f from the graph of f^{-1}.

25.

FIGURE 1.5.20   Graph for Problem 25

26.

FIGURE 1.5.21   Graph for Problem 26

In Problems 27–30, find an inverse function f^{-1} that has the same range as the given function by suitably restricting the domain of f.

27. $f(x) = (5 - 2x)^2$ **28.** $f(x) = 3x^2 + 9$
29. $f(x) = x^2 + 2x + 4$ **30.** $f(x) = -x^2 + 8x$

31. If the functions f and g have inverses, then it can be proved that
$$(f \circ g)^{-1} = g^{-1} \circ f^{-1}.$$
Verify this for $f(x) = x^3$ and $g(x) = 4x + 5$.

32. The equation $y = \sqrt[3]{x} - \sqrt[3]{y}$ defines a one-to-one function $y = f(x)$. Find $f^{-1}(x)$.

In Problems 33–44, obtain the exact value of the given expression. Do not use a calculator.

33. $\arccos\left(-\dfrac{\sqrt{2}}{2}\right)$ **34.** $\cos^{-1}\dfrac{1}{2}$
35. $\arctan(1)$ **36.** $\tan^{-1}\sqrt{3}$
37. $\cot^{-1}(-1)$ **38.** $\sec^{-1}(-1)$
39. $\arcsin\left(-\dfrac{\sqrt{3}}{2}\right)$ **40.** $\operatorname{arccot}(-\sqrt{3})$
41. $\sin\left(\arctan\dfrac{4}{3}\right)$ **42.** $\cos\left(\sin^{-1}\dfrac{2}{5}\right)$
43. $\tan\left(\cot^{-1}\dfrac{1}{2}\right)$ **44.** $\csc\left(\tan^{-1}\dfrac{2}{3}\right)$

In Problems 45–48, evaluate the given expression by means of an appropriate trigonometric identity.

45. $\sin\left(2\sin^{-1}\dfrac{1}{3}\right)$ **46.** $\cos\left(2\cos^{-1}\dfrac{3}{4}\right)$

47. $\sin\left(\arcsin\dfrac{\sqrt{3}}{3}+\arccos\dfrac{2}{3}\right)$

48. $\cos(\tan^{-1}4-\tan^{-1}3)$

In Problems 49–52, write the given expression as an algebraic quantity in x.

49. $\cos(\sin^{-1}x)$ **50.** $\tan(\sin^{-1}x)$

51. $\sec(\tan^{-1}x)$ **52.** $\sin(\sec^{-1}x),\ x\ge 1$

In Problems 53 and 54, graphically verify the identities by a reflection and a vertical shift.

53. $\sin^{-1}x+\cos^{-1}x=\dfrac{\pi}{2}$

54. $\operatorname{arccot}x+\arctan x=\dfrac{\pi}{2}$

55. Prove that $\sec^{-1}x=\cos^{-1}(1/x)$ for $|x|\ge 1$.

56. Prove that $\csc^{-1}x=\sin^{-1}(1/x)$ for $|x|\ge 1$.

57. If $t=\sin^{-1}(-2/\sqrt{5})$, find the exact values of $\cos t$, $\tan t$, $\cot t$, $\sec t$, and $\csc t$.

58. If $\theta=\arctan\tfrac{1}{2}$, find the exact values of $\sin\theta$, $\cos\theta$, $\cot\theta$, $\sec\theta$, and $\csc\theta$.

≡ Calculator/CAS Problems

Most calculators do not have dedicated keys for $\csc^{-1}x$ and $\sec^{-1}x$. In Problems 59 and 60, use a calculator and the identities in Problems 55 and 56 to compute the given quantity.

59. (a) $\sec^{-1}(-\sqrt{2})$ (b) $\csc^{-1}2$

60. (a) $\sec^{-1}(3.5)$ (b) $\csc^{-1}(-1.25)$

61. Use a calculator to verify:
 (a) $\tan(\tan^{-1}1.3)=1.3$ and $\tan^{-1}(\tan 1.3)=1.3$
 (b) $\tan(\tan^{-1}5)=5$ and $\tan^{-1}(\tan 5)=-1.2832$
 Explain why $\tan^{-1}(\tan 5)\ne 5$.

62. Let $x=1.7$ radians. Compare, if possible, the values of $\sin^{-1}(\sin x)$ and $\sin(\sin^{-1}x)$. Explain any differences.

≡ Applications

63. Consider a ladder of length L leaning against a house with a load at point P as shown in FIGURE 1.5.22. The angle β at which the ladder is at the verge of slipping is defined by

$$\dfrac{x}{L}=\dfrac{c}{1+c^2}(c+\tan\beta),$$

where c is the coefficient of friction between the ladder and the ground.

(a) Find β when $c=1$ and the load is at the top of the ladder.

(b) Find β when $c=0.5$ and the load is $\tfrac{3}{4}$ of the way up the ladder.

FIGURE 1.5.22 Ladder in Problem 63

64. An airplane flies west at a constant speed v_1 and a wind blows from the north at a constant speed v_2. The plane's course south of west is given by $\theta=\tan^{-1}(v_2/v_1)$. See FIGURE 1.5.23. Find the course of a plane flying west at 300 km/h if a wind from the north blows at 60 km/h.

FIGURE 1.5.23 Plane in Problem 64

≡ Think About It

In Problems 65 and 66, use a calculator or CAS to obtain the graph of the given function where x is any real number. Explain why the graphs do not violate Theorems 1.5.2(i) and 1.5.2(iii).

65. $f(x)=\sin^{-1}(\sin x)$ **66.** $f(x)=\cos^{-1}(\cos x)$

67. Discuss: Can any periodic function be one-to-one?

68. How are the one-to-one functions $y=f(x)$ shown in FIGURES 1.5.24(a) and 1.5.24(b) related to the inverse functions $y=f^{-1}(x)$? Find at least three explicit functions with this property.

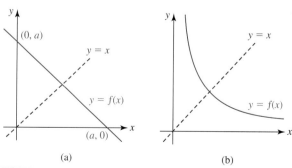

FIGURE 1.5.24 Graph for Problem 68

1.6 Exponential and Logarithmic Functions

■ **Introduction** In the preceding sections we considered functions such as $f(x) = x^2$, that is, a function with a variable base x and constant power or exponent 2. We now review functions such as $f(x) = 2^x$ having a constant base 2 and a variable exponent x.

Definition 1.6.1 Exponential Function

If $b > 0$ and $b \neq 1$, then an **exponential function** $y = f(x)$ is a function of the form

$$f(x) = b^x. \tag{1}$$

The number b is called the **base** and x is called the **exponent**.

> In (1), the base b is restricted to positive numbers in order to guarantee that b^x is a real number. Also, $b = 1$ is of no interest since $f(x) = 1^x = 1$.

The **domain** of an exponential function f defined in (1) is the set of all real numbers $(-\infty, \infty)$.

■ **Exponents** Because the domain of an exponential function (1) is the set of real numbers, the exponent x can be either a rational or an irrational number. For example, if the base $b = 3$ and the exponent x is a *rational number*, for example, $x = \frac{1}{5}$ and $x = 1.4$, then

$$3^{1/5} = \sqrt[5]{3} \quad \text{and} \quad 3^{1.4} = 3^{14/10} = 3^{7/5} = \sqrt[5]{3^7}.$$

The function (1) is also defined for every *irrational number* x. The following procedure illustrates a way of defining a number such as $3^{\sqrt{2}}$. From the decimal representation $\sqrt{2} = 1.414213562\ldots$ we see that the rational numbers

$$1, 1.4, 1.41, 1.414, 1.4142, 1.41421, \ldots$$

> One definition of b^x, for x irrational, is given by
> $$b^x = \lim_{t \to x} b^t,$$
> where t is rational. This is read "b^x is the **limit** of b^t as t approaches x". We will study limits in detail in Chapter 2.

are successively better approximations to $\sqrt{2}$. By using these rational numbers as exponents, we would expect that the numbers

$$3^1, 3^{1.4}, 3^{1.41}, 3^{1.414}, 3^{1.4142}, 3^{1.41421}, \ldots$$

are then successively better approximations to $3^{\sqrt{2}}$. In fact, this can be shown to be true with a precise definition of b^x for an irrational value of x. But on a practical level, we can use the $\boxed{y^x}$ key on a calculator to obtain the approximation 4.728804388 to $3^{\sqrt{2}}$.

■ **Laws of Exponents** Since b^x is defined for all real numbers x when $b > 0$, it can be proved that the **laws of exponents** hold for all real-number exponents. If $a > 0$, $b > 0$ and $x, x_1,$ and x_2 denote real numbers, then

(i) $b^{x_1} \cdot b^{x_2} = b^{x_1 + x_2}$ (ii) $\dfrac{b^{x_1}}{b^{x_2}} = b^{x_1 - x_2}$ (iii) $(b^{x_1})^{x_2} = b^{x_1 x_2}$

(iv) $\dfrac{1}{b^{x_2}} = b^{-x_2}$ (v) $(ab)^x = a^x b^x$ (vi) $\left(\dfrac{a}{b}\right)^x = \dfrac{a^x}{b^x}.$

■ **Graphs** We distinguish two types of graphs for (1) depending on whether the base b satisfies $b > 1$ or $0 < b < 1$. The next example illustrates the graphs of $f(x) = 3^x$ and $f(x) = \left(\frac{1}{3}\right)^x$. Before graphing, we can make some intuitive observations about both functions. Since the bases $b = 3$ and $b = \frac{1}{3}$ are positive, the values of 3^x and $\left(\frac{1}{3}\right)^x$ are positive for every real number x. Moreover, neither 3^x nor $\left(\frac{1}{3}\right)^x$ can be 0 for any x and so the graphs of $f(x) = 3^x$ and $f(x) = \left(\frac{1}{3}\right)^x$ have no x-intercepts. Also, $3^0 = 1$ and $\left(\frac{1}{3}\right)^0 = 1$ means that the graphs of $f(x) = 3^x$ and $f(x) = \left(\frac{1}{3}\right)^x$ have the same y-intercept $(0, 1)$.

EXAMPLE 1 Graphs of Exponential Functions

Graph the functions

(a) $f(x) = 3^x$, (b) $f(x) = \left(\dfrac{1}{3}\right)^x.$

Solution

(a) We first construct a table of some function values corresponding to preselected values of x. As shown in FIGURE 1.6.1(a), we plot the corresponding points obtained from the table

x	-3	-2	-1	0	1	2
$f(x)$	$\frac{1}{27}$	$\frac{1}{9}$	$\frac{1}{3}$	1	3	9

and connect them with a continuous curve. The graph shows that f is an increasing function on the interval $(-\infty, \infty)$.

(b) Proceeding as in part (a), we construct a table of some function values

x	-3	-2	-1	0	1	2
$f(x)$	27	9	3	1	$\frac{1}{3}$	$\frac{1}{9}$

corresponding to preselected values of x. Note, for example, by the laws of exponents $f(-2) = \left(\frac{1}{3}\right)^{-2} = (3^{-1})^{-2} = 3^2 = 9$. As shown in Figure 1.6.1(b), we plot the corresponding points obtained from the table and connect them with a continuous curve. In this case the graph shows that f is a decreasing function on the interval $(-\infty, \infty)$.

Note: Exponential functions with bases satisfying $0 < b < 1$, such as $b = \frac{1}{3}$, are frequently written in an alternative manner. By writing $y = \left(\frac{1}{3}\right)^x$ as $y = (3^{-1})^x$ and using (iii) of the laws of exponents we see that $y = \left(\frac{1}{3}\right)^x$ is the same as $y = 3^{-x}$.

FIGURE 1.6.1 Graphs of functions in Example 1

■ **Horizontal Asymptote** FIGURE 1.6.2 illustrates the two general shapes that the graph of an exponential function $f(x) = b^x$ can have. But there is one more important aspect of all such graphs. Observe in Figure 1.6.2 that for $0 < b < 1$, the function values $f(x)$ approach 0 as x becomes unbounded in the positive direction (the red graph), and for $b > 1$, the function values $f(x)$ approach 0 as x becomes unbounded in the negative direction (the blue graph). In other words, the line $y = 0$ (the x-axis) is a **horizontal asymptote** for both types of exponential graphs.

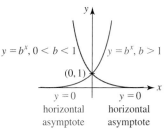

FIGURE 1.6.2 f increasing for $b > 1$; f decreasing for $0 < b < 1$

■ **Properties of an Exponential Function** The following list summarizes some of the important properties of the exponential function f with base b. Reexamine the graphs in Figure 1.6.2 as you read this list.

- The domain of f is the set of real numbers, that is, $(-\infty, \infty)$.
- The range of f is the set of positive real numbers, that is, $(0, \infty)$.
- The y-intercept of f is $(0, 1)$. The graph of f has no x-intercept.
- The function f is increasing on the interval $(-\infty, \infty)$ for $b > 1$ and decreasing on the interval $(-\infty, \infty)$ for $0 < b < 1$.
- The x-axis, that is, $y = 0$, is a horizontal asymptote for the graph of f.
- The function f is one-to-one.

Although the graphs of $y = b^x$ in the case when $b > 1$ all share the same basic shape and all pass through the same point $(0, 1)$, there are subtle differences. The larger the base b the more steeply the graph rises as x increases. In FIGURE 1.6.3 we compare the graphs of $y = 5^x$, $y = 3^x$, $y = 2^x$, and $y = (1.2)^x$ in green, blue, gold, and red, respectively, on the same coordinate axes. We see from its graph that the values of $y = (1.2)^x$ increase slowly as x increases.

The fact that (1) is a one-to-one function follows from the horizontal line test discussed in Section 1.5.

FIGURE 1.6.3 Graphs of $y = b^x$ for $b = 1.2, 2, 3, 5$

■ **The Number e** Most every student of mathematics has heard of, and has likely worked with, the famous irrational number $\pi = 3.141592654\ldots$. In calculus and applied mathematics the irrational number,

$$e = 2.718281828459\ldots \qquad (2)$$

arguably plays a role more important than the number π. The usual definition of the number e is the number that the function $f(x) = (1 + 1/x)^x$ approaches as we let x become large without bound in the positive direction. If we let the arrow symbol $\rightarrow$ represent the word *approach*, then the fact that $f(x) \rightarrow e$ as $x \rightarrow \infty$ is evident in the table of numerical values of f

x	100	1000	10,000	100,000	1,000,000
$(1 + 1/x)^x$	2.704814	2.716924	2.718146	2.718268	2.718280

FIGURE 1.6.4 $y = e$ is a horizontal asymptote of the graph of f

and from the graph in FIGURE 1.6.4. In the figure the horizontal dashed red line $y = e$ is a horizontal asymptote for the graph of f. We also say that e is the *limit* of $f(x) = (1 + 1/x)^x$ as $x \rightarrow \infty$ and write

$$e = \lim_{x \to \infty} \left(1 + \frac{1}{x}\right)^x. \tag{3}$$

You will often see an alternative definition of the number e. If we let $h = 1/x$ in (3), then as $x \rightarrow \infty$ we have simultaneously $h \rightarrow 0$. Hence an equivalent form of (3) is

$$e = \lim_{h \to 0}(1 + h)^{1/h}. \tag{4}$$

■ **The Natural Exponential Function** When the base in (1) is chosen to be $b = e$, the function $f(x) = e^x$ is called the **natural exponential function**. Since $b = e > 1$ and $b = 1/e < 1$, the graphs of $y = e^x$ and $y = e^{-x}$ are given in FIGURE 1.6.5. On the face of it, $f(x) = e^x$ possesses no noticeable graphical characteristic that distinguishes it from, say, the function $f(x) = 3^x$, and has no special properties other than the ones given in the bulleted list on page 49. Questions as to why $f(x) = e^x$ is a "natural" and frankly, the most important exponential function, will be answered in the chapters ahead and in your courses beyond calculus.

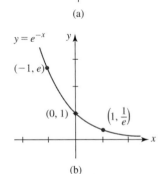

FIGURE 1.6.5 Natural exponential function in (a)

■ **Inverse of the Exponential Function** Since an exponential function $y = b^x$ is one-to-one, we know that it has an inverse function. To find this inverse, we interchange the variables x and y to obtain $x = b^y$. This last formula defines y as a function of x:

- y is that exponent of the base b that produces x.

By replacing the word *exponent* with the word *logarithm*, we can rephrase the preceding line as:

- y is that logarithm of the base b that produces x.

This last line is abbreviated by the notation $y = \log_b x$ and is called the **logarithmic function**.

Definition 1.6.2 Logarithmic Function

The **logarithmic function** with base $b > 0, b \neq 1$, is defined by

$$y = \log_b x \quad \text{if and only if} \quad x = b^y. \tag{5}$$

For $b > 0$ there is no real number y for which b^y can either be 0 or negative. It then follows from $x = b^y$ that $x > 0$. In other words, the **domain** of a logarithmic function $y = \log_b x$ is the set of positive real numbers $(0, \infty)$.

For emphasis, all that is being said in the preceding sentences is:

- The logarithmic expression $y = \log_b x$ and the exponential expression $x = b^y$ are equivalent,

that is, they mean the same thing. As a consequence, within a specific context such as solving a problem, we can use whichever form happens to be more convenient. The following list illustrates several examples of equivalent logarithmic and exponential statements:

Logarithmic Form	Exponential Form
$\log_3 9 = 2$	$9 = 3^2$
$\log_8 2 = \frac{1}{3}$	$2 = 8^{1/3}$
$\log_{10} 0.001 = -3$	$0.001 = 10^{-3}$
$\log_b 5 = -1$	$5 = b^{-1}$

Graphs Because a logarithmic function is the inverse of an exponential function, we can obtain the graph of the former by reflecting the graph of the latter in the line $y = x$. As you inspect the two graphs in FIGURE 1.6.6, remember that the domain $(-\infty, \infty)$ and range $(0, \infty)$ of $y = b^x$ become, in turn, the range $(-\infty, \infty)$ and domain $(0, \infty)$ of $y = \log_b x$. Note that the y-intercept $(0, 1)$ for the exponential function (blue graph) becomes the x-intercept $(1, 0)$ for the logarithmic function (red graph). Also, when the exponential function is reflected in the line $y = x$, the horizontal asymptote $y = 0$ for the graph of $y = b^x$ becomes a vertical asymptote for the graph of $y = \log_b x$. In Figure 1.6.6 we see that for $b > 1$, $x = 0$, which is the equation of the y-axis, is a **vertical asymptote** for the graph of $y = \log_b x$.

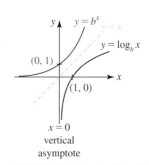

FIGURE 1.6.6 Graph of logarithmic function with base $b > 1$

Properties of a Logarithmic Function The following list summarizes some of the important properties of the logarithmic function $f(x) = \log_b x$:

- The domain of f is the set of positive real numbers, that is, $(0, \infty)$.
- The range of f is the set of real numbers, that is, $(-\infty, \infty)$.
- The x-intercept of f is $(1, 0)$. The graph of f has no y-intercept.
- The function f is increasing on the interval $(0, \infty)$ for $b > 1$ and decreasing on the interval $(0, \infty)$ for $0 < b < 1$.
- The y-axis, that is, $x = 0$, is a vertical asymptote for the graph of f.
- The function f is one-to-one.

We would like to call attention to the third entry in the foregoing list for special emphasis

$$\log_b 1 = 0 \quad \text{since} \quad b^0 = 1. \tag{6}$$

Also,
$$\log_b b = 1 \quad \text{since} \quad b^1 = b. \tag{7}$$

The result in (7) means that in addition to $(1, 0)$, the graph of any logarithmic function (5) with base b also contains the point $(b, 1)$. The equivalence of $y = \log_b x$ and $x = b^y$ also yields two sometimes-useful identities. By substituting $y = \log_b x$ into $x = b^y$, and then $x = b^y$ into $y = \log_b x$ gives

$$x = b^{\log_b x} \quad \text{and} \quad y = \log_b b^y. \tag{8}$$

For example, from (8), $2^{\log_2 10} = 10$ and $\log_3 3^7 = 7$.

Natural Logarithm Logarithms with base $b = 10$ are called **common logarithms** and logarithms with base $b = e$ are called **natural logarithms**. Furthermore, it is customary to write the natural logarithm $\log_e x$ as $\ln x$. The symbol "$\ln x$" is usually read phonetically as "ell-en of x." Since $b = e > 1$, the graph of $y = \ln x$ has the characteristic logarithmic shape shown in red in Figure 1.6.6. For base $b = e$, (5) becomes

$$y = \ln x \quad \text{if and only if} \quad x = e^y. \tag{9}$$

The analogues of (6) and (7) for the natural logarithm are

$$\ln 1 = 0 \quad \text{since} \quad e^0 = 1. \tag{10}$$

$$\ln e = 1 \quad \text{since} \quad e^1 = e. \tag{11}$$

The identities in (8) become

$$x = e^{\ln x} \quad \text{and} \quad y = \ln e^y. \tag{12}$$

For example, from (12), $e^{\ln 25} = 25$.

■ **Laws of Logarithms** The laws of exponents can be restated equivalently as the laws of logarithms. For example, if $M = b^{x_1}$ and $N = b^{x_2}$, then by (5), $x_1 = \log_b M$ and $x_2 = \log_b N$. By (i) of the laws of exponents, $MN = b^{x_1 + x_2}$. Expressed as a logarithm this is $x_1 + x_2 = \log_b MN$. Substituting for x_1 and x_2 gives $\log_b M + \log_b N = \log_b MN$. The remaining parts of the next theorem can be proved in the same manner.

Theorem 1.6.1 Laws of Logarithms

For any base $b > 0$, $b \neq 1$, and positive numbers M and N:

(i) $\log_b MN = \log_b M + \log_b N$

(ii) $\log_b \left(\dfrac{M}{N} \right) = \log_b M - \log_b N$

(iii) $\log_b M^c = c \log_b M$, for c any real number.

EXAMPLE 2 Laws of Logarithms

Simplify and write $\frac{1}{2}\ln 36 + 2\ln 4$ as a single logarithm.

Solution By (iii) of the laws of logarithms we can write

$$\frac{1}{2}\ln 36 + 2\ln 4 = \ln(36)^{1/2} + \ln 4^2 = \ln 6 + \ln 16.$$

Then by (i) of the laws of logarithms,

$$\frac{1}{2}\ln 36 + 2\ln 4 = \ln 6 + \ln 16 = \ln(6 \cdot 16) = \ln 96. \quad \blacksquare$$

EXAMPLE 3 Rewriting Logarithmic Expressions

Use the laws of logarithms to rewrite each expression and evaluate.

(a) $\ln \sqrt{e}$ (b) $\ln 5e$ (c) $\ln \dfrac{1}{e}$

Solution

(a) Since $\sqrt{e} = e^{1/2}$ we have from (iii) of the laws of logarithms:

$$\ln \sqrt{e} = \ln e^{1/2} = \frac{1}{2}\ln e = \frac{1}{2}. \quad \leftarrow \text{from (11), } \ln e = 1$$

(b) From (i) of the laws of logarithms and a calculator:

$$\ln 5e = \ln 5 + \ln e = \ln 5 + 1 \approx 2.6094. \quad \leftarrow \text{from (11), } \ln e = 1$$

(c) From (ii) of the laws of logarithms:

$$\ln \frac{1}{e} = \ln 1 - \ln e = 0 - 1 = -1. \quad \leftarrow \text{from (10) and (11)}$$

Note that (iii) of the laws of logarithms can also be used here:

$$\ln \frac{1}{e} = \ln e^{-1} = (-1)\ln e = -1. \quad \blacksquare$$

EXAMPLE 4 Solving Equations

(a) Solve the logarithmic equation $\ln 2 + \ln(4x - 1) = \ln(2x + 5)$ for x.
(b) Solve the exponential equation $e^{10k} = 7$ for k.

Solution

(a) By (i) of the laws of logarithms, the left-hand side of the equation can be written

$$\ln 2 + \ln(4x - 1) = \ln 2(4x - 1) = \ln(8x - 2).$$

The original equation is then

$$\ln(8x - 2) - \ln(2x + 5) = 0 \quad \text{or} \quad \ln \frac{8x - 2}{2x + 5} = 0.$$

It follows from (9) that

$$\frac{8x - 2}{2x + 5} = e^0 = 1 \quad \text{or} \quad 8x - 2 = 2x + 5.$$

From the last equation we find that $x = \frac{7}{6}$.

(b) We use (9) to rewrite the exponential expression $e^{10k} = 7$ as the logarithmic expression $10k = \ln 7$. Therefore, with the aid of a calculator

$$k = \frac{1}{10} \ln 7 \approx 0.1946. \qquad \blacksquare$$

■ **Change of Base** The graph of $y = 2^x - 5$ is the graph $y = 2^x$ shifted down 5 units. As seen in FIGURE 1.6.7 the graph has an x-intercept. By setting $y = 0$, we see that x is the solution of the equation $2^x - 5 = 0$ or $2^x = 5$. Now a perfectly valid solution is $x = \log_2 5$. But from a computational viewpoint (that is, expressing x as a number), the last answer is not desirable because no calculator has a logarithmic function with base 2. We can compute the answer by changing $\log_2 5$ to the natural logarithm by simply taking the natural log of both sides of the exponential equation $2^x = 5$:

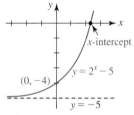

FIGURE 1.6.7 x-intercept of $y = 2^x - 5$

$$\ln 2^x = \ln 5$$
$$x \ln 2 = \ln 5$$

Note: We actually divide the logarithms here →
$$x = \frac{\ln 5}{\ln 2} \approx 2.3219.$$

By the way, since we started with $x = \log_2 5$, the last result also proves the equality $\log_2 5 = \frac{\ln 5}{\ln 2}$. The x-intercept of the graph is then $(\log_2 5, 0) = (\ln 5/\ln 2, 0) \approx (2.32, 0)$.

In general, to convert a logarithm with any base $b > 0$ to the natural logarithm, we first rewrite the logarithmic expression $x = \log_b N$ as an equivalent exponential expression $b^x = N$. Then take the natural logarithm of both sides of the last equality and solve the resulting equation $x \ln b = \ln N$ for x. This yields the general **change of base formula**:

$$\log_b N = \frac{\ln N}{\ln b}. \qquad (13)$$

Exercises 1.6 Answers to selected odd-numbered problems begin on page ANS-6.

≡ Fundamentals

In Problems 1–6, sketch the graph of the given function f. Find the y-intercept and the horizontal asymptote of the graph.

1. $f(x) = \left(\frac{3}{4}\right)^x$
2. $f(x) = \left(\frac{4}{3}\right)^x$
3. $f(x) = -2^x$
4. $f(x) = -2^{-x}$
5. $f(x) = -5 + e^x$
6. $f(x) = 2 + e^{-x}$

In Problems 7–10, find an exponential function $f(x) = b^x$ such that the graph of f passes through the given point.

7. $(3, 216)$
8. $(-1, 5)$
9. $(-1, e^2)$
10. $(2, e)$

In Problems 11–14, use a graph to solve the given inequality for x.

11. $2^x > 16$
12. $e^x \leq 1$

13. $e^{x-2} < 1$ **14.** $\left(\frac{1}{2}\right)^x \geq 8$

In Problems 15 and 16, use $f(-x) = f(x)$ to demonstrate that the given function is even. Sketch the graph of f.

15. $f(x) = e^{x^2}$ **16.** $f(x) = e^{-|x|}$

In Problems 17 and 18, use the graphs obtained in Problems 15 and 16 as an aid in sketching the graph of the given function f.

17. $f(x) = 1 - e^{x^2}$ **18.** $f(x) = 2 + 3e^{-|x|}$

19. Show that $f(x) = 2^x + 2^{-x}$ is an even function. Sketch the graph of f.

20. Show that $f(x) = 2^x - 2^{-x}$ is an odd function. Sketch the graph of f.

In Problems 21 and 22, sketch the graph of the given piecewise-defined function f.

21. $f(x) = \begin{cases} -e^x, & x < 0 \\ -e^{-x}, & x \geq 0 \end{cases}$ **22.** $f(x) = \begin{cases} e^{-x}, & x \leq 0 \\ -e^x, & x > 0 \end{cases}$

In Problems 23–26, rewrite the given exponential expression as an equivalent logarithmic expression.

23. $4^{-1/2} = \frac{1}{2}$ **24.** $9^0 = 1$

25. $10^4 = 10{,}000$ **26.** $10^{0.3010} = 2$

In Problems 27–30, rewrite the given logarithmic expression as an equivalent exponential expression.

27. $\log_2 128 = 7$ **28.** $\log_5 \frac{1}{25} = -2$

29. $\log_{\sqrt{3}} 81 = 8$ **30.** $\log_{16} 2 = \frac{1}{4}$

In Problems 31 and 32, find a logarithmic function $f(x) = \log_b x$ such that the graph of f passes through the given point.

31. $(49, 2)$ **32.** $\left(4, \frac{1}{3}\right)$

In Problems 33–38, find the exact value of the given expression.

33. $\ln e^e$ **34.** $\ln(e^4 e^9)$
35. $10^{\log_{10} 6^2}$ **36.** $25^{\log_5 8}$
37. $e^{-\ln 7}$ **38.** $e^{\frac{1}{2}\ln \pi}$

In Problems 39–42, find the domain of the given function f. Find the x-intercept and the vertical asymptote of the graph. Sketch the graph of f.

39. $f(x) = -\ln x$ **40.** $f(x) = -1 + \ln x$
41. $f(x) = -\ln(x + 1)$ **42.** $f(x) = 1 + \ln(x - 2)$

In Problems 43 and 44, find the domain of the given function f.

43. $f(x) = \ln(9 - x^2)$ **44.** $f(x) = \ln(x^2 - 2x)$

45. Show that $f(x) = \ln|x|$ is an even function. Sketch the graph of f. Find the x-intercepts and the vertical asymptote of the graph.

46. Use the graph obtained in Problem 45 to sketch the graph of $y = \ln|x - 2|$. Find the x-intercepts and the vertical asymptote of the graph.

In Problems 47–50, use the laws of logarithms to rewrite the given expression as one logarithm.

47. $\ln(x^4 - 4) - \ln(x^2 + 2)$ **48.** $\ln\left(\frac{x}{y}\right) - 2\ln x^3 - 4\ln y$

49. $\ln 5 + \ln 5^2 + \ln 5^3 - \ln 5^6$ **50.** $5\ln 2 + 2\ln 3 - 3\ln 4$

In Problems 51–54, use the laws of logarithms so that $\ln y$ contains no products, quotients, or powers.

51. $y = \dfrac{x^{10}\sqrt{x^2 + 5}}{\sqrt[3]{8x^3 + 2}}$ **52.** $y = \sqrt{\dfrac{(2x + 1)(3x + 2)}{4x + 3}}$

53. $y = \dfrac{(x^3 - 3)^5(x^4 + 3x^2 + 1)^8}{\sqrt{x}(7x + 5)^9}$

54. $y = 64x^6\sqrt{x + 1}\sqrt[3]{x^2 + 2}$

In Problems 55 and 56, use the natural logarithm to find x in the domain of the given function for which f takes on the indicated value.

55. $f(x) = 6^x$; $f(x) = 51$ **56.** $f(x) = \left(\frac{1}{2}\right)^x$; $f(x) = 7$

In Problems 57–60, use the natural logarithm to solve for x.

57. $2^{x+5} = 9$ **58.** $4 \cdot 7^{2x} = 9$
59. $5^x = 2e^{x+1}$ **60.** $3^{2(x-1)} = 2^{x-3}$

In Problems 61 and 62, solve for x.

61. $\ln x + \ln(x - 2) = \ln 3$
62. $\ln 3 + \ln(2x - 1) = \ln 4 + \ln(x + 1)$

≡ Mathematical Models

63. Exponential Growth An exponential model for the number of bacteria in a culture at time t is given by $P(t) = P_0 e^{kt}$, where P_0 is the initial population and $k > 0$ is the growth constant.

(a) After 2 h the initial number of bacteria in a culture is observed to have doubled. Find an exponential growth model $P(t)$.

(b) According to the model in part (a), what is the number of bacteria present in the culture after 5 h?

(c) Find the time that it takes the culture to grow to 20 times its initial size.

64. Exponential Decay An exponential model for the amount of a radioactive substance remaining at time t is given by $A(t) = A_0 e^{kt}$, where A_0 is the initial amount and $k < 0$ is the decay constant.

(a) Initially 200 mg of a radioactive substance was present. After 6 h the mass had decreased by 3%. Construct an exponential model for the amount of the decaying substance remaining after t h.

(b) Find the amount remaining after 24 h.
(c) The time at which $A(t) = \frac{1}{2}A_0$ is called the **half-life** of the substance. What is the half-life of the substance in part (a)?

65. **Logistic Growth** A student sick with a flu virus returns to an isolated college campus of 2000 students. The number of students infected with the flu t days after the student's return is predicted by the logistic function

$$P(t) = \frac{2000}{1 + 1999e^{-0.8905t}}.$$

(a) According to this mathematical model, how many students will be infected with the flu after 5 days?
(b) How long will it take for one-half of the student population to become infected?
(c) How many students does the model predict will become infected after a very long period of time?
(d) Sketch a graph of $P(t)$.

66. **Newton's Law of Cooling** If an object or body is placed in a medium (such as air, water, etc.) that is held at a constant temperature T_m and if the initial temperature of the object is T_0, then Newton's law of cooling predicts that the temperature of the object at time t is given by

$$T(t) = T_m + (T_0 - T_m)e^{kt}, k < 0.$$

(a) A cake is removed from an oven where the temperature was 350°F into a kitchen where the temperature is a constant 75°F. One minute later the temperature of the cake is measured to be 300°F. What is the temperature of the cake after 6 minutes?
(b) At what time is the temperature of the cake 80°F?

≡ **Think About It**

67. Discuss: How can the graphs of the given functions be obtained from the graph of $f(x) = \ln x$ by means of a rigid transformation (a shift or a reflection)?
(a) $y = \ln 5x$
(b) $y = \ln \frac{x}{4}$
(c) $y = \ln x^{-1}$
(d) $y = \ln(-x)$

68. (a) Use a graphing utility to obtain the graph of the function $f(x) = \ln(x + \sqrt{x^2 + 1})$.
(b) Show that f is an odd function, that is, $f(-x) = -f(x)$.

1.7 From Words to Functions

Introduction In Chapters 4 and 6 there will be several instances when you will be expected to translate the words that describe a *function* or an *equation* into mathematical symbols.

In this section we focus on problems that involve functions. We begin with a verbal description about the product of two numbers.

EXAMPLE 1 Product of Two Numbers

The sum of two nonnegative numbers is 5. Express the product of one and the square of the other as a function of one of the numbers.

Solution We first represent the two numbers by the symbols x and y and recall that *nonnegative* means that $x \geq 0$ and $y \geq 0$. Using these symbols, the words "the sum . . . is 5" translates into the equation $x + y = 5$; this is *not* the function we are seeking. The word *product* in the second sentence suggests that we use the symbol P to denote the function we want. Now P is the product of one of the numbers, say, x and the square of the other, that is, y^2:

$$P = xy^2. \qquad (1)$$

No, we are not finished because P is supposed to be a "function of *one* of the numbers." We now use the fact that the numbers x and y are related by $x + y = 5$. From this last equation we substitute $y = 5 - x$ into (1) to obtain the desired result:

$$P(x) = x(5 - x)^2. \qquad (2) \blacksquare$$

Here is a symbolic diagram of the analysis of the problem given in Example 1:

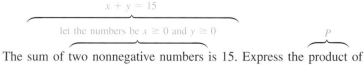

(3)

Notice that the second sentence is vague about which number is squared. This implies that it really does not matter; (1) could also be written as $P = yx^2$. Also, we could have used $x = 5 - y$ in (1) to arrive at $P(y) = (5 - y)y^2$. In a calculus setting it would not have mattered whether we worked with $P(x)$ or with $P(y)$ because by finding *one* of the numbers we automatically find the other from the equation $x + y = 5$. This last equation is called a **constraint**. A constraint not only defines a relationship between the variables x and y but often puts a limitation on how x and y can vary. As we see in the next example, the constraint helps in determining the domain of the function.

EXAMPLE 2 Example 1 Continued

What is the domain of the function $P(x)$ in (2)?

Solution Taken out of the context of the statement of the problem in Example 1, one would have to conclude that since

$$P(x) = x(5 - x)^2 = 25x - 10x^2 + x^3$$

is a polynomial function its domain is the set of real numbers $(-\infty, \infty)$. *But* in the context of the original problem, the numbers were to be nonnegative. From the requirement that $x \geq 0$ and $y = 5 - x \geq 0$ we get $x \geq 0$ and $x \leq 5$, which means that x must satisfy the simultaneous inequality $0 \leq x \leq 5$. Using interval notation, the domain of the product function P in (2) is the closed interval $[0, 5]$. ∎

If we allowed $x > 5$, then $y = 5 - x < 0$, contrary to the assumption that $y > 0$.

Often in problems that require words translated into a function, it is a good idea to sketch a curve or a picture and identify given quantities in your sketch. Keep your sketch simple.

EXAMPLE 3 Amount of Fencing

A rancher intends to mark off a rectangular plot of land that will have an area of 1000 m². The plot will be fenced and divided into two equal portions by an additional fence parallel to two sides. Express the amount of fence used as a function of the length of one side of the plot.

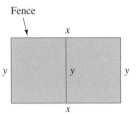

FIGURE 1.7.1 Rectangular plot of land in Example 3

Solution Your drawing should be a rectangle with a line drawn down its middle, similar to that given in FIGURE 1.7.1. As shown in the figure, let $x > 0$ be the length of the rectangular plot of land and let $y > 0$ denote its width. The function we seek is the "amount of fence." If the symbol F represents this amount, then the sum of the lengths of the *five* portions—two horizontal and three vertical—of the fence is

$$F = 2x + 3y. \qquad (4)$$

But the fenced-in land is to have an area of 1000 m², and so x and y must be related by the constraint $xy = 1000$. From the last equation we get $y = 1000/x$, which can be used to eliminate y in (4). Thus, the amount of fence F as a function of the length x is $F(x) = 2x + 3(1000/x)$ or

$$F(x) = 2x + \frac{3000}{x}. \qquad (5)$$

Since x represents a physical dimension that satisfies $xy = 1000$, we conclude that it is positive. But other than that, there is no restriction on x. Thus, unlike the previous example, the function (5) is not defined on a closed interval. The domain of $F(x)$ is the interval $(0, \infty)$. ∎

FIGURE 1.7.2 Rectangle in Example 4

EXAMPLE 4 Area of a Rectangle

A rectangle has two vertices on the x-axis and two vertices on the semicircle whose equation is $y = \sqrt{25 - x^2}$. See FIGURE 1.7.2(a). Express the area of the rectangle as a function of x.

Solution If $(x, y), x > 0, y > 0$, denotes the vertex of the rectangle on the circle in the first quadrant, then as shown in Figure 1.7.2(b) the area A is length × width, or

$$A = (2x) \times y = 2xy. \qquad (6)$$

The equation of the semicircle $y = \sqrt{25 - x^2}$ is the constraint in this problem. We use that equation to eliminate y in (6) and obtain the area of the rectangle as a function of x,

$$A(x) = 2x\sqrt{25 - x^2}. \qquad (7)$$

The implicit domain of (7) is the closed interval $[-5, 5]$, but because we assumed that (x, y) was a point on the semicircle in the first quadrant we must have $x > 0$. Thus the domain of (7) is the interval $(0, 5)$. ∎

EXAMPLE 5 Distance

Express the distance from a point (x, y) in the first quadrant on the circle $x^2 + y^2 = 1$ to the point $(2, 4)$ as a function of x.

Solution Let (x, y) denote a point in the first quadrant on the circle, and let d represent the distance from (x, y) to $(2, 4)$. See FIGURE 1.7.3. Then from the distance formula,

$$d = \sqrt{(x - 2)^2 + (y - 4)^2} = \sqrt{x^2 + y^2 - 4x - 8y + 20}. \qquad (8)$$

The constraint in this problem is the equation of the circle $x^2 + y^2 = 1$. From this equation we can immediately replace $x^2 + y^2$ in (8) by the number 1. Moreover, using the constraint to write $y = \sqrt{1 - x^2}$ allows us to eliminate the symbol y in (8). Thus the distance d as a function of x is:

$$d(x) = \sqrt{21 - 4x - 8\sqrt{1 - x^2}}. \qquad (9)$$

Since (x, y) is a point on the circle in the first quadrant the variable x can range between 0 and 1, that is, the domain of the function in (9) is the open interval $(0, 1)$. ∎

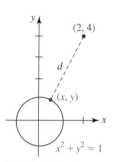

FIGURE 1.7.3 Distance d in Example 5

◀ A point on either the x-axis or the y-axis is not considered to be in any quadrant.

If a word problem involves triangles, you should study the problem carefully and determine whether the Pythagorean Theorem, similar triangles, or right-triangle trigonometry is applicable.

EXAMPLE 6 Length of a Shadow

A tree is planted 30 ft from the base of a streetlamp that is 25 ft tall. Express the length of the tree's shadow as a function of its height.

Solution As shown in FIGURE 1.7.4(a), we let h and s denote the height of the tree and the length of its shadow, respectively. Because the triangles shown Figure 1.7.4(b) are right triangles, we might think of using the Pythagorean Theorem. For this problem, however, the Pythagorean Theorem would lead us astray. The important thing to notice here is that triangles ABC and $AB'C'$ are similar. We then use the fact that the ratios of corresponding sides of similar triangles are equal to write

$$\frac{h}{s} = \frac{25}{s + 30} \quad \text{or} \quad (s + 30)h = 25s.$$

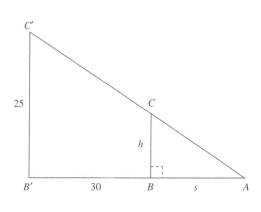

(a)   (b)

FIGURE 1.7.4 Streetlamp and tree in Example 6

By solving the last equation for s in terms of h, we obtain the rational function

$$s(h) = \frac{30h}{25 - h}. \tag{10}$$

It makes physical sense to take the domain of the function in (10) to be defined by $0 \le h < 25$. If $h > 25$, then $s(h)$ is negative, which makes no sense in the physical context of the problem. ∎

EXAMPLE 7 Length of a Ladder

A 10-ft wall stands 5 ft from a building. A ladder, supported by the wall, is to reach from the ground to the building as shown in FIGURE 1.7.5. Express the length of the ladder in terms of the distance x between the base of the wall and the base of the ladder.

Solution Let L denote the length of the ladder. With the variables x and y defined in Figure 1.7.5, we see again that there are two right triangles; the larger triangle has three sides with lengths L, y, and $x + 5$ and the smaller triangle has two sides of lengths x and 10. The ladder is the hypotenuse of the larger right triangle, so by the Pythagorean Theorem,

$$L^2 = (x + 5)^2 + y^2. \tag{11}$$

The right triangles in Figure 1.7.5 are similar because they both contain a right angle and share the common acute angle the ladder makes with the ground. We again use the fact that the ratios of corresponding sides of similar triangles are equal. This enables us to write

$$\frac{y}{x + 5} = \frac{10}{x} \quad \text{so that} \quad y = \frac{10(x + 5)}{x}.$$

Using the last result, (11) becomes

$$L^2 = (x + 5)^2 + \left[\frac{10(x + 5)}{x}\right]^2$$

$$= (x + 5)^2 \left[1 + \frac{100}{x^2}\right]$$

$$= (x + 5)^2 \left[\frac{x^2 + 100}{x^2}\right].$$

Taking the square root gives L as a function of x,

$$L(x) = \frac{x + 5}{x}\sqrt{x^2 + 100}. \tag{12}$$ ∎

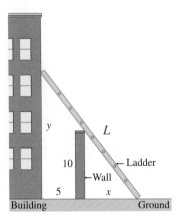

FIGURE 1.7.5 Ladder in Example 7

EXAMPLE 8 Distance

A plane flies at a constant height of 3000 ft over level ground away from an observer on the ground. Express the horizontal distance between the plane and the observer as a function of the angle of elevation of the plane measured by the observer.

Solution As shown in FIGURE 1.7.6, let x be the horizontal distance between the plane and the observer, and let θ denote the angle of elevation. The triangle in the figure is a right triangle. Hence, from right-triangle trigonometry the side opposite θ is related to the side adjacent to θ by $\tan\theta = \text{opp}/\text{adj}$. Therefore,

$$\tan\theta = \frac{3000}{x} \quad \text{or} \quad x(\theta) = 3000\cot\theta, \tag{13}$$

where $0 < \theta < \pi/2$. ∎

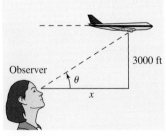

FIGURE 1.7.6 Plane in Example 8

Exercises 1.7 Answers to selected odd-numbered problems begin on page ANS-7.

≡ Fundamentals

In Problems 1–32, translate the words into an appropriate function. Give the domain of the function.

1. The product of two positive numbers is 50. Express their sum as a function of one of the numbers.
2. Express the sum of a nonzero number and its reciprocal as a function of the number.
3. The sum of two nonnegative numbers is 1. Express the sum of the square of one and twice the square of the other as a function of one of the numbers.
4. Let m and n be positive integers. The sum of two nonnegative numbers is S. Express the product of the mth power of one and the nth power of the other as a function of one of the numbers.
5. A rectangle has a perimeter of 200 in. Express the area of the rectangle as a function of the length of one of its sides.
6. A rectangle has an area of 400 in^2. Express the perimeter of the rectangle as a function of the length of one of its sides.
7. Express the area of the rectangle shaded in FIGURE 1.7.7 as a function of x.

FIGURE 1.7.7 Rectangle in Problem 7

8. Express the length of the line segment containing the point $(2, 4)$ shown in FIGURE 1.7.8 as a function of x.

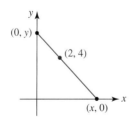

FIGURE 1.7.8 Line segment in Problem 8

9. Express the distance from a point (x, y) on the graph of $x + y = 1$ to the point $(2, 3)$ as a function of x.
10. Express the distance from a point (x, y) on the graph of $y = 4 - x^2$ to the point $(0, 1)$ as a function of x.
11. Express the perimeter of a square as a function of its area A.
12. Express the area of a circle as a function of its diameter d.
13. Express the diameter of a circle as a function of its circumference C.
14. Express the volume of a cube as a function of the area A of its base.
15. Express the area of an equilateral triangle as a function of its height h.
16. Express the area of an equilateral triangle as a function of the length s of one of its sides.
17. A wire of length x is bent into the shape of a circle. Express the area of the circle as a function of x.
18. A wire of length L is cut x units from one end. One piece of the wire is bent into a square and the other piece is bent into a circle. Express the sum of the areas as a function of x.
19. A rancher wishes to enclose a 1000-ft^2 rectangular corral using two different kinds of fence. Along two parallel sides, the fence costs $4 per foot; for the other two parallel sides, the fence costs $1.60 per foot. Express the total cost to enclose the corral as a function of the length of the sides with fence that costs $4 per foot.
20. The frame of a kite consists of six pieces of lightweight plastic. The outer frame of the kite consists of four precut pieces; two pieces of length 2 ft and two pieces of length 3 ft. Express the area of the kite as a function of x, where $2x$ is the length of the horizontal crossbar piece shown in FIGURE 1.7.9.

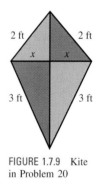

FIGURE 1.7.9 Kite in Problem 20

21. A company wants to construct an open rectangular box with a volume of 450 in^3 so that the length of its base is three times its width. Express the surface area of the box as a function of its width.
22. A conical tank, with vertex down, has a radius of 5 ft and a height of 15 ft. See FIGURE 1.7.10. Water is pumped into the tank. Express the volume of the water as a function of its depth. [*Hint*: The volume of a cone is $V = \frac{1}{3}\pi r^2 h$.]

FIGURE 1.7.10 Conical tank in Problem 22

23. Car A passes point O heading east at a constant rate of 40 mi/h; car B passes the same point 1 hour later heading north at a constant rate of 60 mi/h. Express the distance between the cars as a function of time t, where t is measured starting when car B passes point O. See FIGURE 1.7.11.

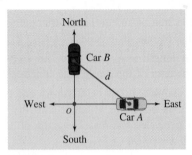

FIGURE 1.7.11 Cars in Problem 23

24. At time $t = 0$ (measured in hours), two planes with a vertical separation of 1 mi pass each other going in opposite directions. See FIGURE 1.7.12. The planes are flying horizontally at rates of 500 mi/h and 550 mi/h.
   (a) Express the horizontal distance between them as a function of t. [*Hint*: Distance = rate × time.]
   (b) Express the diagonal distance between them as a function of t.

FIGURE 1.7.12 Planes in Problem 24

25. The swimming pool shown in FIGURE 1.7.13 is 3 ft deep at the shallow end, 8 ft deep at the deep end, 40 ft long, 30 ft wide, and the bottom is an inclined plane. Water is pumped into the pool. Express the volume of the water in the pool as a function of the height h of the water above the deep end. [*Hint*: The volume will be a piecewise-defined function with domain defined by $0 \leq h \leq 8$.]

FIGURE 1.7.13 Swimming pool in Problem 25

26. U.S. Postal Service regulations for parcel post stipulate that the length plus girth (the perimeter of one end) of a package must not exceed 108 in. Express the volume of the package as a function of the width x shown in FIGURE 1.7.14.

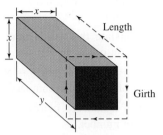

FIGURE 1.7.14 Package in Problem 26

27. Express the height of the balloon shown in FIGURE 1.7.15 as a function of its angle of elevation.

FIGURE 1.7.15 Balloon in Problem 27

28. A long sheet of metal 40 in. wide is made into a V-shaped trough by bending it in the middle along its length. Express the area of the triangular cross section of the trough as a function of the angle θ at the vertex of the V. See FIGURE 1.7.16.

FIGURE 1.7.16 Triangular cross section in Problem 28

29. As shown in FIGURE 1.7.17, a plank is supported by a sawhorse so that one end rests on the ground and the other end rests against a building. Express the length L of the plank as a function of the indicated angle θ. [*Hint*: Use two right triangles.]

FIGURE 1.7.17 Plank in Problem 29

30. A farmer wishes to enclose a pasture in the form of a right triangle using 2000 ft of fencing on hand. See FIGURE 1.7.18. Express the area of the pasture as a function of the indicated angle θ. [*Hint*: Use the symbols in the figure to form $\cot\theta$ and $\csc\theta$.]

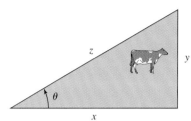

FIGURE 1.7.18 Pasture in Problem 30

31. A statue is placed on a pedestal as shown in FIGURE 1.7.19. Express the viewing angle θ as a function of the distance x from the pedestal.

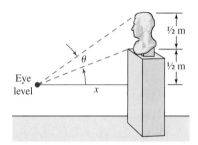

FIGURE 1.7.19 Statue in Problem 31

32. A woman on an island wishes to reach a point R on a straight shore on the mainland from a point P on the island. The point P is 9 mi from the shore and 15 mi from point R. See FIGURE 1.7.20. If the woman rows a boat at a rate of 3 mi/h to a point Q on land, then walks the rest of the way at a rate of 5 mi/h, express the total time it takes the woman to reach point R as a function of the indicated angle θ. [*Hint*: Distance = rate × time.]

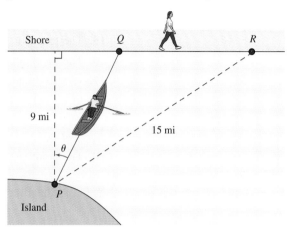

FIGURE 1.7.20 Woman rowing to shore in Problem 32

≡ **Think About It**

33. Suppose the height of the building in Example 7 is 60 ft. What is the domain of the function $L(x)$ given in (12)?

34. In an engineering text, the area of the octagon shown in FIGURE 1.7.21 is given as $A = 3.31r^2$. Show that this formula is actually an approximation to the area; that is, find the exact area A of the octagon as a function of r.

FIGURE 1.7.21 Octagon in Problem 34

Chapter 1 in Review

Answers to selected odd-numbered problems begin on page ANS-7.

A. True/False

In Problems 1–20, indicate whether the given statement is true or false.

1. If f is a function and $f(a) = f(b)$, then $a = b$. _____
2. The function $f(x) = x^5 - 4x^3 + 2$ is an odd function. _____
3. The graph of the function $f(x) = 5x^2\cos x$ is symmetric with respect to the y-axis. _____
4. The graph of $y = f(x + 3)$ is the graph of $y = f(x)$ shifted 3 units to the right. _____
5. The graph of the function $f(x) = \dfrac{1}{x-1} + \dfrac{1}{x-2}$ has no x-intercept. _____
6. An asymptote is a line that the graph of a function approaches but never crosses. _____
7. The graph of a function can have at most two horizontal asymptotes. _____
8. If $f(x) = p(x)/q(x)$ is a rational function and $q(a) = 0$, then the line $x = a$ is a vertical asymptote for the graph of f. _____
9. The function $y = -10\sec x$ has amplitude 10. _____
10. The range of the function $f(x) = 2 + \cos x$ is $[1, 3]$. _____

11. If $f(x) = 1 + x + 2e^x$ is one-to-one, then $f^{-1}(3) = 0$. _____
12. If $\tan(5\pi/4) = -1$, then $\tan^{-1}(-1) = 5\pi/4$. _____
13. No even function can be one-to-one. _____
14. A point of intersection of the graphs of f and f^{-1} must be on the line $y = x$. _____
15. The graph of $y = \sec x$ does not cross the x-axis. _____
16. The function $f(x) = \sin^{-1} x$ is not periodic. _____
17. $y = 10^{-x}$ and $y = (0.1)^x$ are the same function. _____
18. $\ln(e + e) = 1 + \ln 2$ _____
19. $\ln \dfrac{e^b}{e^a} = b - a$ _____
20. The point $(b, 1)$ is on the graph of $f(x) = \log_b x$. _____

B. Fill in the Blanks

In Problems 1–20, fill in the blanks.

1. The domain of the function $f(x) = \sqrt{x + 2}/x$ is _____.
2. If $f(x) = 4x^2 + 7$ and $g(x) = 2x + 3$, then $(f \circ g)(1) =$ _____, $(g \circ f)(1) =$ _____, and $(f \circ f)(1) =$ _____.
3. The vertex of the graph of the quadratic function $f(x) = x^2 + 16x + 70$ is _____.
4. The x-intercepts of the graph of $f(x) = x^2 + 2x - 35$ are _____.
5. The graph of the polynomial function $f(x) = x^3(x - 1)^2(x - 5)$ is tangent to the x-axis at _____ and passes through the x-axis at _____.
6. The range of the function $f(x) = 10/(x^2 + 1)$ is _____.
7. The y-intercept of the graph of $f(x) = (2x - 4)/(5 - x)$ is _____.
8. A rational function whose graph has the horizontal asymptote $y = 1$ and x-intercept $(3, 0)$ is $f(x) =$ _____.
9. The period of the function $y = 2\sin\dfrac{\pi}{3}x$ is _____.
10. The graph of the function $y = \sin(3x - \pi/4)$ is the graph of $f(x) = \sin 3x$ shifted _____ units to the _____.
11. $\sin^{-1}(\sin \pi) =$ _____.
12. If f is a one-to-one function such that $f^{-1}(3) = 1$, then a point on the graph of f is _____.
13. By rigid transformations, the point $(0, 1)$ on the graph of $y = e^x$ is moved to the point _____ on the graph of $y = 4 + e^{x-3}$.
14. $e^{3\ln 10} =$ _____.
15. If $3^x = 5$, then $x =$ _____.
16. If $3e^x = 4e^{-3x}$, then $x =$ _____.
17. If $\log_3 x = -2$, then $x =$ _____.
18. Written as an exponential statement, $\log_9 27 = 1.5$ is equivalent to _____.
19. The inverse of $y = e^x$ is _____.
20. If $f(x) = e^x - 3$, then $f(-\ln 2) =$ _____.

C. Exercises

1. Estimate the function value by referring to the graph of the function $y = f(x)$ in FIGURE 1.R.1.
    (a) $f(-4)$
    (b) $f(-3)$
    (c) $f(-2)$
    (d) $f(-1)$
    (e) $f(0)$
    (f) $f(1)$
    (g) $f(1.5)$
    (h) $f(2)$
    (i) $f(3.5)$
    (j) $f(4)$

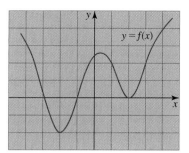

FIGURE 1.R.1 Graph for Problem 1

2. Given that
$$g(t)\begin{cases} t^2, & -1 < t \leq 1 \\ 2t, & t \leq -1 \text{ or } t > 1 \end{cases}$$
find for $0 < a < 1$:

(a) $g(1 + a)$ (b) $g(1 - a)$
(c) $g(1.5 - a)$ (d) $g(a)$
(e) $g(-a)$ (f) $g(2a)$

3. Determine whether the numbers 1, 5, and 8 are in the range of the function
$$f(x) = \begin{cases} 2x, & -2 \leq x < 2 \\ 3, & x = 2 \\ x + 4, & x > 2. \end{cases}$$

4. Suppose $f(x) = \sqrt{x + 4}$, $g(x) = \sqrt{5 - x}$, and $h(x) = x^2$. Find the domain of each of the given functions.

(a) $f \circ h$ (b) $g \circ h$
(c) $f \circ f$ (d) $g \circ g$
(e) $f + g$ (f) f/g

In Problems 5 and 6, compute $\dfrac{f(x + h) - f(x)}{h}$, $h \neq 0$, and simplify.

5. $f(x) = -x^3 + 2x^2 - x + 5$ 6. $f(x) = 1 + 2x - \dfrac{3}{x}$

In Problems 7–16, match the given rational function with one of the graphs (a)–(j).

(a)
FIGURE 1.R.2

(b)
FIGURE 1.R.3

(c)
FIGURE 1.R.4

(d)
FIGURE 1.R.5

(e)
FIGURE 1.R.6

(f)
FIGURE 1.R.7

(g)
FIGURE 1.R.8

(h)
FIGURE 1.R.9

(i)
FIGURE 1.R.10

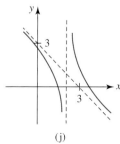
(j)
FIGURE 1.R.11

7. $f(x) = \dfrac{2x}{x^2 + 1}$ 8. $f(x) = \dfrac{x^2 - 1}{x^2 + 1}$

9. $f(x) = \dfrac{2x}{x - 2}$ 10. $f(x) = 2 - \dfrac{1}{x^2}$

11. $f(x) = \dfrac{x}{(x - 2)^2}$ 12. $f(x) = \dfrac{(x - 1)^2}{x - 2}$

13. $f(x) = \dfrac{x^2 - 10}{2x - 4}$ 14. $f(x) = \dfrac{-x^2 + 5x - 5}{x - 2}$

15. $f(x) = \dfrac{2x}{x^3 + 1}$ 16. $f(x) = \dfrac{3}{x^2 + 1}$

In Problems 17 and 18, find the slope of the red line L in each figure.

17. $f(x) = 3^{-(x+1)}$

FIGURE 1.R.12 Graph for Problem 17

18.

FIGURE 1.R.13 Graph for Problem 18

In Problems 19 and 20, suppose $2^t = a$ and $6^t = b$. Use the laws of exponents given in Section 1.6 to find the value of the given quantity.

19. (a) 12^t (b) 3^t (c) 6^{-t}
20. (a) 6^{3t} (b) $2^{-3t}2^{7t}$ (c) 18^t
21. Find a function $f(x) = ae^{kx}$ if $(0, 5)$ and $(6, 1)$ are points on the graph of f.
22. Find a function $f(x) = a10^{kx}$ if $f(3) = 8$ and $f(0) = \tfrac{1}{2}$.
23. Find a function $f(x) = a + b^x$, $0 < b < 1$, if $f(1) = 5.5$ and the graph of f has a horizontal asymptote $y = 5$.
24. Find a function $f(x) = a + \log_3(x - c)$ if $f(11) = 10$ and the graph of f has a vertical asymptote $x = 2$.

In Problems 25–30, match each of the following functions with one of the given graphs.

(a) $y = \ln(x - 2)$ (b) $y = 2 - \ln x$
(c) $y = 2 + \ln(x + 2)$ (d) $y = -2 - \ln(x + 2)$
(e) $y = -\ln(2x)$ (f) $y = 2 + \ln(-x + 2)$

25.

FIGURE 1.R.14 Graph for Problem 25

26.

FIGURE 1.R.15 Graph for Problem 26

27.

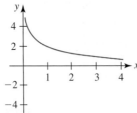

FIGURE 1.R.16 Graph for Problem 27

28.

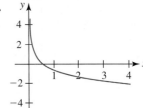

FIGURE 1.R.17 Graph for Problem 28

29.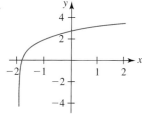

FIGURE 1.R.18 Graph for Problem 29

30.

FIGURE 1.R.19 Graph for Problem 30

31. The width of a rectangular box is three times its length and its height is two times its length.
   (a) Express the volume V of the box as a function of its length l.
   (b) As a function of its width w
   (c) As a function of its height h

32. A closed box in the form of a cube is to be constructed from two different materials. The material for the sides costs 1 cent per square centimeter and the material for the top and bottom costs 2.5 cents per square centimeter. Express the total cost C of construction as a function of the length x of a side.

33. Express the volume V of the box shown in FIGURE 1.R.20 as a function of the indicated angle θ.

FIGURE 1.R.20 Box in Problem 33

34. Consider the circle of radius h with center (h, h) shown in FIGURE 1.R.21. Express the area A of the shaded region as a function of h.

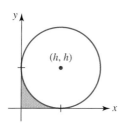

FIGURE 1.R.21 Circle in Problem 34

35. A gutter is to be made from a sheet of metal 30 cm wide by turning up the edges of width 10 cm along each side so that the sides make equal angles ϕ with the vertical. See FIGURE 1.R.22. Express the cross-sectional area of the gutter as a function of the angle ϕ.

FIGURE 1.R.22 Gutter in Problem 35

36. A metal pipe is to be carried horizontally around a right-angled corner from a hallway 8 ft wide into a hallway that is 6 ft wide. See FIGURE 1.R.23. Express the length L of the pipe as a function of the angle θ shown in the figure.

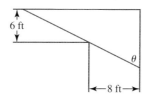

FIGURE 1.R.23 Pipe in Problem 36

37. FIGURE 1.R.24 shows a prism whose parallel faces are equilateral triangles. The rectangular base of the prism is perpendicular to the x-axis and is inscribed within the circle $x^2 + y^2 = 1$. Express the volume V of the prism as a function of x.

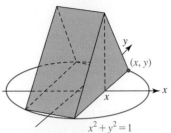

FIGURE 1.R.24 Prism in Problem 37

38. The container shown in FIGURE 1.R.25 consists of an inverted cone (open at its top) attached to the bottom of a right-circular cylinder (open at its top and bottom) of fixed radius R. The container has a fixed volume V. Express the total surface area S of the container as a function of the indicated angle θ. [*Hint*: The lateral surface area of a cone is given by $\pi R \sqrt{R^2 + h^2}$.]

FIGURE 1.R.25 Container in Problem 38

Chapter 2

Limit of a Function

In This Chapter Many topics are included in a typical course in calculus. But the three most fundamental topics in this study are the concepts of *limit*, *derivative*, and *integral*. Each of these concepts deals with functions, which is why we began this text by first reviewing some important facts about functions and their graphs.

Historically, two problems are used to introduce the basic tenets of calculus. These are the *tangent line problem* and the *area problem*. We will see in this and the subsequent chapters that the solutions to both problems involve the limit concept.

2.1 Limits—An Informal Approach
2.2 Limit Theorems
2.3 Continuity
2.4 Trigonometric Limits
2.5 Limits That Involve Infinity
2.6 Limits—A Formal Approach
2.7 The Tangent Line Problem
    Chapter 2 in Review

2.1 Limits—An Informal Approach

Introduction The two broad areas of calculus known as *differential* and *integral calculus* are built on the foundation concept of a *limit*. In this section our approach to this important concept will be intuitive, concentrating on understanding *what* a limit is using numerical and graphical examples. In the next section, our approach will be analytical, that is, we will use algebraic methods to *compute* the value of a limit of a function.

Limit of a Function–Informal Approach Consider the function

$$f(x) = \frac{16 - x^2}{4 + x} \qquad (1)$$

whose domain is the set of all real numbers except -4. Although f cannot be evaluated *at* -4 because substituting -4 for x results in the undefined quantity $0/0$, $f(x)$ can be calculated at any number x that is very *close* to -4. The two tables

x	-4.1	-4.01	-4.001
$f(x)$	8.1	8.01	8.001

x	-3.9	-3.99	-3.999
$f(x)$	7.9	7.99	7.999

(2)

show that as x approaches -4 from either the left or right, the function values $f(x)$ appear to be approaching 8, in other words, when x is near -4, $f(x)$ is near 8. To interpret the numerical information in (1) graphically, observe that for every number $x \neq -4$, the function f can be simplified by cancellation:

$$f(x) = \frac{16 - x^2}{4 + x} = \frac{(4 + x)(4 - x)}{4 + x} = 4 - x.$$

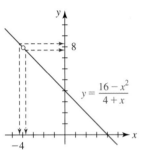

FIGURE 2.1.1 When x is near -4, $f(x)$ is near 8

As seen in FIGURE 2.1.1, the graph of f is essentially the graph of $y = 4 - x$ with the exception that the graph of f has a *hole* at the point that corresponds to $x = -4$. For x sufficiently close to -4, represented by the two arrowheads on the x-axis, the two arrowheads on the y-axis, representing function values $f(x)$, simultaneously get closer and closer to the number 8. Indeed, in view of the numerical results in (2), the arrowheads can be made as *close as we like* to the number 8. We say 8 is the **limit** of $f(x)$ as x approaches -4.

Informal Definition Suppose L denotes a finite number. The notion of $f(x)$ approaching L as x approaches a number a can be defined informally in the following manner.

- If $f(x)$ can be made arbitrarily close to the number L by taking x sufficiently close to but different from the number a, from both the left and right sides of a, then the **limit** of $f(x)$ as x approaches a is L.

Notation The discussion of the limit concept is facilitated by using a special notation. If we let the arrow symbol $\rightarrow$ represent the word *approach*, then the symbolism

$$x \rightarrow a^- \text{ indicates that } x \text{ approaches a number } a \text{ from the } \textbf{left},$$

that is, through numbers that are less than a, and

$$x \rightarrow a^+ \text{ signifies that } x \text{ approaches } a \text{ from the } \textbf{right},$$

that is, through numbers that are greater than a. Finally, the notation

$$x \rightarrow a \text{ signifies that } x \text{ approaches } a \text{ from } \textbf{both sides},$$

in other words, from the left and the right sides of a on a number line. In the left-hand table in (2) we are letting $x \rightarrow -4^-$ (for example, -4.001 is to the left of -4 on the number line), whereas in the right-hand table $x \rightarrow -4^+$.

One-Sided Limits In general, if a function $f(x)$ can be made arbitrarily close to a number L_1 by taking x sufficiently close to, but not equal to, a number a from the *left*, then we write

$$f(x) \rightarrow L_1 \text{ as } x \rightarrow a^- \qquad \text{or} \qquad \lim_{x \rightarrow a^-} f(x) = L_1. \qquad (3)$$

The number L_1 is said to be the **left-hand limit of $f(x)$ as x approaches a**. Similarly, if $f(x)$ can be made arbitrarily close to a number L_2 by taking x sufficiently close to, but not equal to, a number a from the *right*, then L_2 is the **right-hand limit of $f(x)$ as x approaches a** and we write

$$f(x) \to L_2 \text{ as } x \to a^+ \quad \text{or} \quad \lim_{x \to a^+} f(x) = L_2. \tag{4}$$

The quantities in (3) and (4) are also referred to as **one-sided limits**.

■ **Two-Sided Limits** If both the left-hand limit $\lim_{x \to a^-} f(x)$ and the right-hand limit $\lim_{x \to a^+} f(x)$ exist and have a common value L,

$$\lim_{x \to a^-} f(x) = L \quad \text{and} \quad \lim_{x \to a^+} f(x) = L,$$

then we say that L is the **limit of $f(x)$ as x approaches a** and write

$$\lim_{x \to a} f(x) = L. \tag{5}$$

A limit such as (5) is said to be a **two-sided limit**. See FIGURE 2.1.2. Since the numerical tables in (2) suggest that

$$f(x) \to 8 \text{ as } x \to -4^- \quad \text{and} \quad f(x) \to 8 \text{ as } x \to -4^+, \tag{6}$$

we can replace the two symbolic statements in (6) by the statement

$$f(x) \to 8 \text{ as } x \to -4 \quad \text{or equivalently} \quad \lim_{x \to -4} \frac{16 - x^2}{4 + x} = 8. \tag{7}$$

FIGURE 2.1.2 $f(x) \to L$ as $x \to a$ if and only if $f(x) \to L$ as $x \to a^-$ and $f(x) \to L$ as $x \to a^+$

■ **Existence and Nonexistence** Of course a limit (one-sided or two-sided) does not have to exist. But it is important that you keep firmly in mind:

- *The existence of a limit of a function f as x approaches a (from one side or from both sides), does not depend on whether f is defined at a but only on whether f is defined for x near the number a.*

For example, if the function in (1) is modified in the following manner

$$f(x) = \begin{cases} \dfrac{16 - x^2}{4 + x}, & x \neq -4 \\ 5, & x = -4, \end{cases}$$

then $f(-4)$ is defined and $f(-4) = 5$, but still $\lim_{x \to -4} \dfrac{16 - x^2}{4 + x} = 8$. See FIGURE 2.1.3. In general, the two-sided limit $\lim_{x \to a} f(x)$ **does not exist**

- if either of the one-sided limits $\lim_{x \to a^-} f(x)$ or $\lim_{x \to a^+} f(x)$ fails to exist, or
- if $\lim_{x \to a^-} f(x) = L_1$ and $\lim_{x \to a^+} f(x) = L_2$, but $L_1 \neq L_2$.

FIGURE 2.1.3 Whether f is defined at a or is not defined at a has no bearing on the existence of the limit of $f(x)$ as $x \to a$

EXAMPLE 1 A Limit That Exists

The graph of the function $f(x) = -x^2 + 2x + 2$ is shown in FIGURE 2.1.4. As seen from the graph and the accompanying tables, it seems plausible that

$$\lim_{x \to 4^-} f(x) = -6 \quad \text{and} \quad \lim_{x \to 4^+} f(x) = -6$$

and consequently $\lim_{x \to 4} f(x) = -6$.

$x \to 4^-$	3.9	3.99	3.999
$f(x)$	-5.41000	-5.94010	-5.99400

$x \to 4^+$	4.1	4.01	4.001
$f(x)$	-6.61000	-6.06010	-6.00600

Note that in Example 1 the given function is certainly defined at 4, but at no time did we substitute $x = 4$ into the function to find the value of $\lim_{x \to 4} f(x)$.

FIGURE 2.1.4 Graph of function in Example 1

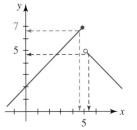

FIGURE 2.1.5 Graph of function in Example 2

EXAMPLE 2 A Limit That Exists

The graph of the piecewise-defined function

$$f(x) = \begin{cases} x^2, & x < 2 \\ -x + 6, & x > 2 \end{cases}$$

is given in FIGURE 2.1.5. Notice that $f(2)$ is not defined, but that is of no consequence when considering $\lim_{x \to 2} f(x)$. From the graph and the accompanying tables,

$x \to 2^-$	1.9	1.99	1.999
$f(x)$	3.61000	3.96010	3.99600

$x \to 2^+$	2.1	2.01	2.001
$f(x)$	3.90000	3.99000	3.99900

we see that when we make x close to 2, we can make $f(x)$ arbitrarily close to 4, and so

$$\lim_{x \to 2^-} f(x) = 4 \quad \text{and} \quad \lim_{x \to 2^+} f(x) = 4.$$

That is, $\lim_{x \to 2} f(x) = 4$. ∎

EXAMPLE 3 A Limit That Does Not Exist

The graph of the piecewise-defined function

$$f(x) = \begin{cases} x + 2, & x \le 5 \\ -x + 10, & x > 5 \end{cases}$$

is given in FIGURE 2.1.6. From the graph and the accompanying tables, it appears that as x approaches 5 through numbers less than 5 that $\lim_{x \to 5^-} f(x) = 7$. Then as x approaches 5 through numbers greater than 5 it appears that $\lim_{x \to 5^+} f(x) = 5$. But since

$$\lim_{x \to 5^-} f(x) \ne \lim_{x \to 5^+} f(x),$$

we conclude that $\lim_{x \to 5} f(x)$ does not exist.

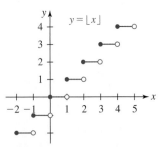

FIGURE 2.1.6 Graph of function in Example 3

$x \to 5^-$	4.9	4.99	4.999
$f(x)$	6.90000	6.99000	6.99900

$x \to 5^+$	5.1	5.01	5.001
$f(x)$	4.90000	4.99000	4.99900

EXAMPLE 4 A Limit That Does Not Exist

Recall, the **greatest integer function** or **floor function** $f(x) = \lfloor x \rfloor$ is defined to be the greatest integer that is less than or equal to x. The domain of f is the set of real numbers $(-\infty, \infty)$. From the graph in FIGURE 2.1.7 we see that $f(n)$ is defined for every integer n; nonetheless, for each integer n, $\lim_{x \to n} f(x)$ does not exist. For example, as x approaches, say, the number 3, the two one-sided limits exist but have different values:

$$\lim_{x \to 3^-} f(x) = 2 \quad \text{whereas} \quad \lim_{x \to 3^+} f(x) = 3. \tag{8}$$

In general, for an integer n,

$$\lim_{x \to n^-} f(x) = n - 1 \quad \text{whereas} \quad \lim_{x \to n^+} f(x) = n. \qquad \blacksquare$$

The greatest integer function was discussed in Section 1.1.

FIGURE 2.1.7 Graph of function in Example 4

EXAMPLE 5 A Right-Hand Limit

From FIGURE 2.1.8 it should be clear that $f(x) = \sqrt{x} \to 0$ as $x \to 0^+$, that is,

$$\lim_{x \to 0^+} \sqrt{x} = 0.$$

It would be incorrect to write $\lim_{x \to 0} \sqrt{x} = 0$ since this notation carries with it the connotation that the limits from the left and from the right exist and are equal to 0. In this case $\lim_{x \to 0^-} \sqrt{x}$ does not exist since $f(x) = \sqrt{x}$ is not defined for $x < 0$. ∎

FIGURE 2.1.8 Graph of function in Example 5

If $x = a$ is a vertical asymptote for the graph of $y = f(x)$, then $\lim_{x \to a} f(x)$ will always fail to exist because the function values $f(x)$ must become unbounded from at least one side of the line $x = a$.

EXAMPLE 6 A Limit That Does Not Exist

A vertical asymptote always corresponds to an infinite break in the graph of a function f. In FIGURE 2.1.9 we see that the y-axis or $x = 0$ is a vertical asymptote for the graph of $f(x) = 1/x$. The tables

$x \to 0^-$	-0.1	-0.01	-0.001
$f(x)$	-10	-100	-1000

$x \to 0^+$	0.1	0.01	0.001
$f(x)$	10	100	1000

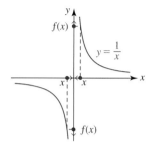

FIGURE 2.1.9 Graph of function in Example 6

clearly show that the function values $f(x)$ become unbounded in absolute value as we get close to 0. In other words, $f(x)$ is not approaching a real number as $x \to 0^-$ nor as $x \to 0^+$. Therefore, neither the left-hand nor the right-hand limit exists as x approaches 0. Thus we conclude that $\lim_{x \to 0} f(x)$ does not exist. ∎

EXAMPLE 7 An Important Trigonometric Limit

To do the calculus of the trigonometric functions $\sin x$, $\cos x$, $\tan x$, and so on, it is important to realize that the variable x is either a real number or an angle measured in radians. With that in mind, consider the numerical values of $f(x) = (\sin x)/x$ as $x \to 0^+$ given in the table that follows.

$x \to 0^+$	0.1	0.01	0.001	0.0001
$f(x)$	0.99833416	0.99998333	0.99999983	0.99999999

It is easy to see that the same results given in the table hold as $x \to 0^-$. Because $\sin x$ is an odd function, for $x > 0$ and $-x < 0$ we have $\sin(-x) = -\sin x$ and as a consequence

$$f(-x) = \frac{\sin(-x)}{-x} = \frac{\sin x}{x} = f(x).$$

As can be seen in FIGURE 2.1.10, f is an even function. The table of numerical values as well as the graph of f strongly suggest the following result:

FIGURE 2.1.10 Graph of function in Example 7

$$\lim_{x \to 0} \frac{\sin x}{x} = 1. \tag{9}$$ ∎

The limit in (9) is a very important result and will be used in Section 3.4. Another trigonometric limit that you are asked to verify as an exercise is given by

$$\lim_{x \to 0} \frac{1 - \cos x}{x} = 0. \tag{10}$$

See Problem 43 in Exercises 2.1. Because of their importance, both (9) and (10) will be proven in Section 2.4.

■ An Indeterminate Form A limit of a quotient $f(x)/g(x)$, where both the numerator and the denominator approach 0 as $x \to a$, is said to have the **indeterminate form 0/0**. The limit (7) in our initial discussion has this indeterminate form. Many important limits, such as (9) and (10), and the limit

$$\lim_{h \to 0} \frac{f(x + h) - f(x)}{h},$$

which forms the backbone of differential calculus, also have the indeterminate form $0/0$.

EXAMPLE 8 An Indeterminate Form

The limit $\lim_{x\to 0} |x|/x$ has the indeterminate form $0/0$, but unlike (7), (9), and (10) this limit fails to exist. To see why, let us examine the graph of the function $f(x) = |x|/x$. For $x \neq 0$, $|x| = \begin{cases} x, & x > 0 \\ -x, & x < 0 \end{cases}$ and so we recognize f as the piecewise-defined function

$$f(x) = \frac{|x|}{x} = \begin{cases} 1, & x > 0 \\ -1, & x < 0. \end{cases} \qquad (11)$$

FIGURE 2.1.11 Graph of function in Example 8

From (11) and the graph of f in FIGURE 2.1.11 it should be apparent that both the left-hand and right-hand limits of f exist and

$$\lim_{x\to 0^-} \frac{|x|}{x} = -1 \quad \text{and} \quad \lim_{x\to 0^+} \frac{|x|}{x} = 1.$$

Because these one-sided limits are different, we conclude that $\lim_{x\to 0} |x|/x$ does not exist. ∎

$\lim_{x\to a}$ NOTES FROM THE CLASSROOM

While graphs and tables of function values may be convincing for determining whether a limit does or does not exist, you are certainly aware that all calculators and computers work only with approximations and that graphs can be drawn inaccurately. A blind use of a calculator can also lead to a false conclusion. For example, $\lim_{x\to 0} \sin(\pi/x)$ is known not to exist, but from the table of values

$x \to 0$	±0.1	±0.01	±0.001
$f(x)$	0	0	0

one would naturally conclude that $\lim_{x\to 0} \sin(\pi/x) = 0$. On the other hand, the limit

$$\lim_{x\to 0} \frac{\sqrt{x^2 + 4} - 2}{x^2} \qquad (12)$$

can be shown to exist and equals $\frac{1}{4}$. See Example 11 in Section 2.2. One calculator gives

$x \to 0$	±0.00001	±0.000001	±0.0000001
$f(x)$	0.200000	0.000000	0.000000

The problem in calculating (12) for x very close to 0 is that $\sqrt{x^2 + 4}$ is correspondingly very close to 2. When subtracting two numbers of nearly equal values on a calculator a loss of significant digits may occur due to round-off error.

Exercises 2.1   Answers to selected odd-numbered problems begin on page ANS-8.

≡ Fundamentals

In Problems 1–14, sketch the graph of the function to find the given limit, or state that it does not exist.

1. $\lim_{x\to 2}(3x + 2)$
2. $\lim_{x\to 2}(x^2 - 1)$
3. $\lim_{x\to 0}\left(1 + \frac{1}{x}\right)$
4. $\lim_{x\to 5}\sqrt{x - 1}$
5. $\lim_{x\to 1}\frac{x^2 - 1}{x - 1}$
6. $\lim_{x\to 0}\frac{x^2 - 3x}{x}$
7. $\lim_{x\to 3}\frac{|x - 3|}{x - 3}$
8. $\lim_{x\to 0}\frac{|x| - x}{x}$
9. $\lim_{x\to 0}\frac{x^3}{x}$
10. $\lim_{x\to 1}\frac{x^4 - 1}{x^2 - 1}$
11. $\lim_{x\to 0} f(x)$ where $f(x) = \begin{cases} x + 3, & x < 0 \\ -x + 3, & x \geq 0 \end{cases}$
12. $\lim_{x\to 2} f(x)$ where $f(x) = \begin{cases} x, & x < 2 \\ x + 1, & x \geq 2 \end{cases}$
13. $\lim_{x\to 2} f(x)$ where $f(x) = \begin{cases} x^2 - 2x, & x < 2 \\ 1, & x = 2 \\ x^2 - 6x + 8, & x > 2 \end{cases}$

14. $\lim_{x \to 0} f(x)$ where $f(x) = \begin{cases} x^2, & x < 0 \\ 2, & x = 0 \\ \sqrt{x} - 1, & x > 0 \end{cases}$

In Problems 15–18, use the given graph to find the value of each quantity, or state that it does not exist.

(a) $f(1)$ (b) $\lim_{x \to 1^+} f(x)$ (c) $\lim_{x \to 1^-} f(x)$ (d) $\lim_{x \to 1} f(x)$

15.

FIGURE 2.1.12 Graph for Problem 15

16.

FIGURE 2.1.13 Graph for Problem 16

17.

FIGURE 2.1.14 Graph for Problem 17

18.

FIGURE 2.1.15 Graph for Problem 18

In Problems 19–28, each limit has the value 0, but some of the notation is incorrect. If the notation is incorrect, give the correct statement.

19. $\lim_{x \to 0} \sqrt[3]{x} = 0$
20. $\lim_{x \to 0} \sqrt[4]{x} = 0$
21. $\lim_{x \to 1} \sqrt{1 - x} = 0$
22. $\lim_{x \to -2^+} \sqrt{x + 2} = 0$
23. $\lim_{x \to 0^-} \lfloor x \rfloor = 0$
24. $\lim_{x \to \frac{1}{2}} \lfloor x \rfloor = 0$
25. $\lim_{x \to \pi} \sin x = 0$
26. $\lim_{x \to 1} \cos^{-1} x = 0$
27. $\lim_{x \to 3^+} \sqrt{9 - x^2} = 0$
28. $\lim_{x \to 1} \ln x = 0$

In Problems 29 and 30, use the given graph to find each limit, or state that it does not exist.

29. (a) $\lim_{x \to -4^+} f(x)$ (b) $\lim_{x \to -2} f(x)$
    (c) $\lim_{x \to 0} f(x)$ (d) $\lim_{x \to 1} f(x)$
    (e) $\lim_{x \to 3} f(x)$ (f) $\lim_{x \to 4} f(x)$

FIGURE 2.1.16 Graph for Problem 29

30. (a) $\lim_{x \to -5} f(x)$ (b) $\lim_{x \to -3^-} f(x)$
    (c) $\lim_{x \to -3^+} f(x)$ (d) $\lim_{x \to -3} f(x)$
    (e) $\lim_{x \to 0} f(x)$ (f) $\lim_{x \to 1} f(x)$

FIGURE 2.1.17 Graph for Problem 30

In Problems 31–34, sketch a graph of a function f with the given properties.

31. $f(-1) = 3, f(0) = -1, f(1) = 0, \lim_{x \to 0} f(x)$ does not exist
32. $f(-2) = 3, \lim_{x \to 0^-} f(x) = 2, \lim_{x \to 0^+} f(x) = -1, f(1) = -2$
33. $f(0) = 1, \lim_{x \to 1^-} f(x) = 3, \lim_{x \to 1^+} f(x) = 3, f(1)$ is undefined, $f(3) = 0$
34. $f(-2) = 2, f(x) = 1, -1 \le x \le 1, \lim_{x \to -1} f(x) = 1, \lim_{x \to 1} f(x)$ does not exist, $f(2) = 3$

≡ Calculator/CAS Problems

In Problems 35–40, use a calculator or CAS to obtain the graph of the given function f on the interval $[-0.5, 0.5]$. Use the graph to conjecture the value of $\lim_{x \to 0} f(x)$, or state that the limit does not exist.

35. $f(x) = \cos \dfrac{1}{x}$
36. $f(x) = x \cos \dfrac{1}{x}$
37. $f(x) = \dfrac{2 - \sqrt{4 + x}}{x}$
38. $f(x) = \dfrac{9}{x}\left[\sqrt{9 - x} - \sqrt{9 + x}\right]$
39. $f(x) = \dfrac{e^{-2x} - 1}{x}$
40. $f(x) = \dfrac{\ln |x|}{x}$

In Problems 41–50, proceed as in Examples 3, 6, and 7 and use a calculator to construct tables of function values. Conjecture the value of each limit, or state that it does not exist.

41. $\lim_{x \to 1} \dfrac{6\sqrt{x} - 6\sqrt{2x - 1}}{x - 1}$
42. $\lim_{x \to 1} \dfrac{\ln x}{x - 1}$
43. $\lim_{x \to 0} \dfrac{1 - \cos x}{x}$
44. $\lim_{x \to 0} \dfrac{1 - \cos x}{x^2}$
45. $\lim_{x \to 0} \dfrac{x}{\sin 3x}$
46. $\lim_{x \to 0} \dfrac{\tan x}{x}$
47. $\lim_{x \to 4} \dfrac{\sqrt{x} - 2}{x - 4}$
48. $\lim_{x \to 3} \left[\dfrac{6}{x^2 - 9} - \dfrac{6\sqrt{x - 2}}{x^2 - 9}\right]$
49. $\lim_{x \to 1} \dfrac{x^4 + x - 2}{x - 1}$
50. $\lim_{x \to -2} \dfrac{x^3 + 8}{x + 2}$

2.2 Limit Theorems

Introduction The intention of the informal discussion in Section 2.1 was to give you an intuitive grasp of when a limit does or does not exist. However, it is neither desirable nor practical, in every instance, to reach a conclusion about the existence of a limit based on a graph or on a table of numerical values. We must be able to evaluate a limit, or discern its non-existence, in a somewhat mechanical fashion. The theorems that we shall consider in this section establish such a means. The proofs of some of these results are given in the *Appendix*.

The first theorem gives two basic results that will be used throughout the discussion of this section.

Theorem 2.2.1 Two Fundamental Limits

(*i*) $\lim_{x \to a} c = c$, where c is a constant

(*ii*) $\lim_{x \to a} x = a$

Although both parts of Theorem 2.2.1 require a formal proof, Theorem 2.2.1(*ii*) is almost tautological when stated in words:

- *The limit of x as x is approaching a is a.*

See the *Appendix* for a proof of Theorem 2.2.1(*i*).

EXAMPLE 1 Using Theorem 2.2.1

(a) From Theorem 2.2.1(*i*),

$$\lim_{x \to 2} 10 = 10 \quad \text{and} \quad \lim_{x \to 6} \pi = \pi.$$

(b) From Theorem 2.1.1(*ii*),

$$\lim_{x \to 2} x = 2 \quad \text{and} \quad \lim_{x \to 0} x = 0. \quad \blacksquare$$

The limit of a constant multiple of a function f is the constant times the limit of f as x approaches a number a.

Theorem 2.2.2 Limit of a Constant Multiple

If c is a constant, then

$$\lim_{x \to a} c f(x) = c \lim_{x \to a} f(x).$$

We can now start using theorems in conjunction with each other.

EXAMPLE 2 Using Theorems 2.2.1 and 2.2.2

From Theorems 2.2.1 (*ii*) and 2.2.2,

(a) $\lim_{x \to 8} 5x = 5 \lim_{x \to 8} x = 5 \cdot 8 = 40$

(b) $\lim_{x \to -2} \left(-\tfrac{3}{2} x\right) = -\tfrac{3}{2} \lim_{x \to -2} x = \left(-\tfrac{3}{2}\right) \cdot (-2) = 3.$ $\blacksquare$

The next theorem is particularly important because it gives us a way of computing limits in an algebraic manner.

Theorem 2.2.3 Limit of a Sum, Product, and Quotient

Suppose a is a real number and $\lim_{x \to a} f(x)$ and $\lim_{x \to a} g(x)$ exist. If $\lim_{x \to a} f(x) = L_1$ and $\lim_{x \to a} g(x) = L_2$, then

(i) $\lim_{x \to a} [f(x) \pm g(x)] = \lim_{x \to a} f(x) \pm \lim_{x \to a} g(x) = L_1 \pm L_2$,

(ii) $\lim_{x \to a} [f(x)g(x)] = \left(\lim_{x \to a} f(x)\right)\left(\lim_{x \to a} g(x)\right) = L_1 L_2$, and

(iii) $\lim_{x \to a} \dfrac{f(x)}{g(x)} = \dfrac{\lim_{x \to a} f(x)}{\lim_{x \to a} g(x)} = \dfrac{L_1}{L_2}$, $L_2 \neq 0$.

Theorem 2.2.3 can be stated in words:

- *If both limits exist, then*
    - (i) *the limit of a sum is the sum of the limits,*
    - (ii) *the limit of a product is the product of the limits, and*
    - (iii) *the limit of a quotient is the quotient of the limits provided the limit of the denominator is not zero.*

Note: If all limits exist, then Theorem 2.2.3 is also applicable to one-sided limits, that is, the symbolism $x \to a$ in Theorem 2.2.3 can be replaced by either $x \to a^-$ or $x \to a^+$. Moreover, Theorem 2.2.3 extends to differences, sums, products, and quotients that involve more than two functions. See the *Appendix* for a proof of Theorem 2.2.3.

EXAMPLE 3 Using Theorem 2.2.3

Evaluate $\lim_{x \to 5}(10x + 7)$.

Solution From Theorems 2.2.1 and 2.2.2, we know that $\lim_{x \to 5} 7$ and $\lim_{x \to 5} 10x$ exist. Hence, from Theorem 2.2.3(*i*),

$$\lim_{x \to 5}(10x + 7) = \lim_{x \to 5} 10x + \lim_{x \to 5} 7$$
$$= 10 \lim_{x \to 5} x + \lim_{x \to 5} 7$$
$$= 10 \cdot 5 + 7 = 57. \qquad \blacksquare$$

▍ **Limit of a Power** Theorem 2.2.3(*ii*) can be used to calculate the limit of a positive integer power of a function. For example, if $\lim_{x \to a} f(x) = L$, then from Theorem 2.2.3(*ii*) with $g(x) = f(x)$,

$$\lim_{x \to a}[f(x)]^2 = \lim_{x \to a}[f(x) \cdot f(x)] = \left(\lim_{x \to a} f(x)\right)\left(\lim_{x \to a} f(x)\right) = L^2.$$

By the same reasoning we can apply Theorem 2.2.3(*ii*) to the general case where $f(x)$ is a factor n times. This result is stated as the next theorem.

Theorem 2.2.4 Limit of a Power

Let $\lim_{x \to a} f(x) = L$ and n be a positive integer. Then

$$\lim_{x \to a}[f(x)]^n = \left[\lim_{x \to a} f(x)\right]^n = L^n.$$

For the special case $f(x) = x$, the result given in Theorem 2.2.4 yields

$$\lim_{x \to a} x^n = a^n. \qquad (1)$$

EXAMPLE 4 Using (1) and Theorem 2.2.3

Evaluate

(a) $\lim_{x \to 10} x^3$ (b) $\lim_{x \to 4} \dfrac{5}{x^2}$.

Solution

(a) From (1),
$$\lim_{x \to 10} x^3 = 10^3 = 1000.$$

(b) From Theorem 2.2.1 and (1) we know that $\lim_{x \to 4} 5 = 5$ and $\lim_{x \to 4} x^2 = 16 \ne 0$. Therefore by Theorem 2.2.3(*iii*),
$$\lim_{x \to 4} \frac{5}{x^2} = \frac{\lim_{x \to 4} 5}{\lim_{x \to 4} x^2} = \frac{5}{4^2} = \frac{5}{16}. \blacksquare$$

EXAMPLE 5 Using Theorem 2.2.3

Evaluate $\lim_{x \to 3}(x^2 - 5x + 6)$.

Solution In view of Theorem 2.2.1, Theorem 2.2.2, and (1) all limits exist. Therefore by Theorem 2.2.3(*i*),
$$\lim_{x \to 3}(x^2 - 5x + 6) = \lim_{x \to 3} x^2 - \lim_{x \to 3} 5x + \lim_{x \to 3} 6 = 3^2 - 5 \cdot 3 + 6 = 0. \blacksquare$$

EXAMPLE 6 Using Theorems 2.2.3 and 2.2.4

Evaluate $\lim_{x \to 1}(3x - 1)^{10}$.

Solution First, we see from Theorem 2.2.3(*i*) that
$$\lim_{x \to 1}(3x - 1) = \lim_{x \to 1} 3x - \lim_{x \to 1} 1 = 2.$$

It then follows from Theorem 2.2.4 that
$$\lim_{x \to 1}(3x - 1)^{10} = \left[\lim_{x \to 1}(3x - 1)\right]^{10} = 2^{10} = 1024. \blacksquare$$

■ **Limit of a Polynomial Function** Some limits can be evaluated by *direct substitution*. We can use (1) and Theorem 2.2.3(*i*) to compute the limit of a general polynomial function. If
$$f(x) = c_n x^n + c_{n-1} x^{n-1} + \cdots + c_1 x + c_0$$

is a polynomial function, then
$$\lim_{x \to a} f(x) = \lim_{x \to a}\left(c_n x^n + c_{n-1} x^{n-1} + \cdots + c_1 x + c_0\right)$$
$$= \lim_{x \to a} c_n x^n + \lim_{x \to a} c_{n-1} x^{n-1} + \cdots + \lim_{x \to a} c_1 x + \lim_{x \to a} c_0$$
$$= c_n a^n + c_{n-1} a^{n-1} + \cdots + c_1 a + c_0. \;\leftarrow \begin{smallmatrix}f \text{ is defined at } x = a \text{ and}\\ \text{this limit is } f(a)\end{smallmatrix}$$

In other words, to evaluate a limit of a polynomial function f as x approaches a real number a, we need only evaluate the function at $x = a$:
$$\lim_{x \to a} f(x) = f(a). \qquad (2)$$

A reexamination of Example 5 shows that $\lim_{x \to 3} f(x)$, where $f(x) = x^2 - 5x + 6$, is given by $f(3) = 0$.

Because a rational function f is a quotient of two polynomials $p(x)$ and $q(x)$, it follows from (2) and Theorem 2.2.3(*iii*) that a limit of a rational function $f(x) = p(x)/q(x)$ can also be found by evaluating f at $x = a$:
$$\lim_{x \to a} f(x) = \lim_{x \to a} \frac{p(x)}{q(x)} = \frac{p(a)}{q(a)}. \qquad (3)$$

Of course we must add to (3) the all-important requirement that the limit of the denominator is not 0, that is, $q(a) \neq 0$.

EXAMPLE 7 Using (2) and (3)

Evaluate $\lim\limits_{x \to -1} \dfrac{3x - 4}{8x^2 + 2x - 2}$.

Solution $f(x) = \dfrac{3x - 4}{8x^2 + 2x - 2}$ is a rational function and so if we identify the polynomials $p(x) = 3x - 4$ and $q(x) = 8x^2 + 2x - 2$, then from (2),

$$\lim_{x \to -1} p(x) = p(-1) = -7 \quad \text{and} \quad \lim_{x \to -1} q(x) = q(-1) = 4.$$

Since $q(-1) \neq 0$ it follows from (3) that

$$\lim_{x \to -1} \frac{3x - 4}{8x^2 + 2x - 2} = \frac{p(-1)}{q(-1)} = \frac{-7}{4} = -\frac{7}{4}. \qquad \blacksquare$$

You should not get the impression that we can *always* find a limit of a function by substituting the number *a directly into the function.*

EXAMPLE 8 Using Theorem 2.2.3

Evaluate $\lim\limits_{x \to 1} \dfrac{x - 1}{x^2 + x - 2}$.

Solution The function in this limit is rational, but if we substitute $x = 1$ into the function we see that this limit has the indeterminate form $0/0$. However, by simplifying *first*, we can then apply Theorem 2.2.3(*iii*):

$$\lim_{x \to 1} \frac{x - 1}{x^2 + x - 2} = \lim_{x \to 1} \frac{x - 1}{(x - 1)(x + 2)} \quad \leftarrow \begin{array}{l}\text{cancellation is valid} \\ \text{provided that } x \neq 1\end{array}$$

$$= \lim_{x \to 1} \frac{1}{x + 2}$$

$$= \frac{\lim\limits_{x \to 1} 1}{\lim\limits_{x \to 1}(x + 2)} = \frac{1}{3}. \qquad \blacksquare$$

◀ If a limit of a rational function has the indeterminate form $0/0$ as $x \to a$, then by the Factor Theorem of algebra $x - a$ must be a factor of both the numerator and the denominator. Factor those quantities and cancel the factor $x - a$.

Sometimes you can tell at a glance when *a limit does not exist.*

Theorem 2.2.5 A Limit That Does Not Exist

Let $\lim\limits_{x \to a} f(x) = L_1 \neq 0$ and $\lim\limits_{x \to a} g(x) = 0$. Then

$$\lim_{x \to a} \frac{f(x)}{g(x)}$$

does not exist.

PROOF We will give an indirect proof of this result based on Theorem 2.2.3. Suppose $\lim\limits_{x \to a} f(x) = L_1 \neq 0$ and $\lim\limits_{x \to a} g(x) = 0$ and suppose further that $\lim\limits_{x \to a}(f(x)/g(x))$ exists and equals L_2. Then

$$L_1 = \lim_{x \to a} f(x) = \lim_{x \to a}\left(g(x) \cdot \frac{f(x)}{g(x)}\right), \quad g(x) \neq 0,$$

$$= \left(\lim_{x \to a} g(x)\right)\left(\lim_{x \to a} \frac{f(x)}{g(x)}\right) = 0 \cdot L_2 = 0.$$

By contradicting the assumption that $L_1 \neq 0$, we have proved the theorem. $\blacksquare$

EXAMPLE 9 Using Theorems 2.2.3 and 2.2.5

Evaluate

(a) $\lim_{x \to 5} \dfrac{x}{x-5}$ (b) $\lim_{x \to 5} \dfrac{x^2 - 10x - 25}{x^2 - 4x - 5}$ (c) $\lim_{x \to 5} \dfrac{x-5}{x^2 - 10x + 25}$.

Solution Each function in the three parts of the example is rational.

(a) Since the limit of the numerator x is 5, but the limit of the denominator $x-5$ is 0, we conclude from Theorem 2.2.5 that the limit does not exist.

(b) Substituting $x = 5$ makes both the numerator and denominator 0, and so the limit has the indeterminate form $0/0$. By the Factor Theorem of algebra, $x-5$ is a factor of both the numerator and denominator. Hence,

$$\lim_{x \to 5} \frac{x^2 - 10x - 25}{x^2 - 4x - 5} = \lim_{x \to 5} \frac{(x-5)^2}{(x-5)(x+1)} \quad \leftarrow \text{cancel the factor } x - 5$$

$$= \lim_{x \to 5} \frac{x-5}{x+1}$$

$$= \frac{0}{6} = 0. \quad \leftarrow \text{limit exists}$$

(c) Again, the limit has the indeterminate form $0/0$. After factoring the denominator and canceling the factors we see from the algebra

$$\lim_{x \to 5} \frac{x-5}{x^2 - 10x + 25} = \lim_{x \to 5} \frac{x-5}{(x-5)^2}$$

$$= \lim_{x \to 5} \frac{1}{x-5}$$

that the limit does not exist since the limit of the numerator in the last expression is now 1 but the limit of the denominator is 0. ∎

Limit of a Root The limit of the nth root of a function is the nth root of the limit whenever the limit exists and has a real nth root. The next theorem summarizes this fact.

Theorem 2.2.6 Limit of a Root

Let $\lim_{x \to a} f(x) = L$ and n be a positive integer. Then

$$\lim_{x \to a} \sqrt[n]{f(x)} = \sqrt[n]{\lim_{x \to a} f(x)} = \sqrt[n]{L},$$

provided that $L \geq 0$ when n is even.

An immediate special case of Theorem 2.2.6 is

$$\lim_{x \to a} \sqrt[n]{x} = \sqrt[n]{a}, \qquad (4)$$

provided $a \geq 0$ when n is even. For example, $\lim_{x \to 9} \sqrt{x} = [\lim_{x \to 9} x]^{1/2} = 9^{1/2} = 3$.

EXAMPLE 10 Using (4) and Theorem 2.2.3

Evaluate $\lim_{x \to -8} \dfrac{x - \sqrt[3]{x}}{2x + 10}$.

Solution Since $\lim_{x \to -8} (2x + 10) = -6 \neq 0$, we see from Theorem 2.2.3(*iii*) and (4) that

$$\lim_{x \to -8} \frac{x - \sqrt[3]{x}}{2x + 10} = \frac{\lim_{x \to -8} x - [\lim_{x \to -8} x]^{1/3}}{\lim_{x \to -8} (2x + 10)} = \frac{-8 - (-8)^{1/3}}{-6} = \frac{-6}{-6} = 1. \quad ∎$$

When a limit of an algebraic function involving radicals has the indeterminate form $0/0$, rationalization of the numerator or the denominator may be something to try.

EXAMPLE 11 Rationalization of a Numerator

Evaluate $\lim\limits_{x \to 0} \dfrac{\sqrt{x^2 + 4} - 2}{x^2}$.

◀ We have seen this limit in (12) in *Notes from the Classroom* at the end of Section 2.1.

Solution Because $\lim\limits_{x \to 0} \sqrt{x^2 + 4} = \sqrt{\lim\limits_{x \to 0}(x^2 + 4)} = 2$ we see by inspection that the given limit has the indeterminate form $0/0$. However, by rationalization of the numerator we obtain

$$\lim_{x \to 0} \frac{\sqrt{x^2 + 4} - 2}{x^2} = \lim_{x \to 0} \frac{\sqrt{x^2 + 4} - 2}{x^2} \cdot \frac{\sqrt{x^2 + 4} + 2}{\sqrt{x^2 + 4} + 2}$$

$$= \lim_{x \to 0} \frac{(x^2 + 4) - 4}{x^2(\sqrt{x^2 + 4} + 2)}$$

$$= \lim_{x \to 0} \frac{x^2}{x^2(\sqrt{x^2 + 4} + 2)} \quad \leftarrow \text{cancel } x\text{'s}$$

$$= \lim_{x \to 0} \frac{1}{\sqrt{x^2 + 4} + 2}. \quad \leftarrow \text{this limit is no longer } 0/0$$

We are now in a position to use Theorems 2.2.3 and 2.2.6:

$$\lim_{x \to 0} \frac{\sqrt{x^2 + 4} - 2}{x^2} = \lim_{x \to 0} \frac{1}{\sqrt{x^2 + 4} + 2}$$

$$= \frac{\lim\limits_{x \to 0} 1}{\sqrt{\lim\limits_{x \to 0}(x^2 + 4)} + \lim\limits_{x \to 0} 2}$$

$$= \frac{1}{2 + 2} = \frac{1}{4}. \quad \blacksquare$$

In case anyone is wondering whether there can be more than one limit of a function $f(x)$ as $x \to a$, we state the last theorem for the record.

Theorem 2.2.7 Existence Implies Uniqueness

If $\lim\limits_{x \to a} f(x)$ exists, then it is unique.

$\lim\limits_{x \to a}$ NOTES FROM THE CLASSROOM

In mathematics it is just as important to be aware of what a definition or a theorem does *not* say as what it says.

(*i*) Property (*i*) of Theorem 2.2.3 does not say that the limit of a sum is *always* the sum of the limits. For example, $\lim\limits_{x \to 0}(1/x)$ does not exist, so

$$\lim_{x \to 0}\left[\frac{1}{x} - \frac{1}{x}\right] \neq \lim_{x \to 0} \frac{1}{x} - \lim_{x \to 0} \frac{1}{x}.$$

Nevertheless, since $1/x - 1/x = 0$ for $x \neq 0$, the limit of the difference exists

$$\lim_{x \to 0}\left[\frac{1}{x} - \frac{1}{x}\right] = \lim_{x \to 0} 0 = 0.$$

(*ii*) Similarly, the limit of a product could exist and yet not be equal to the product of the limits. For example, $x/x = 1$, for $x \neq 0$, and so

$$\lim_{x \to 0}\left(x \cdot \frac{1}{x}\right) = \lim_{x \to 0} 1 = 1$$

but

$$\lim_{x \to 0}\left(x \cdot \frac{1}{x}\right) \neq \left(\lim_{x \to 0} x\right)\left(\lim_{x \to 0} \frac{1}{x}\right)$$

because $\lim\limits_{x \to 0}(1/x)$ does not exist.

(iii) Theorem 2.2.5 does not say that the limit of a quotient fails to exist whenever the limit of the denominator is zero. Example 8 provides a counterexample to that interpretation. However, Theorem 2.2.5 states that a limit of a quotient does not exist whenever the limit of the denominator is zero *and* the limit of the numerator is not zero.

Exercises 2.2
Answers to selected odd-numbered problems begin on page ANS-8.

≡ Fundamentals

In Problems 1–52, find the given limit, or state that it does not exist.

1. $\lim_{x\to -4} 15$
2. $\lim_{x\to 0} \cos \pi$
3. $\lim_{x\to 3}(-4)x$
4. $\lim_{x\to 2}(3x-9)$
5. $\lim_{x\to -2} x^2$
6. $\lim_{x\to 5}(-x^3)$
7. $\lim_{x\to -1}(x^3 - 4x + 1)$
8. $\lim_{x\to 6}(-5x^2 + 6x + 8)$
9. $\lim_{x\to 2} \dfrac{2x+4}{x-7}$
10. $\lim_{x\to 0} \dfrac{x+5}{3x}$
11. $\lim_{t\to 1}(3t-1)(5t^2+2)$
12. $\lim_{t\to -2}(t+4)^2$
13. $\lim_{s\to 7} \dfrac{s^2 - 21}{s+2}$
14. $\lim_{x\to 6} \dfrac{x^2 - 6x}{x^2 - 7x + 6}$
15. $\lim_{x\to -1}(x + x^2 + x^3)^{135}$
16. $\lim_{x\to 2} \dfrac{(3x-4)^{40}}{(x^2-2)^{36}}$
17. $\lim_{x\to 6} \sqrt{2x-5}$
18. $\lim_{x\to 8}(1 + \sqrt[3]{x})$
19. $\lim_{t\to 1} \dfrac{\sqrt{t}}{t^2 + t - 2}$
20. $\lim_{x\to 2} x^2\sqrt{x^2 + 5x + 2}$
21. $\lim_{y\to -5} \dfrac{y^2 - 25}{y + 5}$
22. $\lim_{u\to 8} \dfrac{u^2 - 5u - 24}{u - 8}$
23. $\lim_{x\to 1} \dfrac{x^3 - 1}{x - 1}$
24. $\lim_{t\to -1} \dfrac{t^3 + 1}{t^2 - 1}$
25. $\lim_{x\to 10} \dfrac{(x-2)(x+5)}{(x-8)}$
26. $\lim_{x\to -3} \dfrac{2x+6}{4x^2 - 36}$
27. $\lim_{x\to 2} \dfrac{x^3 + 3x^2 - 10x}{x - 2}$
28. $\lim_{x\to 1.5} \dfrac{2x^2 + 3x - 9}{x - 1.5}$
29. $\lim_{t\to 1} \dfrac{t^3 - 2t + 1}{t^3 + t^2 - 2}$
30. $\lim_{x\to 0} x^3(x^4 + 2x^3)^{-1}$
31. $\lim_{x\to 0^+} \dfrac{(x+2)(x^5-1)^3}{(\sqrt{x}+4)^2}$
32. $\lim_{x\to -2} x\sqrt{x+4}\sqrt[3]{x-6}$
33. $\lim_{x\to 0}\left[\dfrac{x^2 + 3x - 1}{x} + \dfrac{1}{x}\right]$
34. $\lim_{x\to 2}\left[\dfrac{1}{x-2} - \dfrac{6}{x^2 + 2x - 8}\right]$
35. $\lim_{x\to 3^+} \dfrac{(x+3)^2}{\sqrt{x-3}}$
36. $\lim_{x\to 3}(x-4)^{99}(x^2-7)^{10}$
37. $\lim_{x\to 10}\sqrt{\dfrac{10x}{2x+5}}$
38. $\lim_{r\to 1} \dfrac{\sqrt{(r^2 + 3r - 2)^3}}{\sqrt[3]{(5r-3)^2}}$
39. $\lim_{h\to 4}\sqrt{\dfrac{h}{h+5}}\left(\dfrac{h^2 - 16}{h-4}\right)^2$
40. $\lim_{t\to 2}(t+2)^{3/2}(2t+4)^{1/3}$
41. $\lim_{x\to 0^-} \sqrt[5]{\dfrac{x^3 - 64x}{x^2 + 2x}}$
42. $\lim_{x\to 1^+}\left(8x + \dfrac{2}{x}\right)^5$
43. $\lim_{t\to 1}(at^2 - bt)^2$
44. $\lim_{x\to -1}\sqrt{u^2x^2 + 2xu + 1}$
45. $\lim_{h\to 0} \dfrac{(8+h)^2 - 64}{h}$
46. $\lim_{h\to 0} \dfrac{1}{h}[(1+h)^3 - 1]$
47. $\lim_{h\to 0} \dfrac{1}{h}\left(\dfrac{1}{x+h} - \dfrac{1}{x}\right)$
48. $\lim_{h\to 0} \dfrac{\sqrt{x+h} - \sqrt{x}}{h}$ $(x > 0)$
49. $\lim_{t\to 1} \dfrac{\sqrt{t} - 1}{t - 1}$
50. $\lim_{u\to 5} \dfrac{\sqrt{u+4} - 3}{u - 5}$
51. $\lim_{v\to 0} \dfrac{\sqrt{25+v} - 5}{\sqrt{1+v} - 1}$
52. $\lim_{x\to 1} \dfrac{4 - \sqrt{x+15}}{x^2 - 1}$

In Problems 53–60, assume that $\lim_{x\to a} f(x) = 4$ and $\lim_{x\to a} g(x) = 2$. Find the given limit, or state that it does not exist.

53. $\lim_{x\to a}[5f(x) + 6g(x)]$
54. $\lim_{x\to a}[f(x)]^3$
55. $\lim_{x\to a} \dfrac{1}{g(x)}$
56. $\lim_{x\to a} \sqrt{\dfrac{f(x)}{g(x)}}$
57. $\lim_{x\to a} \dfrac{f(x)}{f(x) - 2g(x)}$
58. $\lim_{x\to a} \dfrac{[f(x)]^2 - 4[g(x)]^2}{f(x) - 2g(x)}$
59. $\lim_{x\to a} xf(x)g(x)$
60. $\lim_{x\to a} \dfrac{6x+3}{xf(x) + g(x)}, a \neq -\dfrac{1}{2}$

≡ Think About It

In Problems 61 and 62, use the first result to find the limits in parts (a)–(c). Justify each step in your work citing the appropriate property of limits.

61. $\lim_{x\to 1} \dfrac{x^{100} - 1}{x - 1} = 100$

(a) $\lim_{x\to 1} \dfrac{x^{100} - 1}{x^2 - 1}$ (b) $\lim_{x\to 1} \dfrac{x^{50} - 1}{x - 1}$ (c) $\lim_{x\to 1} \dfrac{(x^{100} - 1)^2}{(x - 1)^2}$

62. $\lim_{x\to 0} \dfrac{\sin x}{x} = 1$

(a) $\lim_{x\to 0} \dfrac{2x}{\sin x}$ (b) $\lim_{x\to 0} \dfrac{1 - \cos^2 x}{x^2}$ (c) $\lim_{x\to 0} \dfrac{8x^2 - \sin x}{x}$

63. Using $\lim_{x\to 0} \dfrac{\sin x}{x} = 1$, show that $\lim_{x\to 0} \sin x = 0$.

64. If $\lim_{x\to 2} \dfrac{2f(x) - 5}{x + 3} = 4$, find $\lim_{x\to 2} f(x)$.

2.3 Continuity

Introduction In the discussion in Section 1.1 on graphing functions, we used the phrase "connect the points with a smooth curve." This phrase invokes the image of a graph that is a nice *continuous* curve—in other words, a curve with no breaks, gaps, or holes in it. Indeed, a continuous function is often described as one whose graph can be drawn without lifting pencil from paper.

In Section 2.2 we saw that the function value $f(a)$ played no part in determining the existence of $\lim_{x \to a} f(x)$. But we did see in Section 2.2 that limits as $x \to a$ of polynomial functions and certain rational functions could be found by simply evaluating the function *at* $x = a$. The reason we can do that in some instances is the fact that the function is *continuous* at a number a. In this section we will see that both the value $f(a)$ and the limit of f as x approaches a number a play major roles in defining the notion of continuity. Before giving the definition, we illustrate in FIGURE 2.3.1 some intuitive examples of graphs of functions that are *not* continuous at a.

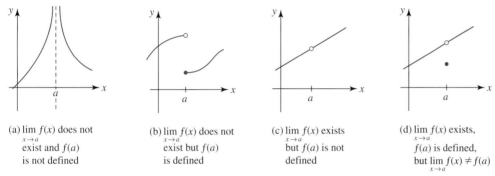

(a) $\lim_{x \to a} f(x)$ does not exist and $f(a)$ is not defined

(b) $\lim_{x \to a} f(x)$ does not exist but $f(a)$ is defined

(c) $\lim_{x \to a} f(x)$ exists but $f(a)$ is not defined

(d) $\lim_{x \to a} f(x)$ exists, $f(a)$ is defined, but $\lim_{x \to a} f(x) \neq f(a)$

FIGURE 2.3.1 Four examples of f *not* continuous at a

Continuity at a Number Figure 2.3.1 suggests the following threefold condition of continuity of a function f at a number a.

Definition 2.3.1 Continuity at a

A function f is said to be **continuous** at a number a if

(*i*) $f(a)$ is defined,   (*ii*) $\lim_{x \to a} f(x)$ exists, and   (*iii*) $\lim_{x \to a} f(x) = f(a)$.

If any one of the three conditions in Definition 2.3.1 fails, then f is said to be **discontinuous** at the number a.

EXAMPLE 1 Three Functions

Determine whether each of the functions is continuous at 1.

(a) $f(x) = \dfrac{x^3 - 1}{x - 1}$ (b) $g(x) = \begin{cases} \dfrac{x^3 - 1}{x - 1}, & x \neq 1 \\ 2, & x = 1 \end{cases}$ (c) $h(x) = \begin{cases} \dfrac{x^3 - 1}{x - 1}, & x \neq 1 \\ 3, & x = 1 \end{cases}$.

Solution

(a) f is discontinuous at 1 since substituting $x = 1$ into the function results in $0/0$. We say that $f(1)$ is not defined and so the first condition of continuity in Definition 2.3.1 is violated.

(b) Because g is defined at 1, that is, $g(1) = 2$, we next determine whether $\lim_{x \to 1} g(x)$ exists. From

$$\lim_{x \to 1} \frac{x^3 - 1}{x - 1} = \lim_{x \to 1} \frac{(x - 1)(x^2 + x + 1)}{x - 1} = \lim_{x \to 1}(x^2 + x + 1) = 3 \qquad (1)$$

◀ Recall from algebra that $a^3 - b^3 = (a - b)(a^2 + ab + b^2)$

we conclude $\lim_{x\to 1} g(x)$ exists and equals 3. Since this value is not the same as $g(1) = 2$, the second condition of Definition 2.3.1 is violated. The function g is discontinuous at 1.

(c) First, $h(1)$ is defined, in this case, $h(1) = 3$. Second, $\lim_{x\to 1} h(x) = 3$ from (1) of part (b). Third, we have $\lim_{x\to 1} h(x) = h(1) = 3$. Thus *all* three conditions in Definition 2.3.1 are satisfied and so the function h is continuous at 1.

The graphs of the three functions are compared in FIGURE 2.3.2.

  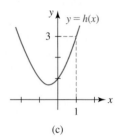

(a)   (b)   (c)

FIGURE 2.3.2 Graphs of functions in Example 1

EXAMPLE 2 Piecewise-Defined Function

Determine whether the piecewise-defined function is continuous at 2.

$$f(x) = \begin{cases} x^2, & x < 2 \\ 5, & x = 2 \\ -x + 6, & x > 2. \end{cases}$$

Solution First, observe that $f(2)$ is defined and equals 5. Next, we see from

$$\left. \begin{array}{l} \lim_{x\to 2^-} f(x) = \lim_{x\to 2^-} x^2 = 4 \\ \lim_{x\to 2^+} f(x) = \lim_{x\to 2^+} (-x + 6) = 4 \end{array} \right\} \text{ implies } \lim_{x\to 2} f(x) = 4$$

that the limit of f as $x \to 2$ exists. Finally, because $\lim_{x\to 2} f(x) \neq f(2) = 5$, it follows from (*iii*) of Definition 2.3.1 that f is discontinuous at 2. The graph of f is shown in FIGURE 2.3.3.

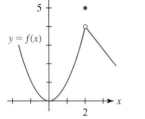

FIGURE 2.3.3 Graph of function in Example 2

■ **Continuity on an Interval** We will now extend the notion of continuity at a number a to **continuity on an interval**.

Definition 2.3.2 Continuity on an Interval

A function f is continuous

(*i*) on an **open interval** (a, b) if it is continuous at every number in the interval; and
(*ii*) on a **closed interval** $[a, b]$ if it is continuous on (a, b) and, in addition,

$$\lim_{x\to a^+} f(x) = f(a) \quad \text{and} \quad \lim_{x\to b^-} f(x) = f(b).$$

If the right-hand limit condition $\lim_{x\to a^+} f(x) = f(a)$ given in (*ii*) of Definition 2.3.1 is satisfied, we say that ***f* is continuous from the right at a**; if $\lim_{x\to b^-} f(x) = f(b)$, then ***f* is continuous from the left at b**.

Extensions of these concepts to intervals such as $[a, b)$, $(a, b]$, (a, ∞), $(-\infty, b)$, $(-\infty, \infty)$, $[a, \infty)$, and $(-\infty, b]$ are made in the expected manner. For example, f is continuous on $[1, 5)$ if it is continuous on the open interval $(1, 5)$ and continuous from the right at 1.

EXAMPLE 3 Continuity on an Interval

(a) As we see from FIGURE 2.3.4(a), $f(x) = 1/\sqrt{1 - x^2}$ is continuous on the open interval $(-1, 1)$ but is not continuous on the closed interval $[-1, 1]$, since neither $f(-1)$ nor $f(1)$ is defined.

(b) $f(x) = \sqrt{1 - x^2}$ is continuous on $[-1, 1]$. Observe from Figure 2.3.4(b) that
$$\lim_{x \to -1^+} f(x) = f(-1) = 0 \text{ and } \lim_{x \to 1^-} f(x) = f(1) = 0.$$

(c) $f(x) = \sqrt{x - 1}$ is continuous on the unbounded interval $[1, \infty)$, because
$$\lim_{x \to a} f(x) = \sqrt{\lim_{x \to a}(x - 1)} = \sqrt{a - 1} = f(a),$$
for any real number a satisfying $a > 1$, and f is continuous from the right at 1 since
$$\lim_{x \to 1^+} \sqrt{x - 1} = f(1) = 0.$$
See Figure 2.3.4(c). ∎

(a)

(b)

(c)

FIGURE 2.3.4 Graphs of functions in Example 3

A review of the graphs in Figures 1.4.1 and 1.4.2 shows that $y = \sin x$ and $y = \cos x$ are continuous on $(-\infty, \infty)$. Figures 1.4.3 and 1.4.5 show that $y = \tan x$ and $y = \sec x$ are discontinuous at $x = (2n + 1)\pi/2, n = 0, \pm 1, \pm 2, \ldots$, whereas Figures 1.4.4 and 1.4.6 show that $y = \cot x$ and $y = \csc x$ are discontinuous at $x = n\pi, n = 0, \pm 1, \pm 2, \ldots$. The inverse trigonometric functions $y = \sin^{-1} x$ and $y = \cos^{-1} x$ are continuous on the closed interval $[-1, 1]$. See Figures 1.5.9 and 1.5.12. The natural exponential function $y = e^x$ is continuous on $(-\infty, \infty)$, whereas the natural logarithmic function $y = \ln x$ is continuous on $(0, \infty)$. See Figures 1.6.5 and 1.6.6.

Continuity of a Sum, Product, and Quotient When two functions f and g are continuous at a number a, then the combinations of functions formed by addition, multiplication, and division are also continuous at a. In the case of division f/g we must, of course, require that $g(a) \neq 0$.

Theorem 2.3.1 Continuity of a Sum, Product, and Quotient

If the functions f and g are continuous at a number a, then the sum $f + g$, the product fg, and the quotient f/g ($g(a) \neq 0$) are continuous at $x = a$.

PROOF OF CONTINUITY OF THE PRODUCT fg As a consequence of the assumption that the functions f and g are continuous at a number a, we can say that both functions are defined at $x = a$, the limits of both functions as x approaches a exist, and
$$\lim_{x \to a} f(x) = f(a) \quad \text{and} \quad \lim_{x \to a} g(x) = g(a).$$
Because the limits exist, we know that the limit of a product is the product of the limits:
$$\lim_{x \to a}(f(x)g(x)) = \left(\lim_{x \to a} f(x)\right)\left(\lim_{x \to a} g(x)\right) = f(a)g(a).$$
The proofs of the remaining parts of Theorem 2.3.1 are obtained in a similar manner. ∎

Since Definition 2.3.1 implies that $f(x) = x$ is continuous at any real number x, we see from successive applications of Theorem 2.3.1 that the functions $x, x^2, x^3, \ldots, x^n$ are also continuous for every x in the interval $(-\infty, \infty)$. Because a polynomial function is just a sum of powers of x, another application of Theorem 2.3.1 shows:

- *A polynomial function f is continuous on $(-\infty, \infty)$.*

Functions, such as polynomials and the sine and cosine, that are continuous for *all* real numbers, that is, on the interval $(-\infty, \infty)$, are said to be **continuous everywhere**. A function that is continuous everywhere is also just said to be **continuous**. Now, if $p(x)$ and $q(x)$ are polynomial functions, it also follows directly from Theorem 2.3.1 that:

- A rational function $f(x) = p(x)/q(x)$ is continuous except at numbers at which the denominator $q(x)$ is zero.

■ **Terminology** A discontinuity of a function f is often given a special name.

- If $x = a$ is a vertical asymptote for the graph of $y = f(x)$, then f is said to have an **infinite discontinuity** at a.

FIGURE 2.3.5 Jump discontinuity at $x = 0$

Figure 2.3.1(a) illustrates a function with an infinite discontinuity at a.

- If $\lim_{x \to a^-} f(x) = L_1$ and $\lim_{x \to a^+} f(x) = L_2$ and $L_1 \ne L_2$, then f is said to have a **finite discontinuity** or a **jump discontinuity** at a.

The function $y = f(x)$ given in FIGURE 2.3.5 has a jump discontinuity at 0, since $\lim_{x \to 0^-} f(x) = -1$ and $\lim_{x \to 0^+} f(x) = 1$. The greatest integer function $f(x) = \lfloor x \rfloor$ has a jump discontinuity at every integer value of x.

- If $\lim_{x \to a} f(x)$ exists but either f is not defined at $x = a$ or $f(a) \ne \lim_{x \to a} f(x)$, then f is said to have a **removable discontinuity** at a.

(a) Not continuous at 1

For example, the function $f(x) = (x^2 - 1)/(x - 1)$ is not defined at $x = 1$ but $\lim_{x \to 1} f(x) = 2$. By *defining* $f(1) = 2$, the new function

$$f(x) = \begin{cases} \dfrac{x^2 - 1}{x - 1}, & x \ne 1 \\ 2, & x = 1 \end{cases}$$

is continuous everywhere. See FIGURE 2.3.6.

(b) Continuous at 1
FIGURE 2.3.6 Removable discontinuity at $x = 1$

■ **Continuity of f^{-1}** The plausibility of the next theorem follows from the fact that the graph of an inverse function f^{-1} is a reflection of the graph of f in the line $y = x$.

Theorem 2.3.2   Continuity of an Inverse Function

If f is a continuous one-to-one function on an interval $[a, b]$, then f^{-1} is continuous on either $[f(a), f(b)]$ or $[f(b), f(a)]$.

The sine function, $f(x) = \sin x$, is continuous on $[-\pi/2, \pi/2]$ and, as noted previously, the inverse of f, $y = \sin^{-1} x$, is continuous on the closed interval $[f(-\pi/2), f(\pi/2)] = [-1, 1]$.

■ **Limit of a Composite Function** The next theorem tells us that if a function f is continuous, then the limit of the function is the function of the limit. The proof of Theorem 2.3.3 is given in the *Appendix*.

Theorem 2.3.3   Limit of a Composite Function

If $\lim_{x \to a} g(x) = L$ and f is continuous at L, then

$$\lim_{x \to a} f(g(x)) = f\left(\lim_{x \to a} g(x)\right) = f(L).$$

Theorem 2.3.3 is useful in proving other theorems. If the function g is continuous at a and f is continuous at $g(a)$, then we see that

$$\lim_{x \to a} f(g(x)) = f\left(\lim_{x \to a} g(x)\right) = f(g(a)).$$

We have just proved that the composite of two continuous functions is continuous.

Theorem 2.3.4 Continuity of a Composite Function

If g is continuous at a number a and f is continuous at $g(a)$, then the composite function $(f \circ g)(x) = f(g(x))$ is continuous at a.

EXAMPLE 4 Continuity of a Composite Function

$f(x) = \sqrt{x}$ is continuous on the interval $[0, \infty)$ and $g(x) = x^2 + 2$ is continuous on $(-\infty, \infty)$. But, since $g(x) \geq 0$ for all x, the composite function

$$(f \circ g)(x) = f(g(x)) = \sqrt{x^2 + 2}$$

is continuous everywhere.

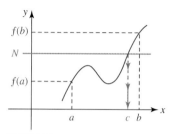

FIGURE 2.3.7 A continuous function f takes on all values between $f(a)$ and $f(b)$

If a function f is continuous on a closed interval $[a, b]$, then, as illustrated in FIGURE 2.3.7, f takes on all values between $f(a)$ and $f(b)$. Put another way, a continuous function f does not "skip" any values.

Theorem 2.3.5 Intermediate Value Theorem

If f denotes a function continuous on a closed interval $[a, b]$ for which $f(a) \neq f(b)$, and if N is any number between $f(a)$ and $f(b)$, then there exists at least one number c between a and b such that $f(c) = N$.

EXAMPLE 5 Consequence of Continuity

The polynomial function $f(x) = x^2 - x - 5$ is continuous on the interval $[-1, 4]$ and $f(-1) = -3, f(4) = 7$. For any number N for which $-3 \leq N \leq 7$, Theorem 2.3.5 guarantees that there is a solution to the equation $f(c) = N$, that is, $c^2 - c - 5 = N$ in $[-1, 4]$. Specifically, if we choose $N = 1$, then $c^2 - c - 5 = 1$ is equivalent to

$$c^2 - c - 6 = 0 \quad \text{or} \quad (c - 3)(c + 2) = 0.$$

Although the latter equation has two solutions, only the value $c = 3$ is between -1 and 4.

The foregoing example suggests a corollary to the Intermediate Value Theorem.

- If f satisfies the hypotheses of Theorem 2.3.5 and $f(a)$ and $f(b)$ have opposite algebraic signs, then there exists a number x between a and b for which $f(x) = 0$.

This fact is often used in locating real zeros of a continuous function f. If the function values $f(a)$ and $f(b)$ have opposite signs, then by identifying $N = 0$, we can say that there is at least one number c in (a, b) for which $f(c) = 0$. In other words, if either $f(a) > 0, f(b) < 0$ or $f(a) < 0, f(b) > 0$, then $f(x)$ has at least one zero c in the interval (a, b). The plausibility of this conclusion is illustrated in FIGURE 2.3.8.

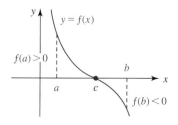

(a) One zero c in (a, b)

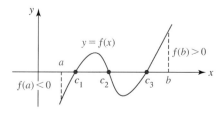

(b) Three zeros c_1, c_2, c_3 in (a, b)

FIGURE 2.3.8 Locating zeros of functions using the Intermediate Value Theorem

Bisection Method

As a direct consequence of the Intermediate Value Theorem, we can devise a means of approximating the zeros of a continuous function to any degree of accuracy. Suppose $y = f(x)$ is continuous on the closed interval $[a, b]$ such that $f(a)$ and $f(b)$ have opposite algebraic signs. Then, as we have just seen, f has a zero in $[a, b]$. Suppose we bisect the interval $[a, b]$ by finding its midpoint $m_1 = (a + b)/2$. If $f(m_1) = 0$, then m_1 is a zero of f and we proceed no further, but if $f(m_1) \neq 0$, then we can say that:

- If $f(a)$ and $f(m_1)$ have opposite algebraic signs, then f has a zero c in $[a, m_1]$.
- If $f(m_1)$ and $f(b)$ have opposite algebraic signs, then f has a zero c in $[m_1, b]$.

FIGURE 2.3.9 The number m_1 is an approximation to the number c

That is, if $f(m_1) \neq 0$, then f has a zero in an interval that is one-half the length of the original interval. See FIGURE 2.3.9. We now repeat the process by bisecting this new interval by finding its midpoint m_2. If m_2 is a zero of f, we stop, but if $f(m_2) \neq 0$, we have located a zero in an interval that is one-fourth the length of $[a, b]$. We continue this process of locating a zero of f in shorter and shorter intervals indefinitely. This method of approximating a zero of a continuous function by a sequence of midpoints is called the **bisection method**. Reinspection of Figure 2.3.9 shows that the error in an approximation to a zero in an interval is less than one-half the length of the interval.

EXAMPLE 6 Zeros of a Polynomial Function

(a) Show that the polynomial function $f(x) = x^6 - 3x - 1$ has a real zero in $[-1, 0]$ and in $[1, 2]$.

(b) Approximate the zero in $[1, 2]$ to two decimal places.

Solution

(a) Observe that $f(-1) = 3 > 0$ and $f(0) = -1 < 0$. This change in sign indicates that the graph of f must cross the x-axis at least once in the interval $[-1, 0]$. In other words, there is at least one zero of f in $[-1, 0]$.

Similarly, $f(1) = -3 < 0$ and $f(2) = 57 > 0$ implies that there is at least one zero of f in the interval $[1, 2]$.

(b) A first approximation to the zero in $[1, 2]$ is the midpoint of the interval:

$$m_1 = \frac{1 + 2}{2} = \frac{3}{2} = 1.5, \qquad \text{error} < \frac{1}{2}(2 - 1) = 0.5.$$

Now since $f(m_1) = f\left(\frac{3}{2}\right) > 0$ and $f(1) < 0$, we know that the zero lies in the interval $\left[1, \frac{3}{2}\right]$.

The second approximation is the midpoint of $\left[1, \frac{3}{2}\right]$:

$$m_2 = \frac{1 + \frac{3}{2}}{2} = \frac{5}{4} = 1.25, \qquad \text{error} < \frac{1}{2}\left(\frac{3}{2} - 1\right) = 0.25.$$

Since $f(m_2) = f\left(\frac{5}{4}\right) < 0$, the zero lies in the interval $\left[\frac{5}{4}, \frac{3}{2}\right]$.

The third approximation is the midpoint of $\left[\frac{5}{4}, \frac{3}{2}\right]$:

$$m_3 = \frac{\frac{5}{4} + \frac{3}{2}}{2} = \frac{11}{8} = 1.375, \qquad \text{error} < \frac{1}{2}\left(\frac{3}{2} - \frac{5}{4}\right) = 0.125.$$

After eight calculations, we find that $m_8 = 1.300781$ with error less than 0.005. Hence, 1.30 is an approximation to the zero of f in $[1, 2]$ that is accurate to two decimal places. The graph of f is given in FIGURE 2.3.10. ∎

FIGURE 2.3.10 Graph of function in Example 6

▶ If we wish the approximation to be accurate to *three* decimal places, we continue until the error becomes less than 0.0005, and so on.

Exercises 2.3 Answers to selected odd-numbered problems begin on page ANS-8.

≡ Fundamentals

In Problems 1–12, determine the numbers, if any, at which the given function f is discontinuous.

1. $f(x) = x^3 - 4x^2 + 7$

2. $f(x) = \dfrac{x}{x^2 + 4}$

3. $f(x) = (x^2 - 9x + 18)^{-1}$

4. $f(x) = \dfrac{x^2 - 1}{x^4 - 1}$

5. $f(x) = \dfrac{x - 1}{\sin 2x}$

6. $f(x) = \dfrac{\tan x}{x + 3}$

7. $f(x) = \begin{cases} x, & x < 0 \\ x^2, & 0 \le x < 2 \\ x, & x > 2 \end{cases}$ 8. $f(x) = \begin{cases} \dfrac{|x|}{x}, & x \ne 0 \\ 1, & x = 0 \end{cases}$

9. $f(x) = \begin{cases} \dfrac{x^2 - 25}{x - 5}, & x \ne 5 \\ 10, & x = 5 \end{cases}$

10. $f(x) = \begin{cases} \dfrac{x-1}{\sqrt{x}-1}, & x \ne 1 \\ \dfrac{1}{2}, & x = 1 \end{cases}$

11. $f(x) = \dfrac{1}{2 + \ln x}$ 12. $f(x) = \dfrac{2}{e^x - e^{-x}}$

In Problems 13–24, determine whether the given function f is continuous on the indicated intervals.

13. $f(x) = x^2 + 1$
    (a) $[-1, 4]$ (b) $[5, \infty)$

14. $f(x) = \dfrac{1}{x}$
    (a) $(-\infty, \infty)$ (b) $(0, \infty)$

15. $f(x) = \dfrac{1}{\sqrt{x}}$
    (a) $(0, 4]$ (b) $[1, 9]$

16. $f(x) = \sqrt{x^2 - 9}$
    (a) $[-3, 3]$ (b) $[3, \infty)$

17. $f(x) = \tan x$
    (a) $[0, \pi]$ (b) $[-\pi/2, \pi/2]$

18. $f(x) = \csc x$
    (a) $(0, \pi)$ (b) $(2\pi, 3\pi)$

19. $f(x) = \dfrac{x}{x^3 + 8}$
    (a) $[-4, -3]$ (b) $(-\infty, \infty)$

20. $f(x) = \dfrac{1}{|x| - 4}$
    (a) $(-\infty, -1]$ (b) $[1, 6]$

21. $f(x) = \dfrac{x}{2 + \sec x}$
    (a) $(-\infty, \infty)$ (b) $[\pi/2, 3\pi/2]$

22. $f(x) = \sin\dfrac{1}{x}$
    (a) $[1/\pi, \infty)$ (b) $[-2/\pi, 2/\pi]$

23.

FIGURE 2.3.11 Graph for Problem 23

(a) $[-1, 3]$ (b) $(2, 4]$

24.

FIGURE 2.3.12 Graph for Problem 24

(a) $[2, 4]$ (b) $[1, 5]$

In Problems 25–28, find values of m and n so that the given function f is continuous.

25. $f(x) = \begin{cases} mx, & x < 4 \\ x^2, & x \ge 4 \end{cases}$

26. $f(x) = \begin{cases} \dfrac{x^2 - 4}{x - 2}, & x \ne 2 \\ m, & x = 2 \end{cases}$

27. $f(x) = \begin{cases} mx, & x < 3 \\ n, & x = 3 \\ -2x + 9, & x > 3 \end{cases}$

28. $f(x) = \begin{cases} mx - n, & x < 1 \\ 5, & x = 1 \\ 2mx + n, & x > 1 \end{cases}$

In Problems 29 and 30, $\lfloor x \rfloor$ denotes the greatest integer not exceeding x. Sketch a graph to determine the points at which the given function is discontinuous.

29. $f(x) = \lfloor 2x - 1 \rfloor$ 30. $f(x) = \lfloor x \rfloor - x$

In Problems 31 and 32, determine whether the given function has a removable discontinuity at the given number a. If the discontinuity is removable, define a new function that is continuous at a.

31. $f(x) = \dfrac{x - 9}{\sqrt{x} - 3}$, $a = 9$ 32. $f(x) = \dfrac{x^4 - 1}{x^2 - 1}$, $a = 1$

In Problems 33–42, use Theorem 2.3.3 to find the given limit.

33. $\lim_{x \to \pi/6} \sin(2x + \pi/3)$ 34. $\lim_{x \to \pi^2} \cos\sqrt{x}$

35. $\lim_{x \to \pi/2} \sin(\cos x)$ 36. $\lim_{x \to \pi/2} (1 + \cos(\cos x))$

37. $\lim_{t \to \pi} \cos\left(\dfrac{t^2 - \pi^2}{t - \pi}\right)$ 38. $\lim_{t \to 0} \tan\left(\dfrac{\pi t}{t^2 + 3t}\right)$

39. $\lim_{t \to \pi} \sqrt{t - \pi + \cos^2 t}$ 40. $\lim_{t \to 1}(4t + \sin 2\pi t)^3$

41. $\lim_{x \to -3} \sin^{-1}\left(\dfrac{x + 3}{x^2 + 4x + 3}\right)$ 42. $\lim_{x \to \pi} e^{\cos 3x}$

In Problems 43 and 44, determine the interval(s) where $f \circ g$ is continuous.

43. $f(x) = \dfrac{1}{\sqrt{x - 1}}$, $g(x) = x + 4$

44. $f(x) = \dfrac{5x}{x - 1}$, $g(x) = (x - 2)^2$

In Problems 45–48, verify the Intermediate Value Theorem for f on the given interval. Find a number c in the interval for the indicated value of N.

45. $f(x) = x^2 - 2x$, $[1, 5]$; $N = 8$
46. $f(x) = x^2 + x + 1$, $[-2, 3]$; $N = 6$
47. $f(x) = x^3 - 2x + 1$, $[-2, 2]$; $N = 1$
48. $f(x) = \dfrac{10}{x^2 + 1}$, $[0, 1]$; $N = 8$

49. Given that $f(x) = x^5 + 2x - 7$, show that there is a number c such that $f(c) = 50$.

50. Given that f and g are continuous on $[a, b]$ such that $f(a) > g(a)$ and $f(b) < g(b)$, show that there is a number c in (a, b) such that $f(c) = g(c)$. [*Hint*: Consider the function $f - g$.]

In Problems 51–54, show that the given equation has a solution in the indicated interval.

51. $2x^7 = 1 - x$, $(0, 1)$
52. $\dfrac{x^2 + 1}{x + 3} + \dfrac{x^4 + 1}{x - 4} = 0$, $(-3, 4)$
53. $e^{-x} = \ln x$, $(1, 2)$
54. $\dfrac{\sin x}{x} = \dfrac{1}{2}$, $(\pi/2, \pi)$

≡ **Calculator/CAS Problems**

In Problems 55 and 56, use a calculator or CAS to obtain the graph of the given function. Use the bisection method to approximate, to an accuracy of two decimal places, the real zeros of f that you discover from the graph.

55. $f(x) = 3x^5 - 5x^3 - 1$
56. $f(x) = x^5 + x - 1$

57. Use the bisection method to approximate the value of c in Problem 49 to an accuracy of two decimal places.

58. Use the bisection method to approximate the solution in Problem 51 to an accuracy of two decimal places.

59. Use the bisection method to approximate the solution in Problem 52 to an accuracy of two decimal places.

60. Suppose a closed right-circular cylinder has a given volume V and surface area S (lateral side, top, and bottom).
    (a) Show that the radius r of the cylinder must satisfy the equation $2\pi r^3 - Sr + 2V = 0$.
    (b) Suppose $V = 3000$ ft^3 and $S = 1800$ ft^2. Use a calculator or CAS to obtain the graph of
    $$f(r) = 2\pi r^3 - 1800r + 6000.$$
    (c) Use the graph in part (b) and the bisection method to find the dimensions of the cylinder corresponding to the volume and surface area given in part (b). Use an accuracy of two decimal places.

≡ **Think About It**

61. Given that f and g are continuous at a number a, prove that $f + g$ is continuous at a.

62. Given that f and g are continuous at a number a and $g(a) \neq 0$, prove that f/g is continuous at a.

63. Let $f(x) = \lfloor x \rfloor$ be the greatest integer function and $g(x) = \cos x$. Determine the points at which $f \circ g$ is discontinuous.

64. Consider the functions
$$f(x) = |x| \quad \text{and} \quad g(x) = \begin{cases} x + 1, & x < 0 \\ x - 1, & x \geq 0. \end{cases}$$
Sketch the graphs of $f \circ g$ and $g \circ f$. Determine whether $f \circ g$ and $g \circ f$ are continuous at 0.

65. **A Mathematical Classic** The **Dirichlet function**
$$f(x) = \begin{cases} 1, & x \text{ rational} \\ 0, & x \text{ irrational} \end{cases}$$
is named after the German mathematician **Johann Peter Gustav Lejeune Dirichlet** (1805–1859). Dirichlet is responsible for the definition of a function as we know it today.
    (a) Show that f is discontinuous at every real number a. In other words, f is a *nowhere continuous function*.
    (b) What does the graph of f look like?
    (c) If r is a positive rational number, show that f is *r*-periodic, that is, $f(x + r) = f(x)$.

2.4 Trigonometric Limits

■ **Introduction** In this section we examine limits that involve trigonometric functions. As the examples in this section will illustrate, computation of trigonometric limits entails both algebraic manipulations and knowledge of some basic trigonometric identities. We begin with some simple limit results that are consequences of continuity.

■ **Using Continuity** We saw in the preceding section that the sine and cosine functions are everywhere continuous. It follows from Definition 2.3.1 that for any real number a,

$$\lim_{x \to a} \sin x = \sin a, \tag{1}$$

$$\lim_{x \to a} \cos x = \cos a. \tag{2}$$

Similarly, for a number a in the domain of the given trigonometric function

$$\lim_{x \to a} \tan x = \tan a, \qquad \lim_{x \to a} \cot x = \cot a, \tag{3}$$

$$\lim_{x \to a} \sec x = \sec a, \qquad \lim_{x \to a} \csc x = \csc a. \tag{4}$$

EXAMPLE 1 Using (1) and (2)

From (1) and (2) we have

$$\lim_{x \to 0} \sin x = \sin 0 = 0 \quad \text{and} \quad \lim_{x \to 0} \cos x = \cos 0 = 1. \tag{5}$$

We will draw on the results in (5) in the following discussion on computing other trigonometric limits. But first, we consider a theorem that is particularly useful when working with trigonometric limits.

Squeeze Theorem The next theorem has many names: **Squeeze Theorem**, **Pinching Theorem**, **Sandwiching Theorem**, **Squeeze Play Theorem**, and **Flyswatter Theorem** are just a few of them. As shown in FIGURE 2.4.1, if the graph of $f(x)$ is "squeezed" between the graphs of two other functions $g(x)$ and $h(x)$ for all x close to a, and if the functions g and h have a common limit L as $x \to a$, it stands to reason that f also approaches L as $x \to a$. The proof of Theorem 2.4.1 is given in the *Appendix*.

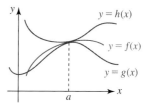

FIGURE 2.4.1 Graph of f squeezed between the graphs g and h

Theorem 2.4.1 Squeeze Theorem

Suppose f, g, and h are functions for which $g(x) \leq f(x) \leq h(x)$ for all x in an open interval that contains a number a, except possibly at a itself. If

$$\lim_{x \to a} g(x) = L \quad \text{and} \quad \lim_{x \to a} h(x) = L,$$

then $\lim_{x \to a} f(x) = L$.

◀ A colleague from Russia said this result was called the **Two Soldiers Theorem** when he was in school. Think about it.

Before applying Theorem 2.4.1, let us consider a trigonometric limit that does not exist.

EXAMPLE 2 A Limit That Does Not Exist

The limit $\lim_{x \to 0} \sin(1/x)$ does not exist. The function $f(x) = \sin(1/x)$ is odd but is not periodic. The graph f oscillates between -1 and 1 as $x \to 0$:

$$\sin \frac{1}{x} = \pm 1 \quad \text{for} \quad \frac{1}{x} = \frac{\pi}{2} + n\pi, \quad n = 0, \pm 1, \pm 2, \ldots.$$

For example, $\sin(1/x) = 1$ for $n = 500$ or $x \approx 0.00064$, and $\sin(1/x) = -1$ for $n = 501$ or $x \approx 0.00063$. This means that near the origin the graph of f becomes so compressed that it appears to be one continuous smear of color. See FIGURE 2.4.2.

FIGURE 2.4.2 Graph of function in Example 2

EXAMPLE 3 Using the Squeeze Theorem

Find the limit $\lim_{x \to 0} x^2 \sin \frac{1}{x}$.

Solution First observe that

$$\lim_{x \to 0} x^2 \sin \frac{1}{x} \neq \left(\lim_{x \to 0} x^2 \right)\left(\lim_{x \to 0} \sin \frac{1}{x} \right)$$

because we have just seen in Example 2 that $\lim_{x \to 0} \sin(1/x)$ does not exist. But for $x \neq 0$ we have $-1 \leq \sin(1/x) \leq 1$. Therefore,

$$-x^2 \leq x^2 \sin \frac{1}{x} \leq x^2.$$

Now if we make the identifications $g(x) = -x^2$ and $h(x) = x^2$, it follows from (1) of Section 2.2 that $\lim_{x \to 0} g(x) = 0$ and $\lim_{x \to 0} h(x) = 0$. Hence, from the Squeeze Theorem we conclude that

$$\lim_{x \to 0} x^2 \sin\frac{1}{x} = 0.$$

In FIGURE 2.4.3 note the small scale on the x- and y-axes.

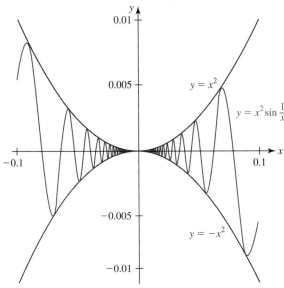

FIGURE 2.4.3 Graph of function in Example 3

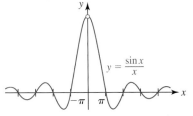

FIGURE 2.4.4 Graph of $f(x) = (\sin x)/x$

■ **An Important Trigonometric Limit** Although the function $f(x) = (\sin x)/x$ is not defined at $x = 0$, the numerical table in Example 7 of Section 2.1 and the graph in FIGURE 2.4.4 suggests that $\lim_{x \to 0} (\sin x)/x$ exists. We are now able to prove this conjecture using the Squeeze Theorem.

Consider a circle centered at the origin O with radius 1. As shown in FIGURE 2.4.5(a), let the shaded region OPR be a sector of the circle with central angle t such that $0 < t < \pi/2$. We see from parts (b), (c), and (d) of Figure 2.4.5 that

$$\text{area of } \triangle OPR \leq \text{area of sector } OPR \leq \text{area of } \triangle OQR. \tag{6}$$

From Figure 2.4.5(b) the height of $\triangle OPR$ is $\overline{OP} \sin t = 1 \cdot \sin t = \sin t$, and so

$$\text{area of } \triangle OPR = \frac{1}{2}\overline{OR} \cdot (\text{height}) = \frac{1}{2} \cdot 1 \cdot \sin t = \frac{1}{2}\sin t. \tag{7}$$

From Figure 2.4.5(d), $\overline{QR}/\overline{OR} = \tan t$ or $\overline{QR} = \tan t$, so that

$$\text{area of } \triangle OQR = \frac{1}{2}\overline{OR} \cdot \overline{QR} = \frac{1}{2} \cdot 1 \cdot \tan t = \frac{1}{2}\tan t. \tag{8}$$

(a) Unit circle   (b) Triangle OPR (c) Sector OPR (d) Right triangle OQR

FIGURE 2.4.5 Unit circle along with two triangles and a circular sector

Finally, the area of a sector of a circle is $\frac{1}{2}r^2\theta$, where r is its radius and θ is the central angle measured in radians. Thus,

$$\text{area of sector } OPR = \frac{1}{2} \cdot 1^2 \cdot t = \frac{1}{2}t. \tag{9}$$

Using (7), (8), and (9) in the inequality (6) gives

$$\frac{1}{2}\sin t < \frac{1}{2}t < \frac{1}{2}\tan t \quad \text{or} \quad 1 < \frac{t}{\sin t} < \frac{1}{\cos t}.$$

From the properties of inequalities, the last inequality can be written

$$\cos t < \frac{\sin t}{t} < 1.$$

We now let $t \to 0^+$ in the last result. Since $(\sin t)/t$ is "squeezed" between 1 and $\cos t$ (which we know from (5) is approaching 1), it follows from Theorem 2.4.1 that $(\sin t)/t \to 1$. While we have assumed $0 < t < \pi/2$, the same result holds for $t \to 0^-$ when $-\pi/2 < t < 0$. Using the symbol x in place of t, we summarize the result:

$$\lim_{x \to 0} \frac{\sin x}{x} = 1. \tag{10}$$

As the following examples illustrate, the results in (1), (2), (3), and (10) are used often to compute other limits. Note that the limit (10) is the indeterminate form $0/0$.

EXAMPLE 4 Using (10)

Find the limit $\lim_{x \to 0} \dfrac{10x - 3\sin x}{x}$.

Solution We rewrite the fractional expression as two fractions with the same denominator x:

$$\lim_{x \to 0} \frac{10x - 3\sin x}{x} = \lim_{x \to 0} \left[\frac{10x}{x} - \frac{3\sin x}{x} \right]$$

$$= \lim_{x \to 0} \frac{10x}{x} - 3 \lim_{x \to 0} \frac{\sin x}{x} \quad \leftarrow \text{since both limits exist, also cancel the } x \text{ in the first expression}$$

$$= \lim_{x \to 0} 10 - 3 \lim_{x \to 0} \frac{\sin x}{x} \quad \leftarrow \text{now use (10)}$$

$$= 10 - 3 \cdot 1$$

$$= 7. \quad \blacksquare$$

EXAMPLE 5 Using the Double-Angle Formula

Find the limit $\lim_{x \to 0} \dfrac{\sin 2x}{x}$.

Solution To evaluate the given limit we make use of the double-angle formula $\sin 2x = 2\sin x \cos x$ of Section 1.4, and the fact the limits exist:

$$\lim_{x \to 0} \frac{\sin 2x}{x} = \lim_{x \to 0} \frac{2\cos x \sin x}{x}$$

$$= 2 \lim_{x \to 0} \left(\cos x \cdot \frac{\sin x}{x} \right)$$

$$= 2 \left(\lim_{x \to 0} \cos x \right) \left(\lim_{x \to 0} \frac{\sin x}{x} \right).$$

From (5) and (10) we know that $\cos x \to 1$ and $(\sin x)/x \to 1$ as $x \to 0$, and so the preceding line becomes

$$\lim_{x \to 0} \frac{\sin 2x}{x} = 2 \cdot 1 \cdot 1 = 2. \quad \blacksquare$$

EXAMPLE 6 Using (5) and (10)

Find the limit $\lim\limits_{x \to 0} \dfrac{\tan x}{x}$.

Solution Using $\tan x = (\sin x)/\cos x$ and the fact that the limits exist we can write

$$\lim_{x \to 0} \frac{\tan x}{x} = \lim_{x \to 0} \frac{(\sin x)/\cos x}{x}$$

$$= \lim_{x \to 0} \frac{1}{\cos x} \cdot \frac{\sin x}{x}$$

$$= \left(\lim_{x \to 0} \frac{1}{\cos x}\right)\left(\lim_{x \to 0} \frac{\sin x}{x}\right)$$

$$= \frac{1}{1} \cdot 1 = 1. \quad \leftarrow \text{from (5) and (10)}$$

∎

■ Using a Substitution We are often interested in limits similar to that considered in Example 5. But if we wish to find, say, $\lim\limits_{x \to 0} \dfrac{\sin 5x}{x}$ the procedure employed in Example 5 breaks down at a practical level since we do not have a readily available trigonometric identity for $\sin 5x$. There is an alternative procedure that allows us to quickly find $\lim\limits_{x \to 0} \dfrac{\sin kx}{x}$, where $k \neq 0$ is any real constant, by simply changing the variable by means of a **substitution**. If we let $t = kx$, then $x = t/k$. Notice that as $x \to 0$ then necessarily $t \to 0$. Thus we can write

$$\lim_{x \to 0} \frac{\sin kx}{x} = \lim_{t \to 0} \frac{\sin t}{t/k} = \lim_{t \to 0}\left(\frac{\sin t}{1} \cdot \frac{k}{t}\right) = k \underbrace{\lim_{t \to 0} \frac{\sin t}{t}}_{\text{this limit is 1 from (10)}} = k.$$

Thus we have proved the general result

$$\lim_{x \to 0} \frac{\sin kx}{x} = k. \tag{11}$$

From (11), with $k = 2$, we get the same result $\lim\limits_{x \to 0} \dfrac{\sin 2x}{x} = 2$ obtained in Example 5.

EXAMPLE 7 Using a Substitution

Find the limit $\lim\limits_{x \to 1} \dfrac{\sin(x - 1)}{x^2 + 2x - 3}$.

Solution Before beginning observe that the limit has the indeterminate form $0/0$ as $x \to 1$. By factoring $x^2 + 2x - 3 = (x + 3)(x - 1)$ the given limit can be expressed as a limit of a product:

$$\lim_{x \to 1} \frac{\sin(x-1)}{x^2 + 2x - 3} = \lim_{x \to 1} \frac{\sin(x-1)}{(x+3)(x-1)} = \lim_{x \to 1}\left[\frac{1}{x+3} \cdot \frac{\sin(x-1)}{x-1}\right]. \tag{12}$$

Now if we let $t = x - 1$, we see that $x \to 1$ implies $t \to 0$. Therefore,

$$\lim_{x \to 1} \frac{\sin(x-1)}{x-1} = \lim_{t \to 0} \frac{\sin t}{t} = 1. \quad \leftarrow \text{from (10)}$$

Returning to (12), we can write

$$\lim_{x \to 1} \frac{\sin(x-1)}{x^2 + 2x - 3} = \lim_{x \to 1}\left[\frac{1}{x+3} \cdot \frac{\sin(x-1)}{x-1}\right]$$

$$= \left(\lim_{x \to 1} \frac{1}{x+3}\right)\left(\lim_{x \to 1} \frac{\sin(x-1)}{x-1}\right)$$

$$= \left(\lim_{x \to 1} \frac{1}{x+3}\right)\left(\lim_{t \to 0} \frac{\sin t}{t}\right)$$

since both limits exist. Thus,

$$\lim_{x\to 1}\frac{\sin(x-1)}{x^2+2x-3} = \left(\lim_{x\to 1}\frac{1}{x+3}\right)\left(\lim_{t\to 0}\frac{\sin t}{t}\right) = \frac{1}{4}\cdot 1 = \frac{1}{4}.$$

EXAMPLE 8 Using a Pythagorean Identity

Find the limit $\lim\limits_{x\to 0}\dfrac{1-\cos x}{x}$.

Solution To compute this limit we start with a bit of algebraic cleverness by multiplying the numerator and denominator by the conjugate factor of the numerator. Next we use the fundamental Pythagorean identity $\sin^2 x + \cos^2 x = 1$ in the form $1 - \cos^2 x = \sin^2 x$:

$$\lim_{x\to 0}\frac{1-\cos x}{x} = \lim_{x\to 0}\frac{1-\cos x}{x}\cdot\frac{1+\cos x}{1+\cos x}$$

$$= \lim_{x\to 0}\frac{1-\cos^2 x}{x(1+\cos x)}$$

$$= \lim_{x\to 0}\frac{\sin^2 x}{x(1+\cos x)}.$$

For the next step we resort back to algebra to rewrite the fractional expression as a product, then use the results in (5):

$$\lim_{x\to 0}\frac{1-\cos x}{x} = \lim_{x\to 0}\frac{\sin^2 x}{x(1+\cos x)}$$

$$= \lim_{x\to 0}\left(\frac{\sin x}{x}\cdot\frac{\sin x}{1+\cos x}\right)$$

$$= \left(\lim_{x\to 0}\frac{\sin x}{x}\right)\cdot\left(\lim_{x\to 0}\frac{\sin x}{1+\cos x}\right).$$

Because $\lim\limits_{x\to 0}(\sin x)/(1+\cos x) = 0/2 = 0$ we have

$$\lim_{x\to 0}\frac{1-\cos x}{x} = 0. \qquad (13)$$

Since the limit in (13) is equal to 0, we can write

$$\lim_{x\to 0}\frac{1-\cos x}{x} = \lim_{x\to 0}\frac{-(\cos x-1)}{x} = (-1)\lim_{x\to 0}\frac{\cos x-1}{x} = 0.$$

Dividing by -1 then gives another important trigonometric limit:

$$\lim_{x\to 0}\frac{\cos x-1}{x} = 0. \qquad (14)$$

FIGURE 2.4.6 shows the graph of $f(x) = (\cos x - 1)/x$. We will use the results in (10) and (14) in Exercises 2.7 and again in Section 3.4.

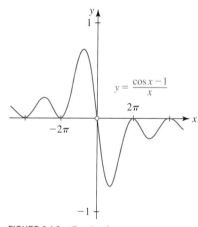

FIGURE 2.4.6 Graph of $f(x) = (\cos x - 1)/x$

Exercises 2.4 Answers to selected odd-numbered problems begin on page ANS-8.

≡ Fundamentals

In Problems 1–36, find the given limit, or state that it does not exist.

1. $\lim\limits_{t\to 0}\dfrac{\sin 3t}{2t}$

2. $\lim\limits_{t\to 0}\dfrac{\sin(-4t)}{t}$

3. $\lim\limits_{x\to 0}\dfrac{\sin x}{4+\cos x}$

4. $\lim\limits_{x\to 0}\dfrac{1+\sin x}{1+\cos x}$

5. $\lim\limits_{x\to 0}\dfrac{\cos 2x}{\cos 3x}$

6. $\lim\limits_{x\to 0}\dfrac{\tan x}{3x}$

7. $\lim\limits_{t\to 0}\dfrac{1}{t\sec t\csc 4t}$

8. $\lim\limits_{t\to 0} 5t\cot 2t$

9. $\lim\limits_{t\to 0}\dfrac{2\sin^2 t}{t\cos^2 t}$

10. $\lim\limits_{t\to 0}\dfrac{\sin^2(t/2)}{\sin t}$

11. $\lim\limits_{t\to 0}\dfrac{\sin^2 6t}{t^2}$

12. $\lim\limits_{t\to 0}\dfrac{t^3}{\sin^2 3t}$

13. $\lim\limits_{x\to 1}\dfrac{\sin(x-1)}{2x-2}$

14. $\lim\limits_{x\to 2\pi}\dfrac{x-2\pi}{\sin x}$

15. $\lim_{x \to 0} \dfrac{\cos x}{x}$

16. $\lim_{\theta \to \pi/2} \dfrac{1 + \sin \theta}{\cos \theta}$

17. $\lim_{x \to 0} \dfrac{\cos(3x - \pi/2)}{x}$

18. $\lim_{x \to -2} \dfrac{\sin(5x + 10)}{4x + 8}$

19. $\lim_{t \to 0} \dfrac{\sin 3t}{\sin 7t}$

20. $\lim_{t \to 0} \sin 2t \csc 3t$

21. $\lim_{t \to 0^+} \dfrac{\sin t}{\sqrt{t}}$

22. $\lim_{t \to 0^+} \dfrac{1 - \cos \sqrt{t}}{\sqrt{t}}$

23. $\lim_{t \to 0} \dfrac{t^2 - 5t \sin t}{t^2}$

24. $\lim_{t \to 0} \dfrac{\cos 4t}{\cos 8t}$

25. $\lim_{x \to 0^+} \dfrac{(x + 2\sqrt{\sin x})^2}{x}$

26. $\lim_{x \to 0} \dfrac{(1 - \cos x)^2}{x}$

27. $\lim_{x \to 0} \dfrac{\cos x - 1}{\cos^2 x - 1}$

28. $\lim_{x \to 0} \dfrac{\sin x + \tan x}{x}$

29. $\lim_{x \to 0} \dfrac{\sin 5x^2}{x^2}$

30. $\lim_{t \to 0} \dfrac{t^2}{1 - \cos t}$

31. $\lim_{x \to 2} \dfrac{\sin(x - 2)}{x^2 + 2x - 8}$

32. $\lim_{x \to 3} \dfrac{x^2 - 9}{\sin(x - 3)}$

33. $\lim_{x \to 0} \dfrac{2 \sin 4x + 1 - \cos x}{x}$

34. $\lim_{x \to 0} \dfrac{4x^2 - 2 \sin x}{x}$

35. $\lim_{x \to \pi/4} \dfrac{1 - \tan x}{\cos x - \sin x}$

36. $\lim_{x \to \pi/4} \dfrac{\cos 2x}{\cos x - \sin x}$

37. Suppose $f(x) = \sin x$. Use (10) and (14) of this section along with (17) of Section 1.4 to find the limit:
$$\lim_{h \to 0} \dfrac{f\left(\dfrac{\pi}{4} + h\right) - f\left(\dfrac{\pi}{4}\right)}{h}.$$

38. Suppose $f(x) = \cos x$. Use (10) and (14) of this section along with (18) of Section 1.4 to find the limit:
$$\lim_{h \to 0} \dfrac{f\left(\dfrac{\pi}{6} + h\right) - f\left(\dfrac{\pi}{6}\right)}{h}.$$

In Problems 39 and 40, use the Squeeze Theorem to establish the given limit.

39. $\lim_{x \to 0} x \sin \dfrac{1}{x} = 0$

40. $\lim_{x \to 0} x^2 \cos \dfrac{\pi}{x} = 0$

41. Use the properties of limits given in Theorem 2.2.3 to show that

   (a) $\lim_{x \to 0} x^3 \sin \dfrac{1}{x} = 0$
   (b) $\lim_{x \to 0} x^2 \sin^2 \dfrac{1}{x} = 0$.

42. If $|f(x)| \le B$ for all x in an interval containing 0, show that $\lim_{x \to 0} x^2 f(x) = 0$.

In Problems 43 and 44, use the Squeeze Theorem to evaluate the given limit.

43. $\lim_{x \to 2} f(x)$ where $2x - 1 \le f(x) \le x^2 - 2x + 3, x \ne 2$

44. $\lim_{x \to 0} f(x)$ where $|f(x) - 1| \le x^2, x \ne 0$

≡ **Think About It**

In Problems 45–48, use an appropriate substitution to find the given limit.

45. $\lim_{x \to \pi/4} \dfrac{\sin x - \cos x}{x - \pi/4}$

46. $\lim_{x \to \pi} \dfrac{x - \pi}{\tan 2x}$

47. $\lim_{x \to 1} \dfrac{\sin(\pi/x)}{x - 1}$

48. $\lim_{x \to 2} \dfrac{\cos(\pi/x)}{x - 2}$

49. Discuss: Is the function
$$f(x) = \begin{cases} \dfrac{\sin x}{x}, & x \ne 0 \\ 1, & x = 0 \end{cases}$$
continuous at 0?

50. The existence of $\lim_{x \to 0} \dfrac{\sin x}{x}$ does not imply the existence of $\lim_{x \to 0} \dfrac{\sin |x|}{x}$. Explain why the second limit fails to exist.

2.5 Limits That Involve Infinity

Introduction In Sections 1.2 and 1.3 we considered some functions whose graphs possessed asymptotes. We will see in this section that vertical and horizontal asymptotes of a graph are defined in terms of limits involving the concept of *infinity*. Recall, the **infinity symbols**, $-\infty$ ("minus infinity") and ∞ ("infinity"), are notational devices used to indicate, in turn, that a quantity becomes unbounded in the negative direction (in the Cartesian plane this means to the left for x and downward for y) and in the positive direction (to the right for x and upward for y).

Although the terminology and notation used when working with $\pm \infty$ is standard, it is nevertheless a bit unfortunate and can be confusing. So let us make it clear at the outset that we are going to consider two kinds of limits. First, we are going to examine

- *infinite limits*.

The words *infinite limit* always refer to *a limit that does not exist* because the function f exhibits unbounded behavior: $f(x) \to -\infty$ or $f(x) \to \infty$. Next, we will consider

- *limits at infinity*.

Some texts use the symbol $+\infty$ and the words *plus infinity* instead of ∞ and *infinity*.

The words *at infinity* mean that we are trying to determine whether a function f possesses a limit when the variable x is allowed to become unbounded: $x \to -\infty$ or $x \to \infty$. Such limits may or may not exist.

▶ Throughout the discussion, bear in mind that $-\infty$ and ∞ do not represent real numbers and should *never* be manipulated arithmetically like a number.

▮ Infinite Limits

The limit of a function f will fail to exist as x approaches a number a whenever the function values increase or decrease without bound. The fact that the function values $f(x)$ increase without bound as x approaches a is denoted symbolically by

$$f(x) \to \infty \text{ as } x \to a \quad \text{or} \quad \lim_{x \to a} f(x) = \infty. \tag{1}$$

If the function values decrease without bound as x approaches a, we write

$$f(x) \to -\infty \text{ as } x \to a \quad \text{or} \quad \lim_{x \to a} f(x) = -\infty. \tag{2}$$

Recall, the use of the symbol $x \to a$ signifies that f exhibits the same behavior—in this instance, unbounded behavior—from both sides of the number a on the x-axis. For example, the notation in (1) indicates that

$$f(x) \to \infty \text{ as } x \to a^- \quad \text{and} \quad f(x) \to \infty \text{ as } x \to a^+.$$

See FIGURE 2.5.1.

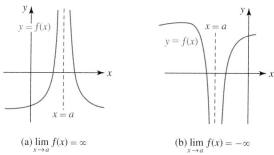

(a) $\lim_{x \to a} f(x) = \infty$ (b) $\lim_{x \to a} f(x) = -\infty$

FIGURE 2.5.1 Two types of infinite limits

Similarly, FIGURE 2.5.2 shows the unbounded behavior of a function f as x approaches a from one side. Note in Figure 2.5.2(c), we cannot describe the behavior of f near a using just one limit symbol.

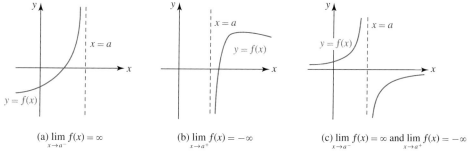

(a) $\lim_{x \to a^-} f(x) = \infty$ (b) $\lim_{x \to a^+} f(x) = -\infty$ (c) $\lim_{x \to a^-} f(x) = \infty$ and $\lim_{x \to a^+} f(x) = -\infty$

FIGURE 2.5.2 Three more types of infinite limits

In general, any limit of the six types

$$\lim_{x \to a^-} f(x) = -\infty, \quad \lim_{x \to a^-} f(x) = \infty,$$
$$\lim_{x \to a^+} f(x) = -\infty, \quad \lim_{x \to a^+} f(x) = \infty, \tag{3}$$
$$\lim_{x \to a} f(x) = -\infty, \quad \lim_{x \to a} f(x) = \infty,$$

is called an **infinite limit**. Again, in each case of (3) we are simply describing in a symbolic manner the behavior of a function f near the number a. *None of the limits in (3) exist.*

In Section 1.3 we reviewed how to identify a vertical asymptote for the graph of a rational function $f(x) = p(x)/q(x)$. We are now in a position to define a vertical asymptote of any function in terms of the limit concept.

Definition 2.5.1 Vertical Asymptote

A line $x = a$ is said to be a **vertical asymptote** for the graph of a function f if at least one of the six statements in (3) is true.

See Figure 1.2.1. ▶

In the review of functions in Chapter 1 we saw that the graphs of rational functions often possess asymptotes. We saw that the graphs of the rational functions $y = 1/x$ and $y = 1/x^2$ were similar to the graphs in Figure 2.5.2(c) and Figure 2.5.1(a), respectively. The y-axis, that is, $x = 0$, is a vertical asymptote for each of these functions. The graphs of

$$y = \frac{1}{x - a} \quad \text{and} \quad y = \frac{1}{(x - a)^2} \qquad (4)$$

are obtained by shifting the graphs of $y = 1/x$ and $y = 1/x^2$ horizontally $|a|$ units. As seen in FIGURE 2.5.3, $x = a$ is a vertical asymptote for the rational functions in (4). We have

$$\lim_{x \to a^-} \frac{1}{x - a} = -\infty \quad \text{and} \quad \lim_{x \to a^+} \frac{1}{x - a} = \infty \qquad (5)$$

and

$$\lim_{x \to a} \frac{1}{(x - a)^2} = \infty. \qquad (6)$$

The infinite limits in (5) and (6) are just special cases of the following general result:

$$\lim_{x \to a^-} \frac{1}{(x - a)^n} = -\infty \quad \text{and} \quad \lim_{x \to a^+} \frac{1}{(x - a)^n} = \infty, \qquad (7)$$

for n an odd positive integer, and

$$\lim_{x \to a} \frac{1}{(x - a)^n} = \infty, \qquad (8)$$

for n an even positive integer. As a consequence of (7) and (8), the graph of a rational function $y = 1/(x - a)^n$ either resembles the graph in Figure 2.5.3(a) for n odd or that in Figure 2.5.3(b) for n even.

For a general rational function $f(x) = p(x)/q(x)$, where p and q have no common factors, it should be clear from this discussion that when q contains a factor $(x - a)^n$, n a positive integer, then the shape of the graph near the vertical line $x = a$ must be either one of those shown in Figure 2.5.3 or its reflection in the x-axis.

(a)

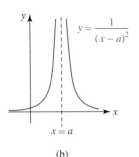

(b)

FIGURE 2.5.3 Graphs of functions in (4)

EXAMPLE 1 Vertical Asymptotes of a Rational Function

Inspection of the rational function

$$f(x) = \frac{x + 2}{x^2(x + 4)}$$

shows that $x = -4$ and $x = 0$ are vertical asymptotes for the graph of f. Since the denominator contains the factors $(x - (-4))^1$ and $(x - 0)^2$ we expect the graph of f near the line $x = -4$ to resemble Figure 2.5.3(a) or its reflection in the x-axis, and the graph near $x = 0$ to resemble Figure 2.5.3(b) or its reflection in the x-axis.

For x close to 0, from either side of 0, it is easily seen that $f(x) > 0$. But, for x close to -4, say $x = -4.1$ and $x = -3.9$, we have $f(x) > 0$ and $f(x) < 0$, respectively. Using the additional information that there is only a single x-intercept $(-2, 0)$, we obtain the graph of f in FIGURE 2.5.4. ∎

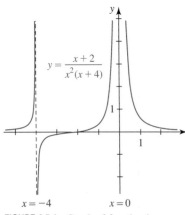

FIGURE 2.5.4 Graph of function in Example 1

EXAMPLE 2 One-Sided Limit

In Figure 1.6.6 we saw that the y-axis, or the line $x = 0$, is a vertical asymptote for the natural logarithmic function $f(x) = \ln x$ since

$$\lim_{x \to 0^+} \ln x = -\infty.$$

The graph of the logarithmic function $y = \ln(x + 3)$ is the graph of $f(x) = \ln x$ shifted 3 units to the left. Thus $x = -3$ is a vertical asymptote for the graph of $y = \ln(x + 3)$ since $\lim_{x \to -3^+} \ln(x + 3) = -\infty$.

EXAMPLE 3 One-Sided Limit

Graph the function $f(x) = \dfrac{x}{\sqrt{x + 2}}$.

Solution Inspection of f reveals that its domain is the interval $(-2, \infty)$ and the y-intercept is $(0, 0)$. From the accompanying table we conclude that f decreases

$x \to -2^+$	-1.9	-1.99	-1.999	-1.9999
$f(x)$	-6.01	-19.90	-63.21	-199.90

without bound as x approaches -2 from the right:

$$\lim_{x \to -2^+} f(x) = -\infty.$$

Hence, the line $x = -2$ is a vertical asymptote. The graph of f is given in FIGURE 2.5.5.

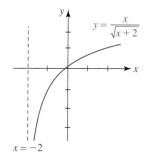

FIGURE 2.5.5 Graph of function in Example 3

■ **Limits at Infinity** If a function f approaches a constant value L as the independent variable x increases without bound ($x \to \infty$) or as x decreases ($x \to -\infty$) without bound, then we write

$$\lim_{x \to -\infty} f(x) = L \quad \text{or} \quad \lim_{x \to \infty} f(x) = L \qquad (9)$$

and say that f possesses a **limit at infinity**. Here are all the possibilities for limits at infinity $\lim_{x \to -\infty} f(x)$ and $\lim_{x \to \infty} f(x)$:

- One limit exists but the other does not,
- Both $\lim_{x \to -\infty} f(x)$ and $\lim_{x \to \infty} f(x)$ exist and equal the same number,
- Both $\lim_{x \to -\infty} f(x)$ and $\lim_{x \to \infty} f(x)$ exist but are different numbers,
- Neither $\lim_{x \to -\infty} f(x)$ nor $\lim_{x \to \infty} f(x)$ exists.

If at least one of the limits exists, say, $\lim_{x \to \infty} f(x) = L$, then the graph of f can be made arbitrarily close to the line $y = L$ as x increases in the positive direction.

Definition 2.5.2 Horizontal Asymptote

A line $y = L$ is said to be a **horizontal asymptote** for the graph of a function f if at least one of the two statements in (9) is true.

In FIGURE 2.5.6 we have illustrated some typical horizontal asymptotes. We note, in conjunction with Figure 2.5.6(d) that, in general, the graph of a function can have at most *two* horizontal asymptotes but the graph of a *rational function* $f(x) = p(x)/q(x)$ can have at most *one*. If the graph of a rational function f possesses a horizontal asymptote $y = L$, then its end behavior is as shown in Figure 2.5.6(c), that is:

$$f(x) \to L \text{ as } x \to -\infty \quad \text{and} \quad f(x) \to L \text{ as } x \to \infty.$$

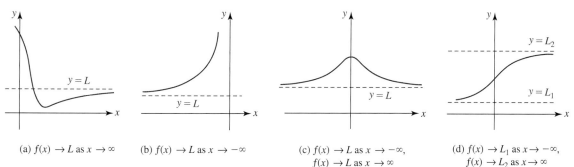

(a) $f(x) \to L$ as $x \to \infty$ (b) $f(x) \to L$ as $x \to -\infty$ (c) $f(x) \to L$ as $x \to -\infty$, $f(x) \to L$ as $x \to \infty$ (d) $f(x) \to L_1$ as $x \to -\infty$, $f(x) \to L_2$ as $x \to \infty$

FIGURE 2.5.6 $y = L$ is a horizontal asymptote in (a), (b), and (c); $y = L_1$ and $y = L_2$ are horizontal asymptotes in (d)

For example, if x becomes unbounded in either the positive or negative direction, the functions in (4) decrease to 0 and we write

$$\lim_{x\to-\infty}\frac{1}{x-a}=0,\ \lim_{x\to\infty}\frac{1}{x-a}=0 \quad\text{and}\quad \lim_{x\to-\infty}\frac{1}{(x-a)^2}=0,\ \lim_{x\to\infty}\frac{1}{(x-a)^2}=0. \quad (10)$$

In general, if r is a positive rational number and if $(x-a)^r$ is defined, then

These results are also true when $x-a$ is replaced by $a-x$, provided $(a-x)^r$ is defined.

$$\lim_{x\to-\infty}\frac{1}{(x-a)^r}=0 \quad\text{and}\quad \lim_{x\to\infty}\frac{1}{(x-a)^r}=0. \quad (11)$$

EXAMPLE 4 Horizontal and Vertical Asymptotes

The domain of the function $f(x)=\dfrac{4}{\sqrt{2-x}}$ is the interval $(-\infty, 2)$. In view of (11) we can write

$$\lim_{x\to-\infty}\frac{4}{\sqrt{2-x}}=0.$$

Note that we cannot consider the limit of f as $x\to\infty$ because the function is not defined for $x\geq 2$. Nevertheless $y=0$ is a horizontal asymptote. Now from infinite limit

$$\lim_{x\to 2^-}\frac{4}{\sqrt{2-x}}=\infty$$

we conclude that $x=2$ is a vertical asymptote for the graph of f. See **FIGURE 2.5.7**. ∎

FIGURE 2.5.7 Graph of function in Example 4

In general, if $F(x)=f(x)/g(x)$, then the following table summarizes the limit results for the forms $\lim_{x\to a} F(x)$, $\lim_{x\to\infty} F(x)$, and $\lim_{x\to -\infty} F(x)$. The symbol L denotes a real number.

limit form: $x\to a, \infty, -\infty$	$\dfrac{L}{\pm\infty}$	$\dfrac{\pm\infty}{L}, L\neq 0$	$\dfrac{L}{0}, L\neq 0$
limit is:	0	infinite	infinite

(12)

Limits of the form $\lim_{x\to\infty} F(x)=\pm\infty$ or $\lim_{x\to-\infty} F(x)=\pm\infty$ are said to be **infinite limits at infinity**. Furthermore, the limit properties given in Theorem 2.2.3 hold by replacing the symbol a by ∞ or $-\infty$ provided the limits exist. For example,

$$\lim_{x\to\infty} f(x)g(x)=\left(\lim_{x\to\infty} f(x)\right)\left(\lim_{x\to\infty} g(x)\right) \quad\text{and}\quad \lim_{x\to\infty}\frac{f(x)}{g(x)}=\frac{\lim_{x\to\infty} f(x)}{\lim_{x\to\infty} g(x)}, \quad (13)$$

whenever $\lim_{x\to\infty} f(x)$ and $\lim_{x\to\infty} g(x)$ exist. In the case of the limit of a quotient we must also have $\lim_{x\to\infty} g(x)\neq 0$.

■ **End Behavior** In Section 1.3 we saw that how a function f behaves when $|x|$ is very large is its **end behavior**. As already discussed, if $\lim_{x\to\infty} f(x)=L$, then the graph of f can be made arbitrarily close to the line $y=L$ for large positive values of x. The graph of a polynomial function,

$$f(x)=a_n x^n + a_{n-1} x^{n-1} + \cdots + a_2 x^2 + a_1 x + a_0,$$

resembles the graph of $y=a_n x^n$ for $|x|$ very large. In other words, for

$$f(x)=a_n x^n \boxed{+\ a_{n-1}x^{n-1}+\cdots+a_1 x+a_0} \quad (14)$$

the terms enclosed in the blue rectangle in (14) are irrelevant when we look at a graph of a polynomial globally—that is, for $|x|$ large. Thus we have

$$\lim_{x\to\pm\infty} a_n x^n = \lim_{x\to\pm\infty}\left(a_n x^n + a_{n-1} x^{n-1} + \cdots + a_1 x + a_0\right), \quad (15)$$

where (15) is either ∞ or $-\infty$ depending on a_n and n. In other words, the limit in (15) is an example of an infinite limit at infinity.

EXAMPLE 5 Limit at Infinity

Evaluate $\lim_{x\to\infty} \dfrac{-6x^4 + x^2 + 1}{2x^4 - x}$.

Solution We cannot apply the limit quotient law in (13) to the given function, since $\lim_{x\to\infty}(-6x^4 + x^2 + 1) = -\infty$ and $\lim_{x\to\infty}(2x^4 - x) = \infty$. However, by dividing the numerator and the denominator by x^4, we can write

$$\lim_{x\to\infty} \frac{-6x^4 + x^2 + 1}{2x^4 - x} = \lim_{x\to\infty} \frac{-6 + \left(\dfrac{1}{x^2}\right) + \left(\dfrac{1}{x^4}\right)}{2 - \left(\dfrac{1}{x^3}\right)}$$

$$= \frac{\lim_{x\to\infty}\left[-6 + \left(\dfrac{1}{x^2}\right) + \left(\dfrac{1}{x^4}\right)\right]}{\lim_{x\to\infty}\left[2 - \left(\dfrac{1}{x^3}\right)\right]} \quad \leftarrow \begin{array}{l}\text{Limit of the numerator}\\\text{and denominator both}\\\text{exist and the limit of}\\\text{the denominator is not}\\\text{zero}\end{array}$$

$$= \frac{-6 + 0 + 0}{2 - 0} = -3.$$

This means the line $y = -3$ is a horizontal asymptote for the graph of the function.

Alternative Solution In view of (14), we can discard all powers of x other than the highest:

discard terms in the blue boxes
↓

$$\lim_{x\to\infty} \frac{-6x^4 \boxed{+ x^2 + 1}}{2x^4 \boxed{- x}} = \lim_{x\to\infty} \frac{-6x^4}{2x^4} = \lim_{x\to\infty} \frac{-6}{2} = -3.$$ ∎

EXAMPLE 6 Infinite Limit at Infinity

Evaluate $\lim_{x\to\infty} \dfrac{1 - x^3}{3x + 2}$.

Solution By (14),

$$\lim_{x\to\infty} \frac{1 - x^3}{3x + 2} = \lim_{x\to\infty} \frac{-x^3}{3x} = -\frac{1}{3}\lim_{x\to\infty} x^2 = -\infty.$$

In other words, the limit does not exist. ∎

EXAMPLE 7 Graph of a Rational Function

Graph the function $f(x) = \dfrac{x^2}{1 - x^2}$.

Solution Inspection of the function f reveals that its graph is symmetric with respect to the y-axis, the y-intercept is $(0, 0)$, and the vertical asymptotes are $x = -1$ and $x = 1$. Now, from the limit

$$\lim_{x\to\infty} f(x) = \lim_{x\to\infty} \frac{x^2}{1 - x^2} = \lim_{x\to\infty} \frac{x^2}{-x^2} = -\lim_{x\to\infty} 1 = -1$$

we conclude that the line $y = -1$ is a horizontal asymptote. The graph of f is given in FIGURE 2.5.8.

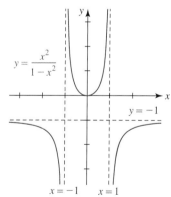

FIGURE 2.5.8 Graph of function in Example 7

Another limit law that holds true for limits at infinity is that the limit of an nth root of a function is the nth root of the limit, whenever the limit exists and the nth root is defined. In symbols, if $\lim_{x\to\infty} g(x) = L$, then

$$\lim_{x\to\infty} \sqrt[n]{g(x)} = \sqrt[n]{\lim_{x\to\infty} g(x)} = \sqrt[n]{L}, \qquad (16)$$

provided $L \geq 0$ when n is even. The result also holds for $x \to -\infty$.

EXAMPLE 8 Limit of a Square Root

Evaluate $\lim_{x \to \infty} \sqrt{\dfrac{2x^3 - 5x^2 + 4x - 6}{6x^3 + 2x}}$.

Solution Because the limit of the rational function inside the radical exists and is positive, we can write

$$\lim_{x \to \infty} \sqrt{\frac{2x^3 - 5x^2 + 4x - 6}{6x^3 + 2x}} = \sqrt{\lim_{x \to \infty} \frac{2x^3 - 5x^2 + 4x - 6}{6x^3 + 2x}} = \sqrt{\lim_{x \to \infty} \frac{2x^3}{6x^3}} = \sqrt{\frac{1}{3}} = \frac{1}{\sqrt{3}}. \quad \blacksquare$$

EXAMPLE 9 Graph with Two Horizontal Asymptotes

Determine whether the graph of $f(x) = \dfrac{5x}{\sqrt{x^2 + 4}}$ has any horizontal asymptotes.

Solution Since the function is not rational, we must investigate the limit of f as $x \to \infty$ and as $x \to -\infty$. First, recall from algebra that $\sqrt{x^2}$ is nonnegative, or more to the point,

$$\sqrt{x^2} = |x| = \begin{cases} x, & x \geq 0 \\ -x, & x < 0. \end{cases}$$

We then rewrite f as

$$f(x) = \frac{\dfrac{5x}{\sqrt{x^2}}}{\dfrac{\sqrt{x^2+4}}{\sqrt{x^2}}} = \frac{\dfrac{5x}{|x|}}{\dfrac{\sqrt{x^2+4}}{\sqrt{x^2}}} = \frac{\dfrac{5x}{|x|}}{\sqrt{1 + \dfrac{4}{x^2}}}.$$

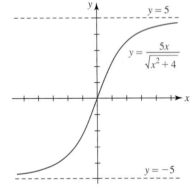

FIGURE 2.5.9 Graph of function in Example 9

The limits of f as $x \to \infty$ and as $x \to -\infty$ are, respectively,

$$\lim_{x \to \infty} f(x) = \lim_{x \to \infty} \frac{\dfrac{5x}{|x|}}{\sqrt{1 + \dfrac{4}{x^2}}} = \lim_{x \to \infty} \frac{\dfrac{5x}{x}}{\sqrt{1 + \dfrac{4}{x^2}}} = \frac{\lim_{x \to \infty} 5}{\sqrt{\lim_{x \to \infty}\left(1 + \dfrac{4}{x^2}\right)}} = \frac{5}{1} = 5,$$

and $\lim_{x \to -\infty} f(x) = \lim_{x \to -\infty} \dfrac{\dfrac{5x}{|x|}}{\sqrt{1 + \dfrac{4}{x^2}}} = \lim_{x \to -\infty} \dfrac{\dfrac{5x}{-x}}{\sqrt{1 + \dfrac{4}{x^2}}} = \dfrac{\lim_{x \to -\infty}(-5)}{\sqrt{\lim_{x \to -\infty}\left(1 + \dfrac{4}{x^2}\right)}} = \dfrac{-5}{1} = -5.$

Thus the graph of f has two horizontal asymptotes $y = 5$ and $y = -5$. The graph of f, which is similar to Figure 2.5.6(d), is given in FIGURE 2.5.9. $\blacksquare$

In the next example we see that the form of the given limit is $\infty - \infty$, but the limit exists and is *not* 0.

EXAMPLE 10 Using Rationalization

Evaluate $\lim_{x \to \infty}\left(x^2 - \sqrt{x^4 + 7x^2 + 1}\right)$.

Solution Because $f(x) = x^2 - \sqrt{x^4 + 7x^2 + 1}$ is an even function (verify that $f(-x) = f(x)$) with domain $(-\infty, \infty)$, if $\lim_{x \to \infty} f(x)$ exists it must be the same as $\lim_{x \to -\infty} f(x)$. We first rationalize the numerator:

$$\lim_{x \to \infty}\left(x^2 - \sqrt{x^4 + 7x^2 + 1}\right) = \lim_{x \to \infty} \frac{\left(x^2 - \sqrt{x^4 + 7x^2 + 1}\right)}{1} \cdot \left(\frac{x^2 + \sqrt{x^4 + 7x^2 + 1}}{x^2 + \sqrt{x^4 + 7x^2 + 1}}\right)$$

$$= \lim_{x \to \infty} \frac{x^4 - (x^4 + 7x^2 + 1)}{x^2 + \sqrt{x^4 + 7x^2 + 1}}$$

$$= \lim_{x \to \infty} \frac{-7x^2 - 1}{x^2 + \sqrt{x^4 + 7x^2 + 1}}.$$

Next, we divide the numerator and denominator by $\sqrt{x^4} = x^2$:

$$\lim_{x\to\infty} \frac{-7x^2-1}{x^2+\sqrt{x^4+7x^2+1}} = \lim_{x\to\infty} \frac{\dfrac{-7x^2}{\sqrt{x^4}} - \dfrac{1}{\sqrt{x^4}}}{\dfrac{x^2+\sqrt{x^4+7x^2+1}}{\sqrt{x^4}}}$$

$$= \lim_{x\to\infty} \frac{-7-\dfrac{1}{x^2}}{1+\sqrt{1+\dfrac{7}{x^2}+\dfrac{1}{x^4}}}$$

$$= \frac{\lim\limits_{x\to\infty}\left(-7-\dfrac{1}{x^2}\right)}{\lim\limits_{x\to\infty} 1 + \sqrt{\lim\limits_{x\to\infty}\left(1+\dfrac{7}{x^2}+\dfrac{1}{x^4}\right)}}$$

$$= \frac{-7}{1+1} = -\frac{7}{2}.$$

With the help of a CAS, the graph of the function f is given in FIGURE 2.5.10. The line $y = -\frac{7}{2}$ is a horizontal asymptote. Note the symmetry of the graph with respect to the y-axis. ■

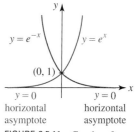

FIGURE 2.5.10 Graph of function in Example 10

When working with functions containing the natural exponential function, the following four limits merit special attention:

$$\lim_{x\to\infty} e^x = \infty, \quad \lim_{x\to-\infty} e^x = 0, \quad \lim_{x\to\infty} e^{-x} = 0, \quad \lim_{x\to-\infty} e^{-x} = \infty. \tag{17}$$

As discussed in Section 1.6 and verified by the second and third limit in (17), $y = 0$ is a horizontal asymptote for the graphs of $y = e^x$ and $y = e^{-x}$. See FIGURE 2.5.11.

FIGURE 2.5.11 Graphs of exponential functions

EXAMPLE 11 Graph with Two Horizontal Asymptotes

Determine whether the graph of $f(x) = \dfrac{6}{1+e^{-x}}$ has any horizontal asymptotes.

Solution Because f is not a rational function, we must examine $\lim\limits_{x\to\infty} f(x)$ and $\lim\limits_{x\to-\infty} f(x)$. First, in view of the third result given in (17) we can write

$$\lim_{x\to\infty} \frac{6}{1+e^{-x}} = \frac{\lim\limits_{x\to\infty} 6}{\lim\limits_{x\to\infty}(1+e^{-x})} = \frac{6}{1+0} = 6.$$

Thus $y = 6$ is a horizontal asymptote. Now, because $\lim\limits_{x\to-\infty} e^{-x} = \infty$ it follows from the table in (12) that

$$\lim_{x\to-\infty} \frac{6}{1+e^{-x}} = 0.$$

Therefore $y = 0$ is a horizontal asymptote. The graph of f is given in FIGURE 2.5.12. ■

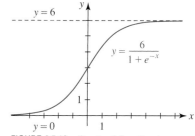

FIGURE 2.5.12 Graph of function in Example 11

▌ Composite Functions Theorem 2.3.3, the limit of a composite function, holds when a is replaced by $-\infty$ or ∞ and the limit exists. For example, if $\lim\limits_{x\to\infty} g(x) = L$ and f is continuous at L, then

$$\lim_{x\to\infty} f(g(x)) = f\left(\lim_{x\to\infty} g(x)\right) = f(L). \tag{18}$$

The limit result in (16) is just a special case of (18) when $f(x) = \sqrt[n]{x}$. The result in (18) also holds for $x \to -\infty$. Our last example illustrates (18) involving a limit at ∞.

EXAMPLE 12 A Trigonometric Function Revisited

In Example 2 of Section 2.4 we saw that $\lim_{x \to 0} \sin(1/x)$ does not exist. However, the limit at infinity, $\lim_{x \to \infty} \sin(1/x)$, exists. By (18) we can write

$$\lim_{x \to \infty} \sin \frac{1}{x} = \sin\left(\lim_{x \to \infty} \frac{1}{x}\right) = \sin 0 = 0.$$

As we see in FIGURE 2.5.13, $y = 0$ is a horizontal asymptote for the graph of $f(x) = \sin(1/x)$. You should compare this graph with that given in Figure 2.4.2. ∎

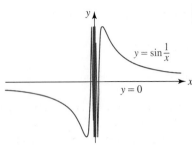

FIGURE 2.5.13 Graph of function in Example 12

Exercises 2.5   Answers to selected odd-numbered problems begin on page ANS-8.

≡ Fundamentals

In Problems 1–24, express the given limit as a number, as $-\infty$, or as ∞.

1. $\lim_{x \to 5^-} \dfrac{1}{x - 5}$

2. $\lim_{x \to 6} \dfrac{4}{(x - 6)^2}$

3. $\lim_{x \to -4^+} \dfrac{2}{(x + 4)^3}$

4. $\lim_{x \to 2^-} \dfrac{10}{x^2 - 4}$

5. $\lim_{x \to 1} \dfrac{1}{(x - 1)^4}$

6. $\lim_{x \to 0^+} \dfrac{-1}{\sqrt{x}}$

7. $\lim_{x \to 0^+} \dfrac{2 + \sin x}{x}$

8. $\lim_{x \to \pi^+} \csc x$

9. $\lim_{x \to \infty} \dfrac{x^2 - 3x}{4x^2 + 5}$

10. $\lim_{x \to \infty} \dfrac{x^2}{1 + x^{-2}}$

11. $\lim_{x \to \infty} \left(5 - \dfrac{2}{x^4}\right)$

12. $\lim_{x \to -\infty} \left(\dfrac{6}{\sqrt[3]{x}} + \dfrac{1}{\sqrt[5]{x}}\right)$

13. $\lim_{x \to \infty} \dfrac{8 - \sqrt{x}}{1 + 4\sqrt{x}}$

14. $\lim_{x \to -\infty} \dfrac{1 + 7\sqrt[3]{x}}{2\sqrt[3]{x}}$

15. $\lim_{x \to \infty} \left(\dfrac{3x}{x + 2} - \dfrac{x - 1}{2x + 6}\right)$

16. $\lim_{x \to \infty} \left(\dfrac{x}{3x + 1}\right)\left(\dfrac{4x^2 + 1}{2x^2 + x}\right)^3$

17. $\lim_{x \to \infty} \sqrt{\dfrac{3x + 2}{6x - 8}}$

18. $\lim_{x \to -\infty} \sqrt[3]{\dfrac{2x - 1}{7 - 16x}}$

19. $\lim_{x \to \infty} (x - \sqrt{x^2 + 1})$

20. $\lim_{x \to \infty} (\sqrt{x^2 + 5x} - x)$

21. $\lim_{x \to \infty} \cos\left(\dfrac{5}{x}\right)$

22. $\lim_{x \to -\infty} \sin\left(\dfrac{\pi x}{3 - 6x}\right)$

23. $\lim_{x \to -\infty} \sin^{-1}\left(\dfrac{x}{\sqrt{4x^2 + 1}}\right)$

24. $\lim_{x \to \infty} \ln\left(\dfrac{x}{x + 8}\right)$

In Problems 25–32, find $\lim_{x \to -\infty} f(x)$ and $\lim_{x \to \infty} f(x)$ for the given function f.

25. $f(x) = \dfrac{4x + 1}{\sqrt{x^2 + 1}}$

26. $f(x) = \dfrac{\sqrt{9x^2 + 6}}{5x - 1}$

27. $f(x) = \dfrac{2x + 1}{\sqrt{3x^2 + 1}}$

28. $f(x) = \dfrac{-5x^2 + 6x + 3}{\sqrt{x^4 + x^2 + 1}}$

29. $f(x) = \dfrac{e^x - e^{-x}}{e^x + e^{-x}}$

30. $f(x) = 1 + \dfrac{2e^{-x}}{e^x + e^{-x}}$

31. $f(x) = \dfrac{|x - 5|}{x - 5}$

32. $f(x) = \dfrac{|4x| + |x - 1|}{x}$

In Problems 33–42, find all vertical and horizontal asymptotes for the graph of the given function. Sketch the graph.

33. $f(x) = \dfrac{1}{x^2 + 1}$

34. $f(x) = \dfrac{x}{x^2 + 1}$

35. $f(x) = \dfrac{x^2}{x + 1}$

36. $f(x) = \dfrac{x^2 - x}{x^2 - 1}$

37. $f(x) = \dfrac{1}{x^2(x - 2)}$

38. $f(x) = \dfrac{4x^2}{x^2 + 4}$

39. $f(x) = \sqrt{\dfrac{x}{x - 1}}$

40. $f(x) = \dfrac{1 - \sqrt{x}}{\sqrt{x}}$

41. $f(x) = \dfrac{x - 2}{\sqrt{x^2 + 1}}$

42. $f(x) = \dfrac{x + 3}{\sqrt{x^2 - 1}}$

In Problems 43–46, use the given graph to find:

(a) $\lim_{x \to 2^-} f(x)$ (b) $\lim_{x \to 2^+} f(x)$
(c) $\lim_{x \to -\infty} f(x)$ (d) $\lim_{x \to \infty} f(x)$

43.

FIGURE 2.5.14 Graph for Problem 43

44.

FIGURE 2.5.15 Graph for Problem 44

45.

FIGURE 2.5.16 Graph for Problem 45

46.

FIGURE 2.5.17 Graph for Problem 46

In Problems 47–50, sketch a graph of a function f that satisfies the given conditions.

47. $\lim_{x \to 1^+} f(x) = -\infty$, $\lim_{x \to 1^-} f(x) = -\infty$, $f(2) = 0$, $\lim_{x \to \infty} f(x) = 0$

48. $f(0) = 1$, $\lim_{x \to -\infty} f(x) = 3$, $\lim_{x \to \infty} f(x) = -2$

49. $\lim_{x \to 2} f(x) = \infty$, $\lim_{x \to -\infty} f(x) = \infty$, $\lim_{x \to \infty} f(x) = 1$

50. $\lim_{x \to 1^-} f(x) = 2$, $\lim_{x \to 1^+} f(x) = -\infty$, $f(\frac{3}{2}) = 0$, $f(3) = 0$, $\lim_{x \to -\infty} f(x) = 0$, $\lim_{x \to \infty} f(x) = 0$

51. Use an appropriate substitution to evaluate
$$\lim_{x \to \infty} x \sin \frac{3}{x}.$$

52. According to Einstein's theory of relativity, the mass m of a body moving with velocity v is $m = m_0/\sqrt{1 - v^2/c^2}$, where m_0 is the initial mass and c is the speed of light. What happens to m as $v \to c^-$?

≡ Calculator/CAS Problems

In Problems 53 and 54, use a calculator or CAS to investigate the given limit. Conjecture its value.

53. $\lim_{x \to \infty} x^2 \sin \frac{2}{x^2}$ **54.** $\lim_{x \to \infty} \left(\cos \frac{1}{x} \right)^x$

55. Use a calculator or CAS to obtain the graph of $f(x) = (1 + x)^{1/x}$. Use the graph to conjecture the values of $f(x)$ as
(a) $x \to -1^+$, (b) $x \to 0$, and (c) $x \to \infty$.

56. (a) A regular n-gon is an n-sided polygon inscribed in a circle; the polygon is formed by n equally spaced points on the circle. Suppose the polygon shown in FIGURE 2.5.18 represents a regular n-gon inscribed in a circle of radius r. Use trigonometry to show that the area $A(n)$ of the n-gon is given by
$$A(n) = \frac{n}{2} r^2 \sin\left(\frac{2\pi}{n}\right).$$

(b) It stands to reason that the area $A(n)$ approaches the area of the circle as the number of sides of the n-gon increases. Use a calculator to compute $A(100)$ and $A(1000)$.

(c) Let $x = 2\pi/n$ in $A(n)$ and note that as $n \to \infty$ then $x \to 0$. Use (10) of Section 2.4 to show that $\lim_{n \to \infty} A(n) = \pi r^2$.

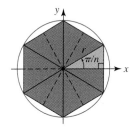

FIGURE 2.5.18 Inscribed n-gon for Problem 56

≡ Think About It

57. (a) Suppose $f(x) = x^2/(x+1)$ and $g(x) = x - 1$. Show that
$$\lim_{x \to \pm \infty} [f(x) - g(x)] = 0.$$

(b) What does the result in part (a) indicate about the graphs of f and g where $|x|$ is large?

(c) If possible, give a name to the function g.

58. Very often students and even instructors will sketch vertically shifted graphs incorrectly. For example, the graphs of $y = x^2$ and $y = x^2 + 1$ are incorrectly drawn in FIGURE 2.5.19(a) but are correctly drawn in Figure 2.5.19(b). Demonstrate that Figure 2.5.19(b) is correct by showing that the horizontal distance between the two points P and Q shown in the figure approaches 0 as $x \to \infty$.

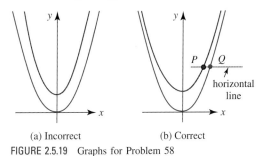

(a) Incorrect (b) Correct

FIGURE 2.5.19 Graphs for Problem 58

2.6 Limits—A Formal Approach

■ **Introduction** In the discussion that follows we will consider an alternative approach to the notion of a limit that is based on analytical concepts rather than on intuitive concepts. A **proof** of the existence of a limit can never be based on one's ability to sketch graphs or on tables of numerical values. Although a good intuitive understanding of $\lim_{x \to a} f(x)$ is sufficient for proceeding with the study of the calculus in this text, an intuitive understanding is admittedly too vague to be of any use in proving theorems. To give a rigorous demonstration of the existence of a limit, or to prove the important theorems of Section 2.2, we must start with a precise definition of a limit.

Limit of a Function Let us try to prove that $\lim_{x \to 2}(2x + 6) = 10$ by elaborating on the following idea: "If $f(x) = 2x + 6$ can be made arbitrarily close to 10 by taking x sufficiently close to 2, from either side but different from 2, then $\lim_{x \to 2} f(x) = 10$." We need to make the concepts of *arbitrarily close* and *sufficiently close* precise. In order to set a standard of arbitrary closeness, let us demand that the distance between the numbers $f(x)$ and 10 be less than 0.1; that is,

$$|f(x) - 10| < 0.1 \quad \text{or} \quad 9.9 < f(x) < 10.1. \tag{1}$$

Then, how close must x be to 2 to accomplish (1)? To find out, we can use ordinary algebra to rewrite the inequality

$$9.9 < 2x + 6 < 10.1$$

as $1.95 < x < 2.05$. Adding -2 across this simultaneous inequality then gives

$$-0.05 < x - 2 < 0.05.$$

Using absolute values and remembering that $x \neq 2$, we can write the last inequality as $0 < |x - 2| < 0.05$. Thus, for an "arbitrary closeness to 10" of 0.1, "sufficiently close to 2" means within 0.05. In other words, if x is a number different from 2 such that its distance from 2 satisfies $|x - 2| < 0.05$, then the distance of $f(x)$ from 10 is guaranteed to satisfy $|f(x) - 10| < 0.1$. Expressed in yet another way, when x is a number different from 2 but in the open interval (1.95, 2.05) on the x-axis, then $f(x)$ is in the interval (9.9, 10.1) on the y-axis.

Using the same example, let us try to generalize. Suppose ε (the Greek letter *epsilon*) denotes an arbitrary *positive number* that is our measure of arbitrary closeness to the number 10. If we demand that

$$|f(x) - 10| < \varepsilon \quad \text{or} \quad 10 - \varepsilon < f(x) < 10 + \varepsilon, \tag{2}$$

then from $10 - \varepsilon < 2x + 6 < 10 + \varepsilon$ and algebra, we find

$$2 - \frac{\varepsilon}{2} < x < 2 + \frac{\varepsilon}{2} \quad \text{or} \quad -\frac{\varepsilon}{2} < x - 2 < \frac{\varepsilon}{2}. \tag{3}$$

Again using absolute values and remembering that $x \neq 2$, we can write the last inequality in (3) as

$$0 < |x - 2| < \frac{\varepsilon}{2}. \tag{4}$$

If we denote $\varepsilon/2$ by the new symbol δ (the Greek letter *delta*), (2) and (4) can be written as

$$|f(x) - 10| < \varepsilon \quad \text{whenever} \quad 0 < |x - 2| < \delta.$$

Thus, for a new value for ε, say $\varepsilon = 0.001$, $\delta = \varepsilon/2 = 0.0005$ tells us the corresponding closeness to 2. For any number x different from 2 in (1.9995, 2.0005),* we can be sure $f(x)$ is in (9.999, 10.001). See **FIGURE 2.6.1**.

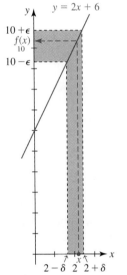

FIGURE 2.6.1 $f(x)$ is in $(10 - \varepsilon, 10 + \varepsilon)$ whenever x is in $(2 - \delta, 2 + \delta), x \neq 2$

A Definition The foregoing discussion leads us to the so-called ε-δ **definition of a limit**.

Definition 2.6.1 Definition of a Limit

Suppose a function f is defined everywhere on an open interval, except possibly at a number a in the interval. Then

$$\lim_{x \to a} f(x) = L$$

means that for every $\varepsilon > 0$, there exists a number $\delta > 0$ such that

$$|f(x) - L| < \varepsilon \quad \text{whenever} \quad 0 < |x - a| < \delta.$$

---

*For this reason, we use $0 < |x - 2| < \delta$ rather than $|x - 2| < \delta$. Keep in mind when considering $\lim_{x \to 2} f(x)$, we do not care about f at 2.

Let $\lim_{x \to a} f(x) = L$ and suppose $\delta > 0$ is the number that "works" in the sense of Definition 2.6.1 for a given $\varepsilon > 0$. As shown in FIGURE 2.6.2(a), every x in $(a - \delta, a + \delta)$, with the possible exception of a itself, will then have an image $f(x)$ in $(L - \varepsilon, L + \varepsilon)$. Furthermore, as in Figure 2.6.2(b), a choice $\delta_1 < \delta$ for the same ε also "works" in that every x not equal to a in $(a - \delta_1, a + \delta_1)$ gives $f(x)$ in $(L - \varepsilon, L + \varepsilon)$. However, Figure 2.6.2(c) shows that choosing a smaller ε_1, $0 < \varepsilon_1 < \varepsilon$, will demand finding a new value of δ. Observe in Figure 2.6.2(c) that x is in $(a - \delta, a + \delta)$ but not in $(a - \delta_1, a + \delta_1)$, and so $f(x)$ is not necessarily in $(L - \varepsilon_1, L + \varepsilon_1)$.

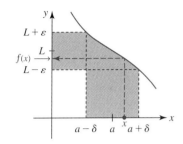

(a) A δ that works for a given ε

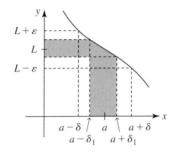

(b) A smaller δ_1 will also work for the same ε

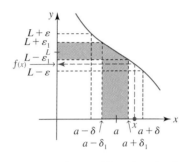

(c) A smaller ε_1 will require a $\delta_1 < \delta$. For x in $(a - \delta, a + \delta)$, $f(x)$ is not necessarily in $(L - \varepsilon_1, L + \varepsilon_1)$

FIGURE 2.6.2 $f(x)$ is in $(L - \varepsilon, L + \varepsilon)$ whenever x is in $(a - \delta, a + \delta), x \neq a$

EXAMPLE 1 Using Definition 2.6.1

Prove that $\lim_{x \to a}(5x + 2) = 17$.

Solution For any arbitrary $\varepsilon > 0$, regardless how small, we wish to find a δ so that

$$|(5x + 2) - 17| < \varepsilon \quad \text{whenever} \quad 0 < |x - 3| < \delta.$$

To do this consider

$$|(5x + 2) - 17| = |5x - 15| = 5|x - 3|.$$

Thus, to make $|(5x + 2) - 17| = 5|x - 3| < \varepsilon$, we need only make $0 < |x - 3| < \varepsilon/5$; that is, choose $\delta = \varepsilon/5$.

Verification If $0 < |x - 3| < \varepsilon/5$, then $5|x - 3| < \varepsilon$ implies

$$|5x - 15| < \varepsilon \quad \text{or} \quad |(5x + 2) - 17| < \varepsilon \quad \text{or} \quad |f(x) - 17| < \varepsilon. \quad \blacksquare$$

EXAMPLE 2 Using Definition 2.6.1

Prove that $\lim_{x \to -4} \dfrac{16 - x^2}{4 + x} = 8$.

◀ We examined this limit in (1) and (2) of Section 2.1.

Solution For $x \neq -4$,

$$\left| \frac{16 - x^2}{4 + x} - 8 \right| = |4 - x - 8| = |-x - 4| = |x + 4| = |x - (-4)|$$

Thus,

$$\left| \frac{16 - x^2}{4 + x} - 8 \right| = |x - (-4)| < \varepsilon$$

whenever we have $0 < |x - (-4)| < \varepsilon$; that is, choose $\delta = \varepsilon$. $\blacksquare$

EXAMPLE 3 A Limit That Does Not Exist

Consider the function

$$f(x) = \begin{cases} 0, & x \leq 1 \\ 2, & x > 1. \end{cases}$$

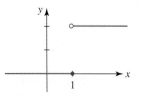

FIGURE 2.6.3 Limit of f does not exist as x approaches 1 in Example 3

We recognize in FIGURE 2.6.3 that f has a jump discontinuity at 1 and so $\lim_{x \to 1} f(x)$ does not exist. However, to *prove* this last fact, we shall proceed indirectly. Assume that the limit exists, namely, $\lim_{x \to 1} f(x) = L$. Then from Definition 2.6.1 we know that for the choice $\varepsilon = \frac{1}{2}$ there must exist a $\delta > 0$ so that

$$|f(x) - L| < \frac{1}{2} \quad \text{whenever} \quad 0 < |x - 1| < \delta.$$

Now to the right of 1, let us choose $x = 1 + \delta/2$. Since

$$0 < \left|1 + \frac{\delta}{2} - 1\right| = \left|\frac{\delta}{2}\right| < \delta$$

we must have

$$\left|f\left(1 + \frac{\delta}{2}\right) - L\right| = |2 - L| < \frac{1}{2}. \tag{5}$$

To the left of 1, choose $x = 1 - \delta/2$. But

$$0 < \left|1 - \frac{\delta}{2} - 1\right| = \left|-\frac{\delta}{2}\right| < \delta$$

implies

$$\left|f\left(1 - \frac{\delta}{2}\right) - L\right| = |0 - L| = |L| < \frac{1}{2}. \tag{6}$$

Solving the absolute-value inequalities (5) and (6) gives, respectively,

$$\frac{3}{2} < L < \frac{5}{2} \quad \text{and} \quad -\frac{1}{2} < L < \frac{1}{2}.$$

Since no number L can satisfy both of these inequalities, we conclude that $\lim_{x \to 1} f(x)$ does not exist. ∎

In the next example we consider the limit of a quadratic function. We shall see that finding the δ in this case requires a bit more ingenuity than in Examples 1 and 2.

EXAMPLE 4 Using Definition 2.6.1

We examined this limit in Example 1 of Section 2.1. ▶ Prove that $\lim_{x \to 4}(-x^2 + 2x + 2) = -6$.

Solution For an arbitrary $\varepsilon > 0$ we must find a $\delta > 0$ so that

$$|-x^2 + 2x + 2 - (-6)| < \varepsilon \quad \text{whenever} \quad 0 < |x - 4| < \delta.$$

Now,

$$\begin{aligned}|-x^2 + 2x + 2 - (-6)| &= |(-1)(x^2 - 2x - 8)| \\ &= |(x + 2)(x - 4)| \\ &= |x + 2||x - 4|. \end{aligned} \tag{7}$$

In other words, we want to make $|x + 2||x - 4| < \varepsilon$. But since we have agreed to examine values of x *near* 4, let us consider only those values for which $|x - 4| < 1$. This last inequality gives $3 < x < 5$ or equivalently $5 < x + 2 < 7$. Consequently we can write $|x + 2| < 7$. Hence from (7),

$$0 < |x - 4| < 1 \quad \text{implies} \quad |-x^2 + 2x + 2 - (-6)| < 7|x - 4|.$$

If we now choose δ to be the minimum of the two numbers, 1 and $\varepsilon/7$, written $\delta = \min\{1, \varepsilon/7\}$ we have

$$0 < |x - 4| < \delta \quad \text{implies} \quad |-x^2 + 2x + 2 - (-6)| < 7|x - 4| < 7 \cdot \frac{\varepsilon}{7} = \varepsilon. \blacksquare$$

The reasoning in Example 4 is subtle. Consequently it is worth a few minutes of your time to reread the discussion immediately following Definition 2.6.1, reexamine

Figure 2.3.2(b), and then think again about why $\delta = \min\{1, \varepsilon/7\}$ is the δ that "works" in the example. Remember, you can pick the ε arbitrarily; think about δ for, say, $\varepsilon = 8$, $\varepsilon = 6$, and $\varepsilon = 0.01$.

One-Sided Limits We state next the definitions of the **one-sided limits**, $\lim_{x \to a^-} f(x)$ and $\lim_{x \to a^+} f(x)$.

Definition 2.6.2 Left-Hand Limit

Suppose a function f is defined on an open interval (c, a). Then
$$\lim_{x \to a^-} f(x) = L$$
means for every $\varepsilon > 0$ there exists a $\delta > 0$ such that
$$|f(x) - L| < \varepsilon \quad \text{whenever} \quad a - \delta < x < a.$$

Definition 2.6.3 Right-Hand Limit

Suppose a function f is defined on an open interval (a, c). Then
$$\lim_{x \to a^+} f(x) = L$$
means for every $\varepsilon > 0$ there exists a $\delta > 0$ such that
$$|f(x) - L| < \varepsilon \quad \text{whenever} \quad a < x < a + \delta.$$

EXAMPLE 5 Using Definition 2.6.3

Prove that $\lim_{x \to 0^+} \sqrt{x} = 0$.

Solution First, we can write
$$|\sqrt{x} - 0| = |\sqrt{x}| = \sqrt{x}.$$
Then, $|\sqrt{x} - 0| < \varepsilon$ whenever $0 < x < 0 + \varepsilon^2$. In other words, we choose $\delta = \varepsilon^2$.

Verification If $0 < x < \varepsilon^2$, then $0 < \sqrt{x} < \varepsilon$ implies
$$|\sqrt{x}| < \varepsilon \quad \text{or} \quad |\sqrt{x} - 0| < \varepsilon. \qquad \blacksquare$$

Limits Involving Infinity The two concepts of **infinite limits**
$$f(x) \to \infty \text{ (or } -\infty) \quad \text{as} \quad x \to a$$
and a **limit at infinity**
$$f(x) \to L \quad \text{as} \quad x \to \infty \text{ (or } -\infty)$$
are formalized in the next two definitions.

Recall, an infinite limit is a limit that does not exist as $x \to a$.

Definition 2.6.4 Infinite Limits

(i) $\lim_{x \to a} f(x) = \infty$ means for each $M > 0$, there exists a $\delta > 0$ such that $f(x) > M$ whenever $0 < |x - a| < \delta$.

(ii) $\lim_{x \to a} f(x) = -\infty$ means for each $M < 0$, there exists a $\delta > 0$ such that $f(x) < M$ whenever $0 < |x - a| < \delta$.

Parts (*i*) and (*ii*) of Definition 2.6.4 are illustrated in FIGURE 2.6.4(a) and Figure 2.6.4(b), respectively. Recall, if $f(x) \to \infty$ (or $-\infty$) as $x \to a$, then $x = a$ is a vertical asymptote for the graph of f. In the case when $f(x) \to \infty$ as $x \to a$, then $f(x)$ can be made larger than any arbitrary positive number (that is, $f(x) > M$) by taking x sufficiently close to a (that is, $0 < |x - a| < \delta$).

 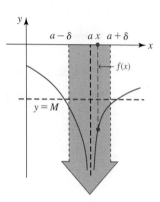

(a) For a given M, whenever $a - \delta < x < a + \delta, x \neq a$, then $f(x) > M$

(b) For a given M, whenever $a - \delta < x < a + \delta, x \neq a$, then $f(x) < M$

FIGURE 2.6.4 Infinite limits as $x \to a$

The four one-sided infinite limits

$$f(x) \to \infty \text{ as } x \to a^-, \quad f(x) \to -\infty \text{ as } x \to a^-$$
$$f(x) \to \infty \text{ as } x \to a^+, \quad f(x) \to -\infty \text{ as } x \to a^+$$

are defined in a manner analogous to that given in Definitions 2.6.2 and 2.6.3.

Definition 2.6.5 Limits at Infinity

(*i*) $\lim_{x \to \infty} f(x) = L$ if for each $\varepsilon > 0$, there exists an $N > 0$ such that $|f(x) - L| < \varepsilon$ whenever $x > N$.

(*ii*) $\lim_{x \to -\infty} f(x) = L$ if for each $\varepsilon > 0$, there exists an $N < 0$ such that $|f(x) - L| < \varepsilon$ whenever $x < N$.

Parts (*i*) and (*ii*) of Definition 2.6.5 are illustrated in FIGURE 2.6.5(a) and Figure 2.6.5(b), respectively. Recall, if $f(x) \to L$ as $x \to \infty$ (or $-\infty$), then $y = L$ is a horizontal asymptote for the graph of f. In the case when $f(x) \to L$ as $x \to \infty$, then the graph of f can be made arbitrarily close to the line $y = L$ (that is, $|f(x) - L| < \varepsilon$) by taking x sufficiently far out on the positive x-axis (that is, $x > N$).

 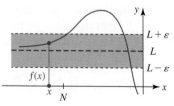

(a) For a given ε, $x > N$ implies $L - \varepsilon < f(x) < L + \varepsilon$

(b) For a given ε, $x < N$ implies $L - \varepsilon < f(x) < L + \varepsilon$

FIGURE 2.6.5 Limits at infinity

EXAMPLE 6 Using Definition 2.6.5(*i*)

Prove that $\lim_{x \to \infty} \dfrac{3x}{x+1} = 3$.

Solution By Definition 2.6.5(*i*), for any $\varepsilon > 0$, we must find a number $N > 0$ such that

$$\left| \dfrac{3x}{x+1} - 3 \right| < \varepsilon \quad \text{whenever} \quad x > N.$$

Now, by considering $x > 0$, we have

$$\left| \dfrac{3x}{x+1} - 3 \right| = \left| \dfrac{-3}{x+1} \right| = \dfrac{3}{x+1} < \dfrac{3}{x} < \varepsilon$$

whenever $x > 3/\varepsilon$. Hence, choose $N = 3/\varepsilon$. For example, if $\varepsilon = 0.01$, then $N = 3/(0.01) = 300$ will guarantee that $|f(x) - 3| < 0.01$ whenever $x > 300$. ∎

■ **Postscript—A Bit of History** After this section you may agree with English philosopher, priest, historian, and scientist William Whewell (1794–1866), who wrote in 1858 that "A limit is a peculiar . . . conception." For many years after the invention of calculus in the seventeenth century, mathematicians argued and debated the nature of a limit. There was an awareness that intuition, graphs, and numerical examples of ratios of vanishing quantities provide at best a shaky foundation for such a fundamental concept. As you will see beginning in the next chapter, the limit concept plays a central role in calculus. The study of calculus went through several periods of increased mathematical rigor beginning with the French mathematician Augustin-Louis Cauchy and continuing later with the German mathematician Karl Wilhelm Weierstrass.

Cauchy

Augustin-Louis Cauchy (1789–1857) was born during an era of upheaval in French history. Cauchy was destined to initiate a revolution of his own in mathematics. For many contributions, but especially for his efforts in clarifying mathematical obscurities, his incessant demand for satisfactory definitions and rigorous proofs of theorems, Cauchy is often called "the father of modern analysis." A prolific writer whose output has been surpassed by only a few, Cauchy produced nearly 800 papers in astronomy, physics, and mathematics. But the same mind that was always open and inquiring in science and mathematics was also narrow and unquestioning in many other areas. Outspoken and arrogant, Cauchy's passionate stands on political and religious issues often alienated him from his colleagues.

Weierstrass

Karl Wilhelm Weierstrass (1815–1897) One of the foremost mathematical analysts of the nineteenth century never earned an academic degree! After majoring in law at the University of Bonn, but concentrating in fencing and beer drinking for four years, Weierstrass "graduated" to real life with no degree. In need of a job, Weierstrass passed a state examination and received a teaching certificate in 1841. During a period of 15 years as a secondary school teacher, his dormant mathematical genius blossomed. Although the quantity of his research publications was modest, especially when compared with that of Cauchy, the quality of these works so impressed the German mathematical community that he was awarded a doctorate, *honoris causa*, from the University of Königsberg and eventually was appointed a professor at the University of Berlin. While there, Weierstrass achieved worldwide recognition both as a mathematician and as a teacher of mathematics. One of his students was Sonja Kowalewski, the greatest female mathematician of the nineteenth century. It was Karl Wilhelm Weierstrass who was responsible for putting the concept of a limit on a firm foundation with the ε-δ definition.

Exercises 2.6

Fundamentals

In Problems 1–24, use Definitions 2.6.1, 2.6.2, or 2.6.3 to prove the given limit result.

1. $\lim\limits_{x\to 5} 10 = 10$
2. $\lim\limits_{x\to -2} \pi = \pi$
3. $\lim\limits_{x\to 3} x = 3$
4. $\lim\limits_{x\to 4} 2x = 8$
5. $\lim\limits_{x\to -1} (x + 6) = 5$
6. $\lim\limits_{x\to 0} (x - 4) = -4$
7. $\lim\limits_{x\to 0} (3x + 7) = 7$
8. $\lim\limits_{x\to 1} (9 - 6x) = 3$
9. $\lim\limits_{x\to 2} \dfrac{2x-3}{4} = \dfrac{1}{4}$
10. $\lim\limits_{x\to 1/2} 8(2x+5) = 48$
11. $\lim\limits_{x\to -5} \dfrac{x^2-25}{x+5} = -10$
12. $\lim\limits_{x\to 3} \dfrac{x^2 - 7x + 12}{2x - 6} = -\dfrac{1}{2}$
13. $\lim\limits_{x\to 0} \dfrac{8x^5 + 12x^4}{x^4} = 12$
14. $\lim\limits_{x\to 1} \dfrac{2x^3 + 5x^2 - 2x - 5}{x^2 - 1} = 7$
15. $\lim\limits_{x\to 0} x^2 = 0$
16. $\lim\limits_{x\to 0} 8x^3 = 0$
17. $\lim\limits_{x\to 0^+} \sqrt{5x} = 0$
18. $\lim\limits_{x\to (1/2)^+} \sqrt{2x-1} = 0$
19. $\lim\limits_{x\to 0^-} f(x) = -1,\ f(x) = \begin{cases} 2x - 1, & x < 0 \\ 2x + 1, & x > 0 \end{cases}$
20. $\lim\limits_{x\to 1^+} f(x) = 3,\ f(x) = \begin{cases} 0, & x \le 1 \\ 3, & x > 1 \end{cases}$
21. $\lim\limits_{x\to 3} x^2 = 9$
22. $\lim\limits_{x\to 2} (2x^2 + 4) = 12$
23. $\lim\limits_{x\to 1} (x^2 - 2x + 4) = 3$
24. $\lim\limits_{x\to 5} (x^2 + 2x) = 35$

25. For $a > 0$, use the identity.
$$|\sqrt{x} - \sqrt{a}| = |\sqrt{x} - \sqrt{a}| \cdot \dfrac{\sqrt{x} + \sqrt{a}}{\sqrt{x} + \sqrt{a}} = \dfrac{|x - a|}{\sqrt{x} + \sqrt{a}}$$
and the fact that $\sqrt{x} \ge 0$ to prove that $\lim\limits_{x\to a} \sqrt{x} = \sqrt{a}$.

26. Prove that $\lim\limits_{x\to 2} (1/x) = \tfrac{1}{2}$. [*Hint*: Consider only those numbers x for which $1 < x < 3$.]

In Problems 27–30, prove that $\lim\limits_{x\to a} f(x)$ does not exist.

27. $f(x) = \begin{cases} 2, & x < 1 \\ 0, & x \ge 1 \end{cases};\ a = 1$
28. $f(x) = \begin{cases} 1, & x \le 3 \\ -1, & x > 3 \end{cases};\ a = 3$
29. $f(x) = \begin{cases} x, & x \le 0 \\ 2 - x, & x > 0 \end{cases};\ a = 0$
30. $f(x) = \dfrac{1}{x};\ a = 0$

In Problems 31–34, use Definition 2.6.5 to prove the given limit result.

31. $\lim\limits_{x\to\infty} \dfrac{5x-1}{2x+1} = \dfrac{5}{2}$
32. $\lim\limits_{x\to\infty} \dfrac{2x}{3x+8} = \dfrac{2}{3}$
33. $\lim\limits_{x\to -\infty} \dfrac{10x}{x-3} = 10$
34. $\lim\limits_{x\to -\infty} \dfrac{x^2}{x^2+3} = 1$

Think About It

35. Prove that $\lim\limits_{x\to 0} f(x) = 0$, where $f(x) = \begin{cases} x, & x \text{ rational} \\ 0, & x \text{ irrational} \end{cases}$.

2.7 The Tangent Line Problem

Introduction In a calculus course you will study many different things, but as mentioned in the introduction to Section 2.1, the subject "calculus" is roughly divided into two broad but related areas known as **differential calculus** and **integral calculus**. The discussion of each of these topics invariably begins with a motivating problem involving the graph of a function. Differential calculus is motivated by the problem

- *Find a tangent line to the graph of a function f,*

whereas integral calculus is motivated by the problem

- *Find the area under the graph of a function f.*

The first problem will be addressed in this section; the second problem will be discussed in Section 5.3.

Tangent Line to a Graph The word *tangent* stems from the Latin verb *tangere*, meaning "to touch." You might remember from the study of plane geometry that a tangent to a circle is a line L that intersects, or touches, the circle in exactly one point P. See FIGURE 2.7.1. It is not quite as easy to define a tangent line to the graph of a function f. The idea of *touching* carries over to the notion of a tangent line to the graph of a function, but the idea of *intersecting the graph in one point* does not carry over.

FIGURE 2.7.1 Tangent line L touches a circle at point P

Suppose $y = f(x)$ is a continuous function. If, as shown in FIGURE 2.7.2, f possesses a line L tangent to its graph at a point P, then what is an equation of this line? To answer this question, we need the coordinates of P and the slope $m_{\tan}$ of L. The coordinates of P pose no difficulty, since a point on the graph of a function f is obtained by specifying a value of x in the domain of f. The coordinates of the point of tangency at $x = a$ are then $(a, f(a))$. Therefore, the problem of finding a tangent line comes down to the problem of finding the slope $m_{\tan}$ of the line. As a means of *approximating* $m_{\tan}$, we can readily find the slopes $m_{\sec}$ of *secant lines* (from the Latin verb *secare*, meaning "to cut") that pass through the point P and any other point Q on the graph. See FIGURE 2.7.3.

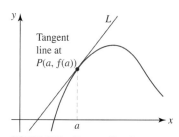

FIGURE 2.7.2 Tangent line L to a graph at point P

■ Slope of Secant Lines
If P has coordinates $(a, f(a))$ and if Q has coordinates $(a + h, f(a + h))$, then as shown in FIGURE 2.7.4, the slope of the secant line through P and Q is

$$m_{\sec} = \frac{\text{change in } y}{\text{change in } x} = \frac{f(a + h) - f(a)}{(a + h) - a}$$

or

$$m_{\sec} = \frac{f(a + h) - f(a)}{h}. \quad (1)$$

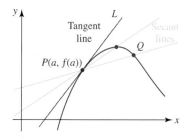

FIGURE 2.7.3 Slopes of secant lines approximate the slope $m_{\tan}$ of L

The expression on the right-hand side of the equality in (1) is called a **difference quotient**. When we let h take on values that are closer and closer to zero, that is, as $h \to 0$, then the points $Q(a + h, f(a + h))$ move along the curve closer and closer to the point $P(a, f(a))$. Intuitively, we expect the secant lines to approach the tangent line L, and that $m_{\sec} \to m_{\tan}$ as $h \to 0$. That is,

$$m_{\tan} = \lim_{h \to 0} m_{\sec}$$

provided this limit exists. We summarize this conclusion in an equivalent form of the limit using the difference quotient (1).

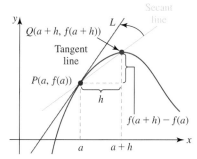

FIGURE 2.7.4 Secant lines swing into the tangent line L as $h \to 0$

Definition 2.7.1 Tangent Line with Slope

Let $y = f(x)$ be continuous at the number a. If the limit

$$m_{\tan} = \lim_{h \to 0} \frac{f(a + h) - f(a)}{h} \quad (2)$$

exists, then the **tangent line** to the graph of f at $(a, f(a))$ is that line passing through the point $(a, f(a))$ with slope $m_{\tan}$.

Just like many of the problems discussed earlier in this chapter, observe that the limit in (2) has the indeterminate form $0/0$ as $h \to 0$.

If the limit in (2) exists, the number $m_{\tan}$ is also called the **slope of the curve** $y = f(x)$ at $(a, f(a))$.

The computation of (2) is essentially a *four-step process*; three of these steps involve only precalculus mathematics: algebra and trigonometry. If the first three steps are done accurately, the fourth step, or the calculus step, *may* be the easiest part of the problem.

Guidelines for Computing (2)

(*i*) Evaluate $f(a)$ and $f(a + h)$.
(*ii*) Evaluate the difference $f(a + h) - f(a)$. Simplify.
(*iii*) Simplify the difference quotient

$$\frac{f(a + h) - f(a)}{h}.$$

(*iv*) Compute the limit of the difference quotient

$$\lim_{h \to 0} \frac{f(a + h) - f(a)}{h}.$$

Note ▶ The computation of the difference $f(a + h) - f(a)$ in step (*ii*) is in most instances the most important step. It is imperative that you simplify this step as much as possible. Here is a tip: In *many* problems involving the computation of (2) you will be able to factor h from the difference $f(a + h) - f(a)$.

EXAMPLE 1 The Four-Step Process

Find the slope of the tangent line to the graph of $y = x^2 + 2$ at $x = 1$.

Solution We use the four-step procedure outlined above with the number 1 playing the part of the symbol a.

(*i*) The initial step is the computation of $f(1)$ and $f(1 + h)$. We have $f(1) = 1^2 + 2 = 3$, and
$$f(1 + h) = (1 + h)^2 + 2$$
$$= (1 + 2h + h^2) + 2$$
$$= 3 + 2h + h^2.$$

(*ii*) Next, from the result in the preceding step the difference is:
$$f(1 + h) - f(1) = 3 + 2h + h^2 - 3$$
$$= 2h + h^2$$
$$= h(2 + h). \leftarrow \text{notice the factor of } h$$

(*iii*) The computation of the difference quotient $\dfrac{f(1 + h) - f(1)}{h}$ is now straightforward. Again, we use the results from the preceding step:
$$\frac{f(1 + h) - f(1)}{h} = \frac{h(2 + h)}{h} = 2 + h. \leftarrow \text{cancel the } h\text{'s}$$

(*iv*) The last step is now easy. The limit in (2) is seen to be
$$m_{\tan} = \lim_{h \to 0} \overset{\text{from the preceding step}}{\overset{\downarrow}{\frac{f(1 + h) - f(1)}{h}}} = \lim_{h \to 0} (2 + h) = 2.$$

The slope of the tangent line to the graph of $y = x^2 + 2$ at $(1, 3)$ is 2. ∎

EXAMPLE 2 Equation of Tangent Line

Find an equation of the tangent line whose slope was found in Example 1.

Solution We know the point of tangency $(1, 3)$ and the slope $m_{\tan} = 2$, and so from the point–slope equation of a line we find
$$y - 3 = 2(x - 1) \quad \text{or} \quad y = 2x + 1.$$

Observe that the last equation is consistent with the x- and y-intercepts of the red line in FIGURE 2.7.5. ∎

FIGURE 2.7.5 Tangent line in Example 2

EXAMPLE 3 Equation of Tangent Line

Find an equation of the tangent line to the graph of $f(x) = 2/x$ at $x = 2$.

Solution We start by using (2) to find $m_{\tan}$ with a identified as 2. In the second of the four steps, we will have to combine two symbolic fractions by means of a common denominator.

(*i*) We have $f(2) = 2/2 = 1$ and $f(2 + h) = 2/(2 + h)$.

(*ii*) $f(2 + h) - f(2) = \dfrac{2}{2 + h} - 1$

$= \dfrac{2}{2 + h} - \dfrac{1}{1} \cdot \dfrac{2 + h}{2 + h} \leftarrow$ a common denominator is $2 + h$

$= \dfrac{2 - 2 - h}{2 + h}$

$= \dfrac{-h}{2 + h}. \leftarrow$ here is the factor of h

(*iii*) The last result is to be divided by h or more precisely $\frac{h}{1}$. We invert and multiply by $\frac{1}{h}$:

$$\frac{f(2+h)-f(2)}{h} = \frac{\frac{-h}{2+h}}{\frac{h}{1}} = \frac{-h}{2+h} \cdot \frac{1}{h} = \frac{-1}{2+h}. \quad \leftarrow \text{cancel the } h\text{'s}$$

(*iv*) From (2) $m_{\tan}$ is

$$m_{\tan} = \lim_{h \to 0} \frac{f(2+h)-f(2)}{h} = \lim_{h \to 0} \frac{-1}{2+h} = -\frac{1}{2}.$$

From $f(2) = 1$ the point of tangency is $(2, 1)$ and the slope of the tangent line at $(2, 1)$ is $m_{\tan} = -\frac{1}{2}$. From the point–slope equation of a line, the tangent line is

$$y - 1 = \frac{1}{2}(x - 2) \quad \text{or} \quad y = -\frac{1}{2}x + 2.$$

The graphs of $y = 2/x$ and the tangent line at $(2, 1)$ are shown in FIGURE 2.7.6.

FIGURE 2.7.6 Tangent line in Example 3

EXAMPLE 4 Slope of Tangent Line

Find the slope of the tangent line to the graph of $f(x) = \sqrt{x-1}$ at $x = 5$.

Solution Replacing a by 5 in (2), we have:

(*i*) $f(5) = \sqrt{5-1} = \sqrt{4} = 2$, and

$$f(5+h) = \sqrt{5+h-1} = \sqrt{4+h}.$$

(*ii*) The difference is

$$f(5+h) - f(5) = \sqrt{4+h} - 2.$$

Because we expect to find a factor of h in this difference, we proceed to rationalize the numerator:

$$f(5+h) - f(5) = \frac{\sqrt{4+h}-2}{1} \cdot \frac{\sqrt{4+h}+2}{\sqrt{4+h}+2}$$

$$= \frac{(4+h)-4}{\sqrt{4+h}+2}$$

$$= \frac{h}{\sqrt{4+h}+2}. \quad \leftarrow \text{here is the factor of } h$$

(*iii*) The difference quotient $\dfrac{f(5+h)-f(5)}{h}$ is then:

$$\frac{f(5+h)-f(5)}{h} = \frac{\frac{h}{\sqrt{4+h}+2}}{h}$$

$$= \frac{h}{h(\sqrt{4+h}+2)}$$

$$= \frac{1}{\sqrt{4+h}+2}.$$

(*iv*) The limit in (2) is

$$m_{\tan} = \lim_{h \to 0} \frac{f(5+h)-f(5)}{h} = \lim_{h \to 0} \frac{1}{\sqrt{4+h}+2} = \frac{1}{\sqrt{4}+2} = \frac{1}{4}.$$

The slope of the tangent line to the graph of $f(x) = \sqrt{x-1}$ at $(5, 2)$ is $\frac{1}{4}$. ∎

The result obtained in the next example should come as no surprise.

FIGURE 2.7.7 Vertical tangent in Example 6

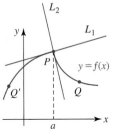

FIGURE 2.7.8 Tangent fails to exist at $(a, f(a))$

FIGURE 2.7.9 Function in Example 7

EXAMPLE 5 Tangent Line to a Line

For any linear function $y = mx + b$, the tangent line to its graph coincides with the line itself. Not unexpectedly then, the slope of the tangent line for any number $x = a$ is

$$m_{\tan} = \lim_{h \to 0} \frac{f(a + h) - f(a)}{h} = \lim_{h \to 0} \frac{m(a + h) + b - (ma + b)}{h} = \lim_{h \to 0} \frac{mh}{h} = \lim_{h \to 0} m = m. \quad \blacksquare$$

▌ **Vertical Tangents** The limit in (2) can fail to exist for a function f at $x = a$ and yet there may be a tangent at the point $(a, f(a))$. The tangent line to a graph can be **vertical**, in which case its slope is undefined. We will consider the concept of vertical tangents in more detail in Section 3.1.

EXAMPLE 6 Vertical Tangent Line

Although we will not pursue the details at this time, it can be shown that the graph of $f(x) = x^{1/3}$ possesses a vertical tangent line at the origin. In FIGURE 2.7.7 we see that the y-axis, that is, the line $x = 0$, is tangent to the graph at the point $(0, 0)$. $\quad \blacksquare$

▌ **A Tangent May Not Exist** The graph of a function f that is continuous at a number a does not have to possess a tangent line at the point $(a, f(a))$. A tangent line will not exist whenever the graph of f has a sharp corner at $(a, f(a))$. FIGURE 2.7.8 indicates what can go wrong when the graph of a function f has a "corner." In this case f is continuous at a, but the secant lines through P and Q approach L_2 as $Q \to P$, and the secant lines through P and Q' approach a different line L_1 as $Q' \to P$. In other words, the limit in (2) fails to exist because the one-sided limits of the difference quotient (as $h \to 0^+$ and as $h \to 0^-$) are different.

EXAMPLE 7 Graph with a Corner

Show that the graph of $f(x) = |x|$ does not have a tangent at $(0, 0)$.

Solution The graph of the absolute-value function in FIGURE 2.7.9 has a corner at the origin. To prove that the graph of f does not possess a tangent line at the origin we must examine

$$\lim_{h \to 0} \frac{f(0 + h) - f(0)}{h} = \lim_{h \to 0} \frac{|0 + h| - |0|}{h} = \lim_{h \to 0} \frac{|h|}{h}.$$

From the definition of absolute value

$$|h| = \begin{cases} h, & h > 0 \\ -h, & h < 0 \end{cases}$$

we see that

$$\lim_{h \to 0^+} \frac{|h|}{h} = \lim_{h \to 0^+} \frac{h}{h} = 1 \quad \text{whereas} \quad \lim_{h \to 0^-} \frac{|h|}{h} = \lim_{h \to 0^-} \frac{-h}{h} = -1.$$

Since the right-hand and left-hand limits are not equal we conclude that the limit (2) does not exist. Even though the function $f(x) = |x|$ is continuous at $x = 0$, the graph of f possesses no tangent at $(0, 0)$. $\quad \blacksquare$

▌ **Average Rate of Change** In different contexts the difference quotient in (1) and (2), or slope of the secant line, is written in terms of alternative symbols. The symbol h in (1) and (2) is often written as Δx and the difference $f(a + \Delta x) - f(a)$ is denoted by Δy, that is, the difference quotient is

$$\frac{\text{change in } y}{\text{change in } x} = \frac{f(a + \Delta x) - f(a)}{(a + \Delta x) - a} = \frac{f(a + \Delta x) - f(a)}{\Delta x} = \frac{\Delta y}{\Delta x}. \qquad (3)$$

Moreover, if $x_1 = a + \Delta x$, $x_0 = a$, then $\Delta x = x_1 - x_0$ and (3) is the same as

$$\frac{f(x_1) - f(x_0)}{x_1 - x_0} = \frac{\Delta y}{\Delta x}. \qquad (4)$$

The slope $\Delta y / \Delta x$ of the secant line through the points $(x_0, f(x_0))$ and $(x_1, f(x_1))$ is called the **average rate of change of the function** f over the interval $[x_0, x_1]$. The limit $\lim_{\Delta x \to 0} \Delta y / \Delta x$ is then called the **instantaneous rate of change of the function** with respect to x at x_0.

Almost everyone has an intuitive notion of speed as a rate at which a distance is covered in a certain length of time. When, say, a bus travels 60 mi in 1 h, the *average speed* of the bus must have been 60 mi/h. Of course, it is difficult to maintain the rate of 60 mi/h for the entire trip because the bus slows down for towns and speeds up when it passes cars. In other words, the speed changes with time. If a bus company's schedule demands that the bus travel the 60 mi from one town to another in 1 h, the driver knows instinctively that he or she must compensate for speeds less than 60 mi/h by traveling at speeds greater than this at other points in the journey. Knowing that the average velocity is 60 mi/h does not, however, answer the question: What is the velocity of the bus at a particular instant?

■ **Average Velocity** In general, the **average velocity** or **average speed** of a moving object is defined by

$$v_{\text{ave}} = \frac{\text{change of distance}}{\text{change in time}}. \tag{5}$$

Consider a runner who finishes a 10-km race in an elapsed time of 1 h 15 min (1.25 h). The runner's average velocity or average speed for the race was

$$v_{\text{ave}} = \frac{10 - 0}{1.25 - 0} = 8 \text{ km/h}.$$

But suppose we now wish to determine the runner's *exact* velocity v at the instant the runner is one-half hour into the race. If the distance run in the time interval from 0 h to 0.5 h is measured to be 5 km, then

$$v_{\text{ave}} = \frac{5}{0.5} = 10 \text{ km/h}.$$

Again, this number is not a measure, or necessarily even a good indicator, of the instantaneous rate v at which the runner is moving 0.5 h into the race. If we determine that at 0.6 h the runner is 5.7 km from the starting line, then the average velocity from 0 h to 0.6 h is $v_{\text{ave}} = 5.7/0.6 = 9.5$ km/h. However, during the time interval from 0.5 h to 0.6 h,

$$v_{\text{ave}} = \frac{5.7 - 5}{0.6 - 0.5} = 7 \text{ km/h}.$$

FIGURE 2.7.10 Runner in a 10-km race

The latter number is a more realistic measure of the rate v. See FIGURE 2.7.10. By "shrinking" the time interval between 0.5 h and the time that corresponds to a measured position close to 5 km, we expect to obtain even better approximations to the runner's velocity at time 0.5 h.

■ **Rectilinear Motion** To generalize the preceding discussion, let us suppose an object, or particle, at point P moves along either a vertical or horizontal coordinate line as shown in FIGURE 2.7.11. Furthermore, let the particle move in such a manner that its position, or coordinate, on the line is given by a function $s = s(t)$, where t represents time. The values of s are directed distances measured from O in units such as centimeters, meters, feet, or miles. When P is either to the right of or above O, we take $s > 0$, whereas $s < 0$ when P is either to the left of or below O. Motion in a straight line is called **rectilinear motion**.

FIGURE 2.7.11 Coordinate lines

If an object, such as a toy car moving on a horizontal coordinate line, is at point P at time t_0 and at point P' at time t_1, then the coordinates of the points, shown in FIGURE 2.7.12, are $s(t_0)$ and $s(t_1)$. By (4) the **average velocity** of the object in the time interval $[t_0, t_1]$ is

$$v_{\text{ave}} = \frac{\text{change in position}}{\text{change in time}} = \frac{s(t_1) - s(t_0)}{t_1 - t_0}. \tag{6}$$

FIGURE 2.7.12 Position of toy car on a coordinate line at two times

EXAMPLE 8 Average Velocity

The height s above ground of a ball dropped from the top of the St. Louis Gateway Arch is given by $s(t) = -16t^2 + 630$, where s is measured in feet and t in seconds. See FIGURE 2.7.13. Find the average velocity of the falling ball between the time the ball is released and the time it hits the ground.

Solution The time at which the ball is released is determined from the equation $s(t) = 630$ or $-16t^2 + 630 = 630$. This gives $t = 0$ s. When the ball hits the ground then $s(t) = 0$ or

FIGURE 2.7.13 Falling ball in Example 8

$-16t^2 + 630 = 0$. The last equation gives $t = \sqrt{315/8} \approx 6.27$ s. Thus from (6) the average velocity in the time interval $[0, \sqrt{315/8}]$ is

$$v_{ave} = \frac{s(\sqrt{351/8}) - s(0)}{\sqrt{351/8} - 0} = \frac{0 - 630}{\sqrt{351/8} - 0} \approx -100.40 \text{ ft/s}.$$

If we let $t_1 = t_0 + \Delta t$, or $\Delta t = t_1 - t_0$, and $\Delta s = s(t_0 + \Delta t) - s(t_0)$, then (6) is equivalent to

$$v_{ave} = \frac{\Delta s}{\Delta t}. \tag{7}$$

This suggests that the limit of (7) as $\Delta t \to 0$ gives the **instantaneous rate of change** of $s(t)$ at $t = t_0$ or the **instantaneous velocity**.

> **Definition 2.7.2** Instantaneous Velocity
>
> Let $s = s(t)$ be a function that gives the position of an object moving in a straight line. Then the **instantaneous velocity** at time $t = t_0$ is
>
> $$v(t_0) = \lim_{\Delta t \to 0} \frac{s(t_0 + \Delta t) - s(t_0)}{\Delta t} = \lim_{\Delta t \to 0} \frac{\Delta s}{\Delta t}, \tag{8}$$
>
> whenever the limit exists.

Note: Except for notation and interpretation, there is no mathematical difference between (2) and (8). Also, the word *instantaneous* is often dropped, and so one often speaks of the *rate of change* of a function or the *velocity* of a moving particle.

EXAMPLE 9 Example 8 Revisited

Find the instantaneous velocity of the falling ball in Example 8 at $t = 3$ s.

Solution We use the same four-step procedure as in the earlier examples with $s = s(t)$ given in Example 8.

(i) $s(3) = -16(9) + 630 = 486$. For any $\Delta t \neq 0$,
$$s(3 + \Delta t) = -16(3 + \Delta t)^2 + 630 = -16(\Delta t)^2 - 96\Delta t + 486.$$

(ii) $s(3 + \Delta t) - s(3) = [-16(\Delta t)^2 - 96\Delta t + 486] - 486$
$$= -16(\Delta t)^2 - 96\Delta t = \Delta t(-16\Delta t - 96)$$

(iii) $\dfrac{\Delta s}{\Delta t} = \dfrac{\Delta t(-16\Delta t - 96)}{\Delta t} = -16\Delta t - 96$

(iv) From (8),
$$v(3) = \lim_{\Delta t \to 0} \frac{\Delta s}{\Delta t} = \lim_{\Delta t \to 0}(-16\Delta t - 96) = -96 \text{ ft/s}. \tag{9}$$

In Example 9, the number $s(3) = 486$ ft is the height of the ball above ground at 3 s. The minus sign in (9) is significant because the ball is moving opposite to the positive or upward direction.

Exercises 2.7

Answers to selected odd-numbered problems begin on page ANS-9.

≡ Fundamentals

In Problems 1–6, sketch the graph of the function and the tangent line at the given point. Find the slope of the secant line through the points that correspond to the indicated values of x.

1. $f(x) = -x^2 + 9, (2, 5); x = 2, x = 2.5$
2. $f(x) = x^2 + 4x, (0, 0); x = -\dfrac{1}{4}, x = 0$
3. $f(x) = x^3, (-2, -8); x = -2, x = -1$
4. $f(x) = 1/x, (1, 1); x = 0.9, x = 1$
5. $f(x) = \sin x, (\pi/2, 1); x = \pi/2, x = 2\pi/3$
6. $f(x) = \cos x, (-\pi/3, \tfrac{1}{2}); x = -\pi/2, x = -\pi/3$

In Problems 7–8, use (2) to find the slope of the tangent line to the graph of the function at the given value of x. Find an equation of the tangent line at the corresponding point.

7. $f(x) = x^2 - 6, x = 3$
8. $f(x) = -3x^2 + 10, x = -1$
9. $f(x) = x^2 - 3x, x = 1$
10. $f(x) = -x^2 + 5x - 3, x = -2$
11. $f(x) = -2x^3 + x, x = 2$
12. $f(x) = 8x^3 - 4, x = \frac{1}{2}$
13. $f(x) = \frac{1}{2x}, x = -1$
14. $f(x) = \frac{4}{x-1}, x = 2$
15. $f(x) = \frac{1}{(x-1)^2}, x = 0$
16. $f(x) = 4 - \frac{8}{x}, x = -1$
17. $f(x) = \sqrt{x}, x = 4$
18. $f(x) = \frac{1}{\sqrt{x}}, x = 1$

In Problems 19 and 20, use (2) to find the slope of the tangent line to the graph of the function at the given value of x. Find an equation of the tangent line at the corresponding point. Before starting, review the limits in (10) and (14) of Section 2.4 and the sum formulas (17) and (18) in Section 1.4.

19. $f(x) = \sin x, x = \pi/6$
20. $f(x) = \cos x, x = \pi/4$

In Problems 21 and 22, determine whether the line that passes through the red point is tangent to the graph of $f(x) = x^2$ at the blue point.

21.
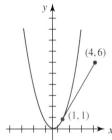
FIGURE 2.7.14 Graph for Problem 21

22.

FIGURE 2.7.15 Graph for Problem 22

23. In FIGURE 2.7.16, the red line is tangent to the graph of $y = f(x)$ at the indicated point. Find an equation of the tangent line. What is the y-intercept of the tangent line?

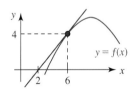
FIGURE 2.7.16 Graph for Problem 23

24. In FIGURE 2.7.17, the red line is tangent to the graph of $y = f(x)$ at the indicated point. Find $f(-5)$.

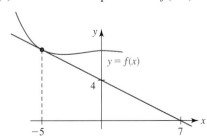
FIGURE 2.7.17 Graph for Problem 24

In Problems 25–28, use (2) to find a formula for $m_{\tan}$ at a general point $(x, f(x))$ on the graph of f. Use the formula for $m_{\tan}$ to determine the points where the tangent line to the graph is horizontal.

25. $f(x) = -x^2 + 6x + 1$
26. $f(x) = 2x^2 + 24x - 22$
27. $f(x) = x^3 - 3x$
28. $f(x) = -x^3 + x^2$

≡ Applications

29. A car travels the 290 mi between Los Angeles and Las Vegas in 5 h. What is its average velocity?

30. Two marks on a straight highway are $\frac{1}{2}$ mi apart. A highway patrol plane observes that a car traverses the distance between the marks in 40 s. Assuming a speed limit of 60 mi/h, will the car be stopped for speeding?

31. A jet airplane averages 920 km/h to fly the 3500 km between Hawaii and San Francisco. How many hours does the flight take?

32. A marathon race is run over a straight 26-mi course. The race begins at noon. At 1:30 P.M. a contestant passes the 10-mi mark and at 3:10 P.M. the contestant passes the 20-mi mark. What is the contestant's average running speed between 1:30 P.M. and 3:10 P.M.?

In Problems 33 and 34, the position of a particle moving on a horizontal coordinate line is given by the function. Use (8) to find the instantaneous velocity of the particle at the indicated time.

33. $s(t) = -4t^2 + 10t + 6, t = 3$
34. $s(t) = t^2 + \frac{1}{5t+1}, t = 0$

35. The height above ground of a ball dropped from an initial altitude of 122.5 m is given by $s(t) = -4.9t^2 + 122.5$, where s is measured in meters and t in seconds.
    (a) What is the instantaneous velocity at $t = \frac{1}{2}$?
    (b) At what time does the ball hit the ground?
    (c) What is the impact velocity?

36. Ignoring air resistance, if an object is dropped from an initial height h, then its height above ground at time $t > 0$ is given by $s(t) = -\frac{1}{2}gt^2 + h$, where g is the acceleration of gravity.
    (a) At what time does the object hit the ground?
    (b) If $h = 100$ ft, compare the impact times for Earth ($g = 32$ ft/s^2), for Mars ($g = 12$ ft/s^2), and for the Moon ($g = 5.5$ ft/s^2).
    (c) Use (8) to find a formula for the instantaneous velocity v at a general time t.
    (d) Using the times found in part (b) and the formula found in part (c), find the corresponding impact velocities for Earth, Mars, and the Moon.

37. The height of a projectile shot from ground level is given by $s = -16t^2 + 256t$, where s is measured in feet and t in seconds.
    (a) Determine the height of the projectile at $t = 2, t = 6, t = 9$, and $t = 10$.
    (b) What is the average velocity of the projectile between $t = 2$ and $t = 5$?
    (c) Show that the average velocity between $t = 7$ and $t = 9$ is zero. Interpret physically.
    (d) At what time does the projectile hit the ground?

(e) Use (8) to find a formula for instantaneous velocity v at a general time t.
(f) Using the time found in part (d) and the formula found in part (e), find the corresponding impact velocity.
(g) What is the maximum height the projectile attains?

38. Suppose the graph shown in FIGURE 2.7.18 is that of position function $s = s(t)$ of a particle moving in a straight line, where s is measured in meters and t in seconds.

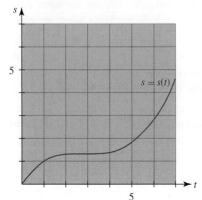

FIGURE 2.7.18 Graph for Problem 38

(a) Estimate the position of the particle at $t = 4$ and at $t = 6$.
(b) Estimate the average velocity of the particle between $t = 4$ and $t = 6$.
(c) Estimate the initial velocity of the particle—that is, its velocity at $t = 0$.
(d) Estimate a time at which the velocity of the particle is zero.
(e) Determine an interval on which the velocity of the particle is decreasing.
(f) Determine an interval on which the velocity of the particle is increasing.

≡ **Think About It**

39. Let $y = f(x)$ be an even function whose graph possesses a tangent line with slope m at $(a, f(a))$. Show that the slope of the tangent line at $(-a, f(a))$ is $-m$. [*Hint*: Explain why $f(-a + h) = f(a - h)$.]

40. Let $y = f(x)$ be an odd function whose graph possesses a tangent line with slope m at $(a, f(a))$. Show that the slope of the tangent line at $(-a, -f(a))$ is m.

41. Proceed as in Example 7 and show that there is no tangent line to graph of $f(x) = x^2 + |x|$ at $(0, 0)$.

Chapter 2 in Review

Answers to selected odd-numbered problems begin on page ANS-9.

A. True/False

In Problems 1–22, indicate whether the given statement is true or false.

1. $\lim\limits_{x \to 2} \dfrac{x^3 - 8}{x - 2} = 12$ _____
2. $\lim\limits_{x \to 5} \sqrt{x - 5} = 0$ _____
3. $\lim\limits_{x \to 0} \dfrac{|x|}{x} = 1$ _____
4. $\lim\limits_{x \to \infty} e^{2x - x^2} = \infty$ _____
5. $\lim\limits_{x \to 0^+} \tan^{-1}\left(\dfrac{1}{x}\right)$ does not exist. _____
6. $\lim\limits_{z \to 1} \dfrac{z^3 + 8z - 2}{z^2 + 9z - 10}$ does not exist. _____
7. If $\lim\limits_{x \to a} f(x) = 3$ and $\lim\limits_{x \to a} g(x) = 0$, then $\lim\limits_{x \to a} f(x)/g(x)$ does not exist. _____
8. If $\lim\limits_{x \to a} f(x)$ exists and $\lim\limits_{x \to a} g(x)$ does not exist, then $\lim\limits_{x \to a} f(x)g(x)$ does not exist. _____
9. If $\lim\limits_{x \to a} f(x) = \infty$ and $\lim\limits_{x \to a} g(x) = \infty$, then $\lim\limits_{x \to a} f(x)/g(x) = 1$. _____
10. If $\lim\limits_{x \to a} f(x) = \infty$ and $\lim\limits_{x \to a} g(x) = \infty$, then $\lim\limits_{x \to a} [f(x) - g(x)] = 0$. _____
11. If f is a polynomial function, then $\lim\limits_{x \to \infty} f(x) = \infty$. _____
12. Every polynomial function is continuous on $(-\infty, \infty)$. _____
13. For $f(x) = x^5 + 3x - 1$ there exists a number c in $[-1, 1]$ such that $f(c) = 0$. _____
14. If f and g are continuous at the number 2, then f/g is continuous at 2. _____
15. The greatest integer function $f(x) = \lfloor x \rfloor$ is not continuous on the interval $[0, 1]$. _____
16. If $\lim\limits_{x \to a^-} f(x)$ and $\lim\limits_{x \to a^+} f(x)$ exist, then $\lim\limits_{x \to a} f(x)$ exists. _____
17. If a function f is discontinuous at the number 3, then $f(3)$ is not defined. _____
18. If a function f is continuous at the number a, then $\lim\limits_{x \to a}(x - a)f(x) = 0$. _____
19. If f is continuous and $f(a)f(b) < 0$, there is a root of $f(x) = 0$ in the interval $[a, b]$. _____

20. The function $f(x) = \begin{cases} \dfrac{x^2 - 6x + 5}{x - 5}, & x \neq 5 \\ 4, & x = 5 \end{cases}$ is discontinuous at 5. _____

21. The function $f(x) = \dfrac{\sqrt{x}}{x + 1}$ has a vertical asymptote at $x = -1$. _____

22. If $y = x - 2$ is a tangent line to the graph of a function $y = f(x)$ at $(3, f(3))$, then $f(3) = 1$. _____

B. Fill in the Blanks

In Problems 1–22, fill in the blanks.

1. $\lim\limits_{x \to 2}(3x^2 - 4x) =$ _____

2. $\lim\limits_{x \to 3}(5x^2)^0 =$ _____

3. $\lim\limits_{t \to \infty} \dfrac{2t - 1}{3 - 10t} =$ _____

4. $\lim\limits_{x \to -\infty} \dfrac{\sqrt{x^2 + 1}}{2x + 1} =$ _____

5. $\lim\limits_{t \to 1} \dfrac{1 - \cos^2(t - 1)}{t - 1} =$ _____

6. $\lim\limits_{x \to 0} \dfrac{\sin 3x}{5x} =$ _____

7. $\lim\limits_{x \to 0^+} e^{1/x} =$ _____

8. $\lim\limits_{x \to 0^-} e^{1/x} =$ _____

9. $\lim\limits_{x \to \infty} e^{1/x} =$ _____

10. $\lim\limits_{x \to -\infty} \dfrac{1 + 2e^x}{4 + e^x} =$ _____

11. $\lim\limits_{x \to \_\_} \dfrac{1}{x - 3} = -\infty$

12. $\lim\limits_{x \to \_\_}(5x + 2) = 22$

13. $\lim\limits_{x \to \_\_} x^3 = -\infty$

14. $\lim\limits_{x \to \_\_} \dfrac{1}{\sqrt{x}} = \infty$

15. If $f(x) = 2(x - 4)/|x - 4|$, $x \neq 4$, and $f(4) = 9$, then $\lim\limits_{x \to 4} f(x) =$ _____.

16. Suppose $x^2 - x^4/3 \leq f(x) \leq x^2$ for all x. Then $\lim\limits_{x \to 0} f(x)/x^2 =$ _____.

17. If f is continuous at a number a and $\lim\limits_{x \to a} f(x) = 10$, then $f(a) =$ _____.

18. If f is continuous at $x = 5$, $f(5) = 2$, and $\lim\limits_{x \to 5} g(x) = 10$, then $\lim\limits_{x \to 5}[g(x) - f(x)] =$ _____.

19. $f(x) = \begin{cases} \dfrac{2x - 1}{4x^2 - 1}, & x \neq \frac{1}{2} \\ 0.5, & x = \frac{1}{2} \end{cases}$ is _____ (continuous/discontinuous) at the number $\frac{1}{2}$.

20. The equation $e^{-x^2} = x^2 - 1$ has precisely _____ roots in the interval $(-\infty, \infty)$.

21. The function $f(x) = \dfrac{10}{x} + \dfrac{x^2 - 4}{x - 2}$ has a removable discontinuity at $x = 2$. To remove the discontinuity, $f(2)$ should be defined to be _____.

22. If $\lim\limits_{x \to -5} g(x) = -9$ and $f(x) = x^2$, then $\lim\limits_{x \to -5} f(g(x)) =$ _____.

C. Exercises

In Problems 1–4, sketch a graph of a function f that satisfies the given conditions.

1. $f(0) = 1, f(4) = 0, f(6) = 0, \lim\limits_{x \to 3^-} f(x) = 2, \lim\limits_{x \to 3^+} f(x) = \infty, \lim\limits_{x \to -\infty} f(x) = 0, \lim\limits_{x \to \infty} f(x) = 2$

2. $\lim\limits_{x \to -\infty} f(x) = 0, f(0) = 1, \lim\limits_{x \to 4^-} f(x) = \infty, \lim\limits_{x \to 4^+} f(x) = \infty, f(5) = 0, \lim\limits_{x \to \infty} f(x) = -1$

3. $\lim\limits_{x \to -\infty} f(x) = 2, f(-1) = 3, f(0) = 0, f(-x) = -f(x)$

4. $\lim\limits_{x \to \infty} f(x) = 0, f(0) = -3, f(1) = 0, f(-x) = f(x)$

In Problems 5–10, state which of the conditions (a)–(j) are applicable to the graph of $y = f(x)$.

(a) $f(a)$ is not defined   (b) $f(a) = L$ (c) f is continuous at $x = a$ (d) f is continuous on $[0, a]$ (e) $\lim\limits_{x \to a^+} f(x) = L$

(f) $\lim\limits_{x \to a} f(x) = L$ (g) $\lim\limits_{x \to a} |f(x)| = \infty$ (h) $\lim\limits_{x \to \infty} f(x) = L$ (i) $\lim\limits_{x \to \infty} f(x) = -\infty$ (j) $\lim\limits_{x \to \infty} f(x) = 0$

5.

6.

FIGURE 2.R.2 Graph for Problem 6

7.

FIGURE 2.R.3 Graph for Problem 7

FIGURE 2.R.1 Graph for Problem 5

8.

FIGURE 2.R.4 Graph for Problem 8

9.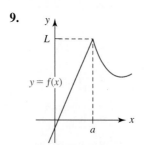

FIGURE 2.R.5 Graph for Problem 9

10.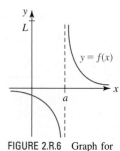

FIGURE 2.R.6 Graph for Problem 10

In Problems 11 and 12, sketch the graph of the given function. Determine the numbers, if any, at which f is discontinuous.

11. $f(x) = |x| + x$

12. $f(x) = \begin{cases} x + 1, & x < 2 \\ 3, & 2 < x < 4 \\ -x + 7, & x > 4 \end{cases}$

In Problems 13–16, determine intervals on which the given function is continuous.

13. $f(x) = \dfrac{x + 6}{x^3 - x}$

14. $f(x) = \dfrac{\sqrt{4 - x^2}}{x^2 - 4x + 3}$

15. $f(x) = \dfrac{x}{\sqrt{x^2 - 5}}$

16. $f(x) = \dfrac{\csc x}{\sqrt{x}}$

17. Find a number k so that

$$f(x) = \begin{cases} kx + 1, & x \le 3 \\ 2 - kx, & x > 3 \end{cases}$$

is continuous at the number 3.

18. Find numbers a and b so that

$$f(x) = \begin{cases} x + 4, & x \le 1 \\ ax + b, & 1 < x \le 3 \\ 3x - 8, & x > 3 \end{cases}$$

is continuous everywhere.

In Problems 19–22, find the slope of the tangent line to the graph of the function at the given value of x. Find an equation of the tangent line at the corresponding point.

19. $f(x) = -3x^2 + 16x + 12$, $x = 2$

20. $f(x) = x^3 - x^2$, $x = -1$

21. $f(x) = \dfrac{-1}{2x^2}$, $x = \dfrac{1}{2}$

22. $f(x) = x + 4\sqrt{x}$, $x = 4$

23. Find an equation of the line that is perpendicular to the tangent line at the point $(1, 2)$ on the graph of $f(x) = -4x^2 + 6x$.

24. Suppose $f(x) = 2x + 5$ and $\varepsilon = 0.01$. Find a $\delta > 0$ that will guarantee that $|f(x) - 7| < \varepsilon$ when $0 < |x - 1| < \delta$. What limit has been proved by finding δ?

Chapter 3

The Derivative

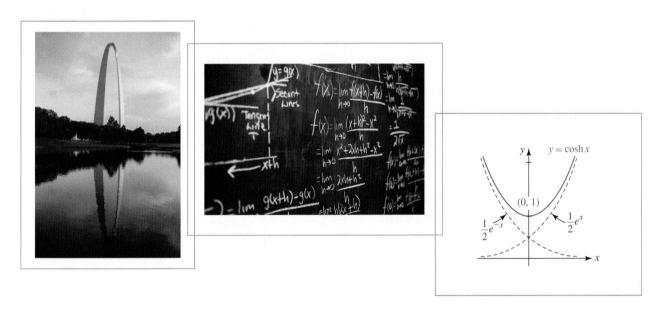

In This Chapter The word *calculus* is a diminutive form of the Latin word *calx*, which means "stone." In ancient civilizations small stones or pebbles were often used as a means of reckoning. Consequently, the word *calculus* can refer to any systematic method of computation. However, over the last several hundred years the connotation of the word *calculus* has evolved to mean that branch of mathematics concerned with the calculation and application of entities known as derivatives and integrals. Thus, the subject known as **calculus** has been divided into two rather broad but related areas: **differential calculus** and **integral calculus**.

In this chapter we will begin our study of differential calculus.

3.1 The Derivative
3.2 Power and Sum Rules
3.3 Product and Quotient Rules
3.4 Trigonometric Functions
3.5 Chain Rule
3.6 Implicit Differentiation
3.7 Derivatives of Inverse Functions
3.8 Exponential Functions
3.9 Logarithmic Functions
3.10 Hyperbolic Functions
Chapter 3 in Review

3.1 The Derivative

Introduction In the last section of Chapter 2 we saw that the tangent line to a graph of a function $y = f(x)$ is the line through a point $(a, f(a))$ with slope given by

$$m_{\tan} = \lim_{h \to 0} \frac{f(a + h) - f(a)}{h}$$

Recall, $m_{\tan}$ is also called the slope of the curve at $(a, f(a))$.

whenever the limit exists. For many functions it is usually possible to obtain a general formula that gives the value of the slope of a tangent line. This is accomplished by computing

$$\lim_{h \to 0} \frac{f(x + h) - f(x)}{h} \tag{1}$$

for *any* x (for which the limit exists). We then substitute a value of x *after* the limit has been found.

A Definition The limit of the difference quotient in (1) defines a function—a function that is *derived* from the original function $y = f(x)$. This new function is called the **derivative function**, or simply the **derivative**, of f and is denoted by f'.

Definition 3.1.1 Derivative

The **derivative** of a function $y = f(x)$ at x is given by

$$f'(x) = \lim_{h \to 0} \frac{f(x + h) - f(x)}{h} \tag{2}$$

whenever the limit exists.

Let us now reconsider Examples 1 and 2 of Section 2.7.

EXAMPLE 1 A Derivative

Find the derivative of $f(x) = x^2 + 2$.

Solution As in the calculation of $m_{\tan}$ in Section 2.7, the process of finding the derivative $f'(x)$ consists of four steps:

(i) $f(x + h) = (x + h)^2 + 2 = x^2 + 2xh + h^2 + 2$

(ii) $f(x + h) - f(x) = [x^2 + 2xh + h^2 + 2] - x^2 - 2 = h(2x + h)$

(iii) $\dfrac{f(x + h) - f(x)}{h} = \dfrac{h(2x + h)}{h} = 2x + h$ ← cancel h's

(iv) $\displaystyle\lim_{h \to 0} \frac{f(x + h) - f(x)}{h} = \lim_{h \to 0} [2x + h] = 2x$.

From step (*iv*) we see that the derivative of $f(x) = x^2 + 2$ is $f'(x) = 2x$. ∎

Observe that the result $m_{\tan} = 2$ obtained in Example 1 of Section 2.7 is obtained by evaluating the derivative $f'(x) = 2x$ at $x = 1$, that is, $f'(1) = 2$.

EXAMPLE 2 Value of the Derivative

For $f(x) = x^2 + 2$, find $f'(-2)$, $f'(0)$, $f'(\tfrac{1}{2})$, and $f'(1)$. Interpret.

Solution From Example 1 we know that the derivative is $f'(x) = 2x$. Hence,

at $x = -2$, $\begin{cases} f(-2) = 6 & \leftarrow \text{point of tangency is } (-2, 6) \\ f'(-2) = -4 & \leftarrow \text{slope of tangent line at } (-2, 6) \text{ is } m = -4 \end{cases}$

at $x = 0$, $\begin{cases} f(0) = 2 & \leftarrow \text{point of tangency is } (0, 2) \\ f'(0) = 0 & \leftarrow \text{slope of tangent line at } (0, 2) \text{ is } m = 0 \end{cases}$

at $x = \frac{1}{2}$, $\begin{cases} f(\frac{1}{2}) = \frac{9}{4} & \leftarrow \text{point of tangency is } (\frac{1}{2}, \frac{9}{4}) \\ f'(\frac{1}{2}) = 1 & \leftarrow \text{slope of tangent line at } (\frac{1}{2}, \frac{9}{4}) \text{ is } m = 1 \end{cases}$

at $x = 1$, $\begin{cases} f(1) = 3 & \leftarrow \text{point of tangency is } (1, 3) \\ f'(1) = 2. & \leftarrow \text{slope of tangent line at } (1, 3) \text{ is } m = 2 \end{cases}$

Recall that a horizontal line has 0 slope. So the fact that $f'(0) = 0$ means that the tangent line is horizontal at $(0, 2)$. ∎

By the way, if you trace back through the four-step process in Example 1, you will find that the derivative of $g(x) = x^2$ is also $g'(x) = 2x = f'(x)$. This makes intuitive sense; since the graph of $f(x) = x^2 + 2$ is a rigid vertical translation or shift of the graph of $g(x) = x^2$ for a given value of x, the points of tangency change but not the slope of the tangent line at the points. For example, at $x = 3$, $g'(3) = 6 = f'(3)$ but the points of tangency are $(3, g(3)) = (3, 9)$ and $(3, f(3)) = (3, 11)$.

EXAMPLE 3 A Derivative

Find the derivative of $f(x) = x^3$.

Solution To calculate $f(x + h)$, we use the Binomial Theorem.

(i) $f(x + h) = (x + h)^3 = x^3 + 3x^2h + 3xh^2 + h^3$

(ii) $f(x + h) - f(x) = [x^3 + 3x^2h + 3xh^2 + h^3] - x^3 = h(3x^2 + 3xh + h^2)$

(iii) $\dfrac{f(x + h) - f(x)}{h} = \dfrac{h[3x^2 + 3xh + h^2]}{h} = 3x^2 + 3xh + h^2$

(iv) $\displaystyle\lim_{h \to 0} \dfrac{f(x + h) - f(x)}{h} = \lim_{h \to 0}[3x^2 + 3xh + h^2] = 3x^2.$

◀ Recall from algebra that $(a + b)^3 = a^3 + 3a^2b + 3ab^2 + b^3$. Now replace a by x and b by h.

The derivative of $f(x) = x^3$ is $f'(x) = 3x^2$. ∎

EXAMPLE 4 Tangent Line

Find an equation of the tangent line to the graph of $f(x) = x^3$ at $x = \frac{1}{2}$.

Solution From Example 3 we have two functions $f(x) = x^3$ and $f'(x) = 3x^2$. As we saw in Example 2, when evaluated at the same number $x = \frac{1}{2}$ these functions give different information:

$f\left(\dfrac{1}{2}\right) = \left(\dfrac{1}{2}\right)^3 = \dfrac{1}{8}$ ← point of tangency is $(\frac{1}{2}, \frac{1}{8})$

$f'\left(\dfrac{1}{2}\right) = 3\left(\dfrac{1}{2}\right)^2 = \dfrac{3}{4}.$ ← slope of tangent at $(\frac{1}{2}, \frac{1}{8})$ is $\frac{3}{4}$

Thus, by the point–slope form of a line, an equation of the tangent line is given by

$$y - \dfrac{1}{8} = \dfrac{3}{4}\left(x - \dfrac{1}{2}\right) \quad \text{or} \quad y = \dfrac{3}{4}x - \dfrac{1}{4}.$$

The graph of the function and the tangent line are given in FIGURE 3.1.1. ∎

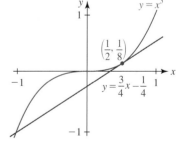

FIGURE 3.1.1 Tangent line in Example 4

EXAMPLE 5 A Derivative

Find the derivative of $f(x) = 1/x$.

Solution In this case you should be able to show that the difference is

$$f(x + h) - f(x) = \dfrac{1}{x + h} - \dfrac{1}{x} = \dfrac{-h}{(x + h)x}.$$ ← add fractions by using a common denominator

Therefore,

$$\lim_{h \to 0} \dfrac{f(x + h) - f(x)}{h} = \lim_{h \to 0} \dfrac{-h}{h(x + h)x}$$

$$= \lim_{h \to 0} \dfrac{-1}{(x + h)x} = \dfrac{-1}{x^2}.$$

The derivative of $f(x) = 1/x$ is $f'(x) = -1/x^2$. ∎

■ **Notation** The following is a list of some of the common **notations** used throughout mathematical literature to denote the derivative of a function:

$$f'(x), \quad \frac{dy}{dx}, \quad y', \quad Dy, \quad D_x y.$$

For a function such as $f(x) = x^2$, we write $f'(x) = 2x$; if the same function is written $y = x^2$, we then utilize $dy/dx = 2x$, $y' = 2x$, or $D_x y = 2x$. We will use the first three notations throughout this text. Of course other symbols are used in various applications. Thus, if $z = t^2$, then

$$\frac{dz}{dt} = 2t \quad \text{or} \quad z' = 2t.$$

The dy/dx notation has its origin in the derivative form of (3) of Section 2.7. Replacing h by Δx and denoting the difference $f(x + h) - f(x)$ by Δy in (2), the derivative is often defined as

$$\frac{dy}{dx} = \lim_{\Delta x \to 0} \frac{f(x + \Delta x) - f(x)}{\Delta x} = \lim_{\Delta x \to 0} \frac{\Delta y}{\Delta x}. \tag{3}$$

EXAMPLE 6 A Derivative Using (3)

Use (3) to find the derivative of $y = \sqrt{x}$.

Solution In the four-step procedure the important algebra takes place in the third step:

(i) $f(x + \Delta x) = \sqrt{x + \Delta x}$

(ii) $\Delta y = f(x + \Delta x) - f(x) = \sqrt{x + \Delta x} - \sqrt{x}$

(iii) $\dfrac{\Delta y}{\Delta x} = \dfrac{f(x + \Delta x) - f(x)}{\Delta x} = \dfrac{\sqrt{x + \Delta x} - \sqrt{x}}{\Delta x}$

$\qquad = \dfrac{\sqrt{x + \Delta x} - \sqrt{x}}{\Delta x} \cdot \dfrac{\sqrt{x + \Delta x} + \sqrt{x}}{\sqrt{x + \Delta x} + \sqrt{x}}$ ← rationalization of numerator

$\qquad = \dfrac{x + \Delta x - x}{\Delta x(\sqrt{x + \Delta x} + \sqrt{x})}$

$\qquad = \dfrac{\Delta x}{\Delta x(\sqrt{x + \Delta x} + \sqrt{x})}$

$\qquad = \dfrac{1}{\sqrt{x + \Delta x} + \sqrt{x}}$

(iv) $\lim_{\Delta x \to 0} \dfrac{\Delta y}{\Delta x} = \lim_{\Delta x \to 0} \dfrac{1}{\sqrt{x + \Delta x} + \sqrt{x}} = \dfrac{1}{\sqrt{x} + \sqrt{x}} = \dfrac{1}{2\sqrt{x}}.$

The derivative of $y = \sqrt{x}$ is $dy/dx = 1/(2\sqrt{x})$. ■

■ **Value of a Derivative** The **value** of the derivative at a number a is denoted by the symbols

$$f'(a), \quad \frac{dy}{dx}\bigg|_{x=a}, \quad y'(a), \quad D_x y \bigg|_{x=a}.$$

EXAMPLE 7 A Derivative

From Example 6, the value of the derivative of $y = \sqrt{x}$ at, say, $x = 9$ is written

$$\frac{dy}{dx}\bigg|_{x=9} = \frac{1}{2\sqrt{x}}\bigg|_{x=9} = \frac{1}{6}.$$

Alternatively, to avoid the clumsy vertical bar we can simply write $y'(9) = \frac{1}{6}$. ■

■ **Differentiation Operators** The process of finding or calculating a derivative is called **differentiation**. Thus differentiation is an operation that is performed on a function $y = f(x)$. The operation of differentiation of a function with respect to the variable x is represented by the symbols d/dx and D_x. These symbols are called **differentiation operators**. For instance, the results in Examples 1, 3, and 6 can be expressed, in turn, as

$$\frac{d}{dx}(x^2 + 2) = 2x, \quad \frac{d}{dx}x^3 = 3x^2, \quad \frac{d}{dx}\sqrt{x} = \frac{1}{2\sqrt{x}}.$$

The symbol

$$\frac{dy}{dx} \quad \text{then means} \quad \frac{d}{dx}y.$$

▌ **Differentiability** If the limit in (2) exists for a given number x in the domain of f, the function f is said to be **differentiable** at x. If a function f is differentiable at every number x in the open intervals (a, b), $(-\infty, b)$, and (a, ∞), then f is **differentiable on the open interval**. If f is differentiable on $(-\infty, \infty)$, then f is said to be **differentiable everywhere**. A function f is **differentiable on a closed interval** $[a, b]$ when f is differentiable on the open interval (a, b), and

$$\begin{aligned} f'_+(a) &= \lim_{h \to 0^+} \frac{f(a + h) - f(a)}{h} \\ f'_-(b) &= \lim_{h \to 0^-} \frac{f(b + h) - f(b)}{h} \end{aligned} \tag{4}$$

both exist. The limits in (4) are called **right-hand** and **left-hand derivatives**, respectively. A function is differentiable on $[a, \infty)$ when it is differentiable on (a, ∞) and has a right-hand derivative at a. A similar definition in terms of a left-hand derivative holds for differentiability on $(-\infty, b]$. Moreover, it can be shown:

- *A function f is differentiable at a number c in an interval (a, b) if and only if $f'_+(c) = f'_-(c)$.* (5)

▌ **Horizontal Tangents** If $y = f(x)$ is continuous at a number a and $f'(a) = 0$, then the tangent line at $(a, f(a))$ is **horizontal**. In Examples 1 and 2 we saw that the value of derivative $f'(x) = 2x$ of the function $f(x) = x^2 + 2$ at $x = 0$ is $f'(0) = 0$. Thus, the tangent line to the graph is horizontal at $(0, f(0))$ or $(0, 0)$. It is left as an exercise (see Problem 7 in Exercises 3.1) to verify by Definition 3.1.1 that the derivative of the continuous function $f(x) = -x^2 + 4x + 1$ is $f'(x) = -2x + 4$. Observe in this latter case that $f'(x) = 0$ when $-2x + 4 = 0$ or $x = 2$. There is a horizontal tangent at the point $(2, f(2)) = (2, 5)$.

▌ **Where f Fails to be Differentiable** A function f fails to have a derivative at $x = a$ if

(*i*) the function f is discontinuous at $x = a$, or
(*ii*) the graph of f has a corner at $(a, f(a))$.

In addition, since the derivative gives slope, f will fail to be differentiable

(*iii*) at a point $(a, f(a))$ at which the tangent line to the graph is vertical.

The domain of the derivative f', defined by (2), is the set of numbers x for which the limit exists. Thus the domain of f' is necessarily a subset of the domain of f.

EXAMPLE 8 Differentiability

(a) The function $f(x) = x^2 + 2$ is differentiable for all real numbers x, that is, the domain of $f'(x) = 2x$ is $(-\infty, \infty)$.
(b) Because $f(x) = 1/x$ is discontinuous at $x = 0$, f is not differentiable at $x = 0$ and consequently not differentiable on any interval containing 0. ∎

(a) Absolute-value function f

(b) Graph of the derivative f'

FIGURE 3.1.2 Graphs of f and f' in Example 9

EXAMPLE 9 Example 7 of Section 2.7 Revisited

In Example 7 of Section 2.7 we saw that the graph of $f(x) = |x|$ possesses no tangent at the origin $(0, 0)$. Thus $f(x) = |x|$ is not differentiable at $x = 0$. But $f(x) = |x|$ is differentiable on the open intervals $(0, \infty)$ and $(-\infty, 0)$. In Example 5 of Section 2.7, we proved that the derivative of a linear function $f(x) = mx + b$ is $f'(x) = m$. Hence, for $x > 0$ we have $f(x) = |x| = x$ and so $f'(x) = 1$. Also, for $x < 0$, $f(x) = |x| = -x$ and so $f'(x) = -1$. Since the derivative of f is a piecewise-defined function,

$$f'(x) = \begin{cases} 1, & x > 0 \\ -1, & x < 0, \end{cases}$$

we can graph it as we would any function. We see in FIGURE 3.1.2(b) that f' is discontinuous at $x = 0$. ∎

In different symbols, what we have shown in Example 9 is that $f'_-(0) = -1$ and $f'_+(0) = 1$. Since $f'_-(0) \neq f'_+(0)$ it follows from (5) that f is not differentiable at 0.

■ **Vertical Tangents** Let $y = f(x)$ be continuous at a number a. If $\lim_{x \to a} |f'(x)| = \infty$, then the graph of f is said to have a **vertical tangent** at $(a, f(a))$. The graphs of many functions with rational exponents possess vertical tangents.

In Example 6 of Section 2.7 we mentioned that the graph of $y = x^{1/3}$ possesses a vertical tangent line at $(0, 0)$. We verify this assertion in the next example.

EXAMPLE 10 Vertical Tangent

It is left as an exercise to prove that the derivative of $f(x) = x^{1/3}$ is given by

$$f'(x) = \frac{1}{3x^{2/3}}.$$

(See Problem 55 in Exercises 3.1.) Although f is continuous at 0, it is clear that f' is not defined at that number. In other words, f is not differentiable at $x = 0$. Moreover, because

$$\lim_{x \to 0^+} f'(x) = \infty \quad \text{and} \quad \lim_{x \to 0^-} f'(x) = \infty$$

we have $|f'(x)| \to \infty$ as $x \to 0$. This is sufficient to say that there is a tangent line at $(0, f(0))$ or $(0, 0)$ and that it is vertical. FIGURE 3.1.3 shows that the tangent lines to the graph on either side of the origin become steeper and steeper as $x \to 0$. ∎

FIGURE 3.1.3 Tangent lines to the graph of the function in Example 10

The graph of a function f can also have a vertical tangent at a point $(a, f(a))$ if f is differentiable only on one side of a, is continuous from the left (right) at a, and either $|f'(x)| \to \infty$ as $x \to a^-$ or $|f'(x)| \to \infty$ as $x \to a^+$.

EXAMPLE 11 One-Sided Vertical Tangent

The function $f(x) = \sqrt{x}$ is not differentiable on the interval $[0, \infty)$ because it is seen from the derivative $f'(x) = 1/(2\sqrt{x})$ that $f'_+(0)$ does not exist. The function $f(x) = \sqrt{x}$ is continuous on $[0, \infty)$ but differentiable on $(0, \infty)$. In addition, because f is continuous at 0 and $\lim_{x \to 0^+} f'(x) = \infty$, there is a one-sided vertical tangent at the origin $(0, 0)$. We see in FIGURE 3.1.4 that the vertical tangent is the y-axis. ∎

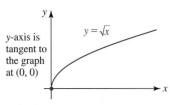

y-axis is tangent to the graph at $(0, 0)$

FIGURE 3.1.4 Vertical tangent in Example 11

Important ▶ The functions $f(x) = |x|$ and $f(x) = x^{1/3}$ are continuous everywhere. In particular, both functions are continuous at 0 but neither are differentiable at that number. In other words, continuity at a number a is not sufficient to guarantee that a function is differentiable at a. However, if a function f is differentiable at a, then f must be continuous at that number. We summarize this last fact in the next theorem.

Theorem 3.1.1 Differentiability Implies Continuity

If f is differentiable at a number a, then f is continuous at a.

PROOF To prove continuity of f at a number a it is sufficient to prove that $\lim_{x \to a} f(x) = f(a)$ or equivalently that $\lim_{x \to a} [f(x) - f(a)] = 0$. The hypothesis is that

$$f'(a) = \lim_{h \to 0} \frac{f(a+h) - f(a)}{h}$$

exists. If we let $x = a + h$, then as $h \to 0$ we have $x \to a$. Thus, the foregoing limit is equivalent to

$$f'(a) = \lim_{x \to a} \frac{f(x) - f(a)}{x - a}.$$

Then we can write

$$\lim_{x \to a} [f(x) - f(a)] = \lim_{x \to a} \frac{f(x) - f(a)}{x - a} \cdot (x - a) \quad \leftarrow \text{multiplication by } \frac{x-a}{x-a} = 1$$

$$= \lim_{x \to a} \frac{f(x) - f(a)}{x - a} \cdot \lim_{x \to a} (x - a) \quad \leftarrow \text{both limits exist}$$

$$= f'(a) \cdot 0 = 0. \qquad \blacksquare$$

■ **Postscript—A Bit of History** It is acknowledged that **Isaac Newton** (1642–1727), an English mathematician and physicist, was the first to set forth many of the basic principles of calculus in unpublished manuscripts on the *method of fluxions*, dated 1665. The word *fluxion* originated from the concept of quantities that "flow"—that is, quantities that change at a certain rate. Newton used the dot notation $\dot{y}$ to represent a fluxion, or as we now know it: the derivative of a function. The symbol $\dot{y}$ never achieved overwhelming popularity among mathematicians and is used today primarily by physicists. For typographical reasons, the so-called "fly-speck notation" has been superseded by the prime notation. Newton attained everlasting fame with the publication of his law of universal gravitation in his monumental treatise *Philosophiae Naturalis Principia Mathematica* in 1687. Newton was also the first to prove, using the calculus and his law of gravitation, Johannes Kepler's three empirical laws of planetary motion and was the first to prove that white light is composed of all colors. Newton was elected to Parliament, was appointed warden of the Royal Mint, and was knighted in 1705. Sir Isaac Newton said about his many accomplishments: "If I have seen farther than others, it is by standing on the shoulders of giants."

Newton

Leibniz

The German mathematician, lawyer, and philosopher **Gottfried Wilhelm Leibniz** (1646–1716) published a short version of his calculus in an article in a periodical journal in 1684. The dy/dx notation for a derivative of a function is due to Leibniz. In fact, it was Leibniz who introduced the word *function* into mathematical literature. But, since it was well known that Newton's manuscripts on the *method of fluxions* dated from 1665, Leibniz was accused of appropriating his ideas from these unpublished works. Fueled by nationalistic prides, a controversy about who was the first to "invent" calculus raged for many years. Historians now agree that both Leibniz and Newton arrived at many of the major premises of calculus independent of each other. Leibniz and Newton are considered the "co-inventors" of the subject.

$\dfrac{d}{dx}$ **NOTES FROM THE CLASSROOM**

(*i*) In the preceding discussion, we saw that the derivative of a function is itself a function that gives the slope of a tangent line. The derivative is, however, *not* an equation of a tangent line. Also, to say that $y - y_0 = f'(x) \cdot (x - x_0)$ is an equation of the tangent at (x_0, y_0) is incorrect. Remember that $f'(x)$ must be evaluated at x_0 *before* it is used in the point–slope form. If f is differentiable at x_0, then an equation of the tangent line at (x_0, y_0) is $y - y_0 = f'(x_0) \cdot (x - x_0)$.

(ii) Although we have emphasized slopes in this section, do not forget the discussion on average rates of change and instantaneous rates of change in Section 2.7. The derivative $f'(x)$ is also the **instantaneous rate of change** of the function $y = f(x)$ with respect to the variable x. More will be said about rates in subsequent sections.

(iii) Mathematicians from the seventeenth to the nineteenth centuries believed that a continuous function *usually* possessed a derivative. (We have noted exceptions in this section.) In 1872 the German mathematician Karl Weierstrass conclusively destroyed this tenet by publishing an example of a function that was *everywhere continuous but nowhere differentiable*.

Exercises 3.1
Answers to selected odd-numbered problems begin on page ANS-10.

≡ Fundamentals

In Problems 1–20, use (2) of Definition 3.1.1 to find the derivative of the given function.

1. $f(x) = 10$
2. $f(x) = x - 1$
3. $f(x) = -3x + 5$
4. $f(x) = \pi x$
5. $f(x) = 3x^2$
6. $f(x) = -x^2 + 1$
7. $f(x) = -x^2 + 4x + 1$
8. $f(x) = \frac{1}{2}x^2 + 6x - 7$
9. $y = (x + 1)^2$
10. $f(x) = (2x - 5)^2$
11. $f(x) = x^3 + x$
12. $f(x) = 2x^3 + x^2$
13. $y = -x^3 + 15x^2 - x$
14. $y = 3x^4$
15. $y = \dfrac{2}{x + 1}$
16. $y = \dfrac{x}{x - 1}$
17. $y = \dfrac{2x + 3}{x + 4}$
18. $f(x) = \dfrac{1}{x} + \dfrac{1}{x^2}$
19. $f(x) = \dfrac{1}{\sqrt{x}}$
20. $f(x) = \sqrt{2x + 1}$

In Problems 21–24, use (2) of Definition 3.1.1 to find the derivative of the given function. Find an equation of the tangent line to the graph of the function at the indicated value of x.

21. $f(x) = 4x^2 + 7x; \quad x = -1$
22. $f(x) = \frac{1}{3}x^3 + 2x - 4; \quad x = 0$
23. $y = x - \dfrac{1}{x}; \quad x = 1$
24. $y = 2x + 1 + \dfrac{6}{x}; \quad x = 2$

In Problems 25–28, use (2) of Definition 3.1.1 to find the derivative of the given function. Find point(s) on the graph of the given function where the tangent line is horizontal.

25. $f(x) = x^2 + 8x + 10$
26. $f(x) = x(x - 5)$
27. $f(x) = x^3 - 3x$
28. $f(x) = x^3 - x^2 + 1$

In Problems 29–32, use (2) of Definition 3.1.1 to find the derivative of the given function. Find point(s) on the graph of the given function where the tangent line is parallel to the given line.

29. $f(x) = \frac{1}{2}x^2 - 1; \quad 3x - y = 1$
30. $f(x) = x^2 - x; \quad -2x + y = 0$
31. $f(x) = -x^3 + 4; \quad 12x + y = 4$
32. $f(x) = 6\sqrt{x} + 2; \quad -x + y = 2$

In Problems 33 and 34, show that the given function is not differentiable at the indicated value of x.

33. $f(x) = \begin{cases} -x + 2, & x \leq 2 \\ 2x - 4, & x > 2 \end{cases}; \quad x = 2$

34. $f(x) = \begin{cases} 3x, & x < 0 \\ -4x, & x \geq 0 \end{cases}; \quad x = 0$

In the proof of Theorem 3.1.1 we saw that an alternative formulation of the derivative of a function f at a is given by

$$f'(a) = \lim_{x \to a} \frac{f(x) - f(a)}{x - a}, \qquad (6)$$

whenever the limit exists. In Problems 35–40, use (6) to compute $f'(a)$.

35. $f(x) = 10x^2 - 3$
36. $f(x) = x^2 - 3x - 1$
37. $f(x) = x^3 - 4x^2$
38. $f(x) = x^4$
39. $f(x) = \dfrac{4}{3 - x}$
40. $f(x) = \sqrt{x}$

41. Find an equation of the tangent line shown in red in FIGURE 3.1.5. What are $f(-3)$ and $f'(-3)$?

FIGURE 3.1.5 Graph for Problem 41

42. Find an equation of the tangent line shown in red in FIGURE 3.1.6. What is $f'(3)$? What is the y-intercept of the tangent line?

FIGURE 3.1.6 Graph for Problem 42

In Problems 43–48, sketch the graph of f' from the graph of f.

43.

FIGURE 3.1.7 Graph for Problem 43

44.

FIGURE 3.1.8 Graph for Problem 44

45.

FIGURE 3.1.9 Graph for Problem 45

46.

FIGURE 3.1.10 Graph for Problem 46

47.

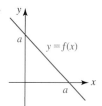

FIGURE 3.1.11 Graph for Problem 47

48.

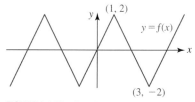

FIGURE 3.1.12 Graph for Problem 48

In Problems 49–54, match the graph of f with a graph of f' from (a)–(f).

(c)   **(d)**

(e)   **(f)**

49.

FIGURE 3.1.13 Graph for Problem 49

50.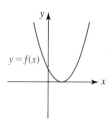

FIGURE 3.1.14 Graph for Problem 50

51.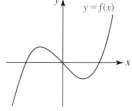

FIGURE 3.1.15 Graph for Problem 51

52.

FIGURE 3.1.16 Graph for Problem 52

53.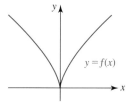

FIGURE 3.1.17 Graph for Problem 53

54.

FIGURE 3.1.18 Graph for Problem 54

≡ **Think About It**

55. Use the alternative definition of the derivative (6) to find the derivative of $f(x) = x^{1/3}$.

[*Hint*: Note that $x - a = (x^{1/3})^3 - (a^{1/3})^3$.]

56. In Examples 10 and 11, we saw, respectively, that the functions $f(x) = x^{1/3}$ and $f(x) = \sqrt{x}$ possessed vertical tangents at the origin $(0, 0)$. Conjecture where the graphs of $y = (x - 4)^{1/3}$ and $y = \sqrt{x + 2}$ may have vertical tangents.

57. Suppose f is differentiable everywhere and has the three properties:

(*i*) $f(x_1 + x_2) = f(x_1)f(x_2)$, (*ii*) $f(0) = 1$, (*iii*) $f'(0) = 1$.

Use (2) of Definition 3.1.1 to show that $f'(x) = f(x)$ for all x.

58. (a) Suppose f is an even differentiable function on $(-\infty, \infty)$. On geometric grounds, explain why $f'(-x) = -f'(x)$; that is, f' is an odd function.
(b) Suppose f is an odd differentiable function on $(-\infty, \infty)$. On geometric grounds, explain why $f'(-x) = f'(x)$; that is, f' is an even function.

59. Suppose f is a differentiable function on $[a, b]$ such that $f(a) = 0$ and $f(b) = 0$. By experimenting with graphs discern whether the following statement is true or false: There is some number c in (a, b) such that $f'(c) = 0$.

60. Sketch graphs of various functions f that have the property $f'(x) > 0$ for all x in $[a, b]$. What do these functions have in common?

≡ **Calculator/CAS Problem**

61. Consider the function $f(x) = x^n + |x|$, where n is a positive integer. Use a calculator or CAS to obtain the graph of f for $n = 1, 2, 3, 4$, and 5. Then use (2) to show that f is not differentiable at $x = 0$ for $n = 1, 2, 3, 4$, and 5. Can you prove this for *any* positive integer n? What is $f'_-(0)$ and $f'_+(0)$ for $n > 1$?

3.2 Power and Sum Rules

Introduction The definition of a derivative

$$f'(x) = \lim_{h \to 0} \frac{f(x + h) - f(x)}{h} \tag{1}$$

has the obvious drawback of being rather clumsy and tiresome to apply. To find the derivative of the polynomial function $f(x) = 6x^{100} + 4x^{35}$ using the above definition we would *only* have to juggle 137 terms in the binomial expansions of $(x + h)^{100}$ and $(x + h)^{35}$. There are more efficient ways of computing derivatives of a function than using the definition each time. In this section, and the sections that follow, we will see that there are shortcuts or general **rules** whereby derivatives of functions such as $f(x) = 6x^{100} + 4x^{35}$ can be obtained, literally, with just a flick of a pencil.

In the last section we saw that the derivatives of the power functions

$$f(x) = x^2, \quad f(x) = x^3, \quad f(x) = \frac{1}{x} = x^{-1}, \quad f(x) = \sqrt{x} = x^{1/2}$$

were, in turn,

See Examples 3, 5, and 6 in Section 3.1.

$$f'(x) = 2x, \quad f'(x) = 3x^2, \quad f'(x) = -\frac{1}{x^2} = -x^{-2}, \quad f'(x) = \frac{1}{2\sqrt{x}} = \frac{1}{2}x^{-1/2}.$$

If the right-hand sides of these four derivatives are written

$$2 \cdot x^{2-1}, \quad 3 \cdot x^{3-1}, \quad (-1) \cdot x^{-1-1}, \quad \frac{1}{2} \cdot x^{\frac{1}{2}-1},$$

we observe that each coefficient (indicated in red) corresponds with the original exponent of x in f and the new exponent of x in f' can be obtained from the old exponent (also indicated in red) by subtracting 1 from it. In other words, the pattern for the derivative of the general power function $f(x) = x^n$ appears to be

$$\underset{\text{decrease exponent by 1}}{\overset{\text{bring down exponent as a multiple}}{(\downarrow) x^{(\uparrow) - 1}}}. \tag{2}$$

Derivative of the Power Function The pattern illustrated in (2) does indeed hold for any real-number exponent n, and we will state it as a formal theorem, but at this point in the course we do not possess the necessary mathematical tools to prove its complete validity. We can, however, readily prove a special case of this power rule; the remaining parts of the proof will be given in the appropriate sections ahead.

Theorem 3.2.1 Power Rule

For any real number n,

$$\frac{d}{dx}x^n = nx^{n-1}. \qquad (3)$$

PROOF We present the proof only in the case when n is a positive integer. To compute (1) for $f(x) = x^n$ we use the four-step method:

(i) $f(x + h) = (x + h)^n = \overbrace{x^n + nx^{n-1}h + \frac{n(n-1)}{2!}x^{n-2}h^2 + \cdots + nxh^{n-1} + h^n}^{\text{general Binomial Theorem}}$

◀ See the *Resource Pages* for a review of the Binomial Theorem.

(ii) $f(x + h) - f(x) = x^n + nx^{n-1}h + \frac{n(n-1)}{2!}x^{n-2}h^2 + \cdots + nxh^{n-1} + h^n - x^n$

$= nx^{n-1}h + \frac{n(n-1)}{2!}x^{n-2}h^2 + \cdots + nxh^{n-1} + h^n$

$= h\left[nx^{n-1} + \frac{n(n-1)}{2!}x^{n-1}h + \cdots + nxh^{n-2} + h^{n-1}\right]$

(iii) $\dfrac{f(x + h) - f(x)}{h} = \dfrac{h\left[nx^{n-1} + \frac{n(n-1)}{2!}x^{n-1}h + \cdots + nxh^{n-2} + h^{n-1}\right]}{h}$

$= nx^{n-1} + \frac{n(n-1)}{2!}x^{n-1}h + \cdots + nxh^{n-2} + h^{n-1}$

(iv) $f'(x) = \lim_{h \to 0} \dfrac{f(x + h) - f(x)}{h}$

$= \lim_{h \to 0}\left[nx^{n-1} + \underbrace{\frac{n(n-1)}{2!}x^{n-1}h + \cdots + nxh^{n-2} + h^{n-1}}_{\text{these terms} \to 0 \text{ as } h \to 0}\right] = nx^{n-1}.$ ∎

EXAMPLE 1 Power Rule

Differentiate

(a) $y = x^7$ (b) $y = x$ (c) $y = x^{-2/3}$ (d) $y = x^{\sqrt{2}}$.

Solution By the Power Rule (3),

(a) with $n = 7$: $\dfrac{dy}{dx} = 7x^{7-1} = 7x^6$,

(b) with $n = 1$: $\dfrac{dy}{dx} = 1x^{1-1} = x^0 = 1$,

(c) with $n = -\dfrac{2}{3}$: $\dfrac{dy}{dx} = \left(-\dfrac{2}{3}\right)x^{(-2/3)-1} = -\dfrac{2}{3}x^{-5/3} = -\dfrac{2}{3x^{5/3}}$,

(d) with $n = \sqrt{2}$: $\dfrac{dy}{dx} = \sqrt{2}x^{\sqrt{2}-1}$. ∎

Observe in part (b) of Example 1 that the result is consistent with the fact that the slope of the line $y = x$ is $m = 1$. See FIGURE 3.2.1.

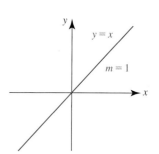

FIGURE 3.2.1 Slope of line $m = 1$ is consistent with $dy/dx = 1$

Theorem 3.2.2 Constant Function Rule

If $f(x) = c$ is a constant function, then $f'(x) = 0$.   (4)

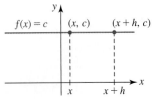

FIGURE 3.2.2 Slope of a horizontal line is 0

PROOF If $f(x) = c$ where c is any real number, then it follows that the difference is $f(x + h) - f(x) = c - c = 0$. Hence from (1),

$$f'(x) = \lim_{h \to 0} \frac{c - c}{h} = \lim_{h \to 0} 0 = 0.$$

Theorem 3.2.2 has an obvious geometric interpretation. As shown in FIGURE 3.2.2, the slope of the horizontal line $y = c$ is, of course, zero. Moreover, Theorem 3.2.2 agrees with (3) in the case when $x \neq 0$ and $n = 0$.

Theorem 3.2.3 Constant Multiple Rule

If c is any constant and f is differentiable at x, then cf is differentiable at x, and

$$\frac{d}{dx} cf(x) = cf'(x). \tag{5}$$

PROOF Let $G(x) = cf(x)$. Then

$$G'(x) = \lim_{h \to 0} \frac{G(x + h) - G(x)}{h} = \lim_{h \to 0} \frac{cf(x + h) - cf(x)}{h}$$

$$= \lim_{h \to 0} c \left[\frac{f(x + h) - f(x)}{h} \right]$$

$$= c \lim_{h \to 0} \frac{f(x + h) - f(x)}{h} = cf'(x). \quad \blacksquare$$

EXAMPLE 2 A Constant Multiple

Differentiate $y = 5x^4$.

Solution From (3) and (5),

$$\frac{dy}{dx} = 5\frac{d}{dx}x^4 = 5(4x^3) = 20x^3.$$

Theorem 3.2.4 Sum and Difference Rules

If f and g are functions differentiable at x, then $f + g$ and $f - g$ are differentiable at x, and

$$\frac{d}{dx}[f(x) + g(x)] = f'(x) + g'(x), \tag{6}$$

$$\frac{d}{dx}[f(x) - g(x)] = f'(x) - g'(x). \tag{7}$$

PROOF OF (6) Let $G(x) = f(x) + g(x)$. Then

$$G'(x) = \lim_{h \to 0} \frac{G(x + h) - G(x)}{h} = \lim_{h \to 0} \frac{[f(x + h) + g(x + h)] - [f(x) + g(x)]}{h}$$

$$= \lim_{h \to 0} \frac{f(x + h) - f(x) + g(x + h) - g(x)}{h} \quad \leftarrow \text{regrouping terms}$$

$$\text{since limits exist,} \atop \text{limit of a sum is} \to \atop \text{the sum of the limits} \quad = \lim_{h \to 0} \frac{f(x + h) - f(x)}{h} + \lim_{h \to 0} \frac{g(x + h) - g(x)}{h}$$

$$= f'(x) + g'(x). \quad \blacksquare$$

Theorem 3.2.4 holds for any finite sum of differentiable functions. For example, if f, g, and h are functions that are differentiable at x, then

$$\frac{d}{dx}[f(x) + g(x) + h(x)] = f'(x) + g'(x) + h'(x).$$

Since $f - g$ can be written as a sum, $f + (-g)$, there is no need to prove (7) since the result follows from a combination of (6) and (5). Hence, we can express Theorem 3.2.4 in words as:

- *The derivative of a sum is the sum of the derivatives.*

■ **Derivative of a Polynomial** Because we now know how to differentiate powers of x and constant multiples of those powers we can easily differentiate sums of those constant multiples. The derivative of a polynomial function is particularly easy to obtain. For example, the derivative of the polynomial function $f(x) = 6x^{100} + 4x^{35}$, mentioned in the introduction to this section, is now readily seen to be $f'(x) = 600x^{99} + 140x^{34}$.

EXAMPLE 3 Polynomial with Six Terms

Differentiate $y = 4x^5 - \frac{1}{2}x^4 + 9x^3 + 10x^2 - 13x + 6$.

Solution Using (3), (5), and (6), we obtain

$$\frac{dy}{dx} = 4\frac{d}{dx}x^5 - \frac{1}{2}\frac{d}{dx}x^4 + 9\frac{d}{dx}x^3 + 10\frac{d}{dx}x^2 - 13\frac{d}{dx}x + \frac{d}{dx}6.$$

Since $\frac{d}{dx}6 = 0$ by (4), we obtain

$$\frac{dy}{dx} = 4(5x^4) - \frac{1}{2}(4x^3) + 9(3x^2) + 10(2x) - 13(1) + 0$$

$$= 20x^4 - 2x^3 + 27x^2 + 20x - 13. \qquad \blacksquare$$

EXAMPLE 4 Tangent Line

Find an equation of a tangent line to the graph of $f(x) = 3x^4 + 2x^3 - 7x$ at the point corresponding to $x = -1$.

Solution From the Sum Rule,

$$f'(x) = 3(4x^3) + 2(3x^2) - 7(1) = 12x^3 + 6x^2 - 7.$$

When evaluated at the same number $x = -1$ the functions f and f' give:

$f(-1) = 8$ ← point of tangency is $(-1, 8)$
$f'(-1) = -13.$ ← slope of tangent at $(-1, 8)$ is -13

The point–slope form gives an equation of the tangent line

$$y - 8 = -13(x - (-1)) \quad \text{or} \quad y = -13x - 5. \qquad \blacksquare$$

■ **Rewriting a Function** In some circumstances, in order to apply a rule of differentiation efficiently it may be necessary to *rewrite* an expression in an alternative form. This alternative form is often the result of some algebraic manipulation or an application of the laws of exponents. For example, we can use (3) to differentiate the following expressions, but first we rewrite them using the laws of exponents ◄ This discussion is worth remembering.

$\boxed{\dfrac{4}{x^2}, \dfrac{10}{\sqrt{x}}, \sqrt{x^3}}$ → rewrite square roots as powers → $\dfrac{4}{x^2}, \dfrac{10}{x^{1/2}}, (x^3)^{1/2},$

then rewrite using negative exponents → $4x^{-2}, 10x^{-1/2}, x^{3/2},$

the derivative of each term using (3) → $\boxed{-8x^{-3}, -5x^{-3/2}, \dfrac{3}{2}x^{1/2}}.$

A function such as $f(x) = (5x + 2)/x^2$ can be rewritten as two fractions

$$f(x) = \frac{5x+2}{x^2} = \frac{5x}{x^2} + \frac{2}{x^2} = \frac{5}{x} + \frac{2}{x^2} = 5x^{-1} + 2x^{-2}.$$

From the last form of f it is now apparent that the derivative f' is

$$f'(x) = 5(-x^{-2}) + 2(-2x^{-3}) = -\frac{5}{x^2} - \frac{4}{x^3}.$$

EXAMPLE 5 Rewriting the Terms of a Function

Differentiate $y = 4\sqrt{x} + \dfrac{8}{x} - \dfrac{6}{\sqrt[3]{x}} + 10$.

Solution Before differentiating we rewrite the first three terms as powers of x:

$$y = 4x^{1/2} + 8x^{-1} - 6x^{-1/3} + 10.$$

Then

$$\frac{dy}{dx} = 4\frac{d}{dx}x^{1/2} + 8\frac{d}{dx}x^{-1} - 6\frac{d}{dx}x^{-1/3} + \frac{d}{dx}10.$$

By the Power Rule (3) and (4), we obtain

$$\frac{dy}{dx} = 4 \cdot \frac{1}{2}x^{-1/2} + 8 \cdot (-1)x^{-2} - 6 \cdot \left(-\frac{1}{3}\right)x^{-4/3} + 0$$

$$= \frac{2}{\sqrt{x}} - \frac{8}{x^2} + \frac{2}{x^{4/3}}.$$

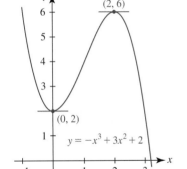

FIGURE 3.2.3 Graph of function in Example 6

EXAMPLE 6 Horizontal Tangents

Find the points on the graph of $f(x) = -x^3 + 3x^2 + 2$ where the tangent line is horizontal.

Solution At a point $(x, f(x))$ on the graph of f where the tangent is horizontal we must have $f'(x) = 0$. The derivative of f is $f'(x) = -3x^2 + 6x$ and the solutions of $f'(x) = -3x^2 + 6x = 0$ or $-3x(x - 2) = 0$ are $x = 0$ and $x = 2$. The corresponding points are then $(0, f(0)) = (0, 2)$ and $(2, f(2)) = (2, 6)$. See FIGURE 3.2.3.

Normal Line A **normal line** at a point P on a graph is one that is perpendicular to the tangent line at P.

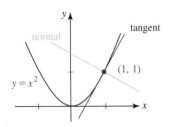

FIGURE 3.2.4 Normal line in Example 7

EXAMPLE 7 Equation of a Normal Line

Find an equation of the normal line to the graph of $y = x^2$ at $x = 1$.

Solution Since $dy/dx = 2x$, we know that $m_{\tan} = 2$ at $(1, 1)$. Thus the slope of the normal line shown in green in FIGURE 3.2.4 is the negative reciprocal of the slope of the tangent line, that is, $m = -\frac{1}{2}$. By the point-slope form of a line, an equation of the normal line is then

$$y - 1 = -\frac{1}{2}(x - 1) \quad \text{or} \quad y = -\frac{1}{2}x + \frac{3}{2}.$$

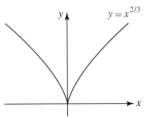

FIGURE 3.2.5 Graph of function in Example 8

EXAMPLE 8 Vertical Tangent

For the power function $f(x) = x^{2/3}$ the derivative is

$$f'(x) = \frac{2}{3}x^{-1/3} = \frac{2}{3x^{1/3}}.$$

Observe that $\lim_{x \to 0^+} f(x) = \infty$ whereas $\lim_{x \to 0^-} f(x) = -\infty$. Since f is continuous at $x = 0$ and $|f'(x)| \to \infty$ as $x \to 0$, we conclude that the y-axis is a vertical tangent at $(0, 0)$. This fact is apparent from the graph in FIGURE 3.2.5.

■ **Cusp** The graph of $f(x) = x^{2/3}$ in Example 8 is said to have a **cusp** at the origin. In general, the graph of a function $y = f(x)$ has a cusp at a point $(a, f(a))$ if f is continuous at a, $f'(x)$ has opposite signs on either side of a, and $|f'(x)| \to \infty$ as $x \to a$.

■ **Higher-Order Derivatives** We have seen that the derivative $f'(x)$ is a function derived from $y = f(x)$. By differentiation of the first derivative, we obtain yet another function called the **second derivative**, which is denoted by $f''(x)$. In terms of the operation symbol d/dx, we define the second derivative with respect to x as the function obtained by differentiating $y = f(x)$ twice in succession:

$$\frac{d}{dx}\left(\frac{dy}{dx}\right).$$

The second derivative is commonly denoted by the symbols

$$f''(x), \quad y'', \quad \frac{d^2y}{dx^2}, \quad \frac{d^2}{dx^2}f(x), \quad D^2, \quad D_x^2.$$

EXAMPLE 9 Second Derivative

Find the second derivative of $y = \dfrac{1}{x^3}$.

Solution We first simplify the function by rewriting it as $y = x^{-3}$. Then by the Power Rule (3), we have

$$\frac{dy}{dx} = -3x^{-4}.$$

The second derivative follows from differentiating the first derivative

$$\frac{d^2y}{dx^2} = \frac{d}{dx}(-3x^{-4}) = -3(-4x^{-5}) = 12x^{-5} = \frac{12}{x^5}. \qquad \blacksquare$$

Assuming that all derivatives exist, we can differentiate a function $y = f(x)$ as many times as we want. The **third derivative** is the derivative of the second derivative; the **fourth derivative** is the derivative of the third derivative; and so on. We denote the third and fourth derivatives by d^3y/dx^3 and d^4y/dx^4 and define them by

$$\frac{d^3y}{dx^3} = \frac{d}{dx}\left(\frac{d^2y}{dx^2}\right) \qquad \text{and} \qquad \frac{d^4y}{dx^4} = \frac{d}{dx}\left(\frac{d^3y}{dx^3}\right).$$

In general, if n is a positive integer, then the **nth derivative** is defined by

$$\frac{d^ny}{dx^n} = \frac{d}{dx}\left(\frac{d^{n-1}y}{dx^{n-1}}\right).$$

Other notations for the first n derivatives are

$$f'(x), \quad f''(x), \quad f'''(x), \quad f^{(4)}(x), \quad \ldots, \quad f^{(n)}(x),$$
$$y', \quad y'', \quad y''', \quad y^{(4)}, \quad \ldots, \quad y^{(n)},$$
$$\frac{d}{dx}f(x), \quad \frac{d^2}{dx^2}f(x), \quad \frac{d^3}{dx^3}f(x), \quad \frac{d^4}{dx^4}f(x), \quad \ldots, \quad \frac{d^n}{dx^n}f(x),$$
$$D, \quad D^2, \quad D^3, \quad D^4, \quad \ldots, \quad D^n,$$
$$D_x, \quad D_x^2, \quad D_x^3, \quad D_x^4, \quad \ldots, \quad D_x^n.$$

Note that the "prime" notation is used to denote only the first three derivatives; after that we use the superscript $y^{(4)}$, $y^{(5)}$, and so on. The **value of the nth derivative** of a function $y = f(x)$ at a number a is denoted by

$$f^{(n)}(a), \quad y^{(n)}(a), \quad \text{and} \quad \left.\frac{d^ny}{dx^n}\right|_{x=a}.$$

EXAMPLE 10 Fifth Derivative

Find the first five derivatives of $f(x) = 2x^4 - 6x^3 + 7x^2 + 5x$.

Solution We have

$$f'(x) = 8x^3 - 18x^2 + 14x + 5$$
$$f''(x) = 24x^2 - 36x + 14$$
$$f'''(x) = 48x - 36$$
$$f^{(4)}(x) = 48$$
$$f^{(5)}(x) = 0.$$

After reflecting a moment, you should be convinced that the $(n + 1)$st derivative of an nth-degree polynomial function is zero.

$\dfrac{d}{dx}$ NOTES FROM THE CLASSROOM

(i) In the different contexts of science, engineering, and business, functions are often expressed in variables other than x and y. Correspondingly we must adapt the derivative notation to the new symbols. For example,

Function	*Derivative*
$v(t) = 32t$	$v'(t) = \dfrac{dv}{dt} = 32$
$A(r) = \pi r^2$	$A'(r) = \dfrac{dA}{dr} = 2\pi r$
$r(\theta) = 4\theta^2 - 3\theta$	$r'(\theta) = \dfrac{dr}{d\theta} = 8\theta - 3$
$D(p) = 800 - 129p + p^2$	$D'(p) = \dfrac{dD}{dp} = -129 + 2p.$

(ii) You may be wondering what interpretation can be given to the higher-order derivatives. If we think in terms of graphs, then f'' gives the slope of tangent lines to the graph of the function f'; f''' gives the slope of the tangent lines to the graph of f'', and so on. In addition, if f is differentiable, then the first-derivative f' gives the instantaneous rate of change of f. Similarly, if f' is differentiable, then f'' gives the instantaneous rate of change of f'.

Exercises 3.2   Answers to selected odd-numbered problems begin on page ANS-10.

≡ Fundamentals

In Problems 1–8, find dy/dx.

1. $y = -18$
2. $y = \pi^6$
3. $y = x^9$
4. $y = 4x^{12}$
5. $y = 7x^2 - 4x$
6. $y = 6x^3 + 3x^2 - 10$
7. $y = 4\sqrt{x} - \dfrac{6}{\sqrt[3]{x^2}}$
8. $y = \dfrac{x - x^2}{\sqrt{x}}$

In Problems 9–16, find $f'(x)$. Simplify.

9. $f(x) = \dfrac{1}{5}x^5 - 3x^4 + 9x^2 + 1$
10. $f(x) = -\dfrac{2}{3}x^6 + 4x^5 - 13x^2 + 8x + 2$
11. $f(x) = x^3(4x^2 - 5x - 6)$
12. $f(x) = \dfrac{2x^5 + 3x^4 - x^3 + 2}{x^2}$

13. $f(x) = x^2(x^2 + 5)^2$ **14.** $f(x) = (x^3 + x^2)^3$
15. $f(x) = (4\sqrt{x} + 1)^2$ **16.** $f(x) = (9 + x)(9 - x)$

In Problems 17–20, find the derivative of the given function.

17. $h(u) = (4u)^3$ **18.** $p(t) = (2t)^{-4} - (2t^{-1})^2$
19. $g(r) = \dfrac{1}{r} + \dfrac{1}{r^2} + \dfrac{1}{r^3} + \dfrac{1}{r^4}$ **20.** $Q(t) = \dfrac{t^5 + 4t^2 - 3}{6}$

In Problems 21–24, find an equation of the tangent line to the graph of the given function at the indicated value of x.

21. $y = 2x^3 - 1$; $x = -1$ **22.** $y = -x + \dfrac{8}{x}$; $x = 2$
23. $f(x) = \dfrac{4}{\sqrt{x}} + 2\sqrt{x}$; $x = 4$ **24.** $f(x) = -x^3 + 6x^2$; $x = 1$

In Problems 25–28, find the point(s) on the graph of the given function at which the tangent line is horizontal.

25. $y = x^2 - 8x + 5$ **26.** $y = \tfrac{1}{3}x^3 - \tfrac{1}{2}x^2$
27. $f(x) = x^3 - 3x^2 - 9x + 2$ **28.** $f(x) = x^4 - 4x^3$

In Problems 29–32, find an equation of the normal line to the graph of the given function at the indicated value of x.

29. $y = -x^2 + 1$; $x = 2$ **30.** $y = x^3$; $x = 1$
31. $f(x) = \tfrac{1}{3}x^3 - 2x^2$; $x = 4$ **32.** $f(x) = x^4 - x$; $x = -1$

In Problems 33–38, find the second derivative of the given function.

33. $y = -x^2 + 3x - 7$ **34.** $y = 15x^2 - 24\sqrt{x}$
35. $y = (-4x + 9)^2$ **36.** $y = 2x^5 + 4x^3 - 6x^2$
37. $f(x) = 10x^{-2}$ **38.** $f(x) = x + \left(\dfrac{2}{x^2}\right)^3$

In Problems 39 and 40, find the indicated higher derivative.

39. $f(x) = 4x^6 + x^5 - x^3$; $f^{(4)}(x)$
40. $y = x^4 - \dfrac{10}{x}$; d^5y/dx^5

In Problems 41 and 42, determine intervals for which $f'(x) > 0$ and intervals for which $f'(x) < 0$.

41. $f(x) = x^2 + 8x - 4$ **42.** $f(x) = x^3 - 3x^2 - 9x$

In Problems 43 and 44, find the point(s) on the graph of f at which $f''(x) = 0$.

43. $f(x) = x^3 + 12x^2 + 20x$ **44.** $f(x) = x^4 - 2x^3$

In Problems 45 and 46, determine intervals for which $f''(x) > 0$ and intervals for which $f''(x) < 0$.

45. $f(x) = (x - 1)^3$ **46.** $f(x) = x^3 + x^2$

An equation containing one or more derivatives of an unknown function $y(x)$ is called a **differential equation**. In Problems 47 and 48, show that the function satisfies the given differential equation.

47. $y = x^{-1} + x^4$; $x^2y'' - 2xy' - 4y = 0$
48. $y = x + x^3 + 4$; $x^2y'' - 3xy' + 3y = 12$
49. Find the point on the graph of $f(x) = 2x^2 - 3x + 6$ at which the slope of the tangent line is 5.

50. Find the point on the graph of $f(x) = x^2 - x$ at which the tangent line is $3x - 9y - 4 = 0$.

51. Find the point on the graph of $f(x) = x^2 - x$ at which the slope of the normal line is 2.

52. Find the point on the graph of $f(x) = \tfrac{1}{4}x^2 - 2x$ at which the tangent line is parallel to the line $3x - 2y + 1 = 0$.

53. Find an equation of the tangent line to the graph of $y = x^3 + 3x^2 - 4x + 1$ at the point where the value of the second derivative is zero.

54. Find an equation of the tangent line to the graph of $y = x^4$ at the point where the value of the third derivative is 12.

≡ Applications

55. The volume V of a sphere of radius r is $V = \tfrac{4}{3}\pi r^3$. Find the surface area S of the sphere if S is the instantaneous rate of change of the volume with respect to the radius.

56. According to the French physician Jean Louis Poiseuille (1799–1869) the velocity v of blood in an artery with a constant circular cross-section radius R is $v(r) = (P/4vl)(R^2 - r^2)$, where P, v, and l are constants. What is the velocity of blood at the value of r for which $v'(r) = 0$?

57. The potential energy of a spring-mass system when the spring is stretched a distance of x units is $U(x) = \tfrac{1}{2}kx^2$, where k is the spring constant. The force exerted on the mass is $F = -dU/dx$. Find the force if the spring constant is 30 N/m and the amount of stretch is $\tfrac{1}{2}$ m.

58. The height s above ground of a projectile at time t is given by
$$s(t) = \tfrac{1}{2}gt^2 + v_0t + s_0,$$
where g, v_0, and s_0 are constants. Find the instantaneous rate of change of s with respect to t at $t = 4$.

≡ Think About It

In Problems 59 and 60, the symbol n represents a positive integer. Find a formula for the given derivative.

59. $\dfrac{d^n}{dx^n} x^n$ **60.** $\dfrac{d^n}{dx^n} \dfrac{1}{x}$

61. From the graphs of f and g in FIGURE 3.2.6, determine which function is the derivative of the other. Explain your choice in words.

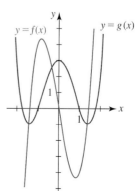

FIGURE 3.2.6 Graphs for Problem 61

62. From the graph of the function $y = f(x)$ given in FIGURE 3.2.7, sketch the graph of f'.

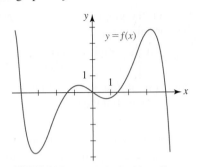

FIGURE 3.2.7 Graph for Problem 62

63. Find a quadratic function $f(x) = ax^2 + bx + c$ such that $f(-1) = -11, f'(-1) = 7$, and $f''(-1) = -4$.

64. The graphs of $y = f(x)$ and $y = g(x)$ are said to be **orthogonal** if the tangent lines to each graph are perpendicular at each point of intersection. Show that the graphs of $y = \frac{1}{8}x^2$ and $y = -\frac{1}{4}x^2 + 3$ are orthogonal.

65. Find the values of b and c so that the graph of $f(x) = x^2 + bx$ possesses the tangent line $y = 2x + c$ at $x = -3$.

66. Find an equation of the line(s) that passes through $\left(\frac{3}{2}, 1\right)$ and is tangent to the graph of $f(x) = x^2 + 2x + 2$.

67. Find the point(s) on the graph of $f(x) = x^2 - 5$ such that the tangent line at the point(s) has x-intercept $(-3, 0)$.

68. Find the point(s) on the graph of $f(x) = x^2$ such that the tangent line at the point(s) has y-intercept $(0, -2)$.

69. Explain why the graph of $f(x) = \frac{1}{5}x^5 + \frac{1}{3}x^3$ has no tangent line with slope -1.

70. Find coefficients A and B so that the function $y = Ax^2 + Bx$ satisfies the differential equation $2y'' + 3y' = x - 1$.

71. Find values of a and b such that the slope of the tangent to the graph of $f(x) = ax^2 + bx$ at $(1, 4)$ is -5.

72. Find the slopes of all the normal lines to the graph of $f(x) = x^2$ that pass through the point $(2, 4)$. [*Hint*: Draw a figure and note that *at* $(2, 4)$ there is only one normal line.]

73. Find a point on the graph of $f(x) = x^2 + x$ and a point on the graph of $g(x) = 2x^2 + 4x + 1$ at which the tangent lines are parallel.

74. Find a point on the graph of $f(x) = 3x^5 + 5x^3 + 2x$ at which the tangent has the least possible slope.

75. Find conditions on the coefficients a, b, and c so that the graph of the polynomial function
$$f(x) = ax^3 + bx^2 + cx + d$$
has exactly one horizontal tangent. Exactly two horizontal tangents. No horizontal tangents.

76. Let f be a differentiable function. If $f'(x) > 0$ for all x in the interval (a, b), sketch possible graphs of f on the interval. Describe in words the behavior of the graph of f on the interval. Repeat if $f'(x) < 0$ for all x in the interval (a, b).

77. Suppose f is a differentiable function such that $f'(x) - f(x) = 0$. Find $f^{(100)}(x)$.

78. The graphs of $y = x^2$ and $y = -x^2 + 2x - 3$ given in FIGURE 3.2.8 show that there are two lines L_1 and L_2 that are simultaneously tangent to both graphs. Find the points of tangency on both graphs. Find an equation of each tangent line.

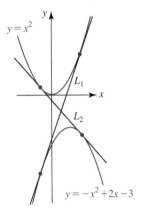

FIGURE 3.2.8 Graphs for Problem 78

≡ **Calculator/CAS Problems**

79. (a) Use a calculator or CAS to obtain the graph of $f(x) = x^4 - 4x^3 - 2x^2 + 12x - 2$.
(b) Evaluate $f''(x)$ at $x = -2, x = -1, x = 0, x = 1, x = 2, x = 3$, and $x = 4$.
(c) From the data in part (b), do you see any relationship between the shape of the graph of f and the algebraic signs of f''?

80. Use a calculator or CAS to obtain the graph of the given functions. By inspection of the graphs indicate where each function may not be differentiable. Find $f'(x)$ at all points where f is differentiable.
(a) $f(x) = |x^2 - 2x|$ (b) $f(x) = |x^3 - 1|$

3.3 Product and Quotient Rules

Introduction So far we know that the derivative of a constant function and a power of x are, in turn:

$$\frac{d}{dx}c = 0 \quad \text{and} \quad \frac{d}{dx}x^n = nx^{n-1}. \quad (1)$$

We also know that for differentiable functions f and g:

$$\frac{d}{dx}cf(x) = cf'(x) \quad \text{and} \quad \frac{d}{dx}[f(x) \pm g(x)] = f'(x) \pm g'(x). \tag{2}$$

Although the results in (1) and (2) allow us to quickly differentiate many algebraic functions (such as polynomials) neither (1) nor (2) are of immediate help in finding derivatives of functions such as $y = x^4\sqrt{x^2 + 4}$ or $y = x/(2x + 1)$. We need additional rules for differentiating products fg and quotients f/g.

Product Rule The rules of differentiation and the derivatives of functions ultimately stem from the definition of the derivative. The Sum Rule in (2), derived in the preceding section, follows from this definition and the fact that the limit of a sum is the sum of the limits whenever the limits exist. We also know that when the limits exist, the limit of a product is the product of the limits. Arguing by analogy, it would then seem plausible that the derivative of a product of two functions is the product of the derivatives. Regrettably, the Product Rule stated next is *not* that simple.

Theorem 3.3.1 Product Rule

If f and g are functions differentiable at x, then fg is differentiable at x, and

$$\frac{d}{dx}[f(x)g(x)] = f(x)g'(x) + g(x)f'(x). \tag{3}$$

PROOF Let $G(x) = f(x)g(x)$. Then by the definition of the derivative along with some algebraic manipulation:

$$G'(x) = \lim_{h \to 0} \frac{G(x + h) - G(x)}{h} = \lim_{h \to 0} \frac{f(x + h)g(x + h) - f(x)g(x)}{h}$$

$$= \lim_{h \to 0} \frac{f(x + h)g(x + h) \overbrace{- f(x + h)g(x) + f(x + h)g(x)}^{\text{zero}} - f(x)g(x)}{h}$$

$$= \lim_{h \to 0} \left[f(x + h)\frac{g(x + h) - g(x)}{h} + g(x)\frac{f(x + h) - f(x)}{h} \right]$$

$$= \lim_{h \to 0} f(x + h) \cdot \lim_{h \to 0} \frac{g(x + h) - g(x)}{h} + \lim_{h \to 0} g(x) \cdot \lim_{h \to 0} \frac{f(x + h) - f(x)}{h}.$$

Because f is differentiable at x, it is continuous there and so $\lim_{h \to 0} f(x + h) = f(x)$. Furthermore, $\lim_{h \to 0} g(x) = g(x)$. Hence the last equation becomes

$$G'(x) = f(x)g'(x) + g(x)f'(x). \qquad \blacksquare$$

The Product Rule is best memorized in words:

- *The first function times the derivative of the second plus the second function times the derivative of the first.*

EXAMPLE 1 Product Rule

Differentiate $y = (x^3 - 2x^2 + 3)(7x^2 - 4x)$.

Solution From the Product Rule (3),

$$\frac{dy}{dx} = \overbrace{(x^3 - 2x^2 + 3)}^{\text{first}} \cdot \overbrace{\frac{d}{dx}(7x^2 - 4x)}^{\text{derivative of second}} + \overbrace{(7x^2 - 4x)}^{\text{second}} \cdot \overbrace{\frac{d}{dx}(x^3 - 2x^2 + 3)}^{\text{derivative of first}}$$

$$= (x^3 - 2x^2 + 3)(14x - 4) + (7x^2 - 4x)(3x^2 - 4x)$$

$$= 35x^4 - 72x^3 + 24x^2 + 42x - 12.$$

Alternative Solution The two terms in the given function could be multiplied out to obtain a fifth-degree polynomial. The derivative can then be gotten using the Sum Rule. ∎

EXAMPLE 2 Tangent Line

Find an equation of the tangent line to the graph of $y = (1 + \sqrt{x})(x - 2)$ at $x = 4$.

Solution Before taking the derivative we rewrite $\sqrt{x}$ as $x^{1/2}$. Then from the Product Rule (3)

$$\frac{dy}{dx} = (1 + x^{1/2})\frac{d}{dx}(x - 2) + (x - 2)\frac{d}{dx}(1 + x^{1/2})$$

$$= (1 + x^{1/2}) \cdot 1 + (x - 2) \cdot \frac{1}{2}x^{-1/2}$$

$$= \frac{3x + 2\sqrt{x} - 2}{2\sqrt{x}}.$$

Evaluating the given function and its derivative at $x = 4$ gives:

$$y(4) = (1 + \sqrt{4})(4 - 2) = 6 \quad \leftarrow \text{point of tangency is } (4, 6)$$

$$\left.\frac{dy}{dx}\right|_{x=4} = \frac{12 + 2\sqrt{4} - 2}{2\sqrt{4}} = \frac{7}{2}. \quad \leftarrow \text{slope of the tangent at } (4, 6) \text{ is } \tfrac{7}{2}$$

By the point–slope form, the tangent line is

$$y - 6 = \frac{7}{2}(x - 4) \quad \text{or} \quad y = \frac{7}{2}x - 8. \qquad \blacksquare$$

Although (3) is stated for only the product of two functions, it can be applied to functions with a greater number of factors. The idea is to group two (or more) functions and treat this grouping as one function. The next example illustrates the technique.

EXAMPLE 3 Product of Three Functions

Differentiate $y = (4x + 1)(2x^2 - x)(x^3 - 8x)$.

Solution We identify the first two factors as the "first function":

$$\frac{dy}{dx} = \overbrace{(4x + 1)(2x^2 - x)}^{\text{first}}\overbrace{\frac{d}{dx}(x^3 - 8x)}^{\text{derivative of second}} + \overbrace{(x^3 - 8x)}^{\text{second}}\overbrace{\frac{d}{dx}(4x + 1)(2x^2 - x)}^{\text{derivative of first}}.$$

Notice that to find the derivative of the first function, we must apply the Product Rule a second time:

$$\frac{dy}{dx} = (4x + 1)(2x^2 - x) \cdot (3x^2 - 8) + (x^3 - 8x) \cdot \overbrace{[(4x + 1)(4x - 1) + (2x^2 - x) \cdot 4]}^{\text{Product Rule again}}$$

$$= (4x + 1)(2x^2 - x)(3x^2 - 8) + (x^3 - 8x)(16x^2 - 1) + 4(x^3 - 8x)(2x^2 - x). \qquad \blacksquare$$

■ **Quotient Rule** The derivative of the quotient of two functions f and g is given next.

Theorem 3.3.2 Quotient Rule

If f and g are functions differentiable at x and $g(x) \neq 0$, then f/g is differentiable at x, and

$$\frac{d}{dx}\left[\frac{f(x)}{g(x)}\right] = \frac{g(x)f'(x) - f(x)g'(x)}{[g(x)]^2}. \tag{4}$$

PROOF Let $G(x) = f(x)/g(x)$. Then

$$G'(x) = \lim_{h \to 0} \frac{G(x+h) - G(x)}{h} = \lim_{h \to 0} \frac{\dfrac{f(x+h)}{g(x+h)} - \dfrac{f(x)}{g(x)}}{h}$$

$$= \lim_{h \to 0} \frac{g(x)f(x+h) - f(x)g(x+h)}{hg(x+h)g(x)}$$

$$= \lim_{h \to 0} \frac{g(x)f(x+h) \overbrace{- g(x)f(x) + g(x)f(x)}^{\text{zero}} - f(x)g(x+h)}{hg(x+h)g(x)}$$

$$= \lim_{h \to 0} \frac{g(x)\dfrac{f(x+h) - f(x)}{h} - f(x)\dfrac{g(x+h) - g(x)}{h}}{g(x+h)g(x)}$$

$$= \frac{\lim_{h \to 0} g(x) \cdot \lim_{h \to 0} \dfrac{f(x+h) - f(x)}{h} - \lim_{h \to 0} f(x) \cdot \lim_{h \to 0} \dfrac{g(x+h) - g(x)}{h}}{\lim_{h \to 0} g(x+h) \cdot \lim_{h \to 0} g(x)}.$$

Since all limits are assumed to exist, the last line is the same as

$$G'(x) = \frac{g(x)f'(x) - f(x)g'(x)}{[g(x)]^2}. \blacksquare$$

In words, the Quotient Rule starts with the denominator:

- *The denominator times the derivative of the numerator minus the numerator times the derivative of the denominator all divided by the denominator squared.*

EXAMPLE 4 Quotient Rule

Differentiate $y = \dfrac{3x^2 - 1}{2x^3 + 5x^2 + 7}$.

Solution From the Quotient Rule (4),

$$\frac{dy}{dx} = \frac{\overbrace{(2x^3 + 5x^2 + 7)}^{\text{denominator}} \cdot \overbrace{\dfrac{d}{dx}(3x^2 - 1)}^{\substack{\text{derivative of} \\ \text{numerator}}} - \overbrace{(3x^2 - 1)}^{\text{numerator}} \cdot \overbrace{\dfrac{d}{dx}(2x^3 + 5x^2 + 7)}^{\substack{\text{derivative of} \\ \text{denominator}}}}{\underbrace{(2x^3 + 5x^2 + 7)^2}_{\substack{\text{square of} \\ \text{denominator}}}}$$

$$= \frac{(2x^3 + 5x^2 + 7) \cdot 6x - (3x^2 - 1) \cdot (6x^2 + 10x)}{(2x^3 + 5x^2 + 7)^2} \leftarrow \text{multiply out numerator}$$

$$= \frac{-6x^4 + 6x^2 + 52x}{(2x^3 + 5x^2 + 7)^2}. \blacksquare$$

EXAMPLE 5 Quotient and Product Rule

Find the points on the graph of $y = \dfrac{(x^2 + 1)(2x^2 + 1)}{3x^2 + 1}$ where the tangent line is horizontal.

Solution We begin with the Quotient Rule and then use the Product Rule when differentiating the numerator:

$$\frac{dy}{dx} = \frac{(3x^2 + 1) \cdot \overbrace{\frac{d}{dx}[(x^2 + 1)(2x^2 + 1)]}^{\text{Product Rule here}} - (x^2 + 1)(2x^2 + 1) \cdot \frac{d}{dx}(3x^2 + 1)}{(3x^2 + 1)^2}$$

$$= \frac{(3x^2 + 1)[(x^2 + 1)4x + (2x^2 + 1)2x] - (x^2 + 1)(2x^2 + 1)6x}{(3x^2 + 1)^2} \quad \leftarrow \text{multiply out numerator}$$

$$= \frac{12x^5 + 8x^3}{(3x^2 + 1)^2}.$$

At a point where the tangent line is horizontal we must have $dy/dx = 0$. The derivative just found can only be 0 when the numerator satisfies

$$12x^5 + 8x^3 = 0 \quad \text{or} \quad x^3(12x^2 + 8) = 0. \tag{5}$$

▶ Of course, values of x that make the numerator zero must *not* simultaneously make the denominator zero.

In (5) because $12x^2 + 8 \neq 0$ for all real numbers x, we must have $x = 0$. Substituting this number into the function gives $y(0) = 1$. The tangent line is horizontal at the y-intercept $(0, 1)$. ■

Postscript—Power Rule Revisited Remember in Section 3.2 we stated that the Power Rule, $(d/dx)x^n = nx^{n-1}$, is valid for all real number exponents n. We are now in a position to prove the rule when the exponent is a negative integer $-m$. Since, by definition, $x^{-m} = 1/x^m$, where m is a positive integer, we can obtain the derivative of x^{-m} by the Quotient Rule and the laws of exponents:

$$\frac{d}{dx}x^{-m} = \frac{d}{dx}\left(\frac{1}{x^m}\right) = \frac{x^m \cdot \frac{d}{dx}1 - 1 \cdot \frac{d}{dx}x^m}{(x^m)^2} = -\frac{mx^{m-1}}{x^{2m}} \overset{\text{subtract exponents}}{=} -mx^{-m-1}.$$

$\frac{d}{dx}$ NOTES FROM THE CLASSROOM

(*i*) The Product and Quotient Rules will usually lead to expressions that demand simplification. If your answer to a problem does not look like the one in the text answer section, you may not have performed sufficient simplifications. Do not be content to simply carry through the mechanics of the various rules of differentiation; it is always a good idea to practice your algebraic skills.

(*ii*) The Quotient Rule is sometimes used when it is not required. Although we could use the Quotient Rule to differentiate functions such as

$$y = \frac{x^5}{6} \quad \text{and} \quad y = \frac{10}{x^3},$$

it is simpler (and faster) to rewrite the functions as $y = \frac{1}{6}x^5$ and $y = 10x^{-3}$ and then use the Constant Multiple and Power Rules:

$$\frac{dy}{dx} = \frac{1}{6}\frac{d}{dx}x^5 = \frac{5}{6}x^4 \quad \text{and} \quad \frac{dy}{dx} = 10\frac{d}{dx}x^{-3} = -30x^{-4}.$$

Exercises 3.3
Answers to selected odd-numbered problems begin on page ANS-10.

≡ Fundamentals

In Problems 1–10, find dy/dx.

1. $y = (x^2 - 7)(x^3 + 4x + 2)$
2. $y = (7x + 1)(x^4 - x^3 - 9x)$
3. $y = \left(4\sqrt{x} + \frac{1}{x}\right)\left(2x - \frac{6}{\sqrt[3]{x}}\right)$
4. $y = \left(x^2 - \frac{1}{x^2}\right)\left(x^3 + \frac{1}{x^3}\right)$

5. $y = \dfrac{10}{x^2 + 1}$

6. $y = \dfrac{5}{4x - 3}$

7. $y = \dfrac{3x + 1}{2x - 5}$

8. $y = \dfrac{2 - 3x}{7 - x}$

9. $y = (6x - 1)^2$

10. $y = (x^4 + 5x)^2$

In Problems 11–20, find $f'(x)$.

11. $f(x) = \left(\dfrac{1}{x} - \dfrac{4}{x^3}\right)(x^3 - 5x - 1)$

12. $f(x) = (x^2 - 1)\left(x^2 - 10x + \dfrac{2}{x^2}\right)$

13. $f(x) = \dfrac{x^2}{2x^2 + x + 1}$

14. $f(x) = \dfrac{x^2 - 10x + 2}{x(x^2 - 1)}$

15. $f(x) = (x + 1)(2x + 1)(3x + 1)$

16. $f(x) = (x^2 + 1)(x^3 - x)(3x^4 + 2x - 1)$

17. $f(x) = \dfrac{(2x + 1)(x - 5)}{3x + 2}$

18. $f(x) = \dfrac{x^5}{(x^2 + 1)(x^3 + 4)}$

19. $f(x) = (x^2 - 2x - 1)\left(\dfrac{x + 1}{x + 3}\right)$

20. $f(x) = (x + 1)\left(x + 1 - \dfrac{1}{x + 2}\right)$

In Problems 21–24, find an equation of the tangent line to the graph of the given function at the indicated value of x.

21. $y = \dfrac{x}{x - 1};\ x = \dfrac{1}{2}$

22. $y = \dfrac{5x}{x^2 + 1};\ x = 2$

23. $y = (2\sqrt{x} + x)(-2x^2 + 5x - 1);\ x = 1$

24. $y = (2x^2 - 4)(x^3 + 5x + 3);\ x = 0$

In Problems 25–28, find the point(s) on the graph of the given function at which the tangent line is horizontal.

25. $y = (x^2 - 4)(x^2 - 6)$

26. $y = x(x - 1)^2$

27. $y = \dfrac{x^2}{x^4 + 1}$

28. $y = \dfrac{1}{x^2 - 6x}$

In Problems 29 and 30, find the point(s) on the graph of the given function at which the tangent line has the indicated slope.

29. $y = \dfrac{x + 3}{x + 1};\ m = -\dfrac{1}{8}$

30. $y = (x + 1)(2x + 5);\ m = -3$

In Problems 31 and 32, find the point(s) on the graph of the given function at which the tangent line has the indicated property.

31. $y = \dfrac{x + 4}{x + 5};$ perpendicular to $y = -x$

32. $y = \dfrac{x}{x + 1};$ parallel to $y = \dfrac{1}{4}x - 1$

33. Find the value of k such that the tangent line to the graph of $f(x) = (k + x)/x^2$ has slope 5 at $x = 2$.

34. Show that the tangent to the graph of $f(x) = (x^2 + 14)/(x^2 + 9)$ at $x = 1$ is perpendicular to the tangent to the graph of $g(x) = (1 + x^2)(1 + 2x)$ at $x = 1$.

In Problems 35–40, f and g are differentiable functions. Find $F'(1)$ if $f(1) = 2$, $f'(1) = -3$, and $g(1) = 6$, $g'(1) = 2$.

35. $F(x) = 2f(x)g(x)$

36. $F(x) = x^2 f(x)g(x)$

37. $F(x) = \dfrac{2g(x)}{3f(x)}$

38. $F(x) = \dfrac{1 + 2f(x)}{x - g(x)}$

39. $F(x) = \left(\dfrac{4}{x} + f(x)\right)g(x)$

40. $F(x) = \dfrac{xf(x)}{g(x)}$

41. Suppose $F(x) = \sqrt{x}\, f(x)$, where f is a differentiable function. Find $F''(4)$ if $f(4) = -16$, $f'(4) = 2$, and $f''(4) = 3$.

42. Suppose $F(x) = xf(x) + xg(x)$, where f and g are differentiable functions. Find $F''(0)$ if $f'(0) = -1$ and $g'(0) = 6$.

43. Suppose $F(x) = f(x)/x$, where f is a differentiable function. Find $F''(x)$.

44. Suppose $F(x) = x^3 f(x)$, where f is a differentiable function. Find $F'''(x)$.

In Problems 45–48, determine intervals for which $f'(x) > 0$ and intervals for which $f'(x) < 0$.

45. $f(x) = \dfrac{5}{x^2 - 2x}$

46. $f(x) = \dfrac{x^2 + 3}{x + 1}$

47. $f(x) = (-2x + 6)(4x + 7)$

48. $f(x) = (x - 2)(4x^2 + 8x + 4)$

≡ Applications

49. The Law of Universal Gravitation states that the force F between two bodies of masses m_1 and m_2 separated by a distance r is $F = km_1 m_2/r^2$, where k is constant. What is the instantaneous rate of change of F with respect to r when $r = \tfrac{1}{2}$ km?

50. The potential energy U between two atoms in a diatomic molecule is given by $U(x) = q_1/x^{12} - q_2/x^6$, where q_1 and q_2 are positive constants and x is the distance between the atoms. The force between the atoms is defined as $F(x) = -U'(x)$. Show that $F\left(\sqrt[6]{2q_1/q_2}\right) = 0$.

51. The **van der Waals equation of state** for an ideal gas is
$$\left(P + \dfrac{a}{V^2}\right)(V - b) = RT,$$
where P is pressure, V is volume per mole, R is the universal gas constant, T is temperature, and a and b are constants depending on the gas. Find dP/dV in the case where T is constant.

52. For a convex lens, the focal length f is related to the object distance p and the image distance q by the **lens equation**
$$\dfrac{1}{f} = \dfrac{1}{p} + \dfrac{1}{q}.$$
Find the instantaneous rate of change of q with respect to p in the case where f is constant. Explain the significance of the negative sign in your answer. What happens to q as p increases?

Think About It

53. **(a)** Graph the rational function $f(x) = \dfrac{2}{x^2+1}$.

    (b) Find all the points on the graph of f such that the normal lines pass through the origin.

54. Suppose $y = f(x)$ is a differentiable function.

    (a) Find dy/dx for $y = [f(x)]^2$.

    (b) Find dy/dx for $y = [f(x)]^3$.

    (c) Conjecture a rule for finding the derivative of $y = [f(x)]^n$, where n is a positive integer.

    (d) Use your conjecture in part (c) to find the derivative of $y = (x^2 + 2x - 6)^{500}$.

55. Suppose $y_1(x)$ satisfies the differential equation $y' + P(x)y = 0$, where P is a known function. Show that $y = u(x)y_1(x)$ satisfies the differential equation

$$y' + P(x)y = f(x)$$

whenever $u(x)$ satisfies $du/dx = f(x)/y_1(x)$.

3.4 Trigonometric Functions

Introduction In this section we develop the derivatives of the six trigonometric functions. Once we have found the derivatives of $\sin x$ and $\cos x$ we can determine the derivatives of $\tan x$, $\cot x$, $\sec x$, and $\csc x$ using the Quotient Rule found in the preceding section. We will see immediately that the derivative of $\sin x$ utilizes the following two limit results

$$\lim_{x \to 0} \frac{\sin x}{x} = 1 \quad \text{and} \quad \lim_{x \to 0} \frac{\cos x - 1}{x} = 0 \qquad (1)$$

found in Section 2.4.

Derivatives of Sine and Cosine To find the derivative of $f(x) = \sin x$ we use the basic definition of the derivative

$$\frac{dy}{dx} = \lim_{h \to 0} \frac{f(x+h) - f(x)}{h} \qquad (2)$$

and the four-step process introduced in Sections 2.7 and 3.1. In the first step we use the sum formula for the sine function,

$$\sin(x_1 + x_2) = \sin x_1 \cos x_2 + \cos x_1 \sin x_2, \qquad (3)$$

but with x and h playing the parts of the symbols x_1 and x_2.

(*i*) $f(x+h) = \sin(x+h) = \sin x \cos h + \cos x \sin h \quad$ ← from (3)

(*ii*) $f(x+h) - f(x) = \sin x \cos h + \cos x \sin h - \sin x \quad$ ← factor $\sin x$ from first and third terms

$\qquad\qquad\qquad\quad = \sin x(\cos h - 1) + \cos x \sin h$

As we see in the next line, we cannot cancel the h's in the difference quotient but we can rewrite the expression to make use of the limit results in (1).

(*iii*) $\dfrac{f(x+h) - f(x)}{h} = \dfrac{\sin x(\cos h - 1) + \cos x \sin h}{h}$

$\qquad\qquad\qquad\quad = \sin x \cdot \dfrac{\cos h - 1}{h} + \cos x \cdot \dfrac{\sin h}{h}$

(*iv*) In this line, the symbol h plays the part of the symbol x in (1):

$$f'(x) = \lim_{h \to 0} \frac{f(x+h) - f(x)}{h} = \sin x \cdot \lim_{h \to 0} \frac{\cos h - 1}{h} + \cos x \cdot \lim_{h \to 0} \frac{\sin h}{h}.$$

From the limit results in (1), the last line is the same as

$$f'(x) = \lim_{h \to 0} \frac{f(x+h) - f(x)}{h} = \sin x \cdot 0 + \cos x \cdot 1 = \cos x.$$

Hence, $\qquad\qquad\qquad \dfrac{d}{dx} \sin x = \cos x. \qquad (4)$

In a similar manner it can be shown that

$$\frac{d}{dx}\cos x = -\sin x. \qquad (5)$$

See Problem 50 in Exercises 3.4.

EXAMPLE 1 Equation of a Tangent Line

Find an equation of the tangent line to the graph of $f(x) = \sin x$ at $x = 4\pi/3$.

Solution From (4) the derivative of $f(x) = \sin x$ is $f'(x) = \cos x$. When evaluated at the same number $x = 4\pi/3$ these functions give:

$$f\left(\frac{4\pi}{3}\right) = \sin\frac{4\pi}{3} = -\frac{\sqrt{3}}{2} \quad \leftarrow \text{point of tangency is } \left(\tfrac{4\pi}{3}, -\tfrac{\sqrt{3}}{2}\right)$$

$$f'\left(\frac{4\pi}{3}\right) = \cos\frac{4\pi}{3} = -\frac{1}{2}. \quad \leftarrow \text{slope of tangent at } \left(\tfrac{4\pi}{3}, -\tfrac{\sqrt{3}}{2}\right) \text{ is } -\tfrac{1}{2}$$

From the point–slope form of a line, an equation of the tangent line is

$$y + \frac{\sqrt{3}}{2} = -\frac{1}{2}\left(x - \frac{4\pi}{3}\right) \quad \text{or} \quad y = -\frac{1}{2}x + \frac{2\pi}{3} - \frac{\sqrt{3}}{2}.$$

The tangent line is shown in red in FIGURE 3.4.1.

FIGURE 3.4.1 Tangent line in Example 1

Other Trigonometric Functions The results in (4) and (5) can be used in conjunction with the rules of differentiation to find the derivatives of the tangent, cotangent, secant, and cosecant functions.

To differentiate $\tan x = \sin x/\cos x$, we use the Quotient Rule:

$$\frac{d}{dx}\frac{\sin x}{\cos x} = \frac{\cos x \dfrac{d}{dx}\sin x - \sin x \dfrac{d}{dx}\cos x}{(\cos x)^2}$$

$$= \frac{\cos x(\cos x) - \sin x(-\sin x)}{(\cos x)^2} = \frac{\overbrace{\cos^2 x + \sin^2 x}^{\text{this equals 1}}}{\cos^2 x}.$$

Using the fundamental Pythagorean identity $\sin^2 x + \cos^2 x = 1$ and the fact that $1/\cos^2 x = (1/\cos x)^2 = \sec^2 x$, the last equation simplifies to

$$\frac{d}{dx}\tan x = \sec^2 x. \qquad (6)$$

The derivative formula for the cotangent

$$\frac{d}{dx}\cot x = -\csc^2 x \qquad (7)$$

is obtained in an analogous fashion and left as an exercise. See Problem 51 in Exercises 3.4.

Now $\sec x = 1/\cos x$. Therefore, we can use the Quotient Rule again to find the derivative of the secant function:

$$\frac{d}{dx}\frac{1}{\cos x} = \frac{\cos x \dfrac{d}{dx}1 - 1 \cdot \dfrac{d}{dx}\cos x}{(\cos x)^2}$$

$$= \frac{0 - (-\sin x)}{(\cos x)^2} = \frac{\sin x}{\cos^2 x}. \qquad (8)$$

By writing $\dfrac{\sin x}{\cos^2 x} = \dfrac{1}{\cos x} \cdot \dfrac{\sin x}{\cos x} = \sec x \tan x$

we can express (8) as

$$\frac{d}{dx}\sec x = \sec x \tan x. \qquad (9)$$

The final result also follows immediately from the Quotient Rule:

$$\frac{d}{dx}\csc x = -\csc x \cot x. \qquad (10)$$

See Problem 52 in Exercises 3.4.

EXAMPLE 2 Product Rule

Differentiate $y = x^2 \sin x$.

Solution The Product Rule along with (4) gives

$$\frac{dy}{dx} = x^2 \frac{d}{dx} \sin x + \sin x \frac{d}{dx} x^2$$

$$= x^2 \cos x + 2x \sin x. \qquad \blacksquare$$

EXAMPLE 3 Product Rule

Differentiate $y = \cos^2 x$.

Solution One way of differentiating this function is to recognize it as a product: $y = (\cos x)(\cos x)$. Then by the Product Rule and (5),

$$\frac{dy}{dx} = \cos x \frac{d}{dx} \cos x + \cos x \frac{d}{dx} \cos x$$

$$= \cos x(-\sin x) + (\cos x)(-\sin x)$$

$$= -2 \sin x \cos x.$$

In the next section we will see that there is an alternative procedure for differentiating a power of a function. $\blacksquare$

EXAMPLE 4 Quotient Rule

Differentiate $y = \dfrac{\sin x}{2 + \sec x}$.

Solution By the Quotient Rule, (4), and (9),

$$\frac{dy}{dx} = \frac{(2 + \sec x)\frac{d}{dx}\sin x - \sin x \frac{d}{dx}(2 + \sec x)}{(2 + \sec x)^2}$$

$$= \frac{(2 + \sec x)\cos x - \sin x (\sec x \tan x)}{(2 + \sec x)^2} \quad \leftarrow \begin{array}{l} \sec x \cos x = 1 \text{ and} \\ \sin x(\sec x \tan x) = \sin^2 x/\cos^2 x \end{array}$$

$$= \frac{1 + 2\cos x - \tan^2 x}{(2 + \sec x)^2}. \qquad \blacksquare$$

EXAMPLE 5 Second Derivative

Find the second derivative of $f(x) = \sec x$.

Solution From (9) the first derivative is

$$f'(x) = \sec x \tan x.$$

To obtain the second derivative we must now use the Product Rule along with (6) and (9):

$$f''(x) = \sec x \frac{d}{dx} \tan x + \tan x \frac{d}{dx} \sec x$$

$$= \sec x (\sec^2 x) + \tan x (\sec x \tan x)$$

$$= \sec^3 x + \sec x \tan^2 x. \qquad \blacksquare$$

For future reference we summarize the derivative formulas introduced in this section.

Theorem 3.4.1 Derivatives of Trigonometric Functions

The derivatives of the six trigonometric functions are

$$\frac{d}{dx}\sin x = \cos x, \qquad \frac{d}{dx}\cos x = -\sin x, \qquad (11)$$

$$\frac{d}{dx}\tan x = \sec^2 x, \qquad \frac{d}{dx}\cot x = -\csc^2 x, \qquad (12)$$

$$\frac{d}{dx}\sec x = \sec x \tan x, \qquad \frac{d}{dx}\csc x = -\csc x \cot x. \qquad (13)$$

$\frac{d}{dx}$ NOTES FROM THE CLASSROOM

When working the problems in Exercises 3.4 you may not get the same answer as given in the answer section in the back of this book. This is because there are so many trigonometric identities that answers can often be expressed in a more compact form. For example, the answer in Example 3:

$$\frac{dy}{dx} = -2\sin x \cos x \quad \text{is the same as} \quad \frac{dy}{dx} = -\sin 2x$$

by the double-angle formula for the sine function. Try to resolve any differences between your answer and the given answer.

Exercises 3.4 Answers to selected odd-numbered problems begin on page ANS-10.

Fundamentals

In Problems 1–12, find dy/dx.

1. $y = x^2 - \cos x$
2. $y = 4x^3 + x + 5\sin x$
3. $y = 1 + 7\sin x - \tan x$
4. $y = 3\cos x - 5\cot x$
5. $y = x\sin x$
6. $y = (4\sqrt{x} - 3\sqrt[3]{x})\cos x$
7. $y = (x^3 - 2)\tan x$
8. $y = \cos x \cot x$
9. $y = (x^2 + \sin x)\sec x$
10. $y = \csc x \tan x$
11. $y = \cos^2 x + \sin^2 x$
12. $y = x^3 \cos x - x^3 \sin x$

In Problems 13–22, find $f'(x)$.

13. $f(x) = (\csc x)^{-1}$
14. $f(x) = \dfrac{2}{\cos x \cot x}$
15. $f(x) = \dfrac{\cot x}{x + 1}$
16. $f(x) = \dfrac{x^2 - 6x}{1 + \cos x}$
17. $f(x) = \dfrac{x^2}{1 + 2\tan x}$
18. $f(x) = \dfrac{2 + \sin x}{x}$
19. $f(x) = \dfrac{\sin x}{1 + \cos x}$
20. $f(x) = \dfrac{1 + \csc x}{1 + \sec x}$
21. $f(x) = x^4 \sin x \tan x$
22. $f(x) = \dfrac{1 + \sin x}{x \cos x}$

In Problems 23–26, find an equation of the tangent line to the graph of the given function at the indicated value of x.

23. $f(x) = \cos x; \quad x = \pi/3$
24. $f(x) = \tan x; \quad x = \pi$
25. $f(x) = \sec x; \quad x = \pi/6$
26. $f(x) = \csc x; \quad x = \pi/2$

In Problems 27–30, consider the graph of the given function on the interval $[0, 2\pi]$. Find the x-coordinates of the point(s) on the graph of the function where the tangent line is horizontal.

27. $f(x) = x + 2\cos x$
28. $f(x) = \dfrac{\sin x}{2 - \cos x}$
29. $f(x) = \dfrac{1}{x + \cos x}$
30. $f(x) = \sin x + \cos x$

In Problems 31–34, find an equation of the normal line to the graph of the given function at the indicated value of x.

31. $f(x) = \sin x; \quad x = 4\pi/3$
32. $f(x) = \tan^2 x; \quad x = \pi/4$
33. $f(x) = x\cos x; \quad x = \pi$
34. $f(x) = \dfrac{x}{1 + \sin x}; \quad x = \pi/2$

In Problems 35 and 36, find the derivative of the given function by first using an appropriate trigonometric identity.

35. $f(x) = \sin 2x$
36. $f(x) = \cos^2 \dfrac{x}{2}$

In Problems 37–42, find the second derivative of the given function.

37. $f(x) = x\sin x$
38. $f(x) = 3x - x^2 \cos x$
39. $f(x) = \dfrac{\sin x}{x}$
40. $f(x) = \dfrac{1}{1 + \cos x}$
41. $y = \csc x$
42. $y = \tan x$

In Problems 43 and 44, C_1 and C_2 are arbitrary real constants. Show that the function satisfies the given differential equation.

43. $y = C_1 \cos x + C_2 \sin x - \frac{1}{2}x\cos x;\quad y'' + y = \sin x$

44. $y = C_1 \dfrac{\cos x}{\sqrt{x}} + C_2 \dfrac{\sin x}{\sqrt{x}};\quad x^2 y'' + xy' + (x^2 - \tfrac{1}{4})y = 0$

≡ Applications

45. When the angle of elevation of the sun is θ, a telephone pole 40 ft high casts a shadow of length s as shown in FIGURE 3.4.2. Find the rate of change of s with respect to θ when $\theta = \pi/3$ radians. Explain the significance of the minus sign in the answer.

FIGURE 3.4.2 Shadow in Problem 45

46. The two ends of a 10-ft board are attached to perpendicular rails, as shown in FIGURE 3.4.3, so that point P is free to move vertically and point R is free to move horizontally.
    (a) Express the area A of triangle PQR as a function of the indicated angle θ.
    (b) Find the rate of change of A with respect to θ.
    (c) Initially the board rests flat on the horizontal rail. Suppose point R is then moved in the direction of point Q, thereby forcing point P to move up the vertical rail. Initially the area of the triangle is 0 ($\theta = 0$), but then it increases for a while as θ increases and then decreases as R approaches Q. When the board is vertical, the area of the triangle is again 0 ($\theta = \pi/2$). Graph the derivative $dA/d\theta$. Interpret this graph to find values of θ for which A is increasing and values of θ for which A is decreasing. Now verify your interpretation of the graph of the derivative by graphing $A(\theta)$.
    (d) Use the graphs in part (c) to find the value of θ for which the area of the triangle is the greatest.

FIGURE 3.4.3 Board in Problem 46

≡ Think About It

47. (a) Find all positive integers n such that
$$\frac{d^n}{dx^n}\sin x = \sin x;\quad \frac{d^n}{dx^n}\cos x = \cos x;$$
$$\frac{d^n}{dx^n}\cos x = \sin x;\quad \frac{d^n}{dx^n}\sin x = \cos x.$$
    (b) Use the results in part (a) as an aid in finding
$$\frac{d^{21}}{dx^{21}}\sin x,\quad \frac{d^{30}}{dx^{30}}\sin x,\quad \frac{d^{40}}{dx^{40}}\cos x,\quad \text{and}\quad \frac{d^{67}}{dx^{67}}\cos x.$$

48. Find two distinct points P_1 and P_2 on the graph of $y = \cos x$ so that the tangent line at P_1 is perpendicular to the tangent line at P_2.

49. Find two distinct points P_1 and P_2 on the graph of $y = \sin x$ so that the tangent line at P_1 is parallel to the tangent line at P_2.

50. Use (1), (2), and the sum formula for the cosine to show that
$$\frac{d}{dx}\cos x = -\sin x.$$

51. Use (4) and (5) and the Quotient Rule to show that
$$\frac{d}{dx}\cot x = -\csc^2 x.$$

52. Use (4) and the Quotient Rule to show that
$$\frac{d}{dx}\csc x = -\csc x \cot x.$$

≡ Calculator/CAS Problems

In Problems 53 and 54, use a calculator or CAS to obtain the graph of the given function. By inspection of the graph indicate where the function may not be differentiable.

53. $f(x) = 0.5(\sin x + |\sin x|)$
54. $f(x) = |x + \sin x|$

55. As shown in FIGURE 3.4.4, a boy pulls a sled on which his little sister is seated. If the sled and girl weigh a total of 70 lb, and if the coefficient of sliding friction of snow-covered ground is 0.2, then the magnitude F of the force (measured in pounds) required to move the sled is
$$F = \frac{70(0.2)}{0.2\sin\theta + \cos\theta},$$
where θ is the angle the tow rope makes with the horizontal.
    (a) Use a calculator or CAS to obtain the graph of F on the interval $[-1, 1]$.
    (b) Find the derivative $dF/d\theta$.
    (c) Find the angle (in radians) for which $dF/d\theta = 0$.
    (d) Find the value of F corresponding to the angle found in part (c).
    (e) Use the graph in part (a) as an aid in interpreting the numbers found in parts (c) and (d).

FIGURE 3.4.4 Sled in Problem 55

3.5 Chain Rule

Introduction As discussed in Section 3.2, the Power Rule

$$\frac{d}{dx}x^n = nx^{n-1}$$

is valid for all real number exponents n. In this section we see that a similar rule holds for the derivative of a power of a function $y = [g(x)]^n$. Before stating the formal result, let us consider an example when n is a positive integer.

Suppose we wish to differentiate

$$y = (x^5 + 1)^2. \tag{1}$$

By writing (1) as $y = (x^5 + 1) \cdot (x^5 + 1)$, we can find the derivative using the Product Rule:

$$\frac{d}{dx}(x^5 + 1)^2 = (x^5 + 1) \cdot \frac{d}{dx}(x^5 + 1) + (x^5 + 1) \cdot \frac{d}{dx}(x^5 + 1)$$

$$= (x^5 + 1) \cdot 5x^4 + (x^5 + 1) \cdot 5x^4$$

$$= 2(x^5 + 1) \cdot 5x^4. \tag{2}$$

Similarly, to differentiate the function $y = (x^5 + 1)^3$, we can write it as $y = (x^5 + 1)^2 \cdot (x^5 + 1)$ and use the Product Rule and the result given in (2):

$$\frac{d}{dx}(x^5 + 1)^3 = \frac{d}{dx}(x^5 + 1)^2 \cdot (x^5 + 1)$$

$$= (x^5 + 1)^2 \cdot \frac{d}{dx}(x^5 + 1) + (x^5 + 1) \cdot \overbrace{\frac{d}{dx}(x^5 + 1)^2}^{\text{we know this from (2)}}$$

$$= (x^5 + 1)^2 \cdot 5x^4 + (x^5 + 1) \cdot 2(x^5 + 1) \cdot 5x^4$$

$$= 3(x^5 + 1)^2 \cdot 5x^4. \tag{3}$$

In like manner, by writing $y = (x^5 + 1)^4$ as $y = (x^5 + 1)^3 \cdot (x^5 + 1)$ we can readily show by the Product Rule and (3) that

$$\frac{d}{dx}(x^5 + 1)^4 = 4(x^5 + 1)^3 \cdot 5x^4. \tag{4}$$

Power Rule for Functions Inspection of (2), (3), and (4) reveals a pattern for differentiating a power of a function g. For example, in (4) we see

$$\underset{\uparrow}{4}(x^5 + 1)^{\underset{\uparrow}{3}} \cdot \underset{\uparrow}{5x^4}$$

bring down exponent as a multiple ↓   ↓ derivative of function inside parentheses
↑ decrease exponent by 1

For emphasis, if we denote a differentiable function by [ ], it appears that

$$\frac{d}{dx}[\ ]^n = n[\ ]^{n-1}\frac{d}{dx}[\ ].$$

The foregoing discussion suggests the result stated in the next theorem.

Theorem 3.5.1   Power Rule for Functions

If n is any real number and $u = g(x)$ is differentiable at x, then

$$\frac{d}{dx}[g(x)]^n = n[g(x)]^{n-1} \cdot g'(x), \tag{5}$$

or equivalently,

$$\frac{d}{dx}u^n = nu^{n-1} \cdot \frac{du}{dx}. \tag{6}$$

Theorem 3.5.1 is itself a special case of a more general theorem, called the **Chain Rule**, which will be presented after we consider some examples of this new power rule.

EXAMPLE 1 Power Rule for Functions

Differentiate $y = (4x^3 + 3x + 1)^7$.

Solution With the identification that $u = g(x) = 4x^3 + 3x + 1$, we see from (6) that

$$\frac{dy}{dx} = \overbrace{7}^{n}\overbrace{(4x^3 + 3x + 1)^6}^{u^{n-1}} \cdot \overbrace{\frac{d}{dx}(4x^3 + 3x + 1)}^{du/dx} = 7(4x^3 + 3x + 1)^6(12x^2 + 3).$$

EXAMPLE 2 Power Rule for Functions

To differentiate $y = 1/(x^2 + 1)$, we could, of course, use the Quotient Rule. However, by rewriting the function as $y = (x^2 + 1)^{-1}$, it is also possible to use the Power Rule for Functions with $n = -1$:

$$\frac{dy}{dx} = (-1)(x^2 + 1)^{-2} \cdot \frac{d}{dx}(x^2 + 1) = (-1)(x^2 + 1)^{-2} 2x = \frac{-2x}{(x^2 + 1)^2}.$$

EXAMPLE 3 Power Rule for Functions

Differentiate $y = \dfrac{1}{(7x^5 - x^4 + 2)^{10}}$.

Solution Write the given function as $y = (7x^5 - x^4 + 2)^{-10}$. Identify $u = 7x^5 - x^4 + 2$, $n = -10$ and use the Power Rule (6):

$$\frac{dy}{dx} = -10(7x^5 - x^4 + 2)^{-11} \cdot \frac{d}{dx}(7x^5 - x^4 + 2) = \frac{-10(35x^4 - 4x^3)}{(7x^5 - x^4 + 2)^{11}}.$$

EXAMPLE 4 Power Rule for Functions

Differentiate $y = \tan^3 x$.

Solution For emphasis, we first rewrite the function as $y = (\tan x)^3$ and then use (6) with $u = \tan x$ and $n = 3$:

$$\frac{dy}{dx} = 3(\tan x)^2 \cdot \frac{d}{dx}\tan x.$$

Recall from (6) of Section 3.4 that $(d/dx)\tan x = \sec^2 x$. Hence,

$$\frac{dy}{dx} = 3\tan^2 x \sec^2 x.$$

EXAMPLE 5 Quotient Rule then Power Rule

Differentiate $y = \dfrac{(x^2 - 1)^3}{(5x + 1)^8}$.

Solution We start with the Quotient Rule followed by two applications of the Power Rule for Functions:

$$\frac{dy}{dx} = \frac{(5x+1)^8 \cdot \overset{\downarrow\text{Power Rule for Functions}}{\frac{d}{dx}(x^2-1)^3} - (x^2-1)^3 \cdot \overset{\downarrow}{\frac{d}{dx}(5x+1)^8}}{(5x+1)^{16}}$$

$$= \frac{(5x+1)^8 \cdot 3(x^2-1)^2 \cdot 2x - (x^2-1)^3 \cdot 8(5x+1)^7 \cdot 5}{(5x+1)^{16}}$$

$$= \frac{6x(5x+1)^8(x^2-1)^2 - 40(5x+1)^7(x^2-1)^3}{(5x+1)^{16}}$$

$$= \frac{(x^2-1)^2(-10x^2+6x+40)}{(5x+1)^9}.$$

EXAMPLE 6 Power Rule then Quotient Rule

Differentiate $y = \sqrt{\dfrac{2x-3}{8x+1}}$.

Solution By rewriting the function as

$$y = \left(\frac{2x-3}{8x+1}\right)^{1/2} \quad \text{we can identify} \quad u = \frac{2x-3}{8x+1}$$

and $n = \frac{1}{2}$. Thus in order to compute du/dx in (6) we must use the Quotient Rule:

$$\frac{dy}{dx} = \frac{1}{2}\left(\frac{2x-3}{8x+1}\right)^{-1/2} \cdot \frac{d}{dx}\left(\frac{2x-3}{8x+1}\right)$$

$$= \frac{1}{2}\left(\frac{2x-3}{8x+1}\right)^{-1/2} \cdot \frac{(8x+1) \cdot 2 - (2x-3) \cdot 8}{(8x+1)^2}$$

$$= \frac{1}{2}\left(\frac{2x-3}{8x+1}\right)^{-1/2} \cdot \frac{26}{(8x+1)^2}.$$

Finally, we simplify using the laws of exponents:

$$\frac{dy}{dx} = \frac{13}{(2x-3)^{1/2}(8x+1)^{3/2}}.$$

■ **Chain Rule** A power of a function can be written as a composite function. If we identify $f(x) = x^n$ and $u = g(x)$, then $f(u) = f(g(x)) = [g(x)]^n$. The Chain Rule gives us a way of differentiating any composition $f \circ g$ of two differentiable functions f and g.

Theorem 3.5.2   Chain Rule

If the function f is differentiable at $u = g(x)$, and the function g is differentiable at x, then the composition $y = (f \circ g)(x) = f(g(x))$ is differentiable at x and

$$\frac{d}{dx}f(g(x)) = f'(g(x)) \cdot g'(x) \tag{7}$$

or equivalently,

$$\frac{dy}{dx} = \frac{dy}{du} \cdot \frac{du}{dx}. \tag{8}$$

PROOF FOR $\Delta u \neq 0$ In this partial proof it is convenient to use the form of the definition of the derivative given in (3) of Section 3.1. For $\Delta x \neq 0$,

$$\Delta u = g(x + \Delta x) - g(x) \tag{9}$$

or $g(x + \Delta x) = g(x) + \Delta u = u + \Delta u$. In addition,

$$\Delta y = f(u + \Delta u) - f(u) = f(g(x + \Delta x)) - f(g(x)).$$

When x and $x + \Delta x$ are in some open interval for which $\Delta u \neq 0$, we can write

$$\frac{\Delta y}{\Delta x} = \frac{\Delta y}{\Delta u} \cdot \frac{\Delta u}{\Delta x}.$$

Since g is assumed to be differentiable, it is continuous. Consequently, as $\Delta x \to 0$, $g(x + \Delta x) \to g(x)$, and so from (9) we see that $\Delta u \to 0$. Thus,

$$\lim_{\Delta x \to 0} \frac{\Delta y}{\Delta x} = \left(\lim_{\Delta x \to 0} \frac{\Delta y}{\Delta u}\right) \cdot \left(\lim_{\Delta x \to 0} \frac{\Delta u}{\Delta x}\right)$$

$$= \left(\lim_{\Delta u \to 0} \frac{\Delta y}{\Delta u}\right) \cdot \left(\lim_{\Delta x \to 0} \frac{\Delta u}{\Delta x}\right). \quad \leftarrow \text{note that } \Delta u \to 0 \text{ in the first term}$$

From the definition of the derivative, (3) of Section 3.1, it follows that

$$\frac{dy}{dx} = \frac{dy}{du} \cdot \frac{du}{dx}. \qquad \blacksquare$$

The assumption that $\Delta u \neq 0$ on some interval does not hold true for every differentiable function g. Although the result given in (7) remains valid when $\Delta u = 0$, the preceding proof does not.

It might help in the understanding of the derivative of a composition $y = f(g(x))$ to think of f as the *outside function* and $u = g(x)$ as the *inside function*. The derivative of $y = f(g(x)) = f(u)$ is then the *product of the derivative of the outside function* (evaluated at the inside function u) *and the derivative of the inside function* (evaluated at x):

derivative of outside function
↓
$$\frac{d}{dx} f(u) = f'(u) \cdot u'. \qquad (10)$$
↑
derivative of inside function

The result in (10) is written in various ways. Since $y = f(u)$, we have $f'(u) = dy/du$, and of course $u' = du/dx$. The product of the derivatives in (10) is the same as (8). On the other hand, if we replace the symbols u and u' in (10) by $g(x)$ and $g'(x)$ we obtain (7).

■ **Proof of the Power Rule for Functions** As noted previously, a power of a function can be written as a composition of $(f \circ g)(x)$ where the outside function is $y = f(x) = x^n$ and the inside function is $u = g(x)$. The derivative of the inside function $y = f(u) = u^n$ is $\frac{dy}{dx} = nu^{n-1}$ and the derivative of the outside function is $\frac{du}{dx}$. The product of these derivatives is then

$$\frac{dy}{dx} = \frac{dy}{du} \cdot \frac{du}{dx} = nu^{n-1} \frac{du}{dx} = n[g(x)]^{n-1} g'(x).$$

This is the Power Rule for Functions given in (5) and (6).

■ **Trigonometric Functions** We obtain the derivatives of the trigonometric functions composed with a differentiable function g as another direct consequence of the Chain Rule. For example, if $y = \sin u$, where $u = g(x)$, then the derivative of y with respect to the variable u is

$$\frac{dy}{du} = \cos u.$$

Hence, (8) gives

$$\frac{dy}{dx} = \frac{dy}{du} \cdot \frac{du}{dx} = \cos u \frac{du}{dx}$$

or equivalently,

$$\frac{d}{dx} \sin[\ ] = \cos[\ ] \frac{d}{dx}[\ ].$$

Similarly, if $y = \tan u$ where $u = g(x)$, then $dy/du = \sec^2 u$ and so

$$\frac{dy}{dx} = \frac{dy}{du} \cdot \frac{du}{dx} = \sec^2 u \frac{du}{dx}.$$

We summarize the Chain Rule results for the six trigonometric functions.

Theorem 3.5.3 Derivatives of Trigonometric Functions

If $u = g(x)$ is a differentiable function, then

$$\frac{d}{dx}\sin u = \cos u \frac{du}{dx}, \qquad \frac{d}{dx}\cos u = -\sin u \frac{du}{dx}, \qquad (11)$$

$$\frac{d}{dx}\tan u = \sec^2 u \frac{du}{dx}, \qquad \frac{d}{dx}\cot u = -\csc^2 u \frac{du}{dx}, \qquad (12)$$

$$\frac{d}{dx}\sec u = \sec u \tan u \frac{du}{dx}, \qquad \frac{d}{dx}\csc u = -\csc u \cot u \frac{du}{dx}. \qquad (13)$$

EXAMPLE 7 Chain Rule

Differentiate $y = \cos 4x$.

Solution The function is $\cos u$ with $u = 4x$. From the second formula in (11) of Theorem 3.5.3 the derivative is

$$\frac{dy}{dx} = \overbrace{-\sin 4x}^{\frac{dy}{du}} \cdot \overbrace{\frac{d}{dx}4x}^{\frac{du}{dx}} = -4\sin 4x.$$

EXAMPLE 8 Chain Rule

Differentiate $y = \tan(6x^2 + 1)$.

Solution The function is $\tan u$ with $u = 6x^2 + 1$. From the first formula in (12) of Theorem 3.5.3 the derivative is

$$\frac{dy}{dx} = \overbrace{\sec^2(6x^2 + 1)}^{\sec^2 u} \cdot \overbrace{\frac{d}{dx}(6x^2 + 1)}^{\frac{du}{dx}} = 12x\sec^2(6x^2 + 1).$$

EXAMPLE 9 Product, Power, and Chain Rule

Differentiate $y = (9x^3 + 1)^2 \sin 5x$.

Solution We first use the Product Rule:

$$\frac{dy}{dx} = (9x^3 + 1)^2 \cdot \frac{d}{dx}\sin 5x + \sin 5x \cdot \frac{d}{dx}(9x^3 + 1)^2$$

followed by the Power Rule (6) and the first formula in (11) of Theorem 3.5.3,

$$\frac{dy}{dx} = (9x^3 + 1)^2 \cdot \overset{\text{from (11)}}{\cos 5x} \cdot \frac{d}{dx}5x + \sin 5x \cdot \overset{\text{from (6)}}{2(9x^3 + 1)} \cdot \frac{d}{dx}(9x^3 + 1)$$

$$= (9x^3 + 1)^2 \cdot 5\cos 5x + \sin 5x \cdot 2(9x^3 + 1) \cdot 27x^2$$

$$= (9x^3 + 1)(45x^3 \cos 5x + 5\cos 5x + 54x^2 \sin 5x).$$

In Sections 3.2 and 3.3 we saw that even though the Sum and Product Rules were stated in terms of two functions f and g, they were applicable to any finite number of differentiable functions. So too, the Chain Rule is stated for the composition of two functions f and g but we can apply it to the composition of three (or more) differentiable functions. In the case of three functions f, g, and h, (7) becomes

$$\frac{d}{dx}f(g(h(x))) = f'(g(h(x))) \cdot \frac{d}{dx}g(h(x))$$

$$= f'(g(h(x))) \cdot g'(h(x)) \cdot h'(x).$$

EXAMPLE 10 Repeated Use of the Chain Rule

Differentiate $y = \cos^4(7x^3 + 6x - 1)$.

Solution For emphasis we first rewrite the given function as $y = [\cos(7x^3 + 6x - 1)]^4$. Observe that this function is the composition $(f \circ g \circ h)(x) = f(g(h(x)))$ where $f(x) = x^4$, $g(x) = \cos x$, and $h(x) = 7x^3 + 6x - 1$. We first apply the Chain Rule in the form of the Power Rule (6) followed by the second formula in (11):

$$\frac{dy}{dx} = 4[\cos(7x^3 + 6x - 1)]^3 \cdot \frac{d}{dx}\cos(7x^3 + 6x - 1) \qquad \leftarrow \text{first Chain Rule: differentiate the power}$$

$$= 4\cos^3(7x^3 + 6x - 1) \cdot \left[-\sin(7x^3 + 6x - 1) \cdot \frac{d}{dx}(7x^3 + 6x - 1)\right] \qquad \leftarrow \text{second Chain Rule: differentiate the cosine}$$

$$= -4(21x^2 + 6)\cos^3(7x^3 + 6x - 1)\sin(7x^3 + 6x - 1).$$

In the final example, the given function is a composition of four functions.

EXAMPLE 11 Repeated Use of the Chain Rule

Differentiate $y = \sin(\tan\sqrt{3x^2 + 4})$.

Solution The function is $f(g(h(k(x))))$, where $f(x) = \sin x$, $g(x) = \tan x$, $h(x) = \sqrt{x}$, and $k(x) = 3x^2 + 4$. In this case we apply the Chain Rule three times in succession

$$\frac{dy}{dx} = \cos(\tan\sqrt{3x^2 + 4}) \cdot \frac{d}{dx}\tan\sqrt{3x^2 + 4} \qquad \leftarrow \text{first Chain Rule: differentiate the sine}$$

$$= \cos(\tan\sqrt{3x^2 + 4}) \cdot \sec^2\sqrt{3x^2 + 4} \cdot \frac{d}{dx}\sqrt{3x^2 + 4} \qquad \leftarrow \text{second Chain Rule: differentiate the tangent}$$

$$= \cos(\tan\sqrt{3x^2 + 4}) \cdot \sec^2\sqrt{3x^2 + 4} \cdot \frac{d}{dx}(3x^2 + 4)^{1/2} \qquad \leftarrow \text{rewrite power}$$

$$= \cos(\tan\sqrt{3x^2 + 4}) \cdot \sec^2\sqrt{3x^2 + 4} \cdot \frac{1}{2}(3x^2 + 4)^{-1/2} \cdot \frac{d}{dx}(3x^2 + 4) \qquad \leftarrow \text{third Chain Rule: differentiate the power}$$

$$= \cos(\tan\sqrt{3x^2 + 4}) \cdot \sec^2\sqrt{3x^2 + 4} \cdot \frac{1}{2}(3x^2 + 4)^{-1/2} \cdot 6x \qquad \leftarrow \text{simplify}$$

$$= \frac{3x\cos(\tan\sqrt{3x^2 + 4}) \cdot \sec^2\sqrt{3x^2 + 4}}{\sqrt{3x^2 + 4}}.$$

You should, of course, become so adept at applying the Chain Rule that you will not have to give a moment's thought as to the number of functions involved in the actual composition.

$\frac{d}{dx}$ NOTES FROM THE CLASSROOM

(*i*) Probably the most common mistake is to forget to carry out the second half of the Chain Rule, namely the derivative of the inside function. This is the du/dx part in

$$\frac{dy}{dx} = \frac{dy}{du}\frac{du}{dx}.$$

For instance, the derivative of $y = (1 - x)^{57}$ is not $dy/dx = 57(1 - x)^{56}$ since $57(1 - x)^{56}$ is only the dy/du part. It might help to consistently use the operation symbol d/dx:

$$\frac{d}{dx}(1 - x)^{57} = 57(1 - x)^{56} \cdot \frac{d}{dx}(1 - x) = 57(1 - x)^{56} \cdot (-1).$$

(*ii*) A less common but probably a worse mistake than the first is to differentiate inside the given function. A student wrote on an examination paper that the derivative of $y = \cos(x^2 + 1)$ was $dy/dx = -\sin(2x)$; that is, the derivative of the cosine is the negative of the sine and the derivative of $x^2 + 1$ is $2x$. Both observations are correct, but how they are put together is incorrect. Bear in mind that the derivative of the inside function is a multiple of the derivative of the outside function. Again, it might help to use the operation symbol d/dx. The correct derivative of $y = \cos(x^2 + 1)$ is the product of two derivatives.

$$\frac{dy}{dx} = -\sin(x^2 + 1) \cdot \frac{d}{dx}(x^2 + 1) = -2x\sin(x^2 + 1).$$

Exercises 3.5   Answers to selected odd-numbered problems begin on page ANS-11.

≡ Fundamentals

In Problems 1–20, find dy/dx.

1. $y = (-5x)^{30}$
2. $y = (3/x)^{14}$
3. $y = (2x^2 + x)^{200}$
4. $y = \left(x - \dfrac{1}{x^2}\right)^5$
5. $y = \dfrac{1}{(x^3 - 2x^2 + 7)^4}$
6. $y = \dfrac{10}{\sqrt{x^2 - 4x + 1}}$
7. $y = (3x - 1)^4(-2x + 9)^5$
8. $y = x^4(x^2 + 1)^6$
9. $y = \sin\sqrt{2x}$
10. $y = \sec x^2$
11. $y = \sqrt{\dfrac{x^2 - 1}{x^2 + 1}}$
12. $y = \dfrac{3x - 4}{(5x + 2)^3}$
13. $y = [x + (x^2 - 4)^3]^{10}$
14. $y = \left[\dfrac{1}{(x^3 - x + 1)^2}\right]^4$
15. $y = x(x^{-1} + x^{-2} + x^{-3})^{-4}$
16. $y = (2x + 1)^3\sqrt{3x^2 - 2x}$
17. $y = \sin(\pi x + 1)$
18. $y = -2\cos(-3x + 7)$
19. $y = \sin^3 5x$
20. $y = 4\cos^2\sqrt{x}$

In Problems 21–38, find $f'(x)$.

21. $f(x) = x^3\cos x^3$
22. $f(x) = \dfrac{\sin 5x}{\cos 6x}$
23. $f(x) = (2 + x\sin 3x)^{10}$
24. $f(x) = \dfrac{(1 - \cos 4x)^2}{(1 + \sin 5x)^3}$
25. $f(x) = \tan(1/x)$
26. $f(x) = x\cot(5/x^2)$
27. $f(x) = \sin 2x \cos 3x$
28. $f(x) = \sin^2 2x \cos^3 3x$
29. $f(x) = (\sec 4x + \tan 2x)^5$
30. $f(x) = \csc^2 2x - \csc 2x^2$
31. $f(x) = \sin(\sin 2x)$
32. $f(x) = \tan\left(\cos\dfrac{x}{2}\right)$
33. $f(x) = \cos(\sin\sqrt{2x + 5})$
34. $f(x) = \tan(\tan x)$
35. $f(x) = \sin^3(4x^2 - 1)$
36. $f(x) = \sec(\tan^2 x^4)$
37. $f(x) = (1 + (1 + (1 + x^3)^4)^5)^6$
38. $f(x) = \left[x^2 - \left(1 + \dfrac{1}{x}\right)^{-4}\right]^2$

In Problems 39–42, find the slope of the tangent line to the graph of the given function at the indicated value of x.

39. $y = (x^2 + 2)^3$; $x = -1$
40. $y = \dfrac{1}{(3x + 1)^2}$; $x = 0$
41. $y = \sin 3x + 4x\cos 5x$; $x = \pi$
42. $y = 50x - \tan^3 2x$; $x = \pi/6$

In Problems 43–46, find an equation of the tangent line to the graph of the given function at the indicated value of x.

43. $y = \left(\dfrac{x}{x + 1}\right)^2$; $x = -\dfrac{1}{2}$
44. $y = x^2(x - 1)^3$; $x = 2$
45. $y = \tan 3x$; $x = \pi/4$
46. $y = (-1 + \cos 4x)^3$; $x = \pi/8$

In Problems 47 and 48, find an equation of the normal line to the graph of the given function at the indicated value of x.

47. $y = \sin\left(\dfrac{\pi}{6x}\right)\cos(\pi x^2)$; $x = \dfrac{1}{2}$
48. $y = \sin^3\dfrac{x}{3}$; $x = \pi$

In Problems 49–52, find the indicated derivative.

49. $f(x) = \sin \pi x$; $f'''(x)$
50. $y = \cos(2x + 1)$; d^5y/dx^5
51. $y = x\sin 5x$; d^3y/dx^3
52. $f(x) = \cos x^2$; $f''(x)$
53. Find the point(s) on the graph of $f(x) = x/(x^2 + 1)^2$ where the tangent line is horizontal. Does the graph of f have any vertical tangents?
54. Determine the values of t at which the instantaneous rate of change of $g(t) = \sin t + \frac{1}{2}\cos 2t$ is zero.
55. If $f(x) = \cos(x/3)$, what is the slope of the tangent line to the graph of f' at $x = 2\pi$?
56. If $f(x) = (1 - x)^4$, what is the slope of the tangent line to the graph of f'' at $x = 2$?

Applications

57. The function $R = (v_0^2/g)\sin 2\theta$ gives the range of a projectile fired at an angle θ from the horizontal with an initial velocity v_0. If v_0 and g are constants, find those values of θ at which $dR/d\theta = 0$.

58. The volume of a spherical balloon of radius r is $V = \frac{4}{3}\pi r^3$. The radius is a function of time t and increases at a constant rate of 5 in/min. What is the instantaneous rate of change of V with respect to t?

59. Suppose a spherical balloon is being filled at a constant rate $dV/dt = 10$ in³/min. At what rate is its radius increasing when $r = 2$ in?

60. Consider a mass on a spring shown in FIGURE 3.5.1. In the absence of damping forces, the displacement (or directed distance) of the mass measured from a position called the **equilibrium position** is given by the function

$$x(t) = x_0\cos\omega t + \frac{v_0}{\omega}\sin\omega t,$$

where $\omega = \sqrt{k/m}$, k is the spring constant (an indicator of the stiffness of the spring), m is the mass (measured in slugs or kilograms), y_0 is the initial displacement of the mass (measured above or below the equilibrium position), v_0 is the initial velocity of the mass, and t is time measured in seconds.

FIGURE 3.5.1 Mass on a spring in Problem 60

(a) Verify that $x(t)$ satisfies the differential equation

$$\frac{d^2x}{dt^2} + \omega^2 x = 0.$$

(b) Verify that $x(t)$ satisfies the initial conditions $x(0) = x_0$ and $x'(0) = v_0$.

Think About It

61. Let F be a differentiable function. What is $\frac{d}{dx}F(3x)$?

62. Let G be a differentiable function. What is $\frac{d}{dx}[G(-x^2)]^2$?

63. Suppose $\frac{d}{du}f(u) = \frac{1}{u}$. What is $\frac{d}{dx}f(-10x + 7)$?

64. Suppose $\frac{d}{dx}f(x) = \frac{1}{1+x^2}$. What is $\frac{d}{dx}f(x^3)$?

In Problems 65 and 66, the symbol n represents a positive integer. Find a formula for the given derivative.

65. $\dfrac{d^n}{dx^n}(1 + 2x)^{-1}$ **66.** $\dfrac{d^n}{dx^n}\sqrt{1 + 2x}$

67. Suppose $g(t) = h(f(t))$, where $f(1) = 3$, $f'(1) = 6$, and $h'(3) = -2$. What is $g'(1)$?

68. Suppose $g(1) = 2, g'(1) = 3, g''(1) - 1, f'(2) = 4$, and $f''(2) = 3$. What is $\dfrac{d^2}{dx^2}f(g(x))\Big|_{x=1}$?

69. Given that f is an odd differentiable function, use the Chain Rule to show that f' is an even function.

70. Given that f is an even differentiable function, use the Chain Rule to show that f' is an odd function.

3.6 Implicit Differentiation

Introduction The graphs of many equations that we study in mathematics are not the graphs of functions. For example, the equation

$$x^2 + y^2 = 4 \tag{1}$$

describes a circle of radius 2 centered at the origin. Equation (1) is not a function, since for any choice of x satisfying $-2 < x < 2$ there corresponds two values of y. See FIGURE 3.6.1(a). Nevertheless, graphs of equations such as (1) can possess tangent lines at various points (x, y). Equation (1) defines *at least* two functions f and g on the interval $[-2, 2]$. Graphically, the obvious functions are the top half and the bottom half of the circle. To obtain formulas for these functions we solve $x^2 + y^2 = 4$ for y in terms of x:

$$y = f(x) = \sqrt{4 - x^2}, \quad \leftarrow \text{upper semicircle} \tag{2}$$

and

$$y = g(x) = -\sqrt{4 - x^2}. \quad \leftarrow \text{lower semicircle} \tag{3}$$

See Figures 3.6.1(b) and (c). We can now find slopes of tangent lines for $-2 < x < 2$ by differentiating (2) and (3) by the Power Rule for Functions.

In this section we will see how the derivative dy/dx can be obtained for (1), as well as for more complicated equations $F(x, y) = 0$, without the necessity of solving the equation for the variable y.

■ Explicit and Implicit Functions A function in which the dependent variable is expressed solely in terms of the independent variable x, namely, $y = f(x)$, is said to be an **explicit function**. For example, $y = \frac{1}{2}x^3 - 1$ is an explicit function. On the other hand, an equivalent equation $2y - x^3 + 2 = 0$ is said to define the function **implicitly**, or y is an **implicit function** of x. We have just seen that the equation $x^2 + y^2 = 4$ defines the two functions $f(x) = \sqrt{4 - x^2}$ and $g(x) = -\sqrt{4 - x^2}$ implicitly.

In general, if an equation $F(x, y) = 0$ defines a function f implicitly on some interval, then $F(x, f(x)) = 0$ is an identity on the interval. The graph of f is a portion or an arc (or all) of the graph of the equation $F(x, y) = 0$. In the case of the functions in (2) and (3), note that both equations

$$x^2 + [f(x)]^2 = 4 \quad \text{and} \quad x^2 + [g(x)]^2 = 4$$

are identities on the interval $[-2, 2]$.

The graph of the equation $x^3 + y^3 = 3xy$ shown in FIGURE 3.6.2(a) is a famous curve called the **Folium of Descartes**. With the aid of a CAS such as *Mathematica* or *Maple*, one of the implicit functions defined by $x^3 + y^3 = 3xy$ is found to be

$$y = \frac{2x}{\sqrt[3]{-4x^3 + 4\sqrt{x^6 - 4x^3}}} + \frac{1}{2}\sqrt[3]{-4x^3 + 4\sqrt{x^6 - 4x^3}}. \tag{4}$$

(a) Not a function

(b) Function

(c) Function

FIGURE 3.6.1 Equation $x^2 + y^2 = 4$ determines at least two functions

The graph of this function is the red arc shown in Figure 3.6.2(b). The graph of another implicit function defined by $x^3 + y^3 = 3xy$ is given in Figure 3.6.2(c).

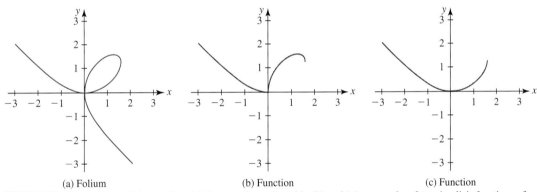

(a) Folium (b) Function (c) Function
FIGURE 3.6.2 The portions of the graph in (a) that are shown in red in (b) and (c) are graphs of two implicit functions of x

■ Implicit Differentiation Do not jump to the conclusion from the preceding discussion that we can always solve an equation $F(x, y) = 0$ for an implicit function of x as we did in (2), (3), and (4). For example, solving an equation such as

$$x^4 + x^2y^3 - y^5 = 2x + y \tag{5}$$

for y in terms of x is more than an exercise in challenging algebra or a lesson in the use of the correct syntax of a CAS. It is *impossible*! Yet (5) may determine several implicit functions on a suitably restricted interval of the x-axis. Nevertheless, we *can* determine the derivative dy/dx by a process known as **implicit differentiation**. This process consists of differentiating both sides of an equation with respect to x, using the rules of differentiation, and then solving for dy/dx. Since we think of y as being determined by the given equation as a differentiable function of x, the Chain Rule, in the form of the Power Rule for Functions, gives the useful result

$$\frac{d}{dx}y^n = ny^{n-1}\frac{dy}{dx}, \tag{6}$$

◂ Although we cannot solve certain equations for an explicit function, it still may be possible to graph the equation with the aid of a CAS. We can then *see* the functions as we did in Figure 3.6.2.

where n is any real number. For example,

$$\frac{d}{dx}x^2 = 2x \quad \text{whereas} \quad \frac{d}{dx}y^2 = 2y\frac{dy}{dx}.$$

Similarly, if y is a function of x, then by the Product Rule,

$$\frac{d}{dx}xy = x\frac{d}{dx}y + y\frac{d}{dx}x = x\frac{dy}{dx} + y,$$

and by the Chain Rule,

$$\frac{d}{dx}\sin 5y = \cos 5y \cdot \frac{d}{dx}5y = 5\cos 5y \frac{dy}{dx}.$$

Guidelines for Implicit Differentiation

(*i*) Differentiate both sides of the equation with respect to x. Use the rules of differentiation and treat y as a differentiable function of x. For powers of the symbol y use (6).
(*ii*) Collect all terms involving dy/dx on the left-hand side of the differentiated equation. Move all other terms to the right-hand side of the equation.
(*iii*) Factor dy/dx from all terms containing this term. Then solve for dy/dx.

In the following examples we shall assume that the given equation determines at least one differentiable implicit function.

EXAMPLE 1 Using Implicit Differentiation

Find dy/dx if $x^2 + y^2 = 4$.

Solution We differentiate both sides of the equation and then utilize (6):

$$\frac{d}{dx}x^2 + \overset{\text{use Power Rule (6) here}}{\frac{d}{dx}y^2} = \frac{d}{dx}4$$

$$2x + 2y\frac{dy}{dx} = 0.$$

Solving for the derivative yields

$$\frac{dy}{dx} = -\frac{x}{y}. \tag{7} \blacksquare$$

As illustrated in (7) of Example 1, implicit differentiation usually yields a derivative that depends on both variables x and y. In our introductory discussion we saw that the equation $x^2 + y^2 = 4$ defines two differentiable implicit functions on the open interval $-2 < x < 2$. The symbolism $dy/dx = -x/y$ represents the derivative of either function on the interval. Note that this derivative clearly indicates that functions (2) and (3) are not differentiable at $x = -2$ and $x = 2$ since $y = 0$ for these values of x. In general, implicit differentiation yields the derivative of any differentiable implicit function defined by an equation $F(x, y) = 0$.

EXAMPLE 2 Slope of a Tangent Line

Find the slopes of the tangent lines to the graph of $x^2 + y^2 = 4$ at the points corresponding to $x = 1$.

Solution Substituting $x = 1$ into the given equation gives $y^2 = 3$ or $y = \pm\sqrt{3}$. Hence, there are tangent lines at $(1, \sqrt{3})$ and $(1, -\sqrt{3})$. Although $(1, \sqrt{3})$ and $(1, -\sqrt{3})$ are points on the graphs of two different implicit functions, indicated by the different colors in FIGURE 3.6.3, (7) of Example 1 gives the correct slope at each number in the interval $(-2, 2)$. We have

$$\left.\frac{dy}{dx}\right|_{(1, \sqrt{3})} = -\frac{1}{\sqrt{3}} \quad \text{and} \quad \left.\frac{dy}{dx}\right|_{(1, -\sqrt{3})} = -\frac{1}{-\sqrt{3}} = \frac{1}{\sqrt{3}}. \quad \blacksquare$$

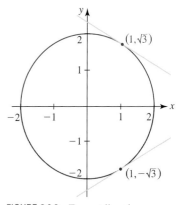

FIGURE 3.6.3 Tangent lines in Example 2 are shown in green

EXAMPLE 3 Using Implicit Differentiation

Find dy/dx if $x^4 + x^2 y^3 - y^5 = 2x + 1$.

Solution In this case, we use (6) and the Product Rule:

$$\underset{\uparrow}{\text{Product Rule here}} \qquad \underset{\uparrow}{\text{Power Rule (6) here}}$$
$$\frac{d}{dx} x^4 + \frac{d}{dx} x^2 y^3 - \frac{d}{dx} y^5 = \frac{d}{dx} 2x + \frac{d}{dx} 1$$

$$4x^3 + x^2 \cdot 3y^2 \frac{dy}{dx} + 2xy^3 - 5y^4 \frac{dy}{dx} = 2 \quad \leftarrow \text{factor } dy/dx \text{ from second and fourth terms}$$

$$(3x^2 y^2 - 5y^4) \frac{dy}{dx} = 2 - 4x^3 - 2xy^3$$

$$\frac{dy}{dx} = \frac{2 - 4x^3 - 2xy^3}{3x^2 y^2 - 5y^4}. \quad \blacksquare$$

■ **Higher Derivatives** Through implicit differentiation we determine dy/dx. By differentiating dy/dx with respect to x we obtain the second derivative d^2y/dx^2. If the first derivative contains y, then d^2y/dx^2 will again contain the symbol dy/dx; we can eliminate that quantity by substituting its known value. The next example illustrates the method.

EXAMPLE 4 Second Derivative

Find d^2y/dx^2 if $x^2 + y^2 = 4$.

Solution From Example 1, we already know that the first derivative is $dy/dx = -x/y$. The second derivative is the derivative of dy/dx, and so by the Quotient Rule:

$$\frac{d^2y}{dx^2} = -\frac{d}{dx}\left(\frac{x}{y}\right) = -\frac{y \cdot 1 - x \cdot \overset{\underset{\downarrow}{\text{substituting for } dy/dx}}{\frac{dy}{dx}}}{y^2} = -\frac{y - x\left(-\frac{x}{y}\right)}{y^2} = -\frac{y^2 + x^2}{y^3}.$$

Noting that $x^2 + y^2 = 4$ permits us to rewrite the second derivative as

$$\frac{d^2y}{dx^2} = -\frac{4}{y^3}. \quad \blacksquare$$

EXAMPLE 5 Chain and Product Rules

Find dy/dx if $\sin y = y \cos 2x$.

Solution From the Chain Rule and Product Rule we obtain

$$\frac{d}{dx} \sin y = \frac{d}{dx} y \cos 2x$$

$$\cos y \cdot \frac{dy}{dx} = y(-\sin 2x \cdot 2) + \cos 2x \cdot \frac{dy}{dx}$$

$$(\cos y - \cos 2x) \frac{dy}{dx} = -2y \sin 2x$$

$$\frac{dy}{dx} = -\frac{2y \sin 2x}{\cos y - \cos 2x}. \quad \blacksquare$$

Postscript—Power Rule Revisited

So far we have proved the Power Rule $(d/dx)x^n = nx^{n-1}$ for all integer exponents n. Implicit differentiation provides a way of proving this rule when the exponent is a rational number p/q, where p and q are integers and $q \neq 0$. In the case $n = p/q$, the function

$$y = x^{p/q} \quad \text{gives} \quad y^q = x^p.$$

Now for $y \neq 0$, implicit differentiation

$$\frac{d}{dx}y^q = \frac{d}{dx}x^p \quad \text{yields} \quad qy^{q-1}\frac{dy}{dx} = px^{p-1}.$$

Solving the last equation for dy/dx and simplifying by the laws of exponents gives

$$\frac{dy}{dx} = \frac{p}{q}\frac{x^{p-1}}{y^{q-1}} = \frac{p}{q}\frac{x^{p-1}}{(x^{p/q})^{q-1}} = \frac{p}{q}\frac{x^{p-1}}{x^{p-p/q}} = \frac{p}{q}x^{p/q-1}.$$

Examination of the last result shows that it is (3) of Section 3.2 with $n = p/q$.

Exercises 3.6   Answers to selected odd-numbered problems begin on page ANS-11.

Fundamentals

In Problems 1–4, assume that y is a differentiable function of x. Find the indicated derivative.

1. $\dfrac{d}{dx}x^2y^4$
2. $\dfrac{d}{dx}\dfrac{x^2}{y^2}$
3. $\dfrac{d}{dx}\cos y^2$
4. $\dfrac{d}{dx}y\sin 3y$

In Problems 5–24, assume that the given equation defines at least one differentiable implicit function. Use implicit differentiation to find dy/dx.

5. $y^2 - 2y = x$
6. $4x^2 + y^2 = 8$
7. $xy^2 - x^2 + 4 = 0$
8. $(y-1)^2 = 4(x+2)$
9. $3y + \cos y = x^2$
10. $y^3 - 2y + 3x^3 = 4x + 1$
11. $x^3y^2 = 2x^2 + y^2$
12. $x^5 - 6xy^3 + y^4 = 1$
13. $(x^2 + y^2)^6 = x^3 - y^3$
14. $y = (x-y)^2$
15. $y^{-3}x^6 + y^6x^{-3} = 2x + 1$
16. $y^4 - y^2 = 10x - 3$
17. $(x-1)^2 + (y+4)^2 = 25$
18. $\dfrac{x+y}{x-y} = x$
19. $y^2 = \dfrac{x-1}{x+2}$
20. $\dfrac{x}{y^2} + \dfrac{y^2}{x} = 5$
21. $xy = \sin(x+y)$
22. $x + y = \cos(xy)$
23. $x = \sec y$
24. $x\sin y - y\cos x = 1$

In Problems 25 and 26, use implicit differentiation to find the indicated derivative.

25. $r^2 = \sin 2\theta; \quad dr/d\theta$
26. $\pi r^2 h = 100; \quad dh/dr$

In Problems 27 and 28, find dy/dx at the indicated point.

27. $xy^2 + 4y^3 + 3x = 0; \quad (1, -1)$
28. $y = \sin xy; \quad (\pi/2, 1)$

In Problems 29 and 30, find dy/dx at the points that correspond to the indicated number.

29. $2y^2 + 2xy - 1 = 0; \quad x = \dfrac{1}{2}$
30. $y^3 + 2x^2 = 11y; \quad y = 1$

In Problems 31–34, find an equation of the tangent line at the indicated point or number.

31. $x^4 + y^3 = 24; \quad (-2, 2)$
32. $\dfrac{1}{x} + \dfrac{1}{y} = 1; \quad x = 3$
33. $\tan y = x; \quad y = \pi/4$
34. $3y + \cos y = x^2; \quad (1, 0)$

In Problems 35 and 36, find the point(s) on the graph of the given equation where the tangent line is horizontal.

35. $x^2 - xy + y^2 = 3$
36. $y^2 = x^2 - 4x + 7$

37. Find the point(s) on the graph of $x^2 + y^2 = 25$ at which the slope of the tangent is $\tfrac{1}{2}$.

38. Find the point where the tangent lines to the graph of $x^2 + y^2 = 25$ at $(-3, 4)$ and $(-3, -4)$ intersect.

39. Find the point(s) on the graph of $y^3 = x^2$ at which the tangent line is perpendicular to the line $y + 3x - 5 = 0$.

40. Find the point(s) on the graph of $x^2 - xy + y^2 = 27$ at which the tangent line is parallel to the line $y = 5$.

In Problems 41–48, find d^2y/dx^2.

41. $4y^3 = 6x^2 + 1$
42. $xy^4 = 5$
43. $x^2 - y^2 = 25$
44. $x^2 + 4y^2 = 16$
45. $x + y = \sin y$
46. $y^2 - x^2 = \tan 2x$
47. $x^2 + 2xy - y^2 = 1$
48. $x^3 + y^3 = 27$

In Problems 49–52, first use implicit differentiation to find dy/dx. Then solve for y explicitly in terms of x and differentiate. Show that the two answers are equivalent.

49. $x^2 - y^2 = x$
50. $4x^2 + y^2 = 1$
51. $x^3y = x + 1$
52. $y\sin x = x - 2y$

In Problems 53–56, determine an implicit function from the given equation such that its graph is the blue curve in the figure.

53. $(y - 1)^2 = x - 2$ **54.** $x^2 + xy + y^2 = 4$

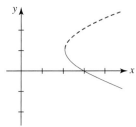

FIGURE 3.6.4 Graph for Problem 53

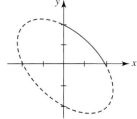

FIGURE 3.6.5 Graph for Problem 54

55. $x^2 + y^2 = 4$ **56.** $y^2 = x^2(2 - x)$

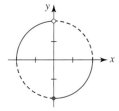

FIGURE 3.6.6 Graph for Problem 55

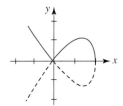

FIGURE 3.6.7 Graph for Problem 56

In Problems 57 and 58, assume that both x and y are differentiable functions of a variable t. Find dy/dt in terms of x, y, and dx/dt.

57. $x^2 + y^2 = 25$ **58.** $x^2 + xy + y^2 - y = 9$

59. The graph of the equation $x^3 + y^3 = 3xy$ is the Folium of Descartes given in Figure 3.6.2(a).

(a) Find an equation of the tangent line at the point in the first quadrant where the Folium intersects the graph of $y = x$.

(b) Find the point in the first quadrant at which the tangent line is horizontal.

60. The graph of $(x^2 + y^2)^2 = 4(x^2 - y^2)$ shown in FIGURE 3.6.8 is called a **lemniscate**.

(a) Find the points on the graph that correspond to $x = 1$.

(b) Find an equation of the tangent line to the graph at each point found in part (a).

(c) Find the points on the graph at which the tangent is horizontal.

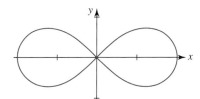

FIGURE 3.6.8 Lemniscate in Problem 60

In Problems 61 and 62, show that the graphs of the given equations are orthogonal at the indicated point of intersection. See Problem 64 in Exercises 3.2.

61. $y^2 = x^3$, $2x^2 + 3y^2 = 5$; (1, 1)
62. $y^3 + 3x^2y = 13$, $2x^2 - 2y^2 = 3x$; (2, 1)

If all the curves of one family of curves $G(x, y) = c_1$, c_1 a constant, intersect orthogonally all the curves of another family $H(x, y) = c_2$, c_2 a constant, then the families are said to be **orthogonal trajectories** of each other. In Problems 63 and 64, show that the families of curves are orthogonal trajectories of each other. Sketch the two families of curves.

63. $x^2 - y^2 = c_1$, $xy = c_2$ **64.** $x^2 + y^2 = c_1$, $y = c_2 x$

≡ Applications

65. A woman drives toward a freeway sign as shown in FIGURE 3.6.9. Let θ be her viewing angle of the sign and let x be her distance (measured in feet) to that sign.

(a) If her eye level is 4 ft from the surface of the road, show that
$$\tan \theta = \frac{4x}{x^2 + 252}.$$

(b) Find the rate at which θ changes with respect to x.

(c) At what distance is the rate in part (b) equal to zero?

FIGURE 3.6.9 Car in Problem 65

66. A jet fighter "loops the loop" in a circle of radius 1 km as shown in FIGURE 3.6.10. Suppose a rectangular coordinate system is chosen so that the origin is at the center of the circular loop. The aircraft releases a missile that flies on a straight-line path that is tangent to the circle and hits a target on the ground whose coordinates are $(2, -2)$.

(a) Determine the point on the circle where the missile was released.

(b) If a missile is released at the point $\left(-\frac{1}{2}, -\frac{\sqrt{3}}{2}\right)$ on the circle, at what point does it hit the ground?

FIGURE 3.6.10 Jet fighter in Problem 66

162 CHAPTER 3 The Derivative

≡ **Think About It**

67. The angle θ ($0 < \theta < \pi$) between two curves is defined to be the angle between their tangent lines at the point P of intersection. If m_1 and m_2 are the slopes of the tangent lines at P, it can be shown that $\tan\theta = (m_1 - m_2)/(1 + m_1 m_2)$. Determine the angle between the graphs of $x^2 + y^2 + 4y = 6$ and $x^2 + 2x + y^2 = 4$ at $(1, 1)$.

68. Show that an equation of the tangent line to the ellipse $x^2/a^2 + y^2/b^2 = 1$ at the point (x_0, y_0) is given by

$$\frac{xx_0}{a^2} + \frac{yy_0}{b^2} = 1.$$

69. Consider the equation $x^2 + y^2 = 4$. Make up another implicit function $h(x)$ defined by this equation for $-2 \leq x \leq 2$ different from the ones given in (2), (3), and Problem 55.

70. For $-1 < x < 1$ and $-\pi/2 < y < \pi/2$, the equation $x = \sin y$ defines a differentiable implicit function.
    (a) Find dy/dx in terms of y.
    (b) Find dy/dx in terms of x.

3.7 Derivatives of Inverse Functions

■ **Introduction** In Section 1.5 we saw that the graphs of a one-to-one function f and its inverse f^{-1} are **reflections** of each other in the line $y = x$. As a consequence, if (a, b) is a point on the graph of f, then (b, a) is a point on the graph of f^{-1}. In this section we will also see that the slopes of tangent lines to the graph of a differentiable function f are related to the slopes of tangents to the graph of f^{-1}.

We begin with two theorems about the continuity of f and f^{-1}.

■ **Continuity of f^{-1}** Although we state the next two theorems without proof, their plausibility follows from the fact that the graph of f^{-1} is a reflection of the graph of f in the line $y = x$.

Theorem 3.7.1 Continuity of an Inverse Function

Let f be a continuous one-to-one function on its domain X. Then f^{-1} is continuous on its domain.

■ **Increasing–Decreasing Functions** Suppose $y = f(x)$ is a function defined on an interval I, and that x_1 and x_2 are any two numbers in the interval such that $x_1 < x_2$. Then from Section 1.3 and Figure 1.3.4 recall that f is said to be

- **increasing** on the interval if $f(x_1) < f(x_2)$, and (1)
- **decreasing** on the interval if $f(x_1) > f(x_2)$. (2)

The next two theorems establish a link between the notions of increasing/decreasing and the existence of an inverse function.

Theorem 3.7.2 Existence of an Inverse Function

Let f be a continuous function and increasing on an interval $[a, b]$. Then f^{-1} exists and is continuous and increasing on $[f(a), f(b)]$.

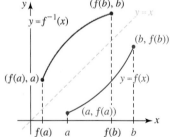

FIGURE 3.7.1 f (blue curve) and f^{-1} (red curve) are continuous and increasing

Theorem 3.7.2 also holds when the word *increasing* is replaced with the word *decreasing* and the interval in the conclusion is replaced by $[f(b), f(a)]$. See FIGURE 3.7.1. In addition, we can conclude from Theorem 3.7.2 that if f is continuous and increasing on an interval $(-\infty, \infty)$, then f^{-1} exists and is continuous and increasing on its domain. Inspection of Figures 1.3.4 and 3.7.1 also shows that if f in Theorem 3.7.2 is a differentiable function on (a, b), then:

f increasing and differentiable means ▶ the tangent lines have positive slope.

- f is increasing on the interval $[a, b]$ if $f'(x) > 0$ on (a, b), and
- f is decreasing on the interval $[a, b]$ if $f'(x) < 0$ on (a, b).

We will prove these statements in the next chapter.

Theorem 3.7.3 Differentiability of an Inverse Function

Suppose f is a differentiable function on an open interval (a, b). If either $f'(x) > 0$ on the interval or $f'(x) < 0$ on the interval, then f is one-to-one. Moreover, f^{-1} is differentiable for all x in the range of f.

EXAMPLE 1 Existence of an Inverse

Prove that $f(x) = 5x^3 + 8x - 9$ has an inverse.

Solution Since f is a polynomial function it is differentiable everywhere, that is, f is differentiable on the interval $(-\infty, \infty)$. Also, $f'(x) = 15x^2 + 8 > 0$ for all x implies that f is increasing on $(-\infty, \infty)$. It follows from Theorem 3.7.3 that f is one-to-one and hence f^{-1} exists. ∎

Derivative of f^{-1} If f is differentiable on an interval I and is one-to-one on that interval, then for a in I the point (a, b) on the graph of f and the point (b, a) on the graph of f^{-1} are mirror images of each other in the line $y = x$. As we see next, the slopes of the tangent lines at (a, b) and (b, a) are also related.

EXAMPLE 2 Derivative of an Inverse

In Example 5 of Section 1.5 we showed that the inverse of the one-to-one function $f(x) = x^2 + 1, x \geq 0$ is $f^{-1}(x) = \sqrt{x - 1}$. At $x = 2$,

$$f(2) = 5 \quad \text{and} \quad f^{-1}(5) = 2.$$

Now from

$$f'(x) = 2x \quad \text{and} \quad (f^{-1})'(x) = \frac{1}{2\sqrt{x-1}}$$

we see $f'(2) = 4$ and $(f^{-1})'(5) = \frac{1}{4}$. This shows that the slope of the tangent to the graph of f at $(2, 5)$ and the slope of the tangent to the graph of f^{-1} at $(5, 2)$ are reciprocals:

$$(f^{-1})'(5) = \frac{1}{f'(2)} \quad \text{or} \quad (f^{-1})'(5) = \frac{1}{f'(f^{-1}(5))}.$$

See FIGURE 3.7.2.

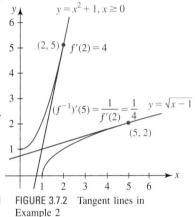

FIGURE 3.7.2 Tangent lines in Example 2

The next theorem shows that the result in Example 2 is no coincidence.

Theorem 3.7.4 Derivative of an Inverse Function

Suppose that f is differentiable on an interval I and $f'(x)$ is never zero on I. If f has an inverse f^{-1} on I, then f^{-1} is differentiable at a number x and

$$\frac{d}{dx} f^{-1}(x) = \frac{1}{f'(f^{-1}(x))}. \tag{3}$$

PROOF As we have seen in (5) of Section 1.5, $f(f^{-1}(x)) = x$ for every x in the domain of f^{-1}. By implicit differentiation and the Chain Rule,

$$\frac{d}{dx} f(f^{-1}(x)) = \frac{d}{dx} x \quad \text{or} \quad f'(f^{-1}(x)) \cdot \frac{d}{dx} f^{-1}(x) = 1.$$

Solving the last equation for $\frac{d}{dx} f^{-1}(x)$ gives (3). ∎

Equation (3) clearly shows that to find the derivative function for f^{-1} we must know $f^{-1}(x)$ explicitly. For a one-to-one function $y = f(x)$ solving the equation $x = f(y)$ for y is sometimes difficult and often impossible. In this case it is convenient to rewrite (3) using different notation. Again by implicit differentiation

$$\frac{d}{dx}x = \frac{d}{dx}f(y) \qquad \text{gives} \qquad 1 = f'(y) \cdot \frac{dy}{dx}.$$

Solving the last equation for dy/dx and writing $dx/dy = f'(y)$ yields

$$\frac{dy}{dx} = \frac{1}{dx/dy}. \qquad (4)$$

If (a, b) is a known point on the graph of f, the result in (4) enables us to evaluate the derivative of f^{-1} at (b, a) without an equation that defines $f^{-1}(x)$.

EXAMPLE 3 Derivative of an Inverse

It was pointed out in Example 1 that the polynomial function $f(x) = 5x^3 + 8x - 9$ is differentiable on $(-\infty, \infty)$ and hence continuous on the interval. Since the end behavior of f is that of the single-term polynomial function $y = 5x^3$ we can conclude that the range of f is also $(-\infty, \infty)$. Moreover, since $f'(x) = 15x^2 + 8 > 0$ for all x, f is increasing on its domain $(-\infty, \infty)$. Hence by Theorem 3.7.3, f has a differentiable inverse f^{-1} with domain $(-\infty, \infty)$. By interchanging x and y, the inverse is defined by the equation $x = 5y^3 + 8y - 9$, but solving this equation for y in terms of x is difficult (it requires the cubic formula). Nevertheless, using $dx/dy = 15y^2 + 8$, the derivative of the inverse function is given by (4):

$$\frac{dy}{dx} = \frac{1}{15y^2 + 8}. \qquad (5)$$

For example, since $f(1) = 4$ we know that $f^{-1}(4) = 1$. Thus, the slope of the tangent line to the graph of f^{-1} at $(4, 1)$ is given by (5):

$$\left.\frac{dy}{dx}\right|_{x=4} = \left.\frac{1}{15y^2 + 8}\right|_{y=1} = \frac{1}{23}. \qquad \blacksquare$$

Read this paragraph a second time. ▶ In Example 3, the derivative of the inverse function can also be obtained directly from $x = 5y^3 + 8y - 9$ using implicit differentiation:

$$\frac{d}{dx}x = \frac{d}{dx}(5y^3 + 8y - 9) \qquad \text{gives} \qquad 1 = 15y^2 \frac{dy}{dx} + 8\frac{dy}{dx}.$$

Solving the last equation for dy/dx gives (5). As a consequence of this observation implicit differentiation can be used to find the derivative of an inverse function with minimum effort. In the discussion that follows we will find the derivatives of the inverse trigonometric functions.

Derivatives of Inverse Trigonometric Functions A review of Figures 1.5.15 and 1.5.17(a) reveals that the inverse tangent and inverse cotangent are differentiable for all x. However, the remaining four inverse trigonometric functions are not differentiable at either $x = -1$ or $x = 1$. We shall confine our attention to the derivations of the derivative formulas for the inverse sine, inverse tangent, and inverse secant and leave the others as exercises.

Inverse Sine: $y = \sin^{-1}x$ if and only if $x = \sin y$, where $-1 \leq x \leq 1$ and $-\pi/2 \leq y \leq \pi/2$. Therefore, implicit differentiation

$$\frac{d}{dx}x = \frac{d}{dx}\sin y \qquad \text{gives} \qquad 1 = \cos y \cdot \frac{dy}{dx}$$

and so

$$\frac{dy}{dx} = \frac{1}{\cos y}. \qquad (6)$$

For the given restriction on the variable y, $\cos y \geq 0$ and so $\cos y = \sqrt{1 - \sin^2 y} = \sqrt{1 - x^2}$. By substituting this quantity in (6), we have shown that

$$\frac{d}{dx}\sin^{-1}x = \frac{1}{\sqrt{1 - x^2}}. \qquad (7)$$

As predicted, note that (7) is not defined at $x = -1$ and $x = 1$. The inverse sine or arcsine function is differentiable on the open interval $(-1, 1)$.

Inverse Tangent: $y = \tan^{-1}x$ if and only if $x = \tan y$, where $-\infty < x < \infty$ and $-\pi/2 < y < \pi/2$. Thus,

$$\frac{d}{dx}x = \frac{d}{dx}\tan y \quad \text{gives} \quad 1 = \sec^2 y \cdot \frac{dy}{dx}$$

or
$$\frac{dy}{dx} = \frac{1}{\sec^2 y}. \tag{8}$$

In view of the identity $\sec^2 y = 1 + \tan^2 y = 1 + x^2$, (8) becomes

$$\frac{d}{dx}\tan^{-1}x = \frac{1}{1 + x^2}. \tag{9}$$

Inverse Secant: For $|x| > 1$ and $0 \le y < \pi/2$ or $\pi/2 < y \le \pi$,

$$y = \sec^{-1}x \quad \text{if and only if} \quad x = \sec y.$$

Differentiating the last equation implicitly gives

$$\frac{dy}{dx} = \frac{1}{\sec y \tan y}. \tag{10}$$

In view of the restrictions on y, we have $\tan y = \pm\sqrt{\sec^2 y - 1} = \pm\sqrt{x^2 - 1}$, $|x| > 1$. Hence, (10) becomes

$$\frac{d}{dx}\sec^{-1}x = \pm\frac{1}{x\sqrt{x^2 - 1}}. \tag{11}$$

We can get rid of the $\pm$ sign in (11) by observing in Figure 1.5.17(b) that the slope of the tangent line to the graph of $y = \sec^{-1}x$ is positive for $x < -1$ and positive for $x > 1$. Thus, (11) is equivalent to

$$\frac{d}{dx}\sec^{-1}x = \begin{cases} -\dfrac{1}{x\sqrt{x^2 - 1}}, & x < -1 \\ \dfrac{1}{x\sqrt{x^2 - 1}}, & x > 1. \end{cases} \tag{12}$$

The result in (12) can be rewritten in a compact form using the absolute value symbol:

$$\frac{d}{dx}\sec^{-1}x = \frac{1}{|x|\sqrt{x^2 - 1}}. \tag{13}$$

The derivative of the composition of an inverse trigonometric function with a differentiable function $u = g(x)$ is obtained from the Chain Rule.

Theorem 3.7.5   Inverse Trigonometric Functions

If $u = g(x)$ is a differentiable function, then

$$\frac{d}{dx}\sin^{-1}u = \frac{1}{\sqrt{1 - u^2}}\frac{du}{dx}, \qquad \frac{d}{dx}\cos^{-1}u = \frac{-1}{\sqrt{1 - u^2}}\frac{du}{dx}, \tag{14}$$

$$\frac{d}{dx}\tan^{-1}u = \frac{1}{1 + u^2}\frac{du}{dx}, \qquad \frac{d}{dx}\cot^{-1}u = \frac{-1}{1 + u^2}\frac{du}{dx}, \tag{15}$$

$$\frac{d}{dx}\sec^{-1}u = \frac{1}{|u|\sqrt{u^2 - 1}}\frac{du}{dx}, \qquad \frac{d}{dx}\csc^{-1}u = \frac{-1}{|u|\sqrt{u^2 - 1}}\frac{du}{dx}. \tag{16}$$

In the formulas in (14) we must have $|u| < 1$, whereas in the formulas in (16) we must have $|u| > 1$.

EXAMPLE 4 Derivative of Inverse Sine

Differentiate $y = \sin^{-1} 5x$.

Solution With $u = 5x$, we have from the first formula in (14),

$$\frac{dy}{dx} = \frac{1}{\sqrt{1 - (5x)^2}} \cdot \frac{d}{dx} 5x = \frac{5}{\sqrt{1 - 25x^2}}.$$

EXAMPLE 5 Derivative of Inverse Tangent

Differentiate $y = \tan^{-1} \sqrt{2x + 1}$.

Solution With $u = \sqrt{2x + 1}$, we have from the first formula in (15),

$$\frac{dy}{dx} = \frac{1}{1 + (\sqrt{2x + 1})^2} \cdot \frac{d}{dx}(2x + 1)^{1/2}$$

$$= \frac{1}{1 + (2x + 1)} \cdot \frac{1}{2}(2x + 1)^{-1/2} \cdot 2$$

$$= \frac{1}{(2x + 2)\sqrt{2x + 1}}.$$

EXAMPLE 6 Derivative of Inverse Secant

Differentiate $y = \sec^{-1} x^2$.

Solution For $x^2 > 1 > 0$, we have from the first formula in (16),

$$\frac{dy}{dx} = \frac{1}{|x^2| \sqrt{(x^2)^2 - 1}} \cdot \frac{d}{dx} x^2$$

$$= \frac{2x}{x^2 \sqrt{x^4 - 1}} = \frac{2}{x\sqrt{x^4 - 1}}. \quad (17)$$

With the aid of a graphing utility we obtain the graph of $y = \sec^{-1} x^2$ given in FIGURE 3.7.3. Notice that (17) gives positive slope for $x > 1$ and negative slope for $x < -1$.

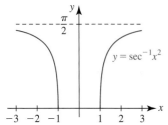

FIGURE 3.7.3 Graph of function in Example 6

EXAMPLE 7 Tangent Line

Find an equation of the tangent line to the graph of $f(x) = x^2 \cos^{-1} x$ at $x = -\frac{1}{2}$.

Solution By the Product Rule and the second formula in (14),

$$f'(x) = x^2 \left(\frac{-1}{\sqrt{1 - x^2}} \right) + 2x \cos^{-1} x.$$

Since $\cos^{-1}(-\frac{1}{2}) = 2\pi/3$, the two functions f and f' evaluated at $x = -\frac{1}{2}$ give:

$$f\left(-\frac{1}{2}\right) = \frac{\pi}{6} \qquad \leftarrow \text{point of tangency is } \left(-\frac{1}{2}, \frac{\pi}{6}\right)$$

$$f'\left(-\frac{1}{2}\right) = -\frac{1}{2\sqrt{3}} - \frac{2\pi}{3}. \quad \leftarrow \text{slope of tangent at } \left(-\frac{1}{2}, \frac{\pi}{6}\right) \text{ is } -\frac{1}{2\sqrt{3}} - \frac{2\pi}{3}$$

By the point–slope form of a line, the unsimplified equation of the tangent line is

$$y - \frac{\pi}{6} = \left(-\frac{1}{2\sqrt{3}} - \frac{2\pi}{3} \right)\left(x + \frac{1}{2} \right).$$

Since the domain of $\cos^{-1} x$ is the interval $[-1, 1]$ the domain of f is $[-1, 1]$. The corresponding range is $[0, \pi]$. FIGURE 3.7.4 was obtained with the aid of a graphing utility.

FIGURE 3.7.4 Tangent line in Example 7.

Exercises 3.7

Answers to selected odd-numbered problems begin on page ANS-11.

≡ Fundamentals

In Problems 1–4, without graphing determine whether the given function f has an inverse.

1. $f(x) = 10x^3 + 8x + 12$
2. $f(x) = -7x^5 - 6x^3 - 2x + 17$
3. $f(x) = x^3 + x^2 - 2x$
4. $f(x) = x^4 - 2x^2$

In Problems 5 and 6, use (3) to find the derivative of f^{-1} at the indicated point.

5. $f(x) = 2x^3 + 8$; $\left(f\left(\tfrac{1}{2}\right), \tfrac{1}{2}\right)$
6. $f(x) = -x^3 - 3x + 7$; $(f(-1), -1)$

In Problems 7 and 8, find f^{-1}. Use (3) to find $(f^{-1})'$ and then verify this result by direct differentiation of f^{-1}.

7. $f(x) = \dfrac{2x+1}{x}$
8. $f(x) = (5x+7)^3$

In Problems 9–12, without finding the inverse, find, at the indicated value of x, the corresponding point on the graph of f^{-1}. Then use (4) to find an equation of the tangent line at this point.

9. $y = \tfrac{1}{3}x^3 + x - 7$; $x = 3$
10. $y = \dfrac{2x+1}{4x-1}$; $x = 0$
11. $y = (x^5 + 1)^3$; $x = 1$
12. $y = 8 - 6\sqrt[3]{x+2}$; $x = -3$

In Problems 13–32, find the derivative of the given function.

13. $y = \sin^{-1}(5x - 1)$
14. $y = \cos^{-1}\left(\dfrac{x+1}{3}\right)$
15. $y = 4\cot^{-1}\dfrac{x}{2}$
16. $y = 2x - 10\sec^{-1}5x$
17. $y = 2\sqrt{x}\tan^{-1}\sqrt{x}$
18. $y = (\tan^{-1}x)(\cot^{-1}x)$
19. $y = \dfrac{\sin^{-1}2x}{\cos^{-1}2x}$
20. $y = \dfrac{\sin^{-1}x}{\sin x}$
21. $y = \dfrac{1}{\tan^{-1}x^2}$
22. $y = \dfrac{\sec^{-1}x}{x}$
23. $y = 2\sin^{-1}x + x\cos^{-1}x$
24. $y = \cot^{-1}x - \tan^{-1}\dfrac{x}{\sqrt{1-x^2}}$
25. $y = \left(x^2 - 9\tan^{-1}\dfrac{x}{3}\right)^3$
26. $y = \sqrt{x - \cos^{-1}(x+1)}$
27. $F(t) = \arctan\left(\dfrac{t-1}{t+1}\right)$
28. $g(t) = \arccos\sqrt{3t+1}$
29. $f(x) = \arcsin(\cos 4x)$
30. $f(x) = \arctan\left(\dfrac{\sin x}{2}\right)$
31. $f(x) = \tan(\sin^{-1}x^2)$
32. $f(x) = \cos(x\sin^{-1}x)$

In Problems 33 and 34, use implicit differentiation to find dy/dx.

33. $\tan^{-1}y = x^2 + y^2$
34. $\sin^{-1}y - \cos^{-1}x = 1$

In Problems 35 and 36, show that $f'(x) = 0$. Interpret the result.

35. $f(x) = \sin^{-1}x + \cos^{-1}x$
36. $f(x) = \tan^{-1}x + \tan^{-1}(1/x)$.

In Problems 37 and 38, find the slope of the tangent line to the graph of the given function at the indicated value of x.

37. $y = \sin^{-1}\dfrac{x}{2}$; $x = 1$
38. $y = (\cos^{-1}x)^2$; $x = 1/\sqrt{2}$

In Problems 39 and 40, find an equation of the tangent line to the graph of the given function at the indicated value of x.

39. $f(x) = x\tan^{-1}x$; $x = 1$
40. $f(x) = \sin^{-1}(x-1)$; $x = \dfrac{1}{2}$

41. Find the points on the graph of $f(x) = 5 - 2\sin x$, $0 \le x \le 2\pi$, at which the tangent line is parallel to the line $y = \sqrt{3}x + 1$.

42. Find all tangent lines to the graph of $f(x) = \arctan x$ that have slope $\tfrac{1}{4}$.

≡ Think About It

43. If f and $(f^{-1})'$ are differentiable, use (3) to find a formula for $(f^{-1})''(x)$.

3.8 Exponential Functions

■ **Introduction** In Section 1.6 we saw that the exponential function $f(x) = b^x$, $b > 0$, $b \ne 1$, is defined for all real numbers, that is, the domain of f is $(-\infty, \infty)$. Inspection of Figure 1.6.2 shows that f is everywhere continuous. It turns out that an exponential function is also differentiable everywhere. In this section we develop the derivative of $f(x) = b^x$.

Derivative of an Exponential Function

To find the derivative of an exponential function $f(x) = b^x$ we will use the definition of the derivative given in (2) of Definition 3.1.1. We first compute the difference quotient

$$\frac{f(x+h) - f(x)}{h} \tag{1}$$

in three steps. For the exponential function $f(x) = b^x$, we have

(i) $f(x+h) = b^{x+h} = b^x b^h$ ← laws of exponents

(ii) $f(x+h) - f(x) = b^{x+h} - b^x = b^x b^h - b^x = b^x(b^h - 1)$ ← laws of exponents and factoring

(iii) $\dfrac{f(x+h) - f(x)}{h} = \dfrac{b^x(b^h - 1)}{h} = b^x \cdot \dfrac{b^h - 1}{h}.$

In the fourth step, the calculus step, we let $h \to 0$ but analogous to the derivatives of $\sin x$ and $\cos x$ in Section 3.4, there is no apparent way of canceling the h in the difference quotient (iii). Nonetheless, the derivative of $f(x) = b^x$ is

$$f'(x) = \lim_{h \to 0} b^x \cdot \frac{b^h - 1}{h}. \tag{2}$$

Because b^x does not depend on the variable h, we can rewrite (2) as

$$f'(x) = b^x \cdot \lim_{h \to 0} \frac{b^h - 1}{h}. \tag{3}$$

Now here are the amazing results. The limit in (3),

$$\lim_{h \to 0} \frac{b^h - 1}{h}, \tag{4}$$

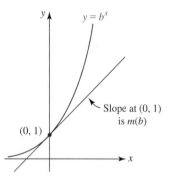

FIGURE 3.8.1 Find a base b so that the slope $m(b)$ of the tangent line at $(0, 1)$ is 1

can be shown to exist for every positive base b. However, as one might expect, we will get a different answer for each base b. So for convenience let us denote the expression in (4) by the symbol $m(b)$. The derivative of $f(x) = b^x$ is then

$$f'(x) = b^x m(b). \tag{5}$$

You are asked to approximate the value of $m(b)$ in the four cases $b = 1.5, 2, 3,$ and 5 in Problems 57–60 of Exercises 3.8. For example, it can be shown that $m(10) \approx 2.302585\ldots$ and as a consequence if $f(x) = 10^x$, then

$$f'(x) = (2.302585\ldots)10^x. \tag{6}$$

We can get a better understanding of what $m(b)$ is by evaluating (5) at $x = 0$. Since $b^0 = 1$, we have $f'(0) = m(b)$. In other words, $m(b)$ is the slope of the tangent line to the graph of $f(x) = b^x$ at $x = 0$, that is, at the y-intercept $(0, 1)$. See FIGURE 3.8.1. Given that we have to calculate a different $m(b)$ for each base b, and that $m(b)$ is likely to be an "ugly" number as in (6), over time the following question arose *naturally*:

- Is there a base b for which $m(b) = 1$? (7)

Derivative of the Natural Exponential Function

To answer the question posed in (7), we must return to the definitions of e given in Section 1.6. Specifically, (4) of Section 1.6,

$$e = \lim_{h \to 0} (1 + h)^{1/h} \tag{8}$$

provides the means for answering the question posed in (7). We know that on an intuitive level, the equality in (8) means that as h gets closer and closer to 0 then $(1 + h)^{1/h}$ can be made arbitrarily close to the number e. Thus for values of h near 0, we have the approximation $(1 + h)^{1/h} \approx e$ and so it follows that $1 + h \approx e^h$. The last expression written in the form

$$\frac{e^h - 1}{h} \approx 1 \tag{9}$$

suggests that

$$\lim_{h \to 0} \frac{e^h - 1}{h} = 1. \tag{10}$$

Since the left-hand side of (10) is $m(e)$ we have the answer to the question posed in (7):

- The base b for which $m(b) = 1$ is $b = e$. (11)

In addition, from (3) we have discovered a wonderfully simple result. The derivative of $f(x) = e^x$ is e^x. In summary,

$$\frac{d}{dx} e^x = e^x. \tag{12}$$

The result in (12) is the same as $f'(x) = f(x)$. Moreover, if $c \neq 0$ is a constant, then the only other nonzero function f in calculus whose derivative is equal to itself is $y = ce^x$ since by the Constant Multiple Rule of Section 3.2

$$\frac{dy}{dx} = \frac{d}{dx} ce^x = c \frac{d}{dx} e^x = ce^x = y.$$

■ **Derivative of $f(x) = b^x$—Revisited** In the preceding discussion we saw that $m(e) = 1$, but left unanswered the question of whether $m(b)$ has an exact value for each $b > 0$. It has. From the identity $e^{\ln b} = b, b > 0$, we can write any exponential function $f(x) = b^x$ in terms of the e base:

$$f(x) = b^x = (e^{\ln b})^x = e^{x(\ln b)}.$$

From the Chain Rule the derivative of b^x is

$$f'(x) = \frac{d}{dx} e^{x(\ln b)} = e^{x(\ln b)} \cdot \frac{d}{dx} x(\ln b) = e^{x(\ln b)}(\ln b).$$

Returning to $b^x = e^{x(\ln b)}$, the preceding line shows that

$$\frac{d}{dx} b^x = b^x (\ln b). \tag{13}$$

Matching the result in (5) with that in (13) we conclude that $m(b) = \ln b$. For example, the derivative of $f(x) = 10^x$ is $f'(x) = 10^x (\ln 10)$. Because $\ln 10 \approx 2.302585$ we see $f'(x) = 10^x (\ln 10)$ is the same as the result in (6).

The Chain Rule forms of the results in (12) and (13) are given next.

Theorem 3.8.1 Derivatives of Exponential Functions

If $u = g(x)$ is a differentiable function, then

$$\frac{d}{dx} e^u = e^u \frac{du}{dx}, \tag{14}$$

and

$$\frac{d}{dx} b^u = b^u (\ln b) \frac{du}{dx}. \tag{15}$$

EXAMPLE 1 Chain Rule

Differentiate

(a) $y = e^{-x}$ (b) $y = e^{1/x^3}$ (c) $y = 8^{5x}$.

Solution

(a) With $u = -x$ we have from (14),

$$\frac{dy}{dx} = e^{-x} \cdot \frac{d}{dx}(-x) = e^{-x}(-1) = -e^{-x}.$$

(b) By rewriting $u = 1/x^3$ as $u = x^{-3}$ we have from (14),

$$\frac{dy}{dx} = e^{1/x^3} \cdot \frac{d}{dx} x^{-3} = e^{1/x^3}(-3x^{-4}) = -3\frac{e^{1/x^3}}{x^4}.$$

(c) With $u = 5x$ we have from (15),

$$\frac{dy}{dx} = 8^{5x} \cdot (\ln 8) \cdot \frac{d}{dx} 5x = 5 \cdot 8^{5x}(\ln 8). \quad \blacksquare$$

EXAMPLE 2 Product and Chain Rule

Find the points on the graph of $y = 3x^2 e^{-x^2}$ where the tangent line is horizontal.

Solution We use the Product Rule along with (14):

$$\frac{dy}{dx} = 3x^2 \cdot \frac{d}{dx} e^{-x^2} + e^{-x^2} \cdot \frac{d}{dx} 3x^2$$
$$= 3x^2(-2xe^{-x^2}) + 6xe^{-x^2}$$
$$= e^{-x^2}(-6x^3 + 6x).$$

FIGURE 3.8.2 Graph of function in Example 2

Since $e^{-x^2} \ne 0$ for all real numbers x, $\frac{dy}{dx} = 0$ when $-6x^3 + 6x = 0$. Factoring the last equation gives $x(x+1)(x-1) = 0$ and so $x = 0$, $x = -1$, and $x = 1$. The corresponding points on the graph of the given function are then $(0, 0)$, $(-1, 3e^{-1})$, and $(1, 3e^{-1})$. The graph of $y = 3x^2 e^{-x^2}$ along with the three tangent lines (in red) are shown in FIGURE 3.8.2. $\blacksquare$

In the next example we recall the fact that an exponential statement can be written in an equivalent logarithmic form. In particular, we use (9) of Section 1.6 in the form

$$y = e^x \quad \text{if and only if} \quad x = \ln y. \tag{16}$$

EXAMPLE 3 Tangent Line Parallel to a Line

Find the point on the graph of $f(x) = 2e^{-x}$ at which the tangent line is parallel to $y = -4x - 2$.

Solution Let $(x_0, f(x_0)) = (x_0, 2e^{-x_0})$ be the unknown point on the graph of $f(x) = 2e^{-x}$ where the tangent line is parallel to $y = -4x - 2$. From the derivative $f'(x) = -2e^{-x}$ the slope of the tangent line at this point is then $f'(x_0) = -2e^{-x_0}$. Since $y = -4x - 2$ and the tangent line are parallel at that point, the slopes are equal:

$$f'(x_0) = -4 \quad \text{or} \quad -2e^{-x_0} = -4 \quad \text{or} \quad e^{-x_0} = 2.$$

FIGURE 3.8.3 Graph of function and lines in Example 3

From (16) the last equation gives $-x_0 = \ln 2$ or $x_0 = -\ln 2$. Hence, the point is $(-\ln 2, 2e^{\ln 2})$. Since $e^{\ln 2} = 2$, the point is $(-\ln 2, 4)$. In FIGURE 3.8.3 the given line is shown in green and the tangent line in red. $\blacksquare$

$\frac{d}{dx}$ NOTES FROM THE CLASSROOM

The numbers e and π are **transcendental** as well as irrational numbers. A transcendental number is one that is *not* a root of a polynomial equation with integer coefficients. For example, $\sqrt{2}$ is irrational but is not transcendental, since it is a root of the polynomial equation $x^2 - 2 = 0$. The number e was proved to be transcendental by the French mathematician **Charles Hermite** (1822–1901) in 1873, whereas π was proved to be transcendental nine years later by the **German mathematician** Ferdinand Lindemann (1852–1939). The latter proof showed conclusively that "squaring a circle" with a rule and a compass was impossible.

Exercises 3.8
Answers to selected odd-numbered problems begin on page ANS-11.

≡ Fundamentals

In Problems 1–26, find the derivative of the given function.

1. $y = e^{-x}$
2. $y = e^{2x+3}$
3. $y = e^{\sqrt{x}}$
4. $y = e^{\sin 10x}$
5. $y = 5^{2x}$
6. $y = 10^{-3x^2}$
7. $y = x^3 e^{4x}$
8. $y = e^{-x}\sin \pi x$
9. $f(x) = \dfrac{e^{-2x}}{x}$
10. $f(x) = \dfrac{xe^x}{x + e^x}$
11. $y = \sqrt{1 + e^{-5x}}$
12. $y = (e^{2x} - e^{-2x})^{10}$
13. $y = \dfrac{2}{e^{x/2} + e^{-x/2}}$
14. $y = \dfrac{e^x + e^{-x}}{e^x - e^{-x}}$
15. $y = \dfrac{e^{7x}}{e^{-x}}$
16. $y = e^{2x} e^{3x} e^{4x}$
17. $y = (e^3)^{x-1}$
18. $y = \left(\dfrac{1}{e^x}\right)^{100}$
19. $f(x) = e^{x^{1/3}} + (e^x)^{1/3}$
20. $f(x) = (2x + 1)^3 e^{-(1-x)^4}$
21. $f(x) = e^{-x}\tan e^x$
22. $f(x) = \sec e^{2x}$
23. $f(x) = e^{x\sqrt{x^2+1}}$
24. $y = e^{\frac{x+2}{x-2}}$
25. $y = e^{e^{x^2}}$
26. $y = e^x + e^{x+e^{-x}}$

27. Find an equation of the tangent line to the graph of $y = (e^x + 1)^2$ at $x = 0$.

28. Find the slope of the normal line to the graph of $y = (x - 1)e^{-x}$ at $x = 0$.

29. Find the point on the graph of $y = e^x$ at which the tangent line is parallel to $3x - y = 7$.

30. Find the point on the graph of $y = 5x + e^{2x}$ at which the tangent line is parallel to $y = 6x$.

In Problems 31 and 32, find the point(s) on the graph of the given function at which the tangent line is horizontal. Use a graphing utility to obtain the graph of each function.

31. $f(x) = e^{-x}\sin x$
32. $f(x) = (3 - x^2)e^{-x}$

In Problems 33–36, find the indicated higher derivative.

33. $y = e^{x^2}$; $\dfrac{d^3 y}{dx^3}$
34. $y = \dfrac{1}{1 + e^{-x}}$; $\dfrac{d^2 y}{dx^2}$
35. $y = \sin e^{2x}$; $\dfrac{d^2 y}{dx^2}$
36. $y = x^2 e^x$; $\dfrac{d^4 y}{dx^4}$

In Problems 37 and 38, C_1 and C_2 are arbitrary real constants. Show that the function satisfies the given differential equation.

37. $y = C_1 e^{-3x} + C_2 e^{2x}$; $y'' + y' - 6y = 0$
38. $y = C_1 e^{-x}\cos 2x + C_2 e^{-x}\sin 2x$; $y'' + 2y' + 5y = 0$

39. If C and k are real constants, show that the function $y = Ce^{kx}$ satisfies the differential equation $y' = ky$.

40. Use Problem 39 to find a function that satisfies the given conditions.
   (a) $y' = -0.01y$ and $y(0) = 100$
   (b) $\dfrac{dP}{dt} - 0.15P = 0$ and $P(0) = P_0$

In Problems 41–46, use implicit differentiation to find dy/dx.

41. $y = e^{x+y}$
42. $xy = e^y$
43. $y = \cos e^{xy}$
44. $y = e^{(x+y)^2}$
45. $x + y^2 = e^{x/y}$
46. $e^x + e^y = y$

47. (a) Sketch the graph of $f(x) = e^{-|x|}$.
    (b) Find $f'(x)$.
    (c) Sketch the graph of f'.
    (d) Is the function differentiable at $x = 0$?

48. (a) Show that the function $f(x) = e^{\cos x}$ is periodic with period 2π.
    (b) Find all points on the graph of f where the tangent is horizontal.
    (c) Sketch the graph of f.

≡ Applications

49. The logistic function
$$P(t) = \dfrac{aP_0}{bP_0 + (a - bP_0)e^{-at}},$$
where a and b are positive constants, often serves as a mathematical model for an expanding but limited population.
    (a) Show that $P(t)$ satisfies the differential equation
    $$\dfrac{dP}{dt} = P(a - bP).$$
    (b) The graph of $P(t)$ is called a **logistic curve** where $P(0) = P_0$ is the initial population. Consider the case when $a = 2$, $b = 1$, and $P_0 = 1$. Find horizontal asymptotes for the graph of $P(t)$ by determining the limits $\lim_{t \to -\infty} P(t)$ and $\lim_{t \to \infty} P(t)$.
    (c) Graph $P(t)$.
    (d) Find the value(s) of t for which $P''(t) = 0$.

50. The **Jenss mathematical model** (1937) represents one of the most accurate empirically devised formulas for predicting the height h (in centimeters) in terms of age t (in years) for preschool-age children (3 months to 6 years):
$$h(t) = 79.04 + 6.39t - e^{3.26 - 0.99t}.$$
    (a) What height does this model predict for a 2-year-old?
    (b) How fast is a 2-year-old increasing in height?
    (c) Use a calculator or CAS to obtain the graph of h on the interval $\left[\frac{1}{4}, 6\right]$.
    (d) Use the graph in part (c) to estimate the age of a preschool-age child who is 100 cm tall.

(b) From the second result in (22),

$$\frac{dy}{dx} = -\text{csch}^2 x^3 \cdot \frac{d}{dx} x^3$$
$$= -\text{csch}^2 x^3 \cdot 3x^2.$$

EXAMPLE 2 Value of a Derivative

Evaluate the derivative of $y = \dfrac{3x}{4 + \cosh 2x}$ at $x = 0$.

Solution From the Quotient Rule,

$$\frac{dy}{dx} = \frac{(4 + \cosh 2x) \cdot 3 - 3x(\sinh 2x \cdot 2)}{(4 + \cosh 2x)^2}.$$

Because $\sinh 0 = 0$ and $\cosh 0 = 1$, we have

$$\left.\frac{dy}{dx}\right|_{x=0} = \frac{15}{25} = \frac{3}{5}.$$

Inverse Hyperbolic Functions Inspection of Figure 3.10.1(a) shows that $y = \sinh x$ is a one-to-one function. That is, for any real number y in the range $(-\infty, \infty)$ of the hyperbolic sine, there corresponds only one real number x in its domain $(-\infty, \infty)$. Hence, $y = \sinh x$ has an inverse function that is written $y = \sinh^{-1} x$. See FIGURE 3.10.3(a). As in our earlier discussion of the inverse trigonometric functions in Section 1.5, this later notation is equivalent to $x = \sinh y$. From Figure 3.10.2(a) it is also seen that $y = \tanh x$ with domain $(-\infty, \infty)$ and range $(-1, 1)$ is also one-to-one and has an inverse $y = \tanh^{-1} x$ with domain $(-1, 1)$ and range $(-\infty, \infty)$. See Figure 3.10.3(c). But from Figures 3.10.1(b) and 3.10.2(c) it is apparent that $y = \cosh x$ and $y = \text{sech}\, x$ are not one-to-one functions and so do not possess inverse functions unless their domains are suitably restricted. Inspection of Figure 3.10.1(b) shows that when the domain of $y = \cosh x$ is restricted to the interval $[0, \infty)$, the corresponding range is $[1, \infty)$. The inverse function $y = \cosh^{-1} x$ then has domain $[1, \infty)$ and range $[0, \infty)$. See Figure 3.10.3(b). The graphs of all the inverse hyperbolic functions along with their domains and ranges are summarized in Figure 3.10.3.

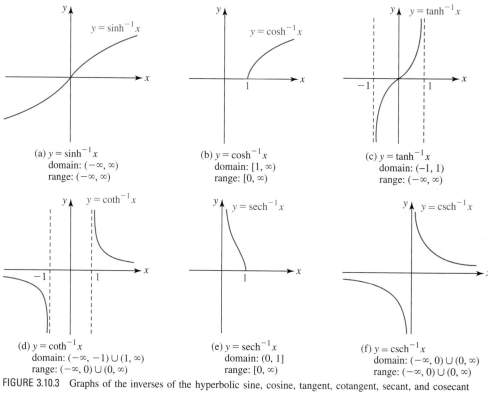

FIGURE 3.10.3 Graphs of the inverses of the hyperbolic sine, cosine, tangent, cotangent, secant, and cosecant

■ **Inverse Hyperbolic Functions as Logarithms** Because all the hyperbolic functions are defined in terms of combinations of e^x, it should not come as any surprise to find that the inverse hyperbolic functions can be expressed in terms of the natural logarithm. For example, $y = \sinh^{-1} x$ is equivalent to $x = \sinh y$, so that

$$x = \frac{e^y - e^{-y}}{2} \quad \text{or} \quad 2x = \frac{e^{2y} - 1}{e^y} \quad \text{or} \quad e^{2y} - 2xe^y - 1 = 0.$$

Because the last equation is quadratic in e^y, the quadratic formula gives

$$e^y = \frac{2x \pm \sqrt{4x^2 + 4}}{2} = x \pm \sqrt{x^2 + 1}. \tag{24}$$

Now the solution corresponding to the minus sign in (24) must be rejected since $e^y > 0$ but $x - \sqrt{x^2 + 1} < 0$. Thus, we have

$$e^y = x + \sqrt{x^2 + 1} \quad \text{or} \quad y = \sinh^{-1} x = \ln(x + \sqrt{x^2 + 1}).$$

Similarly, for $y = \tanh^{-1} x$, $|x| < 1$,

$$x = \tanh y = \frac{e^y - e^{-y}}{e^y + e^{-y}}$$

gives

$$e^y(1 - x) = (1 + x)e^{-y}$$

$$e^{2y} = \frac{1 + x}{1 - x}$$

$$2y = \ln\left(\frac{1 + x}{1 - x}\right)$$

or

$$y = \tanh^{-1} x = \frac{1}{2}\ln\left(\frac{1 + x}{1 - x}\right).$$

We have proved two of the results in the next theorem.

Theorem 3.10.3 Logarithmic Identities

$\sinh^{-1} x = \ln(x + \sqrt{x^2 + 1}) \qquad\qquad \cosh^{-1} x = \ln(x + \sqrt{x^2 - 1}), x \geq 1$ (25)

$\tanh^{-1} x = \frac{1}{2}\ln\left(\frac{1 + x}{1 - x}\right), |x| < 1 \qquad \coth^{-1} x = \frac{1}{2}\ln\left(\frac{x + 1}{x - 1}\right), |x| > 1$ (26)

$\operatorname{sech}^{-1} x = \ln\left(\frac{1 + \sqrt{1 - x^2}}{x}\right), 0 < x \leq 1 \quad \operatorname{csch}^{-1} x = \ln\left(\frac{1}{x} + \frac{\sqrt{1 + x^2}}{|x|}\right), x \neq 0$ (27)

The foregoing identities are a convenient means for obtaining the numerical values of an inverse hyperbolic function. For example, with the aid of a calculator we see from the first result in (25) in Theorem 3.10.3 that when $x = 4$

$$\sinh^{-1} 4 = \ln(4 + \sqrt{17}) \approx 2.0947.$$

■ **Derivatives of Inverse Hyperbolic Functions** To find the derivative of an inverse hyperbolic function, we can proceed in two different ways. For example, if

$$y = \sinh^{-1} x \quad \text{then} \quad x = \sinh y.$$

Using implicit differentiation, we can write

$$\frac{d}{dx}x = \frac{d}{dx}\sinh y$$

$$1 = \cosh y \frac{dy}{dx}.$$

Hence

$$\frac{dy}{dx} = \frac{1}{\cosh y} = \frac{1}{\sqrt{\sinh^2 y + 1}} = \frac{1}{\sqrt{x^2 + 1}}.$$

The foregoing result can be obtained in an alternative manner. We know from Theorem 3.10.3 that

$$y = \ln(x + \sqrt{x^2 + 1}).$$

Therefore, from the derivative of the logarithm, we obtain

$$\frac{dy}{dx} = \frac{1}{x + \sqrt{x^2 + 1}}\left(1 + \frac{1}{2}(x^2 + 1)^{-1/2} \cdot 2x\right) \quad \leftarrow \text{by (3) of Section 3.9}$$

$$= \frac{1}{x + \sqrt{x^2 + 1}} \frac{\sqrt{x^2 + 1} + x}{\sqrt{x^2 + 1}} = \frac{1}{\sqrt{x^2 + 1}}.$$

We have essentially proved the first entry in (28) in the next theorem.

Theorem 3.10.4 Derivatives of Inverse Hyperbolic Functions

If $u = g(x)$ is a differentiable function, then

$$\frac{d}{dx}\sinh^{-1} u = \frac{1}{\sqrt{u^2 + 1}}\frac{du}{dx}, \qquad \frac{d}{dx}\cosh^{-1} u = \frac{1}{\sqrt{u^2 - 1}}\frac{du}{dx}, u > 1, \quad (28)$$

$$\frac{d}{dx}\tanh^{-1} u = \frac{1}{1 - u^2}\frac{du}{dx}, |u| < 1, \qquad \frac{d}{dx}\coth^{-1} u = \frac{1}{1 - u^2}\frac{du}{dx}, |u| > 1, \quad (29)$$

$$\frac{d}{dx}\text{sech}^{-1} u = \frac{-1}{u\sqrt{1 - u^2}}\frac{du}{dx}, 0 < u < 1, \qquad \frac{d}{dx}\text{csch}^{-1} u = \frac{-1}{|u|\sqrt{1 + u^2}}\frac{du}{dx}, u \neq 0. \quad (30)$$

EXAMPLE 3 Derivative of Inverse Hyperbolic Cosine

Differentiate $y = \cosh^{-1}(x^2 + 5)$.

Solution With $u = x^2 + 5$, we have from the second formula in (28),

$$\frac{dy}{dx} = \frac{1}{\sqrt{(x^2 + 5)^2 - 1}} \cdot \frac{d}{dx}(x^2 + 5) = \frac{2x}{\sqrt{x^4 + 10x^2 + 24}}.$$

EXAMPLE 4 Derivative of Inverse Hyperbolic Tangent

Differentiate $y = \tanh^{-1} 4x$.

Solution With $u = 4x$, we have from the first formula in (29),

$$\frac{dy}{dx} = \frac{1}{1 - (4x)^2} \cdot \frac{d}{dx} 4x = \frac{4}{1 - 16x^2}.$$

EXAMPLE 5 Product and Chain Rules

Differentiate $y = e^{x^2} \text{sech}^{-1} x$.

Solution By the Product Rule and the first formula in (30), we have

$$\frac{dy}{dx} = e^{x^2}\underbrace{\left(\frac{-1}{x\sqrt{1 - x^2}}\right)}_{\text{by first formula in (30)}} + \underbrace{2xe^{x^2}\text{sech}^{-1} x}_{\text{by (14) of Section 3.8}}$$

$$= -\frac{e^{x^2}}{x\sqrt{1 - x^2}} + 2xe^{x^2}\text{sech}^{-1} x.$$

$\dfrac{d}{dx}$ NOTES FROM THE CLASSROOM

(*i*) As mentioned in the introduction to this section, the graph of any function of the form $f(x) = k\cosh cx$, k and c constants, is called a **catenary**. The shape assumed by a flexible wire or heavy rope strung between two posts has the basic shape of a graph of a hyperbolic cosine. Furthermore, if two circular rings are held vertically and are not too far apart, then a soap film stretched between the rings will assume a surface having minimum area. The surface is a portion of a **catenoid**, which is the surface obtained by revolving a catenary about the x-axis. See FIGURE 3.10.4.

(*ii*) The similarity between trigonometric and hyperbolic functions extends beyond the derivative formulas and basic identities. If t is an angle measured in radians whose terminal side is OP, then the coordinates of P on a unit *circle* $x^2 + y^2 = 1$ are $(\cos t, \sin t)$. Now, the area of the shaded circular sector shown in FIGURE 3.10.5(a) is $A = \tfrac{1}{2}t$ and so $t = 2A$. In this manner, the *circular functions* $\cos t$ and $\sin t$ can be considered functions of the area A.

You might already know that the graph of the equation $x^2 - y^2 = 1$ is called a *hyperbola*. Because $\cosh t \geq 1$ and $\cosh^2 t - \sinh^2 t = 1$, it follows that the coordinates of a point P on the right-hand branch of the hyperbola are $(\cosh t, \sinh t)$. Furthermore, it can be shown that the area of the hyperbolic sector in Figure 3.10.5(b) is related to the number t by $t = 2A$. Whence we see the origin of the name *hyperbolic function*.

(a) hanging wires

(b) soap film

FIGURE 3.10.4 Catenary in (a); catenoid in (b)

(a) circular sector

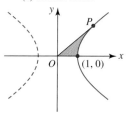
(b) hyperbolic sector

FIGURE 3.10.5 Circle in (a); hyperbola in (b)

Exercises 3.10 Answers to selected odd-numbered problems begin on page ANS-12.

≡ Fundamentals

1. If $\sinh x = -\tfrac{1}{2}$, find the values of the remaining hyperbolic functions.

2. If $\cosh x = 3$, find the values of the remaining hyperbolic functions.

In Problems 3–26, find the derivative of the given function.

3. $y = \cosh 10x$
4. $y = \operatorname{sech} 8x$
5. $y = \tanh \sqrt{x}$
6. $y = \operatorname{csch} \dfrac{1}{x}$
7. $y = \operatorname{sech}(3x - 1)^2$
8. $y = \sinh e^{x^2}$
9. $y = \coth(\cosh 3x)$
10. $y = \tanh(\sinh x^3)$
11. $y = \sinh 2x \cosh 3x$
12. $y = \operatorname{sech} x \coth 4x$
13. $y = x\cosh x^2$
14. $y = \dfrac{\sinh x}{x}$
15. $y = \sinh^3 x$
16. $y = \cosh^4 \sqrt{x}$
17. $f(x) = (x - \cosh x)^{2/3}$
18. $f(x) = \sqrt{4 + \tanh 6x}$
19. $f(x) = \ln(\cosh 4x)$
20. $f(x) = (\ln(\operatorname{sech} x))^2$
21. $f(x) = \dfrac{e^x}{1 + \cosh x}$
22. $f(x) = \dfrac{\ln x}{x^2 + \sinh x}$
23. $F(t) = e^{\sinh t}$
24. $H(t) = e^t e^{\operatorname{csch} t^2}$
25. $g(t) = \dfrac{\sin t}{1 + \sinh 2t}$
26. $w(t) = \dfrac{\tanh t}{(1 + \cosh t)^2}$

27. Find an equation of the tangent line to the graph of $y = \sinh 3x$ at $x = 0$.

28. Find an equation of the tangent line to the graph of $y = \cosh x$ at $x = 1$.

In Problems 29 and 30, find the point(s) on the graph of the given function at which the tangent is horizontal.

29. $f(x) = (x^2 - 2)\cosh x - 2x \sinh x$
30. $f(x) = \cos x \cosh x - \sin x \sinh x$

In Problems 31 and 32, find d^2y/dx^2 for the given function.

31. $y = \tanh x$
32. $y = \operatorname{sech} x$

In Problems 33 and 34, C_1, C_2, C_3, C_4 and k are arbitrary real constants. Show that the function satisfies the given differential equation.

33. $y = C_1 \cosh kx + C_2 \sinh kx$; $y'' - k^2 y = 0$

34. $y = C_1 \cos kx + C_2 \sin kx + C_3 \cosh kx + C_4 \sinh kx$; $y^{(4)} - k^4 y = 0$

In Problems 35–48, find the derivative of the given function.

35. $y = \sinh^{-1} 3x$

36. $y = \cosh^{-1} \dfrac{x}{2}$

37. $y = \tanh^{-1}(1 - x^2)$

38. $y = \coth^{-1} \dfrac{1}{x}$

39. $y = \coth^{-1}(\csc x)$

40. $y = \sinh^{-1}(\sin x)$

41. $y = x \sinh^{-1} x^3$

42. $y = x^2 \operatorname{csch}^{-1} x$

43. $y = \dfrac{\operatorname{sech}^{-1} x}{x}$

44. $y = \dfrac{\coth^{-1} e^{2x}}{e^{2x}}$

45. $y = \ln(\operatorname{sech}^{-1} x)$

46. $y = x \tanh^{-1} x + \ln \sqrt{1 - x^2}$

47. $y = (\cosh^{-1} 6x)^{1/2}$

48. $y = \dfrac{1}{(\tanh^{-1} 2x)^3}$

≡ Applications

49. **(a)** Assume that k, m, and g are real constants. Show that the function

$$v(t) = \sqrt{\dfrac{mg}{k}} \tanh\left(\sqrt{\dfrac{kg}{m}}\, t\right)$$

satisfies the differential equation $m \dfrac{dv}{dt} = mg - kv^2$.

(b) The function v represents the velocity of a falling mass m when air resistance is taken to be proportional to the square of the instantaneous velocity. Find the limiting or **terminal velocity** $v_{\text{ter}} = \lim\limits_{t \to \infty} v(t)$ of the mass.

(c) Suppose a 80-kg skydiver delays opening the parachute until terminal velocity is attained. Determine the terminal velocity if it is known that $k = 0.25$ kg/m.

50. A woman, W, starting at the origin, moves in the direction of the positive x-axis pulling a boat along the curve C, called a **tractrix**, indicated in FIGURE 3.10.6. The boat, initially located on the y-axis at $(0, a)$, is pulled by a rope of constant length a that is kept taut throughout the motion. An equation of the tractrix is given by

$$x = a \ln\left(\dfrac{a + \sqrt{a^2 - y^2}}{y}\right) - \sqrt{a^2 - y^2}.$$

(a) Rewrite this equation using a hyperbolic function.

(b) Use implicit differentiation to show that the equation of the tractrix satisfies the differential equation

$$\dfrac{dy}{dx} = -\dfrac{y}{\sqrt{a^2 - y^2}}.$$

(c) Interpret geometrically the differential equation in part (b).

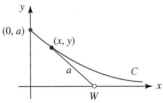

FIGURE 3.10.6 Tractrix in Problem 50

≡ Think About It

In Problems 51 and 52, find the exact numerical value of the given quantity.

51. $\cosh(\ln 4)$

52. $\sinh(\ln 0.5)$

In Problems 53 and 54, express the given quantity as a rational function of x.

53. $\sinh(\ln x)$

54. $\tanh(3 \ln x)$

55. Show that for any positive integer n,

$$(\cosh x + \sinh x)^n = \cosh nx + \sinh nx.$$

Chapter 3 in Review

Answers to selected odd-numbered problems begin on page ANS-13.

A. True/False

In Problems 1–20, indicate whether the given statement is true or false.

1. If $y = f(x)$ is continuous at a number a, then there is a tangent line to the graph of f at $(a, f(a))$. _____

2. If f is differentiable at every real number x, then f is continuous everywhere. _____

3. If $y = f(x)$ has a tangent line at $(a, f(a))$, then f is necessarily differentiable at $x = a$. _____

4. The instantaneous rate of change of $y = f(x)$ with respect to x at x_0 is the slope of the tangent line to the graph at $(x_0, f(x_0))$. _____

5. At $x = -1$, the tangent line to the graph of $f(x) = x^3 - 3x^2 - 9x$ is parallel to the line $y = 2$. _____

6. The derivative of a product is the product of the derivatives. _____

7. A polynomial function has a tangent line at every point on its graph. _____

8. For $f(x) = -x^2 + 5x + 1$ an equation of the tangent line is $f'(x) = -2x + 5$. _____
9. The function $f(x) = x/(x^2 + 9)$ is differentiable on the interval $[-3, 3]$. _____
10. If $f'(x) = g'(x)$, then $f(x) = g(x)$. _____
11. If m is the slope of a tangent line to the graph of $f(x) = \sin x$, then $-1 \le m \le 1$. _____
12. For $y = \tan^{-1} x$, $dy/dx > 0$ for all x. _____
13. $\dfrac{d}{dx} \cos^{-1} x = -\sin^{-1} x$ _____
14. The function $f(x) = x^5 + x^3 + x$ has an inverse. _____
15. If $f'(x) < 0$ on the interval $[2, 8]$, then $f(3) > f(5)$. _____
16. If f is an increasing differentiable function on an interval, then $f'(x)$ is also increasing on the interval. _____
17. The only function for which $f'(x) = f(x)$ is $f(x) = e^x$. _____
18. $\dfrac{d}{dx} \ln|x| = \dfrac{1}{|x|}$ _____
19. $\dfrac{d}{dx} \cosh^2 x = \dfrac{d}{dx} \sinh^2 x$ _____
20. Every inverse hyperbolic function is a logarithm. _____

B. Fill in the Blanks

In Problems 1–20, fill in the blanks.

1. If $y = f(x)$ is a polynomial function of degree 3, then $\dfrac{d^4}{dx^4} f(x) =$ _____.

2. The slope of the tangent line to the graph of $y = \ln|x|$ at $x = -\dfrac{1}{2}$ is _____.

3. The slope of the normal line to the graph of $f(x) = \tan x$ at $x = \pi/3$ is _____.

4. $f(x) = \dfrac{x^{n+1}}{n+1}$, $n \ne -1$, then $f'(x) =$ _____.

5. An equation of the tangent line to the graph of $y = (x + 3)/(x - 2)$ at $x = 0$ is _____.

6. For $f(x) = 1/(1 - 3x)$ the instantaneous rate of change of f' with respect to x at $x = 0$ is _____.

7. If $f'(4) = 6$ and $g'(4) = 3$, then the slope of the tangent line to the graph of $y = 2f(x) - 5g(x)$ at $x = 4$ is _____.

8. If $f(2) = 1$, $f'(2) = 5$, $g(2) = 2$, and $g'(2) = -3$, then $\left.\dfrac{d}{dx} \dfrac{x^2 f(x)}{g(x)}\right|_{x=2} =$ _____.

9. If $g(1) = 2$, $g'(1) = 3$, $g''(1) = -1$, $f'(2) = 4$, and $f''(2) = 3$, then $\left.\dfrac{d^2}{dx^2} f(g(x))\right|_{x=1} =$ _____.

10. If $f'(x) = x^2$, then $\dfrac{d}{dx} f(x^3) =$ _____.

11. If F is a differentiable function, then $\dfrac{d^2}{dx^2} F(\sin 4x) =$ _____.

12. The function $f(x) = \cot x$ is not differentiable on the interval $[0, \pi]$ because _____.

13. The function
$$f(x) = \begin{cases} ax + b, & x \le 3 \\ x^2, & x > 3 \end{cases}$$
is differentiable at $x = 3$ when $a =$ _____ and $b =$ _____.

14. If $f'(x) = \sec^2 2x$, then $f(x) =$ _____.

15. The tangent line to the graph of $f(x) = 5 - x + e^{x-1}$ is horizontal at the point _____.

16. $\dfrac{d}{dx} 2^x =$ _____.

17. $\dfrac{d}{dx} \log_{10} x =$ _____.

18. If $f(x) = \ln|2x - 4|$, the domain of $f'(x)$ is _____.

19. The graph of $y = \cosh x$ is called a _____.

20. $\cosh^{-1} 1 =$ _____.

C. Exercises

In Problems 1–28, find the derivative of the given function.

1. $f(x) = \dfrac{4x^{0.3}}{5x^{0.2}}$
2. $y = \dfrac{1}{x^3 + 4x^2 - 6x + 11}$
3. $F(t) = (t + \sqrt{t^2 + 1})^{10}$
4. $h(\theta) = \theta^{1.5}(\theta^2 + 1)^{0.5}$
5. $y = \sqrt[4]{x^4 + 16} \sqrt[3]{x^3 + 8}$
6. $g(u) = \sqrt{\dfrac{6u - 1}{u + 7}}$
7. $y = \dfrac{\cos 4x}{4x + 1}$
8. $y = 10 \cot 8x$
9. $f(x) = x^3 \sin^2 5x$
10. $y = \tan^2(\cos 2x)$
11. $y = \sin^{-1} \dfrac{3}{x}$
12. $y = \cos x \cos^{-1} x$
13. $y = (\cot^{-1} x)^{-1}$
14. $y = \text{arc sec}(2x - 1)$
15. $y = 2\cos^{-1} x + 2x\sqrt{1 - x^2}$
16. $y = x^2 \tan^{-1} \sqrt{x^2 - 1}$
17. $y = xe^{-x} + e^{-x}$
18. $y = (e + e^2)^x$
19. $y = x^7 + 7^x + 7^\pi + e^{7x}$
20. $y = (e^x + 1)^{-e}$
21. $y = \ln(x\sqrt{4x - 1})$
22. $y = (\ln \cos^2 x)^2$
23. $y = \sinh^{-1}(\sin^{-1} x)$
24. $y = (\tan^{-1} x)(\tanh^{-1} x)$
25. $y = xe^{x \cosh^{-1} x}$
26. $y = \sinh^{-1} \sqrt{x^2 - 1}$
27. $y = \sinh e^{x^3}$
28. $y = (\tanh 5x)^{-1}$

In Problems 29–34, find the indicated derivative.

29. $y = (3x + 1)^{5/2}$; $\dfrac{d^3 y}{dx^3}$
30. $y = \sin(x^3 - 2x)$; $\dfrac{d^2 y}{dx^2}$
31. $s = t^2 + \dfrac{1}{t^2}$; $\dfrac{d^4 s}{dt^4}$
32. $W = \dfrac{v - 1}{v + 1}$; $\dfrac{d^3 W}{dv^3}$
33. $y = e^{\sin 2x}$; $\dfrac{d^2 y}{dx^2}$
34. $f(x) = x^2 \ln x$; $f'''(x)$

35. First use the laws of logarithms to simplify
$$y = \ln \left| \dfrac{(x + 5)^4 (2 - x)^3}{(x + 8)^{10} \sqrt[3]{6x + 4}} \right|,$$
and then find dy/dx.

36. Find dy/dx for $y = 5^{x^2} x^{\sin 2x}$.

37. Given that $y = x^3 + x$ is a one-to-one function, find the slope of the tangent line to the graph of the inverse function at $x = 1$.

38. Given that $f(x) = 8/(1 - x^3)$ is a one-to-one function, find f^{-1} and $(f^{-1})'$.

In Problems 39 and 40, find dy/dx.

39. $xy^2 = e^x - e^y$ **40.** $y = \ln(xy)$

41. Find an equation of a tangent line to the graph of $f(x) = x^3$ that is perpendicular to the line $y = -3x$.

42. Find the point(s) on the graph of $f(x) = \frac{1}{2}x^2 - 5x + 1$ at which
   (a) $f''(x) = f(x)$ and   **(b)** $f''(x) = f'(x)$.

43. Find equations for the lines through $(0, -9)$ that are tangent to the graph of $y = x^2$.

44. (a) Find the x-intercept of the tangent line to the graph of $y = x^2$ at $x = 1$.
   (b) Find an equation of the line with the same x-intercept that is perpendicular to the tangent line in part (a).
   (c) Find the point(s) where the line in part (b) intersects the graph of $y = x^2$.

45. Find the point on the graph of $f(x) = \sqrt{x}$ at which the tangent line is parallel to the secant line through $(1, f(1))$ and $(9, f(9))$.

46. If $f(x) = (1 + x)/x$, what is the slope of the tangent line to the graph of f'' at $x = 2$?

47. Find the x-coordinates of all points on the graph of $f(x) = 2\cos x + \cos 2x$, $0 \le x \le 2\pi$, at which the tangent line is horizontal.

48. Find the point on the graph of $y = \ln 2x$ such that the tangent line passes through the origin.

49. Suppose a series circuit contains a capacitor and a variable resistor. If the resistance at time t is given by $R = k_1 + k_2 t$, where k_1 and k_2 are positive known constants, then the charge $q(t)$ on the capacitor is given by

$$q(t) = E_0 C + (q_0 - E_0 C)\left(\frac{k_1}{k_1 + k_2 t}\right)^{1/Ck_2},$$

where C is a constant called the **capacitance** and $E(t) = E_0$ is the impressed voltage. Show that $q(t)$ satisfies the initial condition $q(0) = q_0$ and the differential equation

$$(k_1 + k_2 t)\frac{dq}{dt} + \frac{1}{C}q = E_0.$$

50. Assume that C_1 and C_2 are arbitrary real constants. Show that the function

$$y = C_1 x + C_2\left[\frac{x}{2}\ln\left(\frac{x-1}{x+1}\right) - 1\right]$$

satisfies the differential equation

$$(1 - x^2)y'' - 2xy' + 2y = 0.$$

In Problems 51 and 52, C_1, C_2, C_3, and C_4 are arbitrary real constants. Show that the function satisfies the given differential equation.

51. $y = C_1 e^{-x} + C_2 e^x + C_3 x e^{-x} + C_4 x e^x$; $y^{(4)} - 2y'' + y = 0$

52. $y = C_1 \cos x + C_2 \sin x + C_3 x \cos x + C_4 x \sin x$; $y^{(4)} + 2y'' + y = 0$

53. (a) Find the points on the graph of $y^3 - y + x^2 - 4 = 0$ corresponding to $x = 2$.
   (b) Find the slopes of the tangent lines at the points found in part (a).

54. Sketch the graph of f' from the graph of f given in FIGURE 3.R.1.

FIGURE 3.R.1 Graph for Problem 54

55. The graph of $x^{2/3} + y^{2/3} = 1$, shown in FIGURE 3.R.2, is called a **hypocycloid**.*
Find equations of the tangent lines to the graph at the points corresponding to $x = \frac{1}{8}$.

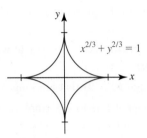

FIGURE 3.R.2 Hypocycloid in Problem 55

56. Find d^2y/dx^2 for the equation in Problem 55.

57. Suppose
$$f(x) = \begin{cases} x^2, & x \leq 0 \\ \sqrt{x}, & x > 0. \end{cases}$$

Find $f'(x)$ for $x \neq 0$. Use the definition of the derivative, (2) of Section 3.1, to determine whether $f'(0)$ exists.

---

*Go to the website **http://mathworld.wolfram.com/Hypocycloid.html** to see various kinds of hypocycloids and their properties.

Chapter 4

Applications of the Derivative

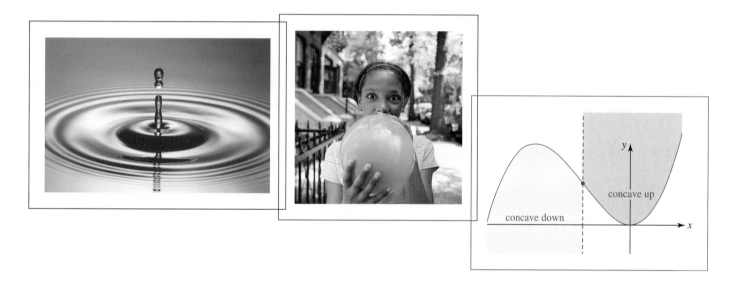

In This Chapter The first and second derivatives of a function f can be used to determine the shape of its graph. If you imagine a graph of a function as a curve that rises and falls, then the high points and low points of the graph, or more precisely, the maximum and minimum values of the function, can be found using the derivative. As we have already seen, the derivative also gives a rate of change. We saw briefly in Section 2.7 that the rate of change with respect to time t of a function that gives the position of a moving object is the velocity of the object.

Finding the maximum and minimum values of a function along with the problem of finding rates of change are two of the central topics of study in this chapter.

4.1 Rectilinear Motion
4.2 Related Rates
4.3 Extrema of Functions
4.4 Mean Value Theorem
4.5 Limits Revisited—L'Hôpital's Rule
4.6 Graphing and the First Derivative
4.7 Graphing and the Second Derivative
4.8 Optimization
4.9 Linearization and Differentials
4.10 Newton's Method
  Chapter 4 in Review

4.1 Rectilinear Motion

Introduction In Section 2.7, the motion of an object in a straight line, either horizontally or vertically, was said to be **rectilinear motion**. A function $s = s(t)$ that gives the coordinate of the object on a horizontal or vertical line is called a **position function**. The variable t represents time and the function value $s(t)$ represents a directed distance, which is measured in centimeters, meters, feet, miles, and so on, from a reference point $s = 0$ on the line. Recall that on a horizontal scale, we take the positive s-direction to be to the right of $s = 0$, and on a vertical scale we take the positive s-direction to be upward.

EXAMPLE 1 Position of a Moving Particle

A particle moves on a horizontal line according to the position function $s(t) = -t^2 + 4t + 3$, where s is measured in centimeters and t in seconds. What is the position of the particle at 0, 2, and 6 seconds?

Solution Substitution into the position function gives
$$s(0) = 3, \quad s(2) = 7, \quad s(6) = -9.$$

As shown in FIGURE 4.1.1, $s(6) = -9 < 0$ means that the position of the particle is to the left of the reference point $s = 0$.

FIGURE 4.1.1 Position of a particle at various times in Example 1

Velocity and Acceleration If the **average velocity** of a body in motion over a time interval of length Δt is
$$\frac{\text{change in position}}{\text{change in time}} = \frac{s(t + \Delta t) - s(t)}{\Delta t},$$
then the instantaneous rate of change, or velocity of the body, is given by
$$v(t) = \lim_{\Delta t \to 0} \frac{s(t + \Delta t) - s(t)}{\Delta t}.$$

Thus, we have the following definition.

Definition 4.1.1 Velocity Function

If $s(t)$ is a position function of an object that moves rectilinearly, then its **velocity function** $v(t)$ at time t is
$$v(t) = \frac{ds}{dt}.$$

The **speed** of the object at time t is $|v(t)|$.

Velocity is measured in centimeters per second (cm/s), meters per second (m/s), feet per second (ft/s), kilometers per hour (km/h), miles per hour (mi/h), and so on.

We can also compute the rate of change of velocity.

Definition 4.1.2 Acceleration Function

If $v(t)$ is the velocity function of an object that moves rectilinearly, then its **acceleration function** $a(t)$ at time t is
$$a(t) = \frac{dv}{dt} = \frac{d^2s}{dt^2}.$$

Typical units for measuring acceleration are meters per second per second (m/s^2), feet per second per second (ft/s^2), miles per hour per hour (mi/h^2), and so on. Often we read units of acceleration literally as "meters per second squared."

■ **Significance of Algebraic Signs** In Section 3.7 we saw that whenever the derivative of a function f is *positive* on an interval I, then f is *increasing* on I. Geometrically, the graph of an increasing function rises as x increases. Similarly, if the derivative of a function f is *negative* on I, then f is *decreasing*, which means its graph goes down as x increases. On a time interval for which $v(t) = s'(t) > 0$, we can say $s(t)$ is increasing. Thus the object is moving to the *right* on a horizontal line or moving *upward* on a vertical line. On the other hand, $v(t) = s'(t) < 0$ implies that $s(t)$ is decreasing and motion is to the *left* on a horizontal line or motion *downward* on a vertical line. See FIGURE 4.1.2. If $a(t) = v'(t) > 0$ on a time interval, then the velocity $v(t)$ of the object is *increasing*, whereas $a(t) = v'(t) < 0$ indicates that the velocity $v(t)$ of the object is *decreasing*. For example, an acceleration of -25 m/s^2 means that the velocity is decreasing by 25 m/s every second. Do not confuse the terms "velocity decreasing" and "velocity increasing" with the concepts of "slowing down" or "speeding up." For example, consider a stone that is dropped from the top of a tall building. The acceleration of gravity is a negative constant, -32 ft/s^2. The negative sign means that the velocity of the stone decreases starting from zero. When the stone hits the ground, its speed $|v(t)|$ is fairly large, but $v(t) < 0$. Specifically, an object that moves rectilinearly on, say, a horizontal line is *slowing down* when $v(t) > 0$ (motion to right) and $a(t) < 0$ (velocity decreasing), or when $v(t) < 0$ (motion to left) and $a(t) > 0$ (velocity increasing). Similarly, an object that moves rectilinearly on a horizontal line is *speeding up* when $v(t) > 0$ (motion to right) and $a(t) > 0$ (velocity increasing) or when $v(t) < 0$ (motion to left) and $a(t) < 0$ (velocity decreasing). In general:

An object that moves rectilinearly

- *is slowing down when its velocity and acceleration have opposite algebraic signs, and*
- *is speeding up when its velocity and acceleration have the same algebraic sign.*

Alternatively, an object is slowing down when its speed $|v(t)|$ is decreasing and speeding up when its speed is increasing.

FIGURE 4.1.2 Significance of the sign of velocity function

EXAMPLE 2 Example 1 Revisited

In Example 1 the velocity and acceleration functions for the particle are, respectively,

$$v(t) = \frac{ds}{dt} = -2t + 4 \quad \text{and} \quad a(t) = \frac{dv}{dt} = -2.$$

At times 0, 2, and 6 s, the velocities are $v(0) = 4$ cm/s, $v(2) = 0$ cm/s, and $v(6) = -8$ cm/s, respectively. Since the acceleration is always negative, the velocity is always decreasing. Notice that $v(t) = 2(-t + 2) > 0$ for $t < 2$ and $v(t) = 2(-t + 2) < 0$ for $t > 2$. If the time t is allowed to be negative as well as positive, then the particle moves to the right for the time interval $(-\infty, 2)$ and moves to the left for the time interval $(2, \infty)$. The motion can be represented by the graph given in FIGURE 4.1.3(a). Since the motion actually takes place *on* the horizontal line, you should envision the movement of a point P that corresponds to the projection of a point on the graph onto the horizontal line. See Figure 4.1.3(b).

(a) $s(t) = -t^2 + 4t + 3$
(b) particle at point P is moving on the s-axis

FIGURE 4.1.3 Representation of the motion of the particle in Example 2

EXAMPLE 8   One-Sided Limit

Evaluate $\lim\limits_{x \to 1^+} \dfrac{\ln x}{\sqrt{x-1}}$.

Solution   The given limit has the indeterminate form $0/0$ at $x = 1$. Hence, by L'Hôpital's Rule,

$$\lim_{x \to 1^+} \frac{\ln x}{\sqrt{x-1}} \overset{h}{=} \lim_{x \to 1^+} \frac{1/x}{\frac{1}{2}(x-1)^{-1/2}} = \lim_{x \to 1^+} \frac{2\sqrt{x-1}}{x} = \frac{0}{1} = 0.$$ ∎

■ **Other Indeterminate Forms**   There are five additional indeterminate forms:

$$\infty - \infty, \quad 0 \cdot \infty, \quad 0^0, \quad \infty^0, \quad \text{and} \quad 1^\infty. \tag{5}$$

By a combination of algebra and a little cleverness we can often convert one of these new limit forms to either $0/0$ or ∞/∞.

■ **The Form $\infty - \infty$**   The next example illustrates a limit that has the indeterminate form $\infty - \infty$. This example should destroy any unwarranted convictions that $\infty - \infty = 0$.

EXAMPLE 9   Indeterminate Form $\infty - \infty$

Evaluate $\lim\limits_{x \to 0^+} \left[\dfrac{3x+1}{\sin x} - \dfrac{1}{x} \right]$.

Solution   We note that $(3x + 1)/\sin x \to \infty$ and $1/x \to \infty$ as $x \to 0^+$. However, after writing the difference as a single fraction, we recognize the form $0/0$:

$$\lim_{x \to 0^+} \left[\frac{3x+1}{\sin x} - \frac{1}{x} \right] = \lim_{x \to 0^+} \frac{3x^2 + x - \sin x}{x \sin x} \quad \leftarrow \text{common denominator}$$

$$\overset{h}{=} \lim_{x \to 0^+} \frac{6x + 1 - \cos x}{x \cos x + \sin x}$$

$$\overset{h}{=} \lim_{x \to 0^+} \frac{6 + \sin x}{-x \sin x + 2 \cos x}$$

$$= \frac{6 + 0}{0 + 2} = 3.$$ ∎

■ **The Form $0 \cdot \infty$**   If

$$f(x) \to 0 \quad \text{and} \quad |g(x)| \to \infty \quad \text{as} \quad x \to a,$$

then $\lim\limits_{x \to a} f(x)g(x)$ has the indeterminate form $0 \cdot \infty$. We can change a limit that has this form to one with the form $0/0$ or ∞/∞ by writing, in turn,

$$f(x)g(x) = \frac{f(x)}{1/g(x)} \quad \text{or} \quad f(x)g(x) = \frac{g(x)}{1/f(x)}.$$

EXAMPLE 10   Indeterminate Form $0 \cdot \infty$

Evaluate $\lim\limits_{x \to \infty} x \sin \dfrac{1}{x}$.

Solution   Since $1/x \to 0$, we have $\sin(1/x) \to 0$ as $x \to \infty$. Hence, the limit has the indeterminate form $0 \cdot \infty$. By writing

$$\lim_{x \to \infty} \frac{\sin(1/x)}{1/x}$$

we now have the form $0/0$. Hence,

$$\lim_{x \to \infty} \frac{\sin(1/x)}{1/x} \overset{h}{=} \lim_{x \to \infty} \frac{(-x^{-2}) \cos(1/x)}{(-x^{-2})}$$

$$= \lim_{x \to \infty} \cos \frac{1}{x} = 1.$$

In the last line we used the fact that $1/x \to 0$ as $x \to \infty$ and $\cos 0 = 1$. ∎

■ **The Forms 0^0, ∞^0 and 1^∞**   Suppose $y = f(x)^{g(x)}$ tends toward 0^0, ∞^0, or 1^∞ as $x \to a$. By taking the natural logarithm of y:

$$\ln y = \ln f(x)^{g(x)} = g(x) \ln f(x)$$

we see that the right-hand side of

$$\lim_{x \to a} \ln y = \lim_{x \to a} g(x) \ln f(x)$$

has the form $0 \cdot \infty$. If it is assumed that $\lim_{x \to a} \ln y = \ln(\lim_{x \to a} y) = L$, then

$$\lim_{x \to a} y = e^L \quad \text{or} \quad \lim_{x \to a} f(x)^{g(x)} = e^L.$$

Of course, the procedure just outlined is applicable to limits involving

$$x \to a^-, \quad x \to a^+, \quad x \to \infty, \quad \text{or} \quad x \to -\infty.$$

EXAMPLE 11 Indeterminate Form 0^0

Evaluate $\lim_{x \to 0^+} x^{1/\ln x}$.

Solution Because $\ln x \to -\infty$ as $x \to 0^+$, it follows from (2) that $1/\ln x \to 0$. Thus the given limit has the indeterminate form 0^0. Now, if we set $y = x^{1/\ln x}$, then

$$\ln y = \frac{1}{\ln x} \ln x = 1.$$

Notice that we do not need L'Hôpital's Rule in this case, since

$$\lim_{x \to 0^+} \ln y = \lim_{x \to 0^+} 1 = 1 \quad \text{or} \quad \ln\left(\lim_{x \to 0^+} y\right) = 1.$$

Hence, $\lim_{x \to 0^+} y = e^1$ or equivalently $\lim_{x \to 0^+} x^{1/\ln x} = e$. ■

EXAMPLE 12 Indeterminate Form 1^∞

Evaluate $\lim_{x \to \infty} \left(1 - \frac{3}{x}\right)^{2x}$.

Solution Because $1 - 3/x \to 1$ as $x \to \infty$ the indeterminate form is 1^∞. If

$$y = \left(1 - \frac{3}{x}\right)^{2x} \quad \text{then} \quad \ln y = 2x \ln\left(1 - \frac{3}{x}\right).$$

Observe that the form of $\lim_{x \to \infty} 2x \ln(1 - 3/x)$ is $\infty \cdot 0$, whereas the form of

$$\lim_{x \to \infty} \frac{2 \ln\left(1 - \frac{3}{x}\right)}{\frac{1}{x}}$$

is $0/0$. Applying (4) to the latter limit and simplifying gives

$$\lim_{x \to \infty} 2 \frac{\ln(1 - 3/x)}{1/x} \stackrel{h}{=} \lim_{x \to \infty} 2 \frac{\frac{3/x^2}{(1 - 3/x)}}{-1/x^2} = \lim_{x \to \infty} \frac{-6}{(1 - 3/x)} = -6.$$

From $\lim_{x \to \infty} \ln y = \ln\left(\lim_{x \to \infty} y\right) = -6$ we conclude that $\lim_{x \to \infty} y = e^{-6}$ or

$$\lim_{x \to \infty} \left(1 - \frac{3}{x}\right)^{2x} = e^{-6}.$$ ■

L'Hôpital

■ **Postscript—A Bit of History** It is questionable whether the French mathematician **Marquis Guillaume François Antoine de L'Hôpital** (1661–1704) discovered the rule that bears his name. The result is probably due to Johann Bernoulli. However, L'Hôpital was the first to publish the rule in his text *Analyse des Infiniment Petits*. The book was published in 1696 and is considered to be the first textbook on calculus.

$f'(x)$ NOTES FROM THE CLASSROOM

(*i*) In the application of L'Hôpital's Rule, students will sometimes misinterpret

$$\lim_{x\to a}\frac{f'(x)}{g'(x)} \quad \text{as} \quad \lim_{x\to a}\frac{d}{dx}\frac{f(x)}{g(x)}.$$

Remember, L'Hôpital's Rule utilizes the *quotient of derivatives* and *not* the *derivative of the quotient*.

(*ii*) Inspect a problem before you leap to its solution. The limit $\lim_{x\to 0}(\cos x)/x$ is the form $1/0$ and, as a consequence, does not exist. Lack of mathematical forethought in writing

$$\lim_{x\to 0}\frac{\cos x}{x} = \lim_{x\to 0}\frac{-\sin x}{1} = 0$$

is an incorrect application of L'Hôpital's Rule. Of course, the "answer" has no significance.

(*iii*) L'Hôpital's Rule is not a cure-all for every indeterminate form. For example, $\lim_{x\to\infty} e^x/e^{x^2}$ is certainly of the form ∞/∞, but

$$\lim_{x\to\infty}\frac{e^x}{e^{x^2}} = \lim_{x\to\infty}\frac{e^x}{2xe^{x^2}}$$

is of no practical help.

Exercises 4.5
Answers to selected odd-numbered problems begin on page ANS-14.

≡ Fundamentals

In Problems 1–40, use L'Hôpital's Rule where appropriate to find the given limit, or state that it does not exist.

1. $\lim_{x\to 0}\dfrac{\cos x - 1}{x}$

2. $\lim_{t\to 3}\dfrac{t^3 - 27}{t - 3}$

3. $\lim_{x\to 1}\dfrac{2x - 2}{\ln x}$

4. $\lim_{x\to 0^+}\dfrac{\ln 2x}{\ln 3x}$

5. $\lim_{x\to 0}\dfrac{e^{2x} - 1}{3x + x^2}$

6. $\lim_{x\to 0}\dfrac{\tan x}{2x}$

7. $\lim_{t\to\pi}\dfrac{5\sin^2 t}{1 + \cos t}$

8. $\lim_{\theta\to 1}\dfrac{\theta^2 - 1}{e^{\theta^2} - e}$

9. $\lim_{x\to 0}\dfrac{6 + 6x + 3x^2 - 6e^x}{x - \sin x}$

10. $\lim_{x\to\infty}\dfrac{3x^2 - 4x^3}{5x + 7x^3}$

11. $\lim_{x\to 0^+}\dfrac{\cot 2x}{\cot x}$

12. $\lim_{x\to 0}\dfrac{\arcsin(x/6)}{\arctan(x/2)}$

13. $\lim_{t\to 2}\dfrac{t^2 + 3t - 10}{t^3 - 2t^2 + t - 2}$

14. $\lim_{r\to -1}\dfrac{r^3 - r^2 - 5r - 3}{(r+1)^2}$

15. $\lim_{x\to 0}\dfrac{x - \sin x}{x^3}$

16. $\lim_{x\to 1}\dfrac{x^2 + 4}{x^2 + 1}$

17. $\lim_{x\to 0}\dfrac{\cos 2x}{x^2}$

18. $\lim_{x\to\infty}\dfrac{2e^{4x} + x}{e^{4x} + 3x}$

19. $\lim_{x\to 1^+}\dfrac{\ln\sqrt{x}}{x - 1}$

20. $\lim_{x\to\infty}\dfrac{\ln(3x^2 + 5)}{\ln(5x^2 + 1)}$

21. $\lim_{x\to 2}\dfrac{e^{x^2} - e^{2x}}{x - 2}$

22. $\lim_{x\to 0}\dfrac{4^x - 3^x}{x}$

23. $\lim_{x\to\infty}\dfrac{x\ln x}{x^2 + 1}$

24. $\lim_{t\to 0}\dfrac{1 - \cosh t}{t^2}$

25. $\lim_{x\to 0}\dfrac{x - \tan^{-1}x}{x^3}$

26. $\lim_{x\to 0}\dfrac{(\sin 2x)^2}{x^2}$

27. $\lim_{x\to\infty}\dfrac{e^x}{x^4}$

28. $\lim_{x\to\infty}\dfrac{e^{1/x}}{\sin(1/x)}$

29. $\lim_{x\to 0}\dfrac{x - \tan^{-1}x}{x - \sin^{-1}x}$

30. $\lim_{t\to 1}\dfrac{t^{1/3} - t^{1/2}}{t - 1}$

31. $\lim_{u\to\pi/2}\dfrac{\ln(\sin u)}{(2u - \pi)^2}$

32. $\lim_{\theta\to\pi/2}\dfrac{\tan\theta}{\ln(\cos\theta)}$

33. $\lim_{x\to -\infty}\dfrac{1 + e^{-2x}}{1 - e^{-2x}}$

34. $\lim_{x\to 0}\dfrac{e^x - x - 1}{2x^2}$

35. $\lim_{r\to 0}\dfrac{r - \cos r}{r - \sin r}$

36. $\lim_{t\to\pi}\dfrac{\csc 7t}{\csc 2t}$

37. $\lim_{x\to 0^+}\dfrac{x^2}{\ln^2(1 + 3x)}$

38. $\lim_{x\to 3}\left(\dfrac{\ln x - \ln 3}{x - 3}\right)^2$

39. $\lim_{x\to 0}\dfrac{3x^2 + e^x - e^{-x} - 2\sin x}{x\sin x}$

40. $\lim_{x\to 8}\dfrac{\sqrt{x+1} - 3}{x^2 - 64}$

In Problems 41–74, identify the given limit as one of the indeterminate forms given in (5). Use L'Hôpital's Rule where appropriate to find the limit, or state that it does not exist.

41. $\lim_{x \to 0} \left(\dfrac{1}{e^x - 1} - \dfrac{1}{x} \right)$

42. $\lim_{x \to 0^+} (\cot x - \csc x)$

43. $\lim_{x \to \infty} x(e^{1/x} - 1)$

44. $\lim_{x \to 0^+} x \ln x$

45. $\lim_{x \to 0^+} x^x$

46. $\lim_{x \to 1^-} x^{1/(1-x)}$

47. $\lim_{x \to 0} \left[\dfrac{1}{x} - \dfrac{1}{\sin x} \right]$

48. $\lim_{x \to 0} \left[\dfrac{1}{x^2} - \dfrac{\cos 3x}{x^2} \right]$

49. $\lim_{t \to 3} \left[\dfrac{\sqrt{t+1}}{t^2 - 9} - \dfrac{2}{t^2 - 9} \right]$

50. $\lim_{x \to 0^+} \left[\dfrac{1}{x} - \dfrac{1}{\ln(x+1)} \right]$

51. $\lim_{\theta \to 0} \theta \csc 4\theta$

52. $\lim_{x \to \pi/2^-} (\sin^2 x)^{\tan x}$

53. $\lim_{x \to \infty} (2 + e^x)^{e^{-x}}$

54. $\lim_{x \to 0^-} (1 - e^x)^{x^2}$

55. $\lim_{t \to \infty} \left(1 + \dfrac{3}{t} \right)^t$

56. $\lim_{h \to 0} (1 + 2h)^{4/h}$

57. $\lim_{x \to 0} x^{(1 - \cos x)}$

58. $\lim_{\theta \to 0} (\cos 2\theta)^{1/\theta^2}$

59. $\lim_{x \to \infty} \dfrac{1}{x^2 \sin^2(2/x)}$

60. $\lim_{x \to 1} (x^2 - 1)^{x^2}$

61. $\lim_{x \to 1} \left[\dfrac{1}{x - 1} - \dfrac{5}{x^2 + 3x - 4} \right]$

62. $\lim_{x \to 0} \left[\dfrac{1}{x^2} - \dfrac{1}{x} \right]$

63. $\lim_{x \to \infty} x^5 e^{-x}$

64. $\lim_{x \to \infty} (x + e^x)^{2/x}$

65. $\lim_{x \to \infty} x \left(\dfrac{\pi}{2} - \arctan x \right)$

66. $\lim_{t \to \pi/4} \left(t - \dfrac{\pi}{4} \right) \tan 2t$

67. $\lim_{x \to \infty} x \tan \left(\dfrac{5}{x} \right)$

68. $\lim_{x \to 0^+} x \ln(\sin x)$

69. $\lim_{x \to -\infty} \left[\dfrac{1}{e^x} - x^2 \right]$

70. $\lim_{x \to 0} (1 + 5 \sin x)^{\cot x}$

71. $\lim_{x \to \infty} \left(\dfrac{3x}{3x + 1} \right)^x$

72. $\lim_{\theta \to \pi/2^-} (\sec^3 \theta - \tan^3 \theta)$

73. $\lim_{x \to 0} (\sinh x)^{\tan x}$

74. $\lim_{x \to 0^+} x^{(\ln x)^2}$

In Problems 75 and 76, find the given limit.

75. $\lim_{x \to 0^+} \dfrac{1}{x} \ln \left(\dfrac{e^x - 1}{x} \right)$

76. $\lim_{x \to \infty} \dfrac{1}{x} \ln \left(\dfrac{e^x - 1}{x} \right)$

≡ Calculator/CAS Problems

In Problems 77 and 78, use a calculator or CAS to obtain the graph of the given function for the value of n on the indicated interval. In each case conjecture the value of $\lim_{x \to \infty} f(x)$.

77. $f(x) = \dfrac{e^x}{x^n}$; $n = 3$ on $[0, 15]$; $n = 4$ on $[0, 20]$;

   $n = 5$ on $[0, 25]$

78. $f(x) = \dfrac{x^n}{e^x}$; $n = 3$ on $[0, 15]$; $n = 4$ on $[0, 15]$;

   $n = 5$ on $[0, 20]$

In Problems 79 and 80, use $n! = 1 \cdot 2 \cdot 3 \cdots (n-1) \cdot n$,

$$\dfrac{d^n}{dx^n} x^n = n!,$$

where n is a positive integer, and L'Hôpital's Rule to find the limit.

79. $\lim_{x \to \infty} \dfrac{x^n}{e^x}$

80. $\lim_{x \to \infty} \dfrac{e^x}{x^n}$

≡ Applications

81. Consider the circle shown in FIGURE 4.5.1.

   (a) If the arc ABC is 5 in. long, express the area A of the shaded blue region as a function of the indicated angle θ. [*Hint*: The area of a circular sector is $\frac{1}{2} r^2 \theta$ and the arc length on a circle is $r\theta$, where θ is measured in radians.]

   (b) Evaluate $\lim_{\theta \to 0} A(\theta)$.

   (c) Evaluate $\lim_{\theta \to 0} dA/d\theta$.

FIGURE 4.5.1 Circle in Problem 81

82. In the absence of damping forces, a mathematical model for the displacement $x(t)$ of a mass on a spring (see Problem 60 in Exercises 3.5) when the system is sinusoidally driven by an external force with amplitude F_0 and frequency $\gamma/2\pi$ is

$$x(t) = \dfrac{F_0}{\omega(\omega^2 - \gamma^2)} (-\gamma \sin \omega t + \omega \sin \gamma t), \quad \gamma \neq \omega,$$

where $\omega/2\pi$ is the frequency of free (undriven) vibrations of the system.

   (a) When $\gamma = \omega$, the spring/mass system is said to be in **pure resonance**, and the displacement of the mass is defined by

$$x(t) = \lim_{\gamma \to \omega} \dfrac{F_0}{\omega(\omega^2 - \gamma^2)} (-\gamma \sin \omega t + \omega \sin \gamma t).$$

   Determine $x(t)$ by finding this limit.

   (b) Use a graphing utility to examine the graph of $x(t)$ found in part (a) in the case where $F_0 = 2$, $\gamma = \omega = 1$. Describe the behavior of the spring/mass system in pure resonance as $t \to \infty$.

83. When an ideal gas expands from pressure p_1 and volume v_1 to pressure p_2 and volume v_2 such that $pv^\gamma = k$ (constant) throughout the expansion, if $\gamma \neq 1$, then the work done is given by

$$W = \dfrac{p_2 v_2 - p_1 v_1}{1 - \gamma}.$$

(a) Show that

$$W = p_1 v_1 \left[\frac{(v_2/v_1)^{1-\gamma} - 1}{1 - \gamma} \right].$$

(b) Find the work done in the case when $pv = k$ (constant) throughout the expansion by letting $\gamma \to 1$ in the expression in part (a).

84. The retina is most sensitive to photons that enter the eye near the center of the pupil, and less sensitive to light that enters near the edge of the pupil. (This phenomenon is known as the **Stiles–Crawford effect** of the first kind.) The percentage σ of photons that reach the photopigments is related to the pupil radius p (measured in mm) by the mathematical model

$$\sigma = \frac{1 - 10^{-0.05p^2}}{0.115p^2} \times 100.$$

See FIGURE 4.5.2.

(a) What percentage of photons reach the photopigments when $p = 2$ mm?

(b) What, according to the formula, is the limiting percentage as the pupil radius tends to zero? Can you explain why it seems to be more than 100%?

FIGURE 4.5.2 Eye in Problem 84

≡ **Think About It**

85. Suppose f has a second derivative. Evaluate

$$\lim_{h \to 0} \frac{f(x + h) - 2f(x) + f(x - h)}{h^2}.$$

86. (a) Use a calculator or CAS to obtain the graph of

$$f(x) = \frac{x \sin x}{x^2 + 1}.$$

(b) From the graph in part (a), conjecture the value of $\lim_{x \to \infty} f(x)$.

(c) Explain why L'Hôpital's Rule does not apply to $\lim_{x \to \infty} f(x)$.

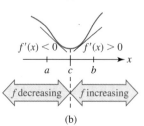

FIGURE 4.6.1 Relative maximum in (a); relative minimum in (b)

4.6 Graphing and the First Derivative

Introduction Knowing that a function does, or does not, possess relative extrema is a great aid in drawing its graph. We saw in Section 4.3 (Theorem 4.3.2) that when a function has a relative extremum it must occur at a critical number. By finding the critical numbers of a function, we have a *list of candidates* for the x-coordinates of the points that correspond to relative extrema. We shall now combine the ideas of the earlier sections of this chapter to devise two tests for determining when a critical number actually is the x-coordinate of a relative extremum.

First Derivative Test Suppose f is continuous on the closed interval $[a, b]$ and differentiable on an open interval (a, b) except possibly at a critical number c within the interval. If $f'(x) > 0$ for all x in (a, c) and $f'(x) < 0$ for all x in (c, b), then the graph of f on the interval (a, b) could be as shown in FIGURE 4.6.1(a); that is, $f(c)$ is a relative maximum. On the other hand, when $f'(x) < 0$ for all x in (a, c) and $f'(x) > 0$ for all x in (c, b), then, as shown in Figure 4.6.1(b), $f(c)$ is a relative minimum. We have demonstrated two special cases of the next theorem.

Theorem 4.6.1 First Derivative Test

Let f be continuous on $[a, b]$ and differentiable on (a, b) except possibly at the critical number c.

(i) If $f'(x)$ changes from positive to negative at c, then $f(c)$ is a relative maximum.
(ii) If $f'(x)$ changes from negative to positive at c, then $f(c)$ is a relative minimum.
(iii) If $f'(x)$ has the same algebraic sign on each side of c, then $f(c)$ is not an extremum.

The conclusions of Theorem 4.6.1 can be summarized in one sentence:

- A function f has a relative extremum at a critical number c where $f'(x)$ changes sign.

FIGURE 4.6.2 illustrates what might be the case when $f'(c)$ *does not* change sign at a critical number c. In Figures 4.6.2(a) and 4.6.2(b) we have shown a horizontal tangent at $(c, f(c))$ and so $f'(c) = 0$ but $f(c)$ is neither a relative maximum nor a relative minimum. In

Figure 4.6.2(c) we have shown a vertical tangent at $(c, f(c))$ and so $f'(c)$ does not exist, but again $f(c)$ is not a relative extremum because $f'(c)$ does not change sign at the critical number c.

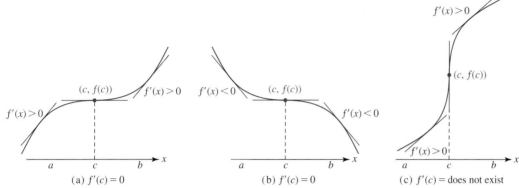

FIGURE 4.6.2 No extremum because $f'(x)$ does not change sign at the critical number c

In the next five examples we will illustrate the usefulness of Theorem 4.6.1 in sketching a graph of a function f by hand. In addition to the calculus:

- Find the derivative of f and factor f' if possible.
- Find the critical numbers of f.
- Apply the First Derivative Test to each critical number.

it also pays to ask:

- What is the domain of f?
- Does the graph of f have any intercepts? ← x-intercepts: Solve $f(x) = 0$; y-intercept: Find $f(0)$
- Does the graph of f have any symmetry? ← determine whether $f(-x) = f(x)$ or $f(-x) = -f(x)$
- Does the graph of f have any asymptotes?

The functions considered in Examples 1 and 2 are polynomials. Notice that these functions consist of both even and odd powers of x; this is sufficient to conclude that the graphs of these functions are not symmetric with respect to either the y-axis or the origin.

EXAMPLE 1   Polynomial Function of Degree 3

Graph $f(x) = x^3 - 3x^2 - 9x + 2$.

Solution The first derivative

$$f'(x) = 3x^2 - 6x - 9 = 3(x + 1)(x - 3) \qquad (1)$$

yields the critical numbers -1 and 3. Now the First Derivative Test is essentially the procedure used in finding the intervals on which f is either increasing or decreasing. We see in FIGURE 4.6.3(a) that $f'(x) > 0$ for $-\infty < x < -1$ and $f'(x) < 0$ for $-1 < x < 3$. In other words, $f'(x)$ changes from positive to negative at -1 and so it follows from part (i) of Theorem 4.6.1 that $f(-1) = 7$ is a relative maximum. Similarly, $f'(x) < 0$ for $-1 < x < 3$ and $f'(x) > 0$ for $3 < x < \infty$. Because $f'(x)$ changes from negative to positive at 3, part (ii) of Theorem 4.6.1 indicates that $f(3) = -25$ is a relative minimum. Now since $f(0) = 2$, the point $(0, 2)$ is the y-intercept for the graph of f. Furthermore, testing the equation $x^3 - 3x^2 - 9x + 2 = 0$ for rational roots reveals that $x = -2$ is a real root. Division by the factor $x + 2$ then gives $(x + 2)(x^2 - 5x + 1) = 0$. The quadratic formula applied to the quadratic factor reveals two additional real solutions:

$$\tfrac{1}{2}(5 - \sqrt{21}) \approx 0.21 \quad \text{and} \quad \tfrac{1}{2}(5 + \sqrt{21}) \approx 4.79.$$

◀ See the *SRM* for a brief review on how to find rational roots of polynomial equations.

The x-intercepts are then $(-2, 0)$, $\left(\tfrac{5}{2} - \tfrac{\sqrt{21}}{2}, 0\right)$, and $\left(\tfrac{5}{2} + \tfrac{\sqrt{21}}{2}, 0\right)$. Putting all this information together leads to the graph given in Figure 4.6.3(b).

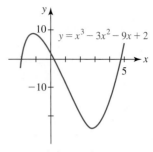

(a) First Derivative Test (b) Note x- and y-intercepts

FIGURE 4.6.3 Graph of function in Example 1

EXAMPLE 2 Polynomial Function of Degree 4

Graph $f(x) = x^4 - 4x^3 + 10$.

Solution The derivative

$$f'(x) = 4x^3 - 12x^2 = 4x^2(x - 3)$$

shows that 0 and 3 are critical numbers. Now, as seen in FIGURE 4.6.4(a), f' has the same negative algebraic sign in the adjacent intervals $(-\infty, 0)$ and $(0, 3)$. Hence, $f(0) = 10$ is *not* an extremum. In this case $f'(0) = 0$ means there is only a horizontal tangent at the y-intercept $(0, f(0)) = (0, 10)$. However, it is evident from the First Derivative Test that $f(3) = -17$ is a relative minimum. Indeed, the information that f is decreasing on the left side and increasing on the right side of the critical number 3 (the graph of f cannot turn back down) allows us to conclude that $f(3) = -17$ is also an *absolute minimum*. Finally, we see that the graph of f has two x-intercepts. With the aid of a calculator or a CAS, the x-intercepts are approximately $(1.61, 0)$ and $(3.82, 0)$.

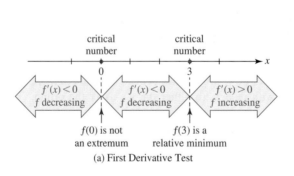

(a) First Derivative Test (b) $f'(0) = 0$ but $f(0) = 10$ is not an extremum

FIGURE 4.6.4 Graph of function in Example 2

EXAMPLE 3 Graph of a Rational Function

Graph $f(x) = \dfrac{x^2 - 3}{x^2 + 1}$.

Solution The following list summarizes some facts that can be discovered about the graph of this rational function f before actually graphing.

y-intercept: $f(0) = -3$, therefore the y-intercept is $(0, -3)$.
x-intercepts: $f(x) = 0$ when $x^2 - 3 = 0$. Thus, $x = -\sqrt{3}$ and $x = \sqrt{3}$. The x-intercepts are $(-\sqrt{3}, 0)$ and $(\sqrt{3}, 0)$.
Symmetry: y-axis, since $f(-x) = f(x)$.
Vertical asymptotes: None, since $x^2 + 1 \neq 0$ for all real numbers.
Horizontal asymptotes: Since the limit at infinity is the indeterminate form ∞/∞ we can use L'Hôpital's Rule to show

$$\lim_{x \to \infty} \frac{x^2 - 3}{x^2 + 1} \stackrel{h}{=} \lim_{x \to \infty} \frac{2x}{2x} = \lim_{x \to \infty} \frac{2}{2} = 1,$$

and so the line $y = 1$ is a horizontal asymptote.

Derivative: The Quotient Rule gives $f'(x) = \dfrac{8x}{(x^2+1)^2}$.

Critical numbers: $f'(x) = 0$ when $x = 0$. Therefore, 0 is the only critical number.

First Derivative Test: See FIGURE 4.6.5(a); $f(0) = -3$ is a relative minimum.

Graph: See Figure 4.6.5(b).

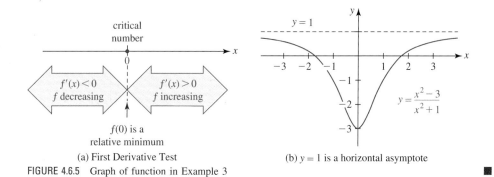

(a) First Derivative Test
(b) $y = 1$ is a horizontal asymptote

FIGURE 4.6.5 Graph of function in Example 3

EXAMPLE 4   Graph with a Vertical Asymptote

Graph $f(x) = x^2 + x - \ln|x|$.

Solution First note that the domain of f is $(-\infty, 0) \cup (0, \infty)$. Then by setting the numerator of the derivative

$$f'(x) = 2x + 1 - \frac{1}{x} = \frac{2x^2 + x - 1}{x} = \frac{(2x-1)(x+1)}{x}$$

equal to zero we see that -1 and $\tfrac{1}{2}$ are critical numbers. Although f is not differentiable at $x = 0$, 0 is not a critical number since 0 is not in the domain of f. In fact, $x = 0$ is a vertical asymptote for $\ln|x|$ and is also a vertical asymptote for the graph of f. We put the critical numbers and 0 on the number line because the sign of the derivative to the left and right of 0 indicates the behavior of f. As seen in FIGURE 4.6.6(a), $f'(x) < 0$ for $-\infty < x < -1$ and $f'(x) > 0$ for $-1 < x < 0$. We conclude that $f(-1) = 0$ is a relative minimum (at the same time $f(-1) = 0$ shows that $x = -1$ is the x-coordinate of an x-intercept). Continuing, $f'(x) < 0$ for $0 < x < \tfrac{1}{2}$ and $f'(x) > 0$ for $\tfrac{1}{2} < x < \infty$ shows that $f(\tfrac{1}{2}) = \tfrac{3}{4} - \ln\tfrac{1}{2} \approx 1.44$ is another relative minimum.

As noted, f is not defined at $x = 0$ and so there is no y-intercept. Finally, there is no symmetry either with respect to the y-axis or with respect to the origin. The graph of the function f is given in Figure 4.6.6(b).

◀ Verify that $f(-x) \neq f(x)$ and $f(-x) \neq -f(x)$.

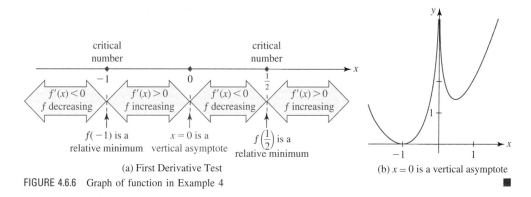

(a) First Derivative Test
(b) $x = 0$ is a vertical asymptote

FIGURE 4.6.6 Graph of function in Example 4

EXAMPLE 5   Graph with a Cusp

Graph $f(x) = -x^{5/3} + 5x^{2/3}$.

Solution The derivative is

$$f'(x) = -\frac{5}{3}x^{2/3} + \frac{10}{3}x^{-1/3} = \frac{5}{3}\frac{(-x+2)}{x^{1/3}}.$$

Notice that f' does not exist at 0 but 0 is in the domain of the function since $f(0) = 0$. The critical numbers are 0 and 2. The First Derivative Test, illustrated in FIGURE 4.6.7(a), shows that $f(0) = 0$ is a relative minimum and that $f(2) = -(2)^{5/3} + 5(2)^{2/3} \approx 4.76$ is a relative maximum. Moreover, since $f'(x) \to \infty$ as $x \to 0^+$ and $f'(x) \to -\infty$ as $x \to 0^-$ there is a cusp at (0, 0). Finally, by writing $f(x) = x^{2/3}(-x + 5)$, we see that $f(x) = 0$ at $x = 0$ and $x = 5$. The x-intercepts are the points (0, 0) and (5, 0). The graph of f is given in Figure 4.6.7(b).

Review the definition of a *cusp* in Section 3.2.

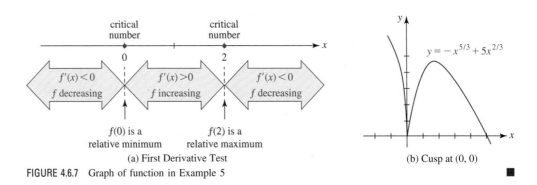

FIGURE 4.6.7 Graph of function in Example 5

It is sometimes convenient to know in advance of graphing, or even with the bother of graphing, whether a relative extremum $f(c)$ is an *absolute* extremum. The next theorem helps a little. You should sketch some graphs and convince yourself of its plausibility.

Theorem 4.6.2 The Sole Critical Number Test

Suppose c is the only critical number of a function f within an interval I. If it is proved that $f(c)$ is a relative extremum, then $f(c)$ is an absolute extremum.

In Example 3, it was shown that $f(0) = 0$ is a relative minimum by the First Derivative Test. We could have also concluded immediately that this function value is an absolute minimum. This fact follows from Theorem 4.6.2 because 0 is the only critical number in the interval $(-\infty, \infty)$.

Exercises 4.6   Answers to selected odd-numbered problems begin on page ANS-15.

≡ Fundamentals

In Problems 1–32, use the First Derivative Test to find the relative extrema of the given function. Graph. Find intercepts when possible.

1. $f(x) = -x^2 + 2x + 1$
2. $f(x) = (x - 1)(x + 3)$
3. $f(x) = x^3 - 3x$
4. $f(x) = \frac{1}{3}x^3 - \frac{1}{2}x^2 + 1$
5. $f(x) = x(x - 2)^2$
6. $f(x) = -x^3 + 3x^2 + 9x - 1$
7. $f(x) = x^3 + x - 3$
8. $f(x) = x^3 + 3x^2 + 3x - 3$
9. $f(x) = x^4 + 4x$
10. $f(x) = (x^2 - 1)^2$
11. $f(x) = \frac{1}{4}x^4 + \frac{4}{3}x^3 + 2x^2$
12. $f(x) = 2x^4 - 16x^2 + 3$
13. $f(x) = -x^2(x - 3)^2$
14. $f(x) = -3x^4 + 8x^3 - 6x^2 - 2$
15. $f(x) = 4x^5 - 5x^4$
16. $f(x) = (x - 2)^2(x + 3)^3$
17. $f(x) = \frac{x^2 + 3}{x + 1}$
18. $f(x) = x + \frac{25}{x}$
19. $f(x) = \frac{1}{x} - \frac{1}{x^3}$
20. $f(x) = \frac{x^2}{x^2 - 4}$
21. $f(x) = \frac{10}{x^2 + 1}$
22. $f(x) = \frac{x^2}{x^4 + 1}$
23. $f(x) = (x^2 - 4)^{2/3}$
24. $f(x) = (x^2 - 1)^{1/3}$
25. $f(x) = x\sqrt{1 - x^2}$
26. $f(x) = x(x^2 - 5)^{1/3}$
27. $f(x) = x - 12x^{1/3}$
28. $f(x) = x^{4/3} + 32x^{1/3}$
29. $f(x) = x^3 - 24 \ln|x|$
30. $f(x) = \frac{\ln x}{x}$
31. $f(x) = (x + 3)^2 e^{-x}$
32. $f(x) = 8x^2 e^{-x^2}$

In Problems 33–36, sketch a graph of a function f whose derivative f' has the given graph.

33.
FIGURE 4.6.8   Graph for Problem 33

34.
FIGURE 4.6.9   Graph for Problem 34

35.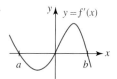
FIGURE 4.6.10   Graph for Problem 35

36.
FIGURE 4.6.11   Graph for Problem 36

In Problems 37 and 38, sketch the graph of f' from the graph of f.

37.
FIGURE 4.6.12   Graph for Problem 37

38.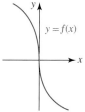
FIGURE 4.6.13   Graph for Problem 38

In Problems 39–42, sketch a graph of a function f that has the given properties.

39. $f(-1) = 0, f(0) = 1$
$f'(3)$ does not exist, $f'(5) = 0$
$f'(x) > 0, x < 3$ and $x > 5$
$f'(x) < 0, 3 < x < 5$

40. $f(0) = 0$
$f'(-1) = 0, f'(0) = 0, f'(1) = 0$
$f'(x) < 0, x < -1, -1 < x < 0$
$f'(x) > 0, 0 < x < 1, x > 1$

41. $f(-x) = f(x)$
$f(2) = 3$
$f'(x) < 0, 0 < x < 2$
$f'(x) > 0, x > 2$

42. $f(1) = -2, f(0) = -1$
$\lim_{x \to 3} f(x) = \infty, f'(4) = 0$
$f'(x) < 0, x < 1$
$f'(x) < 0, x > 4$

In Problems 43 and 44, determine where the slope of the tangent to the graph of the given function has a relative maximum or a relative minimum.

43. $f(x) = x^3 + 6x^2 - x$

44. $f(x) = x^4 - 6x^2$

45. (a) From the graph of $g(x) = \sin 2x$ determine the intervals for which $g(x) > 0$ and the intervals for which $g(x) < 0$.

(b) Find the critical numbers of $f(x) = \sin^2 x$. Use the First Derivative Test and the information in part (a) to find the relative extrema of f.

(c) Sketch the graph of the function f in part (b).

46. (a) Find the critical numbers of $f(x) = x - \sin x$.
(b) Show that f has no relative extrema.
(c) Sketch the graph of f.

≡ Applications

47. The **arithmetic mean**, or **average**, of the n numbers $a_1, a_2, \ldots, a_n$ is given by
$$\bar{x} = \frac{a_1 + a_2 + \cdots + a_n}{n}.$$

(a) Show that $\bar{x}$ is a critical number of the function
$$f(x) = (x - a_1)^2 + (x - a_2)^2 + \cdots + (x - a_n)^2.$$

(b) Show that $f(\bar{x})$ is a relative minimum.

48. When sound passes from one medium to another, some of its energy can be lost because of a difference in the acoustic resistances of the two media. (Acoustic resistance is the product of density and elasticity.) The fraction of energy transmitted is given by
$$T(r) = \frac{4r}{(r + 1)^2},$$
where r is the ratio of the acoustic resistances of the two media.

(a) Show that $T(r) = T(1/r)$. Explain what this means physically.
(b) Use the First Derivative Test to find the relative extrema of T.
(c) Sketch the graph of the function T for $r \geq 0$.

≡ Think About It

49. Find values of a, b, and c such that $f(x) = ax^2 + bx + c$ has a relative maximum 6 at $x = 2$ and the graph of f has y-intercept 4.

50. Find values of a, b, c, and d such that $f(x) = ax^3 + bx^2 + cx + d$ has a relative minimum -3 at $x = 0$ and a relative maximum 4 at $x = 1$.

51. Suppose f is a differentiable function whose graph is symmetric about the y-axis. Prove that $f'(0) = 0$. Does f necessarily have a relative extremum at $x = 0$?

52. Let m and n denote positive integers. Show that $f(x) = x^m(x - 1)^n$ always has a relative minimum.

53. Suppose f and g are differentiable functions and have relative maxima at the same critical number c.

(a) Show that c is a critical number for the functions $f + g, f - g,$ and fg.
(b) Does it follow that the functions $f + g, f - g,$ and fg have relative maxima at c? Prove your assertions or give a counterexample.

4.7 Graphing and the Second Derivative

■ **Introduction** In the discussion that follows, our goal is to relate the concept of the concavity of a graph with the second derivative of a function. The second derivative then provides us another way of testing to see whether a relative extremum of a function f occurs at a critical number.

■ **Concavity** You probably have an *intuitive* idea of what is meant by concavity. FIGURES 4.7.1(a) and 4.7.1 (b) illustrate geometric shapes that are **concave up** and **concave down**, respectively. For example, the Gateway Arch in St. Louis is concave down; the cables between the vertical supports of the Golden Gate Bridge are concave up. Often a shape that is concave up is said to "hold water," whereas a shape that is concave down "spills water." However, the precise definition of concavity is given in terms of the derivative.

(a) "Holds water"

(b) "Spills water"
FIGURE 4.7.1 Concavity

Definition 4.7.1 Concavity

Let f be a differentiable function on an interval (a, b).

(*i*) If f' is an increasing function on (a, b), then the graph of f is **concave up** on the interval.
(*ii*) If f' is a decreasing function on (a, b), then the graph of f is **concave down** on the interval.

In other words, if the slopes of the tangent lines to the graph of f increase (decrease) as x increases on (a, b), then the graph of f is concave up (down) on the interval. If the slopes increase (decrease) as x increases, then this means that the tangent lines are turning counter-clockwise (clockwise) on the interval. The plausibility of Definition 4.7.1 is illustrated in FIGURE 4.7.2. An equivalent way of looking at concavity is also apparent in Figure 4.7.2. The graph of a function f is concave up (down) on an interval if the graph at any point lies above (below) the tangent lines.

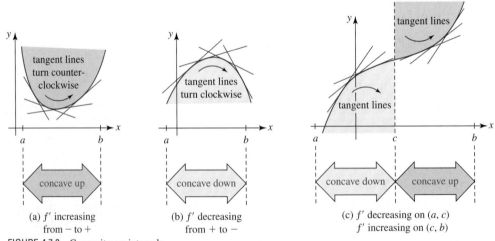

(a) f' increasing from $-$ to $+$
(b) f' decreasing from $+$ to $-$
(c) f' decreasing on (a, c)
f' increasing on (c, b)
FIGURE 4.7.2 Concavity on intervals

■ **Concavity and the Second Derivative** In Theorem 4.4.4 of Section 4.4 we saw that the algebraic sign of the derivative of a function indicates when the function is increasing or decreasing on an interval. Specifically, if the function referred to in the preceding sentence is the derivative f', then we can conclude that the algebraic sign of the derivative of f', that is, f'', indicates when f' is either increasing or decreasing on an interval. For example, if $f''(x) > 0$ on (a, b), then f' increases on (a, b). In view of Definition 4.7.1, if f' increases on (a, b), then the graph of f is concave up on the interval. Therefore, we are led to the following test for concavity.

4.7 Graphing and the Second Derivative

Theorem 4.7.1 Test for Concavity

Let f be a function for which f'' exists on (a, b).

(i) If $f''(x) > 0$ for all x in (a, b), then the graph of f is concave up on (a, b).
(ii) If $f''(x) < 0$ for all x in (a, b), then the graph of f is concave down on (a, b).

EXAMPLE 1 Test for Concavity

Determine the intervals on which the graph of $f(x) = x^3 + \frac{9}{2}x^2$ is concave up and the intervals on which the graph is concave down.

Solution From $f'(x) = 3x^2 + 9x$ we obtain

$$f''(x) = 6x + 9 = 6\left(x + \frac{3}{2}\right).$$

We see that $f''(x) < 0$ when $6\left(x + \frac{3}{2}\right) < 0$ or $x < -\frac{3}{2}$ and that $f''(x) > 0$ when $6\left(x + \frac{3}{2}\right) > 0$ or $x > -\frac{3}{2}$. It follows from Theorem 4.7.1 that the graph of f is concave down on the interval $\left(-\infty, -\frac{3}{2}\right)$ and concave up on $\left(-\frac{3}{2}, \infty\right)$. ∎

■ **Point of Inflection** The graph of the function in Example 1 changes concavity at the point that corresponds to $x = -\frac{3}{2}$. As x increases through $-\frac{3}{2}$, the graph of f changes from concave down to concave up at the point $\left(-\frac{3}{2}, \frac{27}{4}\right)$. A point on the graph of a function where the concavity changes from up to down, or vice versa, is given a special name.

Definition 4.7.2 Point of Inflection

Let f be continuous on an interval (a, b) containing the number c. A point $(c, f(c))$ is a **point of inflection** of the graph of f if there is a tangent line at $(c, f(c))$ and the graph changes concavity at this point.

A reexamination of Example 1 shows that $f(x) = x^3 + \frac{9}{2}x^2$ is continuous at $-\frac{3}{2}$, has a tangent line at $\left(-\frac{3}{2}, \frac{27}{4}\right)$, and changes concavity at that point. Hence, $\left(-\frac{3}{2}, \frac{27}{4}\right)$ is a point of inflection. Also note that $f''\left(-\frac{3}{2}\right) = 0$. See FIGURE 4.7.3(a). We also know that the function $f(x) = x^{1/3}$ is continuous at 0 and possesses a vertical tangent at $(0, 0)$ (see Example 10 of Section 3.1). From $f''(x) = -\frac{2}{9}x^{-5/3}$ it is seen that $f''(x) > 0$ for $x < 0$ and $f''(x) < 0$ for $x > 0$. Hence, $(0, 0)$ is a point of inflection. Note in this case $f''(x) = -\frac{2}{9}x^{-5/3}$ is not defined at $x = 0$. See Figure 4.7.3(b). We have illustrated two cases of the next theorem.

(a) $f''\left(-\frac{3}{2}\right) = 0$

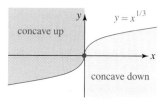
(b) $f''(x)$ does not exist at 0
FIGURE 4.7.3 Points of inflection

Theorem 4.7.2 Point of Inflection

If $(c, f(c))$ is a point of inflection for the graph of a function f, then either $f''(c) = 0$ or $f''(c)$ does not exist.

■ **Second Derivative Test** If c is a critical number of a function $y = f(x)$, and, say, $f''(c) > 0$, then the graph of f is concave up on some interval (a, b) that contains c. Necessarily then, $f(c)$ is a relative minimum. Similarly, $f''(c) < 0$ at a critical value c implies $f(c)$ is a relative maximum. This so-called **Second Derivative Test** is illustrated in FIGURE 4.7.4.

FIGURE 4.7.4 Second Derivative Test

Theorem 4.7.3 Second Derivative Test

Let f be a function for which f'' exists on an interval (a, b) that contains the critical number c.

(i) If $f''(c) > 0$, then $f(c)$ is a relative minimum.
(ii) If $f''(c) < 0$, then $f(c)$ is a relative maximum.
(iii) If $f''(c) = 0$, the test fails and $f(c)$ may or may not be a relative extremum. In this case, use the First Derivative Test.

At this point one might ask, Why do we need another test for relative extrema when we already have the First Derivative Test? If the function f under consideration is a polynomial, it is very easy to compute the second derivative. In using Theorem 4.7.3 we need only determine the algebraic sign of $f''(x)$ *at* the critical number. Contrast this with Theorem 4.6.1 where we must determine the sign of $f'(x)$ at numbers to the right and left of the critical number. If f' is not readily factored, the latter procedure may be somewhat difficult. On the other hand, it may be equally tedious to use Theorem 4.7.3 in the case of some functions that involve products, quotients, powers, and so on. Thus, Theorems 4.6.1 and 4.7.3 both have advantages and disadvantages.

EXAMPLE 2 Second Derivative Test

Graph $f(x) = 4x^4 - 4x^2$.

Solution From $f(x) = 4x^2(x^2 - 1) = 4x^2(x + 1)(x - 1)$ we see that the graph of f has the intercepts $(-1, 0)$, $(0, 0)$, and $(1, 0)$. Furthermore, since f is a polynomial with only even powers, we conclude that its graph is symmetric with respect to the y-axis (even function). Now the first and second derivatives are

$$f'(x) = 16x^3 - 8x = 8x(\sqrt{2}x + 1)(\sqrt{2}x - 1)$$
$$f''(x) = 48x^2 - 8 = 8(\sqrt{6}x + 1)(\sqrt{6}x - 1).$$

From f' we see that the critical numbers of f are 0, $-\sqrt{2}/2$, and $\sqrt{2}/2$. The Second Derivative Test is summarized in the accompanying table.

x	Sign of $f''(x)$	$f(x)$	Conclusion
0	−	0	rel. max.
$\sqrt{2}/2$	+	−1	rel. min.
$-\sqrt{2}/2$	+	−1	rel. min.

Finally, from the factored form of f'' it is seen that $f''(x)$ changes signs at $x = -\sqrt{6}/6$ and at $x = \sqrt{6}/6$. Hence, the graph of f possesses two points of inflection: $\left(-\sqrt{6}/6, -\frac{5}{9}\right)$ and $\left(\sqrt{6}/6, -\frac{5}{9}\right)$. See FIGURE 4.7.5. ∎

FIGURE 4.7.5 Graph of function in Example 2

EXAMPLE 3 Failure of the Second Derivative Test

Consider the simple function $f(x) = x^4 + 1$. From $f'(x) = 4x^3$ we see that 0 is a critical number. But from the second derivative $f''(x) = 12x^2$ we get $f''(0) = 0$. Thus the Second Derivative Test leads to no conclusion. However, from the first derivative $f'(x) = 4x^3$ we see:

$$f'(x) < 0 \quad \text{for} \quad x < 0 \quad \text{and} \quad f'(x) > 0 \quad \text{for} \quad x > 0.$$

This First Derivative Test indicates that $f(0) = 1$ is a relative minimum. FIGURE 4.7.6 shows $f(0) = 1$ is actually an absolute minimum. ∎

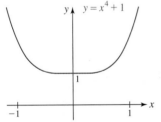

FIGURE 4.7.6 Graph of function in Example 3

EXAMPLE 4 Second Derivative Test

Graph $f(x) = 2\cos x - \cos 2x$.

Solution Because $\cos x$ and $\cos 2x$ are even functions, the graph of f possesses symmetry with respect to the y-axis. Also, $f(0) = 1$ yields the y-intercept $(0, 1)$. Now the first and second derivatives are

$$f'(x) = -2\sin x + 2\sin 2x \quad \text{and} \quad f''(x) = -2\cos x + 4\cos 2x.$$

Using the trigonometric identity $\sin 2x = 2\sin x \cos x$, we can simplify the equation $f'(x) = 0$ to $\sin x(1 - 2\cos x) = 0$. The solutions of $\sin x = 0$ are $0, \pm\pi, \pm 2\pi, \ldots$ and the solutions of $\cos x = \frac{1}{2}$ are $\pm\pi/3, \pm 5\pi/3, \ldots$. But since f is 2π periodic (show this!), it suffices to consider only those critical numbers in $[0, 2\pi]$, namely, $0, \pi/3, \pi, 5\pi/3,$ and 2π. The Second Derivative Test applied to these values is summarized in the accompanying table.

x	Sign of $f''(x)$	$f(x)$	Conclusion
0	$+$	1	rel. min.
$\pi/3$	$-$	$\frac{3}{2}$	rel. max.
π	$+$	-3	rel. min.
$5\pi/3$	$-$	$\frac{3}{2}$	rel. max.
2π	$+$	1	rel. min.

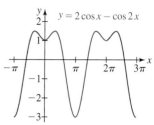

FIGURE 4.7.7 Graph of function in Example 4

The graph of f is the 2π-periodic extension of the blue portion shown in FIGURE 4.7.7 on the interval $[0, 2\pi]$.

$f'(x)$ NOTES FROM THE CLASSROOM

(*i*) If $(c, f(c))$ is a point of inflection, then either $f''(c) = 0$ or $f''(c)$ does not exist. The converse of this statement is not necessarily true. We cannot conclude, simply from the fact that when $f''(c) = 0$ or $f''(c)$ does not exist, that $(c, f(c))$ is a point of inflection. For example, in Example 3 we saw $f''(0) = 0$ for $f(x) = x^4 + 1$. But it is apparent from Figure 4.7.6 that $(0, f(0))$ is not a point of inflection. Also, for $f(x) = 1/x$, we see that $f''(x) = 2/x^3$ is undefined at $x = 0$ and that the graph of f changes concavity at $x = 0$:

$$f''(x) < 0 \quad \text{for} \quad x < 0 \quad \text{and} \quad f''(x) > 0 \quad \text{for} \quad x > 0.$$

However, $x = 0$ is not the x-coordinate of a point of inflection because f is not continuous at 0.

(*ii*) You should not think that the graph of a function *has to have* concavity. There are perfectly good differentiable functions whose graphs possess no concavity. See Problem 60 in Exercises 4.7.

(*iii*) You should be aware that textbooks disagree on the precise definition of a point of inflection. This is nothing to be concerned with but if interested, see Problem 65 in Exercises 4.7.

Exercises 4.7   Answers to selected odd-numbered problems begin on page ANS-16.

≡ Fundamentals

In Problems 1–12, use the second derivative to determine the intervals on which the graph of the given function is concave up and the intervals on which it is concave down.

1. $f(x) = -x^2 + 7x$
2. $f(x) = -(x + 2)^2 + 8$
3. $f(x) = -x^3 + 6x^2 + x - 1$
4. $f(x) = (x + 5)^3$
5. $f(x) = x(x - 4)^3$
6. $f(x) = 6x^4 + 2x^3 - 12x^2 + 3$
7. $f(x) = x^{1/3} + 2x$
8. $f(x) = x^{8/3} - 20x^{2/3}$
9. $f(x) = x + \dfrac{9}{x}$
10. $f(x) = \sqrt{x^2 + 10}$
11. $f(x) = \dfrac{1}{x^2 + 3}$
12. $f(x) = \dfrac{x - 1}{x + 2}$

In Problems 13–16, estimate from the graph of the given function f the intervals on which f' is increasing and the intervals on which f' is decreasing.

13.

FIGURE 4.7.8 Graph for Problem 13

14.

FIGURE 4.7.9 Graph for Problem 14

15.

FIGURE 4.7.10 Graph for Problem 15

16.

FIGURE 4.7.11 Graph for Problem 16

17. Show that the graph of $f(x) = \sec x$ is concave up on those intervals on which $\cos x > 0$, and concave down on those intervals on which $\cos x < 0$.

18. Show that the graph of $f(x) = \csc x$ is concave up on those intervals on which $\sin x > 0$, and concave down on those intervals on which $\sin x < 0$.

In Problems 19–26, use the second derivative to locate all points of inflection.

19. $f(x) = x^4 - 12x^2 + x - 1$ **20.** $f(x) = x^{5/3} + 4x$
21. $f(x) = \sin x$ **22.** $f(x) = \cos x$
23. $f(x) = x - \sin x$ **24.** $f(x) = \tan x$
25. $f(x) = x + xe^{-x}$ **26.** $f(x) = xe^{-x^2}$

In Problems 27–44, use the Second Derivative Test, when applicable, to find the relative extrema of the given function. Graph. Find intercepts and points of inflection when possible.

27. $f(x) = -(2x - 5)^2$ **28.** $f(x) = \frac{1}{3}x^3 - 2x^2 - 12x$
29. $f(x) = x^3 + 3x^2 + 3x + 1$ **30.** $f(x) = \frac{1}{4}x^4 - 2x^2$
31. $f(x) = 6x^5 - 10x^3$ **32.** $f(x) = x^3(x + 1)^2$
33. $f(x) = \dfrac{x}{x^2 + 2}$ **34.** $f(x) = x^2 + \dfrac{1}{x^2}$
35. $f(x) = \sqrt{9 - x^2}$ **36.** $f(x) = x\sqrt{x - 6}$
37. $f(x) = x^{1/3}(x + 1)$ **38.** $f(x) = x^{1/2} - \frac{1}{4}x$
39. $f(x) = \cos 3x$, $[0, 2\pi]$ **40.** $f(x) = 2 + \sin 2x$, $[0, 2\pi]$
41. $f(x) = \cos x + \sin x$, $[0, 2\pi]$
42. $f(x) = 2\sin x + \sin 2x$, $[0, 2\pi]$
43. $f(x) = 2x - x\ln x$ **44.** $f(x) = \ln(x^2 + 2)$

In Problems 45–48, determine whether the given function has a relative extremum at the indicated critical number.

45. $f(x) = \sin x \cos x$; $\pi/4$ **46.** $f(x) = x\sin x$; 0
47. $f(x) = \tan^2 x$; π **48.** $f(x) = (1 + \sin 4x)^3$; $\pi/8$

In Problems 49–52, sketch a graph of a function f that has the given properties.

49. $f(-2) = 0, f(4) = 0$
$f'(3) = 0, f''(1) = 0, f''(2) = 0$
$f''(x) < 0, x < 1, x > 2$
$f''(x) > 0, 1 < x < 2$

50. $f(0) = 5, f(2) = 0$
$f'(2) = 0, f''(3)$ does not exist
$f''(x) > 0, x < 3$
$f''(x) < 0, x > 3$

51. $f(0) = -1, f(\pi/2) > 0$
$f'(x) \geq 0$ for all x
$f''(x) > 0, (2n-1)\dfrac{\pi}{2} < x < (2n+1)\dfrac{\pi}{2}$, n even
$f''(x) < 0, (2n-1)\dfrac{\pi}{2} < x < (2n+1)\dfrac{\pi}{2}$, n odd

52. $f(-x) = -f(x)$
vertical asymptote $x = 2$, $\lim_{x \to \infty} f(x) = 0$
$f''(x) < 0, 0 < x < 2$
$f''(x) > 0, x > 2$

≡ Think About It

53. Find values of a, b, and c such that the graph of $f(x) = ax^3 + bx^2 + cx$ passes through $(-1, 0)$ and has a point of inflection at $(1, 1)$.

54. Find values of a, b, and c such that the graph of $f(x) = ax^3 + bx^2 + cx$ has a horizontal tangent at the point of inflection $(1, 1)$.

55. Use the Second Derivative Test as an aid in graphing $f(x) = \sin(1/x)$. Observe that f is discontinuous at $x = 0$.

56. Show that the graph of a general polynomial function
$$f(x) = a_n x^n + a_{n-1} x^{n-1} + \cdots + a_1 x + a_0, a_n \neq 0$$
can have at most $n - 2$ points of inflection.

57. Let $f(x) = (x - x_0)^n$, where n is a positive integer.
(a) Show that $(x_0, 0)$ is a point of inflection of the graph of f if n is an odd integer.
(b) Show that $(x_0, 0)$ is not a point of inflection of the graph of f but corresponds to a relative minimum when n is an even integer.

58. Prove that the graph of a quadratic polynomial function $f(x) = ax^2 + bx + c, a \neq 0$, is concave upward on the x-axis when $a > 0$ and concave downward on the x-axis when $a < 0$.

59. Let f be a function for which f''' exists on an interval (a, b) that contains the number c. If $f''(c) = 0$ and $f'''(c) \neq 0$, what can be said about $(c, f(c))$?

60. Give an example of a differentiable function whose graph possesses no concavity. Do not think profound thoughts.

61. Prove or disprove: A point of inflection for a function f must occur at a critical value of f'.

62. Without graphing, explain why the graph of $f(x) = 10x^2 - x - 40 + e^x$ cannot have a point of inflection.

63. Prove or disprove: The function
$$f(x) = \begin{cases} 4x^2 - x, & x \leq 0 \\ -x^3, & x > 0 \end{cases}$$
has a point of inflection at $(0, 0)$.

64. Suppose f is a polynomial function of degree 3 and that c_1 and c_2 are distinct critical numbers.
(a) Are $f(c_1)$ and $f(c_2)$ necessarily relative extrema of the function? Prove your assertion.
(b) What do you think is the x-coordinate of the point of inflection for the graph of f? Prove your assertion.

≡ Project

65. Points of Inflection Find other calculus texts and note how they define a point of inflection. Then do some Internet research on the definition of a point of inflection. Write a short paper comparing these definitions. Illustrate your paper with appropriate graphs.

4.8 Optimization

Introduction In science, engineering, and business one is often interested in the maximum and minimum values of functions; for example, a company is naturally interested in maximizing revenue while minimizing cost. The next time you go to a supermarket note that all cans containing, say, 15 oz of food (0.01566569 ft^3) have the same physical appearance. The fact that all cans of a specified volume have the same shape (same radius and height) is no coincidence, since there are specific dimensions that minimize the amount of metal used and, hence, minimize the cost of the construction of the can to a company. In the same vein, many of the so-called economy cars have appearances that are remarkably similar. This is not just a simple matter of one company copying the success of another company, but, rather, for a given volume engineers strive for a design that minimizes the amount of material used.

Playing with Some Numbers Let us begin with a simple problem:

Find two nonnegative numbers whose sum is 5 such that the product of one and the square of the other is as large as possible. (1)

◀ At this point a review of Section 1.7 is strongly recommended.

In Example 1 of Section 1.7 we encountered the problem:

The sum of two nonnegative numbers is 5. Express the product of one and the square of the other as a function of one of the numbers. (2)

A comparison of (1) and (2) shows that (2), where we are simply asked to set up a function, is embedded within the calculus problem (1). The calculus part of (1) requires that we find nonnegative numbers so that the product is a maximum. A review of Examples 1 and 2 of Section 1.7 indicates that the product described in (1) is

$$P = x(5 - x)^2 \quad \text{or} \quad P(x) = 25x - 10x^2 + x^3. \quad (3)$$

The domain of the function $P(x)$ in (3) is the interval $[0, 5]$. This fact came from combining the two inequalities $x \geq 0$ and $y = 5 - x \geq 0$ or the recognition that if x were allowed to be larger than 5, then y would be negative, contrary to our initial assumption. There are an infinite number of pairs of nonnegative real numbers (rational and irrational) that add up to 5. Note that we did not say nonnegative *integers*! For example

Numbers: x, y	Product: $P = xy^2$
1, 4	$P = 1 \cdot 4^2 = 16$
2, 3	$P = 2 \cdot 3^2 = 18$
$\frac{1}{2}, \frac{9}{2}$	$P = \frac{1}{2} \cdot \left(\frac{9}{2}\right)^2 = 10.125$
$\pi, 5 - \pi$	$P = \pi \cdot (5 - \pi)^2 \approx 10.85$

Pairs of numbers, such as $-1, 6$, that add up to 5 are rejected because both numbers must be nonnegative. Now how do we know when we have discovered the numbers x and y that give the largest, that is, the maximum or optimal, value of P? The answer lies in the realization that the domain of the function $P(x)$ is the closed interval $[0, 5]$. We know from Theorem 4.3.3 that the continuous function $P(x)$ has an absolute extremum either at an endpoint of the interval or at a critical number in the open interval $(0, 5)$. From (3) we see that $P'(x) = 25 - 20x + 3x^2 = (3x - 5)(x - 5)$ so that the only critical number in the open interval $(0, 5)$ is $\frac{5}{3}$. The function values $P(0) = 0$ and $P(5) = 0$ obviously represent the absolute minimum product, and so the absolute maximum product is $P\left(\frac{5}{3}\right) = \frac{5}{3}(5) - \left(\frac{5}{3}\right)^2 = \frac{500}{27} \approx 18.52$. In other words, the two numbers are $x = \frac{5}{3}$ and $y = 5 - \frac{5}{3} = \frac{10}{3}$.

Terminology In general, the function that describes the quantity we seek to optimize, by finding its maximum or its minimum value, is called the **objective function**. The function in (3) is the objective function for the problem given in (1). A relationship among the variables in an optimization problem, such as the equation $x + y = 5$ between the numbers x and y in the foregoing discussion, is called a **constraint**. The constraint allows us to eliminate one of the variables in the construction of the objective function, such as $P(x)$ in (3), as well as places a limitation on how the variables such as x and y can actually vary. We saw that the limitations $x \geq 0$ and $y = 5 - x \geq 0$ helped us to infer that the domain of the function $P(x)$ in (3) was the interval $[0, 5]$. You should be aware that the type of word problems considered in this section *may* or *may not* have a constraint.

■ **Suggestions** In the examples and problems that follow either we will be *given* an objective function or we will have to translate the words into mathematical symbols and construct an objective function. These are the kinds of word problems that show off the power of calculus and provide one of many possible answers to the age-old question: What's it good for? While not guaranteeing anything, here are some suggestions to keep in mind when solving an optimization problem. First and foremost:

Develop a positive and analytical attitude. Try to be neat and organized.

Guidelines for Solving Optimization Problems

(*i*) Read the problem slowly, then read it again.
(*ii*) Draw a picture when possible; keep it simple.
(*iii*) Introduce variables (in your picture, if there is one) and note any constraint between the variables.
(*iv*) Using all necessary variables, set up the objective function. If more than one variable is used, then employ the constraint to reduce the function to one variable.
(*v*) Note the interval on which the function is defined. Determine all critical numbers.
(*vi*) If the objective function is continuous and defined on a closed interval $[a, b]$, then test for endpoint extrema. If the desired extremum does not occur at an endpoint, it must occur at a critical number in the open interval (a, b).
(*vii*) If the objective function is defined on an interval that is not closed, then a derivative test should be used at each critical number in that interval.

In our first example, we examine a mathematical model that comes from physics.

EXAMPLE 1 Maximum Range

When air resistance is ignored, the horizontal range R of a projectile is given by

$$R(\theta) = \frac{v_0^2}{g}\sin 2\theta, \tag{4}$$

where v_0 is the constant initial velocity, g is the acceleration due to gravity, and θ is the angle of elevation or departure. Find the maximum range of the projectile.

Solution As a physical model of the problem, let us imagine that the projectile is a cannonball. See **FIGURE 4.8.1**. For angles θ greater than $\pi/2$, the cannon shown in the figure would shoot backward. Thus, it makes physical sense to restrict the function in (4) to the closed interval $[0, \pi/2]$. From

$$\frac{dR}{d\theta} = \frac{v_0^2}{g}2\cos 2\theta$$

we see that $dR/d\theta = 0$ when $\cos 2\theta = 0$ or $2\theta = \pi/2$, and so the only critical number in the open interval $(0, \pi/2)$ is $\pi/4$. Evaluating the function at the endpoints and the critical number gives

$$R(0) = 0, \quad R(\pi/4) = \frac{v_0^2}{g}, \quad R(\pi/2) = 0.$$

FIGURE 4.8.1 Cannonball in Example 1

Since $R(\theta)$ is continuous on the closed interval $[0, \pi/2]$, these values indicate that the minimum range is $R(0) = R(\pi/2) = 0$ and the maximum range is $R(\pi/4) = v_0^2/g$. In other words, to achieve maximum distance, the projectile should be launched at an angle of $45°$ to the horizontal. ∎

If cannonballs in Example 1 are launched with the initial velocity v_0 but with varying angles of elevation θ different from $45°$ then their horizontal ranges are less than $R_{\max} = v_0^2/g$. Inspection of the function in (4) shows that the same horizontal range is attained for complementary angles such as $20°$ and $70°$, and $30°$ and $60°$. See FIGURE 4.8.2. If air resistance is taken into account, all projectiles will fall short of v_0^2/g, even when launched at an angle of elevation of $45°$.

FIGURE 4.8.2 Same range for complementary angles

EXAMPLE 2 Maximum Volume

A 20-ft-long water trough has ends in the form of isosceles triangles with sides that are 4 ft long. Determine the dimension across the top of the triangular end so that the volume of the trough is a maximum. Find the maximum volume.

Solution The trough with the unknown dimension x is shown in FIGURE 4.8.3. The volume V of the trough is

$$V = (\text{area of triangular end}) \times (\text{length}).$$

FIGURE 4.8.3 Water trough in Example 2

From FIGURE 4.8.4 and the Pythagorean Theorem, the area of the triangular end as a function of x is $\frac{1}{2}x\sqrt{16 - x^2/4}$. Consequently, the volume of the trough as a function of x, the objective function, is

$$V(x) = 20 \cdot \left(\frac{1}{2}x\sqrt{16 - \frac{1}{4}x^2}\right) = 5x\sqrt{64 - x^2}.$$

The function $V(x)$ makes sense only on the closed interval $[0, 8]$. (Why?)

Taking the derivative and simplifying yield

$$V'(x) = -10\frac{x^2 - 32}{\sqrt{64 - x^2}}.$$

FIGURE 4.8.4 Triangular end of trough in Example 2

Although $V'(x) = 0$ for $x = \pm 4\sqrt{2}$, the only critical number in the open interval $(0, 8)$ is $4\sqrt{2}$. Since the function $V(x)$ is continuous on $[0, 8]$, we know from Theorem 4.3.3 that $V(0) = V(8) = 0$ must be its absolute minimum. The absolute maximum of $V(x)$ must then occur when the width across the top of the trough is $4\sqrt{2} \approx 5.66$ ft. The maximum volume is $V(4\sqrt{2}) = 160$ ft^3. ∎

Note: Often a problem can be solved in more than one way. In hindsight, you should verify that the solution of Example 2 is slightly "cleaner" if the dimension across the top of the end of the trough is labeled $2x$ rather than x. Indeed, as the next example shows, Example 2 can be solved using an entirely different variable.

EXAMPLE 3 Alternative Solution to Example 2

As shown in FIGURE 4.8.5, we let θ denote the angle between the vertical and one of the sides. From right-triangle trigonometry the height and base of the triangular end are $4\cos\theta$ and $8\sin\theta$, respectively. Expressed as a function of θ, V is $\left(\frac{1}{2} \cdot \text{base} \cdot \text{height}\right) \times (\text{length})$, or

$$\begin{aligned}V(\theta) &= \tfrac{1}{2}(4\cos\theta)(8\sin\theta) \cdot 20 \\ &= 320\sin\theta\cos\theta \\ &= 160(2\sin\theta\cos\theta) \\ &= 160\sin 2\theta, \quad \leftarrow \text{double-angle formula}\end{aligned}$$

FIGURE 4.8.5 Triangular end of trough in Example 3

where $0 \leq \theta \leq \pi/2$. Proceeding as in Example 1, we find that the maximum value $V = 160$ ft^3 occurs at $\theta = \pi/4$. The dimension across the top of the trough, or the base of the isosceles triangle, is $8\sin(\pi/4) = 4\sqrt{2}$ ft. ∎

Problems with Constraints It is often more convenient to set up a function in terms of two variables instead of one. In this case we need to find a relationship between these variables that can be used to eliminate one of the variables from the function under consideration. As discussed in conjunction with (1), this relationship is usually an equation called a **constraint**. The next two examples illustrate this concept.

EXAMPLE 4 Closest Point

Find the point in the first quadrant on the circle $x^2 + y^2 = 1$ that is closest to (2, 4).

Solution Let (x, y), $x > 0$, $y > 0$, denote the point on the circle closest to the point (2, 4). See FIGURE 4.8.6.

As shown in the figure, the distance d between (x, y) and $(2, 4)$ is

$$d = \sqrt{(x-2)^2 + (y-4)^2} \quad \text{or} \quad d^2 = (x-2)^2 + (y-4)^2.$$

Now the point that minimizes the square of the distance d^2 also minimizes the distance d. Let us write $D = d^2$. By expanding $(x - 2)^2$ and $(y - 4)^2$ and using the constraint $x^2 + y^2 = 1$ in the form $y = \sqrt{1 - x^2}$, we find

$$D(x) = x^2 - 4x + 4 + \overbrace{(1 - x^2)}^{y^2} - 8\overbrace{\sqrt{1 - x^2}}^{y} + 16$$
$$= -4x - 8\sqrt{1 - x^2} + 21.$$

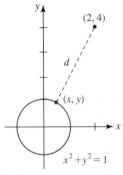

FIGURE 4.8.6 Circle and point in Example 4

Because we have assumed x and y to be positive, the domain of the foregoing function is the open interval (0, 1). However, the solution of the problem will not be affected in any way by assuming that the domain is the closed interval [0, 1].

Differentiation gives

$$D'(x) = -4 - 4(1 - x^2)^{-1/2}(-2x) = \frac{-4\sqrt{1 - x^2} + 8x}{\sqrt{1 - x^2}}.$$

Now $D'(x) = 0$ only if $-4\sqrt{1 - x^2} + 8x = 0$ or $2x = \sqrt{1 - x^2}$. After squaring both sides and simplifying, we find that $\sqrt{5}/5$ is the only critical number in the interval (0, 1). Because $D(x)$ is continuous on [0, 1], we conclude from the function values

$$D(0) = 13, \quad D(\sqrt{5}/5) = 21 - 4\sqrt{5} \approx 12.06, \quad \text{and} \quad D(1) = 17$$

that D and, hence, the distance d are a minimum when $x = \sqrt{5}/5$. Using the constraint $x^2 + y^2 = 1$, we find correspondingly that $y = 2\sqrt{5}/5$. This means $(\sqrt{5}/5, 2\sqrt{5}/5)$ is the point on the circle closest to (2, 4). ∎

EXAMPLE 5 Least Fencing

A rancher intends to mark off a rectangular plot of land that will have an area of 1500 m². The plot will be fenced and divided into two equal portions by an additional fence parallel to two sides. Find the dimensions of the land that require the least amount of fencing.

Solution As shown in FIGURE 4.8.7, we let x and y denote the dimensions of the fenced-in land. The function we wish to minimize is the total amount of fence, that is, the sum of the lengths of the five portions of fence. If we denote this sum by the symbol L, we have

$$L = 2x + 3y. \tag{5}$$

FIGURE 4.8.7 Rectangular plot of land in Example 5

Because the fenced-in land is to have an area of 1500 m², x and y must be related by the requirement that $xy = 1500$. We use this constraint in the form $y = 1500/x$ to eliminate y in (5) and write the objective function L as a function of x:

$$L(x) = 2x + \frac{4500}{x} \tag{6}$$

Since x represents a physical dimension that satisfies $xy = 1500$, we conclude that it is positive. But other than that, there is no restriction on x. Thus, unlike the prior examples, the function we are considering is not defined on a closed interval; $L(x)$ is defined on the unbounded interval $(0, \infty)$.

Setting the derivative

$$L'(x) = 2 - \frac{4500}{x^2}$$

equal to zero and solving for x, we find that the only critical number is $15\sqrt{10}$. Since the second derivative is easy to compute, we shall use the Second Derivative Test. From

$$L''(x) = \frac{9000}{x^3}$$

we observe that $L''(15\sqrt{10}) > 0$. It follows from Theorem 4.7.3 that $L(15\sqrt{10}) = 2(15\sqrt{10}) + 4500/(15\sqrt{10}) = 60\sqrt{10}$ m is the required minimum amount of fencing. Returning to the constraint $y = 1500/x$, we find the corresponding value of y is $10\sqrt{10}$. Therefore, the dimensions of the land should be $15\sqrt{10}$ m $\times$ $10\sqrt{10}$ m. ∎

If an object is moving at a constant rate, then distance, rate, and time are related by *distance = rate $\times$ time*. We shall use this result in our last example in the form

$$\text{time} = \frac{\text{distance}}{\text{rate}}. \tag{7}$$

EXAMPLE 6 Minimum Time

A woman at point P on an island wishes to reach a village located at point S on a straight shore on the mainland. Point P is 9 mi from the closest point Q on the shore and the village at point S is 15 mi from point Q. See FIGURE 4.8.8. If the woman rows a boat at a rate of 3 mi/h to a point R on land, then walks the rest of the way to S at a rate of 5 mi/h, determine where she should land on shore in order to minimize the total time of travel.

Solution As shown in the figure, if x denotes the distance from point Q on shore to the point R where she lands on shore, then by the Pythagorean Theorem, the distance she rows is $\sqrt{81 + x^2}$. The distance she walks is $15 - x$. By (7) the total time of the trip from P to S is

$$T = \text{time rowing} + \text{time walking} \quad \text{or} \quad T(x) = \frac{\sqrt{81 + x^2}}{3} + \frac{15 - x}{5}.$$

FIGURE 4.8.8 Traveling woman in Example 6

Since $0 \le x \le 15$, the function $T(x)$ is defined on the closed interval $[0, 15]$.

The derivative of T is

$$\frac{dT}{dx} = \frac{1}{6}(81 + x^2)^{-1/2}(2x) - \frac{1}{5} = \frac{x}{3\sqrt{81 + x^2}} - \frac{1}{5}.$$

We set this derivative equal to 0 and solve for x:

$$\frac{x}{3\sqrt{81 + x^2}} = \frac{1}{5}$$

$$\frac{x^2}{81 + x^2} = \frac{9}{25}$$

$$16x^2 = 729$$

$$x = \frac{27}{4}.$$

That is, $\frac{27}{4}$ is the only critical number in $[0, 15]$. Since $T(x)$ is continuous on the interval we see from the three function values

$$T(0) = 6 \text{ h}, \quad T\left(\tfrac{27}{4}\right) = 5.4 \text{ h}, \quad \text{and} \quad T(15) \approx 5.83 \text{ h}$$

that the minimum total travel time occurs when $x = \frac{27}{4} = 6.75$. In other words, the woman lands the boat at point R, 6.75 mi from point Q, and then walks the remaining 8.25 mi to point S. ∎

$f'(x)$ NOTES FROM THE CLASSROOM

An observant reader may question at least two aspects of Example 5.

(i) Where did the assumption that the land be divided into two equal portions enter into the solution? In point of fact, it did not. What is important is that the dividing fence be parallel to the two ends. Ask yourself what $L(x)$ would be if this were *not* the case. However, the actual positioning of the dividing fence between the ends is irrelevant as long as it is parallel to them.

(ii) In an applied problem we are naturally interested in only absolute extrema. Therefore, another question might be: Since the function L in (6) is not defined on a closed interval and since the Second Derivative Test does not guarantee absolute extrema, how can we be certain that $L(15\sqrt{10})$ is an absolute minimum? When in doubt, we can always draw a graph. FIGURE 4.8.9 answers the question for $L(x)$. Also, look again at Theorem 4.6.2 in Section 4.6. Because $15\sqrt{10}$ is the *only* critical number in the interval $(0, \infty)$ and because $L(15\sqrt{10})$ was proved to be a relative minimum, Theorem 4.6.2 guarantees that the function value $L(15\sqrt{10}) = 60\sqrt{10}$ is an absolute minimum.

FIGURE 4.8.9 Graph of objective function in Example 5

Exercises 4.8
Answers to selected odd-numbered problems begin on page ANS-17.

≡ Fundamentals

1. Find two nonnegative numbers whose sum is 60 and whose product is a maximum.
2. Find two nonnegative numbers whose product is 50 and whose sum is a minimum.
3. Find a number that exceeds its square by the greatest amount.
4. Let m and n be positive integers. Find two nonnegative numbers whose sum is S such that the product of the mth power of one and the nth power of the other is a maximum.
5. Find two nonnegative numbers whose sum is 1 such that the sum of the square of one and twice the square of the other is a minimum.
6. Find the minimum value of the sum of a nonnegative number and its reciprocal.
7. Find the point(s) on the graph of $y^2 = 6x$ closest to (5, 0). closest to (3, 0).
8. Find the point on the graph of $x + y = 1$ closest to (2, 3).
9. Determine the point on the graph of $y = x^3 - 4x^2$ at which the tangent line has minimum slope.
10. Determine the point on the graph of $y = 8x^2 + 1/x$ at which the tangent line has maximum slope.

In Problems 11 and 12, find the dimensions of the shaded region such that its area is a maximum.

11.

FIGURE 4.8.10 Graph for Problem 11

12.

FIGURE 4.8.11 Graph for Problem 12

13. Find the vertices $(x, 0)$ and $(0, y)$ of the shaded triangular region in FIGURE 4.8.12 so that its area is a minimum.

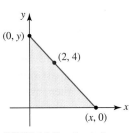

FIGURE 4.8.12 Graph for Problem 13

14. Find the maximum vertical distance d between the graphs of $y = x^2 - 1$ and $y = 1 - x$ for $-2 \leq x \leq 1$. See FIGURE 4.8.13.

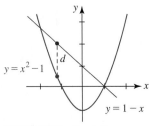

FIGURE 4.8.13 Graph for Problem 14

15. A rancher has 3000 ft of fencing on hand. Determine the dimensions of a rectangular corral that encloses a maximum area.
16. A rectangular plot of land will be fenced into three equal portions by two dividing fences parallel to two sides. See FIGURE 4.8.14. If the area to be enclosed is 4000 m², find the dimensions of the land that require the least amount of fence.

FIGURE 4.8.14 Plot of land in Problem 16

17. If the total fence to be used is 8000 m, find the dimensions of the enclosed land in Figure 4.8.14 that has the greatest area.

18. A rectangular yard is to be enclosed with a fence by attaching it to a house whose width is 40 ft. See FIGURE 4.8.15. The amount of fence to be used is 160 ft. Describe how the fence should be used so that the greatest area is enclosed.

FIGURE 4.8.15 House and yard in Problem 18

19. Solve Problem 18 if the amount of fence to be used is 80 ft.

20. A rancher wishes to build a rectangular corral of 128,000 ft² with one side along a vertical cliff. The fencing along the cliff costs $1.50 per foot, whereas along the other three sides the fencing costs $2.50 per foot. Find the dimensions of the corral so that the cost of fencing is a minimum.

21. An open rectangular box is to be constructed with a square base and a volume of 32,000 cm³. Find the dimensions of the box that require the least amount of material.

22. In Problem 21, find the dimensions of a closed box that require the least amount of material.

23. A box, open at the top, is to be made from a square piece of cardboard by cutting a square out of each corner and turning up the sides. In FIGURE 4.8.16, the white squares are cut out and the cardboard is folded along the dashed lines. Given that the cardboard measures 40 cm on a side, find the dimensions of the box that will give the maximum volume. What is the maximum volume?

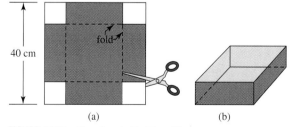

FIGURE 4.8.16 Open box in Problem 23

24. A box, open at the top, is to be made from a rectangular piece of cardboard that is 30 in. long and 20 in. wide. The box can hold itself together by cutting a square out of each corner, cutting on the interior solid lines, and then folding the cardboard on the dashed lines. See FIGURE 4.8.17. Express the volume of the box as a function of the indicated variable x. Find the dimensions of the box that give the maximum volume. What is the maximum volume?

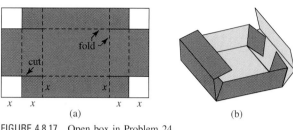

FIGURE 4.8.17 Open box in Problem 24

25. A gutter with a rectangular cross-section is made by bending up equal amounts from the ends of a 30-cm-wide piece of tin. What are the dimensions of the cross-section so that the volume is a maximum?

26. A gutter will be made so that its cross-section is an isosceles trapezoid with dimensions as indicated in FIGURE 4.8.18. Determine the value of θ so that the volume is a maximum.

FIGURE 4.8.18 Gutter in Problem 26

27. Two flagpoles are secured by wires that are attached at a single point between the poles. See FIGURE 4.8.19. Where should the point be located to minimize the amount of wire used?

FIGURE 4.8.19 Flagpoles in Problem 27

28. The running track shown in FIGURE 4.8.20 is to consist of two parallel straight parts and two semicircular parts. The length of the track is to be 2 km. Find the design of the track so that the rectangular plot of land enclosed by the track is a maximum.

FIGURE 4.8.20 Running track in Problem 28

29. A Norman window consists of a rectangle surmounted by a semicircle. Find the dimensions of the window with largest area if its perimeter is 10 m. See FIGURE 4.8.21.

FIGURE 4.8.21 Norman window in Problem 29

30. Rework Problem 29 given that the rectangle is surmounted by an equilateral triangle.

31. A 10-ft wall stands 5 ft away from a building, as shown in FIGURE 4.8.22. Find the length L of the shortest ladder, supported by the wall, that reaches from the ground to the building.

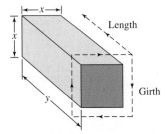

FIGURE 4.8.22 Ladder in Problem 31

32. U.S. Postal Service regulations state that a rectangular box sent by fourth-class mail must satisfy the requirement that its length plus its girth (perimeter of one end) must not exceed 108 in. Given that a box is to be constructed so that it has a square base, find the dimensions of the box that has a maximum volume. See FIGURE 4.8.23.

FIGURE 4.8.23 Box in Problem 32

33. Find the dimensions of the right circular cylinder with greatest volume that can be inscribed in a right circular cone of radius 8 in. and height 12 in. See FIGURE 4.8.24.

FIGURE 4.8.24 Inscribed cylinder in Problem 33

34. Find the maximum length L of a thin board than can be carried horizontally around the right-angle corner shown in FIGURE 4.8.25. [*Hint*: Use similar triangles.]

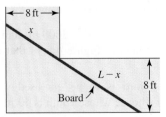

FIGURE 4.8.25 Board in Problem 34

35. A juice can is to be made in the form of a right circular cylinder and have a volume of 32 in^3. See FIGURE 4.8.26. Find the dimensions of the can so that the least amount of material is used in its construction. [*Hint*: Material = total surface area of can = area of top + area of bottom + area of lateral side. If the circular top and bottom covers are removed and the cylinder is cut straight up its side and flattened out, the result is the rectangle shown in Figure 4.8.26(c).]

(a) Circle cylinder   (b) Top and bottom are circular   (c) Lateral side is rectangular

FIGURE 4.8.26 Juice can in Problem 35

36. In Problem 35, suppose that the circular top and bottom are cut from square sheets of metal as shown in FIGURE 4.8.27. If the metal cut from the corners of the square sheet is wasted, then find the dimensions of the can so that the least amount of material (including waste) is used in its construction.

FIGURE 4.8.27 Top and bottom of can in Problem 36

37. Some birds fly more slowly over water than over land. A bird flies at constant rates of 6 km/h over water and 10 km/h over land. Use the information in FIGURE 4.8.28 to find the path the bird should take to minimize the total flying time between the shore of one island and its nest on the shore of another island. [*Hint*: Use *distance* = *rate* × *time*.]

FIGURE 4.8.28 Bird in Problem 37

38. A pipeline is to be constructed from a refinery across a swamp to storage tanks. See FIGURE 4.8.29. The cost of construction is $25,000 per mile over the swamp and $20,000 per mile over land. How should the pipeline be made so that the cost of construction is a minimum?

FIGURE 4.8.29 Pipeline in Problem 38

39. Rework Problem 38 given that the cost per mile across the swamp is twice the cost per mile over land.

40. At midnight ship A is 50 km north of ship B. Ship A is sailing south at 20 km/h and ship B is sailing west at 10 km/h. At what time will the distance between the ships be a minimum?

41. A container for transporting biohazardous waste is made of heavy plastic and is formed by adjoining two hemispheres to the ends of a right circular cylinder as shown in FIGURE 4.8.30. The total volume of the container is to be 30π ft^3. The cost per square foot of the plastic for the ends is one and a half times the cost per square foot of the plastic used in the cylindrical part. Find the dimensions of the container so that the cost of its construction is a minimum. [*Hint*: The volume of a sphere is $\frac{4}{3}\pi r^3$ and its surface area is $4\pi r^2$.]

FIGURE 4.8.30 Container in Problem 41

42. A printed page will have 2-in. margins of white space on the sides and 1-in. margins of white space on the top and bottom. See FIGURE 4.8.31. The area of the printed portion is 32 in^2. Determine the dimensions of the page so that the least amount of paper is used.

FIGURE 4.8.31 Printed page in Problem 42

43. One corner of an 8.5 in. $\times$ 11 in. piece of paper is folded over to the other edge of the paper as shown in FIGURE 4.8.32. Find the width x of the fold so that the length L of the crease is a minimum.

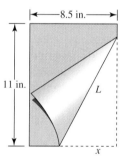

FIGURE 4.8.32 Piece of paper in Problem 43

44. The frame of a kite consists of six pieces of lightweight plastic. As shown in FIGURE 4.8.33, the outer frame of the kite consists of four precut pieces; two pieces of length 2 ft and two pieces of length 3 ft. The remaining crossbar pieces, labeled x in the figure, are to be cut to lengths so that the kite is as large as possible. Find these lengths.

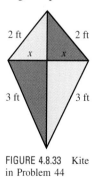

FIGURE 4.8.33 Kite in Problem 44

45. Find the dimensions of the rectangle of greatest area that can be circumscribed about a rectangle of length a and width b. See the red rectangle in FIGURE 4.8.34.

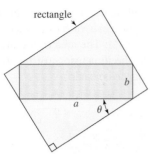

FIGURE 4.8.34 Rectangle in Problem 45

46. A statue is placed on a pedestal as shown in FIGURE 4.8.35. How far should a person stand from the pedestal to maximize the viewing angle θ? [*Hint*: Review the trigonometric identity for $\tan(\theta_2 - \theta_1)$. Also, it suffices to maximize $\tan\theta$ rather than θ. Why?]

FIGURE 4.8.35 Statue in Problem 46

47. A cross-section of a rectangular wooden beam cut from a circular log of diameter d has width x and depth y. See FIGURE 4.8.36. The strength of the beam varies directly as the product of the width and the square of the depth. Find the dimensions of the cross-section of the beam of greatest strength.

FIGURE 4.8.36 Log in Problem 47

48. The container shown in FIGURE 4.8.37 is to be constructed by attaching an inverted cone (open at its top) to the bottom of a right circular cylinder (open at its top and bottom) of radius 5 ft. The container is to have a volume of 100 ft^3. Find the value of the indicated angle so that the total surface area of the container is a minimum. What is the minimum surface area? [*Hint*: See Problem 38 in part C of Chapter 1 in Review.]

FIGURE 4.8.37 Container in Problem 48

≡ **Mathematical Models**

49. The illuminance E due to a light source or intensity I at a distance r from the source is given by $E = I/r^2$. The total illuminance from two light bulbs of intensities $I_1 = 125$ and $I_2 = 216$ is the sum of the illuminances. Find the point P between the two light bulbs 10 m apart at which the total illuminance is a minimum. See FIGURE 4.8.38.

FIGURE 4.8.38 Light bulbs in Problem 49

50. The illuminance E at any point P on the edge of a circular table caused by a light placed directly above its center is given by $E = (i\cos\theta)/r^2$. See FIGURE 4.8.39. Given that the radius of the table is 1 m and $I = 100$, find the height at which the light should be placed so that E is a maximum.

FIGURE 4.8.39 Light and table in Problem 50

51. **Fermat's Principle** in optics states that light travels from point A (in the xy-plane) in one medium to point B in another medium on a path that requires minimum time. Denote the speed of light in the medium that contains point A by c_1 and the speed of light in the medium that contains point B by c_2. Show that the time of travel from A to B is a minimum when the angles θ_1 and θ_2, shown in FIGURE 4.8.40, satisfy **Snell's law**:

$$\frac{\sin\theta_1}{c_1} = \frac{\sin\theta_2}{c_2}.$$

FIGURE 4.8.40 Two mediums in Problem 51

52. Blood is carried throughout the body by the vascular system, which consists of capillaries, veins, arterioles, and arteries. One consideration of the problem of minimizing the energy expended in moving the blood through the various organs is to find an optimum angle θ for *vascular branching* such that the total resistance to the blood along a path from a larger blood vessel to a smaller blood vessel is a minimum. See FIGURE 4.8.41. Use **Poiseuille's law**, which states that the resistance R of a blood vessel of length l and radius r is $R = kl/r^4$, where k is a constant, to show that the total resistance

$$R = k\left(\frac{x}{r_1^4}\right) + k\left(\frac{y}{r_2^4}\right)$$

along the path $P_1 P_2 P_3$ is a minimum when $\cos\theta = r_2^4/r_1^4$. [*Hint*: Express x and y in terms of θ and a.]

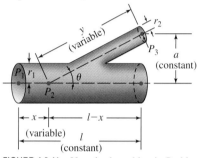

FIGURE 4.8.41 Vascular branching in Problem 52

53. The potential energy between two atoms in a diatomic molecule is given by $U(x) = 2/x^{12} - 1/x^6$. Find the minimum potential energy between the two atoms.

54. The height of a projectile launched with a constant initial velocity v_0 at an angle of elevation θ_0 is given by $y = (\tan\theta_0)x - \left(\dfrac{g}{2v_0^2 \cos^2\theta_0}\right)x^2$, where x is its horizontal displacement measured from the point of launch. Show that the maximum height attained by the projectile is $h = (v_0^2/2g)\sin^2\theta_0$.

55. A beam of length L is embedded in concrete walls as shown in FIGURE 4.8.42. When a constant load w_0 is uniformly distributed along its length, the deflection curve $y(x)$. for the beam is given by

$$y(x) = \frac{w_0 L^2}{24 EI} x^2 - \frac{w_0 L}{12 EI} x^3 + \frac{w_0}{24 EI} x^4,$$

where E and I are constants. (E is **Young's modulus of elasticity** and I is a moment of inertia of a cross-section of the beam.) The deflection curve approximates the shape of the beam.
(a) Determine the maximum deflection of the beam.
(b) Sketch a graph of $y(x)$.

FIGURE 4.8.42 Beam in Problem 55

56. The relationship between the height h and the diameter d of a tree can be approximated by the quadratic expression $h = 137 + ad - bd^2$, where h and d are measured in centimeters, and a and b are positive parameters that depend on the type of tree. See FIGURE 4.8.43.

(a) Suppose a tree attains its maximum height of H centimeters at a diameter of D centimeters. Show that

$$h = 137 + 2\frac{H - 137}{D}d - \frac{H - 137}{D^2}d^2.$$

(b) Suppose a certain tree reaches its maximum possible height (according to the formula) of 15 m at a diameter of 0.8 m. What was the diameter of the tree when the tree was 10 m tall?

FIGURE 4.8.43 Tree in Problem 56

57. The long bones in mammals may be represented as hollow cylindrical tubes, filled with marrow, of outer radius R and inner radius r. Bones should be constructed to be lightweight yet capable of withstanding certain bending moments. In order to withstand a bending moment M, it can be shown that the mass m per unit length of the bone and marrow is given by

$$m = \pi\rho\left[\frac{M}{K(1 - x^4)}\right]^{2/3}\left(1 - \frac{1}{2}x^2\right),$$

where ρ is the density of the bone and K is a positive constant. If $x = r/R$, show that m is a minimum when $r = 0.63R$ (approximately).

58. The rate P (in mg carbon/m³/h) at which photosynthesis takes place for a certain species of phytoplankton is related to the light intensity I (in 10^3 ft-candles) by the function

$$P = \frac{100I}{I^2 + I + 4}.$$

At what light intensity is P the largest?

≡ Think About It

59. **A Mathematical Classic** A person would like to cut a 1-m-long piece of wire into two pieces. One piece will be bent into the shape of a circle and the other into the shape of a square. How should the wire be cut so that the sum of the areas is a maximum?

60. In Problem 59, suppose one piece of wire will be bent into the shape of a circle and the other into the shape of an equilateral triangle. How should the wire be cut so that the sum of the areas is a minimum? A maximum?

61. A conical cup is made from a circular piece of paper of radius R by cutting out a circular sector and then joining the dashed edges shown in FIGURE 4.8.44.

   (a) Determine the value of r indicated in Figure 4.8.44(b) so that the volume of the cup is a maximum.
   (b) What is the maximum volume of the cup?
   (c) Find the central angle θ of the circular sector so that the volume of the conical cup is a maximum.

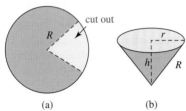

FIGURE 4.8.44 Conical cup in Problem 61

62. The lateral side of a cylinder is to be made from a rectangle of flimsy sheet plastic. Because the plastic material cannot support itself, a thin stiff wire is embedded in the material as shown in FIGURE 4.8.45(a). Find the dimensions of the cylinder of largest volume that can be constructed if the wire has a fixed length L. [*Hint*: There are two constraints in this problem. In Figure 4.8.45(b), the circumference of a circular end of the cylinder is y.]

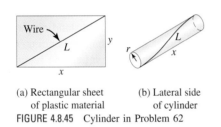

(a) Rectangular sheet of plastic material   (b) Lateral side of cylinder
FIGURE 4.8.45 Cylinder in Problem 62

63. In Problem 27, show that when the optimal amount of wire (the least amount) is used, then the angle θ_1 the wire to the left flagpole makes with the ground is the same as the angle θ_2 the wire to the right flagpole makes with the ground. See Figure 4.8.19.

64. Find an equation of the tangent line L to the graph of $y = 1 - x^2$ at $P(x_0, y_0)$ such that the triangle in the first quadrant bounded by the coordinate axes and L has minimum area. See FIGURE 4.8.46.

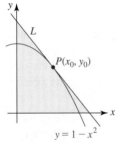

FIGURE 4.8.46 Triangle in Problem 64

Calculator/CAS Problems

65. In a race a woman is required to swim from a floating dock A to the beach and, without stopping, swim from the beach out to another floating dock C. The distances are shown in FIGURE 4.8.47(a). She estimates that she can swim from dock A to the beach at a constant rate of 3 mi/h and out from the beach to dock C at a rate of 2 mi/h. Where should she touch the beach in order to minimize the total swimming time from A to C? Introduce an xy-coordinate system as shown in Figure 4.8.47(b). Use a CAS to find the critical numbers.

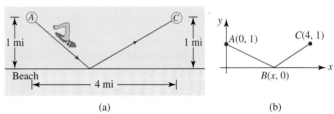

FIGURE 4.8.47 Swimmer in Problem 65

66. A two-story house under construction consists of two structures A and B with rectangular cross-sections and dimensions as indicated in FIGURE 4.8.48. The framing for structure B requires temporary wooden reinforcing buttresses from ground level that rest against structure A as shown.

   (a) Express the length L of a buttress as a function of the indicated angle θ.
   (b) Find $L'(\theta)$.
   (c) Use a calculator or CAS to obtain the graph of $L'(\theta)$ on the interval $(0, \pi/2)$. Use this graph to show that L has only one critical number θ_c in $(0, \pi/2)$. Use this graph to determine the algebraic sign of $L'(\theta)$ for $0 < \theta < \theta_c$, and the algebraic sign of $L'(\theta)$ for $\theta_c < \theta < \pi/2$. What is your conclusion?
   (d) Find the minimum length of a buttress.

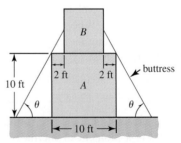

FIGURE 4.8.48 House in Problem 66

67. Consider the three cables shown in FIGURE 4.8.49.

   (a) Express the total length L of the three cables shown in Figure 4.8.49(a) as a function of the length L of the cable AB.
   (b) Use a calculator or CAS to verify that the graph of L has a minimum.
   (c) Find the length of the cable AB so that the total length L of the lengths of the three cables is a minimum.

(d) Express the total length L of the three cables shown in Figure 4.8.49(b) as a function of the length of the cable AB.
(e) Use a calculator or CAS to verify that the graph of L has a minimum.
(f) Use the graph obtained in part (e) or a CAS as an aid in approximating the length of the cable AB that minimizes the function L obtained in part (d).

FIGURE 4.8.49 Cables in Problem 67

≡ Project

68. **Frequency Interference** When the Federal Aviation Administration (FAA) allocates numerous frequencies for an airport radio transmitter quite often nearby transmitters use the same frequencies. As a consequence, the FAA would like to minimize the interference between these transmitters. In FIGURE 4.8.50, the point (x_t, y_t) represents the location of a transmitter whose radio jurisdiction is indicated by the circle C of radius with center at the origin. A second transmitter is located at $(x_i, 0)$ as shown in the figure. In this problem you will develop and analyze a function to find the interference between two transmitters.

(a) The strength of the signal from a transmitter to a point is inversely proportional to the square of the distance between them. Assume that a point (x, y) is located on the upper portion of the circle C as shown in Figure 4.8.50. Express the primary strength of the signal at (x, y) from a transmitter at (x_t, y_t) as a function of x. Express the secondary strength at (x, y) from the transmitter at $(x_i, 0)$ as a function of x. Now define a function $R(x)$ as a quotient of the primary signal strength to the secondary signal strength. $R(x)$ can be thought of as a *signal to noise ratio*. To guarantee that the interference remains small we need to show that the minimum signal to noise ratio is greater than the FAA's minimum threshold of -0.7.

(b) Suppose that $x_t = 760$ m, $y_t = -560$ m, $r = 1.1$ km, and $x_i = 12$ km. Use a CAS to simplify and then plot the graph of $R(x)$. Use the graph to estimate the domain and range of $R(x)$.

(c) Use the graph in part (b) to estimate the value of x where the minimum ratio R occurs. Estimate the value of R at that point. Does this value of R exceed the FAA's minimum threshold?

(d) Use a CAS to differentiate $R(x)$. Use a CAS to find the root of $R'(x) = 0$ and to compute the corresponding value of $R(x)$. Compare your answers here with the estimates in part (c).

(e) What is the point (x, y) on circle C?

(f) We assumed that the point (x, y) was in the top half plane when (x_t, y_t) was in the lower half plane. Explain why this assumption is correct.

(g) Use a CAS to find the value of x where the minimum interference occurs in terms of the symbols x_t, y_t, x_i, and r.

(h) Where is that point that minimizes the signal to noise ratio when the transmitter at (x_t, y_t) is on the x-axis? Give a convincing argument justifying your answer.

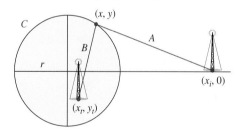

FIGURE 4.8.50 Radio transmitters in Problem 68

4.9 Linearization and Differentials

Introduction We started the discussion of the derivative with the problem of finding the tangent line to the graph of a function $y = f(x)$ at a point $(a, f(a))$. Intuitively, we would expect that a tangent line is very close to the graph of f whenever x is close to the number a. Put in other words, when x is in a small neighborhood of a the function values $f(x)$ are very close to the values of the y-coordinates on the tangent line. Thus, by finding an equation of the tangent line at $(a, f(a))$ we can use that equation to approximate $f(x)$.

An equation of the tangent line shown in red in FIGURE 4.9.1 is given by

$$y - f(a) = f'(a)(x - a) \quad \text{or} \quad y = f(a) + f'(a)(x - a). \tag{1}$$

Using standard functional notation, let us write the last linear equation in (1) as $L(x) = f(a) + f'(a)(x - a)$. This linear function is given a special name.

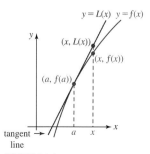

FIGURE 4.9.1 When x is close to a, the value $L(x)$ is close to $f(x)$

Definition 4.9.1 Linearization

If a function $y = f(x)$ is differentiable at a number a, then the function

$$L(x) = f(a) + f'(a)(x - a) \qquad (2)$$

is said to be a **linearization** of f at a. For a number x near a, the approximation

$$f(x) \approx L(x) \qquad (3)$$

is called a **local linear approximation** of f at a.

There is no need to memorize (2); it is simply the point–slope form of the tangent line at $(a, f(a))$.

EXAMPLE 1 Linearization of sin x

Find a linearization of $f(x) = \sin x$ at $a = 0$.

Solution Using $f(0) = 0, f'(x) = \cos x$, and $f'(0) = 1$ the tangent line to the graph of $f(x) = \sin x$ at $(0, 0)$ is $y - 0 = 1 \cdot (x - 0)$. Therefore, the linearization of $f(x) = \sin x$ at $a = 0$ is $L(x) = x$. As seen in FIGURE 4.9.2 the graph of $f(x) = \sin x$ and its linearization at $a = 0$ are nearly indistinguishable near the origin. The local linear approximation $f(x) \approx L(x)$ of f at $a = 0$ is

$$\sin x \approx x. \qquad (4) \quad \blacksquare$$

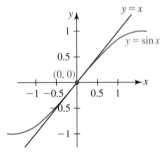

FIGURE 4.9.2 Graph of function and linearization in Example 1

Errors Example 1 reemphasizes something you already know from trigonometry. The local linear approximation (4) shows that the sine of a small angle x (measured in radians) is approximately the same as the angle. For comparison, if we choose $x = 0.1$, then (4) indicates that $f(0.1) \approx L(0.1)$ or $\sin(0.1) \approx 0.1$. For comparison a calculator gives (rounded to five decimal places) $f(0.1) = \sin(0.1) = 0.09983$. Now an error in a calculation is defined by

$$\text{error} = \text{true value} - \text{approximate value}. \qquad (5)$$

However, in practice the

$$\text{relative error} = \frac{\text{error}}{\text{true value}} \qquad (6)$$

is usually more important than the error. Moreover, (relative error) $\cdot$ 100 is called **percentage error**. Thus with the aid of a calculator, the percentage error in the approximation $f(0.1) \approx L(0.1)$ is roughly only 0.2%. Figure 4.9.2 clearly shows that as x moves away from 0, the accuracy of the approximation $\sin x \approx x$ diminishes. For example, for the number 0.9 a calculator gives $f(0.9) = \sin(0.9) = 0.78333$, whereas $L(0.9) = 0.9$. This time the percentage error is about 15%.

We have also seen the result of Example 1 presented in a slightly different manner in Section 2.4. If we divide the local linear approximation $\sin x \approx x$ by x, we get $\dfrac{\sin x}{x} \approx 1$ for values of x near 0. This leads us back to the important trigonometric limit $\lim\limits_{x \to 0} \dfrac{\sin x}{x} = 1$.

EXAMPLE 2 Linearization and Approximation

(a) Find a linearization of $f(x) = \sqrt{x + 1}$ at $a = 3$.
(b) Use a local linear approximation to approximate $\sqrt{3.95}$ and $\sqrt{4.01}$.

Solution
(a) By the Power Rule for Functions, the derivative of f is

$$f'(x) = \frac{1}{2}(x + 1)^{-1/2} = \frac{1}{2\sqrt{x + 1}}.$$

When evaluated at $a = 3$ the two functions give:

$$f(3) = \sqrt{4} = 2 \quad \leftarrow \text{point of tangency is } (3, 2)$$

$$f'(3) = \frac{1}{2\sqrt{4}} = \frac{1}{4}. \quad \leftarrow \text{slope of tangent at } (3, 2) \text{ is } \tfrac{1}{4}$$

Thus, by the point–slope form of a line the linearization of f at $a = 3$ is given by $y - 2 = \tfrac{1}{4}(x - 3)$ or

$$L(x) = 2 + \frac{1}{4}(x - 3). \tag{7}$$

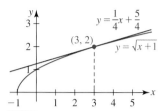

FIGURE 4.9.3 Graphs of function and linearization in Example 2

The graphs of f and L are given in FIGURE 4.9.3. Of course, L can be expressed in the slope–intercept form $L(x) = \tfrac{1}{4}x + \tfrac{5}{4}$ but for computational purposes the form given in (7) is more convenient.

(b) Using (7) from part (a), we have the local linear approximation $f(x) \approx L(x)$ or

$$\sqrt{x + 1} \approx 2 + \frac{1}{4}(x - 3), \tag{8}$$

whenever x is close to 3. Now, setting $x = 2.95$ and $x = 3.01$ in (8) gives, in turn, the approximations:

$$\overbrace{\sqrt{3.95}}^{f(2.95)} \approx \overbrace{2 + \frac{1}{4}(2.95 - 3)}^{L(2.95)} = 2 - \frac{0.05}{4} = 1.9875.$$

and $\overbrace{\sqrt{4.01}}^{f(3.01)} \approx \overbrace{2 + \frac{1}{4}(3.01 - 3)}^{L(3.01)} = 2 + \frac{0.01}{4} = 2.0025.$ ∎

■ **Differentials** The fundamental idea of a linearization of a function was originally couched in the terminology of *differentials*. Suppose $y = f(x)$ is a differentiable function in an open interval containing the number a. If x_1 is a different number on the x-axis, then **increments** Δx and Δy are the differences

$$\Delta x = x_1 - a \quad \text{and} \quad \Delta y = f(x_1) - f(a).$$

But since $x_1 = a + \Delta x$, the **change in the function** is

$$\Delta y = f(a + \Delta x) - f(a).$$

For values of Δx that are close to 0, the difference quotient

$$\frac{f(a + \Delta x) - f(a)}{\Delta x} = \frac{\Delta y}{\Delta x}$$

is an approximation of the value of the derivative of f at a:

$$\frac{\Delta y}{\Delta x} \approx f'(a) \quad \text{or} \quad \Delta y \approx f'(a)\Delta x.$$

The quantities Δx and $f'(a)\Delta x$ are called **differentials** and are denoted by the symbols dx and dy, respectively. That is,

$$\Delta x = dx \quad \text{and} \quad dy = f'(a)\,dx.$$

As shown in FIGURE 4.9.4, for a change dx in x the quantity $dy = f'(a)\,dx$ represents the **change in the linearization** (the *rise* in the tangent line at $(a, f(a))$*. And so when $dx \approx 0$, the change in the function Δy is approximately the same as the change in the linearization dy:

$$\Delta y \approx dy. \tag{9}$$

FIGURE 4.9.4 Geometric interpretations of dx, Δy, and dy

---

*For this reason, the Leibniz notation for the derivative dy/dx has the appearance of a quotient.

Definition 4.9.2 Differentials

The **differential of the independent variable** x is the nonzero number Δx and is denoted by dx; that is,

$$dx = \Delta x. \tag{10}$$

If f is a differentiable function at x, then the **differential of the dependent variable** y is denoted by dy; that is,

$$dy = f'(x)\Delta x = f'(x)\,dx. \tag{11}$$

EXAMPLE 3 Differentials

(a) Find Δy and dy for $f(x) = 5x^2 + 4x + 1$.
(b) Compare the values of Δy and dy for $x = 6$, $\Delta x = dx = 0.02$.

Solution

(a) $\Delta y = f(x + \Delta x) - f(x)$
$= [5(x + \Delta x)^2 + 4(x + \Delta x) + 1] - [5x^2 + 4x + 1]$
$= 10x\Delta x + 4\Delta x + 5(\Delta x)^2.$

Now, since $f'(x) = 10x + 4$, we have from (11) of Definition 4.9.2,

$$dy = (10x + 4)\,dx. \tag{12}$$

By rewriting Δy as $\Delta y = (10x + 4)\Delta x + 5(\Delta x)^2$ and using $dx = \Delta x$, observe that $dy = (10x + 4)\Delta x$ and $\Delta y = (10x + 4)\Delta x + 5(\Delta x)^2$ differ by the amount $5(\Delta x)^2$.

(b) When $x = 6$, $\Delta x = 0.02$:

$$\Delta y = 10(6)(0.02) + 4(0.02) + 5(0.02)^2 = 1.282$$

whereas $\quad dy = (10(6) + 4)(0.02) = 1.28.$

The difference in answers is, of course, $5(\Delta x)^2 = 5(0.02)^2 = 0.002$. ∎

In Example 3 the value $\Delta y = 1.282$ is the *exact* amount by which the function $f(x) = 5x^2 + 4x + 1$ changes as x changes from 6 to 6.02. The differential $dy = 1.28$ represents an *approximation* of the amount by which the function changes. As shown in (9), for a small change or increment Δx in the independent variable, the corresponding change Δy in the dependent variable can be approximated by the differential dy.

∎ **Linear Approximation Revisited** Differentials can be used to approximate the value $f(x + \Delta x)$. From $\Delta y = f(x + \Delta x) - f(x)$, we get

$$f(x + \Delta x) = f(x) + \Delta y.$$

But in view of (9), for a small change in x we can write

$$f(x + \Delta x) \approx f(x) + dy.$$

With $dy = f'(x)\,dx = f'(x)\Delta x$ the preceding line is the same as

$$f(x + \Delta x) \approx f(x) + f'(x)\,dx. \tag{13}$$

We have already seen the formula in (13) in a different guise. If we let $x = a$ and $dx = \Delta x = x - a$, then (13) becomes

$$f(x) \approx f(a) + f'(a)(x - a). \tag{14}$$

The right-hand side of the equality in (14) is recognized as $L(x)$ and (13) becomes $f(x) \approx L(x)$ which is the result given in (3).

EXAMPLE 4 Approximation by Differentials

Use (13) to approximate $(2.01)^3$.

Solution First identify the function $f(x) = x^3$. We wish to calculate the approximate value of $f(x + \Delta x) = (x + \Delta x)^3$ when $x = 2$ and $\Delta x = 0.01$. Now from (11),

$$dy = 3x^2\,dx = 3x^2\Delta x.$$

Thus (13) gives

$$(x + \Delta x)^3 \approx x^3 + 3x^2\Delta x.$$

With $x = 2$ and $\Delta x = 0.01$, the preceding formula gives the approximation

$$(2.01)^3 \approx 2^3 + 3(2)^2(0.01) = 8.12.$$ ∎

EXAMPLE 5 Approximation by Differentials

A side of a cube is measured to be 30 cm with a possible error of ± 0.02 cm. What is the approximate maximum possible error in the volume of the cube?

Solution The volume of a cube is $V = x^3$, where x is the length of one side. If Δx represents the error in the length of one side, then the corresponding error in the volume is

$$\Delta V = (x + \Delta x)^3 - x^3.$$

To simplify matters, we utilize the differential $dV = 3x^2\,dx = 3x^2\Delta x$ as an approximation to ΔV. Thus, for $x = 30$ and $\Delta x = \pm 0.02$ the approximate maximum error is

$$dV = 3(30)^2(\pm 0.02) = \pm 54 \text{ cm}^3.$$ ∎

In Example 5, an error of about 54 cm^3 in the volume for an error of 0.02 cm in the length of a side seems considerable. But, observe, if the relative error (6) is $\Delta V/V$, then the *approximate* relative error is dV/V. When $x = 30$ and $V = (30)^3 = 27{,}000$ the approximate maximum relative error is $\pm 54/27{,}000 = \pm 1/500$, and the maximum percentage error is only about $\pm 0.2\%$.

▌Rules for Differentials The rules for differentiation considered in this chapter can be rephrased in terms of differentials; for example, if $u = f(x)$ and $v = g(x)$ and $y = f(x) + g(x)$, then $dy/dx = f'(x) + g'(x)$. Hence, $dy = [f'(x) + g'(x)]\,dx = f'(x)\,dx + g'(x)\,dx = du + dv$. We summarize the differential equivalents of the Sum, Product, and Quotient Rules:

$$d(u + v) = du + dv \tag{15}$$

$$d(uv) = u\,dv + v\,du \tag{16}$$

$$d(u/v) = \frac{v\,du - u\,dv}{v^2}. \tag{17}$$

As the next example shows, there is little need for memorizing (15), (16), and (17).

EXAMPLE 6 Differential of y

Find dy for $y = x^2\cos 3x$.

Solution To find the differential of a function, we can simply multiply its derivative by dx. Thus, by the Product Rule,

$$\frac{dy}{dx} = x^2(-\sin 3x \cdot 3) + \cos 3x(2x)$$

and so

$$dy = \left(\frac{dy}{dx}\right) \cdot dx = (-3x^2\sin 3x + 2x\cos 3x)\,dx. \tag{18}$$

Alternative Solution Applying (16) gives

$$dy = x^2 d(\cos 3x) + \cos 3x\, d(x^2)$$
$$= x^2(-\sin 3x \cdot 3\,dx) + \cos 3x(2x\,dx). \tag{19}$$

Factoring dx from (19) yields (18). ∎

Exercises 4.9

≡ Fundamentals

In Problems 1–8, find a linearization of the given function at the indicated number.

1. $f(x) = \sqrt{x}$; $a = 9$
2. $f(x) = \dfrac{1}{x^2}$; $a = 1$
3. $f(x) = \tan x$; $a = \pi/4$
4. $f(x) = \cos x$; $a = \pi/2$
5. $f(x) = \ln x$; $a = 1$
6. $f(x) = 5x + e^{x-2}$; $a = 2$
7. $f(x) = \sqrt{1+x}$; $a = 3$
8. $f(x) = \dfrac{1}{\sqrt{3+x}}$; $a = 6$

In Problems 9–16, use a linearization at $a = 0$ to establish the given local linear approximation.

9. $e^x \approx 1 + x$
10. $\tan x \approx x$
11. $(1+x)^{10} \approx 1 + 10x$
12. $(1+2x)^{-3} \approx 1 - 6x$
13. $\sqrt{1-x} \approx 1 - \dfrac{1}{2}x$
14. $\sqrt{x^2 + x + 4} \approx 2 + \dfrac{1}{4}x$
15. $\dfrac{1}{3+x} \approx \dfrac{1}{3} - \dfrac{1}{9}x$
16. $\sqrt[3]{1-4x} \approx 1 - \dfrac{4}{3}x$

In Problems 17–20, use an appropriate result from Problems 1–8 to find an approximation of the given quantity.

17. $(1.01)^{-2}$
18. $\sqrt{9.05}$
19. $10.5 + e^{0.1}$
20. $\ln 0.98$

In Problems 21–24, use an appropriate result from Problems 9–16 to find an approximation of the given quantity.

21. $\dfrac{1}{(1.1)^3}$
22. $(1.02)^{10}$
23. $(0.88)^{1/3}$
24. $\sqrt{4.11}$

In Problems 25–32, use an appropriate function and local linear approximation to find an approximation of the given quantity.

25. $(1.8)^5$
26. $(7.9)^{2/3}$
27. $\dfrac{(0.9)^4}{(0.9)+1}$
28. $(1.1)^3 + 6(1.1)^2$
29. $\cos\left(\dfrac{\pi}{2} - 0.4\right)$
30. $\sin 1°$
31. $\sin 33°$
32. $\tan\left(\dfrac{\pi}{4} + 0.1\right)$

In Problems 33 and 34, find a linearization $L(x)$ of f at the given value of a. Use $L(x)$ to approximate the indicated function value.

33. $a = 1$; $f(1.04)$
34. $a = -2$; $f(-1.98)$

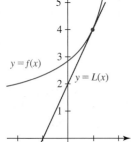

FIGURE 4.9.5 Graph for Problem 33

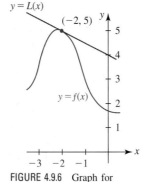

FIGURE 4.9.6 Graph for Problem 34

In Problems 35–42, find Δy and dy.

35. $y = x^2 + 1$
36. $y = 3x^2 - 5x + 6$
37. $y = (x+1)^2$
38. $y = x^3$
39. $y = \dfrac{3x+1}{x}$
40. $y = \dfrac{1}{x^2}$
41. $y = \sin x$
42. $y = -4\cos 2x$

In Problems 43 and 44, complete the following table for each function.

x	Δx	Δy	dy	$\Delta y - dy$
2	1			
2	0.5			
2	0.1			
2	0.01			

43. $y = 5x^2$
44. $y = 1/x$

45. Compute the approximate amount by which the function $f(x) = 4x^2 + 5x + 8$ changes as x changes from:
    (a) 4 to 4.03
    (b) 3 to 2.9.

46. (a) Find an equation of the tangent line to the graph of $f(x) = x^3 + 3x^2$ at $x = 1$.
    (b) Find the y-coordinate of the point on the tangent line in part (a) that corresponds to $x = 1.02$.
    (c) Use (3) to find an approximation to $f(1.02)$. Compare your answer with that of part (b).

47. The area of a circle with radius r is $A = \pi r^2$.
    (a) Given that the radius of a circle changes from 4 cm to 5 cm, find the exact change in the area.
    (b) What is the approximate change in the area?

≡ Applications

48. According to Poiseuille, the resistance R of a blood vessel of length l and radius r is $R = kl/r^4$, where k is a constant. Given that l is constant, find the approximate change in R when r changes from 0.2 mm to 0.3 mm.

49. Many golf balls consist of a spherical cover over a solid core. Find the exact volume of the cover if its thickness is t and the radius of the core is r. [Hint: The volume of a sphere is $V = \frac{4}{3}\pi r^3$. Consider concentric spheres having radii r and $r + \Delta r$.] Use differentials to find an approximation to the volume of the cover. See FIGURE 4.9.7. Find an approximation to the volume of the cover if $r = 0.8$ in. and $t = 0.04$ in.

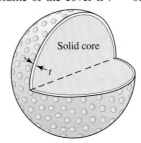

FIGURE 4.9.7 Golf ball in Problem 49

50. A hollow metal pipe is 1.5 m long. Find an approximation to the volume of the metal if the inner radius of the pipe is 2 cm and the thickness of the metal is 0.25 cm. See FIGURE 4.9.8.

FIGURE 4.9.8 Pipe in Problem 50

51. The side of a square is measured to be 10 cm with a possible error of ± 0.3 cm. Use differentials to find an approximation to the maximum error in the area. Find the approximate relative error and the approximate percentage error.

52. An oil storage tank in the form of a circular cylinder has a height of 5 m. The radius is measured to be 8 m with a possible error of ± 0.25 m. Use differentials to estimate the maximum error in the volume. Find the approximate relative error and the approximate percentage error.

53. In the study of some adiabatic processes, the pressure P of a gas is related to the volume V that it occupies by $P = c/V^\gamma$, where c and γ are constants. Show that the approximate relative error in P is proportional to the approximate relative error in V.

54. The range R of a projectile with an initial velocity v_0 and angle of elevation θ is given by $R = (v_0^2/g)\sin 2\theta$, where g is the acceleration of gravity. If v_0 and θ are held constant, then show that the percentage error in the range is proportional to the percentage error in g.

55. Use the formula in Problem 54 to determine the range of a projectile when the initial velocity is 256 ft/s, the angle of elevation is $45°$, and the acceleration of gravity is 32 ft/s². What is the approximate change in the range of the projectile if the initial velocity is increased to 266 ft/s?

56. The acceleration due to gravity g is not constant but changes with altitude. For practical purposes, at the surface of the Earth g is taken to be 32 ft/s², 980 cm/s², or 9.8 m/s².

   (a) From the Law of Universal Gravitation, the force F between a body of mass m_1 and the Earth of mass m_2 is $F = km_1m_2/r^2$, where k is constant and r is the distance to the center of the Earth. Alternatively, Newton's second law of motion implies $F = m_1 g$. Show that $g = km_2/r^2$.
   (b) Use the result from part (a) to show $dg/g = -2dr/r$.
   (c) Let $r = 6400$ km at the surface of the Earth. Use part (b) to show that the approximate value of g at an altitude of 16 km is 9.75 m/s².

57. The acceleration due to gravity g also changes with latitude. The International Geodesy Association has defined g (at sea level) as a function of latitude θ as follows:

$$g = 978.0318(1 + 53.024 \times 10^{-4}\sin^2\theta - 5.9 \times 10^{-6}\sin^2 2\theta),$$

where g is measured in cm/s².

   (a) According to this mathematical model, where is g a minimum? Where is g a maximum?
   (b) What is the value of g at latitude $60°$N?
   (c) What is the approximate change in g as θ changes from $60°$N to $61°$N? [*Hint:* Remember to use radian measure.]

58. The period (in seconds) of a simple pendulum of length L is $T = 2\pi\sqrt{L/g}$, where g is the acceleration due to gravity. Compute the exact change in the period if L is increased from 4 m to 5 m. Then use differentials to find an approximation to the change in the period. Assume $g = 9.8$ m/s².

59. In Problem 58, given that L is fixed at 4 m, find an approximation to the change in the period if the pendulum is moved to an altitude where $g = 9.75$ m/s².

60. Since license plates are all approximately the same size (12 in. across), a computerized optical sensor mounted at the front of car A could register the distance D to car B directly in front of car A by measuring the angle θ subtended by car B's license plate. See FIGURE 4.9.9.

FIGURE 4.9.9 Cars in Problem 60

   (a) Express D as a function of the subtended angle θ.
   (b) Find the distance to the front car if the subtended angle θ is 30 min of an arc (that is, $\frac{1}{2}°$).
   (c) Suppose in part (b) that θ is decreasing at the rate of 2 min of arc per second, and that car A is traveling at a rate of 30 mi/h. At what rate is car B moving?
   (d) Show that the approximate relative error in measuring D is given by

$$\frac{dD}{D} = -\frac{d\theta}{\sin\theta},$$

where $d\theta$ is the approximate error (in radians) in measuring θ. What is the approximate relative error in D in part (b) if the subtended angle θ is measured with a possible error of ± 1 min of arc?

≡ Think About It

61. Suppose that the function $y = f(x)$ is differentiable at a number a. If a polynomial $p(x) = c_1 x + c_0$ has the properties that $p(a) = f(a)$ and $p'(a) = f'(a)$, then show $p(x) = L(x)$, where L is defined in (2).

62. Without appealing to trigonometry, explain why for small values of x, $\cos x = 1$.

63. Suppose a function f and f' are differentiable at a number a and that $L(x)$ is a linearization of f at a. Discuss: If $f''(x) > 0$ for all x in some open interval containing a, will $L(x)$ overestimate or underestimate $f(x)$ for x near a?

64. Suppose $(c, f(c))$ is a point of inflection for the graph of $y = f(x)$ such that $f''(c) = 0$ and suppose further that $L(x)$ is a linearization of f at c. Describe what the graph of $y = f(x) - L(x)$ looks like in a neighborhood of c.

65. The area of a square with side of length x is $A = x^2$. Suppose, as shown in FIGURE 4.9.10, that each side of the square is increased by an amount Δx. In Figure 4.9.10 identify by color the areas ΔA, dA, and $\Delta A - dA$.

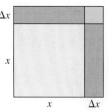

FIGURE 4.9.10 Square in Problem 65

4.10 Newton's Method

Introduction Finding the roots of certain kinds of equations was a problem that captivated mathematicians for centuries. The zeros of a *polynomial* function f of degree 4 or less—that is, the roots of the equation $f(x) = 0$, can always be found by means of an algebraic formula that expresses the unknown x in terms of the coefficients of f. For example, the polynomial equation of degree 2, $ax^2 + bx + c$, $a \neq 0$, can be solved by the quadratic formula. One of the major achievements in the history of mathematics was the proof in the nineteenth century that polynomial equations of degree greater than 4 cannot be solved by means of algebraic formulas, in other words, in terms of radicals. Thus, solving an algebraic equation such as

$$x^5 - 3x^2 + 4x - 6 = 0 \tag{1}$$

poses a quandary unless the fifth-degree polynomial $x^5 - 3x^2 + 4x - 6$ factors. Furthermore, in scientific analyses, one is often asked to find roots of transcendental equations such as

$$2x = \tan x. \tag{2}$$

In the case of problems such as (1) and (2) it is common practice to employ some method that yields an approximation or estimation of the roots. In this section we consider an approximation technique that makes use of the derivative of a function f, or more precisely, a tangent line to the graph of f. This new method is known as **Newton's Method** or the **Newton–Raphson Method**.

Newton's Method Suppose f is differentiable and suppose c represents an unknown real root of $f(x) = 0$; that is, $f(c) = 0$. Let x_0 denote a number that is chosen arbitrarily as a first guess to c. If $f(x_0) \neq 0$, compute $f'(x_0)$ and, as shown in FIGURE 4.10.1(a), construct a tangent to the graph of f at $(x_0, f(x_0))$. If we let $(x_1, 0)$ denote the x-intercept of the tangent line $y - f(x_0) = f'(x_0)(x - x_0)$, then the coordinates $x = x_1$ and $y = 0$ must satisfy this equation. By solving $0 - f(x_0) = f'(x_0)(x_1 - x_0)$ for x_1 we get

$$x_1 = x_0 - \frac{f(x_0)}{f'(x_0)}.$$

(a) One tangent line   (b) Three tangent lines

FIGURE 4.10.1 Successive x-coordinates of x-intercepts of tangent lines approximate the root c

Repeat the procedure at $(x_1, f(x_1))$ and let $(x_2, 0)$ be the x-intercept of the second tangent line $y - f(x_1) = f'(x_1)(x - x_1)$. From $0 - f(x_1) = f'(x_1)(x_2 - x_1)$ we find

$$x_2 = x_1 - \frac{f(x_1)}{f'(x_1)}.$$

Continuing in this fashion, we determine x_{n+1} from x_n using the formula

$$x_{n+1} = x_n - \frac{f(x_n)}{f'(x_n)}. \quad (3)$$

For $n = 0, 1, 2, \ldots$ formula (3) yields a sequence of approximations $x_1, x_2, x_3, \ldots$ to the root c. As suggested in Figure 4.10.1(b), if the terms in the sequence $x_1, x_2, x_3, \ldots$ become closer and closer to c as n increases without bound, that is, $x_n \to c$ as $n \to \infty$, we write $\lim_{x \to \infty} x_n = c$ and say that the sequence **converges** to c.

■ **Graphical Analysis** Before applying (3), it is a good idea to determine the existence and number of real roots of $f(x) = 0$ through graphical means. For example, the irrational number $\sqrt{3}$ can be interpreted as either

- a root of the quadratic equation $x^2 - 3 = 0$ and hence, a zero of the continuous function $f(x) = x^2 - 3$, or
- the x-coordinate of a point of intersection of the graphs of $y = x^2$ and $y = 3$.

Both interpretations are illustrated in FIGURE 4.10.2. Of course, another reason for a graph is to enable us to choose the initial guess x_0 so that it is close to the root c.

Although the actual computation of the number $\sqrt{3}$ is trivial on a calculator, its calculation serves nicely as an introduction to the use of Newton's Method.

(a) x-coordinate of x-intercept

(b) x-coordinate of point of intersection of two graphs

FIGURE 4.10.2 Graphical location of $\sqrt{3}$

EXAMPLE 1 Using Newton's Method

Approximate $\sqrt{3}$ by Newton's Method.

Solution If we define $f(x) = x^2 - 3$, then $f'(x) = 2x$ and (3) becomes

$$x_{n+1} = x_n - \frac{x_n^2 - 3}{2x_n} \quad \text{or} \quad x_{n+1} = \frac{1}{2}\left(x_n + \frac{3}{x_n}\right). \quad (4)$$

From Figure 4.10.2 it seems reasonable to choose $x_0 = 1$ as an initial guess to the value of $\sqrt{3}$. We use (4) and display each calculation to eight decimal places:

$$x_1 = \frac{1}{2}\left(x_0 + \frac{3}{x_0}\right) = \frac{1}{2}(1 + 3) = 2$$

$$x_2 = \frac{1}{2}\left(x_1 + \frac{3}{x_1}\right) = \frac{1}{2}\left(2 + \frac{3}{2}\right) = 1.75$$

$$x_3 = \frac{1}{2}\left(x_2 + \frac{3}{x_2}\right) = \frac{1}{2}\left(\frac{7}{4} + \frac{12}{7}\right) \approx 1.73214286$$

$$x_4 = \frac{1}{2}\left(x_3 + \frac{3}{x_3}\right) \approx 1.73205081$$

$$x_5 = \frac{1}{2}\left(x_4 + \frac{3}{x_4}\right) \approx 1.73205081.$$

The process is continued until we obtain two consecutive approximations x_n and x_{n+1} that agree to the desired number of decimal places. Thus, if we are content with an eight decimal approximation to $\sqrt{3}$, we can stop with x_5 and conclude $\sqrt{3} \approx 1.73205081$. ■

EXAMPLE 2 Approximating a Root of an Equation

Use Newton's Method to approximate the real roots of $x^3 - x + 1 = 0$.

Solution Since the given equation is equivalent to $x^3 = x - 1$ we have graphed the functions $y = x^3$ and $y = x - 1$ in FIGURE 4.10.3. The figure should convince you that the original

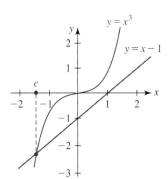

FIGURE 4.10.3 Graphs of functions in Example 2

equation has only one real root c, namely, the x-coordinate of the point of intersection of the two graphs. Now if $f(x) = x^3 - x + 1$, then $f'(x) = 3x^2 - 1$. Hence, (3) is

$$x_{n+1} = x_n - \frac{x_n^3 - x_n + 1}{3x_n^2 - 1} \quad \text{or} \quad x_{n+1} = \frac{2x_n^3 - 1}{3x_n^2 - 1}. \tag{5}$$

If we are interested in three and possibly four decimal place accuracy, we use (5) to compute $x_1, x_2, x_3, \ldots$ until two successive x_n in the sequence agree to four decimal places. Also, Figure 4.10.3 prompts us to make $x_0 = -1.5$ the initial guess. Consequently,

$$x_1 = \frac{2x_0^3 - 1}{3x_0^2 - 1} = \frac{2(-1.5)^3 - 1}{3(-1.5)^2 - 1} \approx -1.3478$$

$$x_2 = \frac{2x_1^3 - 1}{3x_1^2 - 1} \approx -1.3252$$

$$x_3 = \frac{2x_2^3 - 1}{3x_2^2 - 1} \approx -1.3247$$

$$x_4 = \frac{2x_3^3 - 1}{3x_3^2 - 1} \approx -1.3247.$$

Hence, the root of the given equation is approximately $c \approx -1.3247$. ∎

EXAMPLE 3 Approximating a Root of an Equation

Approximate the first positive root of $2x = \tan x$.

Solution FIGURE 4.10.4 shows that there are infinitely many points of intersection of the graphs of $y = 2x$ and $y = \tan x$. The first positive x-coordinate corresponding to a point of intersection is indicated by the letter c in the figure. With $f(x) = 2x - \tan x$ and $f'(x) = 2 - \sec^2 x$, (3) becomes

$$x_{n+1} = x_n - \frac{2x_n - \tan x_n}{2 - \sec^2 x_n}.$$

If a calculator is used in the recursive use of the preceding formula, it is best to express the formula in terms of $\sin x$ and $\cos x$:

$$x_{n+1} = x_n - \frac{2x_n \cos^2 x_n - \sin x_n \cos x_n}{2\cos^2 x_n - 1}. \tag{6}$$

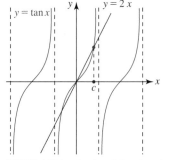

FIGURE 4.10.4 Graphs of functions in Example 3

Since the first vertical asymptote of $y = \tan x$ to the right of the y-axis is $x = \pi/2 \approx 1.57$, it appears from Figure 4.10.4 that the first positive root is near $x_0 = 1$. Using this initial guess, and setting our calculator in radian mode, (6) then yields

$$x_1 \approx 1.310478$$
$$x_2 \approx 1.223929$$
$$x_3 \approx 1.176051$$
$$x_4 \approx 1.165927$$
$$x_5 \approx 1.165562$$
$$x_6 \approx 1.165561$$
$$x_7 \approx 1.165561.$$

We conclude that the first positive root is approximately $c \approx 1.165561$. ∎

Example 3 illustrates the importance of the selection of the initial value x_0. You should verify that the choice $x_0 = \frac{1}{2}$ in (6) leads to a sequence of values $x_1, x_2, x_3, \ldots$ that converges to the one obvious root of $2x = \tan x$, namely, $c = 0$.

▌ Postscript—A Bit of History The problem of finding a formula that expresses the roots of a general nth degree *polynomial* equation $f(x) = 0$ in terms of its coefficients perplexed mathematicians for centuries. We know that in the case of a second-degree, or quadratic, polynomial function $f(x) = ax^2 + bx + c$ where the coefficients a, b, and c are real numbers, the roots c_1 and c_2 of the equation $ax^2 + bx + c = 0$ can be found using the quadratic formula.

The solution of finding roots of a general third-degree, or cubic, polynomial equation is usually attributed to the Italian mathematician **Nicolo Fontana** (1499–1557), also known as Tartaglia the "stammerer." Around 1540, the Italian mathematician **Lodovico Farrari** (1522–1565) discovered an algebraic formula for the roots of the general fourth-degree, or quartic, polynomial equation. Since these formulas are complicated and difficult to use, they are rarely discussed in elementary courses.

Niels Henrik Abel

For the next 284 years no one discovered any formula for roots of polynomial equations of degree 5 or greater. For good reason! In 1824, at age 22, the Norwegian mathematician **Niels Henrik Abel** (1802–1829), was the first to prove that it is *impossible* to find formulas for the roots of general polynomial equations of degrees $n \geq 5$ in terms of their coefficients.

$f'(x)$ NOTES FROM THE CLASSROOM

There are problems with Newton's Method.

(*i*) We must compute $f'(x)$. Needless to say, the form of $f'(x)$ could be formidable when the equation $f(x) = 0$ is complicated.

(*ii*) If the root c of $f(x) = 0$ is near a value for which $f'(x) = 0$, then the denominator in (3) is approaching zero. This necessitates a computation of $f(x_n)$ and $f'(x_n)$ to a high degree of accuracy. A calculation of this kind requires a computer.

(*iii*) It is necessary to find an approximate location of a root $f(x) = 0$ before x_0 is chosen. Attendant to this are the usual difficulties in graphing. But, worse, the iteration of (3) *may not converge* for an imprudent or perhaps blindly chosen x_0. In FIGURE 4.10.5 we see that x_2 is undefined because $f'(x_1) = 0$.

(*iv*) Now some good news. The problems just discussed notwithstanding, the major advantage of Newton's Method is that when it converges to a root c of an equation $f(x) = 0$, it usually does so rather rapidly. It can be shown that under certain conditions Newton's Method converges quadratically, that is, the error at any step in the calculation is not greater than a constant multiple of the square of the error in the preceding step. Roughly this means that the number of places of accuracy can (but not necessarily) double with each step.

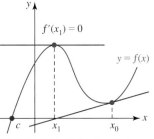

FIGURE 4.10.5 If $f'(x_1) = 0$, then x_2 is undefined

Exercises 4.10 Answers to selected odd-numbered problems begin on page ANS-17.

≡ Fundamentals

In Problems 1–6, determine graphically whether the given equation possesses any real roots.

1. $x^3 = -2 + \sin x$
2. $x^3 - 3x = x^2 - 1$
3. $x^4 + x^2 - 2x + 3 = 0$
4. $\cot x = x$
5. $e^{-x} = x + 2$
6. $e^{-x} - 2\cos x = 0$

In Problems 7–10, use Newton's Method to find an approximation for the given number.

7. $\sqrt{10}$
8. $1 + \sqrt{5}$
9. $\sqrt[3]{4}$
10. $\sqrt[5]{2}$

In Problems 11–16, use Newton's Method, if necessary, to find approximations to all real roots of the given equation.

11. $x^3 = -x + 1$
12. $x^3 - x^2 + 1 = 0$
13. $x^4 + x^2 - 3 = 0$
14. $x^4 = 2x + 1$
15. $x^2 = \sin x$
16. $x + \cos x = 0$

17. Find the smallest positive x-intercept of the graph of $f(x) = 3\cos x + 4\sin x$.

18. Consider the function $f(x) = x^5 + x^2$. Use Newton's Method to approximate the smallest positive number for which $f(x) = 4$.

≡ Applications

19. A cantilever beam 20 ft long with a load of 600 lb at its end is deflected by an amount $d = (60x^2 - x^3)/16{,}000$, where d is measured in inches and x in feet. See FIGURE 4.10.6. Use Newton's Method to approximate the value of x that corresponds to a deflection of 0.01 in.

FIGURE 4.10.6 Beam in Problem 19

20. A vertical solid cylindrical column of fixed radius r that supports its own weight will eventually buckle when its height is increased. It can be proved that the maximum, or critical, height of such a column is $h_{cr} = kr^{2/3}$, where k is a constant and r is measured in meters. Use Newton's Method to approximate the diameter of a column for which $h_{cr} = 10$ m and $k = 35$.

21. A beam of light originating at point P in medium A, whose index of refraction is n_1, strikes the surface of medium B, whose index of refraction is n_2. It can be proved from Snell's Law that the beam is refracted tangent to the surface for the critical angle determined from $\sin\theta_c = n_2/n_1$, $0 < \theta_c < 90°$. For angles of incidence greater than the critical angle, all light is reflected internally to medium A. See FIGURE 4.10.7. If $n_2 = 1$ for air and $n_1 = 1.5$ for glass, use Newton's Method to approximate θ_c in radians.

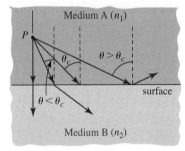

FIGURE 4.10.7 Refraction of light in Problem 21

22. For a suspension bridge, the length s of a cable between two vertical supports whose span is l (horizontal distance) is related to the sag d of the cable by

$$s = l + \frac{8d^2}{3l} - \frac{32d^4}{5l^3}.$$

See FIGURE 4.10.8. If $s = 404$ ft and $l = 400$ ft, use Newton's Method to approximate the sag. Round your answer to one decimal place.* [*Hint*: The root c satisfies $20 < c < 30$.]

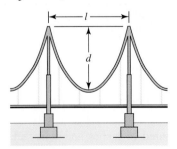

FIGURE 4.10.8 Suspension bridge in Problem 22

---

*The formula for s is itself only an approximation.

23. A rectangular block of steel is hollowed out, making a tub with a uniform thickness t. The dimensions of the tub are shown in FIGURE 4.10.9(a). For the tub to float in water, as shown in Figure 4.10.9(b), the weight of the water displaced must equal the weight of the tub (Archimedes' Principle). If the weight density of water is 62.4 lb/ft^3 and the weight density of the steel is 490 lb/ft^3, then

weight of water displaced = 62.4 × (volume of water displaced)

weight of tub = 490 × (volume of steel in tub).

(a) Show that t satisfies the equation

$$t^3 - 7t^2 + \frac{61}{4}t - \frac{1638}{1225} = 0.$$

(b) Use Newton's Method to approximate the largest positive root of the equation in part (a).

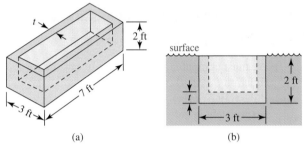

FIGURE 4.10.9 Floating tub in Problem 23

24. A flexible strip of metal 10 ft long is bent into the shape of a circular arc by securing the ends together by means of a cable that is 8 ft long. See FIGURE 4.10.10. Use Newton's Method to approximate the radius r of the circular arc.

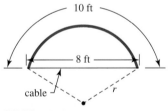

FIGURE 4.10.10 Bent metal strip in Problem 24

25. Two ends of a railroad track L feet long are pushed ℓ feet closer together so that the track bows upward in the arc of a circle of radius R. See FIGURE 4.10.11. The question is, what is the height h above ground of the highest point on the track?

(a) Use Figure 4.10.11 to show that

$$h = \frac{L(1 - \ell/L)^2\theta}{2(1 + \sqrt{1 - (1 - \ell/L)^2\theta^2})},$$

where $\theta > 0$ satisfies $\sin\theta = (1 - \ell/L)\theta$. [*Hint*: In a circular sector, how are the arc length, the radius, and the central angle related?]

(b) If $L = 5280$ ft and $\ell = 1$ ft, use Newton's Method to approximate θ and then solve for the corresponding value of h.

(c) If ℓ/L and θ are very small, then $h \approx L\theta/4$ and $\sin\theta \approx \theta - \frac{1}{6}\theta^3$. Use these two approximations to show that $h \approx \sqrt{3\ell L/8}$. Use this formula with $L = 5280$ ft and $\ell = 1$ ft, and compare with the result in part (b).

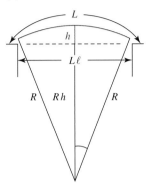

FIGURE 4.10.11 Bowed railroad track in Problem 25

26. At a foundry a metal sphere of radius 2 ft is recast in the form of a rod that is a right circular cylinder 15 ft long surmounted by a hemisphere at one end. The radius r of the hemisphere is the same as the base radius of the cylinder. Use Newton's Method to approximate r.

27. A round but unbalanced wheel of mass M and radius r is connected by a rope and frictionless pulleys to a mass m as shown in FIGURE 4.10.12. O is the center of the wheel and P is its center of mass. If it is released from rest, it can be shown that the angle θ at which the wheel first stops satisfies the equation

$$Mg\frac{r}{2}\sin\theta - mgr\theta = 0,$$

where g is the acceleration due to gravity. Use Newton's Method to approximate θ if the mass of the wheel is four times the mass m.

FIGURE 4.10.12 Unbalanced wheel in Problem 27

28. Two ladders of lengths $L_1 = 40$ ft and $L_2 = 30$ ft are placed against two vertical walls as shown in FIGURE 4.10.13. The height of the point where the ladders cross is $h = 10$ ft.

(a) Show that the indicated height x in the figure can be determined from the equation

$$x^4 - 2hx^3 + (L_1^2 - L_2^2)x^2 - 2h(L_1^2 - L_2^2)x + h^2(L_1^2 - L_2^2) = 0.$$

(b) Use Newton's Method to approximate the solution of the equation in part (a). Why does it make sense to choose $x_0 \geq 10$?

(c) Approximate the distance z between the two walls.

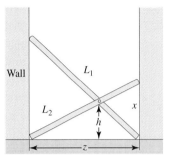

FIGURE 4.10.13 Ladders in Problem 28

≡ Calculator/CAS Problems

In Problems 29 and 30, use a calculator or CAS to obtain the graph of the given function. Use Newton's Method to approximate the roots of $f(x) = 0$ that you discover from the graph.

29. $f(x) = 2x^5 + 3x^4 - 7x^3 + 2x^2 + 8x - 8$

30. $f(x) = 4x^{12} + x^{11} - 4x^8 + 3x^3 + 2x^2 + x - 10$

31. (a) Use a calculator or CAS to obtain the graphs of $f(x) = 0.5x^3 - x$ and $g(x) = \cos x$ on the same coordinate axes.
(b) Use a calculator or CAS to obtain the graph of $y = f(x) - g(x)$, where f and g are as given in part (a).
(c) Use the graphs in part (a) or the graph in part (b) to determine the number of roots of the equation $0.5x^3 - x = \cos x$.
(d) Use Newton's Method to approximate the roots of the equation in part (c).

≡ Think About It

32. Let f be a differentiable function. Show that if $f(x_0) = -f(x_1)$ and $f'(x_0) = f'(x_1)$, then (3) implies $x_2 = x_0$.

33. For the piecewise-defined function

$$f(x) = \begin{cases} -\sqrt{4-x}, & x < 4 \\ \sqrt{x-4}, & x \geq 4 \end{cases}$$

observe that $f(4) = 0$. Show that for any choice of x_0, Newton's Method will fail to converge to the root. [*Hint*: See Problem 32.]

Chapter 4 in Review

Answers to selected odd-numbered problems begin on page ANS-17.

A. True/False

In Problems 1–20, indicate whether the given statement is true or false.

1. If f is increasing on an interval, then $f'(x) > 0$ on the interval. _____
2. A function f has an extremum at c when $f'(c) = 0$. _____
3. A particle moving rectilinearly slows down when the velocity $v(t)$ decreases. _____
4. If the position of a particle moving rectilinearly on a horizontal line is $s(t) = t^2 - 2t$, then the particle is speeding up for $t > 1$. _____
5. If $f''(x) < 0$ for all x in the interval (a, b), then the graph of f is concave down on the interval. _____
6. If $f''(c) = 0$, then $(c, f(c))$ is a point of inflection. _____
7. If $f(c)$ is a relative maximum, then $f'(c) = 0$ and $f'(x) > 0$ for $x < c$ and $f'(x) < 0$ for $x > c$. _____
8. If $f(c)$ is a relative minimum, then $f''(c) > 0$. _____
9. A function f that is continuous on a closed interval $[a, b]$ has both an absolute maximum and an absolute minimum on the interval. _____
10. Every absolute extremum is also a relative extremum. _____
11. If $c > 0$ is a constant and $f(x) = \frac{1}{3}x^3 - cx^2$, then $(c, f(c))$ is a point of inflection. _____
12. $x = 1$ is a critical number of the function $f(x) = \sqrt{x^2 - 2x}$. _____
13. If $f'(x) > 0$ and $g'(x) > 0$ on an interval I, then $f + g$ is increasing on I. _____
14. If $f'(x) > 0$ on an interval I, then $f''(x) > 0$ on I. _____
15. A limit of the form $\infty - \infty$ always has the value 0. _____
16. A limit of the form 1^∞ is always 1. _____
17. A limit of the form ∞/∞ is indeterminate. _____
18. A limit of the form $0/\infty$ is indeterminate. _____
19. If $\lim_{x \to \infty} \frac{f(x)}{g(x)}$ and $\lim_{x \to \infty} \frac{f'(x)}{g'(x)}$ are both of the form ∞/∞, then the first limit does not exist. _____
20. For an indeterminate form, L'Hôpital's Rule states that the limit of a quotient is the same as the derivative of the quotient. _____

B. Fill in the Blanks

In Problems 1–10, fill in the blanks.

1. For a particle moving rectilinearly, acceleration is the first derivative of _____.
2. The graph of a cubic polynomial can have at most _____ point(s) of inflection.
3. An example of a function $y = f(x)$ that is concave up on $(-\infty, 0)$, concave down on $(0, \infty)$, and increasing on $(-\infty, \infty)$ is _____.
4. Two nonnegative numbers whose sum is 8 such that the sum of their squares is a maximum are _____.
5. If f is continuous on $[a, b]$, differentiable on (a, b), and $f(a) = f(b) = 0$, then there exists some c in (a, b) such that $f'(c) =$ _____.
6. $\lim_{x \to \infty} \frac{x^n}{e^x} =$ _____ for every integer n.
7. The sum of a positive number and its reciprocal is always greater than or equal to _____.
8. If $f(1) = 13$ and $f'(x) = 5^{x^2}$, then a linearization of f at $a = 1$ is _____ and $f(1.1) \approx$ _____.

9. If $y = x^2 - x$, then $\Delta y =$ _____.

10. If $y = x^3 e^{-x}$, then $dy =$ _____.

C. Exercises

In Problems 1–4, find the absolute extrema of the given function on the indicated interval.

1. $f(x) = x^3 - 75x + 150$; $[-3, 4]$

2. $f(x) = 4x^2 - \dfrac{1}{x}$; $[\tfrac{1}{4}, 1]$

3. $f(x) = \dfrac{x^2}{x + 4}$; $[-1, 3]$

4. $f(x) = (x^2 - 3x + 5)^{1/2}$; $[1, 3]$

5. Sketch a graph of a continuous function that has the properties:

$$f(0) = 1, \quad f(2) = 3$$
$$f'(0) = 0, \quad f'(2) \text{ does not exist}$$
$$f'(x) > 0, \quad x < 0$$
$$f'(x) > 0, \quad 0 < x < 2$$
$$f'(x) < 0, \quad x > 2.$$

6. Use the first and second derivatives as an aid in comparing the graphs of

$$y = x + \sin x \quad \text{and} \quad y = x + \sin 2x.$$

7. The position of a particle moving on a horizontal line is given by $s(t) = -t^3 + 6t^2$.

(a) Graph the motion on the time interval $[-1, 5]$.
(b) At what time is the velocity function a maximum?
(c) Does this time correspond to the maximum speed?

8. The height above ground of a projectile fired vertically is $s(t) = -4.9t^2 + 14.7t + 49$, where s is measured in meters and t in seconds.

(a) What is the maximum height attained by the projectile?
(b) At what speed does the projectile strike the ground?

9. Suppose f is a polynomial function with zeros of multiplicity 2 at $x = a$ and $x = b$; that is,

$$f(x) = (x - a)^2(x - b)^2 g(x)$$

where g is a polynomial function.

(a) Show that f' has at least three zeros in the closed interval $[a, b]$.
(b) If $g(x)$ is a constant, find the zeros of f' in $[a, b]$.

10. Show that the function $f(x) = x^{1/3}$ does not satisfy the hypothesis of the Mean Value Theorem on the interval $[-1, 8]$, yet a number c can be found in $(-1, 8)$ such that $f'(c) = [f(b) - f(a)]/(b - a)$. Explain.

In Problems 11–14, find the relative extrema of the given function f. Graph.

11. $f(x) = 2x^3 + 3x^2 - 36x$

12. $f(x) = x^5 - \dfrac{5}{3}x^3 + 2$

13. $f(x) = 4x - 6x^{2/3} + 2$

14. $f(x) = \dfrac{x^2 - 2x + 2}{x - 1}$

In Problems 15–18, find the relative extrema and the points of inflection of the given function f. Do not graph.

15. $f(x) = x^4 + 8x^3 + 18x^2$

16. $f(x) = x^6 - 3x^4 + 5$

17. $f(x) = 10 - (x - 3)^{1/3}$

18. $f(x) = x(x - 1)^{5/2}$

In Problems 19–24, match each figure with one or more of the following statements. On the interval corresponding to the portion of the graph of $y = f(x)$ shown:

(a) f has a positive first derivative.
(b) f has a negative second derivative.
(c) The graph of f has a point of inflection.
(d) f is differentiable.

(e) f has a relative extremum.
(f) The slopes of the tangent lines increase as x increases.

19.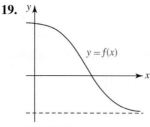

FIGURE 4.R.1 Graph for Problem 19

20.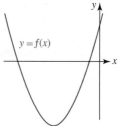

FIGURE 4.R.2 Graph for Problem 20

21.

FIGURE 4.R.3 Graph for Problem 21

22.

FIGURE 4.R.4 Graph for Problem 22

23.

FIGURE 4.R.5 Graph for Problem 23

24.

FIGURE 4.R.6 Graph for Problem 24

25. Let a, b, and c be real numbers. Find the x-coordinate of the point of inflection for the graph of
$$f(x) = (x - a)(x - b)(x - c).$$

26. A triangle is expanding with time. The area of the triangle is increasing at a rate of 15 in^2/min, whereas the length of its base is decreasing at a rate of $\frac{1}{2}$ in./min. At what rate is the altitude of the triangle changing when the altitude is 8 in. and the base is 6 in.?

27. A square is inscribed in a circle of radius r as shown in FIGURE 4.R.7. At what rate is the area of the square increasing at the instant the radius of the circle is 2 in. and increasing at a rate of 4 in./min?

FIGURE 4.R.7 Circle in Problem 27

28. Water drips into a hemispherical tank of radius 10 m at a rate of $\frac{1}{10}$ m^3/min and drips out a hole in the bottom of the tank at a rate of $\frac{1}{5}$ m^3/min. It can be shown that the volume of the water in the tank at t is $V = 10\pi h^2 - (\pi/3)h^3$. See FIGURE 4.R.8.

(a) Is the depth of the water increasing or decreasing?
(b) At what rate is the depth of the water changing when the depth is 5 m?

FIGURE 4.R.8 Tank in Problem 28

29. Two coils that carry the same current produce a magnetic field at point Q on the x-axis of strength

$$B = \frac{1}{2}\mu_0 r_0^2 I\left\{\left[r_0^2 + \left(x + \frac{1}{2}r_0\right)^2\right]^{-3/2} + \left[r_0^2 + \left(x - \frac{1}{2}r_0\right)^2\right]^{-3/2}\right\},$$

where μ_0, r_0, and I are constants. See FIGURE 4.R.9. Show that the maximum value of B occurs at $x = 0$.

FIGURE 4.R.9 Coils in Problem 29

30. A battery with constant emf E and constant internal resistance r is wired in series with a resistor that has resistance R. The current in the circuit is then $I = E/(r + R)$. Find the value of R for which the power $P = RI^2$ dissipated in the external load is a maximum. This is called **impedance matching**.

31. When a hole is punched into the lateral side of a cylindrical tank full of water, the resulting stream hits the ground at a distance x ft from the base, where $x = 2\sqrt{y(h - y)}$. See FIGURE 4.R.10.

   (a) At what point should the hole be punched in the side of the tank so that the stream attains a maximum distance from the base?
   (b) What is the maximum distance?

FIGURE 4.R.10 Leaking tank in Problem 31

32. The area of a circular sector of radius r and arc length s is $A = \frac{1}{2}rs$. See FIGURE 4.R.11. Find the maximum area of a sector enclosed by a perimeter of 60 cm.

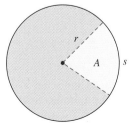

FIGURE 4.R.11 Circular sector in Problem 32

33. A pigpen, attached to a barn, is enclosed using fence on two sides, as shown in FIGURE 4.R.12. The amount of fence to be used is 585 ft. Find the values of x and y indicated in the figure so that the greatest area is enclosed. What is the greatest area?

FIGURE 4.R.12 Pigpen in Problem 33

34. A rancher wants to use 100 m of fence to construct a diagonal fence connecting two existing walls that meet at a right angle. How should this be done so that the area enclosed by the walls and the fence is a maximum?

35. According to **Fermat's Principle**, a ray of light originating at point A and reflected from a plane surface to point B travels on a path requiring the least time. See FIGURE 4.R.13. Assume that the speed of light c as well as h_1, h_2, and d are constants. Show that the time is a minimum when $\tan\theta_1 = \tan\theta_2$. Since $0 < \theta_1 < \pi/2$ and $0 < \theta_2 < \pi/2$, it follows that $\theta_1 = \theta_2$. In other words, the angle of incidence equals the angle of reflection. [*Note*: Figure 4.R.13 is inaccurate on purpose.]

FIGURE 4.R.13 Reflected light rays in Problem 35

36. Determine the dimensions of a right circular cone having minimum volume V that circumscribes a sphere of radius r. See FIGURE 4.R.14. [*Hint*: Use similar triangles.]

FIGURE 4.R.14 Sphere and cone in Problem 36

37. A container in the form of a right circular cylinder has a volume of 100 in^3. The top of the container costs three times as much per unit area as the bottom and the sides. Show that the dimension that gives the least cost of construction is a height that is four times the radius.

38. A box with a cover is to be made from a rectangular piece of cardboard 30 in. long and 15 in. wide by cutting a square out of each corner at one end of the cardboard and cutting a rectangle out of each corner at the other end. The cardboard is then folded on the dashed lines, as shown in FIGURE 4.R.15. Find the dimensions of the box that will give the maximum volume. What is the maximum volume?

(a)   (b)

FIGURE 4.R.15 Box in Problem 38

In Problems 39–48, use L'Hôpital's Rule to find the limit.

39. $\lim\limits_{x \to \sqrt{3}} \dfrac{\sqrt{3} - \tan(\pi/x^2)}{x - \sqrt{3}}$

40. $\lim\limits_{\theta \to 0} \dfrac{10\theta - 5\sin 2\theta}{10\theta - 2\sin 5\theta}$

41. $\lim\limits_{x \to \infty} x\left(\cos\dfrac{1}{x} - e^{2/x}\right)$

42. $\lim\limits_{y \to 0} \left[\dfrac{1}{y} - \dfrac{1}{\ln(y+1)}\right]$

43. $\lim\limits_{t \to 0} \dfrac{(\sin t)^2}{\sin t^2}$

44. $\lim\limits_{x \to 0} \dfrac{\tan(5x)}{e^{3x/2} - e^{-x/2}}$

45. $\lim\limits_{x \to 0^+} (3x)^{-1/\ln x}$

46. $\lim\limits_{x \to 0} (2x + e^{3x})^{4/x}$

47. $\lim\limits_{x \to \infty} \ln\left(\dfrac{x + e^{2x}}{1 + e^{4x}}\right)$

48. $\lim\limits_{x \to 0^+} x(\ln x)^2$

In Problems 49 and 50, use Newton's Method to find the indicated root. Carry out the method until two successive approximations agree to four decimal places.

49. $x^3 - 4x + 2 = 0$, the largest positive root

50. $\left(\dfrac{\sin x}{x}\right)^2 = \dfrac{1}{2}$, the smallest positive root

Chapter 5

Integrals

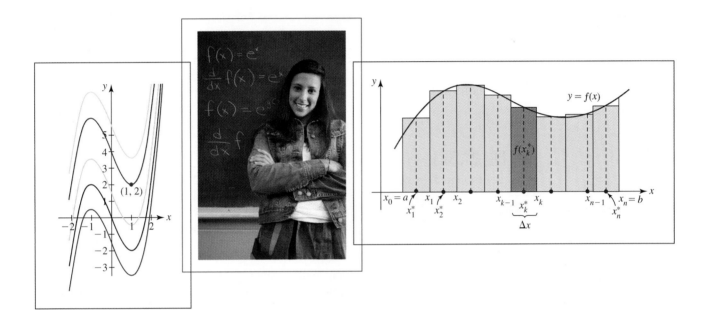

In This Chapter In the last two chapters we have been concerned with the definition, properties, and applications of the derivative. We turn now from differential to integral calculus. Leibniz originally called this second of the two major divisions of calculus, *calculus summatorius.* In 1696, at the persuasion of the Swiss mathematician Johann Bernoulli, Leibniz changed its name to *calculus integralis.* As the original Latin words suggest, the notion of a *sum* plays an important role in the full development of the integral.

In Chapter 2 we saw that the tangent problem leads naturally to the derivative of a function. In the area problem, the motivational problem for integral calculus, we want to find the area bounded by the graph of a function and the *x*-axis. This problem leads to the concept of a *definite integral.*

5.1 The Indefinite Integral
5.2 Integration by the *u*-Substitution
5.3 The Area Problem
5.4 The Definite Integral
5.5 Fundamental Theorem of Calculus
Chapter 5 in Review

5.1 The Indefinite Integral

Introduction In Chapters 3 and 4 we were concerned only with the basic problem:

- Given a function f, find its derivative f'.

In this chapter and in subsequent chapters of this text we shall see that an equally important problem is:

- Given a function f, find a function F whose derivative is f.

In other words, for a given function f, we now think of f as a derivative. We wish to find a function F whose derivative is f, that is, $F'(x) = f(x)$ for all x on some interval. Roughly put, we must do differentiation in reverse.

We begin with a definition.

Definition 5.1.1 Antiderivative

A function F is said to be an **antiderivative** of a function f on some interval I if $F'(x) = f(x)$ for all x in I.

EXAMPLE 1 An Antiderivative

An antiderivative of $f(x) = 2x$ is $F(x) = x^2$, since $F'(x) = 2x$. ∎

There is always more than one antiderivative of a function. For instance, in the foregoing example, $F_1(x) = x^2 - 1$ and $F_2(x) = x^2 + 10$ are also antiderivatives of $f(x) = 2x$, since $F_1'(x) = F_2'(x) = 2x$.

We shall now prove that any antiderivative of f must be of the form $G(x) = F(x) + C$; that is, *two antiderivatives of the same function can differ by at most a constant*. Hence, $F(x) + C$ is *the most general antiderivative* of $f(x)$.

Theorem 5.1.1 Antiderivatives Differ by a Constant

If $G'(x) = F'(x)$ for all x in some interval $[a, b]$, then

$$G(x) = F(x) + C$$

for all x in the interval.

PROOF Suppose we define $g(x) = G(x) - F(x)$. Then, since $G'(x) = F'(x)$, it follows that $g'(x) = G'(x) - F'(x) = 0$ for all x in $[a, b]$. If x_1 and x_2 are any two numbers that satisfy $a \leq x_1 < x_2 \leq b$, it follows from the Mean Value Theorem (Theorem 4.4.2) that a number k exists in the open interval (x_1, x_2) for which

$$g'(k) = \frac{g(x_2) - g(x_1)}{x_2 - x_1} \quad \text{or} \quad g(x_2) - g(x_1) = g'(k)(x_2 - x_1).$$

But $g'(x) = 0$ for all x in $[a, b]$; in particular, $g'(k) = 0$. Hence, $g(x_2) - g(x_1) = 0$ or $g(x_2) = g(x_1)$. Now, by assumption, x_1 and x_2 are any two, but different, numbers in the interval. Since the function values $g(x_1)$ and $g(x_2)$ are the same, we must conclude that the function $g(x)$ is a constant C. Thus, $g(x) = C$ implies $G(x) - F(x) = C$ or $G(x) = F(x) + C$. ∎

The notation $F(x) + C$ represents a *family of functions*; each member has a derivative equal to $f(x)$. Returning to Example 1, the most general antiderivative of $f(x) = 2x$ is the family $F(x) = x^2 + C$. As we see in FIGURE 5.1.1 the graph of an antiderivative of $f(x) = 2x$ is a vertical translation of the graph of x^2.

EXAMPLE 2 Most General Antiderivatives

(a) An antiderivative of $f(x) = 2x + 5$ is $F(x) = x^2 + 5x$ since $F'(x) = 2x + 5$. The most general antiderivative of $f(x) = 2x + 5$ is $F(x) = x^2 + 5x + C$.

(b) An antiderivative of $f(x) = \sec^2 x$ is $F(x) = \tan x$ since $F'(x) = \sec^2 x$. The most general antiderivative of $f(x) = \sec^2 x$ is $F(x) = \tan x + C$. ∎

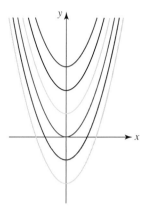

FIGURE 5.1.1 Some members of the family of antiderivatives of $f(x) = 2x$

■ **Indefinite Integral Notation** For convenience, let us introduce a notation for an antiderivative of a function. If $F'(x) = f(x)$, we shall represent the most general antiderivative of f by

$$\int f(x)\, dx = F(x) + C.$$

The symbol $\int$ was introduced by Leibniz and is called an **integral sign**. The notation $\int f(x)\, dx$ is called the **indefinite integral** of $f(x)$ with respect to x. The function $f(x)$ is called the **integrand**. The process of finding an antiderivative is called **antidifferentiation** or **integration**. The number C is called a **constant of integration**. Just as $\frac{d}{dx}(\ )$ denotes the operation of differentiation of $(\ )$ *with respect to x*, the symbolism $\int (\ )\, dx$ denotes the operation of integration of $(\ )$ *with respect to x*.

Differentiation and integration are fundamentally inverse operations. If $\int f(x)\, dx = F(x) + C$, then F is an antiderivative of f, that is, $F'(x) = f(x)$ and so

$$\int F'(x)\, dx = F(x) + C. \qquad (1)$$

Moreover,

$$\frac{d}{dx}\int f(x)\, dx = \frac{d}{dx}(F(x) + C) = F'(x) = f(x) \qquad (2)$$

In words, (1) and (2) are, respectively:

- *An antiderivative of the derivative of a function is that function plus a constant.*
- *The derivative of an antiderivative of a function is that function.*

From this it follows that whenever we obtain the derivative of a function, we get at the same time an integration formula. For example, in view of (1) if

$\dfrac{d}{dx}\dfrac{x^{n+1}}{n+1} = x^n$ then $\displaystyle\int \dfrac{d}{dx}\dfrac{x^{n+1}}{n+1}\, dx = \int x^n\, dx = \dfrac{x^{n+1}}{n+1} + C,$ ◀ This first result is valid only if $n \neq -1$.

$\dfrac{d}{dx}\ln|x| = \dfrac{1}{x}$ then $\displaystyle\int \dfrac{d}{dx}\ln|x|\, dx = \int \dfrac{1}{x}\, dx = \ln|x| + C,$

$\dfrac{d}{dx}\sin x = \cos x$ then $\displaystyle\int \dfrac{d}{dx}\sin x\, dx = \int \cos x\, dx = \sin x + C,$

$\dfrac{d}{dx}\tan^{-1} x = \dfrac{1}{1+x^2}$ then $\displaystyle\int \dfrac{d}{dx}\tan^{-1} x\, dx = \int \dfrac{1}{1+x^2}\, dx = \tan^{-1} x + C.$

In this manner we can construct an integration formula from each derivative formula. TABLE 5.1.1 summarizes *some* of the important derivative formulas for the functions we have studied so far and their integration formula analogues.

TABLE 5.1.1

Differentiation Formula	Integration Formula	Differentiation Formula	Integration Formula
1. $\dfrac{d}{dx}x = 1$	$\int dx = x + C$	10. $\dfrac{d}{dx}\sin^{-1}x = \dfrac{1}{\sqrt{1-x^2}}$	$\int \dfrac{1}{\sqrt{1-x^2}}\,dx = \sin^{-1}x + C$
2. $\dfrac{d}{dx}\dfrac{x^{n+1}}{n+1} = x^n\,(n \neq -1)$	$\int x^n\,dx = \dfrac{x^{n+1}}{n+1} + C$	11. $\dfrac{d}{dx}\tan^{-1}x = \dfrac{1}{1+x^2}$	$\int \dfrac{1}{1+x^2}\,dx = \tan^{-1}x + C$
3. $\dfrac{d}{dx}\ln|x| = \dfrac{1}{x}$	$\int \dfrac{1}{x}\,dx = \ln|x| + C$	12. $\dfrac{d}{dx}\sec^{-1}x = \dfrac{1}{|x|\sqrt{x^2-1}}$	$\int \dfrac{1}{x\sqrt{x^2-1}}\,dx = \sec^{-1}|x| + C$
4. $\dfrac{d}{dx}\sin x = \cos x$	$\int \cos x\,dx = \sin x + C$	13. $\dfrac{d}{dx}b^x = b^x(\ln b),$ $(b>0, b \neq 1)$	$\int b^x\,dx = \dfrac{b^x}{\ln b} + C$
5. $\dfrac{d}{dx}\cos x = -\sin x$	$\int \sin x\,dx = -\cos x + C$	14. $\dfrac{d}{dx}e^x = e^x$	$\int e^x\,dx = e^x + C$
6. $\dfrac{d}{dx}\tan x = \sec^2 x$	$\int \sec^2 x\,dx = \tan x + C$	15. $\dfrac{d}{dx}\sinh x = \cosh x$	$\int \cosh x\,dx = \sinh x + C$
7. $\dfrac{d}{dx}\cot x = -\csc^2 x$	$\int \csc^2 x\,dx = -\cot x + C$	16. $\dfrac{d}{dx}\cosh x = \sinh x$	$\int \sinh x\,dx = \cosh x + C$
8. $\dfrac{d}{dx}\sec x = \sec x \tan x$	$\int \sec x \tan x\,dx = \sec x + C$		
9. $\dfrac{d}{dx}\csc x = -\csc x \cot x$	$\int \csc x \cot x\,dx = -\csc x + C$		

With regard to entry 3 of Table 5.1.1, it is true that the derivative formulas

$$\frac{d}{dx}\ln x = \frac{1}{x}, \qquad \frac{d}{dx}\ln|x| = \frac{1}{x}, \qquad \frac{d}{dx}\frac{\log_b x}{\ln b} = \frac{1}{x}$$

mean that *an* antiderivative of $1/x = x^{-1}$ can be taken to be $\ln x, x > 0$, $\ln|x|, x \neq 0$, or $\log_b x/\ln b, x > 0$. But we write

$$\int \frac{1}{x}\,dx = \ln|x| + C,$$

as the most general and useful result. Also note that only three formulas involving inverse trigonometric functions are given in Table 5.1.1. This is because, in indefinite integral form, the three remaining formulas are redundant. For example, from the derivatives

$$\frac{d}{dx}\sin^{-1}x = \frac{1}{\sqrt{1-x^2}} \quad \text{and} \quad \frac{d}{dx}\cos^{-1}x = \frac{-1}{\sqrt{1-x^2}}$$

it is seen that we could take either

$$\int \frac{1}{\sqrt{1-x^2}}\,dx = \sin^{-1}x + C \quad \text{or} \quad \int \frac{1}{\sqrt{1-x^2}}\,dx = -\cos^{-1}x + C.$$

Similar observations hold for the inverse cotangent and inverse cosecant.

EXAMPLE 3 An Important but Simple Antiderivative

The integration formula in entry 1 in Table 5.1.1 is included for emphasis:

$$\int dx = \int 1 \cdot dx = x + C \quad \text{because} \quad \frac{d}{dx}(x + C) = 1 + 0 = 1.$$

This result can also be obtained from integration formula 2 of Table 5.1.1 with $n = 0$. ∎

It is often necessary to rewrite an integrand $f(x)$ before carrying out the integration.

EXAMPLE 4 Rewriting the Integrand

Evaluate

(a) $\int \dfrac{1}{x^5}\, dx$ and (b) $\int \sqrt{x}\, dx$.

Solution

(a) By rewriting $1/x^5$ as x^{-5} and identifying $n = -5$, we have from integration formula 2 of Table 5.1.1:

$$\int x^{-5}\, dx = \dfrac{x^{-5+1}}{-5+1} + C = -\dfrac{x^{-4}}{4} + C = -\dfrac{1}{4x^4} + C.$$

(b) We first rewrite the radical $\sqrt{x}$ as $x^{1/2}$ and then use integration formula 2 of Table 5.1.1 with $n = \tfrac{1}{2}$:

$$\int x^{1/2}\, dx = \dfrac{x^{3/2}}{3/2} + C = \dfrac{2}{3} x^{3/2} + C.$$ ∎

It should be kept in mind that the *results of integration can always be checked by differentiation*; for example, in part (b) of Example 4:

$$\dfrac{d}{dx}\left(\dfrac{2}{3} x^{3/2} + C\right) = \dfrac{2}{3} \cdot \dfrac{3}{2} x^{3/2 - 1} = x^{1/2} = \sqrt{x}.$$

Some properties of the indefinite integral are given in the next theorem.

Theorem 5.1.2 Properties of the Indefinite Integral

Let $F'(x) = f(x)$ and $G'(x) = g(x)$. Then

(i) $\int k f(x)\, dx = k \int f(x)\, dx = kF(x) + C$, where k is any constant,

(ii) $\int [f(x) \pm g(x)]\, dx = \int f(x)\, dx \pm \int g(x)\, dx = F(x) \pm G(x) + C.$

These properties follow immediately from the properties of the derivative. For example, (ii) is a consequence of the fact that the derivative of a sum is the sum of the derivatives.

Observe in Theorem 5.1.2(ii) that there is no reason to use two constants of integration, since

$$\int [f(x) \pm g(x)]\, dx = (F(x) + c_1) \pm (G(x) + C_2)$$
$$= F(x) \pm G(x) + (C_1 \pm C_2) = F(x) \pm G(x) + C,$$

where we have replaced $C_1 \pm C_2$ by the single constant C.

An indefinite integral of any finite sum of functions can be obtained by integrating each term.

EXAMPLE 5 Using Theorem 5.1.2

Evaluate $\int \left(4x - \dfrac{2}{x} + 5 \sin x \right) dx$

Solution From parts (i) and (ii) of Theorem 5.1.2 we can write this indefinite integral as three integrals:

$$\int \left(4x - \dfrac{2}{x} + 5 \sin x \right) dx = 4 \int x\, dx - 2 \int \dfrac{1}{x}\, dx + 5 \int \sin x\, dx.$$

In view of integration formulas 2, 3, and 5 in Table 5.1.1 we then have

$$\int \left(4x - \frac{2}{x} + 5\sin x\right) dx = 4 \cdot \frac{x^2}{2} - 2 \cdot \ln|x| + 5 \cdot (-\cos x) + C$$
$$= 2x^2 - 2\ln|x| - 5\cos x + C.$$

■ **Using Division** Putting an integrand in a more tractable form may sometimes entail division. The next two examples illustrate the idea.

EXAMPLE 6 Termwise Division

Evaluate $\int \frac{6x^3 - 5}{x} dx$.

If we read the common denominator concept

$$\frac{a}{c} + \frac{b}{c} = \frac{a+b}{c}$$

from right to left we are performing "termwise division."

Solution By termwise division, Theorem 5.1.2, and integration formulas 2 and 3 of Table 5.1.1:

$$\int \frac{6x^3 - 5}{x} dx = \int \left(\frac{6x^3}{x} - \frac{5}{x}\right) dx$$
$$= \int \left(6x^2 - \frac{5}{x}\right) dx = 6 \cdot \frac{x^3}{3} - 5 \cdot \ln|x| + C = 2x^3 - 5\ln|x| + C.$$ ■

For the problem of evaluating $\int f(x)\, dx$, where $f(x) = p(x)/q(x)$ is a rational function, a working rule to keep in mind in this and subsequent sections is summarized next.

Integration of a Rational Function

Suppose $f(x) = p(x)/q(x)$ is a rational function. If the degree of the polynomial function $p(x)$ is greater than or equal to the degree of the degree of the polynomial function $q(x)$, use long division before integration, that is, write

$$\frac{p(x)}{q(x)} = \text{a polynomial} + \frac{r(x)}{q(x)},$$

where the degree of the polynomial $r(x)$ is less than the degree of $q(x)$.

EXAMPLE 7 Long Division

Evaluate $\int \frac{x^2}{1+x^2} dx$.

Solution Because the degree of the numerator of the integrand is equal to the degree of the denominator we perform long division:

$$\frac{x^2}{1+x^2} = 1 - \frac{1}{1+x^2}.$$

From (*ii*) of Theorem 5.1.2 and integration formulas 1 and 11 in Table 5.1.1 we obtain

$$\int \frac{x^2}{1+x^2} dx = \int \left(1 - \frac{1}{1+x^2}\right) dx = x - \tan^{-1} x + C.$$ ■

■ **Differential Equations** In several exercise sets in Chapter 3 you were asked to verify that a given function satisfies a **differential equation**. Roughly, a differential equation is an equation that involves derivatives or the differential of an unknown function. Differential equations are classified by the **order** of the highest derivative appearing in the equation. The goal is to *solve* differential equations. A **first-order differential equation** of the form

$$\frac{dy}{dx} = g(x) \tag{3}$$

can be solved using indefinite integration. From (1) we see that

$$\int \left(\frac{dy}{dx}\right) dx = y.$$

Thus the solution of (3) is the most general antiderivative of g, that is,

$$y = \int g(x)\, dx. \tag{4}$$

EXAMPLE 8 Solving a Differential Equation

Find a function $y = f(x)$ whose graph passes through the point $(1, 2)$ that also satisfies the differential equation $dy/dx = 3x^2 - 3$.

Solution From (3) and (4) it follows that if

$$\frac{dy}{dx} = 3x^2 - 3 \quad \text{then} \quad y = \int (3x^2 - 3)\, dx.$$

That is,

$$y = \int (3x^2 - 3)\, dx = 3 \cdot \frac{x^3}{3} - 3 \cdot x + C$$

or $y = x^3 - 3x + C$. Now when $x = 1$, $y = 2$, so that $2 = 1 - 3 + C$ or $C = 4$. Hence, $y = x^3 - 3x + 4$. Thus out of the family of antiderivatives of $3x^2 - 3$ shown in FIGURE 5.1.2, we see that there is only one whose graph (shown in red) passes through $(1, 2)$. ■

FIGURE 5.1.2 The red curve is the graph of the solution of the problem in Example 8

When solving a differential equation such as $dy/dx = 3x^2 - 3$ in Example 8, the specified side condition that the graph pass through $(1, 2)$, that is, $f(1) = 2$, is called an **initial condition**. It is common practice to write an initial condition such as this as $y(1) = 2$. The solution $y = x^3 - 3x + 4$ that was determined from the family of solutions $y = x^3 - 3x + C$ by the initial condition is called a **particular solution**. The problem of solving (3) subject to an initial condition,

$$\frac{dy}{dx} = g(x), \quad y(x_0) = y_0$$

is called an **initial-value problem**.

We note that an *n*th-order differential equation of the form $d^n y/dx^n = g(x)$ can be solved by integrating the function $g(x)$ in succession n times. In this case the family of solutions will contain n constants of integration.

EXAMPLE 9 Solving a Differential Equation

Find a function $y = f(x)$ such that $\dfrac{d^2 y}{dx^2} = 1$.

Solution We integrate the given differential equation in succession two times. The first integration gives

$$\frac{dy}{dx} = \int \frac{d^2 y}{dx^2}\, dx = \int 1 \cdot dx = x + C_1.$$

The second integration gives $y = f(x)$:

$$y = \int \frac{dy}{dx}\, dx = \int (x + C_1)\, dx = \frac{x^2}{2} + C_1 x + C_2.$$ ■

∫ NOTES FROM THE CLASSROOM

Students often have a more difficult time calculating antiderivatives than they do derivatives. Two words of advice. First, be very, very careful with your algebra—especially the laws of exponents. The second word of advice has been stated previously but it bears repeating: Keep firmly in mind that the *results of indefinite integration can always be checked*. On a quiz or test it is worth a few seconds of your valuable time to check your answer by taking its derivative. You can often do this in your head. For example,

$$\underset{\text{check by differentiation}}{\overset{\text{integration}}{\int x^2 \, dx = \frac{x^3}{3} + C}}$$

Exercises 5.1   Answers to selected odd-numbered problems begin on page ANS-18.

☰ Fundamentals

In Problems 1–30, evaluate the given indefinite integral.

1. $\int 3 \, dx$
2. $\int (\pi^2 - 1) \, dx$
3. $\int x^5 \, dx$
4. $\int 5x^{1/4} \, dx$
5. $\int \frac{1}{\sqrt[3]{x}} \, dx$
6. $\int \sqrt[3]{x^2} \, dx$
7. $\int (1 - t^{-0.52}) \, dt$
8. $\int 10w\sqrt{w} \, dw$
9. $\int (3x^2 + 2x - 1) \, dx$
10. $\int \left(2\sqrt{t} - t - \frac{9}{t^2}\right) dt$
11. $\int \sqrt{x}(x^2 - 2) \, dx$
12. $\int \left(\frac{5}{\sqrt[3]{s^2}} + \frac{2}{\sqrt{s^3}}\right) ds$
13. $\int (4x + 1)^2 \, dx$
14. $\int (\sqrt{x} - 1)^2 \, dx$
15. $\int (4w - 1)^3 \, dw$
16. $\int (5u - 1)(3u^3 + 2) \, du$
17. $\int \frac{r^2 - 10r + 4}{r^3} \, dr$
18. $\int \frac{(x + 1)^2}{\sqrt{x}} \, dx$
19. $\int \frac{x^{-1} - x^{-2} + x^{-3}}{x^2} \, dx$
20. $\int \frac{t^3 - 8t + 1}{(2t)^4} \, dt$
21. $\int (4\sin x - 1 + 8x^{-5}) \, dx$
22. $\int (-3\cos x + 4\sec^2 x) \, dx$
23. $\int \csc x(\csc x - \cot x) \, dx$
24. $\int \frac{\sin t}{\cos^2 t} \, dt$
25. $\int \frac{2 + 3\sin^2 x}{\sin^2 x} \, dx$
26. $\int \left(40 - \frac{2}{\sec \theta}\right) d\theta$
27. $\int (8x + 1 - 9e^x) \, dx$
28. $\int (15x^{-1} - 4\sinh x) \, dx$
29. $\int \frac{2x^3 - x^2 + 2x + 4}{1 + x^2} \, dx$
30. $\int \frac{x^6}{1 + x^2} \, dx$

In Problems 31 and 32, use a trigonometric identity to evaluate the given indefinite integral.

31. $\int \tan^2 x \, dx$
32. $\int \cos^2 \frac{x}{2} \, dx$

In Problems 33–40, verify the given integration result by differentiation and the Chain Rule.

33. $\int \frac{1}{\sqrt{2x + 1}} \, dx = \sqrt{2x + 1} + C$
34. $\int (2x^2 - 4x)^9 (x - 1) \, dx = \frac{1}{40}(2x^2 - 4x)^{10} + C$
35. $\int \cos 4x \, dx = \frac{1}{4}\sin 4x + C$
36. $\int \sin x \cos x \, dx = \frac{1}{2}\sin^2 x + C$
37. $\int x \sin x^2 \, dx = -\frac{1}{2}\cos x^2 + C$
38. $\int \frac{\cos x}{\sin^3 x} \, dx = -\frac{1}{2\sin^2 x} + C$
39. $\int \ln x \, dx = x \ln x - x + C$
40. $\int x e^x \, dx = x e^x - e^x + C$

In Problems 41 and 42, perform the indicated operations.

41. $\frac{d}{dx}\int (x^2 - 4x + 5) \, dx$
42. $\int \frac{d}{dx}(x^2 - 4x + 5) \, dx$

In Problems 43–48, solve the given differential equation.

43. $\dfrac{dy}{dx} = 6x^2 + 9$

44. $\dfrac{dy}{dx} = 10x + 3\sqrt{x}$

45. $\dfrac{dy}{dx} = \dfrac{1}{x^2}$

46. $\dfrac{dy}{dx} = \dfrac{(2+x)^2}{x^5}$

47. $\dfrac{dy}{dx} = 1 - 2x + \sin x$

48. $\dfrac{dy}{dx} = \dfrac{1}{\cos^2 x}$

49. Find a function $y = f(x)$ whose graph passes through the point $(2, 3)$ that also satisfies the differential equation $dy/dx = 2x - 1$.

50. Find a function $y = f(x)$ so that $dy/dx = 1/\sqrt{x}$ and $f(9) = 1$.

51. If $f''(x) = 2x$, find $f'(x)$ and $f(x)$.

52. Find a function f such that $f''(x) = 6, f'(-1) = 2$, and $f(-1) = 0$.

53. Find a function f such that $f''(x) = 12x^2 + 2$ for which the slope of the tangent line to its graph at $(1, 1)$ is 3.

54. If $f^{(n)}(x) = 0$, what is f?

In Problems 55 and 56, the graph of a function f is shown in blue. Of the graphs of functions F, G, and H whose graphs are shown in black, green, and red, respectively, which function is the graph of an antiderivative of f? State your reasoning.

55.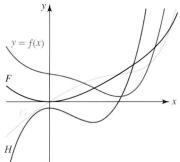

FIGURE 5.1.3 Graphs for Problem 55

56.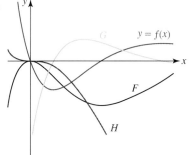

FIGURE 5.1.4 Graphs for Problem 56

≡ Applications

57. A bucket that contains liquid is rotating about a vertical axis at a constant angular velocity ω. The shape of the cross-section of the rotating liquid in the xy-plane is determined from

$$\dfrac{dy}{dx} = \dfrac{\omega^2}{g} x.$$

With coordinate axes as shown in FIGURE 5.1.5, find $y = f(x)$.

FIGURE 5.1.5 Bucket in Problem 57

58. The ends of a beam of length L rest on two supports as shown in FIGURE 5.1.6. With a uniform load on the beam, its shape (or elastic curve) is determined from

$$EIy'' = \dfrac{1}{2}qLx - \dfrac{1}{2}qx^2,$$

where E, I, and q are constants. Find $y = f(x)$ if $f(0) = 0$ and $f'(L/2) = 0$.

FIGURE 5.1.6 Beam in Problem 58

≡ Think About It

In Problems 59 and 60, determine f.

59. $\displaystyle\int f(x)\,dx = \ln|\ln x| + C$

60. $\displaystyle\int f(x)\,dx = x^2 e^x - 2xe^x + 2e^x + C$

61. Find a function f such that $f'(x) = x^2$ and $y = 4x + 7$ is a tangent line to the graph of f.

62. Simplify the expression $e^{4\int dx/x}$ as much as possible.

63. Determine which of the following two results is correct:

$$\int (x+1)^3\,dx = \dfrac{1}{4}(x+1)^4 + C$$

or

$$\int (x+1)^3\,dx = \dfrac{1}{4}x^4 + x^3 + \dfrac{3}{2}x^2 + x + C?$$

64. Given that $\dfrac{d}{dx}\sin \pi x = \pi \cos \pi x$. Find an antiderivative F of $\cos \pi x$ that has the property that $F(\tfrac{3}{2}) = 0$.

5.2 Integration by the *u*-Substitution

Introduction In the last section we discussed the fact that for each formula for the derivative of a function there is a corresponding antiderivative or indefinite integral formula. For example, by interpreting each of the functions

$$x^n \ (n \neq -1), \quad x^{-1}, \quad \text{and} \quad \cos x$$

as a derivative, we find the corresponding "reverse of the derivative" is a family of antiderivatives:

$$\int x^n \, dx = \frac{x^{n+1}}{n+1} + C \ (n \neq -1), \quad \int \frac{1}{x} \, dx = \ln|x| + C, \quad \int \cos x \, dx = \sin x + C. \quad (1)$$

Review Section 4.9 ▶

In the present exposition, we examine the "reverse of the Chain Rule." The concept of a **differential** of a function plays an important role in this discussion. Recall, if $u = g(x)$ is a differentiable function, then its differential is $du = g'(x) \, dx$.

We begin with an example.

Power of a Function If we wish to find a function F such that

$$\int (5x+1)^{1/2} \, dx = F(x) + C,$$

we must have

$$F'(x) = (5x+1)^{1/2}.$$

By reasoning "backward," we could argue that to obtain $(5x+1)^{1/2}$ we must have differentiated $(5x+1)^{3/2}$. It would then seem that we could proceed as in the first formula in (1)—namely, increase the power by 1 and divide by the new power:

$$\int (5x+1)^{1/2} \, dx = \frac{(5x+1)^{3/2}}{3/2} + C = \frac{2}{3}(5x+1)^{3/2} + C. \quad (2)$$

Regrettably the "answer" in (2) does not check, since the Chain Rule, in the form of the Power Rule for Functions, gives

$$\frac{d}{dx}\left[\frac{2}{3}(5x+1)^{3/2} + C\right] = \frac{2}{3} \cdot \frac{3}{2}(5x+1)^{1/2} \cdot 5 = 5(5x+1)^{1/2} \neq (5x+1)^{1/2}. \quad (3)$$

To account for the missing factor of 5 in (2) we use Theorem 5.1.2(*i*) and a little bit of cleverness:

$$\int (5x+1)^{1/2} \, dx = \int (5x+1)^{1/2} \boxed{\frac{1}{5} \cdot 5} \, dx \quad \leftarrow \frac{5}{5} = 1$$

$$= \frac{1}{5} \int \boxed{(5x+1)^{1/2} \, 5} \, dx \quad \leftarrow \text{derivative of } \frac{2}{3}(5x+1)^{3/2}$$

$$= \frac{1}{5} \cdot \frac{2}{3}(5x+1)^{3/2} + C \quad \leftarrow \text{from (3)}$$

$$= \frac{2}{15}(5x+1)^{3/2} + C.$$

You should now verify by differentiation that the last function is indeed an antiderivative of $(5x+1)^{1/2}$.

The key to evaluating indefinite integrals such as

$$\int (5x+1)^{1/2} \, dx, \quad \int \frac{x}{(4x^2+3)^6} \, dx, \quad \text{and} \quad \int \sin 10x \, dx \quad (4)$$

lies in the *recognition* that the integrands in (4),

$$(5x+1)^{1/2}, \quad \frac{x}{(4x^2+3)^6}, \quad \text{and} \quad \sin 10x$$

are the result of differentiating a composite function by the Chain Rule. In order to make this recognition it helps to make a substitution in an indefinite integral.

> **Theorem 5.2.1** *u*-Substitution Rule
>
> If $u = g(x)$ is a differentiable function whose range is an interval I, f is a function continuous on I, and F is an antiderivative of f on I, then
>
> $$\int f(g(x))g'(x)\,dx = \int f(u)\,du. \qquad (5)$$

PROOF By the Chain Rule,

$$\frac{d}{dx}F(g(x)) = F'(g(x))g'(x)$$

and so from the definition of an antiderivative we have,

$$\int F'(g(x))g'(x)\,dx = F(g(x)) + C.$$

Because F is an antiderivative of f, that is, if $F' = f$, the preceding line becomes

$$\int f(g(x))g'(x)\,dx = F(g(x)) + C = F(u) + C = \int F'(u)\,du = \int f(u)\,du. \qquad (6) \blacksquare$$

The interpretation of the result in (6) and its summary in (5) is subtle. In Section 5.1, the symbol dx was used simply as an indicator that we are integrating with respect to the variable x. In (6), we see that it is allowable to interpret dx and du as *differentials*.

Using the *u*-Substitution The basic idea is to be able to recognize an indefinite integral in a variable x (such as those given in (4)) that it is the reverse of the Chain Rule by converting it into a different indefinite integral in the variable u by means of the substitution $u = g(x)$. For convenience we list some guidelines for evaluating $\int f(g(x))g'(x)\,dx$ by carrying out a *u*-substitution.

Guidelines for Using a *u*-Substitution

(*i*) In the integral $\int f(g(x))g'(x)\,dx$ identify the functions $g(x)$ and $g'(x)\,dx$.
(*ii*) Express the integral *entirely* in terms of the symbol u by substituting u and du for $g(x)$ and $g'(x)\,dx$, respectively. In your substitution there should be no x-variables left in the integral.
(*iii*) Carry out the integration with respect to the variable u.
(*iv*) Finally, resubstitute $g(x)$ for the symbol u.

Indefinite Integral of a Power of a Function The derivative of a power of a function was an important special case of the Chain Rule. Recall, if $F(x) = x^{n+1}/(n+1)$, n a real number, $n \neq -1$, and if $u = g(x)$ is a differentiable function, then

$$F(g(x)) = \frac{[g(x)]^{n+1}}{n+1} \quad \text{and} \quad \frac{d}{dx}F(g(x)) = [g(x)]^n g'(x).$$

Hence, Theorem 5.2.1 immediately implies

$$\int [g(x)]^n g'(x)\,dx = \frac{[g(x)]^{n+1}}{n+1} + C. \qquad (7)$$

In terms of the substitutions

$$u = g(x) \quad \text{and} \quad du = g'(x)\,dx,$$

(7) can be summarized in the following manner:

$$\int u^n\,du = \frac{u^{n+1}}{n+1} + C, \quad n \neq -1. \qquad (8)$$

In the next example we evaluate the second of the three indefinite integrals in (4).

EXAMPLE 1 Using (8)

Evaluate $\int \dfrac{x}{(4x^2 + 3)^6} \, dx$.

Solution Let us rewrite the integral as

$$\int (4x^2 + 3)^{-6} x \, dx$$

and make the identifications

$$u = 4x^2 + 3 \quad \text{and} \quad du = 8x \, dx.$$

Now, to get the precise form $\int u^{-6} \, du$ we must adjust the integrand by multiplying and dividing by 8:

$$\int (4x^2 + 3)^{-6} x \, dx = \dfrac{1}{8} \int \overbrace{(4x^2 + 3)^{-6}}^{u^{-6}} \overbrace{(8x \, dx)}^{du} \quad \leftarrow \text{substitution}$$

$$= \dfrac{1}{8} \int u^{-6} \, du \quad \leftarrow \text{now use (8)}$$

$$= \dfrac{1}{8} \cdot \dfrac{u^{-5}}{-5} + C$$

$$= -\dfrac{1}{40}(4x^2 + 3)^{-5} + C. \quad \leftarrow \text{resubstitution}$$

Check by Differentiation: By the Power Rule for Functions,

$$\dfrac{d}{dx}\left[-\dfrac{1}{40}(4x^2 + 3)^{-5} + C \right] = \left(-\dfrac{1}{40}\right)(-5)(4x^2 + 3)^{-6}(8x) = \dfrac{x}{(4x^2 + 3)^6}. \quad \blacksquare$$

EXAMPLE 2 Using (8)

Evaluate $\int (2x - 5)^{11} \, dx$.

Solution If $u = 2x - 5$, then $du = 2 \, dx$. We adjust the integral by multiplying and dividing by 2 to get the correct form of the differential du:

$$\int (2x - 5)^{11} \, dx = \dfrac{1}{2} \int \overbrace{(2x - 5)^{11}}^{u^{11}} \overbrace{(2 \, dx)}^{du} \quad \leftarrow \text{substitution}$$

$$= \dfrac{1}{2} \int u^{11} \, du \quad \leftarrow \text{now use (8)}$$

$$= \dfrac{1}{2} \cdot \dfrac{u^{12}}{12} + C$$

$$= \dfrac{1}{24}(2x - 5)^{12} + C. \quad \leftarrow \text{resubstitution} \quad \blacksquare$$

In Examples 1 and 2 we "fixed up" or adjusted the integrand by multiplying and dividing by a constant in order to obtain the appropriate du. This procedure works fine if you immediately recognize $g(x)$ in $\int f(g(x))g'(x) \, dx$ and that $g'(x) \, dx$ is simply missing an appropriate constant multiple. The next example illustrates a slightly different technique.

EXAMPLE 3 Using (8)

Evaluate $\int \cos^4 x \sin x \, dx$.

Solution For emphasis we rewrite the integrand as $\int (\cos x)^4 \sin x \, dx$. With the identification $u = \cos x$ we get $du = -\sin x \, dx$. Solving for the product $\sin x \, dx$ from the last differential we get $\sin x \, dx = -du$. Then

$$\int (\cos x)^4 \sin x \, dx = \int \overbrace{(\cos x)^4}^{u^4} \overbrace{(\sin x \, dx)}^{-du} \quad \leftarrow \text{substitution}$$

$$= -\int u^4 \, du \quad \leftarrow \text{now use (8)}$$

$$= -\frac{u^5}{5} + C$$

$$= -\frac{1}{5}\cos^5 x + C. \quad \leftarrow \text{resubstitution}$$

You are again encouraged to differentiate the last result.

In the remaining examples in this section we will alternate between the methods employed in Examples 1 and 3.

On a practical level it not always obvious that we are dealing with an integral of the form $\int [g(x)]^n g'(x) \, dx$. As you work more and more problems you will see that integrals are not always what they seem to be on first inspection. For example, you should convince yourself using u-substitutions that the integral $\int \cos^2 x \, dx$ is *not* of the form $\int [g(x)]^n g'(x) \, dx$. In a more general sense, is it not always obvious in $\int f(g(x))g'(x) \, dx$ what functions should be chosen as u and du.

▌ Indefinite Integrals of Trigonometric Functions   If $u = g(x)$ is a differentiable function, then the differentiation formulas

$$\frac{d}{dx}\sin u = \cos u \frac{du}{dx} \quad \text{and} \quad \frac{d}{dx}(-\cos u) = \sin u \frac{du}{dx}$$

yield, in turn, the integration formulas

$$\int \cos u \frac{du}{dx} dx = \sin u + C \tag{9}$$

and

$$\int \sin u \frac{du}{dx} dx = -\cos u + C. \tag{10}$$

Since $du = g'(x) \, dx = \dfrac{du}{dx} dx$, (9) and (10) are, respectively, equivalent to

$$\int \cos u \, du = \sin u + C, \tag{11}$$

$$\int \sin u \, du = -\cos u + C. \tag{12}$$

EXAMPLE 4   Using (11)

Evaluate $\int \cos 2x \, dx$.

Solution   If $u = 2x$, then $du = 2 \, dx$ and $dx = \dfrac{1}{2} du$. Accordingly, we write

$$\int \cos 2x \, dx = \int \cos \overbrace{2x}^{u} \overbrace{(dx)}^{\frac{1}{2}du} \quad \leftarrow \text{substitution}$$

$$= \frac{1}{2}\int \cos u \, du \quad \leftarrow \text{now use (11)}$$

$$= \frac{1}{2}\sin u + C$$

$$= \frac{1}{2}\sin 2x + C. \quad \leftarrow \text{resubstitution}$$

Integration formulas (8), (11), and (12) are the Chain Rule analogues of integration formulas 2, 4, and 5 in Table 5.1.1. In Table 5.2.1 that follows we summarize the Chain Rule analogues of the sixteen integration formulas in Table 5.1.1.

TABLE 5.2.1

Integration Formulas

1. $\int du = u + C$
2. $\int u^n \, du = \dfrac{u^{n+1}}{n+1} + C \quad (n \neq -1)$
3. $\int \dfrac{1}{u} \, du = \ln|u| + C$
4. $\int \cos u \, du = \sin u + C$
5. $\int \sin u \, du = -\cos u + C$
6. $\int \sec^2 u \, du = \tan u + C$
7. $\int \csc^2 u \, du = -\cot u + C$
8. $\int \sec u \tan u \, du = \sec u + C$
9. $\int \csc u \cot u \, du = -\csc u + C$
10. $\int \dfrac{1}{\sqrt{1-u^2}} \, du = \sin^{-1} u + C$
11. $\int \dfrac{1}{1+u^2} \, du = \tan^{-1} u + C$
12. $\int \dfrac{1}{u\sqrt{u^2-1}} \, du = \sec^{-1}|u| + C$
13. $\int b^u \, du = \dfrac{b^u}{\ln b} + C$
14. $\int e^u \, du = e^u + C$
15. $\int \cosh u \, du = \sinh u + C$
16. $\int \sinh u \, du = \cosh u + C$

In other textbooks, formulas such as 3, 10, 11, and 12 in Table 5.2.1 are frequently written with the differential du as the numerator:

$$\int \frac{du}{u}, \quad \int \frac{du}{\sqrt{1-u^2}}, \quad \int \frac{du}{1+u^2}, \quad \int \frac{du}{u\sqrt{u^2-1}}.$$

But since we have found, over the years, that these latter formulas are often subject to misunderstanding in a classroom environment, we prefer the forms given in the table.

EXAMPLE 5 Using Table 5.2.1

Evaluate $\int \sec^2(1 - 4x) \, dx$.

Solution We recognize that the indefinite integral has the form of the integration formula 6 in Table 5.2.1. If $u = 1 - 4x$, then $du = -4 \, dx$. Adjusting the integrand to obtain the correct form of the differential requires multiplying and dividing by -4:

$$\int \sec^2(1 - 4x) \, dx = -\frac{1}{4} \int \sec^2(\overbrace{1 - 4x}^{u})(\overbrace{-4 \, dx}^{du})$$

$$= -\frac{1}{4} \int \sec^2 u \, du \quad \leftarrow \text{formula 6 in Table 5.2.1}$$

$$= -\frac{1}{4} \tan u + C$$

$$= -\frac{1}{4} \tan(1 - 4x) + C.$$

EXAMPLE 6 Using Table 5.2.1

Evaluate $\int \dfrac{x^2}{x^3 + 5}\, dx$.

Solution If $u = x^3 + 5$, then $du = 3x^2\, dx$ and $x^2\, dx = \dfrac{1}{3} du$. Hence,

$$\int \dfrac{x^2}{x^3 + 5}\, dx = \int \dfrac{1}{x^3 + 5}\, (x^2\, dx)$$

$$= \dfrac{1}{3} \int \dfrac{1}{u}\, du$$

$$= \dfrac{1}{3} \ln|u| + C \quad \leftarrow \text{formula 3 in Table 5.2.1}$$

$$= \dfrac{1}{3} \ln|x^3 + 5| + C. \qquad \blacksquare$$

EXAMPLE 7 Rewriting and Using Table 5.2.1

Evaluate $\int \dfrac{1}{1 + e^{-2x}}\, dx$.

Solution The given integral does not look like any of the integration formulas in Table 5.2.1. However, if we multiply the numerator and denominator by e^{2x}, then we have

$$\int \dfrac{1}{1 + e^{-2x}}\, dx = \int \dfrac{e^{2x}}{e^{2x} + 1}\, dx.$$

If $u = e^{2x} + 1$, then $du = 2e^{2x}\, dx$ and so from formula 3 of Table 5.2.1,

$$\int \dfrac{1}{1 + e^{-2x}}\, dx = \dfrac{1}{2} \int \dfrac{1}{e^{2x} + 1}\, (2e^{2x}\, dx)$$

$$= \dfrac{1}{2} \int \dfrac{1}{u}\, du$$

$$= \dfrac{1}{2} \ln|u| + C$$

$$= \dfrac{1}{2} \ln(e^{2x} + 1) + C.$$

Note that the absolute value symbol can be dropped because $e^{2x} + 1 > 0$ for all values of x. $\blacksquare$

EXAMPLE 8 Using Table 5.2.1

Evaluate $\int e^{5x}\, dx$.

Solution Let $u = 5x$ so that $du = 5\, dx$. Then

$$\int e^{5x}\, dx = \dfrac{1}{5} \int e^{5x}(5\, dx)$$

$$= \dfrac{1}{5} \int e^{u}\, du \quad \leftarrow \text{formula 14 in Table 5.2.1}$$

$$= \dfrac{1}{5} e^{u} + C$$

$$= \dfrac{1}{5} e^{5x} + C. \qquad \blacksquare$$

EXAMPLE 9 Using Table 5.2.1

Evaluate $\int \dfrac{e^{4/x}}{x^2}\, dx$.

Solution If we let $u = 4/x$, then $du = (-4/x^2)\, dx$ and $(1/x^2)\, dx = -\dfrac{1}{4} du$.

Again from formula 14 of Table 5.2.1 we see that

$$\int \frac{e^{4/x}}{x^2} dx = \int e^{4/x} \left(\frac{1}{x^2} dx \right)$$
$$= \int e^u \left(-\frac{1}{4} du \right)$$
$$= -\frac{1}{4} \int e^u \, du$$
$$= -\frac{1}{4} e^u + C$$
$$= -\frac{1}{4} e^{4/x} + C.$$

EXAMPLE 10 Using Table 5.2.1

Evaluate $\int \frac{(\tan^{-1} x)^2}{1 + x^2} dx.$

Solution Like Example 7, the given integral at first glance does not resemble any of the formulas in Table 5.2.1. But if we try the u-substitution with $u = \tan^{-1} x$ and $du = \frac{1}{1 + x^2} dx$, then

$$\int \frac{(\tan^{-1} x)^2}{1 + x^2} dx = \int \overbrace{(\tan^{-1} x)^2}^{u} \overbrace{\frac{1}{1 + x^2} dx}^{du}$$
$$= \int u^2 \, du \quad \leftarrow \text{formula 2 in Table 5.2.1}$$
$$= \frac{u^3}{3} + C$$
$$= \frac{1}{3} (\tan^{-1} x)^3 + C.$$

EXAMPLE 11 Using Table 5.2.1

Evaluate $\int \frac{1}{\sqrt{100 - x^2}} dx.$

Solution By factoring 100 from the radical and identifying $u = \frac{1}{10} x$ and $du = \frac{1}{10} dx$, the result is obtained from formula 10 of Table 5.2.1:

$$\int \frac{1}{\sqrt{100 - x^2}} dx = \int \frac{1}{\sqrt{1 - \left(\frac{x}{10}\right)^2}} \left(\frac{1}{10} dx \right)$$
$$= \int \frac{1}{\sqrt{1 - u^2}} du$$
$$= \sin^{-1} u + C$$
$$= \sin^{-1} \frac{x}{10} + C.$$

■ **Three Alternative Formulas** As a matter of convenience, integration formulas 10, 11, and 12 in Table 5.2.1 are extended in the following manner. For $a > 0$,

$$\int \frac{1}{\sqrt{a^2 - u^2}} du = \sin^{-1} \frac{u}{a} + C \qquad (13)$$

$$\int \frac{1}{a^2 + u^2} du = \frac{1}{a} \tan^{-1} \frac{u}{a} + C \qquad (14)$$

$$\int \frac{1}{u\sqrt{u^2 - a^2}} du = \frac{1}{a} \sec^{-1} \left| \frac{u}{a} \right| + C. \qquad (15)$$

For practice you are encouraged to verify these results by differentiation. Observe that the indefinite integral in Example 11 can be quickly evaluated by identifying $u = x$ and $a = 10$ in (13).

■ **Special Trigonometric Integrals** The integration formulas given next, which relate some trigonometric functions with the natural logarithm, occur often enough in practice to merit special attention:

$$\int \tan x \, dx = -\ln|\cos x| + C \qquad (16)$$

$$\int \cot x \, dx = \ln|\sin x| + C \qquad (17)$$

$$\int \sec x \, dx = \ln|\sec x + \tan x| + C \qquad (18)$$

$$\int \csc x \, dx = \ln|\csc x - \cot x| + C. \qquad (19)$$

◀ In tables of integral formulas you will often see (16) written as
$\int \tan x \, dx = \ln|\sec x| + C.$
By the properties of logarithms
$-\ln|\cos x| = \ln|\cos x|^{-1} = \ln|\sec x|.$

To obtain (16) we write

$$\int \tan x \, dx = \int \frac{\sin x}{\cos x} \, dx \qquad (20)$$

and identify $u = \cos x$, $du = -\sin x \, dx$ so that,

$$\int \tan x \, dx = \int \frac{\sin x}{\cos x} \, dx = -\int \frac{1}{\cos x}(-\sin x \, dx)$$

$$= -\int \frac{1}{u} \, du$$

$$= -\ln|u| + C$$

$$= -\ln|\cos x| + C.$$

To obtain (18) we write

$$\int \sec x \, dx = \int \sec x \, \frac{\sec x + \tan x}{\sec x + \tan x} \, dx$$

$$= \int \frac{\sec^2 x + \sec x \tan x}{\sec x + \tan x} \, dx.$$

If we let $u = \sec x + \tan x$, then $du = (\sec x \tan x + \sec^2 x) \, dx$ and so,

$$\int \sec x \, dx = \int \frac{1}{\sec x + \tan x} (\sec^2 x + \sec x \tan x) \, dx$$

$$= \int \frac{1}{u} \, du$$

$$= \ln|u| + C$$

$$= \ln|\sec x + \tan x| + C.$$

Also, each of the formulas (16)–(19) can be written in a general form:

$$\int \tan u \, dx = -\ln|\cos u| + C \qquad (21)$$

$$\int \cot u \, du = \ln|\sin u| + C \qquad (22)$$

$$\int \sec u \, dx = \ln|\sec u + \tan u| + C \qquad (23)$$

and

$$\int \csc u \, du = \ln|\csc u - \cot u| + C. \qquad (24)$$

Useful Identities When working with trigonometric functions it is often necessary to use a trigonometric identity to solve a problem. The half-angle formulas for the cosine and sine in the form

$$\cos^2 x = \frac{1}{2}(1 + \cos 2x) \quad \text{and} \quad \sin^2 x = \frac{1}{2}(1 - \cos 2x) \qquad (25)$$

are particularly useful in problems that require antiderivatives of $\cos^2 x$ and $\sin^2 x$.

EXAMPLE 12 Using a Half-Angle Formula

Evaluate $\int \cos^2 x \, dx$.

Solution It should be verified that the integral is *not* of the form $\int u^2 \, du$. Now using the half-angle formula $\cos^2 x = \frac{1}{2}(1 + \cos 2x)$, we obtain

$$\int \cos^2 x \, dx = \int \frac{1}{2}(1 + \cos 2x) \, dx$$

$$= \frac{1}{2}\left[\int dx + \frac{1}{2}\int \cos 2x (2 \, dx)\right] \quad \leftarrow \text{see Example 4}$$

$$= \frac{1}{2}\left[x + \frac{1}{2}\sin 2x\right] + C$$

$$= \frac{1}{2}x + \frac{1}{4}\sin 2x + C. \qquad \blacksquare$$

Of course, the method illustrated in Example 12 works equally well in finding antiderivatives such as $\int \cos^2 5x \, dx$ and $\int \sin^2 \frac{1}{2}x \, dx$. With x replaced by $5x$ and then x replaced by $\frac{1}{2}x$, the formulas in (25) enable us to write, respectively,

$$\int \cos^2 5x \, dx = \int \frac{1}{2}(1 + \cos 10x) \, dx = \frac{1}{2}x + \frac{1}{20}\sin 10x + C$$

$$\int \sin^2 \frac{1}{2}x \, dx = \int \frac{1}{2}(1 - \cos x) \, dx = \frac{1}{2}x - \frac{1}{2}\sin x + C.$$

We will consider antiderivatives of more complicated powers of trigonometric functions in Section 7.4.

∫ NOTES FROM THE CLASSROOM

The following example illustrates a common, but *totally incorrect*, procedure for evaluating an indefinite integral. Because $2x/2x = 1$,

$$\int (4 + x^2)^{1/2} \, dx = \int (4 + x^2)^{1/2} \frac{2x}{2x} \, dx$$

$$= \frac{1}{2x}\int (4 + x^2)^{1/2} 2x \, dx$$

$$= \frac{1}{2x}\int u^{1/2} \, du$$

$$= \frac{1}{2x} \cdot \frac{2}{3}(4 + x^2)^{3/2} + C.$$

You should verify that differentiation of the latter function does *not* yield $(4 + x^2)^{1/2}$. The mistake is in the first line of the "solution." Variables, in this case $2x$, cannot be brought outside an integral symbol. If $u = x^2 + 4$, then the integrand lacks the function $du = 2x \, dx$; in fact, there is no way of adjusting the problem to fit the form given in (8). With the "tools" we presently have on hand, the integral $\int (4 + x^2)^{1/2} \, dx$ simply cannot be evaluated.

Exercises 5.2

≡ Fundamentals

In Problems 1–50, evaluate the given indefinite integral using an appropriate u-substitution.

1. $\displaystyle\int \sqrt{1-4x}\,dx$
2. $\displaystyle\int (8x+2)^{1/3}\,dx$
3. $\displaystyle\int \frac{1}{(5x+1)^3}\,dx$
4. $\displaystyle\int (7-x)^{49}\,dx$
5. $\displaystyle\int x\sqrt{x^2+4}\,dx$
6. $\displaystyle\int \frac{t}{\sqrt[3]{t^2+9}}\,dt$
7. $\displaystyle\int \sin^5 3x \cos 3x\,dx$
8. $\displaystyle\int \sin 2\theta \cos^4 2\theta\,d\theta$
9. $\displaystyle\int \tan^2 2x \sec^2 2x\,dx$
10. $\displaystyle\int \sqrt{\tan x}\sec^2 x\,dx$
11. $\displaystyle\int \sin 4x\,dx$
12. $\displaystyle\int 5\cos\frac{x}{2}\,dx$
13. $\displaystyle\int (\sqrt{2t}-\cos 6t)\,dt$
14. $\displaystyle\int \sin(2-3x)\,dx$
15. $\displaystyle\int x\sin x^2\,dx$
16. $\displaystyle\int \frac{\cos(1/x)}{x^2}\,dx$
17. $\displaystyle\int x^2 \sec^2 x^3\,dx$
18. $\displaystyle\int \csc^2(0.1x)\,dx$
19. $\displaystyle\int \frac{\csc\sqrt{x}\cot\sqrt{x}}{\sqrt{x}}\,dx$
20. $\displaystyle\int \tan 5v \sec 5v\,dv$
21. $\displaystyle\int \frac{1}{7x+3}\,dx$
22. $\displaystyle\int (5x+6)^{-1}\,dx$
23. $\displaystyle\int \frac{x}{x^2+1}\,dx$
24. $\displaystyle\int \frac{x^2}{5x^3+8}\,dx$
25. $\displaystyle\int \frac{x}{x+1}\,dx$
26. $\displaystyle\int \frac{(x+3)^2}{x+2}\,dx$
27. $\displaystyle\int \frac{1}{x\ln x}\,dx$
28. $\displaystyle\int \frac{1-\sin\theta}{\theta+\cos\theta}\,d\theta$
29. $\displaystyle\int \frac{\sin(\ln x)}{x}\,dx$
30. $\displaystyle\int \frac{1}{x(\ln x)^2}\,dx$
31. $\displaystyle\int e^{10x}\,dx$
32. $\displaystyle\int \frac{1}{e^{4x}}\,dx$
33. $\displaystyle\int x^2 e^{-2x^3}\,dx$
34. $\displaystyle\int \frac{e^{1/x^3}}{x^4}\,dx$
35. $\displaystyle\int \frac{e^{-\sqrt{x}}}{\sqrt{x}}\,dx$
36. $\displaystyle\int \sqrt{e^x}\,dx$
37. $\displaystyle\int \frac{e^x - e^{-x}}{e^x + e^{-x}}\,dx$
38. $\displaystyle\int e^{3x}\sqrt{1+2e^{3x}}\,dx$
39. $\displaystyle\int \frac{1}{\sqrt{5-x^2}}\,dx$
40. $\displaystyle\int \frac{1}{\sqrt{9-16x^2}}\,dx$
41. $\displaystyle\int \frac{1}{1+25x^2}\,dx$
42. $\displaystyle\int \frac{1}{2+9x^2}\,dx$
43. $\displaystyle\int \frac{e^x}{1+e^{2x}}\,dx$
44. $\displaystyle\int \frac{\theta}{\sqrt{1-\theta^4}}\,d\theta$
45. $\displaystyle\int \frac{2x-3}{\sqrt{1-x^2}}\,dx$
46. $\displaystyle\int \frac{x-8}{x^2+2}\,dx$
47. $\displaystyle\int \frac{\tan^{-1}x}{1+x^2}\,dx$
48. $\displaystyle\int \sqrt{\frac{\sin^{-1}x}{1-x^2}}\,dx$
49. $\displaystyle\int \tan 5x\,dx$
50. $\displaystyle\int e^x \cot e^x\,dx$

In Problems 51–56, use the identities in (25) to evaluate the given indefinite integral.

51. $\displaystyle\int \sin^2 x\,dx$
52. $\displaystyle\int \cos^2 \pi x\,dx$
53. $\displaystyle\int \cos^2 4x\,dx$
54. $\displaystyle\int \sin^2 \frac{3}{2}x\,dx$
55. $\displaystyle\int (3-2\sin x)^2\,dx$
56. $\displaystyle\int (1+\cos 2x)^2\,dx$

In Problems 57 and 58, solve the given differential equation.

57. $\dfrac{dy}{dx} = \sqrt[3]{1-x}$
58. $\dfrac{dy}{dx} = \dfrac{(1-\tan x)^5}{\cos^2 x}$

59. Find a function $y = f(x)$ whose graph passes through the point $(\pi, -1)$ that also satisfies $dy/dx = 1 - 6\sin 3x$.

60. Find a function f such that $f''(x) = (1+2x)^5$, $f(0) = 0$, and $f'(0) = 0$.

61. Show that:
    (a) $\displaystyle\int \sin x \cos x\,dx = \frac{1}{2}\sin^2 x + C_1$
    (b) $\displaystyle\int \sin x \cos x\,dx = -\frac{1}{2}\cos^2 x + C_2$
    (c) $\displaystyle\int \sin x \cos x\,dx = -\frac{1}{4}\cos 2x + C_3$.

62. In Problem 61:
    (a) Verify that the derivative of each answer in parts (a), (b), and (c) is $\sin x \cos x$.
    (b) By a trigonometric identity, show how the result in part (b) can be obtained from the answer in part (a).
    (c) By adding the results in parts (a) and (b), obtain the result in part (c).

≡ Applications

63. Consider the plane pendulum, shown in FIGURE 5.2.1, that swings between points A and C. If B is midway between A and C, it can be shown that
$$\frac{dt}{ds} = \sqrt{\frac{L}{g(s_C^2 - s^2)}},$$
where g is the acceleration due to gravity.

(a) If $t(0) = 0$, then show that the time the pendulum takes to travel between points B and P is
$$t(s) = \sqrt{\frac{L}{g}} \sin^{-1}\left(\frac{s}{s_C}\right).$$
(b) Use the result in part (a) to determine the time of travel from B to C.
(c) Use (b) to determine the period T of the pendulum—that is, the time of an oscillation from A to C and back to A.

FIGURE 5.2.1 Pendulum in Problem 63

≡ Think About It

64. Find a function $y = f(x)$ for which $f(\pi/2) = 0$ and $\dfrac{dy}{dx} = \cos^3 x$. [Hint: $\cos^3 x = \cos^2 x \cos x$.]

In Problems 65 and 66, use the identities in (25) to evaluate the given indefinite integral.

65. $\displaystyle\int \cos^4 x \, dx$

66. $\displaystyle\int \sin^4 x \, dx$

In Problems 67 and 68, evaluate the given indefinite integral.

67. $\displaystyle\int \frac{1}{x\sqrt{x^4 - 16}} \, dx$

68. $\displaystyle\int \frac{e^{2x}}{e^x + 1} \, dx$

In Problems 69 and 70, evaluate the given indefinite integral.

69. $\displaystyle\int \frac{1}{1 - \cos x} \, dx$

70. $\displaystyle\int \frac{1}{1 + \sin 2x} \, dx$

In Problems 71–74, evaluate the given indefinite integral. Assume f is a differentiable function.

71. $\displaystyle\int f'(8x) \, dx$

72. $\displaystyle\int xf'(5x^2) \, dx$

73. $\displaystyle\int \sqrt{f(2x)} f'(2x) \, dx$

74. $\displaystyle\int \frac{f'(3x + 1)}{f(3x + 1)} \, dx$

75. Evaluate $\displaystyle\int f''(4x) \, dx$ if $f(x) = \sqrt{x^4 + 1}$.

76. Evaluate $\displaystyle\int \left\{\int \sec^2 3x \, dx\right\} dx$.

5.3 The Area Problem

Introduction As the derivative is motivated by the geometric problem of constructing a tangent to a curve, the historical problem leading to the definition of a definite integral is the problem of finding area. Specifically, we are interested in the following version of this problem:

- Find the area A of a region bounded by the x-axis and the graph of a continuous nonnegative function $y = f(x)$ defined on an interval $[a, b]$.

We shall call the area of this region the **area under the graph** of f on the interval $[a, b]$. The requirement that f be nonnegative on $[a, b]$ means that no portion of its graph on the interval is below the x-axis. See FIGURE 5.3.1.

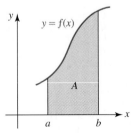

FIGURE 5.3.1 Area under the graph of f on $[a, b]$

Before pursuing the solution of the area problem we need to digress briefly to discuss a helpful notation for a sum of numbers such as

$$1 + 2 + 3 + \cdots + n \quad \text{and} \quad 1^2 + 2^2 + 3^2 + \cdots + n^2.$$

Sigma Notation Let a_k be a real number that depends on an integer k. We denote the sum $a_1 + a_2 + a_3 + \cdots + a_n$ by the symbol $\sum_{k=1}^{n} a_k$; that is,

$$\sum_{k=1}^{n} a_k = a_1 + a_2 + a_3 + \cdots + a_n. \tag{1}$$

Since Σ is the capital Greek letter *sigma*, (1) is called **sigma notation** or **summation notation**. The variable k is called the **index of summation**. Thus,

$$\text{symbol } \Sigma \text{ indicates addition of } a_k \rightarrow \sum_{k=1}^{n} a_k \quad \begin{array}{c} \text{end with this value of } k \\ \downarrow \\ \uparrow \\ \text{start with indicated value of } k \end{array}$$

is the sum of all numbers of the form a_k as k takes on the successive values $k = 1, k = 2, \ldots,$ end and concludes with $k = n$.

EXAMPLE 1 Using Sigma Notation

The sum of the first ten positive even integers

$$2 + 4 + 6 + \cdots + 18 + 20$$

can be written succinctly as $\sum_{k=1}^{10} 2k$. The sum of the first ten positive odd integers

$$1 + 3 + 5 + \cdots + 17 + 19$$

can be written $\sum_{k=1}^{10}(2k - 1)$. ∎

The index of summation need not start at the value $k = 1$; for example,

$$\sum_{k=3}^{5} 2^k = 2^3 + 2^4 + 2^5 \quad \text{and} \quad \sum_{k=0}^{5} 2^k = 2^0 + 2^1 + 2^2 + 2^3 + 2^4 + 2^5.$$

Note that the sum of the first ten odd positive integers in Example 1 can also be written as $\sum_{k=0}^{9}(2k + 1)$. However, in a general discussion we shall always assume that the summation index starts at $k = 1$. This assumption is for convenience rather than necessity. The index of summation is often called a **dummy variable**, since the symbol itself is not important; it is the successive integer values of the index and the corresponding sum that are important. In general,

$$\sum_{k=1}^{n} a_k = \sum_{i=1}^{n} a_i = \sum_{j=1}^{n} a_j = \sum_{m=1}^{n} a_m.$$

For example,

$$\sum_{k=1}^{10} 4^k = \sum_{i=1}^{10} 4^i = \sum_{j=1}^{10} 4^j = 4^1 + 4^2 + 4^3 + \cdots + 4^{10}.$$

Properties The following is a list of some important properties of the sigma notation.

Theorem 5.3.1 Properties of Sigma Notation

For positive integers m and n,

(i) $\sum_{k=1}^{n} ca_k = c \sum_{k=1}^{n} a_k$, where c is any constant

(ii) $\sum_{k=1}^{n} (a_k \pm b_k) = \sum_{k=1}^{n} a_k \pm \sum_{k=1}^{n} b_k$

(iii) $\sum_{k=1}^{n} a_k = \sum_{k=1}^{m} a_k + \sum_{k=m+1}^{n} a_k, \; m < n.$

The proof of formula (i) is an immediate consequence of the distributive law. Of course, (ii) of Theorem 5.3.1 holds for the sum of three of more terms; for example

$$\sum_{k=1}^{n} (a_k + b_k + c_k) = \sum_{k=1}^{n} a_k + \sum_{k=1}^{n} b_k + \sum_{k=1}^{n} c_k.$$

Special Summation Formulas For special kinds of indicated sums, particularly, sums involving positive integer powers of the summation index (such as sums of successive positive integers, successive squares, successive cubes, and so on) it is possible to find a formula that gives the actual numerical value of the sum. For purposes of this section we shall confine our attention to the following four formulas.

Theorem 5.3.2 Summation Formulas

For n a positive integer and c any constant,

(i) $\sum_{k=1}^{n} c = nc$ (ii) $\sum_{k=1}^{n} k = \dfrac{n(n+1)}{2}$

(iii) $\sum_{k=1}^{n} k^2 = \dfrac{n(n+1)(2n+1)}{6}$ (iv) $\sum_{k=1}^{n} k^3 = \dfrac{n^2(n+1)^2}{4}.$

Formulas (i) and (ii) can be easily justified. If c is a constant—that is, independent of the summation index k—then $\sum_{k=1}^{n} c$ means $c + c + c + \cdots + c$. Since there are n c's, we have $\sum_{k=1}^{n} c = n \cdot c$, which is (i) of Theorem 5.3.2. Now, the sum of the first n positive integers can be written as $\sum_{k=1}^{n} k$. If this sum is denoted by the letter S, then

$$S = 1 + 2 + 3 + \cdots + (n-2) + (n-1) + n. \qquad (2)$$

Equivalently, $\quad S = n + (n-1) + (n-2) + \cdots + 3 + 2 + 1. \qquad (3)$

If we add (2) and (3) by adding corresponding first terms, second terms, and so on, then

$$2S = \underbrace{(n+1) + (n+1) + (n+1) + \cdots + (n+1)}_{n \text{ terms of } n+1} = n(n+1).$$

Solving for S gives $S = n(n+1)/2$, which is (ii). You should be able to derive formulas (iii) and (iv) with the hints supplied in Problems 55 and 56 in Exercises 5.3.

EXAMPLE 2 Using Summation Formulas

Find the numerical value of $\sum_{k=1}^{20} (k+5)^2$.

Solution By expanding $(k + 5)^2$ and using (*i*) and (*ii*) of Theorem 5.3.1, we can write

$$\sum_{k=1}^{20}(k+5)^2 = \sum_{k=1}^{20}(k^2 + 10k + 25) \quad \leftarrow \text{squaring the binomial}$$

$$= \sum_{k=1}^{20} k^2 + 10\sum_{k=1}^{20} k + \sum_{k=1}^{20} 25. \quad \leftarrow (i) \text{ and } (ii) \text{ of Theorem 5.3.1}$$

With the identification $n = 20$, it follows from summation formulas (*iii*), (*ii*), and (*i*) of Theorem 5.3.2, respectively, that

$$\sum_{k=1}^{20}(k+5)^2 = \frac{20(21)(41)}{6} + 10\frac{20(21)}{2} + 20 \cdot 25 = 5470. \quad \blacksquare$$

Sigma notation and the foregoing summation formulas will be put to immediate use in the next discussion.

Area of a Triangle Assume for the moment that we do not know a formula for calculating the area A of the right triangle given in FIGURE 5.3.2(a). By superimposing a rectangular coordinate system on the triangle, as shown in Figure 5.3.2(b), we see that the problem is the same as finding the area in the first quadrant bounded by the straight lines $y = (h/b)x$, $y = 0$ (the *x*-axis), and $x = b$. In other words, we wish to find the area under the graph of $y = (h/b)x$ on the interval $[0, b]$.

(a) Right triangle

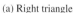

Using rectangles, FIGURE 5.3.3 indicates three different ways of *approximating* the area A. For convenience, let us pursue the procedure hinted at in Figure 5.3.3(b) in greater detail. We begin by dividing the interval $[0, b]$ into n subintervals of equal width $\Delta x = b/n$. If the right endpoint of each of these intervals is denoted by x_k^*, then

$$x_1^* = \Delta x = \frac{b}{n}$$

$$x_2^* = 2\Delta x = 2\left(\frac{b}{n}\right)$$

$$x_3^* = 3\Delta x = 3\left(\frac{b}{n}\right)$$

$$\vdots$$

$$x_n^* = n\Delta x = n\left(\frac{b}{n}\right) = b.$$

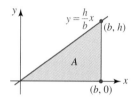

(b) Right triangle in a coordinate system

FIGURE 5.3.2 Find the area A of a right triangle

(a)

(b)

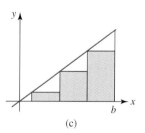
(c)

FIGURE 5.3.3 Approximating the area A using three rectangles

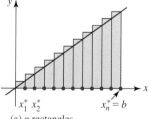

(a) n rectangles

As shown in FIGURE 5.3.4(a), we now construct a rectangle of length $f(x_k^*)$ and width Δx on each of the n subintervals. Since the area of a rectangle is *length* $\times$ *width*, the area of each rectangle is $f(x_k^*)\Delta x$. See Figure 5.3.4(b). The sum of the areas of the n rectangles is an approximation to the number A. We write

$$A \approx f(x_1^*)\Delta x + f(x_2^*)\Delta x + \cdots + f(x_n^*)\Delta x,$$

or in sigma notation,

$$A \approx \sum_{k=1}^{n} f(x_k^*)\Delta x. \qquad (4)$$

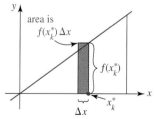

(b) Area of a general rectangle

FIGURE 5.3.4 Area A of the triangle is approximated by the sum of the areas of n rectangles

It seems plausible that we can reduce the error introduced by this method of approximation (the area of each rectangle is larger than the area under the graph on a subinterval $[x_{k-1}, x_k]$) by dividing the interval $[0, b]$ into finer subdivisions. In other words, we expect that a better approximation to A can be obtained by using more and more rectangles ($n \to \infty$) of decreasing widths ($\Delta x \to 0$). Now,

$$f(x) = \frac{h}{b}x, \quad x_k^* = k\left(\frac{b}{n}\right), \quad f(x_k^*) = \frac{h}{n} \cdot k, \quad \text{and} \quad \Delta x = \frac{b}{n},$$

so that with the aid of summation formula (ii) of Theorem 5.3.2, (4) becomes

$$A \approx \sum_{k=1}^{n} \left(\frac{h}{n} \cdot k\right)\frac{b}{n} = \frac{bh}{n^2}\sum_{k=1}^{n} k = \frac{bh}{n^2} \cdot \frac{n(n+1)}{2} = \frac{bh}{2}\left(1 + \frac{1}{n}\right). \quad (5)$$

Finally, by letting $n \to \infty$ on the right-hand side of (5), we obtain the familiar formula for the area of a triangle:

$$A = \frac{1}{2}bh \cdot \lim_{n \to \infty}\left(1 + \frac{1}{n}\right) = \frac{1}{2}bh.$$

The General Problem Now, let us turn from the preceding specific example to the general problem of finding the area A under the graph of a function $y = f(x)$ that is continuous on an interval $[a, b]$. As shown in FIGURE 5.3.5(a), we shall also assume that $f(x) \geq 0$ for all x in the interval $[a, b]$. As suggested in Figure 5.3.5(b), the area A can be approximated by adding the areas of n rectangles that are constructed on the interval. One possible procedure for determining A is summarized as follows:

- Divide the interval $[a, b]$ into n subintervals $[x_{k-1}, x_k]$, where

$$a = x_0 < x_1 < x_2 < \cdots < x_{n-1} < x_n = b,$$

so that each subinterval has the same width $\Delta x = (b - a)/n$. This collection of numbers is called a **regular partition** of the interval $[a, b]$.

- Choose a number x_k^* in each of the n subintervals $[x_{k-1}, x_k]$ and form the n products $f(x_k^*)\Delta x$. Since the area of a rectangle is length $\times$ width, $f(x_k^*)\Delta x$ is the area of the rectangle of length $f(x_k^*)$ and width Δx built up on the kth subinterval $[x_{k-1}, x_k]$. The n numbers $x_1^*, x_2^*, x_3^*, \ldots, x_n^*$ are called **sample points**.

- The sum of the areas of the n rectangles

$$\sum_{k=1}^{n} f(x_k^*)\Delta x = f(x_1^*)\Delta x + f(x_2^*)\Delta x + f(x_3^*)\Delta x + \cdots + f(x_n^*)\Delta x,$$

represents an approximation to the value of the area A under the graph of f on the interval $[a, b]$.

With these preliminaries, we are now in a position to define the concept of area under a graph.

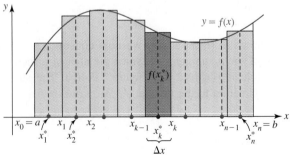

(a) Area A under the graph (b) n rectangles

FIGURE 5.3.5 Find the area A under the graph of f on the interval $[a, b]$

Definition 5.3.1 Area Under a Graph

Let f be continuous on $[a, b]$ and $f(x) \geq 0$ for all x in the interval. We define the **area A under the graph** of f on the interval to be

$$A = \lim_{n \to \infty} \sum_{k=1}^{n} f(x_k^*) \Delta x. \qquad (6)$$

It can be proved that when f is *continuous*, the limit in (6) always exists regardless of the manner used to divide $[a, b]$ into subintervals; that is, the subintervals may or may not be taken of equal width, and the points x_k^* can be chosen quite arbitrarily in the subintervals $[x_{k-1}, x_k]$. However, if the subintervals are not of equal width, then a different kind of limiting process is necessary in (6). We must replace $n \to \infty$ with the requirement that the length of the widest subinterval approach zero.

A Practical Form of (6) To use (6), suppose we choose x_k^* as we did in the discussion of Figure 5.3.4; namely, let x_k^* be the **right endpoint** of each subinterval. Since the width of each of the n subintervals of equal width is $\Delta x = (b-a)/n$, we have

$$x_k^* = a + k\Delta x = a + k\frac{b-a}{n}.$$

Then for $k = 1, 2, \ldots, n$ we have

$$x_1^* = a + \Delta x = a + \frac{b-a}{n}$$

$$x_2^* = a + 2\Delta x = a + 2\left(\frac{b-a}{n}\right)$$

$$x_3^* = a + 3\Delta x = a + 3\left(\frac{b-a}{n}\right)$$

$$\vdots$$

$$x_n^* = a + n\Delta x = a + n\left(\frac{b-a}{n}\right) = b.$$

By substituting $a + k(b-a)/n$ for x_k^* and $(b-a)/n$ for Δx in (6), it follows that the area A is also given by

$$A = \lim_{n \to \infty} \sum_{k=1}^{n} f\left(a + k\frac{b-a}{n}\right) \cdot \frac{b-a}{n}. \qquad (7)$$

We note that since $\Delta x = (b-a)/n$, $n \to \infty$ implies $\Delta x \to 0$.

EXAMPLE 3 Area Using (7)

Find the area A under the graph of $f(x) = x + 2$ on the interval $[0, 4]$.

Solution The area is bounded by the trapezoid indicated in FIGURE 5.3.6(a). By identifying $a = 0$ and $b = 4$, we find

$$\Delta x = \frac{4-0}{n} = \frac{4}{n}.$$

Thus, (7) becomes

$$A = \lim_{n \to \infty} \sum_{k=1}^{n} f\left(0 + k\frac{4}{n}\right)\frac{4}{n} = \lim_{n \to \infty} \frac{4}{n} \sum_{k=1}^{n} f\left(\frac{4k}{n}\right)$$

$$= \lim_{n \to \infty} \frac{4}{n} \sum_{k=1}^{n} \left(\frac{4k}{n} + 2\right)$$

$$= \lim_{n \to \infty} \frac{4}{n}\left[\frac{4}{n}\sum_{k=1}^{n} k + 2\sum_{k=1}^{n} 1\right]. \quad \leftarrow \text{by properties } (i) \text{ and } (ii) \text{ of Theorem 5.3.1}$$

FIGURE 5.3.6 Area under the graph in Example 3

Now, by summation formulas *(ii)* and *(i)* of Theorem 5.3.2, we can write

$$A = \lim_{n\to\infty} \frac{4}{n}\left[\frac{4}{n}\cdot\frac{n(n+1)}{2} + 2n\right]$$

$$= \lim_{n\to\infty}\left[\frac{16}{2}\frac{n(n+1)}{n^2} + 8\right] \quad \leftarrow \text{divide by } n^2$$

$$= \lim_{n\to\infty}\left[8\left(1 + \frac{1}{n}\right) + 8\right]$$

$$= 8\lim_{n\to\infty}\left(1 + \frac{1}{n}\right) + 8\lim_{n\to\infty} 1$$

$$= 8 + 8 = 16 \text{ square units.} \quad \blacksquare$$

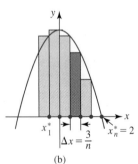

FIGURE 5.3.7 Area under the graph in Example 4

EXAMPLE 4 Area Using (7)

Find the area A under the graph of $f(x) = 4 - x^2$ on the interval $[-1, 2]$.

Solution The area is indicated in FIGURE 5.3.7(a). Since $a = -1$ and $b = 2$, it follows that

$$\Delta x = \frac{2 - (-1)}{n} = \frac{3}{n}.$$

Let us review the steps leading up to (7). The width of each rectangle is given by $\Delta x = (2 - (-1))/n = 3/n$. Now, starting at $x = -1$, the right endpoint of each of the n subintervals is

$$x_1^* = -1 + \frac{3}{n}$$

$$x_2^* = -1 + 2\left(\frac{3}{n}\right) = -1 + \frac{6}{n}$$

$$x_3^* = -1 + 3\left(\frac{3}{n}\right) = -1 + \frac{9}{n}$$

$$\vdots$$

$$x_n^* = -1 + n\left(\frac{3}{n}\right) = 2.$$

The length of each rectangle is then

$$f(x_1^*) = f\left(-1 + \frac{3}{n}\right) = 4 - \left[-1 + \frac{3}{n}\right]^2$$

$$f(x_2^*) = f\left(-1 + \frac{6}{n}\right) = 4 - \left[-1 + \frac{6}{n}\right]^2$$

$$f(x_3^*) = f\left(-1 + \frac{9}{n}\right) = 4 - \left[-1 + \frac{9}{n}\right]^2$$

$$\vdots$$

$$f(x_n^*) = f\left(-1 + \frac{3n}{n}\right) = f(2) = 4 - (2)^2 = 0.$$

The area of the kth rectangle is *length* $\times$ *width*:

$$f(x_k^*)\frac{3}{n} = \left(4 - \left[-1 + k\frac{3}{n}\right]^2\right)\frac{3}{n} = \left(3 + 6\frac{k}{n} - 9\frac{k^2}{n^2}\right)\frac{3}{n}.$$

Adding the areas of the n rectangles gives an approximation to the area under the graph on the interval: $A \approx \sum_{k=1}^{n} f(x_k^*)(3/n)$. As the number n of rectangles increases without bound, we obtain

$$A = \lim_{n\to\infty} \sum_{k=1}^{n} \left(3 + 6\frac{k}{n} - 9\frac{k^2}{n^2}\right)\frac{3}{n}$$

$$= \lim_{n\to\infty} \frac{3}{n} \sum_{k=1}^{n} \left(3 + 6\frac{k}{n} - 9\frac{k^2}{n^2}\right)$$

$$= \lim_{n\to\infty} \frac{3}{n} \left[\sum_{k=1}^{n} 3 + \frac{6}{n}\sum_{k=1}^{n} k - \frac{9}{n^2}\sum_{k=1}^{n} k^2\right].$$

Using summation formulas (*i*), (*ii*), and (*iii*) of Theorem 5.3.2, we get

$$A = \lim_{n\to\infty} \frac{3}{n}\left[3n + \frac{6}{n}\cdot\frac{n(n+1)}{2} - \frac{9}{n^2}\cdot\frac{n(n+1)(2n+1)}{6}\right]$$

$$= \lim_{n\to\infty}\left[9 + 9\left(1 + \frac{1}{n}\right) - \frac{9}{2}\left(1 + \frac{1}{n}\right)\left(2 + \frac{1}{n}\right)\right]$$

$$= 9 + 9 - 9 = 9 \text{ square units.} \qquad \blacksquare$$

■ **Other Choices for x_k^*** There is nothing special about choosing x_k^* to be the right endpoint of each subinterval. We reemphasize that x_k^* can be taken to be any convenient number in $[x_{k-1}, x_k]$. Had we chosen x_k^* to be the **left endpoint** of each subinterval, then

$$x_k^* = a + (k-1)\Delta x = a + (k-1)\frac{b-a}{n}, \quad k = 1, 2, \ldots, n,$$

and (7) would become

$$A = \lim_{n\to\infty} \sum_{k=1}^{n} f\left(a + (k-1)\frac{b-a}{n}\right)\cdot\frac{b-a}{n}. \qquad (8)$$

In Example 4 the corresponding rectangles would be as shown in FIGURE 5.3.8. In this case, we would have $x_k^* = -1 + (k-1)(3/n)$. In Problems 45 and 46 of Exercises 5.3 you will be asked to solve the area problem in Example 4 by choosing x_k^* to be first the left endpoint and then the midpoint of each subinterval $[x_{k-1}, x_k]$. By choosing x_k^* to be the midpoint of each $[x_{k-1}, x_k]$, then

$$x_k^* = a + \left(k - \frac{1}{2}\right)\Delta x, \quad k = 1, 2, \ldots, n. \qquad (9)$$

FIGURE 5.3.8 Rectangles using left endpoints of subintervals

Exercises 5.3

Answers to selected odd-numbered problems begin on page ANS-18.

≡ Fundamentals

In Problems 1–10, expand the indicated sum.

1. $\sum_{k=1}^{5} 3k$
2. $\sum_{k=1}^{5} (2k-3)$
3. $\sum_{k=1}^{4} \frac{2^k}{k}$
4. $\sum_{k=1}^{4} \left(\frac{3}{10}\right)^k$
5. $\sum_{k=1}^{10} \frac{(-1)^k}{2k+5}$
6. $\sum_{k=1}^{10} \frac{(-1)^{k-1}}{k^2}$
7. $\sum_{j=2}^{5} (j^2 - 2j)$
8. $\sum_{m=0}^{4} (m+1)^2$
9. $\sum_{k=1}^{5} \cos k\pi$
10. $\sum_{k=1}^{5} \frac{\sin(k\pi/2)}{k}$

In Problems 11–20, write the given sum using sigma notation.

11. $3 + 5 + 7 + 9 + 11 + 13 + 15$
12. $2 + 4 + 8 + 16 + 32 + 64$
13. $1 + 4 + 7 + 10 + \cdots + 37$
14. $2 + 6 + 10 + 14 + \cdots + 38$
15. $1 - \frac{1}{2} + \frac{1}{3} - \frac{1}{4} + \frac{1}{5}$
16. $-\frac{1}{2} + \frac{2}{3} - \frac{3}{4} + \frac{4}{5} - \frac{5}{6}$
17. $6 + 6 + 6 + 6 + 6 + 6 + 6 + 6$
18. $1 + \sqrt{2} + \sqrt{3} + 2 + \sqrt{5} + \cdots + 3$
19. $\cos\frac{\pi}{p}x - \frac{1}{4}\cos\frac{2\pi}{p}x + \frac{1}{9}\cos\frac{3\pi}{p}x - \frac{1}{16}\cos\frac{4\pi}{p}x$
20. $f'(1)(x-1) - \frac{f''(1)}{3}(x-1)^2 + \frac{f'''(1)}{5}(x-1)^3 - \frac{f^{(4)}(1)}{7}(x-1)^4 + \frac{f^{(5)}(1)}{9}(x-1)^5$

In Problems 21–28, find the numerical value of the given sum.

21. $\sum_{k=1}^{20} 2k$
22. $\sum_{k=0}^{50} (-3k)$

23. $\sum_{k=1}^{10} (k+1)$
24. $\sum_{k=1}^{1000} (2k-1)$
25. $\sum_{k=1}^{6} (k^2+3)$
26. $\sum_{k=1}^{5} (6k^2-k)$
27. $\sum_{p=0}^{10} (p^3+4)$
28. $\sum_{i=1}^{10} (2i^3-5i+3)$

In Problems 29–42, use (7) and Theorem 5.3.2 to find the area under the graph of the given function on the indicated interval.

29. $f(x) = x$, $[0, 6]$
30. $f(x) = 2x$, $[1, 3]$
31. $f(x) = 2x + 1$, $[1, 5]$
32. $f(x) = 3x - 6$, $[2, 4]$
33. $f(x) = x^2$, $[0, 2]$
34. $f(x) = x^2$, $[-2, 1]$
35. $f(x) = 1 - x^2$, $[-1, 1]$
36. $f(x) = 2x^2 + 3$, $[-3, -1]$
37. $f(x) = x^2 + 2x$, $[1, 2]$
38. $f(x) = (x-1)^2$, $[0, 2]$
39. $f(x) = x^3$, $[0, 1]$
40. $f(x) = x^3 - 3x^2 + 4$, $[0, 2]$
41. $f(x) = \begin{cases} 2, & 0 \le x < 1 \\ x+1, & 1 \le x \le 4 \end{cases}$
42. $f(x) = \begin{cases} -x+1, & 0 \le x < 1 \\ x+2, & 1 \le x \le 3 \end{cases}$

43. Sketch the graph of $y = 1/x$ on the interval $[\tfrac{1}{2}, \tfrac{5}{2}]$. By dividing the interval into four subintervals of equal widths, construct rectangles that approximate the area A under the graph on the interval. First use the right endpoint of each subinterval, and then use the left endpoint.

44. Repeat Problem 43 for $y = \cos x$ on the interval $[-\pi/2, \pi/2]$.

45. Rework Example 4 by choosing x_k^* to be the left endpoint of each subinterval. See (8).

46. Rework Example 4 by choosing x_k^* to be the midpoint of each subinterval. See (9).

In Problems 47 and 48, sketch the region whose area A is given by the formula. Do not try to evaluate.

47. $A = \lim_{n \to \infty} \sum_{k=1}^{n} \sqrt{4 - \dfrac{4k^2}{n^2}} \dfrac{2}{n}$
48. $A = \lim_{n \to \infty} \sum_{k=1}^{n} \left(\sin \dfrac{k\pi}{n} \right) \dfrac{\pi}{n}$

≡ Think About It

In Problems 49 and 50, write the given decimal number using sigma notation.

49. 0.11111111
50. 0.3737373737

51. Use summation formula (iii) of Theorem 5.3.2 to find the numerical value of $\sum_{k=21}^{60} k^2$.

52. Write the sum $8 + 7 + 8 + 9 + 10 + 11 + 12$ using sigma notation so that the index of summation starts with $k = 0$. With $k = 1$. With $k = 2$.

53. Solve for $\bar{x}$: $\sum_{k=1}^{n} (x_k - \bar{x})^2 = 0$.

54. (a) Find the value of $\sum_{k=1}^{n} [f(k) - f(k-1)]$. A sum of this form is said to **telescope**.

(b) Use part (a) to find the numerical value of
$$\sum_{k=1}^{400} (\sqrt{k} - \sqrt{k-1}).$$

55. (a) Use part (a) of Problem 54 to show that
$$\sum_{k=1}^{n} [(k+1)^2 - k^2] = -1 + (n+1)^2 = n^2 + 2n.$$

(b) Use the fact that $(k+1)^2 - k^2 = 2k + 1$ to show that
$$\sum_{k=1}^{n} [(k+1)^2 - k^2] = n + 2 \sum_{k=1}^{n} k.$$

(c) Compare the results of parts (a) and (b) to derive summation formula (iii) of Theorem 5.3.2.

56. Show how the pattern illustrated in FIGURE 5.3.9 can be used to infer the summation formula (iv) of Theorem 5.3.2.

FIGURE 5.3.9 Array for Problem 56

57. Derive the formula for the area of the trapezoid given in FIGURE 5.3.10.

FIGURE 5.3.10 Trapezoid in Problem 57

58. In a supermarket, 136 cans are displayed in a triangular stack as shown in FIGURE 5.3.11. How many cans are there in the bottom row of the stack?

FIGURE 5.3.11 Stack of cans in Problem 58

59. Use (7) and the summation formula
$$\sum_{k=1}^{n} k^4 = \frac{n(n+1)(6n^3 + 9n^2 + n - 1)}{30}$$
to find the area under the graph of $f(x) = 16 - x^4$ on $[-2, 2]$.

60. Find the area under the graph of $y = \sqrt{x}$ on $[0, 1]$ by considering the area under the graph of $y = x^2$ on $[0, 1]$. Carry out your ideas.

61. Find the area under the graph of $y = \sqrt[3]{x}$ on $[0, 8]$ by considering the area under the graph of $y = x^3$ on $0 \leq x \leq 2$.

62. (a) Suppose $y = ax^2 + bx + c \geq 0$ on the interval $[0, x_0]$. Show that the area under the graph on $[0, x_0]$ is given by
$$A = a\frac{x_0^3}{3} + b\frac{x_0^2}{2} + cx_0.$$

(b) Use the result in part (a) to find the area under the graph of $y = 6x^2 + 2x + 1$ on the interval $[2, 5]$.

63. A summation formula for the sum of the n terms of a finite geometric sequence $a, ar, ar^2, \ldots, ar^{n-1}$ is given by
$$\sum_{k=1}^{n} ar^{k-1} = a\left(\frac{1-r^n}{1-r}\right).$$
Use this summation formula, (8) of this section, and L'Hôpital's Rule, to find the area under the graph of $y = e^x$ on $[0, 1]$.

64. A Bit of History Everyone knows in a beginning course in physics that the distance a body falls is proportional to the square of the elapsed time. **Galileo Galilei** (1564–1642) was the first to discover this fact. Galileo found that the distance a mass moves down an inclined plane in consecutive time intervals is proportional to a positive odd integer. Hence the total distance s that a mass moves in n seconds, with n a positive integer, is proportional to $1 + 3 + 5 + \cdots + 2n - 1$. Show that this is the same as saying that the total distance a mass moves down an inclined plane is proportional to the square of the elapsed time n.

5.4 The Definite Integral

Introduction In the previous section we saw that the area under the graph of a continuous nonnegative function f on an interval $[a, b]$ was defined as a limit of a sum. We see in this section that the same kind of limiting process leads to the notion of a **definite integral**.

Let $y = f(x)$ be a function defined on a closed interval $[a, b]$. Consider the following four steps.

- Divide the interval $[a, b]$ into n subintervals $[x_{k-1}, x_k]$ of widths $\Delta x_k = x_k - x_{k-1}$, where
$$a = x_0 < x_1 < x_2 < \cdots < x_{n-1} < x_n = b. \quad (1)$$

  The collection of numbers (1) is called a **partition** of the interval and is denoted by P.
- Let $\|P\|$ denote the largest number of the n subinterval widths $\Delta x_1, \Delta x_2, \ldots, \Delta x_n$. The number $\|P\|$ is called the **norm** of the partition P.
- Choose a number x_k^* in each subinterval $[x_{k-1}, x_k]$ as shown in FIGURE 5.4.1. The n numbers $x_1^*, x_2^*, x_3^*, \ldots, x_n^*$ are called **sample points** in these subintervals.
- Form the sum
$$\sum_{k=1}^{n} f(x_k^*) \Delta x_k. \quad (2)$$

FIGURE 5.4.1 Sample point x_k^* in $[x_{k-1}, x_k]$

Sums of the kind given in (2) corresponding to various partitions of $[a, b]$ are known as **Riemann sums** and are named for the famous German mathematician, **Georg Friedrich Bernhard Riemann**.

Although the foregoing procedure looks very similar to the steps leading up to the definition of area under a graph given in Section 5.3, there are some important differences. Observe that a Riemann sum (2) does not require that f be either continuous or nonnegative on the interval $[a, b]$. Thus, (2) does not necessarily represent an approximation to the area under a graph. Keep in mind that "area under a graph" refers to *the area bounded between the graph of a continuous nonnegative function and the x-axis*. As shown in FIGURE 5.4.2, if $f(x) < 0$ for some x in $[a, b]$, a Riemann sum could contain terms $f(x_k^*)\Delta x_k$, where $f(x_{k1}^*) < 0$. In this case the products $f(x_k^*)\Delta x_k$ are numbers that are the negatives of the areas of rectangles drawn below the x-axis.

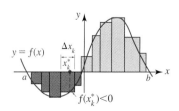

FIGURE 5.4.2 The function f is positive and negative on the interval $[a, b]$

EXAMPLE 1 A Riemann Sum

Compute the Riemann sum for $f(x) = x^2 - 4$ on $[-2, 3]$ with five subintervals determined by $x_0 = -2$, $x_1 = -\frac{1}{2}$, $x_2 = 0$, $x_3 = 1$, $x_4 = \frac{7}{4}$, $x_5 = 3$ and $x_1^* = -1$, $x_2^* = -\frac{1}{4}$, $x_3^* = \frac{1}{2}$, $x_4^* = \frac{3}{2}$, $x_5^* = \frac{5}{2}$. Find the norm of the partition.

Solution FIGURE 5.4.3 shows that the numbers x_k, $k = 0, 1, \ldots, 5$ determine five subintervals $\left[-2, -\frac{1}{2}\right]$, $\left[-\frac{1}{2}, 0\right]$, $[0, 1]$, $\left[1, \frac{7}{4}\right]$, and $\left[\frac{7}{4}, 3\right]$ of the interval $[-2, 3]$ and a sample point x_k^* (in red) within each subinterval.

FIGURE 5.4.3 Five subintervals and sample points in Example 1

Now, evaluate the function f at each sample point and determine the width of each subinterval:

$$f(x_1^*) = f(-1) = -3, \quad \Delta x_1 = x_1 - x_0 = -\frac{1}{2} - (-2) = \frac{3}{2}$$

$$f(x_2^*) = f\left(-\frac{1}{4}\right) = -\frac{63}{16}, \quad \Delta x_2 = x_2 - x_1 = 0 - \left(-\frac{1}{2}\right) = \frac{1}{2}$$

$$f(x_3^*) = f\left(\frac{1}{2}\right) = -\frac{15}{4}, \quad \Delta x_3 = x_3 - x_2 = 1 - 0 = 1$$

$$f(x_4^*) = f\left(\frac{3}{2}\right) = -\frac{7}{4}, \quad \Delta x_4 = x_4 - x_3 = \frac{7}{4} - 1 = \frac{3}{4}$$

$$f(x_5^*) = f\left(\frac{5}{2}\right) = \frac{9}{4}, \quad \Delta x_5 = x_5 - x_4 = 3 - \frac{7}{4} = \frac{5}{4}.$$

The **Riemann sum** for this partition and choice of sample points is then

$$f(x_1^*)\Delta x_1 + f(x_2^*)\Delta x_2 + f(x_3^*)\Delta x_3 + f(x_4^*)\Delta x_4 + f(x_5^*)\Delta x_5$$
$$= (-3)\left(\frac{3}{2}\right) + \left(-\frac{63}{16}\right)\left(\frac{1}{2}\right) + \left(-\frac{15}{4}\right)(1) + \left(-\frac{7}{4}\right)\left(\frac{3}{4}\right) + \left(\frac{9}{4}\right)\left(\frac{5}{4}\right) = -\frac{279}{32} \approx -8.72.$$

Inspection of the values of the five Δx_k shows that the norm of the partition is $\|P\| = \frac{3}{2}$. ∎

For a function f defined on an interval $[a, b]$, there are an infinite number of possible Riemann sums for a given partition P of the interval, since the numbers x_k^* can be chosen arbitrarily in each subinterval $[x_{k-1}, x_k]$.

EXAMPLE 2 Another Riemann Sum

Compute the Riemann sum for the function in Example 1 if the partition of $[-2, 3]$ is the same but the sample points are $x_1^* = -\frac{3}{2}$, $x_2^* = -\frac{1}{8}$, $x_3^* = \frac{3}{4}$, $x_4^* = \frac{3}{2}$, and $x_5^* = 2.1$.

Solution We need only compute f at the new sample points since the numbers Δx_k are the same as before:

$$f(x_1^*) = f\left(-\frac{3}{2}\right) = -\frac{7}{4}$$

$$f(x_2^*) = f\left(-\frac{1}{8}\right) = -\frac{255}{64}$$

$$f(x_3^*) = f\left(\frac{3}{4}\right) = -\frac{55}{16}$$

$$f(x_4^*) = f\left(\frac{3}{2}\right) = -\frac{7}{4}$$

$$f(x_5^*) = f(2.1) = 0.41.$$

The Riemann sum is now

$$f(x_1^*)\Delta x_1 + f(x_2^*)\Delta x_2 + f(x_3^*)\Delta x_3 + f(x_4^*)\Delta x_4 + f(x_5^*)\Delta x_5$$
$$= \left(-\frac{7}{4}\right)\left(\frac{3}{2}\right) + \left(-\frac{255}{64}\right)\left(\frac{1}{2}\right) + \left(-\frac{55}{16}\right)(1) + \left(-\frac{7}{4}\right)\left(\frac{3}{4}\right) + (0.41)\left(\frac{5}{4}\right) \approx -8.85. \blacksquare$$

We are interested in a special kind of limit of (2). If the Riemann sums $\sum_{k=1}^{n} f(x_k^*)\Delta x_k$ are close to a number L for *every* partition P of $[a, b]$ for which the norm $\|P\|$ is close to zero, we then write

$$\lim_{\|P\| \to 0} \sum_{k=1}^{n} f(x_k^*)\Delta x_k = L \tag{3}$$

and say that L is the **definite integral** of f on the interval $[a, b]$. In the following definition we introduce a new symbol for the number L.

Definition 5.4.1 The Definite Integral

Let f be a function defined on a closed interval $[a, b]$. Then the **definite integral of f from a to b**, denoted by $\int_a^b f(x)\, dx$, is defined to be

$$\int_a^b f(x)\, dx = \lim_{\|P\| \to 0} \sum_{k=1}^{n} f(x_k^*)\Delta x_k. \tag{4}$$

If the limit in (4) exists, the function f is said to be **integrable** on the interval. The numbers a and b in the preceding definition are called the **lower** and **upper limits of integration**, respectively. The function f is called the **integrand**. The integral symbol $\int$, as used by Leibniz, is an elongated S for the word *sum*. Also, note that $\|P\| \to 0$ always implies that the number of subintervals n becomes infinite ($n \to \infty$). However, as shown in FIGURE 5.4.4, the fact that $n \to \infty$ does not necessarily imply $\|P\| \to 0$.

FIGURE 5.4.4 Infinite number of subintervals does not imply $\|P\| \to 0$.

■ **Integrability** In the next two theorems we state conditions that are sufficient for a function f to be integrable on an interval $[a, b]$. The proofs of these theorems will not be given.

Theorem 5.4.1 Continuity Implies Integrability

If f is continuous on the closed interval $[a, b]$, then $\int_a^b f(x)\, dx$ exists; that is, f is integrable on the interval.

There are functions that are defined for every value of x in $[a, b]$ for which the limit in (4) does not exist. Also, if the function f is not defined for all values of x in the interval, the definite integral *may* not exist; for example, later on we will see why an integral such as $\int_{-3}^{2}(1/x)\, dx$ does not exist. Notice that $y = 1/x$ is discontinuous at $x = 0$ and is unbounded on the interval. However, one should not conclude from this one example that when a function f has a discontinuity in $[a, b]$, $\int_a^b f(x)\, dx$ necessarily does not exist. Continuity of a function f on $[a, b]$ is *sufficient* but *not necessary* to guarantee the existence of $\int_a^b f(x)\, dx$. The set of functions continuous on $[a, b]$ is a subset of the set of functions that are integrable on the interval.

The next theorem gives another sufficient condition for integrability on $[a, b]$.

Theorem 5.4.2 Sufficient Conditions for Integrability

If a function f is bounded on the closed interval $[a, b]$, that is, if there exists a positive constant B such that $-B \leq f(x) \leq B$ for all x in the interval, and has a finite number of discontinuities in $[a, b]$, then f is integrable on the interval.

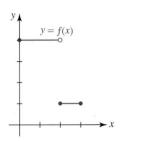

FIGURE 5.4.5 Definite integral of f on $[0, 3]$ exists

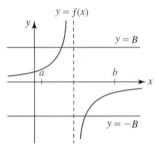

FIGURE 5.4.6 The function f is not bounded on $[a, b]$

When a function f is bounded, its complete graph must lie between two horizontal lines, $y = B$ and $y = -B$. In other words, $|f(x)| \leq B$ for all x in $[a, b]$. The function

$$f(x) = \begin{cases} 4, & 0 \leq x < 2 \\ 1, & 2 \leq x \leq 3 \end{cases}$$

shown in FIGURE 5.4.5 is discontinuous at $x = 2$ but is bounded on $[0, 3]$, since $|f(x)| \leq 4$ for all x in $[0, 3]$. (For that matter, $1 \leq f(x) \leq 4$ for all x in $[0, 3]$ shows that f is bounded on the interval.) It follows from Theorem 5.4.2 that $\int_0^3 f(x)\,dx$ exists. FIGURE 5.4.6 shows the graph of a function f that is unbounded on an interval $[a, b]$. Regardless of how large the number B is chosen, the graph of f cannot be confined to the region between the horizontal lines, $y = B$ and $y = -B$.

▌ **Regular Partition** If it is known that a definite integral exists (say, the integrand f is continuous on $[a, b]$), then:

- *The limit in* (4) *exists for every possible way of partitioning* $[a, b]$ *and for every way of choosing* x_k^* *in the subintervals* $[x_{k-1}, x_k]$.

In particular, by choosing the subintervals of equal width and the sample points to be the right endpoints of the subintervals $[x_{k-1}, x_k]$, that is,

$$\Delta x = \frac{b - a}{n} \quad \text{and} \quad x_k^* = a + k\frac{b - a}{n}, \quad k = 1, 2, \ldots, n,$$

we can write (4) in the alternative manner

$$\int_a^b f(x)\,dx = \lim_{n \to \infty} \sum_{k=1}^n f\left(a + k\frac{b - a}{n}\right)\frac{b - a}{n}. \tag{5}$$

Recall from Section 5.3 that a partition P of $[a, b]$ in which the subintervals have the same width is called a **regular partition**.

▌ **Area** You might conclude that the formulations of $\int_a^b f(x)\,dx$ given in (4) and (5) are exactly the same as (6) and (7) of Section 5.3 for the general case of finding the area under the curve $y = f(x)$ on $[a, b]$. In a way this is correct; however, Definition 5.4.1 is a more general concept, since, as noted before, we are not requiring that f be continuous on $[a, b]$ or that $f(x) \geq 0$ on the interval. Thus, a *definite integral need not be area*. What then is a definite integral? For now, accept the fact that a definite integral is simply a real number. Contrast this with the indefinite integral, which is a function (or a family of functions). Is the area under the graph of a continuous nonnegative function a definite integral? The answer is *yes*.

Theorem 5.4.3 Area As a Definite Integral

If f is continuous on the closed interval $[a, b]$ and $f(x) \geq 0$ for all x in the interval, then the **area A under the graph** on $[a, b]$ is

$$A = \int_a^b f(x)\,dx. \tag{6}$$

EXAMPLE 3 Area as a Definite Integral

Consider the definite integral $\int_{-1}^1 \sqrt{1 - x^2}\,dx$. The integrand is continuous and nonnegative and so the definite integral represents the area under the graph of $f(x) = \sqrt{1 - x^2}$ on the interval $[-1, 1]$. Because the graph of the function f is the upper semicircle of $x^2 + y^2 = 1$, the area under the graph is the shaded region in FIGURE 5.4.7. From geometry we know that the area of a circle of radius r is πr^2, and so with $r = 1$ the area of the semicircle, and hence the value of the definite integral, is

$$\int_{-1}^1 \sqrt{1 - x^2}\,dx = \frac{1}{2}\pi(1)^2 = \frac{1}{2}\pi. \qquad \blacksquare$$

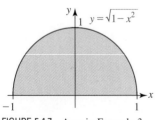

FIGURE 5.4.7 Area in Example 3

We shall return to the question of finding areas by means of the definite integral in Section 6.2.

EXAMPLE 4 Definite Integral Using (5)

Evaluate $\int_{-2}^{1} x^3 \, dx$.

Solution Since $f(x) = x^3$ is continuous on $[-2, 1]$, we know from Theorem 5.4.1 that the definite integral exists. We use a regular partition and the result given in (5). Choosing

$$\Delta x = \frac{1 - (-2)}{n} = \frac{3}{n} \quad \text{and} \quad x_k^* = -2 + k \cdot \frac{3}{n}$$

we have

$$f\left(-2 + \frac{3k}{n}\right) = \left(-2 + \frac{3k}{n}\right)^3 = -8 + 36\left(\frac{k}{n}\right) - 54\left(\frac{k^2}{n^2}\right) + 27\left(\frac{k^3}{n^3}\right).$$

It then follows from (5) and summation formulas (*i*), (*ii*), (*iii*), and (*iv*) of Theorem 5.3.2 that

$$\int_{-2}^{1} x^3 \, dx = \lim_{n \to \infty} \sum_{k=1}^{n} f\left(-2 + \frac{3k}{n}\right) \frac{3}{n}$$

$$= \lim_{n \to \infty} \frac{3}{n} \sum_{k=1}^{n} \left[-8 + 36\left(\frac{k}{n}\right) - 54\left(\frac{k^2}{n^2}\right) + 27\left(\frac{k^3}{n^3}\right)\right]$$

$$= \lim_{n \to \infty} \frac{3}{n} \left[-8n + \frac{36}{n} \cdot \frac{n(n+1)}{2} - \frac{54}{n^2} \cdot \frac{n(n+1)(2n+1)}{6} + \frac{27}{n^3} \cdot \frac{n^2(n+1)^2}{4}\right]$$

$$= \lim_{n \to \infty} \left[-24 + 54\left(1 + \frac{1}{n}\right) - 27\left(1 + \frac{1}{n}\right)\left(2 + \frac{1}{n}\right) + \frac{81}{4}\left(1 + \frac{1}{n}\right)\left(1 + \frac{1}{n}\right)\right]$$

$$= -24 + 54 - 27(2) + \frac{81}{4} = -\frac{15}{4}.$$

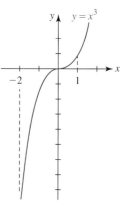

FIGURE 5.4.8 Graph of function in Example 4

FIGURE 5.4.8 shows that we are not considering area under the graph on $[-2, 1]$. ■

EXAMPLE 5 Definite Integral Using (5)

The values of the Riemann sums in Examples 1 and 2 are approximations to the value of the definite integral $\int_{-2}^{3} (x^2 - 4) \, dx$. It is left as an exercise to show that (5) gives

$$\int_{-2}^{3} (x^2 - 4) \, dx = -\frac{25}{3} \approx -8.33.$$

See Problem 16 in Exercises 5.4. ■

Properties of the Definite Integral We examine next some of the important properties of the definite integral defined in (4).

The following two definitions are useful when working with definite integrals.

Definition 5.4.2 Limits of Integration

(*i*) **Equality of Limits** If a is in the domain of f, then

$$\int_{a}^{a} f(x) \, dx = 0. \tag{7}$$

(*ii*) **Reversing Limits** If f is integrable on $[a, b]$, then

$$\int_{b}^{a} f(x) \, dx = -\int_{a}^{b} f(x) \, dx. \tag{8}$$

Definition 5.4.2(*i*) can be motivated by thinking that the area under the graph of f and above a single point a on the x-axis is zero.

In the definition of $\int_{a}^{b} f(x) \, dx$ it was assumed that $a < b$, and so the usual "direction" of definite integration is left to right. Part (*ii*) of Definition 5.4.2 states that reversing this direction, that is, interchanging the limits of integration, results in the negative of the integral.

EXAMPLE 6 Definition 5.4.2

By part (*i*) of Definition 5.4.2,

$$\underset{\substack{\text{limits of integration} \to \\ \text{are the same} \to}}{\int_1^1} (x^3 + 3x)\, dx = 0.$$

EXAMPLE 7 Example 4 Revisited

In Example 4 we saw that $\int_{-2}^{1} x^3\, dx = -\frac{15}{4}$. It follows from part (*ii*) of Definition 5.4.2 that

$$\int_1^{-2} x^3\, dx = -\int_{-2}^{1} x^3\, dx = -\left(-\frac{15}{4}\right) = \frac{15}{4}.$$

In the next theorem we list some of the basic properties of the definite integral. These properties are analogous to the properties of the sigma notation given in Theorem 5.3.1 as well as the properties of the indefinite integral, or antiderivative, discussed in Section 5.1.

Theorem 5.4.4 Properties of the Definite Integral

If f and g are integrable functions on the closed interval $[a, b]$, then

(*i*) $\displaystyle\int_a^b kf(x)\, dx = k\int_a^b f(x)\, dx$, where k is any constant

(*ii*) $\displaystyle\int_a^b [f(x) \pm g(x)]\, dx = \int_a^b f(x)\, dx + \int_a^b g(x)\, dx$.

Theorem 5.4.4(*ii*) extends to any finite sum of integrable functions on the interval $[a, b]$:

$$\int_a^b [f_1(x) + f_2(x) + \cdots + f_n(x)]\, dx = \int_a^b f_1(x)\, dx + \int_a^b f_2(x)\, dx + \cdots + \int_a^b f_n(x)\, dx.$$

The independent variable x in a definite integral is called a **dummy variable** of integration. The value of the integral does not depend on the symbol used. In other words,

$$\int_a^b f(x)\, dx = \int_a^b f(r)\, dr = \int_a^b f(s)\, ds = \int_a^b f(t)\, dt \qquad (9)$$

and so on.

EXAMPLE 8 Example 4 Revisited

From (9), it does not matter what symbol is used as the variable of integration:

$$\int_{-2}^{1} x^3\, dx = \int_{-2}^{1} r^3\, dr = \int_{-2}^{1} s^3\, ds = \int_{-2}^{1} t^3\, dt = -\frac{15}{4}.$$

Theorem 5.4.5 Additive Interval Property

If f is an integrable function on a closed interval containing the numbers a, b, and c, then

$$\int_a^b f(x)\, dx = \int_a^c f(x)\, dx + \int_c^b f(x)\, dx. \qquad (10)$$

It is easy to interpret the additive interval property given in Theorem 5.4.5 in the special case when f is continuous on $[a, b]$ and $f(x) \geq 0$ for all x in the interval. As seen in FIGURE 5.4.9, the area under the graph of f on $[a, c]$ plus the area under the graph on the adjacent interval $[c, b]$ is the same as the area under the graph on the entire interval $[a, b]$.

Note: The conclusion of Theorem 5.4.5 holds when a, b, and c are *any* three numbers in a closed interval. In other words, it is not necessary to have the order $a < c < b$ as shown in Figure 5.4.9. Moreover, the result in (10) extends to any finite number of numbers $a, b, c_1, c_2, \ldots, c_n$ in the interval. For example, for a closed interval containing the numbers a, b, c_1, and c_2,

$$\int_a^b f(x)\, dx = \int_a^{c_1} f(x)\, dx + \int_{c_1}^{c_2} f(x)\, dx + \int_{c_2}^b f(x)\, dx.$$

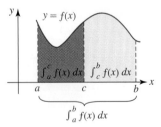

FIGURE 5.4.9 Areas are additive

For a given partition P of an interval $[a, b]$, it stands to reason that

$$\lim_{\|P\| \to 0} \sum_{k=1}^{n} \Delta x_k = b - a, \tag{11}$$

in other words, the limit $\lim_{\|P\| \to 0} \sum_{k=1}^{n} \Delta x_k$ is simply the width of the interval. As a consequence of (11), we have the following theorem.

Theorem 5.4.6 Definite Integral of a Constant

For any constant k,

$$\int_a^b k\, dx = k \int_a^b dx = k(b - a).$$

FIGURE 5.4.10 If $k > 0$, area under the graph is $k(b - a)$

If $k > 0$, then Theorem 5.4.6 implies that $\int_a^b k\, dx$ is simply the area of a rectangle of width $b - a$ and height k. See FIGURE 5.4.10.

EXAMPLE 9 Definite Integral of a Constant

From Theorem 5.4.6,

$$\int_2^8 5\, dx = 5 \int_2^8 dx = 5(8 - 2) = 30.$$

EXAMPLE 10 Using Examples 4 and 9

Evaluate $\int_{-2}^{1} (x^3 + 5)\, dx$.

Solution From Theorem 5.4.4(*ii*) we can write the given integral as two integrals:

$$\int_{-2}^{1} (x^3 + 5)\, dx = \int_{-2}^{1} x^3\, dx + \int_{-2}^{1} 5\, dx.$$

Now, from Example 4 we know that $\int_{-2}^{1} x^3\, dx = -\frac{15}{4}$ and with the help of Theorem 5.4.6 we see that $\int_{-2}^{1} 5\, dx = 5[1 - (-2)] = 15$. Therefore,

$$\int_{-2}^{1} (x^3 + 5)\, dx = \left(-\frac{15}{4}\right) + 15 = \frac{45}{4}.$$

Finally, the following results are not surprising if you interpret the integrals as area.

Theorem 5.4.7 Comparison Properties

Let f and g be integrable functions on the closed interval $[a, b]$.
(*i*) If $f(x) \geq g(x)$ for all x in the interval, then

$$\int_a^b f(x)\, dx \geq \int_a^b g(x)\, dx.$$

(*ii*) If $m \leq f(x) \leq M$ for all x in the interval, then

$$m(b - a) \leq \int_a^b f(x)\, dx \leq M(b - a).$$

Properties (*i*) and (*ii*) of Theorem 5.4.7 are easily understood in terms of area. For (*i*) if we assume $f(x) \geq g(x) \geq 0$ for all x in $[a, b]$, then on the interval the area A_1 under the graph of f is greater than or equal to the area A_2 under the graph of g. Similarly, for (*ii*) if we assume that f is continuous and positive on the interval $[a, b]$, then by the Extreme Value Theorem,

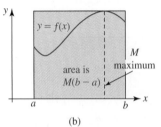

FIGURE 5.4.11 Motivation for part (ii) of Theorem 5.4.7

FIGURE 5.4.12 Definite integral of f on $[a, b]$ gives net signed area

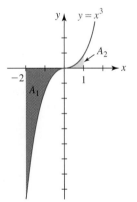

FIGURE 5.4.13 Net signed area in Example 11

f has an absolute minimum $m > 0$ and an absolute maximum $M > 0$ on the interval. The area under the graph $\int_a^b f(x)\,dx$ on the interval is then greater than or equal to the area $m(b - a)$ of the smaller rectangle shown in FIGURE 5.4.11(a) and less than or equal to the area $M(b - a)$ of the larger rectangle shown in Figure 5.4.11(b).

If we let $g(x) = 0$ in (i) of Theorem 5.4.7 and use the fact that $\int_a^b 0\,dx = 0$, we conclude:

- If $f(x) \geq 0$ on $[a, b]$, then $\int_a^b f(x)\,dx \geq 0$. $\qquad(12)$

In like manner by choosing $f(x) = 0$ in (i), it follows that:

- If $g(x) \leq 0$ on $[a, b]$, then $\int_a^b g(x)\,dx \leq 0$. $\qquad(13)$

■ **Net Signed Area** Because the function f in FIGURE 5.4.12 takes on both positive and negative values on $[a, b]$ the definite integral $\int_a^b f(x)\,dx$ does not represent area under the graph of f on the interval. By Theorem 5.4.5, the additive interval property,

$$\int_a^b f(x)\,dx = \int_a^{c_1} f(x)\,dx + \int_{c_1}^{c_2} f(x)\,dx + \int_{c_2}^b f(x)\,dx. \qquad(14)$$

Because $f(x) \geq 0$ on $[a, c_1]$ and $[c_2, b]$ we have

$$\int_a^{c_1} f(x)\,dx = A_1 \quad \text{and} \quad \int_{c_2}^b f(x)\,dx = A_3,$$

where A_1 and A_3 denote the areas under the graph of f on the intervals $[a, c_1]$ and $[c_2, b]$, respectively. But since $f(x) \leq 0$ on $[c_1, c_2]$ we have in view of (13), $\int_{c_1}^{c_2} f(x)\,dx \leq 0$ and so $\int_{c_1}^{c_2} f(x)\,dx$ does not represent area. However, the value of $\int_{c_1}^{c_2} f(x)\,dx$ is the negative of the actual area A_2 bounded between the graph of f and the x-axis on the interval $[c_1, c_2]$. That is, $\int_{c_1}^{c_2} f(x)\,dx = -A_2$. Hence (14) is

$$\int_a^b f(x)\,dx = A_1 + (-A_2) + A_3 = A_1 - A_2 + A_3.$$

We see that the definite integral gives the **net signed area** between the graph of f and the x-axis on the interval $[a, b]$.

EXAMPLE 11 Net Signed Area

The result $\int_{-2}^1 x^3\,dx = -\frac{15}{4}$ obtained in Example 4 can be interpreted as the net signed area between the graph of $f(x) = x^3$ and the x-axis on $[-2, 1]$. Although the observation that

$$\int_{-2}^1 x^3\,dx = \int_{-2}^0 x^3\,dx + \int_0^1 x^3\,dx = -A_1 + A_2 = -\frac{15}{4}$$

does not give us the values of A_1 and A_2, the negative value is consistent with FIGURE 5.4.13 where it is obvious that the area A_1 is larger than A_2. ∎

■ **The Theory** Let f be a function defined on $[a, b]$ and let L denote a real number. The intuitive concept that Riemann sums are close to L whenever the norm $\|P\|$ of a partition P is close to zero can be expressed in a precise manner using the ε-δ symbols introduced in Section 2.6. To say that f is integrable on $[a, b]$, we mean that for every real number $\varepsilon > 0$ there exists a real number $\delta > 0$ such that

$$\left| \sum_{k=1}^n f(x_k^*)\Delta x_k - L \right| < \varepsilon, \qquad(15)$$

whenever P is a partition of $[a, b]$ for which $\|P\| < \delta$ and the x_k^* are numbers in the subintervals $[x_{k-1}, x_k]$, $k = 1, 2, \ldots, n$. In other words,

$$\lim_{\|P\| \to 0} \sum_{k=1}^n f(x_k^*)\Delta x_k$$

exists and is equal to the number L.

■ **Postscript—A Bit of History** Georg Friedrich Bernhard Riemann (1826–1866) born in Hanover, Germany, in 1826, was the son of a Lutheran minister. Although a devout Christian, Riemann was disinclined to follow his father's vocation and abandoned the study of theology at the University of Göttingen in favor of a course of studies in which his genius was obvious: mathematics. It is likely that the concept of Riemann sums grew out of a course on the definite integral that he had taken at the university; this concept reflects his attempt to give a precise mathematical meaning to the definite integral of Newton and Leibniz. After submitting his doctoral dissertation on the foundations of functions of a complex variable to the examining committee at the University of Göttingen, Karl Friedrich Gauss, the "prince of mathematicians," paid Riemann a very rare compliment: "The dissertation offers convincing evidence ... of a creative, active, truly mathematical mind ... of glorious fertile originality."

Riemann

Riemann, like so many other promising scholars of that time, possessed a fragile constitution. He died at age 39 of pleurisy. His original contributions to differential geometry, topology, non-Euclidean geometry, and his bold investigations into the nature of space, electricity, and magnetism, foreshadowed the work of Einstein in the next century.

$\int_a^b$ NOTES FROM THE CLASSROOM

The procedure outlined in (5) has limited utility as a practical means of computing a definite integral. In the next section we will introduce a theorem that enables us to find the number $\int_a^b f(x)\,dx$ in a much easier manner. This important theorem is the bridge between differential and integral calculus.

Exercises 5.4   Answers to selected odd-numbered problems begin on page ANS-19.

≡ Fundamentals

In Problems 1–6, compute the Riemann sum $\sum_{k=1}^n f(x_k^*)\Delta x_k$ for the given partition. Specify $\|P\|$.

1. $f(x) = 3x + 1$, $[0, 3]$, four subintervals; $x_0 = 0, x_1 = 1$, $x_2 = \frac{5}{3}, x_3 = \frac{7}{3}, x_4 = 3$; $x_1^* = \frac{1}{2}, x_2^* = \frac{4}{3}, x_3^* = 2, x_4^* = \frac{8}{3}$

2. $f(x) = x - 4$, $[-2, 5]$, five subintervals; $x_0 = -2, x_1 = -1$, $x_2 = -\frac{1}{2}, x_3 = \frac{1}{2}, x_4 = 3, x_5 = 5$; $x_1^* = -\frac{3}{2}, x_2^* = -\frac{1}{2}$, $x_3^* = 0, x_4^* = 2, x_5^* = 4$

3. $f(x) = x^2$, $[-1, 1]$, four subintervals: $x_0 = -1, x_1 = -\frac{1}{4}$, $x_2 = \frac{1}{4}, x_3 = \frac{3}{4}, x_4 = 1$; $x_1^* = -\frac{3}{4}, x_2^* = 0, x_3^* = \frac{1}{2}$, $x_4^* = \frac{7}{8}$

4. $f(x) = x^2 + 1$, $[1, 3]$, three subintervals; $x_0 = 1, x_1 = \frac{3}{2}$, $x_2 = \frac{5}{2}, x_3 = 3$; $x_1^* = \frac{5}{4}, x_2^* = \frac{7}{4}, x_3^* = 3$

5. $f(x) = \sin x$, $[0, 2\pi]$, three subintervals; $x_0 = 0, x_1 = \pi$, $x_2 = 3\pi/2, x_3 = 2\pi$; $x_1^* = \pi/2, x_2^* = 7\pi/6, x_3^* = 7\pi/4$

6. $f(x) = \cos x$, $[-\pi/2, \pi/2]$, four subintervals; $x_0 = -\pi/2$, $x_1 = -\pi/4$, $x_2 = 0$, $x_3 = \pi/3$, $x_4 = \pi/2$; $x_1^* = -\pi/3$, $x_2^* = -\pi/6, x_3^* = \pi/4, x_4^* = \pi/3$

7. Given $f(x) = x - 2$ on $[0, 5]$, compute the Riemann sum using a partition with five subintervals of equal length. Let x_k^*, $k = 1, 2, \ldots, 5$, be the right endpoint of each subinterval.

8. Given $f(x) = x^2 - x + 1$ on $[0, 1]$, compute the Riemann sum using a partition with three subintervals of equal length. Let x_k^*, $k = 1, 2, 3$, be the left endpoint of each subinterval.

In Problems 9 and 10, let P be a partition of the indicated interval and x_k^* a number in the kth subinterval. Write the given sum as a definite integral on the indicated interval.

9. $\lim\limits_{\|P\|\to 0} \sum\limits_{k=1}^n \sqrt{9 + (x_k^*)^2}\,\Delta x_k$; $[-2, 4]$

10. $\lim\limits_{\|P\|\to 0} \sum\limits_{k=1}^n (\tan x_k^*)\,\Delta x_k$; $[0, \pi/4]$

In Problems 11 and 12, let P be a regular partition of the indicated interval and x_k^* the right endpoint of each subinterval. Write the given sum as a definite integral.

11. $\lim\limits_{n\to\infty} \sum\limits_{k=1}^n \left(1 + \frac{2k}{n}\right)\frac{2}{n}$; $[0, 2]$

12. $\lim\limits_{n\to\infty} \sum\limits_{k=1}^n \left(1 + \frac{3k}{n}\right)^3 \frac{3}{n}$; $[1, 4]$

In Problems 13–18, use (5) and the summation formulas in Theorem 5.3.2 to evaluate the given definite integral.

13. $\int_{-3}^1 x\,dx$

14. $\int_0^3 x\,dx$

15. $\int_1^2 (x^2 - x)\,dx$

16. $\int_{-2}^3 (x^2 - 4)\,dx$

17. $\int_0^1 (x^3 - 1)\, dx$

18. $\int_0^2 (3 - x^3)\, dx$

In Problems 19 and 20, proceed as in Problems 13–18 to obtain the given result.

19. $\int_a^b x\, dx = \frac{1}{2}(b^2 - a^2)$

20. $\int_a^b x^2\, dx = \frac{1}{3}(b^3 - a^3)$

21. Use Problem 19 to evaluate $\int_{-1}^3 x\, dx$.

22. Use Problem 20 to evaluate $\int_{-1}^3 x^2\, dx$.

In Problems 23 and 24, use Theorem 5.4.6 to evaluate the given definite integral.

23. $\int_3^6 4\, dx$

24. $\int_{-2}^5 (-2)\, dx$

In Problems 25–38, use Definition 5.4.2 and Theorems 5.4.4, 5.4.5, and 5.4.6 to evaluate the given definite integral. Use the results obtained in Problems 21 and 22 where appropriate.

25. $\int_4^{-2} \frac{1}{2}\, dx$

26. $\int_5^5 10x^4\, dx$

27. $-\int_3^{-1} 10x\, dx$

28. $\int_{-1}^3 (3x + 1)\, dx$

29. $\int_3^{-1} t^2\, dt$

30. $\int_{-1}^3 (3x^2 - 5)\, dx$

31. $\int_{-1}^3 (-3x^2 + 4x - 5)\, dx$

32. $\int_{-1}^3 6x(x - 1)\, dx$

33. $\int_{-1}^0 x^2\, dx + \int_0^3 x^2\, dx$

34. $\int_{-1}^{1.2} 2t\, dt - \int_3^{1.2} 2t\, dt$

35. $\int_0^4 x\, dx + \int_0^4 (9 - x)\, dx$

36. $\int_{-1}^0 t^2\, dt + \int_0^2 x^2\, dx + \int_2^3 u^2\, du$

37. $\int_0^3 x^3\, dx + \int_3^0 t^3\, dt$

38. $\int_{-1}^{-1} 5x\, dx - \int_3^{-1} (x - 4)\, dx$

In Problems 39–42, evaluate the definite integral using the given information.

39. $\int_2^5 f(x)\, dx$ if $\int_0^2 f(x)\, dx = 6$ and $\int_0^5 f(x)\, dx = 8.5$

40. $\int_1^3 f(x)\, dx$ if $\int_1^4 f(x)\, dx = 2.4$ and $\int_3^4 f(x)\, dx = -1.7$

41. $\int_{-1}^2 [2f(x) + g(x)]\, dx$ if
$\int_{-1}^2 f(x)\, dx = 3.4$ and $\int_{-1}^2 3g(x)\, dx = 12.6$

42. $\int_{-2}^2 g(x)\, dx$ if
$\int_2^{-2} f(x)\, dx = 14$ and $\int_{-2}^2 [f(x) - 5g(x)]\, dx = 24$

In Problems 43 and 44, evaluate the definite integrals

(a) $\int_a^b f(x)\, dx$ (b) $\int_b^c f(x)\, dx$ (c) $\int_c^d f(x)\, dx$

(d) $\int_a^c f(x)\, dx$ (e) $\int_b^d f(x)\, dx$ (f) $\int_a^d f(x)\, dx$

using the information in the given figure.

43.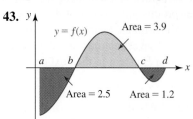

FIGURE 5.4.14 Graph for Problem 43

44.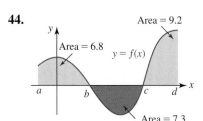

FIGURE 5.4.15 Graph for Problem 44

In Problems 45–48, the given integral represents the area under a graph on an interval. Sketch this region.

45. $\int_{-1}^1 (2x + 3)\, dx$

46. $\int_0^4 (-x^2 + 4x)\, dx$

47. $\int_{2\pi}^{3\pi} \sin x\, dx$

48. $\int_{-2}^0 \sqrt{x + 2}\, dx$

In Problems 49–52, the given integral represents the area under a graph on the interval. Use appropriate formulas from geometry to find the area.

49. $\int_{-2}^4 (x + 2)\, dx$

50. $\int_0^3 |x - 1|\, dx$

51. $\int_0^1 \sqrt{1 - x^2}\, dx$

52. $\int_{-3}^3 (2 + \sqrt{9 - x^2})\, dx$

In Problems 53–56, the given integral represents the net signed area between a graph and the x-axis on an interval. Sketch this region.

53. $\int_0^5 (-2x + 6)\, dx$

54. $\int_{-1}^2 (1 - x^2)\, dx$

55. $\int_{-1/2}^3 \frac{4x}{x + 1}\, dx$

56. $\int_0^{5\pi/2} \cos x\, dx$

In Problems 57–60, the given integral represents the net signed area between a graph and the x-axis on an interval. Use appropriate formulas from geometry to find the net signed area.

57. $\int_{-1}^{4} 2x \, dx$

58. $\int_{0}^{8} \left(\frac{1}{2}x - 2\right) dx$

59. $\int_{-1}^{1} \left(x - \sqrt{1 - x^2}\right) dx$

60. $\int_{-1}^{2} (1 - |x|) \, dx$

In Problems 61–64, the function f is defined to be

$$f(x) = \begin{cases} x, & x \leq 3 \\ 3, & x > 3. \end{cases}$$

Use appropriate formulas from geometry to evaluate the given definite integral.

61. $\int_{-2}^{0} f(x) \, dx$

62. $\int_{-1}^{3} f(x) \, dx$

63. $\int_{-4}^{5} f(x) \, dx$

64. $\int_{0}^{10} f(x) \, dx$

In Problems 65–68, use Theorem 5.4.7 to establish the given inequality.

65. $\int_{-1}^{0} e^x \, dx \leq \int_{-1}^{0} e^{-x} \, dx$

66. $\int_{0}^{\pi/4} (\cos x - \sin x) \, dx \geq 0$

67. $1 \leq \int_{0}^{1} (x^3 + 1)^{1/2} \, dx \leq 1.42$

68. $-2 \leq \int_{0}^{2} (x^2 - 2x) \, dx \leq 0$

In Problems 69 and 70, compare the given two integrals by means of an inequality symbol $\leq$ or $\geq$.

69. $\int_{0}^{1} x^2 \, dx, \quad \int_{0}^{1} x^3 \, dx$

70. $\int_{0}^{1} \sqrt{4 + x^2} \, dx, \quad \int_{0}^{1} \sqrt{4 + x} \, dx$

≡ **Think About It**

71. If f is integrable on the interval $[a, b]$, then so is f^2. Explain why $\int_a^b f^2(x) \, dx \geq 0$.

72. Consider the function defined for all x in the interval $[-1, 1]$:

$$f(x) = \begin{cases} 0, & x \text{ rational} \\ 1, & x \text{ irrational}. \end{cases}$$

Show that f is not integrable on $[-1, 1]$, that is, $\int_{-1}^{1} f(x) \, dx$ does not exist. [*Hint:* The result in (11) may be useful.]

73. Evaluate the definite integral $\int_0^1 \sqrt{x} \, dx$ by using a partition of $[0, 1]$ in which the subintervals $[x_{k-1}, x_k]$ are defined by $[(k-1)^2/n^2, k^2/n^2]$ and choosing x_k^* to be the right endpoint of each subinterval.

74. Evaluate the definite integral $\int_0^{\pi/2} \cos x \, dx$ by using a regular partition of $[0, \pi/2]$ and choosing x_k^* to be the midpoint of each subinterval $[x_{k-1}, x_k]$. Use the known results

(i) $\cos\theta + \cos 3\theta + \cdots + \cos(2n - 1)\theta = \dfrac{\sin 2n\theta}{2\sin\theta}$

(ii) $\lim\limits_{n\to\infty} \dfrac{1}{n\sin(\pi/4n)} = \dfrac{4}{\pi}$.

5.5 Fundamental Theorem of Calculus

Introduction At the end of Section 5.4, we indicated that there is an easier way of evaluating a definite integral than by computing a limit of a sum. This "easier way" is by means of the so-called **Fundamental Theorem of Calculus**. In this section you will see that there are two forms of this important theorem; it is the first form presented next that enables us to evaluate many definite integrals.

Fundamental Theorem of Calculus—First Form In the next theorem we see that the concept of an antiderivative of a continuous function provides the bridge between differential calculus and integral calculus.

Theorem 5.5.1 Fundamental Theorem of Calculus—Antiderivative Form

If f is continuous on an interval $[a, b]$ and F is any antiderivative of f on the interval, then

$$\int_a^b f(x) \, dx = F(b) - F(a). \tag{1}$$

We will present two proofs of Theorem 5.5.1. In the proof given we use the basic premise that a definite integral is a limit of a sum. After we have proved the second form of the Fundamental Theorem of Calculus we will return to Theorem 5.5.1 and present an alternative proof.

▶ **PROOF** If F is an antiderivative of f, then by definition $F'(x) = f(x)$. Since F is differentiable on (a, b), the Mean Value Theorem (Theorem 4.4.2) guarantees that there exists an x_k^* in each subinterval (x_{k-1}, x_k) of the partition P:

$$a = x_0 < x_1 < x_2 < \cdots < x_{n-1} < x_n = b$$

such that

$$F(x_k) - F(x_{k-1}) = F'(x_k^*)(x_k - x_{k-1}) \quad \text{or} \quad F(x_k) - F(x_{k-1}) = f(x_k^*) \Delta x_k.$$

Now, for $k = 1, 2, 3, \ldots, n$ the last result gives

$$F(x_1) - F(a) = f(x_1^*) \Delta x_1$$
$$F(x_2) - F(x_1) = f(x_2^*) \Delta x_2$$
$$F(x_3) - F(x_2) = f(x_3^*) \Delta x_3$$
$$\vdots$$
$$F(b) - F(x_{n-1}) = f(x_n^*) \Delta x_n.$$

If we add the preceding columns,

$$[F(x_1) - F(a)] + [F(x_2) - F(x_1)] + \cdots + [F(b) - F(x_{n-1})] = \sum_{k=1}^{n} f(x_k^*) \Delta x_k$$

we see that all but the two terms in black type on the left-hand side of the equality add to 0 leaving

$$F(b) - F(a) = \sum_{k=1}^{n} f(x_k^*) \Delta x_k. \qquad (2)$$

But $\lim_{\|P\| \to 0} [F(b) - F(a)] = F(b) - F(a)$, and so the limit of (2) as $\|P\| \to 0$ is

$$F(b) - F(a) = \lim_{\|P\| \to 0} \sum_{k=1}^{n} f(x_k^*) \Delta x_k. \qquad (3)$$

From Definition 5.4.1, the right-hand side of (3) is $\int_a^b f(x)\, dx$. ■

The difference $F(b) - F(a)$ in (1) is usually represented by the symbol $F(x) \Big]_a^b$, that is,

$$\underbrace{\int_a^b f(x)\, dx}_{\text{definite integral}} = \underbrace{\left[\int f(x)\, dx \right]_a^b}_{\text{indefinite integral}} = F(x) \Big]_a^b.$$

Since Theorem 5.5.1 indicates that F is *any* antiderivative of f, we may always choose the constant of integration C to be zero. Observe that if $C \neq 0$, then

$$(F(x) + C) \Big]_a^b = (F(b) + C) - (F(a) + C) = F(b) - F(a) = F(x) \Big]_a^b.$$

EXAMPLE 1 Using (1)

In Example 4 of Section 5.4 we resorted to the rather lengthy definition of the definite integral to show that $\int_{-2}^{1} x^3 \, dx = -\frac{15}{4}$. Since $F(x) = \frac{1}{4}x^4$ is an antiderivative of $f(x) = x^3$, we now obtain immediately from (1)

$$\int_{-2}^{1} x^3 \, dx = \frac{x^4}{4} \Big]_{-2}^{1} = \frac{1}{4} - \frac{1}{4}(-2)^4 = \frac{1}{4} - \frac{16}{4} = -\frac{15}{4}. \qquad ■$$

EXAMPLE 2 Using (1)

Evaluate $\int_1^3 x \, dx$.

Solution An antiderivative of $f(x) = x$ is $F(x) = \frac{1}{2}x^2$. Consequently (1) of Theorem 5.5.1 gives

$$\int_1^3 x \, dx = \frac{x^2}{2} \Big]_1^3 = \frac{9}{2} - \frac{1}{2} = 4. \qquad ■$$

EXAMPLE 3 Using (1)

Evaluate $\int_{-2}^{2} (3x^2 - x + 1)\, dx$.

Solution We apply (*ii*) of Theorem 5.1.2 and integration formula 2 of Table 5.1.1 to each term of the integrand and then use the Fundamental Theorem:

$$\int_{-2}^{2} (3x^2 - x + 1)\, dx = \left(x^3 - \frac{x^2}{2} + x\right)\Big]_{-2}^{2}$$
$$= (8 - 2 + 2) - (-8 - 2 - 2) = 20. \quad \blacksquare$$

EXAMPLE 4 Using (1)

Evaluate $\int_{\pi/6}^{\pi} \cos x\, dx$.

Solution An antiderivative of $f(x) = \cos x$ is $F(x) = \sin x$. Therefore,

$$\int_{\pi/6}^{\pi} \cos x\, dx = \sin x \Big]_{\pi/6}^{\pi} = \sin \pi - \sin \frac{\pi}{6} = 0 - \frac{1}{2} = -\frac{1}{2}. \quad \blacksquare$$

■ **Fundamental Theorem of Calculus—Second Form** Suppose f is continuous on an interval $[a, b]$ and so we know that the integral $\int_{a}^{b} f(t)\, dt$ exists. For each x in the interval $[a, b]$, the definite integral

$$g(x) = \int_{a}^{x} f(t)\, dt \qquad (4)$$

represents a single number. In this way, it is seen that (4) is a function with domain $[a, b]$. In FIGURE 5.5.1 we have shown f to be a positive function on $[a, b]$, and so as x varies across the interval we can interpret $g(x)$ as the area under the graph on the interval $[a, x]$. In the second form of the Fundamental Theorem of Calculus we will show that $g(x)$ defined in (4) is a differentiable function.

◀ Keep in mind that a definite integral does not depend on the variable of integration t.

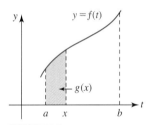

FIGURE 5.5.1 $g(x)$ as area

Theorem 5.5.2 *Fundamental Theorem of Calculus—Derivative Form*

Let f be continuous on $[a, b]$ and let x be any number in the interval. Then $g(x) = \int_{a}^{x} f(t)\, dt$ is continuous on $[a, b]$ and differentiable on (a, b) and

$$g'(x) = f(x). \qquad (5)$$

PROOF FOR $h > 0$ Let x and $x + h$ be in (a, b), where $h > 0$. From the definition of the derivative,

$$g'(x) = \lim_{h \to 0} \frac{g(x + h) - g(x)}{h}. \qquad (6)$$

Using the properties of the definite integral the difference $g(x + h) - g(x)$ can be written

$$g(x + h) - g(x) = \int_{a}^{x+h} f(t)\, dt - \int_{a}^{x} f(t)\, dt$$
$$= \int_{a}^{x+h} f(t)\, dt + \int_{x}^{a} f(t)\, dt \quad \leftarrow \text{by (8) of Section 5.4}$$
$$= \int_{x}^{x+h} f(t)\, dt. \qquad \leftarrow \text{by (10) of Section 5.4}$$

Hence (6) becomes

$$g'(x) = \lim_{h \to 0} \frac{1}{h} \int_{x}^{x+h} f(t)\, dt. \qquad (7)$$

Since f is continuous on the closed interval $[x, x + h]$, we know from the Extreme Value Theorem (Theorem 4.3.1) that f attains a minimum value m and a maximum value M on the interval. Since m and M are constant relative to the integration on the variable t it follows from Theorem 5.4.7(ii) that

$$\int_x^{x+h} m \, dt \le \int_x^{x+h} f(t) \, dt \le \int_x^{x+h} M \, dt. \tag{8}$$

With the aid of Theorem 5.5.1,

$$\int_x^{x+h} m \, dt = mt \Big]_x^{x+h} = m(x + h - x) = mh$$

and

$$\int_x^{x+h} M \, dt = Mt \Big]_x^{x+h} = M(x + h - x) = Mh.$$

Hence the inequality in (8) becomes

$$mh \le \int_x^{x+h} f(t) \, dt \le Mh \quad \text{or} \quad m \le \frac{1}{h} \int_x^{x+h} f(t) \, dt \le M. \tag{9}$$

Because f is continuous on $[x, x + h]$ it stands to reason that $\lim_{h \to 0^+} m = \lim_{h \to 0^+} M = f(x)$. Taking the limit of the second expression in (9) as $h \to 0^+$ gives

$$f(x) \le \lim_{h \to 0^+} \frac{1}{h} \int_x^{x+h} f(t) \, dt \le f(x).$$

This shows that $g'(x)$ exists and from $f(x) \le g'(x) \le f(x)$ we conclude that $g'(x) = f(x)$. Since g is differentiable, it is necessarily continuous. A similar argument holds for $h < 0$. ∎

An alternative, and more traditional, way of writing the result in (5) is

$$\frac{d}{dx} \int_a^x f(t) \, dt = f(x). \tag{10}$$

EXAMPLE 5 Using (10)

From (10),

(a) $\dfrac{d}{dx} \displaystyle\int_{-2}^x t^3 \, dt = x^3$ **(b)** $\dfrac{d}{dx} \displaystyle\int_1^x \sqrt{t^2 + 1} \, dt = \sqrt{x^2 + 1}.$ ∎

EXAMPLE 6 Chain Rule

Find $\dfrac{d}{dx} \displaystyle\int_\pi^{x^3} \cos t \, dt.$

Solution If we identify $g(x) = \int_\pi^x \cos t \, dt$, then the given integral is the composition $g(x^3)$. We carry out the differentiation using the Chain Rule with $u = x^3$:

$$\frac{d}{dx} \int_\pi^{x^3} \cos t \, dt = \frac{d}{du} \left(\int_\pi^u \cos t \, dt \right) \frac{du}{dx}$$

$$= \cos u \cdot \frac{du}{dx} = \cos x^3 \cdot 3x^2$$

$$= 3x^2 \cos x^3. \qquad \blacksquare$$

■ **Alternative Proof of Theorem 5.5.1** It is worthwhile to examine yet another proof of Theorem 5.5.1 using Theorem 5.5.2. For a function f continuous on $[a, b]$, the statement $g'(x) = f(x)$ for $g(x) = \int_a^x f(t) \, dt$ means that $g(x)$ is an antiderivative of the integrand f. If F is any antiderivative of f, we know from Theorem 5.1.1 that $g(x) - F(x) = C$ or

$g(x) = F(x) + C$, where C is any arbitrary constant. Since $g(x) = \int_a^x f(t)\,dt$, it follows for any x in $[a, b]$ that

$$\int_a^x f(t)\,dt = F(x) + C. \tag{11}$$

If we substitute $x = a$ in (11), then

$$\int_a^a f(t)\,dt = F(a) + C$$

implies $C = -F(a)$, since $\int_a^a f(t)\,dt = 0$. Thus, (11) becomes

$$\int_a^x f(t)\,dt = F(x) - F(a).$$

Since the latter equation is valid at $x = b$, we find

$$\int_a^b f(t)\,dt = F(b) - F(a). \qquad \blacksquare$$

■ **Piecewise-Continuous Functions** A function f is said to be **piecewise continuous** on an interval $[a, b]$ if there are at most a finite number of points c_k, $k = 1, 2, \ldots, n$, $(c_{k-1} < c_k)$ at which f has a finite, or jump, discontinuity and f is continuous on each open interval (c_{k-1}, c_k). See FIGURE 5.5.2. If a function f is piecewise continuous on $[a, b]$, it is bounded on the interval, and hence by Theorem 5.4.2, f is integrable on $[a, b]$. A definite integral of a piecewise-continuous function on $[a, b]$ can be evaluated with the help of Theorem 5.4.5:

$$\int_a^b f(x)\,dx = \int_a^{c_1} f(x)\,dx + \int_{c_1}^{c_2} f(x)\,dx + \cdots + \int_{c_n}^b f(x)\,dx$$

FIGURE 5.5.2 Piecewise-continuous function

and by simply treating the integrands of the definite integrals on the right side of the above equation as if they were continuous on the closed intervals $[a, c_1], [c_1, c_2], \ldots, [c_n, b]$.

EXAMPLE 7 Integrating a Piecewise-Continuous Function

Evaluate $\int_{-1}^4 f(x)\,dx$ where

$$f(x) = \begin{cases} x + 1, & -1 \le x < 0 \\ x, & 0 \le x < 2 \\ 3, & 2 \le x \le 4. \end{cases}$$

Solution The graph of the piecewise-continuous function f is given in FIGURE 5.5.3. Now, from the preceding discussion and the definition of f:

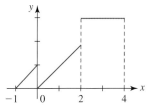

FIGURE 5.5.3 Graph of function in Example 7

EXAMPLE 8 Integrating a Piecewise-Continuous Function

Evaluate $\int_0^3 |x - 2|\,dx$.

Solution From the definition of absolute value,

$$|x - 2| = \begin{cases} x - 2 & \text{if } x - 2 \ge 0 \\ -(x - 2) & \text{if } x - 2 < 0 \end{cases} \quad \text{or} \quad |x - 2| = \begin{cases} x - 2 & \text{if } x \ge 2 \\ -x + 2 & \text{if } x < 2. \end{cases}$$

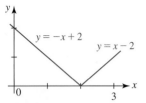

FIGURE 5.5.4 Graph of function in Example 8

The graph of $f(x) = |x - 2|$ is given in FIGURE 5.5.4. Now in view of (10) of Theorem 5.4.5 we can write

$$\int_0^3 |x - 2| \, dx = \int_0^2 |x - 2| \, dx + \int_2^3 |x - 2| \, dx$$

$$= \int_0^2 (-x + 2) \, dx + \int_2^3 (x - 2) \, dx$$

$$= \left(-\frac{1}{2}x^2 + 2x\right)\Big]_0^2 + \left(\frac{1}{2}x^2 - 2x\right)\Big]_2^3$$

$$= (-2 + 4) + \left(\frac{9}{2} - 6\right) - (2 - 4) = \frac{5}{2}. \quad \blacksquare$$

■ **Substitution in a Definite Integral** Recall from Section 5.2 that we sometimes used a substitution as an aid in evaluating an indefinite integral of the form $\int f(g(x))g'(x) \, dx$. Care should be exercised when using a substitution in a definite integral $\int_a^b f(g(x))g'(x) \, dx$, since we can proceed in *two ways*.

Guidelines for Substituting in a Definite Integral

- Evaluate the indefinite integral $\int f(g(x))g'(x) \, dx$ by means of the substitution $u = g(x)$. Resubstitute $u = g(x)$ in the antiderivative and then apply the Fundamental Theorem of Calculus by using the original limits of integration $x = a$ and $x = b$.
- Alternatively, the resubstitution can be avoided by changing the limits of integration to correspond to the value of u at $x = a$ and u at $x = b$. The latter method, which is usually quicker, is summarized in the next theorem.

Theorem 5.5.3 Substitution in a Definite Integral

Let $u = g(x)$ be a function that has a continuous derivative on the interval $[a, b]$, and let f be a function that is continuous on the range of g. If $F'(u) = f(u)$ and $c = g(a), d = g(b)$, then

$$\int_a^b f(g(x))g'(x) \, dx = \int_{g(a)}^{g(b)} f(u) \, du = F(d) - F(c). \qquad (12)$$

PROOF If $u = g(x)$, then $du = g'(x) \, dx$. Therefore,

$$\int_a^b f(g(x))g'(x) \, dx = \int_{g(a)}^{g(b)} f(u) \frac{du}{dx} \, dx = \int_c^d f(u) \, du = F(u)\Big]_c^d = F(d) - F(c). \quad \blacksquare$$

EXAMPLE 9 Substitution in a Definite Integral

Evaluate $\int_0^2 \sqrt{2x^2 + 1} \, x \, dx.$

Solution We shall first illustrate the two procedures outlined in the guidelines preceding Theorem 5.5.3.

(a) To evaluate the indefinite integral $\int \sqrt{2x^2 + 1} \, x \, dx$, we use $u = 2x^2 + 1$ and $du = 4x \, dx$. Thus,

$$\int \sqrt{2x^2 + 1} \, x \, dx = \frac{1}{4} \int \sqrt{2x^2 + 1} \, (4x \, dx) \quad \leftarrow \text{substitution}$$

$$= \frac{1}{4} \int u^{1/2} \, du$$

$$= \frac{1}{4} \frac{u^{3/2}}{3/2} + C$$

$$= \frac{1}{6} (2x^2 + 1)^{3/2} + C. \quad \leftarrow \text{resubstitution}$$

Therefore, by Theorem 5.5.1,

$$\int_0^2 \sqrt{2x^2 + 1}\, x\, dx = \frac{1}{6}(2x^2 + 1)^{3/2} \Big]_0^2$$

$$= \frac{1}{6}[9^{3/2} - 1^{3/2}]$$

$$= \frac{1}{6}[27 - 1] = \frac{13}{3}.$$

(b) If $u = 2x^2 + 1$, then $x = 0$ implies $u = 1$, whereas $x = 2$ gives $u = 9$. Thus, by Theorem 5.5.3,

$$\int_0^2 \sqrt{2x^2 + 1}\, x\, dx = \frac{1}{4} \int_1^9 u^{1/2}\, du \quad \leftarrow \begin{array}{l} \text{integration with} \\ \text{respect to } u \end{array}$$

with u limits indicated on the integral.

$$= \frac{1}{4} \frac{u^{3/2}}{3/2} \Big]_1^9$$

$$= \frac{1}{6}[9^{3/2} - 1^{3/2}] = \frac{13}{3}. \blacksquare$$

When the graph of a function $y = f(x)$ is symmetric with respect to either the y-axis (even function) or the origin (odd function), then the definite integral of f on a symmetric interval $[-a, a]$, that is, $\int_{-a}^{a} f(x)\, dx$, can be evaluated by means of a "shortcut."

Theorem 5.5.4   Even Function Rule

If f is an even integrable function on $[-a, a]$, then

$$\int_{-a}^{a} f(x)\, dx = 2 \int_0^a f(x)\, dx. \tag{13}$$

We will prove the next theorem but leave the proof of Theorem 5.5.4 as an exercise.

Theorem 5.5.5   Odd Function Rule

If f is an odd integrable function on $[-a, a]$, then

$$\int_{-a}^{a} f(x)\, dx = 0. \tag{14}$$

PROOF   Assume f is an odd function. By the additive interval property, Theorem 5.4.5, we have

$$\int_{-a}^{a} f(x)\, dx = \int_{-a}^{0} f(x)\, dx + \int_0^a f(x)\, dx.$$

In the first integral on the right-hand side, let $x = -t$, so that $dx = -dt$, and when $x = -a$ and $x = 0$, then $t = a$ and $t = 0$:

$$\int_{-a}^{a} f(x)\, dx = \int_a^0 f(-t)(-dt) + \int_0^a f(x)\, dx \quad \leftarrow f(-t) = -f(t), f \text{ an odd function}$$

$$= \int_a^0 f(t)\, dt + \int_0^a f(x)\, dx$$

$$= -\int_0^a f(t)\, dt + \int_0^a f(x)\, dx \quad \leftarrow \text{by (8) of Section 5.4}$$

$$= -\int_0^a f(x)\,dx + \int_0^a f(x)\,dx \quad \leftarrow t \text{ was a "dummy variable" of integration}$$

$$= 0.$$

The point in Theorem 5.5.5 is this: When we integrate an odd integrable function f on a symmetric interval $[-a, a]$, there is no need to find an antiderivative of f; the value of the integral is always zero.

Geometric motivations for the results in Theorems 5.5.4 and 5.5.5 are given in FIGURE 5.5.5.

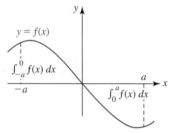

(a) Even function: Value of the definite integral on $[-a, 0]$ is the same as the value on $[0, a]$

(b) Odd function: Value of the definite integral on $[-a, 0]$ is the opposite of the value on $[0, a]$

FIGURE 5.5.5 Even Function Rule in (a); Odd Function Rule in (b)

EXAMPLE 10 Using the Even Function Rule

Evaluate $\int_{-1}^{1} (x^4 + x^2)\,dx$.

Solution The integrand $f(x) = x^4 + x^2$ is a polynomial function with all even powers, and so f is necessarily an even function. Since the interval of integration is the symmetric interval $[-1, 1]$, it follows from Theorem 5.5.4 that we can integrate on $[0, 1]$ and multiply the result by 2:

$$\int_{-1}^{1} (x^4 + x^2)\,dx = 2\int_0^1 (x^4 + x^2)\,dx$$

$$= 2\left(\frac{1}{5}x^5 + \frac{1}{3}x^3\right)\Big]_0^1$$

$$= 2\left(\frac{1}{5} + \frac{1}{3}\right) = \frac{16}{15}.$$

EXAMPLE 11 Using the Odd Function Rule

Evaluate $\int_{-\pi/2}^{\pi/2} \sin x\,dx$.

Solution In this case $f(x) = \sin x$ is an odd function on the symmetric interval $[-\pi/2, \pi/2]$. Thus, by Theorem 5.5.5 we have immediately

$$\int_{-\pi/2}^{\pi/2} \sin x\,dx = 0.$$

NOTES FROM THE CLASSROOM

The antiderivative form of the Fundamental Theorem of Calculus is an extremely important and powerful tool for evaluating definite integrals. Why should we bother with a clumsy limit of a sum when the value of $\int_a^b f(x)\,dx$ can be found by computing $\int f(x)\,dx$ at the two numbers a and b? This is true up to a point—however, it is time to learn another fact of mathematical life. There are continuous functions for which the antiderivative $\int f(x)\,dx$

cannot be expressed in terms of *elementary functions*: sums, products, quotients, and powers of polynomial functions, trigonometric functions, inverse trigonometric functions, logarithmic, and exponential functions. The simple continuous function $f(x) = \sqrt{x^3 + 1}$ possesses no antiderivative that is an elementary function. Although, in view of Theorem 5.4.1 we can say that the definite integral $\int_0^1 \sqrt{x^3 + 1}\, dx$ exists, Theorem 5.5.1 provides no help in finding its value. The integral $\int_0^1 \sqrt{x^3 + 1}\, dx$ is called **nonelementary**. Nonelementary integrals are important and appear in many applications such as probability theory and optics. Here a few more nonelementary integrals:

$$\int \frac{\sin x}{x}\, dx, \quad \int \sin x^2\, dx, \quad \int_0^x e^{-t^2}\, dt, \quad \text{and} \quad \int \frac{e^x}{x}\, dx.$$

See Problems 71 and 72 in Exercises 5.5.

Exercises 5.5   Answers to selected odd-numbered problems begin on page ANS-19.

≡ Fundamentals

In Problems 1–42, use the Fundamental Theorem of Calculus given in Theorem 5.5.1 to evaluate the given definite integral.

1. $\int_3^7 dx$
2. $\int_2^{10} (-4)\, dx$
3. $\int_{-1}^2 (2x + 3)\, dx$
4. $\int_{-5}^4 t^2\, dt$
5. $\int_1^3 (6x^2 - 4x + 5)\, dx$
6. $\int_{-2}^1 (12x^5 - 36)\, dx$
7. $\int_0^{\pi/2} \sin x\, dx$
8. $\int_{-\pi/3}^{\pi/4} \cos\theta\, d\theta$
9. $\int_{\pi/4}^{\pi/2} \cos 3t\, dt$
10. $\int_{1/2}^1 \sin 2\pi x\, dx$
11. $\int_{1/2}^{3/4} \frac{1}{u^2}\, du$
12. $\int_{-3}^{-1} \frac{2}{x}\, dx$
13. $\int_{-1}^1 e^x\, dx$
14. $\int_0^2 (2x - 3e^x)\, dx$
15. $\int_0^2 x(1 - x)\, dx$
16. $\int_3^2 x(x - 2)(x + 2)\, dx$
17. $\int_{-1}^1 (7x^3 - 2x^2 + 5x - 4)\, dx$
18. $\int_{-3}^{-1} (x^2 - 4x + 8)\, dx$
19. $\int_1^4 \frac{x - 1}{\sqrt{x}}\, dx$
20. $\int_2^4 \frac{x^2 + 8}{x^2}\, dx$
21. $\int_1^{\sqrt{3}} \frac{1}{1 + x^2}\, dx$
22. $\int_0^{1/4} \frac{1}{\sqrt{1 - 4x^2}}\, dx$
23. $\int_{-4}^{12} \sqrt{z + 4}\, dz$
24. $\int_0^{7/2} (2x + 1)^{-1/3}\, dx$
25. $\int_0^3 \frac{x}{\sqrt{x^2 + 16}}\, dx$
26. $\int_{-2}^1 \frac{t}{(t^2 + 1)^2}\, dt$
27. $\int_{1/2}^1 \left(1 + \frac{1}{x}\right)^3 \frac{1}{x^2}\, dx$
28. $\int_1^4 \frac{\sqrt[3]{1 + 4\sqrt{x}}}{\sqrt{x}}\, dx$
29. $\int_0^1 \frac{x + 1}{\sqrt{x^2 + 2x + 3}}\, dx$
30. $\int_{-1}^1 \frac{u^3 + u}{(u^4 + 2u^2 + 1)^5}\, du$
31. $\int_0^{\pi/8} \sec^2 2x\, dx$
32. $\int_{\sqrt{\pi/4}}^{\sqrt{\pi/2}} x\csc x^2 \cot x^2\, dx$
33. $\int_{-1/2}^{3/2} (x - \cos\pi x)\, dx$
34. $\int_1^4 \frac{\cos\sqrt{x}}{2\sqrt{x}}\, dx$
35. $\int_0^{\pi/2} \sqrt{\cos x}\, \sin x\, dx$
36. $\int_{\pi/6}^{\pi/3} \sin x \cos x\, dx$
37. $\int_{\pi/6}^{\pi/2} \frac{1 + \cos\theta}{(\theta + \sin\theta)^2}\, d\theta$
38. $\int_{-\pi/4}^{\pi/4} (\sec x + \tan x)^2\, dx$
39. $\int_0^{3/4} \sin^2 \pi x\, dx$
40. $\int_{-\pi/2}^{\pi/2} \cos^2 x\, dx$
41. $\int_1^5 \frac{1}{1 + 2x}\, dx$
42. $\int_{-1}^1 \tan x\, dx$

In Problems 43–48, use the Fundamental Theorem of Calculus given in Theorem 5.5.2 to find the indicated derivative.

43. $\dfrac{d}{dx} \int_0^x t e^t\, dt$
44. $\dfrac{d}{dx} \int_1^x \ln t\, dt$
45. $\dfrac{d}{dt} \int_2^t (3x^2 - 2x)^6\, dx$
46. $\dfrac{d}{dx} \int_x^9 \sqrt[3]{u^2 + 2}\, du$
47. $\dfrac{d}{dx} \int_3^{6x-1} \sqrt{4t + 9}\, dt$
48. $\dfrac{d}{dx} \int_\pi^{\sqrt{x}} \sin t^2\, dt$

In Problems 49 and 50, use the Fundamental Theorem of Calculus given in Theorem 5.5.2 to find $F'(x)$. [*Hint*: Use two integrals.]

49. $F(x) = \int_{3x}^{x^2} \dfrac{1}{t^3 + 1}\, dt$
50. $F(x) = \int_{\sin x}^{5x} \sqrt{t^2 + 1}\, dt$

In Problems 51 and 52, verify the given result by first evaluating the definite integral and then differentiating.

51. $\dfrac{d}{dx}\displaystyle\int_1^x (6t^2 - 8t + 5)\,dt = 6x^2 - 8x + 5$

52. $\dfrac{d}{dt}\displaystyle\int_\pi^t \sin\dfrac{x}{3}\,dx = \sin\dfrac{t}{3}$

53. Consider the function $f(x) = \int_1^x \ln(2t+1)\,dt$. Find the indicated function value.

   (a) $f(1)$ (b) $f'(1)$
   (c) $f''(1)$ (d) $f'''(1)$

54. Suppose $G(x) = \int_a^x f(t)\,dt$ and $G'(x) = f(x)$. Find the given expression.

   (a) $G(x^2)$ (b) $\dfrac{d}{dx}G(x^2)$
   (c) $G(x^3 + 2x)$ (d) $\dfrac{d}{dx}G(x^3 + 2x)$

In Problems 55 and 56, evaluate $\int_{-1}^2 f(x)\,dx$ for the given function f.

55. $f(x) = \begin{cases} -x, & x < 0 \\ x^2, & x \geq 0 \end{cases}$

56. $f(x) = \begin{cases} 2x + 3, & x \leq 0 \\ 3, & x > 0 \end{cases}$

In Problems 57–60, evaluate the definite integral of the given piecewise-continuous function f.

57. $\displaystyle\int_0^3 f(x)\,dx$, where $f(x) = \begin{cases} 4, & 0 \leq x < 2 \\ 1, & 2 \leq x \leq 3 \end{cases}$

58. $\displaystyle\int_0^\pi f(x)\,dx$, where $f(x) = \begin{cases} \sin x, & 0 \leq x < \pi/2 \\ \cos x, & \pi/2 \leq x \leq \pi \end{cases}$

59. $\displaystyle\int_{-2}^2 f(x)\,dx$, where $f(x) = \begin{cases} x^2, & -2 \leq x < -1 \\ 4, & -1 \leq x < 1 \\ x^2, & 1 \leq x \leq 2 \end{cases}$

60. $\displaystyle\int_0^4 f(x)\,dx$, where $f(x) = \lfloor x \rfloor$ is the greatest integer function.

In Problems 61–66, proceed as in Example 8 to evaluate the given definite integral.

61. $\displaystyle\int_{-3}^1 |x|\,dx$

62. $\displaystyle\int_0^4 |2x - 6|\,dx$

63. $\displaystyle\int_{-8}^3 \sqrt{|x| + 1}\,dx$

64. $\displaystyle\int_0^2 |x^2 - 1|\,dx$

65. $\displaystyle\int_{-\pi}^\pi |\sin x|\,dx$

66. $\displaystyle\int_0^\pi |\cos x|\,dx$

In Problems 67–70, proceed as in part (b) of Example 9 and evaluate the given definite integral using the indicated u-substitution.

67. $\displaystyle\int_{1/2}^e \dfrac{(\ln 2t)^5}{t}\,dt;\quad u = \ln 2t$

68. $\displaystyle\int_{\sqrt{2}/2}^1 \dfrac{1}{(\tan^{-1}x)(1+x^2)}\,dx;\quad u = \tan^{-1}x$

69. $\displaystyle\int_0^1 \dfrac{e^{-2x}}{e^{-2x}+1}\,dx;\quad u = e^{-2x}+1$

70. $\displaystyle\int_0^{1/\sqrt{2}} \dfrac{x}{\sqrt{1-x^4}}\,dx;\quad u = x^2$

≡ Applications

71. In applied mathematics some important functions are defined in terms of nonelementary integrals. One such special function is called the **error function** and is defined as
$$\text{erf}(x) = \dfrac{2}{\sqrt{\pi}}\int_0^x e^{-t^2}\,dt.$$

   (a) Show that $\text{erf}(x)$ is an increasing function on the interval $(-\infty, \infty)$.
   (b) Show that the function $y = e^{x^2}[1 + \sqrt{\pi}\,\text{erf}(x)]$ satisfies the differential equation
$$\dfrac{dy}{dx} - 2xy = 2,$$
   and that $y(0) = 1$.

72. Another special function defined by a nonelementary integral is the **sine integral function**
$$\text{Si}(x) = \int_0^x \dfrac{\sin t}{t}\,dt.$$
   The function $\text{Si}(x)$ has an infinite number of relative extrema.

   (a) Find the first four critical numbers for $x > 0$. Use the second derivative test to determine whether these critical numbers correspond to a relative maximum or a relative minimum.
   (b) Use a CAS to obtain the graph of $\text{Si}(x)$. [*Hint:* In *Mathematica* the sine integral function is denoted by SinIntegral[x].]

≡ Think About It

In Problems 73 and 74, let P be a partition of the indicated interval and x_k^* a number in the kth subinterval. Determine the value of the given limit.

73. $\displaystyle\lim_{\|P\|\to 0}\sum_{k=1}^n (2x_k^* + 5)\,\Delta x_k;\quad [-1, 3]$

74. $\displaystyle\lim_{\|P\|\to 0}\sum_{k=1}^n \cos\dfrac{x_k^*}{4}\,\Delta x_k;\quad [0, 2\pi]$

In Problems 75 and 76, let P be a regular partition of the indicated interval and x_k^* a number in the kth subinterval. Establish the given result.

75. $\displaystyle\lim_{n\to\infty}\dfrac{\pi}{n}\sum_{k=1}^n \sin x_k^* = 2;\quad [0, \pi]$

76. $\displaystyle\lim_{n\to\infty}\dfrac{2}{n}\sum_{k=1}^n x_k^* = 0;\quad [-1, 1]$

In Problems 77 and 78, evaluate the given definite integral.

77. $\int_{-1}^{2} \left\{ \int_{1}^{x} 12t^2 \, dt \right\} dx$ **78.** $\int_{0}^{\pi/2} \left\{ \int_{0}^{t} \sin x \, dx \right\} dt$

79. Prove the even function rule, Theorem 5.5.4.

80. Suppose f is an odd function that is defined on the interval $[-4, 4]$. Suppose further that f is differentiable on the interval, $f(-2) = 3.5$, has zeros at -3 and 3, and has critical numbers -2 and 2.

(a) What is $f(0)$?
(b) Sketch a rough graph of f.
(c) Suppose F is a function defined on $[-4, 4]$ by $F(x) = \int_{-3}^{x} f(t) \, dt$. Find $F(-3)$ and $F(3)$.
(d) Sketch a rough graph of F.
(e) Find the critical numbers and points of inflection of F.

81. Determine whether the following reasoning is correct:

$$\int_{-\pi/2}^{\pi/2} \sin^2 t \, dt = -\int_{-\pi/2}^{\pi/2} \sin t (-\sin t \, dt)$$

$$= -\int_{-\pi/2}^{\pi/2} \sqrt{1 - \cos^2 t} \, (-\sin t \, dt) \leftarrow \begin{cases} u = \cos t \\ du = -\sin t \, dt \end{cases}$$

$$= -\int_{0}^{0} \sqrt{1 - u^2} \, du = 0. \leftarrow \begin{cases} \text{Theorem 5.5.3} \\ \text{Definition 5.4.2}(i) \end{cases}$$

82. Compute the derivatives.

(a) $\dfrac{d}{dx} \int_{1}^{2x} \sqrt{t^3 + 7} \, dt$ (b) $\dfrac{d}{dx} \int_{1}^{4} \sqrt{t^3 + 7} \, dt$

≡ **Calculator/CAS Problems**

83. (a) Use a calculator or CAS to obtain the graphs of $f(x) = \cos^3 x$ and $g(x) = \sin^3 x$.
(b) Based on your interpretation of net signed area, use the graphs in part (a) to conjecture the values of $\int_{0}^{2\pi} \cos^3 x \, dx$ and $\int_{0}^{2\pi} \sin^3 x \, dx$.

≡ **Projects**

84. Integration by Darts In this problem we illustrate a method for approximating the area under a graph by "throwing darts." Suppose that we wish to find the area A under the graph of $f(x) = \cos^3(\pi x/2)$ on the interval $[0, 1]$; that is, we wish to approximate $A = \int_{0}^{1} \cos^3(\pi x/2) \, dx$.

If we throw, with no particular attempt at being skillful, a large number of darts, say N, at a 1×1 square target shown in FIGURE 5.5.6 and n darts strike the red-colored region under the graph of $f(x) = \cos^3(\pi x/2)$, then it can be shown that the probability of a dart striking the region is given by the ratio of two areas:

$$\frac{\text{area of region}}{\text{area of square}} = \frac{A}{1}.$$

Moreover, this theoretical probability is approximately the same as the empirical probability n/N:

$$\frac{A}{1} \approx \frac{n}{N} \quad \text{or} \quad A \approx \frac{n}{N}.$$

To simulate the throwing of darts at the target, use a CAS such as *Mathematica* and its random number function to generate a table of N ordered pairs (x, y), $0 < x < 1$, $0 < y < 1$.

(a) Let $N = 50$. Plot the points and the graph of f on the same set of coordinate axes. Use the figure to count the number of hits n. Construct at least 10 different tables of random points and plots. For each plot compute the ratio n/N.
(b) Repeat part (a) for $N = 100$.
(c) Use the CAS to find the exact value of area A and compare this value with the approximations obtained in parts (a) and (b).

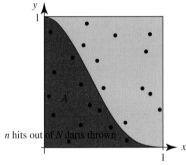

FIGURE 5.5.6 Target in Problem 84

85. Expanding Oil Spill A mathematical model that can be used to determine the time t required for an expanding oil spill to evaporate is given by the formula

$$\frac{RT}{Pv} = \int_{0}^{t} \frac{KA(u)}{V_0} \, du,$$

where $A(u)$ is the area of the spill at time u, RT/Pv is a dimensionless thermodynamic term, K is a mass transfer coefficient, and V_0 is the initial volume of the spill.

(a) Suppose the oil spill is expanding in the form of a circle whose initial radius is r_0. See FIGURE 5.5.7. If the radius r of the spill is increasing at a rate $dr/dt = C$ (in meters per second), solve for t in terms of the other symbols.
(b) Typical values for RT/Pv and K are 1.9×10^6 (for tridecane) and 0.01 mm/s, respectively. If $C = 0.01$ m/s^2, $r_0 = 100$ m, and $V_0 = 10,000$ m^3, determine how long it will take for the oil to evaporate.
(c) Using the result in part (b), determine the final area of the oil spill.

FIGURE 5.5.7 Circular oil spill in Problem 85

86. The Mercator Projection and the Integral of $\sec x$
Roughly, a Mercator map is a representation of a three-dimensional global map onto a two-dimensional surface. See FIGURE 5.5.8. Find and study the article, "Mercator's World Map and the Calculus," Philip M. Tuchinsky, UMAP, Unit 206, Newton, MA, 1978. Write a short report summarizing the article and why **Gerhardus Mercator** (c. 1569) needed the value of the definite integral $\int_0^{\theta_0} \sec x \, dx$ to carry out his constructions.

(a) Globe

(b) Mercator map

FIGURE 5.5.8 Globe and Mercator Projection in Problem 86

Chapter 5 in Review

Answers to selected odd-numbered problems begin on page ANS-19.

A. True/False

In Problems 1–16, indicate whether the given statement is true or false.

1. If $f'(x) = 3x^2 + 2x$, then $f(x) = x^3 + x^2$. _____

2. $\sum_{k=2}^{6} (2k - 3) = \sum_{j=0}^{4} (2j + 1)$ _____

3. $\sum_{k=1}^{40} 5 = \sum_{k=1}^{20} 10$ _____

4. $\int_1^3 \sqrt{t^2 + 7} \, dt = -\int_3^1 \sqrt{t^2 + 7} \, dt$ _____

5. If f is continuous, then $\int_0^1 f(t) \, dt + \int_1^0 f(x) \, dx = 0$. _____

6. If f is integrable, then f is continuous. _____

7. $\int_0^1 (x - x^3) \, dx$ is the area under the graph of $y = x - x^3$ on the interval $[0, 1]$. _____

8. If $\int_a^b f(x) \, dx > 0$, then $\int_a^b f(x) \, dx$ is the area under the graph of f on $[a, b]$. _____

9. If P is a partition of $[a, b]$ into n subintervals, then $n \to \infty$ implies $\|P\| \to 0$. _____

10. If $F'(x) = 0$ for all x, then $F(x) = C$ for all x. _____

11. If f is an odd integrable function on $[-\pi, \pi]$, then $\int_{-\pi}^{\pi} f(x) \, dx = 0$. _____

12. $\int_{-1}^{1} |x| \, dx = 2\int_0^1 x \, dx$ _____

13. $\int \sin x \, dx = \cos x + C$ _____

14. $\int x \cos x \, dx = x \sin x + \cos x + C$ _____

15. $\int_a^b f'(t) \, dt = f(b) - f(a)$ _____

16. The function $F(x) = \int_{-5}^{2x} (t + 4)e^{-t} \, dt$ is increasing on the interval $[-2, \infty)$. _____

B. Fill in the Blanks

In Problems 1–16, fill in the blanks.

1. If G is an antiderivative of a function f, then $G'(x) = $ _____.

2. $\int \dfrac{d}{dx} x^2 \, dx = $ _____.

3. If $\int f(x) \, dx = \dfrac{1}{2}(\ln x)^2 + C$, then $f(x) = $ _____.

4. The value of $\dfrac{d}{dx} \int_3^x \sqrt{t^2 + 5} \, dt$ at $x = 1$ is _____.

5. If g is differentiable, then $\dfrac{d}{dx} \int_{g(x)}^b f(t) \, dt = $ _____.

6. $\dfrac{d}{dx} \int_{5x}^{\sqrt{x}} e^{-t^2} \, dt = $ _____.

7. Using sigma notation, the sum $\dfrac{1}{3} + \dfrac{2}{5} + \dfrac{3}{7} + \dfrac{4}{9} + \dfrac{5}{11}$ can be expressed as _____.

8. The numerical value of $\sum_{k=1}^{15} (3k^2 - 2k)$ is _____.

9. If $u = t^2 + 1$, then the definite integral $\int_2^4 t(t^2 + 1)^{1/3} \, dt$ becomes $\dfrac{1}{2} \int_{\_}^{\_} u^{1/3} \, du$.

10. The area under the graph of $f(x) = 2x$ on the interval $[0, 2]$ is _____, and the net signed area between the graph of $f(x) = 2x$ and the x-axis on $[-1, 2]$ is _____.

11. If the interval $[1, 6]$ is partitioned into four subintervals determined by $x_0 = 1$, $x_1 = 2$, $x_2 = \dfrac{5}{2}$, $x_3 = 5$, and $x_4 = 6$, the norm of the partition is _____.

12. A partition of an interval $[a, b]$ in which all the subintervals have equal width is called a _____ partition.

13. If P is a partition of $[0, 4]$ and x_k^* is a number in the kth subinterval, then $\lim_{\|P\| \to 0} \sum_{k=1}^n \sqrt{x_k^*} \, \Delta x_k$ is the definition of the definite integral _____. By the Fundamental Theorem of Calculus, the value of this definite integral is _____.

14. If $\int_0^6 f(x) \, dx = 11$ and $\int_0^4 f(x) \, dx = 15$, then $\int_4^6 f(x) \, dx = $ _____.

15. $\int_{-1}^1 \left\{ \int_0^x e^{-t} \, dt \right\} dx = $ _____ and $\int_{-1}^1 \dfrac{d}{dx} \left\{ \int_0^x e^{-t} \, dt \right\} dx = $ _____.

16. For $t > 0$, the net signed area $\int_0^t (x^3 - x^2) \, dx = 0$ when $t = $ _____.

C. Exercises

In Problems 1–20, evaluate the given integral.

1. $\int_{-1}^1 (4x^3 - 6x^2 + 2x - 1) \, dx$

2. $\int_1^9 \dfrac{6}{\sqrt{x}} \, dx$

3. $\int (5t + 1)^{100} \, dt$

4. $\int w^2 \sqrt{3w^3 + 1} \, dw$

5. $\int_0^{\pi/4} (\sin 2x - 5 \cos 4x) \, dx$

6. $\int_{\pi^2/9}^{\pi^2} \dfrac{\sin \sqrt{z}}{\sqrt{z}} \, dz$

7. $\int_4^4 (-2x^2 + x^{1/2}) \, dx$

8. $\int_{-\pi/4}^{\pi/4} dx + \int_{-\pi/4}^{\pi/4} \tan^2 x \, dx$

9. $\displaystyle\int \cot^6 8x \csc^2 8x \, dx$

10. $\displaystyle\int \csc 3x \cot 3x \, dx$

11. $\displaystyle\int (4x^2 - 16x + 7)^4 (x - 2) \, dx$

12. $\displaystyle\int (x^2 + 2x - 10)^{2/3}(5x + 5) \, dx$

13. $\displaystyle\int \frac{x^2 + 1}{\sqrt[3]{x^3 + 3x - 16}} \, dx$

14. $\displaystyle\int \frac{x^2 + 1}{x^3 + 3x - 16} \, dx$

15. $\displaystyle\int_0^4 \frac{x}{16 + x^2} \, dx$

16. $\displaystyle\int_0^4 \frac{1}{16 + x^2} \, dx$

17. $\displaystyle\int_0^2 \frac{1}{\sqrt{16 - x^2}} \, dx$

18. $\displaystyle\int_0^2 \frac{x}{\sqrt{16 - x^2}} \, dx$

19. $\displaystyle\int \tan 10x \, dx$

20. $\displaystyle\int \cot 10x \, dx$

21. Suppose $\displaystyle\int_0^5 f(x) \, dx = -3$ and $\displaystyle\int_0^7 f(x) \, dx = 2$. Evaluate $\displaystyle\int_5^7 f(x) \, dx$.

22. Suppose $\displaystyle\int_1^4 f(x) \, dx = 2$ and $\displaystyle\int_4^9 f(x) \, dx = -8$. Evaluate $\displaystyle\int_1^9 f(x) \, dx$.

In Problems 23–28, evaluate the given integral.

23. $\displaystyle\int_0^3 (1 + |x - 1|) \, dx$

24. $\displaystyle\int_0^1 \frac{d}{dt}\left[\frac{10t^4}{(2t^3 + 6t + 1)^2}\right] dt$

25. $\displaystyle\int_{\pi/2}^{\pi/2} \frac{\sin^{10} t}{16t^7 + 1} \, dt$

26. $\displaystyle\int_{-1}^1 t^5 \sin t^2 \, dt$

27. $\displaystyle\int_{-1}^1 \frac{1}{1 + 3x^2} \, dx$

28. $\displaystyle\int_{-2}^2 f(x) \, dx$, where $f(x) = \begin{cases} x^3, & x \leq 0 \\ x^2, & 0 < x \leq 1 \\ x, & x > 1 \end{cases}$

In Problems 29 and 30, find the given limit.

29. $\displaystyle\lim_{n \to \infty} \frac{1 + 2 + 3 + \cdots + n}{n^2}$

30. $\displaystyle\lim_{n \to \infty} \frac{1^2 + 2^2 + 3^2 + \cdots + n^2}{n^3}$

31. A bucket with dimensions (in feet) shown in FIGURE 5.R.1 is filled at a constant rate of $dV/dt = \frac{1}{4}$ ft^3/min. At $t = 0$ the scale reads 31.2 lb. If water weighs 62.4 lb/ft^3, what does the scale read at the end of 8 min? When is the bucket full? [*Hint*: See page RP-2 for the formula for the volume of a frustum of a cone. Also, ignore the weight of the bucket.]

FIGURE 5.R.1 Bucket and scale in Problem 31

32. The **Tower of Hanoi** is a stack of circular disks, each larger than the one above it, set on a pole through holes in the disks' centers. See FIGURE 5.R.2. An ancient king once commanded that such a tower be built of gold disks to the following specifications: Each disk was to be one finger width thick, and the diameter of each disk was to be one finger width larger than the disk above it. The hole through the centers of the disks was to be one finger width in diameter, and the top disk was to be two finger widths in diameter. Assume that a finger width is 1.5 cm and gold weighs 19.3 g/cm^3 and is valued at $14 per gram.

(a) Find a formula for the value of gold in the king's Tower of Hanoi if the tower has n disks.

(b) The usual number of gold disks in a Tower of Hanoi is 64. What is the value of the gold in this tower?

FIGURE 5.R.2 Tower of Hanoi in Problem 32

33. Consider the one-to-one function $f(x) = x^3 + x$ on the interval $[1, 2]$. See FIGURE 5.R.3. Without finding f^{-1}, determine the value of

$$\int_{f(1)}^{f(2)} f^{-1}(x)\, dx.$$

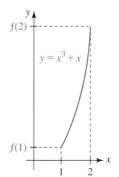

FIGURE 5.R.3 Graph for Problem 33

Chapter 6

Applications of the Integral

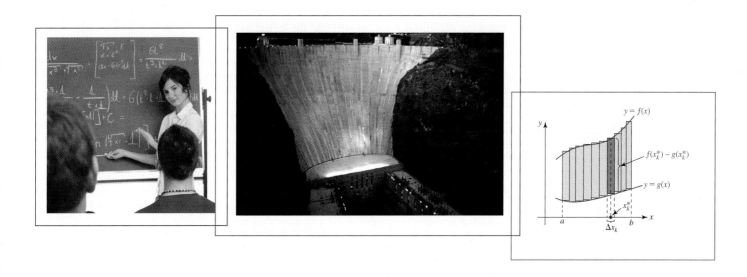

In This Chapter Although we return to the problem of finding areas by definite integration in Section 6.2, you will see in the subsequent sections of this chapter that a definite integral has many other interpretations besides area.

We begin the chapter with an application of the indefinite integral.

- 6.1 Rectilinear Motion Revisited
- 6.2 Area Revisited
- 6.3 Volumes of Solids: Slicing Method
- 6.4 Volumes of Solids: Shell Method
- 6.5 Length of a Graph
- 6.6 Area of a Surface of Revolution
- 6.7 Average Value of a Function
- 6.8 Work
- 6.9 Fluid Pressure and Force
- 6.10 Centers of Mass and Centroids
  Chapter 6 in Review

6.1 Rectilinear Motion Revisited

Introduction We began Chapter 4, *Applications of the Derivative*, with the notion of rectilinear motion. If $s = f(t)$ is the position function of an object moving rectilinearly—that is, in a straight line, then we know

$$\text{velocity} = v(t) = \frac{ds}{dt} \quad \text{and} \quad \text{acceleration} = a(t) = \frac{dv}{dt}.$$

As an immediate consequence of the definition of an antiderivative, the quantities s and v can be written as indefinite integrals

$$s(t) = \int v(t)\, dt \quad \text{and} \quad v(t) = \int a(t)\, dt. \tag{1}$$

By knowing the **initial position** $s(0)$ and the **initial velocity** $v(0)$, we can find specific values of the constants of integration used in (1).

Recall that when a body moves horizontally on a line, the positive direction is to the right. For motion in a vertical line, we take the positive direction to be upward. As shown in **FIGURE 6.1.1**, if an arrow is shot upward from ground level, then **initial conditions** are $s(0) = 0, v(0) > 0$, whereas if the arrow is shot downward from some initial height, say h meters off the ground, then the initial conditions are $s(0) = h, v(0) < 0$. A body that moves in a vertical line close to the surface of the earth, such as the arrow shot upward, is acted upon by the force of gravity. This force causes a body to accelerate. Near the surface of the earth the acceleration due to gravity, $a(t) = -g$, is assumed to be a constant. The magnitude g of this acceleration is approximately

$$32 \text{ ft/s}^2, \quad 9.8 \text{ m/s}^2, \quad \text{or} \quad 980 \text{ cm/s}^2.$$

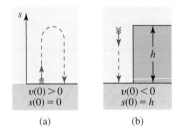

FIGURE 6.1.1 Initial conditions

EXAMPLE 1 Projectile Motion

A projectile is shot vertically upward from ground level with an initial velocity of 49 m/s. What is its velocity at $t = 2$ s? What is the maximum height attained by the projectile? How long is the projectile in the air? What is its impact velocity?

Solution Starting with $a(t) = -9.8$ we obtain by indefinite integration,

$$v(t) = \int (-9.8)\, dt = -9.8t + C_1. \tag{2}$$

From the given initial condition $v(0) = 49$, we see that (2) implies $C_1 = 49$. Hence,

$$v(t) = -9.8t + 49,$$

and so $v(2) = -9.8(2) + 49 = 29.4$ m/s. Notice that $v(2) > 0$ implies the projectile is traveling upward.

Now, the height of the projectile, measured from ground level, is the indefinite integral of the velocity function,

$$s(t) = \int v(t)\, dt = \int (-9.8t + 49)\, dt = -4.9t^2 + 49t + C_2. \tag{3}$$

Since the projectile starts from ground level, $s(0) = 0$ and (3) gives $C_2 = 0$. Hence,

$$s(t) = -4.9t^2 + 49t. \tag{4}$$

When the projectile attains its maximum height, $v(t) = 0$. Solving $-9.8t + 49 = 0$ then gives $t = 5$. From (4) we find the corresponding height to be $s(5) = 122.5$ m.

Finally, to find the time that the projectile hits the ground, we solve $s(t) = 0$ or $-4.9t^2 + 49t = 0$. Writing the latter equation as $-4.9t(t - 10) = 0$, we see the projectile is in the air for 10 s. The impact velocity is $v(10) = -49$ m/s. ∎

When air resistance is ignored, the magnitude of the impact velocity (speed) is the same as the initial upward velocity from ground level. See Problem 32 in Exercises 6.1. This is not true when air resistance is taken into consideration.

EXAMPLE 2 Projectile Motion

A tennis ball is thrown vertically downward from a height of 54 ft with an initial velocity of 8 ft/s. What is its impact velocity if it hits a 6-ft-tall person on the head? See FIGURE 6.1.2.

Solution In this case $a(t) = -32$, $s(0) = 54$, and, since the ball is thrown downward, $v(0) = -8$. Now

$$v(t) = \int (-32) \, dt = -32t + C_1.$$

Using the initial velocity $v(0) = -8$, we find $C_1 = -8$. Therefore,

$$v(t) = -32t - 8.$$

Continuing, we find

$$s(t) = \int (-32t - 8) \, dt = -16t^2 - 8t + C_2.$$

When $t = 0$, we know $s = 54$ and so the last equation implies $C_2 = 54$. Hence,

$$s(t) = -16t^2 - 8t + 54.$$

To determine the time that corresponds to $s = 6$ we solve

$$-16t^2 - 8t + 54 = 6.$$

Simplifying gives $-8(2t - 3)(t + 2) = 0$ and $t = \frac{3}{2}$. The velocity of the ball when it hits the person is then $v(\frac{3}{2}) = -56$ ft/s. ∎

FIGURE 6.1.2 Thrown ball in Example 2

Distance The **total distance** an object travels rectilinearly in a time interval $[t_1, t_2]$ is given by the definite integral

$$\text{total distance} = \int_{t_1}^{t_2} |v(t)| \, dt. \tag{5}$$

The absolute value is necessary in (5), since the object may be moving to the left and hence has negative velocity for some part of the time.

EXAMPLE 3 Distance Traveled

The position function of an object that moves on a coordinate line is $s(t) = t^2 - 6t$, where s is measured in centimeters and t in seconds. Find the distance traveled in the time interval $[0, 9]$.

Solution The velocity function $v(t) = ds/dt = 2t - 6 = 2(t - 3)$ shows that the motion is as indicated in FIGURE 6.1.3; namely, $v < 0$ for $0 \leq t < 3$ (motion to the left) and $v \geq 0$ for $3 \leq t \leq 9$ (motion to the right). Hence, from (5) the distance traveled is

$$\int_0^9 |2t - 6| \, dt = \int_0^3 |2t - 6| \, dt + \int_3^9 |2t - 6| \, dt$$

$$= \int_0^3 -(2t - 6) \, dt + \int_3^9 (2t - 6) \, dt$$

$$= (-t^2 + 6t) \Big]_0^3 + (t^2 - 6t) \Big]_3^9 = 45 \text{ cm}.$$

FIGURE 6.1.3 Representation of the motion of the object in Example 3

Of course, the last result must be consistent with the number obtained by simply counting units in Figure 6.1.3 between $s(0)$ and $s(3)$, and between $s(3)$ and $s(9)$. ∎

Exercises 6.1 Answers to selected odd-numbered problems begin on page ANS-20.

≡ Fundamentals

In Problems 1–6, a body moves in a straight line with velocity $v(t)$. Find the position function $s(t)$.

1. $v(t) = 6$; $s = 5$ when $t = 2$
2. $v(t) = 2t + 1$; $s = 0$ when $t = 1$
3. $v(t) = t^2 - 4t$; $s = 6$ when $t = 3$
4. $v(t) = \sqrt{4t + 5}$; $s = 2$ when $t = 1$

5. $v(t) = -10\cos(4t + \pi/6)$; $s = \frac{5}{4}$ when $t = 0$
6. $v(t) = 2\sin 3t$; $s = 0$ when $t = \pi$

In Problems 7–12, a body moves in a straight line with acceleration $a(t)$. Find $v(t)$ and $s(t)$.

7. $a(t) = -5$; $v = 4$ and $s = 2$ when $t = 1$
8. $a(t) = 6t$; $v = 0$ and $s = -5$ when $t = 2$
9. $a(t) = 3t^2 - 4t + 5$; $v = -3$ and $s = 10$ when $t = 0$
10. $a(t) = (t - 1)^2$; $v = 4$ and $s = 6$ when $t = 1$
11. $a(t) = 7t^{1/3} - 1$; $v = 50$ and $s = 0$ when $t = 8$
12. $a(t) = 100\cos 5t$; $v = -20$ and $s = 15$ when $t = \pi/2$

In Problems 13–18, an object moves in a straight line according to the given position function. If s is measured in centimeters, find the total distance traveled by the object in the indicated time interval.

13. $s(t) = t^2 - 2t$; $[0, 5]$
14. $s(t) = -t^2 + 4t + 7$; $[0, 6]$
15. $s(t) = t^3 - 3t^2 - 9t$; $[0, 4]$
16. $s(t) = t^4 - 32t^2$; $[1, 5]$
17. $s(t) = 6\sin\pi t$; $[1, 3]$
18. $s(t) = (t - 3)^2$; $[2, 7]$

≡ Applications

19. A driver of a car that is traveling in a straight line at a constant 60 mi/h takes his eyes off the road for 2 s. How many feet does the car move in this time?
20. A ball is dropped (released from rest) from a height of 144 ft. How long does it take for the ball to hit the ground? At what speed does it hit the ground?
21. An egg is dropped from the top of a building and hits the ground 4 s from release. How tall is the building?
22. A stone is dropped into a well and the splash is heard 2 s later. If the speed of sound in air is 1080 ft/s, find the depth of the well.
23. An arrow is projected vertically upward from ground level with an initial velocity of 24.5 m/s. How high does it rise?
24. How high would the arrow in Problem 23 rise on the planet Mars where $g = 3.6$ m/s^2?
25. A golf ball is thrown vertically upward from the edge of the roof of a 384-ft-high building with an initial velocity of 32 ft/s. What is the maximum height attained by the ball? At what time does the ball hit the ground?
26. In Problem 25, what is the velocity of the golf ball as it passes an observer in a window that is 256 ft off the ground?
27. A person throws a marshmallow vertically downward with an initial velocity of 16 ft/s from a window that is 102 ft off the ground. If the marshmallow hits a 6-ft-tall person on the head, what is the impact velocity?
28. The person hit on the head in Problem 27 climbs to the top of a 22-ft-high ladder and throws a stone vertically upward with an initial velocity of 96 ft/s. If the stone hits the culprit at the 102-ft level, what is the impact velocity?

≡ Think About It

29. In March 1979, the Voyager 1 space probe photographed an active volcanic eruption on Io, one of the moons of Jupiter. Find the ejection velocity of a rock from the volcano Loki if the rock attains a height of 200 km above the summit of the volcano. On Io the acceleration due to gravity is $g = 1.8$ m/s^2.
30. As shown in FIGURE 6.1.4, at a point 30 ft from a 25-ft-tall street lamp a ball is thrown vertically downward from a height of 25 ft with an initial velocity of 2 ft/s.
    (a) Find the rate at which the shadow of the ball is moving toward the base of the street lamp.
    (b) Find the rate at which the shadow of the ball is moving toward the base of the street lamp at $t = \frac{1}{2}$.

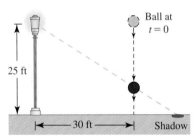

FIGURE 6.1.4 Street lamp in Problem 30

31. If a body is moving rectilinearly with a constant acceleration a and $v = v_0$ when $s = 0$, show that
$$v^2 = v_0^2 + 2as. \quad \left[\text{Hint: } \frac{dv}{dt} = \frac{dv}{ds}\frac{ds}{dt} = \frac{dv}{ds}v.\right]$$

32. Show that, when air resistance is ignored, a projectile shot vertically upward from ground level hits the ground again with a speed equal to the initial velocity v_0.
33. Suppose the acceleration due to gravity on a planet is one-half that on the Earth. Prove that a ball tossed vertically upward from the surface of the planet would attain a maximum height twice that on the Earth when the same initial velocity is used.
34. In Problem 33, suppose the initial velocity of the ball on the planet is v_0 and the initial velocity of the ball on the Earth is $2v_0$. Compare the maximum heights attained. Determine the initial velocity of the ball on the Earth (in terms of v_0) so that the maximum height attained is the same as on the planet.

6.2 Area Revisited

Introduction If a function f takes on both positive and negative values on $[a, b]$, then the definite integral $\int_a^b f(x)\,dx$ does not represent the area under the graph of f on the interval. As we saw in Section 5.4, we can interpret the value of $\int_a^b f(x)\,dx$ as the *net signed area* between the graph of f and the x-axis on the interval $[a, b]$. In this section we investigate two area problems:

- Find the **total area** of a region bounded by the graph of f and the x-axis on an interval $[a, b]$.
- Find the **area of the region** bounded between two graphs on an interval $[a, b]$.

We will see that the first problem is just a special case of the second problem.

Total Area Suppose the function $y = f(x)$ is continuous on the interval $[a, b]$ and that $f(x) < 0$ on $[a, c]$ and $f(x) \geq 0$ on $[c, b]$. The **total area** is the area of the region bounded by the graph of f, the x-axis, and the vertical lines $x = a$ and $x = b$. To find this area we employ the absolute value of the function $y = |f(x)|$, which is nonnegative for all x in $[a, b]$. Recall, $|f(x)|$ is defined in a piecewise manner. For the function f shown in FIGURE 6.2.1(a), $f(x) < 0$ on the interval $[a, c)$ and $f(x) \geq 0$ on the interval $[c, b]$. Thus,

$$|f(x)| = \begin{cases} -f(x), & \text{for } f(x) < 0 \\ f(x), & \text{for } f(x) \geq 0. \end{cases} \quad (1)$$

As shown in Figure 6.2.1(b) the graph of $y = |f(x)|$ on the interval $[a, c)$ is obtained by reflecting that portion of the graph of $y = f(x)$ through the x-axis. On the interval $[c, b]$, where $f(x) \geq 0$, the graphs of $y = f(x)$ and $y = |f(x)|$ are the same. To find the total area $A = A_1 + A_2$ shown in Figure 6.2.1(b) we use the additive interval property of the definite integral along with (1):

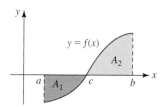

(a) The definite integral of f on $[a, b]$ is not area

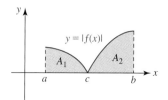

(b) The definite integral of $|f|$ on $[a, b]$ is area

FIGURE 6.2.1 Total area is $A = A_1 + A_2$

◀ See Theorem 5.4.5.

$$\int_a^b |f(x)|\,dx = \int_a^c |f(x)|\,dx + \int_c^b |f(x)|\,dx$$
$$= \int_a^c (-f(x))\,dx + \int_c^b f(x)\,dx$$
$$= A_1 + A_2.$$

We summarize the ideas of the preceding discussion in the following definition.

Definition 6.2.1 Total Area

If $y = f(x)$ is continuous on $[a, b]$, then the **total area** A bounded by its graph and the x-axis on the interval is given by

$$A = \int_a^b |f(x)|\,dx. \quad (2)$$

EXAMPLE 1 Total Area

Find the total area bounded by the graph of $y = x^3$ and the x-axis on $[-2, 1]$.

Solution From (2) we have

$$A = \int_{-2}^1 |x^3|\,dx.$$

In FIGURE 6.2.2 we have compared the graph of $y = x^3$ with the graph of $y = |x^3|$. Since $x^3 < 0$ for $x < 0$, we have on $[-2, 1]$,

$$|f(x)| = \begin{cases} -x^3, & -2 \leq x < 0 \\ x^3, & 0 \leq x \leq 1. \end{cases}$$

(a)

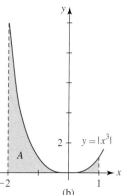

(b)

FIGURE 6.2.2 Graph of function and area in Example 1

Thus, by (2) of Definition 6.2.1 the desired area is

$$A = \int_{-2}^{1} |x^3|\, dx$$

$$= \int_{-2}^{0} |x^3|\, dx + \int_{0}^{1} |x^3|\, dx$$

$$= \int_{-2}^{0} (-x^3)\, dx + \int_{0}^{1} x^3\, dx$$

$$= -\frac{1}{4}x^4 \Big]_{-2}^{0} + \frac{1}{4}x^4 \Big]_{0}^{1}$$

$$= 0 - \left(-\frac{16}{4}\right) + \frac{1}{4} - 0 = \frac{17}{4}.$$

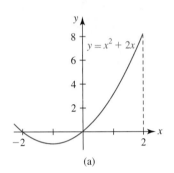

EXAMPLE 2 Total Area

Find the total area bounded by the graph of $y = x^2 + 2x$ and the x-axis on $[-2, 2]$.

Solution The graphs of $y = f(x)$ and $y = |f(x)|$ are given in FIGURE 6.2.3. Now, from Figure 6.2.3(a) we see that on $[-2, 2]$,

$$|f(x)| = \begin{cases} -(x^2 + 2x), & -2 \le x < 0 \\ x^2 + 2x, & 0 \le x \le 2. \end{cases}$$

Therefore, the total area bounded by the graph of f on the interval $[-2, 2]$ and the x-axis is

$$A = \int_{-2}^{2} |x^2 + 2x|\, dx$$

$$= \int_{-2}^{0} |x^2 + 2x|\, dx + \int_{0}^{2} |x^2 + 2x|\, dx$$

$$= \int_{-2}^{0} -(x^2 + 2x)\, dx + \int_{0}^{2} (x^2 + 2x)\, dx$$

$$= \left(-\frac{1}{3}x^3 - x^2\right)\Big]_{-2}^{0} + \left(\frac{1}{3}x^3 + x^2\right)\Big]_{0}^{2}$$

$$= 0 - \left(\frac{8}{3} - 4\right) + \left(\frac{8}{3} + 4\right) - 0 = 8.$$

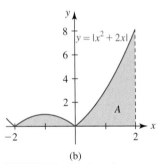

FIGURE 6.2.3 Graphs and area in Example 2

Area Bounded by Two Graphs The foregoing discussion is a special case of the more general problem of finding the **area of the region** bounded between the graphs of two functions f and g and the vertical lines $x = a$ and $x = b$. See FIGURE 6.2.4(a). The area *under* the graph of a continuous nonnegative function $y = f(x)$ on an interval $[a, b]$ can be interpreted as the area of

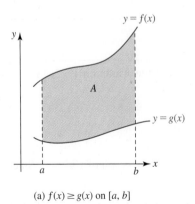

(a) $f(x) \ge g(x)$ on $[a, b]$ (b) Constructing n rectangles between two graphs

FIGURE 6.2.4 Area A bounded between two graphs

the region bounded by the graph of f and the graph of the function $y = 0$ (the x-axis) and the vertical lines $x = a$ and $x = b$.

■ **Building an Integral** Suppose $y = f(x)$ and $y = g(x)$ are continuous on $[a, b]$ and that $f(x) \geq g(x)$ for all x in the interval. Let P be a partition of the interval $[a, b]$ into n subintervals $[x_{k-1}, x_k]$. If we choose a sample point x_k^* in each subinterval, we can then construct n corresponding rectangles that have area

The assumption that $f(x) \geq g(x)$ on the interval means that the graphs of f and g can touch but do not cross each other.

$$A_k = [f(x_k^*) - g(x_k^*)] \Delta x_k.$$

See Figure 6.2.4(b). The area A of the region bounded by the two graphs on the interval $[a, b]$ is approximated by the Riemann sum

$$\sum_{k=1}^{n} A_k = \sum_{k=1}^{n} [f(x_k^*) - g(x_k^*)] \Delta x_k,$$

and this in turn suggests that the area is

$$A = \lim_{\|P\| \to 0} \sum_{k=1}^{n} [f(x_k^*) - g(x_k^*)] \Delta x_k.$$

Since f and g are continuous, so is $f - g$. Hence, the above limit exists and is, by definition, the definite integral

$$A = \int_a^b [f(x) - g(x)]\, dx. \tag{3}$$

Also, (3) applies to regions for which one or both of the functions f and g have negative values. See FIGURE 6.2.5. However, (3) is *not* valid on an interval $[a, b]$ where the graphs of f and g cross each other on the interval. Notice in FIGURE 6.2.6 that g is the upper graph on the intervals (a, c_1) and (c_2, b), whereas f is the upper graph on the interval (c_1, c_2). In the most general case, we have the following definition.

(a) $f(x) > 0$ and $g(x) < 0$ on $[a, b]$

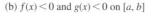

(b) $f(x) < 0$ and $g(x) < 0$ on $[a, b]$

FIGURE 6.2.5 Graphs of f and g can be below the x-axis

FIGURE 6.2.6 Graphs of f and g cross each other on $[a, b]$

Definition 6.2.2 Area Bounded by Two Graphs

If f and g are continuous functions on an interval $[a, b]$, then the **area A of the region bounded by their graphs on the interval** is given by

$$A = \int_a^b |f(x) - g(x)|\, dx. \tag{4}$$

Note that (4) reduces to (2) when $g(x) = 0$ for all x in $[a, b]$. Before using formulas (3) or (4), you are urged to sketch the necessary graphs. If the curves cross on the interval, then as we have seen in Figure 6.2.6 the relative position of the curves changes. In any event, on any subinterval of $[a, b]$ the appropriate integrand is always

$$(\text{upper graph}) - (\text{lower graph}).$$

As in (1), the absolute value of the integrand is given by

$$|f(x) - g(x)| = \begin{cases} -(f(x) - g(x)), & \text{for } f(x) - g(x) < 0 \\ f(x) - g(x), & \text{for } f(x) - g(x) \geq 0. \end{cases} \quad (5)$$

A more practical way of interpreting (5) is to draw the graphs of f and g accurately and visually determine that:

$$|f(x) - g(x)| = \begin{cases} g(x) - f(x), & \text{whenever } g \text{ is the upper graph} \\ f(x) - g(x), & \text{whenever } f \text{ is the upper graph.} \end{cases}$$

In Figure 6.2.6, the area A bounded by the graphs of f and g on $[a, b]$ is

$$A = \int_a^b |f(x) - g(x)|\, dx$$

$$= \int_a^{c_1} |f(x) - g(x)|\, dx + \int_{c_1}^{c_2} |f(x) - g(x)|\, dx + \int_{c_2}^b |f(x) - g(x)|\, dx$$

$$= \int_a^{c_1} \underbrace{[g(x) - f(x)]\, dx}_{g \text{ is the upper graph}} + \int_{c_1}^{c_2} \underbrace{[f(x) - g(x)]\, dx}_{f \text{ is the upper graph}} + \int_{c_2}^b \underbrace{[g(x) - f(x)]\, dx}_{g \text{ is the upper graph}}.$$

EXAMPLE 3 Area Bounded by Two Graphs

Find the area of the region bounded by the graphs of $y = \sqrt{x}$ and $y = x^2$.

Solution As shown in FIGURE 6.2.7, the region in question is located in the first quadrant. Because 0 and 1 are the solutions of the equation $x^2 = \sqrt{x}$, the graphs intersect at the points $(0, 0)$ and $(1, 1)$. In other words, the region lies between the vertical lines $x = 0$ and $x = 1$. Since $y = \sqrt{x}$ is the upper graph on the interval $(0, 1)$, it follows that

$$A = \int_0^1 (\sqrt{x} - x^2)\, dx$$

$$= \left(\frac{2}{3}x^{3/2} - \frac{1}{3}x^3\right)\Big|_0^1$$

$$= \frac{2}{3} - \frac{1}{3} - 0 = \frac{1}{3}. \quad \blacksquare$$

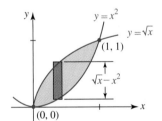

FIGURE 6.2.7 Area in Example 3

EXAMPLE 4 Area Bounded by Two Graphs

Find the area of the region bounded by the graphs of $y = x^2 + 2x$ and $y = -x + 4$ on the interval $[-4, 2]$.

Solution Let us denote the given functions by

$$y_1 = x^2 + 2x \quad \text{and} \quad y_2 = -x + 4.$$

As FIGURE 6.2.8 shows, the graphs cross each other on the interval $[-4, 2]$.

To find the points of intersection we solve the equation $x^2 + 2x = -x + 4$ or $x^2 + 3x - 4 = 0$ and find that $x = -4$ and $x = 1$. The area in question is the sum of the areas $A = A_1 + A_2$:

$$A = \int_{-4}^2 |y_2 - y_1|\, dx = \int_{-4}^1 |y_2 - y_1|\, dx + \int_1^2 |y_2 - y_1|\, dx.$$

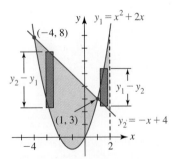

FIGURE 6.2.8 Area in Example 4

But since $y_2 = -x + 4$ is the upper graph on the interval $(-4, 1)$ and $y_1 = x^2 + 2x$ is the upper graph on the interval $(1, 2)$ we can write

$$A = \int_{-4}^{1} [(-x + 4) - (x^2 + 2x)] \, dx + \int_{1}^{2} [(x^2 + 2x) - (-x + 4)] \, dx$$

$$= \int_{-4}^{1} (-x^2 - 3x + 4) \, dx + \int_{1}^{2} (x^2 + 3x - 4) \, dx$$

$$= \left(-\frac{1}{3}x^3 - \frac{3}{2}x^2 + 4x\right)\bigg]_{-4}^{1} + \left(\frac{1}{3}x^3 + \frac{3}{2}x^2 - 4x\right)\bigg]_{1}^{2}$$

$$= \left(-\frac{1}{3} - \frac{3}{2} + 4\right) - \left(\frac{64}{3} - 24 - 16\right) + \left(\frac{8}{3} + 6 - 8\right) - \left(\frac{1}{3} + \frac{3}{2} - 4\right) = \frac{71}{3}. \quad \blacksquare$$

EXAMPLE 5 Area Bounded by Two Graphs

Find the area of the four regions bounded by the graphs of $y = \sin x$ and $y = \cos x$ shown in FIGURE 6.2.9.

Solution There are an infinite number of such regions bounded by the graphs of $y = \sin x$ and $y = \cos x$ and the area of each region is the same. Therefore, we need only find the area of the region on the interval corresponding to the first two positive solutions of the equation $\sin x = \cos x$. By dividing by $\cos x$, a more useful form of the last equation is $\tan x = 1$. The first positive solution is $x = \tan^{-1} 1 = \pi/4$. Then since $\tan x$ is π-periodic, the next positive solution is $x = \pi + \pi/4 = 5\pi/4$. On the interval $(\pi/4, 5\pi/4)$, $y = \sin x$ is the upper graph so the area of the four regions is

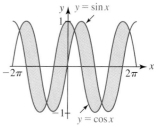

FIGURE 6.2.9 Each of the four regions has the same area in Example 5

$$A = 4 \int_{\pi/4}^{5\pi/4} (\sin x - \cos x) \, dx$$

$$= 4(-\cos x - \sin x) \bigg]_{\pi/4}^{5\pi/4}$$

$$= 4(2\sqrt{2}) = 8\sqrt{2}. \quad \blacksquare$$

In finding the area bounded by two graphs, it is not always convenient to integrate with respect to the variable x.

EXAMPLE 6 Area Bounded by Two Graphs

Find the area of the region bounded by the graphs of $y^2 = 1 - x$ and $2y = x + 2$.

Solution We note that the equation $y^2 = 1 - x$ implicitly defines two functions, $y_2 = \sqrt{1 - x}$ and $y_1 = -\sqrt{1 - x}$ for $x \le 1$. If we define $y_3 = \frac{1}{2}x + 1$, we see from FIGURE 6.2.10 that the height of an element of area on the interval $(-8, 0)$ is $y_3 - y_1$, whereas the height of an element on the interval $(0, 1)$ is $y_2 - y_1$. Thus, if we integrate with respect to x, the desired area is the sum of

$$A_1 = \int_{-8}^{0} (y_3 - y_1) \, dx \quad \text{and} \quad A_2 = \int_{0}^{1} (y_2 - y_1) \, dx.$$

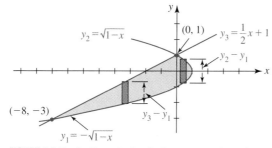

FIGURE 6.2.10 In Example 6 y_3 is the upper graph on the interval $(-8, 0)$; y_2 is the upper graph on the interval $(0, 1)$

Thus the area of the region is the sum of the areas $A = A_1 + A_2$, that is,

$$A = \int_{-8}^{0}\left[\left(\frac{1}{2}x + 1\right) - (-\sqrt{1-x})\right]dx + \int_{0}^{1}[\sqrt{1-x} - (-\sqrt{1-x})]\,dx$$

$$= \int_{-8}^{0}\left(\frac{1}{2}x + 1 + \sqrt{1-x}\right)dx + 2\int_{0}^{1}\sqrt{1-x}\,dx$$

$$= \left(\frac{1}{4}x^2 + x - \frac{2}{3}(1-x)^{3/2}\right)\Big]_{-8}^{0} - \frac{4}{3}(1-x)^{3/2}\Big]_{0}^{1}$$

$$= -\frac{2}{3}\cdot 1^{3/2} - \left(16 - 8 - \frac{2}{3}\cdot 9^{3/2}\right) - \frac{4}{3}\cdot 0 + \frac{4}{3}\cdot 1^{3/2} = \frac{32}{3}. \quad\blacksquare$$

EXAMPLE 7 Alternative Solution to Example 6

The necessity of using two integrals in Example 6 to find the area is avoided by constructing horizontal rectangles and using y as the independent variable. If we define $x_2 = 1 - y^2$ and $x_1 = 2y - 2$, then, as shown in FIGURE 6.2.11, the area of a horizontal element is

$$A_k = [\text{right graph} - \text{left graph}] \cdot \text{width}.$$

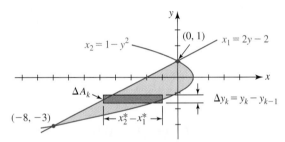

FIGURE 6.2.11 Using y as the variable of integration in Example 7

That is, $A_k = [x_2^* - x_1^*]\Delta y_k,$

where $x_2^* = 1 - (y_k^*)^2,\quad x_1^* = 2y_k^* - 2,\quad$ and $\quad\Delta y_k = y_k - y_{k-1}.$

Summing the rectangles in the positive y-direction leads to

$$A = \lim_{\|P\|\to 0} \sum_{k=1}^{n} [x_2^*(y) - x_1^*(y)]\,\Delta y_k,$$

where $\|P\|$ is the norm of a partition P of the interval on the y-axis defined by $-3 \le y \le 1$. In other words,

$$A = \int_{-3}^{1}(x_2 - x_1)\,dy,$$

where the lower limit -3 and the upper limit 1 are the y-coordinates of the points of intersection $(-8, -3)$ and $(0, 1)$, respectively. Substituting for x_2 and x_1 then gives

$$A = \int_{-3}^{1}[(1 - y^2) - (2y - 2)]\,dy$$

$$= \int_{-3}^{1}(-y^2 - 2y + 3)\,dy$$

$$= \left(-\frac{1}{3}y^3 - y^2 + 3y\right)\Big]_{-3}^{1}$$

$$= \left(-\frac{1}{3} - 1 + 3\right) - \left(\frac{27}{3} - 9 - 9\right) = \frac{32}{3}. \quad\blacksquare$$

$\int_a^b$ NOTES FROM THE CLASSROOM

As mentioned in the introduction, we are going to see different interpretations of the definite integral in this chapter. In each section you will see a variety of definite integrals derived under the paragraph heading *Building an Integral*. Before memorizing such integral formulas you should be aware that the derived result will usually not be applicable to every conceivable geometric or physical situation. For example, as we have seen in Example 7, to find the area of a region in the plane it may be more convenient to integrate with respect to y and you will have to build an entirely different integral. Rather than apply a formula blindly, you should try to understand the process and practice building integrals by analyzing the geometry of each problem.

Exercises 6.2   Answers to selected odd-numbered problems begin on page ANS-20.

≡ Fundamentals

In Problems 1–22, find the total area bounded by the graph of the given function and the x-axis on the indicated interval.

1. $y = x^2 - 1$; $[-1, 1]$
2. $y = x^2 - 1$; $[0, 2]$
3. $y = x^3$; $[-3, 0]$
4. $y = 1 - x^3$; $[0, 2]$
5. $y = x^2 - 3x$; $[0, 3]$
6. $y = -(x + 1)^2$; $[-1, 0]$
7. $y = x^3 - 6x$; $[-1, 1]$
8. $y = x^3 - 3x^2 + 2$; $[0, 2]$
9. $y = (x - 1)(x - 2)(x - 3)$; $[0, 3]$
10. $y = x(x + 1)(x - 1)$; $[-1, 1]$
11. $y = \dfrac{x^2 - 1}{x^2}$; $[\tfrac{1}{2}, 3]$
12. $y = \dfrac{x^2 - 1}{x^2}$; $[1, 2]$
13. $y = \sqrt{x} - 1$; $[0, 4]$
14. $y = 2 - \sqrt{x}$; $[0, 9]$
15. $y = \sqrt[3]{x}$; $[-2, 3]$
16. $y = 2 - \sqrt[3]{x}$; $[-1, 8]$
17. $y = \sin x$; $[-\pi, \pi]$
18. $y = 1 + \cos x$; $[0, 3\pi]$
19. $y = -1 + \sin x$; $[-3\pi/2, \pi/2]$
20. $y = \sec^2 x$; $[0, \pi/3]$
21. $y = \begin{cases} x, & -2 \le x < 0 \\ x^2, & 0 \le x \le 1 \end{cases}$; $[-2, 1]$
22. $y = \begin{cases} x + 2, & -3 \le x < 0 \\ 2 - x^2, & 0 \le x \le 2 \end{cases}$; $[-3, 2]$

In Problems 23–50, find the area of the region bounded by the graphs of the given functions.

23. $y = x$, $y = -2x$, $x = 3$
24. $y = x$, $y = 4x$, $x = 2$
25. $y = x^2$, $y = 4$
26. $y = x^2$, $y = x$
27. $y = x^3$, $y = 8$, $x = -1$
28. $y = x^3$, $y = \sqrt[3]{x}$, first quadrant
29. $y = 4(1 - x^2)$, $y = 1 - x^2$
30. $y = 2(1 - x^2)$, $y = x^2 - 1$
31. $y = x$, $y = 1/x^2$, $x = 3$
32. $y = x^2$, $y = 1/x^2$, $y = 9$, first quadrant
33. $y = -x^2 + 6$, $y = x^2 + 4x$
34. $y = x^2$, $y = -x^2 + 3x$
35. $y = x^{2/3}$, $y = 4$
36. $y = 1 - x^{2/3}$, $y = x^{2/3} - 1$
37. $y = x^2 - 2x - 3$, $y = 2x + 2$, on $[-1, 6]$
38. $y = -x^2 + 4x$, $y = \tfrac{3}{2}x$
39. $y = x^3$, $y = x + 6$, $y = -\tfrac{1}{2}x$
40. $x = y^2$, $x = 0$, $y = 1$
41. $x = -y$, $x = 2 - y^2$
42. $x = y^2$, $x = 6 - y^2$
43. $x = y^2 + 2y + 2$, $x = -y^2 - 2y + 2$
44. $x = y^2 - 6y + 1$, $x = -y^2 + 2y + 1$
45. $y = x^3 - x$, $y = x + 4$, $x = -1$, $x = 1$
46. $x = y^3 - y$, $x = 0$
47. $y = \cos x$, $y = \sin x$, $x = 0$, $x = \pi/2$
48. $y = 2 \sin x$, $y = -x$, $x = \pi/2$
49. $y = 4 \sin x$, $y = 2$, on $[\pi/6, 5\pi/6]$
50. $y = 2 \cos x$, $y = -\cos x$, on $[-\pi/2, \pi/2]$

In Problems 51 and 52, interpret the given definite integral as the area of a region bounded by the graphs of two functions. Sketch two regions that have the area given by the integral.

51. $\displaystyle\int_0^4 \left(\sqrt{x} + x\right) dx$
52. $\displaystyle\int_{-1}^2 \left(\tfrac{1}{2}x^2 + 3 - x\right) dx$

In Problems 53 and 54, interpret the given definite integral as the area of a region bounded by the graphs of two functions on an interval. Evaluate the given integral and sketch the region.

53. $\displaystyle\int_0^2 \left|\dfrac{3}{x+1} - 4x\right| dx$
54. $\displaystyle\int_{-1}^1 |e^x - 2e^{-x}| dx$

In Problems 55–58, use the fact that the area of a circle of radius r is πr^2 to evaluate the given definite integral. Sketch a region whose area is given by the definite integral.

55. $\displaystyle\int_0^3 \sqrt{9 - x^2}\, dx$
56. $\displaystyle\int_{-5}^5 \sqrt{25 - x^2}\, dx$
57. $\displaystyle\int_{-2}^2 \left(1 + \sqrt{4 - x^2}\right) dx$
58. $\displaystyle\int_{-1}^1 \left(2x + 3 - \sqrt{1 - x^2}\right) dx$

59. Set up a definite integral that represents the area of an ellipse $x^2/a^2 + y^2/b^2 = 1$, $a > b > 0$. Use the idea used in Problems 55–58 to evaluate this definite integral.

60. Find the area of the triangle with vertices at (1, 1), (2, 4), and (3, 2).

61. Consider the region bounded by the graphs of $y^2 = -x - 2$, $y = 2$, $y = -2$, and $y = 2(x - 1)$. Compute the area of the region by integrating with respect to x.

62. Compute the area of the region given in Problem 61 by integrating with respect to y.

63. Consider the region bounded by the graphs of $y = 2e^x - 1$, $y = e^x$, and $y = 2$ shown in FIGURE 6.2.12. Express the area of the region as definite integrals first using integration with respect to x and then using integration with respect to y. Choose one of these integral expressions to find the area.

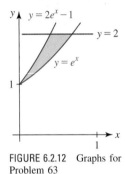

FIGURE 6.2.12 Graphs for Problem 63

≡ Calculator/CAS Problems

64. Use a calculator or CAS to approximate the x-coordinates of the points of intersection of the graphs shown in FIGURE 6.2.13. Find an approximate value of the area of the region.

FIGURE 6.2.13 Graphs for Problem 64

≡ Think About It

65. The line segment between Q and R shown in FIGURE 6.2.14 is tangent to the graph of $y = 1/x$ at point P. Show that the area of triangle QOR is independent of the coordinates of P.

FIGURE 6.2.14 Triangle in Problem 65

66. A trapezoid is bounded by the graphs of $f(x) = Ax + B$, $x = a$, $x = b$, and $x = 0$. Show that the area of the trapezoid is $\dfrac{f(a) + f(b)}{2}(b - a)$.

67. Express the area of the shaded region shown in FIGURE 6.2.15 in terms of the number a. Try to be a little clever.

FIGURE 6.2.15 Graphs for Problem 67

68. Suppose the two swaths of paint shown in FIGURE 6.2.16 are done with one stroke using a paintbrush of width k, $k > 0$, over the interval $[a, b]$. In Figure 6.2.16(b) assume that the painted red region is parallel to the x-axis. Which swath has the greater area? Defend your answer with a solid mathematical demonstration. Can you formulate a general principle?

FIGURE 6.2.16 Swaths of paint in Problem 68

≡ Projects

69. **The Larger Area** The points A and B are on a line and the points C and D are on a line parallel to the first line. The points in FIGURE 6.2.17(a) form a rectangle $ABCD$. The points C and D are moved to the left as shown in Figure 6.2.17(b) in such a manner that $ABC'D'$ forms a parallelogram. Discuss: Which has the larger area, the rectangle $ABCD$ or the parallelogram $ABC'D'$?

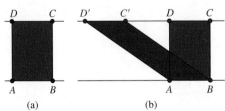

FIGURE 6.2.17 Rectangle and parallelogram in Problem 69

70. **Cavalieri's Principle** Write a short report on Cavalieri's Principle. Discuss Problems 68 and 69 in this report.

6.3 Volumes of Solids: Slicing Method

■ **Introduction** The shape that undoubtedly springs to mind with the words **right cylinder** is the right *circular* cylinder—that is, the usual shape of a tin can. But a right cylinder need not be circular. From geometry, a **right cylinder** is defined as a solid bounded by two congruent plane regions, in parallel planes, and a lateral surface that is generated by a moving line segment that is perpendicular to both planes and whose ends are on the boundaries of the plane regions. When the regions are circles, we obtain the right circular cylinder. If the regions are rectangles, the cylinder is a rectangular parallelepiped. Common to all right cylinders, such as the five shown in FIGURE 6.3.1, is that their volume V is given by the formula:

$$V = B \cdot h, \tag{1}$$

where B denotes the area of a base (that is, the area of one of the plane regions) and h denotes the height of the cylinder (that is, the perpendicular distance between the plane regions).

FIGURE 6.3.1 Five different right cylinders

In this section, we will show how the definite integral can be used to compute the volumes of certain kinds of solids, specifically solids with a known cross-sectional area. Formula (1) will be especially important in the discussion that follows.

■ **Slicing Method** Suppose V is the volume of the solid shown in FIGURE 6.3.2 bounded by planes that are perpendicular to the x-axis at $x = a$ and $x = b$. Furthermore, suppose we know a continuous function $A(x)$ that gives the area of a cross-sectional region that is formed by *slicing* the solid by a plane perpendicular to the x-axis, in other words, a slice is the intersection of the solid and one plane. For example, for $a < x_1 < x_2 < b$ the areas of the cross sections shown in Figure 6.3.2 are $A(x_1)$ and $A(x_2)$. With this is mind, let us imagine slicing the solid into thin slabs by parallel planes (similar to slices of commercially baked bread) so that a slab has thickness or width Δx_k. By using right cylinders to approximate the volumes of these slabs, we can build a definite integral that gives the volume V of the solid.

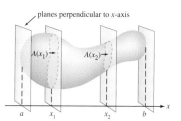

FIGURE 6.3.2 The regions or cross sections have known areas

■ **Building an Integral** Now think of slicing the solid into n slabs. If P is the partition

$$a = x_0 < x_1 < x_2 < \cdots < x_n = b$$

of the interval $[a, b]$ and x_k^* is a sample point in the kth subinterval $[x_{k-1}, x_k]$, then an approximation to the volume of the solid on this subinterval, or slab, is the volume V_k of the right cylinder, which is shown in the enlargement in FIGURE 6.3.3. The area B of the base of the right cylinder is the area $A(x_k^*)$ of the cross section and its height h is Δx_k and so by (1) its volume is

$$V_k = \text{area of base} \cdot \text{height} = A(x_k^*)(x_k - x_{k-1}) = A(x_k^*) \Delta x_k. \tag{2}$$

It follows that the Riemann sum of the volumes $V_k = A(x_k^*) \Delta x_k$ of the n right cylinders,

$$\sum_{k=1}^{n} V_k = \sum_{k=1}^{n} A(x_k^*) \Delta x_k,$$

is an approximation to the volume V of the solid on $[a, b]$. We use the definite integral

$$\lim_{\|P\| \to 0} \sum_{k=1}^{n} A(x_k^*) \Delta x_k = \int_a^b A(x) \, dx$$

as the definition of the volume V of the solid.

A piece of bread is a slab formed by two slices

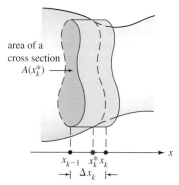

FIGURE 6.3.3 The volume of a right cylinder is an approximation to the volume of a slab

334 CHAPTER 6 Applications of the Integral

> **Definition 6.3.1** Volume by Slicing
>
> Let V be the volume of a solid bounded by planes that are perpendicular to the x-axis at $x = a$ and $x = b$. If $A(x)$ is a continuous function that gives the area of a cross section of the solid formed by a plane perpendicular to the x-axis at any point in the interval $[a, b]$, then the volume of the solid is
>
> $$V = \int_a^b A(x)\,dx. \qquad (3)$$

Bear in mind there is nothing special about the variable x in (3); depending on the geometry and the analysis of the problem we could just as well end up with an integral $\int_c^d A(y)\,dy$.

EXAMPLE 1 Solid with Square Cross Sections

For the solid in FIGURE 6.3.4(a), the cross sections perpendicular to a diameter of a circular base are squares. Given that the radius of the base is 4 ft, find the volume of the solid.

Solution Let the x- and y-axes be as shown in Figure 6.3.4(a), namely, the origin is at the center of the circular base of the solid. In this figure a square cross section is shown perpendicular to the x-axis. Since the base of the solid is a circle we have $x^2 + y^2 = 4^2$. In Figure 6.3.4(b), the dashed line at x_k^* represents the cross section of the solid perpendicular to the x-axis in the subinterval $[x_{k-1}, x_k]$ in a partition of the interval $[-4, 4]$. From this we see that the length of one side of the square cross section is $2y_k^* = 2\sqrt{16 - (x_k^*)^2}$. Thus, the area of a square cross section is

$$A(x_k^*) = \left(2\sqrt{16 - (x_k^*)^2}\right)^2 = 64 - 4(x_k^*)^2.$$

The volume of the approximating right cylinder to the volume of the solid or slab on the subinterval $[x_{k-1}, x_k]$ is

$$V_k = A(x_k^*)\,\Delta x_k = (64 - 4(x_k^*)^2)\,\Delta x_k.$$

Forming the sum $\sum_{k=1}^n V_k$ and taking the limit as $\|P\| \to 0$ gives the definite integral

$$V = \int_{-4}^{4} (64 - 4x^2)\,dx = 64x - \frac{4}{3}x^3 \Big]_{-4}^{4} = \frac{512}{3} - \left(-\frac{512}{3}\right) = \frac{1024}{3}.$$ ∎

(a) Plane perpendicular to x-axis intersects solid in a square

(b) Circular base of solid

FIGURE 6.3.4 Solid in Example 1

■ **Solids of Revolution** If a region R in the xy-plane is revolved about an axis L, it will generate a solid called a **solid of revolution**. See FIGURE 6.3.5.

■ **Disk Method** As just discussed, we can find the volume V of a solid by means of a definite integral whenever we know a function $A(x)$ that gives the area of a cross-sectional region formed by passing a plane through the solid perpendicular to an axis. In the case of finding the volume of a solid of revolution, it is always possible to find $A(x)$; the axis in question is the axis of revolution L. We will see that by slicing the solid by two parallel planes perpendicular to the axis of revolution the volume of the resulting slabs of the solid can be approximated by right *circular* cylinders that are either disks or washers. We will next illustrate building a volume integral using disks.

(a) Region

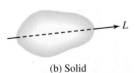

(b) Solid

FIGURE 6.3.5 A solid of revolution is formed by revolving a plane region R about an axis L

■ **Building an Integral** Let R be the region bounded by the graph of a nonnegative continuous function $y = f(x)$, the x-axis, and the vertical lines $x = a$ and $x = b$, as shown in FIGURE 6.3.6. If this region is revolved about the x-axis, let us find the volume V of the resulting solid of revolution.

Let P be a partition of $[a, b]$ and let x_k^* be any number in the kth subinterval $[x_{k-1}, x_k]$ as shown in FIGURE 6.3.7(a). As the rectangular element of width Δx_k and height $f(x_k^*)$ is revolved about the x-axis, it generates a solid disk. Now the cross section of the solid determined by a plane cutting the surface at x_k^* is a circle of radius $r = f(x_k^*)$, and so the area of the cross-

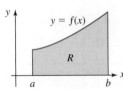

FIGURE 6.3.6 Region to be revolved about the x-axis

sectional region is $A(x_k^*) = \pi[f(x_k^*)]^2$. The volume of the corresponding right-circular cylinder, or solid disk, of radius $r = f(x_k^*)$ and height $h = \Delta x_k$ is $\pi r^2 h$ or

$$V_k = A(x_k^*)\,\Delta x_k = \pi[f(x_k^*)]^2\,\Delta x_k.$$

The Riemann sum

$$\sum_{k=1}^{n} V_k = \sum_{k=1}^{n} A(x_k^*)\,\Delta x_k = \sum_{k=1}^{n} \pi[f(x_k^*)]^2\,\Delta x_k$$

represents an approximation to the volume of the solid shown in Figure 6.3.7(d). This suggests that the volume V of the solid of revolution is given by

$$V = \lim_{\|P\|\to 0} \sum_{k=1}^{n} \pi[f(x_k^*)]^2\,\Delta x_k$$

or
$$V = \int_a^b \pi[f(x)]^2\,dx. \tag{4}$$

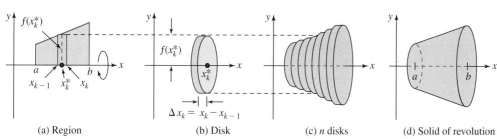

(a) Region (b) Disk (c) n disks (d) Solid of revolution

FIGURE 6.3.7 Revolving the red rectangular element in (a) about the x-axis generates the red circular disk in (b)

If a region R is revolved about some other axis, then (4) may simply not be applicable to the problem of finding the volume of the resulting solid. Rather than apply a formula blindly, you should set up an appropriate integral by carefully analyzing the geometry of each problem. We will examine such a case in Example 6.

EXAMPLE 2 Disk Method

Find the volume V of the solid formed by revolving the region bounded by the graphs of $y = \sqrt{x}$, $y = 0$, and $x = 4$ about the x-axis.

Solution FIGURE 6.3.8(a) shows the region in question. Now, the area of a cross-sectional slice at x_k^* is

$$A(x_k^*) = \pi[f(x_k^*)]^2 = \pi[(x_k^*)^{1/2}]^2 = \pi x_k^*,$$

and so the volume of the corresponding disk shown in Figure 6.3.8(b) is

$$V_k = A(x_k^*)\,\Delta x_k = \pi x_k^*\,\Delta x_k.$$

Hence, the volume of the solid is

$$V = \pi \int_0^4 x\,dx = \pi \frac{1}{2}x^2 \bigg]_0^4 = 8\pi.$$

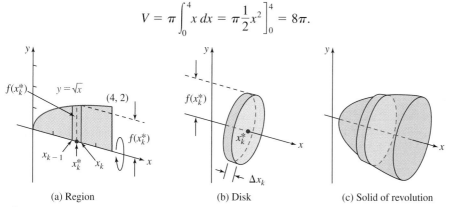

(a) Region (b) Disk (c) Solid of revolution

FIGURE 6.3.8 Region and solid of revolution in Example 2

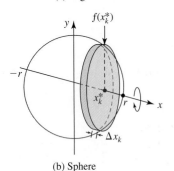

(a) Region

(b) Sphere

FIGURE 6.3.9 Semicircle and sphere in Example 3

EXAMPLE 3 Volume of a Sphere

Show that the volume V of a sphere of radius r is $V = \frac{4}{3}\pi r^3$.

Solution A sphere of radius r can be generated by revolving a semicircle $f(x) = \sqrt{r^2 - x^2}$ about the x-axis. From FIGURE 6.3.9 we see that the area of a cross-sectional region of the solid perpendicular to the x-axis at x_k^* is

$$A(x_k^*) = \pi[f(x_k^*)]^2 = \pi\left(\sqrt{r^2 - (x_k^*)^2}\right)^2 = \pi(r^2 - (x_k^*)^2)$$

and hence, the volume of one disk is

$$V_k = A(x_k^*)\,\Delta x_k = \pi(r^2 - (x_k^*)^2)\,\Delta x_k.$$

Using (4) we see that the volume of the sphere is

$$V = \int_{-r}^{r} \pi(r^2 - x^2)\,dx = \pi\left(r^2 x - \frac{1}{3}x^3\right)\Big|_{-r}^{r} = \pi \frac{2}{3}r^3 - \left(-\pi \frac{2}{3}r^3\right) = \frac{4}{3}\pi r^3. \quad\blacksquare$$

■ **Washer Method** Let the region R bounded by the graphs of the continuous functions $y = f(x)$, $y = g(x)$, and the lines $x = a$ and $x = b$, as shown in FIGURE 6.3.10(a), be revolved about the x-axis. Then the slice perpendicular to the x-axis of the solid of revolution at x_k^* is a circular or annular ring. As the rectangular element of width Δx_k shown in Figure 6.3.10(a) is revolved about the x-axis, it generates a washer. The area of the ring is

$$A(x_k^*) = \text{area of circle} - \text{area of hole}$$
$$= \pi[f(x_k^*)]^2 - \pi[g(x_k^*)]^2 = \pi([f(x_k^*)]^2 - [g(x_k^*)]^2)$$

and the volume V_k of the representative washer shown in Figure 6.3.10(b) is

$$V_k = A(x_k^*)\,\Delta x_k = \pi([f(x_k^*)]^2 - [g(x_k^*)]^2)\,\Delta x_k.$$

Therefore, the volume of the solid is

$$V = \int_a^b \pi([f(x)]^2 - [g(x)]^2)\,dx. \tag{5}$$

Observe that the integral (5) reduces to (4) when $g(x) = 0$.

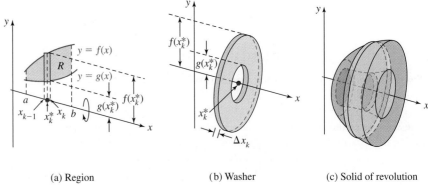

(a) Region   (b) Washer   (c) Solid of revolution

FIGURE 6.3.10 Revolving the red rectangular element in (a) about the x-axis generates the red circular washer in (b)

EXAMPLE 4 Washer Method

Find the volume V of the solid formed by revolving the region bounded by the graphs of $y = x + 2$, $y = x$, $x = 0$, and $x = 3$ about the x-axis.

Solution FIGURE 6.3.11(a) shows the region in question. Now, the area of a cross-sectional region of the solid corresponding to a plane perpendicular to the x-axis at x_k^* is

$$A(x_k^*) = \pi(x_k^* + 2)^2 - (x_k^*)^2 = \pi(4x_k^* + 4).$$

As seen in Figures 6.3.11(a) and (b), a vertical rectangular element of width Δx_k, when revolved about the x-axis, yields a washer having volume

$$V_k = A(x_k^*) \Delta x_k = \pi(4x_k^* + 4) \Delta x_k.$$

The usual summing and limiting process yields the definite integral for the volume V of the solid of revolution:

$$V = \pi \int_0^3 (4x + 4) \, dx = \pi(2x^2 + 4x)\Big]_0^3 = 30\pi.$$

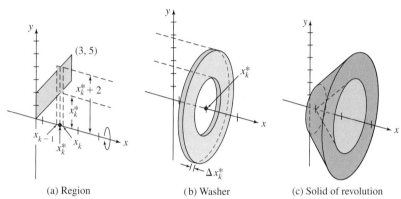

(a) Region   (b) Washer   (c) Solid of revolution

FIGURE 6.3.11 Region and solid of revolution in Example 4

EXAMPLE 5 Integration with Respect to y

Find the volume V of the solid formed by revolving the region bounded by the graphs of $y = \sqrt{x}$ and $y = x$ about the y-axis.

Solution When the horizontal rectangular element in FIGURE 6.3.12(a) is revolved about the y-axis it generates a washer of width Δy_k. The area $A(y_k^*)$ of the annular ring at y_k^* is

$$A(y_k^*) = \text{area of circle} - \text{area of hole}. \quad -1$$

The radius of the circle and the radius of the hole are obtained by solving, in turn, $y = x$ and $y = \sqrt{x}$ for x in terms of y:

$$A(y_k^*) = \pi(y_k^*)^2 - \pi[(y_k^*)^2]^2 = \pi((y_k^*)^2 - (y_k^*)^4).$$

Thus, the volume of a washer is

$$V_k = A(y_k^*) \Delta y_k = \pi((y_k^*)^2 - (y_k^*)^4) \Delta y_k.$$

The usual summing of the V_k and taking the limit of that sum as $\|P\| \to 0$ leads to the definite integral for the volume of the solid:

$$V = \pi \int_0^1 (y^2 - y^4) \, dy = \pi\left(\frac{1}{3}y^3 - \frac{1}{5}y^5\right)\Big]_0^1 = \frac{2}{15}\pi.$$

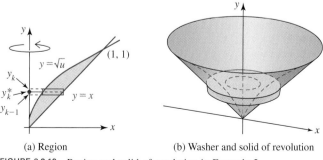

(a) Region   (b) Washer and solid of revolution

FIGURE 6.3.12 Region and solid of revolution in Example 5

Revolution about a Line
The next example shows how to find the volume of a solid of revolution when a region is revolved about an axis that is not a coordinate axis.

EXAMPLE 6 Axis of Revolution not a Coordinate Axis

Find the volume V of the solid that is formed by revolving the region given in Example 2 about the line $x = 4$.

Solution The domed-shaped solid of revolution is shown in FIGURE 6.3.13. From inspection of the figure we see that a horizontal rectangular element of width Δy_k that is perpendicular to the vertical line $x = 4$ generates a solid disk when revolved about that axis. The radius r of that disk is

$$r = \text{(right-most } x\text{-value)} - \text{(left-most } x\text{-value)} = 4 - x_k^*,$$

and so its volume is then

$$V_k = \pi(4 - x_k^*)^2 \Delta y_k.$$

To express x in terms of y we use $y = \sqrt{x}$ to obtain $x_k^* = (y_k^*)^2$. Therefore,

$$V_k = \pi(4 - (y_k^*)^2)^2 \Delta y_k.$$

This leads to the integral

$$V = \pi \int_0^2 (4 - y^2)^2 \, dy$$
$$= \pi \int_0^2 (16 - 8y^2 + y^4) \, dy$$
$$= \pi \left(16y - \frac{8}{3}y^3 + \frac{1}{5}y^5 \right) \Big]_0^2 = \frac{256}{15}\pi.$$

FIGURE 6.3.13 Solid of revolution in Example 6

Exercises 6.3
Answers to selected odd-numbered problems begin on page ANS-20.

≡ Fundamentals

In Problems 1 and 2, use the slicing method to find the volume of the solid if its cross sections perpendicular to a diameter of a circular base are as given. Assume that the radius of the base is 4.

1.
2.

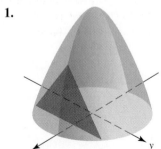

FIGURE 6.3.14 Cross sections are equilateral triangles

FIGURE 6.3.15 Cross sections are semicircles

3. The base of a solid is bounded by the curves $x = y^2$ and $x = 4$ in the xy-plane. The cross sections perpendicular to the x-axis are rectangles for which the height is four times the base. Find the volume of the solid.

4. The base of a solid is bounded by the curve $y = 4 - x^2$ and the x-axis. The cross sections perpendicular to the x-axis are equilateral triangles. Find the volume of the solid.

5. The base of a solid is an isosceles triangle whose base is 4 ft and height is 5 ft. The cross sections perpendicular to the altitude are semicircles. Find the volume of the solid.

6. A hole of radius 1 ft is drilled through the middle of the solid sphere of radius $r = 2$ ft. Find the volume of the remaining solid. See FIGURE 6.3.16.

FIGURE 6.3.16 Hole through sphere in Problem 6

7. The base of a solid is a right isosceles triangle that is formed by the coordinate axes and the line $x + y = 3$. The cross sections perpendicular to the y-axis are squares. Find the volume of the solid.

8. Suppose the pyramid shown in FIGURE 6.3.17 has height h and a square base of area B. Show that the volume of the pyramid is given by $A = \frac{1}{3}hB$. [*Hint*: Let b denote the length of one side of the square base.]

FIGURE 6.3.17 Pyramid in Problem 8

In Problems 9–14, refer to FIGURE 6.3.18. Use the disk or washer method to find the volume of the solid of revolution that is formed by revolving the given region about the indicated line.

9. R_1 about OC
10. R_1 about OA
11. R_2 about OA
12. R_2 about OC
13. R_1 about AB
14. R_2 about AB

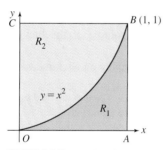

FIGURE 6.3.18 Regions for Problems 9–14

In Problems 15–40, use the disk or washer method to find the volume of the solid of revolution that is formed by revolving the region bounded by the graphs of the given equations about the indicated line or axis.

15. $y = 9 - x^2, y = 0;$ x-axis
16. $y = x^2 + 1, x = 0, y = 5;$ y-axis
17. $y = \frac{1}{x}, x = 1, y = \frac{1}{2};$ y-axis
18. $y = \frac{1}{x}, x = \frac{1}{2}, x = 3, y = 0;$ x-axis
19. $y = (x - 2)^2, x = 0, y = 0;$ x-axis
20. $y = (x + 1)^2, x = 0, y = 0;$ y-axis
21. $y = 4 - x^2, y = 1 - \frac{1}{4}x^2;$ x-axis
22. $y = 1 - x^2, y = x^2 - 1, x = 0,$ first quadrant; y-axis
23. $y = x, y = x + 1, x = 0, y = 2;$ y-axis
24. $x + y = 2, x = 0, y = 0, y = 1;$ x-axis
25. $y = \sqrt{x - 1}, x = 5, y = 0;$ $x = 5$
26. $x = y^2, x = 1;$ $x = 1$
27. $y = x^{1/3}, x = 0, y = 1;$ $y = 2$
28. $x = -y^2 + 2y, x = 0;$ $x = 2$
29. $x^2 - y^2 = 16, x = 5;$ y-axis
30. $y = x^2 - 6x + 9, y = 9 - \frac{1}{2}x^2;$ x-axis
31. $x = y^2, y = x - 6;$ y-axis
32. $y = x^3 + 1, x = 0, y = 9;$ y-axis
33. $y = x^3 - x, y = 0;$ x-axis
34. $y = x^3 + 1, x = 1, y = 0;$ x-axis
35. $y = e^{-x}, x = 1, y = 1;$ $y = 2$
36. $y = e^x, y = 1, x = 2;$ x-axis
37. $y = |\cos x|, y = 0, 0 \leq x \leq 2\pi;$ x-axis
38. $y = \sec x, x = -\pi/4, x = \pi/4, y = 0;$ x-axis
39. $y = \tan x, y = 0, x = \pi/4;$ x-axis
40. $y = \sin x, y = \cos x, x = 0,$ first quadrant; x-axis

≡ Think About It

41. Reread Problems 68–70 in Exercises 6.2 on Cavalieri's Principle. Then show that the circular cylinders in FIGURE 6.3.19 have the same volume.

FIGURE 6.3.19 Cylinders in Problem 41

42. Consider the right circular cylinder of radius a shown in FIGURE 6.3.20. A plane inclined at an angle θ to the base of the cylinder passes through a diameter of the base. Find the volume of the resulting wedge cut from the cylinder when
 (a) $\theta = 45°$
 (b) $\theta = 60°$.

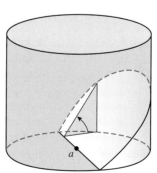

FIGURE 6.3.20 Cylinder and wedge in Problem 42

340 CHAPTER 6 Applications of the Integral

≡ Projects

43. For the Birds A mathematical model for the shape of an egg can be obtained by revolving the region bounded by the graphs of $y = 0$ and the function $f(x) = P(x)\sqrt{1 - x^2}$, where $P(x) = ax^3 + bx^2 + cx + d$ is a cubic polynomial, about the x-axis. For example, an egg of the Common Murre corresponds to $P(x) = -0.07x^3 - 0.02x^2 + 0.2x + 0.56$. FIGURE 6.3.21 shows the graph of f obtained with the aid of a CAS.

(a) Find a general formula for the volume V of an egg based on the mathematical model $f(x) = P(x)\sqrt{1 - x^2}$, where $P(x) = ax^3 + bx^2 + cx + d$. [*Hint*: This problem can be done by hand calculation but it is long and "messy". Use a CAS to carry out the integration.]

(b) Use the formula obtained in part (a) to estimate the volume of an egg of the Common Murre.

(c) An egg of the Red-throated Loon corresponds to $P(x) = -0.06x^3 + 0.04x^2 + 0.1x + 0.54$. Use a calculator or CAS to obtain the graph of f.

(d) Use part (a) to estimate the volume of an egg of the Red-throated Loon.

Common Murre eggs

FIGURE 6.3.21 Model of the shape of the Common Murre egg in Problem 43

44. That Sinking Feeling A wooden spherical ball of radius r is floating on a pond of still water. Let h denote the depth that the ball will sink into the water. See FIGURE 6.3.22.

(a) Show that the volume of the submerged portion of the ball is given by $V = \pi r^2 h - \frac{1}{3}\pi h^3$.

(b) Suppose that the weight density of the ball is denoted by ρ_{ball} and the weight density of the water is ρ_{water} (measured in lb/ft^3). If $r = 3$ in. and $\rho_{ball} = 0.4\rho_{water}$, use Archimedes' Principle—the weight of the ball equals the weight of the water displaced—to determine the approximate depth h that the ball will sink. You will need a calculator or CAS to solve a cubic polynomial equation.

FIGURE 6.3.22 Floating wooden ball in Problem 44

45. Steinmetz Solids The solid formed by two intersecting circular cylinders of radius r whose axes intersect at a right angle is called a **bicylinder** and is a special case of Steinmetz solids. For clarity we have shown one-eighth of the solid in FIGURE 6.3.23.

(a) Find the total volume of the bicylinder illustrated in the figure.

(b) Write a short report on Steinmetz solids.

FIGURE 6.3.23 Intersecting right circular cylinders in Problem 45

6.4 Volumes of Solids: Shell Method

FIGURE 6.4.1 Cylindrical shell

■ **Introduction** In this section we continue the discussion of finding volumes of solids of revolution. But instead of using planes perpendicular to the axis of revolution to slice the solid into slabs whose volume can be approximated by right regular circular cylinders (disks or washers), we will develop a new method for finding volumes of solid of revolution that utilizes circular cylindrical shells. Before building an integral representing this **shell method** we need to find the volume of the general cylindrical shell shown in FIGURE 6.4.1. If, as shown in the figure, r_1 and r_2 denote, respectively, the inner and outer radii of the shell, and h is its height, then its volume is given by the difference

$$\text{volume of outer cylinder} - \text{volume of inner cylinder}$$
$$= \pi r_2^2 h - \pi r_1^2 h = \pi(r_2^2 - r_1^2)h = \pi(r_2 + r_1)(r_2 - r_1)h. \qquad (1)$$

■ **Building an Integral** In Section 6.3 we saw that a rectangular element of area that is perpendicular to an axis of revolution will generate, when revolved, either a circular disk or a circular washer. However, if we were to revolve the rectangular element shown in FIGURE 6.4.2(a) about the y-axis, we generate a hollow shell as shown in Figure 6.4.2(b). To find the volume of the solid shown in Figure 6.4.2(c) we let P denote the arbitrary partition of the interval $[a, b]$:

$$a = x_0 < x_1 < x_2 < \cdots < x_{n-1} < x_n = b.$$

The partition P divides the interval into n subintervals $[x_{k-1}, x_k]$, $k = 1, 2, \ldots, n$, of width $\Delta x_k = x_k - x_{k-1}$. If we identify the outer radius as $r_2 = x_k$ and the inner radius as $r_1 = x_{k-1}$ and define $x_k^* = \frac{1}{2}(x_k + x_{k-1})$, then x_k^* is the midpoint of the subinterval $[x_{k-1}, x_k]$. With the further identification $h = f(x_k^*)$ it follows from (1) that the volume of the representative shell in Figure 6.4.2(b) be written as

$$V_k = \pi(x_k + x_{k-1})(x_k - x_{k-1})h$$
$$= 2\pi \frac{x_k + x_{k-1}}{2} h(x_k - x_{k-1})$$

or
$$V_k = 2\pi x_k^* f(x_k^*) \, \Delta x_k.$$

An approximation to the volume of the solid is given by the Riemann sum

$$\sum_{k=1}^{n} V_k = \sum_{k=1}^{n} 2\pi x_k^* f(x_k^*) \, \Delta x_k. \tag{2}$$

As the norm $\|P\|$ of the partition approaches zero, the limit of (2) is a definite integral that we use as the definition of the volume V of the solid:

$$V = 2\pi \int_a^b x f(x) \, dx. \tag{3}$$

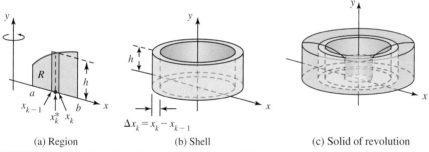

(a) Region   (b) Shell   (c) Solid of revolution

FIGURE 6.4.2 Revolving the red rectangular element in (a) about the y-axis generates the red shell in (b)

As mentioned in the *Notes from the Classroom* at the end of Section 6.2 it is not possible to derive an integral, in this case representing the volume of a solid of revolution, that "works" in every possible case. You are urged again not to memorize a particular formula such as (3). Try to understand the geometric interpretation of the component parts of the integrand. For example, $f(x)$, which represents the height of the rectangle in Figure 6.4.2, could be the difference $f(x) - g(x)$ if the rectangular element is between the graphs of two functions $y = f(x)$ and $y = g(x), f(x) \geq g(x)$. To set up an integral for a given problem without going through a lengthy analysis think of a shell as a circular tin can with its top and bottom removed. To find the volume of the shell, that is, the volume of the metal in the tin can analogy, imagine that the shell is cut straight down its lateral side and flattened out as illustrated in FIGURE 6.4.3(a) and (b). As Figure 6.4.3(c) shows, the volume of the shell is then the volume of a thin rectangular solid:

$$\text{volume} = (\text{length}) \cdot (\text{width}) \cdot (\text{thickness})$$
$$= (\text{circumference of the cylinder}) \cdot (\text{height}) \cdot (\text{thickness})$$
$$= 2\pi r h t. \tag{4}$$

(a) Cut the shell down its side   (b) Flatten it out   (c) The result is a rectangular solid

FIGURE 6.4.3 Finding the volume of a shell

EXAMPLE 1 Using the Shell Method

Reread Example 5 in Section 6.3 before working through this example.

▶ Use the shell method to find the volume V of the solid formed by revolving the region bounded by the graphs of $y = \sqrt{x}$ and $y = x$ about the y-axis.

Solution We solved this problem in Example 5 of Section 6.3. In that example we saw that using a horizontal rectangular element perpendicular to the y-axis of width Δy_k generated a washer when revolved about the y-axis. In contrast, a vertical rectangular element of width Δx_k revolved about the y-axis generates a shell. Using FIGURE 6.4.4(a) we make the identifications in (4) that $r = x_k^*$,

$$h = \text{uppergraph} - \text{lower graph} = \sqrt{x_k^*} - x_k^*,$$

and $t = \Delta x_k$. From the volume of the shell,

$$V_k = 2\pi x_k^* \big(\sqrt{x_k^*} - x_k^*\big)\, \Delta x_k = 2\pi\big((x_k^*)^{3/2} - (x_k^*)^2\big)\, \Delta x_k.$$

we obtain the definite integral getting the volume of the solid:

$$V = 2\pi \int_0^1 (x^{3/2} - x^2)\, dx$$

$$= 2\pi \left(\frac{2}{5} x^{5/2} - \frac{1}{3} x^3\right)\bigg]_0^1 = \frac{2}{15}\pi.$$

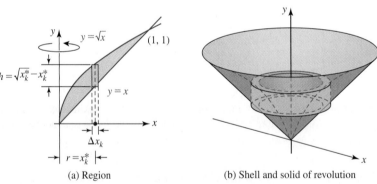

(a) Region   (b) Shell and solid of revolution

FIGURE 6.4.4 Region and solid of revolution in Example 1

It is not always convenient or even possible to use the disk or washer method discussed in the last section to find the volume of a solid of revolution.

EXAMPLE 2 Using the Shell Method

Find the volume V of the solid that is formed by revolving the region bounded by the graph of $y = \sin x^2$ and $y = 0$, $0 \le x \le \sqrt{\pi}$ about the y-axis.

Solution The graph of $y = \sin x^2$ on the indicated interval in FIGURE 6.4.5 was obtained with the help of a CAS.

If we choose to use a horizontal rectangular element to revolve about the y-axis, a washer would be generated. To determine the inner and outer radii of the washer we would have to solve $y = \sin x^2$ for x in terms of y. While this simply leads to x^2 as an inverse sine, this poses the practical problem: We are not in a position to integrate an inverse trigonometric function at this time. Thus we switch our attention to a vertical rectangular element shown in Figure 6.4.5(a). When this element is revolved about the y-axis a shell with radius $r = x_k^*$, height $h = \sin(x_k^*)^2$, and thickness $t = \Delta x_k$ is generated. By (4) the volume of the shell is

$$V_k = 2\pi x_k^* \sin(x_k^*)^2 \, \Delta x_k.$$

Thus, by (3) we have

$$V = 2\pi \int_0^{\sqrt{\pi}} x \sin x^2 \, dx.$$

If we let $u = x^2$, then $du = 2x \, dx$ and $x \, dx = \frac{1}{2} du$. The u-limits of integration are determined from the fact that when $x = 0, u = 0$, and $x = \sqrt{\pi}, u = \pi$. Therefore, the volume of the solid of revolution shown in Figure 6.4.5(b) is

$$V = \pi \int_0^{\pi} \sin u \, du = -\pi \cos u \Big]_0^{\pi} = \pi(1 + 1) = 2\pi.$$

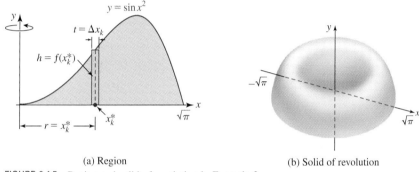

(a) Region (b) Solid of revolution

FIGURE 6.4.5 Region and solid of revolution in Example 2

In the next example we illustrate the shell method when a region is revolved about a line that is not a coordinate axis.

EXAMPLE 3 Axis of Revolution not Coordinate Axis

Find the volume V of the solid that is formed by revolving the region bounded by the graphs of $x = y^2 - 2y$ and $x = 3$ about the line $y = 1$.

Solution In this case a rectangular element of area that is perpendicular to a horizontal line and revolved about the line $y = 1$ would generate a disk. Since the radius of the disk is not measured from the x-axis but from the line $y = 1$, it would be necessary to solve $x = y^2 - 2y$ for y in terms of x. We can avoid this inconvenience by using horizontal elements of area, which then generate shells such as that shown in FIGURE 6.4.6(b). Note that when $x = 3$, the equation $3 = y^2 - 2y$ or equivalently $(y + 1)(y - 3) = 0$, has solutions -1 and 3. Thus, we need only partition the interval $[1, 3]$ on the y-axis. After making the identifications $r = y_k^* - 1$, $h = 3 - x_k^*$, and $t = \Delta y_k$, it follows from (4) that the volume of a shell is

$$\begin{aligned} V_k &= 2\pi(y_k^* - 1)(3 - x_k^*)\Delta y_k \\ &= 2\pi(y_k^* - 1)(3 - (y_k^*)^2 + 2y_k^*)\Delta y_k \\ &= 2\pi(-(y_k^*)^3 + 3(y_k^*)^2 + y_k^* - 3)\Delta y_k. \end{aligned}$$

(a) Region (b) Shell (c) Solid of revolution

FIGURE 6.4.6 Region, shell, and solid of revolution in Example 3

From the last line we see that the volume of the solid is the definite integral

$$V = 2\pi \int_1^3 (-y^3 + 3y^2 + y - 3)\,dy$$

$$= 2\pi \left(-\frac{1}{4}y^4 + y^3 + \frac{1}{2}y^2 - 3y\right)\Big]_1^3$$

$$= 2\pi \left[\left(-\frac{81}{4} + 27 + \frac{9}{2} - 9\right) - \left(-\frac{1}{4} + 1 + \frac{1}{2} - 3\right)\right] = 8\pi. \quad \blacksquare$$

Exercises 6.4 Answers to selected odd-numbered problems begin on page ANS-20.

☰ Fundamentals

In Problems 1–6, refer to FIGURE 6.4.7. Use the shell method to find the volume of the solid of revolution that is formed by revolving the given region about the indicated line.

1. R_1 about OC
2. R_1 about OA
3. R_2 about BC
4. R_2 about OA
5. R_1 about AB
6. R_2 about AB

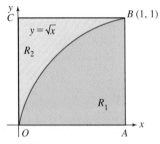

FIGURE 6.4.7 Regions for Problems 1–6

In Problems 7–30, use the shell method to find the volume of the solid of revolution that is formed by revolving the region bounded by the graphs of the given equations about the indicated line or axis.

7. $y = x, x = 0, y = 5$; x-axis
8. $y = 1 - x, x = 0, y = 0$; $y = -2$
9. $y = x^2, x = 0, y = 3$, first quadrant; x-axis
10. $y = x^2, x = 2, y = 0$; y-axis
11. $y = x^2, x = 1, y = 0$; $x = 3$
12. $y = x^2, y = 9$; x-axis
13. $y = x^2 + 4, x = 0, x = 2, y = 2$; y-axis
14. $y = x^2 - 5x + 4, y = 0$; y-axis
15. $y = (x - 1)^2, y = 1$; x-axis
16. $y = (x - 2)^2, y = 4$; $x = 4$
17. $y = x^{1/3}, x = 1, y = 0$; $y = -1$
18. $y = x^{1/3} + 1, y = -x + 1, x = 1$; $x = 1$
19. $y = x^2, y = x$; y-axis
20. $y = x^2, y = x$; $x = 2$
21. $y = -x^3 + 3x^2, y = 0$, first quadrant; y-axis
22. $y = x^3 - x, y = 0$, second quadrant; y-axis
23. $y = x^2 - 2, y = -x^2 + 2, x = 0$, second and third quadrants; y-axis
24. $y = x^2 - 4x, y = -x^2 + 4x$; $x = -1$
25. $x = y^2 - 5y, x = 0$; x-axis
26. $x = y^2 + 2, y = x - 4, y = 1$; x-axis
27. $y = x^3, y = x + 6, x = 0$; y-axis
28. $y = \sqrt{x}, y = \sqrt{1 - x}, y = 0$; x-axis
29. $y = \sin x^2, x = 0, y = 1$; y-axis
30. $y = e^{x^2}, y = 0, x = 0, x = 1$; y-axis

In Problems 31–36, the region in part (a) is revolved about the indicated axis generating the solid given in part (b). Choose between the disk, washer, or shell method to find the volume of the solid of revolution.

31.

(a) (b) Cone

FIGURE 6.4.8 Region and solid for Problem 31

32.

(a) (b) Frustum

FIGURE 6.4.9 Region and solid for Problem 32

33.

(a) (b) Sphere

FIGURE 6.4.10 Region and solid for Problem 33

34.

(a)   (b) Spherical sector

FIGURE 6.4.11 Region and solid for Problem 34

35.

(a)   (b) Prolate spheroid

FIGURE 6.4.12 Region and solid for Problem 35

36.

(a)   (b) Oblate spheroid

FIGURE 6.4.13 Region and solid for Problem 36

≡ **Applications**

37. A cylindrical bucket of radius r that contains a liquid is rotating about the y-axis with a constant angular velocity ω. It can be shown that the surface of the liquid has a parabolic cross-section given by $y = \omega^2 x^2/(2g)$, $-r \leq x \leq r$, where g is the acceleration due to gravity. Use the shell method to find the volume V of the liquid in the rotating bucket given that the height of the bucket is h. See FIGURE 6.4.14.

38. In Problem 37, determine the angular velocity ω for which the fluid will touch the bottom of the bucket. What is the corresponding volume V of the liquid?

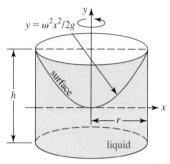

FIGURE 6.4.14 Bucket in Problems 37 and 38

6.5 Length of a Graph

Introduction If a function $y = f(x)$ has a continuous first derivative on an interval $[a, b]$, then its graph is said to be **smooth** and f is called a **smooth function**. As the name implies, a smooth graph has no sharp points. In the discussion that follows we will build an integral formula for the **length** L, or **arc length**, of a smooth graph on an interval $[a, b]$. See FIGURE 6.5.1.

Building an Integral Let f have a smooth graph on $[a, b]$ and let P denote an arbitrary partition of the interval:

$$a = x_0 < x_1 < x_2 < \cdots < x_{n-1} < x_n = b.$$

As usual, let the width of the kth subinterval be given by Δx_k and let $\|P\|$ be the width of the longest subinterval. As shown in 6.5.2(a), we can approximate the length of the graph on each subinterval $[x_{k-1}, x_k]$ by finding the length L_k of the chord between the points $(x_{k-1}, f(x_{k-1}))$ and $(x_k, f(x_k))$ for $k = 1, 2, \ldots, n$. From Figure 6.5.2(b), the Pythagorean Theorem gives the length L_k:

$$L_k = \sqrt{(\Delta x_k)^2 + (\Delta y_k)^2} = \sqrt{(x_k - x_{k-1})^2 + (f(x_k) - f(x_{k-1}))^2}. \quad (1)$$

By the Mean Value Theorem (Section 4.4), we know there exists a number x_k^* in each open subinterval (x_{k-1}, x_k) such that

$$\frac{f(x_k) - f(x_{k-1})}{x_k - x_{k-1}} = f'(x_k^*) \quad \text{or} \quad f(x_k) - f(x_{k-1}) = f'(x_k^*)(x_k - x_{k-1}).$$

Using the last equation, we replace $f(x_k) - f(x_{k-1})$ in (1) and simplify:

$$\begin{aligned} L_k &= \sqrt{(x_k - x_{k-1})^2 + [f'(x_k^*)]^2 (x_k - x_{k-1})^2} \\ &= \sqrt{(x_k - x_{k-1})^2 (1 + [f'(x_k^*)]^2)} \\ &= \sqrt{1 + [f'(x_k^*)]^2} \, (x_k - x_{k-1}) \\ &= \sqrt{1 + [f'(x_k^*)]^2} \, \Delta x_k. \end{aligned}$$

FIGURE 6.5.1 Find the length L of the graph of f on $[a, b]$

(a) n chords

(b) Zoom in on chord on the kth subinterval

FIGURE 6.5.2 Approximating the length of a graph by summing the lengths of chords

The Riemann sum

$$\sum_{k=1}^{n} L_k = \sum_{k=1}^{n} \sqrt{1 + [f'(x_k^*)]^2}\, \Delta x_k$$

represents the length of the polygonal curve joining $(a, f(a))$ and $(b, f(b))$ and gives an approximation to the total length of the graph on $[a, b]$. As $\|P\| \to 0$, we obtain

$$\lim_{\|P\| \to 0} \sum_{k=1}^{n} \sqrt{1 + [f'(x_k^*)]^2}\, \Delta x_k = \int_a^b \sqrt{1 + [f'(x)]^2}\, dx. \tag{2}$$

The foregoing discussion prompts us to use (2) as the definition of the length of the graph on the interval.

Definition 6.5.1 Arc Length

Let f be a function for which f' is continuous on an interval $[a, b]$. Then the **length** L of the graph of $y = f(x)$ on the interval is given by

$$L = \int_a^b \sqrt{1 + [f'(x)]^2}\, dx. \tag{3}$$

The formula for arc length (3) is also written as

$$L = \int_a^b \sqrt{1 + \left(\frac{dy}{dx}\right)^2}\, dx. \tag{4}$$

A graph that has arc length is said to be **rectifiable**.

EXAMPLE 1 Length of a Curve

Find the length of the graph of $y = 4x^{3/2}$ from the origin $(0, 0)$ to the point $(1, 4)$.

Solution The graph of the function on the interval $[0, 1]$ is given in FIGURE 6.5.3. Now,

$$\frac{dy}{dx} = 6x^{1/2}$$

is continuous on the interval. Therefore, it follows from (4) that

$$L = \int_0^1 \sqrt{1 + [6x^{1/2}]^2}\, dx$$

$$= \int_0^1 (1 + 36x)^{1/2}\, dx$$

$$= \frac{1}{36} \int_0^1 (1 + 36x)^{1/2} (36\, dx)$$

$$= \frac{1}{54}(1 + 36x)^{3/2} \bigg]_0^1 = \frac{1}{54}[37^{3/2} - 1] \approx 4.1493.$$

FIGURE 6.5.3 Graph of function in Example 1

■ **Differential of Arc Length** If C is a smooth curve defined by $y = f(x)$, then the arc length between an initial point $(a, f(a))$ and a variable point $(x, f(x))$, where $a \le x \le b$, is given by

$$s(x) = \int_a^x \sqrt{1 + [f'(t)]^2}\, dt, \tag{5}$$

where t represents a dummy variable of integration. The value of the integral in (5) obviously depends on x and so is called the **arc length function**. Then by (10) of Section 5.5, $ds/dx = \sqrt{1 + [f'(x)]^2}$ and, consequently,

$$ds = \sqrt{1 + [f'(x)]^2}\, dx. \tag{6}$$

The latter function is called the **differential of the arc length** and can be used to approximate lengths of curves. With $dy = f'(x)\,dx$, (6) can be written as

$$ds = \sqrt{(dx)^2 + (dy)^2} \quad \text{or} \quad (ds)^2 = (dx)^2 + (dy)^2. \tag{7}$$

FIGURE 6.5.4 shows that the differential ds can be interpreted as the hypotenuse of a right triangle with sides dx and dy.

If (3) is written $L = \int ds$ for brevity and the curve C is defined by $x = g(y)$, $c \le y \le d$, then the last expression in (7) can be used to solve for ds/dy:

$$\frac{ds}{dy} = \sqrt{1 + \left(\frac{dx}{dy}\right)^2} \quad \text{or} \quad ds = \sqrt{1 + \left(\frac{dx}{dy}\right)^2}\,dy.$$

Thus, the y-integration analogue of (4) is

$$L = \int_c^d \sqrt{1 + \left(\frac{dx}{dy}\right)^2}\,dy. \tag{8}$$

See Problems 17 and 18 in Exercises 6.5.

FIGURE 6.5.4 Geometric interpretation of the differential of the arc length

$\int_a^b$ NOTES FROM THE CLASSROOM

The integral in (3) often leads to problems in which specialized techniques of integration are necessary. See Chapter 7. But even with these subsequent procedures, it is not *always* possible to evaluate the indefinite integral $\int \sqrt{1 + [f'(x)]^2}\,dx$ in terms of the familiar elementary functions even for some of the simplest functions such as $y = x^2$. See Problem 45 in Exercises 7.8.

Exercises 6.5 Answers to selected odd-numbered problems begin on page ANS-20.

≡ Fundamentals

In Problems 1–12, find the length of the graph of the given function on the indicated interval. Use a calculator or CAS to obtain the graph.

1. $y = x$; $[-1, 1]$
2. $y = 2x + 1$; $[0, 3]$
3. $y = x^{3/2} + 4$; $[0, 1]$
4. $y = 3x^{2/3}$; $[1, 8]$
5. $y = \frac{2}{3}(x^2 + 1)^{3/2}$; $[1, 4]$
6. $(y + 1)^2 = 4(x + 1)^3$; $[-1, 0]$
7. $y = \frac{1}{3}x^{3/2} - x^{1/2}$; $[1, 4]$
8. $y = \frac{1}{6}x^3 + \frac{1}{2x}$; $[2, 4]$
9. $y = \frac{1}{4}x^4 + \frac{1}{8x^2}$; $[2, 3]$
10. $y = \frac{1}{5}x^5 + \frac{1}{12x^3}$; $[1, 2]$
11. $y = (4 - x^{2/3})^{3/2}$; $[1, 8]$
12. $y = \begin{cases} x - 2, & 2 \le x < 3 \\ (x - 2)^{2/3}, & 3 \le x < 10; \\ \frac{1}{2}(x - 6)^{3/2}, & 10 \le x \le 15 \end{cases}$ $[2, 15]$

In Problems 13–16, set up, but do not evaluate, an integral for the length of the graph of the given function on the indicated interval.

13. $y = x^2$; $[-1, 3]$
14. $y = 2\sqrt{x + 1}$; $[-1, 3]$
15. $y = \sin x$; $[0, \pi]$
16. $y = \tan x$; $[-\pi/4, \pi/4]$

In Problems 17 and 18, use (8) to find the length of the graph of the given equation on the indicated interval.

17. $x = 4 - y^{2/3}$; $[0, 8]$
18. $5x = y^{5/2} + 5y^{-1/2}$; $[4, 9]$
19. Consider the length of the graph of $x^{2/3} + y^{2/3} = 1$ in the first quadrant.
    (a) Show that the use of (3) leads to a discontinuous integrand.
    (b) By assuming that the Fundamental Theorem of Calculus can be used to evaluate the integral obtained in part (a), find the total length of the graph.
20. Set up, but make no attempt to evaluate, an integral that gives the total length of the ellipse $x^2/a^2 + y^2/b^2 = 1$, $a > b > 0$.
21. Given that the circumference of a circle of radius r is $2\pi r$, find the value of the integral

    $$\int_0^1 \frac{1}{\sqrt{1 - x^2}}\,dx.$$

22. Use the differential of the arc length (6) to approximate the length of the graph of $y = \frac{1}{4}x^4$ from $(2, 4)$ to $(2.1, 4.862025)$. [*Hint*: Review (13) of Section 4.9.]

348 CHAPTER 6 Applications of the Integral

6.6 Area of a Surface of Revolution

Introduction As we have seen in Sections 6.3 and 6.4, when the graph of a continuous function $y = f(x)$ on an interval $[a, b]$ is revolved about the x-axis, it generates a solid of revolution. In this section, we are interested in finding the area S of the corresponding surface—that is, a **surface of revolution** on $[a, b]$ as shown in FIGURE 6.6.1(b).

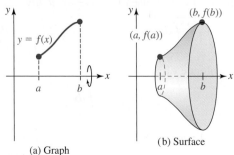

(a) Graph (b) Surface

FIGURE 6.6.1 Surface of revolution

FIGURE 6.6.2 Frustum of a cone

Building an Integral Before building a definite integral for the definition of the area of a surface of revolution, we first need the formula for the lateral area (top and bottom excluded) of a *frustum* of a right circular cone. See FIGURE 6.6.2. If r_1 and r_2 are the radii of the top and bottom and L is the slant height, then the lateral area is given by

$$\pi(r_1 + r_2)L. \quad (1)$$

See Problem 17 in Exercises 6.6. Now suppose $y = f(x)$ is a smooth function and $f(x) \geq 0$ on the interval $[a, b]$. Let P be a partition of the interval:

$$a = x_0 < x_1 < x_2 < \cdots < x_{n-1} < x_n = b.$$

Now, if we connect the points $(x_{k-1}, f(x_{k-1}))$ and $(x_k, f(x_k))$ shown in FIGURE 6.6.3(a) by a chord, we form a trapezoid. When revolved about the x-axis, this trapezoid generates a frustum of a cone with radii $f(x_{k-1})$ and $f(x_k)$. See Figure 6.6.3(b). As shown in cross-section in Figure 6.6.3(c), the slant height can be obtained from the Pythagorean Theorem:

$$\sqrt{(x_k - x_{k-1})^2 + (f(x_k) - f(x_{k-1}))^2}.$$

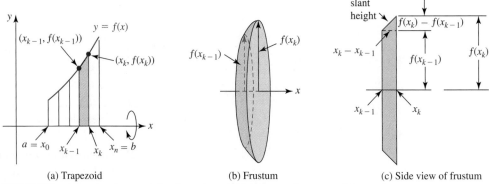

(a) Trapezoid (b) Frustum (c) Side view of frustum

FIGURE 6.6.3 Approximating area of surface of revolution by summing areas of frustums

Thus, from (1) the surface area of this element is

$$S_k = \pi [f(x_k) + f(x_{k-1})] \sqrt{(x_k - x_{k-1})^2 + (f(x_k) - f(x_{k-1}))^2}$$

$$= \pi [f(x_k) + f(x_{k-1})] \sqrt{1 + \left(\frac{f(x_k) - f(x_{k-1})}{x_k - x_{k-1}}\right)^2} (x_k - x_{k-1})$$

$$= \pi [f(x_k) + f(x_{k-1})] \sqrt{1 + \left(\frac{f(x_k) - f(x_{k-1})}{x_k - x_{k-1}}\right)^2} \Delta x_k,$$

where $\Delta x_k = x_k - x_{k-1}$. This last quantity is an approximation to the actual area of the surface of revolution on the subinterval $[x_{k-1}, x_k]$.

Now, as in the discussion of arc length, we invoke the Mean Value Theorem for derivatives to assert that there exists an x_k^* in the open interval (x_{k-1}, x_k) such that

$$f'(x_k^*) = \frac{f(x_k) - f(x_{k-1})}{x_k - x_{k-1}}.$$

The Riemann sum

$$\sum_{k=1}^{n} S_k = \pi \sum_{k=1}^{n} [f(x_k) + f(x_{k-1})] \sqrt{1 + [f'(x_k^*)]^2} \, \Delta x_k$$

is an approximation to the area S on $[a, b]$. This suggests that the surface area S is given by the limit of the Riemann sum:

$$S = \lim_{\|P\| \to 0} \pi \sum_{k=1}^{n} [f(x_k) + f(x_{k-1})] \sqrt{1 + [f'(x_k^*)]^2} \, \Delta x_k. \tag{2}$$

Since we also expect $f(x_{k-1})$ and $f(x_k)$ to approach the common limit $f(x)$ as $\|P\| \to 0$, we have $f(x_k) + f(x_{k-1}) \to 2f(x)$.

The foregoing discussion prompts us to use (2) as the definition of the area of the surface of revolution on the interval.

Definition 6.6.1 Area of a Surface of Revolution

Let f be a function for which f' is continuous and $f(x) \geq 0$ for all x in the interval $[a, b]$. The **area** S of the surface that is obtained by revolving the graph of f on the interval about the x-axis is given by

$$S = 2\pi \int_a^b f(x) \sqrt{1 + [f'(x)]^2} \, dx. \tag{3}$$

EXAMPLE 1 Area of a Surface

Find the area S of the surface that is formed by revolving the graph of $y = \sqrt{x}$ on the interval $[1, 4]$ about the x-axis.

Solution We have $f(x) = x^{1/2}$, $f'(x) = \frac{1}{2}x^{-1/2} = 1/(2\sqrt{x})$, and from (3)

$$S = 2\pi \int_1^4 \sqrt{x} \sqrt{1 + \left(\frac{1}{2\sqrt{x}}\right)^2} \, dx$$

$$= 2\pi \int_1^4 \sqrt{x} \sqrt{1 + \frac{1}{4x}} \, dx$$

$$= 2\pi \int_1^4 \sqrt{x} \sqrt{\frac{4x+1}{4x}} \, dx$$

$$= \pi \int_1^4 \sqrt{4x+1} \, dx$$

$$= \frac{1}{4}\pi \int_1^4 (4x+1)^{1/2}(4\,dx) = \frac{1}{6}\pi(4x+1)^{3/2}\Big]_1^4$$

$$= \frac{1}{6}\pi[17^{3/2} - 5^{3/2}] \approx 30.85.$$

See FIGURE 6.6.4.

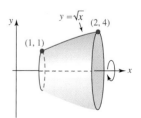

FIGURE 6.6.4 Surface of revolution about x-axis in Example 1

Revolution about y-Axis It can be shown that if the graph of a continuous function $y = f(x)$ on $[a, b]$, $0 \le a < b$, is revolved about the y-axis, then the area S of the resulting surface of revolution is given by

$$S = 2\pi \int_a^b x\sqrt{1 + [f'(x)]^2}\, dx. \tag{4}$$

As in (3), we assume in (4) that $f'(x)$ is continuous on the interval $[a, b]$.

EXAMPLE 2 Area of a Surface

Find the area S of the surface formed by revolving the graph of $y = x^{1/3}$ on the interval $[0, 8]$ about the y-axis.

Solution We have $f'(x) = \frac{1}{3}x^{-2/3}$, so that from (4) it follows that

$$S = 2\pi \int_0^8 x\sqrt{1 + \frac{1}{9}x^{-4/3}}\, dx$$

$$= 2\pi \int_0^8 x\sqrt{\frac{9x^{4/3} + 1}{9x^{4/3}}}\, dx$$

$$= \frac{2}{3}\pi \int_0^8 x^{1/3}\sqrt{9x^{4/3} + 1}\, dx.$$

Let us evaluate the last integral by reviewing the u-substitution method. If we let $u = 9x^{4/3} + 1$, then $du = 12x^{1/3}\, dx$, $dx = \frac{1}{12}x^{-1/3}\, du$, $x = 0$ implies $u = 1$, and $x = 8$ gives $u = 145$. Therefore,

$$S = \frac{1}{18}\pi \int_1^{145} u^{1/2}\, du = \frac{1}{27}\pi u^{3/2}\Big]_1^{145} = \frac{1}{27}\pi(145^{3/2} - 1^{3/2}) \approx 203.04.$$

See **FIGURE 6.6.5**.

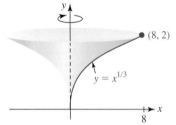

FIGURE 6.6.5 Surface of revolution about y-axis in Example 2

Exercises 6.6
Answers to selected odd-numbered problems begin on page ANS-20.

≡ Fundamentals

In Problems 1–10, find the area of the surface that is formed by revolving each graph on the given interval about the indicated axis.

1. $y = 2\sqrt{x}$, $[0, 8]$; x-axis
2. $y = \sqrt{x + 1}$, $[1, 5]$; x-axis
3. $y = x^3$, $[0, 1]$; x-axis
4. $y = x^{1/3}$, $[1, 8]$; y-axis
5. $y = x^2 + 1$, $[0, 3]$; y-axis
6. $y = 4 - x^2$, $[0, 2]$; y-axis
7. $y = 2x + 1$, $[2, 7]$; x-axis
8. $y = \sqrt{16 - x^2}$, $[0, \sqrt{7}]$; y-axis
9. $y = \frac{1}{4}x^4 + \frac{1}{8x^2}$, $[1, 2]$; y-axis
10. $y = \frac{1}{3}x^3 + \frac{1}{4x}$, $[1, 2]$; x-axis

11. **(a)** The shape of a dish antenna is a parabola revolved about its axis of symmetry and is called a **paraboloid** of revolution. Find the surface area of an antenna of radius r and depth h obtained by revolving the graph of $f(x) = r\sqrt{1 - x/h}$ about the x-axis. See **FIGURE 6.6.6**.

(b) The depth of a dish antenna ranges from 10% to 20% of its radius. If the depth h of the antenna in part (a) is 10% of the radius, show that the surface area of the antenna is approximately the same as the area of a circle of radius r. What is the percentage error in this case?

Dish antennas are paraboloids of revolution

FIGURE 6.6.6 Graph of f in Problem 11

12. The surface formed by two parallel planes cutting a sphere of radius r is called a **spherical zone**. Find the area of the spherical zone shown in FIGURE 6.6.7.

FIGURE 6.6.7 Spherical zone in Problem 12

13. The graph of $y = |x + 2|$ on $[-4, 2]$, shown in FIGURE 6.6.8, is revolved about the x-axis. Find the area S of the surface of revolution.

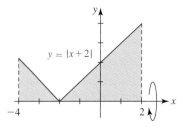

FIGURE 6.6.8 Graph of function in Problem 13

14. Find the area of the surface that is formed by revolving $x^{2/3} + y^{2/3} = a^{2/3}$, $[-a, a]$, about the x-axis.

≡ Think About It

15. Show that the lateral surface area of a right circular cone of radius r and slant height L is $\pi r L$. [*Hint*: A cone cut down its side and flattened forms a circular sector with area $\frac{1}{2}L^2\theta$.]

16. Use Problem 15 to show that the lateral surface area of a right circular cone of radius r and height h is given by $\pi r \sqrt{r^2 + h^2}$. Derive the same result using (3) or (4).

17. Use the result of Problem 15 to derive formula (1). [*Hint*: Consider a complete cone of radius r_2 and slant height L_2. Cut the conical top off. Similar triangles might help.]

18. Show that the surface area of a frustum of a right circular cone of radii r_1 and r_2 and height h is given by $\pi(r_1 + r_2)\sqrt{h^2 + (r_2 - r_1)^2}$.

19. Let $y = f(x)$ be a continuous nonnegative function on $[a, b]$ that has a continuous first derivative on the interval. Prove that if the graph of f is revolved around a horizontal line $y = L$, then the area S of the resulting surface of revolution is given by

$$S = 2\pi \int_a^b |f(x) - L|\sqrt{1 + [f'(x)]^2}\, dx.$$

20. Use the result of Problem 19 to find a definite integral that gives the area of the surface that is formed by revolving $y = x^{2/3}$, $[1, 8]$, about the line $y = 4$. Do not evaluate.

≡ Projects

21. **A View From Space**
    (a) From a spacecraft orbiting the Earth at a distance h from the surface, an astronaut can observe only a portion A_s of the Earth's total surface area A_e. See FIGURE 6.6.9(a). Find a formula for the fractional expression A_s/A_e as a function of h. In Figure 6.6.9(b) we have shown the Earth in cross-section as a circle with center C and radius R. Let the x- and y-axes be as shown and let the y-coordinates of the points B and E be y_B and $y_E = R$, respectively.
    (b) What percentage of the Earth's surface will an astronaut see from a height of 2000 km? Take the radius of the Earth to be $R = 6380$ km.
    (c) At what height h will an astronaut see one-fourth of the Earth's surface?
    (d) What is the limit of A_s/A_e as the height h increases without bound ($h \to \infty$)? Why does the answer make intuitive sense?
    (e) What percentage of the Earth's surface will an astronaut see from the Moon if $h = 3.76 \times 10^5$ km?

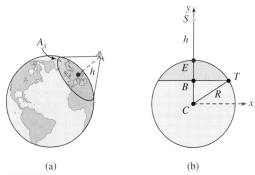

FIGURE 6.6.9 Portion of the Earth's surface in Problem 21

6.7 Average Value of a Function

▌ **Introduction** Every student is aware of averages. If a student takes four examinations in a semester and scores 80%, 75%, 85%, and 92% on them, then his or her average score is

$$\frac{80 + 75 + 85 + 92}{4}$$

352 CHAPTER 6 Applications of the Integral

or 83%. In general, given n numbers $a_1, a_2, \ldots, a_n$, we say that their **arithmetic mean**, or **average**, is

$$\frac{a_1 + a_2 + \cdots + a_n}{n} = \frac{1}{n}\sum_{k=1}^{n} a_k. \tag{1}$$

In this section we shall extend the notion of a discrete average such as (1) to the average of *all* the values of a continuous function f defined over an interval $[a, b]$.

■ **Average of Function Values** Suppose now that we have a continuous function f defined on an interval $[a, b]$. For the arbitrarily chosen numbers $x_i, i = 1, 2, \ldots, n$ such that $a < x_1 < x_2 < \cdots < x_n < b$, then by (1) the average of the set of corresponding function values is

$$\frac{f(x_1) + f(x_2) + \cdots + f(x_n)}{n} = \frac{1}{n}\sum_{k=1}^{n} f(x_k). \tag{2}$$

If we now consider the set of function values $f(x)$ that corresponds to all numbers x in an interval, it should be clear that we cannot use a discrete sum as in (1) since this set of function values is usually an uncountable set. For example, for $f(x) = x^2$ on $[0, 3]$, the values of the function range from a minimum of $f(0) = 0$ to a maximum of $f(3) = 9$. As indicated in FIGURE 6.7.1, we intuitively expect that there exists an average value f_{ave} such that $f(0) \leq f_{\text{ave}} \leq f(3)$.

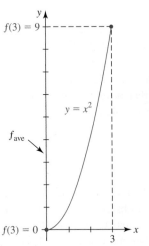

FIGURE 6.7.1 Find the average of all the numbers indicated in red on the y-axis

■ **Building an Integral** Returning to the general case of a continuous function defined on a closed interval $[a, b]$, we let P be a regular partition of the interval into n subintervals of width $\Delta x = (b - a)/n$. If x_k^* is a number chosen in each subinterval, then the average

$$\frac{f(x_k^*) + f(x_k^*) + \cdots + f(x_n^*)}{n}$$

can be written as

$$\frac{f(x_k^*) + f(x_2^*) + \cdots + f(x_n^*)}{\dfrac{b - a}{\Delta x}} \tag{3}$$

since $n = (b - a)/\Delta x$. Rewriting (3) as

$$\frac{1}{b - a}\sum_{k=1}^{n} f(x_k^*)\,\Delta x$$

and taking the limit of this last expression as $\|P\| = \Delta x \to 0$, we obtain the definite integral

$$\frac{1}{b - a}\int_{a}^{b} f(x)\,dx. \tag{4}$$

Because we have assumed that f is continuous on $[a, b]$, let us denote its absolute minimum and absolute maximum on the interval by m and M, respectively. If we multiply the inequality

$$m \leq f(x_k^*) \leq M$$

by $\Delta x > 0$ and sum, we obtain

$$\sum_{k=1}^{n} m\,\Delta x \leq \sum_{k=1}^{n} f(x_k^*)\,\Delta x \leq \sum_{k=1}^{n} M\,\Delta x.$$

Because $\sum_{k=1}^{n} \Delta x = b - a$, the preceding inequality is equivalent to

$$(b - a)m \leq \sum_{k=1}^{n} f(x_k^*)\,\Delta x \leq (b - a)M.$$

And so as $\Delta x \to 0$, it follows that

$$(b - a)m \leq \int_{a}^{b} f(x)\,dx \leq (b - a)M.$$

From the last inequality we conclude that the number from (4) satisfies

$$m \le \frac{1}{b-a}\int_a^b f(x) \le M.$$

By the Intermediate Value Theorem, f takes on all values between m and M. Hence, the number given by (4) actually corresponds to a value of the function on the interval. This prompts us to state the following definition.

Definition 6.7.1 Average Value of a Function

Let $y = f(x)$ be continuous on $[a, b]$. The **average value** of f on the interval is the number

$$f_{\text{ave}} = \frac{1}{b-a}\int_a^b f(x)\, dx. \qquad (5)$$

Although we are interested primarily in continuous functions, Definition 6.7.1 is valid for any integrable function on the interval.

EXAMPLE 1 Finding an Average Value

Find the average value of $f(x) = x^2$ on $[0, 3]$.

Solution From (5) of Definition 6.7.1, we obtain

$$f_{\text{ave}} = \frac{1}{3-0}\int_0^3 x^2\, dx = \frac{1}{3}\left(\frac{1}{3}x^3\right)\Big|_0^3 = 3.$$

It is sometimes possible to determine the value of x in the interval that corresponds to the average value of a function.

EXAMPLE 2 Finding an x Corresponding to f_{ave}

Determine the value of x in the interval $[0, 3]$ that corresponds to the average value f_{ave} of the function $f(x) = x^2$.

Solution Since the function $f(x) = x^2$ is continuous on the closed interval $[0, 3]$, we know from the Intermediate Value Theorem that there exists a number c between 0 and 3 so that

$$f(c) = c^2 = f_{\text{ave}}.$$

But, from Example 1, we know $f_{\text{ave}} = 3$. Thus, the equation $c^2 = 3$ has solutions $c = \pm\sqrt{3}$. As shown in FIGURE 6.7.2, the only solution of this equation in $[0, 3]$ is $c = \sqrt{3}$.

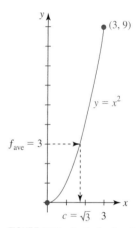

FIGURE 6.7.2 f_{ave} is the function value $f(\sqrt{3})$ in Example 1

■ **Mean Value Theorem for Definite Integrals** The following is an immediate consequence of the foregoing discussion. The result is called the Mean Value Theorem for Integrals.

Theorem 6.7.1 Mean Value Theorem for Integrals

Let $y = f(x)$ be continuous on $[a, b]$. Then there exists a number c in the open interval (a, b) such that

$$f(c)(b-a) = \int_a^b f(x)\, dx. \qquad (6)$$

In the case when $f(x) \ge 0$ for all x in $[a, b]$, Theorem 6.7.1 is readily interpreted in terms of area. The result in (6) simply states that there is some number c in (a, b) for which the area A of a *rectangle* of height $f(c)$ and width $b - a$ shown in FIGURE 6.7.3(a) is the same as the *area A under the graph* indicated in Figure 6.7.3(b).

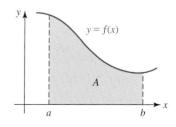

FIGURE 6.7.3 Area A of rectangle is the same as area under the graph on $[a, b]$

EXAMPLE 3 Finding an x Corresponding to f_{ave}

Find the height $f(c)$ of a rectangle so that the area A under the graph of $y = x^2 + 1$ on $[-2, 2]$ is the same as $f(c)[2 - (-2)] = 4f(c)$.

Solution This is basically the same type of problem as illustrated in Example 2. Now, the area under the graph shown in FIGURE 6.7.4(a) is

$$A = \int_{-2}^{2} (x^2 + 1)\,dx = \left(\frac{1}{3}x^3 + x\right)\Big]_{-2}^{2} = \frac{28}{3}.$$

Also, $4f(c) = 4(c^2 + 1)$, so that $4(c^2 + 1) = \frac{28}{3}$ implies $c^2 = \frac{4}{3}$. The solutions $c_1 = 2/\sqrt{3}$ and $c_2 = -2/\sqrt{3}$ are both in the interval $(-2, 2)$. For either number, we see that the height of the rectangle is $f(c_1) = f(c_2) = (\pm 2/\sqrt{3})^2 + 1 = \frac{7}{3}$. The area of the rectangle shown in Figure 6.7.4(b) is $f(c)[2 - (-2)] = \frac{7}{3} \cdot 4 = \frac{28}{3}$.

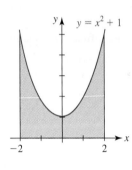
(a) Area under the graph

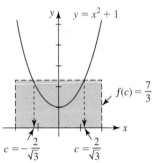
(b) Area of rectangle

FIGURE 6.7.4 Area in (a) is the same as the area in (b) in Example 3

Exercises 6.7 Answers to selected odd-numbered problems begin on page ANS-21.

≡ Fundamentals

In Problems 1–20, find the average value f_{ave} of the given function on the indicated interval.

1. $f(x) = 4x$; $[-3, 1]$
2. $f(x) = 2x + 3$; $[-2, 5]$
3. $f(x) = x^2 + 10$; $[0, 2]$
4. $f(x) = 2x^3 - 3x^2 + 4x - 1$; $[-1, 1]$
5. $f(x) = 3x^2 - 4x$; $[-1, 3]$
6. $f(x) = (x + 1)^2$; $[0, 2]$
7. $f(x) = x^3$; $[-2, 2]$
8. $f(x) = x(3x - 1)^2$; $[0, 1]$
9. $f(x) = \sqrt{x}$; $[0, 9]$
10. $f(x) = \sqrt{5x + 1}$; $[0, 3]$
11. $f(x) = x\sqrt{x^2 + 16}$; $[0, 3]$
12. $f(x) = \left(1 + \frac{1}{x}\right)^{1/3} \frac{1}{x^2}$; $[\frac{1}{2}, 1]$
13. $f(x) = \frac{1}{x^3}$; $[\frac{1}{4}, \frac{1}{2}]$
14. $f(x) = x^{2/3} - x^{-2/3}$; $[1, 4]$
15. $f(x) = \frac{2}{(x + 1)^2}$; $[3, 5]$
16. $f(x) = \frac{(\sqrt{x} - 1)^3}{\sqrt{x}}$; $[4, 9]$
17. $f(x) = \sin x$; $[-\pi, \pi]$
18. $f(x) = \cos 2x$; $[0, \pi/4]$
19. $f(x) = \csc^2 x$; $[\pi/6, \pi/2]$
20. $f(x) = \frac{\sin \pi x}{\cos^2 \pi x}$; $[-\frac{1}{3}, \frac{1}{3}]$

In Problems 21 and 22, find a value of c in the given interval for which $f(c) = f_{ave}$.

21. $f(x) = x^2 + 2x$; $[-1, 1]$
22. $f(x) = \sqrt{x + 3}$; $[1, 6]$

23. The average value of a continuous nonnegative function $y = f(x)$ on the interval $[1, 5]$ is $f_{ave} = 3$. What is the area under the graph on the interval?

24. For $f(x) = 1 - \sqrt{x}$, find a value of b so that $f_{ave} = 0$ on $[0, b]$. Interpret geometrically.

≡ Applications

25. The function $T(t) = 100 + 3t - \frac{1}{2}t^2$ approximates the temperature at t hr past noon on a typical August day in Las Vegas. Find the average temperature between noon and 6 P.M.

26. A company determines that the revenue obtained after the sale of x units of a product is given by $R(x) = 50 + 4x + 3x^2$. Find the average revenue for sales $x = 1$ to $x = 5$. Compare the result with the average $\frac{1}{5}\sum_{k=1}^{5} R(k)$.

27. Let $s(t)$ denote the position of a particle on a horizontal axis as a function of time t. The average velocity $\bar{v}$ during the time interval $[t_1, t_2]$ is $\bar{v} = [s(t_2) - s(t_1)]/(t_2 - t_1)$. Use (5) to show that $v_{ave} = \bar{v}$. [*Hint:* Recall $ds/dt = v$.]

28. In the absence of damping, the position of a mass m on a freely vibrating spring is given by the function $x(t) = A\cos(\omega t + \phi)$, where A, ω, and ϕ are constants. The period of oscillation is $2\pi/\omega$. The potential energy of

the system is $U(x) = \frac{1}{2}kx^2$, where k is the so-called spring constant. The kinetic energy of the system is $K = \frac{1}{2}mv^2$, where $v = dx/dt$. If $\omega^2 = k/m$, show that the average potential energy and the average kinetic energy over one period are the same and that each equals $\frac{1}{4}kA^2$.

29. In physics, the **Impulse-Momentum Theorem** states that the change in momentum of a body in a time interval $[t_0, t_1]$ is $mv_1 - mv_0 = (t_1 - t_0)\overline{F}$, where mv_0 is the initial momentum, mv_1 is the final momentum, and $\overline{F}$ is the average force acting on the body during the interval. Find the change in momentum of a pile driver dropped on a piling between times $t = 0$ and $t = t_1$ if

$$F(t) = k\left[1 - \left(\frac{2t}{t_1} - 1\right)^2\right],$$

where k is a constant.

30. In a small artery the velocity of blood (in cm/s) is given by $v(r) = (P/4\nu l)(R^2 - r^2)$, $0 \leq r \leq R$, where P is blood pressure, ν the viscosity of the blood, l the length of the artery, and R the radius of the artery. Find the average of $v(r)$ on the interval $[0, R]$.

≡ Think About It

31. If $y = f(x)$ is a continuous odd function, then what is f_{ave} on any interval $[-a, a]$?

32. For a linear function $f(x) = ax + b$, $a > 0$, $b > 0$, the average value of the function on $[x_1, x_2]$ is $f_{ave} = aX + b$, where X is some number in the interval. Conjecture the value of X. Prove your assertion.

33. If $y = f(x)$ is a differentiable function, find the average value of f' on the interval $[x, x + h]$, where $h > 0$.

34. Given that n is a positive integer and $a > 1$, show that the average value of $f(x) = (n + 1)x^n$ on the interval $[1, a]$ is $f_{ave} = a^n + a^{n-1} + \cdots + a + 1$.

35. Suppose $y = f(x)$ is a continuous function and f_{ave} is its average value on $[a, b]$. Explain: $\int_a^b [f(x) - f_{ave}]\, dx = 0$.

36. Let $f(x) = \lfloor x \rfloor$ be the greatest integer or floor function. Without integration, what is the average of f on $[0, 1]$? On $[0, 2]$? On $[0, 3]$? On $[0, 4]$? Conjecture the average value of f on the interval $[0, n]$, where n is a positive integer. Prove your assertion.

37. As shown in FIGURE 6.7.5 a chord is drawn randomly between two points on a circle of radius $r = 1$. Discuss: What is the average length of the chords?

FIGURE 6.7.5 Circle in Problem 37

≡ Projects

38. **Human Limbs** The following formula is often used to approximate the surface area S of a human limb:

$$S \approx \text{average circumference} \times \text{length of limb}.$$

(a) As shown in FIGURE 6.7.6, a limb can be considered to be a solid of revolution. For many limbs, $f'(x)$ is small. If $|f'(x)| \leq \varepsilon$ for $a \leq x \leq b$, show that

$$\int_a^b 2\pi f(x)\, dx \leq S \leq \sqrt{1 + \varepsilon^2} \int_a^b 2\pi f(x)\, dx.$$

(b) Show that $\overline{C}L \leq S \leq \sqrt{1 + \varepsilon^2}\, \overline{C}L$, where $\overline{C}$ is the average circumference of the limb over the interval $[a, b]$. Thus, the approximation formula stated above always underestimates S but does well when ε is small (such as for the forearm to wrist).

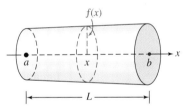

FIGURE 6.7.6 Model of a limb in Problem 38

6.8 Work

■ **Introduction** In physics, when a *constant* force F moves an object a distance d in the same direction of the force, the **work** done is defined to be the product

$$W = Fd. \tag{1}$$

For example, if a 10-lb force moves an object 7 ft in the same direction as the force, then the work done is 70 ft-lb. In this section we shall see how to find the work done by a *variable* force.

Before examining work as a definite integral, let us review some important units.

Units
Commonly used **units** of force, distance, and work are listed in the following table.

Quantity	Engineering system	SI	cgs
Force	pound (lb)	newton (N)	dyne
Distance	foot (ft)	meter (m)	centimeter (cm)
Work	foot-pound (ft-lb)	newton-meter (joule)	dyne-centimeter (erg)

Thus, if a force of 300 N moves an object 15 m, the work done is $W = 300 \cdot 15 = 4500$ N-m or 4500 joules. For comparison, and conversion of one unit to another, we note that

$$1 \text{ N} = 10^5 \text{ dynes} = 0.2247 \text{ lb}$$
$$1 \text{ ft-lb} = 1.356 \text{ joules} = 1.356 \times 10^7 \text{ ergs}.$$

So, for example, 70 ft-lb is equivalent to $70 \times 1.356 = 94.92$ joules, and 4500 joules is equivalent to $4500/1.356 = 3318.584$ ft-lb.

Building an Integral
Now, if $F(x)$ represents a continuous variable force acting across an interval $[a, b]$, then the work is not simply a product as in (1). Suppose P is the partition

$$a = x_0 < x_1 < x_2 < \cdots < x_n = b$$

and Δx_k is the width of the kth subinterval $[x_{k-1}, x_k]$. Let x_k^* denote the sample point chosen arbitrarily in each subinterval. If the width of each $[x_{k-1}, x_k]$ is very small, then, since F is continuous, the function values $F(x)$ cannot vary much across the subinterval. Thus, we can reasonably consider the force that acts over $[x_{k-1}, x_k]$ as the constant $F(x_k^*)$ and the work done from x_{k-1} to x_k is given by the approximation

$$W_k = F(x_k^*) \Delta x_k.$$

An approximation to the total work done from a to b is then given by the Riemann sum

$$\sum_{k=1}^{n} W_k = F(x_1^*) \Delta x_1 + F(x_2^*) \Delta x_2 + \cdots + F(x_n^*) \Delta x_n = \sum_{k=1}^{n} F(x_k^*) \Delta x_k.$$

It is natural to assume that the work done by F over the interval is

$$W = \lim_{\|P\| \to 0} \sum_{k=1}^{n} F(x_k^*) \Delta x_k.$$

We summarize the foregoing discussion with the following definition.

Definition 6.8.1 Work

Let F be continuous on the interval $[a, b]$ and let $F(x)$ represent the force at a number x in the interval. Then the **work** W done by the force in moving an object from a to b is

$$W = \int_a^b F(x)\, dx. \qquad (2)$$

Note: If F is constant, $F(x) = k$ for all x in the interval, then (2) becomes $W = \int_a^b k\, dx = kx\big]_a^b = k(b - a)$, which is consistent with (1).

Spring Problems
Hooke's Law states that, when a spring is stretched (or compressed) beyond its natural length, the restoring force exerted by the spring is directly proportional to the amount of elongation (or compression) x. Thus, in order to stretch a spring x units beyond its natural length, we need to apply the force

$$F(x) = kx, \qquad (3)$$

where k is a constant of proportionality called the **spring constant**. See FIGURE 6.8.1.

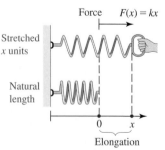

FIGURE 6.8.1 To stretch a spring x units a force $F(x) = kx$ is needed

EXAMPLE 1 Stretching a Spring

It takes a force of 130 N to stretch a spring 50 cm. Find the work done in stretching the spring 20 cm beyond its natural (unstretched) length.

Solution When a force is measured in newtons, distances are commonly expressed in meters. Since $x = 50$ cm $= \frac{1}{2}$ m when $F = 130$ N, (3) becomes $130 = k(\frac{1}{2})$, which implies the spring constant is $k = 260$ N/m. Thus, $F = 260x$. Now, 20 cm $= \frac{1}{5}$ m, so that the work done in stretching the spring by this amount is

$$W = \int_0^{1/5} 260x\, dx = 130x^2 \Big]_0^{1/5} = \frac{26}{5} = 5.2 \text{ joules.} \qquad \blacksquare$$

Note: Suppose the natural length of the spring in Example 1 is 40 cm. An equivalent way of stating the problem is: Find the work done in stretching the spring to a length of 60 cm. Since the elongation is $60 - 40 = 20$ cm $= \frac{1}{5}$ m, we still integrate $F = 260x$ on the interval $[0, \frac{1}{5}]$. However, if the problem were to find the work done in stretching the same spring from 50 cm to 60 cm, we would then integrate on the interval $[\frac{1}{10}, \frac{1}{5}]$. In this situation we are starting from a position where the spring is already stretched 10 cm $(\frac{1}{10} \text{m})$.

■ **Work Done Against Gravity** From the Universal Law of Gravitation, the force between a planet (or satellite) of mass m_1 and a body of mass m_2 is given by

$$F = k\frac{m_1 m_2}{r^2}, \qquad (4)$$

where k is a constant, called the **gravitational constant**, and r is the distance from the center of the planet to the mass m_2. See FIGURE 6.8.2. In lifting the mass m_2 off the surface of a planet of radius R to a height h, the work done can be obtained by using (4) in (2):

$$W = \int_R^{R+h} \frac{km_1 m_2}{r^2} dr = km_1 m_2 \left(-\frac{1}{r}\right)\Big]_R^{R+h} = km_1 m_2 \left(\frac{1}{R} - \frac{1}{R+h}\right). \qquad (5)$$

In SI units, $k = 6.67 \times 10^{-11}$ N $\cdot$ m^2/kg^2. Some masses and values of R are given in the accompanying table.

FIGURE 6.8.2 Lifting a mass m_2 to a height h

Planets	m_1 (in kg)	R (in m)
Venus	4.9×10^{24}	6.2×10^6
Earth	6.0×10^{24}	6.4×10^6
Moon	7.3×10^{22}	1.7×10^6
Mars	6.4×10^{23}	3.3×10^6

EXAMPLE 2 Work Done in Lifting a Payload

The work done in lifting a 5000-kg payload from the surface of the Earth to a height of 30,000 m $(0.03 \times 10^6 \text{ m})$ follows from (5) and the preceding table:

$$W = (6.67 \times 10^{-11})(6.0 \times 10^{24})(5000)\left(\frac{1}{6.4 \times 10^6} - \frac{1}{6.43 \times 10^6}\right)$$

$$\approx 1.46 \times 10^9 \text{ joules.} \qquad \blacksquare$$

■ **Pump Problems** When a liquid that weighs ρ lb/ft^3 is pumped from a tank, the work done in moving a fixed volume or layer of liquid d ft in a vertical direction is

$$W = \text{force} \cdot \text{distance} = (\text{weight per unit volume}) \cdot (\text{volume}) \cdot (\text{distance})$$

or

$$W = \underbrace{\rho \cdot (\text{volume})}_{\text{force}} \cdot d. \qquad (6)$$

In physics the quantity ρ is called the **weight density** of the fluid. For water, $\rho = 62.4$ lb/ft^3 or 9800 N/m^3.

In the next several examples we will use (6) to build the appropriate integral to find the work done in pumping water from a tank.

EXAMPLE 3 Work Done in Pumping Water

A hemispherical tank of radius 20 ft is filled with water to a 15-ft depth. Find the work done in pumping all the water to the top of the tank.

Solution As shown in FIGURE 6.8.3, we let the positive x-axis be directed *downward* and let the origin be at the center-top of the tank. Since the cross-section of the tank is a semicircle, x and y are related by $x^2 + y^2 = (20)^2$, $0 \le x \le 20$. Now suppose the interval $[5, 20]$, corresponding to the water on the x-axis, is partitioned into n subintervals $[x_{k-1}, x_k]$ of width Δx_k. Let x_k^* be any sample point in the kth subinterval and let W_k denote an approximation to the work done by the pump in lifting a circular layer of water of thickness Δx_k to the top of the tank. It follows from (6) that

$$W_k = \underbrace{[62.4\pi(y_k^*)^2 \Delta x_k]}_{\text{force}} \cdot \underbrace{x_k^*}_{\text{distance}},$$

where $(y_k^*)^2 = 400 - (x_k^*)^2$. Hence, the work done by the pump is approximated by the Riemann sum

$$\sum_{k=1}^{n} W_k = \sum_{k=1}^{n} 62.4\pi[400 - (x_k^*)^2] x_k^* \Delta x_k.$$

The work done in pumping all the water to the top of the tank is the limit of this last expression as $\|P\| \to 0$; that is,

$$W = \int_{5}^{20} 62.4\pi(400 - x^2)x\, dx = 62.4\pi\left(200x^2 - \frac{1}{4}x^4\right)\Big]_{5}^{20} \approx 6{,}891{,}869 \text{ ft-lb}.$$

FIGURE 6.8.3 Hemispherical tank in Example 3

It is worth pursuing the analysis of Example 3 for the case where the positive x-axis is taken in the *upward* direction and the origin is at the center-bottom of the tank.

EXAMPLE 4 Alternative Solution to Example 3

With the axes as shown in FIGURE 6.8.4, we see that a circular layer of water must be lifted a distance of $20 - x_k^*$ ft. Since the center of the semicircle is at $(20, 0)$, x and y are now related by $(x - 20)^2 + y^2 = 400$. Hence,

$$W_k = \underbrace{(62.4\pi(y_k^*)^2 \Delta x_k)}_{\text{force}} \cdot \underbrace{(20 - x_k^*)}_{\text{distance}}$$

$$= 62.4\pi[400 - (x - 20)^2](20 - x_k^*)\, \Delta x_k$$

FIGURE 6.8.4 Hemispherical tank in Example 4

and so

$$W = 62.4\pi \int_0^{15} [400 - (x-20)^2](20 - x)\, dx$$
$$= 62.4\pi \int_0^{15} (x^3 - 60x^2 + 800x)\, dx.$$

Note the new limits of integration; this is because the water shown in Figure 6.8.4 corresponds to the interval [0, 15] on the vertical x-axis. You should verify that the value of W in this case is the same as in Example 3. ∎

EXAMPLE 5 Example 3 Revisited

In Example 3, find the work done in pumping all the water to a point 10 ft above the hemispherical tank.

Solution As in Figure 6.8.3, we position the positive x-axis downward. Now, from FIGURE 6.8.5 we see

$$W_k = (62.4\pi(y_k^*)^2\,\Delta x_k) \cdot (10 + x_k^*)$$
$$= 62.4\pi[400 - (x_k^*)^2](10 + x_k^*)\,\Delta x_k.$$

Hence,

$$W = 62.4\pi \int_5^{20} (400 - x^2)(10 + x)\, dx$$
$$= 62.4\pi \int_5^{20} (-x^3 - 10x^2 + 400x + 4000)\, dx$$
$$= 62.4\pi \left(-\frac{1}{4}x^4 - \frac{10}{3}x^3 + 200x^2 + 4000x\right)\Big]_5^{20}$$
$$= 13{,}508{,}063 \text{ ft-lb.}$$ ∎

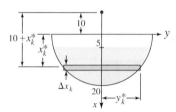

FIGURE 6.8.5 Hemispherical tank in Example 5

■ **Cable Problems** The next example illustrates the fact that when you are calculating the work done in lifting an object by means of a cable (heavy rope or chain), the weight of the cable must be taken into consideration.

EXAMPLE 6 Lifting an Elevator

A cable weighing 6 lb/ft is connected to a construction elevator weighing 1500 lb. Find the work done in lifting the elevator to a height of 500 ft.

Solution Since the weight of the elevator is a constant force, it follows from (1) that the work done in lifting the elevator a distance of 500 ft is simply

$$W_E = (1500)\cdot(500) = 750{,}000 \text{ ft-lb.}$$

The weight of the cable is the variable force. Let W_C denote the work done in lifting the cable. As shown in FIGURE 6.8.6, suppose the positive x-axis is directed upward and the interval [0, 500] is partitioned into n subintervals with lengths Δx_k. At a height of x_k^* ft off the ground, a segment of cable corresponding to the subinterval $[x_{k-1}, x_k]$ weighs $6\Delta x_k$ and must be pulled up an additional $500 - x_k^*$ ft. Hence, we can write

$$(W_C)_k = \underbrace{(6\,\Delta x_k)}_{\text{force}} \cdot \underbrace{(500 - x_k^*)}_{\text{distance}} = (3000 - 6x_k^*)\,\Delta x_k$$

and so

$$W_C = \int_0^{500} (3000 - 6x)\, dx = (3000x - 3x^2)\Big]_0^{500} = 750{,}000 \text{ ft-lb.}$$

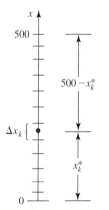

FIGURE 6.8.6 Cable in Example 6

Thus, the total work done in lifting the elevator is

$$W = W_E + W_C = 1{,}500{,}000 \text{ ft-lb.}$$ ∎

FIGURE 6.8.7 Elevator in Examples 6 and 7

EXAMPLE 7 Alternative Solution to Example 6

This is a slightly faster analysis of Example 6. As shown in FIGURE 6.8.7, when the elevator is at a height of x ft, it must be pulled up an additional $500 - x$ ft. The lifting force needed at that height is

$$\underbrace{1500}_{\text{weight of elevator}} + \underbrace{6(500 - x)}_{\text{weight of cable}} = 4500 - 6x.$$

Thus, by (2) the work done is

$$W = \int_0^{500} (4500 - 6x)\, dx = 1,500,000 \text{ ft-lb.} \blacksquare$$

Exercises 6.8   Answers to selected odd-numbered problems begin on page ANS-21.

≡ Fundamentals

1. Find the work done when a 55-lb force moves an object 20 yd in the same direction of the force.

2. A force of 100 N is applied to an object at an angle of 30° measured from the horizontal. If the object moves 8 m horizontally, find the work done by the force.

3. A mass that weighs 10 lb stretches a spring $\frac{1}{2}$ ft. How much will a mass that weighs 8 lb stretch the same spring?

4. A spring has a natural length of 0.5 m. A force of 50 N stretches the spring to a length of 0.6 m.
   (a) What force is needed to stretch the spring x m?
   (b) What force is required to stretch the spring to a length of 1 m?
   (c) How long is the spring when stretched by a force of 200 N?

5. In Problem 4:
   (a) Find the work done in stretching the spring 0.2 m.
   (b) Find the work done in stretching the spring from a length of 1 m to a length of 1.1 m.

6. A force of $F = \frac{3}{2}x$ lb is needed to stretch a 10-in. spring an additional x in.
   (a) Find the work done in stretching the spring to a length of 16 in.
   (b) Find the work done in stretching the spring 16 in.

7. A mass that weighs 10 lb is suspended from a 2-ft spring. The spring is stretched 8 in. and then the mass is removed.
   (a) Find the work done in stretching the spring to a length of 3 ft.
   (b) Find the work done in stretching the spring from a length of 4 ft to a length of 5 ft.

8. A 50-lb force compresses a 15-in.-long spring by 3 in. Find the work done in compressing the spring to a final length of 5 in.

9. Find the work done in lifting a mass of 10,000 kg from the surface of the Earth to a height of 500 km.

10. Find the work done in lifting a mass of 50,000 kg from the surface of the Moon to a height of 200 km.

11. A tank in the form of a right circular cylinder is filled with water. The dimensions of the tank (in feet) are shown in FIGURE 6.8.8. Find the work done in pumping all the water to the top of the tank.

FIGURE 6.8.8 Cylindrical tank in Problem 11

12. A tank in the form of a right circular cone, vertex down, is filled with water to a depth of one-half its height. The dimensions of the tank (in feet) are shown in FIGURE 6.8.9. Find the work done in pumping all the water to the top of the tank. [*Hint*: Assume that the origin is at the vertex of the cone.]

FIGURE 6.8.9 Conical tank in Problem 12

13. For the conical tank in Problem 12, find the work done in pumping all the water to a point 5 ft above the top of the tank.

14. Suppose the cylindrical tank in Problem 11 is horizontal. Find the work done in pumping all the water to a point 2 ft above the top of the tank. [*Hint*: See Problems 55–58 in Exercises 6.2.]

15. A tank has cross sections in the form of isosceles triangles, vertex down. The dimensions of the tank (in feet) are shown in FIGURE 6.8.10. Find the work done in filling the tank with water through a hole in its bottom by a pump located 5 ft below its vertex.

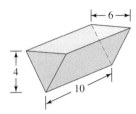

FIGURE 6.8.10 Tank with triangular cross sections in Problem 15

16. A horizontal vat with semicircular cross sections contains oil whose weight density is 80 lb/ft^3. The dimensions of the tank (in feet) are shown in FIGURE 6.8.11. If the depth of the oil is 3 ft, find the work done in pumping all the oil to the top of the tank.

FIGURE 6.8.11 Semicircular vat in Problem 16

17. A 100-ft anchor chain, weighing 20 lb/ft, is hanging vertically over the side of a boat. How much work is performed by pulling in 40 ft of the chain?

18. A ship is anchored in 200 ft of water. In water the ship's anchor weighs 3000 lb and its anchor chain weighs 40 lb/ft. If the anchor chain hangs vertically, how much work is done in pulling in 100 ft of the chain?

19. A bucket of sand weighing 80 lb is lifted vertically by means of a rope and pulley to a height of 65 ft. Find the work done if

   (a) the weight of the rope is negligible and
   (b) the rope weighs $\frac{1}{2}$ lb/ft.

20. A bucket, initially containing 20 ft^3 of water, is lifted vertically from ground level. If the water leaks out at a rate of $\frac{1}{2}$ ft^3 per vertical foot, find the work done in lifting the bucket to a height at which it is empty.

21. The force of attraction between an electron and the nucleus of an atom is inversely proportional to the square of the distance separating them. If the initial distance between nucleus and electron is 1 unit, find the work done by an external force that moves the electron out to a distance four times the initial distance.

22. A rocket weighing 2,500,000 lb when fueled carries a 200,000-lb shuttle orbiter. Assume, in the early stages of the launch, that the rocket burns fuel at a rate of 100 lb/ft.

   (a) Express the total weight of the system in terms of its altitude above the surface of the Earth. See FIGURE 6.8.12.
   (b) Find the work done in lifting the system to an altitude of 1000 ft.

FIGURE 6.8.12 Rocket in Problem 22

23. In thermodynamics, if a gas enclosed in a cylinder expands against a piston so that the volume of the gas changes from v_1 to v_2, then the work done on the piston is given by $W = \int_{v_1}^{v_2} p\, dv$, where p is pressure (force per unit area). See FIGURE 6.8.13. In an adiabatic expansion of an ideal gas, pressure and volume are related by $pv^\gamma = k$, where γ and k are constants. Show that if $\gamma \neq 1$, then

$$W = \frac{p_2 v_2 - p_1 v_1}{1 - \gamma}$$

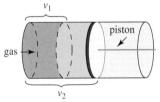

FIGURE 6.8.13 Piston in Problem 23

24. Show that when a body of weight mg is lifted vertically from a point y_1 to a point y_2, $y_2 > y_1$, the work done is the change in potential energy $W = mgy_2 - mgy_1$.

≡ Think About It

25. A person pushes against an immovable wall with a horizontal force of 75 lb. How much work is done?

26. The graph of a variable force F is given in FIGURE 6.8.14. Find the work done by the force in moving a particle from $x = 0$ to $x = 6$.

FIGURE 6.8.14 Graph of force in Problem 26

27. **A Bit of History—A Real Tall Story** In 1977 George Willig, known as the "human fly" or "spider man," scaled the outside of the south tower of the World Trade Center building in New York City to a height of 1350 ft in 3.5 h at a rate of 6.4 ft/min. At the time Willig weighed 165 lb. How much work did George do? (For his efforts, he was fined $1.10, 1 cent for each of the 110 stories of the building.)

28. A bucket containing water weighs 200 lb. As the bucket is lifted by a rope water leaks out of its bottom at a constant rate so that when it reaches a height of 10 ft it weighs 180 lb. Assume that the weight of the rope is negligible. Discuss: Explain why $\frac{200 + 180}{2} \cdot 10 = 1900$ ft-lb is a reasonable approximation to the work done. Without integration, show that the foregoing "approximation" is also the exact value of the work done.

29. As shown in FIGURE 6.8.15, a body of mass m is moved by a horizontal force F on a frictionless surface from a position at x_1 to a position at x_2. At these respective points, the body is moving at velocities v_1 and v_2, where $v_2 > v_1$. Show that the work done by the force is the increase in kinetic energy $W = \frac{1}{2}mv_2^2 - \frac{1}{2}mv_1^2$. [*Hint*: Use Newton's second law, $F = ma$, and express the acceleration a in terms of velocity v. Integrate with respect to time t and make a substitution.]

30. As shown in FIGURE 6.8.16, a bucket containing concrete that is suspended from a cable is pushed horizontally from the vertical by a construction worker. The length of the cable is 30 m and the combined mass m of the bucket and concrete is 550 kg. From the principles of physics it can be shown that the force required to move the bucket x m is given by $F = mg\tan\theta$, where g is the acceleration of gravity. Find the work done by the construction worker in pushing the bucket a horizontal distance of 3 m. [*Hint*: Use (2) and a substitution.]

FIGURE 6.8.16 Bucket in Problem 30

FIGURE 6.8.15 Mass in Problem 29

6.9 Fluid Pressure and Force

■ **Introduction** Everyone has at one time experienced "ear popping" or even pain in the ear when descending in an airplane (or in an elevator) or when diving to the bottom of a swimming pool. These annoying ear sensations are due to a increase in the *pressure* exerted by air or water on mechanisms in the middle ear. Air and water are examples of fluids. In this section we will how the definite integral can be used to find the force exerted by a fluid.

A *fluid* includes liquids (such as water and oil) as well as gases (such as nitrogen).

■ **Force and Pressure** Suppose a *horizontal* flat plate is submerged below the surface of a fluid such as water. The force exerted on the plate by the fluid directly above it, called the **fluid force F**, is defined to be

$$F = \underbrace{(\text{force per unit area})}_{\text{fluid pressure } P} \cdot (\text{area of surface})$$

$$= PA. \qquad (1)$$

If ρ denotes the weight density of the fluid (weight per unit volume) and A is the area of the horizontal plate submerged to a depth h, shown in FIGURE 6.9.1(a), then the **fluid pressure P** on the plate can be expressed in terms of ρ:

$$P = (\text{weight per unit volume}) \cdot (\text{depth}) = \rho h. \qquad (2)$$

Therefore, the fluid force (1) is the same as

$$F = (\text{fluid pressure}) \cdot (\text{area of surface}) = \rho h A. \qquad (3)$$

However, when a *vertical* plate is submerged, the fluid pressure and the fluid force on one side of the plate varies with the depth. See Figure 6.9.1(b). For example, the fluid pressure on a vertical dam is less at the top than at its base.

Before we begin, let us consider a simple example of pressure and force of a horizontally submerged plate.

(a) Horizontal plate   (b) Vertical plate

FIGURE 6.9.1 Fluid pressure and force are constant on a horizontally submerged plate, but fluid pressure and force vary with the depth on a vertically submerged plate

EXAMPLE 1 Pressure and Force

A flat rectangular plate with dimensions 5 ft × 6 ft is submerged horizontally in water at a depth of 10 ft. Determine the pressure and the force exerted on the plate by the water above it.

Solution Recall that the weight density for water is 62.4 lb/ft^3. Hence, by (2) the fluid pressure is
$$P = \rho h = (62.4 \text{ lb/ft}^3) \cdot (10 \text{ ft}) = 624 \text{ lb/ft}^2.$$

Since the surface area of the plate is $A = 30$ ft^2, it follows from (3) that the fluid force on the plate is
$$F = PA = (\rho h)A = (624 \text{ lb/ft}^2) \cdot (30 \text{ ft}^2) = 18{,}720 \text{ lb}. \quad \blacksquare$$

To determine the total force F exerted by a fluid on one side of a vertically submerged flat surface, we employ one form of **Pascal's Principle**:

- *The pressure exerted by a fluid at a depth h is the same in all directions.*

Thus, if a large container with a flat bottom and vertical sidewalls is filled with water to a depth of 10 ft, the pressure of 624 lb/ft^2 at its bottom applies equally to the sidewalls. See FIGURE 6.9.2.

FIGURE 6.9.2 Pressure of 640 lb/ft^2 applies in all directions

■ **Building an Integral** Let the positive x-axis be directed downward with the origin at the surface of a fluid. Suppose a vertical flat plate, bounded by the horizontal lines $x = a$ and $x = b$, is submerged in the fluid as shown in FIGURE 6.9.3(a). Let $w(x)$ be a function that denotes the width of the plate at any number x in $[a, b]$ and let P be any partition of the interval. If x_k^* is a sample point in the kth subinterval $[x_{k-1}, x_k]$, then from (3) with the identifications $h = x_k^*$ and $A = w(x_k^*)\,\Delta x_k$, the force F_k exerted by the fluid on the corresponding rectangular element is approximated by
$$F_k = \rho \cdot x_k^* \cdot w(x_k^*)\,\Delta x_k,$$
where, as before, ρ denotes the weight density of the fluid. Thus, an approximation to the fluid force on one side of the plate is given by the Riemann sum
$$\sum_{k=1}^n F_k = \sum_{k=1}^n \rho x_k^* w(x_k^*)\,\Delta x_k.$$

This suggests that the total fluid force on the plate is
$$F = \lim_{\|P\| \to 0} \sum_{k=1}^n \rho x_k^* w(x_k^*)\,\Delta x_k.$$

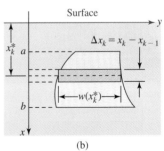

FIGURE 6.9.3 Submerged vertical plate with varying width $w(x)$ on $[a, b]$

Definition 6.9.1 Force Exerted by a Fluid

Let ρ be the weight density of a fluid and let $w(x)$ be a continuous function on $[a, b]$ that describes the width of a vertically submerged plate at a depth x. The **force F** exerted by the fluid on one side of the submerged plate is
$$F = \int_a^b \rho x w(x)\,dx. \quad (4)$$

EXAMPLE 2 Fluid Force

A plate in the shape of an isosceles triangle 3 ft high and 4 ft wide is submerged vertically in water, base downward, with the base 5 ft below the surface. Find the force exerted by the water on one side of the plate.

Solution For convenience, we place the positive x-axis along the axis of symmetry of the triangular plate with the origin at the surface of the water. As indicated in FIGURE 6.9.4, we partition the interval $[2, 5]$ into n subintervals $[x_{k-1}, x_k]$ and choose a point x_k^* in each subinterval. Since the equation of the straight line that contains points $(2, 0)$ and $(5, 2)$ is $y = \frac{2}{3}x - \frac{4}{3}$ we conclude by symmetry that the width of the rectangular element, shown in light red in Figure 6.9.4, is

$$2y_k^* = 2\left(\frac{2}{3}x_k^* - \frac{4}{3}\right).$$

Now $\rho = 62.4$ lb/ft^3 so that the fluid force on that portion of the plate that corresponds to the kth subinterval is approximated by

$$F_k = (62.4) \cdot x_k^* \cdot 2\left(\frac{2}{3}x_k^* - \frac{4}{3}\right)\Delta x_k.$$

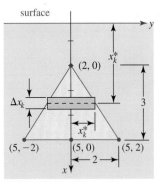

FIGURE 6.9.4 Triangular plate in Example 2

Forming the sum $\sum_{k=1}^{n} F_k$ and taking the limit as $\|P\| \to 0$ give

$$F = \int_2^5 (62.4)2x\left(\frac{2}{3}x - \frac{4}{3}\right) dx$$

$$= (62.4)\frac{4}{3}\int_2^5 (x^2 - 2x)\, dx$$

$$= 83.2\left(\frac{1}{3}x^3 - x^2\right)\bigg]_2^5$$

$$= (83.2) \cdot 18 = 1497.6 \text{ lb}. \quad\blacksquare$$

In problems such as Example 2, the x- and y-axes are placed where convenient. If the y-axis is placed perpendicular to the x-axis at the top of the plate at the point $(2, 0)$, then the four points $(2, 0)$, $(5, -2)$, $(5, 0)$, and $(5, 2)$ in Figure 6.9.4 become $(0, 0)$, $(3, -2)$, $(3, 0)$, and $(3, 2)$, respectively. The equation of the straight line that contains the points $(0, 0)$ and $(3, 2)$ is $y = \frac{2}{3}x$. You should verify then that the force F exerted by the water against the plate is given the definite integral

$$F = (62.4)\frac{4}{3}\int_0^3 x(x + 2)\, dx.$$

EXAMPLE 3 Force of Water Against a Dam

A dam has a vertical rectangular face. Find the force exerted by the water against the vertical face of the dam if the water is h ft deep and l ft wide. See FIGURE 6.9.5(a).

Solution For variety, let us take the positive x-axis pointing upward from the bottom of the rectangular face of the dam as shown in Figure 6.9.5(b). We then divide the interval $[0, h]$ into n subintervals. Suppressing the use of subscripts, the fluid force F_k on that rectangular portion of the plate that corresponds to the kth subinterval, shown in light red in Figure 6.9.5(b), is approximated by

$$F_k = (62.4) \cdot (h - x) \cdot (l\,\Delta x).$$

Here the depth is $h - x$ and the area of the rectangular element is $l\,\Delta x$. Summing these approximations and taking the limit as $\|P\| \to 0$ leads to

$$F = \int_0^h 62.4\,l(h - x)\, dx = \frac{1}{2}(62.4)lh^2. \quad\blacksquare$$

(a) Side view of dam and water

(b) Water against face of dam

FIGURE 6.9.5 Dam in Example 3

In Example 3, if, say, the water is 100 ft deep and 300 ft wide, the fluid force on the face of the dam is then 93,600,000 lb.

Exercises 6.9 Answers to selected odd-numbered problems begin on page ANS-21.

≡ Fundamentals

1. Consider the tanks with flat circular bottoms shown in FIGURE 6.9.6. Each tank is full of water whose weight density is 62.4 lb/ft^3. Find the pressure and force exerted by the water on the bottom of each tank.

FIGURE 6.9.6 Tanks in Problem 1

2. The tanker shown in FIGURE 6.9.7 has a flat bottom and is filled with oil whose weight density is 55 lb/ft^3. The tanker is 350 ft long.
   (a) What is the pressure exerted on the bottom of the tanker by the oil?
   (b) What is the pressure exerted on the bottom of the tanker by the water?
   (c) What is the force exerted on the bottom of the tanker by the oil?
   (d) What is the force exerted on the bottom of the tanker by the water?

FIGURE 6.9.7 Tanker in Problem 2

3. A rectangular swimming pool in the form of a rectangular parallelepiped has dimensions of 30 ft × 15 ft × 9 ft.
   (a) Find the pressure and force exerted on the flat bottom if the pool is filled with water to a depth of 8 ft. See FIGURE 6.9.8.
   (b) Find the force exerted by the water on a vertical sidewall and on a vertical end.

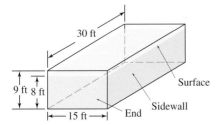

FIGURE 6.9.8 Swimming pool in Problem 3

4. A plate in the shape of an equilateral triangle $\sqrt{3}$ ft on a side is submerged vertically, base downward, with vertex 1 ft below the surface of the water. Find the force exerted by the water on one side of the plate.

5. Find the force on one side of the plate in Problem 4 if it is suspended with base upward 1 ft below the surface of the water.

6. A triangular plate is submerged vertically in water as shown in FIGURE 6.9.9. Find the force exerted by the water on one side of the plate.

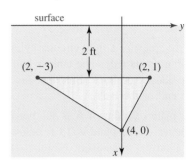

FIGURE 6.9.9 Triangular plate in Problem 6

7. Assuming the positive x-axis is downward, a plate bounded by the parabola $x = y^2$ and the line $x = 4$ is submerged vertically in oil that has weight density 50 lb/ft^3. If the vertex of the parabola is at the surface, find the force exerted by the oil on one side of the plate.

8. Assuming the positive x-axis is downward, a plate bounded by the parabola $x = y^2$ and the line $y = -x + 2$ is submerged vertically in water. If the vertex of the parabola is at the surface, find the force exerted by the water on one side of the plate.

9. A full water trough has vertical ends in the form of trapezoids as shown in FIGURE 6.9.10. Find the force exerted by the water on one end of the trough.

FIGURE 6.9.10 Water trough in Problem 9

10. A full water trough has vertical ends in the form shown in FIGURE 6.9.11. Find the force exerted by the water on one end of the trough.

FIGURE 6.9.11 Water trough in Problem 10

11. A vertical end of a full swimming pool has the shape given in FIGURE 6.9.12. Find the force exerted by the water on this end of the pool.

FIGURE 6.9.12 End of swimming pool in Problem 11

12. A tank in the shape of a right circular cylinder of diameter 10 ft is lying on its side. The tank is half full of oil that has weight density 60 lb/ft^3. Find the force exerted by the oil on one end of the tank.

13. A circular plate of radius 4 ft is submerged vertically so that the center of the plate is 10 ft below the surface of the water. Find the force exerted by the water on one side of the plate. [*Hint*: For simplicity, take the origin to be the center of the plate, positive x-axis downward. Also see Problems 55–58 in Exercises 6.2.]

14. A tank whose ends are in the form of an ellipse $x^2/4 + y^2/9 = 1$ is submerged in a liquid that has weight density ρ so that the end plates are vertical. Find the force exerted by the liquid on one end if its center is 10 ft below the surface of the liquid. [*Hint*: Proceed as in Problem 13 and use the fact that the area of an ellipse $x^2/a^2 + y^2/b^2 = 1$ is πab.]

15. A solid block in the shape of a cube 2 ft on a side is submerged in a large tank of water. The top of the block is horizontal and is 3 ft below the surface of the water. Find the total force on the block (six sides) that is caused by liquid pressure. See FIGURE 6.9.13.

FIGURE 6.9.13 Submerged block in Problem 15

16. In Problem 15, what is the difference between the force on the bottom of the block and the force on the top of the block? This difference is the buoyant force of the water and, by Archimedes' Principle, is equal to the weight of the water displaced. What is the weight of the water displaced by the block?

≡ **Think About It**

17. Consider the rectangular swimming pool shown in FIGURE 6.9.14(a) whose ends are trapezoids. The pool is full of water. By taking the positive x-axis, as shown in Figure 6.9.14(b), find the force exerted by the water on the bottom of the pool. [*Hint*: Express the depth d in terms of x.]

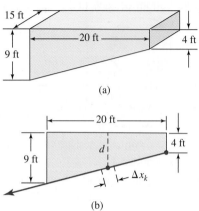

FIGURE 6.9.14 Swimming pool in Problem 17

18. An earthen dam is constructed with dimensions as shown in FIGURE 6.9.15(a). By taking the positive x-axis, as shown in Figure 6.9.15(b), find the force exerted by the water on the slanted wall of the dam.

FIGURE 6.9.15 Dam in Problem 18

19. Analyze Problem 18 with the positive x-axis shown in FIGURE 6.9.16.

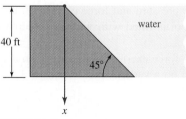

FIGURE 6.9.16 Orientation of x-axis in Problem 19

6.10 Centers of Mass and Centroids

Introduction In this section we consider another application from physics. We use the definite integral to find the mass and center of mass of rods and plane regions. We begin with a review of how to find the center of mass of one- and two-dimensional systems of n discrete or point masses.

One-Dimensional Systems If x denotes the directed distance from the origin O to a mass m, we say that the product mx is the **moment of the mass** about the origin. Some units are summarized in the following table.

Quantity	Engineering system	SI	cgs
Mass	slug	kilogram (kg)	gram (g)
Moment of mass	slug–feet	kilogram–meter	gram–centimeter

Now, for n point masses $m_1, m_2, \ldots, m_n$ at directed distances $x_1, x_2, \ldots, x_n$, respectively, from O, as in FIGURE 6.10.1, we say that

$$m = m_1 + m_2 + \cdots + m_n = \sum_{k=1}^{n} m_k$$

FIGURE 6.10.1 n masses on the x-axes

is the **total mass of the system**, and that

$$M_O = m_1 x_1 + m_2 x_2 + \cdots + m_n x_n = \sum_{k=1}^{n} m_k x_k$$

is the **moment of the system about the origin**. If $\sum_{k=1}^{n} m_k x_k = 0$, the system is said to be in **equilibrium**. See FIGURE 6.10.2. If the system of masses in Figure 6.10.1 is not in equilibrium, there is a point P with coordinate $\bar{x}$ such that

$$\sum_{k=1}^{n} m_k (x_k - \bar{x}) = 0 \quad \text{or} \quad \sum_{k=1}^{n} m_k x_k - \bar{x} \sum_{k=1}^{n} m_k = 0.$$

(a) Seesaw in equilibrium since $m_1 x_1 + m_2 x_2 = 0$

(b) Seesaw not in equilibrium since $m_1 x_1 + m_2 x_2 \neq 0$

FIGURE 6.10.2 Seesaw in equilibrium (a); not in equilibrium (b)

Solving for $\bar{x}$ gives

$$\bar{x} = \frac{M_O}{m} = \frac{\sum_{k=1}^{n} m_k x_k}{\sum_{k=1}^{n} m_k}. \tag{1}$$

The point with coordinate $\bar{x}$ is called the **center of mass** or the **center of gravity** of the system.

Since (1) implies $\bar{x}\left(\sum_{k=1}^{n} m_k\right) = \sum_{k=1}^{n} m_k k_k$, it follows that $\bar{x}$ is the directed distance from the origin to a point at which the total mass of the system can be considered to be concentrated.

◀ In a system in which the acceleration of gravity varies from mass to mass, the center of gravity is not the same as the center of mass.

EXAMPLE 1 Center of Mass of Three Objects

Three bodies of masses 4 kg, 6 kg, and 10 kg are located at $x_1 = -2$, $x_2 = 4$, and $x_3 = 9$, respectively. Distances are measured in meters. Find the center of mass.

Solution From (1),

$$\bar{x} = \frac{4 \cdot (-2) + 6 \cdot 4 + 10 \cdot 9}{4 + 6 + 10} = \frac{106}{20} = 5.3.$$

FIGURE 6.10.3 Center of mass of three point masses

FIGURE 6.10.3 shows that the center of mass $\bar{x}$ is 5.3 m to the right of the origin.

FIGURE 6.10.4 Rod of length L coinciding with x-axis

■ **Building an Integral** Now, let us consider the problem of finding the center of mass of a rod of length L that has a **variable linear density** ρ (mass/unit length measured in slugs/ft, kg/m, or g/cm). We assume that the rod coincides with the x-axis on the interval $[0, L]$, as shown in FIGURE 6.10.4, and that the density is a continuous function $\rho(x)$. After forming a partition P of the interval, we choose a point x_k^* in $[x_{k-1}, x_k]$. The number

$$m_k = \rho(x_k^*) \, \Delta x_k$$

is an approximation to the mass of that portion of the rod on the subinterval. Also, the moment of this element of mass about the origin is approximated by

$$(M_O)_k = x_k^* \rho(x_k^*) \, \Delta x_k.$$

Thus, we conclude that

$$m = \lim_{\|P\| \to 0} \sum_{k=1}^{n} \rho(x_k^*) \, \Delta x_k = \int_0^L \rho(x) \, dx$$

and

$$M_O = \lim_{\|P\| \to 0} \sum_{k=1}^{n} x_k^* \rho(x_k^*) \, \Delta x_k = \int_0^L x\rho(x) \, dx$$

are the **mass of the rod** and its **moment about the origin**, respectively. It then follows from $\bar{x} = M_O/m$ that the center of mass of the rod is given by

$$\bar{x} = \frac{\displaystyle\int_0^L x\rho(x) \, dx}{\displaystyle\int_0^L \rho(x) \, dx}. \tag{2}$$

FIGURE 6.10.5 Rod hanging in balance

As shown in FIGURE 6.10.5, a rod suspended by a string attached to its center of mass would hang in perfect balance.

EXAMPLE 2 Center of Mass of a Rod

A 16-cm-long rod has a linear density, measured in g/cm, given by $\rho(x) = \sqrt{x}$, $0 \leq x \leq 16$. Find its center of mass.

Solution The mass of the rod in grams is

$$m = \int_0^{16} x^{1/2} \, dx = \frac{2}{3} x^{3/2} \Big]_0^{16} = \frac{128}{3}.$$

The moment about the origin (in g-cm) is

$$M_0 = \int_0^{16} x \cdot x^{1/2} \, dx = \frac{2}{5} x^{5/2} \Big]_0^{16} = \frac{2048}{5}.$$

From (2) we find

$$\bar{x} = \frac{2048/5}{128/3} = 9.6.$$

That is, the center of mass $\bar{x}$ of the rod is 9.6 cm from the left end of the rod that coincides with the origin.

Two-Dimensional Systems

For n point masses located in the xy-plane, as indicated in FIGURE 6.10.6, the **center of mass of the system** is defined to be the point $(\bar{x}, \bar{y})$, where

$$\bar{x} = \frac{M_y}{m} = \frac{\sum_{k=1}^{n} m_k x_k}{\sum_{k=1}^{n} m_k} = \frac{\text{moment of system about } y\text{-axis}}{\text{total mass}},$$

$$\bar{y} = \frac{M_x}{m} = \frac{\sum_{k=1}^{n} m_k y_k}{\sum_{k=1}^{n} m_k} = \frac{\text{moment of system about } x\text{-axis}}{\text{total mass}}.$$

FIGURE 6.10.6 n masses in the xy-plane

Lamina

Let us turn now to the problem of finding the center of mass, or balancing point, of a thin two-dimensional smear of matter, or **lamina,** that has a constant density ρ (mass per unit area). See FIGURE 6.10.7. When ρ is constant, the lamina is said to be **homogeneous.**

Building an Integral

As shown in FIGURE 6.10.8(a), let us suppose that the lamina coincides with a region R in the xy-plane bounded by the graph of a continuous nonnegative function $y = f(x)$, the x-axis, and the vertical lines $x = a$ and $x = b$. If P is a partition of the interval $[a, b]$, then the mass of the rectangular element shown in Figure 6.10.8(b) is

$$m_k = \rho \, \Delta A_k = \rho f(x_k^*) \, \Delta x_k,$$

where, in this case, we take x_k^* to be the midpoint of the subinterval $[x_{k-1}, x_k]$ and ρ is the constant density. The moment of this element about the y-axis is

$$(M_y)_k = x_k^* \Delta m_k = x_k^* (\rho \, \Delta A_k) = \rho x_k^* f(x_k^*) \, \Delta x_k.$$

FIGURE 6.10.7 Center of mass of a lamina

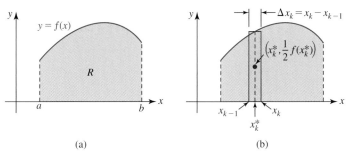

(a)   (b)

FIGURE 6.10.8 Find the center of mass of the region R

Since the density is constant, the center of mass of the element is necessarily at its geometric center $(x_k^*, \tfrac{1}{2}f(x_k^*))$. Hence, the moment of the element about the x-axis is

$$(M_x)_k = \tfrac{1}{2}f(x_k^*)(\rho \, \Delta A_k) = \tfrac{1}{2}\rho [f(x_k^*)]^2 \, \Delta x_k.$$

We conclude that

$$m = \lim_{\|P\|\to 0} \sum_{k=1}^{n} \rho f(x_k^*) \, \Delta x_k = \int_a^b \rho f(x) \, dx,$$

$$M_y = \lim_{\|P\|\to 0} \sum_{k=1}^{n} \rho x_k^* f(x_k^*) \, \Delta x_k = \int_a^b \rho x f(x) \, dx,$$

and

$$M_x = \lim_{\|P\|\to 0} \frac{1}{2} \sum_{k=1}^{n} \rho [f(x_k^*)]^2 \Delta x_k = \frac{1}{2} \int_a^b \rho [f(x)]^2 \, dx.$$

Thus, the coordinates of the center of mass of the lamina are defined to be

$$\bar{x} = \frac{M_y}{m} = \frac{\int_a^b \rho x f(x) \, dx}{\int_a^b \rho f(x) \, dx}, \qquad \bar{y} = \frac{M_x}{m} = \frac{\tfrac{1}{2}\int_a^b \rho [f(x)]^2 \, dx}{\int_a^b \rho f(x) \, dx}. \qquad (3)$$

Centroid

We note that the constant density ρ will cancel in equations (3) for $\bar{x}$ and $\bar{y}$, and that the denominator $\int_a^b f(x)\,dx$ is then the area A of the region R. In other words, the center of mass depends only on the shape of R:

$$\bar{x} = \frac{M_y}{A} = \frac{\int_a^b x f(x)\,dx}{\int_a^b f(x)\,dx}, \qquad \bar{y} = \frac{M_x}{A} = \frac{\frac{1}{2}\int_a^b [f(x)]^2\,dx}{\int_a^b f(x)\,dx}. \qquad (4)$$

To emphasize the distinction, albeit minor, between the physical object, which is the homogeneous lamina, and the geometric object, which is the plane region R, we say the equations in (4) define the coordinates of the **centroid** of the region.

Note: It is important that you understand the result in (4), but it does not pay to memorize the integrals because we have assumed for the sake of discussion that R is bounded by the graph of a function f and the x-axis. R could just as well be the region bounded between the graphs of two functions f and g. See Example 5.

EXAMPLE 3 Centroid of a Region

Find the centroid of the region in the first quadrant bounded by the graph of $y = 9 - x^2$, the x-axis, and the y-axis.

Solution The region is shown in FIGURE 6.10.9. Now, if $f(x) = 9 - x^2$, then

$$A_k = f(x_k^*)\,\Delta x_k$$
$$(M_y)_k = x f(x_k^*)\,\Delta x_k$$

and

$$(M_x)_k = \frac{1}{2} f(x_k^*)(f(x_k^*)\,\Delta x_k) = \frac{1}{2}[f(x_k^*)]^2\,\Delta x_k.$$

Hence,

$$A = \int_0^3 (9 - x^2)\,dx = \left(9x - \frac{1}{3}x^3\right)\bigg|_0^3 = 18$$

$$M_y = \int_0^3 x(9 - x^2)\,dx = \left(\frac{9}{2}x^2 - \frac{1}{4}x^4\right)\bigg|_0^3 = \frac{81}{4}$$

$$M_x = \frac{1}{2}\int_0^3 (9 - x^2)^2\,dx$$

$$= \frac{1}{2}\int_0^3 (81 - 18x^2 + x^4)\,dx$$

$$= \frac{1}{2}\left(81x - 6x^3 + \frac{1}{5}x^5\right)\bigg|_0^3 = \frac{324}{5}.$$

It follows from (4) that the coordinates of the centroid are

$$\bar{x} = \frac{M_y}{A} = \frac{81/4}{18} = \frac{9}{8}, \qquad \bar{y} = \frac{M_x}{A} = \frac{324/5}{18} = \frac{54}{15}. \qquad \blacksquare$$

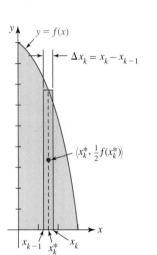

FIGURE 6.10.9 Region in Example 3

EXAMPLE 4 Integration with Respect to y

Find the centroid of the region bounded by the graphs of $x = y^2 + 1$, $x = 0$, $y = 2$, and $y = -2$.

Solution The region is shown in FIGURE 6.10.10. Inspection of the figure suggests that we use horizontal rectangular elements. If $f(y) = y^2 + 1$, then

$$A_k = f(y_k^*)\,\Delta y_k$$
$$(M_x)_k = y_k^* f(y_k^*)\,\Delta y_k$$
$$(M_y)_k = \frac{1}{2} f(y_k^*)(f(y_k^*)\,\Delta y_k) = -\frac{1}{2}[f(y_k^*)]^2\,\Delta y_k$$

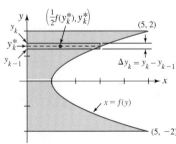

FIGURE 6.10.10 Region in Example 4

and so
$$A = \int_{-2}^{2}(y^2+1)\,dy = \left(\frac{1}{3}y^3 + y\right)\Big]_{-2}^{2} = \frac{28}{3},$$
$$M_x = \int_{-2}^{2} y(y^2+1)\,dy = \left(\frac{1}{4}y^4 + \frac{1}{2}y^2\right)\Big]_{-2}^{2} = 0,$$
$$M_y = \frac{1}{2}\int_{-2}^{2}(y^2+1)^2\,dy = \frac{1}{2}\int_{-2}^{2}(y^4 + 2y^2 + 1)\,dy$$
$$= \frac{1}{2}\left(\frac{1}{5}y^5 + \frac{2}{3}y^3 + y\right)\Big]_{-2}^{2} = \frac{206}{15}.$$

Thus, we have
$$\bar{x} = \frac{M_y}{A} = \frac{206/15}{28/3} = \frac{103}{70}, \quad \bar{y} = \frac{M_x}{A} = \frac{0}{28/3} = 0.$$

As we would expect, since the lamina is symmetric with respect to the x-axis, the centroid is on the axis of symmetry. We also note that the centroid is outside the region. ∎

EXAMPLE 5 Region between Two Graphs

Find the centroid of the region bounded by the graphs of $y = -x^2 + 3$ and $y = x^2 - 2x - 1$.

Solution FIGURE 6.10.11 shows the region in question. We note that the points of intersection of the graphs are $(-1, 2)$ and $(2, -1)$. Now, if $f(x) = -x^2 + 3$ and $g(x) = x^2 - 2x - 1$, then the area of the region is

$$A = \int_{-1}^{2} [f(x) - g(x)]\,dx$$
$$= \int_{-1}^{2} (-2x^2 + 2x + 4)\,dx$$
$$= \left(-\frac{2}{3}x^3 + x^2 + 4x\right)\Big]_{-1}^{2} = 9.$$

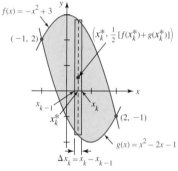

FIGURE 6.10.11 Region in Example 5

Since the coordinates of the midpoint of the indicated element are $(x_k^*, \frac{1}{2}[f(x_k^*) + g(x_k^*)])$, it follows that

$$M_y = \int_{-1}^{2} x[f(x) - g(x)]\,dx$$
$$= \int_{-1}^{2} (-2x^3 + 2x^2 + 4x)\,dx$$
$$= \left(-\frac{1}{2}x^4 + \frac{2}{3}x^3 + 2x^2\right)\Big]_{-1}^{2} = \frac{9}{2},$$

and
$$M_x = \frac{1}{2}\int_{-1}^{2}[f(x) + g(x)][f(x) - g(x)]\,dx$$
$$= \frac{1}{2}\int_{-1}^{2}([f(x)]^2 - [g(x)]^2)\,dx$$
$$= \frac{1}{2}\int_{-1}^{2}[(-x^2+3)^2 - (x^2-2x-1)^2]\,dx$$
$$= \frac{1}{2}\int_{-1}^{2}(4x^3 - 8x^2 - 4x + 8)\,dx$$
$$= \frac{1}{2}\left(x^4 - \frac{8}{3}x^3 - 2x^2 + 8x\right)\Big]_{-1}^{2} = \frac{9}{2}.$$

Thus, the coordinates of the centroid are
$$\bar{x} = \frac{M_y}{A} = \frac{9/2}{9} = \frac{1}{2}, \quad \bar{y} = \frac{M_x}{A} = \frac{9/2}{9} = \frac{1}{2}.$$ ∎

Exercises 6.10   Answers to selected odd-numbered problems begin on page ANS-21.

Fundamentals

In Problems 1–4, find the center of mass of the given system of masses. The mass m_k is located on the x-axis at a point whose directed distance from the origin is x_k. Assume that mass is measured in grams and that distance is measured in centimeters.

1. $m_1 = 2, m_2 = 5; x_1 = 4, x_2 = -2$
2. $m_1 = 6, m_2 = 1, m_3 = 3; x_1 = -\frac{1}{2}, x_2 = -3, x_3 = 8$
3. $m_1 = 10, m_2 = 5, m_3 = 8, m_4 = 7; x_1 = -5, x_2 = 2, x_3 = 6, x_4 = -3$
4. $m_1 = 2, m_2 = \frac{3}{2}, m_3 = \frac{7}{2}, m_4 = \frac{1}{2}; x_1 = 9, x_2 = -4, x_3 = -6, x_4 = -10$
5. Two masses are placed at the ends of a uniform board of negligible mass, as shown in FIGURE 6.10.12. Where should a fulcrum be placed so that the system is in balance? [*Hint:* Although the origin can be placed anywhere, let us agree to choose it halfway between the masses.]

FIGURE 6.10.12 Masses in Problem 5

6. Find the center of mass of the three masses m_1, m_2, and m_3 located at the vertices of the equilateral triangle shown in FIGURE 6.10.13. [*Hint:* First find the center of mass of m_1 and m_2.]

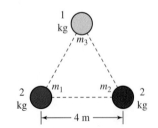

FIGURE 6.10.13 Masses in Problem 6

In Problems 7–14, a rod of linear density $\rho(x)$ kg/m coincides with the x-axis on the interval indicated. Find its center of mass.

7. $\rho(x) = 2x + 1; \quad [0, 5]$
8. $\rho(x) = -x^2 + 2x; \quad [0, 2]$
9. $\rho(x) = x^{1/3}; \quad [0, 1]$
10. $\rho(x) = -x^2 + 1; \quad [0, 1]$
11. $\rho(x) = |x - 3|; \quad [0, 4]$
12. $\rho(x) = 1 + |x - 1|; \quad [0, 3]$
13. $\rho(x) = \begin{cases} x^2, & 0 \le x < 1 \\ 2 - x, & 1 \le x \le 2 \end{cases}; \quad [0, 2]$
14. $\rho(x) = \begin{cases} x, & 0 \le x < 2 \\ 2, & 2 \le x \le 3 \end{cases}; \quad [0, 3]$
15. The density of a 10-ft rod varies as the square of the distance from the left end. Find its center of mass if the density at its center is 12.5 slug/ft.
16. The linear density of a 3-m-long rod varies as the distance from the right end. Find the linear density at the center of the rod if its total mass is 6 kg.

In Problems 17–20, find the center of mass of the given system of masses. The mass m_k is located at the point P_k. Assume that mass is measured in grams and that distance is measured in centimeters.

17. $m_1 = 3, m_2 = 4; P_1 = (-2, 3), P_2 = (1, 2)$
18. $m_1 = 1, m_2 = 3, m_3 = 2; P_1 = (-4, 1), P_2 = (2, 2), P_3 = (5, -2)$
19. $m_1 = 4, m_2 = 8, m_3 = 10; P_1 = (1, 1), P_2 = (-5, 2), P_3 = (7, -6)$
20. $m_1 = 1, m_2 = \frac{1}{2}, m_3 = 4, m_4 = \frac{5}{2}; P_1 = (9, 3), P_2 = (-4, -6), P_3 = (\frac{3}{2}, -1), P_4 = (-2, 10)$

In Problems 21–38, find the centroid of the region bounded by the graphs of the given equations.

21. $y = 2x + 4, y = 0, x = 0, x = 2$
22. $y = x + 1, y = 0, x = 3$
23. $y = x^2, y = 0, x = 1$
24. $y = x^2 + 2, y = 0, x = -1, x = 2$
25. $y = x^3, y = 0, x = 3$
26. $y = x^3, y = 8, x = 0$
27. $y = \sqrt{x}, y = 0, x = 1, x = 4$
28. $x = y^2, x = 1$
29. $y = x^2, y - x = 2$
30. $y = x^2, y = \sqrt{x}$
31. $y = x^3, y = x^{1/3}$, first quadrant
32. $y = 4 - x^2, y = 0, x = 0$, second quadrant
33. $y = 1/x^3, y = 0, x = 1, x = 3$
34. $y = x^2 - 2x + 1, y = -4x + 9$
35. $x = y^2 - 1, y = -1, y = 2, x = -2$
36. $y = x^2 - 4x + 6, y = 0, x = 0, x = 4$
37. $y = 4 - 4x^2, y = 1 - x^2$
38. $y^2 + x = 1, y + x = -1$

In Problems 39 and 40, use symmetry to locate $\bar{x}$ and integration to find $\bar{y}$ of the region bounded by the graphs of the given functions.

39. $y = 1 + \cos x, y = 1, -\pi/2 \le x \le \pi/2$

40. $y = 4\sin x, y = -\sin x, 0 \le x \le \pi$

≡ **Think About It**

41. A theorem due to **Pappus of Alexandria** (c. A.D. 350) states that:

> Let L be an axis in a plane and R a region in the same plane that does not intersect L. When R is revolved about L, the volume V of the resulting solid of revolution is equal to the area A of R times the length of the path traversed by the centroid of R.

(a) As shown in FIGURE 6.10.14, let the region R be bounded by the graphs of $y = f(x)$ and $y = g(x)$. Show that if R is revolved about the x-axis, then $V = (2\pi\bar{y})A$, where A is the area of the region.

(b) What do you think V is given by when the region R is revolved about the y-axis?

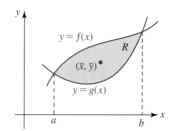

FIGURE 6.10.14 Region in Problem 41

42. Verify the Theorem of Pappus in Problem 41 by revolving the region bounded by $y = x^2 + 1, y = 1, x = 2$ about the x-axis.

43. Use the Theorem of Pappus in Problem 41 to find the volume of the torus shown in FIGURE 6.10.15.

FIGURE 6.10.15 Torus in Problem 43

44. A rod of linear density $\rho(x)$ kg/m coincides with the x-axis on the interval [0, 6]. If $\rho(x) = x(6 - x) + 1$, where would one intuitively expect the center of mass to be? Prove your assertion.

45. Consider the triangular region R in FIGURE 6.10.16. Where do you think the centroid of the triangle is? Think geometrically.

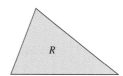

FIGURE 6.10.16 Triangular region in Problem 45

46. Without integration, determine the centroid of the region R shown in FIGURE 6.10.17.

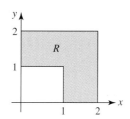

FIGURE 6.10.17 Region in Problem 46

Chapter 6 in Review

Answers to selected odd-numbered problems begin on page ANS-21.

A. True/False

In Problems 1–12, indicate whether the given statement is true or false.

1. When $\int_a^b f(x)\,dx > 0$, the integral gives the area under the graph of $y = f(x)$ on the interval $[a, b]$. _____

2. $\int_0^3 (x - 1)\,dx$ is the area under the graph of $y = x - 1$ on [0, 3]. _____

3. The integral $\int_a^b [f(x) - g(x)]\,dx$ gives the area between the graphs of the continuous functions f and g whenever $f(x) \ge g(x)$ for every x in $[a, b]$. _____

4. The disk and washer methods for finding volumes of solids of revolution are special cases of the slicing method. _____

5. The average value f_{ave} of a continuous function on an interval $[a, b]$ is necessarily a number that satisfies $m \le f_{ave} \le M$, where m and M are the maximum and minimum values of f on the interval, respectively. _____

6. If f and g are continuous on $[a, b]$, then the average value of $f + g$ is $(f + g)_{ave} = f_{ave} + g_{ave}$. _____

7. The center of mass of a pencil with a constant linear density ρ is at its geometric center. _____
8. The center of mass of a lamina that coincides with a plane region R is a point in R at which the lamina would hang in balance. _____
9. The pressure on the flat bottom of a swimming pool is the same as the horizontal pressure on the vertical sidewalls at the same depth. _____
10. Consider a circular tin can with radius 6 in. and a circular reservoir with radius 50 ft. If each has a flat bottom and is filled with water to a depth of 1 ft, then the liquid pressure on the bottom of the reservoir is greater than the pressure on the bottom of the tin can. _____
11. If $s(t)$ is the position function of a body that moves in a straight line, then $\int_{t_1}^{t_2} v(t)\, dt$ is the distance the body moves in the interval $[t_1, t_2]$. _____
12. In the absence of air resistance, when dropped simultaneously from the same height, a cannonball will hit the ground before a marshmallow. _____

B. Fill in the Blanks

In Problems 1–8, fill in the blanks.

1. The unit of work in the SI system of units is _____.
2. To warm up, a 200-lb jogger pushes against a tree for 5 min with a constant force of 60 lb and then runs 2 mi in 10 min. The total work done is _____.
3. The work done by a 100-lb constant force applied at an angle of 60° to the horizontal over a distance of 50 ft is _____.
4. If 80 N of force stretches a spring that is initially 1 m long into a spring that is 1.5 m long, then the spring will measure _____ m long when 100 N of force is applied.
5. The coordinates of the centroid of a region R are $(2, 5)$ and the moment of the region about the x-axis is 30. Hence, the area of R is _____ square units.
6. The weight density of water is _____ lb/ft^3.
7. The graph of a function with a continuous first derivative is said to be _____.
8. A ball dropped from a great height hits the ground in T seconds with a velocity v_{impact}. If its velocity function is $v(t) = -gt$, then the average velocity v_{ave} of the ball for $0 \le t \le T$ in terms of v_{impact} is _____.

C. Exercises

In Problems 1–8, set up the definite integral(s) for the area of the shaded region in each figure.

1.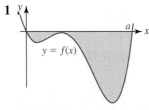

FIGURE 6.R.1 Graph for Problem 1

2.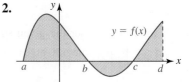

FIGURE 6.R.2 Graph for Problem 2

3.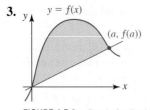

FIGURE 6.R.3 Graph for Problem 3

4.

FIGURE 6.R.4 Graph for Problem 4

5.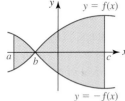

FIGURE 6.R.5 Graph for Problem 5

6.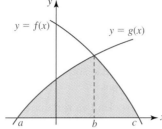

FIGURE 6.R.6 Graph for Problem 6

7.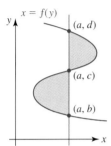

FIGURE 6.R.7 Graph for Problem 7

8.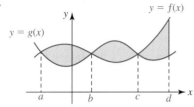

FIGURE 6.R.8 Graph for Problem 8

In Problems 9 and 10, use the definite integral to find the area of the shaded region in terms of a and b.

9.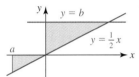

FIGURE 6.R.9 Graph for Problem 9

10.

FIGURE 6.R.10 Graph for Problem 10

In Problems 11–16, consider the region R in FIGURE 6.R.11. Set up the definite integral(s) for the indicated quantity.

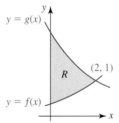

FIGURE 6.R.11 Region for Problems 11–16

11. The centroid of the region
12. The volume of the solid of revolution that is formed by revolving R about the x-axis
13. The volume of the solid of revolution that is formed by revolving R about the y-axis
14. The volume of the solid of revolution that is formed by revolving R about the line $y = -1$
15. The volume of the solid of revolution that is formed by revolving R about the line $x = 2$
16. The volume of the solid with R as its base such that the cross sections of the solid parallel to the y-axis are squares
17. Find the area bounded by the graphs of $y = \sin x$ and $y = \sin 2x$ on the interval $[0, \pi]$.
18. Consider the region bounded by the graphs of $y = e^x$, $y = e^{-x}$, and $x = \ln 2$.
   (a) Find the area of the region.
   (b) Find the volume of the solid of revolution if the region is revolved about the x-axis.

19. Consider the region R bounded by the graphs of $x = y^2$ and $x = \sqrt{2}$. Use the slicing method to find the volume of the solid if the region R is its base and
    (a) cross sections of the solid perpendicular to the x-axis are squares,
    (b) cross sections of the solid perpendicular to the x-axis are circles.
20. Find the volume of the solid of revolution that is formed by revolving the region R bounded by the graphs of $x = 2y - y^2$ and $x = 0$ about the line $y = 3$.
21. A nose cone of a rocket is a right circular cone of height 8 ft and radius 10 ft. The lateral surface is to be covered with canvas except for a section of height 1 ft at the apex of the nose cone. Find the area of the canvas needed.
22. The area under the graph of a continuous nonnegative function $y = f(x)$ on the interval $[-3, 4]$ is 21 square units. What is the average value of the function on the interval?
23. Find the average value of $f(x) = x^{3/2} + x^{1/2}$ on $[1, 4]$.
24. Find a value of x in the interval $[0, 3]$ that corresponds to the average value of the function $f(x) = 2x - 1$.
25. A spring whose unstretched length is $\frac{1}{2}$ m is stretched to a length of 1 m by a force of 50 N. Find the work done in stretching the spring from a length of 1 m to a length of 1.5 m.
26. The work done in stretching a spring 6 in. beyond its natural length is 10 ft-lb. Find the spring constant.
27. A water tank, in the form of a cube that is 10 ft on a side, is filled with water. Find the work done in pumping all the water to a point 5 ft above the tank.
28. A bucket weighing 2 lb contains 30 lb of liquid. As the bucket is raised vertically at a rate of 1 ft/s, the liquid leaks out at a rate of $\frac{1}{4}$ lb/s. Find the work done in lifting the bucket a distance of 5 ft.
29. In Problem 28, find the work done in lifting the bucket to a point where it is empty.
30. In Problem 28, find the work done in lifting the leaking bucket a distance of 5 ft if the rope attached to the bucket weighs $\frac{1}{8}$ lb/ft.
31. A tank on top of a tower 15 ft high consists of a frustum of a cone surmounted by a right circular cylinder. The dimensions (in feet) are given in FIGURE 6.R.12. Find the work done in filling the tank with water from ground level.

FIGURE 6.R.12 Tank in Problem 31

32. A rock is thrown vertically upward from the surface of the Moon with an initial velocity of 44 ft/s.
    (a) If the acceleration of gravity on the Moon is 5.5 ft/s^2, find the maximum height attained. Compare with the Earth.
    (b) On the way down, the rock hits a 6-ft-tall astronaut on the head. What is the impact velocity of the rock?
33. Find the length of the graph of $y = (x - 1)^{3/2}$ from $(1, 0)$ to $(5, 8)$.

34. The linear density of a 6-m-long rod is a linear function of the distance from its left end. The density in the middle of the rod is 11 kg/m and at the right end the density is 17 kg/m. Find the center of mass of the rod.

35. A flat plate, in the form of a quarter-circle, is submerged vertically in oil as shown in FIGURE 6.R.13. If the weight density of the oil is 800 kg/m³, find the force exerted by the oil on one side of the plate.

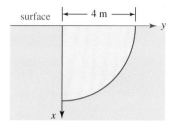

FIGURE 6.R.13 Submerged vertical plate in Problem 35

36. A uniform metal bar of mass 4 kg and length 2m supports two masses, as shown in FIGURE 6.R.14. Where should the wire be attached to the bar so that the system hangs in balance?

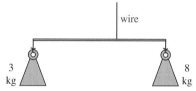

FIGURE 6.R.14 Masses in Problem 36

37. Three masses are suspended from uniform rigid bars of negligible mass as shown in FIGURE 6.R.15. Determine where the indicated wires should be attached so that the entire system hangs in balance.

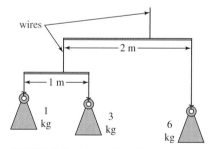

FIGURE 6.R.15 Masses in Problem 37

Chapter 7

Techniques of Integration

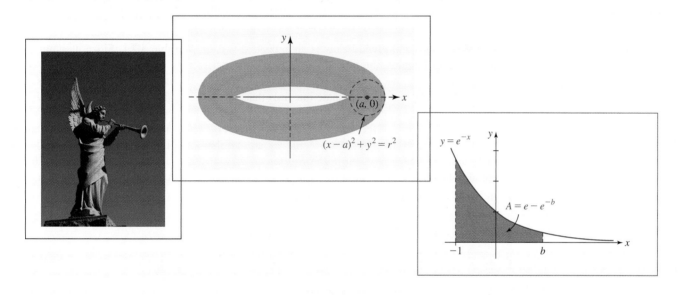

In This Chapter One often encounters an integral that cannot be categorized as a familiar form such as $\int u^n \, du$ or $\int e^u \, du$. For example, it is not possible to evaluate $\int x^2 \sqrt{x+1} \, dx$ by an immediate application of any one of the formulas listed on pages 380–381. However, by applying a **technique of integration**, an integral such as this can sometimes be reduced to one or more of these familiar forms.

- 7.1 Integration—Three Resources
- 7.2 Integration by Substitution
- 7.3 Integration by Parts
- 7.4 Powers of Trigonometric Functions
- 7.5 Trigonometric Substitutions
- 7.6 Partial Fractions
- 7.7 Improper Integrals
- 7.8 Approximate Integration
  Chapter 7 in Review

7.1 Integration—Three Resources

Introduction In this chapter we are going to resume our study of antiderivatives begun in Chapter 5. In that earlier chapter we barely scratched the surface of how to obtain an antiderivative of a function f. Recall, an indefinite integral

$$\int f(x)\, dx = F(x) + C$$

is a family $F(x) + C$ of antiderivatives of the function f, that is, F is related to f by the fact that $F'(x) = f(x)$. In this manner, each time we devise a derivative of a specific function ($\sin x$, $\cos x$, e^x, $\ln x$, and so on) there corresponds an indefinite integral analogue. For example,

$$\frac{d}{dx}\cos x = -\sin x \quad \text{implies} \quad \int \sin x\, dx = -\cos x + C.$$

Tables TABLE 7.1.1 given below is an expanded version of Table 5.2.1. Since it summarizes in indefinite integral notation all the Chain-Rule derivatives of the functions discussed in Chapters 1 and 3, we will refer to the entries in Table 7.1.1 as *familiar* or *basic* forms. The thrust of this chapter is to evaluate integrals that, for the most part, do not fall into any of the forms given in the table.

Table 7.1.1 is just the tip of a rather large iceberg; reference handbooks often contained hundreds of integration formulas. While we are not that ambitious, a more extensive table of integration formulas is given in the *Resource Pages* at the end of this text. As usual, differential notation is used in both of these. If $u = g(x)$ denotes a differentiable function, then the differential of u is the product $du = g'(x)\, dx$.

TABLE 7.1.1

Integration Formulas

Constant Integrands

1. $\int du = u + C$
2. $\int k\, du = ku + C$

Integrands that are Powers

3. $\int u^n\, du = \dfrac{u^{n+1}}{n+1} + C \quad (n \neq -1)$
4. $\int u^{-1}\, du = \int \dfrac{1}{u}\, du = \ln|u| + C$

Exponential Integrands

5. $\int e^u\, du = e^u + C$
6. $\int a^u\, du = \dfrac{1}{\ln a} a^u + C$

Trigonometric Integrands

7. $\int \sin u\, du = -\cos u + C$
8. $\int \cos u\, du = \sin u + C$
9. $\int \sec^2 u\, du = \tan u + C$
10. $\int \csc^2 u\, du = -\cot u + C$
11. $\int \sec u \tan u\, du = \sec u + C$
12. $\int \csc u \cot u\, du = -\csc u + C$
13. $\int \tan u\, du = -\ln|\cos u| + C$
14. $\int \cot u\, du = \ln|\sin u| + C$
15. $\int \sec u\, du = \ln|\sec u + \tan u| + C$
16. $\int \csc u\, du = \ln|\csc u - \cot u| + C$

(continued)

Hyperbolic Integrands

17. $\displaystyle\int \sinh u\, du = \cosh u + C$

18. $\displaystyle\int \cosh u\, du = \sinh u + C$

19. $\displaystyle\int \text{sech}^2 u\, du = \tanh u + C$

20. $\displaystyle\int \text{csch}^2 u\, du = -\coth u + C$

21. $\displaystyle\int \text{sech}\, u\, \tanh u\, du = -\text{sech}\, u + C$

22. $\displaystyle\int \text{csch}\, u\, \coth u\, du = -\text{csch}\, u + C$

Algebraic Integrands

23. $\displaystyle\int \frac{1}{\sqrt{a^2 - u^2}}\, du = \sin^{-1}\frac{u}{a} + C$

24. $\displaystyle\int \frac{1}{a^2 + u^2}\, du = \frac{1}{a}\tan^{-1}\frac{u}{a} + C$

25. $\displaystyle\int \frac{1}{u\sqrt{u^2 - a^2}}\, du = \frac{1}{a}\sec^{-1}\left|\frac{u}{a}\right| + C$

26. $\displaystyle\int \frac{1}{\sqrt{a^2 + u^2}}\, du = \ln\left|u + \sqrt{u^2 + a^2}\right| + C$

27. $\displaystyle\int \frac{1}{\sqrt{u^2 - a^2}}\, du = \ln\left|u + \sqrt{u^2 - a^2}\right| + C$

28. $\displaystyle\int \frac{1}{a^2 - u^2}\, du = \frac{1}{2a}\ln\left|\frac{u + a}{u - a}\right| + C$

29. $\displaystyle\int \frac{1}{u^2 - a^2}\, du = \frac{1}{2a}\ln\left|\frac{u - a}{u + a}\right| + C$

30. $\displaystyle\int \frac{1}{u\sqrt{a^2 - u^2}}\, du = -\frac{1}{a}\ln\left|\frac{a + \sqrt{a^2 - u^2}}{u}\right| + C$

31. $\displaystyle\int \frac{1}{u\sqrt{a^2 + u^2}}\, du = -\frac{1}{a}\ln\left|\frac{a + \sqrt{a^2 + u^2}}{u}\right| + C$

Even though we have designated these integration formulas as *familiar* or *basic* you may not be *that* familiar with some of the formulas, especially 17–22 and 26–31. Because instructors sometimes give short shrift to the hyperbolic functions you are urged to review (or if need be, study for the first time) Section 3.10. Formulas 26–31, which resemble Formulas 23–25, are the indefinite integral forms of the differentiation formulas for the inverse hyperbolic functions combined with the fact that every inverse hyperbolic function is a natural logarithm. See page 183.

▌ **Techniques of Integration** In the sections that follow, the integrals that we are going to examine cannot be categorized as a single familiar form such as $\int u^n\, du$, $\int e^u\, du$, or $\int \sin u\, du$. Nevertheless, Table 7.1.1 is important; as we advance through this chapter we will, of necessity, frequently refer back to it. A wide variety of integrals can be evaluated by performing specific operations on the integrand—called a **technique of integration**—and reducing a given integral to *one or more* of the familiar forms in the table. For example, it is not possible to evaluate $\int \ln x\, dx$ by identifying it with any of the integration formulas in Table 7.1.1. However, we will see in Section 7.3 that by applying a technique of integration, $\int \ln x\, dx$ can be evaluated in a few seconds using the derivative of $\ln x$ along with Formula 1 in the table.

For purposes of review, you are urged to work the problems in Exercises 7.1. By an appropriate u-substitution, each problem can be matched with one of the formulas in Table 7.1.1.

Neither a table, regardless of how large it is, nor techniques of integration, regardless of how powerful they are, is a cure-all for all integration problems. While some integrals, such as $\int e^{x^2}\, dx$, completely defy tables and techniques of integration, others only *appear* to defy these resources. For example, the integral $\int e^{\sin x}\sin 2x\, dx$ does not appear in any tables but it *can* be evaluated by a technique of integration. The problem here is that it is not immediately obvious *which* technique can be applied. There will be times when you will be expected to give some thought to recasting an integrand into a form that is amenable to a technique of integration.

◀ It was pointed out in Section 5.5 (see pages 312–313) that a continuous function f may not have an antiderivative that is an elementary function.

▌ **Technology** A word about technology. If you have not worked with a computer algebra system (CAS) such as *Mathematica*, *Maple*, *Derive*, or *Axiom* you should rectify that deficiency in your background as quickly as possible. A computer algebra system is an extremely sophisticated program designed to perform a wide variety of symbolic mathematical operations such as ordinary algebra, matrix algebra, complex arithmetic, solving polynomial equations, approximating roots of equations, differentiation, integration, graphing equations in two and three dimensions, solving differential equations, manipulation of built-in special functions, and so on. If it is your goal to be a serious student of mathematics, science, or engineering, then an ideal aid to your lecture and laboratory classes (as well as your future career) would be a laptop computer equipped with a program such as *Mathematica*, *Maple*, or MATLAB. Also check the computer labs in your Mathematics and Engineering Departments; the computers therein are undoubtedly equipped with one or more of these programs. Some fundamental command syntax for performing calculus-related operations in *Mathematica* and *Maple* is given in the *Student Resource Manual* that accompanies this text.

As you become adept and comfortable using a CAS, you might be interested in exploring web resource sites such as

http://scienceworld.wolfram.com
http://mathworld.wolfram.com

Wolfram Research is the developer of the computer algebra system *Mathematica*.

Exercises 7.1   Answers to selected odd-numbered problems begin on page ANS-21.

≡ Fundamentals

In Problems 1–32, use a *u*-substitution and Table 7.1.1 to evaluate the given integral.

1. $\int 5^{-5x}\, dx$

2. $\int \dfrac{1}{\sqrt{x}\, e^{\sqrt{x}}}\, dx$

3. $\int \dfrac{\sin\sqrt{1+x}}{\sqrt{1+x}}\, dx$

4. $\int \dfrac{\cos e^{-x}}{e^x}\, dx$

5. $\int \dfrac{x}{\sqrt{25-4x^2}}\, dx$

6. $\int \dfrac{1}{\sqrt{25-4x^2}}\, dx$

7. $\int \dfrac{1}{x\sqrt{4x^2-25}}\, dx$

8. $\int \dfrac{1}{\sqrt{25+4x^2}}\, dx$

9. $\int \dfrac{1}{25+4x^2}\, dx$

10. $\int \dfrac{x}{25+4x^2}\, dx$

11. $\int \dfrac{1}{4x^2-25}\, dx$

12. $\int \dfrac{1}{x\sqrt{4x^2+25}}\, dx$

13. $\int \cot 10x\, dx$

14. $\int x\csc^2 x^2\, dx$

15. $\int \dfrac{6}{(3-5t)^{2.2}}\, dx$

16. $\int x^2 \sqrt{(1-x^3)^5}\, dx$

17. $\int \sec 3x\, dx$

18. $\int 2\csc 2x\, dx$

19. $\int \dfrac{\sin^{-1} x}{\sqrt{1-x^2}}\, dx$

20. $\int \dfrac{1}{(1+x^2)\tan^{-1} x}\, dx$

21. $\int \dfrac{\sin x}{1+\cos^2 x}\, dx$

22. $\int \dfrac{\cos(\ln 9x)}{x}\, dx$

23. $\int \dfrac{x^3}{\cosh^2 x^4}\, dx$

24. $\int \tanh x\, dx$

25. $\int \tan 2x \sec 2x\, dx$

25. $\int \sin x \sin(\cos x)\, dx$

27. $\int \sin x \csc(\cos x) \cot(\cos x)\, dx$

28. $\int \cos x \csc^2(\sin x)\, dx$

29. $\int (1+\tan x)^2 \sec^2 x\, dx$

30. $\int \dfrac{1}{x(\ln x)^2}\, dx$

31. $\int \dfrac{e^{2x}}{1+e^{2x}}\, dx$

32. $\int \dfrac{e^x}{1+e^{2x}}\, dx$

7.2 Integration by Substitution

Introduction In this section we will extend the idea of the ***u*-substitution** introduced in Section 5.2. In Section 5.2 we basically used a *u*-substitution as an aid in recognizing that an integral was actually one of the familiar integration formulas such as $\int u^n\, du$, $\int du/u$, $\int e^u\, du$, and so on. For example, with the substitution $u = \ln x$ and $du = (1/x)\, dx$ we recognize that

$$\int \dfrac{(\ln x)^2}{x}\, dx \quad \text{is the same as} \quad \int u^2\, du.$$

You should verify that the integral $\int x^2\sqrt{2x+1}\, dx$ does not fit any *one* of the 31 integration formulas in Table 7.1.1. Nevertheless, with the aid of a substitution, the integral can be reduced to *several* cases of one of the formulas in Table 7.1.1.

The first example illustrates the general idea.

EXAMPLE 1 Using a u-Substitution

Evaluate $\int x^2 \sqrt{2x+1}\, dx$.

Solution If we let $u = 2x + 1$, then the given integral can be recast entirely in terms of the variable u. To that end, observe that

$$x = \frac{1}{2}(u-1), \qquad dx = \frac{1}{2}\, du,$$

$$x^2 = \frac{1}{4}(u-1)^2 = \frac{1}{4}(u^2 - 2u + 1) \text{ and } \sqrt{2x+1} = u^{1/2}.$$

Substituting these expressions into the given integral yields:

$$\int x^2 \sqrt{2x+1}\, dx = \int \frac{1}{4}(u^2 - 2u + 1)u^{1/2} \frac{1}{2}\, du,$$

that is,

$$\int x^2 \sqrt{2x+1}\, dx = \frac{1}{8}\int (u^2 - 2u + 1)u^{1/2}\, du$$

$$= \frac{1}{8}\int (u^{5/2} - 2u^{3/2} + u^{1/2})\, du \qquad \leftarrow \text{three applications of formula 3 in Table 7.1.1}$$

$$= \frac{1}{8}\int \left(\frac{2}{7}u^{7/2} - \frac{4}{5}u^{5/2} + \frac{2}{3}u^{3/2}\right) + C \qquad \leftarrow \text{now resubstitute for } u$$

$$= \frac{1}{28}(2x+1)^{7/2} - \frac{1}{10}(2x+1)^{5/2} + \frac{1}{12}(2x+1)^{3/2} + C.$$

You should verify that the derivative of the last line actually is $x^2\sqrt{2x+1}$. ∎

The choice of which, if any, substitution to use is not always obvious. Generally, if the integrand contains a power of a function, then it is a good idea to try to let u be that function *or* the power of the function itself. In Example 1, the alternative substitution $u = \sqrt{2x+1}$ or $u^2 = 2x + 1$ leads to the different integral $\frac{1}{4}\int (1 - u^2)^2 u^2\, du$. The latter can be evaluated by expanding the integrand and integrating each term.

EXAMPLE 2 Using a u-Substitution

Evaluate $\int \frac{1}{1+\sqrt{x}}\, dx$.

Solution Let $u = \sqrt{x}$ so that $x = u^2$ and $dx = 2u\, du$. Then

$$\int \frac{1}{1+\sqrt{x}}\, dx = \int \frac{1}{1+u} 2u\, du$$

$$= \int \frac{2u}{1+u}\, du \qquad \leftarrow \text{now use long division}$$

$$= \int \left(2 - \frac{2}{1+u}\right) du \qquad \leftarrow \text{formulas 2 and 4 in Table 7.1.1}$$

$$= 2u - 2\ln|1+u| + C \qquad \leftarrow \text{resubstitute for } u$$

$$= 2\sqrt{x} - 2\ln\left(1 + \sqrt{x}\right) + C.$$

∎

■ **Integrands Containing a Quadratic Expression** If an integrand contains a quadratic expression, $ax^2 + bx + c$, completion of the square may lead to an integral that can be expressed in terms of an inverse trigonometric function or an inverse hyperbolic function. Of course, more complicated integrals can yield other functions as well.

EXAMPLE 3 Completing the Square

Evaluate $\displaystyle\int \frac{x+4}{x^2+6x+18}\,dx$.

Solution After completing the square, the given integral can be written as

$$\int \frac{x+4}{x^2+6x+18}\,dx = \int \frac{x+4}{(x+3)^2+9}\,dx.$$

Now, if $u = x+3$, then $x = u-3$ and $dx = du$. Therefore,

$$\int \frac{x+4}{x^2+6x+18}\,dx = \int \frac{u+1}{u^2+9}\,du \quad \leftarrow \text{termwise division}$$

$$= \int \frac{u}{u^2+9}\,du + \int \frac{1}{u^2+9}\,du$$

$$= \frac{1}{2}\int \frac{2u}{u^2+9}\,du + \int \frac{1}{u^2+9}\,du \quad \leftarrow \begin{array}{l}\text{formulas 4 and 24}\\ \text{in Table 7.1.1}\end{array}$$

$$= \frac{1}{2}\ln(u^2+9) + \frac{1}{3}\tan^{-1}\frac{u}{3} + C$$

$$= \frac{1}{2}\ln[(x+3)^2+9] + \frac{1}{3}\tan^{-1}\frac{x+3}{3} + C$$

$$= \frac{1}{2}\ln(x^2+6x+18) + \frac{1}{3}\tan^{-1}\frac{x+3}{3} + C. \quad \blacksquare$$

The next example illustrates an algebraic substitution in a definite integral.

EXAMPLE 4 A Definite Integral

Evaluate $\displaystyle\int_0^2 \frac{6x+1}{\sqrt[3]{3x+2}}\,dx$.

Solution If $u = 3x+2$, then

$$x = \frac{1}{3}(u-2), \qquad dx = \frac{1}{3}\,du,$$

$$6x+1 = 2(u-2)+1 = 2u-3 \text{ and } \sqrt[3]{3x+2} = u^{1/3}.$$

Since we will change the variable of integration, we must convert the x-limits of integration to u-limits of integration. Observe when $x=0$, $u=2$, and when $x=2$, $u=8$. Therefore, the original integral becomes

$$\int_0^2 \frac{6x+1}{\sqrt[3]{3x+2}}\,dx = \int_2^8 \frac{2u-3}{u^{1/3}}\frac{1}{3}\,du \quad \leftarrow \text{termwise division again}$$

$$= \int_2^8 \left(\frac{2}{3}u^{2/3} - u^{-1/3}\right)du$$

$$= \left(\frac{2}{5}u^{5/3} - \frac{3}{2}u^{2/3}\right)\bigg]_2^8$$

$$= \left(\frac{2}{5}\cdot 2^5 - \frac{3}{2}\cdot 2^2\right) - \left(\frac{2}{5}\cdot 2^{5/3} - \frac{3}{2}\cdot 2^{2/3}\right)$$

$$= \frac{34}{5} - \frac{2}{5}\cdot 2^{5/3} + \frac{3}{2}\cdot 2^{2/3} \approx 7.9112. \quad \blacksquare$$

You are encouraged to rework Example 4 again. The second time use the substitution $u = \sqrt[3]{3x+2}$.

∫ NOTES FROM THE CLASSROOM

(*i*) When working the exercises throughout this chapter, do not be overly disturbed if you do not always obtain the same answer as given in the text. Different techniques applied to the same problem can lead to answers that look different. Remember, two antiderivatives of the same function can differ at most by a constant. Try to resolve any conflicts.

(*ii*) It might also prove helpful at this point to recall that integration of the quotient of two polynomial functions, $p(x)/q(x)$, usually begins with long division if the degree of $p(x)$ is greater than or equal to the degree of $q(x)$. See Example 2.

(*iii*) Look for problems that can be solved by previous methods.

Exercises 7.2   Answers to selected odd-numbered problems begin on page ANS-22.

≡ Fundamentals

In Problems 1–26, use a substitution to evaluate the given integral.

1. $\int x(x+1)^3 \, dx$

2. $\int \dfrac{x^2 - 3}{(x+1)^3} \, dx$

3. $\int (2x+1)\sqrt{x-5} \, dx$

4. $\int (x^2-1)\sqrt{2x+1} \, dx$

5. $\int \dfrac{x}{\sqrt{x-1}} \, dx$

6. $\int \dfrac{x^2}{\sqrt{x+2}} \, dx$

7. $\int \dfrac{x+3}{(3x-4)^{3/2}} \, dx$

8. $\int (x^2+x)\sqrt[3]{x+7} \, dx$

9. $\int \dfrac{\sqrt{x}}{x+1} \, dx$

10. $\int \dfrac{t}{\sqrt{t}+1} \, dt$

11. $\int \dfrac{\sqrt{t}-3}{\sqrt{t}+1} \, dt$

12. $\int \dfrac{\sqrt{r}+3}{r+3} \, dr$

13. $\int \dfrac{x^3}{\sqrt[3]{x^2+1}} \, dx$

14. $\int \dfrac{x^5}{\sqrt[5]{x^2+4}} \, dx$

15. $\int \dfrac{x^2}{(x-1)^4} \, dx$

16. $\int \dfrac{2x+1}{(x+7)^2} \, dx$

17. $\int \sqrt{e^x - 1} \, dx$

18. $\int \dfrac{1}{\sqrt{e^x-1}} \, dx$

19. $\int \sqrt{1-\sqrt{v}} \, dv$

20. $\int \dfrac{\sqrt{w}}{\sqrt{1-\sqrt{w}}} \, dw$

21. $\int \dfrac{\sqrt{1+\sqrt{t}}}{\sqrt{t}} \, dt$

22. $\int \sqrt{t}\sqrt{1+t\sqrt{t}} \, dt$

23. $\int \dfrac{2x+7}{x^2+2x+5} \, dx$

24. $\int \dfrac{6x-1}{4x^2+4x+10} \, dx$

25. $\int \dfrac{2x+5}{\sqrt{16-6x-x^2}} \, dx$

26. $\int \dfrac{4x-3}{\sqrt{11+10x-x^2}} \, dx$

In Problems 27 and 28, use the substitution $u = x^{1/6}$ to evaluate the integral.

27. $\int \dfrac{1}{\sqrt{x}-\sqrt[3]{x}} \, dx$

28. $\int \dfrac{\sqrt[6]{x}}{\sqrt[3]{x}+1} \, dx$

In Problems 29–40, use a substitution to evaluate the given definite integral.

29. $\int_0^1 x\sqrt{5x+4} \, dx$

30. $\int_{-1}^0 x\sqrt[3]{x+1} \, dx$

31. $\int_1^{16} \dfrac{1}{10+\sqrt{x}} \, dx$

32. $\int_4^9 \dfrac{\sqrt{x}-1}{\sqrt{x}+1} \, dx$

33. $\int_2^9 \dfrac{5x-6}{\sqrt[3]{x-1}} \, dx$

34. $\int_{-\sqrt{3}}^0 \dfrac{2x^3}{\sqrt{x^2+1}} \, dx$

35. $\int_0^1 (1-\sqrt{x})^{50} \, dx$

36. $\int_0^4 \dfrac{1}{(1+\sqrt{x})^3} \, dx$

37. $\int_1^8 \dfrac{1}{x^{1/3}+x^{2/3}} \, dx$

38. $\int_1^{64} \dfrac{x^{1/3}}{x^{2/3}+2} \, dx$

39. $\int_0^1 x^2(1-x)^5 \, dx$

40. $\int_0^6 \dfrac{2x+5}{\sqrt{2x+4}} \, dx$

In Problems 41 and 42, use a substitution to establish the given result. Assume $x > 0$.

41. $\int_1^{x^2} \dfrac{1}{t} \, dt = 2\int_1^x \dfrac{1}{t} \, dt$

42. $\int_1^{\sqrt{x}} \dfrac{1}{t} \, dt = \dfrac{1}{2}\int_1^x \dfrac{1}{t} \, dt$

≡ Review of Applications

43. Find the area under the graph of $y = \dfrac{1}{x^{1/3}+1}$ on the interval $[0, 1]$.

44. Find the area bounded by the graph of $y = x^3\sqrt{x+1}$ and the x-axis on the interval $[-1, 1]$.

45. Find the volume of the solid of revolution that is formed by revolving the region bounded by the graphs of $y = \dfrac{1}{\sqrt{x}+1}$, $x = 0$, $x = 4$, and $y = 0$ about the y-axis.

46. Find the volume of the solid of revolution that is formed by revolving the region in Problem 45 about the x-axis.

47. Find the length of the graph of $y = \tfrac{4}{5}x^{5/4}$ on the interval $[0, 9]$.

48. **Bertalanffy's differential equation** is a mathematical model for the growth of an organism in which it is assumed that constructive metabolism (anabolism) of the organism

proceeds at a rate proportional to the surface area, while destructive metabolism (catabolism) proceeds at a rate proportional to the volume. If it is also assumed that surface area is proportional to the two-thirds power of volume and that the organism's weight w is directly proportional to the volume, we can write Bertalanffy's equation as

$$\frac{dw}{dt} = Aw^{2/3} - Bw,$$

where A and B are positive parameters. One can conclude from this equation that the time it takes such an organism to grow from weight w_1 to weight w_2 is given by the definite integral

$$T = \int_{w_1}^{w_2} \frac{1}{Aw^{2/3} - Bw}\, dw.$$

Evaluate this integral. Find an upper limit on how large the organism can grow.

7.3 Integration by Parts

Introduction In this section we are going to develop an important formula that can often be used to integrate the product of two functions. To apply this formula we have to identify one of the functions in the product as a differential. Recall that if $v = g(x)$, then its differential is the function $dv = g'(x)\, dx$.

Integrating Products Since we wish to integrate a product it seems reasonable to begin with the Product Rule of differentiation. If $u = f(x)$ and $v = g(x)$ are differentiable functions, then the derivative of $f(x)g(x)$ is

$$\frac{d}{dx}[f(x)g(x)] = f(x)g'(x) + g(x)f'(x). \tag{1}$$

In turn, integration of both sides of (1),

$$\int \frac{d}{dx}[f(x)g(x)]\, dx = \int f(x)g'(x)\, dx + \int g(x)f'(x)\, dx$$

or

$$f(x)g(x) = \int f(x)g'(x)\, dx + \int g(x)f'(x)\, dx,$$

produces the formula

$$\int f(x)g'(x)\, dx = f(x)g(x) - \int g(x)f'(x)\, dx. \tag{2}$$

The formula in (2) is usually written in terms of the differentials $du = f'(x)\, dx$ and $dv = g'(x)\, dx$:

$$\int u\, dv = uv - \int v\, du. \tag{3}$$

The procedure defined by formula (3) is known as **integration by parts**. The essential idea behind (3) is to evaluate the integral $\int u\, dv$ by means of evaluating another, and it is hoped simpler, integral $\int v\, du$.

Guidelines for Integration by Parts

- The **first step** in the process of integrating by parts consists of choosing and integrating dv in the given integral. As a practical matter, the function dv is *usually* the most complicated factor in the product that can be integrated using one of the basic formulas given in Table 7.1.1.
- The **second step** is the differentiation of the remaining factor u in the given integral. We then form

$$\int u\, dv = uv - \int v\, du.$$
(differentiate u, integrate dv)

- The **third step** is, of course, the evaluation of $\int v\, du$.

Integration problems can sometimes be done by several methods. In the first example, the integral can be evaluated by means of an algebraic substitution (Section 7.2) as well as by integration by parts.

EXAMPLE 1 Using (3)

Evaluate $\int \dfrac{x}{\sqrt{x+1}}\, dx$.

Solution First, we write the integral as

$$\int x(x+1)^{-1/2}\, dx.$$

From this latter form we see that there are several possible choices for the function dv. Of the possible choices for dv,

$$dv = (x+1)^{-1/2}\, dx, \quad dv = x\, dx, \quad \text{or} \quad dv = dx,$$

we choose

$$dv = (x+1)^{-1/2}\, dx \quad \text{and} \quad u = x.$$

Then, by integration Formula 3 in Table 7.1.1, we find

$$v = \int (x+1)^{-1/2}\, dx = 2(x+1)^{1/2}.$$

◂ No constant of integration need be used when integrating dv.

Substituting $v = 2(x+1)^{1/2}$ and $du = dx$ into (3) then gives

$$\int \overbrace{x}^{u}\overbrace{(x+1)^{-1/2}\, dx}^{dv} = \overbrace{x}^{u} \cdot \overbrace{2(x+1)^{1/2}}^{v} - \int \overbrace{2(x+1)^{1/2}}^{v}\, \overbrace{dx}^{du}$$

$$= 2x(x+1)^{1/2} - 2 \cdot \dfrac{2}{3}(x+1)^{3/2} + C \quad \leftarrow \text{we used formula 3 in Table 7.1.1}$$

$$= 2x(x+1)^{1/2} - \dfrac{4}{3}(x+1)^{3/2} + C.$$

Check by Differentiation To verify the preceding result we use the Product Rule:

$$\dfrac{d}{dx}\left(2x(x+1)^{1/2} - \dfrac{4}{3}(x+1)^{3/2} + C\right) = 2x \cdot \dfrac{1}{2}(x+1)^{-1/2} + 2(x+1)^{1/2} - \dfrac{4}{3} \cdot \dfrac{3}{2}(x+1)^{1/2}$$

$$= x(x+1)^{-1/2} + 2(x+1)^{1/2} - 2(x+1)^{1/2}$$

$$= \dfrac{x}{\sqrt{x+1}}. \quad \blacksquare$$

The key to making integration by parts work is to make the "right" choice for the function dv. In the guidelines given prior to Example 1 we stated that dv is usually the most complicated factor in the product that can be immediately integrated by a previously known formula. Yet this cannot be given as a firm rule. Realization that the "right" choice for dv has been made is often based on pragmatic hindsight: Is the second integral $\int v\, du$ less complicated than the first integral $\int u\, dv$? Can we evaluate this second integral? To see what happens when the "wrong" choice is made, let us consider Example 1 again, but this time we select

$$dv = x\, dx \quad \text{and} \quad u = (x+1)^{-1/2}$$

so that

$$v = \dfrac{1}{2}x^2 \quad \text{and} \quad du = -\dfrac{1}{2}(x+1)^{-3/2}\, dx.$$

Applying (3) in this instance gives

$$\int x(x+1)^{-1/2}\, dx = \dfrac{1}{2}x^2(x+1)^{-1/2} + \dfrac{1}{4}\int x^2(x+1)^{-3/2}\, dx.$$

The difficulty here is apparent; the second integral $\int v\, du$ is more complicated than the original $\int u\, dv$. The alternative selection $dv = dx$ also leads to an impasse.

EXAMPLE 2 Using (3)

Evaluate $\int x^3 \ln x\, dx$.

Solution Again there are several possible choices for the function dv:

$$dv = \ln x\, dx, \quad dv = x^3\, dx, \quad \text{or} \quad dv = dx. \tag{4}$$

Although the choice $dv = \ln x\, dx$ is undoubtedly the most complicated factor in the product $x^3 \ln x\, dx$, we reject this choice since it does not match any formula in Table 7.1.1. Of the remaining two functions in (4), the second is the more "complicated." So if we choose

$$dv = x^3\, dx \quad \text{and} \quad u = \ln x,$$

then

$$v = \frac{1}{4}x^4 \quad \text{and} \quad du = \frac{1}{x}\, dx.$$

Hence from (3),

$$\int x^3 \ln x\, dx = \overbrace{\ln x}^{u} \cdot \overbrace{\frac{1}{4}x^4}^{v} - \int \overbrace{\frac{1}{4}x^4}^{v} \cdot \overbrace{\frac{1}{x}}^{du}\, dx \quad \leftarrow \text{simpify the integrand}$$

$$= \frac{1}{4}x^4 \ln x - \frac{1}{4}\int x^3\, dx \quad \leftarrow \text{integrate } x^3$$

$$= \frac{1}{4}x^4 \ln x - \frac{1}{16}x^4 + C.$$

EXAMPLE 3 Using (3)

Evaluate $\int x \tan^{-1} x\, dx$.

Solution The choice $dv = \tan^{-1} x\, dx$ is not a judicious one, since we cannot immediately integrate this function based on a previously known result. So we choose

$$dv = x\, dx \quad \text{and} \quad u = \tan^{-1} x$$

and find

$$v = \frac{1}{2}x^2 \quad \text{and} \quad du = \frac{1}{1+x^2}\, dx.$$

Therefore (3) gives

$$\int \overbrace{(\tan^{-1} x)}^{u} \overbrace{(x\, dx)}^{dv} = \overbrace{(\tan^{-1} x)}^{u} \overbrace{\frac{1}{2}x^2}^{v} - \int \overbrace{\frac{1}{2}x^2}^{v} \cdot \overbrace{\frac{1}{1+x^2}}^{du}\, dx. \quad \leftarrow \text{simplify the integrand}$$

To evaluate the indefinite integral $\int x^2\, dx/(1+x^2)$, we use long division (see Example 7 of Section 5.1). Hence,

$$\int x \tan^{-1} x\, dx = \frac{1}{2}x^2 \tan^{-1} x - \frac{1}{2}\int \left(1 - \frac{1}{1+x^2}\right) dx$$

$$= \frac{1}{2}x^2 \tan^{-1} x - \frac{1}{2}x + \frac{1}{2}\tan^{-1} x + C.$$

▌ **Successive Integrations** A problem may require integration by parts several times in succession. As a rule, integrals of the type

$$\int p(x) \sin kx\, dx, \quad \int p(x) \cos kx\, dx, \quad \text{and} \quad \int p(x) e^{kx}\, dx, \tag{5}$$

where $p(x)$ is a polynomial of degree $n \geq 1$ and k a constant, will require integration by parts n times. Moreover, an integral such as

$$\int x^k (\ln x)^n \, dx, \tag{6}$$

where again n is a positive integer, will also require n applications of (3). The integral in Example 2 is of the form (6) with $k = 3$ and $n = 1$.

EXAMPLE 4 Using (3) Twice in Succession

Evaluate $\int x^2 \cos x \, dx$.

Solution The integral $\int x^2 \cos x \, dx$ is the second of the three forms in (5) with $p(x) = x^2$ and $n = 2$. Consequently we apply (3) twice in succession. In the first integration we use

$$dv = \cos x \, dx \quad \text{and} \quad u = x^2$$

so that

$$v = \sin x \quad \text{and} \quad du = 2x \, dx.$$

Hence (3) becomes

$$\int x^2 \cos x \, dx = x^2 \sin x - 2 \underbrace{\int x \sin x \, dx}_{\text{requires integration by parts}}. \tag{7}$$

The second integral in (7) is the first form in (5) and requires only one integration by parts since the degree of the polynomial $p(x) = x$ is $n = 1$. In this second integral we choose $dv = \sin x \, dx$ and $u = x$:

$$\int x^2 \cos x \, dx = x^2 \sin x - 2\left[x(-\cos x) - \int (-\cos x) \, dx \right]$$

$$= x^2 \sin x + 2x \cos x - 2\int \cos x \, dx \quad \leftarrow \text{formula 8 in Table 7.1.1}$$

$$= x^2 \sin x + 2x \cos x - 2 \sin x + C. \tag{8}$$

The result in (8) can be obtained by a systematic shortcut. If we think of the integral in Example 4 as $\int f(x) g'(x) \, dx$ where $f(x) = x^2$ and $g'(x) = \cos x$, then we can display the derivatives and integrals in an array:

$f(x)$ and its derivatives		$g'(x)$ and its integrals
x^2	$+$	$\cos x$
$2x$	$-$	$\sin x$
2	$+$	$-\cos x$
0		$-\sin x$

We then form the products of the functions joined by the arrows and either add or subtract a product according to the algebraic sign indicated in blue:

$$\int x^2 \cos x \, dx = +x^2(\sin x) - 2x(-\cos x) + 2(-\sin x) + C.$$

The last zero in the derivative column indicates that we need not integrate $g'(x)$ any further; the products from that point on are zero.

This technique for successive integrations by parts works on all integrals of the type shown in (5) and is called **tabular integration**. For an integral such as $\int x^4 e^{-2x} \, dx$ we would pick $f(x) = x^4$ and $g'(x) = e^{-2x}$. You should verify that tabular integration gives

$$\int x^4 e^{-2x} \, dx = +x^4\left(-\frac{1}{2} e^{-2x}\right) - 4x^3\left(\frac{1}{4} e^{-2x}\right) + 12x^2\left(-\frac{1}{8} e^{-2x}\right) - 24x\left(\frac{1}{16} e^{-2x}\right) + 24\left(-\frac{1}{32} e^{-2x}\right) + C$$

$$= -\frac{1}{2} x^4 e^{-2x} - x^3 e^{-2x} - \frac{3}{2} x^2 e^{-2x} - \frac{3}{2} x e^{-2x} - \frac{3}{4} e^{-2x} + C.$$

■ **Solving for Integrals** For certain integrals, one or more applications of integration by parts may result in a situation where the original integral occurs on the right-hand side. In this case the problem of evaluating that integral is completed by *solving* for the original integral. The next example illustrates the technique.

EXAMPLE 5 Solving for the Original Integral

Evaluate $\int \sec^3 x \, dx$.

Solution Inspection of the integral reveals no obvious choice for dv. However, by writing the integrand as the product $\sec^3 x = \sec x \cdot \sec^2 x$, we can identify

$$dv = \sec^2 x \, dx \quad \text{and} \quad u = \sec x$$

so that

$$v = \tan x \quad \text{and} \quad du = \sec x \tan x \, dx.$$

It follows from (3) that

$$\int \sec^3 x \, dx = \sec x \tan x - \int \tan^2 x \sec x \, dx$$

$$= \sec x \tan x - \int (\sec^2 x - 1) \sec x \, dx \quad \leftarrow \text{trig identity for } \tan^2 x$$

$$= \sec x \tan x + \int \sec x \, dx - \int \sec^3 x \, dx$$

$$= \sec x \tan x + \ln|\sec x + \tan x| - \int \sec^3 x \, dx.$$

See (18) in Section 5.2 for the evaluation of the integral $\int \sec x \, dx$. Also, see Formula 15 in Table 7.1.1.

We solve the last equation for $\int \sec^3 x \, dx$ and add a constant of integration:

$$2 \int \sec^3 x \, dx = \sec x \tan x + \ln|\sec x + \tan x|$$

and so

$$\int \sec^3 x \, dx = \frac{1}{2} \sec x \tan x + \frac{1}{2} \ln|\sec x + \tan x| + C. \quad ■$$

Integrals of the type

$$\int e^{ax} \sin bx \, dx \quad \text{and} \quad \int e^{ax} \cos bx \, dx \tag{9}$$

are important in certain aspects of applied mathematics. These integrals require two applications of integration by parts before recovering the original integral on the right-hand side.

EXAMPLE 6 Solving for the Original Integral

Evaluate $\int e^{2x} \cos 3x \, dx$.

Solution If we choose

$$dv = e^{2x} \, dx \quad \text{and} \quad u = \cos 3x,$$

then

$$v = \frac{1}{2} e^{2x} \quad \text{and} \quad du = -3 \sin 3x \, dx.$$

The integration by parts formula (3) then gives

$$\int e^{2x} \cos 3x \, dx = \frac{1}{2} e^{2x} \cos 3x + \frac{3}{2} \int e^{2x} \sin 3x \, dx.$$

We apply integration by parts again to the integral highlighted in color with $dv = e^{2x}\,dx$ and $u = \sin 3x$:

$$\int e^{2x}\cos 3x\,dx = \frac{1}{2}e^{2x}\cos 3x + \frac{3}{2}\left[\frac{1}{2}e^{2x}\sin 3x - \int \frac{1}{2}e^{2x}(3\cos 3x)\,dx\right]$$

$$= \frac{1}{2}e^{2x}\cos 3x + \frac{3}{4}e^{2x}\sin 3x - \frac{9}{4}\int e^{2x}\cos 3x\,dx.$$

Solving the last equation for $\int e^{2x}\cos 3x\,dx$ gives

$$\frac{13}{4}\int e^{2x}\cos 3x\,dx = \frac{1}{2}e^{2x}\cos 3x + \frac{3}{4}e^{2x}\sin 3x.$$

After dividing by $\frac{13}{4}$ and affixing a constant of integration we get

$$\int e^{2x}\cos 3x\,dx = \frac{2}{13}e^{2x}\cos 3x + \frac{3}{13}e^{2x}\sin 3x + C. \qquad \blacksquare$$

In evaluating the integrals $\int e^{ax}\sin bx\,dx$ and $\int e^{ax}\cos bx\,dx$ it does not matter which of the functions are chosen as dv and u. In Example 6 we chose $dv = e^{2x}\,dx$ and $u = \cos 3x$; you are encouraged to rework this example using $dv = \cos 3x\,dx$ and $u = e^{2x}$.

∎ **Definite Integrals** A definite integral can be evaluated using integration by parts in the following manner:

$$\int_a^b f(x)g'(x)\,dx = f(x)g(x)\Big]_a^b - \int_a^b g(x)f'(x)\,dx.$$

For convenience, the foregoing equation is usually written as

$$\int_a^b u\,dv = uv\Big]_a^b - \int_a^b v\,du, \qquad (10)$$

where it is understood the *limits of integration are values of x* and the integrations in the integrals are carried out with respect to the variable x.

EXAMPLE 7 Area Under the Graph

Find the area under the graph of $f(x) = \ln x$ on the interval $[1, e]$.

Solution From FIGURE 7.3.1 we see that $f(x) \geq 0$ for all x in the interval. Hence, the area A is given by the definite integral

$$A = \int_1^e \ln x\,dx.$$

Choosing $\qquad dv = dx \qquad$ and $\qquad u = \ln x,$

then $\qquad\qquad v = x \qquad$ and $\qquad du = \frac{1}{x}\,dx.$

From (10) we have

$$A = x\ln x\Big]_1^e - \int_1^e x\cdot\frac{1}{x}\,dx$$

$$= x\ln x\Big]_1^e - \int_1^e dx$$

$$= x\ln x\Big]_1^e - x\Big]_1^e$$

$$= e\ln e - \ln 1 - e + 1 = 1.$$

Here we have used $\ln e = 1$ and $\ln 1 = 0$. ∎

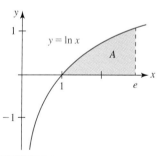

FIGURE 7.3.1 Area under graph in Example 7

Exercises 7.3

Fundamentals

In Problems 1–40, use integration by parts to evaluate the given integral.

1. $\int x\sqrt{x+3}\,dx$
2. $\int \dfrac{x}{\sqrt{2x-5}}\,dx$
3. $\int \ln 4x\,dx$
4. $\int \ln(x+1)\,dx$
5. $\int x\ln 2x\,dx$
6. $\int x^{1/2}\ln x\,dx$
7. $\int \dfrac{\ln x}{x^2}\,dx$
8. $\int \dfrac{\ln x}{\sqrt{x^3}}\,dx$
9. $\int (\ln t)^2\,dt$
10. $\int (t\ln t)^2\,dt$
11. $\int \sin^{-1}x\,dx$
12. $\int x^2\tan^{-1}x\,dx$
13. $\int xe^{3x}\,dx$
14. $\int x^2 e^{5x}\,dx$
15. $\int x^3 e^{-4x}\,dx$
16. $\int x^5 e^x\,dx$
17. $\int x^3 e^{x^2}\,dx$
18. $\int x^5 e^{2x^3}\,dx$
19. $\int t\cos 8t\,dt$
20. $\int x\sinh x\,dx$
21. $\int x^2 \sin x\,dx$
22. $\int x^2\cos\dfrac{x}{2}\,dx$
23. $\int x^3 \cos 3x\,dx$
24. $\int x^4 \sin 2x\,dx$
25. $\int e^x \sin 4x\,dx$
26. $\int e^{-x}\cos 5x\,dx$
27. $\int e^{-2\theta}\cos\theta\,d\theta$
28. $\int e^{\alpha x}\sin\beta x\,dx$
29. $\int \theta\sec\theta\tan\theta\,d\theta$
30. $\int e^{2t}\cos e^t\,dt$
31. $\int \sin x\cos 2x\,dx$
32. $\int \cosh x\cosh 2x\,dx$
33. $\int x^3\sqrt{x^2+4}\,dx$
34. $\int \dfrac{t^5}{(t^3+1)^2}\,dt$
35. $\int \sin(\ln x)\,dx$
36. $\int \cos x\ln(\sin x)\,dx$
37. $\int \csc^3 x\,dx$
38. $\int x\sec^{-1}x\,dx$
39. $\int x\sec^2 x\,dx$
40. $\int x\tan^2 x\,dx$

In Problems 41–46, evaluate the given definite integral.

41. $\displaystyle\int_0^2 x\ln(x+1)\,dx$
42. $\displaystyle\int_0^1 \ln(x^2+1)\,dx$
43. $\displaystyle\int_2^4 xe^{-x/2}\,dx$
44. $\displaystyle\int_{-\pi}^{\pi} e^x\cos x\,dx$
45. $\displaystyle\int_0^1 \tan^{-1}x\,dx$
46. $\displaystyle\int_0^{\sqrt{2}/2} \cos^{-1}x\,dx$

Review of Applications

47. Find the area under the graph of $y = 1 + \ln x$ on the interval $[e^{-1}, 3]$.
48. Find the area bounded by the graph of $y = \tan^{-1}x$ and the x-axis on the interval $[-1, 1]$.
49. The region in the first quadrant bounded by the graphs of $y = \ln x$, $x = 5$, and $y = 0$ is revolved about the x-axis. Find the volume of the solid of revolution.
50. The region in the first quadrant bounded by the graphs of $y = e^x$, $x = 0$, and $y = 3$ is revolved about the y-axis. Find the volume of the solid of revolution.
51. The region in the first quadrant bounded by the graphs of $y = \sin x$ and $y = 0$, $0 \le x \le \pi$, is revolved about the y-axis. Find the volume of the solid of revolution.
52. Find the length of the graph of $y = \ln(\cos x)$ on the interval $[0, \pi/4]$.
53. Find the average value of $f(x) = \tan^{-1}(x/2)$ on the interval $[0, 2]$.
54. A body moves in a straight line with velocity $v(t) = e^{-t}\sin t$, where v is measured in cm/s. Find the position function $s(t)$ if it is known that $s = 0$ when $t = 0$.
55. A body moves in a straight line with acceleration $a(t) = te^{-t}$, where a is measured in cm/s^2. Find the velocity function $v(t)$ and the position function $s(t)$ if $v(0) = 1$ and $s(0) = -1$.
56. A water tank is formed by revolving the region bounded by the graphs of $y = \sin\pi x$ and $y = 0$, $0 \le x \le 1$, about the x-axis, which is taken in the downward direction. The tank is filled to a depth of $\frac{1}{2}$ ft. Determine the work done in pumping all the water to the top of the tank.
57. Find the force caused by liquid pressure on one side of the vertical plate shown in FIGURE 7.3.2. Assume that the plate is submerged in water and that dimensions are in feet.

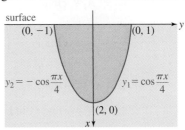

FIGURE 7.3.2 Submerged plate in Problem 57

58. Find the centroid of the region bounded by the graphs of $y = \sin x$, $y = 0$, and $x = \pi/2$.

In Problems 59–62, evaluate the given integral by first using a substitution followed by integration by parts.

59. $\displaystyle\int_1^4 \frac{\tan^{-1}\sqrt{x}}{\sqrt{x}}\,dx$

60. $\displaystyle\int xe^{\sqrt{x}}\,dx$

61. $\displaystyle\int \sin\sqrt{x+2}\,dx$

62. $\displaystyle\int_0^{\pi^2} \cos\sqrt{t}\,dt$

In Problems 63–66, use integration by parts to establish the given **reduction formula**.

63. $\displaystyle\int (\ln x)^n\,dx = x(\ln x)^n - n\int (\ln x)^{n-1}\,dx$

64. $\displaystyle\int \sin^n x\,dx = -\frac{\sin^{n-1}x\cos x}{n} + \frac{n-1}{n}\int \sin^{n-2}x\,dx$

65. $\displaystyle\int \cos^n x\,dx = \frac{\cos^{n-1}x\sin x}{n} + \frac{n-1}{n}\int \cos^{n-2}x\,dx$

66. $\displaystyle\int \sec^n x\,dx = \frac{\sec^{n-2}x\tan x}{n-1} + \frac{n-2}{n-1}\int \sec^{n-2}x\,dx,$

In Problems 67–70, use a reduction formula from Problems 63–66 to evaluate the given integral.

67. $\displaystyle\int \sin^3 x\,dx$

68. $\displaystyle\int \sec^4 x\,dx$

69. $\displaystyle\int \cos^3 10x\,dx$

70. $\displaystyle\int \cos^4 x\,dx$

71. Use Problem 64 to show that for $n \geq 2$,
$$\int_0^{\pi/2} \sin^n x\,dx = \frac{n-1}{n}\int_0^{\pi/2} \sin^{n-2}x\,dx.$$

72. Show how the repeated use of the formula in Problem 71 can be used to obtain the following results.

(a) $\displaystyle\int_0^{\pi/2} \sin^n x\,dx = \frac{\pi}{2}\cdot\frac{1\cdot 3\cdot 5\cdots(n-1)}{2\cdot 4\cdot 6\cdots n},$

n even and $n \geq 2$

(b) $\displaystyle\int_0^{\pi/2} \sin^n x\,dx = \frac{2\cdot 4\cdot 6\cdots(n-1)}{3\cdot 5\cdot 7\cdots n},$

n odd and $n \geq 3$

73. Use part (a) of Problem 72 to evaluate $\displaystyle\int_0^{\pi/2} \sin^8 x\,dx$.

74. Use part (b) of Problem 72 to evaluate $\displaystyle\int_0^{\pi/2} \sin^5 x\,dx$.

≡ **Think About It**

In Problems 75–82, the integration by parts is a bit more challenging. Evaluate the given integral.

75. $\displaystyle\int e^{2x}\tan^{-1}e^x\,dx$

76. $\displaystyle\int (\sin^{-1}x)^2\,dx$

77. $\displaystyle\int \frac{xe^x}{(x+1)^2}\,dx$

78. $\displaystyle\int \frac{x^2 e^x}{(x+2)^2}\,dx$

79. $\displaystyle\int xe^x \sin x\,dx$

80. $\displaystyle\int xe^{-x}\cos 2x\,dx$

81. $\displaystyle\int \ln(x+\sqrt{x^2+1})\,dx$

82. $\displaystyle\int e^{\sin^{-1}x}\,dx$

≡ **Calculator/CAS Problems**

83. (a) Use a calculator or CAS to obtain the graph of $f(x) = 3 + 2\sin^2 x - 5\sin^4 x$.

(b) Find the area under the graph of the function given in part (a) on the interval $[0, 2\pi]$.

84. (a) Use a calculator or CAS to obtain the graphs of $y = x\sin x$ and $y = x\cos x$.

(b) Find the area of the region bounded by the graphs on the interval $[x_1, x_2]$, where x_1 and x_2 are the positive x-coordinates corresponding to the first and second points of intersection of the graphs for $x > 0$.

7.4 Powers of Trigonometric Functions

■ **Introduction** In Section 5.2 we saw how to integrate $\sin^2 x$ and $\cos^2 x$. In this section we see how to integrate higher powers of $\sin x$ and $\cos x$, certain products of powers of $\sin x$ and $\cos x$, and products of powers of $\sec x$ and $\tan x$. The techniques illustrated in this section depend heavily on trigonometric identities.

■ **Integrals of the Form $\int \sin^m x \cos^n x\,dx$** To evaluate integrals of the type

$$\int \sin^m x \cos^n x\,dx, \tag{1}$$

we distinguish two cases.

CASE I: *m* or *n* is an odd positive integer

Let us first assume that $m = 2k + 1$ in (1) is an odd positive integer. Then:

- Begin by splitting off the factor $\sin x$ from $\sin^{2k+1} x$, that is, write $\sin^{2k+1} x = \sin^{2k} x \sin x$, where $2k$ is now even.
- Use the basic Pythagorean identity $\sin^2 x = 1 - \cos^2 x$, to rewrite
$$\sin^{2k} x = (\sin^2 x)^k = (1 - \cos^2 x)^k.$$
- Expand the binomial $(1 - \cos^2 x)^k$.

In this manner we can express the integrand in (1) as a sum of powers of $\cos x$ times $\sin x$. The original integral can then be expressed as a sum of integrals, each having the recognizable form

$$\int \cos^r x \sin x \, dx = -\int \overbrace{(\cos x)^r}^{u^r} \overbrace{(-\sin x \, dx)}^{du} = -\int u^r \, du.$$

If $n = 2k + 1$ is an odd positive integer in (1), then the procedure is the same, except that we write $\cos^{2k+1} x = \cos^{2k} x \cos x$, use $\cos^2 x = 1 - \sin^2 x$, and write the integral as a sum of integrals of the form

$$\int \sin^r x \cos x \, dx = \int \overbrace{(\sin x)^r}^{u^r} \overbrace{(\cos x \, dx)}^{du} = \int u^r \, du.$$

We note that the exponent r need not be an integer.

EXAMPLE 1 Case I of Integral (1)

Evaluate $\int \sin^5 x \cos^2 x \, dx$.

Solution We begin by writing the power of $\sin x$ as $\sin^5 x = \sin^4 x \sin x$:

$$\int \sin^5 x \cos^2 x \, dx = \int \cos^2 x \sin^4 x \sin x \, dx$$

$$= \int \cos^2 x (\sin^2 x)^2 \sin x \, dx \qquad \leftarrow \text{replace } \sin^2 x \text{ by } 1 - \cos^2 x$$

$$= \int \cos^2 x (1 - \cos^2 x)^2 \sin x \, dx$$

$$= \int \cos^2 x (1 - 2\cos^2 x + \cos^4 x) \sin x \, dx \qquad \leftarrow \text{write as three integrals}$$

$$= -\int \overbrace{(\cos x)^2}^{u^2} \overbrace{(-\sin x \, dx)}^{du} + 2\int \overbrace{(\cos x)^4}^{u^4} \overbrace{(-\sin x \, dx)}^{du} - \int \overbrace{(\cos x)^6}^{u^6} \overbrace{(-\sin x \, dx)}^{du}$$

$$= -\frac{1}{3}\cos^3 x + \frac{2}{5}\cos^5 x - \frac{1}{7}\cos^7 x + C. \qquad \blacksquare$$

EXAMPLE 2 Case I of Integral (1)

Evaluate $\int \sin^3 x \, dx$.

Solution As in Example 1 we rewrite the power of $\sin x$ as $\sin^2 x \sin x$:

$$\int \sin^3 x \, dx = \int \sin^2 x \sin x \, dx$$

$$= \int (1 - \cos^2 x) \sin x \, dx$$

$$= \int \sin x \, dx + \int (\cos x)^2 (-\sin x \, dx)$$

$$= -\cos x + \frac{1}{3}\cos^3 x + C. \qquad \blacksquare$$

EXAMPLE 3 Case I of Integral (1)

Evaluate $\int \sin^4 x \cos^3 x \, dx$.

Solution This time we rewrite the power of $\cos x$ as $\cos^2 x \cos x$:

$$\int \sin^4 x \cos^3 x \, dx = \int \sin^4 x \cos^2 x \cos x \, dx$$

$$= \int \sin^4 x (1 - \sin^2 x) \cos x \, dx \qquad \leftarrow \text{write as two integrals}$$

$$= \int \overbrace{(\sin x)^4}^{u^4} \overbrace{(\cos x \, dx)}^{du} - \int \overbrace{(\sin x)^6}^{u^6} \overbrace{(\cos x \, dx)}^{du}$$

$$= \frac{1}{5} \sin^5 x - \frac{1}{7} \sin^7 x + C. \qquad \blacksquare$$

CASE II: *m* and *n* are both even nonnegative integers

When both m and n are even nonnegative integers, the evaluation of (1) relies heavily on the trigonometric identities

$$\sin x \cos x = \frac{1}{2} \sin 2x, \quad \sin^2 x = \frac{1}{2}(1 - \cos 2x), \quad \cos^2 x = \frac{1}{2}(1 + \cos 2x). \qquad (2)$$

We have already seen the last two identities in Section 5.2 as the useful forms for the half-angle formulas for the sine and cosine.

EXAMPLE 4 Using Identities (2) in Integral (1)

Evaluate $\int \sin^2 x \cos^2 x \, dx$.

Solution We will evaluate the integral in two different ways. We begin by using the second and third formulas in (2):

$$\int \sin^2 x \cos^2 x \, dx = \int \frac{1}{2}(1 - \cos 2x) \cdot \frac{1}{2}(1 + \cos 2x) \, dx$$

$$= \frac{1}{4} \int (1 - \cos^2 2x) \, dx$$

$$= \frac{1}{4} \int \left[1 - \frac{1}{2}(1 + \cos 4x) \right] dx \qquad \leftarrow \begin{array}{l} \text{third identity in (2)} \\ \text{with } x \text{ replaced by } 2x \end{array}$$

$$= \frac{1}{4} \int \left(\frac{1}{2} - \frac{1}{2} \cos 4x \right) dx$$

$$= \frac{1}{8} x - \frac{1}{32} \sin 4x + C.$$

Alternative Solution We now use the first formula in (2):

$$\int \sin^2 x \cos^2 x \, dx = \int (\sin x \cos x)^2 \, dx$$

$$= \int \left(\frac{1}{2} \sin 2x \right)^2 dx$$

$$= \frac{1}{4} \int \frac{1}{2}(1 - \cos 4x) \, dx.$$

The remainder of the solution is the same as before. $\blacksquare$

EXAMPLE 5 Using Identities (2) in Integral (1)

Evaluate $\int \cos^4 x \, dx$.

Solution We begin by rewriting $\cos^4 x$ as $(\cos^2 x)^2$ and then use the third identity in (2):

$$\int \cos^4 x \, dx = \int (\cos^2 x)^2 \, dx$$

$$= \int \left[\frac{1}{2}(1 + \cos 2x)\right]^2 dx$$

$$= \frac{1}{4}\int (1 + 2\cos 2x + \cos^2 2x) \, dx \quad \leftarrow \text{use the third formula in (2) a second time}$$

$$= \frac{1}{4}\int \left[1 + 2\cos 2x + \frac{1}{2}(1 + \cos 4x)\right] dx$$

$$= \frac{1}{4}\int \left(\frac{3}{2} + 2\cos 2x + \frac{1}{2}\cos 4x\right) dx$$

$$= \frac{3}{8}x + \frac{1}{4}\sin 2x + \frac{1}{32}\sin 4x + C. \qquad \blacksquare$$

■ **Integrals of the Form $\int \tan^m x \sec^n x \, dx$** To evaluate an integral involving powers of secant and tangent,

$$\int \tan^m x \sec^n x \, dx, \qquad (3)$$

we shall consider three cases. The procedure in the first two cases is similar to Case I for integral (1) in that we break off a factor from the product $\tan^m x \sec^n x$ to serve as part of the differential du.

CASE I: m is an odd positive integer

When $m = 2k + 1$ is an odd positive integer in (3), $2k$ is even. Then:

- Begin by splitting off the factor $\sec x \tan x$ from $\tan^{2k+1} x \sec^n x$, that is, write $\tan^{2k+1} x \sec^n x = \tan^{2k} x \sec^{n-1} x \sec x \tan x$, where $2k$ is now even.
- Use the identity $\tan^2 x = \sec^2 x - 1$ to rewrite

$$\tan^{2k} x = (\tan^2 x)^k = (\sec^2 x - 1)^k.$$

- Expand the binomial $(\sec^2 x - 1)^k$.

In this manner we can express the integrand in (3) as a sum of powers of $\sec x$ times $\sec x \tan x$. The original integral can then be expressed as a sum of integrals, each having the recognizable form

$$\int \overbrace{(\sec x)^r}^{u^r} \overbrace{(\sec x \tan x \, dx)}^{du} = \int u^r \, du.$$

EXAMPLE 6 Case I of Integral (3)

Evaluate $\int \tan^3 x \sec^7 x \, dx$.

Solution By writing $\tan^3 x \sec^7 x = \tan^2 x \sec^6 x \sec x \tan x$, the integral can be written as two integrals that we can evaluate:

$$\int \tan^3 x \sec^7 x \, dx = \int \tan^2 x \sec^6 x \sec x \tan x \, dx$$

$$= \int (\sec^2 x - 1) \sec^6 x \sec x \tan x \, dx$$

$$= \int \overbrace{(\sec x)^8}^{u^8} \overbrace{(\sec x \tan x \, dx)}^{du} - \int \overbrace{(\sec x)^6}^{u^6} \overbrace{(\sec x \tan x \, dx)}^{du}$$

$$= \frac{1}{9} \sec^9 x - \frac{1}{7} \sec^7 x + C. \qquad \blacksquare$$

CASE II: *n* is an even positive integer

Let $n = 2k$ represent an even positive integer in (3). Then:

- Begin by splitting the factor $\sec^2 x$ from $\sec^{2k} x \tan^m x$, that is, write $\sec^{2k} x \tan^m x = \sec^{2(k-1)} x \tan^m x \sec^2 x$.
- Use the identity $\sec^2 x = 1 + \tan^2 x$ to rewrite

$$\sec^{2(k-1)} x = (\sec^2 x)^{k-1} = (1 + \tan^2 x)^{k-1}.$$

- Expand the binomial $(1 + \tan^2 x)^{k-1}$.

In this manner we can express the integrand in (3) as a sum of powers of $\tan x$ times $\sec^2 x$. The original integral can then be expressed as a sum of integrals, each having the recognizable form

$$\int \overbrace{(\tan x)^r}^{u^r} \overbrace{(\sec^2 x \, dx)}^{du} = \int u^r \, du.$$

EXAMPLE 7 Case II of Integral (3)

Evaluate $\int \sqrt{\tan x} \sec^4 x \, dx$.

Solution We rewrite the integrand using $\sec^4 x = \sec^2 x \sec^2 x$:

$$\int \sqrt{\tan x} \sec^4 x \, dx = \int (\tan x)^{1/2} \sec^2 x \sec^2 x \, dx$$

$$= \int (\tan x)^{1/2} (1 + \tan^2 x) \sec^2 x \, dx$$

$$= \int \overbrace{(\tan x)^{1/2}}^{u^{1/2}} \overbrace{(\sec^2 x \, dx)}^{du} + \int \overbrace{(\tan x)^{5/2}}^{u^{5/2}} \overbrace{(\sec^2 x \, dx)}^{du}$$

$$= \frac{2}{3} (\tan x)^{3/2} + \frac{2}{7} (\tan x)^{7/2} + C. \qquad \blacksquare$$

CASE III: *m* is even and *n* is odd

Finally, if *m* is an even positive integer and *n* is an odd positive integer, we write the integrand of (3) in terms of sec *x* and use integration by parts.

EXAMPLE 8 Case III of Integral (3)

Evaluate $\int \tan^2 x \sec x \, dx$.

Solution By writing

$$\int \tan^2 x \sec x \, dx = \int (\sec^2 x - 1) \sec x \, dx$$

$$= \int \sec^3 x \, dx - \int \sec x \, dx$$

we encounter two integrals that were evaluated previously:

$$\int \sec^3 x \, dx = \frac{1}{2} \sec x \tan x + \frac{1}{2} \ln |\sec x + \tan x| + C_1, \qquad (4)$$

$$\int \sec x \, dx = \ln |\sec x + \tan x| + C_2. \qquad (5)$$

◀ For the result in (4), see Example 5 in Section 7.3. For (5), see (18) in Section 5.2.

Subtracting the results in (4) and (5) then yields the desired result:

$$\int \tan^2 x \sec x \, dx = \frac{1}{2} \sec x \tan x - \frac{1}{2} \ln|\sec x + \tan x| + C.$$ ∎

Integrals of the type

$$\int \cot^m x \csc^n x \, dx \qquad (6)$$

are handled in a manner analogous to (3). In this case the identity $\csc^2 x = 1 + \cot^2 x$ is used.

Exercises 7.4   Answers to selected odd-numbered problems begin on page ANS-22.

≡ Fundamentals

In Problems 1–40, evaluate the given indefinite integral. Note that some of the integrals do not, strictly speaking, fall into any of the cases considered in this section. You should be able to evaluate these integrals by previous methods.

1. $\int (\sin x)^{1/2} \cos x \, dx$
2. $\int \cos^4 5x \sin 5x \, dx$
3. $\int \cos^3 x \, dx$
4. $\int \sin^3 4x \, dx$
5. $\int \sin^5 t \, dt$
6. $\int \cos^5 t \, dt$
7. $\int \sin^3 x \cos^3 x \, dx$
8. $\int \sin^5 2x \cos^2 2x \, dx$
9. $\int \sin^4 t \, dt$
10. $\int \cos^6 \theta \, d\theta$
11. $\int \sin^2 x \cos^4 x \, dx$
12. $\int \frac{\cos^3 x}{\sin^2 x} \, dx$
13. $\int \sin^4 x \cos^4 x \, dx$
14. $\int \sin^2 3x \cos^2 3x \, dx$
15. $\int \tan^3 2t \sec^4 2t \, dt$
16. $\int (2 - \sqrt{\tan x})^2 \sec^2 x \, dx$
17. $\int \tan^2 x \sec^3 x \, dx$
18. $\int \tan^2 3x \sec^2 3x \, dx$
19. $\int \tan^3 x (\sec x)^{-1/2} \, dx$
20. $\int \tan^3 \frac{x}{2} \sec^3 \frac{x}{2} \, dx$
21. $\int \tan^3 x \sec^5 x \, dx$
22. $\int \tan^5 x \sec x \, dx$
23. $\int \sec^5 x \, dx$
24. $\int \frac{1}{\cos^4 x} \, dx$
25. $\int \cos^2 x \cot x \, dx$
26. $\int \sin x \sec^7 x \, dx$
27. $\int \cot^{10} x \csc^4 x \, dx$
28. $\int (1 + \csc^2 t)^2 \, dt$
29. $\int \frac{\sec^4(1-t)}{\tan^8(1-t)} \, dt$
30. $\int \frac{\sin^3 \sqrt{t} \cos^2 \sqrt{t}}{\sqrt{t}} \, dt$
31. $\int (1 + \tan x)^2 \sec x \, dx$
32. $\int (\tan x + \cot x)^2 \, dx$
33. $\int \tan^4 x \, dx$
34. $\int \tan^5 x \, dx$
35. $\int \cot^3 t \, dt$
36. $\int \csc^5 t \, dt$
37. $\int (\tan^6 x - \tan^2 x) \, dx$
38. $\int \cot 2x \csc^{5/2} 2x \, dx$
39. $\int x \sin^3 x^2 \, dx$
40. $\int x \tan^8(x^2) \sec^2(x^2) \, dx$

In Problems 41–46, evaluate the given definite integral.

41. $\int_{\pi/3}^{\pi/2} \sin^3 \theta \sqrt{\cos \theta} \, d\theta$
42. $\int_0^{\pi/2} \sin^5 x \cos^5 x \, dx$
43. $\int_0^{\pi} \sin^3 2t \, dt$
44. $\int_{-\pi}^{\pi} \sin^4 x \cos^2 x \, dx$
45. $\int_0^{\pi/4} \tan y \sec^4 y \, dy$
46. $\int_0^{\pi/3} \tan x \sec^{3/2} x \, dx$

In Problems 47–52, use the trigonometric identities

$$\sin mx \cos nx = \frac{1}{2}[\sin(m+n)x + \sin(m-n)x]$$

$$\sin mx \sin nx = \frac{1}{2}[\cos(m-n)x - \cos(m+n)x]$$

$$\cos mx \cos nx = \frac{1}{2}[\cos(m-n)x + \cos(m+n)x]$$

to evaluate the given trigonometric integral.

47. $\int \sin x \cos 2x \, dx$
48. $\int \cos 3x \cos 5x \, dx$

49. $\int \sin 2x \sin 4x \, dx$ **50.** $\int \dfrac{5 - 3\sin 2x}{\sec 6x} \, dx$

51. $\int_0^{\pi/6} \cos 2x \cos x \, dx$ **52.** $\int_0^{\pi/2} \sin \dfrac{3}{2}x \sin \dfrac{1}{2}x \, dx$

53. Show that

$$\int_{-\pi}^{\pi} \sin mx \sin nx \, dx = \begin{cases} 0, & m \neq n \\ \pi, & m = n. \end{cases}$$

54. Evaluate $\int_{-\pi}^{\pi} \sin mx \cos nx \, dx$.

≡ Review of Applications

In Problems 55 and 56, the graphs of $f(x) = \sec^2(x/2)$ (Problem 55) and $f(x) = \sin^2 x$ (Problem 56) are given. Find the volume of the solid of revolution obtained by revolving the region R bounded by the graph of f on the interval $[-\pi/2, \pi/2]$ about the indicated axis.

55. the x-axis;

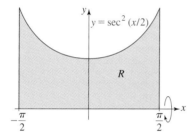

FIGURE 7.4.1 Region in Problem 55

56. the line $y = 1$;

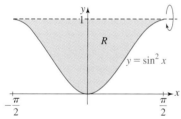

FIGURE 7.4.2 Region in Problem 56

57. Find the area of the region R bounded by the graphs of $y = \sin^3 x$ and $y = \cos^3 x$ on the interval $[-3\pi/4, \pi/4]$. See FIGURE 7.4.3.

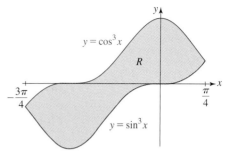

FIGURE 7.4.3 Graphs for Problem 57

58. The graph of the equation $r = |\sin 4\theta \sin \tfrac{1}{2}\theta|$, $0 \leq \theta \leq 2\pi$, encloses a region that is a mathematical model for the shape of a horse chestnut leaf. See FIGURE 7.4.4. We shall see in Chapter 10 that the area A bounded by this graph is given by $A = \tfrac{1}{2}\int_0^{2\pi} r^2 \, d\theta$. Find this area. [*Hint*: Use one of the identities given in the instructions for Problems 47–52.]

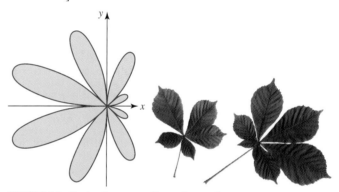

FIGURE 7.4.4 Region in Problem 58 — Horse chestnut leaves

≡ Calculator/CAS Problems

59. Use a calculator or CAS to obtain the graphs of $y = \cos^3 x$, $y = \cos^5 x$, and $y = \cos^7 x$ on the interval $[0, \pi]$. Use the graphs to conjecture the values of the definite integrals

$$\int_0^{\pi} \cos^3 x \, dx, \quad \int_0^{\pi} \cos^5 x \, dx, \quad \text{and} \quad \int_0^{\pi} \cos^7 x \, dx.$$

60. In Problem 59, what do you think is the value of $\int_0^{\pi} \cos^n x \, dx$, where n is a positive odd integer? Prove your conjecture.

7.5 Trigonometric Substitutions

Introduction When an integrand contains integer powers of x and integer powers of

$$\sqrt{a^2 - u^2}, \quad \sqrt{a^2 + u^2}, \quad \text{or} \quad \sqrt{u^2 - a^2}, \, a > 0 \qquad (1)$$

we may be able to evaluate the integral by means of a trigonometric substitution. The three cases we shall consider in this section depend, in turn, on the fundamental Pythagorean identities written in the form:

$$1 - \sin^2\theta = \cos^2\theta, \quad 1 + \tan^2\theta = \sec^2\theta, \quad \text{and} \quad \sec^2\theta - 1 = \tan^2\theta.$$

The procedure for an indefinite integral is similar to the discussion in Sections 5.2 and 7.2:

- Make a substitution in an integral.
- After simplification, carry out the integration with respect to the new variable.
- Return to the original variable by resubstitution.

Before proceeding let us match the type of trigonometric substitution with the radicals in (1).

Guidelines for Trigonometric Substitutions

For integrands containing

- $\sqrt{a^2 - u^2}$, $a > 0$, let $u = a\sin\theta$, where $-\pi/2 \leq \theta \leq \pi/2$.
- $\sqrt{a^2 + u^2}$, $a > 0$, let $u = a\tan\theta$, where $-\pi/2 < \theta < \pi/2$.
- $\sqrt{u^2 - a^2}$, $a > 0$, let $u = a\sec\theta$, where $\begin{cases} 0 \leq \theta < \pi/2, & \text{if } u \geq a \\ \pi/2 < \theta \leq \pi, & \text{if } u \leq -a. \end{cases}$

In each case, the restriction given on the variable θ is precisely the one that accompanies the corresponding inverse trigonometric function. In other words, we want to be able to write $\theta = \sin^{-1}(u/a)$, and so on. Moreover, with the aid of the foregoing identities each of these substitutions yields a perfect square. With the restriction on θ for the substitutions $u = a\sin\theta$ and $u = a\tan\theta$, the square root may be taken without recourse to absolute values. As we shall see, we have to be more careful using the substitution $u = a\sec\theta$.

See Section 1.5 for a review of inverse trigonometric functions.

- If $u = a\sin\theta$, where $-\pi/2 \leq \theta \leq \pi/2$, then
$$\sqrt{a^2 - u^2} = \sqrt{a^2 - a^2\sin^2\theta} = \sqrt{a^2(1 - \sin^2\theta)} = \sqrt{a^2\cos^2\theta} = a\cos\theta.$$

- If $u = a\tan\theta$, where $-\pi/2 < \theta < \pi/2$, then
$$\sqrt{a^2 + u^2} = \sqrt{a^2 + a^2\tan^2\theta} = \sqrt{a^2(1 + \tan^2\theta)} = \sqrt{a^2\sec^2\theta} = a\sec\theta.$$

- If $u = a\sec\theta$, where $0 \leq \theta < \pi/2$ or $\pi/2 < \theta \leq \pi$, then
$$\sqrt{u^2 - a^2} = \sqrt{a^2\sec^2\theta - a^2} = \sqrt{a^2(\sec^2\theta - 1)} = \sqrt{a^2\tan^2\theta} = a|\tan\theta|.$$

When an expression such as $\sqrt{a^2 - u^2}$ appears in the denominator of an integrand, there is the further restriction on the variable θ; in this case $-\pi/2 < \theta < \pi/2$.

After carrying out the integration in θ it is necessary to return to the original variable x. If we construct a reference right triangle, one in which $\sin\theta = u/a$, $\tan\theta = u/a$, or $\sec\theta = u/a$ as shown in FIGURE 7.5.1, then the other trigonometric functions can be readily expressed in terms of u.

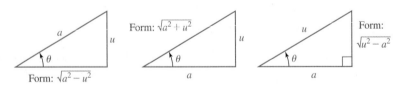

FIGURE 7.5.1 Reference right triangles used to express trigonometric functions in terms of algebraic expression in u and a

In the first two examples, the integrands contain the radical form $\sqrt{a^2 - u^2}$.

EXAMPLE 1 Using a Sine Substitution

Evaluate $\displaystyle\int \frac{x^2}{\sqrt{9 - x^2}}\, dx$.

Solution Identifying $u = x$ and $a = 3$ leads to the substitutions

$$x = 3\sin\theta \quad \text{and} \quad dx = 3\cos\theta\, d\theta,$$

where $-\pi/2 < \theta < \pi/2$. The integral becomes

$$\int \frac{x^2}{\sqrt{9-x^2}} dx = \int \frac{9\sin^2\theta}{\sqrt{9-9\sin^2\theta}} (3\cos\theta\, d\theta) \quad \leftarrow \text{simplify}$$

$$= 9 \int \sin^2\theta\, d\theta.$$

Recall, to evaluate this last trigonometric integral, we make use of the half-angle identity $\sin^2\theta = \frac{1}{2}(1 - \cos 2\theta)$:

$$\int \frac{x^2}{\sqrt{9-x^2}} dx = \frac{9}{2} \int (1 - \cos 2\theta)\, d\theta$$

$$= \frac{9}{2}\theta - \frac{9}{4}\sin 2\theta + C$$

$$= \frac{9}{2}\theta - \frac{9}{2}\sin\theta\cos\theta + C. \quad \leftarrow \text{double-angle formula}$$

In order to express this result back in terms of the variable x, we use $\sin\theta = x/3$ and $\theta = \sin^{-1}(x/3)$. Then from the reference right triangle in FIGURE 7.5.2 we see that $\cos\theta = \sqrt{9-x^2}/3$, and so

$$\int \frac{x^2}{\sqrt{9-x^2}} dx = \frac{9}{2}\sin^{-1}\left(\frac{x}{3}\right) - \frac{1}{2}x\sqrt{9-x^2} + C. \quad \blacksquare$$

FIGURE 7.5.2 Right triangle in Example 1

EXAMPLE 2 Using a Sine Substitution

Evaluate $\int \frac{\sqrt{1-4x^2}}{x} dx.$

Solution With the identifications $u = 2x$, $a = 1$ we let $2x = \sin\theta$, and so $x = \frac{1}{2}\sin\theta$, and $dx = \frac{1}{2}\cos\theta\, d\theta$. Then

$$\int \frac{\sqrt{1-4x^2}}{x} dx = \int \frac{\sqrt{1-\sin^2\theta}}{\frac{1}{2}\sin\theta} \left(\frac{1}{2}\cos\theta\, d\theta\right) \quad \leftarrow \text{simplify}$$

$$= \int \frac{\cos^2\theta}{\sin\theta} d\theta$$

$$= \int \frac{1-\sin^2\theta}{\sin\theta} d\theta \quad \leftarrow \text{use termwise division}$$

$$= \int (\csc\theta - \sin\theta)\, d\theta \quad \leftarrow \text{formulas 16 and 7 in Table 7.1.1}$$

$$= \ln|\csc\theta - \cot\theta| + \cos\theta + C.$$

In FIGURE 7.5.3 we have constructed a right triangle for which $\sin\theta = 2x$ and $\cos\theta = \sqrt{1-4x^2}$. Therefore, $\csc\theta = 1/\sin\theta = 1/(2x)$ and $\cot\theta = \cos\theta/\sin\theta = \sqrt{1-4x^2}/(2x)$. Hence, the result obtained by integrating with respect to θ can be written in terms of x as

$$\int \frac{\sqrt{1-4x^2}}{x} dx = \ln\left|\frac{1-\sqrt{1-4x^2}}{2x}\right| + \sqrt{1-4x^2} + C. \quad \blacksquare$$

FIGURE 7.5.3 Right triangle in Example 2

As a refresher in differentiation skills, you are encouraged to verify periodically that the derivative of the antiderivative you have obtained is the integrand in the original integral. The derivative of the final answer in Example 2,

$$\frac{d}{dx}\left[\ln\left|\frac{1-\sqrt{1-4x^2}}{2x}\right| + \sqrt{1-4x^2} + C\right] = \frac{-1 + 4x^2 + \sqrt{1-4x^2}}{x(-1 + \sqrt{1-4x^2})},$$

is on the face of it not the integrand in Example 2. Use algebra to resolve the difference between this result and the integrand $\sqrt{1-4x^2}/x$.

In the next two examples, the integrands contain an integer power of the radical form $\sqrt{a^2 + u^2}$.

EXAMPLE 3 Using a Tangent Substitution

Evaluate $\displaystyle\int \frac{1}{(4 + x^2)^{3/2}}\, dx$.

Solution Observe that the integrand is an integer power of $\sqrt{4 + x^2}$, since $(4 + x^2)^{3/2} = \left(\sqrt{4 + x^2}\right)^3$. Now, when $u = x$, $x = 2\tan\theta$ and $dx = 2\sec^2\theta\, d\theta$ we have $\sqrt{4 + x^2} = \sqrt{4\sec^2\theta} = 2\sec\theta$ and $(4 + x^2)^{3/2} = 8\sec^3\theta$. Thus,

$$\int \frac{1}{(4 + x^2)^{3/2}}\, dx = \int \frac{1}{8\sec^3\theta}(2\sec^2\theta\, d\theta)$$

$$= \frac{1}{4}\int \cos\theta\, d\theta$$

$$= \frac{1}{4}\sin\theta + C.$$

FIGURE 7.5.4 Right triangle in Example 3

From the triangle in FIGURE 7.5.4, we see that $\sin\theta = x/\sqrt{4 + x^2}$. Hence,

$$\int \frac{1}{(4 + x^2)^{3/2}}\, dx = \frac{1}{4}\frac{x}{\sqrt{4 + x^2}} + C.$$

EXAMPLE 4 Arc Length

Find the length of the graph of $y = \frac{1}{2}x^2 + 3$ on the interval $[0, 1]$.

Solution Recall that the formula for arc length is $L = \displaystyle\int_a^b \sqrt{1 + [f'(x)]^2}\, dx$. Since $dy/dx = x$, we have

$$L = \int_0^1 \sqrt{1 + x^2}\, dx.$$

Now if $u = x$, then we substitute $x = \tan\theta$ and $dx = \sec^2\theta\, d\theta$. The θ-limits of integration in the resulting definite trigonometric integral are obtained from the x-limits in the original integral:

$$x = 0: \quad \theta = \tan^{-1}0 = 0$$

and

$$x = 1: \quad \theta = \tan^{-1}1 = \pi/4.$$

Therefore,

$$L = \int_0^{\pi/4} \sqrt{1 + \tan^2\theta}\, \sec^2\theta\, d\theta$$

$$= \int_0^{\pi/4} \sec^3\theta\, d\theta.$$

The indefinite integral of $\sec^3\theta$ was found in Example 5 of Section 7.3 using integration by parts:

$$L = \left(\frac{1}{2}\sec\theta\tan\theta + \frac{1}{2}\ln|\sec\theta + \tan\theta|\right)\Big]_0^{\pi/4}$$

$$= \frac{1}{2}\sec\frac{\pi}{4}\tan\frac{\pi}{4} + \frac{1}{2}\ln\left|\sec\frac{\pi}{4} + \tan\frac{\pi}{4}\right|$$

$$= \frac{1}{2}\sqrt{2} + \frac{1}{2}\ln(\sqrt{2} + 1) \approx 1.1478.$$

In the next two examples, the integrands contain an integer power of the radical form $\sqrt{u^2 - a^2}$.

EXAMPLE 5 Using a Secant Substitution

Evaluate $\displaystyle\int \frac{\sqrt{x^2 - 16}}{x^4}\, dx$ assuming that $x > 4$.

Solution If we let $u = x$ and $x = 4\sec\theta$, $0 \leq \theta < \pi/2$, and $dx = 4\sec\theta\tan\theta\, d\theta$, the integral becomes

$$\int \frac{\sqrt{x^2 - 16}}{x^4} dx = \int \frac{\sqrt{16\sec^2\theta - 16}}{256\sec^4\theta}(4\sec\theta\tan\theta\, d\theta)$$

$$= \frac{1}{16}\int \frac{\tan^2\theta}{\sec^3\theta} d\theta$$

$$= \frac{1}{16}\int \frac{\sin^2\theta}{\cos^2\theta}\cos^3\theta\, d\theta$$

$$= \frac{1}{16}\int (\sin\theta)^2(\cos\theta\, d\theta)$$

$$= \frac{1}{48}\sin^3\theta + C.$$

The right triangle in FIGURE 7.5.5 was constructed so that $\sec\theta = x/4$ or $\cos\theta = 4/x$ and so we see that $\sin\theta = \sqrt{x^2 - 16}/x$. Because $\sin^3 x = (\sqrt{x^2 - 16})^3/x^3$, it follows that

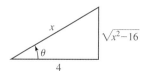

FIGURE 7.5.5 Right triangle in Example 5

$$\int \frac{\sqrt{x^2 - 16}}{x^4} dx = \frac{1}{48}\frac{(x^2 - 16)^{3/2}}{x^3} + C.$$

∎

EXAMPLE 6 A Definite Integral

Evaluate $\displaystyle\int_{-2}^{-1} \frac{\sqrt{x^2 - 1}}{x} dx$.

Solution We identify $u = x$ and $a = 1$ and use the substitutions $x = \sec\theta$ and $dx = \sec\theta\tan\theta\, d\theta$. In this case we must assume that $\pi/2 < \theta \leq \pi$ since the interval of integration indicates that $x \leq -a$, where $-a = -1$. See FIGURE 7.5.6 for a graph of the integrand $f(x) = \sqrt{x^2 - 1}/x$.

As in Example 4 we obtain the θ-limits of integration from the original x-limits of integration:

$$x = -2: \quad \theta = \sec^{-1}(-2) = \frac{2\pi}{3}$$

$$x = -1: \quad \theta = \sec^{-1}(-1) = \pi.$$

FIGURE 7.5.6 Graph of integrand in Example 6

Therefore, $\displaystyle\int_{-2}^{-1} \frac{\sqrt{x^2 - 1}}{x} dx = \int_{2\pi/3}^{\pi} \frac{\sqrt{\sec^2\theta - 1}}{\sec\theta}(\sec\theta\tan\theta\, d\theta)$

$$= \int_{2\pi/3}^{\pi} \sqrt{\tan^2\theta}(\tan\theta\, d\theta).$$

Because $\pi/2 < \theta \leq \pi$, $\tan\theta \leq 0$, $\sqrt{\tan^2\theta} = |\tan\theta| = -\tan\theta$, the last integral becomes

$$\int_{-2}^{-1} \frac{\sqrt{x^2 - 1}}{x} dx = \int_{2\pi/3}^{\pi} \sqrt{\tan^2\theta}(\tan\theta\, d\theta)$$

$$= -\int_{2\pi/3}^{\pi} \tan^2\theta\, d\theta \quad \leftarrow \text{use trig identity}$$

$$= -\int_{2\pi/3}^{\pi} (\sec^2\theta - 1)\, d\theta$$

$$= -(\tan\theta - \theta)\Big]_{2\pi/3}^{\pi}$$

$$= -\left[(0 - \pi) - \left(-\sqrt{3} - \frac{2\pi}{3}\right)\right]$$

$$= \frac{\pi}{3} - \sqrt{3} \approx -0.6849.$$

The negative answer stands to reason, since we see in Figure 7.5.6 that $f(x) \leq 0$ on the interval $[-2, -1]$.

∎

Integrands Containing a Quadratic

A trigonometric substitution can also be used when an integrand contains an integer power of the square root of a quadratic expression $ax^2 + bx + c$. By completion of the square the radical can be expressed as one of the forms:

$$\sqrt{a^2 - u^2}, \quad \sqrt{a^2 + u^2}, \quad \text{or} \quad \sqrt{u^2 - a^2}.$$

If, say, an integrand contains an integer power of

$$\sqrt{x^2 + 4x + 7} = \sqrt{3 + (x + 2)^2},$$

we would identify $u = x + 2$, $a = \sqrt{3}$, and use the substitution $x + 2 = \sqrt{3}\tan\theta$.

EXAMPLE 7 Completing the Square

Evaluate $\displaystyle\int \frac{1}{(x^2 + 8x + 25)^{3/2}}\, dx.$

Solution By completing the square in x, we recognize that the integrand contains an integral power of $a^2 + u^2$,

$$\int \frac{1}{(x^2 + 8x + 25)^{3/2}}\, dx = \int \frac{1}{[9 + (x + 4)^2]^{3/2}}\, dx,$$

where $u = x + 4$ and $a = 3$. Using the substitutions $x + 4 = 3\tan\theta$ and $dx = 3\sec^2\theta\, d\theta$ we find

$$\int \frac{dx}{(x^2 + 8x + 25)^{3/2}} = \int \frac{3\sec^2\theta\, d\theta}{[9 + 9\tan^2\theta]^{3/2}}$$

$$= \frac{1}{9}\int \frac{\sec^2\theta}{\sec^3\theta}\, d\theta$$

$$= \frac{1}{9}\int \cos\theta\, d\theta$$

$$= \frac{1}{9}\sin\theta + C.$$

FIGURE 7.5.7 Right triangle in Example 7

Inspection of the triangle in FIGURE 7.5.7 indicates how to express $\sin\theta$ in terms of x. It follows that

$$\int \frac{dx}{(x^2 + 8x + 25)^{3/2}} = \frac{1}{9}\frac{x + 4}{\sqrt{(x + 4)^2 + 9}} + C = \frac{x + 4}{9\sqrt{x^2 + 8x + 25}} + C. \quad \blacksquare$$

∫ NOTES FROM THE CLASSROOM

Integrals of the form

$$\int \frac{1}{\sqrt{u^2 + a^2}}\, du, \quad \int \frac{1}{\sqrt{u^2 - a^2}}\, du, \quad \text{and} \quad \int \frac{1}{u^2 - a^2}\, du$$

can readily be evaluated by means of trigonometric substitutions. Inspection of Table 7.1.1 of integral formulas shows that each of these integrals is a logarithm. But those with long memories might recognize these are the indefinite integral forms of the differentiation formulas for three of the inverse hyperbolic functions:

$$\frac{d}{dx}\sinh^{-1}\!\left(\frac{u}{a}\right) = \frac{1}{\sqrt{u^2 + a^2}}\frac{du}{dx}, \quad \frac{d}{dx}\cosh^{-1}\!\left(\frac{u}{a}\right) = \frac{1}{\sqrt{u^2 - a^2}}\frac{du}{dx},$$

$$\frac{d}{dx}\frac{1}{a}\tanh^{-1}\!\left(\frac{u}{a}\right) = \frac{1}{a^2 - u^2}\frac{du}{dx}.$$

Every inverse hyperbolic function can be expressed as a natural logarithm. See (25)–(27) in Section 3.10.

Exercises 7.5

Answers to selected odd-numbered problems begin on page ANS-23.

☰ Fundamentals

In Problems 1–38, evaluate the given indefinite integral by a trigonometric substitution where appropriate. You should be able to evaluate some of the integrals without a substitution.

1. $\displaystyle\int \frac{\sqrt{1 - x^2}}{x^2}\, dx$

2. $\displaystyle\int \frac{x^3}{\sqrt{x^2 - 4}}\, dx$

3. $\displaystyle\int \frac{1}{\sqrt{x^2 - 36}}\, dx$

4. $\displaystyle\int \sqrt{3 - x^2}\, dx$

5. $\displaystyle\int x\sqrt{x^2 + 7}\, dx$

6. $\displaystyle\int (1 - x^2)^{3/2}\, dx$

7. $\displaystyle\int x^3 \sqrt{1 - x^2}\, dx$

8. $\displaystyle\int x^3 \sqrt{x^2 - 1}\, dx$

9. $\displaystyle\int \frac{1}{(x^2 - 4)^{3/2}}\, dx$

10. $\displaystyle\int (9 - x^2)^{3/2}\, dx$

11. $\displaystyle\int \sqrt{x^2 + 4}\, dx$

12. $\displaystyle\int \frac{x}{25 + x^2}\, dx$

13. $\displaystyle\int \frac{1}{\sqrt{25 - x^2}}\, dx$

14. $\displaystyle\int \frac{1}{x\sqrt{x^2 - 25}}\, dx$

15. $\displaystyle\int \frac{1}{x\sqrt{16 - x^2}}\, dx$

16. $\displaystyle\int \frac{1}{x^2\sqrt{16 - x^2}}\, dx$

17. $\displaystyle\int \frac{1}{x\sqrt{1 + x^2}}\, dx$

18. $\displaystyle\int \frac{1}{x^2\sqrt{1 + x^2}}\, dx$

19. $\displaystyle\int \frac{\sqrt{1 - x^2}}{x^4}\, dx$

20. $\displaystyle\int \frac{\sqrt{x^2 - 1}}{x^4}\, dx$

21. $\displaystyle\int \frac{x^2}{(9 - x^2)^{3/2}}\, dx$

22. $\displaystyle\int \frac{x^2}{(4 + x^2)^{3/2}}\, dx$

23. $\displaystyle\int \frac{1}{(1 + x^2)^2}\, dx$

24. $\displaystyle\int \frac{x^2}{(x^2 - 1)^2}\, dx$

25. $\displaystyle\int \frac{1}{(4 + x^2)^{5/2}}\, dx$

26. $\displaystyle\int \frac{x^3}{(1 - x^2)^{5/2}}\, dx$

27. $\displaystyle\int \frac{1}{\sqrt{x^2 + 2x + 10}}\, dx$

28. $\displaystyle\int \frac{x}{\sqrt{4x - x^2}}\, dx$

29. $\displaystyle\int \frac{1}{(x^2 + 6x + 13)^2}\, dx$

30. $\displaystyle\int \frac{1}{(11 - 10x - x^2)^2}\, dx$

31. $\displaystyle\int \frac{x - 3}{(5 - 4x - x^2)^{3/2}}\, dx$

32. $\displaystyle\int \frac{1}{(x^2 + 2x)^{3/2}}\, dx$

33. $\displaystyle\int \frac{2x + 4}{x^2 + 4x + 13}\, dx$

34. $\displaystyle\int \frac{1}{4 + (x - 3)^2}\, dx$

35. $\displaystyle\int \frac{x^2}{x^2 + 16}\, dx$

36. $\displaystyle\int \frac{\sqrt{4 - 9x^2}}{x}\, dx$

37. $\displaystyle\int \sqrt{6x - x^2}\, dx$

38. $\displaystyle\int \frac{1}{\sqrt{6x - x^2}}\, dx$

In Problems 39–44, evaluate the given definite integral.

39. $\displaystyle\int_{-1}^{1} \sqrt{4 - x^2}\, dx$

40. $\displaystyle\int_{-1}^{\sqrt{3}} \frac{x^2}{\sqrt{4 - x^2}}\, dx$

41. $\displaystyle\int_{0}^{5} \frac{1}{(x^2 + 25)^{3/2}}\, dx$

42. $\displaystyle\int_{\sqrt{2}}^{2} \frac{1}{x^3\sqrt{x^2 - 1}}\, dx$

43. $\displaystyle\int_{1}^{6/5} \frac{16}{x^4\sqrt{4 - x^2}}\, dx$

44. $\displaystyle\int_{0}^{1/2} x^3(1 + x^2)^{-1/2}\, dx$

In Problems 45 and 46, use integration by parts followed by a trigonometric substitution.

45. $\displaystyle\int x^2 \sin^{-1}x\, dx$

46. $\displaystyle\int x\cos^{-1}x\, dx$

☰ Review of Applications

47. Use a calculator or CAS to obtain the graph of $y = \dfrac{1}{x\sqrt{3 + x^2}}$. Find the area under the graph on the interval $[1, \sqrt{3}]$.

48. Use a calculator or CAS to obtain the graph of $y = x^5\sqrt{1 - x^2}$. Find the area under the graph on the interval $[0, 1]$.

49. Show that the area of a circle given by $x^2 + y^2 = a^2$ is πa^2.

50. Show that the area of an ellipse given by $a^2x^2 + b^2y^2 = a^2b^2$ is πab.

51. The region described in Problem 47 is revolved about the x-axis. Find the volume of the solid of revolution.

52. The region in the first quadrant bounded by the graphs of $y = \dfrac{4}{4 + x^2}$, $x = 2$, and $y = 0$ is revolved about the x-axis. Find the volume of the solid of revolution.

53. The region in the first quadrant bounded by the graphs of $y = x\sqrt{4 + x^2}$, $x = 2$, and $y = 0$ is revolved about the y-axis. Find the volume of the solid of revolution.

54. The region in the first quadrant bounded by the graphs of $y = \dfrac{x}{\sqrt{4 - x^2}}$, $x = 1$, and $y = 0$ is revolved about the y-axis. Find the volume of the solid of revolution.

55. Find the length of the graph of $y = \ln x$ on the interval $[1, \sqrt{3}]$.

56. Find the length of the graph of $y = -\frac{1}{2}x^2 + 2x$ on the interval $[1, 2]$.

57. A woman, W, starting at the origin, moves in the direction of the positive y-axis pulling a mass along the curve C, called a **tractrix**, indicated in FIGURE 7.5.8. The mass, initially located on the x-axis at $(a, 0)$, is pulled by

a rope of constant length a that is kept taut throughout the motion.

(a) Show that the differential equation of the tractrix is
$$\frac{dy}{dx} = -\frac{\sqrt{a^2 - x^2}}{x}.$$

(b) Solve the equation in part (a). Assume that the initial point on the x-axis is $(10, 0)$ and the length of the rope is $a = 10$ ft.

FIGURE 7.5.8 Tractrix in Problem 57

58. The region bounded by the graph of $(x - a)^2 + y^2 = r^2$, $r < a$, is revolved about the y-axis. Find the volume of the solid of revolution or **torus**. See FIGURE 7.5.9.

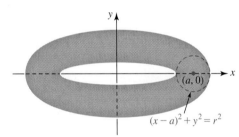

FIGURE 7.5.9 Torus in Problem 58

59. Find the fluid force on one side of the vertical plate shown in FIGURE 7.5.10. Assume that the plate is submerged in water and that dimensions are in feet.

FIGURE 7.5.10 Submerged plate in Problem 59

60. Find the centroid of the region bounded by the graphs of
$$y = \frac{1}{\sqrt{1 + x^2}}, y = 0, x = 0, \text{ and } x = \sqrt{3}.$$

≡ **Think About It**

61. Evaluate the following integrals by an appropriate trigonometric substitution.

(a) $\displaystyle\int \frac{1}{\sqrt{e^{2x} - 1}} dx$ (b) $\displaystyle\int \sqrt{e^{2x} - 1}\, dx$

62. Find the area of the crescent-shaped region shown in yellow in FIGURE 7.5.11. The region, outside the circle of radius a but inside the circle of radius b, $a \neq b$, is called a **lune**.

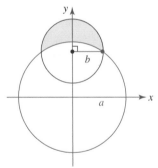

FIGURE 7.5.11 Lune in Problem 62

7.6 Partial Fractions

Introduction When two rational functions, say, $g(x) = 2/(x + 5)$ and $h(x) = 1/(x + 1)$ are added, the terms are combined by means of a common denominator:

$$g(x) + h(x) = \frac{2}{x + 5} + \frac{1}{x + 1} = \frac{2}{x + 5}\left(\frac{x + 1}{x + 1}\right) + \frac{1}{x + 1}\left(\frac{x + 5}{x + 5}\right). \quad (1)$$

Adding numerators on the right-hand side of (1) yields the single rational function

$$f(x) = \frac{3x + 7}{(x + 5)(x + 1)}. \quad (2)$$

Now suppose that we are faced with the problem of integrating the function f. Of course, the solution is obvious: We use the equality of (1) and (2) to write

$$\int \frac{3x + 7}{(x + 5)(x + 1)} dx = \int \left[\frac{2}{x + 5} + \frac{1}{x + 1}\right] dx = 2\ln|x + 5| + \ln|x + 1| + C.$$

This example illustrates a procedure for integrating certain rational functions $f(x) = p(x)/q(x)$. This method consists of reversing the process illustrated in (1), in other words, starting with a rational function, such as (2), break it down into simpler component fractions $g(x) = 2/(x+5)$ and $h(x) = 1/(x+1)$ called **partial fractions**. We then evaluate the integral term-by-term.

◾ **Partial Fractions** The algebraic process for breaking down a rational expression, such as (2), into partial fractions is known as **partial fraction decomposition**. For convenience we will assume that the rational function $f(x) = p(x)/q(x)$, $q(x) \neq 0$, is a **proper fraction** or **proper rational expression**, that is, the degree of $p(x)$ is less than the degree of $q(x)$. We will also assume that the polynomials $p(x)$ and $q(x)$ have no common factors.

In this section, we shall study four cases of partial fraction decomposition.

◾ **Distinct Linear Factors** We state the following fact from algebra without proof. If the denominator $q(x)$ contains a product of n distinct linear factors,

$$(a_1x + b_1)(a_2x + b_2) \cdots (a_nx + b_n),$$

where the a_i and b_i, $i = 1, 2, \ldots, n$ are real numbers, then unique real constants $C_1, C_2, \ldots, C_n$ can be found such that the partial fraction decomposition of $f(x) = p(x)/q(x)$ contains the sum

$$\frac{C_1}{a_1x + b_1} + \frac{C_2}{a_2x + b_2} + \cdots + \frac{C_n}{a_nx + b_n}.$$

In other words, the assumed partial fraction decomposition for f contains one partial fraction for each of the linear factors $a_ix + b_i$.

EXAMPLE 1 Distinct Linear Factors

Evaluate $\displaystyle\int \frac{2x+1}{(x-1)(x+3)}\,dx$.

Solution We make the assumption that the integrand can be written as

$$\frac{2x+1}{(x-1)(x+3)} = \frac{A}{x-1} + \frac{B}{x+3}.$$

Combining the terms of the right-hand member of the equation over a common denominator gives

$$\frac{2x+1}{(x-1)(x+3)} = \frac{A(x+3) + B(x-1)}{(x-1)(x+3)}.$$

Since the denominators are equal, the numerators of the two expressions must be identical:

$$2x + 1 = A(x+3) + B(x-1). \tag{3}$$

Since the last line is an identity, the coefficients of the powers of x are the same

$$2x + 1x^0 = (A+B)x + (3A - B)x^0$$

and therefore,

$$2 = A + B$$
$$1 = 3A - B. \tag{4}$$

By adding the two equations we get $3 = 4A$ and so we find that $A = \frac{3}{4}$. Substituting this value into either equation of (4) then yields $B = \frac{5}{4}$. Hence the desired partial fraction decomposition is

$$\frac{2x+1}{(x-1)(x+3)} = \frac{\frac{3}{4}}{x-1} + \frac{\frac{5}{4}}{x+3}.$$

Therefore,

$$\int \frac{2x+1}{(x-1)(x+3)}\,dx = \int \left[\frac{\frac{3}{4}}{x-1} + \frac{\frac{5}{4}}{x+3}\right] dx = \frac{3}{4}\ln|x-1| + \frac{5}{4}\ln|x+3| + C. \quad \blacksquare$$

▌ **A Shortcut Worth Knowing** If the denominator contains, say, three linear factors, such as in $\dfrac{4x^2 - x + 1}{(x-1)(x+3)(x-6)}$, then the partial fraction decomposition looks like this:

$$\frac{4x^2 - x + 1}{(x-1)(x+3)(x-6)} = \frac{A}{x-1} + \frac{B}{x+3} + \frac{C}{x-6}.$$

By following the same steps as in Example 1, we would find that the analogue of (4) is now three equations in the three unknowns A, B, and C. The point is this: The more linear factors in the denominator, the larger the system of equations we must solve. There is an algebraic procedure worth learning that can cut down on some of the algebra. To illustrate, let us return to the identity (3). Since the equality is true for every value of x, it holds for $x = 1$ and $x = -3$, *the zeros of the denominator*. Setting $x = 1$ in (3) gives $3 = 4A$, from which it follows immediately that $A = \frac{3}{4}$. Similarly, by setting $x = -3$ in (3), we obtain $-5 = (-4)B$ or $B = \frac{5}{4}$.

See the *Notes from the Classroom* at the end of this section for another quick method for determining the constants.

EXAMPLE 2 Area Under a Graph

Find the area A under the graph of $f(x) = \dfrac{1}{x(x+1)}$ on the interval $\left[\frac{1}{2}, 2\right]$.

Solution The area in question is shown in FIGURE 7.6.1. Since $f(x)$ is positive for all x in the interval, the area is the definite integral

$$A = \int_{1/2}^{2} \frac{1}{x(x+1)}\,dx.$$

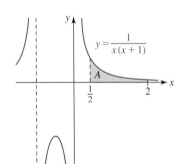

FIGURE 7.6.1 Area under graph in Example 2

Using partial fractions

$$\frac{1}{x(x+1)} = \frac{A}{x} + \frac{B}{x+1} = \frac{A(x+1) + Bx}{x(x+1)}$$

it follows that

$$1 = A(x+1) + Bx. \qquad (5)$$

Following the shortcut discussed prior to this example, we set, in turn, $x = 0$ and $x = -1$ in (5) and obtain $A = 1$ and $B = -1$. Therefore,

$$A = \int_{1/2}^{2} \left[\frac{1}{x} - \frac{1}{x+1}\right] dx = \left(\ln|x| - \ln|x+1|\right)\Big]_{1/2}^{2}$$

$$= \ln\left|\frac{x}{x+1}\right|\,\Big]_{1/2}^{2} = \ln 2 \approx 0.6931. \quad \blacksquare$$

▌ **Repeated Linear Factors** If the denominator of the rational function $f(x) = p(x)/q(x)$ contains a repeated linear factor $(ax + b)^n$, $n > 1$, a and b real numbers, then unique real constants $C_1, C_2, \ldots, C_n$ can be found such that the partial fraction decomposition of f contains the sum

$$\frac{C_1}{ax+b} + \frac{C_2}{(ax+b)^2} + \cdots + \frac{C_n}{(ax+b)^n}.$$

In other words, the assumed partial fraction decomposition for f contains a partial fraction for each power of $ax + b$.

EXAMPLE 3 Repeated Linear Factor

Evaluate $\displaystyle\int \frac{x^2 + 2x + 4}{(x+1)^3}\,dx$.

Solution The decomposition of the integrand contains a partial fraction for each of the three powers of $x + 1$:

$$\frac{x^2 + 2x + 4}{(x + 1)^3} = \frac{A}{x + 1} + \frac{B}{(x + 1)^2} + \frac{C}{(x + 1)^3}.$$

Equating numerators gives

$$x^2 + 2x + 4 = A(x + 1)^2 + B(x + 1) + C = Ax^2 + (2A + B)x + (A + B + C). \quad (6)$$

Note that setting $x = -1$ (the single zero of the denominator) in (6) yields only $C = 3$. But the coefficients of x^2 and x in (6) yield the system of equations

$$1 = A$$
$$2 = 2A + B.$$

From these equations we see that $A = 1$ and $B = 0$. Therefore,

$$\int \frac{x^2 + 2x + 4}{(x + 1)^3} \, dx = \int \left[\frac{1}{x + 1} + \frac{3}{(x + 1)^3} \right] dx$$

$$= \int \left[\frac{1}{x + 1} + 3(x + 1)^{-3} \right] dx$$

$$= \ln|x + 1| - \frac{3}{2}(x + 1)^{-2} + D. \quad \blacksquare$$

When the denominator $q(x)$ contains distinct as well as repeated linear factors, we combine the two cases that we have just considered.

EXAMPLE 4 Repeated Factor and a Distinct Factor

Evaluate $\int \frac{6x - 1}{x^3(2x - 1)} \, dx$.

Solution Since x is a repeated linear factor in the denominator of the integrand, the assumed partial fraction decomposition contains a partial fraction for each of the three powers of x and one partial fraction for the distinct linear factor $2x - 1$:

$$\frac{6x - 1}{x^3(2x - 1)} = \frac{A}{x} + \frac{B}{x^2} + \frac{C}{x^3} + \frac{D}{2x - 1}.$$

After putting the right-hand side over a common denominator we equate numerators:

$$6x - 1 = Ax^2(2x - 1) + Bx(2x - 1) + C(2x - 1) + Dx^3 \quad (7)$$
$$= (2A + D)x^3 + (-A + 2B)x^2 + (-B + 2C)x - C. \quad (8)$$

If we set $x = 0$ and $x = \frac{1}{2}$ in (7), we find $C = 1$ and $D = 16$, respectively. Now, by equating the coefficients of x^3 and x^2 in (8), we get

$$0 = 2A + D$$
$$0 = -A + 2B.$$

Since we know the value of D, the first equation yields $A = -\frac{1}{2}D = -8$. The second then gives $B = \frac{1}{2}A = -4$. Therefore,

$$\int \frac{6x - 1}{x^3(2x - 1)} \, dx = \int \left[-\frac{8}{x} - \frac{4}{x^2} + \frac{1}{x^3} + \frac{16}{2x - 1} \right] dx$$

$$= -8 \ln|x| + 4x^{-1} - \frac{1}{2}x^{-2} + 8 \ln|2x - 1| + E$$

$$= 8 \ln \left| \frac{2x - 1}{x} \right| + 4x^{-1} - \frac{1}{2}x^{-2} + E. \quad \blacksquare$$

The word *irreducible* means that the quadratic expression does not factor over the set of real numbers.

▶ ∎ **Distinct Quadratic Factors** If the denominator of the rational function $f(x) = p(x)/q(x)$ contains a product of n distinct *irreducible* quadratic factors

$$(a_1x^2 + b_1x + c_1)(a_2x^2 + b_2x + c_2)\cdots(a_nx^2 + b_nx + c_n),$$

where the coefficients a_i, b_i, and c_i, $i = 1, 2, \ldots, n$ are real numbers, then unique real constants $A_1, A_2, \ldots, A_n, B_1, B_2, \ldots, B_n$ can be found such that the partial fraction decomposition for f contains the sum

$$\frac{A_1x + B_1}{a_1x^2 + b_1x + c_1} + \frac{A_2x + B_2}{a_2x^2 + b_2x + c_2} + \cdots + \frac{A_nx + B_n}{a_nx^2 + b_nx + c_n}.$$

Analogous to the case where $q(x)$ contains a product of distinct linear factors, the assumed partial fraction decomposition for f contains one partial fraction for each of the quadratic factors $a_ix^2 + b_ix + c_i$.

EXAMPLE 5 Repeated Linear and a Distinct Quadratic

Evaluate $\displaystyle\int \frac{x + 3}{x^4 + 9x^2}\,dx$.

Solution From $x^4 + 9x^2 = x^2(x^2 + 9)$, we see that the problem combines the irreducible quadratic factor $x^2 + 9$ with the repeated linear factor x. Accordingly, the partial fraction decomposition is

$$\frac{x + 3}{x^2(x^2 + 9)} = \frac{A}{x} + \frac{B}{x^2} + \frac{Cx + D}{x^2 + 9}.$$

Proceeding as usual, we find

$$x + 3 = Ax(x^2 + 9) + B(x^2 + 9) + (Cx + D)x^2 \tag{9}$$

$$= (A + C)x^3 + (B + D)x^2 + 9Ax + 9B. \tag{10}$$

Setting $x = 0$ in (9) yields $B = \frac{1}{3}$. Then (10) gives

$$0 = A + C$$
$$0 = B + D$$
$$1 = 9A.$$

From this system we get $A = \frac{1}{9}$, $C = -\frac{1}{9}$, and $D = -\frac{1}{3}$. This gives

$$\int \frac{x + 3}{x^2(x^2 + 9)}\,dx = \int \left[\frac{\frac{1}{9}}{x} + \frac{\frac{1}{3}}{x^2} + \frac{-\frac{1}{9}x - \frac{1}{3}}{x^2 + 9}\right]dx$$

$$= \int \left[\frac{\frac{1}{9}}{x} + \frac{\frac{1}{3}}{x^2} - \frac{1}{18}\frac{2x}{x^2 + 9} - \frac{1}{3}\frac{1}{x^2 + 9}\right]dx$$

$$= \frac{1}{9}\ln|x| - \frac{1}{3}x^{-1} - \frac{1}{18}\ln(x^2 + 9) - \frac{1}{9}\tan^{-1}\frac{x}{3} + E$$

$$= \frac{1}{18}\ln\frac{x^2}{x^2 + 9} - \frac{1}{3}x^{-1} - \frac{1}{9}\tan^{-1}\frac{x}{3} + E. \quad\blacksquare$$

EXAMPLE 6 Distinct Quadratic Factors

Evaluate $\displaystyle\int \frac{4x}{(x^2 + 1)(x^2 + 2x + 3)}\,dx$.

Solution Since each quadratic factor in the denominator of the integrand is irreducible, we write

$$\frac{4x}{(x^2 + 1)(x^2 + 2x + 3)} = \frac{Ax + B}{x^2 + 1} + \frac{Cx + D}{x^2 + 2x + 3}$$

from which we find

$$4x = (Ax + B)(x^2 + 2x + 3) + (Cx + D)(x^2 + 1)$$
$$= (A + C)x^3 + (2A + B + D)x^2 + (3A + 2B + C)x + (3B + D).$$

Since the denominator of the integrand has no real zeros, we compare coefficients of powers of x:

$$0 = A + C$$
$$0 = 2A + B + D$$
$$4 = 3A + 2B + C$$
$$0 = 3B + D.$$

Solving the equations yields $A = 1$, $B = 1$, $C = -1$, and $D = -3$. Therefore,

$$\int \frac{4x}{(x^2 + 1)(x^2 + 2x + 3)}\, dx = \int \left[\frac{x + 1}{x^2 + 1} - \frac{x + 3}{x^2 + 2x + 3}\right] dx.$$

Now, the integral of each term still presents a slight challenge. For the first term in the integrand, we use termwise division to write

$$\frac{x + 1}{x^2 + 1} = \frac{1}{2}\frac{2x}{x^2 + 1} + \frac{1}{x^2 + 1}, \tag{11}$$

and then in the second term we complete the square:

$$\frac{x + 3}{x^2 + 2x + 3} = \frac{x + 1 + 2}{(x + 1)^2 + 2} = \frac{1}{2}\frac{2(x + 1)}{(x + 1)^2 + 2} + \frac{2}{(x + 1)^2 + 2}. \tag{12}$$

In the right-hand members of (11) and (12), we recognize that the integrals of the first and second terms are, respectively, of the forms $\int du/u$ and $\int du/(u^2 + a^2)$. Finally, we obtain

$$\int \frac{4x}{(x^2 + 1)(x^2 + 2x + 3)}\, dx$$

$$= \int \left[\frac{1}{2}\frac{2x}{x^2 + 1} + \frac{1}{x^2 + 1} - \frac{1}{2}\frac{2(x + 1)}{(x + 1)^2 + 2} - \frac{2}{(x + 1)^2 + (\sqrt{2})^2}\right] dx$$

$$= \frac{1}{2}\ln(x^2 + 1) + \tan^{-1}x - \frac{1}{2}\ln[(x + 1)^2 + 2] - \sqrt{2}\tan^{-1}\left(\frac{x + 1}{\sqrt{2}}\right) + E$$

$$= \frac{1}{2}\ln\left(\frac{x^2 + 1}{x^2 + 2x + 3}\right) + \tan^{-1}x - \sqrt{2}\tan^{-1}\left(\frac{x + 1}{\sqrt{2}}\right) + E. \qquad \blacksquare$$

■ **Repeated Quadratic Factors** If the denominator of the rational function $f(x) = p(x)/q(x)$ contains a repeated irreducible quadratic factor $(ax^2 + bx + c)^n$, $n > 1$, where a, b, and c are real numbers, then unique real constants $A_1, A_2, \ldots, A_n, B_1, B_2, \ldots, B_n$ can be found so that the partial fraction decomposition of f contains the sum

$$\frac{A_1x + B_1}{ax^2 + bx + c} + \frac{A_2x + B_2}{(ax^2 + bx + c)^2} + \cdots + \frac{A_nx + B_n}{(ax^2 + bx + c)^n}.$$

That is, the assumed partial fraction decomposition for f contains a partial fraction for each power of $ax^2 + bx + c$.

EXAMPLE 7 Repeated Quadratic Factor

Evaluate $\int \frac{x^2}{(x^2 + 4)^2}\, dx.$

Solution The partial fraction decomposition of the integrand

$$\frac{x^2}{(x^2 + 4)^2} = \frac{Ax + B}{x^2 + 4} + \frac{Cx + D}{(x^2 + 4)^2}$$

leads to
$$x^2 = (Ax + B)(x^2 + 4) + Cx + D$$
$$= Ax^3 + Bx^2 + (4A + C)x + (4B + D)$$

and
$$0 = A$$
$$1 = B$$
$$0 = 4A + C$$
$$0 = 4B + D.$$

From this system we find $A = 0$, $B = 1$, $C = 0$, and $D = -4$. Consequently,

$$\int \frac{x^2}{(x^2 + 4)^2}\, dx = \int \left[\frac{1}{x^2 + 4} - \frac{4}{(x^2 + 4)^2}\right] dx.$$

The integral of the first term is an inverse tangent. However, to evaluate the integral of the second term, we employ the trigonometric substitution $x = 2\tan\theta$:

$$\int \frac{1}{(x^2 + 4)^2}\, dx = \int \frac{2\sec^2\theta\, d\theta}{(4\tan^2\theta + 4)^2}$$
$$= \frac{1}{8}\int \frac{\sec^2\theta}{\sec^4\theta}\, d\theta = \frac{1}{8}\int \cos^2\theta\, d\theta$$
$$= \frac{1}{16}\int (1 + \cos 2\theta)\, d\theta = \frac{1}{16}\left(\theta + \frac{1}{2}\sin 2\theta\right) \quad \leftarrow \text{use double-angle formula here}$$
$$= \frac{1}{16}(\theta + \sin\theta\cos\theta)$$
$$= \frac{1}{16}\left[\tan^{-1}\frac{x}{2} + \frac{x}{\sqrt{x^2 + 4}} \cdot \frac{2}{\sqrt{x^2 + 4}}\right]$$
$$= \frac{1}{16}\left[\tan^{-1}\frac{x}{2} + \frac{2x}{x^2 + 4}\right].$$

Therefore, the original integral is

$$\int \frac{x^2}{(x^2 + 4)^2}\, dx = \frac{1}{2}\tan^{-1}\frac{x}{2} - 4\left[\frac{1}{16}\tan^{-1}\frac{x}{2} + \frac{1}{8}\frac{x}{x^2 + 4}\right] + E$$
$$= \frac{1}{4}\tan^{-1}\frac{x}{2} - \frac{1}{2}\frac{x}{x^2 + 4} + E.$$ ∎

Review Section 5.1.

▶ ▮ **Improper Fractions** In each of the preceding examples the integrand $f(x) = p(x)/q(x)$ was a proper fraction. Recall, when $f(x) = p(x)/q(x)$ is an **improper fraction**, that is, the degree of $p(x)$ is greater than or equal to the degree of $q(x)$, we start with long division.

EXAMPLE 8 Integrand an Improper Fraction

Evaluate $\int \dfrac{x^3 - 2x}{x^2 + 3x + 2}\, dx$.

Solution The integrand is recognized as an improper fraction and we divide the numerator by the denominator:

$$\frac{x^3 - 2x}{x^2 + 3x + 2} = x - 3 + \frac{5x + 6}{x^2 + 3x + 2}.$$

Now since the denominator factors as $x^2 + 3x + 2 = (x + 1)(x + 2)$, we decompose the remainder over the divisor into partial fractions:

See the *SRM* for the command syntax for doing partial fraction decomposition in *Mathematica* and *Maple*.

$$\frac{5x + 6}{x^2 + 3x + 2} = \frac{1}{x + 1} + \frac{4}{x + 2}.$$

With this information, evaluation of the integral is immediate:

$$\int \frac{x^3 - 2x}{x^2 + 3x + 2}\, dx = \int \left[x - 3 + \frac{1}{x + 1} + \frac{4}{x + 2}\right] dx$$
$$= \frac{1}{2}x^2 - 3x + \ln|x + 1| + 4\ln|x + 2| + C.$$ ∎

∫ NOTES FROM THE CLASSROOM

There is another way, called the **cover-up method**, of determining the coefficients in a partial fraction decomposition in the special case when the denominator of the integrand is the product of distinct linear factors:

$$f(x) = \frac{p(x)}{(x - r_1)(x - r_2) \cdots (x - r_n)}.$$

Let us illustrate by means of a specific example. From the foregoing discussion we know there exists unique constants A, B, and C such that

$$\frac{x^2 + 4x - 1}{(x - 1)(x - 2)(x + 3)} = \frac{A}{x - 1} + \frac{B}{x - 2} + \frac{C}{x + 3}. \quad (13)$$

Suppose we multiply both sides of this last expression by $x - 1$, simplify, and then set $x = 1$. Since the coefficients of B and C are zero, we get

$$\left. \frac{x^2 + 4x - 1}{(x - 2)(x + 3)} \right|_{x=1} = A \quad \text{or} \quad A = -1.$$

Written another way,

$$\left. \frac{x^2 + 4x - 1}{\boxed{(x - 1)}(x - 2)(x + 3)} \right|_{x=1} = A$$

where we have shaded or covered up the factor that canceled when the left side of (13) was multiplied by $x - 1$. We *do not evaluate this covered-up factor* at $x = 1$. Now to obtain B and C we simply evaluate the left-hand side of (13) while covering up, in turn, $x - 2$ and $x + 3$:

$$\left. \frac{x^2 + 4x - 1}{(x - 1)\boxed{(x - 2)}(x + 3)} \right|_{x=2} = B \quad \text{or} \quad B = \frac{11}{5}$$

$$\left. \frac{x^2 + 4x - 1}{(x - 1)(x - 2)\boxed{(x + 3)}} \right|_{x=3} = C \quad \text{or} \quad C = -\frac{1}{5}.$$

Thus, we obtain the decomposition

$$\frac{x^2 + 4x - 1}{(x - 1)(x - 2)(x + 3)} = \frac{-1}{x - 1} + \frac{\frac{11}{15}}{x - 2} + \frac{-\frac{1}{5}}{x + 3}.$$

Exercises 7.6
Answers to selected odd-numbered problems begin on page ANS-23.

≡ Fundamentals

In Problems 1–8, write out the appropriate form of the partial fraction decomposition of the given expression. Do not evaluate the coefficients.

1. $\dfrac{x - 1}{x^2 + x}$

2. $\dfrac{9x - 8}{(x - 3)(2x - 5)}$

3. $\dfrac{x^3}{(x - 1)(x + 2)^3}$

4. $\dfrac{2x^2 - 3}{x^3 + 6x^2}$

5. $\dfrac{4}{x^3(x^2 + 3)}$

6. $\dfrac{-x^2 + 3x + 7}{(x + 2)^2(x^2 + x + 1)}$

7. $\dfrac{2x^3 - x}{(x^2 + 9)^2}$

8. $\dfrac{3x^2 - x + 4}{x^4 + 2x^3 + x}$

In Problems 9–42, use partial fractions to evaluate the given integral.

9. $\displaystyle\int \dfrac{1}{x(x - 2)}\,dx$

10. $\displaystyle\int \dfrac{1}{x(2x + 3)}\,dx$

11. $\displaystyle\int \dfrac{x + 2}{2x^2 - x}\,dx$

12. $\displaystyle\int \dfrac{3x + 10}{x^2 + 2x}\,dx$

13. $\displaystyle\int \dfrac{x + 1}{x^2 - 16}\,dx$

14. $\displaystyle\int \dfrac{1}{4x^2 - 25}\,dx$

15. $\displaystyle\int \dfrac{x}{2x^2 + 5x + 2}\,dx$

16. $\displaystyle\int \dfrac{x + 5}{(x + 4)(x^2 - 1)}\,dx$

17. $\displaystyle\int \dfrac{x^2 + 2x - 6}{x^3 - x}\,dx$

18. $\displaystyle\int \dfrac{5x^2 - x + 1}{x^3 - 4x}\,dx$

19. $\displaystyle\int \frac{1}{(x+1)(x+2)(x+3)}\,dx$
20. $\displaystyle\int \frac{1}{(4x^2-1)(x+7)}\,dx$

21. $\displaystyle\int \frac{4t^2+3t-1}{t^3-t^2}\,dt$
22. $\displaystyle\int \frac{2x-11}{x^3+2x^2}\,dx$

23. $\displaystyle\int \frac{1}{x^3+2x^2+x}\,dx$
24. $\displaystyle\int \frac{t-1}{t^4+6t^3+9t^2}\,dt$

25. $\displaystyle\int \frac{2x-1}{(x+1)^3}\,dx$
26. $\displaystyle\int \frac{1}{x^2(x^2-4)^2}\,dx$

27. $\displaystyle\int \frac{1}{(x^2+6x+5)^2}\,dx$

28. $\displaystyle\int \frac{1}{(x^2-x-6)(x^2-2x-8)}\,dx$

29. $\displaystyle\int \frac{x^4+2x^2-x+9}{x^5+2x^4}\,dx$
30. $\displaystyle\int \frac{5x-1}{x(x-3)^2(x+2)^2}\,dx$

31. $\displaystyle\int \frac{x-1}{x(x^2+1)}\,dx$
32. $\displaystyle\int \frac{1}{(x-1)(x^2+3)}\,dx$

33. $\displaystyle\int \frac{x}{(x+1)^2(x^2+1)}\,dx$
34. $\displaystyle\int \frac{x^2}{(x-1)^3(x^2+4)}\,dx$

35. $\displaystyle\int \frac{1}{x^4+5x^2+4}\,dx$
36. $\displaystyle\int \frac{1}{x^4+13x^2+36}\,dx$

37. $\displaystyle\int \frac{1}{x^3-1}\,dx$
38. $\displaystyle\int \frac{81}{x^4+27x}\,dx$

39. $\displaystyle\int \frac{3x^2-x+1}{(x+1)(x^2+2x+2)}\,dx$

40. $\displaystyle\int \frac{4x+12}{(x-2)(x^2+4x+8)}\,dx$

41. $\displaystyle\int \frac{x^2-x+4}{(x^2+4)^2}\,dx$
42. $\displaystyle\int \frac{1}{x^3(x^2+1)^2}\,dx$

In Problems 43 and 44, proceed as in Example 7 to evaluate the given integral.

43. $\displaystyle\int \frac{x^3-2x^2+x-3}{x^4+8x^2+16}\,dx$
44. $\displaystyle\int \frac{x^2}{(x^2+3)^2}\,dx$

In Problems 45 and 46, proceed as in Example 8 to evaluate the given integral.

45. $\displaystyle\int \frac{x^4+3x^2+4}{(x+1)^2}\,dx$
46. $\displaystyle\int \frac{x^5-10x^3}{x^4-10x^2+9}\,dx$

In Problems 47–54, evaluate the given definite integral.

47. $\displaystyle\int_2^4 \frac{1}{x^2-6x+5}\,dx$
48. $\displaystyle\int_0^1 \frac{1}{x^2-4}\,dx$

49. $\displaystyle\int_0^2 \frac{2x-1}{(x+3)^2}\,dx$
50. $\displaystyle\int_1^5 \frac{2x+6}{x(x+1)^2}\,dx$

51. $\displaystyle\int_0^1 \frac{1}{x^3+x^2+2x+2}\,dx$
52. $\displaystyle\int_0^1 \frac{x^2}{x^4+8x^2+16}\,dx$

53. $\displaystyle\int_{-1}^1 \frac{2x^3+5x}{x^4+5x^2+6}\,dx$
54. $\displaystyle\int_1^2 \frac{1}{x^5+4x^4+5x^3}\,dx$

In Problems 55–58, evaluate the given integral by first using the indicated substitution followed by partial fractions.

55. $\displaystyle\int \frac{\sqrt{1-x^2}}{x^3}\,dx;\quad u^2=1-x^2$

56. $\displaystyle\int \sqrt{\frac{x-1}{x+1}}\,dx;\quad u^2=\frac{x-1}{x+1}$

57. $\displaystyle\int \frac{\sqrt[3]{x+1}}{x}\,dx;\quad u^3=x+1$

58. $\displaystyle\int \frac{1}{\sqrt{x}(1+\sqrt[3]{x})^2}\,dx;\quad u^6=x$

≡ Review of Applications

In Problems 59 and 60, find the area under the graph of the given function on the indicated interval. If necessary, use a calculator or CAS to obtain the graph of the function.

59. $y=\dfrac{1}{x^2+2x-3};\quad [2,4]$

60. $y=\dfrac{x^3}{(x^2+1)(x^2+2)};\quad [0,4]$

In Problems 61 and 62, find the area bounded by the graph of the given function and the x-axis on the indicated interval. If necessary, use a calculator or CAS to obtain the graph of the function.

61. $y=\dfrac{x}{(x+2)(x+3)};\quad [-1,1]$

62. $y=\dfrac{3x^3}{x^3-8};\quad [-2,1]$

In Problems 63–66, find the volume of the solid of revolution that is formed by revolving the region bounded in the first quadrant by the graphs of the given equations about the indicated axis. If necessary, use a calculator or CAS to obtain the graph of the given function.

63. $y=\dfrac{2}{x(x+1)},\ x=1,\ x=3,\ y=0;\quad x$-axis

64. $y=\dfrac{1}{\sqrt{(x+1)(x+4)}},\ x=0,\ x=2,\ y=0;\quad x$-axis

65. $y=\dfrac{4}{(x+1)^2},\ x=0,\ x=1,\ y=0;\quad y$-axis

66. $y=\dfrac{8}{(x^2+1)(x^2+4)},\ x=0,\ x=1,\ y=0;\quad y$-axis

≡ Think About It

In Problems 67–70, evaluate the given integral by first making a substitution followed by partial fraction decomposition.

67. $\displaystyle\int \frac{\cos x}{\sin^2 x+3\sin x+2}\,dx$
68. $\displaystyle\int \frac{\sin x}{\cos^2 x-\cos^3 x}\,dx$

69. $\displaystyle\int \frac{e^t}{(e^t+1)^2(e^t-2)}\,dt$
70. $\displaystyle\int \frac{e^{2t}}{(e^t+1)^3}\,dt$

71. Find the length of the graph of $y = e^x$ on the interval $[0, \ln 2]$. [*Hint*: Evaluate the integral by starting with a substitution.]

72. Explain why partial fraction decomposition would be either unnecessary or inappropriate for the given integral. Discuss how these integrals can be evaluated.

(a) $\displaystyle\int \frac{x^3}{(x^2-1)(x^2+1)}\,dx$ (b) $\displaystyle\int \frac{3x+4}{x^2+4}\,dx$

(c) $\displaystyle\int \frac{x}{(x^2+5)^2}\,dx$ (d) $\displaystyle\int \frac{2x^3+5x}{x^4+5x^2+6}\,dx$

73. Although partial fraction decomposition could be used to evaluate

$$\int \frac{x^5}{(x-1)^{10}(x+1)^{10}}\,dx$$

it would require solving 20 equations in 20 unknowns. Evaluate the integral using an easier technique of integration.

74. Why could the answer to Problem 53 be obtained with *no work* whatsoever?

7.7 Improper Integrals

■ **Introduction** Up to now in our study of the definite integral $\int_a^b f(x)\,dx$, it was understood that

- the limits of integration were finite numbers, and
- the function f either was *continuous* on $[a, b]$ or, if discontinuous, was *bounded* on the interval.

When either of these two conditions is dropped, the resulting integral is said to be an **improper integral**. In the discussion that follows, we first consider integrals of functions that are defined and continuous on unbounded intervals, in other words,

- at least one of the limits of integration is ∞ or $-\infty$.

After that, we examine integrals over bounded intervals of functions that become unbounded on an interval. In the latter type of improper integral,

- the integrand f has an *infinite discontinuity* at some number in the interval of integration.

■ **Improper Integrals—Unbounded Intervals** If the integrand f is defined on an unbounded interval, then there are three possible **improper integrals** with infinite limits of integration. Their definitions are summarized as follows:

Definition 7.7.1 Unbounded Intervals

(*i*) If f is continuous on $[a, \infty)$, then

$$\int_a^\infty f(x)\,dx = \lim_{b\to\infty} \int_a^b f(x)\,dx. \qquad (1)$$

(*ii*) If f is continuous on $(-\infty, b]$, then

$$\int_{-\infty}^b f(x)\,dx = \lim_{a\to -\infty} \int_a^b f(x)\,dx. \qquad (2)$$

(*iii*) If f is continuous on $(-\infty, \infty)$, then

$$\int_{-\infty}^\infty f(x)\,dx = \int_{-\infty}^c f(x)\,dx + \int_c^\infty f(x)\,dx. \qquad (3)$$

When the limits in (1) and (2) exist, the integrals are said to **converge**. If the limit fails to exist, the integral is said to **diverge**. In (3) the integral $\int_{-\infty}^\infty f(x)\,dx$ converges provided *both* $\int_{-\infty}^c f(x)\,dx$ and $\int_c^\infty f(x)\,dx$ converge. If either $\int_{-\infty}^c f(x)\,dx$ or $\int_c^\infty f(x)\,dx$ diverges, then the improper integral $\int_{-\infty}^\infty f(x)\,dx$ diverges.

EXAMPLE 1 Using (1)

Evaluate $\int_2^\infty \frac{1}{x^3}\,dx$.

Solution By (1),

$$\int_2^\infty \frac{1}{x^3}\,dx = \lim_{b\to\infty}\int_2^b x^{-3}\,dx = \lim_{b\to\infty}\frac{x^{-2}}{-2}\bigg]_2^b = -\frac{1}{2}\lim_{b\to\infty}(b^{-2}-2^{-2}).$$

Since the limit $\lim_{b\to\infty} b^{-2} = \lim_{b\to\infty}(1/b^2)$ exists,

$$\lim_{b\to\infty}(b^{-2}-2^{-2}) = \lim_{b\to\infty}\left(\frac{1}{b^2}-\frac{1}{4}\right) = 0 - \frac{1}{4} = -\frac{1}{4},$$

the given integral converges, and

$$\int_2^\infty \frac{1}{x^3}\,dx = -\frac{1}{2}\left(-\frac{1}{4}\right) = \frac{1}{8}.$$

EXAMPLE 2 Using (1)

Evaluate $\int_1^\infty x^2\,dx$.

Solution By (1),

$$\int_1^\infty x^2\,dx = \lim_{b\to\infty}\int_1^b x^2\,dx = \lim_{b\to\infty}\frac{1}{3}x^3\bigg]_1^b = \lim_{b\to\infty}\left(\frac{1}{3}b^3 - \frac{1}{3}\right).$$

Since $\lim_{b\to\infty}\left(\frac{1}{3}b^3 - \frac{1}{3}\right) = \infty$ we conclude that the integral diverges.

EXAMPLE 3 Using (3)

Evaluate $\int_{-\infty}^\infty x^2\,dx$.

Solution Since c can be chosen arbitrarily in (3), we pick $c = 1$ and write

$$\int_{-\infty}^\infty x^2\,dx = \int_{-\infty}^1 x^2\,dx + \int_1^\infty x^2\,dx.$$

But, in Example 2 we saw that $\int_1^\infty x^2\,dx$ diverges. This is sufficient to conclude that $\int_{-\infty}^\infty x^2\,dx$ also diverges.

■ **Area** If $f(x) \geq 0$ for all x over $[a, \infty)$, $(-\infty, b]$, or $(-\infty, \infty)$, then each of the integrals in (1), (2), and (3) can be interpreted as area under the graph of f on the interval whenever the integral converges.

EXAMPLE 4 Area

Evaluate $\int_{-1}^\infty e^{-x}\,dx$. Interpret geometrically.

Solution By (1),

$$\int_{-1}^\infty e^{-x}\,dx = \lim_{b\to\infty}\int_{-1}^b e^{-x}\,dx = \lim_{b\to\infty}(-e^{-x})\bigg]_{-1}^b = \lim_{b\to\infty}(e - e^{-b}).$$

Since $\lim_{b\to\infty} e^{-b} = 0$, $\lim_{b\to\infty}(e - e^{-b}) = e$ and so the given integral converges to e. In FIGURE 7.7.1(a) we see that the area under the graph of the positive function $f(x) = e^{-x}$ on the interval $[-1, b]$ is $e - e^{-b}$. But, by taking $b \to \infty$, $e^{-b} \to 0$, and hence, as shown in Figure 7.7.1(b), we can interpret $\int_{-1}^\infty e^{-x}\,dx = e$ as a measure of the area under the graph of f on $[-1, \infty)$.

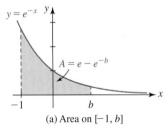

(a) Area on $[-1, b]$

(b) Area on $[-1, \infty)$

FIGURE 7.7.1 Area under the graph in Example 4

EXAMPLE 5 Using (2)

Evaluate $\int_{-\infty}^{0} \cos x \, dx$.

Solution By (2),

$$\int_{-\infty}^{0} \cos x \, dx = \lim_{a \to -\infty} \int_{a}^{0} \cos x \, dx = \lim_{a \to -\infty} \sin x \Big]_{a}^{0} = \lim_{a \to -\infty} (-\sin a).$$

Since $\sin a$ oscillates between -1 and 1, we conclude that $\lim_{a \to -\infty} (-\sin a)$ does not exist. Hence, $\int_{-\infty}^{0} \cos x \, dx$ diverges. ∎

EXAMPLE 6 Using (3)

Evaluate $\int_{-\infty}^{\infty} \dfrac{e^x}{e^x + 1} \, dx$.

Solution Choosing $c = 0$, we can write

$$\int_{-\infty}^{\infty} \frac{e^x}{e^x + 1} \, dx = \int_{-\infty}^{0} \frac{e^x}{e^x + 1} \, dx + \int_{0}^{\infty} \frac{e^x}{e^x + 1} \, dx = I_1 + I_2.$$

First, let us examine I_1:

$$I_1 = \lim_{a \to -\infty} \int_{a}^{0} \frac{e^x}{e^x + 1} \, dx = \lim_{a \to -\infty} \ln(e^x + 1) \Big]_{a}^{0} = \lim_{a \to -\infty} [\ln 2 - \ln(e^a + 1)].$$

Now $e^a + 1 \to 1$ since $e^a \to 0$ as $a \to -\infty$. Therefore, $\ln(e^a + 1) \to \ln 1 = 0$ as $a \to -\infty$. Hence, $I_1 = \ln 2$.

Second, we have

$$I_2 = \lim_{b \to \infty} \int_{0}^{b} \frac{e^x}{e^x + 1} \, dx = \lim_{b \to \infty} \ln(e^x + 1) \Big]_{0}^{b} = \lim_{b \to \infty} [\ln(e^b + 1) - \ln 2].$$

However, $e^b + 1 \to \infty$ as $b \to \infty$, so $\ln(e^b + 1) \to \infty$. Hence, I_2 diverges.

Because *both* I_1 and I_2 do not converge, it follows that the given integral is divergent. ∎

EXAMPLE 7 Using (3)

The improper integral $\int_{-\infty}^{\infty} \dfrac{1}{1 + x^2} \, dx$ converges because

$$\int_{-\infty}^{\infty} \frac{1}{1 + x^2} \, dx = \int_{-\infty}^{0} \frac{1}{1 + x^2} \, dx + \int_{0}^{\infty} \frac{1}{1 + x^2} \, dx = \frac{\pi}{2} + \frac{\pi}{2} = \pi.$$

The result follows from the facts that

$$\int_{-\infty}^{0} \frac{1}{1 + x^2} \, dx = \lim_{a \to -\infty} \int_{a}^{0} \frac{1}{1 + x^2} \, dx = -\lim_{a \to -\infty} \tan^{-1} a = -\left(-\frac{\pi}{2}\right) = \frac{\pi}{2}$$

$$\int_{0}^{\infty} \frac{1}{1 + x^2} \, dx = \lim_{b \to \infty} \int_{0}^{b} \frac{1}{1 + x^2} \, dx = \lim_{b \to \infty} \tan^{-1} b = \frac{\pi}{2}. \quad ∎$$

EXAMPLE 8 Work

In (5) of Section 6.8, we saw that the work done in lifting a mass m_2 off the surface of a planet of mass m_1 to a height h is given by

$$W = \int_{R}^{R+h} \frac{k m_1 m_2}{r^2} \, dr,$$

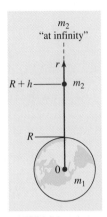

FIGURE 7.7.2 Mass m_2 lifted to "infinity" in Example 8

where R is the radius of the planet. Hence, the amount of work done in lifting m_2 to an unlimited or "infinite distance" from the surface of the planet is

$$W = \int_R^\infty \frac{km_1 m_2}{r^2} dr$$
$$= km_1 m_2 \lim_{b \to \infty} \int_R^b r^{-2} dr$$
$$= km_1 m_2 \lim_{b \to \infty} \left[-\frac{1}{b} + \frac{1}{R} \right]$$
$$= \frac{km_1 m_2}{R}.$$

See FIGURE 7.7.2. From the data in Example 2 of Section 6.8, it follows that the work done in lifting a payload of 5000 kg to an "infinite distance" from the surface of the Earth is

$$W = \frac{(6.67 \times 10^{-11})(6.0 \times 10^{24})(5000)}{6.4 \times 10^6} \approx 3.13 \times 10^{11} \text{ joules.} \qquad \blacksquare$$

Recall, if f is continuous on $[a, b]$, then the definite integral $\int_a^b f(x)\,dx$ exists. Moreover, if $F'(x) = f(x)$, then $\int_a^b f(x)\,dx = F(b) - F(a)$. However, we cannot evaluate an integral such as

$$\int_{-2}^1 \frac{1}{x^2} dx \qquad (4)$$

by the same procedure, since $f(x) = 1/x^2$ possesses an infinite discontinuity in $[-2, 1]$. See FIGURE 7.7.3. In other words, for the integral in (4), the "procedure"

$$-x^{-1} \Big]_{-2}^1 = (-1) - \left(\frac{1}{2}\right) = -\frac{3}{2}$$

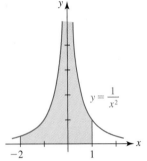

FIGURE 7.7.3 $x = 0$ is a vertical asymptote for the graph of $f(x) = 1/x^2$

is just meaningless scratchings on paper. Thus, we have another type of integral that demands special handling.

▌ Improper Integrals—Infinite Discontinuities An integral $\int_a^b f(x)\,dx$ is also said to be **improper** if f is unbounded on $[a, b]$—that is, if f has an infinite discontinuity at some number in the interval of integration. There are three possible **improper integrals** of this type. Their definitions are summarized in the next definition.

Definition 7.7.2 Infinite Discontinuities

(i) If f is continuous on $[a, b)$ and $|f(x)| \to \infty$ as $x \to b^-$, then

$$\int_a^b f(x)\,dx = \lim_{t \to b^-} \int_a^t f(x)\,dx. \qquad (5)$$

(ii) If f is continuous on $(a, b]$ and $|f(x)| \to \infty$ as $x \to a^+$, then

$$\int_a^b f(x)\,dx = \lim_{s \to a^+} \int_s^b f(x)\,dx. \qquad (6)$$

(iii) If $|f(x)| \to \infty$ as $x \to c$ for some c in (a, b) and f is continuous at all other numbers in $[a, b]$, then

$$\int_a^b f(x)\,dx = \int_a^c f(x)\,dx + \int_c^b f(x)\,dx. \qquad (7)$$

When the limits in (5) and (6) exist, the integrals are said to **converge**. If the limit fails to exist, the integral is said to **diverge**. In (7) the integral $\int_a^b f(x)\,dx$ converges provided *both* $\int_a^c f(x)\,dx$ and $\int_c^b f(x)\,dx$ converge. If either $\int_a^c f(x)\,dx$ or $\int_c^b f(x)\,dx$ diverges, then $\int_a^b f(x)\,dx$ diverges.

EXAMPLE 9 Using (6)

Evaluate $\int_0^4 \frac{1}{\sqrt{x}}\,dx$.

Solution Observe that $f(x) = 1/\sqrt{x} \to \infty$ as $x \to 0^+$, that is, $x = 0$ is a vertical asymptote for the graph of f. Thus, by (6) of Definition 7.7.2,

$$\int_0^4 \frac{1}{\sqrt{x}}\,dx = \lim_{s \to 0^+} \int_s^4 x^{-1/2}\,dx = \lim_{s \to 0^+} 2x^{1/2}\Big]_s^4 = \lim_{s \to 0^+}[4 - 2s^{1/2}].$$

Since $\lim_{s \to 0^+} s^{1/2} = 0$ we have $\lim_{s \to 0^+}[4 - 2s^{1/2}] = 4$. Thus, the given integral converges and

$$\int_0^4 \frac{1}{\sqrt{x}}\,dx = 4.$$

As seen in FIGURE 7.7.4, the number 4 can be regarded as the area under the graph of f on the interval $[0, 4]$.

FIGURE 7.7.4 Area under the graph in Example 9

EXAMPLE 10 Using (6)

Evaluate $\int_0^e \ln x\,dx$.

Solution In this case we know $f(x) = \ln x \to -\infty$ as $x \to 0^+$. Using (6) and integration by parts gives

$$\int_0^e \ln x\,dx = \lim_{s \to 0^+} \int_s^e \ln x\,dx$$

$$= \lim_{s \to 0^+} (x \ln x - x)\Big]_s^e$$

$$= \lim_{s \to 0^+}[(e \ln e - e) - (s \ln s - s)] \quad \leftarrow \ln e = 1$$

$$= \lim_{s \to 0^+} s(1 - \ln s).$$

Now, the last limit has the indeterminate form $0 \cdot \infty$, but if it is written as

$$\lim_{s \to 0^+} \frac{1 - \ln s}{1/s},$$

we recognize the indeterminate form is now ∞/∞. Thus, by L'Hôpital's Rule,

$$\lim_{s \to 0^+} \frac{1 - \ln s}{1/s} \overset{h}{=} \lim_{s \to 0^+} \frac{-1/s}{-1/s^2} = \lim_{s \to 0^+} s = 0.$$

Therefore, the integral converges and $\int_0^e \ln x\,dx = 0$.

The result $\int_0^e \ln x\,dx = 0$ in Example 10 indicates that the net signed area between the graph of $f(x) = \ln x$ and the x-axis on $[0, e]$ is 0. From FIGURE 7.7.5 we see that

$$\int_0^e \ln x\,dx = \int_0^1 \ln x\,dx + \int_1^e \ln x\,dx = -A_1 + A_2 = 0.$$

We saw in Example 7 in Section 7.3 that $\int_1^e \ln x\,dx = 1$, and so $A_1 = A_2 = 1$.

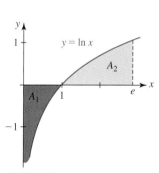

FIGURE 7.7.5 Net signed area in Example 10

EXAMPLE 11 Using (7)

Evaluate $\int_1^5 \frac{1}{(x-2)^{1/3}}\,dx$.

Solution In the interval $[1, 5]$ the integrand has an infinite discontinuity at 2. Consequently, from (7) we write

$$\int_1^5 \frac{1}{(x-2)^{1/3}}\,dx = \int_1^2 (x-2)^{-1/3}\,dx + \int_2^5 (x-2)^{-1/3}\,dx = I_1 + I_2.$$

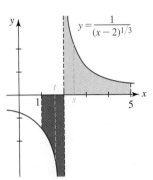

FIGURE 7.7.6 Graph of the integrand in Example 11

Now,
$$I_1 = \lim_{t \to 2^-} \int_1^t (x-2)^{-1/3}\,dx = \lim_{t \to 2^-} \frac{3}{2}(x-2)^{2/3}\bigg]_1^t$$
$$= \frac{3}{2}\lim_{t \to 2^-}[(t-2)^{2/3} - 1] = -\frac{3}{2}.$$

Similarly,
$$I_2 = \lim_{s \to 2^+} \int_s^5 (x-2)^{-1/3}\,dx = \lim_{s \to 2^+} \frac{3}{2}(x-2)^{2/3}\bigg]_s^5$$
$$= \frac{3}{2}\lim_{s \to 2^+}[3^{2/3} - (s-2)^{2/3}] = \frac{3^{5/3}}{2}.$$

Since both I_1 and I_2 converge, the given integral converges and
$$\int_1^5 \frac{dx}{(x-2)^{1/3}} = -\frac{3}{2} + \frac{3^{5/3}}{2} \approx 1.62.$$

Note from FIGURE 7.7.6 that this last number represents a net signed area on the interval $[1, 5]$. ■

EXAMPLE 12 The Integral in (4) Revisited

Evaluate $\displaystyle\int_{-2}^1 \frac{1}{x^2}\,dx$.

Solution This is the integral discussed in (4). Since, in the interval $[-2, 1]$ the integrand has an infinite discontinuity at 0, we write
$$\int_{-2}^1 \frac{1}{x^2}\,dx = \int_{-2}^0 \frac{1}{x^2}\,dx + \int_0^1 \frac{1}{x^2}\,dx = I_1 + I_2.$$

Now, the result
$$I_1 = \int_{-2}^0 \frac{1}{x^2}\,dx = \lim_{t \to 0^-} \int_{-2}^t x^{-2}\,dx = \lim_{t \to 0^-}(-x^{-1})\bigg]_{-2}^t = \lim_{t \to 0^-}\left[-\frac{1}{t} - \frac{1}{2}\right] = \infty$$

indicates there is no need to evaluate $I_2 = \int_0^1 dx/x^2$. The integral $\int_{-2}^1 dx/x^2$ diverges.

∫ **NOTES FROM THE CLASSROOM**

(i) You should verify that $\int_{-\infty}^{\infty} x\,dx = \int_{-\infty}^0 x\,dx + \int_0^{\infty} x\,dx$ diverges since both $\int_{-\infty}^0 x\,dx$ and $\int_0^{\infty} x\,dx$ diverge. A common mistake when working with integrals with doubly infinite limits is to use one limit:
$$\int_{-\infty}^{\infty} x\,dx = \lim_{t \to \infty} \int_{-t}^t x\,dx = \lim_{t \to \infty} \frac{1}{2}x^2\bigg]_{-t}^t = \frac{1}{2}\lim_{t \to \infty}[t^2 - t^2] = 0.$$

Of course, this "answer" is incorrect. Integrals of the type $\int_{-\infty}^{\infty} f(x)\,dx$ require the evaluation of *two independent* limits.

(ii) In our previous work we often wrote without thinking that an integral of a sum is the sum of the integrals:
$$\int_a^b [f(x) + g(x)]\,dx = \int_a^b f(x)\,dx + \int_a^b g(x)\,dx. \tag{8}$$

For improper integrals one should proceed with more caution. For example, the integral $\displaystyle\int_1^{\infty}\left[\frac{1}{x} - \frac{1}{x+1}\right] dx$ converges (see Problem 25 in Exercises 7.7), but
$$\int_1^{\infty}\left[\frac{1}{x} - \frac{1}{x+1}\right] dx \neq \int_1^{\infty} \frac{1}{x}\,dx - \int_1^{\infty} \frac{1}{x+1}\,dx.$$

The property in (8) remains valid for improper integrals whenever both integrals on the right-hand side of the equality converge.

(iii) From examples, problems, and graphs such as Figure 7.7.1, students are often left with the impression that $f(x) \to 0$ as $x \to \infty$ is a necessary condition for the integral $\int_a^\infty f(x)\,dx$ to converge. This is not so. Work Problem 70 when you get to Exercises 9.3.

(iv) It is possible for an integral to have infinite limits of integration *and* an integrand with an infinite discontinuity. To determine whether an integral such as

$$\text{integrand discontinuous at } x = 1 \to \int_1^\infty \frac{1}{x\sqrt{x^2 - 1}}\,dx \leftarrow \text{infinite limit}$$

converges, we break up the integration at some convenient point of continuity of the integrand, say, $x = 2$:

$$\int_1^\infty \frac{1}{x\sqrt{x^2 - 1}}\,dx = \int_1^2 \frac{1}{x\sqrt{x^2 - 1}}\,dx + \int_2^\infty \frac{1}{x\sqrt{x^2 - 1}}\,dx = I_1 + I_2. \qquad (9)$$

I_1 and I_2 are improper integrals; I_1 is of the type given in (6) and I_2 is of the type given in (1). If both I_1 and I_2 converge, then the original integral converges. See Problems 85 and 86 in Exercises 7.7.

(v) The integrand of $\int_a^b f(x)\,dx$ can also have infinite discontinuities at both $x = a$ and $x = b$. In this case the improper integral is defined in a manner analogous to (7). If an integrand f has an infinite discontinuity at several numbers in (a, b), then the improper integral is defined by a natural extension of (7). See Problems 87 and 88 in Exercises 7.7.

(vi) Sometimes strange things happen when working with improper integrals. It is possible to revolve a region with infinite area around an axis and the resulting volume of the solid of revolution can be finite. A very famous problem of this sort is given in Problem 89 of Exercises 7.7.

Exercises 7.7   Answers to selected odd-numbered problems begin on page ANS-24.

≡ Fundamentals

In Problems 1–30, evaluate the given improper integral or show that it diverges.

1. $\int_3^\infty \frac{1}{x^4}\,dx$
2. $\int_{-\infty}^{-1} \frac{1}{\sqrt[3]{x}}\,dx$
3. $\int_1^\infty \frac{1}{x^{0.99}}\,dx$
4. $\int_1^\infty \frac{1}{x^{1.01}}\,dx$
5. $\int_{-\infty}^3 e^{2x}\,dx$
6. $\int_{-\infty}^\infty e^{-x}\,dx$
7. $\int_1^\infty \frac{\ln x}{x}\,dx$
8. $\int_1^\infty \frac{\ln t}{t^2}\,dt$
9. $\int_e^\infty \frac{1}{x(\ln x)^3}\,dx$
10. $\int_e^\infty \ln x\,dx$
11. $\int_{-\infty}^\infty \frac{x}{(x^2 + 1)^{3/2}}\,dx$
12. $\int_{-\infty}^\infty \frac{x}{1 + x^2}\,dx$
13. $\int_{-\infty}^0 \frac{x}{(x^2 + 9)^2}\,dx$
14. $\int_5^\infty \frac{1}{\sqrt[4]{3x + 1}}\,dx$
15. $\int_2^\infty u e^{-u}\,du$
16. $\int_{-\infty}^3 \frac{x^3}{x^4 + 1}\,dx$
17. $\int_{2/\pi}^\infty \frac{\sin(1/x)}{x^2}\,dx$
18. $\int_{-\infty}^\infty t e^{-t^2}\,dt$
19. $\int_{-1}^\infty \frac{1}{x^2 + 2x + 2}\,dx$
20. $\int_{-\infty}^0 \frac{1}{x^2 + 2x + 3}\,dx$
21. $\int_0^\infty e^{-x} \sin x\,dx$
22. $\int_{-\infty}^0 e^x \cos 2x\,dx$
23. $\int_{1/2}^\infty \frac{x + 1}{x^3}\,dx$
24. $\int_0^\infty (e^{-x} - e^{-2x})^2\,dx$
25. $\int_1^\infty \left[\frac{1}{x} - \frac{1}{x + 1}\right]dx$
26. $\int_3^\infty \left[\frac{1}{x} + \frac{1}{x^2 + 9}\right]dx$
27. $\int_2^\infty \frac{1}{x^2 + 6x + 5}\,dx$
28. $\int_{-\infty}^0 \frac{1}{x^2 - 3x + 2}\,dx$
29. $\int_{-\infty}^{-2} \frac{x^2}{(x^3 + 1)^2}\,dx$
30. $\int_0^\infty \frac{1}{e^x + e^{-x}}\,dx$

In Problems 31–52, evaluate the given improper integral or show that it diverges.

31. $\int_0^5 \frac{1}{x}\,dx$
32. $\int_0^8 \frac{1}{x^{2/3}}\,dx$
33. $\int_0^1 \frac{1}{x^{0.99}}\,dx$
34. $\int_0^1 \frac{1}{x^{1.01}}\,dx$
35. $\int_0^2 \frac{1}{\sqrt{2 - x}}\,dx$
36. $\int_1^3 \frac{1}{(x - 1)^2}\,dx$

37. $\int_{-1}^{1} \frac{1}{x^{5/3}} dx$

38. $\int_{0}^{2} \frac{1}{\sqrt[3]{x-1}} dx$

39. $\int_{0}^{2} (x-1)^{-2/3} dx$

40. $\int_{0}^{27} \frac{e^{x^{1/3}}}{x^{2/3}} dx$

41. $\int_{0}^{1} x \ln x \, dx$

42. $\int_{1}^{e} \frac{1}{x \ln x} dx$

43. $\int_{0}^{\pi/2} \tan t \, dt$

44. $\int_{0}^{\pi/4} \frac{\sec^2 \theta}{\sqrt{\tan \theta}} d\theta$

45. $\int_{0}^{\pi} \frac{\sin x}{1 + \cos x} dx$

46. $\int_{0}^{\pi} \frac{\cos x}{\sqrt{1 - \sin x}} dx$

47. $\int_{-1}^{0} \frac{x}{\sqrt{1+x}} dx$

48. $\int_{0}^{3} \frac{1}{x^2 - 1} dx$

49. $\int_{0}^{1} \frac{x^2}{\sqrt{1-x^2}} dx$

50. $\int_{0}^{2} \frac{e^w}{\sqrt{e^w - 1}} dw$

51. $\int_{1}^{3} \frac{1}{\sqrt{3 + 2x - x^2}} dx$

52. $\int_{0}^{1} \left[\frac{1}{\sqrt{x}} + \frac{1}{\sqrt{1-x}} \right] dx$

In Problems 53 and 54, use a substitution to evaluate the given integral.

53. $\int_{12}^{\infty} \frac{1}{\sqrt{x}(x+4)} dx$

54. $\int_{1}^{\infty} \sqrt{x} e^{-\sqrt{x}} dx$

≡ Review of Applications

In Problems 55–58, find the area under the graph of the given function on the indicated interval.

55. $f(x) = \frac{1}{(2x+1)^2}$; $[1, \infty)$

56. $f(x) = \frac{10}{x^2 + 25}$; $(-\infty, 5]$

57. $f(x) = e^{-|x|}$; $(-\infty, \infty)$

58. $f(x) = |x|^3 e^{-x^4}$; $(-\infty, \infty)$

59. Find the area of the region that is bounded by the graphs of $y = 1/\sqrt{x-1}$ and $y = -1/\sqrt{x-1}$ on the interval $[1, 5]$.

60. Consider the region that is bounded by the graphs of $y = 1/\sqrt{x+2}$ and $y = 0$ on the interval $[-2, 1]$.
    (a) Show that the region has finite area.
    (b) Show that the solid of revolution that is formed by revolving the region around the x-axis has infinite volume.

61. Use a calculator or CAS to obtain the graphs of
$$y = \frac{1}{x} \quad \text{and} \quad y = \frac{1}{x(x^2+1)}.$$
Determine whether the area of the region that is bounded by these graphs on the interval $[0, 1]$ is finite.

62. Find the volume of the solid of revolution that is formed by revolving the region bounded by the graphs of $y = xe^{-x}$ and $y = 0$ on $[0, \infty)$ around the x-axis.

63. Find the work done against gravity in lifting a 10,000-kg payload to an infinite distance above the surface of the Moon. [*Hint:* Review page 357 of Section 6.8.]

64. The work done by an external force in moving a test charge q_0 radially from point A to point B in the electric field of a charge q is defined to be:
$$W = -\frac{qq_0}{4\pi e_0} \int_{r_A}^{r_B} \frac{1}{r^2} dr.$$
See FIGURE 7.7.7.
(a) Show that $W = \frac{qq_0}{4\pi e_0} \left(\frac{1}{r_B} - \frac{1}{r_A} \right)$.
(b) Find the work done in bringing the test charge in from an infinite distance to point B.

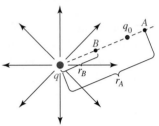

FIGURE 7.7.7 Charge in Problem 64

The **Laplace transform** of a function $y = f(x)$, defined by the integral
$$\mathcal{L}\{f(x)\} = \int_{0}^{\infty} e^{-st} f(t) \, dt,$$
is very useful in some areas of applied mathematics. In Problems 65–72, find the Laplace transform of the given function and state a restriction on s for which the integral converges.

65. $f(x) = 1$
66. $f(x) = x$
67. $f(x) = e^x$
68. $f(x) = e^{-5x}$
69. $f(x) = \sin x$
70. $f(x) = \cos 2x$
71. $f(x) = \begin{cases} 0, & 0 \le x < 1 \\ 1, & x \ge 1 \end{cases}$
72. $f(x) = \begin{cases} 0, & 0 \le x < 3 \\ e^{-x}, & x \ge 3 \end{cases}$

73. **A probability density function** is any nonnegative function f defined on an interval $[a, b]$ for which $\int_a^b f(x) dx = 1$. Verify that for $k > 0$,
$$f(x) = \begin{cases} 0, & x < 0 \\ ke^{-kx}, & x \ge 0 \end{cases}$$
is a probability density function on the interval $(-\infty, \infty)$.

74. Another integral in applied mathematics is the so-called **gamma function**:
$$\Gamma(\alpha) = \int_{0}^{\infty} t^{\alpha-1} e^{-t} dt, \quad x > 0.$$
(a) Show that $\Gamma(\alpha + 1) = \alpha \Gamma(\alpha)$.
(b) Use the result in part (a) to show that
$$\Gamma(n+1) = 1 \cdot 2 \cdot 3 \cdots (n-1) \cdot n = n!,$$
where the symbol $n!$ is read "n factorial." Because of this property, the gamma function is called the **generalized factorial function**.

≡ **Think About It**

In Problems 75–78, determine all values of k such that the given integral is convergent.

75. $\int_1^\infty \dfrac{1}{x^k}\, dx$

76. $\int_{-\infty}^1 x^{2k}\, dx$

77. $\int_0^\infty e^{kx}\, dx$

78. $\int_1^\infty \dfrac{(\ln x)^k}{x}\, dx$

The following is a **Comparison Test** for improper integrals. Suppose f and g are continuous and $0 \le f(x) \le g(x)$ for $x \ge a$. If $\int_a^\infty g(x)\, dx$ converges, then $\int_a^\infty f(x)\, dx$ also converges. In Problems 79–82, use this result to show that the given integral converges.

79. $\int_1^\infty \dfrac{\sin^2 x}{x^2}\, dx$

80. $\int_2^\infty \dfrac{1}{x^3 + 4}\, dx$

81. $\int_0^\infty \dfrac{1}{x + e^x}\, dx$

82. $\int_0^\infty e^{-x^2}\, dx$

In the Comparison Test for improper integrals, if the integral $\int_a^\infty f(x)\, dx$ diverges, then $\int_a^\infty g(x)\, dx$ is divergent. In Problems 83 and 84, use this result to show that the given integral diverges.

83. $\int_1^\infty \dfrac{1 + e^{-2x}}{\sqrt{x}}\, dx$

84. $\int_1^\infty e^{x^2}\, dx$

In Problems 85–88, determine whether the given integral converges or diverges.

85. $\int_1^\infty \dfrac{1}{x\sqrt{x^2 - 1}}\, dx$

86. $\int_{-\infty}^4 \dfrac{1}{(x - 1)^{2/3}}\, dx$

87. $\int_{-1}^1 \dfrac{1}{\sqrt{1 - x^2}}\, dx$

88. $\int_0^2 \dfrac{2x - 1}{\sqrt[3]{x^2 - x}}\, dx$

≡ **Projects**

89. **A Mathematical Classic** The Italian mathematician and physicist **Evangelista Torricelli** (1608–1647) was the first to investigate the interesting properties of the region bounded by the graphs of $y = 1/x$ and $y = 0$ on the interval $[1, \infty)$.

    (a) Show that the region has infinite area.
    (b) Show, however, that the solid of revolution formed by revolving the region about the x-axis has finite volume. The solid shown in FIGURE 7.7.8 is called **Gabriel's horn** or **Torricelli's trumpet**. In some religious traditions Gabriel is held to be the angel of judgment, the destroyer of Sodom and Gomorrah, and so he is frequently identified as the angel who blows the horn to announce the arrival of Judgment Day.
    (c) Use (3) of Section 6.6 to show that the surface area S of the solid of revolution is given by
    $$S = 2\pi \int_1^\infty \dfrac{\sqrt{x^4 + 1}}{x^3}\, dx.$$
    Use the version of the Comparison Test given in Problems 83 and 84 to show that the surface area is infinite.

FIGURE 7.7.8 Gabriel's horn in Problem 89

90. **A Bit of History—Return of the Plague** A study of the Bombay plague epidemic of 1905–06 found that the death rate for that epidemic could be approximated by the mathematical model
$$R = 890\,\text{sech}^2(0.2t - 3.4),$$
where R is the number of deaths per week and t is the time (in weeks) from the onset of the plague.

    (a) What is the peak death rate, and when does it occur?
    (b) Estimate the total number of deaths by computing the integral $\int_{-\infty}^\infty R_0(t)\, dt$.
    (c) Show that more than 99% of the deaths occurred in the first 34 weeks of the epidemic; that is, compare $\int_0^{34} R(t)\, dt$ to the result in part (b).
    (d) Suppose you want to use a "simpler" model for the death rate, of the form
    $$R_0 = \dfrac{a}{t^2 - 2bt + c},$$
    where $c > b^2$. You want this model to have the same peak death rate at the same time as the original model and you also want the total number of deaths, $\int_{-\infty}^\infty R_0(t)\, dt$, to be the same. Find coefficients a, b, and c that satisfy these requirements.
    (e) For the model in part (d), show that less than 89% of the deaths occur in the first 34 weeks of the epidemic.

7.8 Approximate Integration

Introduction Life in mathematics would be extremely pleasant if the antiderivative of every continuous function could be expressed in terms of elementary functions such as polynomial, rational, exponential, or trigonometric functions. As discussed in the *Notes from the Classroom* in Section 5.5 this is not the case. Hence, Theorem 5.5.1 cannot be used to evaluate every definite integral. Sometimes the best we can hope for is an approximation of the value of a definite integral $\int_a^b f(x)\, dx$. In this concluding section of the chapter, we shall consider three such numerical or *approximate integration* procedures.

In the following discussion it will again be useful to interpret the definite integral $\int_a^b f(x)\,dx$ as the area under the graph of f on $[a, b]$. Although continuity of f is essential, there is no actual requirement that $f(x) \geq 0$ on the interval.

Midpoint Rule One way of approximating a definite integral is to proceed in the same manner as we did in the initial discussion about finding the area under a graph—namely, construct rectangular elements under the graph and add their areas. In particular, let us suppose that $y = f(x)$ is continuous on $[a, b]$ and that this interval is divided into n subintervals of equal length $\Delta x = (b - a)/n$. (Recall this is called a *regular partition*.) A simple, but fairly accurate, approximation rule consists of adding the areas of n rectangular elements whose lengths are calculated at the midpoint of each subinterval. See FIGURE 7.8.1(a).

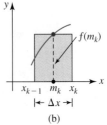

FIGURE 7.8.1 Using n rectangles to approximate the definite integral

Now, if $m_k = (x_{k-1} + x_k)/2$ is the midpoint of a subinterval $[x_{k-1}, x_k]$, then the area of the rectangular element shown in Figure 7.8.1(b) is

$$A_k = f(m_k)\,\Delta x = f\left(\frac{x_{k-1} + x_k}{2}\right)\Delta x.$$

Identifying $a = x_0$ and $b = x_n$ and summing the n areas, we obtain

$$\int_a^b f(x)\,dx \approx f\left(\frac{x_0 + x_1}{2}\right)\Delta x + f\left(\frac{x_1 + x_2}{2}\right)\Delta x + \cdots + f\left(\frac{x_{n-1} + x_n}{2}\right)\Delta x.$$

If we replace Δx by $(b - a)/n$, this midpoint approximation rule can be summarized as follows:

Definition 7.8.1 Midpoint Rule

The **Midpoint Rule** is the approximation $\int_a^b f(x)\,dx \approx M_n$, where

$$M_n = \frac{b - a}{n}\left[f\left(\frac{x_0 + x_1}{2}\right) + f\left(\frac{x_1 + x_2}{2}\right) + \cdots + f\left(\frac{x_{n-1} + x_n}{2}\right)\right]. \quad (1)$$

Since the function $f(x) = 1/x$ is continuous on any interval $[a, b]$ that does not include the origin, we know that $\int_a^b (1/x)\,dx$ exists. For the sake of the next example suspend your knowledge that $\ln|x|$ is an antiderivative of $1/x$.

EXAMPLE 1 Using (1)

Approximate $\int_1^2 (1/x)\,dx$ by the Midpoint Rule for $n = 1$, $n = 2$, and $n = 5$.

Solution As shown in FIGURE 7.8.2(a), the case $n = 1$ is one rectangle in which $\Delta x = 1$. The midpoint of the interval is $m_1 = \frac{3}{2}$ and $f\left(\frac{3}{2}\right) = \frac{2}{3}$. Therefore, from (1),

$$M_1 = 1 \cdot \frac{2}{3} \approx 0.6666.$$

FIGURE 7.8.2 Rectangles in Example 1

When $n = 2$, Figure 7.8.2(b) shows $\Delta x = \frac{1}{2}$, $x_0 = 1$, $x_1 = 1 + \Delta x = \frac{3}{2}$, and $x_2 = 1 + 2\Delta x = 2$. The midpoints of intervals $[1, \frac{3}{2}]$, and $[\frac{3}{2}, 2]$ are, respectively, $m_1 = \frac{5}{4}$ and $m_2 = \frac{7}{4}$ and so $f(\frac{5}{4}) = \frac{4}{5}$ and $f(\frac{7}{4}) = \frac{4}{7}$. Hence, (1) gives

$$M_2 = \frac{1}{2}\left[\frac{4}{5} + \frac{4}{7}\right] \approx 0.6857.$$

Finally, for $n = 5$, $\Delta x = \frac{1}{5}$, $x_0 = 1$, $x_1 = 1 + \Delta x = \frac{6}{5}$, $x_2 = 1 + 2\Delta x = \frac{7}{5}, \ldots, x_5 = 1 + 5\Delta x = 2$. The midpoints of the five subintervals $[1, \frac{6}{5}]$, $[\frac{6}{5}, \frac{7}{5}]$, $[\frac{7}{5}, \frac{8}{5}]$, $[\frac{8}{5}, \frac{9}{5}]$, $[\frac{9}{5}, 2]$ and the corresponding function values are shown in the accompanying table. The information in the table then gives

k	m_k	$f(m_k)$
1	$\frac{11}{10}$	$\frac{10}{11}$
2	$\frac{13}{10}$	$\frac{10}{13}$
3	$\frac{15}{10}$	$\frac{10}{15}$
4	$\frac{17}{10}$	$\frac{10}{17}$
5	$\frac{19}{10}$	$\frac{10}{19}$

$$M_5 = \frac{1}{5}\left[\frac{10}{11} + \frac{10}{13} + \frac{10}{15} + \frac{10}{17} + \frac{10}{19}\right] \approx 0.6919.$$

In other words, $\int_1^2 (1/x)\, dx \approx M_5$ or $\int_1^2 (1/x)\, dx \approx 0.6919$. ∎

■ Error in the Midpoint Rule Suppose $I = \int_a^b f(x)\, dx$ and M_n is an approximation to I using n rectangles. We define the error in the method to be

$$E_n = |I - M_n|.$$

An upper bound for the error can be obtained by means of the next result. The proof is omitted.

Theorem 7.8.1 Error Bound for Midpoint Rule

If there exists a number $M > 0$ such that $|f''(x)| \leq M$ for all x in $[a, b]$, then

$$E_n \leq \frac{M(b - a)^3}{24n^2}. \tag{2}$$

Observe that this upper bound for the error E_n is inversely proportional to n^2. Hence, the accuracy in the method improves as we take more and more rectangles. For example, if the number of rectangles is doubled, the error E_{2n} is less than one-fourth the error bound for E_n. Thus, we see that $\lim_{n \to \infty} M_n = I$.

The next example illustrates how the error bound (2) can be utilized to determine the number of rectangles that will yield a prescribed accuracy.

EXAMPLE 2 Using (2)

Determine a value of n so that (1) will give an approximation to $\int_1^2 (1/x)\, dx$ that is accurate to two decimal places.

Solution The Midpoint Rule will be accurate to two decimal places for those values of n for which the upper bound $M(b-a)^3/24n^2$ for the error is strictly less than 0.005. For $f(x) = 1/x$, we have $f''(x) = 2/x^3$. Since f'' decreases on $[1, 2]$, it follows that $f''(x) \leq f''(1) = 2$ for all x in the interval. Thus, with $M = 2$, $b - a = 1$, we want

◀ If we want accuracy to three decimal places, we use 0.0005, and so on.

$$\frac{2(1)^3}{24n^2} < 0.005 \quad \text{or} \quad n^2 > \frac{50}{3} \approx 16.67.$$

By taking $n \geq 5$ we obtain the desired accuracy. ∎

Example 2 indicates that the third approximation $\int_1^2 (1/x)\, dx \approx 0.6919$ obtained in Example 1 is accurate to two decimal places. By way of comparison, the exact value of the integral, using Theorem 5.5.1,

$$\int_1^2 \frac{1}{x}\, dx = \ln x \Big]_1^2 = \ln 2 - \ln 1 = \ln 2 \approx 0.6931$$

is correct to four decimal places. Thus, for $n = 5$ the error in the method E_n is approximately 0.0012.

Trapezoidal Rule

A more popular method for approximating an integral is based on the plausibility that a better estimate of $\int_a^b f(x)\,dx$ can be obtained by adding the areas of trapezoids instead of the areas of rectangles. See FIGURE 7.8.3(a). The area of the trapezoid shown in Figure 7.8.3(b) is $h(l_1 + l_2)/2$. Thus, the area A_k of the trapezoidal element shown in Figure 7.8.3(c) is

$$A_k = \Delta x \frac{f(x_{k-1}) + f(x_k)}{2}.$$

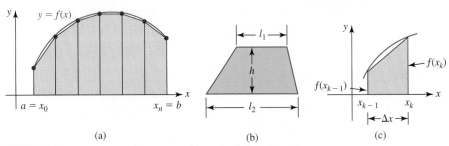

FIGURE 7.8.3 Using n trapezoids to approximate the definite integral

Thus, for a regular partition of the interval $[a, b]$ on which f is continuous, we obtain

$$\int_a^b f(x)\,dx \approx \Delta x \frac{f(x_0) + f(x_1)}{2} + \Delta x \frac{f(x_1) + f(x_2)}{2} + \cdots + \Delta x \frac{f(x_{n-1}) + f(x_n)}{2}.$$

We summarize this new approximation rule in the next definition after we combine like terms and substitute $\Delta x = (b - a)/n$.

Definition 7.8.2 Trapezoidal Rule

The **Trapezoidal Rule** is the approximation $\int_a^b f(x)\,dx \approx T_n$, where

$$T_n = \frac{b - a}{2n}[f(x_0) + 2f(x_1) + 2f(x_2) + \cdots + 2f(x_{n-1}) + f(x_n)]. \tag{3}$$

Error in the Trapezoidal Rule

The error in the method for the Trapezoidal Rule is given by $E_n = |I - T_n|$, where $I = \int_a^b f(x)\,dx$. As the next theorem shows, the error bound for the Trapezoidal Rule is almost the same as that for the Midpoint Rule.

Theorem 7.8.2 Error Bound for Trapezoidal Rule

If there exists a number $M > 0$ such that $|f''(x)| \leq M$ for all x in $[a, b]$, then

$$E_n \leq \frac{M(b - a)^3}{12n^2}. \tag{4}$$

---

k	x_k	$f(x_k)$
0	1	1
1	$\frac{7}{6}$	$\frac{6}{7}$
2	$\frac{4}{3}$	$\frac{3}{4}$
3	$\frac{3}{2}$	$\frac{2}{3}$
4	$\frac{5}{3}$	$\frac{3}{5}$
5	$\frac{11}{6}$	$\frac{6}{11}$
6	2	$\frac{1}{2}$

EXAMPLE 3 Using (4) and (3)

Determine a value of n so that the Trapezoidal Rule will give an approximation to $\int_1^2 (1/x)\,dx$ that is accurate to two decimal places. Approximate the integral.

Solution Using the information in Example 2, we have immediately:

$$\frac{2(1)^3}{12n^2} < 0.005 \quad \text{or} \quad n^2 > \frac{100}{3} \approx 33.33.$$

In this case we take $n \geq 6$ to obtain the desired accuracy. Hence, $\Delta x = \frac{1}{6}$, $x_0 = 1$, $x_1 = 1 + \Delta x = \frac{7}{6}, \ldots, x_6 = 1 + 6\Delta x = 2$. With the information in the accompanying table (3), gives

$$T_6 = \frac{1}{12}\left[1 + 2\left(\frac{6}{7}\right) + 2\left(\frac{3}{4}\right) + 2\left(\frac{2}{3}\right) + 2\left(\frac{3}{5}\right) + 2\left(\frac{6}{11}\right) + \frac{1}{2}\right] \approx 0.6949.$$

EXAMPLE 4 Using (4) and (3)

Approximate $\int_{1/2}^{1} \cos\sqrt{x}\,dx$ by the Trapezoidal Rule so that the error is less than 0.001.

Solution The second derivative of $f(x) = \cos\sqrt{x}$ is

$$f''(x) = \frac{1}{4x}\left(\frac{\sin\sqrt{x}}{\sqrt{x}} - \cos\sqrt{x}\right).$$

For x in the interval $[\frac{1}{2}, 1]$ we have $0 < (\sin\sqrt{x})/\sqrt{x} \leq 1$ and $0 < \cos\sqrt{x} \leq 1$ and consequently $|f''(x)| \leq \frac{1}{4x}$. Therefore, on the interval, $|f''(x)| \leq \frac{1}{2}$. Thus, with $M = \frac{1}{2}$ and $b - a = \frac{1}{2}$, it follows from (4) that we want

$$\frac{\frac{1}{2}\left(\frac{1}{2}\right)^3}{12n^2} < 0.001 \qquad \text{or} \qquad n^2 > \frac{125}{24} \approx 5.21.$$

Hence, to obtain the desired accuracy it suffices to choose $n = 3$ and $\Delta x = \frac{1}{6}$. With the aid of a calculator to obtain the information in the accompanying table, we find the following approximation for $\int_{1/2}^{1} \cos\sqrt{x}\,dx$ from (3):

$$T_3 = \frac{1}{12}\left[\cos\sqrt{\frac{1}{2}} + 2\cos\sqrt{\frac{2}{3}} + 2\cos\sqrt{\frac{5}{6}} + \cos 1\right] \approx 0.3244.$$

k	x_k	$f(x_k)$
0	$\frac{1}{2}$	0.7602
1	$\frac{2}{3}$	0.6848
2	$\frac{5}{6}$	0.6115
3	1	0.5403

Although not obvious from a figure, an improved method of approximating a definite integral $\int_a^b f(x)\,dx$ can be obtained by considering a series of parabolic arcs instead of a series of chords used in the Trapezoidal Rule. It can be proved, under certain conditions, that a parabolic arc passing through *three* specified points will "fit" the graph of f better than a single straight line. See FIGURE 7.8.4. By adding the areas under the parabolic arcs, we obtain an approximation to the integral.

(a) One arc

(b) Three arcs

FIGURE 7.8.4 Fitting a parabolic arc through three consecutive points on the graph of a function

To begin, let us find the area under an arc of a parabola that passes through three points $P_0(x_0, y_0)$, $P_1(x_1, y_1)$, and $P_2(x_2, y_2)$, where $x_0 < x_1 < x_2$ and $x_1 - x_0 = x_2 - x_1 = h$. As shown in FIGURE 7.8.5, this can be done by finding the area under the graph of $y = Ax^2 + Bx + C$ on the interval $[-h, h]$ so that P_0, P_1, and P_2 have coordinates $(-h, y_0)$, $(0, y_1)$, and (h, y_2), respectively. The interval $[-h, h]$ is chosen for simplicity; the area in question does not depend on the location of the y-axis. Using Theorem 5.5.1, the area is

$$\int_{-h}^{h} (Ax^2 + Bx + C)\,dx = \frac{h}{3}(2Ah^2 + 6C). \tag{5}$$

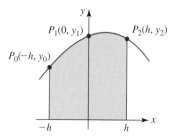

FIGURE 7.8.5 Area under a parabolic arc

But, since the graph is to pass through $(-h, y_0)$, $(0, y_1)$, and (h, y_2), we must have

$$y_0 = Ah^2 - Bh + C \tag{6}$$

$$y_1 = C \tag{7}$$

$$y_2 = Ah^2 + Bh + C. \tag{8}$$

By adding (6) and (8) and using (7), we find $2Ah^2 = y_0 + y_2 - 2y_1$. Thus, (5) can be expressed as

$$\text{area} = \frac{h}{3}(y_0 + 4y_1 + y_2). \tag{9}$$

▌ Simpson's Rule Now suppose that the interval $[a, b]$ is partitioned into n subintervals of equal width $\Delta x = (b - a)/n$, where n is *an even integer*. As shown in FIGURE 7.8.6, on each subinterval $[x_{k-2}, x_k]$ of width $2\Delta x$ we approximate the graph of f by an arc of a parabola through points P_{k-2}, P_{k-1}, and P_k on the graph that corresponds to the endpoints and midpoint of the subinterval. If A_k denotes the area under the parabola on $[x_{k-2}, x_k]$, it follows from (9) that

$$A_k = \frac{\Delta x}{3}[f(x_{k-2}) + 4f(x_{k-1}) + f(x_k)].$$

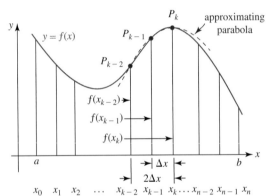

FIGURE 7.8.6 Approximating the function by a parabolic arc

Thus, summing all the A_k gives

$$\int_a^b f(x)\,dx \approx \frac{\Delta x}{3}[f(x_0) + 4f(x_1) + f(x_2)] + \frac{\Delta x}{3}[f(x_2) + 4f(x_3) + f(x_4)] + \cdots + \frac{\Delta x}{3}[f(x_{n-2}) + 4f(x_{n-1}) + f(x_n)].$$

This approximation rule, named after the English mathematician **Thomas Simpson** (1710–1761), is summarized in the following definition.

Definition 7.8.3 Simpson's Rule

Simpson's Rule is the approximation $\int_a^b f(x)\,dx \approx S_n$, where

$$S_n = \frac{b-a}{3n}[f(x_0) + 4f(x_1) + 2f(x_2) + 4f(x_3) + \cdots + 2f(x_{n-2}) + 4f(x_{n-1}) + f(x_n)]. \tag{10}$$

We note again that the integer n in (10) must be even, since each A_k represents the area under a parabolic arc on a subinterval of width $2\Delta x$.

▌ Error in Simpson's Rule If $I = \int_a^b f(x)\,dx$, the next theorem establishes an upper bound for the error in the method $E_n = |I - S_n|$ using an upper bound on the fourth derivative.

Theorem 7.8.3 Error Bound for Simpson's Rule

If there exists a number $M > 0$ such that $|f^{(4)}(x)| \leq M$ for all x in $[a, b]$, then

$$E_n \leq \frac{M(b-a)^5}{180n^4}. \tag{11}$$

EXAMPLE 5 Using (11)

Determine a value of n so that (10) will give an approximation to $\int_1^2 (1/x)\, dx$ that is accurate to two decimal places.

Solution For $f(x) = 1/x$, $f^{(4)}(x) = 24/x^5$ and on $[1, 2]$, $f^{(4)}(x) \leq f^{(4)}(1) = 24$. Thus, with $M = 24$ it follows from (11) that

$$\frac{24(1)^5}{180n^4} < 0.005 \quad \text{or} \quad n^4 > \frac{80}{3} \approx 26.67$$

and so $n > 2.27$. Since n must be an even integer, it suffices to take $n \geq 4$. ∎

EXAMPLE 6 Using (10)

Approximate $\int_1^2 (1/x)\, dx$ by Simpson's Rule for $n = 4$.

Solution When $n = 4$, we have $\Delta x = \frac{1}{4}$. From (10) and the accompanying table we obtain

$$S_4 = \frac{1}{12}\left[1 + 4\left(\frac{4}{5}\right) + 2\left(\frac{2}{3}\right) + 4\left(\frac{4}{7}\right) + \frac{1}{2}\right] \approx 0.6933.$$ ∎

k	m_k	$f(m_k)$
0	1	1
1	$\frac{5}{4}$	$\frac{4}{5}$
2	$\frac{3}{2}$	$\frac{2}{3}$
3	$\frac{7}{4}$	$\frac{4}{7}$
4	2	$\frac{1}{2}$

In Example 6, keep in mind that even though we are using $n = 4$, the definite integral $\int_1^2 (1/x)\, dx$ is being approximated by the area under only two parabolic arcs. Recall that the Midpoint Rule gave $\int_1^2 (1/x)\, dx \approx 0.6919$ with $n = 5$, the Trapezoidal Rule gave $\int_1^2 (1/x)\, dx \approx 0.6949$ with $n = 6$, and 0.6931 is an approximation of the integral correct to four decimal places.

In some applications it may only be possible to obtain numerical values of a quantity $Q(x)$—say, by measurements or by experiment—at specific points in some interval $[a, b]$, and yet it may be necessary to have some idea of the value of the definite integral $\int_a^b Q(x)\, dx$. Even though Q is not defined by means of an explicit formula, we may still be able to apply the Trapezoidal Rule or Simpson's Rule to approximate the integral.

EXAMPLE 7 Area of a Plot of Land

Suppose we wish to find the area of an irregularly shaped piece of land that is bounded between a straight road and the shore of a lake. The boundaries of the land are indicated by the dashed lines in FIGURE 7.8.7(a). Suppose we divide the indicated 1-mi boundary along the road into, say, $n = 8$ subintervals and then, as shown in Figure 7.8.7(b), measure the perpendicular distances from the road to the shore of the lake. We are now in position to approximate the area of the land $A = \int_a^b f(x)\, dx$ by Simpson's Rule. With $b - a = 1$ mi $= 5280$ ft, $\Delta x = (b-a)/n = 5280/8 = 660$, and the identifications $f(x_0) = 83, \ldots, f(x_8) = 28$, (10) gives the following approximation for A:

$$S_8 = \frac{660}{3}[83 + 4(82) + 2(96) + 4(100) + 2(82) + 4(55) + 2(63) + 4(54) + 28]$$

$$= 386{,}540 \text{ ft}^2.$$

Using the fact that 1 acre = 43,560 ft², we see that the land is approximately 8.9 acres.

FIGURE 7.8.7 Lakeside land in Example 7

$\int_a^b$ NOTES FROM THE CLASSROOM

(*i*) The popularity of the Trapezoidal Rule notwithstanding, a direct comparison of the error bounds (2) and (4) shows that the Midpoint Rule is actually more accurate than the Trapezoidal Rule. Specifically, (2) suggests that in some cases the error in the Midpoint Rule can be one-half the error in the Trapezoidal Rule. See Problem 33 in Exercises 7.8.

(*ii*) Under some circumstances the rules considered in the foregoing discussion will give the *exact* value of an integral $\int_a^b f(x)\, dx$. The error bounds (2) and (4) indicate that M_n and T_n will yield the precise value whenever f is a linear function. See Problems 31, 32, and 35 in Exercises 7.8. Simpson's Rule will give the exact value of $\int_a^b f(x)\, dx$ whenever f is a linear, quadratic, or cubic polynomial function. See Problems 34 and 36 in Exercises 7.8.

(*iii*) In general, Simpson's Rule will give greater accuracy than either the Midpoint or the Trapezoidal Rule. So why should we even bother with these other rules? In some instances, the slightly simpler Midpoint and Trapezoidal Rules will yield accuracy that is sufficient for the purpose at hand. Furthermore, the requirement that n must be an even integer in Simpson's Rule may prevent its application to a given problem. Also, to find an error bound for Simpson's Rule, we must compute and then find an upper bound for the fourth derivative. The expression for $f^{(4)}(x)$ can, of course, be very complicated. The error bounds for the other two rules depend on the second derivative.

Exercises 7.8 Answers to selected odd-numbered problems begin on page ANS-24.

≡ Fundamentals

In Problems 1 and 2, compare the exact value of the integral with the approximation obtained from the Midpoint Rule for the indicated value of n.

1. $\int_1^4 (3x^2 + 2x)\, dx;\quad n = 3$ **2.** $\int_0^{\pi/6} \cos x\, dx;\quad n = 4$

In Problems 3 and 4, compare the exact value of the integral with the approximation obtained from the Trapezoidal Rule for the indicated value of n.

3. $\int_1^3 (x^3 + 1)\, dx;\quad n = 4$ **4.** $\int_0^2 \sqrt{x + 1}\, dx;\quad n = 6$

In Problems 5–12, use the Midpoint Rule and the Trapezoidal Rule to obtain an approximation to the given integral for the indicated value of n.

5. $\int_1^6 \dfrac{1}{x}\, dx;\quad n = 5$ **6.** $\int_0^2 \dfrac{1}{3x + 1}\, dx;\quad n = 4$

7. $\int_0^1 \sqrt{x^2 + 1}\, dx;\quad n = 10$ **8.** $\int_1^2 \dfrac{1}{\sqrt{x^3 + 1}}\, dx;\quad n = 5$

9. $\int_0^\pi \dfrac{\sin x}{x + \pi}\, dx;\quad n = 6$ **10.** $\int_0^{\pi/4} \tan x\, dx;\quad n = 3$

11. $\int_0^2 \cos x^2\, dx;\quad n = 6$

12. $\displaystyle\int_0^1 \frac{\sin x}{x}\,dx$; $n = 5$ [Hint: Define $f(0) = 1$.]

In Problems 13 and 14, compare the exact value of the integral with the approximation obtained from Simpson's Rule for the indicated value of n.

13. $\displaystyle\int_0^4 \sqrt{2x+1}\,dx$; $n = 4$ 14. $\displaystyle\int_0^{\pi/2} \sin^2 x\,dx$; $n = 2$

In Problems 15–22, use Simpson's Rule to obtain an approximation to the given integral for the indicated value of n.

15. $\displaystyle\int_{1/2}^{5/2} \frac{1}{x}\,dx$; $n = 4$ 16. $\displaystyle\int_0^5 \frac{1}{x+2}\,dx$; $n = 6$

17. $\displaystyle\int_0^1 \frac{1}{1+x^2}\,dx$; $n = 4$ 18. $\displaystyle\int_{-1}^1 \sqrt{x^2+1}\,dx$; $n = 2$

19. $\displaystyle\int_0^\pi \frac{\sin x}{x+\pi}\,dx$; $n = 6$ 20. $\displaystyle\int_0^1 \cos\sqrt{x}\,dx$; $n = 4$

21. $\displaystyle\int_2^4 \sqrt{x^3+x}\,dx$; $n = 4$ 22. $\displaystyle\int_{\pi/4}^{\pi/2} \frac{1}{2+\sin x}\,dx$; $n = 2$

23. Determine the number of rectangles needed so that an approximation to $\int_{-1}^2 dx/(x+3)$ is accurate to two decimal places.

24. Determine the number of trapezoids needed so that the error in an approximation to $\int_0^{1.5} \sin^2 x\,dx$ is less than 0.0001.

25. Use the Trapezoidal Rule so that an approximation to the area under the graph of $f(x) = 1/(1+x^2)$ on $[0, 2]$ is accurate to two decimal places. [Hint: Examine $f'''(x)$.]

26. The domain of $f(x) = 10^x$ is the set of real numbers and $f(x) > 0$ for all x. Use the Trapezoidal Rule to approximate the area under the graph of f on $[-2, 2]$ with $n = 4$.

27. Using Simpson's Rule, determine n so that the error in approximating $\int_1^3 dx/x$ is less than 10^{-5}. Compare with the n needed in the Trapezoidal Rule to give the same accuracy.

28. Find an upper bound for the error in approximating $\int_0^3 dx/(2x+1)$ by Simpson's Rule with $n = 6$.

In Problems 29 and 30, use the data given in the table and an appropriate rule to approximate the indicated definite integral.

29. $\displaystyle\int_{2.05}^{2.30} f(x)\,dx$;

x	2.05	2.10	2.15	2.20	2.25	2.30
$f(x)$	4.91	4.80	4.66	4.41	3.93	3.58

30. $\displaystyle\int_0^{1.20} f(x)\,dx$;

x	0.0	0.1	0.2	0.4	0.6	0.8	0.9	1.00	1.20
$f(x)$	-0.72	-0.55	-0.16	0.62	0.78	1.34	1.47	1.61	1.51

31. Compare the exact value of the integral $\int_0^4 (2x+5)\,dx$ with the approximation obtained from the Midpoint Rule with $n = 2$ and $n = 4$.

32. Repeat Problem 31 using the Trapezoidal Rule.

33. (a) Find the exact value of the integral $I = \int_{-1}^1 (x^3 + x^2)\,dx$.
    (b) Use the Midpoint Rule with $n = 8$ to find an approximation to I.
    (c) Use the Trapezoidal Rule with $n = 8$ to find an approximation to I.
    (d) Compare the errors $E_8 = |I - M_8|$ and $E_8 = |I - T_8|$.

34. Compare the exact value of the integral $\int_{-1}^3 (x^3 - x^2)\,dx$ with the approximations obtained from Simpson's Rule with $n = 2$ and $n = 4$.

35. Prove that the Trapezoidal Rule will give the exact value of $\int_a^b f(x)\,dx$ when $f(x) = c_1 x + c_0$, with c_0 and c_1 constants. Geometrically, why does this make sense?

36. Prove that Simpson's Rule will give the exact value of $\int_a^b f(x)\,dx$ where $f(x) = c_3 x^3 + c_2 x^2 + c_1 x + c_0$, with $c_0, c_1, c_2,$ and c_3 constants.

37. Use the data given in FIGURE 7.8.8 and Simpson's Rule to find an approximation to the area under the graph of the continuous function f on the interval $[1, 4]$.

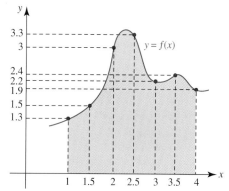

FIGURE 7.8.8 Graph for Problem 37

38. Use the Trapezoidal Rule with $n = 9$ to find an approximation to the area under the graph in FIGURE 7.8.9. Does the Trapezoidal Rule give the exact value of the area?

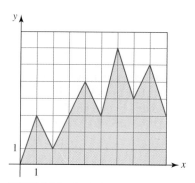

FIGURE 7.8.9 Graph for Problem 38

39. The large irregularly shaped fish pond shown in FIGURE 7.8.10 is filled with water to a uniform depth of 4 ft. Use Simpson's Rule to find an approximation to the number of gallons of water in the tank. Measurements are in feet; the vertical spacing between the horizontal measurements is 1.86 ft. There are 7.48 gal in 1 ft³ of water.

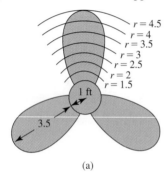

FIGURE 7.8.10 Pond in Problem 39

40. The moment of inertia I of a three-bladed ship's propeller whose dimensions are shown in FIGURE 7.8.11(a) is given by

$$I = \frac{3\rho\pi}{2g} + \frac{3\rho}{g}\int_1^{4.5} r^2 A\, dr,$$

where ρ is the density of the metal, g is the acceleration of gravity, and A is the area of a cross section of the propeller at a distance r ft from the center of the hub. If $\rho = 570$ lb/ft^3 for bronze, use the data in Figure 7.8.11(b) and the Trapezoidal Rule to find an approximation to I.

(a)

r (ft)	1	1.5	2	2.5	3	3.5	4	4.5
A (ft)	0.3	0.50	0.62	0.70	0.60	0.50	0.27	0

(b)

FIGURE 7.8.11 Propeller in Problem 40

≡ **Calculator/CAS Problems**

In Problems 41 and 42, use a calculator or CAS to obtain the graph of the given function. Use Simpson's Rule to approximate the area bounded by the graph of f and the x-axis on the indicated interval. Use $n = 10$.

41. $f(x) = \sqrt[5]{(5^{2.5} - |x|^{2.5})^2}$; $[-5, 5]$

42. $f(x) = 1 + |\sin x|^x$; $[0, 2\pi]$ [*Hint:* Use the graph to discern $f(0)$.]

43. (a) Show that the convergent integral $\int_1^\infty \frac{e^{1/x}}{x^{5/2}}\, dx$ can be written as $\int_0^1 t^{1/2} e^t\, dt$.
  (b) Use the result of part (a) and Simpson's Rule with $n = 4$ to find an approximation to the original improper integral.

44. Use (3) of Section 6.5 and Simpson's Rule with $n = 4$ to find an approximation to the length L of the graph of $y = \frac{1}{3}x^3 + 1$ from the point $(0, 1)$ to $\left(2, \frac{11}{3}\right)$.

45. Use (3) of Section 6.5 and the Trapezoidal Rule with $n = 10$ to find an approximation to the length L of the graph of $y = x^2$ from the origin $(0, 0)$ to the point $(1, 1)$.

46. Use (3) of Section 6.5 and Simpson's Rule with $n = 6$ to find an approximation to the length L of the graph of $y = \ln x$ on the interval $[1, 2]$.

47. Use (3) of Section 6.6 and the Midpoint Rule with $n = 5$ to find an approximation to the area S of the surface that is formed by revolving the graph of $y = \frac{1}{2}x^2$ on the interval $[0, 2]$ about the x-axis.

48. Use Simpson's Rule with $n = 6$ to find an approximation to the area S of the surface that is formed by revolving the graph of $x = y^2 + 1$ for $-1 \le y \le 1$ about the y-axis.

≡ **Think About It**

49. (a) Estimate the length L of the graph given in FIGURE 7.8.12 on the interval $[1, 8]$.
  (b) Explain why using the Trapezoidal Rule with $n = 7$ is not particularly a good idea.

FIGURE 7.8.12 Graph for Problem 49

50. **A Bit of History** The **logarithmic integral function**, Li(x), is defined as the integral

$$\text{Li}(x) = \int_2^x \frac{1}{\ln t}\, dt$$

for $x > 2$. In 1896, the French mathematician **Jacques Hadamard** (1865–1963) and the Belgian mathematician **Charles-Jean de la Vallée Poussin** (1886–1962) independently proved the Prime Number Theorem, which that the number of prime numbers (2, 3, 5, 7, 11, etc.) less than or equal to x, denoted $\pi(x)$, can be approximated by the logarithmic integral meaning that

$$\lim_{x\to\infty} \frac{\pi(x)}{\text{Li}(x)} = 1.$$

  (a) Show that $\pi(x)$ can also be approximated by the function $x/\ln x$ by using L'Hôpital's Rule and the Fundamental Theorem of Calculus to show that

$$\lim_{x\to\infty} \frac{\text{Li}(x)}{x/\ln x} = 1.$$

Since there are an infinite number of primes, Li(x) $\to \infty$ as $x \to \infty$.
  (b) Use Simpson's Rule to approximate Li(100). Compute $x/\ln x$ for $x = 100$. Compare these numbers with the actual number of prime numbers less than 100.

Chapter 7 in Review

Answers to selected odd-numbered problems begin on page ANS-24.

A. True/False

In Problems 1–20, indicate whether the given statement is true or false.

1. Under the change of variable $u = 2x + 3$, the integral $\int_1^5 \dfrac{4x}{\sqrt{2x+3}}\,dx$ becomes $\int_5^{13}(u^{1/2} - 3u^{-1/2})\,du$. _____

2. The trigonometric substitution $u = a\sec\theta$ is appropriate for integrals that contain $\sqrt{a^2 + u^2}$. _____

3. The method of integration by parts is derived from the Product Rule for differentiation. _____

4. $\int_1^e 2x\ln x^2\,dx = e^2 + 1$. _____

5. Partial fractions are not applicable to $\displaystyle\int \dfrac{1}{(x-1)^3}\,dx$. _____

6. A partial fraction decomposition of $x^2/(x+1)^2$ can be found having the form $A/(x+1) + B/(x+1)^2$, where A and B are constants. _____

7. To evaluate $\displaystyle\int \dfrac{1}{(x^2-1)^2}\,dx$, we assume constants A, B, C, and D can be found such that
$$\dfrac{1}{(x^2-1)^2} = \dfrac{Ax+B}{x^2-1} + \dfrac{Cx+D}{(x^2-1)^2}.$$ _____

8. To evaluate $\int x^n e^x\,dx$, n a positive integer, integration by parts is used $n-1$ times. _____

9. To evaluate $\displaystyle\int \dfrac{x}{\sqrt{9-x^2}}\,dx$, it is necessary to use $x = 3\sin\theta$. _____

10. When evaluated, the integral $\int \sin^3 x \cos^2 x\,dx$ can be expressed as a sum of powers of $\cos x$. _____

11. If $\int_a^\infty f(x)\,dx$ and $\int_a^\infty g(x)\,dx$ converge, then $\int_a^\infty [f(x) + g(x)]\,dx$ converges. _____

12. If $\int_a^\infty [f(x) + g(x)]\,dx$ converges, then $\int_a^\infty f(x)\,dx$ converges. _____

13. If f is continuous for all x and $\int_{-\infty}^a f(x)\,dx$ diverges, then $\int_{-\infty}^\infty f(x)\,dx$ diverges. _____

14. The integral $\int_{-\infty}^\infty f(x)\,dx$ is defined by $\displaystyle\lim_{x \to \infty}\int_{-t}^t f(x)\,dx$. _____

15. $\displaystyle\int_{\frac{1}{2}}^1 \dfrac{1}{1+\ln x}\,dx$ is an improper integral. _____

16. $\int_{-1}^1 x^{-3}\,dx = 0$. _____

17. $\int_0^4 x^{-0.999}\,dx$ converges. _____

18. $\int_1^\infty x^{-0.999}\,dx$ diverges. _____

19. $\displaystyle\int_2^\infty \left[\dfrac{e^x}{e^x+1} - \dfrac{e^x}{e^x-1}\right]dx$ diverges, since $\displaystyle\int_2^\infty \dfrac{e^x}{e^x+1}\,dx$ diverges. _____

20. If a positive function f has an infinite discontinuity at a number in $[a, b]$, then the area under the graph on the interval is also infinite. _____

B. Fill in the Blanks

In Problems 1–6, fill in the blanks.

1. $\int_0^\infty e^{-5x}\,dx =$ _____.

2. If $\int_0^\infty e^{-x^2}\,dx = \sqrt{\pi}/2$, then $\int_{-\infty}^\infty e^{-x^2}\,dx =$ _____.

3. If $\int_0^\infty e^{-x^2}\,dx = \sqrt{\pi}/2$, then $\displaystyle\int_0^\infty \dfrac{e^{-x}}{\sqrt{x}}\,dx =$ _____.

4. The integral $\int_1^\infty x^p\,dx$ converges for $p <$ _____ and diverges for $p \geq$ _____.

5. $\int_0^x e^{-2t}\, dt = \int_x^\infty e^{-2t}\, dt$ for $x = $ _____.

6. $\int \sin x \ln(\sin x)\, dx = $ _____.

C. Exercises

In Problems 1–80, use the methods of this chapter, or previous chapters, to evaluate the given integral.

1. $\displaystyle\int \frac{1}{\sqrt{x}+9}\, dx$

2. $\displaystyle\int e^{\sqrt{x+1}}\, dx$

3. $\displaystyle\int \frac{x}{\sqrt{x^2+4}}\, dx$

4. $\displaystyle\int \frac{1}{\sqrt{x^2+4}}\, dx$

5. $\displaystyle\int \frac{1}{(x^2+4)^3}\, dx$

6. $\displaystyle\int \frac{x^2}{x^2+4}\, dx$

7. $\displaystyle\int \frac{x^2+4}{x^2}\, dx$

8. $\displaystyle\int \frac{3x-1}{x(x^2-4)}\, dx$

9. $\displaystyle\int \frac{x-5}{x^2+4}\, dx$

10. $\displaystyle\int \frac{\sqrt[3]{x+27}}{x}\, dx$

11. $\displaystyle\int \frac{(\ln x)^9}{x}\, dx$

12. $\displaystyle\int (\ln 3x)^2\, dx$

13. $\displaystyle\int t \sin^{-1} t\, dt$

14. $\displaystyle\int \frac{\ln x}{(x-1)^2}\, dx$

15. $\displaystyle\int (x+1)^3 (x-2)\, dx$

16. $\displaystyle\int \frac{1}{(x+1)^3(x-2)}\, dx$

17. $\displaystyle\int \ln(x^2+4)\, dx$

18. $\displaystyle\int 8t e^{2t^2}\, dt$

19. $\displaystyle\int \frac{1}{x^4+10x^3+25x^2}\, dx$

20. $\displaystyle\int \frac{1}{x^2+8x+25}\, dx$

21. $\displaystyle\int \frac{x}{x^3+3x^2-9x-27}\, dx$

22. $\displaystyle\int \frac{x+1}{(x^2-x)(x^2+3)}\, dx$

23. $\displaystyle\int \frac{\sin^2 t}{\cos^2 t}\, dt$

24. $\displaystyle\int \frac{\sin^3 \theta}{(\cos \theta)^{3/2}}\, d\theta$

25. $\displaystyle\int \tan^{10} x \sec^4 x\, dx$

26. $\displaystyle\int \frac{x \tan x}{\cos x}\, dx$

27. $\displaystyle\int y \cos y\, dy$

28. $\displaystyle\int x^2 \sin x^3\, dx$

29. $\displaystyle\int (1+\sin^2 t) \cos^3 t\, dt$

30. $\displaystyle\int \frac{\sec^3 \theta}{\tan \theta}\, d\theta$

31. $\displaystyle\int e^w (1+e^w)^5\, dw$

32. $\displaystyle\int (x-1) e^{-x}\, dx$

33. $\displaystyle\int \cot^3 4x\, dx$

34. $\displaystyle\int (3-\sec x)^2\, dx$

35. $\displaystyle\int_0^{\pi/4} \cos^2 x \tan x\, dx$

36. $\displaystyle\int_0^{\pi/3} \sin^4 x \tan x\, dx$

37. $\displaystyle\int \frac{\sin x}{1+\sin x}\, dx$

38. $\displaystyle\int \frac{\cos x}{1+\sin x}\, dx$

39. $\displaystyle\int_0^1 \frac{1}{(x+1)(x+2)(x+3)}\, dx$

40. $\displaystyle\int_{\ln 3}^{\ln 2} \sqrt{e^x+1}\, dx$

41. $\displaystyle\int e^x \cos 3x\, dx$

42. $\displaystyle\int x(x-5)^9\, dx$

43. $\int \cos(\ln t)\, dt$

44. $\int \sec^2 x \ln(\tan x)\, dx$

45. $\int \cos\sqrt{x}\, dx$

46. $\int \dfrac{\cos\sqrt{x}}{\sqrt{x}}\, dx$

47. $\int \cos x \sin 2x\, dx$

48. $\int (\cos^2 x - \sin^2 x)\, dx$

49. $\int \sqrt{x^2 + 2x + 5}\, dx$

50. $\int \dfrac{1}{(8 - 2x - x^2)^{3/2}}\, dx$

51. $\int \tan^5 x \sec^3 x\, dx$

52. $\int \cos^4 \dfrac{x}{2}\, dx$

53. $\int \dfrac{t^5}{1 + t^2}\, dt$

54. $\int \dfrac{1}{\sqrt{1 - x^2}}\, dx$

55. $\int \dfrac{5x^3 + x^2 + 6x + 1}{(x^2 + 1)^2}\, dx$

56. $\int \dfrac{\sqrt{x^2 + 9}}{x^2}\, dx$

57. $\int x \sin^2 x\, dx$

58. $\int (t + 1)^2 e^{3t}\, dt$

59. $\int e^{\sin x} \sin 2x\, dx$

60. $\int e^x \tan^2 e^x\, dx$

61. $\displaystyle\int_0^{\pi/6} \dfrac{\cos x}{\sqrt{1 + \sin x}}\, dx$

62. $\displaystyle\int_0^{\pi/2} \dfrac{1}{\sin x + \cos x}\, dx$

63. $\int \sinh^{-1} t\, dt$

64. $\int x \cot x^2\, dx$

65. $\displaystyle\int_3^8 \dfrac{1}{x\sqrt{x + 1}}\, dx$

66. $\int \dfrac{t + 3}{t^2 + 2t + 1}\, dt$

67. $\int \dfrac{\sec^4 3u}{\cot^{12} 3u}\, du$

68. $\displaystyle\int_0^2 x^5 \sqrt{x^2 + 4}\, dx$

69. $\int \dfrac{3 + \sin x}{\cos^2 x}\, dx$

70. $\int \dfrac{\sin 2x}{5 + \cos^2 x}\, dx$

71. $\int x(1 + \ln x)^2\, dx$

72. $\int x \cos^2 x\, dx$

73. $\int e^x e^{e^x}\, dx$

74. $\int \dfrac{1}{\sqrt{x + 1} - \sqrt{x}}\, dx$

75. $\int \dfrac{2t}{1 + e^{t^2}}\, dt$

76. $\int \cos x \cos 2x\, dx$

77. $\int \dfrac{1}{\sqrt{1 - (5x + 2)^2}}\, dx$

78. $\int (\ln 2x) \ln x\, dx$

79. $\int \cos x \ln|\sin x|\, dx$

80. $\int \ln\left(\dfrac{x + 1}{x - 1}\right) dx$

In Problems 81–92, evaluate the given integral or show that it diverges.

81. $\displaystyle\int_0^3 x(x^2 - 9)^{-2/3}\, dx$

82. $\displaystyle\int_0^5 x(x^2 - 9)^{-2/3}\, dx$

83. $\displaystyle\int_{-\infty}^0 (x + 1)e^x\, dx$

84. $\displaystyle\int_0^\infty \dfrac{e^{2x}}{e^{4x} + 1}\, dx$

85. $\displaystyle\int_3^\infty \dfrac{1}{1 + 5x}\, dx$

86. $\displaystyle\int_0^\infty \dfrac{x}{(x^2 + 4)^2}\, dx$

87. $\displaystyle\int_0^e \ln\sqrt{x}\, dx$

88. $\displaystyle\int_0^{\pi/2} \dfrac{\sec^2 t}{\tan^3 t}\, dt$

89. $\displaystyle\int_0^{\pi/2} \frac{1}{1-\cos x}\,dx$

90. $\displaystyle\int_0^\infty \frac{x}{x+1}\,dx$

91. $\displaystyle\int_0^1 \frac{1}{\sqrt{x}e^{\sqrt{x}}}\,dx$

92. $\displaystyle\int_0^\infty \frac{1}{\sqrt{x}e^{\sqrt{x}}}\,dx$

In Problems 93 and 94, prove the given result.

93. $\displaystyle\int_1^\infty \frac{\sqrt{x}}{(1+x)^2}\,dx = \frac{\pi}{4} + \frac{1}{2}$

94. $\displaystyle\int_0^\infty \frac{1}{\sqrt{x}(x+1)}\,dx = \pi$

In Problems 95 and 96, use the fact that $\displaystyle\int_0^\infty e^{t^2}\,dt = \lim_{x\to\infty}\int_0^x e^{t^2}\,dt = \infty$ to evaluate the given limit.

95. $\displaystyle\lim_{x\to\infty} \frac{x\int_0^x e^{t^2}\,dt}{e^{x^2}}$

96. $\displaystyle\lim_{x\to\infty} \frac{\int_0^x e^{t^2}\,dt}{xe^{x^2}}$

97. Find the area of the region that is bounded by the graphs of $y = e^{-x}$ and $y = e^{-3x}$ on $[0, \infty)$.

98. Consider the region that is bounded by the graphs of $y = 1/\sqrt[3]{1-x}$ and $y = 0$ on the interval $[0, 1]$.

  (a) Find the area of the region.
  (b) Find the volume of the solid of revolution that is formed by revolving the region about the x-axis.
  (c) Find the volume of the solid of revolution that is formed by revolving the region about the line $x = 1$.

99. Consider the graph of $f(x) = (x^2 - 1)/(x^2 + 1)$ given in FIGURE 7.R.1.

  (a) Determine whether the region R_1, which is bounded between the graph of f and its horizontal asymptote, is finite.
  (b) Determine whether the regions R_2 and R_3 have finite areas.

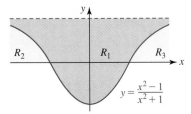

FIGURE 7.R.1 Graph for Problem 99

100. Use Newton's Method to find the number x^* for which the shaded region R in FIGURE 7.R.2 is 99% of the total area under the graph of $y = xe^{-x}$ on $[0, \infty)$.

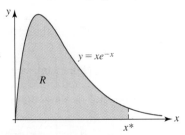

FIGURE 7.R.2 Graph for Problem 100

101. A continuous variable force $f(x)$ acts over the interval $[0, 1]$, where F is measured in Newtons and x in meters. It is determined empirically that

x (m)	0	0.2	0.4	0.6	0.8	1
$F(x)$ (N)	0	50	90	150	210	260

Use an appropriate numerical technique to approximate the work done over the interval.

102. The graph of a variable force F is given in FIGURE 7.R.3.

(a) Use rectangular elements of area to find an approximation to the work done by the force in moving a particle from $x = 1$ to $x = 5$.

(b) Use the Trapezoidal Rule to approximate the work done.

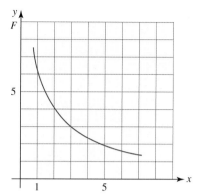

FIGURE 7.R.3 Graph for Problem 102

Chapter 8

First-Order Differential Equations

In This Chapter We are going to study differential equations that have the form $dy/dx = F(x, y)$. These are called *first-order* differential equations. We will examine two solution methods and some applications of these equations. Higher-order differential equations are considered in Chapter 16.

- 8.1 Separable Equations
- 8.2 Linear Equations
- 8.3 Mathematical Models
- 8.4 Solution Curves without a Solution
- 8.5 Euler's Method
  Chapter 8 in Review

8.1 Separable Equations

Introduction In several previous exercise sets you were asked to verify that a given function satisfies a **differential equation (DE)**. Roughly, a differential equation is an equation that involves an unknown function y and one or more of the derivatives of y. Differential equations are classified by the **order** of the highest derivative appearing in the equation. For example, the equation

$$\underset{\uparrow}{\overset{\text{highest-order derivative}}{\frac{d^2y}{dx^2}}} + 4\frac{dy}{dx} + 8y = 0 \tag{1}$$

is an example of a **second-order** differential equation, whereas

$$\frac{dy}{dx} = -\frac{x}{y} \tag{2}$$

is a **first-order** differential equation. Using the "prime" notation the differential equations in (1) and (2) can be written $y'' + 4y' + 8y = 0$ and $y' = -x/y$, respectively. Although the prime notation is easier to write and typeset, the Leibniz notation used in (1) and (2) is often preferred because it clearly displays the independent variable.

Of course, different symbols will often be used. For example,
$$\frac{d^2y}{dt^2} + 4y = \sin 2t$$
is also of a second-order DE.

The exploration of the subject of differential equations usually begins with the study of how to *solve* them. A **solution** of a differential equation is a sufficiently differentiable function $y(x)$, defined explicitly or implicitly, that, when substituted into the equation, reduces it to an identity on some interval. The graph of $y(x)$ is naturally called a **solution curve**.

As mentioned in the chapter opener, in this chapter we are going to study some solution methods and some applications only for *first-order* differential equations. We will make the assumption hereafter that a first-order differential equation can be expressed as

$$\frac{dy}{dx} = F(x, y), \tag{3}$$

Functions of two variables will be discussed in detail in Chapter 13.

where F is a function of two variables x and y. The function F is called the **slope function** and (3) is called the **normal form** of the differential equation. At a point (x, y) on a solution curve of the DE, the value $F(x, y)$ gives the slope of a tangent line.

A Definition We have already solved a simple kind of first-order differential equation in Section 5.1. Recall, the first-order DE,

$$\frac{dy}{dx} = g(x) \tag{4}$$

can be solved by finding the most general antiderivative of g; that is,

$$y = \int g(x)\, dx.$$

For example, a solution of the first-order DE

$$\frac{dy}{dx} = 2x + e^{-3x}$$

is given by

$$y = \int (2x + e^{-3x})\, dx = x^2 - \frac{1}{3}e^{-3x} + C.$$

Equations of the form in (4) are just a special case of a first-order differential equation $dy/dx = F(x, y)$, where the function F can be factored into a product of a function of x times a function of y.

Definition 8.1.1 Separable Differential Equation

A **separable first-order differential equation** is any equation $dy/dx = F(x, y)$ that can be put into the form

$$\frac{dy}{dx} = g(x)f(y). \tag{5}$$

EXAMPLE 1 A Separable DE

The first-order DE

$$\frac{dy}{dx} = -\frac{x}{y} \tag{6}$$

is separable, since the right-hand side of the equality can be rewritten as the product of a function of x times a function of y:

$$\frac{dy}{dx} = \underbrace{-x}_{g(x)} \cdot \underbrace{\frac{1}{y}}_{f(y)}. \qquad \blacksquare$$

Notice that when $f(y) = 1$ in (5) we get (4). Analogous to differential equations of the form (4), we can also solve a separable DE by integration.

Before solving, we rewrite a separable equation in terms of differentials. For example, the equation in (6) can be written alternatively in the differential form

$$y\, dy = -x\, dx.$$

Similarly, by dividing by $f(y)$, (5) can be written as

$$p(y)\, dy = g(x)\, dx,$$

where for notational convenience we have written $p(y) = 1/f(y)$. Now if $y = \phi(x)$ denotes a solution of (5), we must have

$$p(\phi(x))\, \phi'(x) = g(x)$$

and therefore, by integration,

$$\int p(\phi(x))\, \phi'(x)\, dx = \int g(x)\, dx. \tag{7}$$

But $dy = \phi'(x)\, dx$, so (7) is the same as

$$\int p(y)\, dy = \int g(x)\, dx \quad \text{or} \quad H(y) = G(x) + C,$$

where $H(y)$ and $G(x)$ are antiderivatives of $p(y) = 1/f(y)$ and $g(x)$, respectively, and C is a constant. Since C is arbitrary, $H(y) = G(x) + C$ represents a **one-parameter family of solutions**. The parameter is the arbitrary constant C.

Note: There is no need to use two constants in the integration of a separable equation, because if we write $H(y) + C_1 = G(x) + C_2$, then the difference $C_2 - C_1$ can be replaced by a single constant C.

We summarize the discussion.

Guidelines for Solving a Separable DE

(*i*) First, determine whether a first-order equation actually is separable. That is, can the DE be written in the form given in (5)?

(*ii*) If the DE is separable, then rewrite it in differential form:
$$p(y)\, dy = g(x)\, dx.$$

(*iii*) Integrate both sides of the differential form. Integrate the left-hand side with respect to y and the right-hand side with respect to x.

Before illustrating the above solution method, you should be aware that many first-order DEs are not separable. For example, neither of the differential equations
$$\frac{dy}{dx} = x^2 + y^2 \quad \text{and} \quad \frac{dy}{dx} = \sin(x+y)$$
are separable.

EXAMPLE 2 Solving a Separable DE

Solve $\dfrac{dy}{dx} = -\dfrac{x}{y}$.

Solution Rewriting the given equation in differential form
$$y\, dy = -x\, dx$$
and integrating both sides give
$$\int y\, dy = -\int x\, dx \quad \text{or} \quad \frac{y^2}{2} = -\frac{x^2}{2} + C_1.$$

Thus, a one-parameter family of solutions is defined by $x^2 + y^2 = C^2$. Here we have chosen to replace the arbitrary constant $2C_1$ by C^2 because the equation $x^2 + y^2 = C^2$ represents a family of circles centered at the origin of radius $C > 0$. See FIGURE 8.1.1.

The solutions of the differential equation are the functions defined implicitly by the equation $x^2 + y^2 = C^2$, $C > 0$. ∎

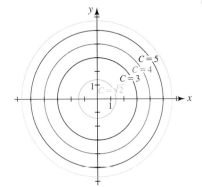

FIGURE 8.1.1 Family of circles in Example 2

■ **Initial-Value Problem** We are often interested in solving a first-order differential equation $dy/dx = F(x, y)$ subject to a prescribed side condition $y(x_0) = y_0$, where x_0 and y_0 are arbitrarily specified real numbers. The problem

Solve: $\quad \dfrac{dy}{dx} = F(x, y)$

Subject to: $\quad y(x_0) = y_0$

is called an **initial-value problem** (IVP). The side condition $y(x_0) = y_0$ is called an **initial condition**. In geometric terms we are seeking at least one solution of the differential equation on an interval I containing x_0 so that a solution curve passes through the point (x_0, y_0). From a practical viewpoint, this often comes down to the problem of determining a specific value of the constant C in a family of solutions.

EXAMPLE 3 An Initial-Value Problem

Solve the initial-value problem $\dfrac{dy}{dx} = -\dfrac{x}{y}$, $y(4) = -3$.

Solution From Example 2, a family of solutions for the given DE is $x^2 + y^2 = C^2$. When $x = 4$, then $y = -3$ so that $16 + 9 = C^2$ gives $C = 5$. Thus the IVP determines $x^2 + y^2 = 25$. Because of its simplicity we can solve the last equation for an explicit function or solution that satisfies the initial condition. Solving for y gives $y = \pm\sqrt{25 - x^2}$. Because the graph must be that of a function and the graph of this function must contain the point $(4, -3)$, we are forced to take the negative square root. In other words, the solution is $y = -\sqrt{25 - x^2}$ defined on the interval $(-5, 5)$. In Figure 8.1.1 the solution curve is the lower semicircle for the circle shown in blue. ∎

The simple first-order differential equation

$$\frac{dy}{dx} = ky, \qquad (8)$$

where k is a constant, has many applications. The equation can be solved by separation of variables.

EXAMPLE 4 Solving a Separable DE

Solve $\dfrac{dy}{dx} = ky$, where $k \neq 0$ is a constant.

Solution We write the differential equation as $\dfrac{1}{y} dy = k\, dx$. Integrating

$$\int \frac{1}{y} dy = k \int dx \qquad \text{gives} \qquad \ln|y| = kx + C_1.$$

Solving for y then yields

$$|y| = e^{kx+C_1} = e^{C_1} e^{kx} \qquad \text{or} \qquad y = \pm e^{kx+C_1} = \pm e^{C_1} e^{kx}.$$

By relabeling the constants $\pm e^{C_1}$ as C, a one-parameter family of solutions is given by $y = Ce^{kx}$. ∎

In order to solve separable differential equations it is obvious that a working knowledge of integration formulas and techniques is imperative. A review of Sections 7.1–7.3 and 7.6 is recommended.

EXAMPLE 5 Solving a Separable DE

Solve $(e^{2y} - y)\dfrac{dy}{dx} = e^y \sin x$.

Solution By rewriting the equation as

$$\frac{e^{2y} - y}{e^y} \frac{dy}{dx} = \sin x \qquad \text{or} \qquad (e^y - ye^{-y})\, dy = \sin x\, dx$$

we see that the equation is separable. From the differential form of the equation,

$$\int (e^y - ye^{-y})\, dy = \int \sin x\, dx,$$

we see that integration by parts must be used to evaluate $\int ye^{-y}\, dy$. The result is

$$e^y + ye^{-y} + e^{-y} = \cos x + C.$$

The last equation defines a solution of the differential equation implicitly. Indeed, it is impossible to solve the last equation for y in terms of x. ∎

EXAMPLE 6 An Initial-Value Problem

Solve $\dfrac{dy}{dx} = y(1 - y), y(0) = \dfrac{1}{3}$.

Solution We rewrite the equation in differential form

$$\frac{1}{y(1-y)} dy = dx \qquad \text{and integrate} \qquad \int \frac{1}{y(1-y)} dy = \int dx.$$

Using partial fractions on the left-hand side of the equality gives

$$\int \left[\frac{1}{y} + \frac{1}{1-y} \right] dy = \int dx$$

$$\ln|y| - \ln|1-y| = x + C_1$$

$$\ln\left|\frac{y}{1-y}\right| = x + C_1$$

$$\left|\frac{y}{1-y}\right| = e^{x+C_1} = e^{C_1}e^x$$

$$\frac{y}{1-y} = C_2 e^x. \quad \leftarrow C_2 = \pm e^{C_1}$$

Solving for y gives

$$y = \frac{C_2 e^x}{1 + C_2 e^x} \quad \text{or} \quad y = \frac{1}{1 + Ce^{-x}}, \qquad (9)$$

where we have replaced $1/C_2$ by C. Substituting $x = 0$ and $y = \frac{1}{3}$ into the last equation then yields $C = 2$. The solution of the initial-value problem is

$$y = \frac{1}{1 + 2e^{-x}}. \qquad (10)$$

Graphs of various members of the family of solutions given in (9) for various positive and negative values of C are illustrated in FIGURE 8.1.2. The graphs shown in color represent solutions of the DE that are defined on the interval $(-\infty, \infty)$. The graph of the solution given in (10) is the blue curve in the figure.

FIGURE 8.1.2 Family of solution curves in Example 6

Exercises 8.1 Answers to selected odd-numbered problems begin on page ANS-25.

≡ Fundamentals

In Problems 1–20, solve the given differential equation by separation of variables.

1. $\dfrac{dy}{dx} = \sin 5x$
2. $\dfrac{dy}{dt} = (t+1)^2$
3. $\dfrac{dy}{dx} = \dfrac{y^3}{x^2}$
4. $\dfrac{dy}{dx} = \dfrac{1}{5y^4}$
5. $\dfrac{dy}{dx} = \left(\dfrac{1+x}{1+y}\right)^2$
6. $\dfrac{dy}{dx} = \sqrt{xy}$
7. $\dfrac{dy}{dx} = \dfrac{1+5x^2}{x^2 \sin y}$
8. $\dfrac{dy}{dx} = y^3 \cos x$
9. $x\dfrac{dy}{dx} = 4y$
10. $\dfrac{dy}{dx} + 2xy = 0$
11. $\dfrac{dy}{dx} = e^{3x+2y}$
12. $e^x y \dfrac{dy}{dx} = e^{-y} + e^{-2x-y}$
13. $\left(\dfrac{y+1}{x}\right)^2 \dfrac{dy}{dx} = y \ln x$
14. $\dfrac{dy}{dx} = \left(\dfrac{2y+3}{4x+5}\right)^2$
15. $\dfrac{dN}{dt} + N = Nte^{t+2}$
16. $\dfrac{dQ}{dt} = k(Q - 70)$
17. $\dfrac{dP}{dt} = 5P - P^2$
18. $\dfrac{dX}{dt} = (10 - X)(50 - X)$
19. $\dfrac{dy}{dx} = \dfrac{xy + 3x - y - 3}{xy - 2x + 4y - 8}$
20. $\dfrac{dy}{dx} = \dfrac{xy + 2y - x - 2}{xy - 3y + x - 3}$

In Problems 21–26, solve the given initial-value problem.

21. $\dfrac{dy}{dx} = \dfrac{1}{(xy)^2}, \quad y(1) = 3$
22. $\dfrac{dy}{dx} = \dfrac{2x + \sec^2 x}{2y}, \quad y(0) = -2$
23. $\dfrac{dx}{dt} = 4(x^2 + 1), \quad x(\pi/4) = 1$
24. $\dfrac{dy}{dx} = \dfrac{y^2 - 1}{x^2 - 1}, \quad y(2) = 2$
25. $x^2 \dfrac{dy}{dx} = y - xy, \quad y(-1) = -1$
26. $\dfrac{dy}{dt} + 2y = 1, \quad y(0) = \dfrac{5}{2}$

In Problems 27 and 28, solve the given initial-value problem. Write the solution as an explicit *algebraic* function $y = f(x)$ (see *Notes From the Classroom* in Section 1.3). You may have to use a trigonometric identity.

27. $\sqrt{1 - x^2}\, \dfrac{dy}{dx} = \sqrt{1 - y^2}, \quad y(0) = \dfrac{\sqrt{3}}{2}$
28. $(1 + x^4)\dfrac{dy}{dx} + x + 4xy^2 = 0, \quad y(1) = 0$

In Problems 29–32, use the concept that $y = k$ on $(-\infty, \infty)$ is a constant function if and only if $dy/dx = 0$ to determine whether the given differential equation possesses constant solutions. Solve the given differential equation. Assume that k is a real number.

29. $x\dfrac{dy}{dx} + 6y = 18$
30. $2\dfrac{dy}{dx} = 5y + 40$
31. $\dfrac{dy}{dx} = y^2 - y - 20$
32. $x\dfrac{dy}{dx} = y^2 + 2y + 4$

In Problems 33 and 34, proceed as in Problems 29–32 to determine whether the given differential equation possesses constant solutions. Solve the given differential equation and then find a solution whose graph passes through the indicated point.

33. $x\dfrac{dy}{dx} = y^2 - y$,

   (a) $(0, 1)$ (b) $(0, 0)$ (c) $\left(\tfrac{1}{2}, \tfrac{1}{2}\right)$

34. $\dfrac{dy}{dx} = y^2 - 9$,

   (a) $(0, 0)$ (b) $(0, 3)$ (c) $\left(\tfrac{1}{3}, 1\right)$

≡ Think About It

35. Without solving, explain why the initial-value problem

$$\dfrac{dy}{dx} = \sqrt{y}, \quad y(x_0) = y_0$$

has no solution for $y_0 < 0$.

36. A solution of a differential equation that is not a member of the family of solutions of the equation is called a **singular solution**. Reexamine Problems 29, 31, 33, and 34 and find any singular solutions. In Example 6, what would be the solution of the IVP if the initial condition were changed to $y(0) = 1$?

37. In Example 3, it was stated that the solution $y = -\sqrt{25 - x^2}$ is defined on the open interval $(-5, 5)$. Why would it be incorrect to say that the solution is defined on the closed interval $[-5, 5]$?

8.2 Linear Equations

Introduction We continue our quest for solutions of first-order DEs by next examining linear equations. Linear differential equations are an especially "friendly" family of differential equations in that, given a linear equation, whether first-order or a higher-order kin, there is always a good possibility that we can find some sort of solution of the equation that we can look at. Nonlinear differential equations, especially equations of order greater than or equal to two, are often impossible to solve in terms of elementary functions.

The technique for solving a linear first-order equation, like a separable equation, consists of integration; but integration only after the original equation has been multiplied by a special function called an *integrating factor*.

A Definition We begin with the definition of a linear first-order equation. As you read the next definition bear in mind the following essential properties:

- In a linear differential equation the dependent variable and its derivative are of the first degree, that is, the power of each term involving the dependent variable is 1, and each coefficient depends at most only on the independent variable.

Definition 8.3.1 Linear Equation

A **linear first-order differential equation** is an equation $dy/dx = F(x, y)$ that can be put into the form

$$a_1(x)\dfrac{dy}{dx} + a_0(x)y = g(x). \tag{1}$$

The functions $a_1(x)$, $a_0(x)$, and $g(x)$ in (1) can of course be constants.

If a first-order DE is not linear, it is said to be **nonlinear**.

EXAMPLE 1 Linear/Nonlinear

(a) By direct comparison with (1) we see that the following differential equations

$$\dfrac{dy}{dx} + 3y = 6 \quad \text{and} \quad x\dfrac{dy}{dx} - 4y = x^6 e^x$$

are linear first-order equations.

(b) The following first-order equations are nonlinear:

power not 1 ↓   coefficient depends on y ↓

$$x\dfrac{dy}{dx} = y^2 \quad \text{and} \quad y\dfrac{dy}{dx} = 2y + \cos x.$$

∎

It is important to note that not every first-order linear differential equation can be solved by the method of separation of variables. The linear equation

$$\frac{dy}{dx} + 2y = x$$

is not separable. Hence, we need a new procedure for solving linear equations.

■ **Standard Form** By dividing (1) by the lead coefficient $a_1(x)$, we obtain the more useful form of a linear equation:

$$\frac{dy}{dx} + P(x)y = f(x). \tag{2}$$

Equation (2) is called the **standard form** of a linear DE (1). We seek solutions of (2) on an interval I for which P and f are continuous. Equation (2) has the property that when multiplied by the function $e^{\int P(x)\,dx}$, the left-hand side of (2) becomes the derivative of the product $e^{\int P(x)\,dx}y$. To see this, observe that the Product Rule and Chain Rule give

$$\frac{d}{dx}\left[e^{\int P(x)\,dx}y\right] = e^{\int P(x)\,dx}\frac{d}{dx}y + y\frac{d}{dx}e^{\int P(x)\,dx}$$

$$= e^{\int P(x)\,dx}\frac{dy}{dx} + e^{\int P(x)\,dx}P(x)y. \tag{3}$$

By the Chain Rule:

$$\frac{d}{dx}e^{\int P(x)\,dx} = e^{\int P(x)\,dx}\frac{d}{dx}\int P(x)\,dx$$

$$= e^{\int P(x)\,dx}P(x)$$

Thus, if we multiply both sides of (2) by $e^{\int P(x)\,dx}$, we get

$$\underbrace{e^{\int P(x)\,dx}\frac{dy}{dx} + e^{\int P(x)\,dx}P(x)y}_{\text{This is } \frac{d}{dx}[e^{\int P(x)\,dx}y]} = e^{\int P(x)\,dx}f(x).$$

By comparing the left-hand side of the last equation with the result in (3), it follows that the last equation is the same as

$$\frac{d}{dx}\left[e^{\int P(x)\,dx}y\right] = e^{\int P(x)\,dx}f(x). \tag{4}$$

The form of equation (4) is the key for solving linear first-order differential equations. We can simply integrate both sides of (4) with respect to x. The function $e^{\int P(x)\,dx}$ that makes this possible is called an **integrating factor** for the DE. The procedure is outlined next.

Guidelines for Solving Linear Equations

(i) Put the given differential equation into the standard form (2); that is, make the coefficient of dy/dx unity by division.

(ii) Identify $P(x)$ (the coefficient of y) and find the integrating factor

$$e^{\int P(x)\,dx}.$$

(iii) Multiply the equation obtained in step (i) by the integrating factor.

(iv) The left-hand side of the equation in step (iii) is the derivative of the integrating factor and the dependent variable:

$$\frac{d}{dx}\left[e^{\int P(x)\,dx}y\right] = e^{\int P(x)\,dx}f(x).$$

(v) Integrate both sides of the equation found in step (iv).

In Section 8.1 we solved the equation $dy/dx - ky = 0$ by separation of variables, but since the DE is linear, we can also solve it by the foregoing procedure.

EXAMPLE 2 Using an Integrating Factor

The linear differential equation

$$\frac{dy}{dx} - ky = 0,$$

k a constant, is already in standard form (2). By identifying $P(x) = -k$, the integrating factor is $e^{\int (-k)dx} = e^{-kx}$ and, after multiplying the equation by this factor, we see that

$$e^{-kx}\frac{dy}{dx} - ke^{-kx}y = 0 \cdot e^{-kx} \quad \text{is the same as} \quad \frac{d}{dx}[e^{-kx}y] = 0.$$

◀ We need not use a constant of integration in computing $e^{\int P(x)dx}$.

Integration of both sides of the last equation with respect to x,

$$\int \frac{d}{dx}[e^{-kx}y]\, dx = \int 0\, dx$$

gives $e^{-kx}y = C$. From this last expression we get the same family of solutions $y = Ce^{kx}$ as in Example 4 of Section 8.1. ∎

Recall, in the discussion of (2) we stated that we seek a solution of a linear equation on an interval I for which P and f are continuous. As you work your way through the next example, note that P and f are both continuous on the interval $(0, \infty)$.

EXAMPLE 3 Solving a Linear DE on an Interval

Solve $x\dfrac{dy}{dx} - 4y = x^6 e^x$.

Solution By dividing by x we get the standard form

$$\frac{dy}{dx} - \frac{4}{x}y = x^5 e^x. \tag{5}$$

From this form we identify $P(x) = -4/x$ and $f(x) = x^5 e^x$ and observe that P and f are continuous for $x > 0$, that is, on $(0, \infty)$. Hence, the integrating factor is

we can use $\ln x$ instead of $\ln|x|$ since $x > 0$
↓
$$e^{-4\int dx/x} = e^{-4\ln x} = e^{\ln x^{-4}} = x^{-4}.$$

◀ The identity $e^{\ln N} = N$ is useful in computing the integrating factor.

Now we multiply (5) by x^{-4},

$$x^{-4}\frac{dy}{dx} - 4x^{-5}y = xe^x \quad \text{and obtain} \quad \frac{d}{dx}[x^{-4}y] = xe^x.$$

Using integration by parts on the right-hand side of

$$\int \frac{d}{dx}[x^{-4}y]\, dx = \int xe^x\, dx$$

yields the solution defined on $(0, \infty)$:

$$x^{-4}y = xe^x - e^x + C \quad \text{or} \quad y = x^5 e^x - x^4 e^x + Cx^4. \quad ∎$$

EXAMPLE 4 Solving a Linear DE on an Interval

Solve $(x^2 - 9)\dfrac{dy}{dx} + xy = 0$.

Solution We write the equation in standard form

$$\frac{dy}{dx} + \frac{x}{x^2 - 9}y = 0 \tag{6}$$

and identify $P(x) = x/(x^2 - 9)$. Although P is continuous on $(-\infty, -3), (-3, 3)$ and on $(3, \infty)$, we shall solve the equation on the first and third intervals. On these intervals, the integrating factor is

$$e^{\int x\,dx/(x^2-9)} = e^{\frac{1}{2}\int 2x\,dx/(x^2-9)} = e^{\frac{1}{2}\ln(x^2-9)} = e^{\ln\sqrt{x^2-9}} = \sqrt{x^2 - 9}.$$

After multiplying the standard form (6) by this factor, we get

$$\frac{d}{dx}\left[\sqrt{x^2 - 9}\,y\right] = 0, \quad \text{and integrating gives} \quad \sqrt{x^2 - 9}\,y = C.$$

For either $(-\infty, -3)$ or $(3, \infty)$, the solution of the equation is $y = \dfrac{C}{\sqrt{x^2 - 9}}$. ∎

EXAMPLE 5 An Initial-Value Problem

Solve the initial-value problem $\dfrac{dy}{dx} + y = x, y(0) = 4$.

Solution The equation is already in standard form, and $P(x) = 1$ and $f(x) = 1$ are continuous on $(-\infty, \infty)$. The integrating factor is $e^{\int dx} = e^x$, and so integrating

$$\frac{d}{dx}[e^x y] = xe^x$$

gives $e^x y = xe^x - e^x + C$. Solving this last equation for y yields the family of solutions

$$y = x - 1 + Ce^{-x}. \tag{7}$$

But from the initial condition we know that $y = 4$ when $x = 0$. Substituting these values in (7) implies $C = 5$. Hence, the solution of the problem on $(-\infty, \infty)$ is $y = x - 1 + 5e^{-x}$ and is the blue curve in FIGURE 8.2.1. ∎

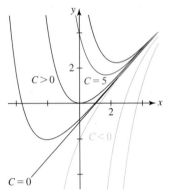

FIGURE 8.2.1 Family of solution curves in Example 5. Solution of the IVP is shown in blue.

It is interesting to observe that as x becomes unbounded in the positive direction, the graphs of *all* members of the family of solutions (7) for $C > 0$ or $C < 0$ are close to the graph of $y = x - 1$, which is shown in black in Figure 8.2.1. Indeed, $y = x - 1$ is the solution of the DE in Example 5 that corresponds to $C = 0$ in (7). This asymptotic behavior is attributable to the fact that the term Ce^{-x} in (7) becomes negligible for increasing values of x, that is, $e^{-x} \to 0$ as $x \to \infty$. We say that Ce^{-x} is a **transient term**. Although this behavior is not a characteristic of all families of solutions of linear equations (see Example 3), the notion of a transient is often important in applied problems.

$\dfrac{dy}{dx} = F(x, y)$ **NOTES FROM THE CLASSROOM**

If we solve (5) on an interval on which P and f are continuous, then it can be proved that a one-parameter family of solutions of the equation yields *all* solutions of the DE defined on the interval. In Example 3, the functions $P(x) = -4/x$ and $f(x) = x^5 e^x$ are continuous on the interval $(0, \infty)$. In this case, *every* solution of $dy/dx - (4/x)y = x^5 e^x$ on $(0, \infty)$ can be obtained from $y = x^5 e^x - x^4 e^x + Cx^4$ for appropriate choices of the constant C. For this reason, the family of solutions $y = x^5 e^x - x^4 e^x + Cx^4$ is called the **general solution** of the differential equation.

Exercises 8.2 Answers to selected odd-numbered problems begin on page ANS-25.

≡ Fundamentals

In Problems 1–22, solve the given linear differential equation.

1. $\dfrac{dy}{dx} = 4y$

2. $\dfrac{dy}{dx} + 2y = 0$

3. $2\dfrac{dy}{dx} + 10y = 1$

4. $x\dfrac{dy}{dx} + 2y = 3$

5. $\dfrac{dy}{dt} + y = e^{3t}$

6. $\dfrac{dy}{dt} = y + e^t$

7. $y' + 3x^2 y = x^2$

8. $y' + 2xy = x^3$

9. $x^2 \dfrac{dy}{dx} + xy = 1$

10. $(1 + x^2) \dfrac{dy}{dx} + xy = 2x$

11. $(1 + e^x) \dfrac{dy}{dx} + e^x y = 0$

12. $(1 - x^3) \dfrac{dy}{dx} = 3x^2 y$

13. $x \dfrac{dy}{dx} - y = x^2 \sin x$

14. $\dfrac{dy}{dx} + y = \cos(e^x)$

15. $\cos x \dfrac{dy}{dx} + (\sin x) y = 1$

16. $\sin x \dfrac{dy}{dx} + (\cos x) y = \sec^2 x$

17. $\dfrac{dy}{dx} + (\cot x) y = 2 \cos x$

18. $\dfrac{dr}{d\theta} + (\sec \theta) r = \cos \theta$

19. $(x + 2)^2 \dfrac{dy}{dx} = 5 - 8y - 4xy$

20. $\dfrac{dP}{dt} + 2tP = P + 4t - 2$

21. $x^2 \dfrac{dy}{dx} + x(x + 2) y = e^x$

22. $x \dfrac{dy}{dx} + (x + 1) y = e^{-x} \sin 2x$

In Problems 23–32, solve the given initial-value problem.

23. $\dfrac{dy}{dx} = x + y$, $y(0) = -4$

24. $\dfrac{dy}{dx} = 2x - 3y$, $y(0) = \tfrac{1}{3}$

25. $x \dfrac{dy}{dx} + y = e^x$, $y(1) = 2$

26. $x \dfrac{dy}{dx} + y = 4x + 1$, $y(1) = 8$

27. $x \dfrac{dy}{dx} - y = 2x^2$, $y(5) = 1$

28. $x(x + 1) \dfrac{dy}{dx} + xy = 1$, $y(1) = 10$

29. $(t + 1) \dfrac{dx}{dt} + x = \ln t$, $x(1) = 10$

30. $y' + (\tan t) y = \cos^2 t$, $y(0) = -1$

31. $L \dfrac{di}{dt} + Ri = E$, $i(0) = i_0$, $L, R,$ and E are constants

32. $\dfrac{dT}{dt} = k(T - T_m)$, $T(0) = T_0$, $k, T_m,$ and T_0 are constants

≡ Calculator/CAS Problems

In Problems 33 and 34, before attempting to solve the given initial-value problem review Problems 71 and 72 in Exercises 5.5.

33. **(a)** Express the solution of the initial-value problem $y' - 2xy = 2$, $y(0) = 1$, in terms of the **error function**

$$\operatorname{erf}(x) = \dfrac{2}{\sqrt{\pi}} \int_0^x e^{-t^2} dt.$$

(b) Use tables or a CAS to calculate $y(2)$. Use a CAS to graph the solution on the interval $(-\infty, \infty)$.

34. The **sine integral function** is defined by

$$\operatorname{Si}(x) = \int_0^x \dfrac{\sin t}{t} dt.$$

(a) Show that the solution of the initial-value problem $x^3 y' + 2x^2 y = 10 \sin x$, $y(1) = 0$ is

$$y = 10 x^{-2} [\operatorname{Si}(t) - \operatorname{Si}(1)].$$

(b) Use tables or a CAS to calculate $y(2)$. Use a CAS to graph the solution on the interval $(0, \infty)$.

≡ Think About It

35. Find a continuous solution of the initial-value problem

$$\dfrac{dy}{dx} + y = f(x), \quad f(x) = \begin{cases} 1, & 0 \le x \le 1 \\ 0, & x > 1 \end{cases}, \quad y(0) = 0.$$

Graph f and the solution of the IVP. [*Hint*: Solve the problem in two parts and use continuity to match the parts of your solution.]

36. Explain why we do not have to use a constant of integration when computing an integrating factor $e^{\int P(x) dx}$ for a linear differential equation.

37. In Example 4 we solved the given differential equation on the intervals $(-\infty, -3)$ and $(3, \infty)$. Find a solution of the DE on the interval $(-3, 3)$.

38. Suppose $P(t)$ represents the population of some animal species present in an environment at time t. If the symbol $\propto$ means "proportional to," in words give a physical interpretation of the mathematical statement

$$\dfrac{dP}{dt} \propto P.$$

39. The following system of differential equations is encountered in the study of a special type of radioactive series of elements:

$$\dfrac{dx}{dt} = -\lambda_1 x$$

$$\dfrac{dy}{dt} = -\lambda_1 x - \lambda_2 y,$$

where λ_1 and λ_2 are constants. Solve the system subject to $x(0) = x_0$, $y(0) = y_0$.

40. The nonlinear first-order differential equation

$$\dfrac{dy}{dx} + \dfrac{1}{x} y = xy^2$$

is a member of a class of nonlinear DEs called **Bernoulli equations**.

(a) Use the substitution $y = u^{-1}$ to show that the given Bernoulli equation becomes

$$\dfrac{du}{dx} - \dfrac{1}{x} u = -x.$$

(b) Find a solution of the given Bernoulli equation by solving the DE in part (a).

41. The differential equation
$$\frac{dy}{dx} = -\frac{1}{x+y}$$
is neither separable nor linear in the variable y. Take the reciprocal of both sides of the equation. Can this new differential equation be solved?

42. Although the differential equation $y'' + y' = x$ is second order it can be solved using the method discussed in this section. Solve the equation by letting $Y = y'$.

43. **(a)** Find a one-parameter family of solutions of the linear equation
$$x\frac{dy}{dx} + 3y = 6x^2.$$

(b) Find the member of the family of solutions in part (a) that satisfies the initial condition $y(-1) = 2$. Give the interval over which this solution is valid.

(c) Find the member of the family of solutions in part (a) that satisfies the initial condition $y(1) = 2$. Give the interval over which this solution is valid.

(d) Find an initial condition so that the corresponding member of the family of solutions in part (a) passes through the origin and is valid on the interval $(-\infty, \infty)$.

8.3 Mathematical Models

■ **Introduction** So far our experience with first-order DEs has been limited to either solving them or verifying that a given function is a solution. But mathematics is a language as well as a tool. As you undoubtedly remember from algebra and Section 1.7, we translate words into mathematics when solving a "word problem." So too, we can interpret words, empirical laws, observations, or simply assumptions, into mathematical terms. When we try to describe something, let us call it a *system*, in mathematical terms we are constructing a *model* of that system. If something in the system changes with time, say, either growing or decreasing at a certain *rate*— and a rate of change is a derivative—then a **mathematical model** of the system may be a differential equation.

In this section we will consider a few simple mathematical models and their solutions.

■ **Population Growth** One of the earliest attempts to model human population growth by means of mathematics was by the English economist **Thomas Malthus** (1776–1834) in 1798. Basically, the idea of the Malthusian model is the assumption that the rate at which a population of a country *grows* or increases is proportional to the total population $P(t)$ of the country at time t. In other words, the more people there are at time t, the more there are going to be in the future. In mathematical terms this assumption can be expressed as

$$\frac{dP}{dt} \propto P \quad \text{or} \quad \frac{dP}{dt} = kP, \tag{1}$$

(proportionality symbol ↓ points to $\propto$)

where k is a constant of proportionality. This simple model, which fails to take into account many factors (immigration and emigration, for example) that can influence human populations to either grow or decline, nevertheless turned out to be fairly accurate in predicting the population of the United States during the years 1790–1860. The differential equation given in (1) is still often used to model, over short intervals of time, the populations of bacteria or small animals.

The constant of proportionality k in (1) can be determined from the solution of the initial-value problem $dP/dt = kP$, $P(t_0) = P_0$ using a subsequent measurement of P at a time $t_1 > t_0$.

EXAMPLE 1 Bacterial Growth

A culture initially has P_0 number of bacteria. At $t = 1$ hr, the number of bacteria present is measured to be $\frac{3}{2}P_0$. If the rate of growth is proportional to the number of bacteria $P(t)$ present at time t, determine the time necessary for the number of bacteria to triple.

Solution We first solve the differential equation in (1) subject to the initial condition $P(0) = P_0$. Then we use the empirical observation that $P(1) = \frac{3}{2}P_0$ to determine the constant of proportionality k. Now we have already see that the equation $dP/dt = kP$ is both separable and linear. From Example 2 of Section 8.2, with the symbols P and t, in turn, playing the parts of y and x, a family of solutions of the DE is

$$P(t) = Ce^{kt}.$$

At $t = 0$, it follows that $P_0 = Ce^0 = C$, so $P(t) = P_0 e^{kt}$. At $t = 1$, we have $\frac{3}{2}P_0 = P_0 e^k$ or $e^k = \frac{3}{2}$. From the last equation $k = \ln\frac{3}{2} = 0.4055$. Thus

$$P(t) = P_0 e^{0.4055t}.$$

To find the time at which the number of bacteria has tripled, we solve $3P_0 = P_0 e^{0.4055t}$ for t. It follows that $0.4055t = \ln 3$, so

$$t = \frac{\ln 3}{0.4055} \approx 2.71 \text{ h}.$$

See FIGURE 8.3.1.

FIGURE 8.3.1 Graph of solution in Example 1

Note: We can write the function $P(t)$ obtained in the preceding example in an alternative form. From the laws of exponents,

$$P(t) = P_0 e^{kt} = P_0 (e^k)^t = P_0 \left(\frac{2}{3}\right)^t,$$

since $e^k = \frac{3}{2}$. This latter solution provides a convenient method for computing $P(t)$ for small positive integral values of t; it also clearly shows the influence of the subsequent experimental observation at $t = 1$ on the solution for all time. We notice too that the actual number of bacteria present initially—that is, at time $t = 0$—is quite irrelevant in finding the time required to triple the number in the culture. The necessary time to triple, say, 100 or even 100,000 bacteria is still approximately 2.7 h.

■ **Radioactive Decay** The nucleus of an atom consists of combinations of protons and neutrons. Many of these combinations of protons and neutrons are unstable, that is, the atoms decay or transmute into the atoms of another substance. Such nuclei are said to be *radioactive*. For example, over time the highly radioactive radium, Ra-226, transmutes into the radioactive gas radon, Rn-222. To model the phenomenon of radioactive decay, it is assumed that the rate dA/dt at which the nuclei of a substance decays is proportional to the amount of the substance (more precisely, the number of nuclei) $A(t)$ remaining at time t:

$$\frac{dA}{dt} \propto A \quad \text{or} \quad \frac{dA}{dt} = kA. \quad (2)$$

The model (2) for decay also occurs in a biological setting, such as determining the time that it takes for 50% of a drug to be eliminated from a body by excretion or metabolism. The point of (1) and (2) is simply this:

- *A single differential equation can serve as a mathematical model for many different phenomena.*

Of course, since equations (1) and (2) are exactly the same, their solutions are exactly the same (namely, Ce^{kt}); the difference is only in the symbols and their interpretation. As shown in FIGURE 8.3.2, the exponential function e^{kt} increases as t increases for $k > 0$ and decreases as t increases for $k < 0$. Thus, problems describing growth (whether of animal populations, bacteria, or even capital) are characterized by a positive value of k, whereas problems involving decay yield a negative k value. Accordingly, we say that k is either a **growth constant** ($k > 0$) or a **decay constant** ($k < 0$).

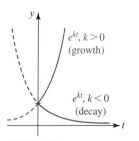

FIGURE 8.3.2 Exponential growth and decay

■ **Half-Life** In physics the **half-life** is a measure of the stability of a radioactive substance. The half-life is simply the time it takes for one-half of the atoms in an initial amount A_0 to disintegrate, or transmute, into the atoms of another element. In terms of the solution $A(t) = Ce^{kt}$ of (2), the half-life of a decaying element is the time t for which $A(t) = \frac{1}{2}A_0$. The longer the half-life of a substance, the more stable it is. For example, the half-life of highly radioactive radium, Ra-226, is about 1700 years. In 1700 years one-half of a given quantity of Ra-226 is transmuted into radon, Rn-222. The most commonly occurring uranium isotope, U-238, has a half-life of approximately 4,500,000,000 years. In about 4.5 billion years, one-half of a quantity of U-238 is transmuted into lead, Pb-206.

■ **Carbon Dating** In the 1940s the chemist Willard Libby devised a method of using radioactive carbon as a means of determining the approximate ages of fossils. The theory of **carbon dating** is based on the fact that the radioactive isotope carbon-14 is produced in the atmosphere by the action of cosmic radiation on nitrogen. The ratio of the amount of C-14 to ordinary carbon in the atmosphere appears to be a constant, and as a consequence the proportionate amount of the isotope present in all living organisms is the same as that in the atmosphere. When an organism dies, the absorption of C-14, by either breathing or eating, ceases. Thus, by comparing the proportionate amount of C-14 present, say, in a fossil with the constant ratio found in the atmosphere, it is possible to obtain a reasonable estimation of its age. The method is based on the knowledge that the half-life of the radioactive C-14 is approximately 5730 years. For his work Libby won the Nobel Prize for chemistry in 1960. Libby's method has been used to date wooden furniture in Egyptian tombs, the woven flax wrappings of the Dead Sea scrolls, and the controversial Shroud of Turin.

EXAMPLE 2 Dating a Fossil

A fossilized bone is found to contain $\frac{1}{1000}$ the original amount of C-14. Determine the age of the fossil.

Solution The starting point is the differential equation $dA/dt = kA$, where $A(t)$ is the amount of C-14 remaining at time t. If A_0 is the initial amount of C-14 in the bone, it follows as in Example 1 that

$$A(t) = A_0 e^{kt}.$$

We can use the fact that $A(5730) = \frac{1}{2}A_0$ to determine the decay constant k. Setting $t = 5730$ in $A(t)$ implies $\frac{1}{2}A_0 = A_0 e^{5730k}$ and so from $5730k = \ln\frac{1}{2} = -\ln 2$ we find that

$$k = -\frac{1}{5730}\ln 2 = -0.00012097.$$

Therefore, $A(t) = A_0 e^{-0.00012097t}$. Now the age of the fossil is determined from the equation $A(t) = \frac{1}{1000}A_0$. That is, $\frac{1}{1000}A_0 = A_0 e^{-0.00012097t}$ and so $-0.00012097t = \ln\frac{1}{1000} = -\ln 1000$ yields

$$t = \frac{\ln 1000}{0.00012097} \approx 57{,}103 \text{ years.} \qquad \blacksquare$$

The date found in Example 2 is really at the border of accuracy for this method. The usual carbon-14 technique is limited to about 9 half-lives of the isotope or about 50,000 years. One reason is that the chemical analysis needed to obtain an accurate measurement of the remaining C-14 becomes somewhat formidable around the point of $\frac{1}{1000}A_0$. Also, this analysis demands the destruction of a rather large sample of the specimen. If this measurement is accomplished indirectly, based on the actual radioactivity of the specimen, then it is very difficult to distinguish between the radiation from the fossil and the normal background radiation.

In recent developments geologists have shown that in some cases, dates determined by carbon dating may be off by as much as 3500 years. One conjecture for this possible error is the fact that carbon-14 levels in the air are known to vary with time. These same scientists have devised another dating technique based on the fact that living organisms ingest traces of uranium. By measuring the relative amounts of uranium and thorium (the isotope into which the uranium decays) and by knowing the half-lives of these elements, scientists can determine the age of a fossil. The advantage of this method is that it can date fossils up to 500,000 years; the disadvantage is that it is effective mostly on marine fossils. Another isotopic technique, using potassium-40 and argon-40, when applicable, can give dates of several million years. See Problem 37 in Exercises 8.3. Nonisotopic methods based on the use of amino acids are also sometimes possible.

■ **Cooling** Newton's law of cooling states that the rate at which the temperature $T(t)$ changes in a cooling body is proportional to the difference between the temperature in the body and the constant temperature T_m of the surrounding medium; that is,

$$\frac{dT}{dt} = k(T - T_m), \tag{3}$$

where k is a constant of proportionality.

EXAMPLE 3   Cooling Cake

When a cake is removed from a baking oven, its temperature is measured at 300°F. Three minutes later its temperature is 200°F. Determine the temperature of the cake at any time after leaving the oven if the room temperature is 70°F.

Solution   We identify the temperature of the room (70°F) as T_m. To find the temperature of the cake at time t, we must solve the initial-value problem

$$\frac{dT}{dt} = k(T - 70), \quad T(0) = 300$$

and determine the value of k so that $T(3) = 200$. The DE is both separable and linear. Assuming $T > 70$, it follows by separation of variables that

$$\frac{1}{T - 70} dT = k\, dt$$

$$\int \frac{1}{T - 70} dT = \int k\, dt$$

$$\ln(T - 70) = kt + C_1$$

$$T - 70 = Ce^{kt} \quad \leftarrow C = e^{C_1}$$

$$T = 70 + Ce^{kt}.$$

When $t = 0$, $T = 300$ so that $300 = 70 + C$ gives $C = 230$ and therefore $T(t) = 70 + 230e^{kt}$. From $T(3) = 200$ we find $e^{3k} = \frac{13}{23}$ and so, to four decimal places, a calculator gives

$$k = \frac{1}{3} \ln \frac{13}{23} = -0.1902.$$

Thus, $T(t) = 70 + 230e^{-0.1902t}$. The graph of T along with some calculated values are given in FIGURE 8.3.3.   ■

t (minutes)	$T(t)$
20.1	75°
21.3	74°
22.8	73°
24.9	72°
28.6	71°
32.3	70.5°

(b)

FIGURE 8.3.3   Graph of solution in Example 3

■ **Mixtures**   The mixing of two fluids sometimes gives rise to a linear first-order differential equation. In the next example we consider the mixture of two salt solutions of different concentrations.

EXAMPLE 4   Mixing a Salt Solution

Initially 50 lb of salt is dissolved in a large tank holding 300 gal of water. A brine solution is pumped into the tank at a rate of 3 gal/min, and the well-stirred solution is then pumped out at the same rate. See FIGURE 8.3.4. If the concentration of the solution entering is 2 lb/gal, determine the amount of salt in the tank at time t. How much salt is present after 50 min? After a long time?

Solution   Let $A(t)$ be the amount of salt (in pounds) in the tank at time t. For problems of this sort, the net rate at which $A(t)$ changes is given by

$$\frac{dA}{dt} = \begin{pmatrix} \text{rate of} \\ \text{substance entering} \end{pmatrix} - \begin{pmatrix} \text{rate of} \\ \text{substance leaving} \end{pmatrix} = R_1 - R_2. \tag{4}$$

FIGURE 8.3.4   Mixing tank in Example 4

Now the rate at which the salt enters the tank is, in pounds per minute,

$$R_1 = (3 \text{ gal/min}) \cdot (2 \text{ lb/gal}) = 6 \text{ lb/min},$$

whereas the rate at which salt is leaving is

$$R_2 = (3 \text{ gal/min}) \cdot \left(\frac{A}{300} \text{ lb/gal}\right) = \frac{A}{100} \text{ lb/min}.$$

Thus, equation (4) becomes

$$\frac{dA}{dt} = 6 - \frac{A}{100} \quad \text{or} \quad \frac{dA}{dt} + \frac{1}{100}A = 6. \tag{5}$$

We solve the last equation subject to the initial condition $A(0) = 50$.

Since the integrating factor is $e^{t/100}$, we can write (5) as

$$\frac{d}{dt}[e^{t/100}A] = 6e^{t/100}$$

and therefore $e^{t/100}A = 600e^{t/100} + C$ or $A = 600 + Ce^{-t/100}$. When $t = 0, A = 50$, so we find that $C = -550$. Finally, we obtain

$$A(t) = 600 - 550e^{-t/100}. \tag{6}$$

At $t = 50$ we find $A(50) = 266.41$ lb. Also, as $t \to \infty$ it is seen from (6) and FIGURE 8.3.5 that $A \to 600$. Of course, this is what we would expect; over a long period of time the number of pounds of salt in the solution must be $(300 \text{ gal})(2 \text{ lb/gal}) = 600$ lb. ∎

t (minutes)	$A(t)$
50	266.41
100	397.67
150	477.27
200	525.57
300	572.62
400	589.93

(b)

FIGURE 8.3.5 Graph of solution in Example 4

In Example 4 we assumed that the rate at which the solution was pumped in was the same as the rate at which the solution was pumped out. However, this need not be the case; the mixed brine solution could be pumped out at a rate faster or slower than the rate at which the other solution is pumped in. For example, if the well-stirred solution is pumped out at the slower rate of 2 gal/min, then the solution is accumulating at a rate of $(3 - 2)$ gal/min $= 1$ gal/min. After t min there are $300 + t$ gal of brine in the tank. The rate at which the salt is leaving is then

$$R_2 = (2 \text{ gal/min}) \cdot \left(\frac{A}{300 + t} \text{lb/gal}\right) = \frac{2A}{300 + t} \text{ lb/min}.$$

Equation (4) in this case becomes

$$\frac{dA}{dt} = 6 - \frac{2A}{300 + t} \quad \text{or} \quad \frac{dA}{dt} + \frac{2}{300 + t}A = 6. \tag{7}$$

Inspection of the last equation shows that it is linear. We leave its solution as an exercise. See Problems 18–20 in Exercises 8.3.

■ **Newton's Second Law of Motion** To construct a mathematical model of the motion of a body moving in a force field, the usual starting point is Newton's second law of motion. Recall, **Newton's first law of motion** states that a body will either remain at rest or will continue to move with a constant velocity unless acted upon by an external force. In each of these two cases, this is equivalent to saying that when the sum of the forces $F = \Sigma F_k$—that is, the net or resultant force—acting on the body is zero, then the acceleration a of the body is zero. **Newton's second law of motion** indicates that when the net force acting on a body is *not* zero, then the net force is proportional to its acceleration a, or more precisely, $F = ma$, where m is the mass of the body.

■ **Falling Bodies and Air Resistance** It has been established empirically that when a body moves through a resistive medium such as air (or water), the retarding force due to the medium, called the **drag force**, acts in the direction opposite to that of the motion and is proportional to a power of the body's velocity, that is, kv^α. Here k is a constant of proportionality and α is constant in the range $1 \leq \alpha \leq 2$. Roughly, for slow speeds we take $\alpha = 1$. Now suppose a falling body of mass m encounters air resistance proportional to its instantaneous velocity v. If we take, in this circumstance, the positive direction to be oriented downward, then the net force acting on the mass is given by $mg - kv$ where the weight mg of the body is a force acting in the positive direction and air resistance is a force acting in the opposite or upward direction. See FIGURE 8.3.6. Now since v is related to acceleration a by $dv/dt = a$, Newton's second law becomes $F = ma = m\,dv/dt$. By equating the net force to this form of Newton's second law, we obtain a linear differential equation for the velocity v of the body at time t,

$$m\frac{dv}{dt} = mg - kv, \quad (8)$$

where k is a positive constant of proportionality.

For high-speed motion—such as a skydiver who is falling before the parachute is opened—it is usually assumed that the air resistance is proportional to the square of the instantaneous velocity, in other words, $\alpha = 2$. If the positive direction is again taken to be downward, then a model for the velocity v of a falling body is given by the nonlinear differential equation

$$m\frac{dv}{dt} = mg - kv^2, \quad (9)$$

where k is a positive constant of proportionality. See Problem 22 in Exercises 8.3.

FIGURE 8.3.6 Forces acting on falling body of mass m

EXAMPLE 5 Velocity of a Falling Body

Solve (8) subject to the initial condition $v(0) = v_0$.

Solution Equation (8) is linear and has the standard form

$$\frac{dv}{dt} + \frac{k}{m}v = g. \quad (10)$$

Multiplying (10) by the integrating factor $e^{kt/m}$ enables us to write the equation as

$$\frac{d}{dt}\left[e^{kt/m}v\right] = g\,e^{kt/m}.$$

Integrating and solving for v then gives $v(t) = mg/k + Ce^{-kt/m}$. The initial condition $v(0) = v_0$ implies $C = v_0 - mg/k$ and so the velocity function for the falling body is

$$v(t) = \frac{mg}{k} + \left(v_0 - \frac{mg}{k}\right)e^{-kt/m}. \quad (11)\ \blacksquare$$

Two observations are in order about the solution in Example 5. If we desire the position function $s(t)$ of the falling body, then it is a simple matter of integrating the equation

$$\frac{ds}{dt} = v(t),$$

where $v(t)$ is given in (11). See Problem 21 in Exercises 8.3. Also, because of the air resistance, the solution (11) clearly shows that the velocity of a body that falls a long distance does not increase indefinitely. Because the term $(v_0 - mg/k)e^{-kt/m}$ in (11) is *transient* (see page 448), we see that $v(t) \to mg/k$ as $t \to \infty$. This limiting value of the velocity $v_{\text{ter}} = mg/k$ is called the **terminal velocity** of the body. It is left as an exercise to find $v(t)$ and v_{ter} when the mathematical model for the velocity is given by (9). See Problem 22 in Exercises 8.3.

Exercises 8.3   Answers to selected odd-numbered problems begin on page ANS-25.

≡ Fundamentals

1. The population of a certain community is known to increase at a rate proportional to the number of people present at time t. If the population has doubled in 5 years, how long will it take to triple? To quadruple?

2. Suppose it is known that the population of the community in Problem 1 is 10,000 after 3 years. What was the initial population? What will the population be in 10 years?

3. The population of a town grows at a rate proportional to the population at time t. Its initial population of 500

increases by 15% in 10 years. What will the population be in 30 years?

4. The population of bacteria in a culture grows at a rate proportional to the number of bacteria present at time t. After 3 h it is observed that there are 400 bacteria present. After 10 h there are 2000 bacteria present. What was the initial number of bacteria?

5. The radioactive isotope of lead, Pb-209, decays at a rate proportional to the amount present at time t and has a half-life of 3.3 h. If 1 g of lead is present initially, how long will it take for 90% of the lead to decay?

6. Initially there were 100 mg of a radioactive substance present. After 6 h the mass decreased by 3%. If the rate of decay is proportional to the amount of the substance present at time t, find the amount remaining after 24 h.

7. Determine the half-life of the radioactive substance described in Problem 6.

8. Show that the half-life of a radioactive substance is, in general,

$$t = \frac{(t_2 - t_1)\ln 2}{\ln(A_1/A_2)},$$

where $A_1 = A(t_1)$ and $A_2 = A(t_2)$, $t_1 < t_2$.

9. When a vertical beam of light passes through a transparent substance, the rate at which its intensity I decreases is proportional to $I(t)$, where t represents the thickness of the medium (in feet). In clear seawater, the intensity 3 ft below the surface is 25% of the initial intensity I_0 of the incident beam. What is the intensity of the beam 15 ft below the surface?

10. When interest is compounded continuously, the amount of money increases at a rate proportional to the amount S present at time t: $dS/dt = rS$, where r is the annual rate of interest.
    (a) Find the amount of money accrued at the end of 5 years when $5000 is deposited in a savings account drawing $5\frac{3}{4}\%$ annual interest compounded continuously.
    (b) In how many years will the initial sum deposited be doubled?
    (c) Use a calculator to compare the number obtained in part (a) with the value

    $$S = 5000\left(1 + \frac{0.0575}{4}\right)^{5(4)}.$$

    This value represents the amount accrued when interest is compounded quarterly.

11. In a piece of burned wood, or charcoal, it was found that 85.5% of the C-14 had decayed. Use the information in Example 2 to determine the approximate age of the wood. (It is precisely these data that archaeologists used to date prehistoric paintings in a cave in Lascaux, France.)

12. A thermometer is taken from an inside room to the outside where the air temperature is 5°F. After 1 min the thermometer reads 55°F, and after 5 min the reading is 30°F. What is the initial temperature of the room?

13. A thermometer is removed from a room where the air temperature is 70°F to the outside where the temperature is 10°F. After $\frac{1}{2}$ min the thermometer reads 50°F. What is the reading at $t = 1$ min? How long will it take for the thermometer to reach 15°F?

14. Equation (3) also holds when an object absorbs heat from the surrounding medium. If a small metal bar whose initial temperature is 20°C is dropped into a container of boiling water, how long will it take for the bar to reach 90°C if it is known that its temperature increased 2° in 1 s? How long will it take the bar to reach 98°C?

15. A tank contains 200 L of fluid in which 30 g of salt is dissolved. Brine containing 1 g of salt per liter is then pumped into the tank at a rate of 4 L/min; the well-mixed solution is pumped out at the same rate. Find the number of grams of salt $A(t)$ in the tank at time t.

16. Solve Problem 15 assuming pure water is pumped into the tank.

17. A large tank is filled with 500 gal of pure water. Brine containing 2 lb of salt per gallon is pumped into the tank at a rate of 5 gal/min. The well-mixed solution is pumped out at the same rate. Find the number of pounds of salt $A(t)$ in the tank at time t.

18. Solve the differential equation (7) subject to the initial condition $A(0) = 50$ lb.

19. Reread the discussion following Example 4. Then solve Problem 17 under the assumption that the solution is pumped out at a faster rate of 10 gal/min. Determine when the tank is empty.

20. A large tank is partially filled with 100 gal of fluid in which 10 lb of salt is dissolved. Brine containing $\frac{1}{2}$ lb of salt per gallon is pumped into the tank at a rate of 6 gal/min. The well-mixed solution is then pumped out at a slower rate of 4 gal/min. Find the number of pounds of salt in the tank after 30 min.

21. Reread the discussion following Example 5. Then find the position function $s(t)$ for the falling body in Example 5. Since the positive direction was assumed to be downward, assume that $s(0) = 0$.

22. Solve the differential equation (9) subject to $v(0) = v_0$. Express the velocity function $v(t)$ in terms of the hyperbolic tangent function. With the aid of Figure 3.10.2(a) determine the terminal velocity v_{ter} of a falling body.

≡ Additional Mathematical Models

23. The rate at which a drug disseminates into the bloodstream is governed by the differential equation

$$\frac{dX}{dt} = A - BX,$$

where A and B are positive constants. The function $X(t)$ describes the concentration of the drug in the bloodstream at time t. Find $X(t)$. What is the limiting value of $X(t)$ as $t \to \infty$? At what time is the concentration one-half this limiting value? Assume that $X(0) = 0$.

24. Suppose a cell is suspended in a solution containing a solute of constant concentration C_s. Suppose further that the cell has constant volume V and that the area of its permeable membrane is the constant A. By **Fick's law**, the rate of change of its mass m is directly proportional to the area A and the difference $C_s - C(t)$, where $C(t)$ is the concentration of the solute inside the cell at time t. Find $C(t)$ if $m = VC(t)$ and $C(0) = C_0$. See FIGURE 8.3.7.

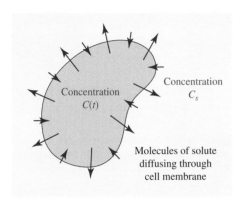

FIGURE 8.3.7 Cell in Problem 24

25. A heart pacemaker, shown in FIGURE 8.3.8, consists of a battery, a capacitor, and the heart as a resistor. When the switch S is at P, the capacitor charges; when S is at Q, the capacitor discharges, sending an electrical stimulus to the heart. During this time, the voltage E applied to the heart is given by the linear differential equation

$$\frac{dE}{dt} = -\frac{1}{RC}E, \quad t_1 < t < t_2,$$

where R and C are constants. Determine $E(t)$ if $E(t_1) = E_0$. (Of course, the opening and closing of the switch are periodic in time to simulate the natural heartbeat.)

FIGURE 8.3.8 Heart pacemaker in Problem 25

26. In a series circuit that contains only a resistor and an inductor, Kirchhoff's second law states that the sum of the voltage drop across the inductor ($L(di/dt)$) and the voltage drop across the resistor (iR) is the same as the impressed voltage (E) on the circuit. See FIGURE 8.3.9. Thus, we obtain the differential equation for the current $i(t)$:

$$L\frac{di}{dt} + Ri = E,$$

FIGURE 8.3.9 Series circuit in Problem 26

where L and R are constants known as the inductance and the resistance, respectively. Determine the current $i(t)$ if E is 12 volts, the inductance is $\frac{1}{2}$ henry, the resistance is 10 ohms, and $i(0) = 0$.

27. A 30-volt battery is connected to a series circuit in which the inductance is 0.1 henry and the resistance is 50 ohms. Find the current $i(t)$ if $i(0) = 0$. Determine the behavior of the current for large values of time. (See Problem 26.)

28. Suppose a water tank has the form of a right-circular cylinder. If water is allowed to drain under the influence of gravity through a hole in the bottom of the tank, then the height h of water at time t is given by the nonlinear differential equation

$$\frac{dh}{dt} = -c\frac{A_h}{A_w}\sqrt{2gh},$$

where A_w and A_h are the cross-sectional areas of the water and the hole, respectively, and c is a friction/contraction factor at the hole. See FIGURE 8.3.10.

(a) Solve the equation if the initial height of the water is 20 ft and $A_w = 50$ ft^2 and $A_h = \frac{1}{4}$ ft^2.
(b) If $c = 1$, at what time is the tank empty?
(c) How long would it take the tank to empty if the friction/contraction factor is $c = 0.6$?

FIGURE 8.3.10 Tank in Problem 28

29. Around 1840, the Belgian mathematician–biologist P. F. Verhulst was concerned with mathematical models for predicting human population of countries. One of the equations he studied was

$$\frac{dP}{dt} = P(a - bP),$$

where $a > 0, b > 0$. This differential equation is now known as the **logistic equation**; the graph of a solution of the DE is known as a **logistic curve**. Show that a solution of this DE subject to the initial condition $P(0) = P_0$ is

$$P(t) = \frac{aP_0}{bP_0 + (a - bP_0)e^{-at}}.$$

30. The population $P(t)$ at time t in a suburb of a large city is modeled by the initial-value problem

$$\frac{dP}{dt} = P(10^{-1} - 10^{-7}P), \quad P(0) = 5000,$$

where t is measured in months. Find $P(t)$ and determine the limiting value of the population over a long period of time. At what time will the population be equal to one-half of this limiting value?

31. Suppose a student carrying a flu virus returns to an isolated college campus of 1000 students. If it is assumed that the rate at which the virus spreads is proportional not only to the number x of infected students but also to the number $1000 - x$ not infected, then a mathematical model for the number of infected students is

$$\frac{dx}{dt} = kx(1000 - x),$$

where $k > 0$ is a constant of proportionality, and t is time measured from the day the student returns to campus. If $x(0) = 1$ and if it is observed that $x(4) = 50$, then according to this model, how many students are infected after 6 days? Sketch a graph of the solution curve.

32. When two chemicals A and B are combined a compound C is formed. The resulting second-order reaction between the two chemicals is modeled by the differential equation

$$\frac{dX}{dt} = k(250 - X)(40 - X),$$

where $X(t)$ denotes the number of grams of the compound C present at time t.

(a) Determine $X(t)$ if it is known that $X(0) = 0$ g and $X(10) = 30$ g.
(b) How much of the compound C is present at 15 min?
(c) The amounts of chemicals A and B remaining at time t are $50 - \frac{1}{5}X$ and $32 - \frac{4}{5}X$, respectively. How many grams of the compound C is formed as $t \to \infty$? How many grams of the chemicals A and B remain as $t \to \infty$?

≡ Think About It

33. A rocket is shot vertically upward from the ground with an initial velocity v_0. See FIGURE 8.3.11. If the positive direction is taken to be upward, then in the absence of air resistance, the differential equation for the velocity v after fuel burnout is

$$v\frac{dv}{dy} = -\frac{k}{y^2},$$

where k is a positive constant.

(a) Solve the differential equation.
(b) If $k = gR^2$ and $g = 32$ ft/s^2, $R = 4000$ mi, use a calculator to show that the "escape velocity" of a rocket is approximately $v_0 = 25{,}000$ mi/h.

FIGURE 8.3.11 Rocket in Problem 33

34. Suppose a sphere of ice melts at a rate proportional to its surface area. Determine the volume V of the sphere at time t.

35. In a model for the growth of tissue, let $A(t)$ be the area of the tissue culture at time t. See FIGURE 8.3.12. Since the majority of cell divisions take place on the peripheral portion of the tissue, the number of cells on the periphery is proportional to $\sqrt{A(t)}$. If it is assumed that the rate of growth of the area is jointly proportional to $\sqrt{A(t)}$ and $M - A(t)$, then a mathematical model for A is given by

$$\frac{dA}{dt} = k\sqrt{A}(M - A),$$

where M is the final area of the tissue when growth is completed.

(a) Solve the differential equation by separation of variables. [*Hint*: Use a substitution as in Section 7.2 to carry out the integration with respect to A.]
(b) Find $\lim_{t \to \infty} A(t)$.

FIGURE 8.3.12 Tissue growth in Problem 35

≡ Projects

36. **A Mathematical Classic—Time of Death** The following problem occurs in almost all texts on differential equations.

A dead body was found within a closed room of a house where the temperature was a constant 70°F. A measurement of the core temperature of the body at the time of its discovery was found to be 85°F. A second measurement, one hour later, showed that the core temperature of the body was 80°F. Use the fact that if $t = 0$ corresponds to the time of death, then the core temperature of the body at that time was 98.6°F. Determine how many hours elapsed before the body was found. [*Hint*: Let $t_1 > 0$ denote the time that the body was discovered.]

37. **Potassium/Argon Dating** The mineral potassium, whose chemical symbol is K, is the eighth most abundant element in the Earth's crust, making up about 2% of it by weight, and one of its naturally occurring isotopes, K-40, is radioactive. The radioactive decay of K-40 is more complex than that of carbon-14 because each of its atoms decays through one of two different nuclear decay reactions into one of two different substances: the mineral calcium-40 (Ca-40) or the gas argon-40 (Ar-40). Dating methods have been developed using both of these decay products. In each case, the age of a sample is calculated using the ratio of two numbers: the amount of the *parent* isotope K-40 in the sample and the amount of the *daughter* isotope (Ca-40 or Ar-40) in the sample that is **radiogenic**, in other words, the substance that originates from the decay of the parent isotope after the formation of the rock.

The amount of K-40 in a sample is easy to calculate. K-40 comprises 1.17% of naturally occurring potassium, and this small percentage is distributed quite uniformly, so that the mass of K-40 in the sample is just 1.17% of the total mass of potassium in the sample, which can be measured. But for several reasons it is complicated, and sometimes problematic, to determine how much of the Ca-40 in a sample is radiogenic. In contrast, when an igneous rock is formed by volcanic activity, all of the argon (and other) gas previously trapped in the rock is driven away by the intense heat. At the moment when the rock cools and solidifies, the gas trapped inside the rock has the same composition as the atmosphere. There are three stable isotopes of argon, and in the atmosphere they occur in the following relative abundances: 0.063% Ar-38, 0.337% Ar-36, and 99.60% Ar-40. Of these, just one, Ar-36, is not created radiogenically by the decay of any element, so any Ar-40 in excess of $99.60/(0.337) = 295.5$ times the amount of Ar-36 must be radiogenic. So the amount of radiogenic Ar-40 in the sample can be determined from the amounts of Ar-38 and Ar-36 in the sample, which can be measured.

Assuming that we have a sample of rock for which the amount of K-40 and the amount of radiogenic Ar-40 have been determined, how can we calculate the age of the rock? Let $P(t)$ be the amount of K-40, $A(t)$ the amount of radiogenic Ar-40, and $C(t)$ the amount of radiogenic Ca-40 in the sample as functions of time t in years since the formation of the rock. Then a mathematical model for the decay of K-40 is the system of linear first-order differential equations

$$\frac{dA}{dt} = k_A P$$

$$\frac{dC}{dt} = k_C P$$

$$\frac{dP}{dt} = -(k_A + k_C)P,$$

where $k_A = 0.581 \times 10^{-10}$ and $k_C = 4.962 \times 10^{-10}$.

(a) Find a formula for $P(t)$. What is the half-life of K-40?

(b) Show that

$$A(t) = \frac{k_A}{k_A + k_C} P(t)(e^{(k_A + k_C)t} - 1).$$

(c) After a very long time (that is, let $t \to \infty$), what percentage of the K-40 originally present in the sample decays to Ar-40? What percentage decays to Ca-40?

(d) Show that the age t of the rock as a function of the present amounts $P(t)$ of K-40 and $A(t)$ of radiogenic Ar-40 in the sample is

$$t = \frac{1}{k_A + k_C} \ln\left[\frac{A(t)}{P(t)}\left(\frac{k_A + k_C}{k_A}\right) + 1\right].$$

(e) Suppose it is found that each gram of a rock sample contains 8.6×10^{-7} grams of radiogenic Ar-40 and 5.3×10^{-6} grams of K-40. How old is the rock?

8.4 Solution Curves without a Solution

Introduction Most differential equations cannot be solved. Perhaps the last sentence should be balanced by saying that *many* differential equations possess solutions, but the problem is finding them. When we say that a solution of a DE exists we do not mean that there also exists a method for finding it in the sense of being able to exhibit an exact solution, namely, a solution either given by an explicit function or as a function defined implicitly. It may be that the best we can do is to analyze a DE *qualitatively* or *numerically*.

In this section we shall examine two ways of analyzing a first-order DE qualitatively. We begin with a fundamental concept: A derivative dy/dx gives slope.

A First-Order DE Defines Slope Because a solution $y(x)$ of a first-order differential equation $dy/dx = F(x, y)$ is a differentiable function on some interval I, it must also be continuous on I. Thus, the corresponding solution curve on I has no breaks and must possess a tangent line at each point $(x, y(x))$. The slope of the tangent line at $(x, y(x))$ on a solution curve is the value of the first derivative dy/dx at this point, and this we know from the **slope function** F of the

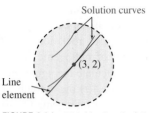

FIGURE 8.4.1 Neighborhood of the point (3, 2)

differential equation: $F(x, y(x))$. Now suppose that (x, y) represents any point in the xy-plane at which the function F is defined. The slope function F assigns a value $F(x, y)$ to the point; the value is the slope of a line. A short line segment, called a **line element**, is drawn through (x, y) with slope $F(x, y)$. For example, consider the equation $dy/dx = x - y$, where $F(x, y) = x - y$. At the point $(3, 2)$, for example, the slope of a line element is $F(3, 2) = 1$. As shown in FIGURE 8.4.1, a solution curve that passes through $(3, 2)$ does so tangent to the line element; a different solution curve that passes close to $(3, 2)$ will have a similar shape in a *small* neighborhood of the point.

▌ **Direction Fields** Now suppose we systematically evaluate $F(x, y)$ over a rectangular grid of points in the xy-plane and draw a line element at each point where F is evaluated. The collection of all these line elements is called a **direction field** or a **slope field** of the differential equation $dy/dx = F(x, y)$. Visually, the direction field suggests the appearance or shape of a family of solution curves of the differential equation and consequently it may be possible to see certain qualitative aspects (for example, increasing, decreasing, and concavity) of a solution curve. A single solution curve that wends its way through the direction field must follow the flow pattern of the field; it is tangent to a line element when it intersects a point in the grid.

▌ **Solution Curves without a Solution** Sketching a direction field by hand is straightforward, but time consuming; it is probably one of those tasks about which an argument can be made for doing it once in a lifetime, but it is overall most efficiently carried out by means of computer software. FIGURE 8.4.2 was obtained using a software direction field application with $dy/dx = x - y$ and a 5×5 rectangular region, where points in that region have a vertical and horizontal separation of $\frac{1}{2}$ unit, that is, at points (mh, nh), $h = \frac{1}{2}$, m and n integers such that $-10 \le m \le 10$, $-10 \le n \le 10$.

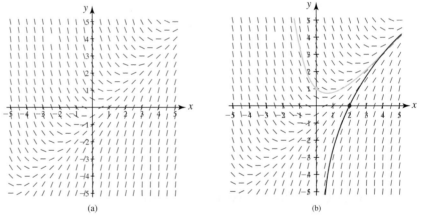

FIGURE 8.4.2 Direction field in (a); solution curves of DE superimposed on direction field in (b)

In Figure 8.4.2(a), notice that at any point along the line $y = x$, slopes are $F(x, x) = 0$ and so the line elements are horizontal. Moreover, the line $y = x$ splits the plane into two regions, above the line $(y > x)$ the line elements have negative slope, whereas below the line $(y < x)$ the line elements have positive slope. Reading left to right, imagine a solution curve starting at a point in the second quadrant, moving downward, becoming flat as it passes through the line $y = x$ and then moving upward into the first quadrant—in other words, its shape would be concave upward. We have seen the family of solutions of this DE in Example 5 of Section 8.2. You should compare the sample graphs in Figure 8.2.1 with the direction field in Figure 8.4.2(a). In Figure 8.4.2(b) we have given the two solution curves corresponding to the solutions of $dy/dx = x - y$ that pass through $(0, 1)$ (in green) and $(2, 0)$ (in red).

EXAMPLE 1 Direction Field

Sketch the direction field for $\dfrac{dy}{dx} = -\dfrac{x}{y}$.

Solution This is the differential equation in Example 2 in Section 8.1.

(a) If we use a grid of points (x, y) with integer coordinates for $-5 \le x \le 5$, $-5 \le y \le 5$ then it is straight forward to compute the slopes of the line elements in the four quadrants by hand. For $x > 0, y > 0$ (first quadrant) the slopes $F(x, y) = -x/y$ are given in the following table.

$F(x, y)$	$y = 1$	$y = 2$	$y = 3$	$y = 4$	$y = 5$
$x = 1$	-1	$-\frac{1}{2}$	$-\frac{1}{3}$	$-\frac{1}{4}$	$-\frac{1}{5}$
$x = 2$	-2	-1	$-\frac{2}{3}$	$-\frac{1}{2}$	$-\frac{2}{5}$
$x = 3$	-3	$-\frac{3}{2}$	-1	$-\frac{3}{4}$	$-\frac{3}{5}$
$x = 4$	-4	-2	$-\frac{4}{3}$	-1	$-\frac{4}{5}$
$x = 5$	-5	$-\frac{5}{2}$	$-\frac{5}{3}$	$-\frac{5}{4}$	-1

For example, $F(3, 4) = -\frac{3}{4}$ is the slope of a line element at $(3, 4)$ and is given in red in the table at the intersection of the row labeled $x = 3$ and the column labeled $y = 4$.

Since the algebraic sign of the quotient x/y at (x, y), $x > 0, y > 0$ is the same as at (x, y), $x < 0, y < 0$ the slopes at the corresponding points in the third quadrant are the same as the slopes in the first quadrant. Similarly, it is easy to see that the slopes of the line segments in the second and fourth quadrants are the negatives of the slopes in the table. Drawing line elements through the points with the slopes determined from the table yields the direction field in FIGURE 8.4.3(a).

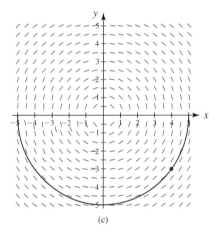

(a) (b) (c)

FIGURE 8.4.3 Direction fields in Example 1

(b) With the aid of a CAS and the grid of points again defined by (mh, nh), $h = \frac{1}{2}$, m and n integers, $-10 \le m \le 10, -10 \le n \le 10$, we get the direction field in Figure 8.4.3(b). Visually the flow of the field is circular. In Figure 8.4.3(c) we have superimposed the solution curve (in red) of the IVP,

$$\frac{dy}{dx} = -\frac{x}{y}, \qquad y(4) = -3$$

obtained in Example 3 of Section 8.1 over the computer-generated direction field. ■

◂ By using more gridpoints we get a better idea of the shapes of solution curves.

Of course the main purpose of constructing a direction field is to be able to obtain a rough sketch of a solution curve when it is impossible to solve a DE exactly.

EXAMPLE 2 Direction Field

Use a direction field to describe an approximate solution curve for the initial-value problem

$$\frac{dy}{dx} = \sin x + \cos y, \qquad y(-4) = 4.$$

Solution We mark the initial point $(-4, 4)$ in the computer-generated direction field in FIGURE 8.4.4(a); moving to the left and to the right we try to draw a curve as long as possible that contains the initial point. When we move to the right of the initial point we see that the line segments almost immediately force a graph downward (roughly for $-3 < x < -0.5$), and then upward as a graph crosses the y-axis (roughly for $-0.5 < x < 2.5$) and finally followed by another downward movement (roughly for $x > 2.5$). The solution curve just described has the approximate shape shown in red in Figure 8.4.4(b).

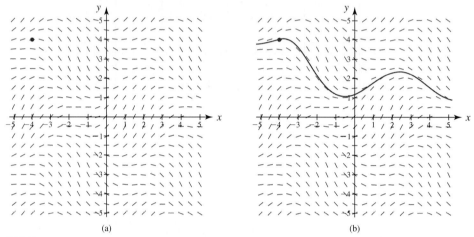

FIGURE 8.4.4 Direction field and approximate solution curve in Example 2

■ **Phase Portraits and Stability** Interpretation of the derivative dy/dx as a function that gives slope played the key role in the construction of direction fields. In the discussion that follows next, we will employ another telling property of the first derivative, namely, if $y(x)$ is a differentiable function, and if $dy/dx > 0$ (or $dy/dx < 0$) for all x in an interval I, then $y(x)$ is increasing (or decreasing) on I.

■ **Autonomous DEs** A first-order differential equation in which the independent variable x does not appear explicitly, is said to be an **autonomous differential equation**. Hence, an autonomous first-order differential equation is one whose normal form is

$$\frac{dy}{dx} = f(y). \tag{1}$$

We shall assume throughout that f and its derivative f' are continuous functions of x on some interval I. The differential equations

$$\overset{f(y)}{\underset{\downarrow}{}}\qquad\qquad\overset{F(x,y)}{\underset{\downarrow}{}}$$

$$\frac{dy}{dx} = 1 + y^2 \qquad \text{and} \qquad \frac{dy}{dx} = 2xy$$

are autonomous and nonautonomous, respectively. Many first-order differential equations encountered in applications are autonomous and of the form (1). All but one of the mathematical models in Section 8.3 are autonomous; equation (7) of that section is nonautonomous. Of course different symbols in Section 8.3 are playing the part of x and y in the current discussion.

■ **Critical Points** The zeros of the function f in (1) are of special interest. We say that a real number c is a **critical point** of the autonomous differential equation (1) if it is a zero of f, that is, $f(c) = 0$. Critical points are also called **equilibrium points** and **stationary points**. Moreover, substituting $y = c$ into (1) makes both sides of the equation zero, so we see that:

- *If c is a critical point of* (1) *then* $y(x) = c$ *is a constant solution of the autonomous equation.*

A constant solution $y(x) = c$ of (1) is called an **equilibrium solution**, and:

- *Equilibrium solutions are the only constant solutions of* (1).

We can tell when a nonconstant solution $y(x)$ of (1) is increasing or decreasing by determining the algebraic sign of the derivative dy/dx; this we do by identifying the intervals over which $f(y)$ is positive or negative.

EXAMPLE 3   Autonomous First-Order DE

Inspection of the differential equation

$$\frac{dy}{dx} = y(a - by) \qquad (2)$$

◀ This is the DE in Problem 29, Exercises 8.3. Here the symbol y plays the part of P and x plays the part of t.

$a > 0, b > 0$, shows that it is autonomous. From $f(y) = y(a - by) = 0$ we also see that 0 and a/b are critical points of the equation. By putting these two numbers on a vertical number line, we divide the line into three intervals determined by the inequalities:

$$-\infty < x < 0, \qquad 0 < x < a/b, \qquad a/b < x < \infty.$$

The arrows on the line shown in FIGURE 8.4.5 indicate the algebraic sign of $f(y) = y(a - by)$ on these intervals, and whether a solution $y(x)$ is increasing or decreasing. The following table explains the figure.

Interval	Sign of $f(y)$	$y(x)$	Arrow
$(-\infty, 0)$	minus	decreasing	points down
$(0, a/b)$	plus	increasing	points up
$(a/b, \infty)$	minus	decreasing	points down

The equilibrium solutions of the DE are $y = 0$ and $y = a/b$.   ■

FIGURE 8.4.5 Two critical points determine three intervals in Example 3

Figure 8.4.5 is called a **one-dimensional phase portrait**, or simply a **phase portrait**, of the differential equation $dy/dx = y(a - by)$. The vertical line or y-axis is called a **phase line**. A phase portrait such as this can also be interpreted in terms of motion of a moving particle. If we imagine that $y(x)$ denotes the position of a particle at *time* x on a vertical line whose positive y-direction is upward, then the rate of change dy/dx represents the velocity of the particle. Positive velocity indicates motion upward and negative velocity indicates that the particle is moving downward. If a particle is placed at a critical point, then it must remain there for all time. Whence the origin of the alternative name *stationary point*.

■ **Solution Curves without the Solution** Without solving an autonomous differential equation, we can usually say a great deal about its solution curves. Relating this back to the first topic of this section, note that a direction field of an autonomous differential equation (1) is independent of x, so at any point on a line parallel to the x-axis all slopes are the same. Thus if $y(x)$ is a solution of (1), then any horizontal translation $y(x - k)$, k a constant, is also a solution. Since the function f in (1) is independent of the variable x, we may consider it defined for $-\infty < x < \infty$ or $0 \le x < \infty$. Also, since f and its derivative f' are continuous functions of x on some interval I, it can be proved that in some horizontal strip or region R in the xy-plane corresponding to I, through any point (x_0, y_0) in R there passes only one solution curve of (1). See FIGURE 8.4.6(a). For the sake of discussion let us suppose that (1) possesses exactly two critical points c_1 and c_2 and that $c_1 < c_2$. The graphs of the equilibrium solutions $y(x) = c_1$ and $y(x) = c_2$ are horizontal lines, and these lines partition the region R into three subregions R_1, R_2, and R_3 as illustrated in Figure 8.4.6(b). Without proof, here are some conclusions that we can draw about a nonconstant solution $y(x)$ of (1):

- If (x_0, y_0) is in a subregion R_i, $i = 1, 2, 3$, and $y(x)$ is a solution whose graph passes through this point, then $y(x)$ remains in the subregion for all x. As illustrated in

(a) Region R

(b) Subregions R_1, R_2, and R_3 of R

FIGURE 8.4.6 Two equilibrium solutions determine three subregions in the plane

Figure 8.4.6(b) the solution $y(x)$ in R_2 is bounded below by c_1 and above by c_2, that is, $c_1 < y(x) < c_2$ for all x. The solution curve stays within R_2 for all x because the graph of a nonconstant solution of (1) cannot cross the graph of an equilibrium solution.

- By continuity of f we must then have either $f(y) > 0$ or $f(y) < 0$ for all y in a subregion R_i, $i = 1, 2, 3$. In other words, $f(y)$ cannot change signs in a subregion.
- Since $dy/dx = f(y(x))$ is either positive or negative in a subregion, a solution $y(x)$ is either increasing or decreasing in a subregion R_i, $i = 1, 2, 3$. Therefore $y(x)$ cannot be oscillatory, nor can it have a relative extremum (maximum or minimum).
- If $y(x)$ is *bounded above* by a critical point c_1 (as in subregion R_1 where $y(x) < c_1$ for all x), then the graph of $y(x)$ must approach the graph of the equilibrium solution $y(x) = c_1$ either as $x \to \infty$ or as $x \to -\infty$. If $y(x)$ is *bounded*, that is, bounded above and below by two consecutive critical points (as in subregion R_2 where $c_1 < y(x) < c_2$ for all x), then the graph of $y(x)$ must approach the graphs of the equilibrium solutions $y(x) = c_1$ and $y(x) = c_2$, one as $x \to \infty$ and the other as $x \to -\infty$. If $y(x)$ is *bounded below* by a critical point (as in subregion R_3 where $c_2 < y(x)$ for all x), then the graph of $y(x)$ must approach the graph of the equilibrium solution $y(x) = c_2$ either as $x \to \infty$ or as $x \to -\infty$.

With the foregoing facts in mind let us reexamine the differential equation in Example 3.

EXAMPLE 4 Example 3 Revisited

The three intervals determined on the y-axis or phase line by the critical points 0 and a/b now correspond in the xy-plane to three subregions defined by:

$$R_1: \quad -\infty < y < 0, \qquad R_2: \quad 0 < y < a/b, \qquad R_3: \quad a/b < y < \infty,$$

where $-\infty < x < \infty$. The phase portrait in Figure 8.4.5 tells us that $y(x)$ is decreasing in R_1, increasing in R_2, and decreasing in R_3. If $y(0) = y_0$ is an initial value, then in R_1, R_2, and R_3 we have, respectively:

(*i*) For $y_0 < 0$, $y(x)$ is bounded above. Since $y(x)$ is decreasing, $y(x)$ decreases without bound for increasing x and so $y(x) \to 0$ as $x \to -\infty$. This means the negative x-axis, $y = 0$, is a horizontal asymptote for a solution curve.

(*ii*) For $0 < y_0 < a/b$, $y(x)$ is bounded. Since $y(x)$ is increasing, $y(x) \to a/b$ as $x \to \infty$, and $y(x) \to 0$ as $x \to -\infty$. The two lines $y = 0$ and $y = a/b$ are horizontal asymptotes for any solution curve starting in this subregion.

(*iii*) For $y_0 > a/b$, $y(x)$ is bounded below. Since $y(x)$ is decreasing, $y(x) \to a/b$ as $x \to \infty$. This means $y = a/b$ is a horizontal asymptote for a solution curve.

In **FIGURE 8.4.7**, the original phase portrait is reproduced to the left of the xy-plane in which the subregions R_1, R_2, and R_3 are shaded. The graphs of the equilibrium solutions $y = a/b$ and $y = 0$ are shown in Figure 8.4.7 as dashed lines; the solid graphs represent typical graphs of $y(x)$ illustrating the three cases just discussed. ■

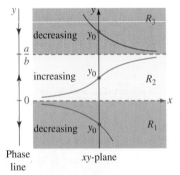

FIGURE 8.4.7 Phase portrait and solution curves in each of the three subregions in Example 4

In a subregion such as R_1 in Example 4, where $y(x)$ is decreasing and unbounded below, we must necessarily have $y(x) \to -\infty$. Do *not* interpret this last statement to mean $y(x) \to -\infty$ as $x \to \infty$; we could have $y(x) \to -\infty$ as $x \to a$, where $a > 0$ is a finite number that depends on the initial condition $y(x_0) = y_0$. Thinking in dynamic terms, $y(x)$ could "blow up" in finite time; or thinking graphically, $y(x)$ could have a vertical asymptote at $x = a > 0$. A similar remark holds true for the subregion R_3. The next example illustrates these concepts.

EXAMPLE 5 Solution Curves

The autonomous equation $dy/dx = (y - 1)^2$ possesses the single critical point 1 and hence has only one constant solution $y(x) = 1$. From the phase portrait in **FIGURE 8.4.8(a)**, we conclude that a nonconstant solution $y(x)$ is an increasing function in the two subregions defined by $-\infty < y < 1$ and $1 < y < \infty$, where $-\infty < x < \infty$. For an initial condition $y(0) = y_0 < 1$, a solution $y(x)$ is increasing and bounded above by 1, so $y(x) \to 1$ as $x \to \infty$; for $y(0) = y_0 > 1$, a solution $y(x)$ is increasing and unbounded.

You should verify by separation of variables that a one-parameter family of solutions of the differential equation is $y(x) = 1 - 1/(x + C)$. For a given initial condition, say, $y(0) = -1 < 1$, we find $C = \frac{1}{2}$ and $y(x) = 1 - 1/(x + \frac{1}{2})$. Observe $x = -\frac{1}{2}$ is a vertical asymptote and $y(x) \to -\infty$ as $x \to -\frac{1}{2}^+$. See Figure 8.4.8(b). For a different initial condition $y(0) = 2 > 1$, we find $C = -1$ and $y(x) = 1 - 1/(x - 1)$. The last function has a vertical asymptote at $x = 1$ and we see in Figure 8.4.8(c) that $y(x) \to \infty$ as $x \to 1^-$. The solution curves are the portions of the graphs in Figures 8.4.8(b) and 8.4.8(c) shown in blue. As predicted by the phase portrait in Figure 8.4.8(a), for the solution curve in Figure 8.4.8(b), $y(x) \to 1$ as $x \to \infty$, whereas for the solution curve in Figure 8.4.8(c), $y(x) \to \infty$ as $x \to 1^-$.

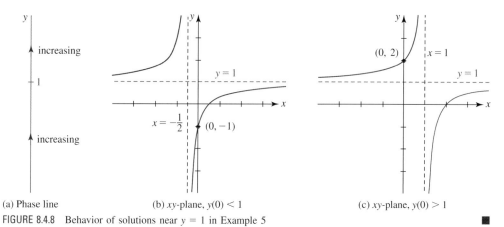

(a) Phase line   (b) xy-plane, $y(0) < 1$ (c) xy-plane, $y(0) > 1$

FIGURE 8.4.8 Behavior of solutions near $y = 1$ in Example 5

■ **Attractors and Repellers** Suppose $y(x)$ is a nonconstant solution of the autonomous differential equation (1) and that c is a critical point of the DE. There are basically three types of behavior $y(x)$ can exhibit near c. In **FIGURE 8.4.9** we have placed c on four vertical phase lines. When both arrowheads on either side of the blue dot labeled c point toward c, as in Figure 8.4.9(a), all solutions $y(x)$ of (1) that start from an initial point (x_0, y_0) sufficiently near c exhibit the asymptotic behavior $\lim_{x \to \infty} y(x) = c$. For this reason the critical point c is said to be **asymptotically stable**. Using a physical analogy, a solution that starts near c is like a charged particle that, over time, is drawn to a particle of opposite charge, so c is often referred to as an **attractor**. When both arrowheads on either side of the blue dot labeled c point away from c, as in Figure 8.4.9(b), all solutions $y(x)$ of (1) that start from an initial point (x_0, y_0) move away from c as x increases. In this case the critical point c is said to be **unstable**. An unstable critical point is called a **repeller**, for obvious reasons. The critical point illustrated in Figures 8.4.9(c) and 8.4.9(d) is neither an attractor nor a repeller. But since c exhibits characteristics of both an attractor *and* a repeller—that is a solution starting from an initial point (x_0, y_0) sufficiently near c is attracted to c from one side and repelled from the other side—we say that the critical point c is **semi-stable**. In Example 3, the critical point a/b is asymptotically stable (an attractor) and the critical point 0 is unstable (a repeller). The critical point 1 in Example 5 is semi-stable.

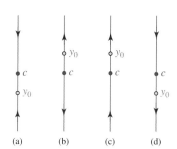

FIGURE 8.4.9 Critical point is: an attractor in (a), a repeller in (b), and semi-stable in (c) and (d)

Exercises 8.4   Answers to selected odd-numbered problems begin on page ANS-25.

≡ **Fundamentals**

In Problems 1–8, use the given computer-generated direction field to sketch an approximate solution curve for the indicated differential equation that passes through each of the given points.

1. $\dfrac{dy}{dx} = \dfrac{x}{y}$

   (a) $y(0) = 3$ (b) $y(3) = 3$
   (c) $y\left(-\frac{3}{2}\right) = 2$ (d) $y(-2) = -3$

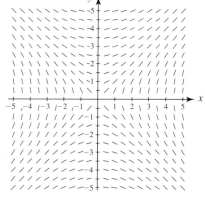

FIGURE 8.4.10 Direction field for Problem 1

2. $\dfrac{dy}{dx} = e^{-0.01xy^3}$

   (a) $y(-4) = 0$ (b) $y(3) = -2$
   (c) $y(0) = -2$ (d) $y(0) = 1$

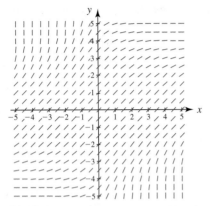

FIGURE 8.4.11 Direction field for Problem 2

3. $\dfrac{dy}{dx} = 1 - xy$

   (a) $y(0) = 0$ (b) $y(-1) = 0$
   (c) $y(2) = 0$ (d) $y(0) = -4$

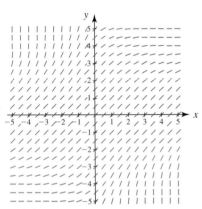

FIGURE 8.4.12 Direction field for Problem 3

4. $\dfrac{dy}{dx} = (\sin x)\cos y$

   (a) $y(0) = 0$ (b) $y(0) = 2$
   (c) $y(-3) = 0$ (d) $y(4) = 0$

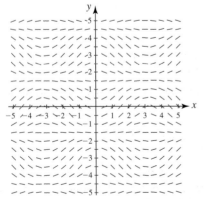

FIGURE 8.4.13 Direction field for Problem 4

5. $\dfrac{dy}{dx} = \dfrac{1}{2}x - \dfrac{1}{2}y^2$

   (a) $y(0) = -2$ (b) $y(0) = 2$
   (c) $y(-1) = 0$ (d) $y(-4) = 0$

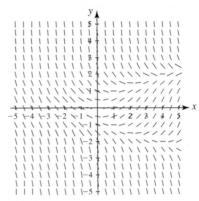

FIGURE 8.4.14 Direction field for Problem 5

6. $\dfrac{dy}{dx} = e^{-\sin y}$

   (a) $y(0) = 0$ (b) $y(0) = 2$
   (c) $y(-2) = 0$ (d) $y(4) = 0$

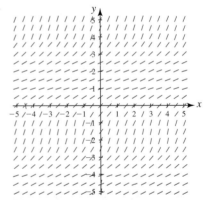

FIGURE 8.4.15 Direction field for Problem 6

7. $\dfrac{dy}{dx} = y\sin x$

   (a) $y(0) = 1$ (b) $y(-3) = -2$
   (c) $y(4) = 1$ (d) $y(2) = 2$

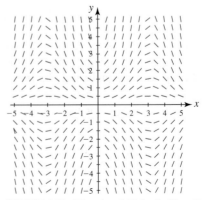

FIGURE 8.4.16 Direction field for Problem 7

8. $\dfrac{dy}{dx} = x^2 - y^2$

(a) $y(-2) = 0$ (b) $y(0) = -2$
(c) $y(0) = 0$ (d) $y(3) = 0$

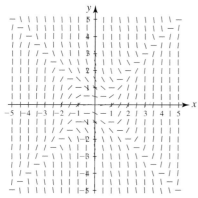

FIGURE 8.4.17 Direction field for Problem 8

In Problems 9–12, use computer software to obtain a direction field for the given differential equation. By hand, sketch an approximate solution curve that passes through each of the given points.

9. $\dfrac{dy}{dx} = \dfrac{1}{y}$

(a) $y(0) = 3$
(b) $y(2) = -2$

10. $\dfrac{dy}{dx} = x + y$

(a) $y(0) = -1$
(b) $y(3) = 0$

11. $\dfrac{dy}{dx} = \dfrac{1}{5}x^2 + y$

(a) $y(0) = 1$
(b) $y(4) = 0$

12. $\dfrac{dy}{dx} = y - 2\cos \pi x$

(a) $y(0) = 0$
(b) $y(2) = -4$

In Problems 13–20, find the critical points and phase portrait of the given autonomous first-order differential equation. Classify each critical point as asymptotically stable (attractor), unstable (repeller), or semi-stable.

13. $\dfrac{dy}{dx} = y^2 - 3y$

14. $\dfrac{dy}{dx} = y^2 - y^3$

15. $\dfrac{dy}{dx} = (y - 2)^2$

16. $\dfrac{dy}{dx} = 10 + 3y - y^2$

17. $\dfrac{dy}{dx} = y^2(4 - y^2)$

18. $\dfrac{dy}{dx} = y(2 - y)(4 - y)$

19. $\dfrac{dy}{dx} = y \ln(y + 2)$

20. $\dfrac{dy}{dx} = \dfrac{ye^y - 9y}{e^y}$

In Problems 21 and 22, consider the given autonomous first-order differential equation and the initial condition $y(0) = y_0$. By hand, sketch a graph of a typical solution $y(x)$ when y_0 has the values:

(a) $y_0 > 1$
(c) $-1 < y_0 < 0$
(b) $0 < y_0 < 1$
(d) $y_0 < -1$

21. $\dfrac{dy}{dx} = y - y^3$

22. $\dfrac{dy}{dx} = y^2 - y^4$

In Problems 23 and 24, consider the autonomous differential equation $dy/dx = f(y)$, where the graph of f is given. Use the graph to locate the critical points of each differential equation. Sketch a phase portrait of each equation. By hand, sketch typical solution curves in the subregions of the xy-plane determined by the graphs of the equilibrium solutions.

23.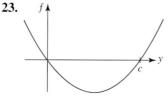

FIGURE 8.4.18 Graph for Problem 23

24.

FIGURE 8.4.19 Graph for Problem 24

≡ Applications

25. We saw in Section 8.3 that the *linear* autonomous differential equation

$$m\dfrac{dv}{dt} = mg - kv$$

is a mathematical model for the velocity of a falling body when air resistance is taken into account. Using only a phase portrait, determine the terminal velocity of the falling body.

26. We also saw in Section 8.3 that the *nonlinear* autonomous differential equation

$$m\dfrac{dv}{dt} = mg - kv^2$$

is a mathematical model for the velocity of a falling body when air resistance is taken into account. Using only a phase portrait, determine the terminal velocity of the falling body.

27. In Problem 26 of Exercises 8.3, we saw that the current $i(t)$ in a series circuit is described by

$$L\dfrac{di}{dt} + Ri = E.$$

If the inductance L, resistance R, and impressed voltage E are positive constants, show that as $t \to \infty$ the current obeys Ohm's law that $E = iR$.

28. When two chemicals are combined, the rate at which a new compound is formed is governed by the differential equation

$$\dfrac{dX}{dt} = k(\alpha - X)(\beta - X),$$

where $k > 0$ is a constant of proportionality and $\beta > \alpha > 0$. Here $X(t)$ denotes the number of grams of the new compound formed in time t.

(a) Describe the behavior of X as $t \to \infty$.
(b) Consider the case when $\alpha = \beta$. What is the behavior of X as $t \to \infty$ if $X(0) < \alpha$? From the phase portrait of the differential equation, can you predict the behavior of X as $t \to \infty$ if $X(0) > \alpha$?
(c) Verify that an explicit solution of the differential equation in the case when $k = 1$ and $\alpha = \beta$ is $X(t) = \alpha - 1/(t + c)$. Find a solution satisfying $X(0) = \alpha/2$. Find a solution satisfying $X(0) = 2\alpha$. Graph these two solutions. Does the behavior of the solutions as $t \to \infty$ agree with your answers to part (b)?

≡ **Think About It**

29. For a differential equation $dy/dx = F(x, y)$, any member of the family of curves $F(x, y) = c$, where c is a constant, is called an **isocline** of the equation. In a direction field of the DE $dy/dx = x^2 + y^2$, what is true about the line segments at points on the isocline $dy/dx = x^2 + y^2 = 1$? Identify the isoclines of the differential equation $dy/dx = x + y$.

30. For a differential equation $dy/dx = F(x, y)$, a curve in the plane defined by $F(x, y) = 0$ is called a **nullcline** of the equation. In a direction field of the DE $dy/dx = x^2 + y^2 - 1$, what is true about the line segments at points on a nullcline? Identify the nullclines of the differential equation $dy/dx = x^2 - y^2$ and indicate them in the direction field given in Figure 8.4.17.

31. The number 0 is a critical point of the autonomous differential equation $dy/dt = y^n$, where n is a positive integer. For what values of n is 0 asymptotically stable? Unstable? Repeat for the equation $dy/dx = -y^n$.

8.5 Euler's Method

Introduction We turn now from the visualization methods examined in the preceding section to a numerical method. By using the DE we are able to construct a simple procedure for obtaining approximations to the numerical values of the y-coordinates of points on a solution curve.

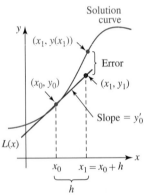

FIGURE 8.5.1 Approximating a point on a solution curve by a point on the tangent line

Euler's Method One of the simplest techniques for approximating a solution of a first-order initial-value problem

$$y' = F(x, y), \quad y(x_0) = y_0 \tag{1}$$

is known as **Euler's method**, or the **method of tangent lines**. This technique uses the fact that the derivative of a function $y(x)$ at a number x_0 determines a linearization of $y(x)$ at $x = x_0$:

$$L(x) = y_0 + y'(x_0)(x - x_0).$$

Recall from Section 4.9 that the linearization of $y(x)$ at x_0 is simply an equation of the tangent line to the graph of $y = y(x)$ at the point (x_0, y_0). We now let h be a positive increment on the x-axis, as shown in FIGURE 8.5.1. Then for $x_1 = x_0 + h$ we have

$$L(x_1) = y_0 + y'(x_0)(x_0 + h - x_0) = y_0 + hy'_0,$$

where $y'_0 = y'(x_0) = F(x_0, y_0)$. Letting $y_1 = L(x_1)$ we get

$$y_1 = y_0 + hF(x_0, y_0).$$

The point (x_1, y_1), which is seen in Figure 8.5.1 to be a point on the tangent line, is an approximation to the point $(x_1, y(x_1))$ on the actual solution curve; that is, $L(x_1) \approx y(x_1)$ or $y_1 \approx y(x_1)$ is an approximation of $y(x)$ at x_1. Of course, the accuracy of the approximation depends heavily on the size of the increment h. Usually we must choose this **step size** to be *reasonably small*. If we repeat the process, using (x_1, y_1) and the new slope $F(x_1, y_1)$ as the new starting point, then we obtain the approximation

$$y(x_2) = y(x_0 + 2h) = y(x_1 + h) \approx y_2 = y_1 + hF(x_1, y_1).$$

In general, it follows that

$$y_{n+1} = y_n + hF(x_n, y_n), \tag{2}$$

where $x_n = x_0 + nh$.

In the next example we apply Euler's method (2) to a differential equation for which we know an explicit solution; in this way we can compare the estimated values y_n with the true values $y(x_n)$.

EXAMPLE 1 Euler's Method

Consider the initial-value problem $y' = 0.2xy$, $y(1) = 1$. Use Euler's method to approximate $y(1.5)$ using first $h = 0.1$ and then $h = 0.05$.

Solution With the identification that $F(x, y) = 0.2xy$, (2) becomes

$$y_{n+1} = y_n + h(0.2x_n y_n).$$

Then for $x_0 = 1$, $y_0 = 1$, and $h = 0.1$ we find

$$y_1 = y_0 + (0.1)(0.2x_0 y_0) = 1 + (0.1)[0.2(1)(1)] = 1.02,$$

which is an estimate to the value of $y(1.1)$. However, if we use $h = 0.05$, it takes *two* steps to reach $x = 1.1$. From

$$y_1 = 1 + (0.05)[0.2(1)(1)] = 1.01$$
$$y_2 = 1.01 + (0.05)[0.2(1.05)(1.01)] = 1.020605$$

we have $y_1 \approx y(1.05)$ and $y_2 \approx y(1.1)$. The remainder of the calculations were carried out using computer software. The results are summarized in TABLES 8.5.1 and 8.5.2. Each entry is rounded to four decimal places. Observe that it takes 5 steps with $h = 0.1$ and 10 steps with $h = 0.05$ to get to $x = 1.5$.

TABLE 8.5.1 Euler's Method with $h = 0.1$

x_n	y_n	True Value	Absolute Error	% Relative Error
1.00	1.0000	1.0000	0.0000	0.00
1.10	1.0200	1.0212	0.0012	0.12
1.20	1.0424	1.0450	0.0025	0.24
1.30	1.0675	1.0714	0.0040	0.37
1.40	1.0952	1.1008	0.0055	0.50
1.50	1.1259	1.1331	0.0073	0.64

TABLE 8.5.2 Euler's Method with $h = 0.05$

x_n	y_n	True Value	Absolute Error	% Relative Error
1.00	1.0000	1.0000	0.0000	0.00
1.05	1.0100	1.0103	0.0003	0.03
1.10	1.0206	1.0212	0.0006	0.06
1.15	1.0318	1.0328	0.0009	0.09
1.20	1.0437	1.0450	0.0013	0.12
1.25	1.0562	1.0579	0.0016	0.16
1.30	1.0694	1.0714	0.0020	0.19
1.35	1.0833	1.0857	0.0024	0.22
1.40	1.0980	1.1008	0.0028	0.25
1.45	1.1133	1.1166	0.0032	0.29
1.50	1.1295	1.1331	0.0037	0.32

In Example 1, the true values in the tables were calculated from the known solution $y = e^{0.1(x^2-1)}$. Also, **absolute error** is defined to be

$$|\text{true value} - \text{approximation}|.$$

◀ Verify this solution by solving the DE by separation of variables.

The **relative error** and **percentage relative error** are, in turn,

$$\frac{\text{absolute error}}{|\text{true value}|} \quad \text{and} \quad \frac{\text{absolute error}}{|\text{true value}|} \times 100.$$

It is apparent by comparing Tables 8.5.1 and 8.5.2 that the accuracy of the approximations improves as the step size h decreases. Also, we see that even though the percentage relative error is growing, it does not appear to be that bad. But you should not be deceived by one example. Watch what happens in the next example, when we simply change the coefficient of the right side of the differential equation from 0.2 to 2.

EXAMPLE 2 Comparison of Exact/Approximate Values

Use the Euler method to approximate $y(1.5)$ for the solution of the initial-value problem $y' = 2xy$, $y(1) = 1$.

Solution You should verify that the exact solution of the IVP is now $y = e^{x^2-1}$. Proceeding as in Example 1, we obtain the results shown in Tables 8.5.3 and 8.5.4.

In this case, with a step size $h = 0.1$, a 16% relative error in the calculation of the approximation to $y(1.5)$ is totally unacceptable. At the expense of doubling the number of calculations, a slight improvement in accuracy is obtained by halving the step size to $h = 0.05$.

TABLE 8.5.3 Euler's Method with $h = 0.1$

x_n	y_n	True Value	Absolute Error	% Relative Error
1.00	1.0000	1.0000	0.0000	0.00
1.10	1.2000	1.2337	0.0337	2.73
1.20	1.4640	1.5527	0.0887	5.71
1.30	1.8154	1.9937	0.1784	8.95
1.40	2.2874	2.6117	0.3244	12.42
1.50	2.9278	3.4904	0.5625	16.12

TABLE 8.5.4 Euler's Method with $h = 0.05$

x_n	y_n	True Value	Absolute Error	% Relative Error
1.00	1.0000	1.0000	0.0000	0.00
1.05	1.1000	1.1079	0.0079	0.72
1.10	1.2155	1.2337	0.0182	1.47
1.15	1.3492	1.3806	0.0314	2.27
1.20	1.5044	1.5527	0.0483	3.11
1.25	1.6849	1.7551	0.0702	4.00
1.30	1.8955	1.9937	0.0982	4.93
1.35	2.1419	2.2762	0.1343	5.90
1.40	2.4311	2.6117	0.1806	6.92
1.45	2.7714	3.0117	0.2403	7.98
1.50	3.1733	3.4904	0.3171	9.08

$$\frac{dy}{dx} = F(x, y)$$ **NOTES FROM THE CLASSROOM**

Euler's method is just one of many different ways a solution of a differential equation can be approximated. Although attractive in its simplicity, Euler's method is seldom used in serious calculations. We have introduced this topic simply to give you a first taste of numerical methods. You will delve into greater detail and examine methods that give significantly greater accuracy in a formal course in differential equations.

Exercises 8.5 Answers to selected odd-numbered problems begin on page ANS-26.

≡ Fundamentals

In Problems 1 and 2, use Euler's method (2) to obtain a four-decimal approximation to the indicated value. Carry out the recursion of (2) by hand, first using $h = 0.1$ and then $h = 0.05$.

1. $\dfrac{dy}{dx} = 2x - 3y + 1, y(1) = 5; \quad y(1.2)$

2. $\dfrac{dy}{dx} = x + y^2, y(0) = 0; \quad y(0.2)$

In Problems 3 and 4, use Euler's method to obtain a four-decimal approximation of the indicated value. First use $h = 0.1$ and then use $h = 0.05$. Find an explicit solution for each initial-value problem and then construct tables similar to Tables 8.5.1 and 8.5.2.

3. $y' = y, y(0) = 1; \quad y(1.0)$
4. $y' = 4x - 2y, y(0) = 2; \quad y(0.5)$

In Problems 5–10, use Euler's method to obtain a four-decimal approximation of the indicated value. First use $h = 0.1$ and then use $h = 0.05$.

5. $y' = e^{-y}, y(0) = 0; \quad y(0.5)$
6. $y' = x^2 + y^2, y(0) = 1; \quad y(0.5)$
7. $y' = (x - y)^2, y(0) = 0.5; \quad y(0.5)$
8. $y' = xy + \sqrt{y}, y(0) = 1; \quad y(0.5)$
9. $y' = xy^2 - \dfrac{y}{x}, y(1) = 1; \quad y(1.5)$
10. $y' = y - y^2, y(0) = 0.5; \quad y(0.5)$

Chapter 8 in Review

Answers to selected odd-numbered problems begin on page ANS-26.

A. True/False

In Problems 1–4, indicate whether the given statement is true or false.
1. The differential equation $dy/dx = x + xy$ is both separable and linear. _____
2. The differential equation $dy/dx = \sin y$ is nonlinear. _____
3. $y = 0$ is a solution of the initial-value problem $dy/dx = x^2 y$, $y(0) = 0$. _____
4. A solution of the differential equation $dy/dx = x^2 y^2 + 4$ is increasing on $(-\infty, \infty)$. _____

B. Fill in the Blanks

In Problems 1–8, fill in the blanks.
1. A one-parameter family of solutions of $dy/dx = 1 - 6x + 12e^{3x}$ is _____.
2. The order of the differential equation $(y'')^3 + y^4 = 1$ is _____.
3. An integrating factor for the linear equation $dy/dx - y = e^{3x}$ is _____.
4. In the direction field of the differential equation $dy/dx = x^2 - y^2$, the slope of a line element at $(2, 4)$ is _____.
5. The time that it takes for a substance decaying through radioactivity to go from its initial amount A_0 to $\frac{1}{2}A_0$ is called its _____.
6. If an initial population P_0 of bacteria doubles in 2 h, then the number of bacteria present after 32 h is _____.
7. If $P(t) = P_0 e^{0.16t}$ gives the population in an environment at time t, then $P(t)$ satisfies the initial-value problem _____.
8. Give an example of a first-order differential equation that is both separable and linear. _____

C. Exercises

In Problems 1–10, solve the given differential equation.

1. $\sin x \dfrac{dy}{dx} + (\cos x) y = 0$
2. $\dfrac{dx}{dt} + x = e^{-t} \cos 2t$
3. $t \dfrac{dy}{dt} - 5y = t$
4. $\dfrac{y}{x^2} \dfrac{dy}{dx} + e^{2x^3 + y^2} = 0$
5. $(x^2 + 4) \dfrac{dy}{dx} = 2x - 8xy$
6. $y \sec^2 x \dfrac{dy}{dx} = y^2 + 1$
7. $\dfrac{dy}{dx} = 2x\sqrt{1 - y^2}$
8. $(e^x + e^{-x}) \dfrac{dy}{dx} = y^2$
9. $y' - 2y = x(e^{3x} - e^{2x})$
10. $\dfrac{dy}{dx} + y = \sqrt{1 + e^x}$

In Problems 11–20, solve the given initial-value problems.

11. $\dfrac{dP}{dt} = 0.05P$, $P(0) = 1000$
12. $\dfrac{dA}{dt} = -0.015A$, $A(0) = 5$
13. $t \dfrac{dy}{dt} + y = t^4 \ln t$, $y(1) = 0$
14. $x \dfrac{dy}{dx} = 10y$, $y(1) = -3$
15. $\dfrac{dy}{dx} = 2y + y^2$, $y(0) = 3$
16. $\dfrac{dy}{dx} = y(10 - 2y)$, $y(0) = 7$
17. $\dfrac{dy}{dx} = 1 + y^2$, $y(\pi/3) = -1$
18. $x \dfrac{dy}{dx} = y^2 - 1$, $y(2) = 2$
19. $\dfrac{dy}{dx} = -8x^3 y^2$, $y(0) = \dfrac{1}{2}$
20. $\dfrac{dy}{dx} = e^{x-y}$, $y(0) = 1$

In Problems 21 and 22, find a function whose graph passes through the given point and has the indicated slope at a point (x, y) on its graph.

21. $(0, 2)$; $2x/3y^3$

22. $(0, 1)$; $x + y$

23. If P_0 is the initial population of a community, show that if the population P is modeled by $dP/dt = kP$, then

$$\left(\frac{P_1}{P_0}\right)^{t_2} = \left(\frac{P_2}{P_0}\right)^{t_1},$$

where $P_1 = P(t_1)$ and $P_2 = P(t_2)$, $t_1 < t_2$.

24. A metal bar is taken out of a furnace whose temperature is 150°C and put into a tank of water whose temperature is maintained at a constant 30°C. After $\frac{1}{4}$ h in the tank, the temperature of the bar is 90°C. What is the temperature of the bar in $\frac{1}{2}$ h? In 1 h?

25. When forgetfulness is taken into account, the rate at which a person can memorize a subject is given by the differential equation

$$\frac{dA}{dt} = k_1(M - A) - k_2 A,$$

where k_1 and k_2 are positive constants, $A(t)$ is the amount of material memorized in time t, M is the total amount to be memorized, and $M - A$ is the amount remaining to be memorized.

 (a) Solve for $A(t)$ if $A(0) = 0$.
 (b) Find the limiting value of A as $t \to \infty$ and interpret the result.
 (c) Graph the solution.

26. Suppose a series circuit contains a capacitor and a variable resistor. If the resistance at time t is given by $R = k_1 + k_2 t$, where k_1 and k_2 are positive known constants, then the charge q on the capacitor is described by the first-order differential equation

$$(k_1 + k_2 t)\frac{dq}{dt} + \frac{1}{C}q = E(t),$$

where C is a constant called the **capacitance** and $E(t)$ is the **impressed voltage**. Show that if $E(t) = E_0$ and $q(0) = q_0$ are constants, then

$$q(t) = E_0 C + (q_0 - E_0 C)\left(\frac{k_1}{k_1 + k_2 t}\right)^{1/Ck_2}.$$

27. The differential equation $dP/dt = (k \cos t)P$, where k is a positive constant, is often used as a model of a population that undergoes yearly seasonal fluctuations.

 (a) Solve for $P(t)$ if $P(0) = P_0$.
 (b) Use a calculator or CAS to obtain the graph of the function found in part (a).

28. A projectile is shot vertically into the air with an initial velocity of v_0 ft/s. Assuming that air resistance is proportional to the square of the instantaneous velocity, the motion is described by the pair of differential equations:

$$m\frac{dv}{dt} = -mg - kv^2, \quad k > 0$$

positive y-axis up and origin at ground level so that $v = v_0$ at $y = 0$, and

$$m\frac{dv}{dt} = mg - kv^2, \quad k > 0$$

positive y-axis down and origin at the maximum height so that $v = 0$ at $y = h$. See FIGURE 8.R.1. The first and second equations describe the motion of the projectile when rising and falling, respectively. Prove that the impact velocity v_i is less than the initial velocity v_0. [*Hint*: By the Chain Rule, $dv/dt = v\, dv/dy$.]

FIGURE 8.R.1 Initial and impact velocities in Problem 28

29. (a) Use a CAS to obtain the direction field for the differential equation $dy/dx = e^{-x} - 3y$ using a 3×3 rectangular grid with points (mh, nh), $h = 0.25$, $-12 \leq m \leq 12$, $-12 \leq n \leq 12$.
    (b) On the direction field, sketch by hand a solution curve that corresponds to each of the initial conditions: $y(0) = 1, y(-2) = 0, y(-1) = -2$.
    (c) Based on the direction field and the solution curves, form a conjecture about the behavior of all solutions $y(x)$ as $x \to \pm\infty$.

30. Construct an autonomous differential equation $dy/dx = f(y)$ whose phase portrait is consistent with part (a) of FIGURE 8.R.2. With part (b) of Figure 8.R.2.

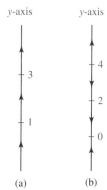

FIGURE 8.R.2 Phase portraits for Problem 30

31. Consider the autonomous differential equation $dy/dx = f(y)$, where
$$f(y) = -0.5y^3 - 1.7y + 3.4.$$

It is seen in FIGURE 8.R.3 that the function $f(y)$ has one zero. Without attempting to solve the differential equation for $y(x)$, estimate the value of $\lim_{x \to \infty} y(x)$.

FIGURE 8.R.3 Phase portrait for Problem 31

32. Use Euler's method with step size $h = 0.1$ to approximate $y(1.2)$, where $y(x)$ is the solution of the initial-value problem $dy/dx = 1 + x\sqrt{y}$, $y(1) = 9$.

33. Two curves are said to be **orthogonal** at a point if and only if their tangent lines L_1 and L_2 are perpendicular at the point of intersection. Show that the curves defined by $y = x^3$ and $x^2 + 3y^2 = 4$ are orthogonal at $(-1, -1)$ and $(1, 1)$.

34. When all the curves of one family of curves $F(x, y, C_1) = 0$ intersect orthogonally all the curves of another family $G(x, y, C_2) = 0$, then the families are said to be **orthogonal trajectories** of each another.

    (a) Find the differential equations of the families $xy = C_1$ and $y^2 - x^2 = C_2$. Show that the two families of curves are orthogonal trajectories of each other.

    (b) Sketch the graphs of some members of each family in part (a) on the same coordinate axis.

Chapter 9

Sequences and Series

In This Chapter Everyday experience gives one an intuitive feeling for the notion of a sequence. The words *sequence of events* or *sequence of numbers* suggest an arrangement whereby the events E or numbers n are set down in some order: $E_1, E_2, E_3, \ldots$ or $n_1, n_2, n_3, \ldots$.

Every student of mathematics is also familiar with the fact that any real number can be written as a decimal. For example, the rational number $\frac{1}{3} = 0.333\ldots$, where the mysterious three dots (an ellipsis) signify that the digit 3 repeats forever. This means that the decimal $0.333\ldots$ is an infinite sum or the *infinite series*

$$\frac{3}{10} + \frac{3}{100} + \frac{3}{1000} + \frac{3}{10{,}000} + \cdots.$$

In this chapter we will see that the concepts of sequence and infinite series are related.

9.1 Sequences
9.2 Monotonic Sequences
9.3 Series
9.4 Integral Test
9.5 Comparison Tests
9.6 Ratio and Root Tests
9.7 Alternating Series
9.8 Power Series
9.9 Representing Functions by Power Series
9.10 Taylor Series
9.11 Binomial Series
    Chapter 9 in Review

9.1 Sequences

Introduction If the domain of a function f is the set of positive integers, then the elements $f(n)$ in the range can be arranged in an order corresponding to increasing values of n:

$$f(1), f(2), f(3), \ldots, f(n), \ldots$$

In the discussion that follows we consider only functions whose domain is the set of positive integers and whose range elements are real numbers.

EXAMPLE 1 Function with Domain the Positive Integers

If n is a positive integer, then the first several elements in the range of the function $f(n) = (1 + 1/n)^n$ are

$$f(1) = 2, \quad f(2) = \frac{9}{4}, \quad f(3) = \frac{64}{27}, \ldots.$$

A function whose domain is the entire set of positive integers is given a special name.

Definition 9.1.1 Sequence

A **sequence** is a function whose domain is the set of positive integers.

Some texts use the words *infinite sequence*. When the domain of the function is a finite subset of the set of positive integers, we get a *finite sequence*. All the sequences in this chapter will be infinite.

Notation and Terms Instead of the customary function notation $f(n)$, a sequence is usually denoted by either $\{a_n\}$ or $\{a_n\}_{n=1}^{\infty}$. The integer n is sometimes called the **index** of a_n. The **terms** of the sequence are formed by letting the index n take on the values $1, 2, 3, \ldots$; the number a_1 is the *first term*, a_2 is the *second term*, and so on. The number a_n is called the *n*th *term* or the **general term** of the sequence. Thus, $\{a_n\}$ is equivalent to

$$a_1, a_2, a_3, \ldots, a_n, \ldots \leftarrow \text{numbers in the range}$$
$$\uparrow \ \uparrow \ \uparrow \qquad \uparrow$$
$$1 \ \ 2 \ \ 3 \qquad n \quad \leftarrow \text{numbers in the domain}$$

For example, the sequence defined in Example 1 would be written $\{(1 + 1/n)^n\}$.

In some circumstances it is convenient to take the first term of a sequence to be a_0 and the sequence is then

$$a_0, a_1, a_2, a_3, \ldots, a_n, \ldots.$$

EXAMPLE 2 Terms of a Sequence

Write out the first four terms of the sequences

(a) $\left\{\dfrac{1}{2^n}\right\}$ \qquad (b) $\{n^2 + n\}$ \qquad (c) $\{(-1)^n\}$.

Solution By substituting $n = 1, 2, 3, 4$ in the respective general term of each sequence, we obtain

(a) $\dfrac{1}{2}, \dfrac{1}{4}, \dfrac{1}{8}, \dfrac{1}{16}, \ldots$ \qquad (b) $2, 6, 12, 20, \ldots$ \qquad (c) $-1, 1, -1, 1, \ldots$

Convergent Sequence For the sequence in part (a) of Example 2, we see that as the index n becomes progressively larger, the values $a_n = \dfrac{1}{2^n}$ do not increase without bound. Indeed, we see that as $n \to \infty$, the terms

$$\frac{1}{2}, \frac{1}{4}, \frac{1}{8}, \frac{1}{16}, \frac{1}{32}, \frac{1}{64}, \ldots$$

approach the limiting value 0. We say that the sequence $\{\frac{1}{2^n}\}$ **converges** to 0. In contrast, the terms of the sequences in parts (b) and (c) do not approach a limiting value as $n \to \infty$. In general we have the following definition.

Definition 9.1.2 Convergent Sequence

A sequence $\{a_n\}$ is said to **converge** to a real number L if for every $\varepsilon > 0$ there exists a positive integer N such that

$$|a_n - L| < \varepsilon \text{ whenever } n > N. \tag{1}$$

The number L is called the **limit** of the sequence.

◀ Compare this definition with the wording in Definition 2.6.5.

If a sequence $\{a_n\}$ converges, then its limit L is unique.

■ **Convergent Sequence** If $\{a_n\}$ is a convergent sequence, (1) means that the terms a_n can be made arbitrarily close to L for n sufficiently large. We indicate that a sequence converges to a number L by writing

$$\lim_{n \to \infty} a_n = L.$$

When $\{a_n\}$ does not converge, that is, when $\lim_{n \to \infty} a_n$ does not exist, we say that the sequence **diverges**.

FIGURE 9.1.1 illustrates several ways in which a sequence $\{a_n\}$ can converge to a number L. Parts (a), (b), (c), and (d) of Figure 9.1.1 show that for four different convergent sequences $\{a_n\}$, *all but a finite* number of the terms a_n are in the interval $(L - \varepsilon, L + \varepsilon)$. The terms of the sequence $\{a_n\}$ that are in $(L - \varepsilon, L + \varepsilon)$ for $n > N$ are represented by red dots in the figure.

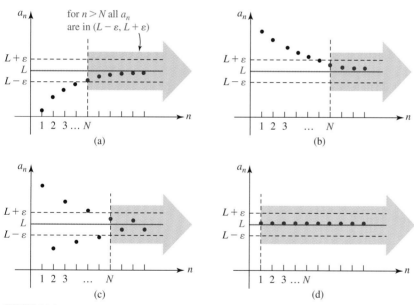

FIGURE 9.1.1 Four ways a sequence can converge to L

EXAMPLE 3 Convergent Sequence

Use Definition 9.1.2 to prove that the sequence $\{1/\sqrt{n}\}$ converges to 0.

Solution Intuitively, we can see from the terms

$$1, \frac{1}{\sqrt{2}}, \frac{1}{\sqrt{3}}, \frac{1}{2}, \frac{1}{\sqrt{5}}, \ldots$$

that as the index n increases without bound the terms approach the limiting value 0. To prove convergence, we start by assuming that $\varepsilon > 0$ is given. Since the terms of the sequence are positive, the inequality $|a_n - 0| < \varepsilon$ is the same as

$$\frac{1}{\sqrt{n}} < \varepsilon.$$

This is equivalent to $\sqrt{n} > 1/\varepsilon$ or $n > 1/\varepsilon^2$. Hence, we need only choose N to be the first positive integer greater than or equal to $1/\varepsilon^2$. For instance, if we choose $\varepsilon = 0.01$, then $|1/\sqrt{n} - 0| = 1/\sqrt{n} < 0.01$ whenever $n > 10{,}000$. That is, we choose $N = 10{,}000$. ∎

In practice, to determine whether a sequence $\{a_n\}$ converges or diverges, we work directly with $\lim_{n\to\infty} a_n$ and proceed as we would in the examination of $\lim_{x\to\infty} f(x)$. If a_n either increases or decreases without bound as $n \to \infty$, then $\{a_n\}$ is necessarily divergent and we write, respectively,

$$\lim_{n\to\infty} a_n = \infty \quad \text{or} \quad \lim_{n\to\infty} a_n = -\infty. \tag{2}$$

In the first case in (2) we say that $\{a_n\}$ **diverges to infinity** and in the second, $\{a_n\}$ **diverges to negative infinity**. A sequence may diverge in a manner other than that given in (2). The next example illustrates two sequences; each diverges in a different way.

EXAMPLE 4 Divergent Sequences

(a) The sequence $\{n^2 + n\}$ diverges to infinity, since $\lim_{n\to\infty}(n^2 + n) = \infty$.

(b) The sequence $\{(-1)^n\}$ is divergent since $\lim_{n\to\infty}(-1)^n$ does not exist. The general term of the sequence does not approach a single constant as $n \to \infty$; as can be seen in part (c) of Example 2 the term $(-1)^n$ alternates between 1 and -1 as $n \to \infty$. ∎

EXAMPLE 5 Determining Convergence

Determine whether the sequence $\left\{\dfrac{3n(-1)^n}{n+1}\right\}$ converges or diverges.

Solution By dividing the numerator and denominator of the general term by n we obtain

$$\lim_{n\to\infty} \frac{3n(-1)^n}{n+1} = \lim_{n\to\infty} \frac{3(-1)^n}{1 + 1/n}.$$

Although $3/(1 + 1/n) \to 3$ as $n \to \infty$, the foregoing limit still does not exist. Because of the factor $(-1)^n$, we see that as $n \to \infty$,

$$a_n \to 3, \quad n \text{ even}, \quad \text{and} \quad a_n \to -3, \quad n \text{ odd}.$$

The sequence diverges. ∎

A sequence, such as those in part (b) of Example 4 and in Example 5, for which

$$\lim_{n\to\infty} a_{2n} = L \quad \text{and} \quad \lim_{n\to\infty} a_{2n+1} = -L,$$

$L \neq 0$, is said to **diverge by oscillation**.

■ **Constant Sequence** A sequence of constants

$$c, c, c, \ldots$$

is written $\{c\}$. Common sense tells us that this sequence converges and that its limit is c. See Figure 9.1.1(d). For example, the sequence $\{\pi\}$ converges to π.

In determining the limit of a sequence it is often useful to replace the discrete variable n by a continuous variable x. If f is a function such that $f(x) \to L$ as $x \to \infty$ and the values of f at the positive integers, $f(1), f(2), f(3), \ldots$, agree with the terms $a_1, a_2, a_3, \ldots$ of $\{a_n\}$, that is,

$$f(1) = a_1, \quad f(2) = a_2, \quad f(3) = a_3, \ldots,$$

then necessarily the sequence $\{a_n\}$ converges to the number L. The plausibility of this result is illustrated in FIGURE 9.1.2.

FIGURE 9.1.2 If $f(x) \to L$ as $x \to \infty$, then $f(n) = a_n \to L$ as $n \to \infty$

Theorem 9.1.1 Limit of a Sequence

Suppose $\{a_n\}$ is a sequence and f is a function such that $f(n) = a_n$ for $n \geq 1$. If

$$\lim_{x \to \infty} f(x) = L \quad \text{then} \quad \lim_{n \to \infty} a_n = L. \quad (3)$$

EXAMPLE 6 Using L'Hôpital's Rule

Show that the sequence $\{(n+1)^{1/n}\}$ converges.

Solution If we define $f(x) = (x+1)^{1/x}$, then we recognize that $\lim_{x\to\infty} f(x)$ has the indeterminate form ∞^0 as $x \to \infty$. Hence, by L'Hôpital's Rule,

◀ See Section 4.5 for a review on how to handle the form ∞^0.

$$\lim_{x\to\infty} \ln f(x) = \lim_{x\to\infty} \frac{\ln(x+1)}{x} \stackrel{h}{=} \lim_{x\to\infty} \frac{\frac{1}{x+1}}{1} = \lim_{x\to\infty} \frac{1}{x+1} = 0.$$

This shows $\lim_{x\to\infty} \ln f(x) = \ln\left[\lim_{x\to\infty} f(x)\right] = 0$ and that $\lim_{x\to\infty} f(x) = e^0 = 1$. Thus, by (3) we have $\lim_{n\to\infty}(n+1)^{1/n} = e^0 = 1$. The sequence converges to 1. ∎

EXAMPLE 7 Convergent Sequence

Show that the sequence $\left\{\dfrac{n(4n+1)(5n+3)}{6n^3+2}\right\}$ converges.

Solution If $f(x) = \dfrac{x(4x+1)(5x+3)}{6x^3+2} = \dfrac{20x^3+17x^2+3x}{6x^3+2}$, then $\lim_{x\to\infty} f(x)$ has the indeterminate form ∞/∞. By L'Hôpital's Rule,

$$\lim_{x\to\infty}\frac{x(4x+1)(5x+3)}{6x^3+2} = \lim_{x\to\infty}\frac{20x^3+17x^2+3x}{6x^3+2}$$

$$\stackrel{h}{=} \lim_{x\to\infty}\frac{60x^2+34x+3}{18x^2}$$

$$\stackrel{h}{=} \lim_{x\to\infty}\frac{120x+34}{36x}$$

$$\stackrel{h}{=} \lim_{x\to\infty}\frac{120}{36} = \frac{10}{3}.$$

From (3) of Theorem 9.1.1, the given sequence converges to $\frac{10}{3}$. ∎

EXAMPLE 8 Determining Convergence

Determine whether the sequence $\left\{\sqrt{\dfrac{n}{9n+1}}\right\}$ converges.

Solution It follows either by using L'Hôpital's Rule or by dividing the numerator and denominator by x that $x/(9x+1) \to \frac{1}{9}$ as $x \to \infty$. Thus, we can write

$$\lim_{n\to\infty}\sqrt{\frac{n}{9n+1}} = \sqrt{\lim_{n\to\infty}\frac{n}{9n+1}} = \sqrt{\frac{1}{9}} = \frac{1}{3}.$$

The sequence converges to $\frac{1}{3}$. ∎

■ **Properties** The following **properties** of sequences are analogous to those given in Theorems 2.2.1, 2.2.2, and 2.2.3.

Theorem 9.1.2 Limit of a Sequence

Let $\{a_n\}$ and $\{b_n\}$ be convergent sequences. If $\lim_{n\to\infty} a_n = L_1$ and $\lim_{n\to\infty} b_n = L_2$, then

(i) $\lim_{n\to\infty} c = c$, c a real number

(ii) $\lim_{n\to\infty} ka_n = k \lim_{n\to\infty} a_n = kL_1$, k a real number

(iii) $\lim_{n\to\infty} (a_n + b_n) = \lim_{n\to\infty} a_n + \lim_{n\to\infty} b_n = L_1 + L_2$

(vi) $\lim_{n\to\infty} a_n b_n = \lim_{n\to\infty} a_n \cdot \lim_{n\to\infty} b_n = L_1 \cdot L_2$

(v) $\lim_{n\to\infty} \dfrac{a_n}{b_n} = \dfrac{\lim_{n\to\infty} a_n}{\lim_{n\to\infty} b_n} = \dfrac{L_1}{L_2}$, $L_2 \neq 0$.

EXAMPLE 9 Determining Convergence

Determine whether the sequence $\left\{\dfrac{2 - 3e^{-n}}{6 + 4e^{-n}}\right\}$ converges.

Solution Observe that $2 - 3e^{-n} \to 2$ and $6 + 4e^{-n} \to 6 \neq 0$ as $n \to \infty$. According to Theorem 9.1.2(v), we have

$$\lim_{n\to\infty} \frac{2 - 3e^{-n}}{6 + 4e^{-n}} = \frac{\lim_{n\to\infty}(2 - 3e^{-n})}{\lim_{n\to\infty}(6 + 4e^{-n})} = \frac{2}{6} = \frac{1}{3}.$$

The sequence converges to $\tfrac{1}{3}$. ∎

Review Section 1.6, especially Figure 1.6.2.

The first of the next two theorems should seem believable based on your knowledge of the behavior of the exponential function. Recall, for $0 < b < 1$, $b^x \to 0$ as $x \to \infty$, whereas for $b > 1$, $b^x \to \infty$ as $x \to \infty$.

Theorem 9.1.3 Sequences of the Form $\{r^n\}$

Suppose r is a nonzero constant. The sequence $\{r^n\}$ converges to 0 if $|r| < 1$ and diverges if $|r| > 1$.

Theorem 9.1.4 Sequences of the Form $\{1/n^r\}$

The sequence $\left\{\dfrac{1}{n^r}\right\}$ converges to 0 for r any positive rational number.

EXAMPLE 10 Applications of Theorems 9.1.3 and 9.1.4

(a) The sequence $\{e^{-n}\}$ converges to 0 by Theorem 9.1.3, since $e^{-n} = \left(\dfrac{1}{e}\right)^n$ and $r = 1/e < 1$.

(b) The sequence $\left\{\left(\dfrac{3}{2}\right)^n\right\}$ diverges by Theorem 9.1.3, since $r = \tfrac{3}{2} > 1$.

(c) The sequence $\left\{\dfrac{4}{n^{5/2}}\right\}$ converges to 0 by Theorem 9.1.2 (ii) and Theorem 9.1.4, since $r = \tfrac{5}{2}$ is a positive rational number. ∎

EXAMPLE 11 Determining Convergence

From Theorem 9.1.2(iii) and Theorem 9.1.4 we see that the sequence $\left\{10 + \dfrac{4}{n^{3/2}}\right\}$ converges to 10. ∎

■ **Recursively Defined Sequence** As the following example indicates, a sequence can be defined by specifying the first term a_1 together with a rule for obtaining the subsequent terms from the preceding terms. In this case the sequence is said to be defined **recursively**. The defining rule is called a **recursion formula**. See Problems 59 and 60 in Exercises 9.1. Newton's Method, given in (3) in Section 4.10, is an example of a recursively defined sequence.

EXAMPLE 12 A Sequence Defined Recursively

Suppose a sequence is defined recursively by $a_{n+1} = 3a_n + 4$, where $a_1 = 2$. Then by letting $n = 1, 2, 3, \ldots$ we obtain

$$a_2 = 3a_1 + 4 = 3(2) + 4 = 10$$
$$a_3 = 3a_2 + 4 = 3(10) + 4 = 34$$
$$a_4 = 3a_3 + 4 = 3(34) + 4 = 106$$

(this number is given as 2)

and so on.

■ **Squeeze Theorem** The following theorem is the sequence equivalent of Theorem 2.4.1.

Theorem 9.1.5 Squeeze Theorem

Suppose $\{a_n\}$, $\{b_n\}$, and $\{c_n\}$ are sequences such that
$$a_n \leq c_n \leq b_n$$
for all values of n larger than some index N (that is, $n > N$). If $\{a_n\}$ and $\{b_n\}$ converge to a common limit L, then $\{c_n\}$ also converges to L.

■ **Factorial** Before presenting an example illustrating Theorem 9.1.5 we need to review a symbol that occurs frequently in this chapter. If n is a positive integer, the symbol $n!$, read "n factorial," is the product of the first n positive integers:

$$n! = 1 \cdot 2 \cdot 3 \cdots (n-1) \cdot n. \qquad (4)$$

For example, $5! = 1 \cdot 2 \cdot 3 \cdot 4 \cdot 5 = 120$. An important property of the factorial is given by

$$n! = (n-1)!n.$$

To see this, consider the case when $n = 6$:

$$6! = 1 \cdot 2 \cdot 3 \cdot 4 \cdot 5 \cdot 6 = \overbrace{(1 \cdot 2 \cdot 3 \cdot 4 \cdot 5)}^{5!} 6 = 5!6.$$

Stated in a slightly different manner, the property $n! = (n-1)!n$ is equivalent to

$$(n+1)! = n!(n+1). \qquad (5)$$

One last point, for purposes of convenience and to ensure that the formula $n! = (n-1)!n$ is valid when $n = 1$, we define $0! = 1$.

EXAMPLE 13 Determining Convergence

Determine whether the sequence $\left\{\dfrac{2^n}{n!}\right\}$ converges.

Solution The convergence or divergence of the given sequence is not obvious since $2^n \to \infty$ and $n! \to \infty$ as $n \to \infty$. Even though the limit form of $\lim_{n\to\infty}(2^n/n!)$ is ∞/∞ we cannot use L'Hôpital's Rule since we have studied no function $f(x) = x!$. We can, however, use Theorem 9.1.5 by algebraically manipulating the general term of the sequence. In view of (4), the general term can be written

$$\frac{2^n}{n!} = \frac{\overbrace{2 \cdot 2 \cdot 2 \cdot 2 \cdots 2}^{n \text{ factor of } 2}}{1 \cdot 2 \cdot 3 \cdot 4 \cdots n} = \overbrace{\frac{2}{1} \cdot \frac{2}{2} \cdot \frac{2}{3} \cdot \frac{2}{4} \cdots \frac{2}{3}}^{n \text{ fractions}}$$

From the preceding line we obtain the inequality

$$0 \leq \frac{2^n}{n!} = \underbrace{\frac{2}{1} \cdot \frac{2}{2} \cdot \frac{2}{3} \cdot \frac{2}{4} \cdots \frac{2}{n}}_{n \text{ fractions}} \leq 2 \cdot 1 \cdot \underbrace{\frac{2}{3} \cdot \frac{2}{3} \cdots \frac{2}{3}}_{n-2 \text{ fractions}} = 2\left(\frac{2}{3}\right)^{n-2} \quad (6)$$

The $n - 2$ fractions of $\frac{2}{3}$ on the right-hand side of (6) results from the fact that after the second factor in the product of n fractions, 3 is the smallest denominator that makes $\frac{2}{3}$ larger than $\frac{2}{4}$, larger than $\frac{2}{5}$, and so on down to the last factor $\frac{2}{n}$. By the laws of exponents (6) is the same as

$$0 \leq \frac{2^n}{n!} \leq \frac{9}{2}\left(\frac{2}{3}\right)^n \quad \text{or} \quad a_n \leq c_n \leq b_n,$$

where we now identify the sequences $\{a_n\} = \{0\}$, $\{b_n\} = \{\frac{9}{2}(\frac{2}{3})^n\}$, and $\{c_n\} = \{2^n/n!\}$. The sequence $\{a_n\}$ is a sequence of 0's and so converges to 0. The sequence $\{b_n\} = \{\frac{9}{2}(\frac{2}{3})^n\}$ also converges to 0 by invoking Theorem 9.1.2(ii) and Theorem 9.1.3 with $r = \frac{2}{3} < 1$. Thus by Theorem 9.1.5, $\{c_n\} = \{2^n/n!\}$ must also converge to 0. ∎

The result $\lim_{n \to \infty} \frac{2^n}{n!} = 0$ shows that $n!$ grows much faster than 2^n as $n \to \infty$. For example, for $n = 10$, $2^{10} = 1024$, whereas $10! = 3,628,800$.

The sequence in the preceding example can also be defined recursively. For $n = 1$, $a_1 = 2^1/1! = 2$. Then by (5) and the laws of exponents,

$$a_{n+1} = \frac{2^{n+1}}{(n+1)!} = \frac{2 \cdot 2^n}{(n+1) \cdot n!} = \frac{2}{n+1} \cdot \underset{\underset{\text{this is } a_n}{\downarrow}}{\frac{2^n}{n!}}.$$

Thus $\{2^n/n!\}$ is the same as

$$a_{n+1} = \frac{2}{n+1}a_n, \quad a_1 = 2. \quad (7)$$

We can use the recursion formula in (7) as an alternative means of finding the limit L of the sequence $\{2^n/n!\}$. Since the sequence was shown to be convergent we have $\lim_{n \to \infty} a_n = L$. This last statement is also equivalent to $\lim_{n \to \infty} a_{n+1} = L$. By letting $n \to \infty$ in (7) and using the properties of limits we can write

$$\lim_{n \to \infty} a_{n+1} = \lim_{n \to \infty}\left(\frac{2}{n+1}a_n\right) = \left(\lim_{n \to \infty}\frac{2}{n+1}\right) \cdot \left(\lim_{n \to \infty} a_n\right). \quad (8)$$

From the last line we see that $L = 0 \cdot L$, which implies that the limit of the sequence is $L = 0$.

Our last theorem for this section is an immediate consequence of Theorem 9.1.5.

Theorem 9.1.6 Sequence of Absolute Values

If the sequence $\{|a_n|\}$ converges to 0, then $\{a_n\}$ converges to 0.

PROOF By the definition of absolute value, $|a_n| = a_n$ if $a_n \geq 0$ and $|a_n| = -a_n$ if $a_n < 0$. It follows that

$$-|a_n| \leq a_n \leq |a_n|. \quad (9)$$

By assumption $\{|a_n|\}$ converges to 0 and so $\lim_{n \to \infty} |a_n| = 0$. From the inequality (9) and Theorem 9.1.5 we conclude that $\lim_{n \to \infty} a_n = 0$. Therefore $\{a_n\}$ converges to 0. ∎

EXAMPLE 14 Using Theorem 9.1.6

The sequence $\left\{\frac{(-1)^n}{\sqrt{n}}\right\}$ converges to zero since we have already shown in Example 3 that the sequence of absolute values $\{|(-1)^n/\sqrt{n}|\} = \{1/\sqrt{n}\}$ converges to 0. ∎

Exercises 9.1

Answers to selected odd-numbered problems begin on page ANS-26.

≡ Fundamentals

In Problems 1–10, list the first four terms of the sequence whose general term is a_n.

1. $a_n = \dfrac{1}{2n+1}$
2. $a_n = \dfrac{3}{4n-2}$
3. $a_n = \dfrac{(-1)^n}{n}$
4. $a_n = \dfrac{(-1)^n n^2}{n+1}$
5. $a_n = 10^n$
6. $a_n = 10^{-n}$
7. $a_n = 2n!$
8. $a_n = (2n)!$
9. $a_n = \sum_{k=1}^{n} \dfrac{1}{k}$
10. $a_n = \sum_{k=1}^{n} 2^{-k}$

In Problems 11–14, use Definition 9.1.2 to show that each sequence converges to the given number L.

11. $\left\{\dfrac{1}{n}\right\}$; $L = 0$
12. $\left\{\dfrac{1}{n^2}\right\}$; $L = 0$
13. $\left\{\dfrac{n}{n+1}\right\}$; $L = 1$
14. $\left\{\dfrac{e^n + 1}{e^n}\right\}$; $L = 1$

In Problems 15–46, determine whether the given sequence converges. If the sequence converges, then find its limit.

15. $\left\{\dfrac{10}{\sqrt{n+1}}\right\}$
16. $\left\{\dfrac{1}{n^{3/2}}\right\}$
17. $\left\{\dfrac{1}{5n+6}\right\}$
18. $\left\{\dfrac{4}{2n+7}\right\}$
19. $\left\{\dfrac{3n-2}{6n+1}\right\}$
20. $\left\{\dfrac{n}{1-2n}\right\}$
21. $\{20(-1)^{n+1}\}$
22. $\left\{\left(-\dfrac{1}{3}\right)^n\right\}$
23. $\left\{\dfrac{n^2-1}{2n}\right\}$
24. $\left\{\dfrac{7n}{n^2+1}\right\}$
25. $\{ne^{-n}\}$
26. $\{n^3 e^{-n}\}$
27. $\left\{\dfrac{\sqrt{n+1}}{n}\right\}$
28. $\left\{\dfrac{n}{\sqrt{n+1}}\right\}$
29. $\{\cos n\pi\}$
30. $\{\sin n\pi\}$
31. $\left\{\dfrac{\ln n}{n}\right\}$
32. $\left\{\dfrac{e^n}{\ln(n+1)}\right\}$
33. $\left\{\dfrac{5-2^{-n}}{7+4^{-n}}\right\}$
34. $\left\{\dfrac{2^n}{3^n+1}\right\}$
35. $\left\{\dfrac{e^n+1}{e^n}\right\}$
36. $\left\{4 + \dfrac{3^n}{2^n}\right\}$
37. $\left\{n\sin\left(\dfrac{6}{n}\right)\right\}$
38. $\left\{\left(1 - \dfrac{5}{n}\right)^n\right\}$
39. $\left\{\dfrac{e^n - e^{-n}}{e^n + e^{-n}}\right\}$
40. $\left\{\dfrac{\pi}{4} - \arctan(n)\right\}$
41. $\{n^{2/(n+1)}\}$
42. $\{10^{(n+1)/n}\}$
43. $\left\{\ln\left(\dfrac{4n+1}{3n-1}\right)\right\}$
44. $\left\{\dfrac{\ln n}{\ln 3n}\right\}$
45. $\{\sqrt{n+1} - \sqrt{n}\}$
46. $\{\sqrt{n}(\sqrt{n+1} - \sqrt{n})\}$

In Problems 47–52, find a formula for the general term a_n of the sequence. Determine whether the given sequence converges. If the sequence converges, then find its limit.

47. $\dfrac{2}{1}, \dfrac{4}{3}, \dfrac{6}{5}, \dfrac{8}{7}, \ldots$
48. $1 + \dfrac{1}{2}, \dfrac{1}{2} + \dfrac{1}{3}, \dfrac{1}{3} + \dfrac{1}{4}, \dfrac{1}{4} + \dfrac{1}{5}, \ldots$
49. $3, -5, 7, -9, \ldots$
50. $-2, 2, -2, 2, \ldots$
51. $2, \dfrac{2}{3}, \dfrac{2}{9}, \dfrac{2}{27}, \ldots$
52. $\dfrac{1}{1\cdot 4}, \dfrac{1}{2\cdot 8}, \dfrac{1}{3\cdot 16}, \dfrac{1}{4\cdot 32}, \ldots$

In Problems 53–56, for the given recursively defined sequence, write the next four terms after the indicated initial term(s).

53. $a_{n+1} = \dfrac{1}{2} a_n$, $a_1 = -1$
54. $a_{n+1} = 2a_n - 1$, $a_1 = 2$
55. $a_{n+1} = \dfrac{a_n}{a_{n-1}}$, $a_1 = 1$, $a_2 = 3$
56. $a_{n+1} = 2a_n - 3a_{n-1}$, $a_1 = 2$, $a_2 = 4$

In Problems 57 and 58, the recursively defined sequence is known to converge for a given initial value $a_1 > 0$. Assume $\lim_{n \to \infty} a_n = L$, and proceed as in (8) of this section to find the limit L of the sequence.

57. $a_{n+1} = \dfrac{1}{4} a_n + 6$
58. $a_{n+1} = \dfrac{1}{2}\left(a_n + \dfrac{5}{a_n}\right)$

In Problems 59 and 60, find a recursion formula that defines the given sequence.

59. $\left\{\dfrac{5^n}{n!}\right\}$
60. $\sqrt{3}, \sqrt{3 + \sqrt{3}}, \sqrt{3 + \sqrt{3 + \sqrt{3}}}, \ldots$

In Problems 61–64, use the Squeeze Theorem to establish convergence of the given sequence.

61. $\left\{\dfrac{\sin^2 n}{4^n}\right\}$
62. $\left\{\sqrt{16 + \dfrac{1}{n^2}}\right\}$
63. $\left\{\dfrac{\ln n}{n(n+2)}\right\}$
64. $\left\{\dfrac{n!}{n^n}\right\}$ $\left[\text{Hint: } a_n = \dfrac{1}{n}\left(\dfrac{2}{n} \cdot \dfrac{3}{n} \cdot \dfrac{4}{n} \cdots \dfrac{n}{n}\right).\right]$

65. Show that for any real number x, the sequence $\{(1 + x/n)^n\}$ converges to e^x.

66. The sequence

$$\left\{1 + \frac{1}{2} + \frac{1}{3} + \cdots + \frac{1}{n} - \ln n\right\}$$

is known to converge to a number γ called **Euler's constant**. Calculate the first 10 terms of the sequence.

≡ Applications

67. A ball is dropped from an initial height of 15 ft onto a concrete slab. Each time it bounces, it reaches a height of $\frac{2}{3}$ its preceding height. What height does it reach on its third bounce? On its nth bounce? See FIGURE 9.1.3.

FIGURE 9.1.3 Bouncing ball in Problem 67

68. A ball, falling from a great height, travels 16 ft during the first second, 48 ft during the second, 80 ft during the third, and so on. How far does the ball travel during the sixth second?

69. A patient takes 15 mg of a drug each day. Suppose 80% of the drug accumulated is excreted each day by bodily functions. Write out the first six terms of the sequence $\{A_n\}$, where A_n is the amount of the drug present in the patient's body immediately after the nth dose.

70. One dollar is deposited in a savings account that pays an annual rate of interest r. If no money is withdrawn, what is the amount accrued in the account after the first, second, and third years?

71. Each person has two parents. Determine how many great-great-great-grandparents each person has.

72. The recursively defined sequence

$$p_{n+1} = 3p_n - \frac{p_n^2}{400}, \quad p_0 = 450,$$

is called a **discrete logistic equation**. Such a sequence is often used to model a population p_n in an environment; here p_0 is the initial population in the environment. Find the **carrying capacity** $K = \lim_{n \to \infty} p_n$ of the environment. Compute the next nine terms of the sequence and show that these terms oscillate around K.

≡ Think About It

73. Consider the sequence $\{a_n\}$ whose first four terms are

$$1, \quad 1 + \frac{1}{2}, \quad 1 + \frac{1}{2 + \frac{1}{2}}, \quad 1 + \frac{1}{2 + \frac{1}{2 + \frac{1}{2}}}, \ldots.$$

(a) With $a_1 = 1$, find a recursion formula that defines the sequence.
(b) What are the fifth and sixth terms of the sequence?
(c) The sequence $\{a_n\}$ is known to converge. Find the limit of the sequence.

74. Conjecture the limit of the convergent sequence
$\sqrt{3}, \sqrt{3\sqrt{3}}, \sqrt{3\sqrt{3\sqrt{3}}}, \ldots.$

75. If the sequence $\{a_n\}$ converges, does the sequence $\{a_n^2\}$ diverge? Defend your answer with sound mathematics.

76. In FIGURE 9.1.4 the square shown in red is 1 unit on a side. A second blue square is constructed inside the first square by connecting the midpoints of the first one. A third green square is constructed by connecting the midpoints of the sides of the second square, and so on.

(a) Find a formula for the area A_n of the nth inscribed square.
(b) Consider the sequence $\{S_n\}$, where $S_n = A_1 + A_2 + \cdots + A_n$. Calculate the numerical values of the first 10 terms of this sequence.
(c) Make a conjecture about the convergence of $\{S_n\}$.

FIGURE 9.1.4 Embedded squares in Problem 76

≡ Projects

77. **A Mathematical Classic** Consider an equilateral triangle with sides of length 1 as shown in FIGURE 9.1.5(a). As shown in Figure 9.1.5(b), on each of the three sides of the first triangle, another equilateral triangle is constructed with sides of length $\frac{1}{3}$. As indicated in Figures 9.1.5(c) and 9.1.5(d), this construction is continued: equilateral triangles are constructed on the sides of each previously new triangle such that the length of the sides of the new triangles is $\frac{1}{3}$ the length of the sides of the previous triangle. Let the perimeter of the first figure be P_1, the perimeter of the second figure P_2, and so on.

(a) Find the values of P_1, P_2, P_3, and P_4.
(b) Find a formula for the perimeter P_n of the nth figure.
(c) What is $\lim_{n \to \infty} P_n$? The perimeter of the snowflake-like region obtained by letting $n \to \infty$ is called a **Koch snowflake curve** and was invented in 1904 by the Swedish mathematician **Helge von Koch** (1870–1924). The Koch curve plays a part in the theory of **fractals**.

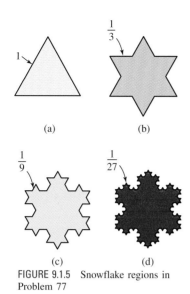

FIGURE 9.1.5 Snowflake regions in Problem 77

Discern the pattern of the solution of this problem and complete the following table.

	Start	After each month											
		1	2	3	4	5	6	7	8	9	10	11	12
Adult pairs	1	1	2	3	5	8	13	21					
Baby pairs	0	1	1	2	3	5	8	13					
Total pairs	1	2	3	5	8	13	21	34					

78. **A Bit of History—How Many Rabbits?** Besides its famous leaning bell tower, the city of Pisa, Italy, is also noted as the birthplace of **Leonardo Pisano**, aka **Leonardo Fibonacci** (1170–1250). Fibonacci was the first in Europe to introduce the Hindu–Arabic place-valued decimal system and the use of Arabic numerals. His book *Liber Abacci*, published in 1202, is basically a text on how to do arithmetic in this decimal system. But in Chapter 12 of *Liber Abacci*, Fibonacci poses and solves the following problem on the reproduction of rabbits:

How many pairs of rabbits will be produced in a year beginning with a single pair, if in every month each pair bears a new pair which become productive from the second month on?

79. Write out five terms, after the initial two, of the sequence defined recursively by $F_{n+1} = F_n + F_{n-1}$, $F_1 = 1, F_2 = 1$. Reexamine Problem 78.

80. **Golden Ratio** If the recursion formula in Problem 79 is divided by F_n, then

$$\frac{F_{n+1}}{F_n} = 1 + \frac{F_{n+1}}{F_n}.$$

If we define $a_n = F_{n+1}/F_n$, then the sequence $\{a_n\}$ is defined recursively by

$$a_n = 1 + \frac{1}{a_{n-1}}, \quad a_1 = 1, n \geq 2.$$

The sequence $\{a_n\}$ is known to converge to the **golden ratio** $\phi = \lim_{n \to \infty} a_n$.

(a) Find ϕ.

(b) Write a short report on the significance of the number ϕ. Include in your report the relationship between the number ϕ and the shape of the multichambered nautilus shell. See the photo in the Chapter 9 Opener on page 475.

9.2 Monotonic Sequences

Introduction In the preceding section we showed that a sequence $\{a_n\}$ converged by finding $\lim_{n \to \infty} a_n$. However, it is not always easy or even possible to determine whether a sequence $\{a_n\}$ converges by seeking the exact value of $\lim_{n \to \infty} a_n$. For example, does the sequence

$$\left\{ 1 + \frac{1}{2} + \frac{1}{3} + \cdots + \frac{1}{n} - \ln n \right\}$$

converge? It turns out that this sequence can be shown to converge, but not by using the basic ideas of the last section. In this section we consider a special type of sequence that can be proved convergent without finding the value of $\{a_n\}$.

We begin with a definition.

Definition 9.2.1 Monotonic Sequence

A sequence $\{a_n\}$ is said to be

(i) **increasing** if $a_{n+1} > a_n$ for all $n \geq 1$,
(ii) **nondecreasing** if $a_{n+1} \geq a_n$ for all $n \geq 1$,
(iii) **decreasing** if $a_{n+1} < a_n$ for all $n \geq 1$,
(iv) **nonincreasing** if $a_{n+1} \leq a_n$ for all $n \geq 1$.

If a sequence $\{a_n\}$ is one of the above types, then it is said to be **monotonic**.

In other words, sequences of the type

$$a_1 < a_2 < a_3 < \cdots < a_n < a_{n+1} < \cdots$$
$$a_1 > a_2 > a_3 > \cdots > a_n > a_{n+1} > \cdots,$$

are increasing and decreasing, respectively. Whereas

$$a_1 \leq a_2 \leq a_3 \leq \cdots \leq a_n \leq a_{n+1} \leq \cdots$$
$$a_1 \geq a_2 \geq a_3 \geq \cdots \geq a_n \geq a_{n+1} \geq \cdots,$$

are nondecreasing and nonincreasing sequences, respectively. The notions of *nondecreasing* and *nonincreasing* allow some adjacent terms in a sequence to be equal.

EXAMPLE 1 Monotonic/Not Monotonic

(a) The three sequences

$$4, 6, 8, 10, \ldots \qquad 1, \frac{1}{2}, \frac{1}{4}, \frac{1}{8}, \ldots \qquad \text{and} \qquad 5, 5, 4, 4, 4, 3, 3, 3, 3, \ldots$$

are monotonic. They are, respectively, increasing, decreasing, and nonincreasing.

(b) The sequence $-1, \frac{1}{2}, -\frac{1}{3}, \frac{1}{4}, -\frac{1}{5}, \ldots$ is not monotonic. ■

It is not always evident whether a sequence is increasing, decreasing, and so on. The following guidelines illustrate some of the ways that monotonicity can be demonstrated.

Guidelines for Demonstrating Monotonicity

(i) Form a **function** $f(x)$ such that $f(n) = a_n$. If $f'(x) > 0$, then $\{a_n\}$ is increasing. If $f'(x) < 0$, then $\{a_n\}$ is decreasing.
(ii) Form the **ratio** a_{n+1}/a_n where $a_n > 0$ for all n. If $a_{n+1}/a_n > 1$ for all n, then $\{a_n\}$ is increasing. If $a_{n+1}/a_n < 1$ for all n, then $\{a_n\}$ is decreasing.
(iii) Form the **difference** $a_{n+1} - a_n$. If $a_{n+1} - a_n > 0$ for all n, then $\{a_n\}$ is increasing. If $a_{n+1} - a_n < 0$ for all n, then $\{a_n\}$ is decreasing.

EXAMPLE 2 A Monotonic Sequence

Show that $\left\{\dfrac{n}{e^n}\right\}$ is a monotonic sequence.

Solution If we define $f(x) = x/e^x$, then $f(n) = a_n$. Now,

$$f'(x) = \frac{1-x}{e^x} < 0$$

for $x > 1$ implies that f is decreasing on $[1, \infty)$. Thus it follows that

$$f(n+1) = a_{n+1} < f(n) = a_n.$$

By Definition 9.2.1 the given sequence is decreasing.

Alternative Solution From the ratio

$$\frac{a_{n+1}}{a_n} = \frac{n+1}{e^{n+1}} \cdot \frac{e^n}{n} = \frac{n+1}{ne} = \frac{1}{e} + \frac{1}{ne} \leq \frac{1}{e} + \frac{1}{e} = \frac{2}{e} < 1$$

we see that $a_{n+1} < a_n$ for $n \geq 1$. This shows the sequence is decreasing. ∎

EXAMPLE 3 A Monotonic Sequence

The sequence $\left\{\dfrac{2n+1}{n+1}\right\}$ or $\dfrac{3}{2}, \dfrac{5}{3}, \dfrac{7}{4}, \dfrac{9}{5}, \ldots$ *appears* to be increasing. From

$$a_{n+1} - a_n = \frac{2n+3}{n+2} - \frac{2n+1}{n+1} = \frac{1}{(n+2)(n+1)} > 0$$

we conclude $a_{n+1} > a_n$ for all $n \geq 1$. This proves the sequence is increasing. ∎

Definition 9.2.2 Bounded Sequence

(*i*) A sequence $\{a_n\}$ is said to be **bounded above** if there is a positive number M such that $a_n \leq M$ for all n.

(*ii*) A sequence $\{a_n\}$ is said to be **bounded below** if there is a positive number m such that $a_n \geq m$ for all n.

(*iii*) A sequence $\{a_n\}$ is said to be **bounded** if it is bounded above and bounded below.

Of course, if a sequence $\{a_n\}$ is not bounded, then it is said to be **unbounded**. An unbounded sequence is divergent. The Fibonacci sequence (see Problems 78 and 79 in Exercises 9.1)

$$1, 1, 2, 3, 5, 8, 13, 21, \ldots$$

is nondecreasing and is an example of an unbounded sequence.

The sequence $1, \frac{1}{2}, \frac{1}{4}, \frac{1}{8}, \ldots$ in Example 1 is bounded since $0 \leq a_n \leq 1$ for all n. Any number smaller than a lower bound m of a sequence is also a lower bound and any number greater than an upper bound M is an upper bound; in other words the numbers m and M in Definition 9.2.2 are not unique. For the sequence $1, \frac{1}{2}, \frac{1}{4}, \frac{1}{8}, \ldots$ it is equally true that $-2 \leq a_n \leq 2$ for all $n \geq 1$.

EXAMPLE 4 A Bounded Sequence

The sequence $\left\{\dfrac{2n+1}{n+1}\right\}$ is bounded above by 2, since the inequality

$$\frac{2n+1}{n+1} \leq \frac{2n+2}{n+1} = \frac{2(n+1)}{n+1} = 2$$

shows that $a_n \leq 2$ for $n \geq 1$. Moreover,

$$a_n = \frac{2n+1}{n+1} \geq 0$$

for $n \geq 1$ shows that the sequence is bounded below by 0. Thus, $0 \leq a_n \leq 2$ for all n implies that the sequence is bounded. ∎

◂ Indeed, from Example 3 we see that the terms of the sequence are bounded below by the first term of the sequence.

The next result will be useful in subsequent sections of this chapter.

Theorem 9.2.1 Sufficient Condition for Convergence

A bounded monotonic sequence $\{a_n\}$ converges.

The existence of a *least upper bound*, that is, an upper bound that is smaller than all other upper bounds for the sequence, is one of the basic axioms in mathematics. It is called the **completeness property** of the real number system.

PROOF We will prove the theorem in the case of a nondecreasing sequence. By assumption, $\{a_n\}$ is bounded and so $m \leq a_n \leq M$ for all n. In turn, this means the infinite set of terms $S = \{a_1, a_2, a_3, \ldots, a_n, \ldots\}$ is bounded above and therefore has a least or smallest upper bound L. The sequence actually converges to L. For $\varepsilon > 0$ we know that $L - \varepsilon < L$, and consequently $L - \varepsilon$ is not an upper bound for S (there are no upper bounds smaller than the least upper bound). Hence, there exists a positive integer N such that $a_N > L - \varepsilon$. But, since $\{a_n\}$ is nondecreasing,

$$L - \varepsilon \leq a_N \leq a_{N+1} \leq a_{N+2} \leq a_{N+3} \leq \cdots \leq L + \varepsilon.$$

It follows that for $n > N$, $L - \varepsilon \leq a_n \leq L + \varepsilon$ or $|a_n - L| < \varepsilon$. From Definition 9.1.2 we conclude that $\lim_{n \to \infty} a_n = L$. ∎

EXAMPLE 5 Bounded and Monotonic

The sequence $\left\{ \dfrac{2n+1}{n+1} \right\}$ was shown to be monotonic (Example 3) and bounded (Example 4). Hence, by Theorem 9.2.1 the sequence is convergent. ∎

EXAMPLE 6 Determining Convergence

Show that the sequence $\left\{ \dfrac{1 \cdot 3 \cdot 5 \cdots (2n-1)}{2 \cdot 4 \cdot 6 \cdots (2n)} \right\}$ converges.

Solution First, the ratio

$$\frac{a_{n+1}}{a_n} = \frac{1 \cdot 3 \cdot 5 \cdots (2n-1)(2n+1)}{2 \cdot 4 \cdot 6 \cdots (2n)(2n+2)} \cdot \frac{2 \cdot 4 \cdot 6 \cdots (2n)}{1 \cdot 3 \cdot 5 \cdots (2n-1)} = \frac{2n+1}{2n+2} < 1$$

shows that $a_{n+1} < a_n$ for all n. The sequence is monotonic since it is decreasing. Next, from the inequality

Why is the product $\dfrac{1}{2} \cdot \dfrac{3}{4} \cdot \dfrac{5}{6} \cdot \dfrac{7}{8} \cdots \dfrac{2n-1}{2n}$ less than 1?

$$0 < \frac{1 \cdot 3 \cdot 5 \cdots (2n-1)}{2 \cdot 4 \cdot 6 \cdots (2n)} = \frac{1}{2} \cdot \frac{3}{4} \cdot \frac{5}{6} \cdot \frac{7}{8} \cdots \frac{2n-1}{2n} < 1$$

we see that the sequence is bounded. It follows from Theorem 9.2.1 that the sequence is convergent. ∎

Theorem 9.2.1 is handy in proving that a sequence $\{a_n\}$ converges, that is, $\lim_{n \to \infty} a_n = L$, but the theorem does not provide us with the specific number L. But the next example shows how to determine L when the sequence is defined recursively.

EXAMPLE 7 Determining Convergence

Show that the sequence $\{a_n\}$ defined by the recursion formula $a_{n+1} = \tfrac{1}{4} a_n + 6$, $a_1 = 1$, converges.

This can be proved using a method called *mathematical induction*.

Solution First, the sequence $\{a_n\}$ is bounded. It can be proved that $a_n < 8$, for all n. This fact is suggested by calculating a_n for $n = 1, 2, 3, \ldots$

$$a_2 = \frac{1}{4} a_1 + 6 = \frac{1}{4}(1) + 6 = \frac{25}{4} = 6.25 < 8$$

$$a_3 = \frac{1}{4} a_2 + 6 = \frac{1}{4}\left(\frac{25}{4}\right) + 6 = \frac{121}{16} = 7.5625 < 8$$

$$a_4 = \frac{1}{4} a_3 + 6 = \frac{1}{4}\left(\frac{121}{16}\right) + 6 = \frac{505}{64} = 7.890625 < 8$$

$$\vdots$$

Because $a_n > 0$ for all n, we have $0 < a_n < 8$ for all n. Thus $\{a_n\}$ is bounded.

Next, we will show that the sequence $\{a_n\}$ is monotonic. Because $a_n < 8$ necessarily $\tfrac{3}{4} a_n < \tfrac{3}{4} \cdot 8 = 6$. Therefore, from the recursion formula,

$$a_{n+1} = \frac{1}{4} a_n + 6 > \frac{1}{4} a_n + \frac{3}{4} a_n = a_n.$$

This shows that $a_{n+1} > a_n$ for all n, and so the sequence is increasing.

Since $\{a_n\}$ is bounded and monotonic it follows from Theorem 9.2.1 that the sequence converges. Because we must have $\lim_{n\to\infty} a_n = L$ and $\lim_{n\to\infty} a_{n+1} = L$ the limit of the sequence can be determined from the recursion formula:

$$\lim_{n\to\infty} a_{n+1} = \lim_{n\to\infty}\left(\frac{1}{4}a_n + 6\right)$$

$$\lim_{n\to\infty} a_{n+1} = \frac{1}{4}\lim_{n\to\infty} a_n + 6$$

$$L = \frac{1}{4}L + 6.$$

By solving the last equation for L we find $\frac{3}{4}L = 6$ or $L = 8$. ∎

∑ NOTES FROM THE CLASSROOM

(*i*) Every convergent sequence $\{a_n\}$ is necessarily bounded. See Problem 31 in Exercises 9.2. But it does not follow that every bounded sequence is convergent. You will be asked to supply an example that illustrates this last statement in Problem 30 of Exercises 9.2.

(*ii*) Some sequences $\{a_n\}$ do not exhibit monotonic behavior until some point on in the sequence, that is, until the index satisfies $n \geq N$, where N is some positive integer. For example, the terms of the sequence $\{5^n/n!\}$ for $n = 1, 2, 3, 4, 5, 6, \ldots$ are:

$$5, \frac{25}{2}, \frac{125}{6}, \frac{625}{24}, \frac{625}{24}, \frac{3125}{144}, \ldots. \tag{1}$$

To see better what is happening in (1), let us approximate the terms using numbers rounded to two decimals:

$$5, 12.5, 20.83, 26.04, 26.04, 21.70, \ldots. \tag{2}$$

In (2) we see that the first four terms of $\{5^n/n!\}$ obviously increase, but starting with the *fourth term* the terms appear to turn to nonincreasing. This can be proven from a recursively defined version of the sequence. Proceeding as we did in obtaining the recurrence formula in (7) in Section 9.1, $\{5^n/n!\}$ is the same as $a_{n+1} = \frac{5}{n+1}a_n$, $a_1 = 5$. Since $\frac{5}{n+1} \leq 1$ for $n \geq 4$ we see that $a_{n+1} \leq a_n$, that is, $\{5^n/n!\}$ is nonincreasing only for $n \geq 4$. In like manner, it is easily shown that $\{100^n/n!\}$ eventually becomes nonincreasing only when $n \geq 99$. By taking the limit of the recursion formula as $n \to \infty$, as in Example 7, we can show that both $\{5^n/n!\}$ and $\{100^n/n!\}$ converge to 0.

Exercises 9.2   Answers to selected odd-numbered problems begin on page ANS-26.

≡ Fundamentals

In Problems 1–12, determine whether the given sequence is monotonic. If so, state whether it is increasing, decreasing, nondecreasing, or nonincreasing.

1. $\left\{\dfrac{n}{3n+1}\right\}$
2. $\left\{\dfrac{10+n}{n}\right\}$
3. $\{(-1)^n\sqrt{n}\}$
4. $\{(n-1)(n-2)\}$
5. $\left\{\dfrac{e^n}{n}\right\}$
6. $\left\{\dfrac{e^n}{n^5}\right\}$
7. $\left\{\dfrac{2^n}{n!}\right\}$
8. $\left\{\dfrac{2^{2n}(n!)^2}{(2n)!}\right\}$
9. $\left\{n + \dfrac{1}{n}\right\}$
10. $\{n^2 + (-1)^n n\}$
11. $\{(\sin 1)(\sin 2)\cdots(\sin n)\}$
12. $\left\{\ln\left(\dfrac{n+2}{n+1}\right)\right\}$

In Problems 13–24, use Theorem 9.2.1 to show that the given sequence converges.

13. $\left\{\dfrac{4n-1}{5n+2}\right\}$
14. $\left\{\dfrac{6-4n^2}{1+n^2}\right\}$
15. $\left\{\dfrac{3^n}{1+3^n}\right\}$
16. $\{n5^{-n}\}$

17. $\{e^{1/n}\}$ **18.** $\left\{\dfrac{n!}{n^n}\right\}$

19. $\left\{\dfrac{n!}{1\cdot 3\cdot 5\cdots(2n-1)}\right\}$ **20.** $\left\{\dfrac{2\cdot 4\cdot 6\cdots(2n)}{1\cdot 3\cdot 5\cdots(2n+1)}\right\}$

21. $\{\tan^{-1} n\}$ **22.** $\left\{\dfrac{\ln(n+3)}{n+3}\right\}$

23. $(0.8), (0.8)^2, (0.8)^3, \ldots$

24. $\sqrt{3}, \sqrt{\sqrt{3}}, \sqrt{\sqrt{\sqrt{3}}}, \ldots$

In Problems 25 and 26, use Theorem 9.2.1 to show that the recursively defined sequence converges. Find the limit of the sequence.

25. $a_{n+1} = \dfrac{1}{2}a_n + 5,\ a_1 = 1$ **26.** $a_{n+1} = \sqrt{2 + a_n},\ a_1 = 0$

27. Express
$$\sqrt{7},\ \sqrt{7\sqrt{7}},\ \sqrt{7\sqrt{7\sqrt{7}}},\ \ldots$$
as a recursively defined sequence $\{a_n\}$. Use the fact that the sequence is bounded, $0 < a_n < 7$ for all n, to show that $\{a_n\}$ is increasing. Find the limit of the sequence.

28. Use Theorem 9.2.1 to show that the recursively defined sequence
$$a_{n+1} = \left(1 - \dfrac{1}{n^2}\right)a_n,\quad a_1 = 2, a_2 = 1, n \geq 2$$
is bounded and monotonic and hence converges. Explain why the recursion formula is no help in finding the limit of the sequence.

≡ **Applications**

29. Certain studies in fishery management hold that the size of an undisturbed fish population changes from one year to the next in accordance with the formula
$$p_{n+1} = \dfrac{bp_n}{a + p_n},\quad n \geq 0,$$
where $p_n > 0$ is the population after n years, and a and b are positive parameters that depend on the species and its environment. Suppose that a population size p_0 is introduced in year 0.

(a) Use the recursion formula to show that the only possible limit values for the sequence $\{p_n\}$ are 0 and $b - a$.

(b) Show that $p_{n+1} < (b/a)p_n$.

(c) Use the result in part (b) to show that if $a > b$, then the population dies out: that is, $\lim_{n\to\infty} p_n = 0$.

(d) Now assume $a < b$. Show that if $0 < p_0 < b - a$, then the sequence $\{p_n\}$ is increasing and bounded above by $b - a$. Show that if $0 < b - a < p_0$, then the sequence $\{p_n\}$ is decreasing and bounded below by $b - a$. Conclude that $\lim_{n\to\infty} p_n = b - a$ for any $p_0 > 0$. [*Hint*: Examine $|b - a - p_{n+1}|$, which is the distance between p_{n+1} and $b - a$.]

≡ **Think About It**

30. Give an example of a bounded sequence that is not convergent.

31. Show that every convergent sequence $\{a_n\}$ is bounded. [*Hint*: Since $\{a_n\}$ is convergent, it follows from Definition 9.1.2 that there exists an N such that $|a_n - L| < 1$ whenever $n > N$.]

32. Show that $\left\{\int_1^n e^{-t^2} dt\right\}$ converges. [*Hint*: For $x > 1$, $e^{-x^2} \leq e^{-x}$.]

33. **A Mathematical Classic** Prove that the sequence
$$\left\{1 + \dfrac{1}{2} + \dfrac{1}{3} + \cdots + \dfrac{1}{n} - \ln n\right\}$$
is bounded and monotonic and hence convergent. The limit of the sequence is denoted by γ and is called **Euler's constant** after the noted Swiss mathematician **Leonhard Euler** (1707–1783). From Problem 66 of Exercises 9.1, $\gamma \approx 0.5772\ldots$. [*Hint*: First prove the inequality
$$\dfrac{1}{2} + \dfrac{1}{3} + \cdots + \dfrac{1}{n-1} + \dfrac{1}{n} < \ln n < 1 + \dfrac{1}{2} + \dfrac{1}{3} + \cdots + \dfrac{1}{n-1}$$
by considering the area under the graph of $y = 1/x$ on the interval $[1, n]$.

9.3 Series

■ **Introduction** The concept of a *series* is closely related to the concept of a *sequence*. If $\{a_n\}$ is the sequence $a_1, a_2, a_3, \ldots, a_n, \ldots$, then the sum of the terms

$$a_1 + a_2 + a_3 + \cdots + a_n + \cdots \qquad (1)$$

is called an **infinite series** or simply a **series**. The a_k, $k = 1, 2, 3, \ldots$, are called the **terms** of the series and a_n is called the **general term**. We write (1) compactly using summation notation as

$$\sum_{k=1}^{\infty} a_k \qquad \text{or for convenience} \qquad \sum a_k.$$

The question we seek to answer in this and the next several sections is:

- *When does an infinite series of constants "add up" to a number?*

EXAMPLE 1 An Infinite Series

In the opening remarks to this chapter we noted that the decimal representation for the rational number $\frac{1}{3}$ is, in fact, an infinite series

$$0.333\cdots = \frac{3}{10} + \frac{3}{10^2} + \frac{3}{10^3} + \cdots = \sum_{k=1}^{\infty} \frac{3}{10^k}.\qquad\blacksquare$$

Intuitively, we expect that $\frac{1}{3}$ is the sum of the series $\sum_{k=1}^{\infty} \frac{3}{10^k}$. But, just as intuitively, we expect that an infinite series such as

$$100 + 1000 + 10{,}000 + 100{,}000 + \cdots$$

where the terms are becoming larger and larger, has no sum. In other words, we do not expect the latter series to "add up" or *converge* to any number. The concept of convergence of an infinite series is defined in terms of the convergence of a special kind of sequence.

■ **Sequence of Partial Sums**   Associated with every infinite series $\sum a_k$, there is a **sequence of partial sums** $\{S_n\}$ whose terms are defined by

$$\begin{aligned}
S_1 &= a_1 \\
S_2 &= a_1 + a_2 \\
S_3 &= a_1 + a_2 + a_3 \\
&\vdots \\
S_n &= a_1 + a_2 + a_3 + \cdots + a_n \\
&\vdots
\end{aligned}$$

The general term $S_n = a_1 + a_2 + \cdots + a_n = \sum_{k=1}^{n} a_k$ of this sequence is called the ***n*th partial sum** of the series.

EXAMPLE 2 An Infinite Series

The sequence of partial sums $\{S_n\}$ for the series $\sum_{k=1}^{\infty} \frac{3}{10^k}$ is

$$\begin{aligned}
S_1 &= \frac{3}{10} = 0.3 \\
S_2 &= \frac{3}{10} + \frac{3}{10^2} = 0.33 \\
S_3 &= \frac{3}{10} + \frac{3}{10^2} + \frac{3}{10^3} = 0.333 \\
&\vdots \\
S_n &= \frac{3}{10} + \frac{3}{10^2} + \frac{3}{10^3} + \cdots + \frac{3}{10^n} = 0.\overbrace{333\ldots 3}^{n\ 3's} \\
&\vdots
\end{aligned}$$

$\blacksquare$

In Example 2, when n is very large, S_n will give a good approximation to $\frac{1}{3}$, and so it seems reasonable to write

$$\frac{1}{3} = \lim_{n\to\infty} S_n = \lim_{n\to\infty} \sum_{k=1}^{n} \frac{3}{10^k} = \sum_{k=1}^{\infty} \frac{3}{10^k}.$$

This leads to the following definition.

492 CHAPTER 9 Sequences and Series

> **Definition 9.3.1** Convergent Series
>
> An infinite series $\sum_{k=1}^{\infty} a_k$ is said to be **convergent** if its sequence of partial sums $\{S_n\} = \{\sum_{k=1}^{n} a_k\}$ converges; that is
>
> $$\lim_{n\to\infty} S_n = \lim_{n\to\infty} \sum_{k=1}^{n} a_k = S.$$
>
> The number S is said to be the **sum** of the series. If $\lim_{n\to\infty} S_n$ does not exist, then the series is said to be **divergent**.

EXAMPLE 3 Using the Sequence of Partial Sums

Show that the series $\sum_{k=1}^{\infty} \dfrac{1}{(k+4)(k+5)}$ is convergent.

Solution By partial fractions the general term a_n of the series can be written as

$$a_n = \frac{1}{n+4} - \frac{1}{n+5}.$$

Thus, the nth partial sum of the series is

$$S_n = \left[\frac{1}{5} - \frac{1}{6}\right] + \left[\frac{1}{6} - \frac{1}{7}\right] + \left[\frac{1}{7} - \frac{1}{8}\right] + \cdots + \left[\frac{1}{n+3} - \frac{1}{n+4}\right] + \left[\frac{1}{n+4} - \frac{1}{n+5}\right]$$

$$= \frac{1}{5} \underbrace{- \frac{1}{6} + \frac{1}{6} - \frac{1}{7} + \frac{1}{7} - \frac{1}{8} + \cdots + \frac{1}{n+3} - \frac{1}{n+4} + \frac{1}{n+4}}_{0} - \frac{1}{n+5}$$

$$= \frac{1}{5} - \frac{1}{n+5}.$$

From the last line we see that $\lim_{n\to\infty} 1/(n+5) = 0$, and so

$$\lim_{n\to\infty} S_n = \lim_{n\to\infty}\left[\frac{1}{5} - \frac{1}{n+5}\right] = \frac{1}{5} - 0 = \frac{1}{5}.$$

Hence, the series converges and we write

$$\sum_{k=1}^{\infty} \frac{1}{(k+4)(k+5)} = \frac{1}{5}. \qquad \blacksquare$$

Telescoping Series Because of the manner in which the general term of the sequence of partial sums "collapses" to two terms, the series in Example 3 is said to be a **telescoping series**. See Problems 11–14 in Exercises 9.3.

Geometric Series Another type of series that can be proven convergent or divergent directly from its sequence of partial sums has the form

$$a + ar + ar^2 + \cdots + ar^{n-1} + \cdots = \sum_{k=1}^{\infty} ar^{k-1}, \qquad (2)$$

where $a \neq 0$ and r are fixed real numbers. A series of the form (2) is called a **geometric series**. Note in (2) that each term after the first is obtained by multiplying the preceding term by r. The number r is called the **common ratio** and, as we see in the next theorem, its magnitude determines whether a geometric series converges or diverges.

Theorem 9.3.1 Sum of a Geometric Series

(*i*) If $|r| < 1$, then a geometric series converges and its sum is

$$\sum_{k=1}^{\infty} ar^{k-1} = \frac{a}{1-r}, \quad a \neq 0.$$

(*ii*) If $|r| \geq 1$, then a geometric series diverges.

PROOF The proof of Theorem 9.3.1 will be given in two parts. In each part we assume $a \neq 0$. We begin with the case that $|r| = 1$. For $r = 1$, the series is

$$\sum_{k=1}^{\infty} a = a + a + a + \cdots$$

and so the nth partial sum $S_n = \overbrace{a + a + \cdots + a}^{n\ a's}$ is simply $S_n = na$. In this case, $\lim_{n \to \infty} S_n = a \cdot \lim_{n \to \infty} n = \infty$. Thus the series diverges. For $r = -1$, the series is

$$\sum_{k=1}^{\infty} a(-1)^{k-1} = a + (-a) + a + (-a) + \cdots$$

and so the sequence of partial sums is

$$S_1, S_2, S_3, S_4, S_5, S_6, \ldots \quad \text{or} \quad a, 0, a, 0, a, 0, \ldots,$$

which is divergent.

Next we consider the case $|r| \neq 1$, which means that $|r| < 1$ or $|r| > 1$. Consider the general term of the sequence of partial sums of (2):

$$S_n = a + ar + ar^2 + \cdots + ar^{n-1}. \tag{3}$$

Multiplying both sides of (3) by r gives

$$rS_n = ar + ar^2 + ar^3 + \cdots + ar^n. \tag{4}$$

We then subtract (4) from (3) and solve for S_n:

$$S_n - rS_n = a - ar^n$$
$$(1 - r)S_n = a(1 - r^n)$$
$$S_n = \frac{a(1 - r^n)}{1 - r}, \quad r \neq 1. \tag{5}$$

Now, from Theorem 9.1.3 we know that $\lim_{n \to \infty} r^n = 0$ for $|r| < 1$. Consequently,

$$\lim_{n \to \infty} S_n = \lim_{n \to \infty} \frac{a(1 - r^n)}{1 - r} = \frac{a}{1 - r}, \quad |r| < 1.$$

If $|r| > 1$, then $\lim_{n \to \infty} r^n$ does not exist and so the limit of (5) fails to exist. ∎

EXAMPLE 4 Geometric Series

(a) In the geometric series

$$\sum_{k=1}^{\infty} \left(-\frac{1}{3}\right)^{k-1} = 1 - \frac{1}{3} + \frac{1}{9} - \frac{1}{27} + \cdots$$

we identify $a = 1$ and the common ratio $r = -\frac{1}{3}$. Since $|r| = |-\frac{1}{3}| = \frac{1}{3} < 1$, the series converges. From Theorem 9.3.1 the sum of the series is then

$$\sum_{k=1}^{\infty} \left(-\frac{1}{3}\right)^{k-1} = \frac{1}{1 - \left(-\frac{1}{3}\right)} = \frac{3}{4}.$$

(b) The common ratio in the geometric series

$$\sum_{k=1}^{\infty} 5\left(\frac{3}{2}\right)^{k-1} = 5 + \frac{15}{2} + \frac{45}{4} + \frac{135}{8} + \cdots$$

is $r = \frac{3}{2}$. The series diverges because $r = \frac{3}{2} > 1$.

Every rational number p/q, where p and $q \neq 0$ are integers, can be expressed either as a terminating decimal or as a repeating decimal. Thus, the series $\sum_{k=1}^{\infty} \frac{3}{10^k}$ in Example 1 converges since it is a geometric series with $r = \frac{1}{10} < 1$. With $a = \frac{3}{10}$ we find

$$\sum_{k=1}^{\infty} \frac{3}{10^k} = \frac{\frac{3}{10}}{1 - \frac{1}{10}} = \frac{\frac{3}{10}}{\frac{9}{10}} = \frac{3}{9} = \frac{1}{3}.$$

In general:

- *Every repeating decimal is a convergent geometric series.*

EXAMPLE 5 Rational Number

Express the repeating decimal $0.121212\ldots$ as a quotient of integers.

Solution We first write the given number as a geometric series

$$0.121212\ldots = \frac{12}{100} + \frac{12}{10{,}000} + \frac{12}{1{,}000{,}000} + \cdots$$

$$= \frac{12}{10^2} + \frac{12}{10^4} + \frac{12}{10^6} + \cdots$$

and make the identifications $a = \frac{12}{100}$ and $r = \frac{1}{10^2} = \frac{1}{100}$. By Theorem 9.3.1 the series converges since $r = \frac{1}{100} < 1$ and its sum is

$$0.121212\ldots = \frac{\frac{12}{100}}{1 - \frac{1}{100}} = \frac{\frac{12}{100}}{\frac{99}{100}} = \frac{12}{99} = \frac{4}{33}.$$

EXAMPLE 6 Watch the Bouncing Ball

If a ball is dropped from a height of s ft above the ground, then the time t it takes to reach the ground is related to s by $s = \frac{1}{2}gt^2$. In other words, it takes the ball $t = \sqrt{2s/g}$ s to reach the ground. Suppose the ball always rebounds to a certain fixed fraction $\beta\,(0 < \beta < 1)$ of its prior height. Find a formula for the time T it takes for the ball to come to rest. See FIGURE 9.3.1.

Solution The time to fall from a height of s ft to the ground is: $\sqrt{2s/g}$; the time to rise βs ft and then fall βs ft to the ground is: $2\sqrt{2\beta s/g}$; the time to rise $\beta(\beta s)$ ft and then fall $\beta(\beta s)$ ft to the ground is: $2\sqrt{2\beta^2 s/g}$; and so on. Thus, the total time T is given by the infinite series

$$T = \sqrt{2s/g} + 2\sqrt{2\beta s/g} + 2\sqrt{2\beta^2 s/g} + \cdots + 2\sqrt{2\beta^n s/g} + \cdots$$

$$= \sqrt{2s/g}\left[1 + 2\sum_{k=1}^{\infty}(\sqrt{\beta})^k\right].$$

Because $0 < \beta < 1$, the series $\sum_{k=1}^{\infty}(\sqrt{\beta})^k$ is a convergent geometric series with $a = \sqrt{\beta}$ and $r = \sqrt{\beta}$. Consequently, from Theorem 9.3.1

$$T = \sqrt{2s/g}\left[1 + 2\frac{\sqrt{\beta}}{1 - \sqrt{\beta}}\right] \quad \text{or} \quad T = \sqrt{2s/g}\left[\frac{1 + \sqrt{\beta}}{1 - \sqrt{\beta}}\right].$$

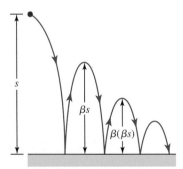

FIGURE 9.3.1 Bouncing ball in Example 6

Stroboscopic photo of a bouncing basketball

■ **Harmonic Series** One of the most famous series is also an example of a divergent series. The **harmonic series** is the sum of the reciprocals of the positive integers:

$$1 + \frac{1}{2} + \frac{1}{3} + \cdots + \frac{1}{n} + \cdots = \sum_{k=1}^{\infty} \frac{1}{k}. \tag{6}$$

◀ Remember this series. It will be important in the subsequent sections of this chapter.

The general term of the sequence of partial sums for (6) is given by

$$S_n = 1 + \frac{1}{2} + \frac{1}{3} + \cdots + \frac{1}{n}.$$

Thus,
$$S_{2n} = 1 + \frac{1}{2} + \frac{1}{3} + \cdots + \frac{1}{n} + \frac{1}{n+1} + \frac{1}{n+2} + \cdots + \frac{1}{2n}$$

$$= S_n + \frac{1}{n+1} + \frac{1}{n+2} + \cdots + \frac{1}{2n}$$

$$\geq S_n + \underbrace{\frac{1}{2n} + \frac{1}{2n} + \cdots + \frac{1}{2n}}_{n \text{ terms of } \frac{1}{2n}} = S_n + n \cdot \frac{1}{2n} = S_n + \frac{1}{2}.$$

The inequality $S_{2n} \geq S_n + \frac{1}{2}$ implies that the sequence of partial sums for the harmonic series is unbounded. To see this, we observe that

$$S_2 \geq S_1 + \frac{1}{2} = 1 + \frac{1}{2} = \frac{3}{2}$$

$$S_4 \geq S_2 + \frac{1}{2} \geq \frac{3}{2} + \frac{1}{2} = 2$$

$$S_8 \geq S_4 + \frac{1}{2} \geq 2 + \frac{1}{2} = \frac{5}{2}$$

$$S_{16} \geq S_8 + \frac{1}{2} \geq \frac{5}{2} + \frac{1}{2} = 3$$

and so on. Hence, we conclude that the harmonic series is divergent.

■ **A Consequence of Convergence** If a_n and S_n are the general terms of a series and the corresponding sequence of partial sums, respectively, then from the subtraction

$$S_n - S_{n-1} = (a_1 + a_2 + \cdots + a_{n-1} + a_n) - (a_1 + a_2 + \cdots + a_{n-1}) = a_n$$

we see that $a_n = S_n - S_{n-1}$. Now, if the series $\sum a_k$ converges to a number S, we have $\lim_{n \to \infty} S_n = S$ and $\lim_{n \to \infty} S_{n-1} = S$. This implies that

$$\lim_{n \to \infty} a_n = \lim_{n \to \infty} (S_n - S_{n-1}) = S - S = 0.$$

We have established the next theorem.

Theorem 9.3.2 Necessary Condition for Convergence

If the series $\sum_{k=1}^{\infty} a_k$ converges, then $\lim_{n \to \infty} a_n = 0$.

■ **Test for a Divergent Series** Theorem 9.3.2 simply states that if an infinite series converges, it is necessary that the *n*th, or general, term of the series approach zero. Equivalently, we conclude:

- *If the nth term a_n of an infinite series **does not** approach zero as $n \to \infty$, then the series **does not** converge.*

We formalize this result as a test for divergence.

> **Theorem 9.3.3** *n*th Term Test for Divergence
>
> If $\lim_{n\to\infty} a_n \neq 0$, then the series $\sum_{k=1}^{\infty} a_k$ diverges.

Theorem 9.3.3 immediately corroborates part (*ii*) of the proof of Theorem 9.3.1, namely, a geometric series $\sum_{k=1}^{\infty} ar^{k-1}$, $a \neq 0$, diverges when $r = \pm 1$. For example, when $r = 1$, $\lim_{n\to\infty} ar^{n-1} = \lim_{n\to\infty} a \neq 0$.

EXAMPLE 7 Divergent Series

(a) Consider the series $\sum_{k=1}^{\infty} \dfrac{4k-1}{5k+3}$. From

$$\lim_{n\to\infty} a_n = \lim_{n\to\infty} \frac{4n-1}{5n+3} = \lim_{n\to\infty} \frac{4 - \dfrac{1}{n}}{5 + \dfrac{3}{n}} = \frac{4}{5} \neq 0$$

it follows from Theorem 9.3.3 that the series diverges.

(b) Consider the series

$$\sum_{k=1}^{\infty} (-1)^{k-1} = 1 - 1 + 1 - 1 + \cdots.$$

Since $\lim_{n\to\infty} a_n = \lim_{n\to\infty} (-1)^{n-1}$ does not exist, we can assert that $\lim_{n\to\infty} a_n \neq 0$. The series diverges by Theorem 9.3.3.?   ∎

You are encouraged to read (and remember) (*iii*) of *Notes From the Classroom* at this time. We state the following three theorems without proof.

> **Theorem 9.3.4** Constant Multiple of a Series
>
> If c is any nonzero constant, then series $\sum_{k=1}^{\infty} a_k$ and $\sum_{k=1}^{\infty} ca_k$ both converge or both diverge.

> **Theorem 9.3.5** Sum of Two Convergent Series
>
> If $\sum_{k=1}^{\infty} a_k$ and $\sum_{k=1}^{\infty} b_k$ converge to S_1 and S_2, respectively, then
>
> (*i*) $\sum_{k=1}^{\infty} (a_k + b_k)$ converges to $S_1 + S_2$, and
>
> (*ii*) $\sum_{k=1}^{\infty} (a_k - b_k)$ converges to $S_1 - S_2$.

Theorem 9.3.5 indicates that when $\sum_{k=1}^{\infty} a_k$ and $\sum_{k=1}^{\infty} b_k$ converge, then

$$\sum_{k=1}^{\infty} (a_k \pm b_k) = \sum_{k=1}^{\infty} a_k \pm \sum_{k=1}^{\infty} b_k.$$

> **Theorem 9.3.6** Sum of a Convergent and a Divergent Series
>
> If $\sum_{k=1}^{\infty} a_k$ converges and $\sum_{k=1}^{\infty} b_k$ diverges, then $\sum_{k=1}^{\infty} (a_k + b_k)$ diverges.

EXAMPLE 8 Sum of Two Convergent Series

With the aid of Theorem 9.3.1, we see that the geometric series $\sum_{k=1}^{\infty} \left(\frac{1}{2}\right)^{k-1}$ and $\sum_{k=1}^{\infty} \left(\frac{1}{3}\right)^{k-1}$ converge to 2 and $\frac{3}{2}$, respectively. Hence, from Theorem 9.3.5, the series $\sum_{k=1}^{\infty} \left[\left(\frac{1}{2}\right)^{k-1} - \left(\frac{1}{3}\right)^{k-1}\right]$ converges and

$$\sum_{k=1}^{\infty}\left[\left(\frac{1}{2}\right)^{k-1} - \left(\frac{1}{3}\right)^{k-1}\right] = \sum_{k=1}^{\infty}\left(\frac{1}{2}\right)^{k-1} - \sum_{k=1}^{\infty}\left(\frac{1}{3}\right)^{k-1} = 2 - \frac{3}{2} = \frac{1}{2}.$$

EXAMPLE 9 Sum of Two Series

From Example 3 we know that $\sum_{k=1}^{\infty} \dfrac{1}{(k+4)(k+5)}$ converges. Since $\sum_{k=1}^{\infty} \dfrac{1}{k}$ is the divergent harmonic series, it follows from Theorem 9.3.6 that the series

$$\sum_{k=1}^{\infty}\left[\frac{1}{(k+4)(k+5)} + \frac{1}{k}\right]$$

diverges.

$\sum$ NOTES FROM THE CLASSROOM

(*i*) The nth term of the sequence of partial sums of the harmonic series is often denoted by $H_n = \sum_{k=1}^{n}(1/k)$. The terms of the sequence $H_1 = 1$, $H_2 = \frac{3}{2}$, $H_3 = \frac{11}{6}$, ... are called **harmonic numbers**. See Problem 71 in Exercises 9.3.

(*ii*) When written in terms of summation notation, a geometric series may not be immediately recognizable, or if it is, the values of a and r may not be apparent. For example, to see whether $\sum_{n=3}^{\infty} 4\left(\frac{1}{2}\right)^{n+2}$ is a geometric series it is a good idea to write out two or three terms:

$$\sum_{n=3}^{\infty} 4\left(\frac{1}{2}\right)^{n+2} = \overbrace{4\left(\frac{1}{2}\right)^5}^{a} + \overbrace{4\left(\frac{1}{2}\right)^6}^{ar} + \overbrace{4\left(\frac{1}{2}\right)^7}^{ar^2} + \cdots.$$

From the right side of the last equality, we can make the identifications $a = 4\left(\frac{1}{2}\right)^5$ and $r = \frac{1}{2} < 1$. Consequently, the sum of the series is $\dfrac{4\left(\frac{1}{2}\right)^5}{1 - \frac{1}{2}} = \dfrac{1}{4}$. If desired, although there is no real need to do this, we can express $\sum_{n=3}^{\infty} 4\left(\frac{1}{2}\right)^{n+2}$ in the more familiar form $\sum_{k=1}^{\infty} ar^{k-1}$ by letting $k = n - 2$. The result is

$$\sum_{n=3}^{\infty} 4\left(\frac{1}{2}\right)^{n+2} = \sum_{k=1}^{\infty} 4\left(\frac{1}{2}\right)^{k+4} = \sum_{k=1}^{\infty} \overbrace{4\left(\frac{1}{2}\right)^5}^{a} \overbrace{\left(\frac{1}{2}\right)^{k-1}}^{r^{k-1}}.$$

(*iii*) Note carefully how Theorems 9.3.2 and 9.3.3 are stated. Specifically, Theorem 9.3.3 *does not* say if $\lim_{n\to\infty} a_n = 0$, then $\sum a_k$ converges. In other words, $\lim_{n\to\infty} a_n = 0$ is not *sufficient* to guarantee that $\sum a_k$ converges. In fact, if $\lim_{n\to\infty} a_n = 0$, the series may either converge or diverge. For example, in the harmonic series $\sum_{k=1}^{\infty}(1/k)$, $a_n = 1/n$ and $\lim_{n\to\infty}(1/n) = 0$, but the series diverges.

(iv) When determining convergence, it is possible, and sometimes convenient, to delete or ignore the first several terms of a series. In other words, the infinite series $\sum_{k=1}^{\infty} a_k$ and $\sum_{k=N}^{\infty} a_k$, $N > 1$, differ by at most a finite number of terms and are either both convergent or both divergent. Of course, deleting the first $N - 1$ terms of a convergent series usually does affect the sum of the series.

Exercises 9.3
Answers to selected odd-numbered problems begin on page ANS-27.

Fundamentals

In Problems 1–10, write out the first four terms in each series.

1. $\sum_{k=1}^{\infty} \dfrac{2k+1}{k}$
2. $\sum_{k=1}^{\infty} \dfrac{2^k}{k}$
3. $\sum_{k=1}^{\infty} \dfrac{(-1)^{k-1}}{k(k+1)}$
4. $\sum_{k=1}^{\infty} \dfrac{(-1)^{k+1}}{k3^k}$
5. $\sum_{n=0}^{\infty} \dfrac{n+1}{n!}$
6. $\sum_{n=1}^{\infty} \dfrac{(2n)!}{n^2+1}$
7. $\sum_{m=1}^{\infty} \dfrac{2 \cdot 4 \cdot 6 \cdots (2m)}{1 \cdot 3 \cdot 5 \cdots (2m-1)}$
8. $\sum_{m=1}^{\infty} \dfrac{1 \cdot 3 \cdot 5 \cdots (2m-1)}{m!}$
9. $\sum_{j=3}^{\infty} \dfrac{\cos j\pi}{2j+1}$
10. $\sum_{i=5}^{\infty} i \sin \dfrac{i\pi}{2}$

In Problems 11–14, proceed as in Example 3 to find the sum of the given telescoping series.

11. $\sum_{k=1}^{\infty} \dfrac{1}{k(k+1)}$
12. $\sum_{k=1}^{\infty} \dfrac{1}{(k+1)(k+2)}$
13. $\sum_{k=1}^{\infty} \dfrac{1}{4k^2-1}$
14. $\sum_{k=1}^{\infty} \dfrac{1}{k^2+7k+12}$

In Problems 15–24, determine whether the given geometric series converges or diverges. If convergent, find the sum of the series.

15. $\sum_{k=1}^{\infty} 3\left(\dfrac{1}{5}\right)^{k-1}$
16. $\sum_{k=1}^{\infty} 10\left(\dfrac{3}{4}\right)^{k-1}$
17. $\sum_{k=1}^{\infty} \dfrac{(-1)^{k-1}}{2^{k-1}}$
18. $\sum_{k=1}^{\infty} \pi^k \left(\dfrac{1}{3}\right)^{k-1}$
19. $\sum_{r=1}^{\infty} 5^r 4^{-r}$
20. $\sum_{s=1}^{\infty} (-3)^s 7^{-s}$
21. $\sum_{n=1}^{\infty} 1000(0.9)^n$
22. $\sum_{n=1}^{\infty} \dfrac{(1.1)^n}{1000}$
23. $\sum_{k=0}^{\infty} \dfrac{1}{(\sqrt{3}-\sqrt{2})^k}$
24. $\sum_{k=0}^{\infty} \left(\dfrac{\sqrt{5}}{1+\sqrt{5}}\right)^k$

In Problems 25–30, write each repeating decimal number as a quotient of integers.

25. $0.222\ldots$
26. $0.555\ldots$
27. $0.616161\ldots$
28. $0.393939\ldots$
29. $1.314314\ldots$
30. $0.5262626\ldots$

In Problems 31 and 32, find the sum of the given series.

31. $\sum_{k=1}^{\infty} \left[\left(\dfrac{1}{3}\right)^{k-1} + \left(\dfrac{1}{4}\right)^{k-1}\right]$
32. $\sum_{k=1}^{\infty} \dfrac{2^k - 1}{4^k}$

In Problems 33–42, show that the given series is divergent.

33. $\sum_{k=1}^{\infty} 10$
34. $\sum_{k=1}^{\infty} (5k+1)$
35. $\sum_{k=1}^{\infty} \dfrac{k}{2k+1}$
36. $\sum_{k=1}^{\infty} \dfrac{k^2+1}{k^2+2k+3}$
37. $\sum_{k=1}^{\infty} (-1)^k$
38. $\sum_{k=1}^{\infty} \ln\left(\dfrac{k}{3k+1}\right)$
39. $\sum_{k=1}^{\infty} \dfrac{10}{k}$
40. $\sum_{k=1}^{\infty} \dfrac{1}{6k}$
41. $\sum_{k=1}^{\infty} \left[\dfrac{1}{2^{k-1}} + \dfrac{1}{k}\right]$
42. $\sum_{k=1}^{\infty} k \sin \dfrac{1}{k}$

In Problems 43–46, determine the values of x for which the given series converges.

43. $\sum_{k=1}^{\infty} \left(\dfrac{x}{2}\right)^{k-1}$
44. $\sum_{k=1}^{\infty} \left(\dfrac{1}{x}\right)^{k-1}$
45. $\sum_{k=1}^{\infty} (x+1)^k$
46. $\sum_{k=0}^{\infty} 2^k x^{2k}$

Applications

47. A ball is dropped from an initial height of 15 ft onto a concrete slab. Each time the ball bounces, it reaches a height of $\frac{2}{3}$ its preceding height. Use geometric series to determine the distance the ball travels before it comes to rest.

48. In Problem 47 determine the time it takes for the ball to come to rest.

49. To eradicate agricultural pests (such as the Medfly), sterilized male flies are released into the general population at regular time intervals. Let N_0 be the number of flies released each day and let s be the proportion that survive a given day. Of the original N_0 sterilized males, $N_0 s^n$ will survive for n successive weeks. Hence, the total number of such males that survive n weeks after the program has begun is $N_0 + N_0 s + N_0 s^2 + \cdots + N_0 s^n$. What does this sum approach as $n \to \infty$? Suppose $s = 0.9$ and 10,000 sterilized males are needed to control the

population in a certain area. Determine the number that should be released each day.

50. In some circumstances the amount of a drug that will accumulate in a patient's body after a long period of time is $A_0 + A_0 e^{-k} + A_0 e^{-2k} + \cdots$, where $k > 0$ is a constant and A_0 is the daily dose of the drug. Find the sum of the series.

51. A patient takes 15 mg of a drug each day. If 80% of the drug accumulated is excreted each day by bodily functions, how much of the drug will accumulate after a long period of time, that is, as $n \to \infty$? (Assume that the measurement of the accumulation is made immediately after each dose. See Problem 69 in Exercises 9.1.)

52. A force is applied to a particle, which moves in a straight line, in such a fashion that after each second the particle moves only one-half the distance it moved in the preceding second. If the particle moves 20 cm in the first second, how far will it move?

≡ Think About It

53. Suppose the sequence $\{a_n\}$ converges to a number $L \neq 0$. Explain why the series $\sum_{k=1}^{\infty} a_k$ diverges.

54. Determine whether the series
$$\frac{1}{1.1} + \frac{1}{1.11} + \frac{1}{1.111} + \cdots$$
converges or diverges.

55. Determine whether the sum of two divergent series is necessarily divergent.

56. Consider the series $\sum_{k=1}^{\infty} \frac{1}{k^2}$. Since $k^2 = k \cdot k$, the nth partial sum of the series is
$$S_n = \frac{1}{1 \cdot 1} + \frac{1}{2 \cdot 2} + \frac{1}{3 \cdot 3} + \cdots + \frac{1}{n \cdot n}.$$
Explain why the following inequalities are true and why they can be used to prove that the given series converges:
$$0 < S_n < 1 + \frac{1}{1 \cdot 2} + \frac{1}{2 \cdot 3} + \cdots + \frac{1}{(n-1) \cdot n}$$
or
$$0 < S_n < 1 + \left(\frac{1}{1} - \frac{1}{2}\right) + \left(\frac{1}{2} - \frac{1}{3}\right) + \cdots + \left(\frac{1}{n-1} - \frac{1}{n}\right).$$

57. Find the sum of the series
$$\frac{1+9}{25} + \frac{1+27}{125} + \frac{1+81}{625} + \cdots.$$

58. Find the sum of the series
$$\sum_{k=1}^{\infty} \left(\int_{k}^{k+1} x e^{-x} \, dx \right).$$

59. Find all values of x in $(-\pi/2, \pi/2)$ for which
$$\lim_{n \to \infty} \left(\frac{1}{1 - \tan x} - \sum_{k=0}^{n} \tan^k x \right) = 0.$$

60. Show that if $\lim_{n \to \infty} f(n+1) = L$, where L is a number, then
$$\sum_{k=1}^{\infty} [f(k+1) - f(k)] = L - f(1).$$

61. Determine whether $\sum_{n=1}^{\infty} \left(\sum_{k=1}^{n} \frac{1}{k} \right)$ converges or diverges.

62. Show that the series $\sum_{k=1}^{\infty} \frac{1}{\sqrt{k}}$ is divergent by showing that $S_n \geq \sqrt{n}$.

63. We saw that the harmonic series $\sum_{k=1}^{\infty} \frac{1}{k}$ diverges since the general term S_n of the sequence of partial sums can be made as large as we like by taking n to be sufficiently large ($S_n \to \infty$ as $n \to \infty$). Nevertheless, the harmonic series diverges *very slowly*.

    (a) Use the graph of $f(x) = 1/x$ for $x \geq 1$ to establish the inequality
    $$\ln(n+1) < 1 + \frac{1}{2} + \frac{1}{3} + \frac{1}{4} + \cdots + \frac{1}{n} < 1 + \ln n.$$

    (b) Use a calculator and the inequality in part (a) to estimate the value of n for which $S_n \geq 10$. Estimate the value of n for which $S_n \geq 100$.

64. In Problem 77 in Exercises 9.1 we considered the perimeters of the regions bounded by the Koch curves shown in Figure 9.1.5. In part (c) of the problem you should have shown that the perimeter of the limiting region is infinite. In this problem we consider the *areas* of the sequence of figures. Let the area of the first figure be A_1, the area of the second figure A_2, and so on.

    (a) Using the fact that the area of an equilateral triangle with sides of length s is $\frac{1}{4}\sqrt{3}s^2$, find the values of $A_1, A_2, A_3,$ and A_4.

    (b) Show that the area of the nth figure is
    $$A_n = \frac{1}{20}\sqrt{3}\left[8 - 3\left(\frac{4}{9}\right)^{n-1} \right].$$

    (c) What is $\lim_{n \to \infty} A_n$?

≡ Projects

65. **A Bit of History—Death by Bread** In 1972, an outbreak of methylmercury poisoning in Iraq resulted in 459 deaths among 6530 cases of poisoning admitted to hospitals. The outbreak was caused by the consumption of homemade bread prepared from wheat that had been treated with a methylmercury fungicide. The first symptoms of *parasthesia* (loss of sensation at the mouth, hands, and feet) begin to occur when the accumulated level of mercury reaches 25 mg. Symptoms of *ataxia* (loss of coordination in gait) begin to occur at 55 mg, *dysarthia* (slurred speech) at 90 mg, and deafness at 170 mg. Death becomes a possibility when the accumulated mercury level exceeds 200 mg. It was estimated that a typical loaf of bread made from the contaminated wheat contained

Homemade breads

1.4 mg of mercury. It is also estimated that the body removes only about 0.9% of the accumulated mercury each day.

(a) Suppose that a person receives a dosage d of mercury each day, and that the body removes a fraction p of the accumulated mercury each day. Find a formula for L_n, the accumulated level after eating on the nth day, and a formula for the limiting level, $\lim_{n\to\infty} L_n$.

(b) Using $d = 1.4$ and $p = 0.009$, find the limiting level of mercury and determine on which days the various symptoms begin to occur.

(c) What would the daily dose have to be in order for death to become possible by the 100th day? (Use $p = 0.009$.)

66. A Bit of History—Zeno's Paradox The Greek philosopher Zeno of Elea (c. 490 BC) was a disciple of the pre-Socratic philosopher Permenides who held that change or motion was an illusion. Of the paradoxes Zeno advanced in support of this philosophy, the most famous is his argument that Achilles, known for his ability to run fast, could not overcome a moving tortoise. The usual form of the story goes something like this:

> Achilles starts from point S and at exactly the same instant a tortoise starts from a point A in front of S. After a certain amount of time Achilles reaches the tortoise's starting point A, but during this time the tortoise has advanced to a new point B. During the time it takes Achilles to reach B, the tortoise has moved ahead again to a new point C. Continuing in this manner, forever, Achilles can never catch up to the tortoise.

See FIGURE 9.3.2. Use infinite series to resolve this apparent paradox. Make the assumption that each moves with a constant speed. It may help to make up reasonable values for the tortoise's head start and for the two speeds.

FIGURE 9.3.2 Achilles and the tortoise in Problem 66

67. Prime Numbers Write a short report in which you define a prime number. In the report include a proof on whether the series of the reciprocal of primes,

$$\sum_{n=1}^{\infty} \frac{1}{p_n} = \frac{1}{2} + \frac{1}{3} + \frac{1}{5} + \frac{1}{7} + \frac{1}{11} + \cdots$$

converges or diverges.

68. Length of a Zigzag Path In FIGURE 9.3.3(a), the blue triangle ABC is an isosceles right triangle. The line segment AP_1 is perpendicular to BC, the line segment P_1P_2 is perpendicular to AC, and so on. Find the length of the red zigzag path $AP_1P_2P_3\ldots$.

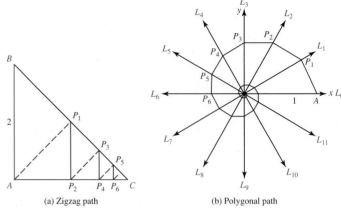

FIGURE 9.3.3 Zigzag and polygonal paths in Problems 68 and 69

69. Length of a Polygonal Path In Figure 9.3.3(b), there are twelve blue rays emanating from the origin and the angle between each pair of consecutive rays is 30°. The line segment AP_1 is perpendicular to ray L_1, the line segment P_1P_2 is perpendicular to ray L_2, and so on. Find the length of the red polygonal path $AP_1P_2P_3\ldots$.

70. An Improper Integral At the end of Section 7.7 we left dangling the question of whether $f(x) \to 0$ as $x \to \infty$ is a necessary requirement for the convergence of an improper integral $\int_a^{\infty} f(x)\, dx$. Here is the answer. Observe that the function f whose graph is given in FIGURE 9.3.4 does *not* approach 0 as $x \to \infty$. Show that $\int_0^{\infty} f(x)\, dx$ converges.

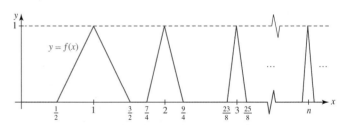

FIGURE 9.3.4 Graph for Problem 70

71. A Stacking Problem Take time out from doing your homework and perform an experiment. You will need a supply of n identical rectangular objects, let us say, the objects are books, but they could also be boards, playing cards, dominoes, and so on. Assume that the length of each book is L. Here is a rough statement of the problem:

> How far can a stack of n books extend over the edge of a table without falling over?

Intuitively the stack should *not* fall provided its center of mass stays above the tabletop. Using the stacking rule illustrated in FIGURE 9.3.5, observe that the book shown in Figure 9.3.5(a) achieves its maximum overhang $d_1 = L/2$ when its center of mass is placed directly at the edge of the table.

(a) Compute the overhangs d_2, d_3, and d_4 from the edge of the table for the stack of books in Figures 9.3.5(b),

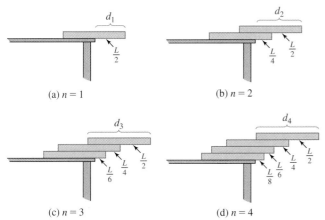

(a) $n = 1$ (b) $n = 2$ (c) $n = 3$ (d) $n = 4$

FIGURE 9.3.5 Method of stacking books in Problem 71

9.3.5(c), and 9.3.5(d), respectively. Then use (1) of Section 6.10 to show that the center of mass of each stack is at the edge of the table. [*Hint*: For n books put the x-axis along the horizontal tabletop with the origin O at the left edge of the first, or bottom, book in the stack.]

(b) What does the value of d_4 in part (a) indicate about the fourth, or top, book in the stack?

(c) Following the pattern of stacking indicated in Figure 9.3.5, for n books the overhang of the first book from the edge of the table would be $L/2n$, the overhang for the second book from the edge of the first book would be $L/2(n-1)$, the overhang for the third book from the edge of the second book would be $L/2(n-2)$, and so on. Find a formula for d_n, the overhang of n books from the edge of the table. Show that the center of mass of the stack of n books is at the edge of the table.

(d) Use the formula d_n for the overhang found in part (c) and find the smallest value of n so that the overhang of n books stacked in the manner described in part (c) is greater than twice the length of one book.

(e) In theory, using the stacking rule in part (c), is there any limitation on the number of books in a stack?

72. A Mathematical Classic—The Trains and the Fly At a specified time two trains T_1 and T_2, 20 mi apart on the same track, start on a collision course at a rate of 10 mph. Suppose that at the precise instant the trains start a fly leaves the front of train T_1, flies at a rate of 20 mph in a straight line to the front of the engine of train T_2, then flies back to T_1 at 20 mph, then back to T_2, and so on. Use geometric series to find the total distance traversed by the fly when the trains collide (and the fly is squashed). Then use common sense to find the total distance the fly flies. See FIGURE 9.3.6.

FIGURE 9.3.6 Trains and fly in Problem 72

9.4 Integral Test

Introduction Unless $\sum_{k=1}^{\infty} a_k$ is a telescoping series or a geometric series it is a difficult, if not futile, task to prove convergence or divergence directly from the sequence of partial sums. However, it is usually possible to determine whether a series converges or diverges by means of a *test* that utilizes only the terms of the series. In this and the next two sections we will examine five such tests that are applicable to infinite series of *positive terms*.

Integral Test The first test that we shall consider relates the concepts of convergence and divergence of an improper integral to convergence and divergence of an infinite series.

Theorem 9.4.1 Integral Test

Suppose $\sum_{k=1}^{\infty} a_k$ is a series of positive terms and f is a continuous function that is nonnegative and decreasing on $[1, \infty)$ such that $f(k) = a_k$ for $k \geq 1$.

(*i*) If $\int_1^\infty f(x)\, dx$ converges, then $\sum_{k=1}^{\infty} a_k$ converges.

(*ii*) If $\int_1^\infty f(x)\, dx$ diverges, then $\sum_{k=1}^{\infty} a_k$ diverges.

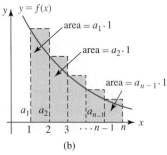

FIGURE 9.4.1 Rectangles in proof of Theorem 9.4.1

$f'(x) < 0$ on the interval $[3, \infty)$.

PROOF If the graph of f is given as in FIGURE 9.4.1, then by considering the areas of the rectangles shown in the figure, we see that

$$0 \leq a_2 + a_3 + a_4 + \cdots + a_n \leq \int_1^n f(x)\, dx \leq a_1 + a_2 + a_3 + \cdots + a_{n-1}$$

or

$$S_n - a_1 \leq \int_1^n f(x)\, dx \leq S_{n-1}.$$

From the inequality $S_n - a_1 \leq \int_1^n f(x)\, dx$, it is apparent that $\lim_{n\to\infty} S_n$ exists whenever $\lim_{n\to\infty} \int_1^n f(x)\, dx$ exists. On the other hand, from the inequality $S_{n-1} \geq \int_1^n f(x)\, dx$, we conclude that $\lim_{n\to\infty} S_{n-1}$ fails to exist whenever $\int_1^\infty f(x)\, dx$ diverges. ∎

EXAMPLE 1 Using the Integral Test

Test for convergence $\sum_{k=1}^{\infty} \dfrac{1}{1+k^2}$.

Solution The function $f(x) = 1/(1+x^2)$ is continuous, nonnegative, and decreasing for $x \geq 1$ such that $f(k) = a_k$ for $k \geq 1$. From

$$\int_1^\infty \frac{1}{1+x^2}\, dx = \lim_{b\to\infty} \int_1^b \frac{1}{1+x^2}\, dx$$

$$= \lim_{b\to\infty} \tan^{-1} x \Big]_1^b$$

$$= \lim_{b\to\infty} (\tan^{-1} b - \tan^{-1} 1) \quad \leftarrow \tan^{-1} 1 = \pi/4$$

$$= \lim_{b\to\infty} \left(\tan^{-1} b - \frac{\pi}{4}\right) \quad \leftarrow \text{see Figure 1.5.15}$$

$$= \frac{\pi}{2} - \frac{\pi}{4} = \frac{\pi}{4}$$

we see that the improper integral is convergent. From Theorem 9.4.1(i) we conclude that the given series also converges. ∎

In the Integral Test, if the positive-term series is of the form $\sum_{k=N}^{\infty} a_k$, we then use

$$\int_N^\infty f(x)\, dx \text{ where } f(k) = a_k.$$

EXAMPLE 2 Using the Integral Test

Test for convergence $\sum_{k=3}^{\infty} \dfrac{\ln k}{k}$.

Solution The function $f(x) = (\ln x)/x$ satisfies the hypotheses of the Integral Test on the interval $[3, \infty)$. Now,

$$\int_3^\infty \frac{\ln x}{x}\, dx = \lim_{b\to\infty} \int_3^b \frac{\ln x}{x}\, dx$$

$$= \lim_{b\to\infty} \frac{1}{2}(\ln x)^2 \Big]_3^b$$

$$= \lim_{b\to\infty} \frac{1}{2}[(\ln b)^2 - (\ln 3)^2] = \infty$$

shows that the improper integral diverges. It follows from Theorem 9.4.1(ii) that the given series also diverges. ∎

***p*-Series** The Integral Test is particularly useful on any series of the form

$$\sum_{k=1}^{\infty} \frac{1}{k^p} = 1 + \frac{1}{2^p} + \frac{1}{3^p} + \cdots, \qquad (1)$$

where p is any fixed real number. The infinite series (1) is called the **p-series** or **hyperharmonic series**. The next theorem indicates the values of p for which the p-series converges (diverges).

Theorem 9.4.2 Convergence of p-Series

The p-series $\sum_{k=1}^{\infty} \frac{1}{k^p}$ converges if $p > 1$ and diverges if $p \leq 1$.

PROOF We distinguish four cases: $p > 1$, $p = 1$, $0 < p < 1$, and $p \leq 0$. In the first and third cases we use the Integral Test with $f(x) = 1/x^p = x^{-p}$.

(i) If $p > 1$, then $p - 1 > 0$ and so

$$\int_1^{\infty} x^{-p} \, dx = \lim_{b \to \infty} \frac{x^{-p+1}}{-p+1} \Big]_1^b = \frac{1}{1-p} \lim_{b \to \infty} \left[\frac{1}{b^{p-1}} - 1 \right] = \frac{1}{1-p}[0-1] = \frac{1}{p-1}.$$

The p-series is convergent by Theorem 9.4.1(i).

(ii) If $p = 1$, then we recognize the p-series as the divergent harmonic series.

(iii) If $0 < p < 1$, then $-p + 1 > 0$ and so

$$\int_1^{\infty} x^{-p} \, dx = \lim_{b \to \infty} \frac{x^{-p+1}}{-p+1} \Big]_1^b = \frac{1}{1-p} \lim_{b \to \infty} [b^{-p+1} - 1] = \infty.$$

The p-series is divergent by Theorem 9.4.1(ii).

(iv) Finally, if $p \leq 0$, then $-p \geq 0$ and so $\lim_{n \to \infty}(1/n^p) = \lim_{n \to \infty} n^{-p} \neq 0$. The p-series is divergent by the nth term test, Theorem 9.3.3. ■

EXAMPLE 3 p-Series

(a) From Theorem 9.4.2, the p-series $\sum_{k=1}^{\infty} \frac{1}{\sqrt{k}} = \sum_{k=1}^{\infty} \frac{1}{k^{1/2}}$ diverges, since $p = \frac{1}{2} < 1$.

(b) From Theorem 9.4.2, the p-series $\sum_{k=1}^{\infty} \frac{1}{k^2}$ converges, since $p = 2 > 1$. ■

$\sum$ NOTES FROM THE CLASSROOM

(i) When applying the Integral Test, you should be aware that the value of the convergent improper integral $\int_1^{\infty} f(x) \, dx$ is not related to the actual sum of the corresponding infinite series. Thus, the series in Example 1 *does not* converge to $\pi/4$. See Problem 36 in Exercises 9.4.

(ii) The results of the Integral Test for $\sum_{k=n}^{\infty} a_k$ hold even if the continuous nonnegative function f does not begin to decrease until $x \geq N \geq n$. For the series $\sum_{k=1}^{\infty} (\ln k)/k$ the function $f(x) = (\ln x)/x$ decreases on the interval $[3, \infty)$. Nonetheless, in the Integral Test we may use $\int_1^{\infty} (\ln x \, dx)/x$.

Exercises 9.4 Answers to selected odd-numbered problems begin on page ANS-27.

≡ **Fundamentals**

In Problems 1–30, determine whether the given series converges or diverges. Use the Integral Test where appropriate.

1. $\sum_{k=1}^{\infty} \frac{1}{k^{1.1}}$

2. $\sum_{k=1}^{\infty} \frac{1}{k^{0.99}}$

3. $1 + \frac{1}{2\sqrt{2}} + \frac{1}{3\sqrt{3}} + \cdots$

4. $\frac{1}{100} + \frac{1}{100\sqrt{2}} + \frac{1}{100\sqrt{3}} + \cdots$

5. $\sum_{k=1}^{\infty} \dfrac{1}{2k+7}$

6. $\sum_{k=1}^{\infty} \dfrac{k}{3k+1}$

7. $\sum_{k=1}^{\infty} \dfrac{1}{1+5k^2}$

8. $\sum_{k=3}^{\infty} \dfrac{k}{k^2+5}$

9. $\sum_{k=1}^{\infty} ke^{-k^2}$

10. $\sum_{k=1}^{\infty} \dfrac{e^{1/k}}{k^2}$

11. $\sum_{k=1}^{\infty} \dfrac{k}{e^k}$

12. $\sum_{k=2}^{\infty} k^2 e^{-k}$

13. $\sum_{k=2}^{\infty} \dfrac{1}{k \ln k}$

14. $\sum_{k=2}^{\infty} \dfrac{k}{\ln k}$

15. $\sum_{k=2}^{\infty} \dfrac{10}{k(\ln k)^2}$

16. $\sum_{k=2}^{\infty} \dfrac{1}{k\sqrt{\ln k}}$

17. $\sum_{k=1}^{\infty} \dfrac{\arctan k}{1+k^2}$

18. $\sum_{k=1}^{\infty} \dfrac{k}{1+k^4}$

19. $\sum_{k=1}^{\infty} \dfrac{1}{\sqrt{1+k}}$

20. $\sum_{k=1}^{\infty} \dfrac{1}{\sqrt{1+k^2}}$

21. $\sum_{n=1}^{\infty} \dfrac{n}{(n^2+1)^3}$

22. $\sum_{n=2}^{\infty} \dfrac{1}{(4n+1)^{3/2}}$

23. $\sum_{k=1}^{\infty} k \sin\left(\dfrac{1}{k}\right)$

24. $\sum_{k=1}^{\infty} \ln\left(1+\dfrac{1}{3}k\right)$

25. $\sum_{k=1}^{\infty} \dfrac{1}{k(k+1)}$

26. $\sum_{k=1}^{\infty} \dfrac{2k+1}{k(k+1)}$

27. $\sum_{k=1}^{\infty} \dfrac{1}{(k+1)(k+2)}$

28. $\sum_{k=1}^{\infty} \dfrac{1}{k(k^2+1)}$

29. $\sum_{k=1}^{\infty} \dfrac{2}{e^k + e^{-k}}$

30. $\sum_{k=0}^{\infty} \dfrac{-1}{\sqrt{e^{3k}}}$

In Problems 31–34, without doing any work determine whether the given series converges or diverges. State your reasons.

31. $\sum_{k=1}^{\infty} \left(\dfrac{2}{k} + \dfrac{3}{k^2}\right)$

32. $\sum_{k=1}^{\infty} \left(5k^{-1.6} - 10k^{-1.1}\right)$

33. $\sum_{k=1}^{\infty} \left(\dfrac{1}{k^2} + \dfrac{1}{2^k}\right)$

34. $\sum_{k=1}^{\infty} \dfrac{1+4\sqrt{k}}{k^2}$

In Problems 35 and 36, determine the values of p for which the given series converges.

35. $\sum_{k=2}^{\infty} \dfrac{1}{k(\ln k)^p}$

36. $\sum_{k=3}^{\infty} \dfrac{1}{k \ln k [\ln(\ln k)]^p}$

≡ **Think About It**

37. Determine the values of p for which the series

$$\sum_{k=2}^{\infty} k^p \ln k$$

is convergent.

38. Suppose that f is a continuous function that is positive and decreasing for $x \geq 1$ such that $f(k) = a_k$ for $k \geq 1$. Show that

$$\int_1^{n+1} f(x)\,dx \leq \sum_{k=1}^n a_k \leq a_1 + \int_1^n f(x)\,dx.$$

39. Show that

$$\dfrac{\pi}{4} \leq \sum_{k=1}^{\infty} \dfrac{1}{1+k^2} \leq \dfrac{1}{2} + \dfrac{\pi}{4}.$$

40. The harmonic series $\sum_{k=1}^{\infty}(1/k)$ was shown to be divergent because the sequence of partial sums diverges. Recall from page 495 that $S_n = \sum_{k=1}^n (1/k) \to \infty$ as $n \to \infty$.

   (a) Use the result of Problem 38 to estimate the sum of the first 10 billion terms of the harmonic series.

   (b) How many terms of the harmonic series are necessary to guarantee that $S_n \geq 100$?

41. Let S denote the sum of a positive-term series $\sum_{k=1}^{\infty} a_k$ and S_n the general term in its sequence of partial sums. Define the **remainder**, or the error made when S is approximated by S_n, to be

$$R_n = S - S_n = a_{n+1} + a_{n+2} + a_{n+3} + \cdots.$$

Suppose that f is a continuous function that is positive and decreasing for $x \geq 1$ such that $f(k) = a_k$ for $k \geq 1$ and that $\int_1^{\infty} f(x)\,dx$ converges. Show that

$$\int_{n+1}^{\infty} f(x)\,dx \leq R_n \leq \int_n^{\infty} f(x)\,dx.$$

42. The sum S of the convergent p-series $\sum_{k=1}^{\infty}(1/k^2)$ is known to be $\pi^2/6$. Use Problem 41 to determine n so that S_n will give an approximation to S that is accurate to three decimal places.

9.5 Comparison Tests

Introduction It is often possible to determine convergence or divergence of a positive-term series $\sum a_k$ by *comparing* its terms with the terms of a *test series* $\sum b_k$ that is known to be either convergent or divergent. In this section we will consider two comparison tests for convergence and divergence.

Direct Comparison Test The proof of the next test will utilize two important properties of sequences. Recall from Section 9.2 that if a sequence is bounded and monotonic it must converge. Also if the terms of a sequence become unbounded then it diverges. We apply these results to the sequence of partial sums of a series.

Theorem 9.5.1 Direct Comparison Test

Suppose $\sum_{k=1}^{\infty} a_k$ and $\sum_{k=1}^{\infty} b_k$ are series of positive terms.

(i) If $\sum_{k=1}^{\infty} b_k$ converges and $a_k \leq b_k$ for every positive integer k, then $\sum_{k=1}^{\infty} a_k$ converges.

(ii) If $\sum_{k=1}^{\infty} b_k$ diverges and $a_k \geq b_k$ for every positive integer k, then $\sum_{k=1}^{\infty} a_k$ diverges.

PROOF Let $a_k > 0$ and $b_k > 0$ for $k = 1, 2, \ldots$ and let

$$S_n = a_1 + a_2 + \cdots + a_n \quad \text{and} \quad T_n = b_1 + b_2 + \cdots + b_n$$

be the general terms of the sequences of partial sums for $\sum a_k$ and $\sum b_k$, respectively.

(i) If $\sum b_k$ is a convergent series for which $a_k \leq b_k$, then $S_n \leq T_n$. Since $\lim_{n \to \infty} T_n$ exists, $\{S_n\}$ is a bounded increasing sequence and, hence, convergent by Theorem 9.2.1. Therefore, $\sum a_k$ is convergent.

(ii) If $\sum b_k$ diverges and $a_k > b_k$, then $S_n > T_n$. Since T_n increases without bound, so does S_n. Hence, $\sum a_k$ is divergent. ∎

In general, if $\sum c_k$ and $\sum d_k$ are two series for which $c_k \leq d_k$ for all k, we say that the series $\sum c_k$ is **dominated** by the series $\sum d_k$. Thus, for positive-term series, parts (i) and (ii) of Theorem 9.5.1 can be restated in the following manner:

- A series $\sum a_k$ is convergent if it is dominated by a convergent series $\sum b_k$.
- A series $\sum a_k$ diverges if it dominates a divergent series $\sum b_k$.

The next two examples illustrate the method. Of course, it goes without saying that to come up with a test series $\sum b_k$ you must be familiar with some series that converge and some that diverge.

◂ It might be a good idea at this point to review the notion of p-series in Section 9.4.

EXAMPLE 1 Using the Direct Comparison Test

Test for convergence $\sum_{k=1}^{\infty} \dfrac{k}{k^3 + 4}$.

Solution We observe that by decreasing the denominator in the general terms we obtain a larger fraction:

$$\frac{k}{k^3 + 4} \leq \frac{k}{k^3} = \frac{1}{k^2}.$$

Because the given series is dominated by the convergent p-series $\sum_{k=1}^{\infty} (1/k^2)$, it follows from Theorem 9.5.1(i) that the given series is also convergent. ∎

EXAMPLE 2 Using the Direct Comparison Test

Test for convergence $\sum_{k=1}^{\infty} \dfrac{\ln(k + 2)}{k}$.

Solution Since $\ln(k + 2) > 1$ for $k \geq 1$, we have

$$\frac{\ln(k + 2)}{k} > \frac{1}{k}.$$

In this case the given series has been shown to dominate the divergent harmonic series $\sum_{k=1}^{\infty} (1/k)$. Hence, by Theorem 9.5.1(ii) the given series diverges. ∎

■ **Limit Comparison Test** Another kind of comparison test involves taking the limit of the ratio of the general term of a series $\sum a_k$ to the general term of a test series $\sum b_k$ that is known to be convergent or divergent.

> **Theorem 9.5.2** Limit Comparison Test
>
> Suppose $\sum_{k=1}^{\infty} a_k$ and $\sum_{k=1}^{\infty} b_k$ are series of positive terms. If
> $$\lim_{n \to \infty} \frac{a_n}{b_n} = L,$$
> where L is finite and $L > 0$, then the two series are either both convergent or both divergent.

PROOF Since $\lim_{n \to \infty} a_n/b_n = L > 0$, we can choose n so large, say $n \geq N$ for some positive integer N, that
$$\frac{1}{2}L \leq \frac{a_n}{b_n} \leq \frac{3}{2}L.$$
Since $a_n > 0$, the inequality implies that $a_n \leq \frac{3}{2}Lb_n$ for $n \geq N$. If $\sum_{k=1}^{\infty} b_k$ converges, it follows from the Direct Comparison Test that $\sum_{k=1}^{\infty} a_k$ and, therefore, $\sum_{k=1}^{\infty} a_k$ is convergent. Furthermore, since $\frac{1}{2}Lb_n \leq a_n$ for $n \geq N$, we see that if $\sum_{k=1}^{\infty} b_k$ diverges, then $\sum_{k=1}^{\infty} a_k$ and $\sum_{k=1}^{\infty} a_k$ diverge. ■

The Limit Comparison Test is often applicable to series $\sum a_k$ for which the Direct Comparison Test is inconvenient.

EXAMPLE 3 Using the Limit Comparison Test

You should convince yourself that it is difficult to apply the Direct Comparison Test to the series $\sum_{k=1}^{\infty} \frac{1}{k^3 - 5k^2 + 1}$. However, we know that $\sum_{k=1}^{\infty}(1/k^3)$ is a convergent p-series ($p = 3 > 1$). Hence, with
$$a_n = \frac{1}{n^3 - 5n^2 + 1} \quad \text{and} \quad b_n = \frac{1}{n^3}$$
we have
$$\lim_{n \to \infty} \frac{a_n}{b_n} = \lim_{n \to \infty} \frac{n^3}{n^3 - 5n^2 + 1} = 1.$$
From Theorem 9.5.2, it follows that the given series converges. ■

If the general term a_n of a series $\sum a_k$ is a quotient of either rational powers of n or roots of polynomials in n, it is possible to discern the general term b_n of a test series $\sum b_k$ by examining the "degree behavior" of a_n for large values of n. In other words, to find a candidate for b_n we need only examine the quotient of the *highest powers of n* in the numerator and denominator of a_n.

EXAMPLE 4 Using the Limit Comparison Test

Test for convergence $\sum_{k=1}^{\infty} \frac{k}{\sqrt[3]{8k^5 + 7}}$.

Solution For large values of n, the general term of the series $a_n = n/\sqrt[3]{8n^5 + 7}$ "behaves like" a constant multiple of
$$\frac{n}{\sqrt[3]{n^5}} = \frac{n}{n^{5/3}} = \frac{1}{n^{2/3}}.$$

Thus, we try the divergent p-series $\sum_{k=1}^{\infty} \dfrac{1}{k^{2/3}}$ as a test series:

$$\lim_{n\to\infty} \frac{a_n}{b_n} = \lim_{n\to\infty} \frac{\dfrac{n}{\sqrt[3]{8n^5+7}}}{\dfrac{1}{n^{2/3}}}$$

$$= \lim_{n\to\infty} \left(\frac{n^5}{8n^5+7}\right)^{1/3} = \left(\frac{1}{8}\right)^{1/3} = \frac{1}{2}.$$

Thus, from Theorem 9.5.2 the given series diverges. ∎

$\sum$ NOTES FROM THE CLASSROOM

(*i*) The hypotheses in the Direct Comparison Test can also be weakened, giving a stronger theorem. For a series with positive terms, it is only required that $a_k \le b_k$ or $a_k \ge b_k$ for k sufficiently large and not for all positive integers.

(*ii*) In the application of the Direct Comparison Test, it is often easy to reach a point where the given series is dominated by a divergent series. For example,

$$\frac{1}{5^k + \sqrt{k}} \le \frac{1}{\sqrt{k}}$$

is certainly true and $\sum_{k=1}^{\infty} \dfrac{1}{\sqrt{k}}$ diverges. This kind of reasoning proves nothing about the series $\sum_{k=1}^{\infty} \dfrac{1}{5^k + \sqrt{k}}$. Indeed, the last series converges. Why? Similarly, no conclusion can be reached by showing that a given series dominates a convergent series.

The following table summarizes the **Direct Comparison Test**. Let $\sum a_k$ be a series of positive terms and $\sum b_k$ a series that we know either converges or diverges (a test series).

Comparison of terms	Test Series $\sum b_k$	Conclusion about $\sum a_k$
$a_k \le b_k$	converges	converges
$a_k \le b_k$	diverges	none
$a_k \ge b_k$	diverges	diverges
$a_k \ge b_k$	converges	none

Exercises 9.5
Answers to selected odd-numbered problems begin on page ANS-27.

≡ Fundamentals

In Problems 1–14, use the Direct Comparison Test to determine whether the given series converges.

1. $\sum_{k=1}^{\infty} \dfrac{1}{(k+1)(k+2)}$

2. $\sum_{k=1}^{\infty} \dfrac{1}{k^2+5}$

3. $\sum_{k=2}^{\infty} \dfrac{1}{\sqrt{k}-1}$

4. $\sum_{k=2}^{\infty} \dfrac{2k^2+1}{k^3-k}$

5. $\sum_{k=2}^{\infty} \dfrac{1}{\ln k}$

6. $\sum_{k=3}^{\infty} \dfrac{\ln k}{k^5}$

7. $\sum_{k=1}^{\infty} \dfrac{1+3^k}{2^k}$

8. $\sum_{k=1}^{\infty} \dfrac{1+8^k}{3+10^k}$

9. $\sum_{k=1}^{\infty} \dfrac{2+\sin k}{\sqrt[3]{k^4+1}}$

10. $\sum_{k=2}^{\infty} \dfrac{2k+1}{k\ln k}$

11. $\sum_{j=1}^{\infty} \dfrac{j + e^{-j}}{5^j(j+9)}$

12. $\sum_{i=1}^{\infty} \dfrac{ie^{-i}}{i+1}$

13. $\sum_{k=1}^{\infty} \dfrac{\sqrt{k+1} - \sqrt{k}}{k}$

14. $\dfrac{1}{1 \cdot 3} + \dfrac{1}{2 \cdot 9} + \dfrac{1}{3 \cdot 27} + \dfrac{1}{4 \cdot 81} + \cdots$

In Problems 15–28, use the Limit Comparison Test to determine whether the given series converges.

15. $\sum_{k=1}^{\infty} \dfrac{1}{2k+7}$

16. $\sum_{k=1}^{\infty} \dfrac{1}{10 + \sqrt{k}}$

17. $\sum_{n=2}^{\infty} \dfrac{1}{n\sqrt{n^2 - 1}}$

18. $\sum_{n=1}^{\infty} \dfrac{1}{\sqrt{(n+1)(n+2)}}$

19. $\sum_{n=1}^{\infty} \dfrac{n^2 - n + 2}{3n^5 + n^2}$

20. $\sum_{n=2}^{\infty} \dfrac{n}{(4n+1)^{3/2}}$

21. $\sum_{k=1}^{\infty} \dfrac{\sqrt{k+1}}{\sqrt[3]{64k^9 + 40}}$

22. $\sum_{k=2}^{\infty} \dfrac{5k^2 - k}{2k^3 + 2k^2 - 8}$

23. $\sum_{k=2}^{\infty} \dfrac{k + \ln k}{k^3 + 2k - 1}$

24. $\sum_{k=1}^{\infty} \dfrac{10}{e^k - 2}$

25. $\sum_{k=1}^{\infty} \sin\!\left(\dfrac{1}{k}\right)$

26. $\sum_{k=1}^{\infty} \left(1 - \cos\!\left(\dfrac{1}{k}\right)\right)$

27. $\sum_{k=1}^{\infty} \left(\dfrac{1}{2} + \dfrac{1}{2k}\right)^k$

28. $\dfrac{1}{2 \cdot 3} + \dfrac{2}{3 \cdot 4} + \dfrac{3}{4 \cdot 5} + \dfrac{4}{5 \cdot 6} + \cdots$

In Problems 29–40, use any appropriate test to determine whether the given series converges.

29. $\sum_{k=1}^{\infty} \dfrac{k}{100\sqrt{k^2 + 1}}$

30. $\sum_{k=1}^{\infty} \dfrac{1}{k + \sqrt{k}}$

31. $\sum_{k=1}^{\infty} \ln\!\left(5 + \dfrac{k}{5}\right)$

32. $\sum_{k=1}^{\infty} \ln\!\left(1 + \dfrac{1}{3^k}\right)$

33. $\sum_{k=1}^{\infty} \dfrac{k}{(k^2 + 1)^2}$

34. $\sum_{k=2}^{\infty} \dfrac{k}{\sqrt{k-1}\sqrt[3]{k^2 - 2}}$

35. $\sum_{k=1}^{\infty} \dfrac{1}{9 + \sin^2 k}$

36. $\sum_{k=1}^{\infty} \dfrac{3^k}{3^{2k} - 1}$

37. $\sum_{k=1}^{\infty} \dfrac{2}{2 + k 2^k}$

38. $\sum_{k=1}^{\infty} \dfrac{2}{2 + k 2^{-k}}$

39. $\sum_{k=2}^{\infty} \ln\!\left(1 + \dfrac{1}{k}\right)$

40. $\sum_{k=1}^{\infty} \dfrac{(0.9)^k}{k}$

≡ Think About It

41. Reread (ii) of Notes from the Classroom on page 507 and then discuss the reasons why the following statement is true:

If $a_k > 0$ for all k and Σa_k converges, then Σa_k^2 converges.

42. Suppose that p and q are polynomial functions with no common factors of degrees n and m, respectively, and that $p(x)/q(x) > 0$ for $x > 0$. Discuss: Under what conditions will the series $\sum_{k=1}^{\infty} p(k)/q(k)$ converge?

43. Discuss whether the following statement is true or false:

If $a_k < b_k$ for all k and Σb_k converges, then Σa_k converges.

44. Show that if the series Σa_k of positive terms converges, then $\Sigma \ln(1 + a_k)$ converges.

In Problems 45 and 46, determine whether the given series converges.

45. $\sum_{k=1}^{\infty} \dfrac{1}{k^{1 + 1/k}}$

46. $\sum_{k=1}^{\infty} \dfrac{1}{1 + 2 + 3 + \cdots + k}$

47. The decimal representation of a positive real number is an infinite series:

$$0.a_1 a_2 a_3 a_4 \ldots = \dfrac{a_1}{10} + \dfrac{a_2}{10^2} + \dfrac{a_3}{10^3} + \dfrac{a_4}{10^4} + \cdots,$$

where a_i represents one of the 10 nonnegative integers $0, 1, 2, \ldots, 9$. Prove that the series of the form

$$\dfrac{a_1}{10} + \dfrac{a_2}{10^2} + \dfrac{a_3}{10^3} + \dfrac{a_4}{10^4} + \cdots = \sum_{k=1}^{\infty} \dfrac{a_k}{10^k}$$

is always convergent.

≡ Project

48. **How Big Is Infinity?** The Integral Test can be used to verify that $\sum_{k=1}^{\infty} \dfrac{1}{k^{1.0001}}$ converges, whereas $\sum_{k=2}^{\infty} \dfrac{1}{k \ln k}$ diverges. However, with the aid of a CAS we see from the graphs of $y = 1/x^{1.0001}$ (in red) and $y = 1/(x \ln x)$ (in blue) in FIGURE 9.5.1 that

$$\dfrac{1}{k \ln k} < \dfrac{1}{k^{1.0001}}$$

for $2 \le k \le 15{,}000$. Indeed, the foregoing inequality is true for $2 \le k \le 99{,}999{,}999 \times 10^{99}$. So why doesn't $\sum_{k=2}^{\infty} \dfrac{1}{k \ln k}$ converge by the Direct Comparison Test?

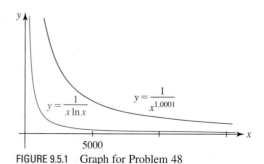

FIGURE 9.5.1 Graph for Problem 48

9.6 Ratio and Root Tests

Introduction In this section, as in the last, the tests that we shall consider are applicable to infinite series of *positive terms*.

Ratio Test The first of these tests employs a limit of the ratio of the $(n + 1)$st term to the nth term of the series. This test is especially useful when a_k involves factorials, kth powers of a constant, and sometimes, kth powers of k.

Theorem 9.6.1 Ratio Test

Suppose $\sum_{k=1}^{\infty} a_k$ is a series of positive terms such that

$$\lim_{n \to \infty} \frac{a_{n+1}}{a_n} = L.$$

(i) If $L < 1$, the series is convergent.
(ii) If $L > 1$, or if $L = \infty$, the series is divergent.
(iii) If $L = 1$, the test is inconclusive.

PROOF

(i) Let r be a positive number such that $0 \le L \le r \le 1$. For n sufficiently large, say $n \ge N$ for some positive integer N, $a_{n+1}/a_n < r$; that is $a_{n+1} < ra_n$, $n \ge N$. The last inequality implies

$$a_{N+1} < ra_N$$
$$a_{N+2} < ra_{N+1} < a_N r^2$$
$$a_{N+3} < ra_{N+2} < a_N r^3,$$

and so on. Thus, the series $\sum_{k=N+1}^{\infty} a_k$ converges by comparison with the convergent geometric series $\sum_{k=1}^{\infty} a_N r^k$. Since $\sum_{k=1}^{\infty} a_k$ differs from $\sum_{k=N+1}^{\infty} a_k$ by at most a finite number of terms, we conclude that the former series also converges.

(ii) Let r be a finite number such that $1 < r < L$. Then for n sufficiently large, say $n \ge N$ for some positive integer N, $a_{n+1}/a_n > r$ or $a_{n+1} > ra_n$. For $r > 1$ this last inequality implies $a_{n+1} > a_n$ and so $\lim_{n \to \infty} a_n \ne 0$. From Theorem 9.3.3 we conclude that $\sum_{k=1}^{\infty} a_k$ diverges. ∎

In the case when $L = 1$, we must apply another test to the series to determine its convergence or divergence.

EXAMPLE 1 Using the Ratio Test

Test for convergence $\sum_{k=1}^{\infty} \frac{5^k}{k!}$.

Solution We identify $a_n = 5^n/n!$ and so $a_{n+1} = 5^{n+1}/(n + 1)!$. We next form the quotient of a_{n+1} and a_n, simplify, and then take the limit as $n \to \infty$:

$$\lim_{n \to \infty} \frac{a_{n+1}}{a_n} = \lim_{n \to \infty} \frac{5^{n+1}}{(n+1)!} \cdot \frac{n!}{5^n}$$

$$= \lim_{n \to \infty} 5 \frac{n!}{(n+1)!}$$

$$= \lim_{n \to \infty} 5 \frac{n!}{n!(n+1)}$$

$$= \lim_{n \to \infty} \frac{5}{n+1} = 0.$$

◀ Review the properties of the factorial in Section 9.1. See (4) and (5) in that section.

Since $L = 0 < 1$, it follows from Theorem 9.6.1(i) that the series is convergent. ∎

EXAMPLE 2 Using the Ratio Test

Test for convergence $\sum_{k=1}^{\infty} \dfrac{k^k}{k!}$.

Solution In this case we have $a_n = n^n/n!$ and $a_{n+1} = (n+1)^{n+1}/(n+1)!$. Then

$$\lim_{n\to\infty} \frac{a_{n+1}}{a_n} = \lim_{n\to\infty} \frac{(n+1)^{n+1}}{(n+1)!} \cdot \frac{n!}{n^n}$$

$$= \lim_{n\to\infty} \frac{(n+1)^{n+1}}{n+1} \cdot \frac{1}{n^n}$$

$$= \lim_{n\to\infty} \left(\frac{n+1}{n}\right)^n$$

$$= \lim_{n\to\infty} \left(1 + \frac{1}{n}\right)^n = e. \quad \leftarrow \text{This limit is (3) of Section 1.6.}$$

Since $L = e > 1$, it follows from Theorem 9.6.1(*ii*) that the series is divergent. ∎

Root Test If the terms of a series $\sum a_k$ consist of only kth powers, then the following test, which involves taking the nth root of the nth term, may be applicable.

Theorem 9.6.2 Root Test

Suppose $\sum_{k=1}^{\infty} a_k$ is a series of positive terms such that

$$\lim_{n\to\infty} \sqrt[n]{a_n} = \lim_{n\to\infty} (a_n)^{1/n} = L.$$

(*i*) If $L < 1$, the series is convergent.
(*ii*) If $L > 1$, or if $L = \infty$, the series is divergent.
(*iii*) If $L = 1$, the test is inconclusive.

The proof of the Root Test is very similar to the proof of the Ratio Test and will not be given.

EXAMPLE 3 Using the Root Test

Test for convergence $\sum_{k=1}^{\infty} \left(\dfrac{5}{k}\right)^k$.

Solution We first identify $a_n = (5/n)^n$, and then compute the limit as $n \to \infty$ of the nth root of a_n:

$$\lim_{n\to\infty} \left[\left(\frac{5}{n}\right)^n\right]^{1/n} = \lim_{n\to\infty} \frac{5}{n} = 0.$$

Since $L = 0 < 1$, we conclude from Theorem 9.6.2(*i*) that the series converges. ∎

Σ NOTES FROM THE CLASSROOM

(*i*) The Ratio Test will always give the inconclusive case when applied to a *p*-series. Try it on the series $\sum_{k=1}^{\infty} 1/k^2$ and see what happens.

(*ii*) The tests examined in this and the preceding two sections tell us when a series has a sum, but none of these tests gives so much as even a clue as to what the actual sum is. But knowing that a series converges, we can now add up five, a hundred, or a thousand terms on a computer to obtain an approximation of the sum.

Exercises 9.6

≡ Fundamentals

In Problems 1–16, use the Ratio Test to determine whether the given series converges.

1. $\sum_{k=1}^{\infty} \frac{1}{k!}$
2. $\sum_{k=1}^{\infty} \frac{2^k}{k!}$
3. $\sum_{k=1}^{\infty} \frac{k!}{1000^k}$
4. $\sum_{k=1}^{\infty} k\left(\frac{2}{3}\right)^k$
5. $\sum_{j=1}^{\infty} \frac{j^{10}}{(1.1)^j}$
6. $\sum_{j=1}^{\infty} \frac{1}{j^5(0.99)^j}$
7. $\sum_{n=1}^{\infty} \frac{4^{n-1}}{n3^{n-2}}$
8. $\sum_{n=1}^{\infty} \frac{n^3 2^{n+3}}{7^{n-1}}$
9. $\sum_{k=1}^{\infty} \frac{k!}{(2k)!}$
10. $\sum_{k=1}^{\infty} \frac{(2k)!}{k!(2k)^k}$
11. $\sum_{k=1}^{\infty} \frac{99^k(k^3+1)}{k^2 10^{2k}}$
12. $\sum_{k=1}^{\infty} \frac{k!}{e^{k^2}}$
13. $\sum_{k=1}^{\infty} \frac{5^k}{k^k}$
14. $\sum_{k=1}^{\infty} \frac{k! 3^k}{k^k}$
15. $\sum_{k=1}^{\infty} \frac{1 \cdot 3 \cdot 5 \cdots (2k-1)}{k!}$
16. $\sum_{k=1}^{\infty} \frac{k!}{2 \cdot 4 \cdot 6 \cdots (2k)}$

In Problems 17–24, use the Root Test to determine whether the given series converges.

17. $\sum_{k=1}^{\infty} \frac{1}{k^k}$
18. $\sum_{k=1}^{\infty} \left(\frac{ke}{k+1}\right)^k$
19. $\sum_{k=2}^{\infty} \left(\frac{k}{\ln k}\right)^k$
20. $\sum_{k=2}^{\infty} \frac{1}{(\ln k)^k}$
21. $\sum_{k=1}^{\infty} \left(\frac{k}{k+1}\right)^{k^2}$
22. $\sum_{k=1}^{\infty} \left(1 - \frac{2}{k}\right)^{k^2}$
23. $\sum_{k=1}^{\infty} \frac{6^{2k+1}}{k^k}$
24. $\sum_{k=1}^{\infty} \frac{k^k}{e^{k+1}}$

In Problems 25–32, use any appropriate test to determine whether the given series converges.

25. $\sum_{k=1}^{\infty} \frac{k^2+k}{k^3+2k+1}$
26. $\sum_{k=1}^{\infty} \left(\frac{3k}{2k+1}\right)^k$
27. $\sum_{n=1}^{\infty} \frac{e^{1/n}}{n^2}$
28. $\sum_{n=1}^{\infty} \frac{n^2+n}{e^n}$
29. $\sum_{k=1}^{\infty} \frac{5^k k!}{(k+1)!}$
30. $\sum_{k=1}^{\infty} \frac{3}{2^k+k}$
31. $\sum_{k=0}^{\infty} \frac{2^k}{3^k+4^k}$
32. $\frac{1}{3} + \frac{2}{4} + \frac{3}{5} + \frac{4}{6} + \cdots$

In Problems 33 and 34, use the Ratio Test to determine the nonnegative values of p for which the given series converges.

33. $\sum_{k=1}^{\infty} kp^k$
34. $\sum_{k=1}^{\infty} k^2 \left(\frac{2}{p}\right)^k$

In Problems 35 and 36, determine all real values of p for which the gives series converges.

35. $\sum_{k=1}^{\infty} \frac{k^p}{k!}$
36. $\sum_{k=2}^{\infty} \frac{\ln k}{k^p}$

37. In Problems 78 and 79 of Exercises 9.1 we saw that the Fibonacci sequence $\{F_n\}$,

$$1, 1, 2, 3, 5, 8, \ldots,$$

is defined by the recursion formula $F_{n+1} = F_n + F_{n-1}$, where $F_1 = 1, F_2 = 1$.

(a) Verify that the general term of the sequence is

$$F_n = \frac{1}{\sqrt{5}}\left(\frac{1+\sqrt{5}}{2}\right)^n - \frac{1}{\sqrt{5}}\left(\frac{1-\sqrt{5}}{2}\right)^n$$

by showing that this result satisfies the recursion formula.

(b) Use the general term in part (a) to calculate $F_1, F_2, F_3, F_4,$ and F_5.

38. Let F_n be the general term of the Fibonacci sequence given in Problem 37. Show that

$$\lim_{n \to \infty} \frac{F_{n+1}}{F_n} = \frac{1+\sqrt{5}}{2}.$$

39. Explain how the result in Problem 38 proves that the series

$$\frac{1}{1} + \frac{1}{1} + \frac{1}{2} + \frac{1}{3} + \frac{1}{5} + \frac{1}{8} + \cdots = \sum_{n=1}^{\infty} \frac{1}{F_n}$$

converges.

40. **A Bit of History** In 1985 William Gosper used the following identity to compute the first 17 million digits of π:

$$\frac{1}{\pi} = \frac{2\sqrt{2}}{9801} \sum_{n=0}^{\infty} (1103 + 26,390n)\frac{(4n)!}{(n!)^4(4 \cdot 99)^{4n}}.$$

This identity was discovered in 1920 by the Indian mathematician **Srinivasa Ramanujan** (1887–1920). Ramanujan was noted for his remarkable insights in handling exceedingly complex algebraic manipulations and calculations.

(a) Verify that the infinite series converges.
(b) How many correct decimal places of π does the first term of the series yield?
(c) How many correct decimal places of π do the first two terms of the series yield?

9.7 Alternating Series

Introduction In the last three sections, we considered tests for convergence that were applicable only to series with positive terms. In the present discussion we consider series in which the terms alternate between positive and negative numbers, that is, series having either form

$$a_1 - a_2 + a_3 - a_4 + \cdots + (-1)^{n+1}a_n + \cdots = \sum_{k=1}^{\infty}(-1)^{k+1}a_k \qquad (1)$$

or

$$-a_1 + a_2 - a_3 + a_4 - \cdots + (-1)^n a_n + \cdots = \sum_{k=1}^{\infty}(-1)^k a_k, \qquad (2)$$

A geometric series such as $\sum_{k=1}^{\infty}(-\frac{1}{3})^{k-1} = 1 - \frac{1}{3} + \frac{1}{9} - \frac{1}{27} + \cdots$ is an alternating series. See Example 4 in Section 9.3.

where $a_k > 0$ for $k = 1, 2, 3, \ldots$. The series in (1) and (2) are said to be **alternating series**. We have already encountered a special type of alternating series in Section 9.3, but in this section we will examine properties of general alternating series and tests for their convergence. Because the series (2) is just a multiple of (1), we will confine our discussion to the latter series.

EXAMPLE 1 Alternating Series

The series

$$1 - \frac{1}{2} + \frac{1}{3} - \frac{1}{4} + \cdots = \sum_{k=1}^{\infty} \frac{(-1)^{k+1}}{k}$$

and

$$\frac{\ln 2}{4} - \frac{\ln 3}{8} + \frac{\ln 4}{16} - \frac{\ln 5}{32} + \cdots = \sum_{k=2}^{\infty}(-1)^k \frac{\ln k}{2^k}$$

are examples of alternating series. ∎

Alternating Series Test The first series in Example 1, $1 - \frac{1}{2} + \frac{1}{3} - \frac{1}{4} + \cdots$, is called the **alternating harmonic series**. Although the harmonic series

$$\sum_{k=1}^{\infty} \frac{1}{k} = 1 + \frac{1}{2} + \frac{1}{3} + \frac{1}{4} + \cdots$$

is divergent, the introduction of positive and negative terms in the sequence of partial sums for the alternating harmonic series is sufficient to produce a convergent series. We will prove that $\sum_{k=1}^{\infty} \frac{(-1)^{k+1}}{k}$ converges by means of the next test.

Theorem 9.7.1 Alternating Series Test

If $\lim_{n\to\infty} a_n = 0$ and $0 < a_{k+1} \leq a_k$ for every positive integer k, then the alternating series $\sum_{k=1}^{\infty}(-1)^{k+1}a_k$ converges.

The condition $0 < a_{k+1} \leq a_k$ means that
$a_1 \geq a_2 \geq a_3 \geq \cdots \geq a_k \geq a_{k+1} \geq \cdots$

PROOF Consider the partial sums that contain $2n$ terms:

$$S_{2n} = a_1 - a_2 + a_3 - a_4 + \cdots + a_{2n-1} - a_{2n} \qquad (3)$$
$$= (a_1 - a_2) + (a_3 - a_4) + \cdots + (a_{2n-1} - a_{2n}).$$

Since the assumption $0 < a_{k+1} \leq a_k$ implies $a_k - a_{k+1} \geq 0$ for $k = 1, 2, 3, \ldots$ we have

$$S_2 \leq S_4 \leq S_6 \leq \cdots \leq S_{2n} \leq \cdots.$$

Thus, the sequence $\{S_{2n}\}$, whose general term S_{2n} contains an even number of terms of the series, is a monotonic sequence. Rewriting (3) as

$$S_{2n} = a_1 - (a_2 - a_3) - \cdots - a_{2n}$$

shows that $S_{2n} < a_1$ for every positive integer n. Hence, $\{S_{2n}\}$ is bounded. By Theorem 9.2.1 it follows that $\{S_{2n}\}$ converges to a limit S. Now,

$$S_{2n+1} = S_{2n} + a_{2n+1}$$

implies that $\lim_{n\to\infty} S_{2n+1} = \lim_{n\to\infty} S_{2n} + \lim_{n\to\infty} a_{2n+1} = S + 0 = S$. This shows that the sequence of partial sums $\{S_{2n+1}\}$, whose general term S_{2n+1} contains an odd number of terms, also converges to S. Because both $\{S_{2n}\}$ and $\{S_{2n+1}\}$ converge to S, we conclude that $\{S_n\}$ converges to S. ∎

EXAMPLE 2 Alternating Harmonic Series

Show that the alternating harmonic series $\sum_{k=1}^{\infty} \frac{(-1)^{k+1}}{k}$ converges.

Solution With the identification $a_n = 1/n$ we have immediately

$$\lim_{n\to\infty} a_n = \lim_{n\to\infty} \frac{1}{n} = 0.$$

Moreover, since

$$\frac{1}{k+1} \leq \frac{1}{k}$$

for $k \geq 1$ we have $0 < a_{k+1} \leq a_k$. If follows from Theorem 9.7.1 that the alternating harmonic series converges. ∎

EXAMPLE 3 Divergent Alternating Series

The alternating series $\sum_{k=1}^{\infty} (-1)^{k+1} \frac{2k+1}{3k-1}$ diverges, since

$$\lim_{n\to\infty} a_n = \lim_{n\to\infty} \frac{2n+1}{3n-1} = \frac{2}{3}.$$

This last result indicates that

$$\lim_{n\to\infty} (-1)^{n+1} \frac{2n+1}{3n-1}$$

does not exist. Recall from Theorem 9.3.2 that it is necessary that the latter limit be 0 for the convergence of the series. ∎

Although showing that $a_{k+1} \leq a_k$ may seem a straightforward task, this is often not the case.

EXAMPLE 4 Using the Alternating Series Test

Test for convergence $\sum_{k=1}^{\infty} (-1)^{k+1} \frac{\sqrt{k}}{k+1}$.

Solution In order to show that the terms of the series satisfy the condition $a_{k+1} \leq a_k$, let us consider the function $f(x) = \sqrt{x}/(x+1)$ for which $f(k) = a_k$. From the derivative, we see that

$$f'(x) = -\frac{x-1}{2\sqrt{x}(x+1)^2} < 0 \quad \text{for} \quad x > 1,$$

and hence, the function f decreases for $x > 1$. Thus, $a_{k+1} \leq a_k$ is true for $k \geq 1$. Moreover, L'Hôpital's Rule shows that

$$\lim_{x\to\infty} f(x) = 0 \quad \text{and so} \quad \lim_{n\to\infty} f(n) = \lim_{n\to\infty} a_n = 0.$$

Hence, the given series converges by the Alternating Series Test. ∎

Approximating the Sum of an Alternating Series

Suppose the alternating series $\sum_{k=1}^{\infty}(-1)^{k+1}a_k$ converges to a number S. The partial sums

$$S_1 = a_1, \quad S_2 = a_1 - a_2, \quad S_3 = a_1 - a_2 + a_3, \quad S_4 = a_1 - a_2 + a_3 - a_4, \ldots$$

can be represented on a number line as shown in FIGURE 9.7.1. The sequence $\{S_n\}$ converges in the manner illustrated in Figure 9.1.1(c); that is, the terms S_n get closer to S as $n \to \infty$ although they oscillate on either side of S. As indicated in Figure 9.7.1, the even-numbered partial sums are less than S and the odd-numbered partial sums are greater than S. Roughly, the even-numbered partial sums increase to the number S, and, in turn, the odd-numbered partial sums decrease to S. Because of this, *the sum S of the series must lie between consecutive partial sums S_n and S_{n+1}*:

$$S_n \leq S \leq S_{n+1}, \quad \text{if } n \text{ is even}, \tag{4}$$

and

$$S_{n+1} \leq S \leq S_n, \quad \text{if } n \text{ is odd}. \tag{5}$$

Now (4) yields $0 \leq S - S_n \leq S_{n+1} - S_n$ for n even, and (5) implies $0 \leq S_n - S \leq S_n - S_{n+1}$ for n odd. Thus, in either case $|S_n - S| \leq |S_{n+1} - S_n|$.

But $S_{n+1} - S_n = a_{n+1}$ for n even and $S_{n+1} - S_n = -a_{n+1}$ for n odd. Thus, $|S_n - S| \leq a_{n+1}$ for all n. We state this result as our next theorem.

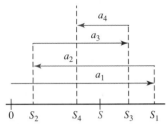

FIGURE 9.7.1 Partial sums on the number line

Theorem 9.7.2 Error Bound for Alternating Series

Suppose the alternating series $\sum_{k=1}^{\infty}(-1)^{k+1}a_k$, $a_k > 0$, converges to a number S. If S_n is the nth partial sum of the series and $a_{k+1} \leq a_k$ for all k, then

$$|S_n - S| \leq a_{n+1}$$

for all n.

Theorem 9.7.2 is useful in approximating the sum of a convergent alternating series. It states that the **error** $|S_n - S|$ between the nth partial sum and the series is less than the absolute value of the $(n + 1)$st term of the series.

EXAMPLE 5 Approximating the Sum of a Series

Approximate the sum of the convergent series $\sum_{k=1}^{\infty}\dfrac{(-1)^{k+1}}{(2k)!}$ to four decimal places.

Solution First, we note that $a_n = 1/(2n)!$. Theorem 9.7.2 indicates that we must have

$$a_{n+1} = \frac{1}{(2n+2)!} < 0.00005$$

in order to approximate the sum of the series to four decimal places. Now from

$$n = 1, \quad a_2 = \frac{1}{4!} \approx 0.041667$$

$$n = 2, \quad a_3 = \frac{1}{6!} \approx 0.001389$$

$$n = 3, \quad a_4 = \frac{1}{8!} \approx 0.000025 < 0.00005$$

we see that $|S_3 - S| \leq a_4 < 0.00005$. Therefore,

$$S_3 = \frac{1}{2!} - \frac{1}{4!} + \frac{1}{6!} \approx 0.4597$$

has the desired accuracy.

■ **Absolute and Conditional Convergence** A series containing mixed signs such as

$$\frac{2}{3} + \left(\frac{2}{3}\right)^2 - \left(\frac{2}{3}\right)^3 - \left(\frac{2}{3}\right)^4 + \left(\frac{2}{3}\right)^5 + \left(\frac{2}{3}\right)^6 - - + + \cdots \qquad (6)$$

is not strictly of the form given in (1) and so is not classified as an alternating series. Theorem 9.7.1 is not applicable to such a series. Nonetheless, we will see that the series (6) is convergent *because* the series of absolute values

$$\frac{2}{3} + \left(\frac{2}{3}\right)^2 + \left(\frac{2}{3}\right)^3 + \left(\frac{2}{3}\right)^4 + \left(\frac{2}{3}\right)^5 + \left(\frac{2}{3}\right)^6 + \cdots \qquad (7)$$

◀ Peek ahead and read the two sentences immediately following Example 7.

is convergent (a geometric series with $r = \frac{2}{3} < 1$). The series (6) is an example of a series that is **absolutely convergent**.

In the next definition we are letting the symbol $\sum_{k=1}^{\infty} a_k$ represent *any* series—the terms a_k could alternate as in (1) or contain mixed signs—the signs can follow some rule (as in (6)) or not.

Definition 9.7.1 Absolute Convergence

A series $\sum_{k=1}^{\infty} a_k$ is said to be **absolutely convergent** if the series of absolute values $\sum_{k=1}^{\infty} |a_k|$ converges.

EXAMPLE 6 Absolute Convergence

The alternating series $\sum_{k=1}^{\infty} \frac{(-1)^{k+1}}{1 + k^2}$ is absolutely convergent, since the series of absolute values

$$\sum_{k=1}^{\infty} \left|\frac{(-1)^{k+1}}{1 + k^2}\right| = \sum_{k=1}^{\infty} \frac{1}{1 + k^2}$$

was shown to be convergent by the Integral Test in Example 1 of Section 9.4. ∎

Definition 9.7.2 Conditional Convergence

A series $\sum_{k=1}^{\infty} a_k$ is said to be **conditionally convergent** if $\sum_{k=1}^{\infty} a_k$ converges but the series of absolute values $\sum_{k=1}^{\infty} |a_k|$ diverges.

EXAMPLE 7 Conditional Convergence

In Example 2 we saw that the alternating harmonic series $\sum_{k=1}^{\infty} \frac{(-1)^{k+1}}{k}$ is convergent. But taking the absolute value of each term gives the divergent harmonic series $\sum_{k=1}^{\infty} \frac{1}{k}$. Thus, $\sum_{k=1}^{\infty} \frac{(-1)^{k+1}}{k}$ is conditionally convergent. ∎

The next result shows that every absolutely convergent series is also convergent. It is for this reason that the series in (6) converges.

Theorem 9.7.3 Absolute Convergence Implies Convergence

If $\sum_{k=1}^{\infty} |a_k|$ converges, then $\sum_{k=1}^{\infty} a_k$ converges.

PROOF If we define $c_k = a_k + |a_k|$, then $c_k \leq 2|a_k|$. Since $\sum |a_k|$ converges, it follows from the Comparison Test that $\sum c_k$ converges. Furthermore, $\sum(c_k - |a_k|)$ converges, since both $\sum c_k$ and $\sum |a_k|$ converge. But

$$\sum_{k=1}^{\infty} a_k = \sum_{k=1}^{\infty} (c_k - |a_k|).$$

Therefore, $\sum a_k$ converges. ∎

Note that $\sum |a_k|$ is a series of positive terms, and so the tests of the preceding section can be utilized to determine whether a series converges absolutely.

EXAMPLE 8 Absolute Convergence Implies Convergence

The series

$$\sum_{k=1}^{\infty} \frac{\sin k}{k^2} = \frac{\sin 1}{1} + \frac{\sin 2}{4} + \frac{\sin 3}{9} + \frac{\sin 4}{16} + \cdots$$

contains positive and negative terms since

$$\sin 1 > 0, \quad \sin 2 > 0, \quad \sin 3 > 0, \quad \sin 4 < 0, \quad \sin 5 < 0, \quad \sin 6 < 0,$$

and so on. From trigonometry we know that $|\sin k| \leq 1$ for all k. Therefore,

$$\left|\frac{\sin k}{k^2}\right| \leq \frac{1}{k^2}$$

for all k. By the Direct Comparison Test, Theorem 9.5.1, the series $\sum_{k=1}^{\infty} \left|\frac{\sin k}{k^2}\right|$ converges because it is dominated by the convergent p-series $\sum_{k=1}^{\infty} \frac{1}{k^2}$. Hence, $\sum_{k=1}^{\infty} \frac{\sin k}{k^2}$ is absolutely convergent, and so by Theorem 9.7.3 it converges. ∎

■ **Ratio and Root Tests** The following modified forms of the Ratio Test and the Root Test can be applied directly to an alternating series.

Theorem 9.7.4 Ratio Test

Suppose $\sum_{k=1}^{\infty} a_k$ is a series of nonzero terms such that:

$$\lim_{n \to \infty} \left|\frac{a_{n+1}}{a_n}\right| = L.$$

(*i*) If $L < 1$, the series is absolutely convergent.
(*ii*) If $L > 1$, or if $L = \infty$, the series is divergent.
(*iii*) If $L = 1$, the test is inconclusive.

EXAMPLE 9 Using the Ratio Test

Test for convergence $\sum_{k=1}^{\infty} \frac{(-1)^{k+1} 2^{2k-1}}{k3^k}$.

Solution With $a_n = (-1)^{n+1} 2^{2n-1}/(n3^n)$, we see that

$$\lim_{n \to \infty} \left|\frac{a_{n+1}}{a_n}\right| = \lim_{n \to \infty} \left|\frac{(-1)^{n+2} 2^{2n+1}}{(n+1)3^{n+1}} \cdot \frac{n3^n}{(-1)^{n+1} 2^{2n-1}}\right|$$

$$= \lim_{n \to \infty} \frac{4n}{3(n+1)} = \frac{4}{3}.$$

Since $L = \frac{4}{3} > 1$, we see from Theorem 9.7.4(*ii*) that the alternating series diverges. ∎

Theorem 9.7.5 Root Test

Suppose $\sum_{k=1}^{\infty} a_k$ is a series such that:
$$\lim_{n \to \infty} \sqrt[n]{|a_n|} = \lim_{n \to \infty} |a_n|^{1/n} = L.$$

(*i*) If $L < 1$, the series is absolutely convergent.
(*ii*) If $L > 1$, or if $L = \infty$, the series is divergent.
(*iii*) If $L = 1$, the test is inconclusive.

■ **Rearrangement of Terms** When working with a *finite* series of terms such as
$$a_1 - a_2 + a_3 - a_4 + a_5 - a_6, \qquad (8)$$
any rearrangement of the order of terms, such as
$$-a_2 + a_1 - a_4 + a_3 - a_6 + a_5$$
or
$$(a_1 - a_2) + (a_3 - a_4) + (a_5 - a_6)$$
has the same sum as the original (8). This kind of carefree manipulation of terms does not carry over to *infinite* series:

- If the terms of a conditionally convergent series are written in a different order, the new series may diverge or converge to an entirely different number.

Indeed, it can be proved that by a suitable rearrangement of its terms, a conditionally convergent series can be made to converge to any prescribed real number r.

In contrast, a rearrangement of the terms of an absolutely convergent series does not effects its sum:

- If a series $\sum a_k$ is absolutely convergent, then the terms of the series can be rearranged in any manner and the resulting series will converge to the same number as the original series.

For example, the geometric series $1 - \frac{1}{3} + \frac{1}{9} - \frac{1}{27} + \cdots$ is absolutely convergent and its sum is $\frac{3}{4}$. The rearrangement $-\frac{1}{3} + \frac{1}{1} - \frac{1}{27} + \frac{1}{9} - \cdots$ of the geometric series is *not* a geometric series, nevertheless the rearranged series converges and its sum is $\frac{3}{4}$. See Problems 53–56 in Exercises 9.7.

∑ NOTES FROM THE CLASSROOM

(*i*) The conclusion of Theorem 9.7.1 remains true when the hypothesis "$a_{k+1} \leq a_k$ for every positive k" is replaced with the statement "$a_{k+1} \leq a_k$ for k sufficiently large." For the alternating series $\sum_{k=1}^{\infty} (-1)^{k+1} (\ln k)/k^{1/3}$, it is readily shown by the procedure used in Example 4 that $a_{k+1} \leq a_k$ for $k \geq 21$. Moreover, $\lim_{n \to \infty} a_n = 0$. Hence, the series converges by the Alternating Series Test.
(*ii*) If the series of absolute values $\sum |a_k|$ is found to be divergent, then no conclusion can be drawn concerning the convergence or divergence of the series $\sum a_k$.

Exercises 9.7 Answers to selected odd-numbered problems begin on page ANS-27.

≡ **Fundamentals**

In Problems 1–14, use the Alternating Series Test to determine whether the given series converges.

1. $\sum_{k=1}^{\infty} \frac{(-1)^{k+1}}{k+2}$

2. $\sum_{k=1}^{\infty} \frac{(-1)^{k-1}}{\sqrt{k}}$

3. $\sum_{k=1}^{\infty} (-1)^{k-1} \frac{k}{k+1}$

4. $\sum_{k=1}^{\infty} (-1)^k \frac{k}{k^2+1}$

5. $\sum_{k=1}^{\infty} (-1)^{k+1} \frac{k^2+2}{k^3}$

6. $\sum_{k=1}^{\infty} (-1)^{k+1} \frac{3k-1}{k+5}$

7. $\sum_{k=1}^{\infty}(-1)^{k+1}\left(\frac{1}{k}+\frac{1}{3^k}\right)$

8. $\sum_{k=1}^{\infty}(-1)^{k+1}\frac{k+1}{4^k}$

9. $\sum_{n=1}^{\infty}(-1)^{n+1}\frac{4\sqrt{n}}{2n+1}$

10. $\sum_{n=1}^{\infty}(-1)^{n-1}\frac{\sqrt[3]{n}}{n+1}$

11. $\sum_{n=2}^{\infty}(\cos n\pi)\frac{\sqrt{n+1}}{n+2}$

12. $\sum_{k=2}^{\infty}(-1)^k\frac{\sqrt{k^2+1}}{k^3}$

13. $\sum_{k=2}^{\infty}(-1)^k\frac{k}{\ln k}$

14. $\sum_{k=2}^{\infty}\frac{(-1)^k}{\ln k}$

In Problems 15–34, determine whether the given series is absolutely convergent, conditionally convergent, or divergent.

15. $\sum_{k=1}^{\infty}\frac{(-1)^{k+1}}{2k+1}$

16. $\sum_{k=1}^{\infty}\frac{(-1)^{k-1}}{\sqrt{k+5}}$

17. $\sum_{k=1}^{\infty}(-1)^{k+1}\left(\frac{2}{3}\right)^k$

18. $\sum_{k=1}^{\infty}(-1)^{k+1}\frac{2^{2k}}{3^k}$

19. $\sum_{k=1}^{\infty}(-1)^k\frac{k}{5^k}$

20. $\sum_{k=1}^{\infty}(-1)^k(k2^{-k})^2$

21. $\sum_{k=1}^{\infty}\frac{(-1)^k}{k!}$

22. $\sum_{k=1}^{\infty}(-1)^k\frac{(k!)^2}{(2k)!}$

23. $\sum_{k=1}^{\infty}(-1)^{k+1}\frac{k!}{100^k}$

24. $\sum_{k=1}^{\infty}(-1)^{k-1}\frac{5^{2k-3}}{10^{k+2}}$

25. $\sum_{k=1}^{\infty}(-1)^{k-1}\frac{k}{1+k^2}$

26. $\sum_{k=1}^{\infty}(-1)^{k+1}\frac{k}{1+k^4}$

27. $\sum_{k=1}^{\infty}\cos k\pi$

28. $\sum_{k=1}^{\infty}\frac{\sin\left(\frac{2k+1}{2}\pi\right)}{\sqrt{k+1}}$

29. $\sum_{k=1}^{\infty}(-1)^{k-1}\sin\left(\frac{1}{k}\right)$

30. $\sum_{k=1}^{\infty}\frac{(-1)^{k-1}}{k^2}\sin\left(\frac{1}{k}\right)$

31. $\sum_{k=1}^{\infty}(-1)^k\left[\frac{1}{k+1}-\frac{1}{k}\right]$

32. $\sum_{k=1}^{\infty}(-1)^k\left[\sqrt{k+1}-\sqrt{k}\right]$

33. $\sum_{k=1}^{\infty}(-1)^k\left(\frac{2k}{k+50}\right)^k$

34. $\sum_{k=1}^{\infty}(-1)^{k+1}\frac{6^{3k}}{k^k}$

In Problems 35 and 36, approximate the sum of the convergent series to the indicated number of decimal places.

35. $\sum_{k=1}^{\infty}\frac{(-1)^{k+1}}{(2k-1)!}$; five

36. $\sum_{k=1}^{\infty}\frac{(-1)^{k+1}}{k!}$; three

In Problems 37 and 38, find the smallest positive integer n so that S_n approximates the sum of the convergent series to the indicated number of decimal places.

37. $\sum_{k=1}^{\infty}\frac{(-1)^{k+1}}{k^3}$; two

38. $\sum_{k=1}^{\infty}\frac{(-1)^{k+1}}{\sqrt{k}}$; three

In Problems 39 and 40, approximate the sum of the convergent series so that the error is less than the indicated amount.

39. $1-\frac{1}{4^2}+\frac{1}{4^3}-\frac{1}{4^4}+\cdots$; 10^{-3}

40. $1-\frac{2}{5^2}+\frac{3}{5^3}-\frac{4}{5^4}+\cdots$; 10^{-4}

In Problems 41 and 42, estimate the error in using the indicated partial sum as an approximation to the sum of the convergent series.

41. $\sum_{k=1}^{\infty}\frac{(-1)^{k+1}}{k}$; S_{100}

42. $\sum_{k=1}^{\infty}\frac{(-1)^{k+1}}{k2^k}$; S_6

In Problems 43–48, state why the Alternating Series Test is not applicable to the given series. Determine whether the series converges.

43. $\sum_{k=1}^{\infty}\frac{\sin(k\pi/6)}{\sqrt{k^4+1}}$

44. $\sum_{k=1}^{\infty}\frac{100+(-1)^k 2^k}{3^k}$

45. $1-\frac{1}{2}-\frac{1}{4}+\frac{1}{8}+\frac{1}{16}--++\cdots$

46. $\frac{1}{1}-\frac{1}{4}-\frac{1}{9}+\frac{1}{16}+\frac{1}{25}+\frac{1}{36}---+++\cdots$

47. $\frac{2}{1}-\frac{1}{1}+\frac{2}{2}-\frac{1}{2}+\frac{2}{3}-\frac{1}{3}+\frac{2}{4}-\frac{1}{4}+\cdots$

[Hint: Consider the partial sums S_{2n} for $n=1,2,3,\ldots$]

48. $\frac{1}{2}+\frac{1}{2}-\frac{1}{3}-\frac{1}{3}-\frac{1}{3}+\frac{1}{4}+\frac{1}{4}+\frac{1}{4}+\frac{1}{4}----\cdots$

In Problems 49–52, determine whether the given series converges.

49. $1-1+1-1+\cdots$

50. $(1-1)+(1-1)+(1-1)+\cdots$

51. $1+(-1+1)+(-1+1)+\cdots$

52. $1+(-1+1)+(-1+1-1)+\cdots$

≡ **Think About It**

53. Reread the discussion just prior to *Notes from the Classroom* in this section. Then explain why the following statement is true:

    If a positive-term series $\sum a_k$ is convergent, then the terms of the series can be rearranged in any manner and the resulting series will converge to the same number as the original series.

54. Suppose S is the sum of the convergent alternating harmonic series $1-\frac{1}{2}+\frac{1}{3}-\frac{1}{4}+\frac{1}{5}-\frac{1}{6}+\cdots$.

    Show that the rearrangement of the series

    $$1-\frac{1}{2}-\frac{1}{4}+\frac{1}{3}-\frac{1}{6}-\frac{1}{8}+\frac{1}{5}-\frac{1}{10}-\frac{1}{12}+\frac{1}{7}-\frac{1}{14}\cdots$$
    $$=\left(1-\frac{1}{2}\right)-\frac{1}{4}+\left(\frac{1}{3}-\frac{1}{6}\right)-\frac{1}{8}+\left(\frac{1}{5}-\frac{1}{10}\right)-\frac{1}{12}$$
    $$+\left(\frac{1}{7}-\frac{1}{14}\right)-\cdots,$$

    gives $\frac{1}{2}S=\frac{1}{2}-\frac{1}{4}+\frac{1}{6}-\frac{1}{8}+\cdots$.

55. Use $S=1-\frac{1}{2}+\frac{1}{3}-\frac{1}{4}+\frac{1}{5}-\frac{1}{6}+\cdots$ and the result of Problem 54 in the form

    $$\frac{1}{2}S=0+\frac{1}{2}+0-\frac{1}{4}+0+\frac{1}{6}+0-\frac{1}{8}+\cdots$$

to show that the sum of another rearrangement of the terms of the alternating harmonic series is

$$\frac{3}{2}S = 1 + \frac{1}{3} - \frac{1}{2} + \frac{1}{5} + \frac{1}{7} - \frac{1}{4} + \cdots.$$

56. The series $1 - \frac{1}{3} + \frac{1}{9} - \frac{1}{27} + \cdots$ is an absolutely convergent geometric series. Show that its rearrangement $-\frac{1}{3} + \frac{1}{1} - \frac{1}{27} + \frac{1}{9} - \cdots$ is convergent. Try the Ratio Test and Root Test. [*Hint*: Examine $3^{k+(-1)^k}$, $k = 0, 1, 2, \ldots$.]

57. If $\sum a_k$ is absolutely convergent, prove that $\sum a_k^2$ converges. [*Hint*: For n sufficiently large, $|a_n| < 1$. Why?]

58. Give an example of a convergent series $\sum a_k$ for which $\sum a_k^2$ diverges.

59. Give an example of a convergent series $\sum a_k$ for which $\sum a_k^2$ converges.

60. Give an example of a divergent series $\sum a_k$ for which $\sum a_k^2$ converges.

61. Explain why the series

$$e^{-x}\sin x + e^{-2x}\sin 2x + e^{-3x}\sin 3x + \cdots$$

converges for every positive value of x.

9.8 Power Series

Introduction In applied mathematics it is common to work with infinite series of functions,

$$\sum_{k=0}^{\infty} c_k u_k(x) = c_0 u_0(x) + c_1 u_1(x) + c_2 u_2(x) + \cdots. \quad (1)$$

The coefficients c_k are constants depending on k and the functions $u_k(x)$ could be various kinds of polynomials or even the sine and cosine functions. When the variable x is specified, say, $x = 1$, then the series reduces to a series of constants. The convergence of a series such as (1) will, of course, depend on the variable x, the series usually converging for some values of x while diverging for other values. In this and the next section we consider infinite series (1) where the functions $u_k(x)$ are the polynomials $(x - a)^k$. We will study the properties of such series and show how to determine the values of x for which the series converge.

Power Series A series containing nonnegative integral powers of $(x - a)^k$,

$$\sum_{k=0}^{\infty} c_k(x - a)^k = c_0 + c_1(x - a) + c_2(x - a)^2 + \cdots + c_n(x - a)^n + \cdots, \quad (2)$$

is called a **power series in $x - a$**. The power series (2) is said to be **centered at a** or have **center a**. An important special case of (2), when $a = 0$,

$$\sum_{k=0}^{\infty} c_k x^k = c_0 + c_1 x + c_2 x^2 + \cdots + c_n x^n + \cdots, \quad (3)$$

is called a **power series in x**. The power series in (3) is centered at 0. A problem we face in this section is:

• *Find the values of x for which a power series converges.*

Observe that (2) and (3) converge to c_0 when $x = a$ and $x = 0$, respectively.

◀ It is conventient to define $(x - a)^0 = 1$ and $x^0 = 1$ even when $x = a$ and $x = 0$, respectively.

EXAMPLE 1 Power Series Centered at 0

The power series in x where the coefficients $c_k = 1$ for all k,

$$\sum_{k=0}^{\infty} x^k = 1 + x + x^2 + \cdots + x^n + \cdots,$$

is recognized as a geometric series with the common ratio $r = x$. By Theorem 9.3.1, the series converges for those values of x that satisfy $|x| < 1$ or $-1 < x < 1$. The series diverges for $|x| \geq 1$, that is, for $x \leq -1$ or $x \geq 1$. ∎

In general, the Ratio Test, as stated in Theorem 9.7.4, is especially helpful in finding the values of x for which a power series converges. The Root Test, in the form of Theorem 9.7.5, is also useful but to a lesser extent.

EXAMPLE 2 Interval of Convergence

Find the interval of convergence for $\sum_{k=0}^{\infty} \frac{x^k}{2^k(k+1)^2}$.

Solution With the identification that $a_n = x^n/(2^n(n+1)^2)$ we use the Ratio Test, Theorem 9.7.4,

$$\lim_{n\to\infty} \left| \frac{a_{n+1}}{a_n} \right| = \lim_{n\to\infty} \left| \frac{x^{n+1}}{2^{n+1}(n+2)^2} \cdot \frac{2^n(n+1)^2}{x^n} \right|$$

$$= \lim_{n\to\infty} \left(\frac{n+1}{n+2}\right)^2 \frac{|x|}{2} \quad \leftarrow \text{divide the numerator and denominator of the first term by } n$$

$$= \frac{|x|}{2} \cdot \lim_{n\to\infty} \left(\frac{1+1/n}{1+2/n}\right)^2 = \frac{|x|}{2}.$$

From part (*i*) of Theorem 9.7.4, we have absolute convergence whenever this limit is strictly less than 1. Thus, the series is absolutely convergent for those values of x that satisfy $|x|/2 < 1$ or $|x| < 2$. Since the absolute-value inequality $|x| < 2$ is equivalent to $-2 < x < 2$, we see the given series will converge for any number x in the open interval $(-2, 2)$. However, if $|x|/2 = 1$, or $|x| = 2$, or when $x = 2$ or $x = -2$, then the Ratio Test gives no information. We must perform separate checks of the given series for convergence at these endpoints. Substituting 2 for x the series becomes

$$\sum_{k=1}^{\infty} \frac{1}{(k+1)^2},$$

which is convergent by direct comparison with the convergent *p*-series $\sum_{k=1}^{\infty}(1/k^2)$. Similarly, substituting -2 for x yields

$$\sum_{k=1}^{\infty} \frac{(-1)^k}{(k+1)^2},$$

which is convergent by the Alternating Series Test, Theorem 9.7.1. We conclude that the given series converges for all x in the closed interval $[-2, 2]$. The series diverges for $x < -2$ and $x > 2$, or equivalently, for $|x| > 2$. ∎

FIGURE 9.8.1 The set of numbers x for which the series in Example 2 converges is shown in blue

■ **Interval of Convergence** In FIGURE 9.8.1 we have illustrated in blue the set $[-2, 2]$ of all real numbers x for which the series in Example 2 convergences and in red the set $(-\infty, -2) \cup (2, \infty)$ of numbers x for which the series diverges. The set of numbers for which the series converges is an interval centered at 0 (the center of the series). As shown in the figure, the radius of this interval is $R = 2$. In general, the set of *all* real numbers x for which a power series $\Sigma c_k(x - a)^k$ converges is said to be its **interval of convergence**. The center of the interval of convergence is the center a of the series. The radius R of the interval of convergence is called the **radius of convergence**.

The next theorem, given without proof, summarizes all the possible ways a power series can converge.

Theorem 9.8.1 Convergence of a Power Series

For a power series $\sum_{k=0}^{\infty} c_k(x-a)^k$ exactly one of the following is true:

(*i*) The series converges only at the *single number* $x = a$.
(*ii*) The series converges absolutely for *all real numbers* x.
(*iii*) The series converges absolutely for the *numbers* x *in a finite interval* $(a - R, a + R)$, $R > 0$, and diverges for numbers in the set $(-\infty, a - R) \cup (a + R, \infty)$. At an endpoint of the finite interval, $x = a - R$ or $x = a + R$, the series may converge absolutely, converge conditionally, or diverge.

Of course in (*ii*) and (*iii*), when the power series converges absolutely at a number x, we know, by Theorem 9.7.3, that it converges. In (*i*) of Theorem 9.8.1 the interval of convergence consists of the singleton set $\{a\}$ and we say that the series has **radius of convergence** $R = 0$. In (*ii*) of Theorem 9.8.1 the interval of convergence is $(-\infty, \infty)$ and the series has

radius of convergence $R = \infty$. Finally, in *(iii)* of Theorem 9.8.1, there are four possibilities for the interval of convergence with **radius of convergence** $R > 0$:

$$(a - R, a + R), \quad [a - R, a + R], \quad (a - R, a + R], \quad \text{or} \quad [a - R, a + R).$$

See FIGURE 9.8.2.

As in Example 1, if $R > 0$, we must handle the question of convergence at an endpoint $x = a \pm R$ by substituting these numbers into the given series and then either *recognizing* the resulting series as convergent or divergent or by *testing* the resulting series for convergence by an appropriate test other than the Ratio Test. Remember:

- *The Ratio Test is always inconclusive at an endpoint* $x = a \pm R$.

FIGURE 9.8.2 Possible finite intervals of convergence with $R > 0$

EXAMPLE 3 Interval of Convergence

Find the interval of convergence for $\sum_{k=0}^{\infty} \dfrac{x^k}{k!}$.

Solution By the Ratio Test, Theorem 9.7.4, we have

$$\lim_{n\to\infty} \left|\frac{a_{n+1}}{a_n}\right| = \lim_{n\to\infty} \left|\frac{x^{n+1}}{(n+1)!} \cdot \frac{n!}{x^n}\right| = \lim_{n\to\infty} \frac{n!}{(n+1)!}|x| = \lim_{n\to\infty} \frac{|x|}{n+1}.$$

Since $\lim_{n\to\infty} |x|/(n+1) = 0$ for any choice of x, the series converges absolutely for every real number. Thus, the interval of convergence is $(-\infty, \infty)$ and the radius of convergence is $R = \infty$. ∎

EXAMPLE 4 Interval of Convergence

Find the interval of convergence for $\sum_{k=1}^{\infty} \dfrac{(x-5)^k}{k3^k}$.

Solution By the Ratio Test, Theorem 9.7.4, we have

$$\lim_{n\to\infty} \left|\frac{a_{n+1}}{a_n}\right| = \lim_{n\to\infty} \left|\frac{(x-5)^{n+1}}{(n+1)3^{n+1}} \cdot \frac{n3^n}{(x-5)^n}\right|$$

$$= \lim_{n\to\infty} \left(\frac{n}{n+1}\right) \frac{|x-5|}{3}$$

$$= \lim_{n\to\infty} \left(\frac{1}{1+1/n}\right) \frac{|x-5|}{3} = \frac{|x-5|}{3}.$$

The series converges absolutely if $|x-5|/3 < 1$ or $|x-5| < 3$. This absolute-value inequality yields the open interval $(2, 8)$. At $x = 2$ and $x = 8$, the endpoints of the interval, we obtain, in turn,

$$\sum_{k=1}^{\infty} \frac{(-1)^k}{k} \quad \text{and} \quad \sum_{k=1}^{\infty} \frac{1}{k}.$$

The first series is a multiple of the alternating harmonic series and so is convergent, the second series is the divergent harmonic series. Consequently, the interval of convergence is $[2, 8)$. The radius of convergence is $R = 3$. The series diverges if $x < 2$ or $x \geq 8$. See FIGURE 9.8.3. ∎

The first series is

$$-1 + \tfrac{1}{2} - \tfrac{1}{3} + \cdots$$

or $(-1)[1 - \tfrac{1}{2} + \tfrac{1}{3} + \cdots]$

◂ The series in the brackets is the convergent alternating harmonic series.

FIGURE 9.8.3 Interval of convergence (blue) in Example 4

EXAMPLE 5 Interval of Convergence

Find the interval of convergence for $\sum_{k=1}^{\infty} k!(x + 10)^k$.

Solution From the Ratio Test,

$$\lim_{n\to\infty} \left|\frac{a_{n+1}}{a_n}\right| = \lim_{n\to\infty} \left|\frac{(n+1)!(x+10)^{n+1}}{n!(x+10)^n}\right|$$

$$= \lim_{n\to\infty} (n+1)|x+10|$$

we see that the limit as $n \to \infty$ can only exist if $|x + 10| = 0$, namely, when $x = -10$. Thus,

$$\lim_{n\to\infty} \left|\frac{a_{n+1}}{a_n}\right| = \begin{cases} \infty, & x \neq -10 \\ 0, & x = -10. \end{cases}$$

The series diverges for every real number x, except $x = -10$. At $x = -10$, we obtain a convergent series consisting of all zeros. The interval of convergence is the set $\{10\}$ and the radius of convergence is $R = 0$. ∎

Exercises 9.8
Answers to selected odd-numbered problems begin on page ANS-27.

≡ Fundamentals

In Problems 1–24, use the Ratio Test to find the interval and radius of convergence of the given power series.

1. $\sum_{k=1}^{\infty} \frac{(-1)^k}{k} x^k$
2. $\sum_{k=1}^{\infty} \frac{x^k}{k^2}$
3. $\sum_{k=1}^{\infty} \frac{2^k}{k} x^k$
4. $\sum_{k=0}^{\infty} \frac{5^k}{k!} x^k$
5. $\sum_{k=1}^{\infty} \frac{(x-3)^k}{k^3}$
6. $\sum_{k=1}^{\infty} \frac{(x+7)^k}{\sqrt{k}}$
7. $\sum_{k=1}^{\infty} \frac{(-1)^k}{10^k}(x-5)^k$
8. $\sum_{k=1}^{\infty} \frac{k}{(k+2)^2}(x-4)^k$
9. $\sum_{k=0}^{\infty} k! 2^k x^k$
10. $\sum_{k=0}^{\infty} \frac{k-1}{k^{2k}} x^k$
11. $\sum_{k=1}^{\infty} \frac{(3x-1)^k}{k^2+k}$
12. $\sum_{k=0}^{\infty} \frac{(4x-5)^k}{3^k}$
13. $\sum_{k=2}^{\infty} \frac{x^k}{\ln k}$
14. $\sum_{k=2}^{\infty} \frac{(-1)^k x^k}{k \ln k}$
15. $\sum_{k=1}^{\infty} \frac{k^2}{3^{2k}}(x+7)^k$
16. $\sum_{k=1}^{\infty} k^3 2^{4k}(x-1)^k$
17. $\sum_{k=1}^{\infty} \frac{2^{5k}}{5^{2k}} \left(\frac{x}{3}\right)^k$
18. $\sum_{k=1}^{\infty} \frac{1000^k}{k^k} x^k$
19. $\sum_{k=0}^{\infty} \frac{(-3)^k}{(k+1)(k+2)}(x-1)^k$
20. $\sum_{k=1}^{\infty} \frac{3^k}{(-2)^k k(k+1)}(x+5)^k$
21. $\sum_{k=1}^{\infty} \frac{(-1)^{k+1}}{(k!)^2} \left(\frac{x-2}{3}\right)^k$
22. $\sum_{k=0}^{\infty} \frac{(6-x)^{k+1}}{\sqrt{2k+1}}$
23. $\sum_{k=0}^{\infty} \frac{(-1)^k}{9^k} x^{2k+1}$
24. $\sum_{k=1}^{\infty} \frac{5^k}{(2k)!} x^{2k}$

In Problems 25-28, use the Root Test to find the interval and radius of convergence of the given power series.

25. $\sum_{k=2}^{\infty} \frac{x^k}{(\ln k)^k}$
26. $\sum_{k=1}^{\infty} (k+1)^k (x+1)^k$
27. $\sum_{k=1}^{\infty} \left(\frac{4}{3}\right)^k (x+3)^k$
28. $\sum_{k=1}^{\infty} \left(\frac{k}{k+1}\right)^{k^2} (x-e)^k$

In Problems 29 and 30, find the radius of convergence of the given power series.

29. $\sum_{k=1}^{\infty} \frac{k!}{1 \cdot 3 \cdot 5 \cdots (2k-1)} \left(\frac{x}{2}\right)^k$
30. $\sum_{k=2}^{\infty} \frac{1 \cdot 3 \cdot 5 \cdots (2k-3)}{3^k k!} (x-1)^k$

In Problems 31-38, the given series is not a power series. Nonetheless, find all values of x for which the given series converges.

31. $\sum_{k=1}^{\infty} \frac{1}{x^k}$
32. $\sum_{k=1}^{\infty} \frac{7^k}{x^{2k}}$
33. $\sum_{k=1}^{\infty} \left(\frac{x+1}{x}\right)^k$
34. $\sum_{k=1}^{\infty} \frac{1}{2^k} \left(\frac{x}{x+2}\right)^k$
35. $\sum_{k=0}^{\infty} \left(\frac{x^2+2}{6}\right)^{k^2}$
36. $\sum_{k=1}^{\infty} \frac{k!}{(kx)^k}$
37. $\sum_{k=0}^{\infty} e^{kx}$
38. $\sum_{k=0}^{\infty} k! e^{-kx^2}$

39. Find all values of x in $[0, 2\pi]$ for which $\sum_{k=1}^{\infty} \left(\frac{2}{\sqrt{3}}\right)^k \sin^k x$ converges.

40. Show that $\sum_{k=1}^{\infty} (\sin kx)/k^2$ converges for all real values of x.

≡ Calculator/CAS Problems

41. In Problems 71 and 72 of Exercises 5.5 we pointed out that some important functions in applied mathematics are defined in terms of nonelementary integrals. Some of these special functions of applied mathematics are also defined by infinite series. The power series

$$J_0(x) = \sum_{k=0}^{\infty} \frac{(-1)^k}{2^{2k}(k!)^2} x^{2k}$$

is called the **Bessel function of order 0**.

(a) The domain of the function $J_0(x)$ is its interval of convergence. Find the domain.

(b) The value of $J_0(x)$ is defined to be the sum of the series for x in its domain:

$$J_0(x) = \lim_{n \to \infty} S_n(x),$$

where
$$S_n(x) = \sum_{k=0}^{n} \frac{(-1)^k}{2^{2k}(k!)^2} x^{2k}$$

is the general term of the sequence of partial sums. Use a calculator or CAS and graph the partial sums $S_0(x), S_1(x), S_2(x), S_3(x)$, and $S_4(x)$.

(c) There are various kinds of Bessel functions of differing orders; $J_0(x)$ is a special case of a more general function $J_\nu(x)$ called the **Bessel function of the first kind of order** ν. Bessel functions are built-in functions in computer algebra systems such as *Mathematica* and *Maple*. Use a CAS to obtain the graph of $J_0(x)$ and compare it with the graphs of the partial sums in part (b). [*Hint*: In *Mathematica* $J_0(x)$ is denoted by BesselJ[0, x].]

9.9 Representing Functions by Power Series

■ **Introduction** For each x in its interval of convergence a power series $\sum c_k(x - a)^k$ converges to a single number. For this reason, a power series is itself a function, which we denote as f, whose *domain* is its interval of convergence. Then for each x in the interval of convergence we define the corresponding element in the *range* of the function, the value $f(x)$, as the sum of the series:

$$f(x) = c_0 + c_1(x - a) + c_2(x - a)^2 + \cdots = \sum_{k=0}^{\infty} c_k(x - a)^k.$$

The next two theorems, which will be stated without proof, answer some of the fundamental questions about differentiability, integrability, and continuity of a function f defined by a power series.

■ **Differentiation of a Power Series** The function f defined by a power series $\sum c_k(x - a)^k$ is differentiable.

Theorem 9.9.1 Differentiation of a Power Series

If $f(x) = \sum_{k=0}^{\infty} c_k(x - a)^k$ converges on an interval $(a - R, a + R)$ for which the radius of convergence R is either positive or ∞, then f is differentiable at each x in $(a - R, a + R)$, and

$$f'(x) = \sum_{k=1}^{\infty} k c_k (x - a)^{k-1}. \tag{1}$$

The radius of convergence R of (1) is the same as that of the original series.

The result in (1) simply states that a power series can be differentiated *term-by-term* as we would for a polynomial function:

$$\begin{aligned} f'(x) &= \frac{d}{dx} c_0 + \frac{d}{dx} c_1(x - a) + \frac{d}{dx} c_2(x - a)^2 + \cdots + \frac{d}{dx} c_n(x - a)^n + \cdots \\ &= c_1 + 2c_2(x - a) + 3c_3(x - a)^2 + \cdots + nc_n(x - a)^{n-1} + \cdots = \sum_{k=1}^{\infty} kc_k(x - a)^{k-1}. \end{aligned} \tag{2}$$

Since (1) is a power series with radius of convergence R, we can apply Theorem 9.9.1 to f' defined in (2). That is, we can say f' is differentiable at each x in $(a - R, a + R)$ and f'' is given by

$$f''(x) = 2c_2 + 3 \cdot 2c_3(x - a) + \cdots + n(n - 1)c_n(x - a)^{n-2} + \cdots = \sum_{k=2}^{\infty} k(k - 1) c_k(x - a)^{k-2}.$$

Continuing in this manner, it follows that:

- *A function f defined by a power series on $(a - R, a + R)$, $R > 0$, or on $(-\infty, \infty)$ possesses derivatives of all orders in the interval.*

The radius of convergence R of each differentiated series is the same as that of the original series. Moreover, since differentiability implies continuity we also have the result:

- *A function f defined by a power series on $(a - R, a + R)$, $R > 0$, or on $(-\infty, \infty)$, is continuous at each x in the interval.*

■ **Integration of a Power Series** As in (1), the process of integration of a power series can be carried out term-by-term:

$$\begin{aligned} \int f(x)\, dx &= \int c_0(x - a)^0\, dx + \int c_1(x - a)\, dx + \int c_2(x - a)^2\, dx + \cdots + \int c_n(x - a)^n\, dx + \cdots \\ &= c_0(x - a) + \frac{c_1}{2}(x - a)^2 + \frac{c_2}{3}(x - a)^3 + \cdots + \frac{c_n}{n + 1}(x - a)^{n+1} + \cdots + C \\ &= \sum_{k=0}^{\infty} \frac{c_k}{k + 1}(x - a)^{k+1} + C. \end{aligned}$$

This result is summarized in the next theorem.

Theorem 9.9.2 Integration of a Power Series

If $f(x) = \sum_{k=0}^{\infty} c_k(x-a)^k$ converges on an interval $(a-R, a+R)$ for which the radius of convergence R is either positive or ∞, then

$$\int f(x)\, dx = \sum_{k=0}^{\infty} \frac{c_k}{k+1}(x-a)^{k+1} + C. \tag{3}$$

The radius of convergence R of (3) is the same as that of the original series.

Since the function $f(x) = \sum_{k=0}^{\infty} c_k(x-a)^k$ is continuous, its definite integral exists and is defined by

$$\int_\alpha^\beta f(x)\, dx = \sum_{k=0}^{\infty} c_k \left(\int_\alpha^\beta (x-a)^k\, dx \right)$$

for any numbers α and β in $(a-R, a+R)$, $R > 0$, or in $(-\infty, \infty)$ if $R = \infty$.

It is recommended that you read this paragraph several times. ▶ In Theorems 9.9.1 and 9.9.2 it was stated that if the function $f(x) = \sum_{k=0}^{\infty} c_k(x-a)^k$ has radius of convergence $R > 0$ or $R = \infty$, then the series obtained by forming $f'(x)$ and $\int f(x)\, dx$ have the same radius of convergence R. This does *not* mean that the power series defining $f(x)$, $f'(x)$, and $\int f(x)\, dx$ have the same intervals of convergence. This is not as bad as it sounds. If the radius of convergence of the series defining $f(x)$, $f'(x)$, and $\int f(x)\, dx$ is $R > 0$, then the intervals of convergence can differ only at the endpoints of the interval. As a rule, by differentiating a function defined by a power series with radius of convergence $R > 0$ we *may lose* convergence at an endpoint of the interval. By integrating a function defined by a power series with radius of convergence $R > 0$ we *may gain* convergence at an endpoint of the interval.

EXAMPLE 1 Interval of Convergence

For the function f defined by $f(x) = \sum_{k=1}^{\infty} \frac{x^k}{k}$, find the intervals of convergence of

(a) $f'(x)$ (b) $\int f(x)\, dx$.

Solution It is readily shown from the Ratio Test that the interval of convergence of the power series that defines f is $[-1, 1)$.

(a) The derivative

$$f'(x) = \sum_{k=1}^{\infty} \frac{d}{dx}\frac{x^k}{k} = \sum_{k=1}^{\infty} x^{k-1} = 1 + x + x^2 + x^3 + \cdots \tag{4}$$

is recognized as a geometric series whose interval of convergence is $(-1, 1)$. The differentiated series (4) has lost convergence at the left endpoint of the interval of convergence for f.

(b) The integral of f is

$$\int f(x)\, dx = \sum_{k=1}^{\infty} \int \frac{x^k}{k}\, dx = \sum_{k=1}^{\infty} \frac{x^{k+1}}{k(k+1)} + C. \tag{5}$$

At $x = -1$ and $x = 1$, the series in (5) become, respectively,

$$\sum_{k=1}^{\infty} \frac{(-1)^{k+1}}{k(k+1)} \quad \text{and} \quad \sum_{k=1}^{\infty} \frac{1}{k(k+1)}.$$

◀ *The first series converges by the Alternating Series Test; the second converges by the Direct Comparison Test (the series is dominated by the convergent p-series $\sum 1/k^2$.)*

Because both series converge, the interval of convergence of (5) is $[-1, 1]$. In this instance, the integrated series (5) has gained convergence at the right endpoint of the interval of convergence for f. ∎

■ **Power Series Representation of a Function** It is often possible to express a *known* or *given* function f (such as e^x or $\tan^{-1} x$) as the sum of a power series on some interval. In this case we then say that the series is a **power series representation of** f on the interval.

The next example is important because it leads to many other results.

EXAMPLE 2 Representing a Function by a Power Series

Find a power series representation of $\dfrac{1}{1-x}$ centered at 0.

Solution Recall that a geometric series converges to $a/(1-r)$ if $|r| < 1$:

$$\frac{a}{1-r} = a + ar + ar^2 + \cdots + ar^{n-1} + \cdots.$$

Identifying $a = 1$ and $r = x$, we see that

$$\frac{1}{1-x} = 1 + x + x^2 + x^3 + \cdots + x^n + \cdots = \sum_{k=0}^{\infty} x^k. \tag{6}$$

The series converges for $|x| < 1$. The interval of convergence is $(-1, 1)$. In FIGURE 9.9.1 we have displayed the graph of $y = 1/(1-x)$ in blue along with the graphs of the partial sums $S_2(x)$, $S_5(x)$, $S_8(x)$, and $S_9(x)$ of the power series (6). When inspecting this figure, pay attention only to the interval $(-1, 1)$; the series does not represent the function outside this interval.

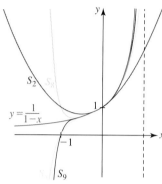

FIGURE 9.9.1 Graphs of partial sums in Example 2

By replacing x by $-x$ in (6), we obtain a power series representation for the function $1/(1 + x)$:

$$\frac{1}{1+x} = 1 - x + x^2 - x^3 + \cdots + (-1)^n x^n + \cdots = \sum_{k=0}^{\infty} (-1)^k x^k. \tag{7}$$

The series (7) converges for $|-x| < 1$ or $|x| < 1$. The interval of convergence is again $(-1, 1)$.

Many known functions can be represented by an infinite series by some sort of manipulation of the series in (6) and (7). For example, we could multiply the series by a power of x, we could replace x with another variable, or perhaps we could combine replacement of x by another variable with the process of integration (or differentiation), and so on.

EXAMPLE 3 Representing a Function by a Power Series

Find a power series representation of $\dfrac{1}{1 + 3x}$ centered at 0.

Solution By simply replacing the symbol x by $3x$ in (7) we get

$$\frac{1}{1+3x} = 1 - 3x + (3x)^2 - (3x)^3 + \cdots + (-1)^n(3x)^n + \cdots = \sum_{k=0}^{\infty} (-1)^k 3^k x^k.$$

This series converges when $|-3x| < 1$ or $|x| < \tfrac{1}{3}$. The interval of convergence is $\left(-\tfrac{1}{3}, \tfrac{1}{3}\right)$.

EXAMPLE 4 Representing a Function by a Power Series

Find a power series representation of $\dfrac{1}{5 - x}$ centered at 0.

Solution By factoring 5 from the denominator,

$$\frac{1}{5-x} = \frac{1}{5\left(1 - \dfrac{x}{5}\right)} = \frac{1}{5} \cdot \frac{1}{1 - \dfrac{x}{5}},$$

we are in a position to use (6). Replacing the symbol x in (6) with $x/5$ we get

$$\frac{1}{5-x} = \frac{1}{5} \cdot \frac{1}{1 - \dfrac{x}{5}} = \frac{1}{5}\left[1 + \frac{x}{5} + \left(\frac{x}{5}\right)^2 + \left(\frac{x}{5}\right)^3 + \cdots\right]$$

or

$$\frac{1}{5-x} = \frac{1}{5}\sum_{k=0}^{\infty}\left(\frac{x}{5}\right)^k = \sum_{k=0}^{\infty} \frac{1}{5^{k+1}} x^k.$$

This series converges for $|x/5| < 1$ or $|x| < 5$. The interval of convergence is $(-5, 5)$.

EXAMPLE 5 Power Series Centered at 3

Find a power series representation of $\dfrac{1}{1+x}$ centered at 3.

Solution Since the center of the power is to be 3, we want the power series to contain only powers of $x - 3$. To that end, we subtract and add 3 in the denominator:

$$\frac{1}{1+x} = \frac{1}{1+x-3+3} = \frac{1}{4+(x-3)}.$$

From this point on, we proceed as in Example 4, namely, we factor 4 from the denominator and use (7) with x replaced by $(x-3)/4$:

$$\frac{1}{1+x} = \frac{1}{4+(x-3)}$$

$$= \frac{1}{4} \cdot \frac{1}{1+\dfrac{x-3}{4}}$$

$$= \frac{1}{4}\left[1 - \frac{x-3}{4} + \left(\frac{x-3}{4}\right)^2 - \left(\frac{x-3}{4}\right)^3 + \cdots\right]$$

or

$$\frac{1}{1+x} = \frac{1}{4}\sum_{k=0}^{\infty}(-1)^k\left(\frac{x-3}{4}\right)^k = \sum_{k=0}^{\infty}\frac{(-1)^k}{4^{k+1}}(x-3)^k.$$

This series converges for $|(x-3)/4| < 1$ or $|x-3| < 4$. Solving the last inequality shows that the interval of convergence is $(-1, 7)$. ∎

EXAMPLE 6 Differentiation of a Power Series

Term-by-term differentiation of (7) yields a power series representation of $1/(1+x)^2$ on the interval $(-1, 1)$:

$$\frac{d}{dx}\frac{1}{1+x} = \frac{d}{dx}1 - \frac{d}{dx}x + \frac{d}{dx}x^2 - \frac{d}{dx}x^3 + \cdots + (-1)^n\frac{d}{dx}x^n + \cdots$$

yields

$$\frac{-1}{(1+x)^2} = -1 + 2x - 3x^2 + \cdots + (-1)^n nx^{n-1} + \cdots \quad \leftarrow \text{multiply both sides by } -1$$

or

$$\frac{1}{(1+x)^2} = 1 - 2x + 3x^2 + \cdots + (-1)^{n+1}nx^{n-1} + \cdots = \sum_{k=1}^{\infty}(-1)^{k+1}kx^{k-1}. \quad \blacksquare$$

EXAMPLE 7 Integration of a Power Series

Find a power series representation of $\ln(1+x)$ on $(-1, 1)$.

Solution We first introduce a dummy variable of integration by substituting $x = t$ in (7):

$$\frac{1}{1+t} = 1 - t + t^2 - t^3 + \cdots + (-1)^n t^n + \cdots.$$

Then, for any x within the interval $(-1, 1)$,

$$\int_0^x \frac{1}{1+t}\,dt = \int_0^x dt - \int_0^x t\,dt + \int_0^x t^2\,dt - \cdots + (-1)^n\int_0^x t^n\,dt + \cdots$$

$$= t\Big]_0^x - \frac{1}{2}t^2\Big]_0^x + \frac{1}{3}t^3\Big]_0^x - \cdots + (-1)^n\frac{1}{n+1}t^{n+1}\Big]_0^x + \cdots$$

$$= x - \frac{x^2}{2} + \frac{x^3}{3} - \cdots + (-1)^n\frac{x^{n+1}}{n+1} + \cdots.$$

But

$$\int_0^x \frac{1}{1+t}\,dt = \ln(1+t)\Big]_0^x = \ln(1+x) - \ln 1 = \ln(1+x)$$

and so

$$\ln(1 + x) = x - \frac{x^2}{2} + \frac{x^3}{3} - \cdots + (-1)^n \frac{x^{n+1}}{n+1} + \cdots = \sum_{k=0}^{\infty} \frac{(-1)^k}{k+1} x^{k+1}. \qquad (8) \blacksquare$$

Notice that the interval of convergence series in (8) is now $(-1, 1]$, that is, we have picked up convergence at $x = 1$. By setting $x = 1$ in (8), the series on the right-hand side of the equality is the convergent alternating harmonic series; on the left-hand side we get $\ln 2$. Thus, we have discovered the sum S of the alternating harmonic series:

$$\ln 2 = 1 - \frac{1}{2} + \frac{1}{3} - \frac{1}{4} + \cdots. \qquad (9)$$

EXAMPLE 8 Approximating a Value of ln x

Approximate $\ln(1.2)$ to four decimal places.

Solution Substituting $x = 0.2$ in (8) gives

$$\ln(1.2) = 0.2 - \frac{(0.2)^2}{2} + \frac{(0.2)^3}{3} - \frac{(0.2)^4}{4} + \frac{(0.2)^5}{5} - \frac{(0.2)^6}{6} + \cdots \qquad (10)$$

$$= 0.2 - 0.02 + 0.00267 - 0.0004 + 0.000064 - 0.00001067 + \cdots$$

$$\approx 0.1823. \qquad (11) \blacksquare$$

If the sum of the series (10) in Example 8 is denoted by S, then we know from Theorem 9.7.2 that $|S_n - S| \leq a_{n+1}$. The number given in (11) is accurate to four decimal places, since, for the fifth partial sum of (10),

$$|S_5 - S| \leq 0.00001067 < 0.00005.$$

■ **Arithmetic of Power Series** Two power series $f(x) = \sum b_k(x - a)^k$ and $g(x) = \sum c_k(x - a)^k$ can be combined by the arithmetic operations of addition, multiplication, and division. We can compute $f(x) + g(x)$ and $f(x)g(x)$ as in the addition and multiplication of two polynomials: We collect terms by like powers of $x - a$. At every point at which the power series defining f and g converge absolutely, the series

$$f(x) + g(x) = (b_0 + c_0) + (b_1 + c_1)(x - a) + (b_2 + c_2)(x - a)^2 + \cdots \qquad (12)$$

and $$f(x)g(x) = b_0 c_0 + (b_0 c_1 + b_1 c_0)(x - a) + (b_0 c_2 + b_1 c_1 + b_2 c_0)(x - a)^2 + \cdots \qquad (13)$$

converge absolutely. Similarly, for $c_0 \neq 0$ we can compute $f(x)/g(x)$ by long division:

$$\begin{array}{r} \frac{b_0}{c_0} + \frac{b_1 c_0 - b_0 c_1}{c_0^2}(x - a) + \cdots \quad \leftarrow \text{quotient} \\ c_0 + c_1(x - a) + \cdots \overline{)\, b_0 + b_1(x - a) + \cdots} \\ b_0 + \frac{b_0 c_1}{c_0}(x - a) + \cdots \\ \hline 0 + \frac{b_1 c_0 - b_0 c_1}{c_0}(x - a) + \cdots \\ \vdots \end{array} \qquad (14)$$

◂ Of course, do not memorize (12), (13), and (14); just carry out the algebra as you would for two polynomials.

The division is valid in *some* neighborhood of the center a of the two series.

We can sometimes use the arithmetic operations just illustrated along with previously known results to obtain a power series representation of a function.

EXAMPLE 9 Addition of Power Series

Find a power series representation of $\dfrac{4x}{x^2 + 2x - 3}$ centered at 0.

Solution To start we decompose the function into partial fractions

$$\frac{4x}{x^2 + 2x - 3} = \frac{3}{3 + x} - \frac{1}{1 - x}.$$

We then factor 3 from the denominator of the first partial fraction and use (7) with x replaced by $x/3$:

$$\frac{3}{3+x} = \frac{1}{1+\frac{x}{3}} = 1 - \frac{x}{3} + \frac{x^2}{3^2} - \frac{x^3}{3^3} + \cdots = \sum_{k=0}^{\infty} \frac{(-1)^k}{3^k} x^k. \quad (15)$$

This series converges for $|x/3| < 1$ or $|x| < 3$. The interval of convergence for (15) is $(-3, 3)$. Now from (6) we know

$$\frac{1}{1-x} = 1 + x + x^2 + x^3 + \cdots = \sum_{k=0}^{\infty} x^k \quad (16)$$

converges for $|x| < 1$. The interval of convergence for (16) is $(-1, 1)$. Finally, the addition of (15) and (16) yields the following power series representation for the given function:

$$\frac{4x}{x^2 + 2x - 3} = \frac{3}{3+x} - \frac{1}{1-x} = -\frac{4}{3}x - \frac{8}{9}x^2 - \frac{28}{27}x^3 - \cdots = \sum_{k=1}^{\infty} \left(\frac{(-1)^k}{3^k} - 1 \right) x^k. \quad (17)$$

The series (17) converges for all x common to (that is, the intersection of) the intervals $(-3, 3)$ and $(-1, 1)$, that is, for all x in $(-1, 1)$. ∎

The result in (17) can also be gotten by multiplying two power series.

EXAMPLE 10 Example 9 Revisited

If we rewrite the function in Example 9 as a product

$$\frac{4x}{x^2 + 2x - 3} = -\frac{4}{3}x \cdot \frac{1}{1 + \frac{x}{3}} \cdot \frac{1}{1-x}$$

and then use (15) and (16), it follows that

$$\frac{4x}{x^2 + 2x - 3} = -\frac{4}{3}x \cdot \left(1 - \frac{x}{3} + \frac{x^2}{3^2} - \frac{x^3}{3^3} + \cdots \right) \cdot (1 + x + x^2 + x^3 + \cdots)$$

$$= -\frac{4}{3}x \cdot \left[1 + 1\left(1 - \frac{1}{3}\right)x + \left(1 - \frac{1}{3} + \frac{1}{3^2}\right)x^2 + \cdots \right]$$

$$= -\frac{4}{3}x - \frac{8}{9}x^2 - \frac{28}{27}x^3 - \cdots. \quad ∎$$

Exercises 9.9 Answers to selected odd-numbered problems begin on page ANS-27.

≡ **Fundamentals**

In Problems 1–8, use (6) and (7) to find a power series representation, centered at 0, of the given function. Give the interval of convergence.

1. $\dfrac{1}{3-x}$
2. $\dfrac{1}{4+x}$
3. $\dfrac{1}{1+2x}$
4. $\dfrac{1}{5+2x}$
5. $\dfrac{1}{1+x^2}$
6. $\dfrac{x}{1+x^2}$
7. $\dfrac{1}{4+x^2}$
8. $\dfrac{4}{4-x^2}$

In Problems 9–14, use differentiation of an appropriate series from Problems 1–8 to find a power series representation, centered at 0, of the given function. Give the interval of convergence.

9. $\dfrac{1}{(3-x)^2}$
10. $\dfrac{1}{(1+2x)^2}$
11. $\dfrac{1}{(5+2x)^3}$
12. $\dfrac{1}{(4+x)^3}$
13. $\dfrac{x}{(1+x^2)^2}$
14. $\dfrac{1-x^2}{(1+x^2)^2}$

In Problems 15–20, use integration of an appropriate series from Problems 1–8 to find a power series representation, centered at 0, of the given function. Give the interval of convergence.

15. $\tan^{-1} x$
16. $\tan^{-1}(x/2)$
17. $\ln(1+x^2)$
18. $\ln(5+2x)$
19. $\ln(4+x)$
20. $\ln\left(\dfrac{3+x}{3-x}\right)$

In Problems 21–28, use (6), (7), or previous results to find a power series representation, centered at 0, of the given function. Give the interval of convergence.

21. $\dfrac{1-x}{1+2x}$

22. $\dfrac{3-x}{1-x}$

23. $\dfrac{x^2}{(1+x)^3}$

24. $\dfrac{x^3}{8+2x}$

25. $x\ln(1+x^2)$

26. $x^2\tan^{-1}x$

27. $\displaystyle\int_0^x \tan^{-1}t\,dt$

28. $\displaystyle\int_0^x \ln(1+t^2)\,dt$

In Problems 29–32, proceed as in Example 5 and find a power series representation, centered at the given number a, of the given function. Give the interval of convergence.

29. $\dfrac{1}{1-x};\quad a=6$

30. $\dfrac{1}{x};\quad a=-2$

31. $\dfrac{x}{2+x};\quad a=-1$

32. $\dfrac{x-2}{x-1};\quad a=2$

In Problems 33 and 34, proceed as in Example 9 and use partial fractions to find a power series representation, centered at 0, of the given function. Give the interval of convergence.

33. $\dfrac{7x}{x^2+x-12}$

34. $\dfrac{3}{x^2-x-2}$

In Problems 35 and 36, proceed as in Example 10 and use multiplication of power series to find the first four nonzero terms of a power series representation, centered at 0, for the given function.

35. $\dfrac{1}{(2-x)(1-x)}$

36. $\dfrac{x}{(1+2x)(1+x^2)}$

In Problems 37 and 38, find the domain of the given function.

37. $f(x) = \dfrac{x}{3} - \dfrac{x^2}{2\cdot 3^2} + \dfrac{x^3}{3\cdot 3^3} - \dfrac{x^4}{4\cdot 3^4} + \cdots$

38. $f(x) = 1 + 2x + \dfrac{4x^2}{1\cdot 2} + \dfrac{8x^2}{1\cdot 2\cdot 3} + \cdots$

In Problems 39–44, use power series to approximate the given quantity to four decimal places.

39. $\ln(1.1)$

40. $\tan^{-1}(0.2)$

41. $\displaystyle\int_0^{1/2} \dfrac{1}{1+x^3}\,dx$

42. $\displaystyle\int_0^{1/3} \dfrac{x}{1+x^4}\,dx$

43. $\displaystyle\int_0^{0.3} x\tan^{-1}x\,dx$

44. $\displaystyle\int_0^{1/2} \tan^{-1}x^2\,dx$

45. Use Problem 15 to show that
$$\dfrac{\pi}{4} = 1 - \dfrac{1}{3} + \dfrac{1}{5} - \dfrac{1}{7} + \cdots.$$

46. The series in Problem 45 is known to converge very slowly. Show this by finding the smallest positive integer n so that S_n approximates $\pi/4$ to four decimal places.

In Problems 47 and 48, show that the function defined by the power series satisfies the given differential equation.

47. $y = \displaystyle\sum_{k=1}^{\infty} \dfrac{(-1)^{k+1}}{k}x^k;\quad (x+1)y'' + y' = 0$

48. $J_0(x) = \displaystyle\sum_{k=0}^{\infty} \dfrac{(-1)^k}{2^{2k}(k!)^2}x^{2k};\quad xy'' + y' + xy = 0$

≡ Think About It

49. (a) If $f(x) = \displaystyle\sum_{k=0}^{\infty} \dfrac{x^k}{k!}$, then show that $f'(x) = f(x)$ for all x in $(-\infty, \infty)$.

(b) What function has the property that its first derivative equals the function? Conjecture what function is represented by the power series in part (a).

50. (a) If $f(x) = \displaystyle\sum_{k=0}^{\infty} \dfrac{(-1)^k}{(2k+1)!}x^{2k+1}$, then show that $f''(x) = -f(x)$ for all x in $(-\infty, \infty)$.

(b) What functions have the property that their second derivative equals the negative of the function? Conjecture what function is represented by the power series in part (a). Note that the powers of x in the power series are odd positive integers.

9.10 Taylor Series

Introduction Suppose $\sum c_k(x-a)^k$ is a power series centered at a that has an interval of convergence with a nonzero radius of convergence R. Then, as we saw in the preceding section, within the interval of convergence a power series is a continuous function that possesses derivatives of all orders. We also touched on the idea of using a power series to *represent* a given function (such as $1/(1+x)$) on an interval. In this section we are going to expand upon the notion of representing a function by a power series. The basic problem is:

- Suppose we are given a function f that possesses derivatives of all orders on an open interval I. Can we find a power series that **represents** f on I?

In slightly different words: Can we **expand** an infinitely differentiable function (such as $f(x) = \sin x$, $f(x) = \cos x$, or $f(x) = e^x$) into a power series $\sum c_k(x-a)^k$ that converges to the correct function value $f(x)$ for all x in some open interval $(a-R, a+R)$, where R is either $R > 0$ or $R = \infty$?

■ **Taylor Series for a Function f** Before answering the question in the last paragraph, let us simply make the *assumption* that an infinitely differentiable function f on an interval $(a - R, a + R)$ can be represented by a power series $\sum c_k(x - a)^k$ on that interval. It is then relatively easy to determine what the coefficients c_k must be. Repeated differentiation of

$$f(x) = c_0 + c_1(x - a) + c_2(x - a)^2 + c_3(x - a)^3 + \cdots + c_n(x - a)^n + \cdots \quad (1)$$

yields

$$f'(x) = c_1 + 2c_2(x - a) + 3c_3(x - a)^2 + \cdots \quad (2)$$

$$f''(x) = 2c_2 + 3 \cdot 2c_3(x - a) + \cdots \quad (3)$$

$$f'''(x) = 3 \cdot 2 \cdot 1 c_3 + \cdots, \quad (4)$$

and so on. By evaluating (1), (2), (3), and (4) at $x = a$, we find that

$$f(a) = c_0, \quad f'(a) = 1!c_1, \quad f''(a) = 2!c_2, \quad \text{and} \quad f'''(a) = 3!c_3,$$

respectively. In general, we see that $f^{(n)}(a) = n!c_n$, or

$$c_n = \frac{f^{(n)}(a)}{n!}, n \geq 0. \quad (5)$$

When $n = 0$ we interpret the *zero*th derivative as $f(a)$ and $0! = 1$. Substituting (5) in (1) yields the results summarized in the next theorem.

Theorem 9.10.1 Form of a Power Series

If a function f possesses a power series representation $f(x) = \sum c_k(x - a)^k$ on an interval $(a - R, a + R)$, then the coefficients must be $c_k = f^{(k)}(a)/k!$.

In other words, if a function f has a power series representation centered at a then it must look like this:

$$f(x) = f(a) + \frac{f'(a)}{1!}(x - a) + \frac{f''(a)}{2!}(x - a)^2 + \frac{f'''(a)}{3!}(x - a)^3 + \cdots = \sum_{k=0}^{\infty} \frac{f^{(k)}(a)}{k!}(x - a)^k. \quad (6)$$

The series in (6) is called the **Taylor series of f at a**, or **centered at a**. The Taylor series centered at $a = 0$,

$$f(x) = f(0) + \frac{f'(0)}{1!}x + \frac{f''(0)}{2!}x^2 + \frac{f'''(0)}{3!}x^3 + \cdots = \sum_{k=0}^{\infty} \frac{f^{(k)}(0)}{k!}x^k \quad (7)$$

is called the **Maclaurin series of f**.

The question posed in the introduction can now be rephrased as:

- *Can we expand an infinitely differentiable function f into a Taylor series* (6)?

It would appear that the answer is yes—by simply calculating the coefficients as dictated by the formula (5). Unfortunately, the concept of expanding a given infinitely differentiable function f in a Taylor series is not that simple. You must bear in mind that (5) and (6) were obtained under the assumption that f was represented by a power series centered at a. If we do not know *a priori* that an infinitely differentiable function f has a power series representation, then we must look upon a power series obtained from either (6) or (7) as a *formal* result, in other words, a power series that is simply **generated** by the function f. We do not know whether the series generated in this manner converges or, even if it does, whether it converges to $f(x)$.

EXAMPLE 1 Taylor Series of $\ln x$

Find the Taylor series of $f(x) = \ln x$ centered at $a = 1$. Find its interval of convergence.

Solution The function f, its derivatives, and their values at 1 are:

$$f(x) = \ln x \qquad\qquad f(1) = 0$$
$$f'(x) = \frac{1}{x} \qquad\qquad f'(1) = 1$$
$$f''(x) = -\frac{1}{x^2} \qquad\qquad f''(1) = -1$$
$$f'''(x) = \frac{1 \cdot 2}{x^3} \qquad\qquad f'''(1) = 2!$$
$$\vdots \qquad\qquad\qquad \vdots$$
$$f^{(n)}(x) = (-1)^{n-1}\frac{(n-1)!}{x^n} \qquad f^{(n)}(1) = (-1)^{n-1}(n-1)!$$

Since $(n-1)!/n! = 1/n$, $n \geq 1$, (6) yields

$$(x-1) - \frac{1}{2}(x-1)^2 + \frac{1}{3}(x-1)^3 - \cdots = \sum_{k=1}^{\infty} \frac{(-1)^{k-1}}{k}(x-1)^k. \qquad (8)$$

The Ratio Test,

$$\lim_{n\to\infty}\left|\frac{a_{n+1}}{a_n}\right| = \lim_{n\to\infty}\left|\frac{(-1)^n(x-1)^{n+1}}{n+1} \cdot \frac{n}{(-1)^{n-1}(x-1)^n}\right|$$

$$= \lim_{n\to\infty}\frac{n}{n+1}|x-1| = |x-1|,$$

shows that the series (8) converges for $|x-1| < 1$ or on the interval $(0, 2)$. At the endpoints $x = 0$ and $x = 2$, the series

$$-\sum_{k=1}^{\infty}\frac{1}{k} \qquad \text{and} \qquad \sum_{k=1}^{\infty}\frac{(-1)^{k-1}}{k}$$

are divergent and convergent, respectively. The interval of convergence for this series is $(0, 2]$. The radius of convergence is $R = 1$. ∎

Notice in Example 1 that we did not write the equality

$$\ln x = \sum_{k=1}^{\infty}\frac{(-1)^{k-1}}{k}(x-1)^k.$$

At this point it has not been established that the series given in (8) represents $\ln x$ on the interval $(0, 2]$.

■ Taylor's Theorem It is apparent from (5) that to have a Taylor series centered at a, it is necessary that a function f must possess derivatives of all orders that are defined at a. Thus, for example, $f(x) = \ln x$ does not possess a Maclaurin series, because $f(x) = \ln x$ and all its derivatives are undefined at 0. Moreover, it is important to note that even if a function f possesses derivatives of all orders and generates a Taylor series convergent on some interval, it is possible that the series does not represent f on the interval, that is, the series does not converge to $f(x)$ at every x in the interval. See Problem 63 in Exercises 9.10. The fundamental question of whether a Taylor series represents the function that generates it can be resolved by means of **Taylor's Theorem**.

Theorem 9.10.2 Taylor's Theorem

Let f be a function such that $f^{(n+1)}(x)$ exists for every x in an interval containing the number a. Then for all x in the interval,

$$f(x) = P_n(x) + R_n(x),$$

where
$$P_n(x) = f(a) + \frac{f'(a)}{1!}(x-a) + \cdots + \frac{f^{(n)}(a)}{n!}(x-a)^n \qquad (9)$$

(continued)

is called the *n*th-degree **Taylor polynomial of f at a**, and

$$R_n(x) = \frac{f^{(n+1)}(c)}{(n+1)!}(x-a)^{n+1} \qquad (10)$$

is called the **Lagrange form of the remainder**. The number c lies between a and x.

There are several forms of the remainder. This form is due to the French mathematician **Joseph Louis Lagrange** (1736–1813).

Since the proof of this theorem would deflect us from the main thrust of our discussion, it is given in the *Appendix*. The importance of Theorem 9.10.2 lies in the fact that the Taylor polynomials $P_n(x)$ are the partial sums of the Taylor series (6). The remainder is defined as

$$R_n(x) = f(x) - P_n(x) \quad \text{and so} \quad P_n(x) = f(x) - R_n(x). \qquad (11)$$

If $\lim_{n\to\infty} P_n(x) = f(x)$, then the function f is the sum of the Taylor series it generates. But from (11) we see that

$$\lim_{n\to\infty} P_n(x) = f(x) - \lim_{n\to\infty} R_n(x)$$

and so if we can somehow show that $R_n(x) \to 0$ as $n \to \infty$, then the sequence of partial sums converges to $f(x)$. We summarize the result.

Theorem 9.10.3 Convergence of a Taylor Series

Suppose f is a function that possesses derivatives of all orders on an interval centered at the number a. If

$$\lim_{n\to\infty} R_n(x) = 0$$

for every x in the interval, then the Taylor series generated by f converges to $f(x)$,

$$f(x) = \sum_{k=0}^{\infty} \frac{f^{(k)}(a)}{k!}(x-a)^k.$$

In practice, the proof that the remainder $R_n(x)$ approaches zero as $n \to \infty$ often depends on the fact that

$$\lim_{n\to\infty} \frac{|x|^n}{n!} = 0. \qquad (12)$$

This latter result follows from applying Theorem 9.3.2 to the series $\sum_{m=1}^{\infty} x^k/k!$, which is known to be absolutely convergent for all real numbers. (See Example 3 in Section 9.8.)

EXAMPLE 2 Example 1 Revisited

Prove that the series (8) represents $f(x) = \ln x$ on the interval $(0, 2]$.

Solution In the solution for Example 1 we saw that the *n*th derivative of $f(x) = \ln x$ is given by

$$f^{(n)}(x) = \frac{(-1)^{n-1}(n-1)!}{x^n}.$$

From $f^{(n+1)}(c) = \dfrac{(-1)^n n!}{c^{n+1}}$ we get from (10),

$$|R_n(x)| = \frac{|f^{(n+1)}(c)|}{(n+1)!}|x-1|^{n+1} = \left|\frac{(-1)^n n!}{c^{n+1}(n+1)!} \cdot (x-1)^{n+1}\right| = \frac{1}{n+1}\left|\frac{x-1}{c}\right|^{n+1},$$

where c is some number in the interval $(0, 2]$ between 1 and x.

If $1 \le x \le 2$, then $0 < x - 1 \le 1$. Since $1 < c < x$, we must have $0 < x - 1 \le 1 < c$ and, consequently, $(x-1)/c < 1$. Hence,

$$|R_n(x)| \le \frac{1}{n+1} \quad \text{and} \quad \lim_{n\to\infty} R_n(x) = 0.$$

In the case where $0 < x < 1$, we can also show that $\lim_{n \to \infty} R_n(x) = 0$. We omit the proof. Hence,

$$\ln x = (x-1) - \frac{1}{2}(x-1)^2 + \frac{1}{3}(x-1)^3 - \cdots = \sum_{k=1}^{\infty} \frac{(-1)^{k-1}}{k}(x-1)^k$$

for all values of x in the interval $(0, 2]$. ∎

EXAMPLE 3 Maclaurin Series Representation of cos x

Find the Maclaurin series of $f(x) = \cos x$. Prove that the Maclaurin series represents $\cos x$ for all x.

Solution We first find the Maclaurin series generated by $f(x) = \cos x$:

$$\begin{array}{ll} f(x) = \cos x & f(0) = 1 \\ f'(x) = -\sin x & f'(0) = 0 \\ f''(x) = -\cos x & f''(0) = -1 \\ f'''(x) = \sin x & f'''(0) = 0 \end{array}$$

and so on. From (7) we obtain the power series

$$1 - \frac{x^2}{2!} + \frac{x^4}{4!} - \frac{x^6}{6!} + \cdots = \sum_{k=0}^{\infty} \frac{(-1)^k}{(2k)!} x^{2k}. \tag{13}$$

The Ratio Test shows that (13) converges absolutely for all real values of x, in other words, the interval of convergence is $(-\infty, \infty)$. Now in order to prove that $\cos x$ is represented by the series (13), we must show that $\lim_{n \to \infty} R_n(x) = 0$. To this end, we note that the derivatives of f satisfy

$$|f^{(n+1)}(x)| = \begin{cases} |\sin x|, & n \text{ even} \\ |\cos x|, & n \text{ odd}. \end{cases}$$

In either case, $|f^{(n+1)}(c)| \leq 1$ for any real number c, and so by (10),

$$|R_n(x)| = \frac{|f^{(n+1)}(c)|}{(n+1)!}|x|^{n+1} \leq \frac{|x|^{n+1}}{(n+1)!}.$$

In view of (12), we have for any fixed but arbitrary choice of x,

$$\lim_{n \to \infty} \frac{|x|^{n+1}}{(n+1)!} = 0.$$

But $\lim_{n \to \infty} |R_n(x)| = 0$ implies that $\lim_{n \to \infty} R_n(x) = 0$. Therefore,

$$\cos x = 1 - \frac{x^2}{2!} + \frac{x^4}{4!} - \frac{x^6}{6!} + \cdots + (-1)^n \frac{x^{2n}}{(2n)!} + \cdots$$

is a valid representation of $\cos x$ for every real number x. ∎

EXAMPLE 4 Taylor Series Representation of sin x

Find the Taylor series of $f(x) = \sin x$ centered at $a = \pi/3$. Prove that the Taylor series represents $\sin x$ for all x.

Solution We have

$$\begin{array}{ll} f(x) = \sin x & f\left(\dfrac{\pi}{3}\right) = \dfrac{\sqrt{3}}{2} \\[4pt] f'(x) = \cos x & f'\left(\dfrac{\pi}{3}\right) = \dfrac{1}{2} \\[4pt] f''(x) = -\sin x & f''\left(\dfrac{\pi}{3}\right) = -\dfrac{\sqrt{3}}{2} \\[4pt] f'''(x) = -\cos x & f'''\left(\dfrac{\pi}{3}\right) = -\dfrac{1}{2} \end{array}$$

and so on. Hence, the Taylor series centered at $\pi/3$ generated by $\sin x$ is

$$\frac{\sqrt{3}}{2} + \frac{1}{2 \cdot 1!}\left(x - \frac{\pi}{3}\right) - \frac{\sqrt{3}}{2 \cdot 2!}\left(x - \frac{\pi}{3}\right)^2 - \frac{1}{2 \cdot 3!}\left(x - \frac{\pi}{3}\right)^3 + \cdots. \tag{14}$$

Again, from the Ratio Test it follows that (14) converges absolutely for all real values of x, that is, its interval of convergence is $(-\infty, \infty)$. To show that

$$\sin x = \frac{\sqrt{3}}{2} + \frac{1}{2 \cdot 1!}\left(x - \frac{\pi}{3}\right) - \frac{\sqrt{3}}{2 \cdot 2!}\left(x - \frac{\pi}{3}\right)^2 - \frac{1}{2 \cdot 3!}\left(x - \frac{\pi}{3}\right)^3 + \cdots$$

for every real x, we note that, as in the preceding example, $|f^{(n+1)}(c)| \le 1$. This implies that

$$|R_n(x)| \le \frac{|x - \pi/3|^{n+1}}{(n+1)!}$$

from which we see, with the help of (12), that $\lim_{n \to \infty} R_n(x) = 0$. ∎

We summarize some important Maclaurin series representations and their intervals of convergence:

Maclaurin Series	Interval of Convergence	
$e^x = 1 + x + \dfrac{x^2}{2!} + \dfrac{x^3}{3!} + \cdots = \displaystyle\sum_{k=0}^{\infty} \dfrac{x^k}{k!}$	$(-\infty, \infty)$	(15)
$\cos x = 1 - \dfrac{x^2}{2!} + \dfrac{x^4}{4!} - \dfrac{x^6}{6!} + \cdots = \displaystyle\sum_{k=0}^{\infty} \dfrac{(-1)^k}{(2k)!} x^{2k}$	$(-\infty, \infty)$	(16)
$\sin x = x - \dfrac{x^3}{3!} + \dfrac{x^5}{5!} - \dfrac{x^7}{7!} + \cdots = \displaystyle\sum_{k=0}^{\infty} \dfrac{(-1)^k}{(2k+1)!} x^{2k+1}$	$(-\infty, \infty)$	(17)
$\tan^{-1} x = x - \dfrac{x^3}{3} + \dfrac{x^5}{5} - \dfrac{x^7}{7} + \cdots = \displaystyle\sum_{k=0}^{\infty} \dfrac{(-1)^k}{2k+1} x^{2k+1}$	$[-1, 1]$	(18)
$\cosh x = 1 + \dfrac{x^2}{2!} + \dfrac{x^4}{4!} + \dfrac{x^6}{6!} + \cdots = \displaystyle\sum_{k=0}^{\infty} \dfrac{x^{2k}}{(2k)!}$	$(-\infty, \infty)$	(19)
$\sinh x = x + \dfrac{x^3}{3!} + \dfrac{x^5}{5!} + \dfrac{x^7}{7!} + \cdots = \displaystyle\sum_{k=0}^{\infty} \dfrac{x^{2k+1}}{(2k+1)!}$	$(-\infty, \infty)$	(20)
$\ln(1+x) = x - \dfrac{x^2}{2} + \dfrac{x^3}{3} - \dfrac{x^4}{4} + \cdots = \displaystyle\sum_{k=0}^{\infty} \dfrac{(-1)^k}{k+1} x^{k+1}$	$[-1, 1]$	(21)

You are asked to demonstrate the validity of the representations (15), (17), (19), and (20) as exercises. See Problems 51–54 in Exercises 9.10.

Also, you are encouraged to look hard at the series given in (16)–(20). Then answer the question in Problem 61 of Exercise 9.10.

■ **Some Graphs of Taylor Polynomials** In Example 3 we saw that the Taylor series of $f(x) = \cos x$ at $a = 0$ represents the function for all x, since $\lim_{n \to \infty} R_n(x) = 0$. It is always of interest to see graphically how partial sums of the Taylor series, which are the Taylor polynomials defined in (9), converge to the function. In FIGURE 9.10.1(a) the graphs of the Taylor polynomials

$$P_0(x) = 1, \quad P_2(x) = 1 - \frac{1}{2!}x^2, \quad P_4(x) = 1 - \frac{1}{2!}x^2 + \frac{1}{4!}x^4,$$

and

$$P_{10}(x) = 1 - \frac{1}{2!}x^2 + \frac{1}{4!}x^4 - \frac{1}{6!}x^6 + \frac{1}{8!}x^8 - \frac{1}{10!}x^{10}$$

are compared with the graph of $f(x) = \cos x$ shown in blue.

A comparison of numerical values is given in Figure 9.10.1(b).

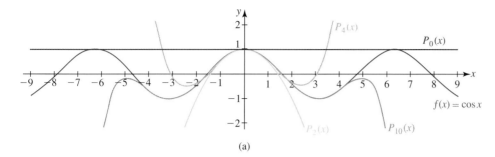

FIGURE 9.10.1 Taylor polynomials P_0, P_2, P_4, and P_{10} for $\cos x$

■ **Approximations** When the value of x is close to the center a ($x \approx a$) of a Taylor series the Taylor polynomial $P_n(x)$ of a function f at a can be used to approximate the function value $f(x)$. The error in this approximation is given by

$$|R_n(x)| = |f(x) - P_n(x)|.$$

EXAMPLE 5 Approximation Using a Taylor Polynomial

Approximate $e^{-0.2}$ by a Taylor polynomial $P_3(x)$. Determine the accuracy of the approximation.

Solution Because the value $x = -0.2$ is close to zero, we use the Taylor polynomial of $f(x) = e^x$ at $a = 0$:

$$P_3(x) = f(0) + \frac{f'(0)}{1!}x + \frac{f''(0)}{1!}x^2 + \frac{f'''(0)}{3!}x^3.$$

It follows from

$$f(x) = f'(x) = f''(x) = f'''(x) = e^x$$
$$f(0) = f'(0) = f''(0) = f'''(0) = 1$$

that

$$P_3(x) = 1 + x + \frac{1}{2}x^2 + \frac{1}{6}x^3.$$

This polynomial is the fourth partial sum of the series given in (15). Now,

$$P_3(-0.2) = 1 + (-0.2) + \frac{1}{2}(-0.2)^2 + \frac{1}{6}(-0.2)^3 \approx 0.8187$$

and so,

$$e^{-0.2} \approx 0.8187. \qquad (22)$$

Now, from (10) we can write

$$|R_3(x)| = \frac{e^c}{4!}|x|^4 < \frac{|x|^4}{4!}$$

since $-0.2 < c < 0$ and $e^c < 1$. The inequality,

$$|R_3(-0.2)| < \frac{|-0.2|^4}{24} < 0.0001$$

implies that the result in (22) is accurate to three decimal places.

In FIGURE 9.10.2 we have compared the graphs of the Taylor polynomials of $f(x) = e^x$ centered at $a = 0$:

$$P_1(x) = 1 + x, \qquad P_2(x) = 1 + x + \frac{1}{2}x^2, \quad \text{and} \quad P_3(x) = 1 + x + \frac{1}{2}x^2 + \frac{1}{6}x^3.$$

Notice in Figures 9.10.2(b) and 9.10.2(c) the graphs of the Taylor polynomials $P_2(x)$ and $P_3(x)$ are indistinguishable from the graph of $y = e^x$ in a small neighborhood of $x = 0.2$.

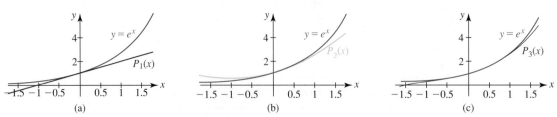

FIGURE 9.10.2 Graphs of Taylor polynomials in Example 5

In the *Notes from the Classroom* in Section 5.5 we introduced the notion of **nonelementary integrals**, namely, an integral such as $\int \sin x^2 \, dx$, where $\sin x^2$ does not possess an antiderivative in the form of an elementary function. Taylor series can be an aid when working with nonelementary integrals. For example, the Maclaurin series obtained by replacing x by x^2 in (17) converges for $-\infty < x < \infty$, and so by Theorem 9.9.2,

$$\int \sin x^2 \, dx = \int \left(x^2 - \frac{x^6}{3!} + \frac{x^{10}}{5!} - \frac{x^{14}}{7!} + \cdots \right) dx$$

$$= \frac{x^3}{3} - \frac{x^7}{7 \cdot 3!} + \frac{x^{11}}{11 \cdot 5!} - \frac{x^{15}}{15 \cdot 7!} + \cdots + C. \qquad (23)$$

EXAMPLE 6 Approximation Using a Taylor Series

Approximate $\int_0^1 \sin x^2 \, dx$ to three decimal places.

Solution From (23) we see immediately that

$$\int_0^1 \sin x^2 \, dx = \left. \frac{x^3}{3} - \frac{x^7}{7 \cdot 3!} + \frac{x^{11}}{11 \cdot 5!} - \frac{x^{15}}{15 \cdot 7!} + \cdots \right]_0^1$$

$$= \frac{1}{3} - \frac{1}{7 \cdot 3!} + \frac{1}{11 \cdot 5!} - \frac{1}{15 \cdot 7!} + \cdots. \qquad (24)$$

By the error-bound theorem for alternating series, Theorem 9.7.2, the fourth term in the series (24) satisfies

$$a_4 = \frac{1}{15 \cdot 7!} \approx 0.000013 < 0.0005.$$

Therefore, the approximation

$$\int_0^1 \sin x^2 \, dx \approx \frac{1}{3} - \frac{1}{7 \cdot 3!} + \frac{1}{11 \cdot 5!} \approx 0.3103$$

is accurate to three decimal places.

Limits A power series representation of a function can sometimes be useful in computing limits. For example, in Section 2.4 we resorted to a subtle geometric argument to prove that $\lim_{x \to 0} \frac{\sin x}{x} = 1$. But if we use (17) and division by x we see immediately that

$$\lim_{x \to 0} \frac{\sin x}{x} = \lim_{x \to 0} \frac{x - \dfrac{x^3}{3!} + \dfrac{x^5}{5!} - \cdots}{x} = \lim_{x \to 0} \left(1 - \overbrace{\frac{x^2}{3!} + \frac{x^4}{5!} - \cdots}^{\text{limit of each of these terms is 0}} \right) = 1.$$

EXAMPLE 7 Calculating a Limit

Evaluate $\lim_{x \to 0} \dfrac{x - \tan^{-1} x}{x^3}$.

Solution Observe that the limit has the indeterminate form $0/0$. If you review Problem 25 in Exercises 4.5, you might recall evaluating this limit by L'Hôpital's Rule. But in view of (18), we can write

$$\lim_{x \to 0} \frac{x - \tan^{-1} x}{x^3} = \lim_{x \to 0} \frac{x - \left(x - \dfrac{x^3}{3} + \dfrac{x^5}{5} - \cdots\right)}{x^3} \quad \substack{\leftarrow \text{ also see Problem 15 in Exercises 9.9}\\ \leftarrow \text{ for the power series representation of } \tan^{-1} x}$$

$$= \lim_{x \to \infty} \frac{\dfrac{x^3}{3} - \dfrac{x^5}{5} + \cdots}{x^3} \quad \leftarrow \text{factor } x^3 \text{ from numerator and cancel}$$

$$= \lim_{x \to 0} \left(\frac{1}{3} - \frac{x^2}{5} + \cdots\right) = \frac{1}{3}. \quad \blacksquare$$

■ **Using the Arithmetic of Power Series** In Section 9.9 we discussed the arithmetic of power series, that is, power series can basically be manipulated arithmetically like polynomials. In the case where the power series representations $f(x) = \sum b_k(x - a)^k$ and $g(x) = \sum c_k(x - a)^k$ converge on the same open interval $(a - R, a + R)$ for $R > 0$ or $(-\infty, \infty)$ for $R = \infty$, the power series representations for $f(x) + g(x)$ and $f(x)g(x)$ can be obtained, in turn, by adding the series and multiplying the series. The sum and product converge on the same interval. If we divide the power series of f by the power series of g, then the quotient represents $f(x)/g(x)$ in some neighborhood of a.

EXAMPLE 8 Maclaurin Series of tan x

Find the first three nonzero terms of the Maclaurin series of $f(x) = \tan x$.

Solution From (16) and (17) we can write

$$\tan x = \frac{\sin x}{\cos x} = \frac{x - \dfrac{x^3}{3!} + \dfrac{x^5}{5!} - \dfrac{x^7}{7!} + \cdots}{1 - \dfrac{x^2}{2!} + \dfrac{x^4}{4!} - \dfrac{x^6}{6!} + \cdots}$$

Then by long division

$$\begin{array}{r}
x + \frac{1}{3}x^3 + \frac{2}{15}x^5 + \cdots \\
1 - \frac{1}{2}x^2 + \frac{1}{24}x^4 - \cdots \overline{\smash{\big)}\,x - \frac{1}{6}x^3 + \frac{1}{120}x^5 - \cdots} \\
\underline{x - \frac{1}{2}x^3 + \frac{1}{24}x^5 - \cdots} \\
\frac{1}{3}x^3 - \frac{1}{30}x^5 + \cdots \\
\underline{\frac{1}{3}x^3 - \frac{1}{6}x^5 + \cdots} \\
\frac{2}{15}x^5 + \cdots \\
\underline{\frac{2}{15}x^5 + \cdots} \\
\vdots
\end{array}$$

Hence, we have

$$\tan x = x + \frac{1}{3}x^3 + \frac{2}{15}x^5 + \cdots. \quad \blacksquare$$

Of course, the last result could also be obtained using (7). See Problem 11 in Exercises 9.10. After working through Example 8 you are encouraged to read (*ii*) in the *Notes from the Classroom*.

■ **Taylor Polynomials—Redux** In Section 4.9 we introduced the notion of a **local linear approximation** of f at a given by $f(x) \approx L(x)$, where

$$L(x) = f(a) + f'(a)(x - a). \tag{25}$$

This equation represents the tangent line to the graph of f at $x = a$. Because it is a linear polynomial, another appropriate symbol for (25) is

$$P_1(x) = f(a) + f'(a)(x - a). \tag{26}$$

The equation is now recognized as the first-degree Taylor polynomial of f at a. The idea behind (25) is that the tangent line can be used to approximate the value of $f(x)$ when x is in a small neighborhood of a. But, since most graphs have concavity and a tangent line does not, it makes sense to expect a polynomial of higher degree would provide a better approximation to $f(x)$ in the sense that its graph would stay near the graph of f over a larger interval containing a. Notice that (26) has the properties that P_1 and its first derivative agree with f and its first derivative at $x = a$:

$$P_1(a) = f(a) \quad \text{and} \quad P_1'(a) = f'(a).$$

If we want a quadratic polynomial function

$$P_2(x) = c_0 + c_1(x - a) + c_2(x - a)^2$$

to have the analogous properties, namely,

$$P_2(a) = f(a), \quad P_2'(a) = f'(a), \quad \text{and} \quad P_2''(a) = f''(a),$$

then, following a procedure similar to (1)–(5), it is seen that P_2 must be

$$P_2(x) = f(a) + \frac{f'(a)}{1!}(x - a) + \frac{f''(a)}{2!}(x - a)^2. \tag{27}$$

$P_n(x)$ is the polynomial of degree n defined in (9).

Graphically, this means that the graph of f and the graph of P_2 have the same tangent line and the same concavity at $x = a$. Of course, (27) is recognized as the second-degree Taylor polynomial. We say that $f(x) \approx P_2(x)$ is a **local quadratic approximation of f at a**. Continuing in this manner we then build up to $f(x) \approx P_n(x)$ which is a **local nth degree approximation of f at a**. With this discussion in mind, you should now look more closely at the graphs of $f(x) = \cos x$, P_0, P_2, P_4, and P_{10} near $x = 0$ in Figure 9.10.1(a) and the approximations in Figure 9.10.1(b). Also reinspect Figure 9.10.2.

■ **Postscript—A Bit of History** Theorem 9.10.2 is named in honor of the English mathematician **Brook Taylor** (1685–1731), who published this result in 1715. However, the formula in (6) was discovered by Johann Bernoulli about 20 years earlier. The series in (7) is named after the Scottish mathematician and former student of Isaac Newton, **Colin Maclaurin** (1698–1746). It is not clear why Maclaurin's name is associated with this series.

Σ NOTES FROM THE CLASSROOM

(*i*) The Taylor series method of finding a power series for a function and then proving that the series represents the function has one big and obvious drawback. Obtaining a general expression for the nth derivative for most functions is nearly impossible. Thus, we are often limited to finding just the first few coefficients c_n.

(*ii*) It is easy to pass over the significance of the results in (6) and (7). Suppose we wish to find the Maclaurin series for $f(x) = 1/(2 - x)$. We can, of course, use (7)–and you are asked to do so in Problem 1 in Exercises 9.10. On the other hand, you should also recognize, from Examples 3–5 of Section 9.9, that a power series representation of f can be obtained utilizing geometric series. The point is:

> • *The representation is unique. Thus, on its interval of convergence, a power series representing a function, regardless of how it is obtained, **is** the Taylor or Maclaurin series of that function.*

Exercises 9.10   Answers to selected odd-numbered problems begin on page ANS-28.

≡ **Fundamentals**

In Problems 1–10, use (7) to find the Maclaurin series for the given function.

1. $f(x) = \dfrac{1}{2-x}$
2. $f(x) = \dfrac{1}{1+5x}$
3. $f(x) = \ln(1+x)$
4. $f(x) = \ln(1+2x)$
5. $f(x) = \sin x$
6. $f(x) = \cos 2x$
7. $f(x) = e^x$
8. $f(x) = e^{-x}$
9. $f(x) = \sinh x$
10. $f(x) = \cosh x$

In Problems 11 and 12, use (7) to find the first four nonzero terms of the Maclaurin series for the given function.

11. $f(x) = \tan x$
12. $f(x) = \sin^{-1} x$

In Problems 13–24, use (6) to find the Taylor series for the given function centered at the indicated value of a.

13. $f(x) = \dfrac{1}{1+x},\ a = 4$
14. $f(x) = \sqrt{x},\ a = 1$
15. $f(x) = \dfrac{1}{x},\ a = 1$
16. $f(x) = \dfrac{1}{x},\ a = -5$
17. $f(x) = \sin x,\ a = \pi/4$
18. $f(x) = \sin x,\ a = \pi/2$
19. $f(x) = \cos x,\ a = \pi/3$
20. $f(x) = \cos x,\ a = \pi/6$
21. $f(x) = e^x,\ a = 1$
22. $f(x) = e^{-2x},\ a = \dfrac{1}{2}$
23. $f(x) = \ln x,\ a = 2$
24. $f(x) = \ln(x+1),\ a = 2$

In Problems 25–32, use previous results, methods, or problems to find the Maclaurin series for the given function.

25. $f(x) = e^{-x^2}$
26. $f(x) = x^2 e^{-3x}$
27. $f(x) = x \cos x$
28. $f(x) = \sin x^3$
29. $f(x) = \ln(1-x)$
30. $f(x) = \ln\left(\dfrac{1+x}{1-x}\right)$
31. $f(x) = \sec^2 x$
32. $f(x) = \ln(\cos x)$

In Problems 33 and 34, use Maclaurin series as an aid in evaluating the given limit.

33. $\displaystyle\lim_{x \to 0} \dfrac{x^3}{x - \sin x}$
34. $\displaystyle\lim_{x \to 0} \dfrac{1 + x - e^x}{1 - \cos x}$

In Problems 35 and 36, use addition of the Maclaurin series for e^x and e^{-x} to find the Maclaurin series for the given function.

35. $f(x) = \cosh x$
36. $f(x) = \sinh x$

In Problems 37 and 38, use multiplication to find the first five nonzero terms of the Maclaurin series for the given function.

37. $f(x) = \dfrac{e^x}{1-x}$
38. $f(x) = e^x \sin x$

In Problems 39 and 40, use division to find the first five nonzero terms of the Maclaurin series for the given function.

39. $f(x) = \dfrac{e^x}{\cos x}$
40. $f(x) = \sec x$

In Problems 41 and 42, establish the indicated value of the given definite integral.

41. $\displaystyle\int_0^1 e^{-x^2}\, dx = 1 - \dfrac{1}{3} + \dfrac{1}{10} - \dfrac{1}{42} + \cdots$
42. $\displaystyle\int_0^1 \dfrac{\sin x}{x}\, dx = 1 - \dfrac{1}{3 \cdot 3!} + \dfrac{1}{5 \cdot 5!} - \dfrac{1}{7 \cdot 7!} + \cdots$

In Problems 43–46, find the sum of the given series.

43. $1 - \dfrac{1}{3} + \dfrac{1}{5} - \dfrac{1}{7} + \cdots$
44. $\dfrac{1}{2!} - \dfrac{1}{3!} + \dfrac{1}{4!} - \dfrac{1}{5!} + \cdots$
45. $1 - \dfrac{\pi^2}{2!} + \dfrac{\pi^4}{4!} - \dfrac{\pi^6}{6!} + \cdots$
46. $\pi - \dfrac{\pi^3}{3!} + \dfrac{\pi^5}{5!} - \dfrac{\pi^7}{7!} + \cdots$

In Problems 47–50, approximate the given quantity using the Taylor polynomial $P_n(x)$ for the indicated values of n and a. Determine the accuracy of the approximation.

47. $\sin 46°,\ n = 2, a = \pi/4$ [Hint: Convert 46° to radian measure.]
48. $\cos 29°,\ n = 2, a = \pi/6$
49. $e^{0.3},\ n = 4, a = 0$
50. $\sinh(0.1),\ n = 3, a = 0$
51. Prove that the series obtained in Problem 5 represents $\sin x$ for every real value of x.
52. Prove that the series obtained in Problem 7 represents e^x for every real value of x.
53. Prove that the series obtained in Problem 9 represents $\sinh x$ for every real value of x.
54. Prove that the series obtained in Problem 10 represents $\cosh x$ for every real value of x.

≡ **Applications**

55. In leveling a long roadway of length L, an allowance must be made for the curvature of the Earth.
    (a) Show that the leveling correction y indicated in FIGURE 9.10.3 is $y = R \sec(L/R) - R$, where R is the radius of the Earth measured in miles.
    (b) If $P_2(x)$ is the second-degree Taylor polynomial for $f(x) = \sec x$ at $a = 0$, use $\sec x \approx P_2(x)$ for x close to zero to show that the approximate leveling correction is $y \approx L^2/(2R)$.
    (c) Find the number of inches of leveling correction needed for 1 mi of roadway. Use $R = 4000$ mi.
    (d) If we use $\sec x \approx P_4(x)$, then show that the leveling correction is
    $$y \approx \dfrac{L^2}{2R} + \dfrac{5L^4}{24R^3}.$$

Redo the calculation in part (c) using the last formula.

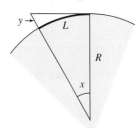

FIGURE 9.10.3 Earth in Problem 55

56. A wave of length L is traveling left to right across water of depth d (in feet), as illustrated in FIGURE 9.10.4. A mathematical model relating the speed v of the wave to L and d is

$$v = \sqrt{\frac{gL}{2\pi}\tanh\left(\frac{2\pi d}{L}\right)}.$$

(a) For deep water show that $v \approx \sqrt{gL/2\pi}$.
(b) Use (7) to find the first three nonzero terms of the Maclaurin series for $f(x) = \tanh x$. Show that when d/L is small, $v \approx \sqrt{gd}$. In other words, in shallow water the speed of a wave is independent of wave length.

FIGURE 9.10.4 Wave in Problem 56

≡ **Think About It**

In Problems 57 and 58, find two ways, other than using (7), of finding the Maclaurin series representation of the given function.

57. $f(x) = \sin^2 x$ **58.** $f(x) = \sin x \cos x$

59. Without using (6), find the Taylor series for the function $f(x) = (x + 1)^2 e^x$ centered at $a = -1$. [*Hint*: $e^x = e^{x+1-1}$.]

60. Discuss: Does $f(x) = \cot x$ possess a Maclaurin series representation?

61. Explain why it stands to reason that the Maclaurin series (16) and (17) for $\cos x$ and $\sin x$ contain only even powers of x and only odd powers of x, respectively. Then reinspect the Maclaurin series in (18), (19), and (20) and make an observation.

62. Suppose it is desired to compute $f^{(10)}(0)$ for $f(x) = x^4 \sin x^2$. Of course, one could use the brute force approach: Use the Product Rule and when the tenth derivative is (eventually) obtained set x equal to 0. Think of a more clever way of determining the value of this derivative.

≡ **Projects**

63. A Mathematical Classic The function

$$f(x) = \begin{cases} e^{-1/x^2}, & x \neq 0 \\ 0, & x = 0 \end{cases}$$

appears in almost every calculus text. The function f is continuous and possesses derivatives of all orders at every value of x.

(a) Use a calculator or CAS to obtain the graph of f.
(b) Use (7) to find the Maclaurin series for f. You will have to use the definition of the derivative to compute $f'(0), f''(0), \ldots$. For example,

$$f'(0) = \lim_{\Delta x \to 0} \frac{f(0 + \Delta x) - f(0)}{\Delta x}.$$

It might help to use $t = \Delta x$ and to recall L'Hôpital's Rule. Show that the Maclaurin series for f converges for every x. Does the series represent the function f that generated it?

9.11 Binomial Series

∎ **Introduction** Most students of mathematics have a familiarity with binomial expansion in the two cases:

$$(1 + x)^2 = 1 + 2x + x^2$$
$$(1 + x)^3 = 1 + 3x + 3x^2 + x^3.$$

In general, if m is a positive integer, then

$$(1 + x)^m = 1 + mx + \frac{m(m-1)}{2!}x^2 + \cdots + \frac{m(m-1)(m-2)\cdots(m-n+1)}{n!}x^n \quad (1)$$
$$+ \cdots + mx^{m-1} + x^m.$$

The expansion of $(1 + x)^m$ in (1) is called the **Binomial Theorem**. Using summation notation, (1) is written

$$(1 + x)^m = \sum_{k=0}^{m}\binom{m}{k}x^k, \quad (2)$$

where the symbol $\binom{m}{k}$ is defined as

$$\underset{\substack{\text{for convenience this}\\ \text{term is defined to be 1}}}{\downarrow}$$

$$\binom{m}{0} = 1, \quad k = 0 \quad \text{and} \quad \binom{m}{k} = \frac{m(m-1)(m-2)\cdots\overset{\overset{(m-k+1)\,=\,(m-(k-1))}{\downarrow}}{(m-k+1)}}{k!}, \quad k \geq 1.$$

These numbers are called the **binomial coefficients**. For example, when $m = 3$, the four binomial coefficients are

$$\binom{3}{0} = 1, \quad \binom{3}{1} = \frac{3}{1} = 3, \quad \binom{3}{2} = \frac{3(3-1)}{2} = 3, \quad \binom{3}{3} = \frac{3(3-1)(3-2)}{6} = 1.$$

Although (2) has the appearance of a series, it is a finite sum consisting of $m + 1$ terms that ends with x^m. In this section we will see that when (1) is extended to powers m other than positive integers the result is an infinite series.

◀ The extension of the **Binomial Theorem** (m a positive integer) to **binomial series** (m fractional and negative real numbers) was first given by Isaac Newton in 1665.

▌**Binomial Series** Now suppose $f(x) = (1 + x)^r$, where r represents any real number. From

$$\begin{aligned}
f(x) &= (1 + x)^r \\
f'(x) &= r(1 + x)^{r-1} \\
f''(x) &= r(r-1)(1 + x)^{r-2} \\
f'''(x) &= r(r-1)(r-2)(1 + x)^{r-3} \\
&\vdots \\
f^{(n)}(x) &= r(r-1)\cdots(r-n+1)(1+x)^{r-n}
\end{aligned} \qquad \begin{aligned}
f(0) &= 1 \\
f'(0) &= r \\
f''(0) &= r(r-1) \\
f'''(0) &= r(r-1)(r-2) \\
&\vdots \\
f^{(n)}(0) &= r(r-1)\cdots(r-n+1)
\end{aligned}$$

we see that the Maclaurin series generated by f is

$$\begin{aligned}
\sum_{k=0}^{\infty} \frac{f^{(k)}(0)}{k!} x^k &= 1 + rx + \frac{r(r-1)}{2!}x^2 + \frac{r(r-1)(r-2)}{3!}x^3 + \cdots + \frac{r(r-1)\cdots(r-n+1)}{n!}x^n + \cdots \\
&= 1 + \sum_{k=1}^{\infty} \frac{r(r-1)\cdots(r-k+1)}{k!} x^k \\
&= \sum_{k=0}^{\infty} \binom{r}{k} x^k. \qquad (3)
\end{aligned}$$

The power series given in (3) is called the **binomial series**. Note that (3) terminates only when r is a positive integer; in this case (3) reduces to (1). From the Ratio Test, the version given in Theorem 9.7.4,

$$\begin{aligned}
\lim_{n\to\infty} \left|\frac{a_{n+1}}{a_n}\right| &= \lim_{n\to\infty} \left| \frac{r(r-1)\cdots(r-n+1)(r-n)x^{n+1}}{(n+1)!} \cdot \frac{n!}{r(r-1)\cdots(r-n+1)x^n} \right| \\
&= \lim_{n\to\infty} \frac{|r-n|}{n+1}|x| \\
&= \lim_{n\to\infty} \frac{\left|\frac{r}{n}-1\right|}{1+\frac{1}{n}}|x| = |x|
\end{aligned}$$

we conclude that the binomial series (3) converges for $|x| < 1$, or $-1 < x < 1$, and diverges for $|x| > 1$, that is, for $x > 1$ or $x < -1$. Convergence at the endpoints $x = \pm 1$ depends on the value of r.

Of course it is no big surprise to learn that the series (3) represents the function f that generated it. We state this as a formal theorem.

> **Theorem 9.11.1** Binomial Series
>
> If $|x| < 1$, then for any real number r,
>
> $$(1 + x)^r = \sum_{k=0}^{\infty} \binom{r}{k} x^k, \qquad (4)$$
>
> where
>
> $$\binom{r}{0} = 1, k = 0, \quad \text{and} \quad \binom{r}{k} = \frac{r(r-1)(r-2)\cdots(r-k+1)}{k!}, \quad k \geq 1.$$

EXAMPLE 1 Representing a Function by a Binomial Series

Find a power series representation for $f(x) = \sqrt{1+x}$.

Solution By rewriting f as $f(x) = (1+x)^{1/2}$ we identify $r = \frac{1}{2}$. It then follows from (4) that for $|x| < 1$,

$$\sqrt{1+x} = 1 + \binom{\frac{1}{2}}{1}x + \binom{\frac{1}{2}}{2}x^2 + \binom{\frac{1}{2}}{3}x^3 + \cdots + \binom{\frac{1}{2}}{n}x^n + \cdots$$

$$= 1 + \frac{1}{2}x + \frac{\frac{1}{2}(\frac{1}{2}-1)}{2!}x^2 + \frac{\frac{1}{2}(\frac{1}{2}-1)(\frac{1}{2}-2)}{3!}x^3 + \cdots$$

$$+ \frac{\frac{1}{2}(\frac{1}{2}-1)(\frac{1}{2}-2)\cdots(\frac{1}{2}-n+1)}{n!}x^n + \cdots$$

$$= 1 + \frac{1}{2}x - \frac{1}{2^2 2!}x^2 + \frac{1 \cdot 3}{2^3 3!}x^3 + \cdots + (-1)^{n-1}\frac{1 \cdot 3 \cdot 5 \cdots (2n-3)}{2^n n!}x^n + \cdots.$$

The last line can be written using summation notation as

$$\sqrt{1+x} = 1 + \frac{1}{2}x + \sum_{k=2}^{\infty} (-1)^{k-1} \frac{1 \cdot 3 \cdot 5 \cdots (2k-3)}{2^k k!} x^k. \quad \blacksquare$$

Suppose the function in Example 1 had been $f(x) = \sqrt{4+x}$. In order to get the binomial series representation of f we would have to rewrite the function in the form $(1+x)^r$ by factoring the 4 out of the radical, that is,

$$f(x) = \sqrt{4+x} = \sqrt{4\left(1 + \frac{1}{4}x\right)^{1/2}} = 2\left(1 + \frac{1}{4}x\right)^{1/2}.$$

We can now use (4) where the symbol x is replaced by $x/4$. The resulting series would then converge for $|x/4| < 1$ or $|x| < 4$.

EXAMPLE 2 A Formula from Physics

In Einstein's theory of relativity, the mass of a particle moving at a velocity v relative to an observer is given by

$$m = \frac{m_0}{\sqrt{1 - v^2/c^2}}, \qquad (5)$$

where m_0 is the rest mass and c is the speed of light.

Many of the results from classical physics do not hold for particles, such as electrons, which may move close to the speed of light. Kinetic energy is no longer $K = \frac{1}{2}m_0 v^2$ but is

$$K = mc^2 - m_0 c^2. \qquad (6)$$

If we identify $r = -\frac{1}{2}$ and $x = -v^2/c^2$ in (5), we have $|x| < 1$, since no particle can surpass the speed of light. Hence, (6) can be written:

$$K = \frac{m_0 c^2}{\sqrt{1+x}} - m_0 c^2$$

$$= m_0 c^2 [(1+x)^{-1/2} - 1]$$

$$= m_0 c^2 \left[\left(1 - \frac{1}{2}x + \frac{3}{8}x^2 - \frac{5}{16}x^3 + \cdots \right) - 1 \right] \quad \leftarrow \text{now substitute for } x$$

$$= m_0 c^2 \left[\frac{1}{2}\left(\frac{v^2}{c^2}\right) + \frac{3}{8}\left(\frac{v^4}{c^4}\right) + \frac{5}{16}\left(\frac{v^6}{c^6}\right) + \cdots \right]. \tag{7}$$

In the everyday world where v is very much smaller than c, terms beyond the first in (7) are negligible. This leads to the well-known classical result

$$K \approx m_0 c^2 \left[\frac{1}{2}\left(\frac{v^2}{c^2}\right) \right] = \frac{1}{2} m_0 v^2. \qquad \blacksquare$$

Σ NOTES FROM THE CLASSROOM

As we come to the end of our discussion of infinite series you probably have a strong impression that divergent series are worthless. Not quite so. Mathematicians hate to see anything go to waste. Divergent series are used in a theory known as *asymptotic representations of functions*. It goes something like this; a divergent series of the form

$$a_0 + a_1/x + a_2/x^2 + \cdots$$

is an **asymptotic representation** of a function f if

$$\lim_{n \to \infty} x^n [f(x) - S_n(x)] = 0,$$

where $S_n(x)$ is the $(n + 1)$st partial sum of the divergent series. Some important functions in applied mathematics are defined in this manner.

Exercises 9.11   Answers to selected odd-numbered problems begin on page ANS-28.

☰ Fundamentals

In Problems 1–10, use (4) to find the first four terms of a power series representation of the given function. Give the radius of convergence.

1. $f(x) = \sqrt[3]{1 + x}$

2. $f(x) = \sqrt{1 - x}$

3. $f(x) = \sqrt{9 - x}$

4. $f(x) = \dfrac{1}{\sqrt{1 + 5x}}$

5. $f(x) = \dfrac{1}{\sqrt{1 + x^2}}$

6. $f(x) = \dfrac{x}{\sqrt[3]{1 - x^2}}$

7. $f(x) = (4 + x)^{3/2}$

8. $f(x) = \dfrac{x}{\sqrt{(1 + x)^5}}$

9. $f(x) = \dfrac{x}{(2 + x)^2}$

10. $f(x) = x^2(1 - x^2)^{-3}$

In Problems 11 and 12, explain why the error in the given approximation is less than the indicated amount. [*Hint*: Review Theorem 9.7.2.]

11. $(1 + x)^{1/3} \approx 1 + \dfrac{x}{3}; \quad \dfrac{1}{9}x^2, \; x > 0$

12. $(1 + x^2)^{-1/2} \approx 1 - \dfrac{x^2}{2} + \dfrac{3}{8}x^4; \quad \dfrac{5}{16}x^6$

13. Find a power series representation for $\sin^{-1} x$ using

$$\sin^{-1} x = \int_0^x \frac{1}{\sqrt{1 - t^2}} \, dt.$$

14. (a) Show that the length of one-quarter of the ellipse $x^2/a^2 + y^2/b^2 = 1$ is given by $L = aE(k)$, where $E(k)$ is

$$E(k) = \int_0^{\pi/2} \sqrt{1 - k^2 \sin^2 \theta} \, d\theta$$

and $k^2 = (a^2 - b^2)/a^2 < 1$. This integral is called the **complete elliptic integral of the second kind**.

(b) Show that

$$L = a\frac{\pi}{2} - \frac{a}{2}\frac{\pi}{4}k^2 - \frac{a}{8}\frac{3\pi}{16}k^4 - \cdots.$$

15. In FIGURE 9.11.1 a hanging cable is supported at points A and B and carries a uniformly distributed load (such as the floor of a bridge). If $y = (4d/l^2)x^2$ is the equation of the cable, show that its length is given by

$$s = l + \frac{8d^2}{3l} - \frac{32d^4}{5l^3} + \cdots.$$

See Problem 22 in Exercises 4.10.

uniform load distributed horizontally

FIGURE 9.11.1 Hanging cable in Problem 15

16. Approximate the following integrals to three decimal places.

(a) $\int_0^{0.2} \sqrt{1 + x^3}\, dx$

(b) $\int_0^{1/2} \sqrt[3]{1 + x^4}\, dx$

17. By the law of cosines the potential at point A in FIGURE 9.11.2 due to a unit charge at point B is $1/R = (1 - 2xr + r^2)^{-1/2}$, where $x = \cos\theta$. The expression $(1 - 2xr + r^2)^{-1/2}$ is said to be the **generating function** for the **Legendre polynomials** $P_k(x)$, since

$$(1 - 2xr + r^2)^{-1/2} = \sum_{k=0}^{\infty} P_k(x) r^k.$$

Use (4) to find $P_0(x)$, $P_1(x)$, and $P_2(x)$.

FIGURE 9.11.2 Unit charge at point B in Problem 17

18. (a) Suppose

$$f(x) = 1 + rx + \frac{r(r-1)}{2!} x^2 + \cdots + \frac{r(r-1)\cdots(r-n+1)}{n!} x^n + \cdots$$

for $|x| < 1$. Find $f'(x)$ and $xf'(x)$.

(b) Show that

$$(n+1)\frac{r(r-1)\cdots(r-n)}{(n+1)!} + n\frac{r(r-1)\cdots(r-n+1)}{n!}$$
$$= r\frac{r(r-1)\cdots(r-n+1)}{n!}.$$

(c) Show that $f'(x) + xf'(x) = rf(x)$.

(d) Solve the first-order differential equation

$$(1 + x)f'(x) = rf(x)$$

subject to $f(0) = 1$.

In Problems 19 and 20, use (4) to find a power series representation in $x - 1$ of the given function. [*Hint*: $1 + x = 2 + (x - 1)$.]

19. $f(x) = \sqrt{1 + x}$

20. $f(x) = (1 + x)^{-2}$

Chapter 9 in Review

Answers to selected odd-numbered problems begin on page ANS-28.

A. True/False

In Problems 1–30, indicate whether the given statement is true or false.

1. The sequence $\left\{\dfrac{(-1)^n n}{2n + 1}\right\}$ converges. _____

2. Every bounded sequence converges. _____

3. If a sequence is not monotonic, it is not convergent. _____

4. The sequence $\left\{\dfrac{10^n}{2^{n^2}}\right\}$ is not monotonic. _____

5. If $a_n \leq B$ for all n and $a_{n+1}/a_n \geq 1$ for all n, then $\{a_n\}$ converges. _____

6. $\lim\limits_{n\to\infty} \dfrac{|x|^n}{n!} = 0$ for every value of x. _____

7. If $\{a_n\}$ is a convergent sequence, then $\sum a_k$ always converges. _____

8. $0.999999\ldots = 1$ _____

9. If $\sum a_k = \frac{3}{2}$, then $a_n \to 0$ as $n \to \infty$. _____

10. If $a_n \to 0$ as $n \to \infty$, then $\sum a_k$ converges. _____

11. If $\sum a_k^2$ converges, then $\sum a_k$ converges. _____

12. If $\sum a_k$ converges and $\sum b_k$ diverges, then $\sum (a_k + b_k)$ diverges. _____

13. $\sum\limits_{k=1}^{\infty} \dfrac{1}{k^p}$ converges for $p = 1.0001$. _____

14. The series $\dfrac{2}{1} + \dfrac{2}{2} + \dfrac{2}{3} + \dfrac{2}{4} + \cdots$ diverges. _____

15. If $\sum |a_k|$ diverges, then $\sum a_k$ diverges. _____

16. If $\sum a_k$, $a_k > 0$, converges, then $\sum (-1)^{k+1} a_k$ converges. _____

17. If $\sum (-1)^{k+1} a_k$ converges absolutely, then $\sum (-1)^{k+1} \dfrac{a_k}{k}$ converges. _____

18. If $\sum b_k$ converges and $a_k \geq b_k$ for every positive integer k, then $\sum a_k$ converges. _____
19. If $\lim_{n\to\infty} \left|\dfrac{a_{n+1}}{a_n}\right| = 1$, then $\sum a_k$ converges absolutely. _____
20. Every power series has a nonzero radius of convergence. _____
21. A power series converges absolutely at every number x in its interval of convergence. _____
22. A power series $\sum c_k x^k$ with an interval of convergence $[-R, R]$, $R > 0$, is an infinitely differentiable function within $(-R, R)$. _____
23. If a power series $\sum c_k x^k$ converges for $-1 < x < 1$ and is convergent at $x = 1$, then the series must also converge at $x = -1$. _____
24. If the power series $\sum a_k x^k$, $a_k > 0$, has the interval of convergence $[-R, R)$, $R > 0$, then the series converges conditionally, but not absolutely, at $x = -R$. _____
25. Since $\int_0^\infty e^{-x}\, dx = 1$, the series $\sum_{k=0}^\infty e^{-k}$ also converges to 1. _____
26. The series $1 + \dfrac{1}{2^2} - \dfrac{1}{3^2} - \dfrac{1}{4^2} + \dfrac{1}{5^2} + \dfrac{1}{6^2} - - + + \cdots$ converges. _____
27. $f(x) = \ln x$ cannot be represented by a Maclaurin series. _____
28. If the power series $\sum c_k(x - 4)^k$ diverges at $x = 7$, the series necessarily diverges at $x = 9$. _____
29. If the sequence $\{\sum_{k=1}^n a_k\}$ converges to 10, then $\sum_{k=1}^\infty a_k = 10$. _____
30. If $f(x) = \sum_{k=1}^\infty c_{2k-1} x^{2k-1}$ is the Maclaurin series of a function f, then $f^{(4)}(0) = 0$. _____

B. Fill in the Blanks

In Problems 1–12, fill in the blanks.

1. If $\{a_n\}$ converges to 4 and $\{b_n\}$ converges to 5, then $\{a_n b_n\}$ converges to _____, $\{a_n + b_n\}$ converges to _____, $\{a_n/b_n\}$ converges to _____, and $\{a_n^2\}$ converges to _____.
2. The sequence $\{\tan^{-1} n\}$ converges to _____.
3. To approximate the sum of the alternating series $\sum_{k=1}^\infty \dfrac{(-1)^{k+1}}{10^k}$ to four decimal places, we need only use the _____ th partial sum.
4. The sum of the series $\sum_{k=0}^\infty 4\left(\dfrac{2}{3}\right)^k$ is _____.
5. If n is an integer, $1 \leq n \leq 9$, then $0.nnn\ldots =$ _____ and so as a quotient of integers, $2.444444\ldots =$ _____.
6. The series $\sum_{k=1}^\infty [\tan^{-1} k - \tan^{-1}(k + 1)]$ converges to _____.
7. The power series $\sum_{k=0}^\infty \dfrac{x^k}{k!}$ represents the function $f(x)$ _____ for all x.
8. The binomial series representation of $f(x) = (4 + x)^{1/2}$ has the radius of convergence _____.
9. The geometric series $\sum_{k=1}^\infty \left(\dfrac{5}{x}\right)^{k-1}$ converges for the following values of x: _____.
10. If $e^x = \sum_{k=0}^\infty \dfrac{x^k}{k!}$ for all real numbers x, then a power series for $e^{-x^3} =$ _____.
11. The interval of convergence of the power series $x - \dfrac{x^2}{2} + \dfrac{x^3}{3} - \dfrac{x^4}{4} + \cdots$ is _____.
12. If $\sum_{k=1}^n a_k = 8 - 3\left(1 - \dfrac{1}{2^n}\right)$, then $\sum_{k=1}^\infty a_k =$ _____.

C. Exercises

In Problems 1–12, determine whether the given series converges or diverges.

1. $\sum_{k=1}^{\infty} \dfrac{k}{(k^2+1)^2}$
2. $\sum_{k=1}^{\infty} \dfrac{1}{1+e^{-k}}$
3. $\sum_{k=1}^{\infty} \pi^{-k}$
4. $\sum_{k=0}^{\infty} \dfrac{1}{(\ln 2.5)^k}$
5. $\sum_{k=1}^{\infty} \dfrac{\sqrt{k}\ln k}{k^4+4}$
6. $\sum_{k=1}^{\infty} \dfrac{\sin k}{k^{3/2}}$
7. $\sum_{k=2}^{\infty} \dfrac{k}{\sqrt[3]{k^6-4k}}$
8. $\sum_{k=2}^{\infty} \dfrac{1}{k\sqrt{\ln k}}$
9. $\sum_{k=1}^{\infty} \dfrac{1+(-1)^k}{\sqrt{k}}$
10. $\sum_{k=1}^{\infty} \dfrac{(k^2)!}{(k!)^2}$
11. $\sum_{k=1}^{\infty} \dfrac{1}{3k^2+4k+6}$
12. $\sum_{k=1}^{\infty} \ln\left(\dfrac{3k}{k+1}\right)$

In Problems 13 and 14, find the sum of the given convergent series.

13. $\sum_{k=1}^{\infty} \dfrac{(-1)^{k-1}+3}{(1.01)^{k-1}}$
14. $\sum_{k=1}^{\infty} \dfrac{1}{k^2+11k+30}$

In Problems 15–18, find the interval of convergence of the given power series.

15. $\sum_{k=1}^{\infty} \dfrac{3^k}{k^3} x^k$
16. $\sum_{k=1}^{\infty} \dfrac{k}{4^k}(2x-1)^k$
17. $\sum_{k=1}^{\infty} k!(x+5)^k$
18. $\sum_{k=2}^{\infty} \dfrac{(2x)^k}{\ln k}$

19. Find the radius of convergence for the power series
$$\sum_{k=1}^{\infty} \dfrac{2\cdot 5\cdot 8\cdots(3k-1)}{3\cdot 7\cdot 11\cdots(4k-1)} x^k.$$

20. Find the values of x for which $\sum_{k=1}^{\infty}(\cos x)^k$ converges.

21. For $|\alpha|>1$, find the sum of the series
$$\dfrac{1}{\alpha}+\dfrac{1}{\alpha^2}+\dfrac{1}{\alpha^3}+\cdots.$$

22. Determine whether the following argument is valid. If
$$S = 1+2+4+8+\cdots, \quad \text{then} \quad 2S = 2+4+8+\cdots = S-1.$$
Solving $2S = S-1$ gives $S = -1$.

In Problems 23–26, find by any method, the first three nonzero terms of the Maclaurin series for the given function.

23. $f(x) = \dfrac{1}{\sqrt[3]{1+x^5}}$
24. $f(x) = \dfrac{x}{2-x}$
25. $f(x) = \sin x \cos x$
26. $f(x) = \int_0^x e^{t^2}\,dt$

27. Find the Taylor series for $f(x) = \cos x$ with center $a = \pi/2$.

28. Prove that the series in Problem 25 represents the function by showing that $R_n(x) \to 0$ as $n \to \infty$.

29. A large convention of free-spending mathematicians from out of town contributes $3 million to the economy of the city of San Francisco. It is estimated that each resident of the city spends $\tfrac{2}{3}$ of his or her income in the city. Thus, of the amount of the money brought in by the convention, $3(\tfrac{2}{3}) = \$2$ million is spent by the people of San Francisco in the city. Of this last amount, $\tfrac{2}{3}$ is spent in the city, and so on. How much, in the long run, do the residents of San Francisco spend in their city as a result of the convention?

30. If P dollars are invested at an annual rate of interest r compounded annually, the return S after m years is $S = (1+r)^m$. The **Rule of 70**, which is often used by loan officers and stock analysts, says that the time required to double an investment earning an annual rate of interest r is approximately $70/(100r)$ years. For example, money invested at an annual rate of 5% takes approximately $79/100(0.05) = 14$ years to double.
   (a) Show that the true doubling time is $\ln 2/\ln(1+r)$.
   (b) Use the Maclaurin series for $\ln(1+r)$ to derive the Rule of 70.
   (c) Use the first three terms of the Maclaurin series for $\ln(1+r)$ to approximate that interest rate for which the Rule of 70 gives the true doubling time.

Chapter 10

Conics and Polar Coordinates

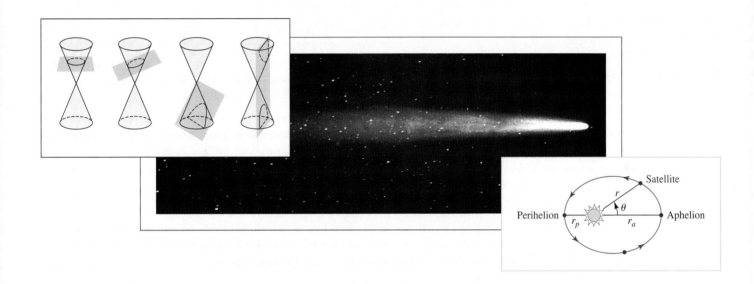

In This Chapter A rectangular or Cartesian equation is not the only, and often not the most convenient, way of describing a curve in the plane. In this chapter we shall consider two additional means by which a curve can be represented. One of the two approaches utilizes an entirely new kind of coordinate system.

We begin by reviewing the notion of a conic section.

10.1 Conic Sections
10.2 Parametric Equations
10.3 Calculus and Parametric Equations
10.4 Polar Coordinate System
10.5 Graphs of Polar Equations
10.6 Calculus in Polar Coordinates
10.7 Conic Sections in Polar Coordinates
    Chapter 10 in Review

10.1 Conic Sections

Hypatia

■ **Introduction** **Hypatia** is the first woman in the history of mathematics about whom we have considerable knowledge. Born in 370 CE, in Alexandria, she was renowned as a mathematician and philosopher. Among her writings is *On the Conics of Apollonius*, which popularized Apollonius' (200 BCE) work on curves that can be obtained by intersecting a double-napped cone with a plane: the circle, parabola, ellipse, and hyperbola. See FIGURE 10.1.1. With the close of the Greek period, interest in conic sections waned; after Hypatia the study of these curves was neglected for over 1000 years.

> When the plane passes through the vertex of the cone we get a *degenerate conic*: a point, a pair of lines, or a single line.

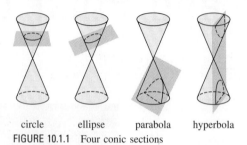

circle   ellipse   parabola   hyperbola

FIGURE 10.1.1 Four conic sections

In the seventeenth century, Galileo showed that in the absence of air resistance the path of a projectile follows a parabolic arc. At about the same time Johannes Kepler hypothesized that the orbits of planets about the Sun are ellipses with the Sun at one focus. This was later verified by Isaac Newton, using the methods of the newly developed calculus. Kepler also experimented with the reflecting properties of parabolic mirrors; these investigations sped the development of the reflecting telescope. The Greeks had known little of these practical applications. They had studied the conics for their beauty and fascinating properties. Rather than using a cone, we shall see in this section how the parabola, ellipse, and hyperbola are defined by means of distance. Using a rectangular coordinate system and the distance formula, we obtain equations for the conics. Each of these equations will be in the form of a quadratic equation in variables x and y:

$$Ax^2 + Bxy + Cy^2 + Dx + Ey + F = 0, \qquad (1)$$

where A, B, C, D, E, and F are constants. The **standard form** of a circle with center (h, k) and radius r,

$$(x - h)^2 + (y - k)^2 = r^2, \qquad (2)$$

is a special case of (1). Equation (2) is a direct result of the definition of a circle:

- A **circle** is defined to be the set of all points $P(x, y)$ in the coordinate plane that are a given fixed distance r, called the **radius**, from a given fixed point (h, k), called the **center**.

In a similar manner, we use the distance formula to obtain equations for the parabola, ellipse, and hyperbola.

The graph of a quadratic function $y = ax^2 + bx + c$, $a \neq 0$, is a parabola. However, not every parabola is the graph of a function of x. In general, a parabola is defined in the following manner.

Definition 10.1.1   Parabola

A **parabola** is the set of all points $P(x, y)$ in the plane that are equidistant from a fixed line L, called the **directrix**, and a fixed point F, called the **focus**.

The line through the focus perpendicular to the directrix is called the **axis** of the parabola. The point of intersection of the parabola and the axis is called the **vertex** of the parabola.

■ **Equation of a Parabola** To describe a parabola analytically, let us assume for the sake of discussion that the directrix L is the horizontal line $y = -p$ and that the focus is $F(0, p)$. Using Definition 10.1.1 and FIGURE 10.1.2, we see that $d(F, P) = d(P, Q)$ is the same as

$$\sqrt{x^2 + (y - p)^2} = y + p.$$

Squaring both sides and simplifying lead to

$$x^2 = 4py. \qquad (3)$$

FIGURE 10.1.2 Parabola with vertex $(0, 0)$ and focus on the y-axis

We say that (3) is the **standard form** for the equation of a parabola with focus $F(0, p)$ and directrix $y = -p$. In like manner, if the directrix and focus are $x = -p$ and $F(p, 0)$, respectively, we find that the standard form for the equation of the parabola is

$$y^2 = 4px. \qquad (4)$$

Although we assumed that $p > 0$ in Figure 10.1.2, this, of course, need not be the case. FIGURE 10.1.3 summarizes information about equations (3) and (4).

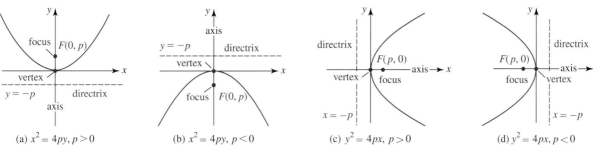

(a) $x^2 = 4py, p > 0$ (b) $x^2 = 4py, p < 0$ (c) $y^2 = 4px, p > 0$ (d) $y^2 = 4px, p < 0$

FIGURE 10.1.3 Pictorial summary of equations (3) and (4)

EXAMPLE 1 Focus and Directrix

Find the focus and directrix of the parabola whose equation is $y = x^2$.

Solution Comparing the equation $y = x^2$ with (3) enables us to identify the coefficient of y, $4p = 1$ and so $p = \frac{1}{4}$. Hence, the focus of the parabola is $\left(0, \frac{1}{4}\right)$ and its directrix is the horizontal line $y = -\frac{1}{4}$. The familiar graph, along with the focus and directrix, is given in FIGURE 10.1.4. ∎

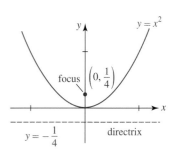

FIGURE 10.1.4 Graph of equation in Example 1

By knowing the basic parabolic shape, all we need to know to sketch a *rough* graph of either equations (3) or (4) is the fact that the graph passes through its vertex $(0, 0)$ and the direction in which the parabola opens. To add more accuracy to the graph it is convenient to use the number p determined by the standard-form equation to plot two additional points. Note that if we choose $y = p$ in (3), then $x^2 = 4p^2$ implies $x = \pm 2p$. Thus $(2p, p)$ and $(-2p, p)$ lie on the graph of $x^2 = 4py$. Similarly, the choice $x = p$ in (2) yields the points $(p, 2p)$ and $(p, -2p)$ on the graph of $y^2 = 4px$. The *line segment* through the focus with endpoints $(2p, p)$, $(-2p, p)$ for equations with standard form (3), and $(p, 2p)$, $(p, -2p)$ for equations with standard form (4) is called the **focal chord**. For example, in Figure 10.1.4, if we choose $y = \frac{1}{4}$, then $x^2 = \frac{1}{4}$ implies $x = \pm\frac{1}{2}$. Endpoints of the horizontal focal chord for $y = x^2$ are $\left(-\frac{1}{2}, \frac{1}{4}\right)$ and $\left(\frac{1}{2}, \frac{1}{4}\right)$.

◀ Graphing tip for equations (3) and (4).

EXAMPLE 2 Finding an Equation of a Parabola

Find the equation in standard form of the parabola with directrix $x = 2$ and focus $(-2, 0)$. Graph.

Solution In FIGURE 10.1.5 we have graphed the directrix and the focus. We see from their placement that the equation we seek is of the form $y^2 = 4px$. Since $p = -2$, the parabola opens to the left and so

$$y^2 = 4(-2)x \qquad \text{or} \qquad y^2 = -8x.$$

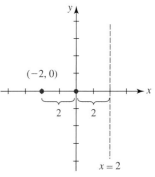

FIGURE 10.1.5 Directrix and focus in Example 2

As mentioned in the discussion preceding this example, if we substitute $x = p = -2$ into the equation $y^2 = -8x$ we can find two points on its graph. From $y^2 = -8(-2) = 16$ we get

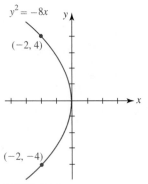

FIGURE 10.1.6 Graph of parabola in Example 2

$y = \pm 4$. As shown in FIGURE 10.1.6, the graph passes through $(0, 0)$ as well as through the endpoints $(-2, -4)$ and $(-2, 4)$ of the focal chord. ∎

Vertex Translated to (h, k) In general, the **standard form** for the equation of a parabola with vertex (h, k) is given by either

$$(x - h)^2 = 4p(y - k) \tag{5}$$

or
$$(y - k)^2 = 4p(x - h). \tag{6}$$

The parabolas defined by these equations are identical in shape to the parabolas defined by equations (3) and (4) because equations (5) and (6) represent rigid transformations (shifts up, down, left, and right) of the graphs of (3) and (4). For example, the parabola $(x + 1)^2 = 8(y - 5)$ has vertex $(-1, 5)$. Its graph is the graph of $x^2 = 8y$ shifted horizontally one unit to the left followed by an upward vertical shift of five units.

For each of the equations, (3) and (4) or (5) and (6), the *distance* from the vertex to the focus, as well as the distance from the vertex to the directrix, is $|p|$.

EXAMPLE 3 Find Everything

Find the vertex, focus, axis, directrix, and graph of the parabola

$$y^2 - 4y - 8x - 28 = 0. \tag{7}$$

Solution In order to write the equation in one of the standard forms we complete the square in y:

$$y^2 - 4y + 4 = 8x + 28 + 4 \quad \leftarrow \text{add 4 to both sides}$$
$$(y - 2)^2 = 8(x + 4).$$

Comparing the last equation with (6) we conclude that the vertex is $(-4, 2)$ and that $4p = 8$ or $p = 2$. From $p = 2 > 0$, the parabola opens to the right and the focus is 2 units to the right of the vertex at $(-2, 2)$. The directrix is the vertical line 2 units to the left of the vertex, $x = -6$. Knowing that the parabola opens to the right from the point $(-4, 2)$ also tells us that the graph has intercepts. To find the x-intercept we set $y = 0$ in (7) and find immediately that $x = -\frac{7}{2}$. The x-intercept is $\left(-\frac{7}{2}, 0\right)$. To find the y-intercepts we set $x = 0$ in (7) and find from the quadratic formula that $y = 2 \pm 4\sqrt{2}$ or $y \approx 7.66$ and $y \approx -3.66$. The y-intercepts are $(0, 2 - 4\sqrt{2})$ and $(0, 2 + 4\sqrt{2})$. Putting all this information together we get the graph in FIGURE 10.1.7. ∎

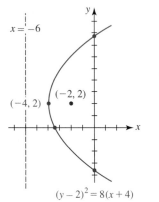

FIGURE 10.1.7 Graph of equation in Example 3

The ellipse is defined as follows.

> **Definition 10.1.2** Ellipse
>
> An **ellipse** is the set of points $P(x, y)$ in the plane such that the sum of the distances between P and two fixed points F_1 and F_2 is a constant. The fixed points F_1 and F_2 are called **foci** (plural for **focus**). The midpoint of the line segment joining F_1 and F_2 is called the **center** of the ellipse.

If P is a point on the ellipse and $d_1 = d(F_1, P)$, $d_2 = d(F_2, P)$ are distances from the foci to P, then Definition 10.1.2 asserts that

$$d_1 + d_2 = k, \tag{8}$$

where $k > 0$ is a constant.

On a practical level, (8) can be used to sketch an ellipse. FIGURE 10.1.8 shows that if a string of length k is attached to a paper by two tacks, then an ellipse can be traced by inserting a pencil against the string and moving it in such a manner that the string remains taut.

FIGURE 10.1.8 A way to draw an ellipse

Equation of an Ellipse For convenience, let us choose $k = 2a$ and put the foci on the x-axis with coordinates $F_1(-c, 0)$ and $F_2(c, 0)$. See FIGURE 10.1.9. It follows from (8) that

$$\sqrt{(x + c)^2 + y^2} + \sqrt{(x - c)^2 + y^2} = 2a. \tag{9}$$

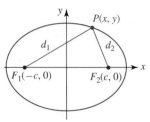

FIGURE 10.1.9 Ellipse with center $(0, 0)$ and foci on the x-axis

Squaring (9), simplifying, and squaring again yields

$$(a^2 - c^2)x^2 + a^2y^2 = a^2(a^2 - c^2). \tag{10}$$

Referring to Figure 10.1.9, we see that the points F_1, F_2, and P form a triangle. Because the sum of the lengths of any two sides of a triangle is greater than the remaining side, we must have $2a > 2c$ or $a > c$. Hence, $a^2 - c^2 > 0$. When we let $b^2 = a^2 - c^2$, then (8) becomes $b^2x^2 + a^2y^2 = a^2b^2$. Dividing this last equation by a^2b^2 gives

$$\frac{x^2}{a^2} + \frac{y^2}{b^2} = 1. \tag{11}$$

Equation (11) is called the **standard form** of the equation of an ellipse centered at $(0, 0)$ with foci $(-c, 0)$ and $(c, 0)$, where c is defined by $b^2 = a^2 - c^2$ and $a > b > 0$.

If the foci are placed on the y-axis, then a repetition of the above analysis leads to

$$\frac{x^2}{b^2} + \frac{y^2}{a^2} = 1. \tag{12}$$

Equation (12) is called the **standard form** of the equation of an ellipse centered at $(0, 0)$ with foci $(0, -c)$ and $(0, c)$, where c is defined by $b^2 = a^2 - c^2$ and $a > b > 0$.

■ Major and Minor Axes The **major axis** of an ellipse is the line segment through its center, containing the foci, and with endpoints on the ellipse. For an ellipse with standard equation (11), the major axis is horizontal whereas, for (12) the major axis is vertical. The line segment through the center, perpendicular to the major axis, and with endpoints on the ellipse is called the **minor axis**. The two endpoints of the major axis are called the **vertices** of the ellipse. For (11) the vertices are the x-intercepts. Setting $y = 0$ in (11) gives $x = \pm a$. The vertices are then $(-a, 0)$ and $(a, 0)$. For (12) the vertices are the y-intercepts $(0, -a)$ and $(0, a)$. For equation (11), the endpoints of the minor axis are $(0, -b)$ and $(0, b)$; for (12) the endpoints are $(-b, 0)$ and $(b, 0)$. For either (11) or (12), the **length of the major axis** is $a - (-a) = 2a$; the length of the minor axis is $2b$. Since $a > b$, the major axis of an ellipse is always longer than its minor axis.

A summary of all this information for equations (11) and (12) is given in FIGURE 10.1.10.

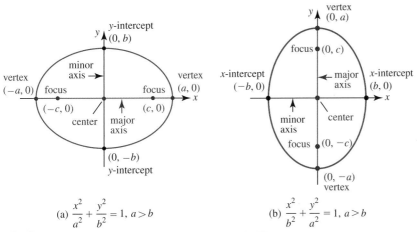

FIGURE 10.1.10 Pictorial summary of equations (11) and (12)

EXAMPLE 4 Vertices, Foci, Graph

Find the vertices and foci of the ellipse whose equation is $9x^2 + 3y^2 = 27$. Graph.

Solution By dividing both sides of the equality by 27 the standard form of the equation is

$$\frac{x^2}{3} + \frac{y^2}{9} = 1.$$

We see that $9 > 3$ and so we identify the equation with (12). From $a^2 = 9$ and $b^2 = 3$, we see that $a = 3$ and $b = \sqrt{3}$. The major axis is vertical with endpoints or vertices $(0, -3)$

and (0, 3). The minor axis is horizontal with endpoints $(-\sqrt{3}, 0)$ and $(\sqrt{3}, 0)$. Of course, the vertices are also the *y*-intercepts and the endpoints of the minor axis are the *x*-intercepts. Now, to find the foci we use $b^2 = a^2 - c^2$ or $c^2 = a^2 - b^2$ to write $c = \sqrt{a^2 - b^2}$. With $a = 3, b = \sqrt{3}$, we get $c = \sqrt{6}$. Hence, the foci are on the *y*-axis at $(0, -\sqrt{6})$ and $(0, \sqrt{6})$. The graph is given in FIGURE 10.1.11. ∎

Center Translated to (h, k) When the center is at (h, k), the standard form for the equation of an ellipse is either

$$\frac{(x - h)^2}{a^2} + \frac{(y - k)^2}{b^2} = 1 \qquad (13)$$

or

$$\frac{(x - h)^2}{b^2} + \frac{(y - k)^2}{a^2} = 1. \qquad (14)$$

The ellipses defined by these equations are identical in shape to the ellipses defined by equations (11) and (12) since equations (13) and (14) represent rigid transformations of the graphs of (11) and (12). For example, the graph of the ellipse

$$\frac{(x - 1)^2}{9} + \frac{(y + 3)^2}{16} = 1$$

with center $(1, -3)$ is the graph of $x^2/9 + y^2/16 = 1$ shifted horizontally 1 unit to the right followed by a downward vertical shift of 3 units.

It is not a good idea to memorize formulas for the vertices and foci of an ellipse with center (h, k). Everything is the same as before, *a, b,* and *c* are positive, $a > b, a > c$, and $c^2 = a^2 - b^2$. You can locate vertices, foci, and endpoints of the minor axis using the fact that *a* is the distance from the center to a vertex, *b* is the distance from the center to an endpoint on the minor axis, and *c* is the distance from the center to a focus.

EXAMPLE 5 Find Everything

Find the vertices and foci of the ellipse $4x^2 + 16y^2 - 8x - 96y + 84 = 0$. Graph.

Solution To write the given equation in one of the standard forms (13) or (14) we complete the square in *x* and in *y*. To do this, recall we want the coefficients of the quadratic terms x^2 and y^2 to be 1. Factoring 4 from the *x* terms and 16 from the *y* terms gives

$$4(x^2 - 2x + 1) + 16(y^2 - 6y + 9) = -84 + 4 \cdot 1 + 16 \cdot 9$$

or $4(x - 1)^2 + 16(y - 3)^2 = 64$. The last equation yields the standard form

$$\frac{(x - 1)^2}{16} + \frac{(y - 3)^2}{4} = 1. \qquad (15)$$

In (15) we identify $a^2 = 16$ or $a = 4$, $b^2 = 4$ or $b = 2$, and $c^2 = a^2 - b^2 = 12$, or $c = 2\sqrt{3}$. The major axis is horizontal and lies on the horizontal line $y = 3$ passing through the center $(1, 3)$. This is the red horizontal dashed line segment in FIGURE 10.1.12. By measuring $a = 4$ units to the left and then to the right of the center along the line $y = 3$ we arrive at the vertices $(-3, 3)$ and $(5, 3)$. By measuring $b = 2$ units both down and up the vertical line $x = 1$ through the center we arrive at the endpoints $(1, 1)$ and $(1, 5)$ of the minor axis. The minor axis is the black dashed vertical line segment in Figure 10.1.12. Finally, by measuring $c = 2\sqrt{3}$ units to the left and right of the center along $y = 3$ we obtain the foci $(1 - 2\sqrt{3}, 3)$ and $(1 + 2\sqrt{3}, 3)$. ∎

The definition of a hyperbola is basically the same as the definition of the ellipse with only one exception: The word *sum* is replaced by the word *difference*.

FIGURE 10.1.12 Ellipse in Example 5

Definition 10.1.3 Hyperbola

A **hyperbola** is the set of points $P(x, y)$ in the plane such that the difference of the distances between P and two fixed points F_1 and F_2 is constant. The fixed points F_1 and F_2 are called **foci** (plural for **focus**). The midpoint of the line segment joining points F_1 and F_2 is called the **center** of the hyperbola.

If P is a point on the hyperbola, then

$$|d_1 - d_2| = k, \tag{16}$$

where $d_1 = d(F_1, P)$ and $d_2 = d(F_2, P)$. Proceeding as for the ellipse, we place the foci on the x-axis at $F_1(-c, 0)$ and $F_2(c, 0)$ as shown in FIGURE 10.1.13 and choose the constant k to be $2a$ for algebraic convenience. As illustrated in the figure the graph of a hyperbola consists of two **branches**.

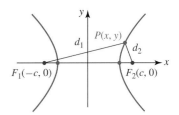

FIGURE 10.1.13 Hyperbola with center $(0, 0)$ and foci on the x-axis

■ **Hyperbola with Center (0, 0)** Applying the usual distance formula and algebra to (16) yields the **standard form** of the equation of a hyperbola centered at $(0, 0)$ with foci $(-c, 0)$ and $(c, 0)$,

$$\frac{x^2}{a^2} - \frac{y^2}{b^2} = 1 \tag{17}$$

When the foci lie on the y-axis, the **standard form** of the equation of a hyperbola centered at $(0, 0)$ with foci $(0, -c)$ and $(0, c)$ is

$$\frac{y^2}{a^2} - \frac{x^2}{b^2} = 1. \tag{18}$$

In both (17) and (18), c is defined by $b^2 = c^2 - a^2$ and $c > a$.

For the hyperbola (unlike the ellipse) bear in mind that in (17) and (18) there is no relationship between the relative sizes of a and b; rather, a^2 is always the denominator of the *positive term* and the intercepts *always* have $\pm a$ as a coordinate.

■ **Transverse and Conjugate Axes** The line segment with endpoints on the hyperbola and lying on the line through the foci is called the **transverse axis**; its endpoints are called the **vertices** of the hyperbola. For the hyperbola described by equation (17), the transverse axis lies on the x-axis. Therefore, the coordinates of the vertices are the x-intercepts. Setting $y = 0$ gives $x^2/a^2 = 1$, or $x = \pm a$. Thus, as shown in FIGURE 10.1.14 the vertices are $(-a, 0)$ and $(a, 0)$; the **length of the transverse axis** is $2a$. Notice that by setting $y = 0$ in (18), we get $-y^2/b^2 = 1$ or $y^2 = -b^2$, which has no real solutions. Hence the graph of any equation in that form has no y-intercepts. Nonetheless, the numbers $\pm b$ are important. The line segment through the center of the hyperbola perpendicular to the transverse axis and with endpoints $(0, -b)$ and $(0, b)$ is called the **conjugate axis**. Similarly, the graph of an equation in standard form (18) has no x-intercepts. The conjugate axis for (18) is the line segment with endpoints $(-b, 0)$ and $(b, 0)$.

This information for equations (17) and (18) is summarized in Figure 10.1.14.

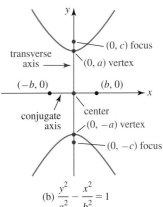

FIGURE 10.1.14 Pictorial summary of equations (17) and (18)

■ **Asymptotes** Every hyperbola possesses a pair of slant asymptotes that pass through its center. These asymptotes are indicative of end behavior, and as such are an invaluable aide in sketching the graph of a hyperbola. Solving (17) for y in terms of x gives

$$y = \pm \frac{b}{a} x \sqrt{1 - \frac{a^2}{x^2}}.$$

As $x \to -\infty$ or as $x \to \infty$, $a^2/x^2 \to 0$, and thus $\sqrt{1 - a^2/x^2} \to 1$. Therefore, for large values of $|x|$, points on the graph of the hyperbola are close to the points on the lines

$$y = \frac{b}{a} x \quad \text{and} \quad y = -\frac{b}{a} x. \tag{19}$$

By a similar analysis we find that the slant asymptotes for (18) are

$$y = \frac{a}{b} x \quad \text{and} \quad y = -\frac{a}{b} x. \tag{20}$$

Each pair of asymptotes intersect at the origin, which is the center of the hyperbola. Note, too, in FIGURE 10.1.15(a) that the asymptotes are simply the *extended diagonals* of a rectangle of width $2a$ (the length of the transverse axis) and height $2b$ (the length of the conjugate axis); in Figure 10.1.15(b) the asymptotes are the extended diagonals of a rectangle of width $2b$ and height $2a$.

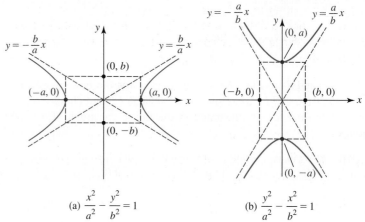

(a) $\dfrac{x^2}{a^2} - \dfrac{y^2}{b^2} = 1$ (b) $\dfrac{y^2}{a^2} - \dfrac{x^2}{b^2} = 1$

FIGURE 10.1.15 Hyperbolas (17) and (18) with slant asymptotes

We recommend that you *do not* memorize the equations in (19) and (20). There is an easy method for obtaining the asymptotes of a hyperbola. For example, since $y = \pm \dfrac{b}{a} x$ is equivalent to

$$\frac{x^2}{a^2} = \frac{y^2}{b^2} \quad \text{or} \quad \frac{x^2}{a^2} - \frac{y^2}{b^2} = 0. \tag{21}$$

Note that the last equation in (21) factors as the difference of two squares:

$$\left(\frac{x}{a} - \frac{y}{b}\right)\left(\frac{x}{a} + \frac{y}{b}\right) = 0.$$

▶ Setting each factor equal to zero and solving for y gives an equation of an asymptote. Equation (21) is simply the left-hand side of the standard form of the equation of a hyperbola given in (17). In like manner, to obtain the asymptotes for (18) just replace 1 by 0 in the standard form, factor $y^2/a^2 - x^2/b^2 = 0$, and solve for y.

This is a mnemonic, or memory device. It has no geometric significance.

EXAMPLE 6 Vertices, Foci, Asymptotes, Graph

Find the vertices, foci, and asymptotes of the hyperbola $9x^2 - 25y^2 = 225$. Graph.

Solution We first put the equation into standard form by dividing both sides of the equality by 225:

$$\frac{x^2}{25} - \frac{y^2}{9} = 1. \tag{22}$$

From this equation we see that $a^2 = 25$ and $b^2 = 9$, and so $a = 5$ and $b = 3$. Therefore, the vertices are $(-5, 0)$ and $(5, 0)$. Since $b^2 = c^2 - a^2$ implies $c^2 = a^2 + b^2$, we have $c^2 = 34$, and so the foci are $(-\sqrt{34}, 0)$ and $(\sqrt{34}, 0)$. To find the slant asymptotes we use the standard form (22) with 1 replaced by 0:

$$\frac{x^2}{25} - \frac{y^2}{9} = 0 \quad \text{factors as} \quad \left(\frac{x}{5} - \frac{y}{3}\right)\left(\frac{x}{5} + \frac{y}{3}\right) = 0.$$

Setting each factor equal to zero and solving for y gives the asymptotes $y = \pm 3x/5$. We plot the vertices and graph the two lines through the origin. Both branches of the hyperbola must become arbitrarily close to the asymptotes as $x \to \pm\infty$. See FIGURE 10.1.16.

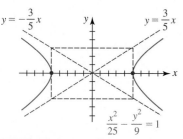

FIGURE 10.1.16 Hyperbola in Example 6

■ **Center Translated to (h, k)** When the center of the hyperbola is (h, k) the **standard form** analogues of equations (17) and (18) are, in turn,

$$\frac{(x-h)^2}{a^2} - \frac{(y-k)^2}{b^2} = 1 \tag{23}$$

and
$$\frac{(y-k)^2}{a^2} - \frac{(x-h)^2}{b^2} = 1. \tag{24}$$

As in (17) and (18) the numbers a^2, b^2, and c^2 are related by $b^2 = c^2 - a^2$.

You can locate vertices and foci using the fact that a is the distance from the center to a vertex, and c is the distance from the center to a focus. The slant asymptotes for (23) can be obtained by factoring

$$\frac{(x-h)^2}{a^2} - \frac{(y-k)^2}{b^2} = 0 \quad \text{as} \quad \left(\frac{x-h}{a} - \frac{y-k}{b}\right)\left(\frac{x-h}{a} + \frac{y-k}{b}\right) = 0.$$

Similarly, the asymptotes for (24) can be obtained from factoring $\dfrac{(y-k)^2}{a^2} - \dfrac{(x-h)^2}{b^2} = 0$, setting each factor equal to zero and solving for y in terms of x. As a check on your work, remember that (h, k) must be a point that lies on each asymptote.

EXAMPLE 7 Find Everything

Find the center, vertices, foci, and asymptotes of the hyperbola $4x^2 - y^2 - 8x - 4y - 4 = 0$. Graph.

Solution Before completing the square in x and y, we factor 4 from the two x-terms and factor -1 from the two y-terms so that the leading coefficient in each expression is 1. Then we have

$$4(x^2 - 2x + 1) - (y^2 + 4y + 4) = 4 + 4 \cdot 1 + (-1) \cdot 4$$
$$4(x-1)^2 - (y+2)^2 = 4$$
$$\frac{(x-1)^2}{1} - \frac{(y+2)^2}{4} = 1.$$

We see now that the center is $(1, -2)$. Since the term in the standard form involving x has the positive coefficient, the transverse axis is horizontal along the line $y = -2$, and we identify $a = 1$ and $b = 2$. The vertices are 1 unit to the left and to the right of the center at $(0, -2)$ and $(2, -2)$, respectively. From $b^2 = c^2 - a^2$ we have $c^2 = a^2 + b^2 = 5$, and so $c = \sqrt{5}$. Hence, the foci are $\sqrt{5}$ units to the left and the right of the center $(1, -2)$ at $(1 - \sqrt{5}, -2)$ and $(1 + \sqrt{5}, -2)$.

To find the asymptotes, we solve

$$\frac{(x-1)^2}{1} - \frac{(y+2)^2}{4} = 0 \quad \text{or} \quad \left(x - 1 - \frac{y+2}{2}\right)\left(x - 1 + \frac{y+2}{2}\right) = 0$$

for y. From $y + 2 = \pm 2(x - 1)$ we find that the asymptotes are $y = -2x$ and $y = 2x - 4$. Observe that by substituting $x = 1$, both equations give $y = -2$, which means that both lines pass through the center. We then locate the center, plot the vertices, and graph the asymptotes. As shown in FIGURE 10.1.17, the graph of the hyperbola passes through the vertices and becomes closer and closer to the asymptotes as $x \to \pm\infty$. ■

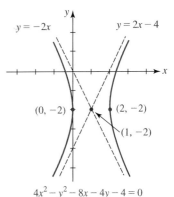

FIGURE 10.1.17 Hyperbola in Example 7

▌ Eccentricity
Associated with each conic section is a number e called its **eccentricity**.

Definition 10.1.4   Eccentricity

The **eccentricity** of an ellipse and a hyperbola is

$$e = \frac{c}{a}.$$

Of course, you must bear in mind that for an ellipse $c = \sqrt{a^2 - b^2}$ and for a hyperbola $c = \sqrt{a^2 + b^2}$. From the inequalities $0 < \sqrt{a^2 - b^2} < a$ and $0 < a < \sqrt{a^2 + b^2}$, we see, in turn, that

- the eccentricity of an ellipse satisfies $0 < e < 1$, and
- the eccentricity of a hyperbola satisfies $e > 1$.

The eccentricity of a parabola will be discussed in Section 10.7.

EXAMPLE 8 Eccentricity

Determine the eccentricity of
  (a) the ellipse in Example 5,   (b) the hyperbola in Example 7.

Solution
  (a) In the solution of Example 5 we found that $a = 4$ and $c = 2\sqrt{3}$. Hence, the eccentricity of the ellipse is $e = (2\sqrt{3})/4 = \sqrt{3}/2 \approx 0.87$.
  (b) In Example 7 we found that $a = 1$ and $c = \sqrt{5}$. Hence, the eccentricity of the hyperbola is $e = \sqrt{5}/1 \approx 2.23$.

Eccentricity is an indicator of the shape of an ellipse or a hyperbola. If $e \approx 0$, then $c = \sqrt{a^2 - b^2} \approx 0$ and consequently $a \approx b$. This means the ellipse is nearly circular. On the other hand, if $e \approx 1$, then $c = \sqrt{a^2 - b^2} \approx a$ and so $b \approx 0$. This means that each focus is close to a vertex and so the ellipse is elongated. See FIGURE 10.1.18. The shapes of a hyperbola in the two extreme cases $e \approx 1$ and e much bigger than 1 are illustrated in FIGURE 10.1.19.

(a) Rays emitted at focus are reflected as parallel rays

(b) Incoming rays reflected to focus

FIGURE 10.1.20 Parabolic reflecting surface

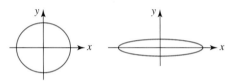

(a) e close to zero   (b) e close to 1
FIGURE 10.1.18 Effect of eccentricity on the shape of an ellipse

(a) e close to 1   (b) e much greater than 1
FIGURE 10.1.19 Effect of eccentricity on the shape of a hyperbola

■ **Applications** The parabola has many properties that make it suitable for certain applications. Reflecting surfaces are often designed to take advantage of a reflection property of parabolas. Such surfaces, called **paraboloids**, are three-dimensional and are formed by rotating a parabola about its axis. As illustrated in FIGURE 10.1.20, rays of light (or electronic signals) from a point source located at the focus of a parabolic reflecting surface will be reflected along lines parallel to the axis. This is the idea behind the design of searchlights, some flashlights, and on-location satellite dishes. Conversely, if the incoming rays of light are parallel to the axis of a parabola, they will be reflected off the surface along lines passing through the focus. Beams of light from a distant object such as a galaxy are essentially parallel, and so when these beams enter a reflecting telescope they are reflected by the parabolic mirror to the focus, where a camera is usually placed to capture the image over time. A parabolic home satellite dish operates on the same principle as the reflecting telescope; the digital signal from a TV satellite is captured at the focus of the dish by a receiver.

Ellipses have a reflection property analogous to the parabola. It can be shown that if a light or sound source is placed at one focus of an ellipse, then all rays or waves will be reflected off the ellipse to the other focus. See FIGURE 10.1.21. For example, if a ceiling is elliptical with two foci on (or near) the floor, but considerably distant from each other, then anyone whispering at one focus will be heard at the other. Some famous "whispering galleries"

TV satellite dish

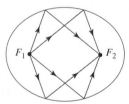

FIGURE 10.1.21 Reflection property of an ellipse

200 inch reflecting telescope at Mt. Palomar

are the Statuary Hall at the Capitol in Washington, DC, the Mormon Tabernacle in Salt Lake City, and St. Paul's Cathedral in London.

Using his Law of Universal Gravitation, Isaac Newton was the first to prove Kepler's first law of planetary motion: The orbit of each planet about the Sun is an ellipse with the Sun at one focus.

EXAMPLE 9 Eccentricity of Earth's Orbit

The perihelion distance of the Earth (the least distance between the Earth and the Sun) is approximately 9.16×10^7 mi, and its aphelion distance (the greatest distance between the Earth and the Sun) is approximately 9.46×10^7 mi. What is the eccentricity of Earth's elliptical orbit?

Solution Let us assume that the orbit of the Earth is as shown in FIGURE 10.1.22. From the figure we see that

$$a - c = 9.16 \times 10^7$$
$$a + c = 9.46 \times 10^7.$$

Solving this system of equations gives $a = 9.31 \times 10^7$ and $c = 0.15 \times 10^7$. Thus, the eccentricity $e = c/a$ is

$$e = \frac{0.15 \times 10^7}{9.31 \times 10^7} \approx 0.016.$$

Statuary Hall in Washington, DC

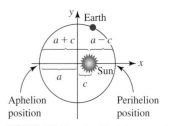
FIGURE 10.1.22 Graphical interpretation of data in Example 9

The orbits of seven of the planets have eccentricities less than 0.1 and, hence, the orbits are not far from circular. Mercury is an exception. Many of the asteroids and comets have highly eccentric orbits. The orbit of the asteroid Hildago is one of the most eccentric, with $e = 0.66$. Another notable case is the orbit of Comet Halley. See Problem 79 in Exercises 10.1.

The hyperbola has several important applications involving sounding techniques. In particular, several navigational systems utilize hyperbolas as follows. Two fixed radio transmitters at a known distance from each other transmit synchronized signals. The difference in reception times by a navigator determines the difference $2a$ of the distances from the navigator to the two transmitters. This information locates the navigator somewhere on the hyperbola with foci at the transmitters and fixed difference in distances from the foci equal to $2a$. By using two sets of signals obtained from a single master station paired with each of two second stations, the long-range navigation system LORAN locates a ship or plane at the intersection of two hyperbolas. See FIGURE 10.1.23.

There are many other applications of the hyperbola. As shown in FIGURE 10.1.24(a), a plane flying at a supersonic speed parallel to level ground leaves a hyperbolic sonic "footprint" on the ground. Like the parabola and ellipse, a hyperbola also possesses a reflecting property. The Cassegrain reflecting telescope shown in Figure 10.1.24(b) utilizes a convex hyperbolic secondary mirror to reflect a ray of light back through a hole to an eyepiece (or camera) behind the parabolic primary mirror. This telescope construction makes use of the fact that

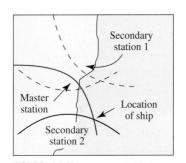
FIGURE 10.1.23 The idea behind LORAN

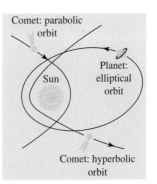

(a) Sonic footprint (b) Cassegrain telescope (c) Orbits around the Sun

FIGURE 10.1.24 Applications of hyperbolas

Exercises 10.1

Answers to selected odd-numbered problems begin on page ANS-29.

Fundamentals

In Problems 1–14, find the vertex, focus, directrix, and axis of the given parabola. Graph the parabola.

1. $y^2 = 4x$
2. $y^2 = \dfrac{7}{2}x$
3. $x^2 = -16y$
4. $x^2 = \dfrac{1}{10}y$
5. $(y-1)^2 = 16x$
6. $(y+3)^2 = -8(x+2)$
7. $(x+5)^2 = -4(y+1)$
8. $(x-2)^2 + y = 0$
9. $y^2 + 12y - 4x + 16 = 0$
10. $x^2 + 6x + y + 11 = 0$
11. $x^2 + 5x - \dfrac{1}{4}y + 6 = 0$
12. $x^2 - 2x - 4y + 17 = 0$
13. $y^2 - 8y + 2x + 10 = 0$
14. $y^2 - 4y - 4x + 3 = 0$

In Problems 15–22, find an equation of the parabola that satisfies the given conditions.

15. Focus, $(0, 7)$, directrix $y = -7$
16. Focus $(-4, 0)$, directrix $x = 4$
17. Focus $\left(\dfrac{5}{2}, 0\right)$, vertex $(0, 0)$
18. Focus $(0, -10)$, vertex $(0, 0)$
19. Focus $(1, -7)$, directrix $x = -5$
20. Focus $(2, 3)$, directrix $y = -3$
21. Vertex $(0, 0)$, through $(-2, 8)$, axis along the y-axis
22. Vertex $(0, 0)$, through $\left(1, \dfrac{1}{4}\right)$, axis along the x-axis

In Problems 23 and 24, find the x- and y-intercepts of the given parabola.

23. $(y+4)^2 = 4(x+1)$
24. $(x-1)^2 = -2(y-1)$

In Problems 25–38, find the center, foci, vertices, endpoints of the minor axis, and eccentricity of the given ellipse. Graph the ellipse.

25. $x^2 + \dfrac{y^2}{16} = 1$
26. $\dfrac{x^2}{25} + \dfrac{y^2}{9} = 1$
27. $9x^2 + 16y^2 = 144$
28. $2x^2 + y^2 = 4$
29. $\dfrac{(x-1)^2}{49} + \dfrac{(y-3)^2}{36} = 1$
30. $\dfrac{(x+1)^2}{25} + \dfrac{(y-2)^2}{36} = 1$
31. $(x+5)^2 + \dfrac{(y+2)^2}{16} = 1$
32. $\dfrac{(x-3)^2}{64} + \dfrac{(y+4)^2}{81} = 1$
33. $4x^2 + \left(y + \dfrac{1}{2}\right)^2 = 4$
34. $36(x+2)^2 + (y-4)^2 = 72$
35. $25x^2 + 9y^2 - 100x + 18y - 116 = 0$
36. $9x^2 + 5y^2 + 18x - 10y - 31 = 0$
37. $x^2 + 3y^2 + 18y + 18 = 0$
38. $12x^2 + 4y^2 - 24x - 4y + 1 = 0$

In Problems 39–48, find an equation of the ellipse that satisfies the given conditions.

39. Vertices $(\pm 5, 0)$, foci $(\pm 3, 0)$
40. Vertices $(\pm 9, 0)$, foci $(\pm 2, 0)$
41. Vertices $(-3, -3), (5, -3)$, endpoints of minor axis $(1, -1), (1, -5)$
42. Vertices $(1, -6), (1, 2)$, endpoints of minor axis $(-2, -2), (4, -2)$
43. Foci $\left(\pm\sqrt{2}, 0\right)$, length of minor axis 6
44. Foci $\left(0, \pm\sqrt{5}\right)$, length of major axis 16
45. Foci $(0, \pm 3)$, passing through $\left(-1, 2\sqrt{2}\right)$
46. Vertices $(\pm 5, 0)$, passing through $\left(\sqrt{5}, 4\right)$
47. Center $(1, 3)$, one focus $(1, 0)$, one vertex $(1, -1)$
48. Endpoints of major axis $(2, 4), (13, 4)$, one focus $(4, 4)$

In Problems 49–62, find the center, foci, vertices, asymptotes, and eccentricity of the given hyperbola. Graph the hyperbola.

49. $\dfrac{x^2}{16} - \dfrac{y^2}{25} = 1$
50. $\dfrac{x^2}{4} - \dfrac{y^2}{4} = 1$
51. $y^2 - 5x^2 = 20$
52. $9x^2 - 16y^2 + 144 = 0$
53. $\dfrac{(x-5)^2}{4} - \dfrac{(y+1)^2}{49} = 1$
54. $\dfrac{(x+2)^2}{10} - \dfrac{(y+4)^2}{25} = 1$
55. $\dfrac{(y-4)^2}{36} - x^2 = 1$
56. $\dfrac{\left(y-\dfrac{1}{4}\right)^2}{4} - \dfrac{(x+3)^2}{9} = 1$
57. $25(x-3)^2 - 5(y-1)^2 = 125$
58. $10(x+1)^2 - 2\left(y-\dfrac{1}{2}\right)^2 = 100$
59. $5x^2 - 6y^2 - 20x + 12y - 16 = 0$
60. $16x^2 - 25y^2 - 256x - 150y + 399 = 0$
61. $4x^2 - y^2 - 8x + 6y - 4 = 0$
62. $2y^2 - 9x^2 - 18x + 20y + 5 = 0$

In Problems 63–70, find an equation of the hyperbola that satisfies the given conditions.

63. Foci $(0, \pm 4)$, one vertex $(0, -2)$
64. Foci $(0, \pm 3)$, one vertex $\left(0, -\frac{3}{2}\right)$
65. Center $(1, -3)$, one focus $(1, -6)$, one vertex $(1, -5)$
66. Vertices $(2, 5)$, $(2, -1)$, one focus $(2, 7)$
67. Center $(-1, 3)$, one vertex $(-1, 4)$, passing through $\left(-5, 3 + \sqrt{5}\right)$
68. Center $(3, -5)$, one vertex $(3, -2)$, passing through $(1, -1)$
69. Center $(2, 4)$, one vertex $(2, 5)$, one asymptote $2y - x - 6 = 0$
70. Eccentricity $\sqrt{10}$, endpoints of conjugate axis $(-5, 4)$, $(-5, 10)$

≡ Applications

71. A large spotlight is designed so that a cross section through its axis is a parabola and the light source is at the focus. Find the position of the light source if the spotlight is 4 ft across at the opening and 2 ft deep.
72. A reflecting telescope has a parabolic mirror that is 20 ft across at the top and 4 ft deep at the center. Where should the eyepiece be located?
73. Suppose that two towers of a suspension bridge are 350 ft apart and the vertex of the parabolic cable is tangent to the road midway between the towers. If the cable is 1 ft above the road at a point 20 ft from the vertex, find the height of the towers above the road.
74. Two 75-ft towers of a suspension bridge with a parabolic cable are 250 ft apart. The vertex of the parabola is tangent to the road midway between the towers. Find the height of the cable above the roadway at a point 50 ft from one of the towers.
75. Assume that the water gushing from the end of a horizontal pipe follows a parabolic arc with vertex at the end of the pipe. The pipe is 20 m above the ground. At a point 2 m below the end of the pipe, the horizontal distance from the water to a vertical line through the end of the pipe is 4 m. See FIGURE 10.1.25. Where does the water strike the ground?

FIGURE 10.1.25 Pipe in Problem 75

76. A dart thrower releases a dart 5 ft above the ground. The dart is thrown horizontally and follows a parabolic path. It hits the ground $10\sqrt{10}$ ft from the dart thrower. At a distance of 10 ft from the dart thrower, how high should a bull's-eye be placed in order for the dart to hit it?

77. The orbit of the planet Mercury is an ellipse with the Sun at one focus. The length of the major axis of this orbit is 72 million mi and the length of the minor axis is 70.4 million mi. What is the least distance (perihelion) between Mercury and the Sun? What is the greatest distance (aphelion)?
78. What is the eccentricity of the orbit of Mercury in Problem 77?
79. The orbit of comet Halley is an ellipse whose major axis is 3.34×10^9 mi long, and whose minor axis is 8.5×10^8 mi long. What is the eccentricity of the comet's orbit?
80. A satellite orbits the Earth in an elliptical path with the center of the Earth at one focus. It has a minimum altitude of 200 mi and a maximum altitude of 1000 mi above the surface of the Earth. If the radius of the Earth is 4000 mi, what is an equation of the satellite's orbit?
81. A semielliptical archway has a vertical major axis. The base of the arch is 10 ft across and the highest part of the arch is 15 ft. Find the height of the arch above the point on the base of the arch 3 ft from the center.
82. Suppose that a room is constructed on a flat elliptical base by rotating a semiellipse 180° about its major axis. Then, by the reflection property of the ellipse, anything whispered at one focus will be distinctly heard at the other focus. If the height of the room is 16 ft and the length is 40 ft, find the location of the whispering and listening posts.

≡ Think About It

83. The graph of the ellipse $x^2/4 + (y - 1)^2/9 = 1$ is shifted 4 units to the right. What are the center, foci, vertices, and endpoints of the minor axis for the shifted graph?
84. The graph of the ellipse $(x - 1)^2/9 + (y - 4)^2 = 1$ is shifted 5 units to the left and 3 units up. What are the center, foci, vertices, and endpoints of the minor axis for the shifted graph?
85. The hyperbolas
$$\frac{x^2}{a^2} - \frac{y^2}{b^2} = 1 \quad \text{and} \quad \frac{y^2}{b^2} - \frac{x^2}{a^2} = 1$$
are said to be **conjugates** of each other.
   (a) Find the equation of the hyperbola that is conjugate to
$$\frac{x^2}{25} - \frac{y^2}{144} = 1.$$
   (b) Discuss how the graphs of conjugate hyperbolas are related.
86. A **rectangular hyperbola** is one for which the asymptotes are perpendicular.
   (a) Show that $y^2 - x^2 + 5y + 3x = 1$ is a rectangular hyperbola.
   (b) Which of the hyperbolas given in Problems 49–62 are rectangular?
87. It can be shown that a ray of light emanating from one focus of a hyperbola will be reflected back along the line from the

opposite focus. See FIGURE 10.1.26. A light ray from the left focus of the hyperbola $x^2/16 - y^2/20 = 1$ strikes the hyperbola at $(-6, -5)$. Find an equation of the reflected ray.

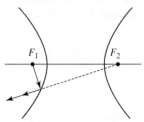

FIGURE 10.1.26 Reflecting property in Problem 87

88. An **oval** is an approximation to an ellipse consisting of arcs from symmetrically placed pairs of circles of different radii, each small circle being tangent to a large circle at two transition points as indicated in FIGURE 10.1.27. Architects in the Renaissance and Baroque periods commonly used ovals because they are simpler to construct than ellipses. In this problem, let the small circles be centered at $(\pm a, 0)$, $a > 0$, with radius r, and let the large circles be centered at $(0, \pm b)$, $b > 0$, with radius R. Also, let $(\pm A, 0)$, $A > 0$, and $(0, \pm B)$, $B > 0$, be the points of intersection of the oval with the x- and y-axes.

(a) Express R in terms of a, b, and r.
(b) Show that $A > B$. This shows that the "major axis" of the oval is always in line with the centers of the small circles, and that the "minor axis" of the oval is always in line with the centers of the large circles. [*Hint:* Show that $A - B = a + b - \sqrt{a^2 + b^2}$.]

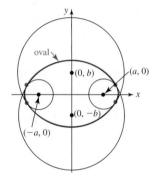

FIGURE 10.1.27 Oval in Problem 88

10.2 Parametric Equations

Introduction A rectangular or Cartesian equation is not the only, and often not the most convenient, way of describing a curve in the coordinate plane. In this section we will consider a different way of representing a curve that is important in the many applications of calculus.

Curvilinear Motion Let us begin with an example. The motion of a particle along a curve, in contrast to a straight line, is called **curvilinear motion**. If it is assumed that a golf ball is hit off the ground, perfectly straight (no hook or slice), and that its path stays in a coordinate plane, then its motion is governed by the fact that its acceleration in the x- and y-directions satisfies

$$a_x = 0, \qquad a_y = -g, \qquad (1)$$

where g is the acceleration due to gravity and $a_x = d^2x/dt^2$, $a_y = d^2y/dt^2$. At $t = 0$ we take $x = 0$, $y = 0$, and the x- and y-components of the initial velocity v_0 to be

$$v_0 \cos\theta_0 \quad \text{and} \quad v_0 \sin\theta_0, \qquad (2)$$

respectively. See FIGURE 10.2.1. Taking two antiderivatives of each equation in (1), we see from the initial conditions (2) that the x- and y-coordinates of the golf ball at time t are given by

$$x = (v_0 \cos\theta_0)t, \qquad y = -\frac{1}{2}gt^2 + (v_0 \sin\theta_0)t, \qquad (3)$$

FIGURE 10.2.1 Fore!

where θ_0 is the launch angle, v_0 is its initial velocity, and $g = 32$ ft/s^2. These equations, which give the golf ball's position in the coordinate plane at time t, are said to be **parametric equations**. The third variable t in (3) is called a **parameter** and is restricted to some interval I, in this case I is defined by $0 \leq t \leq T$, where $t = 0$ gives the origin $(0, 0)$, and $t = T$ is the time the ball hits the ground.

The idea in (3), that is, representing x and y in an ordered pair (x, y) by functions of a third variable t, is used to *define* a curve.

Definition 10.2.1 Plane Curve

If f and g are continuous functions defined on a common interval I, then $x = f(t), y = g(t)$ are called **parametric equations** and t is called a **parameter**. The set C of ordered pairs $(f(t), g(t))$ as t varies over I is called a **plane curve**.

It is also common practice to refer to the set of equations $x = f(t), y = g(t)$, for t in I, as a **parameterization** for C. Hereafter, we will refer to a plane curve C as a **curve**, a **parametric curve**, or as a **parameterized curve**. The **graph** of a parametric curve C is the set of all points (x, y) in the coordinate plane corresponding to the ordered pairs $(f(t), g(t))$. For simplicity, we shall not belabor the distinction between a *parametric curve* and a *graph of a curve*.

EXAMPLE 1 Parametric Curve

Graph the curve C that has the parametric equations

$$x = t^2, \quad y = t^3, \quad -1 \leq t \leq 2.$$

Solution As shown in the accompanying table, for any choice of t in the interval $[-1, 2]$, we obtain a single ordered pair (x, y). By connecting the points with a curve, we obtain the graph in FIGURE 10.2.2.

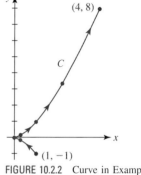

FIGURE 10.2.2 Curve in Example 1

t	-1	$-\dfrac{1}{2}$	0	$\dfrac{1}{2}$	1	$\dfrac{3}{2}$	2
x	1	$\dfrac{1}{4}$	0	$\dfrac{1}{4}$	1	$\dfrac{9}{4}$	4
y	-1	$-\dfrac{1}{8}$	0	$\dfrac{1}{8}$	1	$\dfrac{27}{8}$	2

In Example 1, if we think in terms of motion and t as time, then as t increases from -1 to 2, a point P defined as (t^2, t^3) starts from $(1, -1)$, advances up the lower branch of the curve to the origin $(0, 0)$, passes to the upper branch, and finally stops at $(4, 8)$. In general, a parameter t need have no relation to time. As we plot points corresponding to *increasing values* of the parameter, a curve C is traced out by $(f(t), g(t))$ in a certain *direction* indicated by the arrowheads on the curve in Figure 10.2.2. This direction is called the **orientation** of the curve C.

When the interval I over which f and g are defined is a closed interval $[a, b]$, we say that $(f(a), g(a))$ is the **initial point** of the curve C and that $(f(b), g(b))$ is its **terminal point**. In Example 1, $(1, -1)$ and $(4, 8)$ are the initial and terminal points of C, respectively. If the terminal point is the same as the initial point, that is,

$$(f(a), g(a)) = (f(b), g(b)),$$

then C is a **closed curve**. If C is closed but does not cross itself, then it is called a **simple closed curve**. In FIGURE 10.2.3, A and B represent the initial and terminal points, respectively.

The next example illustrates a simple closed curve.

(a) Plane curve

(b) Simple closed curve

(c) Closed but not simple

FIGURE 10.2.3 Some plane curves

EXAMPLE 2 A Parameterization of a Circle

Find a parameterization for the circle $x^2 + y^2 = a^2$.

Solution The circle has center at the origin and radius $a > 0$. If t represents the central angle, that is, an angle with vertex at the origin and initial side coinciding with the positive x-axis, then as shown in FIGURE 10.2.4 the equations

$$x = a\cos t, \quad y = a\sin t, \quad 0 \leq t \leq 2\pi \qquad (4)$$

give every point P on the circle. For example, at $t = \pi/2$ we get $x = 0$ and $y = a$, in other words, the point is $(0, a)$. The initial point corresponds to $t = 0$ and is $(a, 0)$; the terminal point corresponds to $t = 2\pi$ and is also $(a, 0)$. Since the initial and terminal points are the same, this

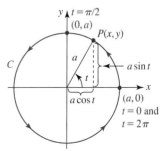
FIGURE 10.2.4 Circle in Example 2

proves the obvious, that the curve C defined by the parametric equations (4) is a closed curve. Note the orientation of C in Figure 10.2.4; as t increases from 0 to 2π, the point $P(x, y)$ traces out C in a counterclockwise direction. ■

In Example 2, the upper *semicircle* $x^2 + y^2 = a^2$, $0 \leq y \leq a$, is defined parametrically by restricting the parameter t to the interval $[0, \pi]$,

$$x = a\cos t, \quad y = a\sin t, \quad 0 \leq t \leq \pi.$$

Observe that when $t = \pi$, the terminal point is now $(-a, 0)$. On the other hand, if we wish to describe *two* complete counterclockwise revolutions around the circle, we again modify the parameter interval by writing

$$x = a\cos t, \quad y = a\sin t, \quad 0 \leq t \leq 4\pi.$$

■ **Eliminating the Parameter** Given a set of parametric equations, we sometimes desire to *eliminate* or *clear* the parameter to obtain a rectangular equation for the curve. To eliminate the parameter in (4), we simply square x and y and add the two equations:

$$x^2 + y^2 = a^2\cos^2 t + a^2\sin^2 t \quad \text{implies} \quad x^2 + y^2 = a^2$$

since $\sin^2 t + \cos^2 t = 1$. There is no unique way of eliminating a parameter.

EXAMPLE 3 Eliminating the Parameter

(a) From the first equation in (3) we have $t = x/(v_0\cos\theta_0)$. Substituting this into the second equation then gives

$$y = -\frac{g}{2(v_0\cos\theta_0)^2}x^2 + (\tan\theta_0)x.$$

Since v_0, θ_0, and g are constants, the last equation has the form $y = ax^2 + bx$ and so the trajectory of any projectile launched at the angle $0 < \theta_0 < \pi/2$ is a parabolic arc.

(b) In Example 1, we can eliminate the parameter from $x = t^2$, $y = t^3$ by solving the second equation for t in terms of y and then substituting in the first equation. We find

$$t = y^{1/3} \quad \text{and so} \quad x = (y^{1/3})^2 = y^{2/3}.$$

The curve shown in Figure 10.2.2 is only a portion of the graph of $x = y^{2/3}$. For $-1 \leq t \leq 2$ we have correspondingly $-1 \leq y \leq 8$. Thus, a rectangular equation for the curve in Example 1 is given by $x = y^{2/3}$, $-1 \leq y \leq 8$. ■

A curve C can have more than one parameterization. For example, an alternative parameterization for the circle in Example 2 is

▶ A curve C can have many different parameterizations.

$$x = a\cos 2t, \quad y = a\sin 2t, \quad 0 \leq t \leq \pi.$$

Note that the parameter interval is now $[0, \pi]$. We see that as t increases from 0 to π, the new angle $2t$ increases from 0 to 2π.

EXAMPLE 4 Alternative Parameterizations

Consider the curve C that has the parametric equations $x = t$, $y = 2t^2$, $-\infty < t < \infty$. We can eliminate the parameter by using $t = x$ and substituting in $y = 2t^2$. This gives the rectangular equation $y = 2x^2$, which we recognize as a parabola. Moreover, since $-\infty < t < \infty$ is equivalent to $-\infty < x < \infty$, the point $(t, 2t^2)$ traces out the complete parabola $y = 2x^2$, $-\infty < x < \infty$.

An alternative parameterization of C is given by $x = t^3/4$, $y = t^6/8$, $-\infty < t < \infty$. Using $t^3 = 4x$ and substituting in $y = t^6/8 = (t^3 \cdot t^3)/8$ gives $y = (4x)^2/8 = 2x^2$. Moreover, $-\infty < t < \infty$ implies $-\infty < t^3 < \infty$ and so $-\infty < x < \infty$. ■

We note in Example 4 that a point on C need not correspond to the same value of the parameter in each set of parametric equations for C. For example, $(1, 2)$ is obtained for $t = 1$ in $x = t$, $y = 2t^2$, but $t = \sqrt[3]{4}$ yields $(1, 2)$ in $x = t^3/4$, $y = t^6/8$.

EXAMPLE 5   Example 4 Revisited

One has to be careful when working with parametric equations. Eliminating the parameter in $x = t^2$, $y = 2t^4$, $-\infty < t < \infty$, would seem to yield the same parabola $y = 2x^2$ as in Example 4. However, this is *not* the case because for any value of t, $t^2 \geq 0$ and so $x \geq 0$. In other words, the last set of equations is a parametric representation of only the right-hand branch of the parabola, that is, $y = 2x^2$, $0 \leq x < \infty$. ∎

◀ You should proceed with caution when eliminating the parameter.

EXAMPLE 6   Eliminating the Parameter

Consider the curve C defined parametrically by

$$x = \sin t, \quad y = \cos 2t, \quad 0 \leq t \leq \pi/2.$$

Eliminate the parameter and obtain a rectangular equation for C.

Solution   Using the double angle formula $\cos 2t = \cos^2 t - \sin^2 t$, we can write

$$\begin{aligned} y &= \cos^2 t - \sin^2 t \\ &= (1 - \sin^2 t) - \sin^2 t \\ &= 1 - 2\sin^2 t \quad \leftarrow \text{substitute } \sin t = x \\ &= 1 - 2x^2. \end{aligned}$$

Now the curve C described by the parametric equations does not consist of the complete parabola, that is, $y = 1 - 2x^2$, $-\infty < x < \infty$. See FIGURE 10.2.5(a). For $0 \leq t \leq \pi/2$ we have $0 \leq \sin t \leq 1$ and $-1 \leq \cos 2t \leq 1$. This means that C is only that portion of the parabola for which the coordinates of a point $P(x, y)$ satisfy $0 \leq x \leq 1$ and $-1 \leq y \leq 1$. The curve C, along with its orientation, is shown in Figure 10.2.5(b). A rectangular equation for C is $y = 1 - 2x^2$ with the restricted domain $0 \leq x \leq 1$. ∎

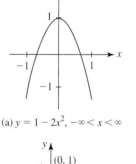

(a) $y = 1 - 2x^2$, $-\infty < x < \infty$

(b) $x = \sin t$, $y = \cos 2t$,
$0 \leq t \leq \pi/2$

FIGURE 10.2.5   Curve C in Example 6

■ **Intercepts**   We can get intercepts of a curve C without finding its rectangular equation. For instance, in Example 6 we can find the x-intercept by finding the value of t in the parameter interval for which $y = 0$. The equation $\cos 2t = 0$ yields $2t = \pi/2$ so that $t = \pi/4$. The corresponding point at which C crosses the x-axis is $(\sqrt{2}, 0)$. Similarly, the y-intercept of C is found by solving $x = 0$ for t. From $\sin t = 0$ we immediately conclude $t = 0$ and so the y-intercept is $(0, 1)$.

■ **Applications of Parametric Equations**   Cycloidal curves were a popular topic of study by mathematicians in the seventeenth century. Suppose a point $P(x, y)$, marked on a circle of radius a, is at the origin when its diameter lies along the y-axis. As the circle rolls along the x-axis, the point P traces out a curve C that is called a **cycloid**. See FIGURE 10.2.6.

(a) Circle rolling on x-axis

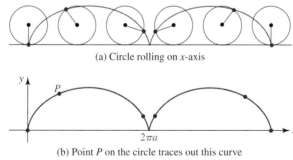

(b) Point P on the circle traces out this curve
FIGURE 10.2.6   Cycloid

Two problems were extensively studied in the seventeenth century. Consider a flexible (frictionless) wire fixed at points A and B and a bead free to slide down the wire starting at P. See FIGURE 10.2.7. Is there a particular shape of the wire so that, regardless of where the bead starts, the time to slide down the wire to B will be the same? Also, what would the shape of the wire be so that the bead slides from P to B in the shortest time? The so-called **tautochrone** (same time) and **brachistochrone** (least time) were shown to be an inverted half-arch of a cycloid.

FIGURE 10.2.7   Sliding bead

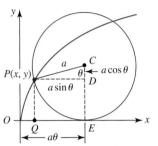

FIGURE 10.2.8 In Example 7 the angle θ is the parameter for the cycloid

EXAMPLE 7 Parameterization of a Cycloid

Find a parameterization for the cycloid shown in Figure 10.2.6(b).

Solution A circle of radius a whose diameter initially lies along the y-axis rolls along the x-axis without slipping. We take as a parameter the angle θ (in radians) through which the circle has rotated. The point $P(x, y)$ starts at the origin, which corresponds to $\theta = 0$. As the circle rolls through an angle θ, its distance from the origin is the arc $PE = \overline{OE} = a\theta$. From FIGURE 10.2.8 we then see that the x-coordinate of P is

$$x = \overline{OE} - \overline{QE} = a\theta - a\sin\theta.$$

Now the y-coordinate of P is seen to be

$$y = \overline{CE} - \overline{CD} = a - a\cos\theta.$$

Hence, parametric equations for the cycloid are

$$x = a\theta - a\sin\theta, \quad y = a - a\cos\theta.$$

As shown in Figure 10.2.6(a), one arch of a cycloid is generated by one rotation of the circle and corresponds to the parameter interval $0 \leq \theta \leq 2\pi$. ∎

Parameterizations of Rectangular Curves A curve C described by a continuous function $y = f(x)$ can always be parameterized by letting $x = t$. Parametric equations for C are then

$$x = t, \quad y = f(t). \tag{5}$$

For example, one cycle of the graph of the sine function $y = \sin x$ can be parameterized by $x = t, y = \sin t, 0 \leq t \leq 2\pi$.

Smooth Curves A curve C, given parametrically by

$$x = f(t), \quad y = g(t), \quad a \leq t < b,$$

is said to be **smooth** if f' and g' are continuous on $[a, b]$ and not simultaneously zero on (a, b). A curve C is said to be **piecewise smooth** if the interval $[a, b]$ can be divided into subintervals such that C is smooth on each subinterval. The curves in Examples 2, 3, and 6 are smooth; the curves in Examples 1 and 7 are piecewise smooth.

FIGURE 10.2.9 Circular helix

FIGURE 10.2.10 Circular Helicoid

DNA is a double helix

Helical antenna

$\dfrac{d}{d\theta}$ **NOTES FROM THE CLASSROOM**

In this section we have focused on **plane curves**, curves C defined parametrically in two dimensions. In the study of multivariable calculus you will see curves and surfaces in three dimensions that are defined by means of parametric equations. For example, a **space curve** C consists of a set of ordered triples $(f(t), g(t), h(t))$, where f, g, and h are defined on a common interval. Parametric equations for C are $x = f(t), y = g(t), z = h(t)$. For example, the **circular helix** such as shown in FIGURE 10.2.9 is a space curve whose parametric equations are

$$x = a\cos t, \quad y = a\cos t, \quad z = bt, \quad t \geq 0. \tag{6}$$

Surfaces in three dimensions can be represented by a set of parametric equations involving *two* parameters, $x = f(u, v), y = g(u, v), z = h(u, v)$. For example, the **circular helicoid** shown in FIGURE 10.2.10 arises from the study of minimal surfaces and is defined by the set of parametric equations similar to those in (6):

$$x = u\cos v, \quad y = u\sin v, \quad z = bv,$$

where b is a constant. The circular helicoid has a circular helix as its boundary. You might recognize the helicoid as the model for the rotating curved blade in machinery such as post hole diggers, ice augers, and snow blowers.

Exercises 10.2

≡ Fundamentals

In Problems 1 and 2, fill in the table for the given set of parametric equations.

1. $x = 2t + 1, y = t^2 + t$

t	−3	−2	−1	0	1	2	3
x							
y							

2. $x = \cos t, y = \sin^2 t$

t	0	$\pi/6$	$\pi/4$	$\pi/3$	$\pi/2$	$5\pi/6$	$7\pi/4$
x							
y							

In Problems 3–10, graph the curve that has the given set of parametric equations.

3. $x = t - 1, y = 2t - 1; \quad -1 \leq t \leq 5$
4. $x = 3t, y = t^2 - 1; \quad -2 \leq t \leq 3$
5. $x = \sqrt{t}, y = 5 - t; \quad t \geq 0$
6. $x = 3 + 2\sin t, y = 4 + \sin t; \quad -\pi/2 \leq t \leq \pi/2$
7. $x = 4\cos t, y = 4\sin t; \quad -\pi/2 \leq t \leq \pi/2$
8. $x = t^3 + 1, y = t^2 - 1; \quad -2 \leq t \leq 2$
9. $x = e^t, y = e^{3t}; \quad 0 \leq t \leq \ln 2$
10. $x = -e^t, y = e^{-t}; \quad t \geq 0$

In Problems 11–16, eliminate the parameter from the given set of parametric equations and obtain a rectangular equation that has the same graph.

11. $x = t^2, y = t^4 + 3t^2 - 1$
12. $x = t^3 + t + 4, y = -2t^3 - 2t$
13. $x = -\cos 2t, y = \sin t; \quad -\pi/4 \leq t \leq \pi/4$
14. $x = e^t, y = \ln t; \quad t > 0$
15. $x = t^3, y = 3\ln t; \quad t > 0$
16. $x = \tan t, y = \sec t; \quad -\pi/2 < t < \pi/2$

In Problems 17–22, graphically show the difference between the given curves.

17. $y = x$ and $x = \sin t, y = \sin t$
18. $y = x^2$ and $x = -\sqrt{t}, y = t, t \geq 0$
19. $y = \frac{1}{4}x^2 - 1$ and $x = 2t, y = t^2 - 1, -1 \leq t \leq 2$
20. $y = -x^2$ and $x = e^t, y = -e^{2t}, t \geq 0$
21. $x^2 - y^2 = 1$ and $x = \cosh t, y = \sinh t$
22. $y = 2x - 2$ and $x = t^2 - 1, y = 2t^2 - 4$

In Problems 23–26, graphically show the difference between the given curves. Assume $a > 0, b > 0$.

23. $x = a\cos t, y = a\sin t, \quad 0 \leq t \leq \pi$
    $x = a\sin t, y = a\cos t, \quad 0 \leq t \leq \pi$
24. $x = a\cos t, y = b\sin t, a > b, \quad \pi \leq t \leq 2\pi$
    $x = a\sin t, y = b\cos t, a > b, \quad \pi \leq t \leq 2\pi$
25. $x = a\cos t, y = a\sin t, \quad -\pi/2 \leq t \leq \pi/2$
    $x = a\cos 2t, y = a\sin 2t, \quad -\pi/2 \leq t \leq \pi/2$
26. $x = a\cos\frac{t}{2}, y = a\sin\frac{t}{2}, \quad 0 \leq t \leq \pi$
    $x = a\cos\left(-\frac{t}{2}\right), y = a\sin\left(-\frac{t}{2}\right), \quad -\pi \leq t \leq 0$

In Problems 27 and 28, graph the curve that has the given parametric equations.

27. $x = 1 + 2\cosh t, y = 2 + 3\sinh t$
28. $x = -3 + 3\cos t, y = 5 + 5\sin t$

In Problems 29–34, determine whether the given set of parametric equations has the same graph as the rectangular equation $xy = 1$.

29. $x = \dfrac{1}{2t + 1}, y = 2t + 1$ 30. $x = t^{1/2}, y = t^{-1/2}$
31. $x = \cos t, y = \sec t$
32. $x = t^2 + 1, y = (t^2 + 1)^{-1}$
33. $x = e^{-2t}, y = e^{2t}$ 34. $x = t^3, y = t^{-3}$

≡ Applications

35. As shown in FIGURE 10.2.11, a piston is attached by means of a rod of length L to a circular crank mechanism of radius r. Parameterize the coordinates of the point P in terms of the angle ϕ.

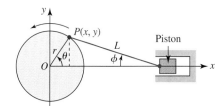

FIGURE 10.2.11 Crank mechanism in Problem 35

36. A point Q traces out a circular path of radius r and a point P moves in the manner shown in FIGURE 10.2.12. If R is constant, find parametric equations of the path traced by P. This curve is called an **epitrochoid**. (Those knowledgeable about automobiles might recognize the

curve traced by P as the shape of the rotor housing in the rotary or Wankel engine.)

FIGURE 10.2.12 Epitrochoid in Problem 36

37. A circular spool wound with thread has its center at the origin. The radius of the spool is a. The end of the thread P, starting from $(a, 0)$, is unwound while the thread is kept taut. See FIGURE 10.2.13. Find parametric equations of the path traced by the point P if the thread PR is tangent to the circular spool at R. The curve is called an **involute** of a circle.

FIGURE 10.2.13 Involute of a circle in Problem 37

38. Imagine a small circle of radius a rolling around inside and on a larger circle of radius $b > a$. A point P on the smaller circle generates a curve called a **hypocycloid**. Use FIGURE 10.2.14 to show that parametric equations of a hypocycloid are

$$x = (b-a)\cos\theta + a\cos\frac{b-a}{a}\theta$$
$$y = (b-a)\sin\theta - a\sin\frac{b-a}{a}\theta.$$

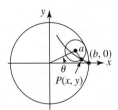

FIGURE 10.2.14 Hypocycloid in Problem 38

39. (a) Use the equations in Problem 38 to show that parametric equations of a **hypocycloid of four cusps** are

$$x = b\cos^3\theta, \quad y = b\sin^3\theta.$$

(b) Use a graphing utility to obtain the graph of the curve in part (a).

(c) Eliminate the parameter and obtain a rectangular equation for the hypocycloid of four cusps.

40. Use FIGURE 10.2.15 to show that parametric equations of an **epicycloid** are given by

$$x = (a+b)\cos\theta - a\cos\frac{a+b}{a}\theta$$
$$y = (a+b)\sin\theta - a\sin\frac{a+b}{a}\theta.$$

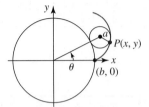

FIGURE 10.2.15 Epicycloid in Problem 40

41. (a) Use the equations in Problem 40 to show that parametric equations of an **epicycloid of three cusps** are

$$x = 4a\cos\theta - a\cos 4\theta, \quad y = 4a\sin\theta - a\sin 4\theta.$$

(b) Use a graphing utility to obtain the graph of the curve in part (a).

42. **A Mathematical Classic**

(a) Consider a circle of radius a, which is tangent to the x-axis at the origin O. Let B be a point on a horizontal line through $(0, 2a)$ and let the line segment OB cut the circle at point A. As shown in FIGURE 10.2.16, the projection of AB on the vertical gives the line segment BP. Find parametric equations of the path traced by the point P as A varies around the circle. The curve is called the **Witch of Agnesi**. No, the curve has nothing to do with witches and goblins. This curve, called *versoria*, which is Latin for a kind of rope, was included in a text on analytic geometry written in 1748 by the Italian mathematician **Maria Gaetana Agnesi** (1718–1799). This text proved to be so popular that it was soon translated into English. The translator confused *versoria* with the Italian word *versiera*, which means *female goblin*. In English, *female goblin* became a *witch*.

(b) In part (a) eliminate the parameter and show that the curve has the rectangular equation

$$y = 8a^3/(x^2 + 4a^2).$$

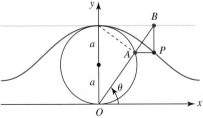

FIGURE 10.2.16 Witch of Agnesi in Problem 42

☰ Calculator/CAS Problems

In Problems 43–48, use a calculator or CAS to obtain the graph of the given set of parametric equations.

43. $x = 4\sin 2t$, $y = 2\sin t$; $0 \le t \le 2\pi$
44. $x = 6\cos 3t$, $y = 4\sin 2t$; $0 \le t \le 2\pi$
45. $x = 6\cos 4t$, $y = 4\sin t$; $0 \le t \le 2\pi$
46. $x = \cos t + t\sin t$, $y = \sin t - t\cos t$; $0 \le t \le 3\pi$
47. $x = t^3 - 4t + 1$, $y = t^4 - 4t^2$; $-5 \le t \le 5$
48. $x = t^5 - t + 1$, $y = t^3 + 2t - 1$; $-3 \le t \le 6$

☰ Think About It

49. Show that parametric equations for a line through (x_1, y_1) and (x_2, y_2) are

    $x = x_1 + (x_2 - x_1)t$, $y = y_1 + (y_2 - y_1)t$, $-\infty < t < \infty$.

    What do these equations represent when $0 \le t \le 1$?

50. (a) Use the result of Problem 49 to find parametric equations of the line through $(-2, 5)$ and $(4, 8)$.
    (b) Eliminate the parameter in part (a) to obtain a rectangular equation for the line.
    (c) Find parametric equations for the line segment with $(-2, 5)$ as the initial point and $(4, 8)$ as the terminal point.

51. A skier hits a mogul on a slope and is launched horizontally into the air with an initial velocity of 75 ft/s. As shown in FIGURE 10.2.17 the slope falls away from the horizontal at an angle of 33°. Use the equations in (3) to determine how far down the slope she will land. [*Hint:* Observe the *x*- and *y*-axes in Figure 10.2.1 are in standard position (to the right and upward, respectively). In Figure 10.2.17 assume the origin is the point where the skier is launched into the air.]

FIGURE 10.2.17   Skier in Problem 51

☰ Projects

52. **Butterfly Curve**   The graph of the set of parametric equations

    $$x = \sin t\left(e^{\cos t} - 2\cos 4t + \sin^5 \frac{1}{12}t\right),$$
    $$y = \cos t\left(e^{\cos t} - 2\cos 4t + \sin^5 \frac{1}{12}t\right)$$

    is said to be a **butterfly curve**. FIGURE 10.2.18 consists of seven colored portions of the curve corresponding to different parameter intervals. Experiment with a CAS to determine these parameter intervals. Use the CAS to generate more colored portions and then combine all the colored curves on one set of coordinate axes.

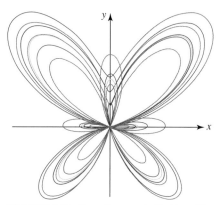

FIGURE 10.2.18   Butterfly curve in Problem 52

53. The curve in Figure 10.2.18 is one of two curves known as a butterfly curve. Write a short report that discusses both kinds of curves.

54. **Bézier Curves**   Most computer graphing applications plot parametric equations in addition to graphs of functions. All graphic calculators can plot parametric equations by repeatedly calculating a point on the curve and then plotting it. In this project we introduce some special parametric curves called **Bézier curves**, which are commonly used in computer-aided design (CAD), in computer drawing programs, and in the mathematical representation of different fonts for many laser printers. A *cubic* Bézier curve is specified by four control points in the plane—for example,

    $P_0(p_0, q_0)$, $P_1(p_1, q_1)$, $P_2(p_2, q_2)$,   and $P_3(p_3, q_3)$.

    The curve starts at the first point for the value $t = 0$, ends at the last point for $t = 1$, and roughly "heads toward" the middle points for parameter values between 0 and 1. Artists and engineering designers can move the control points to adjust the end locations and the shape of the parametric curve. The cubic Bézier curve for these four control points has the following parametric equations

    $$x = p_0(1-t)^3 + 3p_1(1-t)^2 t + 3p_2(1-t)t^2 + p_3 t^3$$
    $$y = q_0(1-t)^3 + 3q_1(1-t)^2 t + 3q_2(1-t)t^2 + q_3 t^3,$$

    where $0 \le t \le 1$. Several Bézier curves can be pieced together continuously by making the last control point on one curve the first control point on the next curve. Equivalently, piecewise parametric equations can be constructed. For example, the next piece can be represented by

    $$x = p_3(2-t)^3 + 3p_4(2-t)^2(t-1)$$
    $$\qquad\qquad + 3p_5(2-t)(t-1)^2 + p_6(t-1)^3$$
    $$y = q_3(2-t)^3 + 3q_4(2-t)^2(t-1)$$
    $$\qquad\qquad + 3q_5(2-t)(t-1)^2 + q_6(t-1)^3,$$

where $1 \leq t \leq 2$.

In (a)–(f), use a graphing utility to obtain the graph of the piecewise continuous Bézier curve associated with the given control points.

(a) $P_0(5, 1)$, $P_1(1, 30)$, $P_2(50, 28)$, $P_3(55, 5)$
(b) $P_0(32, 1)$, $P_1(85, 25)$, $P_2(1, 30)$, $P_3(40, 3)$
(c) $P_0(10, 5)$, $P_1(16, 4)$, $P_2(25, 28)$, $P_3(30, 30)$, $P_4(18, 1)$, $P_5(40, 18)$, $P_6(16, 20)$
(d) $P_0(55, 50)$, $P_1(45, 40)$, $P_2(38, 20)$, $P_3(50, 20)$, $P_4(60, 20)$, $P_5(63, 35)$, $P_6(45, 32)$
(e) $P_0(30, 30)$, $P_1(40, 5)$, $P_2(12, 12)$, $P_3(45, 10)$, $P_4(58, 10)$, $P_5(66, 31)$, $P_6(25, 30)$
(f) $P_0(48, 20)$, $P_1(20, 15)$, $P_2(20, 50)$, $P_3(48, 45)$, $P_4(28, 47)$, $P_5(28, 18)$, $P_6(48, 20)$, $P_7(48, 36)$, $P_8(52, 32)$, $P_9(40, 32)$

In (g)–(i), experiment with the locations for control points to obtain piecewise continuous Bézier curves approximating the indicated shape or object. Give the final control points chosen, and sketch the resulting parametric curve.

(g) Historically the letter "S" has been one of the most difficult to represent mathematically. Use two or three Bézier curve pieces to draw a letter "S" in some simple font style.
(h) The long cross-section of an egg is not quite an ellipse because one end is more pointed than the other. Use several Bézier curve pieces to represent an approximation for the shape of an egg.
(i) Give a curve approximating the shape of the Greek letter epsilon, ε, using as few pieces as possible.

Finish this project by writing a short report that discusses linear, quadratic, and nth degree Bézier curves. Include a discussion on the early history of Bézier curves; for example, what was the contribution of Pierre Bézier?

10.3 Calculus and Parametric Equations

Introduction As with graphs of functions $y = f(x)$, we can obtain useful information about a curve C defined parametrically by examining the derivative dy/dx.

Slope Let $x = f(t)$ and $y = g(t)$ be parametric equations of a smooth curve C. The **slope** of the tangent line at a point $P(x, y)$ on C is given by dy/dx. To calculate this derivative, we use the form of the derivative given in (3) of Section 3.1:

$$\frac{dy}{dx} = \lim_{\Delta x \to 0} \frac{\Delta y}{\Delta x}.$$

For an increment Δt, the increments in x and y are, respectively

$$\Delta x = f(t + \Delta t) - f(t) \quad \text{and} \quad \Delta y = g(t + \Delta t) - g(t)$$

and so

$$\frac{\Delta y}{\Delta x} = \frac{\frac{\Delta y}{\Delta t}}{\frac{\Delta x}{\Delta t}} = \frac{\frac{g(t + \Delta t) - g(t)}{\Delta t}}{\frac{f(t + \Delta t) - f(t)}{\Delta t}}.$$

Therefore,

$$\frac{dy}{dx} = \lim_{\Delta t \to 0} \frac{\Delta y}{\Delta x} = \frac{\lim_{\Delta t \to 0} \Delta y/\Delta t}{\lim_{\Delta t \to 0} \Delta x/\Delta t} = \frac{dy/dt}{dx/dt},$$

when the limit of the denominator is not zero. This parametric form of the derivative is summarized in the next theorem.

Theorem 10.3.1 Slope of a Tangent Line

If $x = f(t)$, $y = g(t)$ define a smooth curve C, then the **slope of the tangent line** at a point $P(x, y)$ on C is

$$\frac{dy}{dx} = \frac{dy/dt}{dx/dt} = \frac{g'(t)}{f'(t)}, \qquad (1)$$

provided that $f'(t) \neq 0$.

EXAMPLE 1 Tangent Line

Find an equation of the tangent line to the curve $x = t^2 - 4t - 2$, $y = t^5 - 4t^3 - 1$ at the point corresponding to $t = 1$.

Solution We first find the slope dy/dx of the tangent line. Since

$$\frac{dx}{dt} = 2t - 4 \quad \text{and} \quad \frac{dy}{dt} = 5t^4 - 12t^2$$

it follows from (1) that

$$\frac{dy}{dx} = \frac{dy/dt}{dx/dt} = \frac{5t^4 - 12t^2}{2t - 4}.$$

Thus, at $t = 1$ we have

$$\left.\frac{dy}{dx}\right|_{t=1} = \frac{-7}{-2} = \frac{7}{2}.$$

By substituting $t = 1$ back into the original parametric equations, we find the point of tangency to be $(-5, -4)$. Hence, an equation of the tangent line at that point is

$$y - (-4) = \frac{7}{2}(x - (-5)) \quad \text{or} \quad y = \frac{7}{2}x + \frac{27}{2}.$$

With the aid of a CAS we obtain the curve given in FIGURE 10.3.1.

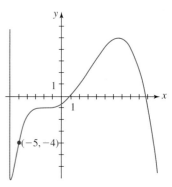

FIGURE 10.3.1 Curve in Example 1

■ **Horizontal and Vertical Tangents** At a point (x, y) on a curve C at which $dy/dt = 0$ and $dx/dt \neq 0$, the tangent line is necessarily **horizontal** because $dy/dx = 0$ at that point. On the other hand, at a point at which $dx/dt = 0$ and $dy/dt \neq 0$, the tangent line is **vertical**. When both dy/dt and dx/dt are zero at a point, we can draw no immediate conclusion about the tangent line.

EXAMPLE 2 Graph of a Parametric Curve

Graph the curve that has the parametric equations $x = t^2 - 4$, $y = t^3 - 3t$.

Solution *x-intercepts*: $y = 0$ implies $t(t^2 - 3) = 0$ at $t = 0$, $t = -\sqrt{3}$, and $t = \sqrt{3}$.
y-intercepts: $x = 0$ implies $t^2 - 4 = 0$ at $t = -2$ and $t = 2$.
Horizontal tangents: $\frac{dy}{dt} = 3t^2 - 3$; $\frac{dy}{dt} = 0$ implies $3(t^2 - 1) = 0$ at $t = -1$ and $t = 1$. Note that $dx/dt \neq 0$ at $t = -1$ and $t = 1$.
Vertical tangents: $\frac{dx}{dt} = 2t$; $\frac{dx}{dt} = 0$ implies $2t = 0$ at $t = 0$. Note that $dy/dt \neq 0$ at $t = 0$.

The points (x, y) on the curve corresponding to these values of the parameter are summarized in the accompanying table.

t	-2	$-\sqrt{3}$	-1	0	1	$\sqrt{3}$	2
x	0	-1	-3	-4	-3	-1	0
y	-2	0	2	0	-2	0	2

From the table we see that: the x-intercepts are $(-1, 0)$ and $(-4, 0)$, the y-intercepts are $(0, -2)$ and $(0, 2)$, the points of horizontal tangency are $(-3, 2)$ and $(-3, -2)$, the point of vertical tangency is $(-4, 0)$. A curve plotted through these points, consistent with the orientation and tangent information, is illustrated in FIGURE 10.3.2.

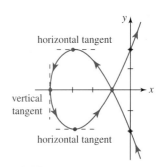

FIGURE 10.3.2 Curve in Example 2

The graph of a differentiable function $y = f(x)$ can have only one tangent line at a point on its graph. In contrast, since a curve C defined parametrically may not be the graph of a function, it is possible that such a curve may have more than one tangent line at a point.

FIGURE 10.3.3 Tangent lines in Example 3

EXAMPLE 3 Two Tangent Lines at a Point

In the table in Example 2 we see that for $t = -\sqrt{3}$ and $t = \sqrt{3}$ we get the single point $(-1, 0)$. As can be seen in Figure 10.3.2 this means the curve intersects itself at $(-1, 0)$. Now, from $x = t^2 - 4$, $y = t^3 - 3t$ we get

$$\frac{dy}{dx} = \frac{3t^2 - 3}{2t}$$

and

$$\left.\frac{dy}{dx}\right|_{t=-\sqrt{3}} = -\sqrt{3} \quad \text{and} \quad \left.\frac{dy}{dx}\right|_{t=\sqrt{3}} = \sqrt{3}.$$

Hence, we conclude that there are two tangent lines at $(-1, 0)$:

$$y = -\sqrt{3}(x + 1) \quad \text{and} \quad y = \sqrt{3}(x + 1).$$

See FIGURE 10.3.3. ∎

☐ **Higher-Order Derivatives** Higher-order derivatives can be found in exactly the same manner as dy/dx. Suppose (1) is written as

$$\frac{d}{dx}(\ ) = \frac{d(\ )/dt}{dx/dt}. \tag{2}$$

If $y' = dy/dx$ is a differentiable function of t, it follows from (2) by replacing ( ) by y' that

$$\frac{d^2y}{dx^2} = \frac{d}{dx}y' = \frac{dy'/dt}{dx/dt}. \tag{3}$$

Similarly, if $y'' = d^2y/dx^2$ is a differentiable function of t, then the third derivative is

$$\frac{d^3y}{dx^3} = \frac{d}{dx}y'' = \frac{dy''/dt}{dx/dt}, \tag{4}$$

and so on.

EXAMPLE 4 Third Derivative

Find d^3y/dx^3 for the curve given by $x = 4t + 6$, $y = t^2 + t - 2$.

Solution To compute the third derivative we must first find the first and second derivatives. From (2) the first derivative is

$$\frac{dy}{dx} = \frac{dy/dt}{dx/dt} = \frac{2t + 1}{4} = y'.$$

Then using (3) and (4) we find the second and third derivatives:

$$\frac{d^2y}{dx^2} = \frac{dy'/dt}{dx/dt} = \frac{\frac{1}{2}}{4} = \frac{1}{8} = y''$$

$$\frac{d^3y}{dx^3} = \frac{dy''/dt}{dx/dt} = \frac{0}{4} = 0.$$

∎

Inspection of Example 4 shows that the curve has a horizontal tangent at $t = -\frac{1}{2}$ or $(4, -\frac{9}{4})$. Furthermore, since $d^2y/dx^2 > 0$ for all t, the graph of the curve is concave upward at every point. Verify this by graphing the curve.

☐ **Length of a Curve** In Section 6.5 we were able to find the length L of the graph of a smooth function $y = f(x)$ by means of a definite integral. We can now generalize the result given in (3) of that section to curves defined parametrically.

■ **Building an Integral** Suppose $x = f(t)$, $y = g(t)$, $a \le t \le b$, are parametric equations of a smooth curve C that does not intersect itself for $a < t < b$. If P is a partition of $[a, b]$ given by the numbers

$$a = t_0 < t_1 < t_2 < \cdots < t_{n-1} < t_n = b,$$

then, as shown in FIGURE 10.3.4, it seems reasonable that C can be approximated by a polygonal path through the points $Q_k(f(t_k), g(t_k))$, $k = 0, 1, \ldots, n$. Denoting the length of the line segment through Q_{k-1} and Q_k by L_k we write the approximate length of C as

$$\sum_{k=1}^{n} L_k, \tag{5}$$

where $\qquad L_k = \sqrt{[f(t_k) - f(t_{k-1})]^2 + [g(t_k) - g(t_{k-1})]^2}.$

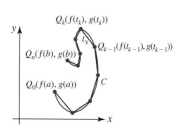

FIGURE 10.3.4 Approximating the length of C (blue) by the length of a polygonal path (red)

Now, since f and g have continuous derivatives, the Mean Value Theorem (see Section 4.4) asserts that there exist numbers u_k^* and v_k^* in the interval (t_{k-1}, t_k) such that

$$f(t_k) - f(t_{k-1}) = f'(u_k^*)(t_k - t_{k-1}) = f'(u_k^*)\Delta t_k \tag{6}$$

and $\qquad g(t_k) - g(t_{k-1}) = g'(v_k^*)(t_k - t_{k-1}) = g'(v_k^*)\Delta t_k. \tag{7}$

Using (6) and (7) in (5) and simplifying yield

$$\sum_{k=1}^{n} L_k = \sum_{k=1}^{n} \sqrt{[f'(u_k^*)]^2 + [g'(v_k^*)]^2}\, \Delta t_k. \tag{8}$$

By taking $\|P\| \to 0$ in (8), we obtain a formula for the length of a smooth curve. Notice that the limit of the sum in (8) is not the usual definition of a definite integral, since we are dealing with two numbers (u_k^* and v_k^*) rather than one in the interval (t_{k-1}, t_k). Nevertheless, it *can* be shown rigorously that the formula given in the next theorem results from (8) by taking $\|P\| \to 0$.

Theorem 10.3.2 Arc Length

If $x = f(t)$ and $y = g(t)$, $a \le t \le b$, define a smooth curve C that does not intersect itself for $a < t < b$, then the length L of C is

$$L = \int_a^b \sqrt{[f'(t)]^2 + [g'(t)]^2}\, dt = \int_a^b \sqrt{\left(\frac{dx}{dt}\right)^2 + \left(\frac{dy}{dt}\right)^2}\, dt. \tag{9}$$

Alternatively, (9) can be obtained using (1). If the curve defined by $x = f(t)$, $y = g(t)$, $a \le t \le b$, can also be represented by an explicit function $y = F(x)$, $x_0 \le x \le x_1$, then by changing variables of integration and using $f(a) = x_0$, $g(b) = x_1$, (3) of Section 6.5 becomes

$$L = \int_{x_0}^{x_1} \sqrt{1 + \left(\frac{dy}{dx}\right)^2}\, dx = \int_a^b \sqrt{1 + \left(\frac{f'(t)}{g'(t)}\right)^2}\, g'(t)\, dt = \int_a^b \sqrt{[f'(t)]^2 + [g'(t)]^2}\, dt.$$

EXAMPLE 5 Length of a Curve

Find the length of the curve given by $x = 4t$, $y = t^2$, $0 \le t \le 2$.

Solution: Since $f'(t) = 4$ and $g'(t) = 2t$, (9) gives

$$L = \int_0^2 \sqrt{16 + 4t^2}\, dt = 2\int_0^2 \sqrt{4 + t^2}\, dt.$$

With the trigonometric substitution $t = 2\tan\theta$, the last integral becomes

$$L = 8\int_0^{\pi/4} \sec^3\theta\, d\theta.$$

Integration by parts leads to (see Example 5, Section 7.3)

$$L = \left[4\sec\theta\tan\theta + 4\ln|\sec\theta + \tan\theta|\right]_0^{\pi/4} = 4\sqrt{2} + 4\ln(\sqrt{2} + 1) \approx 9.1823. \quad \blacksquare$$

Exercises 10.3   Answers to selected odd-numbered problems begin on page ANS-32.

Fundamentals

In Problems 1–6, find the slope of the tangent line at the point corresponding to the indicated value of the parameter.

1. $x = t^3 - t^2, y = t^2 + 5t; \quad t = -1$
2. $x = 4/t, y = 2t^3 - t + 1; \quad t = 2$
3. $x = \sqrt{t^2 + 1}, y = t^4; \quad t = \sqrt{3}$
4. $x = e^{2t}, y = e^{-4t}; \quad t = \ln 2$
5. $x = \cos^2\theta, y = \sin\theta; \quad \theta = \pi/6$
6. $x = 2\theta - 2\sin\theta, y = 2 - 2\cos\theta; \quad \theta = \pi/4$

In Problems 7 and 8, find an equation of the tangent line to the given curve at the point corresponding to the indicated value of the parameter.

7. $x = t^3 + 3t, y = 6t^2 + 1; \quad t = -1$
8. $x = 2t + 4, y = t^2 + \ln t; \quad t = 1$

In Problems 9 and 10, find an equation of the tangent line to the given curve at the indicated point.

9. $x = t^2 + t, y = t^2; \quad (2, 4)$
10. $x = t^4 - 9, y = t^4 - t^2; \quad (0, 6)$
11. What is the slope of the tangent line to the curve given by $x = 4\sin 2t, y = 2\cos t, 0 \le t \le 2\pi$, at the point $(2\sqrt{3}, 1)$?
12. A curve C has parametric equations $x = t^2, y = t^3 + 1$. At what point on C is the tangent line given by $y + 3x - 5 = 0$?
13. A curve C has parametric equations $x = 2t - 5$, $y = t^2 - 4t + 3$. Find an equation of the tangent line to C that is parallel to the line $y = 3x + 1$.
14. Verify that the curve given by $x = -2/\pi + \cos\theta$, $y = -2\theta/\pi + \sin\theta, -\pi \le \theta \le \pi$, intersects itself. Find equations of tangent lines at the point of intersection.

In Problems 15–18, determine the points on the given curve at which the tangent line is either horizontal or vertical. Graph the curve.

15. $x = t^3 - t, y = t^2$
16. $x = \frac{1}{8}t^3 + 1, y = t^2 - 2t$
17. $x = t - 1, y = t^3 - 3t^2$
18. $x = \sin t, y = \cos 3t, 0 \le t \le 2\pi$

In Problems 19–22, find dy/dx, d^2y/dx^2, and d^3y/dx^3.

19. $x = 3t^2, y = 6t^3$
20. $x = \cos t, y = \sin t$
21. $x = e^{-t}, y = e^{2t} + e^{3t}$
22. $x = \frac{1}{2}t^2 + t, y = \frac{1}{2}t^2 - t$

23. Use d^2y/dx^2 to determine the intervals of the parameter for which the curve in Problem 16 is concave upward and the intervals for which it is concave downward.
24. Use d^2y/dx^2 to determine whether the curve given by $x = 2t + 5, y = 2t^3 + 6t^2 + 4t$ has any points of inflection.

In Problems 25–30, find the length of the given curve.

25. $x = \frac{5}{3}t^3 + 2, y = 4t^3 + 6; \quad 0 \le t \le 2$
26. $x = \frac{1}{3}t^3, y = \frac{1}{2}t^2; \quad 0 \le t \le \sqrt{3}$
27. $x = e^t \sin t, y = e^t \cos t; \quad 0 \le t \le \pi$
28. One arch of the cycloid:
$$x = a(\theta - \sin\theta), y = a(1 - \cos\theta); \quad 0 \le \theta \le 2\pi$$
29. One arch of the hypocycloid of four cusps:
$$x = b\cos^3\theta, y = b\sin^3\theta; \quad 0 \le \theta \le \pi/2$$
30. One arch of the epicycloid of three cusps:
$$x = 4a\cos\theta - a\cos 4\theta, y = 4a\sin\theta - a\sin 4\theta, 0 \le \theta \le 2\pi/3$$

Calculator/CAS Problems

31. Consider the curve $x = t^2 - 4t - 2, y = t^5 - 4t^3 - 1$ in Example 1.
    (a) Use a calculator to find an approximation of the y-coordinate of the y-intercept shown in Figure 10.3.1.
    (b) Use Newton's Method to approximate the x-coordinates of the three x-intercepts shown in Figure 10.3.1.

Think About It

32. Let C be a curve described by $y = f(x)$, where F is a continuous nonnegative function on $x_1 \le x \le x_2$. Show that if C is given parametrically by $x = f(t), y = g(t), a \le t \le b$, f' and g continuous, then the **area** under the graph of C is $\int_a^b g(t)f'(t)dt$.
33. Use Problem 32 to show that the area under one arch of the cycloid in Figure 10.2.6(b) is three times the area of the circle.

10.4 Polar Coordinate System

Introduction So far we have used the **rectangular** or **Cartesian coordinate system** to specify a point P or describe a curve C in the plane. We can regard this system as a grid of horizontal and vertical lines. The coordinates (a, b) of a point P are determined by the intersection of two lines: One line $x = a$ is perpendicular to the horizontal reference line called the x-axis, and the other $y = b$ is perpendicular to the vertical reference line called the y-axis. See FIGURE 10.4.1(a). Another system for locating points in the plane is the **polar coordinate system**.

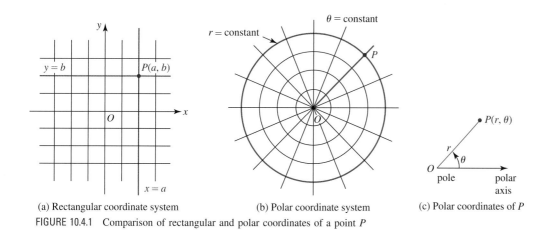

(a) Rectangular coordinate system   (b) Polar coordinate system   (c) Polar coordinates of P

FIGURE 10.4.1 Comparison of rectangular and polar coordinates of a point P

Polar Coordinates To set up a **polar coordinate system**, we use a system of circles centered at a point O, called the **pole**, and straight lines or rays emanating from O. We take as a reference axis a horizontal half-line directed to the right of the pole and call it the **polar axis**. By specifying a directed (signed) distance r from O and an angle θ whose initial side is the polar axis and whose terminal side is the ray OP, we label the point P by (r, θ). We say that the ordered pair (r, θ) are the **polar coordinates** of P. See Figures 10.4.1(b) and 10.4.1(c).

Although the measure of the angle θ can be either in degrees or radians, in calculus radian measure is used almost exclusively. Consequently, we shall use only radian measure in this discussion.

In the polar coordinate system we adopt the following conventions.

Definition 10.4.1 Conventions in Polar Coordinates

(i) Angles $\theta > 0$ are measured counterclockwise from the polar axis, whereas angles $\theta < 0$ are measured clockwise.
(ii) To graph a point $(-r, \theta)$, where $-r < 0$, measure $|r|$ units along the ray $\theta + \pi$.
(iii) The coordinates of the pole O are $(0, \theta)$, where θ is any angle.

EXAMPLE 1 Plotting Polar Points

Plot the points whose polar coordinates are given.

(a) $(4, \pi/6)$ (b) $(2, -\pi/4)$ (c) $(-3, 3\pi/4)$

Solution
(a) Measure 4 units along the ray $\pi/6$ as shown in FIGURE 10.4.2(a).
(b) Measure 2 units along the ray $-\pi/4$. See Figure 10.4.2(b).
(c) Measure 3 units along the ray $3\pi/4 + \pi = 7\pi/4$. Equivalently, we can measure 3 units along the ray $3\pi/4$ extended *backward* through the pole. Note carefully in Figure 10.4.2(c) that the point $(-3, 3\pi/4)$ is not in the same quadrant as the terminal side of the given angle.

FIGURE 10.4.2 Points in polar coordinates in Example 1

In contrast to the rectangular coordinate system, the description of a point in polar coordinates is not unique. This is an immediate consequence of the fact that

$$(r, \theta) \quad \text{and} \quad (r, \theta + 2n\pi), \ n \text{ an integer},$$

are equivalent. To compound the problem, negative values of r can be used.

EXAMPLE 2 Equivalent Polar Points

The following coordinates are some alternative representations of the point $(2, \pi/6)$:

$$(2, 13\pi/6), \quad (2, -11\pi/6), \quad (-2, 7\pi/6), \quad (-2, -5\pi/6).$$ ∎

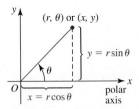

FIGURE 10.4.3 Relating polar and rectangular coordinates

■ **Conversion of Polar Coordinates to Rectangular** By superimposing a rectangular coordinate system on a polar coordinate system, as shown in FIGURE 10.4.3, we can convert a polar description of a point to rectangular coordinates by using

$$x = r\cos\theta, \quad y = r\sin\theta. \tag{1}$$

These conversion formulas hold true for any values of r and θ in an equivalent polar representation of (r, θ).

EXAMPLE 3 Polar to Rectangular

Convert $(2, \pi/6)$ in polar coordinates to rectangular coordinates.

Solution With $r = 2$, $\theta = \pi/6$, we have from (1),

$$x = 2\cos\frac{\pi}{6} = 2\left(\frac{\sqrt{3}}{2}\right) = \sqrt{3}$$

$$y = 2\sin\frac{\pi}{6} = 2\left(\frac{1}{2}\right) = 1.$$

Thus, $(2, \pi/6)$ is equivalent to $\left(\sqrt{3}, 1\right)$ in rectangular coordinates. ∎

■ **Conversion of Rectangular Coordinates to Polar** It should be evident from Figure 10.4.3 that x, y, r, and θ are also related by

$$r^2 = x^2 + y^2, \quad \tan\theta = \frac{y}{x}. \tag{2}$$

The equations in (2) are used to convert the rectangular coordinates (x, y) to the polar coordinates (r, θ).

EXAMPLE 4 Rectangular to Polar

Convert $(-1, 1)$ in rectangular coordinates to polar coordinates.

Solution With $x = -1$, $y = 1$, we have from (2)

$$r^2 = 2 \quad \text{and} \quad \tan\theta = -1.$$

Now, $r^2 = 2$ or $r = \pm\sqrt{2}$, and two of many angles that satisfy $\tan\theta = -1$ are $3\pi/4$ and $7\pi/4$. From FIGURE 10.4.4 we see that two polar representations for $(-1, 1)$ are

$$\left(\sqrt{2}, 3\pi/4\right) \quad \text{and} \quad \left(-\sqrt{2}, 7\pi/4\right).$$ ∎

FIGURE 10.4.4 Point in Example 4

In Example 4, observe that we cannot pair just *any* angle θ and *any* value r that satisfy (2); these solutions must also be consistent with (1). Because the points $\left(-\sqrt{2}, 3\pi/4\right)$ and $\left(\sqrt{2}, 7\pi/4\right)$ lie in the fourth quadrant, they are not polar representations of the second-quadrant point $(-1, 1)$.

There are instances in calculus when a rectangular equation must be expressed as a polar equation $r = f(\theta)$. The next example shows how to do this using the conversion formulas in (1).

EXAMPLE 5 Rectangular Equation to Polar Equation

Find a polar equation that has the same graph as the circle $x^2 + y^2 = 8x$.

Solution Substituting $x = r\cos\theta$, $y = r\sin\theta$ into the given equation we find

$$r^2\cos^2\theta + r^2\sin^2\theta = 8r\cos\theta$$
$$r^2(\cos^2\theta + \sin^2\theta) = 8r\cos\theta \quad \leftarrow \cos^2\theta + \sin^2\theta = 1$$
$$r(r - 8\cos\theta) = 0.$$

The last equation implies that $r = 0$ or $r = 8\cos\theta$. Since $r = 0$ determines only the pole O, we conclude that a polar equation of the circle is $r = 8\cos\theta$. Note that the circle $x^2 + y^2 = 8x$ passes through the origin since $x = 0$ and $y = 0$ satisfy the equation. Relative to the polar equation $r = 8\cos\theta$ of the circle, the origin or pole corresponds to the polar coordinates $(0, \pi/2)$. ∎

EXAMPLE 6 Rectangular Equation to Polar Equation

Find a polar equation that has the same graph as the parabola $x^2 = 8(2 - y)$.

Solution We replace x and y in the given equation by $x = r\cos\theta$, $y = r\sin\theta$ and solve for r in terms of θ:

$$r^2\cos^2\theta = 8(2 - r\sin\theta)$$
$$r^2(1 - \sin^2\theta) = 16 - 8r\sin\theta$$
$$r^2 = r^2\sin^2\theta - 8r\sin\theta + 16 \quad \leftarrow \text{right side is a perfect square}$$
$$r^2 = (r\sin\theta - 4)^2$$
$$r = \pm(r\sin\theta - 4).$$

Solving for r gives two equations,

$$r = \frac{4}{1 + \sin\theta} \quad \text{or} \quad r = \frac{-4}{1 - \sin\theta}.$$

Now recall that, by convention (*ii*) of Definition 10.4.1, (r, θ) and $(-r, \theta + \pi)$ represent the same point. You should verify that if (r, θ) is replaced by $(-r, \theta + \pi)$ in the second of these two equations we obtain the first equation. In other words, the equations are equivalent and so we may simply take the polar equation of the parabola to be $r = 4/(1 + \sin\theta)$. ∎

In the last example we express a polar equation $r = f(\theta)$ as a rectangular equation using (1) and (2).

EXAMPLE 7 Polar Equation to Rectangular Equation

Find a rectangular equation that has the same graph as the polar equation $r^2 = 9\cos 2\theta$.

Solution First, we use the trigonometric identity for the cosine of a double angle:

$$r^2 = 9(\cos^2\theta - \sin^2\theta). \quad \leftarrow \cos 2\theta = \cos^2\theta - \sin^2\theta \qquad (3)$$

Now from (1) we can write $\cos\theta = x/r$ and $\sin\theta = y/r$, and from (2) we have $r^2 = x^2 + y^2$. Therefore,

$$\cos^2\theta = \frac{x^2}{r^2} = \frac{x^2}{x^2 + y^2} \quad \text{and} \quad \sin^2\theta = \frac{y^2}{r^2} = \frac{y^2}{x^2 + y^2}.$$

Substituting r^2, $\cos^2\theta$, and $\sin^2\theta$ into (3) then gives

$$x^2 + y^2 = 9\left(\frac{x^2}{x^2 + y^2} - \frac{y^2}{x^2 + y^2}\right) \quad \text{or} \quad (x^2 + y^2)^2 = 9(x^2 - y^2). \quad ∎$$

The next section will be devoted to graphing polar equations.

Exercises 10.4

≡ **Fundamentals**

In Problems 1–6, plot the point with the given polar coordinates.
1. $(3, \pi)$
2. $(2, -\pi/2)$
3. $(-\frac{1}{2}, \pi/2)$
4. $(-1, \pi/6)$
5. $(-4, -\pi/6)$
6. $(\frac{2}{3}, 7\pi/4)$

In Problems 7–12, find alternative polar coordinates that satisfy
(a) $r > 0, \theta < 0$
(b) $r > 0, \theta > 2\pi$
(c) $r < 0, \theta > 0$
(d) $r < 0, \theta < 0$

for each point with the given polar coordinates.

7. $(2, 3\pi/4)$
8. $(5, \pi/2)$
9. $(4, \pi/3)$
10. $(3, \pi/4)$
11. $(1, \pi/6)$
12. $(3, 7\pi/6)$

In Problems 13–18, find the rectangular coordinates for each point with the given polar coordinates.

13. $(\frac{1}{2}, 2\pi/3)$
14. $(-1, 7\pi/4)$
15. $(-6, -\pi/3)$
16. $(\sqrt{2}, 11\pi/6)$
17. $(4, 5\pi/4)$
18. $(-5, \pi/2)$

In Problems 19–24, find polar coordinates that satisfy
(a) $r > 0, -\pi < \theta \le \pi$
(b) $r < 0, -\pi < \theta \le \pi$

for each point with the given rectangular coordinates.

19. $(-2, -2)$
20. $(0, -4)$
21. $(1, -\sqrt{3})$
22. $(\sqrt{6}, \sqrt{2})$
23. $(7, 0)$
24. $(1, 2)$

In Problems 25–30, sketch the region on the plane that consists of points (r, θ) whose polar coordinates satisfy the given conditions.

25. $2 \le r < 4, 0 \le \theta \le \pi$
26. $2 < r \le 4$
27. $0 \le r \le 2, -\pi/2 \le \theta \le \pi/2$
28. $r \ge 0, \pi/4 < \theta < 3\pi/4$
29. $-1 \le r \le 1, 0 \le \theta \le \pi/2$
30. $-2 \le r < 4, \pi/3 \le \theta \le \pi$

In Problems 31–40, find a polar equation that has the same graph as the given rectangular equation.

31. $y = 5$
32. $x + 1 = 0$
33. $y = 7x$
34. $3x + 8y + 6 = 0$
35. $y^2 = -4x + 4$
36. $x^2 - 12y - 36 = 0$
37. $x^2 + y^2 = 36$
38. $x^2 - y^2 = 1$
39. $x^2 + y^2 + x = \sqrt{x^2 + y^2}$
40. $x^3 + y^3 - xy = 0$

In Problems 41–52, find a rectangular equation that has the same graph as the given polar equation.

41. $r = 2\sec\theta$
42. $r\cos\theta = -4$
43. $r = 6\sin 2\theta$
44. $2r = \tan\theta$
45. $r^2 = 4\sin 2\theta$
46. $r^2 \cos 2\theta = 16$
47. $r + 5\sin\theta = 0$
48. $r = 2 + \cos\theta$
49. $r = \dfrac{2}{1 + 3\cos\theta}$
50. $r(4 - \sin\theta) = 10$
51. $r = \dfrac{5}{3\cos\theta + 8\sin\theta}$
52. $r = 3 + 3\sec\theta$

≡ **Think About It**

53. How would you express the distance d between two points (r_1, θ_1) and (r_2, θ_2) in terms of their polar coordinates?

54. You know how to find a rectangular equation of a line through two points with rectangular coordinates. How would you find a polar equation of a line through two points with polar coordinates (r_1, θ_1) and (r_2, θ_2)? Carry out your ideas by finding a polar equation of the line through $(3, 3\pi/4)$ and $(1, \pi/4)$. Find the polar coordinates of the x- and y-intercepts of the line.

55. In rectangular coordinates the x-intercepts of the graph of a function $y = f(x)$ are determined from the solutions of the equation $f(x) = 0$. In the next section we will graph polar equations $r = f(\theta)$. What is the significance of the solutions of the equation $f(\theta) = 0$?

10.5 Graphs of Polar Equations

■ **Introduction** The graph of a polar equation $r = f(\theta)$ is the set of points P with *at least* one set of polar coordinates that satisfies the equation. Since it is most likely that your classroom does not have a polar coordinate grid, to facilitate graphing and discussion of graphs of a polar equation $r = f(\theta)$, we will, as in the preceding section, superimpose a rectangular coordinate over the polar coordinate system.

We begin with some simple polar graphs.

EXAMPLE 1 A Circle Centered at Origin

Graph $r = 3$.

Solution Since θ is not specified, the point $(3, \theta)$ lies on the graph of $r = 3$ for any value of θ and is 3 units from the origin. We see in FIGURE 10.5.1 that the graph is the circle of radius 3 centered at the origin.

Alternatively, we know from (2) of Section 10.4 that $r = \pm\sqrt{x^2 + y^2}$ so that $r = 3$ yields the familiar rectangular equation $x^2 + y^2 = 3^2$ of a circle of radius 3 centered at the origin. ∎

■ **Circles Centered at the Origin** In general, if a is any nonzero constant, the polar graph of

$$r = a \tag{1}$$

is a circle of radius $|a|$ with center at the origin.

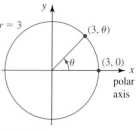

FIGURE 10.5.1 Circle in Example 1

EXAMPLE 2 A Line Through the Origin

Graph $\theta = \pi/4$.

Solution Since r is not specified, the point $(r, \pi/4)$ lies on the graph for any value of r. If $r > 0$, then this point lies on the half-line in the first quadrant; if $r < 0$ then the point lies on the half-line in the third quadrant. For $r = 0$, the point $(0, \pi/4)$ is the pole or origin. Therefore, the polar graph of $\theta = \pi/4$ is the entire line through the origin that makes an angle of $\pi/4$ with the polar axis or positive x-axis. See FIGURE 10.5.2. ∎

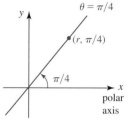

FIGURE 10.5.2 Line in Example 2

■ **Lines Through the Origin** In general, if α is any nonzero real constant, the polar graph of

$$\theta = \alpha \tag{2}$$

is a line through the origin that makes an angle of α radians with the polar axis.

EXAMPLE 3 A Spiral

Graph $r = \theta$.

Solution As $\theta \geq 0$ increases, r increases and the points (r, θ) wind around the pole in a counterclockwise manner. This is illustrated by the blue portion of the graph in FIGURE 10.5.3. The red portion of the graph is obtained by plotting points for $\theta < 0$.

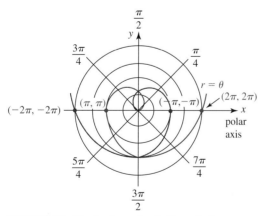

FIGURE 10.5.3 Graph of equation in Example 3 ∎

■ **Spirals** Many graphs in polar coordinates are given special names. The graph in Example 3 is a special case of

$$r = a\theta, \tag{3}$$

where a is a constant. A graph of this equation is called a **spiral of Archimedes**. A polar equation of the form

$$r = ae^{b\theta} \tag{4}$$

is called a **logarithmic spiral**. The curve that describes the multichambered nautilus shell is an example of a logarithmic spiral. See Problems 31 and 32 in Exercises 10.5.

Half of multichambered Nautilus shell

Symmetries of a snowflake

In addition to basic point plotting, symmetry can often be utilized to graph a polar equation.

■ **Symmetry** As illustrated in **FIGURE 10.5.4**, a polar graph can have three types of symmetry. A polar graph is **symmetric with respect to the *y*-axis** if whenever (r, θ) is a point on the graph,

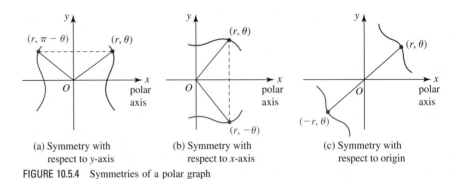

(a) Symmetry with respect to *y*-axis
(b) Symmetry with respect to *x*-axis
(c) Symmetry with respect to origin

FIGURE 10.5.4 Symmetries of a polar graph

$(r, \pi - \theta)$ is also a point on the graph. A polar graph is **symmetric with respect to the *x*-axis** if whenever (r, θ) is a point on the graph, $(r, -\theta)$ is also a point on the graph. Finally, a polar graph is **symmetric with respect to the origin** if whenever (r, θ) is on the graph, $(-r, \theta)$ is also a point on the graph.

We have the following tests for symmetries of a polar graph.

Tests for Symmetry of the Graph of a Polar Equation

The graph of a polar equation is symmetric with respect to:

- the ***y*-axis** if replacing (r, θ) by $(r, \pi - \theta)$ results in the same equation; (5)
- the ***x*-axis** if replacing (r, θ) by $(r, -\theta)$ results in the same equation; (6)
- the **origin** if replacing (r, θ) by $(-r, \theta)$ results in the same equation. (7)

> In rectangular coordinates the description of a point is unique. Hence, in rectangular coordinates if a test for a particular type of symmetry fails, then we can definitely say that the graph does not possess that symmetry.

▶ Because the polar description of a point is not unique, the graph of a polar equation may still have a particular type of symmetry even though the test for it may fail. For example, if replacing (r, θ) by $(r, -\theta)$ fails to give the original polar equation, the graph of that equation may still possess symmetry with respect to the *x*-axis. Therefore, if one of the replacement tests in (5)–(7) fails to give the same polar equation, the best we can say is "no conclusion."

EXAMPLE 4 Graphing a Polar Equation

Graph $r = 1 - \cos\theta$.

Solution One way of graphing this equation is to plot a few well-chosen points corresponding to $0 \leq \theta \leq 2\pi$. As the following table shows

θ	0	$\pi/4$	$\pi/2$	$3\pi/4$	π	$5\pi/4$	$3\pi/2$	$7\pi/4$	2π
r	0	0.29	1	1.71	2	1.71	1	0.29	0

as θ advances from $\theta = 0$ to $\theta = \pi/2$, r increases from $r = 0$ (the origin) to $r = 1$. See

FIGURE 10.5.5(a). As θ advances from $\theta = \pi/2$ to $\theta = \pi$, r continues to increase from $r = 1$ to its maximum value of $r = 2$. See Figure 10.5.5(b). Then, for $\theta = \pi$ to $\theta = 3\pi/2$, r begins to decrease from $r = 2$ to $r = 1$. For $\theta = 3\pi/2$ to $\theta = 2\pi$, r continues to decrease and we end up again at the origin $r = 0$. See Figures 10.5.5(c) and (d).

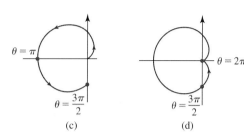

FIGURE 10.5.5 Graph of equation in Example 4

By taking advantage of symmetry we could have simply plotted points for $0 \leq \theta \leq \pi$. From the trigonometric identity for the cosine function $\cos(-\theta) = \cos\theta$ it follows from (6) that the graph of $r = 1 - \cos\theta$ is symmetric with respect to the x-axis. We can obtain the complete graph of $r = 1 - \cos\theta$ by reflecting in the x-axis that portion of the graph given in Figure 10.5.5(b). ∎

■ **Cardioids** The polar equation in Example 4 is a member of a family of equations that all have a "heart-shaped" graph that passes through the origin. A graph of any polar equation of the form

$$r = a \pm a\sin\theta \quad \text{or} \quad r = a \pm a\cos\theta \tag{8}$$

is called a **cardioid**. The only difference in the graph of these four equations is their symmetry with respect to the y-axis ($r = a \pm a\sin\theta$) or symmetry with respect to the x-axis ($r = a \pm a\cos\theta$). In FIGURE 10.5.6 we have assumed that $a > 0$.

By knowing the basic shape and orientation of a cardioid, you can obtain a quick and accurate graph by plotting the four points corresponding to $\theta = 0$, $\theta = \pi/2$, $\theta = \pi$, and $\theta = 3\pi/2$. The graphs of $r = a \pm a\sin\theta$ are symmetric with respect to the y-axis and the graphs of $r = a \pm a\cos\theta$ are symmetric with respect to the x-axis.

■ **Limaçons** Cardioids are special cases of polar curves known as **limaçons**:

$$r = a \pm b\sin\theta \quad \text{or} \quad r = a \pm b\cos\theta. \tag{9}$$

The shape of a limaçon depends on the relative magnitudes of a and b. Let us assume that $a > 0$ and $b > 0$. For $0 < a/b < 1$, we get a **limaçon with an interior loop** as shown in FIGURE 10.5.7(a). When $a = b$ or equivalently $a/b = 1$ we get a **cardioid**. For $1 < a/b < 2$, we get a **dimpled limaçon** as shown in Figure 10.5.7(b). For $a/b \geq 2$, the curve is called a **convex limaçon**. See Figure 10.5.7(c).

(a)

(b)

(c)

(d)

FIGURE 10.5.6 Cardioids

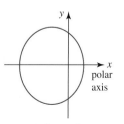

(a) Limaçon with interior loop (b) Dimpled limaçon (c) Convex limaçon

FIGURE 10.5.7 Three kinds of limaçons: for $0 < a/b < 1$ we get (a); for $1 < a/b < 2$ we get (b); for $a/b \geq 2$ we get (c)

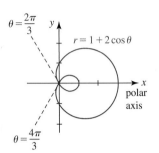

FIGURE 10.5.8 Graph of equation in Example 6

EXAMPLE 5 A Limaçon

The graph of $r = 3 - \sin\theta$ is a convex limaçon, since $a = 3$, $b = 1$, and $a/b = 3 > 2$. The graph of this equation is similar to that in Figure 10.5.7(c) except that the graph is symmetric with respect to the y-axis. ∎

EXAMPLE 6 A Limaçon

The graph of $r = 1 + 2\cos\theta$ is a limaçon with an interior loop, since $a = 1$, $b = 2$, and $a/b = \frac{1}{2} < 1$. For $\theta \geq 0$, notice in FIGURE 10.5.8 the limaçon starts at $\theta = 0$ or $(3, 0)$. The graph passes through the y-axis at $(1, \pi/2)$ and then enters the origin $(r = 0)$ for the first angle for which $r = 0$ or $1 + 2\cos\theta = 0$ or $\cos\theta = -\frac{1}{2}$. This implies that $\theta = 2\pi/3$. At $\theta = \pi$, the curve passes through $(-1, \pi)$. The remainder of the graph can then be completed using the fact that it is symmetric with respect to the x-axis. ∎

■ **Tangents to the Graph at the Origin** In Example 6, the lines $\theta = 2\pi/3$ and $\theta = 4\pi/3$ shown in red in Figure 10.5.8, where the graph of $r = 1 + 2\cos\theta$ enters and exits the origin, respectively, are actually tangent to the graph at the origin. In general, if $r = 0$ for $\theta = \theta_0$ and $dr/d\theta \neq 0$ when $\theta = \theta_0$, then the graph of $r = f(\theta)$ is tangent to the line $\theta = \theta_0$ at the origin. We will prove this in the next section.

EXAMPLE 7 A Rose Curve

Graph $r = 2\cos 2\theta$.

Solution Because

$$\cos 2(\pi - \theta) = \cos 2\theta \quad \text{and} \quad \cos(-2\theta) = \cos 2\theta$$

we conclude by (5) and (6) of the tests for symmetry that the graph is symmetric with respect to both the x- and the y-axes. A moment of reflection should convince you that we need only consider $0 \leq \theta \leq \pi/2$. Using the data in the following table, we see that the dashed portion of the graph given in FIGURE 10.5.9 is that completed by symmetry. The graph is called a **rose curve with four petals**.

θ	0	$\pi/12$	$\pi/6$	$\pi/4$	$\pi/3$	$5\pi/12$	$\pi/2$
r	2	$\sqrt{3}$	1	0	-1	$-\sqrt{3}$	-2

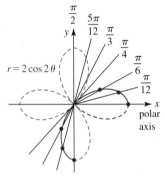

FIGURE 10.5.9 Graph of equation in Example 7

Note from the table that $r = 0$ and $dr/d\theta = -4\sin 2\theta \neq 0$ for $\theta = \pi/4$. Hence, the graph is tangent to the line $\theta = \pi/4$ at the origin. ∎

■ **Rose Curves** In general, if n is a positive integer, then the graphs of

$$r = a\sin n\theta \quad \text{or} \quad r = a\cos n\theta, \quad n \geq 2 \qquad (10)$$

are called **rose curves**, although as you can see in FIGURE 10.5.10 the curve looks more like a daisy. We note that the number of **petals** or **loops** of the curve is:

- n when n is odd, and
- $2n$ when n is even.

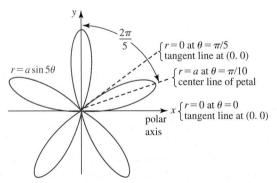

FIGURE 10.5.10 Rose curve with 5 petals

To graph a rose curve we can start by graphing one petal. To begin, we find an angle θ for which r is a maximum. This gives the *center line* of the petal. We then find corresponding values of θ for which the rose curve enters the origin ($r = 0$). To complete the graph we use the fact that the center lines of the petals are spaced $2\pi/n$ radians ($360/n$ degrees) apart if n is odd, and $2\pi/2n = \pi/n$ radians ($180/n$ degrees) apart if n is even. In Figure 10.5.10 we have drawn the graph of $r = a\sin 5\theta$, $a > 0$. The center line of the petal in the first quadrant is determined from the solution of

$$a = a\sin 5\theta \quad \text{or} \quad 1 = \sin 5\theta.$$

The last equation implies that $5\theta = \pi/2$ or $\theta = \pi/10$. The spacing between the center lines of the five petals is $2\pi/5$ radians ($72°$). Also, $r = 0$, or $\sin 5\theta = 0$, for $5\theta = 0$ and $5\theta = \pi$. Since $dr/d\theta = 5a\cos 5\theta \neq 0$ for $\theta = 0$ and $\theta = \pi/5$ the graph of the petal in the first quadrant is tangent to those lines at the origin.

In Example 5 in Section 10.4 we saw that the polar equation $r = 8\cos\theta$ is equivalent to the rectangular equation $x^2 + y^2 = 8x$. By completing the square in x in the rectangular equation, we recognize

$$(x - 4)^2 + y^2 = 16$$

as a circle of radius 4 centered at $(4, 0)$ on the x-axis. Polar equations such as $r = 8\cos\theta$ or $r = 8\sin\theta$ are circles and are also special cases of rose curves. See FIGURE 10.5.11.

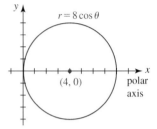

FIGURE 10.5.11 Graph of equation $r = 8\cos\theta$

■ **Circles with Centers on an Axis** When $n = 1$ in (10) we get

$$r = a\sin\theta \quad \text{or} \quad r = a\cos\theta, \qquad (11)$$

which are polar equations of circles passing through the origin with diameters $|a|$ and with centers $(a/2, 0)$ on the x-axis ($r = a\cos\theta$), or with centers $(0, a/2)$ on the y-axis ($r = a\sin\theta$). FIGURE 10.5.12 illustrates the graphs of the equations in (11) in the cases when $a > 0$ and $a < 0$.

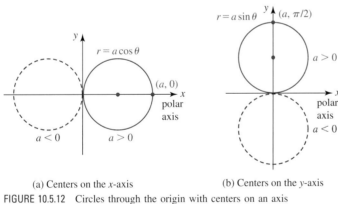

FIGURE 10.5.12 Circles through the origin with centers on an axis

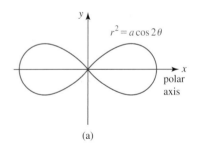

■ **Lemniscates** If n is a positive integer, the graphs of

$$r^2 = a\cos 2\theta \quad \text{or} \quad r^2 = a\sin 2\theta, \qquad (12)$$

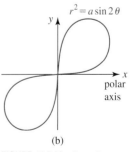

where $a > 0$, are called **lemniscates**. By (7) of the tests for symmetry you can see the graphs of both of the equations in (12) are symmetric with respect to the origin. Moreover, by (6) of the tests for symmetry, the graph of $r^2 = a\cos 2\theta$ is symmetric with respect to the x-axis. FIGURE 10.5.13 shows typical graphs of the equations $r^2 = a\cos 2\theta$ and $r^2 = a\sin 2\theta$, respectively.

FIGURE 10.5.13 Lemniscates

■ **Points of Intersection** In rectangular coordinates we can find the points (x, y) where the graphs of two functions $y = f(x)$ and $y = g(x)$ intersect by equating the y values. The real solutions of the equation $f(x) = g(x)$ correspond to *all* the x-coordinates of the points where the graphs intersect. In contrast, problems may arise in polar coordinates when we try the same method to determine where the graphs of two polar equations $r = f(\theta)$ and $r = g(\theta)$ intersect.

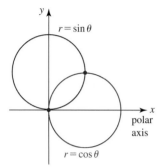

FIGURE 10.5.14 Intersecting circles in Example 8

EXAMPLE 8 Intersecting Circles

In FIGURE 10.5.14 we see that the circles $r = \sin\theta$ and $r = \cos\theta$ have two points of intersection. By equating the r values, the equation $\sin\theta = \cos\theta$ leads to $\theta = \pi/4$. Substituting this value into either equation yields $r = \sqrt{2}/2$. Thus, we have found only a single polar point $(\sqrt{2}/2, \pi/4)$ where the graphs intersect. From the figure, it is apparent that the graphs also intersect at the origin. But the problem here is that the origin or pole is $(0, \pi/2)$ on the graph of $r = \cos\theta$ but is $(0, 0)$ on the graph of $r = \sin\theta$. This situation is analogous to the curves reaching the same point at different times. ■

Rotation of Polar Graphs   In Section 1.2 we saw that if $y = f(x)$ is the rectangular equation of a function, then the graphs of $y = f(x - c)$ and $y = f(x + c)$, $c > 0$, are obtained by *shifting* the graph of f horizontally c units to the right and to the left, respectively. In contrast, if $r = f(\theta)$ is a polar equation, then the graphs of $f(\theta - \gamma)$ and $f(\theta + \gamma)$, where $\gamma > 0$, can be obtained by *rotating* the graph of f by an amount γ. Specifically:

- The graph of $r = f(\theta - \gamma)$ is the graph of $r = f(\theta)$ rotated *counterclockwise* about the origin by an amount γ.
- The graph of $r = f(\theta + \gamma)$ is the graph of $r = f(\theta)$ rotated *clockwise* about the origin by an amount γ.

For example, the graph of the cardioid $r = a(1 + \cos\theta)$ is shown in Figure 10.5.6(a). The graph of $r = a(1 + \cos(\theta - \pi/2))$ is the graph of $r = a(1 + \cos\theta)$ rotated counterclockwise about the origin by an amount $\pi/2$. Its graph then must be that given in Figure 10.5.6(c). This makes sense, because the sum formula of the cosine gives

See the identities in (18) of Section 1.4.

$$r = a[1 + \cos(\theta - \pi/2)] = a[1 + \cos\theta\cos(\pi/2) + \sin\theta\sin(\pi/2)] = a(1 + \sin\theta).$$

Similarly, rotating $r = a(1 + \cos\theta)$ counterclockwise about the origin by an amount π gives

$$r = a[1 + \cos(\theta - \pi)] = a[1 + \cos\theta\cos\pi + \sin\theta\sin\pi] = a(1 - \cos\theta).$$

Now look again at Figure 10.5.13. From

$$r^2 = a\cos 2\left(\theta - \frac{\pi}{4}\right) = a\cos\left(2\theta - \frac{\pi}{2}\right) = a\sin 2\theta$$

we see that the graph of the lemniscate in Figure 10.5.13(b) is the graph in Figure 10.5.13(a) rotated counterclockwise about the origin by an amount $\pi/4$.

EXAMPLE 9 Rotated Polar Graph

Graph $r = 1 + 2\sin(\theta + \pi/4)$.

Solution The graph of the given equation is the graph of the limaçon $r = 1 + 2\sin\theta$ rotated clockwise about the origin by an amount $\pi/4$. In FIGURE 10.5.15 the blue graph is that of $r = 1 + 2\sin\theta$ and the red graph is the rotated graph. ■

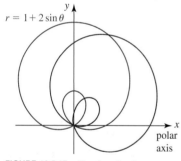

FIGURE 10.5.15 Graphs of polar equations in Example 9

$\dfrac{d}{d\theta}$ NOTES FROM THE CLASSROOM

(*i*) The four-petal rose curve obtained in Example 7 is obtained by plotting r for θ-values satisfying $0 \le \theta < 2\pi$. See FIGURE 10.5.16. Do not assume this is true for every rose curve. Indeed, the five-petal rose curve discussed in Figure 10.5.10 is obtained using θ-values satisfying $0 \le \theta < \pi$. In general, a rose curve $r = a\sin n\theta$ or $r = a\cos n\theta$ is traced out exactly once for $0 \le \theta < 2\pi$ if n is even and once for $0 \le \theta < \pi$ if n is odd. Observations such as these will be important in the next section.

(*ii*) Example 8 illustrates one of several frustrating difficulties of working in polar coordinates:

A point can be on the graph of a polar equation even though its coordinates do not satisfy the equation.

This is something that cannot happen in rectangular coordinates. For example, you should verify that $(2, \pi/2)$ is an alternative polar description of the point $(-2, 3\pi/2)$. Moreover, verify that $(-2, 3\pi/2)$ is a point on the graph of $r = 1 + 3\sin\theta$ by showing that the coordinates satisfy the equation. However, note that the alternative coordinates $(2, \pi/2)$ do not satisfy the equation.

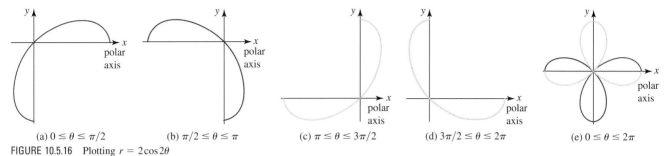

(a) $0 \leq \theta \leq \pi/2$ (b) $\pi/2 \leq \theta \leq \pi$ (c) $\pi \leq \theta \leq 3\pi/2$ (d) $3\pi/2 \leq \theta \leq 2\pi$ (e) $0 \leq \theta \leq 2\pi$

FIGURE 10.5.16   Plotting $r = 2\cos 2\theta$

Exercises 10.5
Answers to selected odd-numbered problems begin on page ANS-33.

≡ Fundamentals

In Problems 1–30, identify by name the graph of the given polar equation. Then sketch the graph of the equation.

1. $r = 6$
2. $r = -1$
3. $\theta = \pi/3$
4. $\theta = 5\pi/6$
5. $r = 2\theta, \theta \leq 0$
6. $r = 3\theta, \theta \geq 0$
7. $r = 1 + \cos\theta$
8. $r = 5 - 5\sin\theta$
9. $r = 2(1 + \sin\theta)$
10. $2r = 1 - \cos\theta$
11. $r = 1 - 2\cos\theta$
12. $r = 2 + 4\sin\theta$
13. $r = 4 - 3\sin\theta$
14. $r = 3 + 2\cos\theta$
15. $r = 4 + \cos\theta$
16. $r = 4 - 2\sin\theta$
17. $r = \sin 2\theta$
18. $r = 3\sin 4\theta$
19. $r = 3\cos 3\theta$
20. $r = 2\sin 3\theta$
21. $r = \cos 5\theta$
22. $r = 2\sin 9\theta$
23. $r = 6\cos\theta$
24. $r = -2\cos\theta$
25. $r = -3\sin\theta$
26. $r = 5\sin\theta$
27. $r^2 = 4\sin 2\theta$
28. $r^2 = 4\cos 2\theta$
29. $r^2 = -25\cos 2\theta$
30. $r^2 = -9\sin 2\theta$

In Problems 31 and 32, the graph of the given equation is a spiral. Sketch its graph.

31. $r = 2^\theta, \theta \geq 0$ (logarithmic)
32. $r\theta = \pi, \theta > 0$ (hyperbolic)

In Problems 33–38, find an equation of the given polar graph.

33.

FIGURE 10.5.17   Graph for Problem 33

34.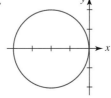

FIGURE 10.5.18   Graph for Problem 34

35.

FIGURE 10.5.19   Graph for Problem 35

36.

FIGURE 10.5.20   Graph for Problem 36

37.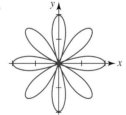

FIGURE 10.5.21   Graph for Problem 37

38.

FIGURE 10.5.22   Graph for Problem 38

In Problems 39–42, find the points of intersection of the graphs of the given pair of polar equations.

39. $r = 2$
   $r = 4\sin\theta$

40. $r = \sin\theta$
   $r = \sin 2\theta$

41. $r = 1 - \cos\theta$
   $r = 1 + \cos\theta$

42. $r = 3 - 3\cos\theta$
   $r = 3\cos\theta$

In Problems 43 and 44, use the fact that $r = f(\theta)$ and $-r = f(\theta + \pi)$ describe the same curve as an aid in finding the points of intersection of the given pair of polar equations.

43. $r = 3$
   $r = 6\sin 2\theta$

44. $r = \cos 2\theta$
   $r = 1 + \cos\theta$

≡ Calculator/CAS Problems

45. Use a graphing utility to obtain the graph of the **bifolium** $r = 4\sin\theta\cos^2\theta$ and the circle $r = \sin\theta$ on the same axes. Find all points of intersection of the graphs.

46. Use a graphing utility to verify that the cardioid $r = 1 + \cos\theta$ and the lemniscate $r^2 = 4\cos\theta$ intersect at four points. Find these points of intersection of the graphs.

In Problems 47 and 48, the graphs of the equations (a)–(d) represent a rotation of the graph of the given equation. Try sketching these graphs by hand. If you have difficulties, then use a calculator or CAS.

47. $r = 1 + \sin\theta$
    (a) $r = 1 + \sin(\theta - \pi/2)$
    (b) $r = 1 + \sin(\theta + \pi/2)$
    (c) $r = 1 + \sin(\theta - \pi/6)$
    (d) $r = 1 + \sin(\theta + \pi/4)$

48. $r = 2 + 4\cos\theta$
    (a) $r = 2 + 4\cos(\theta + \pi/6)$
    (b) $r = 2 + 4\cos(\theta - 3\pi/2)$
    (c) $r = 2 + 4\cos(\theta + \pi)$
    (d) $r = 2 + 4\cos(\theta - \pi/8)$

In Problems 49–52, use a calculator or CAS, if necessary, to match the given graph with the appropriate polar equation in (a)–(d).

(a) $r = 2\cos\dfrac{3\theta}{2}$, $0 \leq \theta \leq 4\pi$

(b) $r = 2\cos\dfrac{\theta}{5}$, $0 \leq \theta \leq 5\pi$

(c) $r = 2\sin\dfrac{\theta}{4}$, $0 \leq \theta \leq 8\pi$

(d) $r = 2\sin\dfrac{\theta}{2}$, $0 \leq \theta \leq 4\pi$

49.

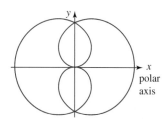

FIGURE 10.5.23 Graph for Problem 49

50.

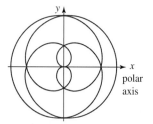

FIGURE 10.5.24 Graph for Problem 50

51.

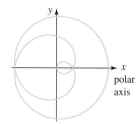

FIGURE 10.5.25 Graph for Problem 51

52.

FIGURE 10.5.26 Graph for Problem 52

53. Use a CAS to obtain graphs of the polar equation $r = a + \cos\theta$ for $a = 0, \frac{1}{4}, \frac{1}{2}, \frac{3}{4}, 1, \frac{5}{4}, \ldots, 3$.

54. Identify all the curves in Problem 53. What happens to the graphs as $a \to \infty$?

≡ Think About It

In Problems 55–58, identify the symmetries if the given pair of points are on the graph of $r = f(\theta)$.

55. $(r, \theta), (-r, \pi - \theta)$

56. $(r, \theta), (r, \theta + \pi)$

57. $(r, \theta), (-r, \theta + 2\pi)$

58. $(r, \theta), (-r, -\theta)$

In Problems 59 and 60, let $r = f(\theta)$ be a polar equation. Interpret the given result geometrically.

59. $f(-\theta) = f(\theta)$ (even function)

60. $f(-\theta) = -f(\theta)$ (odd function)

61. (a) What is the difference between the circles $r = -4$ and $r = 4$?
    (b) What is the difference between the lines through the origin $\theta = \pi/6$ and $\theta = 7\pi/6$?

62. **A Bit of History** The Italian **Galileo Galilei** (1564–1642) is remembered for his many discoveries and innovations in the fields of astronomy and physics. With a reflecting telescope of his own design he was the first to discover the moons of Jupiter. Through his observations of the planet Venus and sun spots, Galileo eventually came to support the controversial opinion of Nicolaus Copernicus that the planets revolved around the Sun. Galileo's empirical work on gravity predates the contributions of Isaac Newton. He was the first to perform scientific studies to determine the acceleration of gravity. By measuring the time it takes metal balls to roll down an inclined plane, Galileo was able to calculate the speed of each ball and from those observations concluded that the distance s a ball moved is related to time t by $s = \frac{1}{2}gt^2$, where g is the acceleration due to gravity.

Suppose several metal balls are released simultaneously from a common point and allowed to slide down frictionless inclined planes at various angles, each ball accelerating because of gravity. See FIGURE 10.5.27. Show that at any instant, all the balls lie on a common circle whose topmost point is the point of release. Galileo was able to show this without the benefit of either rectangular or polar coordinates.

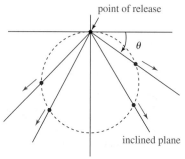

FIGURE 10.5.27 Inclined planes in Problem 62

10.6 Calculus in Polar Coordinates

Introduction In this section we will answer three standard calculus problems in the polar coordinate system.

- What is the slope of a tangent line to a polar graph?
- What is the area bounded by a polar graph?
- What is the length of a polar graph?

We begin with the tangent line problem.

Slope of a Tangent to a Polar Graph Somewhat surprisingly, the slope of a tangent line to the graph of a polar equation $r = f(\theta)$ is *not* the derivative $dr/d\theta = f'(\theta)$. The slope of a tangent line is still dy/dx. To find this latter derivative, we use $r = f(\theta)$ along with $x = r\cos\theta$, $y = r\sin\theta$ to write parametric equations of the curve:

$$x = f(\theta)\cos\theta, \quad y = f(\theta)\sin\theta. \tag{1}$$

Then from (1) of Section 10.3 and the Product Rule,

$$\frac{dy}{dx} = \frac{dy/d\theta}{dx/d\theta} = \frac{f(\theta)\cos\theta + f'(\theta)\sin\theta}{-f(\theta)\sin\theta + f'(\theta)\cos\theta}.$$

This result is summarized in the next theorem.

Theorem 10.6.1 Slope of Tangent Line

If f is a differentiable function of θ, then the **slope of the tangent line** to the graph of $r = f(\theta)$ at a point (r, θ) on the graph is

$$\frac{dy}{dx} = \frac{dy/d\theta}{dx/d\theta} = \frac{f(\theta)\cos\theta + f'(\theta)\sin\theta}{-f(\theta)\sin\theta + f'(\theta)\cos\theta}, \tag{2}$$

provided $dx/d\theta \neq 0$.

Formula (2) in Theorem 10.6.1 is presented "for the record"; do not memorize it. To find dy/dx in polar coordinates simply form the parametric equations $x = f(\theta)\cos\theta$, $y = f(\theta)\sin\theta$ and then use the parametric form of the derivative.

EXAMPLE 1 Slope

Find the slope of the tangent line to the graph of $r = 4\sin 3\theta$ at $\theta = \pi/6$.

Solution From the parametric equations $x = 4\sin 3\theta \cos\theta$, $y = 4\sin 3\theta \sin\theta$ we find

$$\frac{dy}{dx} = \frac{dy/d\theta}{dx/d\theta} = \frac{4\sin 3\theta \cos\theta + 12\cos 3\theta \sin\theta}{-4\sin 3\theta \sin\theta + 12\cos 3\theta \cos\theta}$$

and so

$$\left.\frac{dy}{dx}\right|_{\theta=\pi/6} = -\sqrt{3}.$$

The graph of the equation, which we recognize as a rose curve with three petals, and the tangent line are illustrated in FIGURE 10.6.1.

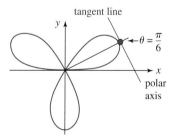

FIGURE 10.6.1 Tangent line in Example 1

EXAMPLE 2 Equation of Tangent Line

Find a rectangular equation of the tangent line in Example 1.

Solution At $\theta = \pi/6$ the parametric equations $x = 4\sin 3\theta \cos\theta$, $y = 4\sin 3\theta \sin\theta$ yield, respectively, $x = 2\sqrt{3}$ and $y = 2$. The rectangular coordinates of the point of tangency are $(2\sqrt{3}, 2)$. Using the slope found in Example 1, the point-slope form gives an equation of the red tangent line shown in Figure 10.6.1:

$$y - 2 = -\sqrt{3}(x - 2\sqrt{3}) \quad \text{or} \quad y = -\sqrt{3}x + 8.$$

We can get a polar equation of the line in Example 2 by replacing x and y in the rectangular equation by $x = r\cos\theta$, $y = r\sin\theta$ and solving for r:

$$r = \frac{8}{\sin\theta + \sqrt{3}\cos\theta}.$$

EXAMPLE 3 Horizontal and Vertical Tangents

Find the points on the graph of $r = 3 - 3\sin\theta$ at which the tangent line is horizontal and the points at which the tangent line is vertical.

Solution Recall from Section 10.3 that a horizontal tangent occurs at a point for which $dy/d\theta = 0$ and $dx/d\theta \neq 0$, whereas a vertical tangent occurs at a point for which $dx/d\theta = 0$ and $dy/d\theta \neq 0$. Now, from the parametric equations

$$x = (3 - 3\sin\theta)\cos\theta, \qquad y = (3 - 3\sin\theta)\sin\theta$$

we get

$$\frac{dx}{d\theta} = (3 - 3\sin\theta)(-\sin\theta) + \cos\theta(-3\cos\theta)$$

$$= -3\sin\theta + 3\sin^2\theta - 3\cos^2\theta$$

$$= -3 - 3\sin\theta + 6\sin^2\theta$$

$$= 3(2\sin\theta + 1)(\sin\theta - 1),$$

$$\frac{dy}{d\theta} = (3 - 3\sin\theta)\cos\theta + \sin\theta(-3\cos\theta)$$

$$= 3\cos\theta(1 - 2\sin\theta).$$

From these derivatives we see that:

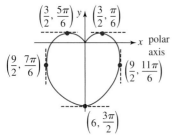

$$\frac{dy}{d\theta} = 0 \left(\frac{dx}{d\theta} \neq 0\right) \text{ at } \theta = \frac{\pi}{6},\ \theta = \frac{5\pi}{6}, \text{ and } \theta = \frac{3\pi}{2},$$

$$\frac{dx}{d\theta} = 0 \left(\frac{dy}{d\theta} \neq 0\right) \text{ at } \theta = \frac{7\pi}{6}, \text{ and } \theta = \frac{11\pi}{6}.$$

Thus, there are

horizontal tangents at: $(\tfrac{3}{2}, \pi/6), (\tfrac{3}{2}, 5\pi/6), (6, 3\pi/2)$,
vertical tangents at: $(\tfrac{9}{2}, 7\pi/6), (\tfrac{9}{2}, 11\pi/6)$.

FIGURE 10.6.2 Horizontal and vertical tangent lines in Example 3

These points, along with the tangent lines, are shown in FIGURE 10.6.2. ■

Tangents to the Graph at the Origin In the preceding section we stated that, in general, if $r = 0$ and $dr/d\theta = f'(\theta) \neq 0$ when $\theta = \theta_0$, then the graph of $r = f(\theta)$ is tangent to the line $\theta = \theta_0$ at the origin. This fact follows from (2). If $r = f(\theta)$ is a differentiable function of θ for which $f(\theta_0) = 0$ and $f'(\theta_0) \neq 0$, then at $\theta = \theta_0$, (2) gives

$$\frac{dy}{dx} = \frac{f(\theta_0)\cos\theta_0 + f'(\theta_0)\sin\theta_0}{-f(\theta_0)\sin\theta_0 + f'(\theta_0)\cos\theta_0} = \frac{f'(\theta_0)\sin\theta_0}{f'(\theta_0)\cos\theta_0} = \tan\theta_0.$$

In the last expression we recognize that $\tan\theta_0$ is the slope of the tangent line $\theta = \theta_0$.

Note in Example 3 that $r = 3 - 3\sin\theta = 0$ at $\theta = \pi/2$. But since both the numerator $dy/d\theta$ and the denominator $dx/d\theta$ in (2) are 0 at $\theta = \pi/2$ we do not draw any conclusion about the tangent line at the origin $(0, \pi/2)$.

Area of Region The problem of finding the area of a region bounded by polar graphs is not quite as straightforward as it was in Section 6.2. As we shall see in the ensuing discussion, in place of a rectangle we use a sector of a circle, such as shown in FIGURE 10.6.3. Since the area A of a circular sector is proportional to the central angle θ, measured in radians, and since the area of the complete circle is πr^2 we have

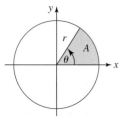

FIGURE 10.6.3 Area A of circular section

$$\frac{A}{\pi r^2} = \frac{\theta}{2\pi} \quad \text{or} \quad A = \frac{1}{2}r^2\theta. \tag{3}$$

Building an Integral

Suppose $r = f(\theta)$ is a nonnegative continuous function on the interval $[\alpha, \beta]$, where $0 \leq \alpha \leq \beta < 2\pi$. To find the area A of the region shown in FIGURE 10.6.4(a) that is bounded by the graph of f and the rays $\theta = \alpha$ and $\theta = \beta$, we start by forming a partition P of $[\alpha, \beta]$:

$$\alpha = \theta_0 < \theta_1 < \theta_2 < \cdots < \theta_n = \beta.$$

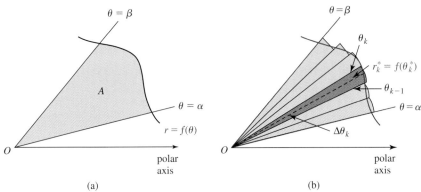

(a) (b)

FIGURE 10.6.4 Area A of a region bounded by a polar graph and two rays

If θ_k^* denotes a sample point in the kth subinterval $[\theta_{k-1}, \theta_k]$, then by (3) the area of the circular sector of radius $r_k = f(\theta_k^*)$ indicated in Figure 10.6.4(b) is

$$A_k = \frac{1}{2}[f(\theta_k^*)]^2 \Delta\theta_k,$$

where $\Delta\theta_k = \theta_k - \theta_{k-1}$ is its central angle. In turn, the Riemann sum

$$\sum_{k=1}^{n} \frac{1}{2}[f(\theta_k^*)]^2 \Delta\theta_k$$

gives an approximation to A. The area A is then given by the limit as $\|P\| \to 0$:

$$A = \lim_{\|P\| \to 0} \sum_{k=1}^{n} \frac{1}{2}[f(\theta_k^*)]^2 \Delta\theta_k.$$

Theorem 10.6.2 Area in Polar Coordinates

If $r = f(\theta)$ is a nonnegative continuous function on $[\alpha, \beta]$, then the **area** bounded by its graph and the rays $\theta = \alpha$ and $\theta = \beta$ is given by

$$A = \int_\alpha^\beta \frac{1}{2}[f(\theta)]^2 \, d\theta = \frac{1}{2} \int_\alpha^\beta r^2 \, d\theta. \tag{4}$$

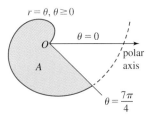

FIGURE 10.6.5 Area in Example 4

EXAMPLE 4 Area Bounded by a Spiral

Find the area of the region that is bounded by the spiral $r = \theta$, $\theta \geq 0$, between the rays $\theta = 0$ and $\theta = 7\pi/4$.

Solution From (4), the area of the shaded region shown in FIGURE 10.6.5 is

$$A = \frac{1}{2} \int_0^{7\pi/4} \theta^2 \, d\theta = \frac{1}{6}\theta^3 \Big]_0^{7\pi/4} = \frac{343}{384}\pi^3 \approx 27.70. \blacksquare$$

EXAMPLE 5 Area Bounded by a Rose Curve

Find the area of one petal of the rose curve $r = 2\cos 5\theta$.

Solution As shown in FIGURE 10.6.6 the rose curve has five petals. Because of symmetry we will find the area of one-half of one petal and multiply the result by 2. By setting $r = 0$ and solving

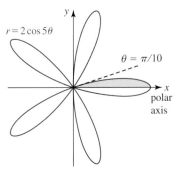

FIGURE 10.6.6 One-half of the area in Example 5

$2\cos 5\theta = 0$ we get $5\theta = \pi/2$ or $\theta = \pi/10$. In other words, the curve enters the origin tangent to the line $\theta = \pi/10$. From (4) the area A of the shaded half-petal in Figure 10.6.6 is then

$$A = \frac{1}{2}\int_0^{\pi/10} (2\cos 5\theta)^2 d\theta$$

$$= 2\int_0^{\pi/10} \cos^2 5\theta\, d\theta$$

$$= 2\int_0^{\pi/10} \frac{1}{2}(1 + \cos 10\theta)\, d\theta$$

$$= \theta + \frac{1}{10}\sin 10\theta \Big]_0^{\pi/10}$$

$$= \frac{\pi}{10}.$$

The area of one petal is then $2(\pi/10) = \pi/5$.

The half-angle formulas:

$$\cos^2\theta = \frac{1}{2}(1 + \cos 2\theta)$$

$$\sin^2\theta = \frac{1}{2}(1 - \cos 2\theta)$$

will be useful in this section.

Of course, the area of each petal in Example 5 is the same and so the area enclosed by the complete five-petal rose curve is $5(\pi/5) = \pi$.

A word of caution. When working problems of the sort in Example 5 be careful with the limits of integration. Do not assume that the area enclosed by the complete five-petal rose curve can be obtained from (4) by integrating on the interval $[0, 2\pi]$. In other words, the area is not $\frac{1}{2}\int_0^{2\pi} (2\cos 5\theta)^2\, d\theta$. This is because the complete curve is obtained for $0 \le \theta \le \pi$. See (i) in *Notes from the Classroom* in Section 10.5.

EXAMPLE 6 Area Bounded Between Two Graphs

Find the area of the region that is common to the interiors of the cardioid $r = 2 - 2\cos\theta$ and the limaçon $r = 2 + \cos\theta$.

Solution Inspection of FIGURE 10.6.7 shows that we need two integrals. Solving the given equations simultaneously:

$$2 - 2\cos\theta = 2 + \cos\theta \quad \text{or} \quad \cos\theta = 0$$

yields $\theta = \pi/2$ so that a point of intersection is $(2, \pi/2)$. By symmetry, it follows that

$$A = 2\left\{\frac{1}{2}\int_0^{\pi/2} (2 - 2\cos\theta)^2\, d\theta + \frac{1}{2}\int_{\pi/2}^{\pi} (2 + \cos\theta)^2\, d\theta\right\}$$

$$= 4\int_0^{\pi/2} (1 - 2\cos\theta + \cos^2\theta)\, d\theta + \int_{\pi/2}^{\pi} (4 + 4\cos\theta + \cos^2\theta)\, d\theta$$

$$= 4\int_0^{\pi/2}\left[1 - 2\cos\theta + \frac{1}{2}(1 + \cos 2\theta)\right] d\theta + \int_{\pi/2}^{\pi}\left[4 + 4\cos\theta + \frac{1}{2}(1 + \cos 2\theta)\right] d\theta$$

$$= 4\left[\frac{3}{2}\theta - 2\sin\theta + \frac{1}{4}\sin 2\theta\right]_0^{\pi/2} + \left[\frac{9}{2}\theta + 4\sin\theta + \frac{1}{4}\sin 2\theta\right]_{\pi/2}^{\pi}$$

$$= \frac{21}{4}\pi - 12 \approx 4.49.$$

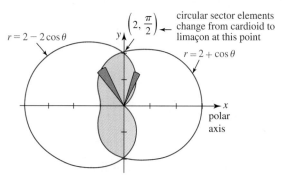

FIGURE 10.6.7 Area in Example 6

■ **Area Bounded by Two Graphs** The area A of the region shown in FIGURE 10.6.8 can be found by subtracting areas. If f and g are continuous on $[\alpha, \beta]$ and $f(\theta) \geq g(\theta)$ on the interval, then the area bounded by the graphs of $r = f(\theta)$, $r = g(\theta)$, $\theta = \alpha$, and $\theta = \beta$ is

$$A = \frac{1}{2}\int_{\alpha}^{\beta} [f(\theta)]^2 \, d\theta - \frac{1}{2}\int_{\alpha}^{\beta} [g(\theta)]^2 \, d\theta.$$

Written as a single integral, the area is given by

$$A = \frac{1}{2}\int_{\alpha}^{\beta} ([f(\theta)]^2 - [g(\theta)]^2) \, d\theta. \tag{5}$$

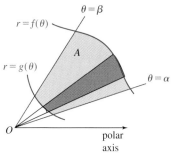

FIGURE 10.6.8 Area of region bounded between two graphs

EXAMPLE 7 Area Bounded by Two Graphs

Find the area of the region in the first quadrant that is outside the circle $r = 1$ and inside the rose curve $r = 2\sin 2\theta$.

Solution Solving the two equations simultaneously:

$$1 = 2\sin 2\theta \quad \text{or} \quad \sin 2\theta = \frac{1}{2}$$

implies that $2\theta = \pi/6$ and $2\theta = 5\pi/6$. Thus, two points of intersection in the first quadrant are $(1, \pi/12)$ and $(1, 5\pi/12)$. The area in question is shown in FIGURE 10.6.9. From (5),

$$A = \frac{1}{2}\int_{\pi/12}^{5\pi/12} [(2\sin 2\theta)^2 - 1^2] \, d\theta$$

$$= \frac{1}{2}\int_{\pi/12}^{5\pi/12} [4\sin^2 2\theta - 1] \, d\theta$$

$$= \frac{1}{2}\int_{\pi/12}^{5\pi/12} \left[4\left(\frac{1 - \cos 4\theta}{2}\right) - 1\right] d\theta$$

$$= \frac{1}{2}\left[\theta - \frac{1}{2}\sin 4\theta\right]_{\pi/12}^{5\pi/12} = \frac{\pi}{6} + \frac{\sqrt{3}}{4} \approx 0.96. \qquad \blacksquare$$

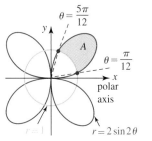

FIGURE 10.6.9 Area in Example 7

■ **Arc Length for Polar Graphs** We have seen that if $r = f(\theta)$ is the equation of a curve C in polar coordinates, then parametric equations of C are

$$x = f(\theta)\cos\theta, \quad y = f(\theta)\sin\theta, \quad \alpha \leq \theta \leq \beta.$$

If f has a continuous derivative, then it is a straightforward matter to derive a formula for arc length in polar coordinates. Since

$$\frac{dx}{d\theta} = f'(\theta)\cos\theta - f(\theta)\sin\theta, \quad \frac{dy}{d\theta} = f'(\theta)\sin\theta + f(\theta)\cos\theta,$$

straightforward algebra shows that

$$\left(\frac{dx}{d\theta}\right)^2 + \left(\frac{dy}{d\theta}\right)^2 = [f(\theta)]^2 + [f'(\theta)]^2 = r^2 + \left(\frac{dr}{d\theta}\right)^2.$$

The next result then follows from (9) of Section 10.3.

Theorem 10.6.3 Length of a Polar Graph

Let f be a function for which f' is continuous on an interval $[\alpha, \beta]$. Then the **length** L of the graph $r = f(\theta)$ on the interval is

$$L = \int_{\alpha}^{\beta} \sqrt{r^2 + \left(\frac{dr}{d\theta}\right)^2} \, d\theta. \tag{6}$$

EXAMPLE 8 Length of a Cardioid

Find the length of the cardioid $r = 1 + \cos\theta$ for $0 \leq \theta \leq \pi$.

Solution The graph of $r = 1 + \cos\theta$ for $0 \leq \theta \leq \pi$ is shown in blue in FIGURE 10.6.10. Now, $dr/d\theta = -\sin\theta$ so that

$$r^2 + \left(\frac{dr}{d\theta}\right)^2 = (1 + 2\cos\theta + \cos^2\theta) + \sin^2\theta = 2 + 2\cos\theta$$

and

$$\sqrt{r^2 + \left(\frac{dr}{d\theta}\right)^2} = \sqrt{2 + 2\cos\theta} = \sqrt{2}\sqrt{1 + \cos\theta}.$$

Hence, from (6) the length of the blue portion of the graph in Figure 10.6.10 is:

$$L = \sqrt{2}\int_0^\pi \sqrt{1 + \cos\theta}\, d\theta.$$

To evaluate this integral, we employ the half-angle formula for the cosine in the form $\cos^2(\theta/2) = \frac{1}{2}(1 + \cos\theta)$ or $1 + \cos\theta = 2\cos^2(\theta/2)$. The length of the graph for $0 \leq \theta \leq \pi$ is given by

$$L = 2\int_0^\pi \cos\frac{\theta}{2}\, d\theta = 4\sin\frac{\theta}{2}\Big]_0^\pi = 4\sin\frac{\pi}{2} = 4.\quad\blacksquare$$

You are encouraged to read the following *Notes from the Classroom*.

$\dfrac{d}{d\theta}$ NOTES FROM THE CLASSROOM

It is easy to make an error in the limits of integration in the area and arc length integrals (4) and (6). In Example 8, we can use symmetry to see that the length of the complete cardioid, that is $r = 1 + \cos\theta$ for $0 \leq \theta \leq 2\pi$ is $2(4) = 8$ units, but yet this is *not* given by (6) by integrating on the interval $0 \leq \theta \leq 2\pi$:

$$L = 2\int_0^{2\pi} \cos(\theta/2)\, d\theta. \tag{7}$$

Think about why an incorrect answer is obtained from (7) and then work Problems 45 and 46 in Exercises 10.6.

Exercises 10.6 Answers to selected odd-numbered problems begin on page ANS-33.

≡ Fundamentals

In Problems 1–6, find the slope of the tangent line at the indicated value of θ.

1. $r = \theta; \quad \theta = \pi/2$
2. $r = 1/\theta; \quad \theta = 3$
3. $r = 4 - 2\sin\theta; \quad \theta = \pi/3$
4. $r = 1 - \cos\theta; \quad \theta = 3\pi/4$
5. $r = \sin\theta; \quad \theta = \pi/6$
6. $r = 10\cos\theta; \quad \theta = \pi/4$

In Problems 7 and 8, find the points on the graph of the given equation at which the tangent line is horizontal and the points at which the tangent line is vertical.

7. $r = 2 + 2\cos\theta$
8. $r = 1 - \sin\theta$

In Problems 9 and 10, find a rectangular equation of the tangent line at the indicated point.

9. $r = 4\cos 3\theta$
10. $r = 1 + 2\cos\theta$

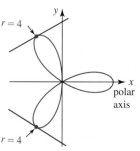

FIGURE 10.6.11 Graph for Problem 9

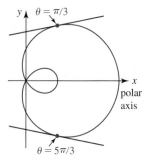

FIGURE 10.6.12 Graph for Problem 10

In Problems 11–16, find a polar equation of each tangent line to the polar graph at the origin.

11. $r = -2\sin\theta$
12. $r = 3\cos\theta$
13. $r = 1 + \sqrt{2}\sin\theta$
14. $r = 1 - 2\sin\theta$
15. $r = 2\cos 5\theta$
16. $r = 2\sin 2\theta$

In Problems 17–24, find the area of the region that is bounded by the graph of the given polar equation.

17. $r = 2\sin\theta$
18. $r = 10\cos\theta$
19. $r = 4 + 4\cos\theta$
20. $r = 1 - \sin\theta$
21. $r = 3 + 2\sin\theta$
22. $r = 2 + \cos\theta$
23. $r = 3\sin 2\theta$
24. $r = \cos 3\theta$

In Problems 25–30, find the area of the region that is bounded by the graph of the given polar equation and the indicated rays.

25. $r = 2\theta, \theta \geq 0, \theta = 0, \theta = 3\pi/2$
26. $r\theta = \pi, \theta > 0, \theta = \pi/2, \theta = \pi$
27. $r = e^\theta, \theta = 0, \theta = \pi$
28. $r = 10e^{-\theta}, \theta = 1, \theta = 2$
29. $r = \tan\theta, \theta = 0, \theta = \pi/4$
30. $r\sin\theta = 5, \theta = \pi/6, \theta = \pi/3$

In Problems 31 and 32, the graph is that of the polar equation $r = 1 + 2\cos\theta$. Find the area of the shaded region.

31.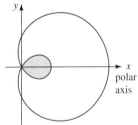

FIGURE 10.6.13 Region for Problem 31

32.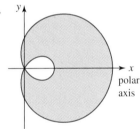

FIGURE 10.6.14 Region for Problem 32

In Problems 33–38, find the area of the described region.

33. Outside the circle $r = 1$ and inside the rose curve $r = 2\cos 3\theta$
34. Common to the interiors of the circles $r = \cos\theta$ and $r = \sin\theta$
35. Inside the circle $r = 5\sin\theta$ and outside the limaçon $r = 3 - \sin\theta$
36. Common to the interiors of the graphs of the equations in Problem 35
37. Inside the cardioid $r = 4 - 4\cos\theta$ and outside the circle $r = 6$
38. Common to the interiors of the graphs of the equations in Problem 37

In Problems 39–44, find the length of the curve for the indicated values of θ.

39. $r = 3, 0 \leq \theta \leq 2\pi$
40. $r = 6\cos\theta$, complete graph
41. $r = e^{\theta/2}, 0 \leq \theta \leq 4$
42. $r = \theta, 0 \leq \theta \leq 1$
43. $r = 3 - 3\cos\theta$, complete graph
44. $r = \sin^3(\theta/3), 0 \leq \theta \leq \pi$

Think About It

45. Consider the lemniscate $r^2 = 9\cos 2\theta$.
    (a) Explain why the area of the region bounded by the graph is not given by the integral $\frac{1}{2}\int_0^{2\pi} 9\cos 2\theta\, d\theta$.
    (b) Using an appropriate integral, find the area of the region bounded by the graph.

46. In Example 8, explain why the length of the complete cardioid $r = 1 + \cos\theta, 0 \leq \theta \leq 2\pi$, is not given by the integral $2\int_0^{2\pi}\cos(\theta/2)\,d\theta$. Then reexamine Problem 43 and explain why there are no difficulties in integrating on the interval $[0, 2\pi]$.

47. Sketch the region common to the interiors of the graphs of $r = \sin 2\theta$ and $r = \cos 2\theta$. Find the area of this region.

48. The area of the region that is bounded by the graph of $r = 1 + \cos\theta$ is $3\pi/2$. What can you say about the areas bounded by the graphs of $r = 1 - \cos\theta$, $r = 1 + \sin\theta$, and $r = 1 - \sin\theta$? Justify your answer without calculating the areas using (4).

49. Is the area of the region bounded by the graph of $r = 2(1 + \cos\theta)$ equal to twice the area of the region bounded by the graph of $r = 1 + \cos\theta$?

50. Find the area of the shaded region in FIGURE 10.6.15. Each circle has radius 1.

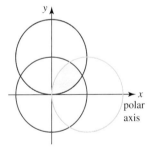

FIGURE 10.6.15 Intersecting circles in Problem 50

Projects

51. **Kepler's Second Law** In polar coordinates the **angular momentum** of a moving particle of mass m is defined to be $L = mr^2\, d\theta/dt$. Assume that the polar coordinates of a planet of mass m are (r_1, θ_1) and (r_2, θ_2) at times $t = a$ and $t = b, a < b$, respectively. Since the gravitational force acting on the planet is a central force, the angular momentum L of the planet is a constant. Use this fact to show that the area A swept out by r is $A = L(b - a)/2m$.

When the Sun is taken to be at the origin, this equation proves Kepler's second law of planetary motion:

- A line joining a planet with the Sun sweeps out equal areas in equal time intervals.

See FIGURE 10.6.16.

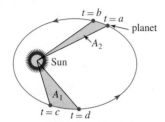

$A_1 = A_2$ when $b - a = d - c$

FIGURE 10.6.16 Orbit of planet in Problem 51

52. **A Bit of History—For the Record** Prior to iPods, MP3 players, and CDs, music was obtained by playing a *record*. In the years 1960–1990 the popular format was the LP (long-playing) record that revolved on a record player at the rate of 33 revolutions per minute.* Although they can now be found in stores specializing in collectible objects, many of us still have collections of these large black vinyl 33-rpm records stored in boxes in the garage. Sound was encoded on these disks by mechanical means along a continuous groove. When a record was played, a needle started at a point near the outer edge of the disk and traversed the groove up to a point near its center. How long is the groove of a record? Suppose a record plays for 20 min at 33 revolutions per minute. As the record plays, the needle goes from an outer radius R_o to an inner radius R_i. See FIGURE 10.6.17. Assume that the groove of the record is a spiral that can be described by

---

*The 33s actually revolved at $33\frac{1}{3}$ revolutions per minute.

a polar equation of the form $r = R_o - k\theta$, where k is a constant and θ is measured in radians.

(a) Express k in terms of R_o, R_i, and N, where N is the number of revolutions completed by the record.

(b) Show that the length L of the record groove is given by
$$L = \frac{1}{k}\int_{R_i}^{R_o} \sqrt{k^2 + u^2}\, du.$$

(c) Use a binomial series to establish the approximation
$$\sqrt{k^2 + u^2} \approx u\left[1 + \frac{1}{2}\left(\frac{k}{u}\right)^2\right].$$

(d) In part (b), use the result obtained in part (c) to show that the length L of the record groove is given by the approximation
$$L \approx \pi N(R_i + R_o) + \frac{R_o - R_i}{4\pi N} \ln\frac{R_o}{R_i}.$$

(e) Use the result in part (d) to approximate the length L if $R_o = 6$ in. and $R_i = 2.5$ in.

(f) Use an appropriate substitution to evaluate the integral in part (b) using the specified values of R_o and R_i given in part (e). Compare this answer with that obtained in part (e).

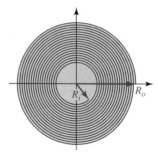

FIGURE 10.6.17 LP record in Problem 52

10.7 Conic Sections in Polar Coordinates

Introduction In the first section of this chapter we derived equations for the parabola, ellipse, and hyperbola using the distance formula in rectangular coordinates. By using polar coordinates and the concept of eccentricity, we can now give one general definition of a conic section that encompasses all three curves.

Definition 10.7.1 Conic Section

Let L be a fixed line in the plane, and let F be a point not on the line. A **conic section** is the set of points P in the plane for which the distance from P to F divided by the distance from P to L is a constant.

The fixed line L is called a **directrix**, and the point F is a **focus**. The fixed constant is the **eccentricity** e of the conic. As FIGURE 10.7.1 shows, the point P lies on the conic if and only if

$$\frac{d(P, F)}{d(P, Q)} = e, \qquad (1)$$

where Q denotes the foot of the perpendicular from P to L. In (1), if

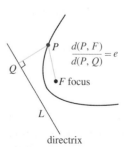

FIGURE 10.7.1 Geometric interpretation of (1)

- $e = 1$, the conic is a **parabola**,
- $0 < e < 1$, the conic is an **ellipse**, and
- $e > 1$, the conic is a **hyperbola**.

FIGURE 10.7.2 Polar coordinate interpretation of (2)

■ **Polar Equations of Conics** Equation (1) is readily interpreted using polar coordinates. Suppose the focus F is placed at the pole and the directrix L is d units ($d > 0$) to the *left* of F perpendicular to the extended polar axis. We see from FIGURE 10.7.2 that (1) written as $d(P, F) = ed(P, Q)$ is the same as

$$r = e(d + r\cos\theta) \quad \text{or} \quad r - er\cos\theta = ed. \tag{2}$$

Solving for r yields

$$r = \frac{ed}{1 - e\cos\theta}. \tag{3}$$

To see that (3) yields the familiar equations of the conics, let us superimpose a rectangular coordinate system on the polar coordinate system with origin at the pole and the positive x-axis coinciding with the polar axis. We then express the first equation in (2) in rectangular coordinates and simplify:

$$\pm\sqrt{x^2 + y^2} = ex + ed$$
$$x^2 + y^2 = e^2x^2 + 2e^2dx + e^2d^2$$
$$(1 - e^2)x^2 - 2e^2dx + y^2 = e^2d^2. \tag{4}$$

Choosing $e = 1$, (4) becomes

$$-2dx + y^2 = d^2 \quad \text{or} \quad y^2 = 2d\left(x + \frac{d}{2}\right),$$

which is an equation in standard form of a parabola whose axis is the x-axis, vertex is at $(-d/2, 0)$ and, consistent with the placement of F, whose focus is at the origin.

It is a good exercise in algebra to show that (2) yields standard form equations of an ellipse in the case $0 < e < 1$ and a hyperbola in the case $e > 1$. See Problem 43 in Exercises 10.7. Thus, depending on the value of e, the polar equation (3) can have three possible graphs as shown in FIGURE 10.7.3.

If we place the directrix L to the *right* of the focus F at the origin in our derivation of the polar equation (3), then the resulting equation would be $r = ed/(1 + e\cos\theta)$. When the directrix L is chosen parallel to the polar axis, that is, horizontal, then the equation of the conic is found to be either $r = ed/(1 - e\sin\theta)$ (directrix below the origin) or $r = ed/(1 + e\sin\theta)$ (directrix above the origin). A summary of the preceding discussion is given in the next theorem.

(a) $e = 1$

(b) $0 < e < 1$

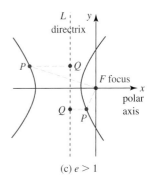

(c) $e > 1$

FIGURE 10.7.3 Graphs of equation (3); directrix L to the left of F

Theorem 10.7.1 Polar Equations of Conics

Any polar equation of the form

$$r = \frac{ed}{1 \pm e\cos\theta} \tag{5}$$

or

$$r = \frac{ed}{1 \pm e\sin\theta} \tag{6}$$

is a **conic section** with focus at the origin and directrix d units from the origin and either perpendicular (in the case of (5)) or parallel (in the case of (6)) to the x-axis. The conic is a parabola if $e = 1$, an ellipse if $0 < e < 1$, and a hyperbola if $e > 1$.

EXAMPLE 1 Identifying Conics

Identify each of the following conics

(a) $r = \dfrac{2}{1 - 2\sin\theta}$ (b) $r = \dfrac{3}{4 + \cos\theta}$.

Solution

(a) A term-by-term comparison of the given equation with the polar form $r = ed/(1 - e\sin\theta)$ enables us to make the identification $e = 2$. Hence the conic is a hyperbola.

(b) In order to identify the conic section, we divide the numerator and the denominator of the given equation by 4. This puts the equation into the form

$$r = \frac{\tfrac{3}{4}}{1 + \tfrac{1}{4}\cos\theta}.$$

Then by comparison with $r = ed/(1 + e\cos\theta)$ we see that $e = \tfrac{1}{4}$. Hence the conic is an ellipse. ∎

Graphs A rough graph of a conic defined by (5) or (6) can be obtained by knowing the orientation of its axis, finding the x- and y-intercepts, and finding the vertices. In the cases of equations (5) and (6) we have, respectively:

- the two vertices of the **ellipse** or a **hyperbola** occur at $\theta = 0$ and $\theta = \pi$; the vertex of a **parabola** can occur at only one of the values: $\theta = 0$ or $\theta = \pi$.
- the two vertices of an **ellipse** or a **hyperbola** occur at $\theta = \pi/2$ and $\theta = 3\pi/2$; the vertex of a **parabola** can occur at only one of the values: $\theta = \pi/2$ or $\theta = 3\pi/2$.

EXAMPLE 2 Graphing a Conic

Graph $r = \dfrac{4}{3 - 2\sin\theta}$.

Solution By writing the equation as $r = \dfrac{\tfrac{4}{3}}{1 - \tfrac{2}{3}\sin\theta}$ we see that the eccentricity is $e = \tfrac{2}{3}$ and so the conic is an ellipse. Moreover, because the equation is of the form given in (6), we know that the directrix is parallel to the x-axis. Now in view of the discussion preceding this example, we obtain:

$$\text{vertices: } (4, \pi/2), \left(\tfrac{4}{5}, 3\pi/2\right)$$

$$x\text{-intercepts: } \left(\tfrac{4}{3}, 0\right), \left(\tfrac{4}{3}, \pi\right).$$

As seen in FIGURE 10.7.4 the major axis of the ellipse lies along the y-axis. ∎

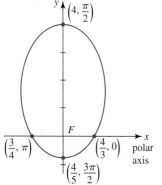

FIGURE 10.7.4 Graph of polar equation in Example 2

EXAMPLE 3 Graphing a Conic

Graph $r = \dfrac{1}{1 - \cos\theta}$.

Solution Inspection of the equation shows that it is of the form given in (5) with $e = 1$. Hence, the conic section is a parabola whose directrix is perpendicular to the x-axis. Since r is undefined at $\theta = 0$, the vertex of the parabola occurs at $\theta = \pi$:

$$\text{vertex: } \left(\tfrac{1}{2}, \pi\right)$$

$$y\text{-intercepts: } (1, \pi/2), (1, 3\pi/2).$$

As seen in FIGURE 10.7.5 the axis of symmetry of the parabola lies along the x-axis. ∎

FIGURE 10.7.5 Graph of polar equation in Example 3

EXAMPLE 4 Graphing a Conic

Graph $r = \dfrac{2}{1 + 2\cos\theta}$.

Solution From (5) we see that $e = 2$ and so the conic section is a hyperbola whose directrix is perpendicular to the x-axis. The vertices occur at $\theta = 0$ and at $\theta = \pi$:

$$\text{vertices: } \left(\tfrac{2}{3}, 0\right), (-2, \pi)$$
$$y\text{-intercepts: } (2, \pi/2), (2, 3\pi/2).$$

As seen in FIGURE 10.7.6 the transverse axis of the hyperbola lies along the x-axis.

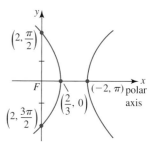

FIGURE 10.7.6 Graph of polar equation in Example 4

Rotated Conics We saw in Section 10.5 that graphs of $r = f(\theta - \gamma)$ and $r = f(\theta + \gamma)$, $\gamma > 0$, are rotations of the graph of the polar equation $r = f(\theta)$ about the origin by an amount γ. Thus,

$$\left.\begin{array}{l} r = \dfrac{ed}{1 \pm e\cos(\theta - \gamma)} \\ r = \dfrac{ed}{1 \pm e\sin(\theta - \gamma)} \end{array}\right\} \begin{array}{l}\text{conics rotated}\\ \text{counterclockwise}\\ \text{about the origin}\end{array} \qquad \left.\begin{array}{l} r = \dfrac{ed}{1 \pm e\cos(\theta + \gamma)} \\ r = \dfrac{ed}{1 \pm e\sin(\theta + \gamma)} \end{array}\right\} \begin{array}{l}\text{conics rotated}\\ \text{clockwise}\\ \text{about the origin}\end{array}$$

EXAMPLE 5 Rotated Conic

In Example 2 we saw that the graph of $r = \dfrac{4}{3 - 2\sin\theta}$ is an ellipse with major axis along the y-axis. This is the blue graph in FIGURE 10.7.7. The graph of $r = \dfrac{4}{3 - 2\sin(\theta - 2\pi/3)}$ is the red graph in Figure 10.7.7 and is a counter-clockwise rotation of the blue graph by the amount $2\pi/3$ (or $120°$) about the origin. The major axis of the red graph lies along the line $\theta = 7\pi/6$.

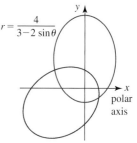

FIGURE 10.7.7 Graphs of polar equations in Example 5

Applications Equations of the type in (5) and (6) are well suited to describe a closed orbit of a satellite around the Sun (Earth or Moon) since such an orbit is an ellipse with the Sun (Earth or Moon) at one focus. Suppose that an equation of the orbit is given by $r = ed/(1 - e\cos\theta)$, $0 < e < 1$, and r_p is the value of r at perihelion (perigee or perilune) and r_a is the value of r at aphelion (apogee or apolune). These are the points in the orbit, occurring on the x-axis, at which the satellite is closest and farthest, respectively, from the Sun (Earth or Moon). See FIGURE 10.7.8. It is left as an exercise to show that the eccentricity e of the orbit is related to r_p and r_a by

$$e = \dfrac{r_a - r_p}{r_a + r_p}. \tag{7}$$

See Problem 44 in Exercises 10.7.

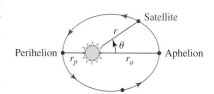

FIGURE 10.7.8 Orbit of satellite around the Sun

EXAMPLE 6 Finding a Polar Equation of an Orbit

Find a polar equation of the orbit of the planet Mercury around the Sun if $r_p = 2.85 \times 10^7$ mi and $r_a = 4.36 \times 10^7$ mi.

Solution From (7), the eccentricity of Mercury's orbit is

$$e = \dfrac{4.36 \times 10^7 - 2.85 \times 10^7}{4.36 \times 10^7 + 2.85 \times 10^7} = 0.21.$$

Hence, $\qquad r = \dfrac{0.21d}{1 - 0.21\cos\theta}.$

Mercury is the closest planet to the Sun

All we need do now is to solve for the quantity $0.21d$. To do this we use the fact that aphelion occurs at $\theta = 0$:

$$4.36 \times 10^7 = \frac{0.21d}{1 - 0.21}.$$

Solving the last equation for the quantity $0.21d$ yields $0.21d = 3.44 \times 10^7$. Hence, a polar equation of Mercury's orbit is

$$r = \frac{3.44 \times 10^7}{1 - 0.21\cos\theta}.$$

Exercises 10.7
Answers to selected odd-numbered problems begin on page ANS-33.

≡ Fundamentals

In Problems 1–10, determine the eccentricity, identify the conic section, and sketch its graph.

1. $r = \dfrac{2}{1 + \cos\theta}$
2. $r = \dfrac{2}{2 - \sin\theta}$
3. $r = \dfrac{15}{4 - \cos\theta}$
4. $r = \dfrac{5}{2 - 2\sin\theta}$
5. $r = \dfrac{4}{1 + 2\sin\theta}$
6. $r = \dfrac{12}{6 + 2\sin\theta}$
7. $r = \dfrac{18}{3 + 6\cos\theta}$
8. $r = \dfrac{6\sec\theta}{\sec\theta - 1}$
9. $r = \dfrac{10}{5 + 4\sin\theta}$
10. $r = \dfrac{2}{2 + 5\cos\theta}$

In Problems 11–14, determine the eccentricity e of the given conic. Then convert the polar equation to a rectangular equation and verify that $e = c/a$.

11. $r = \dfrac{6}{1 + 2\sin\theta}$
12. $r = \dfrac{10}{2 - 3\cos\theta}$
13. $r = \dfrac{12}{3 - 2\cos\theta}$
14. $r = \dfrac{2\sqrt{3}}{\sqrt{3} + \sin\theta}$

In Problems 15–20, find a polar equation of the conic with focus at the origin that satisfies the given conditions.

15. $e = 1$, directrix $x = 3$
16. $e = \frac{3}{2}$, directrix $y = 2$
17. $e = \frac{2}{3}$, directrix $y = -2$
18. $e = \frac{1}{2}$, directrix $x = 4$
19. $e = 2$, directrix $x = 6$
20. $e = 1$, directrix $y = -2$

21. Find a polar equation of the conic in Problem 15 if the graph is rotated clockwise about the origin by an amount $2\pi/3$.

22. Find a polar equation of the conic in Problem 16 if the graph is rotated counterclockwise about the origin by an amount $\pi/6$.

In Problems 23–28, find a polar equation of the parabola with focus at the origin and the given vertex.

23. $(\frac{3}{2}, 3\pi/2)$
24. $(2, \pi)$
25. $(\frac{1}{2}, \pi)$
26. $(2, 0)$
27. $(\frac{1}{4}, 3\pi/2)$
28. $(\frac{3}{2}, \pi/2)$

In Problems 29–32, find the polar coordinates of the vertex or vertices of the given rotated conic.

29. $r = \dfrac{4}{1 + \cos(\theta - \pi/4)}$
30. $r = \dfrac{5}{3 + 2\cos(\theta - \pi/3)}$
31. $r = \dfrac{10}{2 - \sin(\theta + \pi/6)}$
32. $r = \dfrac{6}{1 + 2\sin(\theta + \pi/3)}$

≡ Applications

33. A communications satellite is 12,000 km above the Earth at its apogee. The eccentricity of its orbit is 0.2. Use (7) to find the perigee distance.

34. Find a polar equation $r = ed/(1 - e\cos\theta)$ of the orbit of the satellite in Problem 33.

35. Find a polar equation of the orbit of the Earth around the Sun if $r_p = 1.47 \times 10^8$ km and $r_a = 1.52 \times 10^8$ km.

36. (a) The eccentricity of the elliptical orbit of Comet Halley is 0.97 and the length of the major axis of its orbit is 3.34×10^9 mi. Find a polar equation of its orbit of the form $r = ed/(1 - e\cos\theta)$.
    (b) Use the equation in part (a) to obtain r_p and r_a for the orbit of Comet Halley.

≡ Calculator/CAS Problems

The orbital characteristics (eccentricity, perigee, and major axis) of a satellite near the Earth gradually degrade over time due to many small forces acting on the satellite other than the gravitational force of the Earth. These forces include atmospheric drag, the gravitational attractions of the Sun and the Moon, and magnetic forces. Approximately once a month tiny rockets are activated for a few seconds in order to "boost" the orbital characteristics back into the desired range. Rockets are turned on longer to a major change in the orbit of a satellite. The most fuel-efficient way to move from an inner orbit to an outer orbit, called a **Hohmann transfer**, is to add velocity in the direction of flight at the time the satellite reaches perigee on the inner orbit, follow the Hohmann transfer ellipse halfway around to its apogee, and add velocity again to achieve the outer orbit. A similar process (subtracting velocity at apogee on the outer orbit and subtracting velocity at perigee on the Hohmann transfer orbit) moves a satellite from an outer orbit to an inner orbit.

In Problems 37–40, use a calculator or CAS to superimpose the graphs of the given three polar equations on the same coordinate axes. Print out your result and use a colored pencil to trace out the Hohmann transfer.

37. Inner orbit $r = \dfrac{24}{1 + 0.2\cos\theta}$,

    Hohmann transfer $r = \dfrac{32}{1 + 0.6\cos\theta}$,

    outer orbit $r = \dfrac{56}{1 + 0.3\cos\theta}$

38. Inner orbit $r = \dfrac{5.5}{1 + 0.1\cos\theta}$,

    Hohmann transfer $r = \dfrac{7.5}{1 + 0.5\cos\theta}$,

    outer orbit $r = \dfrac{13.5}{1 + 0.1\cos\theta}$

39. Inner orbit $r = 9$,

    Hohmann transfer $r = \dfrac{15.3}{1 + 0.7\cos\theta}$,

    outer orbit $r = 51$

40. Inner orbit $r = \dfrac{73.5}{1 + 0.05\cos\theta}$,

    Hohmann transfer $r = \dfrac{77}{1 + 0.1\cos\theta}$,

    outer orbit $r = \dfrac{84.7}{1 + 0.01\cos\theta}$

In Problems 41 and 42, use a calculator or CAS to superimpose the graphs of the given two polar equations on the same coordinate axes.

41. $r = \dfrac{4}{4 + 3\cos\theta}$; $r = \dfrac{4}{4 + 3\cos(\theta - \pi/3)}$

42. $r = \dfrac{2}{1 - \sin\theta}$; $r = \dfrac{2}{1 - \sin(\theta + 3\pi/4)}$

≡ Think About It

43. Show that (2) yields standard form equations of an ellipse in the case $0 < e < 1$ and a hyperbola in the case $e > 1$.

44. Use the equation $r = ed/(1 - e\cos\theta)$ to derive the result in (7).

Chapter 10 in Review

Answers to selected odd-numbered problems begin on page ANS-34.

A. True/False

In Problems 1–26, indicate whether the given statement is true or false.

1. For a parabola, the distance from the vertex to the focus is the same as the distance from the vertex to the directrix. _____
2. The minor axis of an ellipse bisects the major axis. _____
3. The asymptotes of $x^2/a^2 - y^2/a^2 = 1$ are perpendicular. _____
4. The y-intercepts of the graph of $x^2/a^2 - y^2/b^2 = 1$ are $(0, b)$ and $(0, -b)$. _____
5. The point $(-2, 5)$ is on the ellipse $x^2/8 + y^2/50 = 1$. _____
6. The graphs of $y = x^2$ and $y^2 - x^2 = 1$ have at most two points in common. _____
7. If for all values of θ the points $(-r, \theta)$ and $(r, \theta + \pi)$ are on the graph of the polar equation $r = f(\theta)$, then the graph is symmetric with respect to the origin. _____
8. The graph of the curve $x = t^2$, $y = t^4 + 1$ is the same as the graph of $y = x^2 + 1$. _____
9. The graph of the curve $x = t^2 + t - 12$, $y = t^3 - 7t$ crosses the y-axis at $(0, 6)$. _____
10. $(3, \pi/6)$ and $(-3, -5\pi/6)$ are polar coordinates of the same point. _____
11. Rectangular coordinates of a point in the plane are unique. _____
12. The graph of the rose curve $r = 5\sin 6\theta$ has six "petals." _____
13. The point $(4, 3\pi/2)$ is not on the graph of $r = 4\cos 2\theta$, since its coordinates do not satisfy the equation. _____
14. The eccentricity of a parabola is $e = 1$. _____
15. The transverse axis of the hyperbola $r = 5/(2 + 3\cos\theta)$ lies along the x-axis. _____
16. The graph of the ellipse $r = 90/(15 - \sin\theta)$ is nearly circular. _____
17. The rectangular coordinates of the point $\left(-\sqrt{2}, 5\pi/4\right)$ in polar coordinates are $(1, 1)$. _____
18. The graph of the polar equation $r = -5\sec\theta$ is a line. _____

19. The terminal side of the angle θ is always in the same quadrant as the point (r, θ). _____
20. The slope of the tangent to the graph of $r = e^\theta$ at $\theta = \pi/2$ is -1. _____
21. The graphs of the cardioids $r = 3 + 3\cos\theta$ and $r = -3 + 3\cos\theta$ are the same. _____
22. The area bounded by $r = \cos 2\theta$ is $2\int_{-\pi/4}^{\pi/4}\cos^2 2\theta\, d\theta$. _____
23. The area bounded by $r = 2\sin 3\theta$ is $6\int_0^{\pi/3}\sin^2 3\theta\, d\theta$. _____
24. The area bounded by $r = 1 - 2\cos\theta$ is $\frac{1}{2}\int_0^{2\pi}(1 - 2\cos\theta)^2\, d\theta$. _____
25. The area bounded by $r^2 = 36\cos 2\theta$ is $18\int_0^{2\pi}\cos 2\theta\, d\theta$. _____
26. The θ-coordinate of a point of intersection of the graphs of the polar equations $r = f(\theta)$ and $r = g(\theta)$ must satisfy the equation $f(\theta) = g(\theta)$. _____

B. Fill in the Blanks

In Problems 1–22, fill in the blanks.

1. $y = 2x^2$, focus _____
2. $\dfrac{x^2}{4} - \dfrac{y^2}{12} = 1$, foci _____
3. $4x^2 + 5(y + 3)^2 = 20$, center _____
4. $25y^2 - 4x^2 = 100$, asymptotes _____
5. $8(y + 3) = (x - 1)^2$, directrix _____
6. $\dfrac{(x + 1)^2}{36} + \dfrac{(y + 7)^2}{16} = 1$, vertices _____
7. $x = y^2 + 4y - 6$, vertex _____
8. $x^2 - 2y^2 = 18$, length of conjugate axis _____
9. $(x - 4)^2 - (y + 1)^2 = 4$, endpoints of transverse axis _____
10. $\dfrac{(x - 3)^2}{7} + \dfrac{(y + 3/2)^2}{8} = 1$, equation of line containing major axis _____
11. $25x^2 + y^2 - 200x + 6y + 384 = 0$, center _____
12. $(x + 1)^2 + (y + 8)^2 = 100$, x-intercepts _____
13. $y^2 - (x - 2)^2 = 1$, y-intercepts _____
14. $y^2 - y + 3x = 3$, slope of tangent line at $(1, 1)$ _____
15. $x = t^3, y = 4t^3$, name of rectangular graph _____
16. $x = t^2 - 1, y = t^3 + t + 1$, y-intercepts _____
17. $r = -2\cos\theta$, name of polar graph _____
18. $r = 2 + \sin\theta$, name of polar graph _____
19. $r = \sin 3\theta$, tangents to the graph at the origin _____
20. $r = \dfrac{1}{2 + 5\sin\theta}$, eccentricity _____
21. $r = \dfrac{10}{1 - \sin\theta}$, focus _____ and vertex _____
22. $r = \dfrac{12}{2 + \cos\theta}$, center _____, foci _____, vertices _____

C. Exercises

1. Find an equation of the line that is normal to the graph of the curve $x = t - \sin t$, $y = 1 - \cos t, 0 \le t \le 2\pi$, at $t = \pi/3$.
2. Find the length of the curve given in Problem 1.
3. Find the points on the graph of the curve $x = t^2 + 4, y = t^3 - 9t^2 + 2$ at which the tangent line is parallel to $6x + y = 8$.
4. Find the points on the graph of the curve $x = t^2 + 1, y = 2t$ at which the tangent line passes through $(1, 5)$.

5. Consider the rectangular equation $y^2 = 4x^2(1 - x^2)$.
   (a) Explain why it is necessary that $|x| \leq 1$.
   (b) If $x = \sin t$, then $|x| \leq 1$. Find parametric equations that have the same graph as the given rectangular equation.
   (c) Using parametric equations, find the points on the graph of the rectangular equation at which the tangent is horizontal.
   (d) Sketch the graph of the rectangular equation.

6. Find the area of the region that is outside the circle $r = 4\cos\theta$ and inside the limaçon $r = 3 + \cos\theta$.

7. Find the area of the region that is common to the interiors of the circle $r = 3\sin\theta$ and the cardioid $r = 1 + \sin\theta$.

8. In polar coordinates, sketch the region whose area A is described by $A = \int_0^{\pi/2}(25 - 25\sin^2\theta)\,d\theta$.

9. Find (a) a rectangular equation and (b) a polar equation of the tangent line to the graph of $r = 2\sin 2\theta$ at $\theta = \pi/4$.

10. Determine the rectangular coordinates of the vertices of the ellipse whose polar equation is $r = 2/(2 - \sin\theta)$.

In Problems 11 and 12, find a rectangular equation that has the same graph as the given polar equation

11. $r = \cos\theta + \sin\theta$ 12. $r = \sec\theta - 5\cos\theta$

In Problems 13 and 14, find a polar equation that has the same graph as the given rectangular equation

13. $2xy = 5$ 14. $(x^2 + y^2 - 2x)^2 = 9(x^2 + y^2)$

15. Find a polar equation for the set of points that are equidistant from the origin (pole) and the line $r = -\sec\theta$.

16. Find a polar equation of the hyperbola with focus at the origin, vertices (in rectangular coordinates) $(0, -\frac{4}{3})$ and $(0, -4)$, and eccentricity 2.

In Problems 17 and 18, write an equation of the given polar graph.

17.
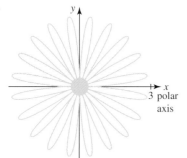

FIGURE 10.R.1 Graph for Problem 17

18.
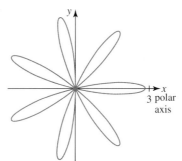

FIGURE 10.R.2 Graph for Problem 18

19. Find an equation of the hyperbola that has asymptotes $3y = 5x$ and $3y = -5x$ and vertices $(0, 10)$ and $(0, -10)$.

20. Find a rectangular equation of the tangent line to the graph of $r = 1/(1 + \cos\theta)$ at $\theta = \pi/2$.

21. The folium of Descartes, first discussed in Section 3.6, has the rectangular equation $x^3 + y^3 = 3axy$, where $a > 0$ is a constant. Use the substitution $y = tx$ to find parametric equations for the curve. See FIGURE 10.R.3.

22. Use the parametric equations found in Problem 21 to find the points on the folium of Descartes where the tangent line is horizontal. See Figure 10.R.3.

23. (a) Find a polar equation for the folium of Descartes in Problem 21.
    (b) Use the polar equation to find the area of the shaded loop in the first quadrant in Figure 10.R.3. [*Hint*: Let $u = \tan\theta$.]

24. Use the parametric equations found in Problem 21 to show that the folium of Descartes has the slant asymptote $x + y + a = 0$. This is the red dashed line in Figure 10.R.3. [*Hint*: Consider what happens to x, y, and $x + y$ as $r \to -1$.]

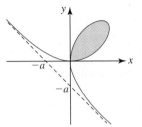

FIGURE 10.R.3 Graph for Problems 21–24

25. The graph of $r = 2\sin(\theta/3)$ given in FIGURE 10.R.4 resembles a limaçon with an interior loop. Find the area of the interior loop.

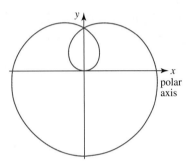

FIGURE 10.R.4 Graph for Problem 25

26. Find the area of the shaded region in FIGURE 10.R.5. Each circle has radius 1.

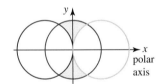

FIGURE 10.R.5 Graph for Problem 26

In Problems 27 and 28, the graph of the given polar equation is rotated about the origin by the indicated amount.

(a) Find a polar equation of the new graph.
(b) Find a rectangular equation for the new graph.

27. $r = 2\cos\theta$; counterclockwise, $\pi/4$
28. $r = 1/(1 + \cos\theta)$; clockwise, $\pi/6$
29. A satellite revolves around the planet Jupiter in an elliptical orbit with the center of the planet at one focus. The length of the major axis of the orbit is 10^9 m and the length of the minor axis is 6×10^8 m. Find the minimum distance between the satellite and the center of Jupiter. What is the maximum distance?
30. Find the width w of each petal of the rose curve $r = \cos 2\theta$. One petal is shown in FIGURE 10.R.6.

(a)   (b)

FIGURE 10.R.6 Graph for Problem 30

Appendix

Proofs of Selected Theorems

■ Section 2.2

PROOF OF THEOREM 2.2.1(*i*): Let $\varepsilon > 0$ be given. To prove (*i*) we must find a $\delta > 0$ so that

$$|c - c| < \varepsilon \quad \text{whenever} \quad 0 < |x - a| < \delta.$$

Since $|c - c| = 0$, the preceding line is equivalent to

$$\varepsilon > 0 \quad \text{whenever} \quad 0 < |x - a| < \delta.$$

The last statement is always true for any choice of $\delta > 0$. ∎

PROOF OF THEOREM 2.2.3(*i*): Let $\varepsilon > 0$ be given. To prove (*i*) we must find a $\delta > 0$ so that

$$|f(x) + g(x) - L_1 - L_2| < \varepsilon \quad \text{whenever} \quad 0 < |x - a| < \delta.$$

Since $\lim_{x \to a} f(x) = L_1$ and $\lim_{x \to a} g(x) = L_2$, we know there exist numbers $\delta_1 > 0$ and $\delta_2 > 0$ for which

$$|f(x) - L_1| < \frac{\varepsilon}{2} \quad \text{whenever} \quad 0 < |x - a| < \delta_1, \tag{1}$$

and

$$|g(x) - L_2| < \frac{\varepsilon}{2} \quad \text{whenever} \quad 0 < |x - a| < \delta_2. \tag{2}$$

Now, if we choose δ to be the smallest number in the set of positive numbers $\{\delta_1, \delta_2\}$, then (1) and (2) *both* hold and so

$$|f(x) + g(x) - L_1 - L_2| = |f(x) - L_1 + g(x) - L_2|$$
$$\leq |f(x) - L_1| + |g(x) - L_2|$$
$$< \frac{\varepsilon}{2} + \frac{\varepsilon}{2} = \varepsilon,$$

whenever $0 < |x - a| < \delta$. ∎

PROOF OF THEOREM 2.2.3(*ii*): By the triangle inequality,

$$|f(x)g(x) - L_1 L_2| = |f(x)g(x) - f(x)L_2 + f(x)L_2 - L_1 L_2|$$
$$\leq |f(x)g(x) - f(x)L_2| + |f(x)L_2 - L_1 L_2|$$
$$= |f(x)||g(x) - L_2| + |L_2||f(x) - L_1| \tag{3}$$
$$\leq |f(x)||g(x) - L_2| + (1 + |L_2|)|f(x) - L_1|.$$

Since $\lim_{x \to a} f(x) = L_1$ and $\lim_{x \to a} g(x) = L_2$, we know there exist numbers $\delta_1 > 0, \delta_2 > 0, \delta_3 > 0$ such that $|f(x) - L_1| < 1$ or

$$|f(x)| < 1 + |L_1| \quad \text{whenever} \quad 0 < |x - a| < \delta_1, \tag{4}$$

$$|f(x) - L_1| < \frac{\varepsilon/2}{1 + |L_2|} \quad \text{whenever} \quad 0 < |x - a| < \delta_2, \tag{5}$$

and

$$|g(x) - L_2| < \frac{\varepsilon/2}{1 + |L_1|} \quad \text{whenever} \quad 0 < |x - a| < \delta_3. \tag{6}$$

Hence, if we choose δ to be the smallest number in the set of positive numbers $\{\delta_1, \delta_2, \delta_3\}$, then we have from (3), (4), (5), and (6),

$$|f(x)g(x) - L_1L_2| < (1 + |L_1|) \cdot \frac{\varepsilon/2}{1 + |L_1|} + (1 + |L_2|) \cdot \frac{\varepsilon/2}{1 + |L_2|} = \frac{\varepsilon}{2} + \frac{\varepsilon}{2} = \varepsilon. \quad \blacksquare$$

PROOF OF THEOREM 2.2.3(iii): We will first prove that

$$\lim_{x \to a} \frac{1}{g(x)} = \frac{1}{L_2}, \quad L_2 \neq 0.$$

Consider
$$\left| \frac{1}{g(x)} - \frac{1}{L_2} \right| = \frac{|g(x) - L_2|}{|L_2||g(x)|}. \quad (7)$$

Since $\lim_{x \to a} g(x) = L_2$, there exists a $\delta_1 > 0$ such that

$$|g(x) - L_2| < \frac{|L_2|}{2} \quad \text{whenever} \quad 0 < |x - a| < \delta_1.$$

For these values of x, the inequality

$$|L_2| = |g(x) - (g(x) - L_2)| \leq |g(x)| + |g(x) - L_2| < |g(x)| + \frac{|L_2|}{2}$$

gives
$$|g(x)| > \frac{|L_2|}{2} \quad \text{and} \quad \frac{1}{|g(x)|} < \frac{2}{|L_2|}.$$

Thus, from (7),

$$\left| \frac{1}{g(x)} - \frac{1}{L_2} \right| < \frac{2}{|L_2|^2} |g(x) - L_2|. \quad (8)$$

Now for $\varepsilon > 0$ there exists a $\delta_2 > 0$ such that

$$|g(x) - L_2| < \frac{|L_2|^2}{2} \varepsilon \quad \text{whenever} \quad 0 < |x - a| < \delta_2.$$

By choosing δ to be the smallest number in the set of positive numbers $\{\delta_1, \delta_2\}$, it follows from (8) that

$$\left| \frac{1}{g(x)} - \frac{1}{L_2} \right| < \varepsilon \quad \text{whenever} \quad 0 < |x - a| < \delta.$$

We conclude the proof using Theorem 2.2.3(ii):

$$\lim_{x \to a} \frac{f(x)}{g(x)} = \lim_{x \to a} \frac{1}{g(x)} \cdot f(x) = \lim_{x \to a} \frac{1}{g(x)} \cdot \lim_{x \to a} f(x) = \frac{L_1}{L_2}. \quad \blacksquare$$

■ **Section 2.3**

PROOF OF THEOREM 2.3.3: To prove the theorem we must find a $\delta > 0$ so that

$$|f(g(x)) - f(L)| < \varepsilon \quad \text{whenever} \quad 0 < |x - a| < \delta.$$

To this end we first use the fact that f is continuous at L, in other words, $\lim_{u \to L} f(u) = f(L)$. This means for a given $\varepsilon > 0$ there exists a $\delta_1 > 0$ such that

$$|f(u) - f(L)| < \varepsilon \quad \text{whenever} \quad |u - L| < \delta_1.$$

Now if $u = g(x)$, then the last line is

$$|f(g(x)) - f(L)| < \varepsilon \quad \text{whenever} \quad |g(x) - L| < \delta_1.$$

Also from the assumption that $\lim_{x \to a} g(x) = L$, we know there exists a $\delta > 0$ such that

$$|g(x) - L| < \delta_1 \quad \text{whenever} \quad 0 < |x - a| < \delta.$$

We now combine the last two results. That is, whenever $0 < |x - a| < \delta$, then $|g(x) - L| < \delta_1$; but whenever $|g(x) - L| < \delta_1$, then necessarily $|f(g(x)) - f(L)| < \varepsilon$. $\blacksquare$

Section 2.4

PROOF OF THEOREM 2.4.1: We assume that $g(x) \leq f(x) \leq h(x)$ for all x in an open interval that contains the number a (except possibly at a itself) and that $\lim_{x \to a} g(x) = L$ and $\lim_{x \to a} h(x) = L$. Let $\varepsilon > 0$ be given. Then there exist numbers $\delta_1 > 0$ and $\delta_2 > 0$ such that $|g(x) - L| < \varepsilon$ whenever $0 < |x - a| < \delta_1$ and $|h(x) - L| < \varepsilon$ whenever $0 < |x - a| < \delta_2$. That is,

$$L - \varepsilon < g(x) < L + \varepsilon \quad \text{whenever} \quad 0 < |x - a| < \delta_1$$
$$L - \varepsilon < h(x) < L + \varepsilon \quad \text{whenever} \quad 0 < |x - a| < \delta_2.$$

Also there must exist $\delta_3 > 0$ such that

$$g(x) \leq f(x) \leq h(x) \quad \text{whenever} \quad 0 < |x - a| < \delta_3.$$

If δ is taken to be the smallest number in the set of positive numbers $\{\delta_1, \delta_2, \delta_3\}$, then for $0 < |x - a| < \delta$ we have

$$L - \varepsilon < g(x) \leq f(x) \leq h(x) < L + \varepsilon$$

or equivalently $|f(x) - L| < \varepsilon$. This means $\lim_{x \to a} f(x) = L$. ∎

Section 9.10

PROOF OF THEOREM 9.10.2 Let x be a fixed number in the interval $(a - r, a + r)$ and let the difference between $f(x)$ and the nth degree Taylor polynomial of f at a be denoted by

$$R_n(x) = f(x) - P_n(x).$$

For any t in the interval $[a, x]$ we define

$$F(t) = f(x) - f(t) - \frac{f'(t)}{1!}(x - t) - \frac{f''(t)}{2!}(x - t)^2 - \cdots - \frac{f^{(n)}(t)}{n!}(x - t)^n - \frac{R_n(x)}{(x - a)^{n+1}}(x - t)^{n+1}. \quad (9)$$

With x held constant we differentiate F with respect to t using the Product and Power Rules:

$$F'(t) = -f'(t) + \left[f'(t) - \frac{f''(t)}{1!}(x - t) \right] + \left[\frac{f''(t)}{1!}(x - t) - \frac{f'''(t)}{2!}(x - t)^2 \right] + \cdots$$
$$+ \left[\frac{f^{(n)}(t)}{(n - 1)!}(x - t)^{n-1} - \frac{f^{(n+1)}(t)}{n!}(x - t)^n \right] + \frac{R_n(x)(n + 1)}{(x - a)^{n+1}}(x - t)^n,$$

for all t in the open interval (a, x). Since the last sum telescopes, we obtain

$$F'(t) = -\frac{f^{(n+1)}(t)}{n!}(x - t)^n + \frac{R_n(x)(n + 1)}{(x - a)^{n+1}}(x - t)^n. \quad (10)$$

Now it is evident from (9) that F is continuous on $[a, x]$ and that

$$F(x) = f(x) - f(x) - 0 - \cdots - 0 = 0.$$

Furthermore, $\quad F(a) = f(x) - P_n(x) - R_n(x) = 0.$

Thus, $F(t)$ satisfies the hypotheses of Rolle's Theorem (Theorem 4.4.1) on $[a, x]$ and so there exists a number c between a and x for which $F'(c) = 0$. From (10) we obtain

$$R_n(x) = \frac{f^{(n+1)}(c)}{(n + 1)!}(x - a)^{n+1}.$$ ∎

Review of Algebra

Integers

$\{\ldots, -4, -3, -2, -1, 0, 1, 2, 3, 4, \ldots\}$

Positive Integers (Natural Numbers)

$\{1, 2, 3, 4, 5, \ldots\}$

Nonnegative Integers (Whole Numbers)

$\{0, 1, 2, 3, 4, 5, \ldots\}$

Rational Numbers

A rational number is a number of the form p/q, where p and $q \neq 0$ are integers

Irrational Numbers

An irrational number is a number that cannot be written in the form p/q, where p and $q \neq 0$ are integers

Real Numbers

The set R of real numbers is the union of the sets of rational and irrational numbers

Laws of Exponents

$$a^m a^n = a^{m+n}, \quad \frac{a^m}{a^n} = a^{m-n}$$

$$(a^m)^n = a^{mn}, \quad (ab)^n = a^n b^n$$

$$\left(\frac{a}{b}\right)^n = \frac{a^n}{b^n}, \quad a^0 = 1, a \neq 0$$

Negative Exponent

$$a^{-n} = \frac{1}{a^n}, \quad n > 0$$

Radical

$a^{1/n} = \sqrt[n]{a}$, $n > 0$ an integer

Rational Exponents and Radicals

$$a^{m/n} = (a^m)^{1/n} = (a^{1/n})^m$$

$$a^{m/n} = \sqrt[n]{a^m} = (\sqrt[n]{a})^m$$

$$\sqrt[n]{ab} = \sqrt[n]{a}\sqrt[n]{b}$$

$$\sqrt[n]{\frac{a}{b}} = \frac{\sqrt[n]{a}}{\sqrt[n]{b}}$$

Quadratic Formula

Roots of a quadratic equation $ax^2 + bx + c = 0$, $a \neq 0$, are

$$x = \frac{-b \pm \sqrt{b^2 - 4ac}}{2a}$$

Binomial Expansions

$(a + b)^2 = a^2 + 2ab + b^2$

$(a + b)^3 = a^3 + 3a^2 b + 3ab^2 + b^3$

$(a + b)^4 = a^4 + 4a^3 b + 6a^2 b^2 + 4ab^3 + b^4$

$(a + b)^5 = a^5 + 5a^4 b + 10a^3 b^2 + 10a^2 b^3 + 5ab^4 + b^5$

Pascal's Triangle

Coefficients in the expansion of $(a + b)^n$ follow the pattern:

$$\begin{array}{c} 1 \\ 1 \quad 1 \\ 1 \quad 2 \quad 1 \\ 1 \quad 3 \quad 3 \quad 1 \\ 1 \quad 4 \quad 6 \quad 4 \quad 1 \\ \vdots \end{array}$$

Each number in the interior of this array is the sum of the two numbers directly above it:

$$\begin{array}{c} 1 \quad 4 \quad 6 \quad 4 \quad 1 \\ 1 \quad 5 \quad 10 \quad 10 \quad 5 \quad 1 \end{array}$$

The last row are the coefficients in the expansion of $(a + b)^5$.

Factorization Formulas

$a^2 - b^2 = (a - b)(a + b)$

$a^3 - b^3 = (a - b)(a^2 + ab + b^2)$

$a^3 + b^3 = (a + b)(a^2 - ab + b^2)$

$a^4 - b^4 = (a - b)(a + b)(a^2 + b^2)$

Definition of Absolute Value

$$|a| = \begin{cases} a & \text{if } a \text{ is nonnegative } (a \geq 0) \\ -a & \text{if } a \text{ is negative } (a < 0) \end{cases}$$

Properties of Inequalities

If $a > b$ and $b > c$, then $a > c$.

If $a < b$, then $a + c < b + c$.

If $a < b$, then $ac < bc$ for $c > 0$.

If $a < b$, then $ac > bc$ for $c < 0$.

Formulas from Geometry

Area A, Circumference C, Volume V, Surface Area S

RECTANGLE

$A = lw$, $C = 2l + 2w$

PARALLELOGRAM

$A = bh$

TRAPEZOID
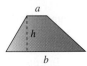

$A = \frac{1}{2}(a+b)h$

RIGHT TRIANGLE
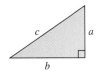

Pythagorean Theorem:
$c^2 = a^2 + b^2$

TRIANGLE
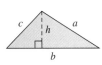

$A = \frac{1}{2}bh$, $C = a+b+c$

EQUILATERAL TRIANGLE

$h = \frac{\sqrt{3}}{2}s$, $A = \frac{\sqrt{3}}{4}s^2$

CIRCLE

$A = \pi r^2$, $C = 2\pi r$

CIRCULAR RING

$A = \pi(R^2 - r^2)$

CIRCULAR SECTOR

$A = \frac{1}{2}r^2\theta$, $s = r\theta$

ELLIPSE

$A = \pi ab$

ELLIPSOID

$V = \frac{4}{3}\pi abc$

SPHERE

$V = \frac{4}{3}\pi r^3$, $S = 4\pi r^2$

RIGHT CYLINDER	RIGHT CIRCULAR CYLINDER	RECTANGULAR PARALLELEPIPED
$V = Bh$, B area of base	$V = \pi r^2 h$, $S = 2\pi r h$ (lateral side)	$V = lwh$, $S = 2(hl + lw + hw)$
CONE	RIGHT CIRCULAR CONE	FRUSTUM OF A CONE
$V = \frac{1}{3} Bh$, B area of base	$V = \frac{1}{3}\pi r^2 h$, $S = \pi r \sqrt{r^2 + h^2}$	$V = \frac{1}{3}\pi h(r_1^2 + r_1 r_2 + r_2^2)$

Graphs and Functions

To Find Intercepts

y-intercepts: Set $x = 0$ in an equation and solve for y
x-intercepts: Set $y = 0$ in an equation and solve for x

Polynomial Functions

$$f(x) = a_n x^n + a_{n-1} x^{n-1} + \cdots + a_1 x + a_0,$$

where n is a nonnegative integer

Linear Function

$f(x) = ax + b, a \neq 0$

Graph of a linear function is a straight line.

Equation forms of lines:

Point-Slope: $y - x_0 = m(x - x_0)$,
Slope-Intercept: $y = mx + b$,

where m is slope

Quadratic Function

$$f(x) = ax^2 + bx + c, a \neq 0$$

Graph of a quadratic function is a parabola.

Vertex (h, k) of a Parabola

Complete the square in x for $f(x) = ax^2 + bx + c$ to obtain $f(x) = a(x - h)^2 + k$. Alternatively, compute the coordinates

$$\left(-\frac{b}{2a}, f\left(-\frac{b}{2a}\right)\right).$$

Even and Odd Functions

Even: $f(-x) = f(x)$; Symmetry of graph: y-axis
Odd: $f(-x) = -f(x)$; Symmetry of graph: origin

Rigid Transformations

Graph of $y = f(x)$ for $c > 0$:
$y = f(x) + c$, shifted up c units
$y = f(x) - c$, shifted down c units
$y = f(x + c)$, shifted left c units
$y = f(x - c)$, shifted right c units
$y = f(-x)$, reflection in y-axis
$y = -f(x)$, reflection in x-axis

Rational Function

$$f(x) = \frac{p(x)}{q(x)} = \frac{a_n x^n + \cdots + a_1 x + a_0}{b_m x^m + \cdots + b_1 x + b_0},$$

where $p(x)$ and $q(x)$ are polynomial functions

Asymptotes

If the polynomial functions $p(x)$ and $q(x)$ have no common factors, then the graph of a rational function

$$f(x) = \frac{p(x)}{q(x)} = \frac{a_n x^n + \cdots + a_1 x + a_0}{b_m x^m + \cdots + b_1 x + b_0}$$

has a

Vertical asymptote:

$x = a$ when $q(a) = 0$,

Horizontal asymptote:

$y = a_n/b_m$ when $n = m$, and $y = 0$ when $n < m$,

Slant asymptote:

$y = ax + b$ when $n = m + 1$.

The graph has no horizontal asymptote when $n > m$. A slant asymptote is found by division.

Power Function

$f(x) = x^n$,

where n is any real number

Review of Trigonometry

Unit Circle Definition of Sine and Cosine

$y = \sin\theta$
$x = \cos\theta$

Other Trigonometric Functions

$$\tan\theta = \frac{y}{x} = \frac{\sin\theta}{\cos\theta}, \quad \cot\theta = \frac{x}{y} = \frac{\cos\theta}{\sin\theta}$$

$$\sec\theta = \frac{1}{x} = \frac{1}{\cos\theta}, \quad \csc\theta = \frac{1}{y} = \frac{1}{\sin\theta}$$

Conversion Formulas

$1 \text{ degree} = \dfrac{\pi}{180}$ radians

$1 \text{ radian} = \dfrac{180}{\pi}$ degrees

Right Triangle Definition of Sine and Cosine

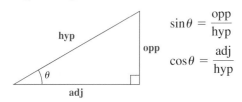

$$\sin\theta = \frac{\text{opp}}{\text{hyp}}$$

$$\cos\theta = \frac{\text{adj}}{\text{hyp}}$$

Other Trigonometric Functions

$$\tan\theta = \frac{\text{opp}}{\text{adj}}, \quad \cot\theta = \frac{\text{adj}}{\text{opp}}$$

$$\sec\theta = \frac{\text{hyp}}{\text{adj}}, \quad \csc\theta = \frac{\text{hyp}}{\text{opp}}$$

Signs of Sine and Cosine

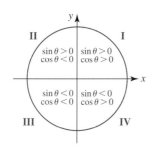

Values of Sine and Cosine for Special Angles

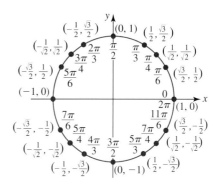

Bounds for Sine and Cosine Functions

$-1 \leq \sin x \leq 1 \quad \text{and} \quad -1 \leq \cos x \leq 1$

Periodicity of Trigonometric Functions

$\sin(x + 2\pi) = \sin x, \quad \cos(x + 2\pi) = \cos x$
$\sec(x + 2\pi) = \sec x, \quad \csc(x + 2\pi) = \csc x$
$\tan(x + \pi) = \tan x, \quad \cot(x + \pi) = \cot x$

Cofunction Identities

$$\sin\left(\frac{\pi}{2} - x\right) = \cos x$$

$$\cos\left(\frac{\pi}{2} - x\right) = \sin x$$

$$\tan\left(\frac{\pi}{2} - x\right) = \cot x$$

Pythagorean Identities

$\sin^2 x + \cos^2 x = 1$
$1 + \tan^2 x = \sec^2 x$
$1 + \cot^2 x = \csc^2 x$

Even/Odd Identities

Even	Odd
$\cos(-x) = \cos x$	$\sin(-x) = -\sin x$
$\sec(-x) = \sec x$	$\csc(-x) = -\csc x$
	$\tan(-x) = -\tan x$
	$\cot(-x) = -\cot x$

Sum Formulas

$\sin(x_1 + x_2) = \sin x_1 \cos x_2 + \cos x_1 \sin x_2$

$\cos(x_1 + x_2) = \cos x_1 \cos x_2 - \sin x_1 \sin x_2$

$\tan(x_1 + x_2) = \dfrac{\tan x_1 + \tan x_2}{1 - \tan x_1 \tan x_2}$

Difference Formulas

$\sin(x_1 - x_2) = \sin x_1 \cos x_2 - \cos x_1 \sin x_2$

$\cos(x_1 - x_2) = \cos x_1 \cos x_2 + \sin x_1 \sin x_2$

$\tan(x_1 - x_2) = \dfrac{\tan x_1 - \tan x_2}{1 + \tan x_1 \tan x_2}$

Double-Angle Formulas

$\sin 2x = 2 \sin x \cos x$

$\cos 2x = \cos^2 x - \sin^2 x$

Alternative Double-Angle Formulas for Cosine

$\cos 2x = 1 - 2\sin^2 x$

$\cos 2x = 2\cos^2 x - 1$

Half-Angle Formulas as Used in Calculus

$\sin^2 x = \tfrac{1}{2}(1 - \cos 2x)$

$\cos^2 x = \tfrac{1}{2}(1 + \cos 2x)$

Law of Sines

$\dfrac{\sin \alpha}{a} = \dfrac{\sin \beta}{b} = \dfrac{\sin \gamma}{c}$

Law of Cosines

$a^2 = b^2 + c^2 - 2bc \cos \alpha$

$b^2 = a^2 + c^2 - 2ac \cos \beta$

$c^2 = a^2 + b^2 - 2ab \cos \gamma$

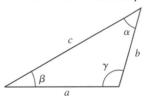

Inverse Trigonometric Functions

$y = \sin^{-1} x$ if and only if $x = \sin y$, $\quad -\pi/2 \leq y \leq \pi/2$

$y = \cos^{-1} x$ if and only if $x = \cos y$, $\quad 0 \leq y \leq \pi$

$y = \tan^{-1} x$ if and only if $x = \tan y$, $\quad -\pi/2 < y < \pi/2$

Cycles for Sine, Cosine, and Tangent

sine

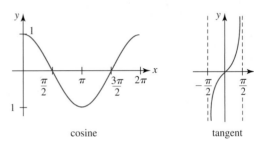

cosine   tangent

Exponential and Logarithmic Functions

The Number e
$e = 2.718281828459...$

Definitions of the Number e
$$e = \lim_{x \to \infty} \left(1 + \frac{1}{x}\right)^x$$
$$e = \lim_{h \to 0} (1 + h)^{1/h}$$

Exponential Function
$f(x) = b^x, \; b > 0, b \neq 1$

Natural Exponential Function
$f(x) = e^x$

Logarithmic Function
$f(x) = \log_b x, \quad x > 0$
where $y = \log_b x$ is equivalent to $x = b^y$

Natural Logarithmic Function
$f(x) = \log_e x = \ln x, \quad x > 0$
where $y = \ln x$ is equivalent to $x = e^y$

Laws of Logarithms
$\log_b MN = \log_b M + \log_b N$
$\log_b \dfrac{M}{N} = \log_b M - \log_b N$
$\log_b M^c = c \log_b M$

Properties of Logarithms
$\log_b b = 1, \quad \log_b 1 = 0$
$\log_b b^x = x, \quad b^{\log_b x} = x$

Change from Base b to Base e
$\log_b x = \dfrac{\ln x}{\ln b}$

Hyperbolic Functions
$$\sinh x = \frac{e^x - e^{-x}}{2}, \quad \cosh x = \frac{e^x + e^{-x}}{2}$$
$$\tanh x = \frac{\sinh x}{\cosh x}, \quad \coth x = \frac{\cosh x}{\sinh x}$$
$$\operatorname{sech} x = \frac{1}{\cosh x}, \quad \operatorname{csch} x = \frac{1}{\sinh x}$$

Inverse Hyperbolic Functions as Logarithms
$\sinh^{-1} x = \ln\left(x + \sqrt{x^2 + 1}\right)$
$\cosh^{-1} x = \ln\left(x + \sqrt{x^2 - 1}\right), \; x \geq 1$
$\tanh^{-1} x = \dfrac{1}{2} \ln\left(\dfrac{1 + x}{1 - x}\right), \; |x| < 1$
$\coth^{-1} x = \dfrac{1}{2} \ln\left(\dfrac{x + 1}{x - 1}\right), \; |x| > 1$
$\operatorname{sech}^{-1} x = \ln\left(\dfrac{1 + \sqrt{1 - x^2}}{x}\right), \; 0 < x \leq 1$
$\operatorname{csch}^{-1} x = \ln\left(\dfrac{1}{x} + \dfrac{\sqrt{1 + x^2}}{|x|}\right), \; x \neq 0$

Even/Odd Identities
Even	Odd
$\cosh(-x) = \cosh x$	$\sinh(-x) = -\sinh x$

Additional Identities
$\cosh^2 x - \sinh^2 x = 1$
$1 - \tanh^2 x = \operatorname{sech}^2 x$
$\coth^2 x - 1 = \operatorname{csch}^2 x$
$\sinh(x_1 \pm x_2) = \sinh x_1 \cosh x_2 \pm \cosh x_1 \sinh x_2$
$\cosh(x_1 \pm x_2) = \cosh x_1 \cosh x_2 \pm \sinh x_1 \sinh x_2$
$\sinh 2x = 2 \sinh x \cosh x$
$\cosh 2x = \cosh^2 x + \sinh^2 x$
$\sinh^2 x = \tfrac{1}{2}(-1 + \cosh 2x)$
$\cosh^2 x = \tfrac{1}{2}(1 + \cosh 2x)$

Differentiation

Rules

1. Constant: $\dfrac{d}{dx}c = 0$

2. Constant Multiple: $\dfrac{d}{dx}cf(x) = cf'(x)$

3. Sum: $\dfrac{d}{dx}[f(x) \pm g(x)] = f'(x) \pm g'(x)$

4. Product: $\dfrac{d}{dx}f(x)g(x) = f(x)g'(x) + g(x)f'(x)$

5. Quotient: $\dfrac{d}{dx}\dfrac{f(x)}{g(x)} = \dfrac{g(x)f'(x) - f(x)g'(x)}{[g(x)]^2}$

6. Chain: $\dfrac{d}{dx}f(g(x)) = f'(g(x))g'(x)$

7. Power: $\dfrac{d}{dx}x^n = nx^{n-1}$

8. Power: $\dfrac{d}{dx}[g(x)]^n = n[g(x)]^{n-1}g'(x)$

Functions

Trigonometric:

9. $\dfrac{d}{dx}\sin x = \cos x$

10. $\dfrac{d}{dx}\cos x = -\sin x$

11. $\dfrac{d}{dx}\tan x = \sec^2 x$

12. $\dfrac{d}{dx}\cot x = -\csc^2 x$

13. $\dfrac{d}{dx}\sec x = \sec x \tan x$

14. $\dfrac{d}{dx}\csc x = -\csc x \cot x$

Inverse trigonometric:

15. $\dfrac{d}{dx}\sin^{-1}x = \dfrac{1}{\sqrt{1-x^2}}$

16. $\dfrac{d}{dx}\cos^{-1}x = -\dfrac{1}{\sqrt{1-x^2}}$

17. $\dfrac{d}{dx}\tan^{-1}x = \dfrac{1}{1+x^2}$

18. $\dfrac{d}{dx}\cot^{-1}x = -\dfrac{1}{1+x^2}$

19. $\dfrac{d}{dx}\sec^{-1}x = \dfrac{1}{|x|\sqrt{x^2-1}}$

20. $\dfrac{d}{dx}\csc^{-1}x = -\dfrac{1}{|x|\sqrt{x^2-1}}$

Hyperbolic:

21. $\dfrac{d}{dx}\sinh x = \cosh x$

22. $\dfrac{d}{dx}\cosh x = \sinh x$

23. $\dfrac{d}{dx}\tanh x = \text{sech}^2 x$

24. $\dfrac{d}{dx}\coth x = -\text{csch}^2 x$

25. $\dfrac{d}{dx}\text{sech}\, x = -\text{sech}\, x \tanh x$

26. $\dfrac{d}{dx}\text{csch}\, x = -\text{csch}\, x \coth x$

Inverse hyperbolic:

27. $\dfrac{d}{dx}\sinh^{-1}x = \dfrac{1}{\sqrt{x^2+1}}$

28. $\dfrac{d}{dx}\cosh^{-1}x = \dfrac{1}{\sqrt{x^2-1}}$

29. $\dfrac{d}{dx}\tanh^{-1}x = \dfrac{1}{1-x^2}$

30. $\dfrac{d}{dx}\coth^{-1}x = \dfrac{1}{1-x^2}$

31. $\dfrac{d}{dx}\text{sech}^{-1}x = -\dfrac{1}{x\sqrt{1-x^2}}$

32. $\dfrac{d}{dx}\text{csch}^{-1}x = -\dfrac{1}{|x|\sqrt{x^2+1}}$

Exponential:

33. $\dfrac{d}{dx}e^x = e^x$

34. $\dfrac{d}{dx}b^x = b^x(\ln b)$

Logarithmic:

35. $\dfrac{d}{dx}\ln|x| = \dfrac{1}{x}$

36. $\dfrac{d}{dx}\log_b x = \dfrac{1}{x(\ln b)}$

Integration Formulas

Basic Forms

1. $\int u\, dv = uv - \int v\, du$

2. $\int u^n\, du = \dfrac{1}{n+1} u^{n+1} + C,\ n \neq -1$

3. $\int \dfrac{du}{u} = \ln|u| + C$

4. $\int e^u\, du = e^u + C$

5. $\int a^u\, du = \dfrac{1}{\ln a} a^u + C$

6. $\int \sin u\, du = -\cos u + C$

7. $\int \cos u\, du = \sin u + C$

8. $\int \sec^2 u\, du = \tan u + C$

9. $\int \csc^2 u\, du = -\cot u + C$

10. $\int \sec u \tan u\, du = \sec u + C$

11. $\int \csc u \cot u\, du = -\csc u + C$

12. $\int \tan u\, du = -\ln|\cos u| + C$

13. $\int \cot u\, du = \ln|\sin u| + C$

14. $\int \sec u\, du = \ln|\sec u + \tan u| + C$

15. $\int \csc u\, du = \ln|\csc u - \cot u| + C$

16. $\int \dfrac{du}{\sqrt{a^2 - u^2}} = \sin^{-1}\dfrac{u}{a} + C$

17. $\int \dfrac{du}{a^2 + u^2} = \dfrac{1}{a}\tan^{-1}\dfrac{u}{a} + C$

18. $\int \dfrac{du}{u\sqrt{u^2 - a^2}} = \dfrac{1}{a}\sec^{-1}\left|\dfrac{u}{a}\right| + C$

19. $\int \dfrac{du}{a^2 - u^2} = \dfrac{1}{2a}\ln\left|\dfrac{u+a}{u-a}\right| + C$

20. $\int \dfrac{du}{u^2 - a^2} = \dfrac{1}{2a}\ln\left|\dfrac{u-a}{u+a}\right| + C$

Forms Involving $\sqrt{a^2 + u^2}$

21. $\int \sqrt{a^2 + u^2}\, du = \dfrac{u}{2}\sqrt{a^2 + u^2} + \dfrac{a^2}{2}\ln|u + \sqrt{a^2 + u^2}| + C$

22. $\int u^2\sqrt{a^2 + u^2}\, du = \dfrac{u}{8}(a^2 + 2u^2)\sqrt{a^2 + u^2} - \dfrac{a^4}{8}\ln|u + \sqrt{a^2 + u^2}| + C$

23. $\int \dfrac{\sqrt{a^2 + u^2}}{u}\, du = \sqrt{a^2 + u^2} - a\ln\left|\dfrac{a + \sqrt{a^2 + u^2}}{u}\right| + C$

24. $\int \dfrac{\sqrt{a^2 + u^2}}{u^2}\, du = -\dfrac{\sqrt{a^2 + u^2}}{u} + \ln|u + \sqrt{a^2 + u^2}| + C$

25. $\int \dfrac{du}{\sqrt{a^2 + u^2}} = \ln|u + \sqrt{a^2 + u^2}| + C$

26. $\int \dfrac{u^2\, du}{\sqrt{a^2 + u^2}} = \dfrac{u}{2}\sqrt{a^2 + u^2} - \dfrac{a^2}{2}\ln|u + \sqrt{a^2 + u^2}| + C$

27. $\int \dfrac{du}{u\sqrt{a^2 + u^2}} = -\dfrac{1}{a}\ln\left|\dfrac{\sqrt{a^2 + u^2} + a}{u}\right| + C$

28. $\int \dfrac{du}{u^2\sqrt{a^2 + u^2}} = -\dfrac{\sqrt{a^2 + u^2}}{a^2 u} + C$

29. $\int \dfrac{du}{(a^2 + u^2)^{3/2}} = \dfrac{u}{a^2\sqrt{a^2 + u^2}} + C$

Form Involving $\sqrt{a^2 - u^2}$

30. $\int \sqrt{a^2 - u^2}\, du = \dfrac{u}{2}\sqrt{a^2 - u^2} + \dfrac{a^2}{2}\sin^{-1}\dfrac{u}{a} + C$

31. $\int u^2\sqrt{a^2 - u^2}\, du = \dfrac{u}{8}(2u^2 - a^2)\sqrt{a^2 - u^2} + \dfrac{a^4}{8}\sin^{-1}\dfrac{u}{a} + C$

32. $\int \dfrac{\sqrt{a^2 - u^2}}{u}\, du = \sqrt{a^2 - u^2} - a\ln\left|\dfrac{a + \sqrt{a^2 - u^2}}{u}\right| + C$

33. $\int \dfrac{\sqrt{a^2 - u^2}}{u^2}\, du = -\dfrac{1}{u}\sqrt{a^2 - u^2} - \sin^{-1}\dfrac{u}{a} + C$

34. $\int \dfrac{u^2\, du}{\sqrt{a^2 - u^2}} = -\dfrac{u}{2}\sqrt{a^2 - u^2} + \dfrac{a^2}{2}\sin^{-1}\dfrac{u}{a} + C$

35. $\int \dfrac{du}{u\sqrt{a^2 - u^2}} = -\dfrac{1}{a}\ln\left|\dfrac{a + \sqrt{a^2 - u^2}}{u}\right| + C$

36. $\int \dfrac{du}{u^2\sqrt{a^2-u^2}} = -\dfrac{1}{a^2 u}\sqrt{a^2-u^2} + C$

37. $\int (a^2-u^2)^{3/2}\,du = -\dfrac{u}{8}(2u^2-5a^2)\sqrt{a^2-u^2}$
$+ \dfrac{3a^4}{8}\sin^{-1}\dfrac{u}{a} + C$

38. $\int \dfrac{du}{(a^2-u^2)^{3/2}} = \dfrac{u}{a^2\sqrt{a^2-u^2}} + C$

Forms Involving $\sqrt{u^2 - a^2}$

39. $\int \sqrt{u^2-a^2}\,du = \dfrac{u}{2}\sqrt{u^2-a^2}$
$- \dfrac{a^2}{2}\ln|u+\sqrt{u^2-a^2}| + C$

40. $\int u^2\sqrt{u^2-a^2}\,du = \dfrac{u}{8}(2u^2-a^2)\sqrt{u^2-a^2}$
$- \dfrac{a^4}{8}\ln|u+\sqrt{u^2-a^2}| + C$

41. $\int \dfrac{\sqrt{u^2-a^2}}{u}\,du = \sqrt{u^2-a^2} - a\cos^{-1}\dfrac{a}{u} + C$

42. $\int \dfrac{\sqrt{u^2-a^2}}{u^2}\,du = -\dfrac{\sqrt{u^2-a^2}}{u}$
$+ \ln|u+\sqrt{u^2-a^2}| + C$

43. $\int \dfrac{du}{\sqrt{u^2-a^2}} = \ln|u+\sqrt{u^2-a^2}| + C$

44. $\int \dfrac{u^2\,du}{\sqrt{u^2-a^2}} = \dfrac{u}{2}\sqrt{u^2-a^2}$
$+ \dfrac{a^2}{2}\ln|u+\sqrt{u^2-a^2}| + C$

45. $\int \dfrac{du}{u^2\sqrt{u^2-a^2}} = \dfrac{\sqrt{u^2-a^2}}{a^2 u} + C$

46. $\int \dfrac{du}{(u^2-a^2)^{3/2}} = -\dfrac{u}{a^2\sqrt{u^2-a^2}} + C$

Forms Involving $a + bu$

47. $\int \dfrac{u\,du}{a+bu} = \dfrac{1}{b^2}(a+bu - a\ln|a+bu|) + C$

48. $\int \dfrac{u^2\,du}{a+bu} = \dfrac{1}{2b^3}[(a+bu)^2 - 4a(a+bu)$
$+ 2a^2\ln|a+bu|] + C$

49. $\int \dfrac{du}{u(a+bu)} = \dfrac{1}{a}\ln\left|\dfrac{u}{a+bu}\right| + C$

50. $\int \dfrac{du}{u^2(a+bu)} = -\dfrac{1}{au} + \dfrac{b}{a^2}\ln\left|\dfrac{a+bu}{u}\right| + C$

51. $\int \dfrac{u\,du}{(a+bu)^2} = \dfrac{a}{b^2(a+bu)} + \dfrac{1}{b^2}\ln|a+bu| + C$

52. $\int \dfrac{du}{u(a+bu)^2} = \dfrac{1}{a(a+bu)} - \dfrac{1}{a^2}\ln\left|\dfrac{a+bu}{u}\right| + C$

53. $\int \dfrac{u^2\,du}{(a+bu)^2} = \dfrac{1}{b^3}\left(a+bu - \dfrac{a^2}{a+bu} - 2a\ln|a+bu|\right) + C$

54. $\int u\sqrt{a+bu}\,du = \dfrac{2}{15b^2}(3bu-2a)(a+bu)^{3/2} + C$

55. $\int \dfrac{u\,du}{\sqrt{a+bu}} = \dfrac{2}{3b^2}(bu-2a)\sqrt{a+bu} + C$

56. $\int \dfrac{u^2\,du}{\sqrt{a+bu}} = \dfrac{2}{15b^3}(8a^2+3b^2u^2-4abu)\sqrt{a+bu} + C$

57. $\int \dfrac{du}{u\sqrt{a+bu}} = \dfrac{1}{\sqrt{a}}\ln\left|\dfrac{\sqrt{a+bu}-\sqrt{a}}{\sqrt{a+bu}+\sqrt{a}}\right| + C$, if $a > 0$
$= \dfrac{2}{\sqrt{-a}}\tan^{-1}\sqrt{\dfrac{a+bu}{-a}} + C$, if $a < 0$

58. $\int \dfrac{\sqrt{a+bu}}{u}\,du = 2\sqrt{a+bu} + a\int \dfrac{du}{u\sqrt{a+bu}}$

59. $\int \dfrac{\sqrt{a+bu}}{u^2}\,du = -\dfrac{\sqrt{a+bu}}{u} + \dfrac{b}{2}\int \dfrac{du}{u\sqrt{a+bu}}$

60. $\int u^n\sqrt{a+bu}\,du = \dfrac{2u^n(a+bu)^{3/2}}{b(2n+3)}$
$- \dfrac{2na}{b(2n+3)}\int u^{n-1}\sqrt{a+bu}\,du$

61. $\int \dfrac{u^n\,du}{\sqrt{a+bu}} = \dfrac{2u^n\sqrt{a+bu}}{b(2n+1)} - \dfrac{2na}{b(2n+1)}\int \dfrac{u^{n-1}\,du}{\sqrt{a+bu}}$

62. $\int \dfrac{du}{u^n\sqrt{a+bu}} = -\dfrac{\sqrt{a+bu}}{a(n-1)u^{n-1}}$
$- \dfrac{b(2n-3)}{2a(n-1)}\int \dfrac{du}{u^{n-1}\sqrt{a+bu}}$

Trigonometric Forms

63. $\int \sin^2 u\,du = \dfrac{1}{2}u - \dfrac{1}{4}\sin 2u + C$

64. $\int \cos^2 u\,du = \dfrac{1}{2}u + \dfrac{1}{4}\sin 2u + C$

65. $\int \tan^2 u\,du = \tan u - u + C$

66. $\int \cot^2 u\,du = -\cot u - u + C$

67. $\int \sin^3 u\,du = -\dfrac{1}{3}(2+\sin^2 u)\cos u + C$

68. $\int \cos^3 u\,du = \dfrac{1}{3}(2+\cos^2 u)\sin u + C$

69. $\int \tan^3 u\,du = \dfrac{1}{2}\tan^2 u + \ln|\cos u| + C$

70. $\int \cot^3 u\,du = -\dfrac{1}{2}\cot^2 u - \ln|\sin u| + C$

71. $\int \sec^3 u\,du = \dfrac{1}{2}\sec u \tan u + \dfrac{1}{2}\ln|\sec u + \tan u| + C$

72. $\int \csc^3 u \, du = -\frac{1}{2} \csc u \cot u + \frac{1}{2} \ln|\csc u - \cot u| + C$

73. $\int \sin^n u \, du = -\frac{1}{n} \sin^{n-1} u \cos u + \frac{n-1}{n} \int \sin^{n-2} u \, du$

74. $\int \cos^n u \, du = \frac{1}{n} \cos^{n-1} u \sin u + \frac{n-1}{n} \int \cos^{n-2} u \, du$

75. $\int \tan^n u \, du = \frac{1}{n-1} \tan^{n-1} u - \int \tan^{n-2} u \, du$

76. $\int \cot^n u \, du = \frac{-1}{n-1} \cot^{n-1} u - \int \cot^{n-2} u \, du$

77. $\int \sec^n u \, du = \frac{1}{n-1} \tan u \sec^{n-2} u + \frac{n-2}{n-1} \int \sec^{n-2} u \, du$

78. $\int \csc^n u \, du = \frac{-1}{n-1} \cot u \csc^{n-2} u + \frac{n-2}{n-1} \int \csc^{n-2} u \, du$

79. $\int \sin au \sin bu \, du = \frac{\sin(a-b)u}{2(a-b)} - \frac{\sin(a+b)u}{2(a+b)} + C$

80. $\int \cos au \cos bu \, du = \frac{\sin(a-b)u}{2(a-b)} + \frac{\sin(a+b)u}{2(a+b)} + C$

81. $\int \sin au \cos bu \, du = -\frac{\cos(a-b)u}{2(a-b)} - \frac{\cos(a+b)u}{2(a+b)} + C$

82. $\int u \sin u \, du = \sin u - u \cos u + C$

83. $\int u \cos u \, du = \cos u + u \sin u + C$

84. $\int u^n \sin u \, du = -u^n \cos u + n \int u^{n-1} \cos u \, du$

85. $\int u^n \cos u \, du = u^n \sin u - n \int u^{n-1} \sin u \, du$

86. $\int \sin^n u \cos^m u \, du = -\frac{\sin^{n-1} u \cos^{m+1} u}{n+m}$
$\qquad + \frac{n-1}{n+m} \int \sin^{n-1} u \cos^m u \, du$
$\qquad = \frac{\sin^{n+1} u \cos^{m-1} u}{n+m}$
$\qquad + \frac{m-1}{n+m} \int \sin^n u \cos^{m-2} u \, du$

87. $\int \frac{du}{1 - \sin au} = \frac{1}{a} \tan\left(\frac{\pi}{4} + \frac{au}{2}\right) + C$

88. $\int \frac{du}{1 + \sin au} = -\frac{1}{a} \tan\left(\frac{\pi}{4} - \frac{au}{2}\right) + C$

89. $\int \frac{u \, du}{1 - \sin au} = \frac{u}{a} \tan\left(\frac{\pi}{4} + \frac{au}{2}\right)$
$\qquad + \frac{2}{a^2} \ln\left|\sin\left(\frac{\pi}{4} - \frac{au}{2}\right)\right| + C$

Inverse Trigonometric Forms

90. $\int \sin^{-1} u \, du = u \sin^{-1} u + \sqrt{1 - u^2} + C$

91. $\int \cos^{-1} u \, du = u \cos^{-1} u - \sqrt{1 - u^2} + C$

92. $\int \tan^{-1} u \, du = u \tan^{-1} u - \frac{1}{2} \ln(1 + u^2) + C$

93. $\int u \sin^{-1} u \, du = \frac{2u^2 - 1}{4} \sin^{-1} u + \frac{u\sqrt{1 - u^2}}{4} + C$

94. $\int u \cos^{-1} u \, du = \frac{2u^2 - 1}{4} \cos^{-1} u - \frac{u\sqrt{1 - u^2}}{4} + C$

95. $\int u \tan^{-1} u \, du = \frac{u^2 + 1}{2} \tan^{-1} u - \frac{u}{2} + C$

96. $\int u^n \sin^{-1} u \, du = \frac{1}{n+1}\left[u^{n+1} \sin^{-1} u \right.$
$\qquad \left. - \int \frac{u^{n+1} \, du}{\sqrt{1 - u^2}}\right], \quad n \neq -1$

97. $\int u^n \cos^{-1} u \, du = \frac{1}{n+1}\left[u^{n+1} \cos^{-1} u \right.$
$\qquad \left. + \int \frac{u^{n+1} \, du}{\sqrt{1 - u^2}}\right], \quad n \neq -1$

98. $\int u^n \tan^{-1} u \, du = \frac{1}{n+1}\left[u^{n+1} \tan^{-1} u \right.$
$\qquad \left. - \int \frac{u^{n+1} \, du}{1 + u^2}\right], \quad n \neq -1$

Exponential and Logarithmic Forms

99. $\int u e^{au} \, du = \frac{1}{a^2}(au - 1) e^{au} + C$

100. $\int u^n e^{au} \, du = \frac{1}{a} u^n e^{au} - \frac{n}{a} \int u^{n-1} e^{au} \, du$

101. $\int e^{au} \sin bu \, du = \frac{e^{au}}{a^2 + b^2}(a \sin bu - b \cos bu) + C$

102. $\int e^{au} \cos bu \, du = \frac{e^{au}}{a^2 + b^2}(a \cos bu + b \sin bu) + C$

103. $\int \ln u \, du = u \ln u - u + C$

104. $\int \frac{1}{u \ln u} \, du = \ln|\ln u| + C$

105. $\int u^n \ln u \, du = \frac{u^{n+1}}{(n+1)^2}[(n+1) \ln u - 1] + C$

106. $\int u^m \ln^n u \, du = \frac{u^{m+1} \ln^n u}{m+1}$
$\qquad - \frac{n}{m+1} \int u^m \ln^{n-1} u \, du, \quad m \neq -1$

107. $\int \ln(u^2 + a^2)\, du = u\ln(u^2 + a^2) - 2u + 2a\tan^{-1}\dfrac{u}{a} + C$

108. $\int \ln|u^2 - a^2|\, du = u\ln|u^2 - a^2| - 2u + a\ln\left|\dfrac{u+a}{u-a}\right| + C$

109. $\int \dfrac{du}{a + be^u} = \dfrac{u}{a} - \dfrac{1}{a}\ln|a + be^u| + C$

Hyperbolic Forms

110. $\int \sinh u\, du = \cosh u + C$

111. $\int \cosh u\, du = \sinh u + C$

112. $\int \tanh u\, du = \ln(\cosh u) + C$

113. $\int \coth u\, du = \ln|\sinh u| + C$

114. $\int \operatorname{sech} u\, du = \tan^{-1}(\sinh u) + C$

115. $\int \operatorname{csch} u\, du = \ln|\tanh\tfrac{1}{2}u| + C$

116. $\int \operatorname{sech}^2 u\, du = \tanh u + C$

117. $\int \operatorname{csch}^2 u\, du = -\coth u + C$

118. $\int \operatorname{sech} u \tanh u\, du = -\operatorname{sech} u + C$

119. $\int \operatorname{csch} u \coth u\, du = -\operatorname{csch} u + C$

Forms Involving $\sqrt{2au - u^2}$

120. $\int \sqrt{2au - u^2}\, du = \dfrac{u-a}{2}\sqrt{2au - u^2} + \dfrac{a^2}{2}\cos^{-1}\left(\dfrac{a-u}{a}\right) + C$

121. $\int u\sqrt{2au - u^2}\, du = \dfrac{2u^2 - au - 3a^2}{6}\sqrt{2au - u^2} + \dfrac{a^3}{2}\cos^{-1}\left(\dfrac{a-u}{a}\right) + C$

122. $\int \dfrac{\sqrt{2au - u^2}}{u}\, du = \sqrt{2au - u^2} + a\cos^{-1}\left(\dfrac{a-u}{a}\right) + C$

123. $\int \dfrac{\sqrt{2au - u^2}}{u^2}\, du = -\dfrac{2\sqrt{2au - u^2}}{u} - \cos^{-1}\left(\dfrac{a-u}{a}\right) + C$

124. $\int \dfrac{du}{\sqrt{2au - u^2}} = \cos^{-1}\left(\dfrac{a-u}{a}\right) + C$

125. $\int \dfrac{u\, du}{\sqrt{2au - u^2}} = -\sqrt{2au - u^2} + a\cos^{-1}\left(\dfrac{a-u}{a}\right) + C$

126. $\int \dfrac{u^2\, du}{\sqrt{2au - u^2}} = -\dfrac{(u + 3a)}{2}\sqrt{2au - u^2} + \dfrac{3a^2}{2}\cos^{-1}\left(\dfrac{a-u}{a}\right) + C$

127. $\int \dfrac{du}{u\sqrt{2ua - u^2}} = -\dfrac{\sqrt{2au - u^2}}{au} + C$

Some Definite Integrals

128. $\displaystyle\int_0^{\pi/2} \sin^{2n} x\, dx = \int_0^{\pi/2} \cos^{2n} x\, dx$
$= \dfrac{\pi}{2}\dfrac{1\cdot 3\cdot 5\cdots(2n-1)}{2\cdot 4\cdot 6\cdots 2n},\ n = 1, 2, 3, \ldots$

129. $\displaystyle\int_0^{\pi/2} \sin^{2n+1} x\, dx = \int_0^{\pi/2} \cos^{2n+1} x\, dx$
$= \dfrac{2\cdot 4\cdot 6\cdots 2n}{1\cdot 3\cdot 5\cdots(2n+1)},\ n = 1, 2, 3, \ldots$

Answers to Test Yourself

1. false
2. true
3. false
4. true
5. 12
6. -243
7. $\dfrac{3x^3 + 8x}{\sqrt{x^2 + 4}}$
8. $2\left(x + \tfrac{3}{2}\right)^2 + \tfrac{1}{2}$
9. (a) 0, 7 (b) $-1 + \sqrt{6}, -1 - \sqrt{6}$ (c) 1 (d) 1
10. (a) $(5x + 1)(2x - 3)$ (b) $x^2(x + 3)(x - 5)$ (c) $(x - 3)(x^2 + 3x + 9)$ (d) $(x - 2)(x + 2)(x^2 + 4)$
11. false
12. false
13. true
14. $6; -6$
15. $-a + 5$
16. (a), (b), (d), (e), (g), (h), (i), (l)
17. (i) (d); (ii) (c); (iii) (a); (iv) (b)
18. (a) $-2 < x < 2$; (b) $|x| < 2$
19. ⟵─┤++++[─⟶
    $\quad\ -1\quad\ 3$
20. $(-\infty, -2) \cup \left(\tfrac{8}{3}, \infty\right)$
21. $(-\infty, -5] \cup [3, \infty)$
22. $(-\infty, -2) \cup [0, 1]$
23. fourth
24. $(5, -7)$
25. $-12; 9$
26. (a) $(1, -5)$ (b) $(-1, 5)$ (c) $(-1, -5)$
27. $(-2, 0), (0, -4), (0, 4)$
28. second and fourth
29. $x = 6 \ $ or $\ x = -4$
30. $x^2 + y^2 = 25$
31. $d(P_1, P_2) + d(P_2, P_3) = d(P_1, P_3)$
32. (c)
33. false
34. -27
35. 8
36. $\tfrac{2}{3}; (-9, 0); (0, 6)$
37. $y = -5x + 3$
38. $y = 2x - 14$
39. $y = -\dfrac{1}{3}x + 3$
40. $y = -\dfrac{5}{8}x$
41. $x - \sqrt{3}y + 4\sqrt{3} - 7 = 0$
42. (i) (g); (ii) (e); (iii) (h); (iv) (a); (v) (b); (vi) (f); (vii) (d); (viii) (c)
43. false
44. false
45. $4\pi/3$
46. 15
47. 0.23
48. $\cos t = -\dfrac{2\sqrt{2}}{3}$
49. $\sin\theta = \tfrac{3}{5}; \cos\theta = \tfrac{4}{5}; \tan\theta = \tfrac{3}{4}; \cot\theta = \tfrac{4}{3}; \sec\theta = \tfrac{5}{4}; \csc\theta = \tfrac{5}{3}$
50. $b = 10\tan\theta, c = 10\sec\theta$
51. $k = 10\ln 5$
52. $4 = 64^{1/3}$
53. $\log_b 125$
54. approximately 2.3347
55. 1000
56. true

Answers to Selected Odd-Numbered Problems

Exercises 1.1, Page 8

1. 24; 2; 8; 35,
3. 0; 1; 2; $\sqrt{6}$
5. $-\dfrac{3}{2}$; 0; $\dfrac{3}{2}$; $\sqrt{2}$
7. $-2x^2 + 3x$; $-8a^2 + 6a$; $-2a^4 + 3a^2$; $-50x^2 - 15x$; $-8a^2 - 2a + 1$; $-2x^2 - 4xh - 2h^2 + 3x + 3h$
9. $-2, 2$
11. $[\frac{1}{2}, \infty)$
13. $(-\infty, 1)$
15. $\{x | x \neq 0, x \neq 3\}$
17. $\{x | x \neq 5\}$
19. $(-\infty, \infty)$
21. $[-5, 5]$
23. $(-\infty, 0] \cup [5, \infty)$
25. $(-2, 3]$
27. not a function
29. function
31. domain: $[-4, 4]$; range: $[0, 5]$
33. domain: $[1, 9]$; range: $[1, 6]$
35. $(8, 0), (0, -4)$
37. $(\frac{3}{2}, 0), (\frac{5}{2}, 0), (0, 15)$
39. $(-1, 0), (2, 0), (0, 0)$
41. $(0, -\frac{1}{4})$
43. $(-2, 0), (2, 0), (0, 3)$
45. 0; -3.4; 0.3; 2; 3.8; 2.9; (0, 2)
47. 3.6; 2; 3.3; 4.1; 2; -4.1; $(-3.2, 0), (2.3, 0), (3.8, 0)$
49. $f_1(x) = \sqrt{x+5}, f_2(x) = -\sqrt{x+5}$; $[-5, \infty)$
51. (a) 2; 6; 120; 5040 (c) 5; 42
    (d) $(n+1)(n+2)(n+3)$

Exercises 1.2, Page 18

1. $-2x + 13$; $6x - 3$; $-8x^2 - 4x + 40$; $\dfrac{2x+5}{-4x+8}, x \neq 2$
3. $\dfrac{x^2+x+1}{x(x+1)}$; $\dfrac{x^2-x-1}{x(x+1)}$; $\dfrac{1}{x+1}$; $\dfrac{x^2}{x+1}, x \neq 0, x \neq -1$
5. $2x^2 + 5x - 7$; $-x + 1$; $x^4 + 5x^3 - x^2 - 17x + 12$; $\dfrac{x+3}{x+4}, x \neq 1, x \neq -4$
7. the interval $[1, 2]$
9. the interval $[1, 2)$
11. $3x + 16$; $3x + 4$
13. $x^6 + 2x^5 + x^4$; $x^6 + x^4$
15. $\dfrac{3x+3}{x}$; $\dfrac{3}{3+x}$
17. $(-\infty, -1] \cup [1, \infty)$
19. $[-\sqrt{5}, \sqrt{5}]$
21. $128x^9$; $\dfrac{1}{4x^9}$
23. $36x^2 - 36x + 15$
25. $-2x + 9$
27. $f(x) = 2x^2 - x, g(x) = x^2$
29. $(-2, 3), (3, -2)$
31. $(-8, 1), (-3, -4)$
33. $(-6, 2), (-1, -3)$
35. $(2, 1), (-3, -4)$

37. (a)
    (b)
    (c) (d)
    (e) (f)

39. (a) (b)
    (c) (d)
    (e) (f)

41. (a) (b)

(c) (d)

(e) (f)

(g) (h)

43. $y = (x - 1)^3 + 5$ **45.** $y = -(x + 7)^4$
47.   **49.** $10, 8, -1, 2, 0$

51.   **53.** $y = 2 - 3U(x - 2) + U(x - 3)$

55.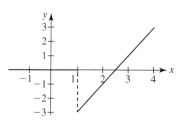

Exercises 1.3, Page 28

1. $y = \dfrac{2}{3}x + \dfrac{4}{3}$ **3.** $y = 2$

5. $y = -x + 3$

7. $\dfrac{3}{4}$; $(-4, 0), (0, 3)$; 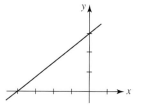   **9.** $\dfrac{2}{3}$; $\left(\dfrac{9}{2}, 0\right), (0, -3)$;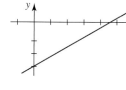

11. $y = -2x + 7$ **13.** $y = -3x - 2$

15. $y = -4x + 11$ **17.** $f(x) = \dfrac{1}{2}x + \dfrac{11}{2}$

19. $y = x + 3$

21. (a) $(0, 0), (-5, 0)$ (b) $y = \left(x + \dfrac{5}{2}\right)^2 - \dfrac{25}{4}$

(c) $\left(-\dfrac{5}{2}, -\dfrac{25}{4}\right)$; $x = -\dfrac{5}{2}$ (d)

(e) $\left[-\dfrac{25}{4}, \infty\right)$ (f) $\left[-\dfrac{5}{2}, \infty\right)$; $\left(-\infty, -\dfrac{5}{2}\right]$

23. (a) $(-1, 0), (3, 0), (0, 3)$ (b) $y = -(x - 1)^2 + 4$

(c) $(1, 4)$; $x = 1$ (d)

(e) $(-\infty, 4]$ (f) $(-\infty, 1]$; $[1, \infty)$

25. (a) $(1, 0), (2, 0), (0, 2)$ (b) $y = \left(x - \dfrac{3}{2}\right)^2 - \dfrac{1}{4}$

(c) $\left(\dfrac{3}{2}, -\dfrac{1}{4}\right)$; $x = \dfrac{3}{2}$ (d)

(e) $\left[-\dfrac{1}{4}, \infty\right)$ (f) $\left[\dfrac{3}{2}, \infty\right)$; $\left(-\infty, \dfrac{3}{2}\right]$

27. graph is shifted horizontally 10 units to the right
29. graph is compressed vertically, followed by a reflection in the x-axis, followed by a horizontal shift of 4 units to the left, followed by a vertical shift of 9 units upward
31. graph is shifted horizontally 6 units to the left, followed by a vertical shift of 4 units downward

33.

35.

37.

39.

41.

43. (f)

45. (e)

47. (b)

49. asymptotes: $x = -\frac{3}{2}$, $y = 2$; intercepts: $(\frac{9}{4}, 0)$, $(0, -3)$;

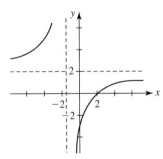

51. asymptotes: $x = 1$, $y = 0$; intercepts: $(0, 1)$;

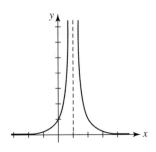

53. asymptotes: $x = -1$, $x = 1$, $y = 0$; intercepts: $(0, 0)$;

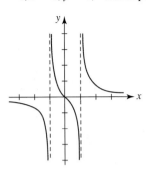

55. asymptotes: $x = 0$, $y = -1$; intercepts: $(-1, 0)$, $(1, 0)$;

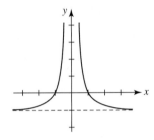

57. asymptotes: $x = 0$, $y = x$; intercepts: $(-3, 0)$, $(3, 0)$;

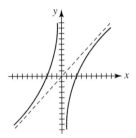

59. asymptotes: $x = -2$, $y = x - 2$; intercepts: $(0, 0)$;

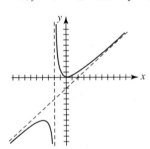

61. asymptotes: $x = 1$, $y = x - 1$; intercepts: $(-1, 0)$, $(3, 0)$, $(0, 3)$;

63. -1 is in the range of f, but 2 is not in the range of f

65. $T_F = \frac{9}{5}T_C + 32$

67. 1680; approximately 35.3 years

69. $t = 0$ and $t = 6$;

Exercises 1.4, Page 35

1.

3.

5.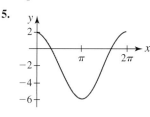

7. amplitude: 4; period: 2;
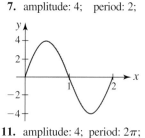

9. amplitude: 3; period: 1;

11. amplitude: 4; period: 2π;
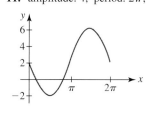

13. amplitude: 1; period: 3π;
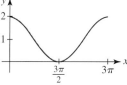

15. $y = -3\sin x$

17. $y = 1 - 3\cos x$

19. $y = 3\sin 2x$

21. $y = \dfrac{1}{2}\cos \pi x$

23. $y = -\sin \pi x$

25. amplitude: 1; period: 2π; phase shift: $\pi/6$;
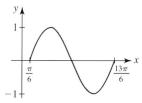

27. amplitude: 1; period: 2π; phase shift: $\pi/4$;
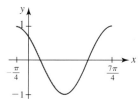

29. amplitude: 4; period: π; phase shift: $3\pi/4$;
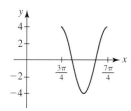

31. amplitude: 3; period: 4π; phase shift: $2\pi/3$;
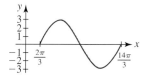

33. amplitude: 4; period: 6; phase shift: 1;
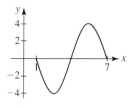

35. $y = 5\sin\left(\pi x - \dfrac{\pi}{2}\right)$

37. $(\pi/2, 0)$; $(\pi/2 + 2n\pi, 0)$, where n is an integer

39. $(n, 0)$, where n is an integer

41. $((2n + 1)\pi, 0)$, where n is an integer

43. $(\pi/4 + n\pi, 0)$, where n is an integer

45. period: 1; x-intercepts: $(n, 0)$, where n is an integer; asymptotes: $x = \tfrac{1}{2}(2n + 1)$, where n is an integer;

47. period: $\dfrac{\pi}{2}$; x-intercepts: $\left(\tfrac{1}{4}(2n + 1)\pi, 0\right)$, where n is an integer; asymptotes: $x = n\pi/2$, where n is an integer;
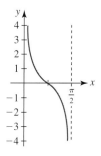

49. period: 2π; x-intercepts: $(\pi/2 + 2n\pi, 0)$, where n is an integer; asymptotes: $x = 3\pi/2 + 2n\pi$, where n is an integer;

51. period: 1; x-intercepts: $(\frac{1}{4} + n, 0)$, where n is an integer; asymptotes: $x = n$, where n is an integer;

53. period: 2; asymptotes: $x = n$, where n is an integer;

55. period: $2\pi/3$; asymptotes: $x = n\pi/3$, where n is an integer;

57.

59. (a) 978.0309 cm/s²   (b) 983.21642 cm/s²
    (c) 980.61796 cm/s²

Exercises 1.5, Page 46

1. because $f(0) = 1$ and $f(5) = 1$ **3.** not one-to-one
5. one-to-one   **7.** one-to-one
9. $f^{-1}(x) = \sqrt[3]{\dfrac{x-7}{3}}$ **11.** $f^{-1}(x) = \dfrac{2-x}{1-x}$
15. domain: $[0, \infty)$; range: $[-2, \infty)$
17. domain: $(-\infty, 0) \cup (0, \infty)$; range: $(-\infty, -3) \cup (-3, \infty)$
19. $(20, 2)$ **21.** $x = 12$
23.

25.

27. $f(x) = (5 - 2x)^2, x \geq \frac{5}{2}$; $f^{-1}(x) = \frac{1}{2}(5 - \sqrt{x})$
29. $f(x) = x^2 + 2x + 4, x \geq -1$; $f^{-1}(x) = -1 + \sqrt{x-3}$
33. $3\pi/4$ **35.** $\pi/4$
37. $3\pi/4$ **39.** $-\pi/3$
41. $\dfrac{4}{5}$ **43.** 2
45. $4\sqrt{2}/9$ **47.** $\sqrt{3}(2 + \sqrt{10})/9$
49. $\sqrt{1 - x^2}$ **51.** $\sqrt{1 + x^2}$
57. $\cos t = \sqrt{5}/5$, $\tan t = -2$, $\cot t = -\dfrac{1}{2}$, $\sec t = \sqrt{5}$, $\csc t = -\sqrt{5}/2$
63. (a) $\pi/4$ (b) 0.942 radian $\approx 53.97°$

Exercises 1.6, Page 53

1. $(0, 1)$; $y = 0$;

3. $(0, -1)$; $y = 0$;

5. $(0, -4)$; $y = -5$;

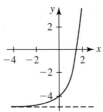

7. $f(x) = 6^x$

9. $f(x) = e^{-2x}$ **11.** $x > 4$
13. $x < 2$ **15.**

17.

19.

21.

23. $-\dfrac{1}{2} = \log_4 \dfrac{1}{2}$

25. $4 = \log_{10} 10{,}000$

27. $2^7 = 128$

29. $(\sqrt{3})^8 = 81$

31. $f(x) = \log_7 x$

33. e

35. 36

37. $\dfrac{1}{7}$

39. $(0, \infty)$; $(1, 0)$; $x = 0$;

41. $(-1, \infty)$; $(0, 0)$; $x = -1$;

43. the interval $(-3, 3)$

45. $(-1, 0), (1, 0)$; $x = 0$;

47. $\ln(x^2 - 2)$

49. 0

51. $10 \ln x + \dfrac{1}{2}\ln(x^2 + 5) - \dfrac{1}{3}\ln(8x^3 + 2)$

53. $5\ln(x^3 - 3) + 8\ln(x^4 + 3x^2 + 1) - \dfrac{1}{2}\ln x - 9\ln(7x + 5)$

55. $\log_6 51 = \dfrac{\ln 51}{\ln 6} \approx 2.1944$

57. $-5 + \dfrac{\ln 9}{\ln 2} \approx 1.8301$

59. $\dfrac{1 + \ln 2}{-1 + \ln 5} \approx 2.7782$

61. 3

63. (a) $P(t) = P_0 e^{0.3466t}$ **(b)** $5.66 P_0$ **(c)** 8.64 h

65. (a) 82 **(b)** 8.53 days **(c)** 2000 **(d)**

Exercises 1.7, Page 59

1. $S(x) = x + \dfrac{50}{x}$; $(0, \infty)$

3. $S(x) = 3x^2 - 4x + 2$; $[0, 1]$

5. $A(x) = 100x - x^2$; $[0, 100]$

7. $A(x) = 2x - \dfrac{1}{2}x^2$; $[0, 4]$

9. $d(x) = \sqrt{2x^2 + 8}$; $(-\infty, \infty)$

11. $P(A) = 4\sqrt{A}$; $(0, \infty)$

13. $d(C) = C/\pi$; $(0, \infty)$

15. $A(h) = \dfrac{1}{\sqrt{3}} h^2$; $(0, \infty)$

17. $A(x) = \dfrac{1}{4\pi} x^2$; $(0, \infty)$

19. $C(x) = 8x + \dfrac{3200}{x}$; $(0, \infty)$

21. $S(w) = 3w^2 + \dfrac{1200}{w}$; $(0, \infty)$

23. $d(t) = 20\sqrt{13t^2 + 8t + 4}$; $(0, \infty)$

25. $V(h) = \begin{cases} 120h^2, & 0 \le h < 5 \\ 1200h - 3000, & 5 \le h \le 8 \end{cases}$; $[0, 8]$

27. $h(\theta) = 300 \tan \theta$; $(0, \pi/2)$

29. $L(\theta) = 3\csc\theta + 4\sec\theta$; $(0, \pi/2)$

31. $\theta(x) = \tan^{-1}(1/x) - \tan^{-1}(1/2x)$; $(0, \infty)$

Chapter 1 in Review, Page 61

A. 1. false
3. true
5. false
7. true
9. false
11. true
13. true
15. true
17. true
19. true

B. 1. $[-2, 0) \cup (0, \infty)$
3. $(-8, 6)$
5. $(1, 0)$; $(0, 0), (5, 0)$
7. $\left(0, -\dfrac{4}{5}\right)$
9. 6
11. 0
13. $(3, 5)$
15. $\log_3 5 = \dfrac{\ln 5}{\ln 3}$
17. $\dfrac{1}{9}$
19. $y = \ln x$

C. 1. (a) 3 **(b)** 0 **(c)** -2 **(d)** 0 **(e)** 2.5
  (f) 2 **(g)** 1 **(h)** 0 **(i)** 3 **(j)** 4

3. 1 and 8 are in the range; 5 is not in the range

5. $-3x^2 + 4x - 3xh - h^2 + 2h - 1$

7. (f)
9. (d)
11. (h)
13. (c)
15. (b)
17. $\dfrac{3^{1-h} - 3}{h}$

19. (a) ab **(b)** b/a **(c)** $1/b$

21. $f(x) = 5e^{(-\frac{1}{6}\ln 5)x} = 5e^{-0.2682x}$
23. $f(x) = 5 + \left(\dfrac{1}{2}\right)^x$

25. (b)
27. (d)

29. (c)

31. (a) $V = 6l^3$ (b) $V = \frac{2}{9}w^3$ (c) $V = \frac{3}{4}h^3$

33. $V(\theta) = 360 + 75\cot\theta$

35. $A(\phi) = 100\cos\phi + 50\sin 2\phi$ **37.** $V(x) = 2\sqrt{3}(1 - x^2)$

Exercises 2.1, Page 72

1. 8
3. does not exist
5. 2
7. does not exist
9. 0
11. 3
13. 0
15. (a) 1 (b) -1 (c) 2 (d) does not exist
17. (a) 2 (b) -1 (c) -1 (d) -1
19. correct
21. $\lim_{x\to 1^-}\sqrt{1-x} = 0$
23. $\lim_{x\to 0^+}\lfloor x\rfloor = 0$
25. correct
27. $\lim_{x\to 3^-}\sqrt{9-x^2} = 0$
29. (a) -1 (b) 0 (c) -3 (d) -2 (e) 0 (f) 1
35. does not exist
37. $-\frac{1}{4}$
39. -2
41. -3
43. 0
45. $\frac{1}{3}$
47. $\frac{1}{4}$
49. 5

Exercises 2.2, Page 80

1. 15
3. -12
5. 4
7. 4
9. $-\frac{8}{5}$
11. 14
13. $\frac{28}{9}$
15. -1
17. $\sqrt{7}$
19. does not exist
21. -10
23. 3
25. 60
27. 14
29. $\frac{1}{5}$
31. $-\frac{1}{8}$
33. 3
35. does not exist
37. 2
39. $\frac{128}{3}$
41. -2
43. $a^2 - 2ab + b^2$
45. 16
47. $-1/x^2$
49. $\frac{1}{2}$
51. $\frac{1}{5}$
53. 32
55. $\frac{1}{2}$
57. does not exist
59. $8a$

Exercises 2.3, Page 86

1. none
3. 3 and 6
5. $n\pi/2, n = 0, \pm 1, \pm 2,\ldots$
7. 2
9. none
11. e^{-2}
13. (a) continuous (b) continuous
15. (a) continuous (b) continuous
17. (a) not continuous (b) not continuous
19. (a) continuous (b) not continuous
21. (a) not continuous (b) not continuous
23. (a) not continuous (b) continuous
25. $m = 4$
27. $m = 1; n = 3$
29. discontinuous at $n/2$, where n is an integer;

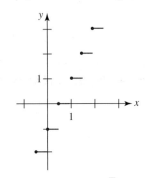

31. define $f(9) = 6$
33. $\frac{\sqrt{3}}{2}$
35. 0
37. 1
39. 1
41. $-\pi/6$
43. $(-3, \infty)$
45. $c = 4$
47. $c = 0, c = \pm\sqrt{2}$
55. $-1.22, -0.64, 1.34$
57. 2.21
59. 0.78

Exercises 2.4, Page 93

1. $\frac{3}{2}$
3. 0
5. 1
7. 4
9. 0
11. 36
13. $\frac{1}{2}$
15. does not exist
17. 3
19. $\frac{3}{7}$
21. 0
23. -4
25. 4
27. $\frac{1}{2}$
29. 5
31. $\frac{1}{6}$
33. 8
35. $\sqrt{2}$
37. $\frac{\sqrt{2}}{2}$
43. 3

Exercises 2.5, Page 102

1. $-\infty$
3. ∞
5. ∞
7. ∞
9. $\frac{1}{4}$
11. 5
13. $-\frac{1}{4}$
15. $\frac{5}{2}$

17. $\dfrac{1}{\sqrt{2}}$ **19.** 0

21. 1   **23.** $-\pi/6$

25. $-4; 4$ **27.** $-\dfrac{2}{\sqrt{3}}, \dfrac{2}{\sqrt{3}}$

29. $-1; 1$ **31.** $-1; 1$

33. VA: none;   HA: $y = 0$;

35. VA: $x = -1$;   HA: none;

37. VA: $x = 0, x = 2$;   HA: $y = 0$;

39. VA: $x = 1$;   HA: $y = 1$;

41. VA: none;   HA: $y = -1, y = 1$;

43. (a) 2   (b) $-\infty$ (c) 0   (d) 2
45. (a) $-\infty$ (b) -1 (c) ∞ (d) 0
51. 3

Exercises 2.6, Page 110

1. choose $\delta = \varepsilon$ **3.** choose $\delta = \varepsilon$
5. choose $\delta = \varepsilon$ **7.** choose $\delta = \varepsilon/3$
9. choose $\delta = 2\varepsilon$ **11.** choose $\delta = \varepsilon$
13. choose $\delta = \varepsilon/8$ **15.** choose $\delta = \sqrt{\varepsilon}$

17. choose $\delta = \varepsilon^2/5$ **19.** choose $\delta = \varepsilon/2$
21. choose $\delta = \min\{1, \varepsilon/7\}$ **23.** choose $\delta = \sqrt{\varepsilon}$
25. choose $\delta = \sqrt{a\varepsilon}$ **31.** choose $N = 7/(4\varepsilon)$
33. choose $N = -30/\varepsilon$

Exercises 2.7, Page 116

1. -4.5;

3. 7;

5. $\dfrac{3\sqrt{3} - 6}{\pi}$;

7. $m_{\tan} = 6$; $y = 6x - 15$

9. $m_{\tan} = -1$; $y = -x - 1$
11. $m_{\tan} = -23$; $y = -23x + 32$
13. $m_{\tan} = -\dfrac{1}{2}$; $y = -\dfrac{1}{2}x - 1$
15. $m_{\tan} = 2$; $y = 2x + 1$
17. $m_{\tan} = \dfrac{1}{4}$; $y = \dfrac{1}{4}x + 1$
19. $m_{\tan} = \dfrac{\sqrt{3}}{2}$; $y = \dfrac{\sqrt{3}}{2}x - \dfrac{\sqrt{3}\pi}{12} + \dfrac{1}{2}$
21. not a tangent line   **23.** $y = x - 2$; $(0, -2)$
25. $m_{\tan} = -2x + 6$; $(3, 10)$
27. $m_{\tan} = 3x^2 - 3$; $(-1, 2), (1, -2)$
29. 58 mi/h   **31.** 3.8 h
33. -14
35. (a) -4.9 m/s   (b) 5 s   (c) -49 m/s
37. (a) 448 ft;   960 ft;   1008 ft;   960 ft
   (b) 144 ft/s   (d) 16 s   (e) $-32t + 256$
   (f) -256 ft/s   (g) 1024 ft

Chapter 2 in Review, Page 118

A. 1. true   **3.** false
5. false   **7.** true
9. false   **11.** false
13. true   **15.** true
17. false   **19.** true
21. false

B. 1. 4   **3.** $-\dfrac{1}{5}$
5. 0   **7.** ∞

9. 1
11. 3^-
13. $-\infty$
15. -2
17. 10
19. continuous
21. 9

C. 5. (a), (e), (f), (h)
7. (c), (h)
9. (b), (c), (d), (e), (f)

11. ; continuous everywhere

13. $(-\infty, -1), (-1, 0), (0, 1), (1, \infty)$
15. $(-\infty, -\sqrt{5}), (\sqrt{5}, \infty)$
17. $\dfrac{1}{6}$
19. $y = 4x + 24$
21. $y = 8x - 6$
23. $y = \dfrac{1}{2}x + \dfrac{3}{2}$

Exercises 3.1, Page 128

1. 0
3. -3
5. $6x$
7. $-2x + 4$
9. $2x + 2$
11. $3x^2 + 1$
13. $-3x^2 + 30x - 1$
15. $-2/(x + 1)^2$
17. $5/(x + 4)^2$
19. $-1/(2x^{3/2})$
21. $y = -x - 4$
23. $y = 2x - 2$
25. $(-4, -6)$
27. $(1, -2), (-1, 2)$
29. $x; \left(3, \dfrac{7}{2}\right)$
31. $-3x^2;\ (2, -4), (-2, 12)$
33. $f'_+(2) = 2$ but $f'_-(2) = -1$
35. $20a$
37. $3a^2 - 8a$
39. $4/(3 - a)^2$
41. $y = \dfrac{1}{2}x + 3;\ f(-3) = \dfrac{3}{2};\ f'(-3) = \dfrac{1}{2}$

43.
45.
47.
49. (e)
51. (b)
53. (a)

Exercises 3.2, Page 136

1. 0
3. $9x^8$
5. $14x - 4$
7. $2x^{-1/2} + 4x^{-5/3}$
9. $x^4 - 12x^3 + 18x$
11. $20x^4 - 20x^3 - 18x^2$
13. $6x^5 + 40x^3 + 50x$
15. $16 + 4/\sqrt{x}$
17. $192u^2$
19. $-1/r^2 - 2/r^3 - 3/r^4 - 4/r^5$
21. $y = 6x + 3$
23. $y = \dfrac{1}{4}x + 5$
25. $(4, -11)$
27. $(3, -25), (-1, 7)$
29. $y = \dfrac{1}{4}x - \dfrac{7}{2}$
31. $x = 4$
33. -2
35. 32
37. $60/x^4$
39. $1440x^2 + 120x$
41. $(-4, \infty), (-\infty, -4)$
43. $(-4, 48)$
45. $(1, \infty), (-\infty, 1)$
49. $(2, 8)$
51. $\left(\dfrac{1}{4}, -\dfrac{3}{16}\right)$
53. $y = -7x$
55. $S = 4\pi r^2$
57. -15 N

Exercises 3.3, Page 142

1. $5x^4 - 9x^2 + 4x - 28$
3. $8x^{-7/3} - 4x^{-5/6} + 12^{1/2}$
5. $-20x/(x^2 + 1)^2$
7. $-17/(5 - 2x)^2$
9. $72x - 12$
11. $(2x^5 + x^2 - 40x - 12)/x^4$
13. $(x^2 + 2x)/(2x^2 + x + 1)^2$
15. $18x^2 + 22x + 6$
17. $(6x^2 + 8x - 3)/(3x + 2)^2$
19. $(2x^3 + 8x^2 - 6x - 8)/(x + 3)^2$
21. $y = -4x + 1$
23. $y = 7x - 1$
25. $(0, 24), (\sqrt{5}, -1), (-\sqrt{5}, -1)$
27. $(0, 0), \left(-1, \dfrac{1}{2}\right), \left(1, \dfrac{1}{2}\right)$
29. $\left(3, \dfrac{3}{2}\right), \left(-5, \dfrac{1}{2}\right)$
31. $(-4, 0), (-6, 2)$
33. $k = -21$
35. -28
37. $\dfrac{11}{3}$
39. -30
41. $\dfrac{13}{2}$
43. $(x^2 f''(x) - 2xf'(x) + 2f(x))/x^3$
45. $f'(x) > 0$ on $(-\infty, 0) \cup (0, 1)$; $f'(x) < 0$ on $(1, 2) \cup (2, \infty)$
47. $f'(x) > 0$ on $\left(-\infty, \dfrac{5}{8}\right)$; $f'(x) < 0$ on $\left(\dfrac{5}{8}, \infty\right)$
49. $-16\,km_1 m_2$
51. $-\dfrac{RT}{(V - b)^2} + \dfrac{2a}{V^3}$

Exercises 3.4, Page 147

1. $2x + \sin x$
3. $7\cos x - \sec^2 x$
5. $x\cos x + \sin x$
7. $(x^3 - 2)\sec^2 x + 3x^2 \tan x$
9. $x^2 \sec x \tan x + 2x \sec x + \sec^2 x$
11. 0
13. $\cos x$
15. $\dfrac{-x\csc^2 x - \csc^2 x - \cot x}{(x + 1)^2}$
17. $\dfrac{-2x^2 \sec^2 x + 4x \tan x + 2x}{(1 + 2\tan x)^2}$
19. $\dfrac{1}{1 + \cos x}$
21. $x^4 \sin x \sec^2 x + x^4 \sin x + 4x^3 \sin x \tan x$
23. $y = -\dfrac{\sqrt{3}}{2}x + \dfrac{1}{2} + \dfrac{\sqrt{3}\pi}{6}$
25. $y = \dfrac{2}{3}x + \dfrac{2}{\sqrt{3}} - \dfrac{\pi}{9}$
27. $\pi/6, 5\pi/6$
29. $\pi/2$
31. $y = 2x - \dfrac{\sqrt{3}}{2} - \dfrac{8\pi}{3}$
33. $y = x - 2\pi$
35. $2(\cos^2 x - \sin^2 x) = 2\cos 2x$
37. $2\cos x - x\sin x$
39. $\dfrac{-x^2 \sin x - 2x\cos x + 2\sin x}{x^3}$
41. $\csc x \cot^2 x + \csc^3 x$
45. $-\dfrac{160}{3}$; as the angle of elevation increases, the length s of the shadow decreases

53. not differentiable at $0, \pm\pi, \pm 2\pi, \ldots$

55. **(b)** $-\dfrac{14(0.2\cos\theta - \sin\theta)}{(0.2\sin\theta + \cos\theta)^2}$ **(c)** 0.1974 radians
  (d) approximately 13.7281
  (e) The minimum force required to pull the sled is about 13.73 lb when θ is approximately 0.1974 radians or 11.31°.

Exercises 3.5, Page 155

1. $-150(-3x)^{29}$
3. $200(2x^2 + x)^{199}(4x + 1)$
5. $-4(x^3 - 2x^2 + 7)^{-5}(3x^2 - 2x)$
7. $-2(3x - 1)^3(-2x + 9)^4(27x - 59)$
9. $\dfrac{\cos\sqrt{2x}}{\sqrt{2x}}$
11. $\dfrac{2x}{\sqrt{x^2 - 1}(x^2 + 1)^{3/2}}$
13. $10(1 + 6x(x^2 - 4)^2)(x + (x^2 - 4)^3)^9$
15. $\dfrac{5x^{14} + 9x^{13} + 13x^{12}}{(x^2 + x + 1)^5}$
17. $\pi\cos(\pi x + 1)$
19. $15\sin^2 5x \cos 5x$
21. $-3x^5\sin x^3 + 3x^2\cos x^3$
23. $10(2 + x\sin 3x)^9(3x\cos 3x + \sin 3x)$
25. $-x^{-2}\sec^2(1/x)$
27. $-3\sin 2x \sin 3x + 2\cos 2x \cos 3x$
29. $5(\sec 4x + \tan 2x)^4(4\sec 4x \tan 4x + 2\sec^2 2x)$
31. $2\cos 2x \cos(\sin 2x)$
33. $-(2x + 5)^{-1/2}\cos\sqrt{2x + 5}\sin(\sin\sqrt{2x + 5})$
35. $24x\sin^2(4x^2 - 1)\cos(4x^2 - 1)$
37. $360x^2(1 + x^3)^3(1 + (1 + x^3)^4)^4(1 + (1 + (1 + x^3)^4)^5)^5$
39. -54
41. -7
43. $y = -8x - 3$
45. $y = 6x - 1 - \dfrac{3\pi}{2}$
47. $y = \dfrac{\sqrt{6}}{4} + \dfrac{12}{\pi(2\sqrt{2} + 3\sqrt{6})}\left(x - \dfrac{1}{2}\right)$
49. $-\pi^3\cos\pi x$
51. $-125x\cos 5x - 75\sin 5x$
53. $(\sqrt{3}/3, 3\sqrt{3}/16), (-\sqrt{3}/3, -3\sqrt{3}/16)$; no
55. $\dfrac{1}{18}$
57. If $0 \le \theta \le \pi$, then $\theta = \pi/4$ or $\theta = 3\pi/4$.
59. $dr/dt = 5/(8\pi)$ in/min

Exercises 3.6, Page 160

1. $4x^2y^3\dfrac{dy}{dx} + 2xy^4$
3. $-2y\sin y^2\dfrac{dy}{dx}$
5. $\dfrac{1}{2y - 2}$
7. $\dfrac{2x - y^2}{2xy}$
9. $\dfrac{2x}{3 - \sin y}$
11. $\dfrac{4x - 3x^2y^2}{2x^3y - 2y}$
13. $\dfrac{x^2 - 4x(x^2 + y^2)^5}{y^2 + 4y(x^2 + y^2)^5}$
15. $\dfrac{2x^4y^4 + 3y^{10} - 6x^9y}{6xy^9 - 3x^{10}}$
17. $\dfrac{1 - x}{y + 4}$
19. $\dfrac{3}{2y(x + 2)^2}$
21. $\dfrac{\cos(x + y) - y}{x - \cos(x + y)}$
23. $\cos y \cot y$
25. $\dfrac{\cos 2\theta}{r}$
27. $-\dfrac{2}{5}$
29. $-\dfrac{1}{3}$ and $-\dfrac{2}{3}$
31. $y = \dfrac{8}{3}x + \dfrac{22}{3}$
33. $y = \dfrac{1}{2}x - \dfrac{1}{2} + \dfrac{\pi}{4}$
35. $(1, 2), (-1, -2)$
37. $(-\sqrt{5}, 2\sqrt{5}), (\sqrt{5}, -2\sqrt{5})$
39. $(8, 4)$
41. $\dfrac{y^3 - 2x^2}{y^5}$
43. $\dfrac{-25}{y^3}$
45. $\dfrac{-\sin y}{(1 - \cos y)^3}$
47. $\dfrac{-2}{(y - x)^3}$
49. $\dfrac{2x - 1}{2\sqrt{x^2 - x}}, -\dfrac{2x - 1}{2\sqrt{x^2 - x}}$
51. $\dfrac{-2x - 3}{x^4}$
53. $y = 1 - \sqrt{x - 2}$
55. $y = \begin{cases} \sqrt{4 - x^2}, & -2 \le x < 0 \\ -\sqrt{4 - x^2}, & 0 \le x < 2 \end{cases}$
57. $\dfrac{dy}{dt} = -\dfrac{x}{y}\dfrac{dx}{dt}$
59. **(a)** $y = -x + 3$ **(b)** $(\sqrt[3]{2}, \sqrt[3]{4})$
65. **(b)** $\dfrac{4(252 - x^2)}{(x^2 + 252)^2 + 16x^2}$ **(c)** $x = 6\sqrt{7} \approx 15.87$ ft

Exercises 3.7, Page 167

1. $f'(x) > 0$ for all x shows that f is increasing on $(-\infty, \infty)$. It follows from Theorem 3.7.3 that f is one-to-one.
3. $f(0) = 0, f(1) = 0$ implies f is not one-to-one.
5. $\dfrac{2}{3}$
7. $(f^{-1})'(x) = -1/(x - 2)^2$
9. $(5, 3);\ y = \dfrac{1}{10}x + \dfrac{5}{2}$
11. $(8, 1);\ y = \dfrac{1}{60}x + \dfrac{13}{15}$
13. $\dfrac{5}{\sqrt{1 - (5x - 1)^2}}$
15. $\dfrac{-8}{4 + x^2}$
17. $\dfrac{1}{1 + x} + \dfrac{\tan^{-1}\sqrt{x}}{\sqrt{x}}$
19. $\dfrac{2(\cos^{-1}2x + \sin^{-1}2x)}{\sqrt{1 - 4x^2}(\cos^{-1}2x)^2}$
21. $\dfrac{-2x}{(1 + x^4)(\tan^{-1}x^2)^2}$
23. $\dfrac{2 - x}{\sqrt{1 - x^2}} + \cos^{-1}x$
25. $3\left(x^2 - 9\tan^{-1}\dfrac{x}{3}\right)^2\left(2x - \dfrac{27}{9 + x^2}\right)$
27. $\dfrac{1}{t^2 + 1}$
29. $\dfrac{-4\sin 4x}{|\sin 4x|}$
31. $\dfrac{2x\sec^2(\sin^{-1}x^2)}{\sqrt{1 - x^4}}$
33. $\dfrac{2x(1 + y^2)}{1 - 2y - 2y^3}$
35. $\sin^{-1}x + \cos^{-1}x = $ constant
37. $\sqrt{3}/3$
39. $y = \dfrac{2 + \pi}{4}x - \dfrac{1}{2}$
41. $(5\pi/6, 4), (7\pi/6, 6)$

Exercises 3.8, Page 171

1. $-e^{-x}$
3. $\dfrac{e^{\sqrt{x}}}{2\sqrt{x}}$
5. $5^{2x}(2\ln 5)$
7. $x^2 e^{4x}(3 + 4x)$
9. $\dfrac{-e^{-2x}(2x + 1)}{x^2}$
11. $-\dfrac{5}{2}(1 + e^{-5x})^{-1/2}e^{-5x}$
13. $-\dfrac{e^{x/2} - e^{-x/2}}{(e^{x/2} + e^{-x/2})^2}$
15. $8e^{8x}$
17. $3e^{3x-3}$
19. $\dfrac{1}{3}x^{-2/3}e^{x^{1/3}} + \dfrac{1}{3}e^{x/3}$

21. $\sec^2 e^x - e^{-x}\tan e^x$

23. $\dfrac{e^{x\sqrt{x^2+1}}(2x^2+1)}{\sqrt{x^2+1}}$

25. $2xe^{x^2}e^{e^{x^2}}$

27. $y = 4x + 4$

29. $(\ln 3, 3)$

31. $x = \pi/4 + n\pi,\ n = 0, \pm 1, 2, \ldots$

33. $4e^{x^2}(2x^3 + 3x)$

35. $4e^{2x}\cos e^{2x} - 4e^{4x}\sin e^{2x}$

41. $\dfrac{e^{x+y}}{1 - e^{x+y}}$

43. $\dfrac{-ye^{xy}\sin e^{xy}}{1 + xe^{xy}\sin e^{xy}}$

45. $\dfrac{-y^2 + ye^{x/y}}{2y^3 + xe^{x/y}}$

47. (a)   (b) $f'(x) = \begin{cases} e^x, & x > 0 \\ -e^{-x}, & x < 0 \end{cases}$

(c)   (d) no

37. $\dfrac{2}{x^3}$

39. $\dfrac{2 - 2\ln|x|}{x^2}$

43. $\dfrac{y}{2xy^2 - x}$

45. $\dfrac{y - xy}{2xy^2 + x}$

47. $\dfrac{2x - x^2 y - y^3}{x^3 + xy^2 - 2y}$

49. $x^{\sin x}\left[\dfrac{\sin x}{x} + (\cos x)\ln x\right]$

51. $x(x-1)^x\left[\dfrac{1}{x} + \dfrac{x}{x-1} + \ln(x-1)\right]$

53. $\dfrac{\sqrt{(2x+1)(3x+2)}}{4x+3}\left[\dfrac{1}{2x+1} + \dfrac{3/2}{3x+2} - \dfrac{4}{4x+3}\right]$

55. $\dfrac{(x^3-1)^5(x^4+3x^3)^4}{(7x+5)^9}\left[\dfrac{15x^2}{x^3-1} + \dfrac{16x^3+36x^2}{x^4+3x^3} - \dfrac{63}{7x+5}\right]$

57. $y = 3x - 2$

59. $(e^{-1}, e^{-e^{-1}})$;

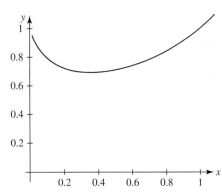

65. (b) one interval is $(\pi, 2\pi)$ **67.** $4 - 4\ln 4 \approx -1.55$

Exercises 3.10, Page 185

1. $\cosh x = \sqrt{5}/2,\ \tanh x = -\sqrt{5}/5,\ \coth x = -\sqrt{5}$, $\operatorname{sech} x = 2\sqrt{5}/5,\ \operatorname{csch} x = -2$

3. $10\sinh 10x$

5. $\dfrac{1}{2}x^{-1/2}\operatorname{sech}^2\sqrt{x}$

7. $-6(3x-1)\operatorname{sech}(3x-1)^2\tanh(3x-1)^2$

9. $-3\sinh 3x\operatorname{csch}^2(\cosh 3x)$

11. $3\sinh 2x\sinh 3x + 2\cosh 2x\cosh 3x$

13. $2x^2\sinh x^2 + \cosh x^2$

15. $3\sinh^2 x\cosh x$

17. $\dfrac{2}{3}(x - \cosh x)^{-1/3}(1 - \sinh x)$ **19.** $4\tanh 4x$

21. $\dfrac{e^x + 1}{(1 + \cosh x)^2}$

23. $e^{\sinh t}\cosh t$

25. $\dfrac{\cos t + \cos t\sinh 2t - 2\sin t\cosh 2t}{(1 + \sinh 2t)^2}$

27. $y = 3x$

29. $(0, -2), (-2, 2\cosh 2 - 4\sinh 2), (2, 2\cosh 2 - 4\sinh 2)$

31. $-2\operatorname{sech}^2 x\tanh x$

35. $\dfrac{3}{\sqrt{9x^2+1}}$

37. $\dfrac{-2x}{1-(1-x^2)^2}$

39. $\sec x$

41. $\dfrac{3x^3}{\sqrt{x^6+1}} + \sinh^{-1}x^3$

43. $-\dfrac{1}{x^2\sqrt{1-x^2}} - \dfrac{\operatorname{sech}^{-1}x}{x^2}$

45. $\dfrac{-1}{x\sqrt{1-x^2}\operatorname{sech}^{-1}x}$

47. $\dfrac{3}{\sqrt{\cosh^{-1}6x}\sqrt{36x^2-1}}$

49. (b) $v_{\text{ter}} = \sqrt{mg/k}$ (c) 56 m/s

49. (b) $P = 0, P = 2$ (c)   (d) $t = 0$

61. $f'(0) = 0$

Exercises 3.9, Page 177

1. $\dfrac{10}{x}$

3. $\dfrac{1}{2x}$

5. $\dfrac{4x^3 + 6x}{x^4 + 3x^2 + 1}$

7. $3x + 6x\ln x$

9. $\dfrac{1 - \ln x}{x^2}$

11. $\dfrac{1}{x(x+1)}$

13. $\tan x$

15. $\dfrac{-1}{x(\ln x)^2}$

17. $\dfrac{1 + \ln x}{x\ln x}$

19. $\dfrac{1}{4x\sqrt{\ln\sqrt{x}}}$

21. $\dfrac{2}{t} + \dfrac{2t}{t^2 + 2}$

23. $\dfrac{1}{x+1} + \dfrac{1}{x+2} - \dfrac{1}{x+3}$

25. $y = x - 1$

27. 4

29. -8

31. (e, e^{-1})

33. $\dfrac{1}{\sqrt{x^2 - 1}}$

35. $\sec x$

Chapter 3 in Review, Page 186

A. 1. false 3. false
5. true 7. true
9. true 11. true
13. false 15. true
17. false 19. true

B. 1. 0 3. $-\dfrac{1}{4}$
5. $y = -\dfrac{5}{4}x - \dfrac{3}{2}$ 7. -3
9. 23
11. $-16F'(\sin 4x)\sin 4x + 16F''(\sin 4x)\cos^2 x$
13. $a = 6;\ b = -9$ 15. $(1, 5)$
17. $\dfrac{1}{x(\ln 10)}$ 19. catenary

C. 1. $0.08x^{-0.9}$
3. $10(t + \sqrt{t^2 + 1})^9(1 + t(t^2 + 1)^{-1/2})$
5. $x^2(x^4 + 16)^{1/4}(x^3 + 8)^{-2/3} + x^3(x^4 + 16)^{-3/4}(x^3 + 8)^{1/3}$
7. $-\dfrac{16x\sin 4x + 4\sin 4x + 4\cos 4x}{(4x + 1)^2}$
9. $10x^3 \sin 5x \cos 5x + 3x^2 \sin^2 5x$
11. $\dfrac{-3}{|x|\sqrt{x^2 - 9}}$ 13. $\dfrac{1}{(\cot^{-1}x)^2(1 + x^2)}$
15. $\dfrac{-4x^2}{\sqrt{1 - x^2}}$ 17. $-xe^{-x}$
19. $7x^6 + 7^x(\ln 7) + 7e^{7x}$ 21. $\dfrac{1}{x} + \dfrac{2}{4x - 1}$
23. $\dfrac{1}{\sqrt{(\sin^{-1}x)^2 + 1}\sqrt{1 - x^2}}$
25. $e^{x\cosh^{-1}x}\left[\dfrac{x^2}{\sqrt{x^2 - 1}} + x\cosh^{-1}x + 1\right]$
27. $3x^2 e^{x^3}\cosh e^{x^3}$ 29. $\dfrac{405}{8\sqrt{1 + 3x}}$
31. $\dfrac{120}{t^6}$ 33. $4e^{\sin 2x}(\cos^2 2x - \sin 2x)$
35. $\dfrac{4}{x + 5} - \dfrac{3}{2 - x} - \dfrac{10}{x + 8} - \dfrac{2}{6x + 4}$
37. $\dfrac{1}{4}$ 39. $\dfrac{e^x - y^2}{2xy + e^y}$
41. $y = \dfrac{1}{3}x - \dfrac{2}{27},\ y = \dfrac{1}{3}x + \dfrac{2}{27}$ 43. $y = 6x - 9,\ y = -6x - 9$
45. $(4, 2)$ 47. $0, 2\pi/3, \pi, 4\pi/3, 2\pi$
53. (a) $(2, 0), (2, -1), (2, 1)$ (b) $4, -2, -2$
55. $y = \sqrt{3}x - \dfrac{\sqrt{3}}{2},\ y = -\sqrt{3}x + \dfrac{\sqrt{3}}{2}$

Exercises 4.1, Page 195

1. $-1, 19;\ -2, 18;\ 2, 18;\ 8, 8$
3. $18, 6;\ -23, 1;\ 23, 1;\ 18, -6$
5. $-\dfrac{15}{4}, 0;\ 17, 2;\ 17, 2;\ -128, -2$

7. $1, \dfrac{1}{2};\ 1 - \pi, 1;\ \pi - 1, 1;\ 0, \pi^2$
9. (a) $-6, 6$ (b) $-8, 8$
11. (a) $-6\sqrt{2}, 6\sqrt{2}$ (b) 15 (c) $-4, 8$
13. slowing down on the time intervals $(-\infty, -3), (0, 3)$; speeding up on the time intervals $(-3, 0), (3, \infty)$
15. $v(t) = 2t, a(t) = 2$; slowing down on the time interval $(-1, 0)$; speeding up on the time interval $(0, 3)$;

17. $v(t) = 2t - 4, a(t) = 2$; slowing down on the time interval $(-1, 2)$; speeding up on the time interval $(2, 5)$;

19. $v(t) = 6t^2 - 12t, a(t) = 12t - 12$; slowing down on the time intervals $(-2, 0), (1, 2)$; speeding up on the time intervals $(0, 1), (2, 3)$;

21. $v(t) = 12t^3 - 24t^2, a(t) = 36t^2 - 48t$;

23. $v(t) = 1 - 2t^{-1/2}, a(t) = t^{-3/2}$;

25. $v(t) = \dfrac{\pi}{2}\cos\dfrac{\pi}{2}t, a(t) = -\left(\dfrac{\pi}{2}\right)^2\sin\dfrac{\pi}{2}t$;

27. $v(t) = e^{-t}(-t^3 + 3t^2), a(t) = e^{-t}(t^3 - 6t^2 + 6t)$;

29.

positive	negative
zero	zero
positive	positive
positive	negative
negative	negative
negative	positive

; slowing down on the time intervals $(a, b), (d, e), (f, g)$; speeding up on the time intervals $(c, d), (e, f)$

31. (a) $v > 0$ on $\left[0, \dfrac{3}{2}\right), v < 0$ on $\left(\dfrac{3}{2}, \dfrac{1}{4}(6 + \sqrt{42})\right]$
(b) 42 ft
33. $64\sqrt{2}$ ft/s; 16 ft/s^2
35. $-8\sqrt{\pi}$ ft/s; the y-coordinate is decreasing

Exercises 4.2, Page 200

1. $\dfrac{dV}{dt} = 3x^2 \dfrac{dx}{dt}$
3. $8\sqrt{3}$ cm²/h
5. $\dfrac{4}{3}$ in/h
7. $\dfrac{dx}{dt} = s\cos\theta \dfrac{d\theta}{dt} + \sin\theta \dfrac{ds}{dt}$
9. -6 or 6
11. $\dfrac{4}{9}$ cm²/h
13. (a) 1 ft/s (b) 4 ft/s
15. $-\dfrac{1}{\sqrt{2}}$ ft/min
19. 17 knots
21. $-\dfrac{5}{4}$ ft/s
23. 15 rad/h
25. -360 mi/h
27. $\dfrac{8\pi}{9}$ km/min
29. (a) $500\sqrt{3}$ mi/h (b) 500 mi/h
31. $\dfrac{5}{32\pi}$ m/min
33. (a) $-\dfrac{1}{4\pi}$ ft/min (b) $-\dfrac{1}{12\pi}$ ft/min
   (c) approximately -0.0124 ft/min
35. (a) $\dfrac{\sqrt{3}}{10}$ ft/min (c) $\dfrac{165\sqrt{3}}{4} \approx 71.45$ min; 0.035 ft/min
39. $-\dfrac{1}{3}$ in²/min
41. 668.7 ft/min
43. $\dfrac{dR}{dt} = \dfrac{R^2}{R_1^2} \dfrac{dR_1}{dt} + \dfrac{R^2}{R_2^2} \dfrac{dR_2}{dt}$
45. (a) increases (b) approximately 2.8% per day
47. (a) 24,000 kg km/h² (b) 2,023,100 kg km/h²

Exercises 4.3, Page 209

1. (a) abs. max. $f(2) = -2$, abs. min. $f(-1) = -5$
   (b) abs. max. $f(7) = 3$, abs. min. $f(3) = -1$
   (c) no extrema
   (d) abs. max. $f(4) = 0$, abs. min. $f(1) = -3$
3. (a) abs. max. $f(4) = 0$, abs. min. $f(2) = -4$
   (b) abs. max. $f(1) = f(3) = -3$, abs. min. $f(2) = -4$
   (c) abs. min. $f(2) = -4$
   (d) abs. max. $f(5) = 5$
5. (a) no extrema
   (b) abs. max. $f(\pi/4) = 1$, abs. min. $f(-\pi/4) = -1$
   (c) abs. max. $f(\pi/3) = \sqrt{3}$, abs. min. $f(0) = 0$
   (d) no extrema
7. $\dfrac{3}{2}$
9. $-1, 6$
11. $\dfrac{4}{3}, 2$
13. 1
15. $\dfrac{3}{4}$
17. $-2, -\dfrac{11}{7}, 1$
19. $2n\pi$, n an integer
21. 2
23. abs. max. $f(3) = 9$, abs. min. $f(1) = 5$
25. abs. max. $f(8) = 4$, abs. min. $f(0) = 0$
27. abs. max. $f(0) = 2$, abs. min. $f(-3) = -79$
29. abs. max. $f(3) = 8$, abs. min. $f(-4) = -125$
31. abs. max. $f(2) = 16$, abs. min. $f(0) = f(1) = 0$
33. abs. max. $f(\pi/6) = f(5\pi/6) = f(7\pi/6) = f(11\pi/6) = \dfrac{3}{2}$,
    abs. min. $f(\pi/2) = f(3\pi/2) = -3$
35. abs. max. $f(\pi/8) = f(3\pi/8) = f(5\pi/8) = f(7\pi/8) = 5$,
    abs. min. $f(0) = f(\pi/4) = f(\pi/2) = f(3\pi/4) = f(\pi) = 3$
37. endpoint abs. max. $f(3) = 3$, rel. max. $f(0) = 0$,
    abs. min. $f(-1) = f(1) = -1$
39. (a) c_1, c_3, c_4, c_{10}
    (b) $c_2, c_5, c_6, c_7, c_8, c_9$
    (c) abs. min. $f(c_7)$, endpoint abs. max. $f(b)$
    (d) rel. max. $f(c_3), f(c_5), f(c_9)$, rel. min. $f(c_2), f(c_4), f(c_7), f(c_{10})$
41. (a) $s(t) \geq 0$ only for $0 \leq t \leq 20$ (b) $s(10) = 1600$
53. (b) $0, \pi/3, \pi, 5\pi/3, 2\pi$
    (c) abs. max. $f(\pi) = 3$, abs. min. $f(\pi/3) = f(5\pi/3) = -\dfrac{3}{2}$

Exercises 4.4, Page 215

1. $c = 0$
3. $f(-3) = 0$ but $f(-2) \neq f(-3)$
5. $c = -\dfrac{2}{3}$
7. $c = -\pi/2, \pi/2,$ or $3\pi/2$
9. f is not differentiable on the interval
11. $f(a) \neq 0$ and $f(b) = 0$, so $f(a) \neq f(b)$
13. $c = 3$
15. $c = \sqrt{13}$
17. f is not continuous on the interval
19. $c = \dfrac{9}{4}$
21. $c = 1 - \sqrt{6}$
23. f is not continuous on $[a, b]$
25. f increasing on $[0, \infty)$; f decreasing on $(-\infty, 0]$
27. f increasing on $[-3, \infty)$; f decreasing on $(-\infty, -3]$
29. f increasing on $(-\infty, 0]$ and $[2, \infty)$; f decreasing on $[0, 2]$
31. f increasing on $[3, \infty)$; f decreasing on $(-\infty, 0]$ and $[0, 3]$
33. f decreasing on $(-\infty, 0]$ and $[0, \infty)$
35. f increasing on $(-\infty, -1]$ and $[1, \infty)$; f decreasing on $[-1, 0)$ and $(0, 1]$
37. f increasing on $[-2, 2]$; f decreasing on $[-2\sqrt{2}, -2]$ and $[2, 2\sqrt{2}]$
39. f increasing on $(-\infty, 0]$; f decreasing on $[0, \infty)$
41. f increasing on $(-\infty, 1]$ and $[3, \infty)$; f decreasing on $[1, 3]$
43. f increasing on $[-\pi/2 + 2n\pi, \pi/2 + 2n\pi]$; f decreasing on $[\pi/2 + 2n\pi, 3\pi/2 + 2n\pi]$, where n is an integer
45. f increasing on $[0, \infty)$; f decreasing on $(-\infty, 0]$
47. f is increasing on $(-\infty, \infty)$
49. If the motorist travels at the speed limit, he will have gone no more than 65 mi.
61. $c \approx 0.3451$ radian

Exercises 4.5, Page 222

1. 0
3. 2
5. $\dfrac{2}{3}$
7. 10
9. -6
11. $\dfrac{1}{2}$
13. $\dfrac{7}{5}$
15. $\dfrac{1}{6}$
17. does not exist
19. $\dfrac{1}{2}$

21. $2e^4$
23. 0
25. $\frac{1}{3}$
27. ∞
29. -2
31. $-\frac{1}{8}$
33. -1
35. does not exist
37. $\frac{1}{9}$
39. 3
41. $\infty - \infty$; $-\frac{1}{2}$
43. $0 \cdot \infty$; 1
45. 0^0; 1
47. $\infty - \infty$; 0
49. $\infty - \infty$; $\frac{1}{24}$
51. $0 \cdot \infty$; $\frac{1}{4}$
53. ∞^0; 1
55. 1^∞; e^3
57. 0^0; 1
59. denominator is $0 \cdot \infty$; $\frac{1}{4}$
61. $\infty - \infty$; $\frac{1}{5}$
63. $0 \cdot \infty$; 0
65. $0 \cdot \infty$; 1
67. $0 \cdot \infty$; 5
69. $\infty - \infty$; does not exist
71. 1^∞; $e^{-1/3}$
73. 0^0; 1
75. $\frac{1}{2}$
79. 0
81. (a) $A(\theta) = 25 \dfrac{\theta - \frac{1}{2}\sin 2\theta}{\theta^2}$ (b) 0 (c) $\dfrac{50}{3}$
83. (b) $p_1 v_1 \ln(v_2/v_1)$

Exercises 4.6, Page 228

1. rel. max. $f(1) = 2$;

3. rel. max. $f(-1) = 2$, rel. min. $f(1) = -2$;

5. rel. max. $f(\frac{2}{3}) = \frac{32}{27}$, rel. min. $f(2) = 0$;

7. no extrema;

9. rel. min. $f(-1) = -3$;

11. rel. min. $f(0) = 0$;

13. rel. max. $f(0) = f(3) = 0$, rel. min. $f(\frac{3}{2}) = -\frac{81}{16}$;

15. rel. max. $f(0) = 0$, rel. min. $f(1) = -1$;

17. rel. max. $f(-3) = -6$, rel. min. $f(1) = 2$;

19. rel. max. $f(\sqrt{3}) = \dfrac{2\sqrt{3}}{9}$, rel. min. $f(-\sqrt{3}) = -\dfrac{2\sqrt{3}}{9}$;

21. rel. max. $f(0) = 10$;

23. rel. max. $f(0) = \sqrt[3]{16}$, rel. min. $f(-2) = f(2) = 0$;

25. rel. max. $f\left(\frac{\sqrt{2}}{2}\right) = \frac{1}{2}$, rel. min. $f\left(-\frac{\sqrt{2}}{2}\right) = -\frac{1}{2}$;

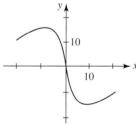

27. rel. max. $f(-8) = 16$, rel. min. $f(8) = -16$;

29. rel. min. $f(2) \approx -8.64$;

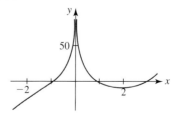

31. rel. min. $f(-3) = 0$, rel. max. $f(-1) = 4e$;

33.

35.

37.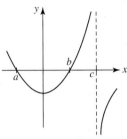

43. rel. min. $f'(-2) = -13$

45. (a) $(n\pi, \pi/2 + n\pi), (\pi/2 + n\pi, \pi + n\pi)$, n an integer
(b) $n\pi/2$, n an integer; rel. max. is $f(-\pi/2) = f(\pi/2) = \cdots 1$, rel. min. is $f(0) = f(\pi) = \cdots 0$
(c)

Exercises 4.7, Page 233

1. concave down on $(-\infty, \infty)$
3. concave up on $(-\infty, 2)$; concave down on $(2, \infty)$
5. concave up on $(-\infty, 2)$ and $(4, \infty)$; concave down on $(2, 4)$
7. concave up on $(-\infty, 0)$; concave down on $(0, \infty)$
9. concave up on $(0, \infty)$; concave down on $(-\infty, 0)$
11. concave up on $(-\infty, -1)$ and $(1, \infty)$; concave down on $(-1, 1)$
13. approximate answers: f' increasing on $(-2, 2)$; f' decreasing on $(-\infty, -2)$ and $(2, \infty)$
15. approximate answers: f' increasing on $(-\infty, -1)$ and $(3, \infty)$; f' decreasing on $(-1, 3)$
19. $(-\sqrt{2}, -21 - \sqrt{2}), (\sqrt{2}, -21 + \sqrt{2})$
21. $(n\pi, 0)$, n an integer
23. $(n\pi, n\pi)$, n an integer
25. $(2, 2 + 2e^{-2})$
27. rel. max. $f(\tfrac{5}{2}) = 0$;

29. point of inflection: $(-1, 0)$;

31. rel. max. $f(-1) = 4$, rel. min. $f(1) = -4$; points of inflection: $(0, 0), \left(-\tfrac{\sqrt{2}}{2}, \tfrac{7\sqrt{2}}{4}\right), \left(\tfrac{\sqrt{2}}{2}, -\tfrac{7\sqrt{2}}{4}\right)$;

33. rel. max. $f(\sqrt{2}) = \tfrac{\sqrt{2}}{4}$, rel. min. $f(-\sqrt{2}) = -\tfrac{\sqrt{2}}{4}$; points of inflection: $(0, 0), \left(-\sqrt{6}, -\tfrac{\sqrt{6}}{8}\right), \left(\sqrt{6}, \tfrac{\sqrt{6}}{8}\right)$;

35. rel. max. $f(0) = 3$;

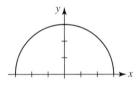

37. rel. min. $f\left(-\tfrac{1}{4}\right) = -3/4^{4/3}$; points of inflection: $(0, 0), (1/2, 3/2^{4/3})$;

39. rel. max. $f(2\pi/3) = f(4\pi/3) = 1$, rel. min. $f(\pi/3) = f(\pi) = f(5\pi/3) = -1$; points of inflection: $(\pi/6, 0), (\pi/2, 0), (5\pi/6, 0), (7\pi/6, 0), (9\pi/6, 0), (11\pi/6, 0)$;

41. rel. max. $f(\pi/4) = \sqrt{2}$, rel. max. $f(5\pi/4) = -\sqrt{2}$; points of inflection: $(3\pi/4, 0), (7\pi/4, 0)$;

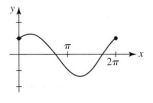

43. rel. max. $f(e) = e$;

45. rel. max. $f(\pi/4) = \frac{1}{2}$ **47.** rel. min. $f(\pi) = 0$

Exercises 4.8, Page 240

1. 30 and 30
3. $\frac{1}{2}$
5. $\frac{1}{3}$ and $\frac{2}{3}$
7. $(2, 2\sqrt{3}), (2, -2\sqrt{3}); (0, 0)$
9. $\left(\frac{4}{3}, -\frac{128}{27}\right)$
11. base $\frac{3}{2}$, height 1
13. $(4, 0)$ and $(0, 8)$
15. 750 ft by 750 ft
17. 2000 m by 1000 m
19. yard should be rectangle 40 ft long and 20 ft wide
21. base 40 cm by 40 cm, height 20 cm
23. base $\frac{80}{3}$ cm by $\frac{80}{3}$ cm, height $\frac{20}{3}$ cm; max. vol. $\frac{128,000}{27}$ cm^3
25. height $\frac{15}{2}$ cm, width 15 cm
27. 10 ft from the flag pole on the right in Figure 4.8.19
29. radius of circular portion $10/(4 + \pi)$ m, width $20/(4 + \pi)$ m, height of rectangular portion $10/(4 + \pi)$ m
31. $L \approx 20.81$ ft
33. radius $16/3$, height 4
35. radius $\sqrt[3]{16/\pi}$, height $2\sqrt[3]{16/\pi}$
37. fly to the point 17.75 km from the nest
39. minimum cost when $x = \dfrac{4}{\sqrt{3}}$
41. $r = \sqrt[3]{9}, h = 2\sqrt[3]{9}$
43. minimum length when $x = 6.375$ in.
45. square with length of side $(a + b)/\sqrt{2}$
47. length of cross section $\sqrt{3}d/3$, width of cross section $\sqrt{6}d/3$
49. $\frac{50}{11}$ m from bulb with illumincance I_1
53. $-\dfrac{1}{8}$
55. (a) $w_0 L^4/384EI$
(b)

65. She should swim from A to a point B about 3.18 mi down the beach from the point on the beach closest to A, and then proceed directly to C.
67. (a) $L = x + 2\sqrt{4 + (4 - x)^2}$
(c) $x = 4 - \dfrac{2}{3}\sqrt{3}$
(d) $L = x + \sqrt{1 + (4 - x)^2} + \sqrt{4 + (4 - x)^2}$
(f) $x \approx 3.1955$

Exercises 4.9, Page 252

1. $L(x) = 3 + \dfrac{1}{6}(x - 9)$ **3.** $L(x) = 1 + 2\left(x - \dfrac{\pi}{4}\right)$
5. $L(x) = x - 1$ **7.** $L(x) = 2 + \dfrac{1}{4}(x - 3)$

17. 0.98   **19.** 11.6
21. 0.7   **23.** 0.96
25. 16   **27.** 0.325
29. 0.4   **31.** $\dfrac{1}{2} + \dfrac{\sqrt{3}\pi}{120} \approx 0.5453$
33. $L(x) = 4 + 2(x - 1)$;   4.08
35. $\Delta y = 2x\,\Delta x + (\Delta x)^2$; $dy = 2x\,dx$
37. $\Delta y = 2(x + 1)\Delta x + (\Delta x)^2$; $dy = 2(x + 1)\,dx$
39. $\Delta y = -\dfrac{\Delta x}{x(x + \Delta x)}$; $dy = -\dfrac{1}{x^2}dx$
41. $\Delta y = \cos x \sin \Delta x + \sin x (\cos \Delta x - 1)$; $dy = \cos x\,dx$
43.

x	Δx	Δy	dy	$\Delta y - dy$
2	1	25	20	5
2	0.5	11.25	10	1.25
2	0.1	2.05	2	0.05
2	0.01	0.2005	0.2	0.0005

45. (a) 1.11   (b) -2.9 **47.** (a) 9π cm^2 (b) 8π cm^2
49. exact volume is $\Delta V = \frac{4}{3}\pi(3r^2 t + 3rt^2 + t^3)$; approximate volume is $dV = 4\pi r^2 t$, where $t = \Delta r$; $(0.1024)\pi$ in^3
51. ± 6 cm^2; ± 0.06; $\pm 6\%$ **55.** 2048 ft;   160 ft
57. (a) minimum at the equator ($\theta = 0°$);   maximum at the North Pole ($\theta = 90°$ N)
(b) 981.9169 cm/s^2 (c) 0.07856 cm/s^2
59. 0.0102 s

Exercises 4.10, Page 257

1. one real root   **3.** no real roots
5. one real root   **7.** 3.1623
9. 1.5874   **11.** 0.6823
13. ± 1.1414 **15.** 0, 0.8767
17. 2.4981   **19.** 1.6560 ft
21. 0.7297   **23.** (b) 0.0915 ft
25. (b) 0.33711, 44.494   (c) 44.497
27. 1.8955 radians   **29.** 1.0000, -1.2494, -2.6638
31. (d) 1.4645

Chapter 4 in Review, Page 260

A. 1. false   **3.** false
5. true   **7.** false
9. true   **11.** true
13. true   **15.** false
17. true   **19.** false

B. 1. the velocity function   **3.** $y = \tan^{-1} x$
5. 0   **7.** 2
9. $2x\Delta x - \Delta x + (\Delta x)^2$

C. 1. abs. max. $f(-3) = 348$, abs. min. $f(4) = -86$
3. abs. max. $f(3) = \frac{9}{7}$, abs. min. $f(0) = 0$

7. max. vel. $v(2) = 12$, max. speed $|v(-1)| = |v(5)| = 15$;

9. (b) $a, b, (a + b)/2$

11. rel. max. $f(-3) = 81$,
    rel. min. $f(2) = -44$;

13. rel. max. $f(0) = 2$,
    rel. min. $f(1) = 0$;

15. rel. min. $f(0) = 0$, points of inflection: $(-3, 27), (-1, 11)$
17. point of inflection: $(3, 10)$
19. (c), (d)
21. (c), (d), (e)
23. (c)
25. $(a + b + c)/3$
27. $32 \text{ in}^2/\text{min}$
31. $y = \tfrac{1}{2}h$; max. distance is h
33. $x = 195 \text{ ft}, y = 390 \text{ ft}; \ 57{,}037.5 \text{ ft}^2$
39. $8\sqrt{3}\pi/9$
41. -2
43. 1
45. e^{-1}
47. $-\infty$
49. 1.6751

Exercises 5.1, Page 274

1. $3x + C$
3. $\tfrac{1}{6}x^6 + C$
5. $\tfrac{3}{2}x^{2/3} + C$
7. $t - \tfrac{25}{12}t^{0.48} + C$
9. $x^3 + x^2 - x + C$
11. $\tfrac{2}{7}x^{7/2} - \tfrac{4}{3}x^{3/2} + C$
13. $\tfrac{16}{3}x^3 + 4x^2 + x + C$
15. $16w^4 - 16w^3 + 6w^2 - w + C$
17. $\ln|r| + 10r^{-1} - 2r^{-2} + C$
19. $-\tfrac{1}{2}x^{-2} + \tfrac{1}{3}x^{-3} - \tfrac{1}{4}x^{-4} + C$
21. $-4\cos x - x - 2x^{-4} + C$
23. $-\cot x + \csc x + C$
25. $-2\cot x + 3x + C$
27. $4x^2 + x - 9e^x + C$
29. $x^2 - x + 5\tan^{-1}x + C$
31. $\tan x - x + C$
41. $x^2 - 4x + 5$
43. $2x^3 + 9x + C$
45. $-x^{-1} + C$
47. $x - x^2 - \cos x + C$
49. $y = x^2 - x + 1$
51. $f'(x) = x^2 + C_1$; $f(x) = \tfrac{1}{3}x^3 + C_1x + C_2$
53. $f(x) = x^4 + x^2 - 3x + 2$
55. G
57. $y = \tfrac{\omega^2}{2g}x^2$

Exercises 5.2, Page 285

1. $-\tfrac{1}{6}(1 - 4x)^{3/2} + C$
3. $-\tfrac{1}{10}(5x + 1)^{-2} + C$
5. $\tfrac{1}{3}(x^2 + 4)^{3/2} + C$
7. $\tfrac{1}{18}\sin^6 3x + C$
9. $\tfrac{1}{6}\tan^3 2x + C$
11. $-\tfrac{1}{4}\cos 4x + C$
13. $\tfrac{1}{3}(2t)^{3/2} - \tfrac{1}{6}\sin 6t + C$
15. $-\tfrac{1}{2}\cos x^2 + C$
17. $\tfrac{1}{3}\tan x^3 + C$
19. $-2\csc\sqrt{x} + C$
21. $\tfrac{1}{7}\ln|7x + 3| + C$
23. $\tfrac{1}{2}\ln(x^2 + 1) + C$
25. $x - \ln|x + 1| + C$
27. $\ln|\ln x| + C$
29. $-\cos(\ln x) + C$
31. $\tfrac{1}{10}e^{10x} + C$
33. $-\tfrac{1}{6}e^{-2x^3} + C$
35. $-2e^{-\sqrt{x}} + C$
37. $\ln(e^x + e^{-x}) + C$
39. $\sin^{-1}\!\left(\tfrac{x}{\sqrt{5}}\right) + C$
41. $\tfrac{1}{5}\tan^{-1}5x + C$
43. $\tan^{-1}e^x + C$
45. $-2\sqrt{1 - x^2} - 3\sin^{-1}x + C$
47. $\tfrac{1}{2}(\tan^{-1}x)^2 + C$
49. $-\tfrac{1}{5}\ln|\cos 5x| + C$
51. $\tfrac{1}{2}x - \tfrac{1}{4}\sin 2x + C$
53. $\tfrac{1}{2}x + \tfrac{1}{16}\sin 8x + C$
55. $11x + 12\cos x - \sin 2x + C$
57. $-\tfrac{3}{4}(1 - x)^{4/3} + C$
59. $y = x + 2\cos 3x + 1 - \pi$
63. (b) $\tfrac{1}{2}\pi\sqrt{L/g}$ (c) $2\pi\sqrt{L/g}$

Exercises 5.3, Page 293

1. $3 + 6 + 9 + 12 + 15$
3. $\tfrac{2}{1} + \tfrac{2^2}{2} + \tfrac{2^3}{3} + \tfrac{2^4}{4}$
5. $-\tfrac{1}{7} + \tfrac{1}{9} - \tfrac{1}{11} + \tfrac{1}{13} - \tfrac{1}{15} + \tfrac{1}{17} - \tfrac{1}{19} + \tfrac{1}{21} - \tfrac{1}{23} + \tfrac{1}{25}$
7. $(2^2 - 4) + (3^2 - 6) + (4^2 - 8) + (5^2 - 10)$
9. $-1 + 1 - 1 + 1 - 1$
11. $\sum_{k=1}^{7}(2k + 1)$
13. $\sum_{k=0}^{12}(3k + 1)$
15. $\sum_{k=1}^{5}\tfrac{(-1)^{k+1}}{k}$
17. $\sum_{k=1}^{8}6$
19. $\sum_{k=1}^{4}\tfrac{(-1)^{k+1}}{k^2}\cos\tfrac{k\pi}{p}x$
21. 420
23. 65
25. 109
27. 3069
29. 18
31. 28
33. $\tfrac{8}{3}$
35. $\tfrac{4}{3}$
37. $\tfrac{16}{3}$
39. $\tfrac{1}{4}$
41. $\tfrac{25}{2}$
43. $\tfrac{77}{60}, \tfrac{25}{12}$

45. 9

47.

25. 1

27. $\dfrac{65}{4}$

29. $\sqrt{6} - \sqrt{3}$

31. $\dfrac{1}{2}$

33. 1

35. $\dfrac{2}{3}$

Exercises 5.4, Page 303

37. $\dfrac{4\pi + 6}{(\pi + 2)(\pi + 3)}$

39. $\dfrac{3}{8} + \dfrac{1}{4\pi}$

1. $\dfrac{33}{2}$; 1

3. $\dfrac{189}{256}$; $\dfrac{3}{4}$

41. $\dfrac{1}{2}\ln\dfrac{11}{3}$

43. xe^x

5. $\dfrac{1}{4}(3 - \sqrt{2})\pi$; π

7. 5

45. $(3t^2 - 2t)^6$

47. $6\sqrt{24x + 5}$

9. $\displaystyle\int_{-2}^{4}\sqrt{9 + x^2}\,dx$

11. $\displaystyle\int_{0}^{2}(1 + x)\,dx$

49. $\dfrac{2x}{x^6 + 1} - \dfrac{3}{27x^3 + 1}$

53. (a) 0 (b) $\ln 3$ (c) $\dfrac{2}{3}$ (d) $-\dfrac{4}{9}$

13. -4

15. $\dfrac{5}{6}$

55. $\dfrac{19}{6}$

57. 9

17. $-\dfrac{3}{4}$

21. 4

59. $\dfrac{38}{3}$

61. 5

23. 12

25. -3

27. 40

29. $-\dfrac{28}{3}$

63. 22

65. 4

67. $\dfrac{1}{6}(1 + \ln 2)^6$

69. $\dfrac{1}{2}\ln\left(\dfrac{2}{1 + e^{-2}}\right)$

31. -32

33. $\dfrac{28}{3}$

35. 36

37. 0

Chapter 5 in Review, Page 316

39. 2.5

41. 11

A. 1. false

3. true

43. (a) -2.5 (b) 3.9 (c) -1.2 (d) 1.4 (e) 2.7 (f) 0.2

5. true

7. true

45.

47.

9. false

11. true

13. false

15. true

B. 1. $f(x)$

3. $\dfrac{\ln x}{x}$

49. 18

51. $\dfrac{9}{4}\pi$

5. $-f(g(x))g'(x)$

7. $\displaystyle\sum_{k=1}^{5}\dfrac{k}{2k + 1}$

53.

55.

9. $\displaystyle\int_{5}^{17}$

11. $\dfrac{5}{2}$

13. $\displaystyle\int_{0}^{4}\sqrt{x}\,dx$; $\dfrac{16}{3}$

15. $2 + e^{-1} - e$; $e - e^{-1}$

57. 15

59. $-\dfrac{\pi}{2}$

C. 1. -6

3. $\dfrac{1}{505}(5t + 1)^{101} + C$

61. -2

63. $\dfrac{5}{2}$

5. $\dfrac{1}{2}$

7. 0

69. $\geq$

9. $-\dfrac{1}{56}\cot^7 8x + C$

11. $\dfrac{1}{40}(4x^2 - 16x + 7)^5 + C$

Exercises 5.5, Page 313

13. $\dfrac{1}{2}(x^3 + 3x - 16)^{2/3} + C$

15. $\dfrac{1}{2}\ln 2$

1. 4

3. 12

5. 46

7. 1

17. $\dfrac{\pi}{6}$

19. $-\dfrac{1}{10}\ln|\cos 10x| + C$

9. $-\dfrac{1}{3} - \dfrac{\sqrt{2}}{6}$

11. $\dfrac{2}{3}$

21. 5

23. $\dfrac{11}{2}$

13. $e - e^{-1}$

15. $-\dfrac{2}{3}$

25. 0

27. $\dfrac{2}{3\sqrt{3}}\pi$

17. $-\dfrac{28}{3}$

19. $\dfrac{8}{3}$

29. $\dfrac{1}{2}$

21. $\dfrac{\pi}{12}$

23. $\dfrac{128}{3}$

31. 156 lb; approximately 20 min

33. $\dfrac{51}{4}$

Exercises 6.1, Page 323

1. $s(t) = 6t - 7$
3. $s(t) = \frac{1}{3}t^3 - 2t^2 + 15$
5. $s(t) = -\frac{5}{2}\sin(4t + \pi/6) + \frac{5}{2}$
7. $v(t) = -5t + 9$; $s(t) = -\frac{5}{2}t^2 + 9t - \frac{9}{2}$
9. $v(t) = t^3 - 2t^2 + 5t - 3$; $s(t) = \frac{1}{4}t^4 - \frac{2}{3}t^3 + \frac{5}{2}t^2 - 3t + 10$
11. $v(t) = \frac{21}{4}t^{4/3} - t - 26$; $s(t) = \frac{9}{4}t^{7/3} - \frac{1}{2}t^2 - 26t - 48$
13. 17 cm
15. 34 cm
17. 24 cm
19. $\frac{1}{30}$ mi = 176 ft
21. 256 ft
23. 30.625 m
25. 400 ft; 6 s
27. -80 ft/s

Exercises 6.2, Page 331

1. $\frac{4}{3}$
3. $\frac{81}{4}$
5. $\frac{9}{2}$
7. $\frac{11}{2}$
9. $\frac{11}{4}$
11. $\frac{11}{6}$
13. 2
15. $\frac{3}{4}(2^{4/3} + 3^{4/3})$
17. 4
19. 2π
21. $\frac{7}{3}$
23. $\frac{27}{2}$
25. $\frac{32}{3}$
27. $\frac{81}{4}$
29. 4
31. $\frac{10}{3}$
33. $\frac{64}{3}$
35. $\frac{128}{5}$
37. $\frac{118}{3}$
39. 22
41. $\frac{9}{2}$
43. $\frac{8}{3}$
45. 8
47. $2\sqrt{2} - 2$
49. $4\sqrt{3} - 4\pi/3$
53. $7 + 3\ln\frac{3}{4} \approx 6.1370$
55. $9\pi/4$
57. $4 + 2\pi$
59. πab
61. $\frac{52}{3}$
63. $A = \int_0^{\ln\frac{3}{2}}(e^x - 1)\,dx + \int_{\ln\frac{3}{2}}^{\ln 2}(2 - e^x)\,dx$,

$A = \int_1^2 \left[\ln y - \ln\frac{1}{2}(y + 1)\right] dy$; $\ln\frac{32}{27} \approx 0.1699$

Exercises 6.3, Page 338

1. $\frac{256\sqrt{3}}{3}$
3. 128
5. $10\pi/3$
7. 9
9. $\pi/2$
11. $4\pi/5$
13. $\pi/6$
15. $1296\pi/5$
17. $\pi/2$
19. $32\pi/5$
21. 32π
23. $7\pi/3$
25. $256\pi/15$
27. $3\pi/5$
29. 36π
31. $500\pi/3$
33. $16\pi/105$
35. $\pi\left(2e^{-1} - \frac{1}{2}e^{-2} - \frac{1}{2}\right)$
37. π^2
39. $\frac{1}{4}(4\pi - \pi^2)$

Exercises 6.4, Page 344

1. $4\pi/5$
3. $\pi/6$
5. $8\pi/15$
7. $250\pi/3$
9. $36\sqrt{3}\pi/5$
11. $3\pi/2$
13. 16π
15. $8\pi/5$
17. $21\pi/10$
19. $\pi/6$
21. $243\pi/10$
23. 4π
25. $625\pi/6$
27. $248\pi/15$
29. $\frac{1}{2}(\pi^2 - 2\pi)$
31. $\frac{1}{3}\pi r^2 h$
33. $\frac{4}{3}\pi r^3$
35. $\frac{4}{3}\pi ab^2$
37. $V = \pi r^2 h - \frac{\pi\omega^2 r^4}{4g}$

Exercises 6.5, Page 347

1. $2\sqrt{2}$
3. $\frac{1}{27}(13^{3/2} - 8) \approx 1.4397$
5. 45
7. $\frac{10}{3}$
9. $\frac{4685}{288} \approx 16.2674$
11. 9
13. $\int_{-1}^{3}\sqrt{1 + 4x^2}\,dx$
15. $\int_0^{\pi}\sqrt{1 + \cos^2 x}\,dx$
17. $\frac{1}{27}(40^{3/2} - 8) \approx 9.0734$
19. (b) 6
21. $\pi/2$

Exercises 6.6, Page 350

1. $208\pi/3$
3. $\frac{\pi}{27}(10^{3/2} - 1) \approx 3.5631$
5. $\frac{\pi}{6}(37^{3/2} - 1) \approx 117.3187$
7. $100\sqrt{5}\pi$
9. $253\pi/20$
11. (a) $(\pi r/6h^2)[(r^2 + 4h^2)^{3/2} - r^3]$
    (b) approximately $0.99\% < 1\%$
13. $20\sqrt{2}\pi$

Exercises 6.7, Page 354

1. -4
3. $\dfrac{34}{3}$
5. 3
7. 0
9. 2
11. $\dfrac{61}{9}$
13. 24
15. $\dfrac{1}{12}$
17. 0
19. $3\sqrt{3}/\pi$
21. $-1 + \dfrac{2\sqrt{3}}{3} \approx 0.1547$
23. 12
25. $103°$
29. $2kt_1/3$

Exercises 6.8, Page 360

1. 3300 ft-lb
3. $\dfrac{2}{5}$ ft
5. (a) 10 joules (b) 27.5 joules
7. (a) 7.5 ft-lb (b) 37.5 ft-lb
9. 453.1×10^8 joules
11. 127,030.9 ft-lb
13. 45,741.6 ft-lb
15. 57,408 ft-lb
17. 64,000 ft-lb
19. (a) 5200 ft-lb (b) 6256.25 ft-lb
21. $3k/4$, where k is constant of proportionality

Exercises 6.9, Page 365

1. (a) 196,000 N/m²; 4,900,000π N
   (b) 196,000 N/m²; 784,000π N
   (c) 196,000 N/m²; 19,600,000π N
3. (a) 499.2 lb/ft²; 244,640 lb (b) 59,904 lb; 29,952 lb
5. 129.59 lb
7. 1280 lb
9. 3660.8 lb
11. 13,977.6 lb
13. 9984π lb
15. 5990.4 lb

Exercises 6.10, Page 372

1. $-\dfrac{2}{7}$
3. $-\dfrac{13}{30}$
5. 1
7. $\dfrac{115}{36}$
9. $\dfrac{4}{7}$
11. $\dfrac{19}{15}$
13. $\dfrac{11}{10}$
15. $\dfrac{15}{2}$
17. $\bar{x} = -\dfrac{2}{7}, \bar{y} = \dfrac{17}{7}$
19. $\bar{x} = \dfrac{17}{11}, \bar{y} = -\dfrac{20}{11}$
21. $\bar{x} = \dfrac{10}{9}, \bar{y} = \dfrac{28}{9}$
23. $\bar{x} = \dfrac{3}{4}, \bar{y} = \dfrac{3}{10}$
25. $\bar{x} = \dfrac{12}{5}, \bar{y} = \dfrac{54}{7}$
27. $\bar{x} = \dfrac{93}{35}, \bar{y} = \dfrac{45}{56}$
29. $\bar{x} = \dfrac{1}{2}, \bar{y} = \dfrac{8}{5}$
31. $\bar{x} = \dfrac{16}{35}, \bar{y} = \dfrac{16}{35}$
33. $\bar{x} = \dfrac{3}{2}, \bar{y} = \dfrac{121}{540}$
35. $\bar{x} = -\dfrac{7}{10}, \bar{y} = \dfrac{7}{8}$
37. $\bar{x} = 0, \bar{y} = 2$
39. $\bar{x} = 0, \bar{y} = \dfrac{1}{8}(\pi + 8)$

Chapter 6 in Review, Page 373

A. 1. false
3. true
5. true
7. true
9. true
11. false

B. 1. joule
3. 2500 ft-lb
5. 6
7. smooth

C. 1. $-\displaystyle\int_0^a f(x)\, dx$
3. $\displaystyle\int_0^a \left[f(x) - \dfrac{f(a)}{a}x\right] dx$
5. $-\displaystyle\int_a^b 2f(x)\, dx + \displaystyle\int_b^c 2f(x)\, dx$
7. $\displaystyle\int_b^c [a - f(y)]\, dy + \displaystyle\int_c^d [f(y) - a]\, dy$
9. $\dfrac{1}{4}a^2 + b^2$
11. $\bar{x} = \dfrac{\displaystyle\int_0^2 x[f(x) - g(x)]\, dx}{\displaystyle\int_0^2 [f(x) - g(x)]\, dx}, \ \bar{y} = \dfrac{\dfrac{1}{2}\displaystyle\int_0^2 ([f(x)]^2 - [g(x)]^2)\, dx}{\displaystyle\int_0^2 [f(x) - g(x)]\, dx}$
13. $2\pi \displaystyle\int_0^2 x[f(x) - g(x)]\, dx$
15. $2\pi \displaystyle\int_0^2 (2 - x)[f(x) - g(x)]\, dx$
17. $\dfrac{5}{2}$
19. (a) 4 (b) π
21. $\dfrac{315\sqrt{41}}{16}\pi$ ft² $\approx$ 396.03 ft²
23. $\dfrac{256}{45}$
25. 37.5 joules
27. 624,000 ft-lb
29. 2040 ft-lb
31. 691,612.83 ft-lb
33. $\dfrac{1}{27}(40^{3/2} - 8) \approx 9.07$
35. 17,066.7 N
37. $\dfrac{3}{4}$ m from the left along the 1-m bar and $\dfrac{6}{5}$ m from the left along the 2-m bar

Exercises 7.1, Page 382

1. $-\dfrac{5^{-5x}}{5\ln 5} + C$
3. $-2\cos\sqrt{1 + x} + C$
5. $-\dfrac{1}{4}\sqrt{25 - 4x^2} + C$
7. $\dfrac{1}{5}\sec^{-1}\left|\dfrac{2}{5}x\right| + C$
9. $\dfrac{1}{10}\tan^{-1}\left(\dfrac{2}{5}x\right) + C$
11. $\dfrac{1}{20}\ln\left|\dfrac{2x - 5}{2x + 5}\right| + C$
13. $\dfrac{1}{10}\ln|\sin 10x| + C$
15. $(3 - 5t)^{-1.2} + C$
17. $\dfrac{1}{3}\ln|\sec 3x + \tan 3x| + C$
19. $\dfrac{1}{2}(\sin^{-1}x)^2 + C$
21. $-\tan^{-1}(\cos x) + C$
23. $\dfrac{1}{4}\tanh x^4 + C$
25. $\dfrac{1}{2}\sec 2x + C$
27. $\csc(\cos x) + C$
29. $\dfrac{1}{3}(1 + \tan x)^3 + C$
31. $\dfrac{1}{2}\ln(1 + e^{2x}) + C$

Exercises 7.2, Page 385

1. $\dfrac{1}{5}(x+1)^5 - \dfrac{1}{4}(x+1)^4 + C$

3. $\dfrac{4}{5}(x-5)^{5/2} + \dfrac{22}{3}(x-5)^{3/2} + C$

5. $\dfrac{2}{3}(x-1)^{3/2} + 2(x-1)^{1/2} + C$

7. $\dfrac{2}{9}(3x-4)^{1/2} - \dfrac{26}{9}(3x-4)^{-1/2} + C$

9. $2\sqrt{x} - 2\tan^{-1}\sqrt{x} + C$

11. $(\sqrt{t}+1)^2 - 10(\sqrt{t}+1) + 8\ln(\sqrt{t}+1) + C$

13. $\dfrac{3}{10}(x^2+1)^{5/3} - \dfrac{3}{4}(x^2+1)^{2/3} + C$

15. $-\dfrac{1}{x-1} - \dfrac{1}{(x-1)^2} - \dfrac{1}{3(x-1)^3} + C$

17. $2\sqrt{e^x-1} - 2\tan^{-1}\sqrt{e^x-1} + C$

19. $\dfrac{4}{5}(1-\sqrt{v})^{5/2} - \dfrac{4}{3}(1-\sqrt{v})^{3/2} + C$

21. $\dfrac{4}{3}(1+\sqrt{t})^{3/2} + C$

23. $\ln(x^2+2x+5) + \dfrac{5}{2}\tan^{-1}\left(\dfrac{x+1}{2}\right) + C$

25. $-2\sqrt{16-6x-x^2} - \sin^{-1}\left(\dfrac{x+3}{5}\right) + C$

27. $2x^{1/2} + 3x^{1/3} + 6x^{1/6} + 6\ln|x^{1/6}-1| + C$

29. $\dfrac{506}{375}$

31. $6 + 20\ln\dfrac{11}{14}$

33. $\dfrac{177}{2}$

35. $\dfrac{1}{1326}$

37. $3 + 3\ln\dfrac{2}{3}$

39. $\dfrac{1}{168}$

43. $-\dfrac{3}{2} + 3\ln 2$

45. $\dfrac{32\pi}{3} - 4\pi\ln 3$

47. $\dfrac{232}{15}$

Exercises 7.3, Page 392

1. $\dfrac{2}{3}x(x+3)^{3/2} - \dfrac{4}{15}(x+3)^{5/2} + C$

3. $x\ln 4x - x + C$

5. $\dfrac{1}{2}x^2\ln 2x - \dfrac{1}{4}x^2 + C$

7. $-x^{-1}\ln x - x^{-1} + C$

9. $t(\ln t)^2 - 2t\ln t + 2t + C$

11. $x\sin^{-1}x + \sqrt{1-x^2} + C$

13. $\dfrac{1}{3}xe^{3x} - \dfrac{1}{9}e^{3x} + C$

15. $-\dfrac{1}{4}x^3e^{-4x} - \dfrac{3}{16}x^2e^{-4x} - \dfrac{3}{32}xe^{-4x} - \dfrac{3}{128}e^{-4x} + C$

17. $\dfrac{1}{2}x^2e^{x^2} - \dfrac{1}{2}e^{x^2} + C$

19. $\dfrac{1}{8}t\sin 8t + \dfrac{1}{64}\cos 8t + C$

21. $-x^2\cos x + 2x\sin x + 2\cos x + C$

23. $\dfrac{1}{3}x^3\sin 3x + \dfrac{1}{3}x^2\cos 3x - \dfrac{2}{9}x\sin 3x - \dfrac{2}{27}\cos 3x + C$

25. $\dfrac{1}{17}e^x(\sin 4x - 4\cos 4x) + C$

27. $\dfrac{1}{5}e^{-2\theta}(\sin\theta - 2\cos\theta) + C$

29. $\theta\sec\theta - \ln|\sec\theta + \tan\theta| + C$

31. $\dfrac{1}{3}\cos x\cos 2x + \dfrac{2}{3}\sin x\sin 2x + C$

33. $\dfrac{1}{3}x^2(x^2+4)^{3/2} - \dfrac{2}{15}(x^2+4)^{5/2} + C$

35. $\dfrac{1}{2}x\sin(\ln x) - \dfrac{1}{2}x\cos(\ln x) + C$

37. $-\dfrac{1}{2}\csc x\cot x + \dfrac{1}{2}\ln|\csc x - \cot x| + C$

39. $x\tan x + \ln|\cos x| + C$

41. $\dfrac{3}{2}\ln 3$

43. $-12e^{-2} + 8e^{-1}$

45. $\dfrac{\pi}{4} - \dfrac{1}{2}\ln 2$

47. $3\ln 3 + e^{-1}$

49. $5\pi(\ln 5)^2 - 10\pi\ln 5 + 8\pi$

51. $2\pi^2$

53. $\dfrac{\pi}{4} - \dfrac{1}{2}\ln 2$

55. $v(t) = -te^{-t} - e^{-t} + 2;\quad s(t) = te^{-t} + 2e^{-t} + 2t - 3$

57. $(124.8)\cdot\dfrac{8(\pi-2)}{\pi^2} \approx 115.48\,\text{lb}$

59. $4\tan^{-1}2 - \pi/2 - \ln\dfrac{5}{2}$

61. $-2\sqrt{x+2}\cos\sqrt{x+2} + 2\sin\sqrt{x+2} + C$

67. $-\dfrac{1}{3}\sin^2 x\cos x - \dfrac{2}{3}\cos x + C$

69. $\dfrac{1}{30}\cos^2 10x\sin 10x + \dfrac{1}{15}\sin 10x + C$

73. $\dfrac{35\pi}{256}$

83. (b) $\dfrac{17\pi}{4}$

Exercises 7.4, Page 398

1. $\dfrac{2}{3}(\sin x)^{3/2} + C$

3. $\sin x - \dfrac{1}{3}\sin^3 x + C$

5. $-\cos t + \dfrac{2}{3}\cos^3 t - \dfrac{1}{5}\cos^5 t + C$

7. $\dfrac{1}{4}\sin^4 x - \dfrac{1}{6}\sin^6 x + C$

9. $\dfrac{3}{8}t - \dfrac{1}{4}\sin 2t + \dfrac{1}{32}\sin 4t + C$

11. $\dfrac{1}{16}x - \dfrac{1}{64}\sin 4x + \dfrac{1}{48}\sin^3 2x + C$

13. $\dfrac{3}{128}x - \dfrac{1}{128}\sin 4x + \dfrac{1}{1024}\sin 8x + C$

15. $\dfrac{1}{8}\tan^4 2t + \dfrac{1}{12}\tan^6 2t + C$

17. $\dfrac{1}{4}\tan x\sec^3 x - \dfrac{1}{8}\sec x\tan x - \dfrac{1}{8}\ln|\sec x + \tan x| + C$

19. $\dfrac{2}{3}(\sec x)^{3/2} + 2(\sec x)^{-1/2} + C$

21. $\dfrac{1}{7}\sec^7 x - \dfrac{1}{5}\sec^5 x + C$

23. $\dfrac{1}{4}\tan x\sec^3 x + \dfrac{3}{8}\sec x\tan x + \dfrac{3}{8}\ln|\sec x + \tan x| + C$

25. $\ln|\sin x| + \dfrac{1}{2}\cos^2 x + C$

27. $-\dfrac{1}{11}\cot^{11}x - \dfrac{1}{13}\cot^{13}x + C$

29. $\dfrac{1}{7\tan^7(1-t)} + \dfrac{1}{5\tan^5(1-t)} + C$

31. $\frac{1}{2}\sec x \tan x + 2\sec x + \frac{1}{2}\ln|\sec x + \tan x| + C$

33. $\frac{1}{3}\tan^3 x - \tan x + x + C$ **35.** $-\frac{1}{2}\csc^2 t - \ln|\sin t| + C$

37. $\frac{1}{5}\tan^5 x - \frac{1}{3}\tan^3 x + C$ **39.** $-\frac{1}{2}\cos x^2 + \frac{1}{6}\cos^3 x^2 + C$

41. $\frac{25\sqrt{2}}{168}$ **43.** 0

45. $\frac{3}{4}$ **47.** $-\frac{1}{6}\cos 3x + \frac{1}{2}\cos x + C$

49. $\frac{1}{4}\sin 2x - \frac{1}{12}\sin 6x + C$ **51.** $\frac{5}{12}$

55. $\frac{16\pi}{3}$ **57.** $\frac{5\sqrt{2}}{3}$

Exercises 7.5, Page 405

1. $-\sin^{-1}x - \frac{\sqrt{1-x^2}}{x} + C$ **3.** $\ln\left|\frac{x + \sqrt{x^2 - 36}}{6}\right| + C$

5. $\frac{1}{3}(x^2 + 7)^{3/2} + C$

7. $-\frac{1}{3}(1 - x^2)^{3/2} + \frac{1}{5}(1 - x^2)^{5/2} + C$

9. $-\frac{x}{4\sqrt{x^2 - 4}} + C$

11. $\frac{1}{2}x\sqrt{x^2 + 4} + 2\ln x\left|\frac{\sqrt{x^2 + 4} + x}{2}\right| + C$

13. $\sin^{-1}\left(\frac{x}{5}\right) + C$

15. $\frac{1}{4}\ln\left|\frac{4 - \sqrt{16 - x^2}}{x}\right| + C$

17. $\ln\left|\frac{\sqrt{x^2 + 1} - 1}{x}\right| + C$ **19.** $-\frac{(1 - x^2)^{3/2}}{3x^3} + C$

21. $\frac{x}{\sqrt{9 - x^2}} - \sin^{-1}\left(\frac{x}{3}\right) + C$

23. $\frac{1}{2}\tan^{-1}x + \frac{x}{2(1 + x^2)} + C$

25. $\frac{x}{16\sqrt{4 + x^2}} - \frac{x^3}{48(4 + x^2)^{3/2}} + C$

27. $\ln\left|\frac{\sqrt{x^2 + 2x + 10} + x + 1}{3}\right| + C$

29. $\frac{1}{16}\tan^{-1}\left(\frac{x + 3}{2}\right) + \frac{x + 3}{8(x^2 + 6x + 13)} + C$

31. $\frac{-5x - 1}{9\sqrt{5 - 4x - x^2}} + C$ **33.** $\ln(x^2 + 4x + 13) + C$

35. $x - 4\tan^{-1}\left(\frac{x}{4}\right) + C$

37. $\frac{9}{2}\sin^{-1}\left(\frac{x - 3}{3}\right) + \frac{1}{2}(x - 3)\sqrt{9 - (x - 3)^2} + C$

39. $\frac{2\pi}{3} + \sqrt{3}$ **41.** $\frac{\sqrt{2}}{50}$

43. $2\sqrt{3} - \frac{172}{81}$

45. $\frac{1}{3}x^3\sin^{-1}x + \frac{1}{3}\sqrt{1 - x^2} - \frac{1}{9}(1 - x^2)^{3/2} + C$

47. $\frac{1}{\sqrt{3}}\ln\left(\frac{\sqrt{2} - 1}{2 - \sqrt{3}}\right)$ **51.** $\frac{\pi\sqrt{3}}{9}\left(\sqrt{3} - 1 - \frac{\pi}{12}\right)$

53. $12\pi\sqrt{2} - 4\pi\ln(\sqrt{2} + 1)$

55. $2 - \sqrt{2} - \ln(\sqrt{6} - \sqrt{3})$

57. (b) $y = -10\ln\left(\frac{10 - \sqrt{100 - x^2}}{x}\right) - \sqrt{100 - x^2}$

59. $15.6\pi \approx 49.01$ lb

Exercises 7.6, Page 413

1. $\frac{A}{x} + \frac{B}{x + 1}$

3. $\frac{A}{x - 1} + \frac{B}{x + 2} + \frac{C}{(x + 2)^2} + \frac{D}{(x + 2)^3}$

5. $\frac{A}{x} + \frac{B}{x^2} + \frac{C}{x^3} + \frac{Dx + E}{x^2 + 3}$

7. $\frac{Ax + B}{x^2 + 9} + \frac{Cx + D}{(x^2 + 9)^2}$

9. $-\frac{1}{2}\ln|x| + \frac{1}{2}\ln|x - 2| + C$

11. $-2\ln|x| + \frac{5}{2}\ln|2x - 1| + C$

13. $\frac{5}{8}\ln|x - 4| + \frac{3}{8}\ln|x + 4| + C$

15. $-\frac{1}{6}\ln|2x + 1| + \frac{2}{3}\ln|x + 2| + C$

17. $6\ln|x| - \frac{7}{2}\ln|x + 1| - \frac{3}{2}\ln|x - 1| + C$

19. $\frac{1}{2}\ln|x + 1| - \ln|x + 2| + \frac{1}{2}\ln|x + 3| + C$

21. $-2\ln|t| - t^{-1} + 6\ln|t - 1| + C$

23. $\ln|x| - \ln|x + 1| + (x + 1)^{-1} + C$

25. $-2(x + 1)^{-1} + \frac{3}{2}(x + 1)^{-2} + C$

27. $-\frac{1}{32}\ln|x + 1| - \frac{1}{16}(x + 1)^{-1} + \frac{1}{32}\ln|x + 5|$
$\quad - \frac{1}{16}(x + 5)^{-1} + C$

29. $-\frac{19}{16}\ln|x| - \frac{19}{8}x^{-1} + \frac{11}{8}x^{-2} - \frac{3}{2}x^{-3} + \frac{35}{16}\ln|x + 2| + C$

31. $-\ln|x| + \frac{1}{2}\ln(x^2 + 1) + \tan^{-1}x + C$

33. $\frac{1}{2}(x + 1)^{-1} + \frac{1}{2}\tan^{-1}x + C$

35. $\frac{1}{3}\tan^{-1}x - \frac{1}{6}\tan^{-1}\left(\frac{x}{2}\right) + C$

37. $\frac{1}{3}\ln|x - 1| - \frac{1}{6}\ln|x^2 + x + 1| - \frac{1}{\sqrt{3}}\tan^{-1}\left(\frac{2x + 1}{\sqrt{3}}\right) + C$

39. $5\ln|x + 1| - \ln(x^2 + 2x + 2) - 7\tan^{-1}(x + 1) + C$

41. $\frac{1}{2(x^2 + 4)} + \frac{1}{2}\tan^{-1}\left(\frac{x}{2}\right) + C$

43. $\frac{1}{2}\ln(x^2 + 4) - \frac{11}{16}\tan^{-1}\left(\frac{x}{2}\right) + \frac{5x + 12}{8(x^2 + 4)} + C$

45. $\frac{1}{3}x^3 - x^2 + 6x - 10\ln|x + 1| - 8(x + 1)^{-1} + C$

47. $-\frac{1}{2}\ln 3$

49. $2\ln\frac{5}{3} - \frac{14}{15}$

51. $\frac{1}{6}\ln\frac{8}{3} + \frac{1}{3\sqrt{2}}\tan^{-1}\left(\frac{1}{\sqrt{2}}\right) + C$ 53. 0

55. $\frac{1}{4}\ln\left|\frac{1+\sqrt{1-x^2}}{1-\sqrt{1-x^2}}\right| - \frac{\sqrt{1-x^2}}{2x^2} + C$

57. $3(x+1)^{1/3} + \ln|(x+1)^{1/3} - 1| -$
$\frac{1}{2}\ln|(x+1)^{2/3} + (x+1)^{1/3} + 1| - \sqrt{3}\tan^{-1}\left(\frac{2(x+1)^{1/3}+1}{\sqrt{3}}\right) + C$

59. $\frac{1}{4}\ln\frac{15}{7} \approx 0.191$

61. $7\ln 2 - 8\ln 3 + 3\ln 4 \approx 0.222$

63. $8\pi\ln\frac{2}{3} + \frac{11\pi}{3} \approx 1.329$

65. $8\pi\ln 2 - 4\pi \approx 4.854$

Exercises 7.7, Page 421

1. $\frac{1}{81}$
3. diverges
5. $\frac{1}{2}e^6$
7. diverges
9. $\frac{1}{2}$
11. 0
13. $-\frac{1}{18}$
15. $3e^{-2}$
17. 1
19. $\frac{\pi}{2}$
21. $\frac{1}{2}$
23. 4
25. $\ln 2$
27. $\frac{1}{4}\ln\frac{7}{3}$
29. $\frac{1}{21}$
31. diverges
33. 100
35. $2\sqrt{2}$
37. diverges
39. 6
41. $-\frac{1}{4}$
43. diverges
45. diverges
47. $-\frac{4}{3}$
49. $\frac{\pi}{4}$
51. $\frac{\pi}{2}$
53. $\frac{\pi}{6}$
55. $\frac{1}{6}$
57. 2
59. 8
61. $\frac{1}{2}\ln 2$
63. 2.86×10^{10} joules
65. $\frac{1}{s}, s > 0$
67. $\frac{1}{s-1}, s > 1$
69. $\frac{1}{s^2+1}, s > 0$
71. $\frac{e^{-s}}{s}, s > 0$

Exercises 7.8, Page 430

1. 78; $M_3 = 77.25$
3. 22; $T_3 = 22.5$
5. 1.7564; 1.8667
7. 1.1475; 1.1484
9. 0.4393; 0.4228
11. 0.4470; 0.4900
13. $\frac{26}{3}$; $S_4 = 8.6611$
15. 1.6222
17. 0.7854
19. 0.4339
21. 11.1053
23. $n \geq 8$
25. 1.11
27. Simpson's Rule: $n \geq 26$; Trapezoidal Rule: $n \geq 366$
29. Trapezoidal Rule gives 1.10
31. For $n = 2$ and $n = 4$, the Midpoint Rule gives the exact value of the integral: 36
33. (a) $\frac{2}{3}$ (b) $M_8 = \frac{21}{22}$ (c) $T_8 = \frac{11}{16}$
(d) $E_8 = \frac{1}{96}$ for Midpoint Rule and $E_8 = \frac{1}{48}$ for Trapezoidal Rule. The error for the Midpoint Rule is one-half the error for the Trapezoidal Rule.
37. 7.0667
39. approximately 4975 gal
41. 41.4028
43. (b) 1.2460
45. 1.4804
47. 14.9772

Chapter 7 in Review, Page 433

A. 1. true
3. true
5. true
7. false
9. false
11. true
13. true
15. false
17. true
19. false

B. 1. $\frac{1}{5}$
3. $\sqrt{\pi}$
5. $\ln\sqrt{2}$

C. 1. $2\sqrt{x} - 18\ln(\sqrt{x} + 9) + C$
3. $(x^2 + 4)^{1/2} + C$
5. $\frac{3}{256}\tan^{-1}\left(\frac{x}{2}\right) + \frac{x}{32(x^2+4)} + \frac{x}{32(x^2+4)^2} - \frac{x^3}{128(x^2+4)^2} + C$
7. $x - \frac{4}{x} + C$
9. $\frac{1}{2}\ln(x^2 + 4) - \frac{5}{2}\tan^{-1}\left(\frac{x}{2}\right) + C$
11. $\frac{1}{10}(\ln x)^{10} + C$
13. $\frac{1}{2}t^2\sin^{-1}t - \frac{1}{4}\sin^{-1}t + \frac{1}{4}t\sqrt{1-t^2} + C$
15. $\frac{1}{5}(x+1)^5 - \frac{3}{4}(x+1)^4 + C$
17. $x\ln(x^2 + 4) - 2x + 4\tan^{-1}\left(\frac{x}{2}\right) + C$
19. $-\frac{2}{125}\ln|x| - \frac{1}{25}x^{-1} + \frac{2}{125}\ln|x+5| - \frac{1}{25}(x+5)^{-1} + C$
21. $-\frac{1}{12}\ln|x+3| - \frac{1}{2}(x+3)^{-1} + \frac{1}{12}\ln|x-3| + C$

23. $\tan t - t + C$

25. $\frac{1}{13}\tan^{13}t + \frac{1}{11}\tan^{11}t + C$

27. $y\sin y + \cos y + C$

29. $\sin t - \frac{1}{5}\sin^5 t + C$

31. $\frac{1}{6}(1 + e^w)^6 + C$

33. $-\frac{1}{8}\csc^2 4x - \frac{1}{4}\ln|\sin 4x| + C$

35. $\frac{1}{4}$

37. $\sec x - \tan x + x + C$

39. $\frac{5}{2}\ln 2 - \frac{3}{2}\ln 3$

41. $\frac{1}{10}e^x(\cos 3x + 3\sin 3x) + C$

43. $\frac{1}{2}t\cos(\ln t) + \frac{1}{2}t\sin(\ln t) + C$

45. $2\sqrt{x}\sin\sqrt{x} + 2\cos\sqrt{x} + C$

47. $-\frac{2}{3}\cos^3 x + C$

49. $\frac{1}{2}(x+1)\sqrt{x^2+2x+5} + 2\ln\left|\frac{\sqrt{x^2+2x+5}+x+1}{2}\right| + C$

51. $\frac{1}{7}\tan^7 x - \frac{2}{5}\sec^5 x + \frac{1}{3}\sec^3 x + C$

53. $\frac{1}{4}t^4 - \frac{1}{2}t^2 + \frac{1}{2}\ln(1+t^2) + C$

55. $\frac{5}{2}\ln(x^2+1) + \tan^{-1}x - \frac{1}{2}(x^2+1)^{-1} + C$

57. $\frac{1}{4}x^2 - \frac{1}{4}x\sin 2x - \frac{1}{8}\cos 2x + C$

59. $2(\sin x)e^{\sin x} - 2e^{\sin x} + C$

61. $\sqrt{6} - 2$

63. $t\sinh^{-1}t - \sqrt{t^2+1} + C$

65. $\ln\frac{3}{2}$

67. $\frac{1}{39}\tan^{13}3u + \frac{1}{45}\tan^{15}3u + C$

69. $3\tan x + \sec x + C$

71. $\frac{1}{2}x^2(1+\ln x)^2 - \frac{1}{2}x^2(1+\ln x) + \frac{1}{4}x^2 + C$

73. $e^{e^x} + C$

75. $t^2 - \ln(1+e^{t^2}) + C$

77. $\frac{1}{5}\sin^{-1}(5x+2) + C$

79. $(\sin x)\ln|\sin x| - \sin x + C$

81. $\frac{3}{2}\sqrt[3]{9}$

83. 0

85. diverges

87. 0

89. diverges

91. $2 - 2e^{-1}$

95. $\frac{1}{2}$

97. $\frac{2}{3}$

99. (a) 2π (b) the areas are infinite

101. 126 joules

Exercises 8.1, Page 444

1. $y = -\frac{1}{5}\cos 5x + C$

3. $y^{-2} = 2x^{-1} + C$

5. $y + y^2 + \frac{1}{3}y^3 = x + x^2 + \frac{1}{3}x^3 + C$

7. $\cos y = x^{-1} - 5x + C$

9. $y = Cx^4$

11. $-3e^{-2y} = 2e^{3x} + C$

13. $\frac{1}{3}x^3\ln x - \frac{1}{9}x^3 = \frac{1}{2}y^2 + 2y + \ln|y| + C$

15. $\ln|N| = te^{t+2} - e^{t+2} - t + C$

17. $P = \dfrac{5}{1 + Ce^{-5t}}$

19. $(y+3)^5 e^x = C(x+4)^5 e^y$

21. $y^3 = -3x^{-1} + 30$

23. $x = \tan\left(4t - \frac{3}{4}\pi\right)$

25. $y = \dfrac{e^{-(1+1/x)}}{x}$

27. $y = \frac{1}{2}x + \frac{\sqrt{3}}{2}\sqrt{1-x^2}$

29. $y = 3$

31. $y = -4, y = 5$

33. (a) $y = \dfrac{1}{1+Cx}$ (b) $y = 0$ (c) $y = \dfrac{1}{1+2x}$

Exercises 8.2, Page 448

1. $y = Ce^{4x}$

3. $y = \frac{1}{10} + Ce^{-5x}$

5. $y = \frac{1}{4}e^{3t} + Ce^{-t}$

7. $y = \frac{1}{3} + Ce^{-x^3}$

9. $y = \dfrac{\ln x}{x} + \dfrac{C}{x}$

11. $y = \dfrac{C}{1+e^x}$

13. $y = -x\cos x + Cx$

15. $y = \sin x + C\cos x$

17. $y = \sin x + C\csc x$

19. $y = \frac{5}{3}(x+2)^{-1} + C(x+2)^{-4}$

21. $y = \dfrac{e^x}{2x^2} + C\dfrac{e^{-x}}{x^2}$

23. $y = -x - 1 - 3e^x$

25. $y = \dfrac{e^x + 2 - e}{x}$

27. $y = 2x^2 - \dfrac{49}{5}x$

29. $(t+1)x = t\ln t - t + 21$

31. $i = \dfrac{E}{R} + \left(i_0 - \dfrac{E}{R}\right)e^{-Rt/L}$

33. (a) $y = e^{x^2}[1 + \sqrt{\pi}\operatorname{erf}(x)]$ (b) $y(2) = 150.92$;

Exercises 8.3, Page 455

1. 7.9 years; 10 years

3. 760

5. approximately 11 h

7. 136.5 h

9. $0.00098 I_0$

11. 15,600 years

13. $36.67°$; approximately 3.06 min

15. $A(t) = 200 - 170e^{-t/50}$

17. $A(t) = 1000 - 1000e^{-t/100}$

19. 100 min

21. $s(t) = \dfrac{mg}{k}t - \dfrac{m}{k}\left(v_0 - \dfrac{mg}{k}\right)e^{-kt/m} + \dfrac{mv_0}{k} - \dfrac{m^2 g}{k^2}$

23. $X(t) = \dfrac{A}{B} - \dfrac{A}{B}e^{-Bt}$; $X(t) \to \dfrac{A}{B}$ as $t \to \infty$; $t = (\ln 2)/B$

25. $E(t) = E_0 e^{-(t-t_1)/RC}$

27. $i(t) = \dfrac{3}{5} - \dfrac{3}{5}e^{-500t}$; $i(t) \to \dfrac{3}{5}$ as $t \to \infty$

31. 276

Exercises 8.4, Page 465

13. 0 is asymptotically stable, 3 is unstable

15. 2 is semi-stable

17. -2 is unstable, 0 is semi-stable; 2 is asymptotically stable

19. -1 is asymptotically stable, 0 is unstable

21.

25. mg/k

27. $i = E/R$ is an equilibrium solution and E/R is asymptotically stable

Exercises 8.5, Page 470

1. $y_2 = 2.9800$, $y_4 = 3.1151$

3. $y_{10} = 2.5937$, $y_{20} = 2.6533$; $y = e^x$

5. $y_5 = 0.4198$, $y_{10} = 0.4124$

7. $y_5 = 0.5639$, $y_{10} = 0.5565$

9. $y_5 = 1.2194$, $y_{10} = 1.2696$

Chapter 8 in Review, Page 471

A. 1. true   **3.** true

B. 1. $y = x - 3x^2 + 4e^{3x} + C$ **3.** e^{-x}

5. half life   **7.** $dP/dt = 0.16P$, $P(0) = P_0$

C. 1. $y = C \csc x$ **3.** $y = -\frac{1}{4}t + Ct^5$

5. $y = \frac{1}{4} + C(x^2 + 4)^{-4}$ **7.** $y = \sin(x^2 + C)$

9. $y = xe^{3x} - e^{3x} - \frac{1}{2}x^2 e^{2x} + Ce^{2x}$

11. $P(t) = 1000e^{0.05t}$

13. $y = \frac{1}{25}t^{-1} + \frac{1}{25}t^4(-1 + 5\ln t)$ **15.** $y = \dfrac{6}{5e^{-2x} - 3}$

17. $y = \tan(x - 7\pi/12)$ **19.** $y = \dfrac{1}{2(1 + x^4)}$

21. $3y^4 = 4x^2 + 48$

25. (a) $A(t) = \dfrac{k_1 M}{k_1 + k_2}(1 - e^{-(k_1+k_2)t})$

(b) $A \to \dfrac{k_1 M}{k_1 + k_2}$ as $t \to \infty$, the material will never be completely memorized

(c)

27. (a) $P(t) = P_0 e^{k \sin t}$ **(b)**

31. 1.3214

Exercises 9.1, Page 483

1. $\dfrac{1}{3}, \dfrac{1}{5}, \dfrac{1}{7}, \dfrac{1}{9}, \ldots$ **3.** $-1, \dfrac{1}{2}, -\dfrac{1}{3}, \dfrac{1}{4}, \ldots$

5. 10, 100, 1000, 10000, …   **7.** 2, 4, 12, 48, …

9. $1, 1 + \dfrac{1}{2}, 1 + \dfrac{1}{2} + \dfrac{1}{3}, 1 + \dfrac{1}{2} + \dfrac{1}{3} + \dfrac{1}{4}, \ldots$

15. 0   **17.** 0

19. $\dfrac{1}{2}$ **21.** sequence diverges

23. sequence diverges   **25.** 0

27. 0   **29.** sequence diverges

31. 0   **33.** $\dfrac{5}{7}$

35. 1   **37.** 6

39. 1   **41.** 1

43. $\ln \dfrac{4}{3}$ **45.** 0

47. $\left\{\dfrac{2n}{2n - 1}\right\}$, converges to 1

49. $\{(-1)^{n+1}(2n + 1)\}$, diverges   **51.** $\left\{\dfrac{2}{3^{n-1}}\right\}$, converges to 0

53. $-\dfrac{1}{2}, -\dfrac{1}{4}, -\dfrac{1}{8}, -\dfrac{1}{16}, \ldots$ **55.** $3, 1, \dfrac{1}{3}, \dfrac{1}{3}, \ldots$

57. 8   **59.** $a_{n+1} = \dfrac{5}{n+1} a_n$, $a_1 = 5$

61. converges to 0   **63.** converges to 0

67. $\dfrac{40}{9}$ ft; $15\left(\dfrac{2}{3}\right)^n$ ft

69. 15, 18, 18.6, 18.72, 18.744, 18.7488, …

71. 32

Exercises 9.2, Page 489

1. increasing   **3.** not monotonic

5. increasing   **7.** nonincreasing

9. increasing   **11.** not monotonic

13. bounded and increasing   **15.** bounded and increasing

17. bounded and decreasing   **19.** bounded and decreasing

21. bounded and increasing   **23.** bounded and decreasing

25. 10   **27.** 7

Exercises 9.3, Page 498

1. $3 + \frac{5}{2} + \frac{7}{3} + \frac{9}{4} + \cdots$
3. $\frac{1}{2} - \frac{1}{6} + \frac{1}{12} - \frac{1}{20} + \cdots$
5. $1 + 2 + \frac{3}{2} + \frac{2}{3} + \cdots$
7. $2 + \frac{8}{3} + \frac{16}{5} + \frac{128}{35} + \cdots$
9. $-\frac{1}{7} + \frac{1}{9} - \frac{1}{11} + \frac{1}{13} - \cdots$
11. 1
13. $\frac{1}{2}$
15. $\frac{15}{4}$
17. $\frac{2}{3}$
19. diverges
21. 9000
23. diverges
25. $\frac{2}{9}$
27. $\frac{61}{99}$
29. $\frac{1313}{999}$
31. $\frac{17}{6}$
43. $-2 < x < 2$
45. $-2 < x < 0$
47. 75 ft
49. $\frac{N_0}{1-s}$; 1000
51. 18.75 mg

Exercises 9.4, Page 503

1. converges
3. converges
5. diverges
7. converges
9. converges
11. converges
13. diverges
15. converges
17. converges
19. diverges
21. converges
23. diverges
25. converges
27. converges
29. converges
31. diverges
33. converges
35. converges for $p > 1$, diverges for $p \leq 1$

Exercises 9.5, Page 507

1. converges
3. diverges
5. diverges
7. diverges
9. converges
11. converges
13. converges
15. diverges
17. converges
19. converges
21. converges
23. converges
25. diverges
27. converges
29. diverges
31. diverges
33. converges
35. diverges
37. converges
39. diverges

Exercises 9.6, Page 511

1. converges
3. diverges
5. converges
7. diverges
9. converges
11. converges
13. converges
15. diverges
17. converges
19. diverges
21. converges
23. converges
25. diverges
27. converges
29. diverges
31. converges
33. converges for $0 \leq p < 1$
35. converges for all real values of p
39. use the Ratio Test

Exercises 9.7, Page 517

1. converges
3. diverges
5. converges
7. converges
9. converges
11. converges
13. diverges
15. conditionally convergent
17. absolutely convergent
19. absolutely convergent
21. absolutely convergent
23. divergent
25. conditionally convergent
27. divergent
29. conditionally convergent
31. absolutely convergent
33. divergent
35. 0.84147
37. 5
39. 0.9492
41. less than $\frac{1}{101} \approx 0.009901$
43. the series contains mixed algebraic signs but the signs do not alternate; converges
45. the algebraic signs do not alternate; converges
47. $a_{k+1} \leq a_k$ is not satisfied for k sufficiently large. The sequence of partial sums $\{S_{2n}\}$ is the same as the sequence of partial sums for the harmonic series. This implies that the series diverges.
49. diverges
51. converges

Exercises 9.8, Page 522

1. $(-1, 1]$; 1
3. $[-\frac{1}{2}, \frac{1}{2})$; $\frac{1}{2}$
5. $[2, 4]$; 1
7. $(-5, 15)$; 10
9. $\{0\}$; 0
11. $[0, \frac{2}{3}]$; $\frac{1}{3}$
13. $[-1, 1)$; 1
15. $(-16, 2)$; 9
17. $(-\frac{75}{32}, \frac{75}{32})$; $\frac{75}{32}$
19. $[\frac{2}{3}, \frac{4}{3}]$; $\frac{1}{3}$
21. $(-\infty, \infty)$; ∞
23. $(-3, N)$; 3
25. $(-\infty, \infty)$; ∞
27. $(-\frac{15}{4}, -\frac{9}{4})$; $\frac{3}{4}$
29. 4
31. $x > 1$ or $x < -1$
33. $x < -\frac{1}{2}$
35. $-2 < x < 2$
37. $x < 0$
39. $0 \leq x < \pi/3, 2\pi/3 < x < 4\pi/3, 5\pi/3 < x \leq 2\pi$
41. (a) $(-\infty, \infty)$

Exercises 9.9, Page 528

1. $\sum_{k=0}^{\infty} \frac{x^k}{3^{k+1}}$; $(-3, 3)$
3. $\sum_{k=0}^{\infty} (-1)^k 2^k x^k$; $(-\frac{1}{2}, \frac{1}{2})$

5. $\sum_{k=0}^{\infty} (-1)^k x^{2k}$; $(-1, 1)$ 7. $\sum_{k=0}^{\infty} \frac{(-1)^k}{4^{k+1}} x^{2k}$; $(-2, 2)$ 21. $\sum_{k=0}^{\infty} \frac{e}{k!}(x - 1)^k$

9. $\sum_{k=1}^{\infty} \frac{k}{3^{k+1}} x^{k-1}$; $(-3, 3)$ 23. $\ln 2 + \sum_{k=1}^{\infty} \frac{(-1)^{k+1}}{k 2^k}(x - 2)^k$ 25. $\sum_{k=0}^{\infty} \frac{(-1)^k}{k!} x^{2k}$

11. $\sum_{k=2}^{\infty} \frac{(-1)^k k(k - 1) 2^{k-3}}{5^{k+1}} x^{k-2}$; $\left(-\frac{5}{2}, \frac{5}{2}\right)$ 27. $\sum_{k=0}^{\infty} \frac{(-1)^k}{(2k)!} x^{2k+1}$ 29. $\sum_{k=1}^{\infty} \frac{-1}{k} x^k$

13. $\sum_{k=1}^{\infty} (-1)^{k+1} k x^{2k-1}$; $(-1, 1)$ 31. $1 + x^2 + \frac{2}{3}x^4 + \frac{17}{45}x^6 + \cdots$ 33. 6

15. $\sum_{k=0}^{\infty} \frac{(-1)^k}{2k + 1} x^{2k+1}$; $[-1, 1]$ 35. $\sum_{k=0}^{\infty} \frac{x^{2k}}{(2k)!}$

17. $\sum_{k=0}^{\infty} \frac{(-1)^k}{k + 1} x^{2k+2}$; $[-1, 1]$ 37. $1 + 2x + \frac{5}{2}x^2 + \frac{8}{3}x^3 + \frac{65}{24}x^4 + \cdots$

19. $\ln 4 + \sum_{k=0}^{\infty} \frac{(-1)^k}{(k + 1) 4^{k+1}} x^{k+1}$; $(-4, 4]$ 39. $1 + x + x^2 + \frac{2}{3}x^3 + \frac{1}{2}x^4 + \cdots$

43. $\frac{\pi}{4}$ 45. -1

21. $1 + \frac{3}{2} \sum_{k=1}^{\infty} (-1)^k (2x)^k$; $\left(-\frac{1}{2}, \frac{1}{2}\right)$ 47. 0.71934; four decimal places

23. $\frac{1}{2} \sum_{k=2}^{\infty} (-1)^k k(k - 1) x^k$; $(-1, 1)$ 49. 1.34983; four decimal places

55. **(c)** $y = 7.92$ in.   **(d)** $y = 7.92000021$ in.

25. $\sum_{k=0}^{\infty} \frac{(-1)^k}{k + 1} x^{2k+3}$; $[-1, 1]$

Exercises 9.11, Page 543

27. $\sum_{k=0}^{\infty} \frac{(-1)^k}{(2k + 1)(2k + 2)} x^{2k+2}$; $[-1, 1]$

1. $1 + \frac{1}{3}x - \frac{1 \cdot 2}{3^2 \cdot 2!} x^2 + \frac{1 \cdot 2 \cdot 5}{3^3 \cdot 3!} x^3 - \cdots$; 1

29. $\sum_{k=0}^{\infty} \frac{(-1)^{k+1}}{5^{k+1}} (x - 6)^k$; $(1, 11)$

3. $3 - \frac{3}{2 \cdot 9} x - \frac{3 \cdot 1}{2^2 \cdot 2! \cdot 9^2} x^2 - \frac{3 \cdot 1 \cdot 3}{2^3 \cdot 3! \cdot 9^3} x^3 - \cdots$; 9

31. $-1 + 2 \sum_{k=0}^{\infty} (-1)^k (x + 1)^{k+1}$; $(-2, 0)$

5. $1 - \frac{1}{2} x^2 + \frac{1 \cdot 3}{2^2 \cdot 2!} x^4 - \frac{1 \cdot 3 \cdot 5}{2^3 \cdot 3!} x^6 + \cdots$; 1

33. $\sum_{k=1}^{\infty} \left[\frac{(-1)^k}{4^k} - \frac{1}{3^k} \right] x^k$; $(-3, 3)$

7. $8 + \frac{8 \cdot 3}{2 \cdot 4} x + \frac{8 \cdot 3 \cdot 1}{2^2 \cdot 2! \cdot 4^2} x^2 - \frac{8 \cdot 3 \cdot 1}{2^3 \cdot 3! \cdot 4^3} x^3 + \cdots$; 4

35. $\frac{1}{2} + \frac{3}{4}x + \frac{7}{8}x^2 + \frac{15}{16}x^3 + \cdots$ 37. $(-3, 3]$

9. $\frac{1}{4}x - \frac{2}{4 \cdot 2} x^2 + \frac{2 \cdot 3}{4 \cdot 2! \cdot 2^2} x^3 - \frac{2 \cdot 3 \cdot 4}{4 \cdot 3! \cdot 2^3} x^4 + \cdots$; 2

39. 0.0953   41. 0.4854

11. $|S_2 - S| < a_3 = \frac{1}{9} x^2$

43. 0.0088

13. $x + \sum_{k=1}^{\infty} \frac{1 \cdot 3 \cdot 5 \cdots (2k - 1)}{2^k k!(2k + 1)} x^{2k+1}$

Exercises 9.10, Page 539

17. $P_0(x) = 1$, $P_1(x) = x$, $P_2(x) = \frac{1}{2}(3x^2 - 1)$

1. $\sum_{k=0}^{\infty} \frac{x^k}{2^{k+1}}$ 3. $\sum_{k=0}^{\infty} \frac{(-1)^k}{k + 1} x^{k+1}$

19. $\sqrt{2} + \frac{\sqrt{2}}{2^2}(x - 1) - \frac{\sqrt{2}}{2^4 \cdot 2!}(x - 1)^2 + \frac{\sqrt{2} \cdot 1 \cdot 3}{2^6 \cdot 3!}(x - 1)^3 - \cdots$

5. $\sum_{k=0}^{\infty} \frac{(-1)^k}{(2k + 1)!} x^{2k+1}$ 7. $\sum_{k=0}^{\infty} \frac{x^k}{k!}$

Chapter 9 in Review, Page 544

9. $\sum_{k=0}^{\infty} \frac{x^{2k+1}}{(2k + 1)!}$

A. 1. false   **3.** false

11. $x + \frac{1}{3}x^3 + \frac{2}{15}x^5 + \frac{17}{315}x^7 + \cdots$

5. true   **7.** false

9. true   **11.** false

13. $\sum_{k=0}^{\infty} \frac{(-1)^k}{5^{k+1}}(x - 4)^k$ 15. $\sum_{k=0}^{\infty} (-1)^k (x - 1)^k$

13. true   **15.** false

17. true   **19.** false

21. false   **23.** false

17. $\frac{\sqrt{2}}{2} + \frac{\sqrt{2}}{2}\left(x - \frac{\pi}{4}\right) - \frac{\sqrt{2}}{2 \cdot 2!}\left(x - \frac{\pi}{4}\right)^2 - \frac{\sqrt{2}}{2 \cdot 3!}\left(x - \frac{\pi}{4}\right)^3 + \cdots$

25. false   **27.** true

29. true

19. $\frac{1}{2} - \frac{\sqrt{3}}{2}\left(x - \frac{\pi}{3}\right) - \frac{1}{2 \cdot 2!}\left(x - \frac{\pi}{3}\right)^2 + \frac{\sqrt{3}}{2 \cdot 3!}\left(x - \frac{\pi}{3}\right)^3 + \cdots$

B. 1. 20; 9; $\frac{4}{5}$; 16   **3.** 4

5. $n/9$; $22/9$

7. e^x

9. $x < -5$ or $x > 5$

11. $(-1, 1]$

C. 1. converges

3. converges

5. converges

7. diverges

9. diverges

11. converges

13. $\dfrac{61{,}004}{201}$

15. $\left[-\tfrac{1}{3}, \tfrac{1}{3}\right]$

17. $\{-5\}$

19. $\dfrac{4}{3}$

21. $\dfrac{1}{\alpha - 1}$

23. $1 - \dfrac{1}{3}x^5 + \dfrac{2}{9}x^{10} - \cdots$

25. $x - \dfrac{2}{3}x^3 + \dfrac{2}{15}x^5 - \cdots$

27. $\displaystyle\sum_{k=0}^{\infty} \dfrac{(-1)^{k+1}}{(2k+1)!}(x - \pi/2)^{2k+1}$

29. $6 million

Exercises 10.1, Page 558

1. vertex: $(0, 0)$; focus: $(1, 0)$; directrix: $x = -1$; axis: $y = 0$;

$y^2 = 4x$

3. vertex: $(0, 0)$; focus: $(0, -4)$; directrix: $y = 4$; axis: $x = 0$;

$x^2 = -16y$

5. vertex: $(0, 1)$; focus: $(4, 1)$; directrix: $x = -4$; axis: $y = 1$;

$(y - 1)^2 = 16x$

7. vertex: $(-5, -1)$; focus: $(-5, -2)$; directrix: $y = 0$; axis: $x = -5$;

$(x + 5)^2 = -4(y + 1)$

9. vertex: $(-5, -6)$; focus: $(-4, -6)$; directrix: $x = -6$; axis: $y = -6$;

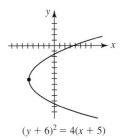

$(y + 6)^2 = 4(x + 5)$

11. vertex: $\left(-\tfrac{5}{2}, -1\right)$; focus: $\left(-\tfrac{5}{2}, -\tfrac{15}{16}\right)$; directrix: $y = -\tfrac{17}{16}$; axis: $x = -\tfrac{5}{2}$;

$\left(x + \tfrac{5}{2}\right)^2 = \tfrac{1}{4}(y + 1)$

13. vertex: $(3, 4)$; focus: $\left(\tfrac{5}{2}, 4\right)$; directrix: $x = \tfrac{7}{2}$; axis: $y = 4$;

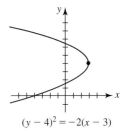

$(y - 4)^2 = -2(x - 3)$

15. $x^2 = 28y$

17. $y^2 = 10x$

19. $(y + 7)^2 = 12(x + 2)$

21. $x^2 = \dfrac{1}{2}y$

23. $(3, 0), (0, -2), (0, -6)$

25. center: $(0, 0)$; foci: $(0, \pm\sqrt{15})$; vertices: $(0, \pm 4)$; minor axis endpoints: $(\pm 1, 0)$; eccentricity: $\tfrac{\sqrt{15}}{4}$;

$\dfrac{x^2}{1} + \dfrac{y^2}{16} = 1$

27. center: $(0, 0)$; foci: $(\pm\sqrt{7}, 0)$; vertices: $(\pm 4, 0)$; minor axis endpoints: $(0, \pm 3)$; eccentricity: $\tfrac{\sqrt{7}}{4}$

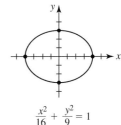

$\dfrac{x^2}{16} + \dfrac{y^2}{9} = 1$

29. center: $(1, 3)$; foci: $(1 \pm \sqrt{13}, 3)$; vertices: $(-6, 3), (8, 3)$; minor axis endpoints: $(1, -3), (1, 9)$; eccentricity: $\tfrac{\sqrt{13}}{7}$;

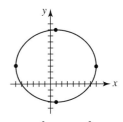

$\dfrac{(x - 1)^2}{49} + \dfrac{(y - 3)^2}{36} = 1$

31. center: $(-5, -2)$; foci: $(-5, -2 \pm \sqrt{15})$; vertices: $(-5, -6)$, $(-5, 2)$; minor axis endpoints: $(-6, -2)$, $(-4, -2)$; eccentricity: $\frac{\sqrt{15}}{4}$;

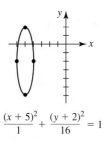

$$\frac{(x+5)^2}{1} + \frac{(y+2)^2}{16} = 1$$

33. center: $\left(0, -\frac{1}{2}\right)$; foci: $\left(0, -\frac{1}{2} \pm \sqrt{3}\right)$; vertices: $\left(0, -\frac{5}{2}\right), \left(0, \frac{3}{2}\right)$; minor axis endpoints: $\left(-1, -\frac{1}{2}\right), \left(1, -\frac{1}{2}\right)$; eccentricity: $\frac{\sqrt{3}}{2}$;

$$\frac{x^2}{1} + \frac{(y+\frac{1}{2})^2}{4} = 1$$

35. center: $(2, -1)$; foci: $(2, -5), (2, 3)$; vertices: $(2, -6), (2, 4)$; minor axis endpoints: $(-1, -1), (5, -1)$; eccentricity: $\frac{4}{5}$;

$$\frac{(x-2)^2}{9} + \frac{(y+1)^2}{25} = 1$$

37. center: $(0, -3)$; foci: $(\pm \sqrt{6}, -3)$; vertices: $(-3, -3), (3, -3)$; minor axis endpoints: $(0, -3 \pm \sqrt{3})$; eccentricity: $\frac{\sqrt{6}}{3}$;

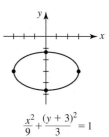

$$\frac{x^2}{9} + \frac{(y+3)^2}{3} = 1$$

39. $\frac{x^2}{25} + \frac{y^2}{16} = 1$

41. $\frac{(x-1)^2}{16} + \frac{(y+3)^2}{4} = 1$

43. $\frac{x^2}{11} + \frac{y^2}{9} = 1$

45. $\frac{x^2}{3} + \frac{y^2}{12} = 1$

47. $\frac{(x-1)^2}{7} + \frac{(y-3)^2}{16} = 1$

49. center: $(0,0)$; foci: $(\pm \sqrt{41}, 0)$; vertices: $(\pm 4, 0)$; asymptotes: $y = \pm \frac{5}{4}x$; eccentricity: $\frac{\sqrt{41}}{4}$;

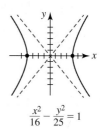

$$\frac{x^2}{16} - \frac{y^2}{25} = 1$$

51. center: $(0, 0)$; foci: $(0, \pm 2\sqrt{6})$; vertices: $(0, \pm 2\sqrt{5})$; asymptotes: $y = \pm \sqrt{5}x$; eccentricity: $\sqrt{\frac{6}{5}}$;

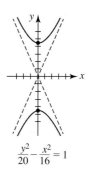

$$\frac{y^2}{20} - \frac{x^2}{16} = 1$$

53. center: $(5, -1)$; foci: $(5 \pm \sqrt{53}, -1)$; vertices: $(3, -1), (7, -1)$; asymptotes: $y = -1 \pm \frac{7}{2}(x - 5)$; eccentricity: $\frac{\sqrt{53}}{2}$;

$$\frac{(x-5)^2}{4} - \frac{(y+1)^2}{49} = 1$$

55. center: $(0, 4)$; foci: $(0, 4 \pm \sqrt{37})$; vertices: $(0, -2), (0, 10)$; asymptotes: $y = 4 \pm 6x$; eccentricity: $\frac{\sqrt{37}}{6}$;

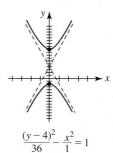

$$\frac{(y-4)^2}{36} - \frac{x^2}{1} = 1$$

57. center: $(3, 1)$; foci: $(3 \pm \sqrt{30}, 1)$; vertices: $(3 \pm \sqrt{5}, 1)$; asymptotes: $y = 1 \pm \sqrt{5}(x - 3)$; eccentricity: $\sqrt{6}$;

$$\frac{(x-3)^2}{5} - \frac{(y-1)^2}{25} = 1$$

59. center: $(2, 1)$; foci: $(2 \pm \sqrt{11}, 1)$; vertices: $(2 \pm \sqrt{6}, 1)$; asymptotes: $y = 1 \pm \sqrt{\frac{5}{6}}(x - 2)$; eccentricity: $\sqrt{\frac{11}{6}}$;

$$\frac{(x-2)^2}{6} - \frac{(y-1)^2}{5} = 1$$

61. center: $(1, 3)$; foci: $\left(1, 3 \pm \frac{\sqrt{5}}{2}\right)$; vertices: $(1, 2), (1, 4)$; asymptotes: $y = 3 \pm 2(x - 1)$; eccentricity: $\frac{\sqrt{5}}{2}$;

$$\frac{(y-3)^2}{1} + \frac{(x-1)^2}{\frac{1}{4}} = 1$$

63. $\dfrac{y^2}{4} - \dfrac{x^2}{12} = 1$ **65.** $\dfrac{(y+3)^2}{4} - \dfrac{(x-1)^2}{5} = 1$

67. $(y - 3)^2 - \dfrac{(x+1)^2}{4} = 1$ **69.** $(y - 4)^2 - \dfrac{(x-2)^2}{4} = 1$

71. at the focus 6 in. from the vertex

73. 76.5625 ft

75. 12.65 m from the point on the ground directly beneath the end of the pipe

77. least distance is 28.5 million mi; greatest distance is 43.5 million mi

79. approximately 0.97   **81.** 12 ft

Exercises 10.2, Page 564

1.

t	-3	-2	-1	0	1	2	3
x	-5	-3	-1	1	3	5	7
y	6	2	0	0	2	6	12

3.   **5.**

7.   **9.**

11. $y = x^2 + 3x - 1, x \geq 0$

13. $x = -1 + 2y^2, -1 \leq x \leq 0$ **15.** $y = \ln x, x > 0$

17.

19.

21.

23.

25.

27.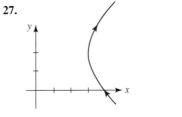

29. yes   **31.** no
33. no
35. $x = \pm\sqrt{r^2 - L^2\sin^2\phi},\ y = L\sin\phi$
37. $x = a(\cos\theta + \theta\sin\theta),\ y = a(\sin\theta - \theta\cos\theta)$
39. (b) 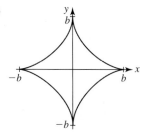   (c) $x^{2/3} + y^{2/3} = b^{2/3}$

41. (b)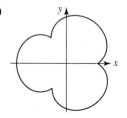

Exercises 10.3, Page 572

1. $\dfrac{3}{5}$ **3.** 24

5. -1 **7.** $y = -2x - 1$

9. $y = \dfrac{4}{3}x + \dfrac{4}{3}$ **11.** $\dfrac{\sqrt{3}}{4}$

13. $y = 3x - 7$

15. horizontal tangent at $(0,0)$, vertical tangent at $\left(-\dfrac{2}{3\sqrt{3}}, \dfrac{1}{3}\right)$ and $\left(\dfrac{2}{3\sqrt{3}}, \dfrac{1}{3}\right)$;

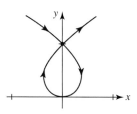

17. horizontal tangents at $(-1, 0)$ and $(1, -4)$, no vertical tangents;

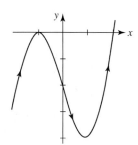

19. $3t;\ 1/(2t);\ -1/(12t^3)$
21. $-2e^{3t} - 3e^{4t};\ 6e^{4t} + 12e^{5t};\ -24e^{5t} - 60e^{6t}$
23. concave up for $0 < t < 2$, concave down for $t < 0$ and $t > 2$

25. $\dfrac{104}{3}$ **27.** $\sqrt{2}(e^\pi - 1)$

29. $\dfrac{3}{2}|b|$

31. (a) -0.6551 (b) $-5.9991, 1.0446, 9.7361$

Exercises 10.4, Page 576

1. 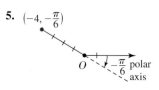   **3.**

$(3, \pi)$, O polar axis

$\left(-\dfrac{1}{2}, \dfrac{\pi}{2}\right)$ polar axis

5. $\left(-4, -\dfrac{\pi}{6}\right)$

7. (a) $(2, -5\pi/4)$ (b) $(2, 11\pi/4)$
    (c) $(-2, 7\pi/4)$ (d) $(-2, -\pi/4)$
9. (a) $(4, -5\pi/3)$ (b) $(4, 7\pi/3)$
    (c) $(-4, 4\pi/3)$ (d) $(-4, -2\pi/3)$
11. (a) $(1, -11\pi/6)$ (b) $(1, 13\pi/6)$
    (c) $(-1, 7\pi/6)$ (d) $(-1, -5\pi/6)$
13. $\left(-\dfrac{1}{4}, \dfrac{\sqrt{3}}{4}\right)$ **15.** $(-3, 3\sqrt{3})$
17. $(-2\sqrt{2}, -2\sqrt{2})$
19. (a) $(2\sqrt{2}, -3\pi/4)$ (b) $(-2\sqrt{2}, \pi/4)$
21. (a) $(2, -\pi/3)$ (b) $(-2, 2\pi/3)$
23. (a) $(7, 0)$ (b) $(-7, \pi)$

25.

27. 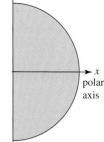   **29.**

31. $r = 5\csc\theta$ **33.** $\theta = \tan^{-1}7$
35. $r = 2/(1 + \cos\theta)$ **37.** $r = 6$
39. $r = 1 - \cos\theta$ **41.** $x = 2$
43. $(x^2 + y^2)^3 = 144x^2y^2$ **45.** $(x^2 + y^2)^2 = 8xy$
47. $x^2 + y^2 + 5y = 0$ **49.** $8x^2 - 12x - y^2 + 4 = 0$
51. $3x + 8y = 5$

Exercises 10.5, Page 583

1. circle;

3. line through origin;

5. spiral;

7. cardioid;

9. cardioid;
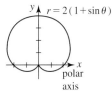

11. limaçon with an interior loop;

13. dimpled limaçon;

15. convex limaçon;

17. rose curve;

19. rose curve;

21. rose curve;
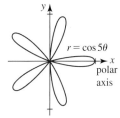

23. circle with center on x-axis;
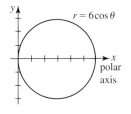

25. circle with center on y-axis;

27. lemniscate;

29. lemniscate;

31.
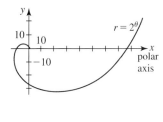

33. $r = \dfrac{5}{2}$

35. $r = 4 - 3\cos\theta$

37. $r = 2\cos 4\theta$

39. $(2, \pi/6), (2, 5\pi/6)$

41. $(1, \pi/2), (1, 3\pi/2)$, origin

43. $(3, \pi/12), (3, 5\pi/12), (3, 13\pi/12), (3, 17\pi/12), (3, -\pi/12), (3, -5\pi/12), (3, -13\pi/12), (3, -17\pi/12)$

45. $(0, 0), \left(\tfrac{\sqrt{3}}{2}, \pi/3\right), \left(\tfrac{\sqrt{3}}{2}, 2\pi/3\right)$ 49. (d)

51. (b)

Exercises 10.6, Page 590

1. $-2/\pi$

3. $\dfrac{\sqrt{3} - 2}{2\sqrt{3} - 1}$

5. $\sqrt{3}$

7. horizontal tangent at $(3, \pi/3)$ and $(3, 5\pi/3)$, vertical tangent at $(4, 0), (1, 2\pi/3)$, and $(1, 4\pi/3)$

9. $y = \dfrac{1}{\sqrt{3}}x + \dfrac{8}{\sqrt{3}}, \; y = -\dfrac{1}{\sqrt{3}}x - \dfrac{8}{\sqrt{3}}$

11. $\theta = 0$ 13. $\theta = 5\pi/4, \theta = 7\pi/4$

15. $\theta = \pi/10, \theta = 3\pi/10, \theta = \pi/2, \theta = 7\pi/10, \theta = 9\pi/10$

17. π 19. 24π

21. 11π 23. $\dfrac{9}{2}\pi$

25. $\dfrac{9}{4}\pi^3$ 27. $\dfrac{1}{4}(e^{2\pi} - 1)$

29. $\dfrac{1}{8}(4 - \pi)$ 31. $\pi - \dfrac{3\sqrt{3}}{2}$

33. $\dfrac{1}{6}(2\pi + 3\sqrt{3})$ 35. $\pi + 6\sqrt{3}$

37. $18\sqrt{3} - 4\pi$ 39. 6π

41. $\sqrt{5}(e^2 - 1)$ 43. 24

Exercises 10.7, Page 596

1. $e = 1$; parabola;
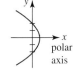

3. $e = \dfrac{1}{4}$; ellipse;
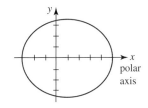

5. $e = 2$;   hyperbola

7. $e = 2$;   hyperbola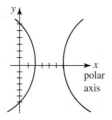

9. $e = \frac{4}{5}$; ellipse;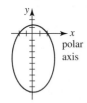

11. $e = 2$; $\dfrac{(y-4)^2}{4} - \dfrac{x^2}{12} = 1$

13. $e = \frac{2}{3}$; $\dfrac{\left(x - \frac{24}{5}\right)^2}{\frac{1296}{25}} + \dfrac{y^2}{\frac{144}{5}} = 1$ **15.** $r = \dfrac{3}{1 + \cos\theta}$

17. $r = \dfrac{4}{3 - 2\sin\theta}$ **19.** $r = \dfrac{12}{1 + 2\cos\theta}$

21. $r = \dfrac{3}{1 + \cos(\theta + 2\pi/3)}$ **23.** $r = \dfrac{3}{1 - \sin\theta}$

25. $r = \dfrac{1}{1 - \cos\theta}$ **27.** $r = \dfrac{1}{2 - 2\sin\theta}$

29. vertex: $(2, \pi/4)$

31. vertices: $(10, \pi/3)$ and $\left(\frac{10}{3}, 4\pi/3\right)$

33. $r_p = 8000$ km   **35.** $r = \dfrac{1.495 \times 10^8}{1 - 0.0167\cos\theta}$

Chapter 10 in Review, Page 597

A. 1. true   **3.** true
5. true   **7.** false

9. true   **11.** true
13. false   **15.** true
17. true   **19.** false
21. true   **23.** true
25. false

B. 1. $\left(0, \frac{1}{8}\right)$ **3.** $(0, -3)$
5. $y = -5$ **7.** $(-10, -2)$
9. $(2, -1), (6, -1)$ **11.** $(4, -3)$
13. $(0, \sqrt{5}), (0, -\sqrt{5})$ **15.** line through origin
17. circle through origin   **19.** $\theta = 0, \theta = \pi/3, \theta = 2\pi/3$
21. $(0, 0), (5, 3\pi/2)$

C. 1. $y = -\dfrac{\sqrt{3}}{3}x + \dfrac{\sqrt{3}\pi}{9}$ **3.** $(8, -26)$

5. (b) $x = \sin t, y = \sin 2t, 0 \le t \le 2\pi$

 (c) $\left(\frac{\sqrt{2}}{2}, 1\right), \left(\frac{\sqrt{2}}{2}, -1\right), \left(-\frac{\sqrt{2}}{2}, 1\right), \left(-\frac{\sqrt{2}}{2}, -1\right)$

 (d)

7. $5\pi/4$

9. (a) $x + y = 2\sqrt{2}$ **(b)** $r = 2\sqrt{2}/(\cos\theta + \sin\theta)$
11. $x^2 + y^2 = x + y$ **13.** $r^2 = 5\csc 2\theta$
15. $r = 1/(1 - \cos\theta)$ **17.** $r = 3\sin 10\theta$

19. $\dfrac{y^2}{100} - \dfrac{x^2}{36} = 1$ **21.** $x = \dfrac{3at}{1 + t^3}, y = \dfrac{3at^2}{1 + t^3}$

23. (a) $r = \dfrac{3a\cos\theta\sin\theta}{\cos^3\theta + \sin^3\theta}$ **(b)** $\dfrac{3}{2}a^2$

25. $\pi - \dfrac{3\sqrt{3}}{2}$

27. (a) $r = 2\cos(\theta - \pi/4)$ **(b)** $x^2 + y^2 = \sqrt{2}x + \sqrt{2}y$
29. 10^8 m; 9×10^8 m

INDEX

A

Abel, Niels Henrik, 257
Absolute convergence:
  definition of, 515
  implies convergence, 515
Absolute error, 469
Absolute extrema:
  definition of, 206
  finding, 208
Absolute maximum and minimum, 206, 732
Absolute value:
  definition of, 6, RP-1
  function, 6
Absolute value of a function, 325, 328
Absolute-value function, 6
Acceleration:
  due to gravity (Earth), 37
  due to gravity (Moon), 376
  function, 192
Achilles and the tortoise, 500
Addition:
  of functions, 11
Additive interval property, 300
Agnesi, Maria Gaetana, 566
Air resistance, 186, 467
Algebraic function, 27, 30
Algebraic sign of a definite integral, 302
Alternating harmonic series, 512
Alternating series:
  definition of, 512
  error bound for, 514
  remainder for, 512
Alternating series test, 512
Amplitude, 32
Analyse des Infiniment Petits, 222
Antiderivative of a function, 268
Antidifferentiation, 269
Aphelion, 557, 595
Approximate integration, 423
Approximate relative error, 251
Approximation:
  by differentials, 250–251
  by Euler's method, 468–470
  by interval bisection, 86
  local linear, 250, 538
  local nth degree, 538
  local quadratic, 538
  by Midpoint Rule, 424
  by Newton's Method, 255
  by Riemann sums, 307
  by Simpson's Rule, 428
  to sum of an alternating series, 514
  by Taylor polynomials, 535, 538
  by Trapezoidal Rule, 426
Archimedean spiral, 577
Archimedes' Principle, 258, 340
Arc length, 345–346
Arc length function, 346
Arccosine function, 43
Arcsine function, 41–42
Arctangent function, 44
Area:
  bounded by two graphs, 326–330
  bounded by two polar graphs, 589
  of a circular sector, RP-2
  of an ellipse, RP-2
  of an equilateral triangle, RP-2
  net signed, 302
  in polar coordinates, 587, 589
  of a region, 326–328
    of a surface of revolution, 348–350
  of a triangle, 289–290
  under a graph, 298
Area problem, 67, 110, 286
Area under a graph:
  as a definite integral, 298
  as a limit of a sum, 291, 293
Arithmetic combinations of functions, 11
Arithmetic mean, 229, 352
Arithmetic of power series, 527
Asymptotes:
  horizontal, 26, 97, RP-4
  of a hyperbola, 553–554
  slant, 26, RP-4
  vertical, 26, 96, RP-4
Asymptotically stable critical point, 465
Asymptotic representation, 543
Asymptotically stable critical point, 465
Attractor, 465
Autonomous first-order differential
    equation, 462
Average, 229, 352
Average rate of change, 114
Average speed, 115
Average value of a function, 353
Average velocity, 115, 192
Axis of a parabola, 23, 548
Axis, polar, 573

B

Base:
   of an exponential function, 48, 50
   of a logarithmic function, 50, 51
Bernoulli, Johann, 222, 267
Bernoulli equations, 449
Bertalanffy's differential equation, 385
Bessel function of the first kind of order ν, 522
Bessel function of order 0, 522
Bézier, Pierre, 568
Bézier curves, 567
Bicylinder, 340, 762
Bifolium, 583
Binomial coefficients, 541, RP-1
Binomial series, 541–542
Binomial theorem, 2, 123, 131, 540, RP-1
Bisection method, 86
Bombay plague epidemic, 423
Bounded above, 487
Bounded below, 487
Bounded function, 297
Bounded sequence, 487
Boxcar function, 19
Brachistochrone, 563
Brain weight, 204
Branches:
   of the graph of a hyperbola, 553
   of the graph of a rational function, 26
Butterfly curve, 567

C

Cable problems, 359
Capacitance, 189, 472
Carbon dating, 452
Cardioid, 579
Carrying capacity, 484
Cartesian coordinate system, 4
Cassegrain reflecting telescope, 557
Catenary, 179, 185
Catenoid, 185
Cauchy, Augustin-Louis, 109
Cavalieri's Principle, 332, 339
Ceiling function, 7, 10
Celsius, relationship to Fahrenheit
       scale, 29
Celsius, relationship to Kelvin scale, 29
Center:
   of a circle, 548
   of an ellipse, 550, 552
   of curvature,
   of gravity, 367
   of a power series, 519
Center of mass:
   of a lamina, 369
   of a rod, 368
   of a system of point masses, 367, 369
Centroid:
   of a plane region, 370
Chain Rule:
   for ordinary derivatives, 151, 156
Change of base, 53

Change in the function, 21, 111, 249
Change in the linearization, 249
Change of variables:
   in a definite integral, 310
Change in x, 21, 111, 114
Change in y, 21, 111, 114
Circle:
   center of, 548
   definition of, 548
   in polar coordinates, 577, 581
   radius of, 548
   in rectangular coordinates, 548
Circular functions, 185
Circular helicoid, 564, 841
Circular helix, 564
Clearing the parameter, 562
Closed curve, 561
Comet Halley, 557, 559, 596
Common logarithm, 51
Common Murre egg, model for, 340
Common ratio, 492
Comparison property of definite integrals, 312
Comparison test:
   for improper integrals, 423
   for series, 505
Complete elliptic integral of second kind, 543
Completeness of the real number system, 488
Completing the square, 22, 404
Composition of functions:
   definition of, 13
   domain of, 14
Compressions of a graph, 16
Computer algebra systems, 381–382
Concave down, 230
Concave up, 230
Concavity:
   definition of, 230
   and the second derivative, 230
   test for, 231
Conditional convergence, 515
Cone:
   double-napped, 548
Conic sections, 548, 592
Conic sections in polar coordinates:
   definition of, 592
   equations of, 593
   graphs of, 594
   rotated, 595
Conjugate axis of a hyperbola, 553
Conjugate hyperbolas, 559
Constant function, 3, 20
Constant Function Rule, 131
Constant of integration, 269
Constant Multiple Rule, 134
Constant term of a polynomial, 20
Constraint, 56, 236
Continuity:
   on a closed interval, 82
   of a composite function, 84
   definition of, 81
   everywhere, 83
   from the left, 82

from the right, 82
of an inverse function, 84, 162
nowhere, 88
at a number, 81
on an open interval, 82
of a polynomial function, 83
of a product, 83
of a quotient, 83
of a rational function, 84
of a sum, 83
Continuous everywhere, 83
Continuous function, 84
Convergence:
absolute, 515
conditional, 515
of an improper integral, 415, 418
of an infinite series, 492
interval of, 520
radius of, 520
of a sequence, 477
Convergent improper integral, 415, 418
Convergent infinite series, 492
Convergent sequence, 255, 477
Conversion of coordinates:
polar to rectangular, 574
rectangular to polar, 574
Convex limaçon, 579
Coordinates:
polar, 573
rectangular, 4
Corner, 125
Cosecant function:
derivative of, 146, 147, 148, 153
domain of, 31
graph of, 32
period of, 32
Cosine function:
cycle of, 30
derivative of, 145, 147, 153
domain of, 30
graph of, 30
Maclaurin series of, 533–534
period of, 30
zeros of, 31
Cotangent function:
derivative of, 145, 147, 153
domain of, 31
graph of, 32
period of, 32
Cover-up method, 423
Critical number, 206
Critical point, 207, 462
Cubic function, 21
Curve(s):
closed, 561
initial point of, 561
length of, 571
orientation of, 561
parametric equations of, 561
piecewise smooth, 564
plane, 561
simple closed, 561

smooth, 564
space, 564
terminal point of, 561
Curvilinear motion in the plane, 560
Cusp, 135
Cycle, 30, 32, RP-6
Cycloid:
definition of, 563
parametric equations of, 564
Cycloidal curves, 563–564
Cylinder:
right, 333
Cylindrical shells, method of, 340–341

D

Dead Sea scrolls, 452
Death by bread, 499
Decay constant, 54, 451
Decimal:
as an infinite series, 475, 491, 494
repeating, 494
Decreasing function, 22, 162, 213
Decreasing sequence, 486
Definite integral(s):
as area, 298
definition of, 297
dummy variable in, 301
equality of limits of integration, 299
even function rule for, 311
integrand of, 297
lower limit of integration, 297
odd function rule for, 311
properties of, 300–301
reversing limits of integration, 299
substitution in, 310
sufficient conditions for existence of, 297
upper limit of integration, 297
Definition of a limit:
functions of one variable, 68, 104
Degenerate conic, 548
Degree of a polynomial function, 20
de la Vallée Poussin, Charles-Jean, 432
Deleting terms of an infinite series, 498
Delta (δ), 104
Delta (Δ), 21
Dependent variable, 2
Derivative(s):
applications of, 191
Chain Rule of, 151
of a constant function, 131
of a constant multiple of a function, 132
definition of, 122, 128
of a difference, 132
domain of, 125
of exponential functions, 168–169
fourth, 135
higher-order, 135
of hyperbolic functions, 180–181
of implicit functions, 157–158
of an integral, 307, 308
of inverse functions, 163

Derivative(s) (*continued*)
  of inverse hyperbolic functions, 184
  of inverse trigonometric functions, 164–165
  left-hand, 125
  of logarithmic functions, 172–173
  notation for, 124
  *n*th, 135
  in parametric form, 568
  of a polynomial, 133
  of power series, 523
  of powers of a function, 149
  of powers of *x*, 131
  of a product, 139
  of a quotient, 140
  as a rate of change, 114, 116
  related to continuity, 126–127
  right-hand, 125
  second, 135
  of a sum, 132–133
  third, 135
  of trigonometric functions, 147, 153
  value of, 124, 135
Difference formulas of trigonometry, 35, RP-6
Difference quotient, 111
Difference Rule of differentiation, 132
Difference of two cubes, 81, RP-1
Difference of two squares, RP-1
Differentiability:
  and continuity, 126
Differentiable:
  at *x*, 125
  on a closed interval, 125
  everywhere, 125
  failure to be, 125
  function of one variable, 125
  on an open interval, 125
Differential(s):
  of arc length, 346–347
  calculus, 68, 121
  definition of, 250
  of dependent variable, 250
  of independent variable, 250
  rules for, 251
  of *x*, 250
  of *y*, 250
Differential equation:
  autonomous, 462
  definition of, 137, 440
  first-order, 273, 440
  general solution of, 448
  initial-value problem for, 442
  integrating factor for, 446
  linear first-order, 445
  as a mathematical model, 450
  nonlinear, 445
  normal form of, 440
  notation, 124
  one-parameter family of solutions of, 441
  order of, 282
  second-order, 440,
  separable, 441

  singular solution of, 445
  solution of, 440
Differentiation:
  definition of, 124
  formulas, table of, RP-9
  implicit, 156–159
  logarithmic, 175
  operators, 124–125
Dimpled limaçon, 579
Direct Comparison Test, 505, 507
Direction field, 460
Directrix:
  of a conic section, 592
  of a parabola, 548
Dirichlet, Johann Peter Gustav Lejeune, 88
Dirichlet function, 88
Discontinuity:
  finite, 84
  infinite, 84
  jump, 84
  removable, 84
Discontinuous function, 81
Discrete logistic equation, 484
Disk method, 334
Distance:
  between two polar points, 576
Distance traveled by an object moving rectilinearly, 323
Divergence:
  of an improper integral, 415, 418
  of an infinite series, 492
  to infinity, 478
  to negative infinity, 478
  by oscillation, 478
  of a sequence, 477
Domain of a function:
  of an arithmetic combination, 11
  of a composition of two functions, 14
  definition of, 2
  implicit (natural), 3
  of a polynomial, 12
Dominated series, 505
Double angle formulas of trigonometry, 35, RP-6
Double-napped cone, 548
Downward concavity, 230
Drag force, 454
Dummy variable:
  of integration, 300
  of summation, 287
Dyne, 356

E

e:
  as a limit, 50, RP-7
  the number, 49, RP-7
Earth:
  Mercator projection, 316
  view from space, 351
Eccentricity:
  of asteroid Hidalgo, 557
  of Comet Halley, 557, 559, 596
  of conic sections, 555, 593

definition of, 555
of Earth's orbit, 557
of planet Mercury, 595
Elementary functions, 313
Eliminating the parameter, 562
Ellipse:
center of, 550, 552
definition of, 550
eccentricity of, 555, 593
equations of, 551–552
foci of, 550
major axis of, 551
minor axis of, 551
in polar coordinates, 593
standard forms of, 551–552
vertices of, 551
Elliptic integral of second kind, 543
End behavior of a function, 4, 24–25, 98
Endpoint extremum, 205, 208
Endpoints of the interval of convergence, 521, 524
Epicycloid, 566
Epicycloid of three cusps, 566
Epitrochoid, 565
Epsilon-delta (ε-δ) definition of a limit, 104–105
Equality of limits of integration, 299
Equation:
of a circle, 548
of an ellipse, 551, 552
of a hyperbola, 553, 555
of a parabola, 23, 549, 550
Equation(s) of a line:
point-slope, 22
slope-intercept, 22
Equilibrium point, 463
Equilibrium position, 156
Equilibrium solution, 463
Equilibrium of a system, 367
Error(s):
in approximate integration methods, 425, 426, 429
between alternating series and its nth partial sum, 514
definition of, 248, 514
percentage, 248, 469
relative, 248, 469
Error bound:
for alternating series, 514
for bisection method, 86
for Midpoint Rule, 425
for Simpson's Rule, 429
for Trapezoidal Rule, 426
Error function, 314, 449
Escape velocity, 458
Euler, Leonhard, 490
Euler's constant, 484, 490
Euler's method, 468
Even function, 17
Even function rule, 311
Existence:
of a derivative, 125
of an inverse function, 38, 162
of a limit, 69, 77
Explicit function, 157
Exponential decay, 54

Exponential function:
with base b, 48
derivative of, 168–169
domain of, 48
graphs of, 49–50
horizontal asymptote for, 49
integral of, 270, 280, 380
natural, 50
properties of, 49
Exponential growth, 54
Exponents, laws of, 48
Extended Mean Value Theorem, 217
Extrema:
absolute, 204
endpoint, 205
on a closed interval, 205
relative, 205–206
Extreme Value Theorem, 205

F

Factor theorem, 27, 77
Factorial, 9, 481
Factorial function, 9
Factorization formulas, RP-1
Fahrenheit, relationship to Celsius scale, 29
Falling bodies and air resistance, 454–455
Familiar integration formulas, 380
Family of functions, 269
Farrari, Lodovico, 257
Fermat's principle, 244, 264
Fibonacci, Leonardo, 485
Fibonacci's sequence, 485, 511
Fick's Law, 457
Finite discontinuity, 86
First Derivative Test, 224
First-order differential equation, 273, 440
Floor function, 7, 70
Flow of blood in an artery, 137
Fluid force, 362
Fluid pressure, 362
Fluxion, 127
Flyswatter Theorem, 89
Focal chord, of a parabola, 549
Focus:
of an ellipse, 550, 592–593
of a hyperbola, 553, 592–593
of a parabola, 548, 592–593
Folium of Descartes, 157
Fontana, Nicolo, 257
Force:
exerted by a fluid, 362–364
Formulas from geometry, RP-2, RP-3
Four-step process, 111, 122
Fourth derivative, 135
Fractals, 484
Fraction:
improper, 412
proper, 407
Fractional part of x function, 20
Frequency interference, 247
Frustum of a right circular cone, 344, 348, RP-2

Function(s):
　absolute extrema of, 204
　absolute value, 6
　absolute value of a, 325, 328
　acceleration, 192
　algebraic, 27, 30
　arithmetic combination of, 11
　average value of, 353
　Bessel, 522
　bounded on an interval, 297–298
　boxcar, 19
　ceiling, 7, 10
　composition of, 13
　constant, 3, 20
　continuity of, 81–85
　as a correspondence, 2
　critical number of, 206
　critical point of, 729
　cubic, 20
　decreasing, 22, 162, 213
　defined by nonelementary integrals, 314
　definition of, 2
　dependent variable of, 2
　derivative, 122
　difference of, 11
　differentiable, 125
　discontinuous, 81
　domain of, 2–3, 11
　end behavior of, 4, 24, 98
　endpoint extrema of, 205, 208
　error, 314, 449
　even, 17, 24
　explicit, 157
　exponential, 48–50
　extrema of, 204–206
　factorial, 9
　floor, 7, 70
　fraction part, 20
　graph of, 4
　greatest integer, 7, 70
　Heaviside, 19
　hyperbolic, 179
　implicit, 157
　implicit domain of, 3
　increasing, 22, 162, 213
　independent variable of, 2
　integer, 10
　integrable, 297
　inverse, 38
　inverse hyperbolic, 182–184
　inverse trigonometric, 41–45
　limit of, 68–69, 104–105
　linear, 20
　logarithmic, 50–53
　as a mathematical model, 7
　maximum of, 204–206
　minimum of, 204–206
　natural domain of, 3
　natural exponential, 50
　natural logarithmic, 51
　objective, 235
　odd, 17, 24
　one-to-one, 38
　piecewise continuous, 309
　piecewise-defined, 5–6
　polynomial, 12, 20
　position, 192, 322
　postage stamp, 7
　power, 10
　power series representation of, 529
　product of, 11
　quadratic, 20, 23
　quotient of, 11
　range of, 2
　rational, 12, 20
　of a real variable, 2
　reciprocal, 3
　relative extrema of, 205–206
　sine integral, 314, 449
　single-term polynomial, 12
　smooth, 345
　square root, 3
　squaring, 2
　sum of, 11
　symmetry of graphs of, 17
　transcendental, 27, 30
　trigonometric, 30–35
　undefined, 3
　unit step, 19
　value of, 2
　velocity, 116, 192
　vertical line test for, 4
　x-intercepts of, 5
　y-intercepts of, 5
　zero of, 5, 25
　zero polynomial, 20
Fundamental Theorem of calculus:
　antiderivative form, 305
　derivative form, 307, 308

G

g, 322
Gabriel's horn, 423
Galileo Galilei, 295, 584
Gamma function, 422
Gateway arch, 178–179
Gauss, Johann Karl Friedrich, 303
General solution of a differential
　　equation, 448
General term:
　of a sequence, 476
　of a series, 490
Generalized factorial function, 422
Generating function, 544
Geometric formulas, RP-2
Geometric series:
　common ratio of, 492
　definition of, 492
　sum of, 493
Global behavior of a function, 4
Global extrema, 204
Golden ratio, 485
Gosper, William, 511

Graph(s):
  asymptotes of, 26, 96, 97
  branches of, 26
  compression of, 16
  of conic sections, 549, 551, 553, 594
  cycle of, 30
  of exponential functions, 48–50
  of a function, 4–6, 683, RP-4
  hole in, 27
  of hyperbolic functions, 179–180
  intercepts of, 5, 25, 26, 49, 51
  of inverse hyperbolic functions, 182–183
  of inverse trigonometric functions, 41, 43, 44, 45
  length of, 345
  of logarithmic functions, 51
  nonrigid transformations of, 16
  of polar equations, 577, 579, 580, 581
  reflections, 15, 39–40
  rigid transformations of, 14–15
  shifted, 14, RP-4
  stretching of, 16
  symmetry of, 17
  of trigonometric functions, 30, 32
Gravitational constant, 357
Greatest integer function, 7
Growth constant, 54, 451
Growth and decay, 54–55, 450–453

H

Half-angle formulas of trigonometry, 35, RP-6
Half-life, 55, 451
Halley's Comet, 557, 559, 596
Hardamard, Jacques, 432
Harmonic numbers, 497
Harmonic series, 495
Heaviside, Oliver, 19
Heaviside function, 19
Helicoid, circular, 564
Helix:
  circular, 564
Hermite, Charles, 170
Higher-order derivatives, 135
Hildago, eccentricity of, 557
Hohmann transfer, 596
Hole in a graph, 27
Homogeneous lamina, 369
Hooke's Law, 356
Horizontal asymptote, 26, 49, 97
Horizontal compression, 16
Horizontal line, 21–22, 131–132
Horizontal Line Test, 38
Horizontal tangent, 125, 569
Horse chestnut leaves, mathematical model for, 399
How big is infinity, 508
Human limb, mathematical model for, 355
Hydrostatic pressure and force, 362–364
Hypatia, 548
Hyperbola:
  asymptotes of, 553–555
  branches of, 553
  center of, 553, 554
  conjugate, 559
  conjugate axis of, 553
  definition of, 553
  eccentricity of, 555, 593
  equations of, 553, 554–555
  foci of, 553
  in polar coordinates, 593
  rectangular, 559
  standard forms of, 553, 554–555
  transverse axis of, 553
  vertices of, 553
Hyperbolic functions:
  definitions of, 179, RP-7
  derivatives of, 180–181
  graphs of, 179–180
  identities for, 180, RP-7
  integrals of, 270, 280, 381, RP-12
  inverses of, 182
Hyperharmonic series:
  convergence of, 503
  definition of, 503
Hypocycloid, 190
Hypocycloid of four cusps, 566

I

Image of x under f, 2
Impedance matching, 263
Implicit differentiation:
  ordinary, 157–158
Implicit domain of a function, 3
Implicit function, 157
Impressed voltage, 472
Improper fraction, 412
Improper integral:
  convergence of, 415, 418, 421
  definition of, 415, 418, 421
  divergence of, 415, 418
Impulse-Momentum Theorem, 355
Increasing function, 22, 162, 213
Increasing sequence, 486
Indefinite integral:
  definition of, 269
  notation for, 269
  properties of, 271
Independent variable, 2
Indeterminate forms:
  of the type ∞/∞, 216
  of the type $\infty - \infty$, 220
  of the type $0/0$, 71, 216
  of the type $0^0, \infty^0, 1^\infty$, 221
  of the type $0 \cdot \infty$, 220
Index of summation, 287
Inequalities, properties of, RP-1
Infinite discontinuity, 84, 425
Infinite limits, 95, 107
Infinite limits at infinity, 99
Infinite sequence, 476
Infinite series:
  absolute convergence of, 515
  addition of two, 496
  alternating, 512

Infinite series (*continued*)
   alternating harmonic, 512
   alternating series test for, 512
   binomial, 541–542
   conditional convergence of, 515
   constant multiple of, 496
   convergent, 492
   definition of, 490
   deleting terms of, 498
   direct comparison test for, 504–505
   divergent, 492
   dominated, 505
   general term of, 490
   geometric, 492–493
   harmonic, 495
   hyperharmonic, 502–503
   integral test for, 501
   limit comparison test for, 506
   Maclaurin, 530
   nth partial sum of, 491
   nth term test for divergence, 496
   p-series, 502–503
   power, 519
   ratio test for, 509, 516
   rearrangement of terms, 517
   root test for, 510, 517
   sum of, 492
   Taylor, 530
   telescoping, 492
   terms of, 490
Infinity, 94
Infinity, limit at, 97
Infinity symbols, 94
Inflection, point of, 239
Initial condition, 273, 322, 442
Initial point:
   of a parametric curve, 561
Initial position, 322
Initial-value problem, 273, 442
Initial velocity, 322
Input of a function, 2
Instantaneous rate of change, 114, 116, 128
Instantaneous velocity, 116
Integer function, 10
Integrability, sufficient conditions for, 297
Integrable, 297
Integral(s):
   algebraic sign of, 302
   as areas, 298, 325, 326–327
   calculus, 121, 267
   change of variable in, 310
   definite, 297
   improper, 415, 418
   indefinite, 269
   Mean Value Theorem, 353
   as net signed area, 302
   nonelementary, 313, 536
   properties of, 271, 300–301
   sign, 269
   tables of, 270, 280, 380–381, RP-9

Integral notation, 269, 297
Integral tables, use of, 380
Integral test, 501
Integrand, 269, 297
Integrating factor, 446
Integration:
   by algebraic substitutions, 382
   approximate, 423
   by computer algebra systems, 381
   constant of, 269
   of a constant, 301
   by darts, 315
   dummy variable of, 300
   of exponential functions, 270, 280, 380
   formulas, 270, 280, 380, RP-9
   of hyperbolic functions, 270, 280, 380, RP-9
   indefinite, 269
   limits of, 297
   by partial fractions, 406
   by parts, 386
   of power series, 523
   of powers of functions, 277
   of powers of trigonometric functions, 393
   of rational functions, 272, 406–407, 412
   reversing the direction of, 299
   reversing the limits of, 299,
   by tables, 380–381, RP-9
   of trigonometric functions, 269–270, 279–280, 380
   by trigonometric substitutions, 399
   by u-substitutions, 282, 382
Intercept form of a line, 21
Intercepts, 5, 24, 26, 31, 49, 51
Intermediate Value Theorem, 85
Intersection of sets, 11
Interval of convergence, 520
Inverse function:
   continuity of, 84, 162
   definition of, 38
   derivative of, 163
   existence of, 162
   graph of, 39–40
   methods for finding, 39
   notation for, 38
   properties of, 39
   reflective property of, 39
Inverse hyperbolic functions:
   definitions of, 183
   derivatives of, 183–184
   graphs of, 182
   as logarithms, 183, RP-6
Inverse trigonometric functions:
   definitions of, 42, 43, 44, 45, RP-6
   derivatives of, 164–165
   integrals yielding, 270, 280, 381
   properties of, 44
Involute of a circle, 566
Irrational number, RP-1
Irreducible quadratic, 410
Isocline, 468
Iverson, Kenneth E., 7
IVP, 442

J

Jenss model, 171
Joule, 356
Jump discontinuity, 84

K

Kepler, Johannes, 127, 557
Kepler's laws of planetary motion:
   first law, 557
   second law, 591
Kinetic energy, 542
Koch, Helge von, 484
Koch, snowflake curve, 484–485

L

Lagrange, Joseph Louis, 532
Lagrange form of remainder, 532
Lamina:
   center of mass of, 369
   centroid of, 370
   definition of, 369
   homogeneous, 369
   mass of, 369
   moments of, 369
Laplace transform, 422
Law of cosines, RP-6
Law of sines, RP-6
Law of Universal Gravitation, 143, 253, 557
Laws of exponents, 48, RP-1
Laws of logarithms, 52
Leading coefficient of a polynomial, 20
Least upper bound, 488
Left endpoint, 293
Left-hand derivative, 125
Left-hand limit, 69, 107
Legendre polynomials, generating function, 544
Leibniz, Gottfried Wilhelm, 127, 267
Leibniz notation:
   for the derivative, 127
   for an integral, 269
Lemniscate, 161, 581
Length:
   of a graph of a function, 346
   of the major axis of an ellipse, 551
   of a parametric curve, 571
   of a polar graph, 589
   of a polygonal path, 345
   of a record groove, 592
   of the transverse axis of a hyperbola, 553
Lens equation, 143
Less than, properties of, RP-1
L'Hôpital, Guillaume Francois Antoine de, 217, 222
L'Hôpital's Rule, 217
Libby, Willard, 452
Liber Abacci, 485
Limaçon:
   convex, 579
   dimpled, 579
   equations of, 579
   with interior loop, 579

Limit(s):
   of a composite function, 84, 101
   of a constant, 74
   of a constant multiple of a function, 74
   ε-δ definition of, 104
   of a difference of functions, 75
   existence of, 69
   indeterminate form ∞/∞, 216
   indeterminate form $0/0$, 71, 216
   infinite, 95, 107
   infinite at infinity, 98
   at infinity, 97, 108
   informal definition, 68
   of integration, 297, 299
   left-hand, 69, 107
   nonexistence of, 69, 77, 98
   notation, 68–69
   of nth root of a function, 78, 99
   of nth root of x, 78
   one-sided, 68, 107
   of a polynomial function, 76
   of a power of a function, 75
   of a product of functions, 75
   of a quotient of functions, 75
   of a rational function, 76–77
   right-hand, 69, 107
   of a sequence, 477
   of a sum of functions, 75
   of trigonometric functions, 71, 88
   two-sided, 69
   uniqueness of, 79
   using a substitution, 92
   of x, 74
Limit comparison test, 506
Lindemann, Ferdinand, 170
Linear approximation formula, 248
Linear density, 368
Linear depreciation, 29
Linear equation:
   in two variables, 22
Linear factors, 407
Linear ordinary differential equation:
   first-order, 445
Linearization:
   of a function of one variable, 248
   as a tangent line, 247
Line element, 460
Line(s):
   element, 460
   horizontal, 21
   normal, 134
   parallel, 22
   perpendicular, 22
   point-slope equation of, 21
   rise of, 21
   run of, 21
   slope of, 21
   slope-intercept equation of, 21
   tangent, 111, 122
   vertical, 21
Local extrema, 206

Local linear approximation:
   for a function of one variable, 248, 538
Local nth degree approximation, 538
Local quadratic approximation, 538
Logarithmic differentiation, 175
Logarithmic function:
   definition of, 50
   derivative of, 172–173
   domain of, 50
   graph of, 51
   natural, 51
   properties of, 51
   vertical asymptote for, 51
Logarithmic integral function, 432
Logarithmic spiral, 577
Logistic curve, 171, 457
Logistic differential equation, 171, 457
Logistic function, 55, 171
Logistic growth, 55
LORAN, 557
Lower limit of integration, 297
Lower semicircle, 6
Lune, 406

M

Maclaurin, Colin, 538
Maclaurin series:
   of cosine function, 534
   definition of, 530
   of exponential function, 534
   of hyperbolic cosine function, 534
   of hyperbolic sine function, 534
   of inverse tangent function, 534
   of sine function, 534
Malthus, Thomas, 450
Major axis of an ellipse, 551
Mass:
   center of, 367, 368
   of a lamina, 369
   of a rod, 368
Mathematical model:
   definition of, 7
   as a differential equation, 450
   as a function, 7
Maximum:
   absolute, 204
   relative, 206
Mean Value Theorem:
   for definite integrals, 353
   for derivatives, 211
Medfly population control, 498
Mercator, Gerhardus, 316
Mercator projection, 316
Mercury, the planet, 595
Method of partial fractions, 406
Method of tangent lines, 468
Methylmercury poisoning, 499
Midpoint:
   of an interval, 293
Midpoint Rule, 424

Minimum:
   absolute, 204
   relative, 206
Minor axis of an ellipse, 551
Mirror image, 15
Mixtures, 453
Moment(s):
   about the origin, 368
Momentum:
   linear, 204
Monotonic sequence, 486
Mormon tabernacle, 557

N

n factorial, 9,
Natural domain of a function, 3
Natural exponential function:
   definition of, 50
   derivative of, 169
   graph of, 50
   integral of, 270, 280, 380
   inverse of, 50
   Maclaurin series of, 534
Natural numbers, RP-1
Natural logarithm, 51
Natural logarithmic function:
   definition of, 51
   derivative of, 172–173
   integrals yielding, 270, 280, 380–381
   laws of, 52
Net signed area, 302
Newton, Isaac, 127
Newton (unit of force), 356
Newton-meter (unit of work), 356
Newton-Raphson method, 255
Newton's first law of motion, 454
Newton's Law of cooling, 55, 453
Newton's Law of Universal Gravitation, 253, 557
Newton's Method, 254–255
Newton's second law of motion, 198, 454
Nondecreasing sequence, 486
Nonelementary integral, 313, 536
Nonincreasing sequence, 486
Nonlinear differential equation, 445
Nonnegative numbers, RP-1
Nonrigid transformation, 14, 16
Norm:
   of a partition, 295
Normal form of a differential equation, 440
Normal line:
   to a graph, 134
Notation:
   for derivatives, 124
   for differentiation operators, 125
   for higher derivatives, 135
   for left-hand derivative, 125
   for right-hand derivative, 125
   for the value of a derivative, 124, 135
Nowhere continuous function, 88
Nowhere differentiable function, 128
nth degree Taylor polynomial, 531–532, 534–536

nth derivative:
   definition of, 135
   value of, 135
nth partial sum, 491
nth term test for divergence, 496
Nullcine, 468
Numerical integration:
   Midpoint Rule, 424
   Simpson's Rule, 426
   Trapezoidal Rule, 428

O

Objective function, 235
Oblate spheroid, 345
Odd function, 17
Odd function rule, 311
One-dimensional phase portrait, 463
One-sided derivative, 125
One-sided limits, 69, 109
One-to-one function, 38
One-parameter family of solutions, 441
Operator, differentiation, 125
Optimization problems, 235
Order of a derivative, 135
Order of a differential equation, 273
Order of integration, definite integral, 299
Orientation of a parametric curve, 561, 656,
Origin, symmetry with respect to, 17, 578
Orthogonal curves, 474
Orthogonal graphs, 138
Orthogonal trajectories, 161, 474
Oscillation, divergence by, 478
Output of a function, 2
Oval, 560

P

p-series, 503
Pappus, of Alexandria, 373
Pappus, Theorem of, 373
Parabola:
   axis of, 23, 548
   definition of, 548
   directrix of, 548
   eccentricity of, 593
   equations of, 23, 549, 550
   focal chord of, 549
   focus of, 548
   in polar coordinates, 593
   standard forms of, 23, 549, 550
   vertex of, 23, 548, RP-4
Paraboloid, 556
Paraboloid of revolution, 350
Parallel lines, 22
Parameter, 560–561
Parameter interval, 560, 562
Parametric curve:
   change of parameter, 562
   closed, 561
   definition of, 561,
   eliminating the parameter, 562

   initial point of, 561
   orientation of, 561,
   simple closed, 561
   terminal point of, 561
   2-space, 561
Partial fraction decomposition, 407
Partial fractions, 407
Partial sum of a series, 491
Particular solution, 273
Partition:
   of an interval, 295
   norm of, 295
   regular, 298
   sample points in, 295
Parts, integration by, 386
Pascal's Principle, 363
Pascal's triangle, RP-1
Percentage error, 248, 469
Perihelion, 557, 595
Period, 30, 32, RP-4
Perpendicular lines, 22
Petals (loops) of a rose curve, 580
Phase line, 463
Phase portrait, 463
Phase shift, 34
Piecewise-continuous function, 309
Piecewise-defined function, 5–6
Piecewise smooth curve, 564
Pinching Theorem, 89
Pisano, Leonardo, 485
Plane curve, 561, 564
Point of inflection, 231
Point-slope equation of a line, 21
Points of intersection of polar graphs, 581
Poiseuille, Jean Louis, 137, 210
Poiseuille's Law, 245
Polar axis, 573
Polar coordinate system, 574
Polar coordinates:
   and arc length, 589
   and area bounded by two polar graphs,
      588–589
   area in, 586–587
   conic sections in, 592
   conventions in, 573
   graphs in, 576, 593
   of a point, 574
   relation to rectangular coordinates, 574
   rotation of graphs in, 582, 595
   slope of tangent line in, 585
   symmetry in, 578
   tangent line to graph at the origin, 580, 586
Polar equation:
   of a cardioid, 579
   of a circle, 577, 581
   of a conic, 593
   graph of, 576
   of a lemniscate, 581
   of a limaçon, 579
   of a line through origin, 577
   of a rose curve, 580
   of a spiral, 577

Polar graph, rotation of, 582
Pole, 573
Polynomial function:
   constant, 20
   constant term of, 20
   continuity of, 83
   cubic, 20
   degree of, 20
   end behavior of, 24
   leading coefficient of, 20
   linear, 20
   of one variable, 20
   quadratic, 21
   x-intercepts of, 25
   zero, 20
   zeros of, 25
Position function, 115, 192
Positive numbers, RP-1
Positive term series, 501
Postage stamp function, 7
Potassium/Argon dating, 459
Power function, 10, RP-4
Power Rule, 131
Power Rule for Functions, 149
Power series:
   arithmetic of, 527
   center of, 519
   continuity of, 523
   convergence of, 520–521
   definition of, 519
   differentiation of, 523
   divergence of, 520
   as a function, 523
   integration of, 523–524
   interval of convergence, 520
   radius of convergence, 520–521
   representations of a function, 524
   in x, 519
   in $x - a$, 519
Pressure of a fluid, 362
Prime numbers, 500
Probability density function, 422
Product of functions, 11
Product Law of limits, 75
Product Rule of differentiation, 139
Prolate spheroid, 345
Proper fraction, 407
Proper rational expression, 407
Properties of the definite integral, 299–301
Properties of exponential functions, 49
Properties of inequalities, RP-1
Properties of inverse functions, 30
Properties of the inverse trigonometric functions, 44
Properties of logarithmic functions, 51
Properties of sequences, 480
p-series:
   convergence of, 503
   definition of, 503
   divergence of, 503
Pump problems, 357
Pure resonance, 223
Pythagorean identities, 35, RP-5

Q

Quadratic formula, RP-1
Quadratic function, 20, 23
Quadratic inequality, 3
Quadric surfaces:
   paraboloid of revolution, 362
Quotient of functions, 11
Quotient Rule of differentiation, 140

R

Radicand, 3
Radioactive decay, 54, 451
Radiogenic, 459
Radius of convergence, 520–521
Ramanujan, Srinivasa, 511
Range of a function, 2
Rate of change:
   average, 114
   derivative as, 128
   instantaneous, 116, 128
Rates, related, 196
Rational exponents and radicals, RP-1
Rational function:
   asymptotes of graphs, 26
   graph with a hole, 27
   of one variable, 12, 26
   intercepts of, 26
Rational number, RP-1
Ratio Test, 509, 516
Real numbers, RP-1
Real-valued function of a real variable, 2
Rearrangement of the terms of a series, 517
Reciprocal function, 3
Rectangular coordinate system, 4
Rectangular hyperbola, 559
Rectifiable graph, 346
Rectilinear motion, 115, 192
Recursion formula, 481
Recursively defined sequence, 481
Reduction formula, 393
Reflecting property:
   of an ellipse, 556
   of a parabola, 556
Reflection of a graph, 15, 39
Regular partition, 290
Related rates, 196
Relative error, 248, 469
Relative extrema:
   definition of, 206
   First Derivative Test for, 224
   Second Derivative Test for, 231
Removable discontinuity, 84
Repeated zero, 25
Repeller, 465
Restricted domains, 40
Return of the plague, 423
Reversing the limits of integration, 299
Reversing the order of integration, 299
Review:
   of algebra, RP-1
   of differentiation rules and formulas, RP-8

of exponential and logarithmic functions, RP-7
of geometric formulas, RP-2, RP-3
of graphs of functions, RP-4
of integration formulas, RP-9
of trigonometry, RP-4, RP-5
Revolution, solids of, 334, 340
Riemann, Georg Friedrich Bernhard, 295, 303
Riemann sum:
 for a definite integral, 295
Right cylinder, 333
Right endpoint, 291
Right-hand derivative, 125
Right-hand limit, 69, 107
Rigid transformation, 14
Rise of a line, 21
Rolle, Michel, 214
Rolle's Theorem, 210
Root of an equation, 5,
Root Test, 510, 517
Rose curves, 580
Rotation of a polar graph, 582, 595
Rule of correspondence, 2
Rule of 70, 546
Rules for differentials, 251
Rules of differentiation:
 chain, 151
 constant function, 131
 constant multiple, 132
 difference, 132
 sum, 132
 power, 130–131, 149
 product, 139
 quotient, 140
Run of a line, 21

S

Sample point, 295
Sandwiching Theorem, 89
Secant function:
 derivative of, 146, 147, 153
 domain of, 31
 graph of, 32
 period of, 32
Secant line, 111
Second derivative:
 ordinary, 135
Second Derivative Test, 231
Second-order differential equation:
 definition of, 440,
Semicircles, 6
Semistable critical point, 465
Separable differential equation, 441
Sequence(s):
 of absolute values, 482
 bounded, 487
 bounded above, 487
 bounded below, 487
 constant, 478
 convergent, 476–477
 decreasing, 486
 definition of, 476
 divergent, 477
 divergent to infinity, 478
 divergent to negative infinity, 478
 divergent by oscillation, 478
 Fibonacci, 485, 511
 finite, 476
 general term of, 476
 increasing, 486
 index, 476
 infinite, 476
 limit of, 477, 479
 monotonic, 486
 nondecreasing, 486
 nonincreasing, 486
 nth term of, 476
 of partial sums, 491
 properties of, 480
 recursively defined, 481
 squeeze theorem for, 481
 terms of, 476
 unbounded, 487
Series:
 absolute convergence of, 515
 alternating, 512
 alternating harmonic, 512
 binomial, 541–542
 conditional convergence of, 515
 convergent, 492
 definition of, 490
 divergent, 492
 general term of, 490
 geometric, 492–493
 harmonic, 495
 hyperharmonic, 502–503
 infinite, 490
 Maclaurin, 530
 nth partial sum of, 491
 p-series, 502–503
 power, 519
 sum of, 492
 Taylor, 530
 telescoping, 492
 terms of, 490
Shell method, 340
Shifted graphs, 14
Shroud of Turin, 452
Sigma notation:
 definition of, 287
 properties of, 287–288
Signal to noise ratio, 247
Simple interest, 29
Simple zero, 25
Simpson, Thomas, 428
Simpson's Rule, 428
Sine function:
 cycle of, 30
 derivative of, 144, 147, 153
 domain of, 30
 graph of, 30
 Maclaurin series of, 534
 period of, 30
 zeros of, 31

Sine integral function, 314, 449
Single-term polynomial function, 12
Single-valued correspondence, 2
Singular solution, 445
Slant asymptote, 26
Slicing method, 333–334
Slope(s):
   of a curve, 111
   field, 460
   function, 440, 459
   of a line, 21
   of parallel lines, 22
   of perpendicular lines, 22
   of tangent lines, 111, 122
   undefined, 21
Slope-intercept equation of line, 21
Slope of a tangent line:
   to the graph of a function, 111, 122
   to the graph of a polar equation, 585
   to the graph of a polar equation at the origin, 580, 586
   to a parametric curve, 568
Slowing down, 193
Smooth curve, 345, 564
Smooth function, 345
Smooth graph, 345
Snell's Law, 244
Snowflake region, 484–485
Sole critical number test, 228
Solid of revolution, 334, 340–341
Solid of revolution about a line, 343
Solution curve, 440
Solution of a differential equation, 440
Space curve:
   definition of, 564
Speed, 192
Speeding up, 193
Spherical sector, 345
Spherical zone, 351
Spirals:
   of Archimedes, 577
   logarithmic, 577
Spring constant, 356
Spring problems, 356
Square root function, 3
Squaring function, 2
Squeeze Theorem, 89, 481
Squeeze Play Theorem, 89
Stacking problem, 500
Standard form:
   of a circle, 548
   of an equation of a ellipse, 551, 552
   of an equation of a hyperbola, 553, 554–55
   of an equation of a parabola, 23, 549–550
   of an equation of a sphere, 540
   of a first-order linear differential equation, 446
Stationary point, 463
Statuary hall, 557
Steinmetz solid, 340
Step functions, 7, 19
Step size, Euler's method, 468

Stiles–Crawford effect, 224
Stretches of a graph, 16
Subinterval, 289–291, 295
Substitutions in an integral:
   definite, 310
   indefinite, 277
Sum:
   of a binomial series, 542
   of a convergent and a divergent series, 498
   of functions, 11
   of a geometric series, 493
   of an infinite series, 492
Sum and difference formulas, 35
Sum formulas of trigonometry, 35, RP-6
Sum Law of limits, 75
Summation formulas, 288
Summation notation:
   index of, 287
   properties of, 288
Sum Rule of differentiation, 132
Surface(s) of revolution:
   area of, 349
   definition of, 348–349
Surface(s):
   area of, 349,
   of revolution, 348
Symbols for ordinary derivatives, 124, 135
Symmetry:
   of graphs of functions, 17, 24
   of polar graphs, 578
   tests for, 17

T

Table of integrals, 270, 280, 380–381, RP-9
Tabular integration, 389
Tangent function:
   derivative of, 145, 147, 153
   graph of, 32
   Maclaurin series of, 537
   period of, 32
Tangent line:
   definition of, 111
   horizontal, 125, 569
   at the origin (pole), 580
   to a parametric curve, 568,
   to a polar curve, 580, 585, 586
   slope of, 111, 568
   vertical, 113, 126, 569
Tangent line problem, 110
Tartaglia, 257
Tautochrone, 563
Taylor, Brook, 538
Taylor polynomials:
   approximation with, 535–536
   definition of, 531–532
Taylor series:
   center of, 530
   convergence of, 532
   definition of, 530
Taylor's Theorem, 531–532
Techniques of integration, 379

Telescoping series, 492
Telescoping sum, 294
Term:
    of a sequence, 476
    of a series, 490
Terminal point:
    of a parametric curve, 561
Terminal velocity, 186, 455, 467
Termwise division, 272
Test for concavity, 231
Test for decreasing function, 213
Test for even and odd functions, 17
Test for increasing function, 213
Tests for convergence and divergence of a series:
    Alternating Series Test, 512
    Direct Comparison Test, 504–505
    Integral Test, 501
    Limit Comparison Test, 506
    nth-term Test for Divergence, 496
    Ratio Test, 509, 516
    Root Test, 510, 517
Theorem of Pappus, 373
Third derivative, 135
Time of death, 458
Torricelli, Evangelista, 423
Torricelli's trumpet, 423
Torus, 373, 406
Total area, 325
Total distance, 323
Total mass of a system, 367
Tower of Hanoi, 319
Tractrix, 186, 405
Trains and fly problem, 501
Transcendental function, 27, 30
Transcendental number, 170
Transformation:
    nonrigid, 14, 16
    rigid, 14
Transient term, 448, 455
Translation of graphs of functions, 14
Transverse axis of a hyperbola, 553
Trapezoidal Rule, 426
Trigonometric functions:
    amplitudes of, 32
    asymptotes of, 32
    continuity of, 88–89
    cycle of, 30, 32
    derivatives of, 144–147, 153
    domains of, 30, 31
    graphs of, 30, 32
    integration of, 270, 280, 380
    inverse, 41–45
    limits of, 88–89, 91, 92, 93
    periodicity of, 30, 32
    range, 30, 31
    zeros of, 31
Trigonometric identities, 35, RP-5, RP-6
Trigonometric integrals, 393
Trigonometric substitutions, 399
Two-sided limit, 69
Two soldiers theorem, 89

U

Unbounded sequence, 487
Undefined function at a number, 3
Undefined slope, 21
Unit circle, RP-5
Unit step function, 19
Units of force, distance, and work, 356
Universal Law of Gravitation, 357
Unstable critical point, 465
Upper limit of integration, 297
Upper semicircle, 6
Upward concavity, 230
u-Substitution Rule, 277, 382

V

Value of a derivative, 124, 135
Value of a function, 2
Values of $\sin x$ and $\cos x$, RP-5
Van der Waals equation of state, 143
Variable:
    change of, 310
    dependent, 2
    dummy, 287, 300
    independent, 2
    of integration, 269, 277, 399
    linear density, 368
Vascular branching, 245
Velocity of a falling body, 115–116
Velocity function, 192
Verhulst, P. F., 457
Vertex of a parabola, 23, 548, 550, RP-4
Vertices:
    of an ellipse, 551
    of a hyperbola, 553
Vertical asymptote, 26, 51, 96
Vertical compression, 16
Vertical line, 21–22
Vertical line test for a function, 4
Vertical stretch of a graph, 16
Vertical tangent, 114, 126, 569
View from space, 351
Volume:
    of a cone, RP-3
    using cylindrical shells, 340–343
    using disks, 334–335
    of a frustum of a cone, RP-3
    by slicing, 333–334
    of a solid of revolution, 333, 340
    using washers, 336

W

Wankel engine, 565–566
Washer method, 336
Wave speed, mathematical model for, 540
Weierstrass, Karl Wilhelm, 109, 128
Weight density, 357
Whewell, William, 109
Whispering galleries, 556–557
Willig, George, 362
Witch of Agnesi, 566

Work:
   as a definite integral, 356
   done against gravity, 357
   done by a constant force, 356
   done by a variable force, 356

X

x-axis, 4
x-increment, 249
x-intercept, 5, 25, 26

Y

y-axis, 4
y-axis, symmetry with respect to, 18
y-increment, 249
y-intercept, 5, 26
Young's modulus of elasticity, 245

Z

Zeno of Elea, 500
Zeno's paradox, 500
Zero(s):
   approximation of, 86
   of a function, 5
   of multiplicity m, 25
   repeated, 25
   simple, 25
   of sine and cosine functions, 31
Zero polynomial, 20
Zone of a sphere, 351

Photo Credits

Table of Contents

Page xi (top) © Pavel Kapish/ShutterStock, Inc.; **page xi (bottom)** Polka Dot Images/age footstock; **page xii (top)** © Jose Gil/ShutterStock, Inc.; **page xii (middle)** © coko/ShutterStock, Inc.; **page xii (bottom)** © RubberBall/age footstock; **page xiii (top)** © Kim Steele/Alamy Images; **page xiii (middle)** © Joy Brown/ShutterStock, Inc.; **page xiii (bottom)** © siloto/ShutterStock, Inc.; **page xiv (top)** © Suzanne Tucker/ShutterStock, Inc.; **page xiv (bottom)** Courtesy of The Observatories of the Carnegie Institution of Washington

Chapter 1

Opener (left) © Pavel Kapish/ShutterStock, Inc.; **Opener (middle)** © Design Pics/age footstock; **page 2** © PhotoCreate/ShutterStock, Inc.; **page 15** Courtesy of Joanna Lee, Flickr.com

Chapter 2

Opener (left) © Polka Dot Images/age fotostock

Chapter 3

Opener (left) © Jose Gil/ShutterStock, Inc.; **Opener (middle)** © ImageSource/age footstock; **page 179** © Marcy J. Levinson/ShutterStock, Inc.

Chapter 4

Opener (left) © coko/ShutterStock, Inc.; **Opener (middle)** © Corbis/age fotostock

Chapter 5

Opener (middle) © RubberBall/age footstock

Chapter 6

Opener (left) © Tomasz Trojanowski/ShutterStock, Inc.; **Opener (middle)** © Kim Steele/Alamy Images; **page 333** © Photoeuphoria/Dreamstime.com; **page 340** Courtesy of Vernon Byrd/U.S. Fish and Wildlife Service; **page 350** © Sergei Chumakov/ShutterStock, Inc.

Chapter 7

Opener (left) © Joy Brown/ShutterStock, Inc.; **page 399** © SlavaK/ShutterStock, Inc.

Chapter 8

Opener (left) Courtesy of Christine Myaskovsky; **Opener (middle)** © siloto/ShutterStock, Inc.

Chapter 9

Opener (middle left) © E.A. Janes/age footstock; **Opener (middle right)** © Suzanne Tucker/ShutterStock, Inc.; **page 494** Courtesy of Michael Maggs; **page 499** © Andrjuss/ShutterStock, Inc.

Chapter 10

Opener (middle) Courtesy of The Observatories of the Carnegie Institution of Washington; **page 556 (left)** © Dennis Donohue/ShutterStock, Inc.; **page 556 (right)** Courtesy of the Palomar Observatory, California Institute of Technology; **page 557** © Architect of the Capital; **page 564** © Jan Kliciak/ShutterStock, Inc.; **page 577** © Mircea Bezergheanu/ShutterStock, Inc.; **page 578** © Jgroup/Dreamstime.com; **page 595** Courtesy of Mariner 10, Astrogeology Team, and USGS

Unless otherwise indicated, all photographs and illustrations are under copyright of Jones and Bartlett Publishers, LLC. Portrait of mathematicians were illustrated by Diana Coe.

Review of Trigonometry

Unit Circle Definition of Sine and Cosine

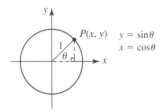

$y = \sin\theta$
$x = \cos\theta$

Other Trigonometric Functions

$\tan\theta = \dfrac{y}{x} = \dfrac{\sin\theta}{\cos\theta}, \quad \cot\theta = \dfrac{x}{y} = \dfrac{\cos\theta}{\sin\theta}$

$\sec\theta = \dfrac{1}{x} = \dfrac{1}{\cos\theta}, \quad \csc\theta = \dfrac{1}{y} = \dfrac{1}{\sin\theta}$

Conversion Formulas

1 degree $= \dfrac{\pi}{180}$ radians

1 radian $= \dfrac{180}{\pi}$ degrees

Right Triangle Definition of Sine and Cosine

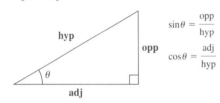

$\sin\theta = \dfrac{\text{opp}}{\text{hyp}}$

$\cos\theta = \dfrac{\text{adj}}{\text{hyp}}$

Other Trigonometric Functions

$\tan\theta = \dfrac{\text{opp}}{\text{adj}}, \quad \cot\theta = \dfrac{\text{adj}}{\text{opp}}$

$\sec\theta = \dfrac{\text{hyp}}{\text{adj}}, \quad \csc\theta = \dfrac{\text{hyp}}{\text{opp}}$

Signs of Sine and Cosine

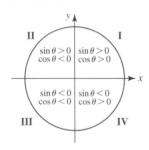

Values of Sine and Cosine for Special Angles

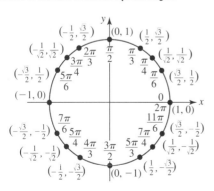

Bounds for Sine and Cosine Functions

$-1 \le \sin x \le 1$ and $-1 \le \cos x \le 1$

Periodicity of Trigonometric Functions

$\sin(x + 2\pi) = \sin x, \quad \cos(x + 2\pi) = \cos x$
$\sec(x + 2\pi) = \sec x, \quad \csc(x + 2\pi) = \csc x$
$\tan(x + \pi) = \tan x, \quad \cot(x + \pi) = \cot x$

Cofunction Identities

$\sin\left(\dfrac{\pi}{2} - x\right) = \cos x$

$\cos\left(\dfrac{\pi}{2} - x\right) = \sin x$

$\tan\left(\dfrac{\pi}{2} - x\right) = \cot x$

Pythagorean Identities

$\sin^2 x + \cos^2 x = 1$
$1 + \tan^2 x = \sec^2 x$
$1 + \cot^2 x = \csc^2 x$

Even/Odd Identities

Even
$\cos(-x) = \cos x$
$\sec(-x) = \sec x$

Odd
$\sin(-x) = -\sin x$
$\csc(-x) = -\csc x$
$\tan(-x) = -\tan x$
$\cot(-x) = -\cot x$

Sum Formulas

$\sin(x_1 + x_2) = \sin x_1 \cos x_2 + \cos x_1 \sin x_2$
$\cos(x_1 + x_2) = \cos x_1 \cos x_2 - \sin x_1 \sin x_2$
$\tan(x_1 + x_2) = \dfrac{\tan x_1 + \tan x_2}{1 - \tan x_1 \tan x_2}$

Difference Formulas

$\sin(x_1 - x_2) = \sin x_1 \cos x_2 - \cos x_1 \sin x_2$
$\cos(x_1 - x_2) = \cos x_1 \cos x_2 + \sin x_1 \sin x_2$
$\tan(x_1 - x_2) = \dfrac{\tan x_1 - \tan x_2}{1 + \tan x_1 \tan x_2}$

Double-Angle Formulas

$\sin 2x = 2 \sin x \cos x$
$\cos 2x = \cos^2 x - \sin^2 x$

Alternative Double-Angle Formulas for Cosine

$\cos 2x = 1 - 2\sin^2 x$
$\cos 2x = 2\cos^2 x - 1$

Half-Angle Formulas as Used in Calculus

$\sin^2 x = \tfrac{1}{2}(1 - \cos 2x)$
$\cos^2 x = \tfrac{1}{2}(1 + \cos 2x)$

Law of Sines

$\dfrac{\sin\alpha}{a} = \dfrac{\sin\beta}{b} = \dfrac{\sin\gamma}{c}$

Law of Cosines

$a^2 = b^2 + c^2 - 2bc\cos\alpha$
$b^2 = a^2 + c^2 - 2ac\cos\beta$
$c^2 = a^2 + b^2 - 2ab\cos\gamma$

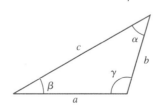

Inverse Trigonometric Functions

$y = \sin^{-1} x$ if and only if $x = \sin y, \quad -\pi/2 \le y \le \pi/2$
$y = \cos^{-1} x$ if and only if $x = \cos y, \quad 0 \le y \le \pi$
$y = \tan^{-1} x$ if and only if $x = \tan y, \quad -\pi/2 < y < \pi/2$

Cycles for Sine, Cosine, and Tangent

sine

tangent